THE
WRITER'S
HANDBOOK
2009

Barry Turner has worked on both sides of publishing, as a journalist and author, editor and marketing director. He started his career as a teacher before joining *The Observer* and then moving on to radio and television. His first book, a study of British politics in the early twentieth century, was published in 1970. He has written over twenty books including *A Place in the Country*, which inspired a television series, and a best-selling biography of the actor, Richard Burton. For many years he wrote on travel for *The Times* and now reviews and serializes books for the paper. Among his recent books are *Countdown to Victory* on the last months of World War II and *Suez 1956: The First Oil War*. As founding editor of *The Writer's Handbook* he has taken this annual reference title through to its twenty-second edition. For twelve years, Barry has edited the annual *Statesman's Yearbook*. He is a founder and current chairman of the National Academy of Writing at Birmingham City University.

Remember to use your FREE access to The Writer's Handbook Online

1. Visit **www.thewritershandbook.com**

2. Register using your unique online access code
(this can be found on the sticker on the cover of your book)

3. Explore the great Writer's Resources on offer!

Register with the site and you will have access to:

- The entire directory of *The Writer's Handbook* – accessible by browse and quick searches

- Extra resource material – this includes a list of useful websites, top tips and FAQs for Writers

- Book industry newsfeeds

- Publishing insights – includes a glossary of publishing and advice from an editor

- Extra articles from previous editions of *The Writer's Handbook* which offer a unique insight into how the book industry has changed over the years

- A blog written by industry insiders and creative writing gurus

www.thewritershandbook.com

THE
WRITER'S
HANDBOOK
2009

EDITOR

BARRY TURNER

MACMILLAN

First published 2008 by
MACMILLAN PUBLISHERS LTD
Houndmills, Basingstoke, Hampshire RG21 6XS and
175 Fifth Avenue, New York, N.Y. 10010
Companies and representatives throughout the world

ISBN-13: 978-0230-57323-9
ISBN-10: 0230-57323-1

Inclusion in *The Writer's Handbook* is entirely at the editor's discretion. Macmillan Publishers Ltd makes no
recommendation whatsoever by inclusion or omission of any agency, publisher or organization.

While every effort has been made to ensure all of the information contained in this publication is
correct and accurate, the publisher cannot accept any responsibility for any omission or errors that may
occur or for any consequences arising therefrom.

This book is printed on paper suitable for recycling and made from fully
managed and sustained forest sources. Logging, pulping and manufacturing
processes are expected to conform to the environmental regulations of the
country of origin.

A catalogue record for this book is available from the British Library.

A catalog record for this book is available from the Library of Congress.

10 9 8 7 6 5 4 3 2 1
17 16 15 14 13 12 11 10 09 08

Credits
Editor *Barry Turner*
Assistant Editor *Jill Fenner*
Editorial Assistant *Kenneth Hadley*
Poetry Editor *Chris Hamilton-Emery*
Contributors   *Kim Fletcher*
                Clare Hodder
                Nicola Monaghan
                Ali Muirden
                Jon Reed
Tax and Finance Adviser *Ian Spring*

Printed and bound in Great Britain by
Cromwell Press Ltd, Trowbridge, Wiltshire

www.writershandbook.com

Contents

vii Foreword
Ion Trewin

Articles

1 The publishing state we're in
Barry Turner

4 To speak on your behalf – Finding and working with agent
Nicola Monaghan

9 Sell yourself online – New ways of promoting your book
Jon Reed

14 Bringing in the money
Barry Turner

17 The sound of words – Audio books
Ali Muirden

20 Reflections on the lives of others – The British love affair with biography
Barry Turner

25 The re-engineering of UK Poetry Ltd
Chris Hamilton-Emery

34 The worst of times, the best of times – The fortunes of the freelance journalist
Kim Fletcher

39 Page to screen – The journey from novel to screenplay
Barry Turner

42 Getting it together – Johnson's dictionary
Barry Turner

44 English as Johnson understood it

46 Clearing permissions
Clare Hodder

52 Settling accounts – Tax and the writer
Ian Spring

Directory

64 UK Publishers

153 Irish Publishers

160 European Publishers

172 US Publishers

191 Other International Publishers

199 Poetry

199 *Poetry Presses*

208 *Poetry Magazines*

219 *Organizations of Interest to Poets*

225 Small Presses

238 Audio Books

242 UK Packagers

247 Electronic Publishing and Other Services

249 UK Literary Agents

274 UK Literary Scouts

276 PR Consultants

279 Irish Literary Agents

280 US Literary Agents

293 Book Clubs

294 Magazines

372 National Newspapers

379 Regional Newspapers

393 News Agencies

394 US Media Contacts in the UK

396 Television and Radio

413 European Television Companies

417 Film, TV and Radio Producers

438 Theatre Producers

453 UK and Irish Writers' Courses

469 US Writers' Courses

Includes FREE online access to **www.thewritershandbook.com**

477 Writers' Circles and Workshops

482 Bursaries, Fellowships and Grants

488 Prizes

526 Professional Associations and Societies

552 Literary Societies

572 Arts Councils and Regional Offices

574 Library Services

601 Picture Libraries

628 Press Cuttings Agencies

629 Festivals

642 Useful Websites

649 Miscellany

652 Index of Entries

686 Subject Index

Includes FREE online access to **www.thewritershandbook.com**

Foreword

Moving around, as I do, the offices of publishers, publicists, independent film companies and the like, the one book I notice most frequently on desks and obviously much used – its yellow cover unmistakeable – is *The Writer's Handbook*. I'm not just saying this because Barry Turner asked me to write a Foreword to the 2009 edition. What these sightings demonstrate is that even with the Internet, with Google and the rest, for authors, publishers, producers, etc, having one handy, chunky reference book containing everything you could possibly need to know, and updated annually, demonstrates yet again that the book as a means of purveying information still has an important, no vital, place in our lives.

I asked my assistant to keep a record for a week as to how she used her copy. Photographic libraries – she needed some ideas when searching for a picture; German publishing – she knew only that the firm was in Munich. Next day and closer to home – the name of Blackpool's evening newspaper? – it is the *Gazette*. Does Ireland have literary agents? – yes! Is there a book prize specializing in folklore? – by golly, there is: the Katharine Briggs award, worth £200. Finally an author wanted to identify a postgraduate course in playwriting – it took but a minute to find the full details of the University of Birmingham's course that has now been going for twenty years.

Writing in all its forms is unique to the individual concerned. Thus everyone who uses *The Writer's Handbook* does so in different ways. As I was about to sit down to write this piece my phone rang; it was a friend, an established biographer and novelist. He was calling from holiday in Ghent in Belgium and urgently needed help. His mobile had just pinged with a message from a British gardening magazine wanting a feature. The connection, however, had been lost; the recall button didn't help. I reached for that familiar yellow covered friend – it took but a minute to run through Britain's gardening magazines (I had no idea there were so many) before he and I identified the one that had probably called him.

Next day he rang back to thank me. The editor had told him that with time at a premium she had been on the verge of trying someone else, but my friend's prompt response meant that he now had a valuable commission. He wanted to thank me. I said, thank *The Writer's Handbook*.

Ion Trewin

Ion Trewin is the Administrator of the Man Booker Prizes, having been Editor in Chief of Weidenfeld & Nicolson and before that Literary Editor of The Times.

The publishing state we're in

Barry Turner

The book trade is a mystery wrapped in an enigma, as Churchill might have observed. Among publishers and booksellers, the mood can be downbeat, even gloomy. Yet there are more books being published than ever before, more books being sold than ever before and more, many more, aspiring authors than ever before. A YouGov poll revealed that 10 per cent of respondents put a literary career at the top of their list of chosen professions, way ahead of any other first choice. Oh yes, and there are more people reading than ever before, 17 per cent up in three decades. So what is the problem? The answer is, in one highly emotive word – change.

A changing industry

You might think that by now book people would be used to change. There has been enough of it in recent years, starting with the end of price fixing through to the boom in online sales. But all this has happened within familiar contours, with the book as a recognizably solid object which readers want to own and keep. Now we are moving into new territory. A computer literate generation is increasingly accustomed to reading on screen.

The Big Three

The digitization of libraries is well under way. Within ten years Google plans to offer up to thirty-two million titles on its database. Rivals Amazon and Microsoft are also digitizing hundreds of thousands of books. Bill Gates predicts that he will soon be able to offer students an affordable hand-held device containing all the books they need for any particular course. Eventually, readers will be able to follow the lead of music fans and simply digitize their own libraries. The revolution extends to everyday pleasure reading. In the next year or two hand-held electronic readers containing up to two hundred titles will be commonplace. Move on a short time for e-books to be available in a multiplicity of formats, from mobile phones and iPods to MP3 players.

A rapidly evolving market

Two questions immediately spring to mind. If all the world is going digital, what future is there for the printed book? And if the printed book is on the way out, how will anyone in the trade – publishers, booksellers and, not least, writers – make their living?

We start with a few words of comfort. History tells us that when a new medium of communication or entertainment comes along, however exciting and user friendly, it rarely, if ever, does away with older rivals. On the contrary, competition can stimulate a rejuvenating burst of creativity. It happened when live theatre was threatened by the movies and again when television led the assault on the cinema. Now, television is having to reinvent itself as computer games and other laptop diversions bid for audiences. The same could happen to the printed book. It all depends on how the book trade adapts to a rapidly evolving market.

Thinking digital

After a wobbly start when all things digital were seen as an enemy to be quashed before it became too powerful, it is beginning to dawn on publishers that the new economy is more opportunity than threat. It is still early days but publishers' websites are becoming increasingly sophisticated with specialized sites attracting like-minded readers and book buyers. Booksellers too are beginning to use the Net to discover niche markets.

Leading the way, it hardly needs saying, is Amazon. Is it really only five years since the received wisdom in the book trade was of the imminent collapse of Amazon? Now it is the third largest bookseller with 11 per cent of total sales. With consumers converting enthusiastically to online shopping, Amazon has the edge over the high street, not least in the range of titles on offer: 1.3 million as against, say, 50,000 in a large chain bookshop.

Authors are heard to complain that while online sales and aggressive marketing by the chains (three for the price of two and so on) are building revenue and profits, the emphasis on top titles by instantly recognizable names is to favour quantity over quality. A quick glance at front-counter displays of leading bookshops would seem to support this thesis. Dumb books for dumb people, might be the appropriate slogan for certain retailers.

New opportunities for new writers

But the fears of mid-list authors, those who stop well short of the bestseller list, need to be put into context. The mid-list consists of authors on the way down as well as authors on the way up. The loudest groans come from writers who find they are going out of fashion. To blame the Internet or the decline of the independent bookshop or the state of the weather is too often a face-saving excuse for something that was going to happen anyway. For talented newcomers who need to get a foothold in the market, the signs are more encouraging. One of the most impressive book trade statistics to appear recently is that a quarter of Amazon's sales come from outside its top 100,000 titles.

Changing tact

Those on the lower rungs can also take solace from the growing power of niche websites. As for the high street retailers, it is true that their marketing has been driven by bestsellers at the expense of medium sellers. But this may soon change as the chains recognize the error of trying to compete with the supermarkets by heavy discounting on lead titles. It is a battle they cannot win. But they can score profit points by catering for a more discriminating public who value informed and helpful sales staff offering a range of titles that give credence to basic intelligence. Waterstones.com guide on how to get published, its New Voices promotion and the renewed support for local publishing are all promising signs of the times.

An exciting time for publishing

But the biggest opportunity for high street booksellers to raise profits and their image is surely in the area of e-book retailing. Francis Bennett and Michael Holdsworth, whose Booksellers Association-sponsored report *Embracing the Digital Age*, was published last November, warns that the industry must prepare itself for the new or risk losing to an outside player 'who may be bigger than the industry itself'.

> **⌐⊸ Top tips**
>
> - Keep up to date with what's going on in the industry; after all, the more informed you are about the state of the industry, the more equipped you'll be to negotiate with your agent or publisher
> - Visit *The Writer's Handbook* website for popular industry-focused news feeds

Publishers are picking up speed in digitizing their back list and it will not be long before new titles are available in e-format simultaneously with hard print publication. Penguin and Pan, for example, have recently announced that they will be publishing some of their front list simultaneously within the next few months. Booksellers, as Bennett and Holdsworth acknowledge, have been much slower at coming into the game. They urge uniform standards for all forms of digital marketing. Says Holdsworth, 'We want a situation where someone can go into a bookshop with their Blackberry, ask for a title by a particular author and for the bookseller to be able to explain what is available and to go through the formats.'

This is the level of thinking that should excite authors, particularly those who are still to make their breakthrough. The publishers and booksellers who are unable to face up to change will go out of business. For the rest, the tantalizing prospect is of an expanding market for a wider range of titles. Writers should wish them every success.

To speak on your behalf

Nicola Monaghan advises on finding and working with an agent

The novice writer, sending out speculative proposals to agents from information in a book such as this, may often pick her targets quite randomly. In a talk I attended as a student, Carole Blake suggested that one of the reasons Blake Friedmann receive huge numbers of unsolicited submissions was their place alphabetically in the listings. For an unpublished writer, getting an agent, *any* agent, is often the only driving force, with many other considerations being ignored. I've heard a number of unpublished writers suggest that it might be precocious of them to think above and beyond this, that they would be satisfied if any industry professional took an interest in their work; but I believe the writer–agent relationship is a crucial one. If you are good enough to get an agent at all, then you really should think about how to go about getting the right agent.

The right agent

The right agent for you can be many things. Perhaps you wish to write a blockbuster thriller and get a huge advance, in which case you may want to know who the big hitters are. The best way to do this is by consulting the trade press: publications like *The Bookseller*, which report the significant deals. If you are a literary writer, you may feel that this isn't the way to go and look for someone who will build your career more gradually. Explosions onto the scene can be quite unsettling and lead to an unfavourable backlash towards the author. Bigger deals also often tie a writer into producing two or three books to deadline, pressure that can be a deadening force on one's creative energy.

The right fit is crucial

A step away from a pin in *The Writer's Handbook* is some research into who represents fiction similar to yours. There are many ways to find this information. Many of the agents in this book provide names of their clients, so it's possible to get a feel for the type of person they represent and imagine whether you would fit in. This may also feel precocious – adding yourself to a list of names that includes Margaret Atwood or the estate of T.S. Eliot – but I believe seeing yourself up there is part of the battle. Another place to look is the acknowledgements pages of books that you've loved, or that you feel have particular synergies with your work. Usually, a writer will thank at least her agent and editor and, indeed, you may well find these are the first two names mentioned.

Of course, this approach doesn't always target your work as appropriately as you might like. Often, an agent has clients on her books for historical reasons, or has inherited them from elsewhere, without necessarily having a real passion for the work that comes with them. Sometimes it's more random than that. A good while ago, I was writing edgy fantasy stories similar to those of a writer I admire, and I looked up his agent. When I approached him, it turned out this wasn't really his area of expertise. He felt it was easy to represent someone writing in this genre who had an established career, but that selling a first novel of this type, at that time, was close to impossible. The genre turned out to be a passing phase for me and, in fact, it was this very same agent who came to represent me when I changed tack and wrote *The Killing Jar*.

Getting noticed

Often, a large part of the challenge as a new writer is getting someone actually to read your work. To understand why this is, I think you need to see what's happening from an agent's point of view. Submissions come flying through the door at a rate of hundreds a month. It is absolutely impossible for an agent to read everything and, as a result, often the ways of working through unsolicited manuscripts are quite brutal. An agent once told me he rejects proposals straight away if they come with anything more than a simple rubber band for binding, or if there are spelling mistakes in the cover letter. Most of those left are then rejected after reading just one paragraph of the prose. Agents *are* looking for the next big thing. They're also looking for any excuse they can find to send back your manuscript so they can get on with their other work.

Professionalism

The way you present yourself is crucial. Always use a simple font, such as Arial, white A4 paper and double spacing. Don't add clips or pins and don't staple sheets together. Keep your cover letter simple. 'Here is my writing, I hope you like it' has always worked well for me. You can include a couple of sentences summarizing the kind of book you are proposing, how many words you have written, basic factual information, but be wary of adding much else. An agent doesn't want to know what your friends and family think about your writing, for example, although it may be worth mentioning if Jim Crace has seen and liked it. Enclose the sample of prose after the cover letter, not the synopsis. The latter are often rather boring documents, and it is notoriously difficult to produce a really good one. Besides, what an agent really wants to know is whether or not you can write well.

Making good contacts

Aside from presentation, there are ways to give your proposal a better chance of being read. One of the most effective is to know the agent you are sending it to or, at least, have had some kind of introduction. This may seem impossible if you live outside London and don't work in publishing, but it's not nearly as difficult as it sounds. I knew no one in the publishing industry when I was looking for an agent, but I was working in the City, and I knew one thing: contacts are everything. I came to the conclusion that my best strategy was to do a Creative Writing MA. I knew that it was likely to be good for my writing to follow a course and I hoped I would meet agents, or people who knew agents. I got much more out of the MA than I ever anticipated but, importantly, I did reach the networks I was hoping to. Soon, I found I was getting somewhere and being asked for full or partial manuscripts. Obviously, this is not the only route. There are industry fairs and book events. An increasing phenomenon is the writing conference, specifically set up for new writers to meet those with established careers and to seek advice, as well as for professional agents and editors.

Handling rejection and staying motivated

If you are rejected by an agent, try not to take it personally. It's not personal for her, it's business, and there is only so much time for reading submissions. One agent I know gets over 200 emails a day from his established clients, which doesn't leave much time for the slush pile. It also doesn't leave time for any kind of feedback loop. If you are rejected by someone, don't waste time asking why. The answer won't be forthcoming unless it was clear from the letter and you might just get your name noted and remembered for the wrong reasons. Do take encouraging rejections seriously, though. Agents do not have the time to write personalized letters or emails saying how good they thought the writing was, and how it somehow just fell short of what they needed, unless they really mean it.

> **⚷ Key advice**
>
> - Never use an agent who asks for money up front
> - Find an agent who represents writers similar to you, as she will have the best contacts and expertise
> - Persevere

Don't let rejections drag you down or cause you to give up either. Remember that picture of the busy agent reading just the first paragraph, in between responding to 200 emails, and send your work out again. If you've queried, say, 30 agencies and you are still getting the same response then you should question your writing, but it's not an agent's job to help you get it right. Take a course or join a writing group. There are people you can pay to read your work too. I've done this, more than once, and found it an invaluable help in working out the strengths and weaknesses of an entire manuscript.

Working with your agent

Editorial guidance

That said, if an agent feels there are enough positives in your manuscript, she may well be prepared to work with you on getting it into a publishable shape. That is certainly what happened to me. Your novel doesn't even need to be complete and, in fact, many agents prefer to work with new writers at an earlier stage. I had just ten thousand words written of *The Killing Jar* when I was signed up, and it was a great help for me to have editorial guidance as I put together the entire narrative. My agent taught me so much about structure and novel writing; things that I felt were terribly hard to learn from textbooks or teaching.

Agents with passion

The most important quality of an agent is an absolute passion for your work. If your agent isn't your biggest fan, perhaps aside from family members, then I'd suggest something is going wrong. In practice, this usually happens somewhat organically. The marketplace is not easy, especially for new writers. An agent has to invest her time in working with a new writer, with no promise of any kind of return unless the book gets sold. (An agent who asks you for any kind of payment up front should be avoided.) All this means it's highly unlikely you'll be taken on by anyone who doesn't have that passion for your work. If you want to find the perfect match, you can do much worse than follow the advice of Miss Snark, the scathing New York agent with an anonymous blog: *write well and query widely*.

The writer–agent relationship

So you find your dream agent and she is interested in your work. What happens next? How is she likely to contact you to tell you of her interest? Will she want to meet and discuss your work? The answers to these questions vary widely, depending on the agent, your work and your personal circumstances. My agent rang me, and he suggested we meet for lunch; but I know writers who have never met their agent, and plenty for whom the first point of contact and ongoing discussion for some time was by email. However, it is most likely that at some point you will meet. That first meeting can be stressful and slightly surreal. Suddenly the relationship has changed. There was a time when you would have mowed gardens or stacked shelves for the chance to meet her. Now you are her prospective client. Whilst there's still a need to impress with your manner and personality, you've done much of the work already or, at least, your writing has. Your prospective agent might even be slightly nervous of meeting you. You may be treated to lunch at somewhere rather nice. This switch in the power balance can be somewhat unsettling.

One of my students came up with the perfect analogy for this first meeting. '*So it's like a blind date then?*' And this applies in so many ways. This is an important point in the career of an aspiring writer, and yet there's little you can do to influence the outcome. Much is already clear from your writing and the discussions you've had beforehand. The rest is down to the way you come across in aspects that are outside your control, and there are times when the relationship simply doesn't work. As with a date, or with a job interview, or anything else really, I believe your very best bet is to be yourself. After all, if you are completely true to yourself and you don't get on, it doesn't bode well for an ongoing relationship. If you fake that first 'date' you'll be found out in the end; there's no chance of a happy 'marriage'.

Trusting your agent

Once you've established a relationship, there are no guarantees. I remember my agent telling me that things might not work out, despite the fact he was prepared to help and advise me as I wrote the novel; and I appreciated his honesty. Here is where there's a difficult line to tread. Your agent may well want various changes to the manuscript you submitted and you might not agree with all of them. Do you roll over and play dead, changing anything that's suggested to you as you strive for publication? Well, of course not. You need to keep your artistic integrity, remember what your work is about and make your own decisions about changes. But I would urge the beginning writer to listen very hard to anything the agent suggests. She has years of experience of the publishing industry and understands how books work, and what sells. You do not know better! In my experience, I have made one change in my entire career at the suggestion of an agent or editor that I have regretted. These people are professionals and are almost always right.

Getting published

You probably won't be given much in the way of promises, or paperwork, until you have a product that is ready to be submitted to editors. Given that agents need to have and keep the respect of the contacts they have at publishing houses, they will want something to be as good as it can before submission. When your agent is happy with your work, you will probably be given a short contract to sign. This isn't especially complicated, and it's easily severable by either party, but it is the basis of your relationship, outlining what she will do for you and how much you pay for that service. You will pay a commission, usually 15 per cent for home territories and 20 per cent for overseas, though it varies, and there is some talk of these figures rising in the future. You may also have to pay for photocopying, or purchase of copies of your books that your agent has to undertake in order to promote you.

The deal

So, let's say you've got this far. Your agent sends out your manuscript and gets you a deal. You probably don't give up your day job, although you might for a while. The editorial work begins again, this time in earnest, with an editor at the publishing house. Officially, this is the end of an agent's commitment to you. In practice, a good agent will stay involved. She may read updates of the manuscript and offer advice. She may help you in your battles over cover proofs; I have certainly called mine in when I've needed an extra voice. She may accompany you to marketing meetings at the publishing house, come up with ideas to help you promote yourself, give you pointers to setting up a website or blog. She will use her contacts and knowledge of the industry to help give your book an extra push. After all, it's in her interest too that it sells lots of copies. Of course, she will also attend any launch parties held in your honour, time and other commitments permitting.

A friend of mine, drunk at his own launch party, once told me that the only important thing about an agent was that she was someone you could go to lunch with. On reflection, I think he had something there. Your agent should be someone you can relax with and have fun, someone who treats you well. Ultimately, it works best if your agent is someone you can be friends with because, although this is a business relationship, writing is a very personal art. Emotions run deep over your work; I've heard a number of writers call a manuscript their 'baby'. And you wouldn't leave care of your baby to just anyone, would you?

Nicola Monaghan is the National Academy of Writing Fellow at Birmingham City University. Described as the rising young star of British mainstream fiction, she is the author of The Killing Jar, *published in 2006 and winner of the Authors' Club Best First Novel award and a Betty Trask Award. Her latest novel,* Starfishing, *was published in March 2008.*

Sell yourself online

Jon Reed explores new ways of promoting your book

The new marketing

Do you have a blog? No? Put this book down immediately and go and start one! Are you on Facebook or MySpace? Go and join now. Feeling more creative? How about starting a podcast, or creating a YouTube video?

When it comes to book promotion, it's never been so easy and so cost-effective to run your own online marketing campaign. Using social media – free tools used to create and share content online – you can connect with your readers and build an army of fans in no time. That's just as well, because the bad news is that most publishers are not going to do this for you.

In 2006, UK businesses spent 11.4 per cent of their advertising budgets online, according to the Internet Advertising Bureau. In publishing it was 2.4 per cent. Trade publishers spend only £33m on above-the-line advertising; and of that only £800,000 goes on online advertising (Steve Hatch, 'Communication Breakdown', *The Deal: The Official Magazine of the London Book Fair*, Spring 2008, pp. 6–10).

This is way behind where most consumers are. People are switching off from traditional marketing channels – TV, magazines, newspapers, radio – and turning on to the Internet. We're so used to buying books online with a couple of clicks, isn't that where they should be marketed too? Millions of us are on Facebook and MySpace, watching videos on YouTube, reading and writing blogs, and listening to podcasts. Instead of shouting marketing messages at people who aren't interested, we need to reach people who are already looking for what we have to offer. This is permission-based, opt-in, conversational marketing. It's a shift from megaphone marketing to Martini marketing: people are accessing content anytime, anyplace, anywhere – and that's where you need to be. Go where your market is.

This is the age of social media, where anyone with something to say, a little talent and an Internet connection can create and share content. Is this a threat to traditional media such as book publishing? Not at all. We still want to read books. It's how books are marketed that has to change. And you can be part of it. Word of mouth has always been hugely important in selling books – but you don't have to be Richard & Judy or Oprah to use it.

Steve Weber, author of *Plug Your Book*, says: 'Internet social networking has handed authors their most powerful tool since the invention of paper. In the Networked Age, the stock of gatekeepers is going down, and the power of authors and readers is soaring' (p. 16). Many authors are already ahead of publishers when it comes to online marketing. Authors have always had a role to play in marketing their own books, and the more successful ones are proactive about it. Because social media is a personal medium where authenticity matters, it is more effective coming from you than from your publisher anyway. But your publisher can – and should – help you do it.

The new tools

Even if you write in a niche area, social media gives you a global reach, and your niche suddenly becomes much larger: it's the 'Long Tail' of publishing. Here's how you can join in.

Blogging

A blog is a little like an online diary, with entries (or 'posts') in reverse chronological order. You can be up and running in minutes using free software such as WordPress or Blogger. You might want to start a blog that's built around you as an author; or around your book. If you choose the latter, make sure it has the same title as your book and is designed to match the jacket.

It does require some commitment to keep the conversation going and the content fresh. But it's cheap, it enhances your online presence, it builds loyalty among your readers, and it gives you a global reach. It's also an obvious medium for writers.

Blogs are great for search engines: Google loves sites with lots of incoming links and regularly updated content. And if you're writing about anything topical – such as a literary festival you attended, or a recent news story relevant to your subject – people are likely to be searching for keywords that will lead them to your blog – and your book.

If you don't want to start your own blog, you can also tap into the conversation by going on a blog tour. It's likely that there are existing blogs where there is a conversation taking place about your subject area or genre of writing. Engage with those communities. Comment on postings. Better yet, email the blogger and see if you can write a post yourself as a 'guest blogger'. Make sure you link back to your own website or blog.

Another method of tapping into the blogosphere is 'blogger outreach'. Your publisher could send review copies to specific, targeted blogs, in the hope that they'll mention the book. Some blogs attract very large readerships, and if people are talking about your book online, it also increases your search engine rankings. This is an approach that worked for Stormhoek wine – a little-known South African winery that got itself talked about, boosted its search engine rankings, and became successful by giving away bottles of wine to bloggers. Publishers are often wary of giving things away for free – but the marketing value can be enormous. Wine is a social object. So is a book.

What if you're an aspiring author who doesn't yet have a book to plug online? A blog is still useful for getting your writing out there. Some books even started as blogs. Most notoriously 'Belle de Jour: Diary of a London Call Girl', which became two books published by Orion and a UK TV series called *Secret Diary of a Call*

Girl starring Billie Piper. But you don't have to sell yourself to sell your book. In 2007 Judith O'Reilly started her blog 'Wife in the North' about downshifting from London to rural Northumberland. Six weeks later, she landed a £70,000 publishing deal with Penguin. These are the exceptions – but don't underestimate the power of blogging. Like Lily Allen on MySpace, it is possible to build up an audience online before breaking into traditional media.

Podcasting

A podcast is really just an audio blog – audio files that people can subscribe to. They can also be video files. The word is a combination of 'iPod' and 'broadcast' – though you don't need an iPod or other device to listen to a podcast. About half of podcast listeners listen from their computers. If you have something to say, or some interesting people to interview about your subject matter, you might consider a podcast. You may even want to produce a series of audio extracts from your book, in discussion with your publisher. You don't have to do an episode a week for ever; you could also do a limited run podcast of 4–6 episodes leading up to the publication of your book.

Your publisher might be able to let you have a hand-held recording device to start your own podcast, and do some post-production for you. You can also download free audio-editing software such as Audacity (http://audacity.sourceforge.net) and do it yourself. Steve Parks, author of *How to be an Entrepreneur* among other books for budding entrepreneurs, is one author who has used podcasting successfully, and has created an online community, 'Flying Startups', around his books. He involves his community in the podcast by interviewing them. You could also include voicemail comments from listeners in your podcast.

Video

Online video is seen by some as the next big thing. With a digital video camera, some basic editing skills and software, and a free account on the video sharing site YouTube, you can have a more direct connection with your readers. You may not top the charts on YouTube – although Richard Wiseman, author of *Quirkology,* did, with his 'colour changing card trick' video; but you will engage your readers and provide another way for them to find you. You should, of course, use your video on your own blog or website.

Social networking

Online social networking sites such as Facebook, MySpace and Bebo have really taken off in the last year. I would recommend having a personal profile rather than one for your book, though you might want to use your book jacket as your profile picture, especially in the first month or so after it's published. But I think profiles are for people, and groups are for books or products. On Facebook, you can use a group to create a fan club for your book (or your blog or podcast) and as a way to stay in touch with readers. You can start discussion forums, post links, pictures and videos, and send messages to members. It's a ready-made, opted-in mailing list that you can use to stay in touch with your community. And everyone on the list has sought you out and chosen to be there.

Social bookmarking

Have you seen links on some blogs to Digg, StumbleUpon, Reddit or del.icio.us? These are social bookmarking sites, where readers can save links or highlight articles they like, and share them with like-minded people. Digg is, essentially, an online newspaper where the front page is chosen by its readers, according to the popularity of articles. You can also add links to 'Share on Facebook', 'email this' and so on to help get the word out.

Virtual worlds

Virtual worlds such as Second Life (SL) are harder to get into, but have their rewards. SL is, in some ways, just another social network – but you have to learn how to use the 3D graphical interface: how to walk, talk and fly. It looks like a video game, but isn't. The 'characters' are in fact 'avatars': graphical representations of real people, logging on from around the world. There are shops, businesses and universities. Sweden has an embassy. There is a real economy: Linden dollars, the in-world currency, can be exchanged for US dollars.

SL has communities for writers and would-be writers, and plenty of book-related activities. There are bookshops, literary magazines and workshops. A few major publishers, including Random House, have their own islands where they host book launches and author readings. Many more small independent publishers, including Cinnamon Press, Bluechrome and Snowbooks, cluster around Book Island, Publishing Island and Cookie Island. The first Second Life Book Fair (SLBF) took place in 2007. SL's own in-world TV channel, the Second Life Cable Network (SLCN), has a regular chat show with author avatars, *Meet the Author*, hosted by Adele Ward (Jilly Kidd in SL), one of the SLBF organizers.

Getting started

- Familiarize yourself with the new online marketing tools. Read blogs, listen to podcasts, watch videos and join social networking sites such as Facebook and MySpace.
- Start a blog, get your publisher to start one for you, use 'blogger outreach', or go on a blog tour. Make sure your publisher links to your blog from their website.
- Include a sales link on your own blog or website. Use an Amazon affiliate account and/or link to the ordering page on your publisher's site. Many of your other social media activities will be aimed at getting people to visit your website and this link.
- Think about starting a podcast or recording a video. Can your publisher help with equipment or production?
- Include links to social bookmarking sites underneath each of your blog postings, such as 'Digg this', 'add to del.icio.us' and 'Share on Facebook'.
- Don't go for the hard sell. Engage your readers with useful, interesting content on your blog, podcast, video or Facebook group. Think 'infotainment' rather than 'infomercial'.
- Include calls to action such as 'visit my website', 'sign up to my mailing list',

'get a free copy of my book by ...'

- Measure your results. Track the numbers of readers, listeners, viewers or group members, and how many people 'click through' to order your book.
- Involve your publisher, and get them to help.
- Have a go – and have fun!

You are the brand

Finally, don't forget that you, the author, are the brand. No one really cares who the publisher is. Marketing guru and prolific author Seth Godin, writing on his own blog, says: 'Authors are brands. Some are billion-dollar brands, some are tiny ones. The web is custom made for authors but, so far, it's largely going unused' (http://sethgodin.typepad.com).

Help your book stand out from the crowd by using the Internet. It's the ultimate in word-of-mouth marketing. Join the conversation. Start a blog. Become a brand.

Jon Reed is a publishing consultant who specializes in social media. He previously worked in publishing for ten years, including as publishing director for McGraw-Hill. His blog and podcast can be found at www.publishingtalk.eu

References

Steve Parks, *How to be an Entrepreneur.* Prentice Hall, 2006.
Steve Weber, *Plug Your Book.* Weber Books, 2007.
Richard Wiseman, *Quirkology.* Pan Macmillan, 2007.

Bringing in the money

Dear Editor

With my first novel accepted for publication, the future looks promising. But I am realistic enough to know that it will be some time before I earn a decent living from my fiction. What can I do meanwhile to keep the bank manager happy?

I assume that you are looking for ways to make money that involve writing. But while it is self-evidently true that the more writing you do the better, just bear in mind that exclusive attention to the muse is not the sole requirement for literary excellence. Many successful authors have found inspiration and a living wage in jobs that have no obvious connection with writing.

A recent survey on earnings from published work is the giveaway here. Outside the golden realms of the bestseller, the average writer's income is a mere £4,000. But this is not to say that they are suffering abject poverty. Rather, it is an indication of the number of part-timers in the writing trade. I'll come back to that after examining the alternatives to a 'proper' job.

Keeping the wheels of industry turning

It is one of the clichés of capitalism that it would operate more efficiently if business people found it easier to communicate, that is, to write and speak plain English. As practised in company reports, training manuals and customer service brochures, Business English is made impenetrable by convoluted sentences, an overload of jargon and a failure to understand basic grammar. You, the professional writer, can do something about this.

Starting locally, check out those companies that need help and can afford to pay for it. Then write to a senior executive offering your services. I know of one young author who got into – and much enjoyed – speech writing and another who was hired to write a company history.

Putting a life into words

Celebrities who have made a lot of money inevitably reach a stage in their career when they want to put their name to a book. If they are not themselves fluent writers, and few are, they or their publisher will be on the lookout for a ghost-writer, one who can transform a motley collection of memories and experiences into a sellable product.

Never underestimate the challenge. To write convincingly, the ghost must be able to adopt the personality of the subject. This calls for the skills of an actor as much as for the skills of an author.

Don't expect much in the way of a credit. For one of my early ghosting jobs, I was reduced to a mention in the copyright line. But if you are prepared to be effacing – and remember to check the small print in the contract (a share of any subsidiary rights should come your way) – ghosting can bring substantial rewards.

Delving into the archives

The stars of the book chains, in fiction and non-fiction, are under pressure to deliver more of the same, preferably at an accelerating rate. One way for them to save time for writing is to engage researchers to work the libraries or the Internet for essential information. The strongest demand is at the popular end of non-fiction – few well-known historians are without backup from one or more fact finders – but novelists too are hungry for the sort of knowledge that adds authority to their stories.

Cutting it short

There is a market for freelance editors who can reduce overweight volumes to manageable proportions. Shorter versions of the classics are popular with young readers, and many talking books are abridgements of the original texts.

Crossing the linguistic boundary

Are you fluent in a second language? In a country notorious for its linguistic limitations, the ability to master another tongue, combined with a sensitivity to the nuances of language, adds up to a market value for imaginative translation.

Publishers have been slow to adapt. Traditionally underplaying continental literature, they have only recently come to realize that good stories travel when the translator can match the creativity of the original author. But take heart. The day of the supposedly bilingual wordsmith, working laboriously with two dictionaries, is thankfully over.

Breaking into journalism

Newcomers tend to get overexcited by the prospect of seeing their bylines splashed across the feature pages of the national press. It all looks so easy. As newspapers get fatter someone has to find the words to fill the extra pages. And then there are all those magazines cramming the newsagents' shelves.

But take heed. Editors work to a frenetic schedule, which is why they are inclined to stick with the freelancers they know and trust to deliver the goods. They rarely have the time or the inclination to nurture unproven talent. Specialist magazines welcome relevant expertise, but they are likely to assume they are doing you a favour by showcasing your work. Your notion that they might pay, let alone pay adequately, for your contribution may be treated by them as an impertinence.

That said, if you are well informed on a subject of interest to the general reader and able to write to fit the space (it is amazing how many authors find it near impossible to restrict themselves to a thousand or fifteen hundred words) you are in with a chance. Explore your own life for triggers to set you off on popular topics. Parenting and personal relationships are a good place to start. Just don't rely on freelance journalism to provide a regular income.

Spoken word

If you enjoy appearing in front of an audience, you might offer yourself as an occasional speaker (for a modest fee, of course) to writing groups or to luncheon clubs with a literary bent. You have only to go down well once or twice for word of mouth recommendations to get around. The invitations to repeat your performance will then start rolling in. Think of it as a trial run for all those festival appearances you will be asked to make once your name is really well known.

State aid

Check out the bursaries, grants and other support services available from such bodies as the Arts Council. There is a full list on pages 572–3.

The stark alternatives

Writers, like actors and others in irregular employment, may have no choice but to adopt a secondary occupation, if only for the short term. But, however unappealing it is to be diverted from the prime objective, it is surely comforting to know that all experience is grist to the literary mill. One author of my acquaintance was the only applicant for the job of nightwatchman on a Thames barge. He made good money, enjoyed perfect quiet and, supported by an inexhaustible supply of teabags, was able to compose a series of river adventures for young readers.

On a more conventional note, part-time teaching finds favour, not least because it affords long breaks which can be put to creative use. Working as an assistant in a bookshop or, better still, as a publisher's rep concentrates the mind on what can or cannot sell in the high street. The experience could save you a lot of heartache on the way to bestsellerdom.

And think of all the interesting characters you will meet. As life studies they have more reader value than the introspection of writers exclusively preoccupied with their trade. Writing about the problems of being a writer, the only subject they know about, has strictly limited appeal. E.M. Forster may have been overstating it when he said that a writer should live and read for thirty years, then stop both and start writing. But far better that than the reverse.

Barry Turner

The sound of words

Audio books are entering a new age. Ali Muirden,
Macmillan Digital Audio Publisher, explains

There has been a brisk and rather nippy 'wind of change' in the world of audio books in the last few years and the reason for this exciting new era can be summed up in that little word of three syllables: digital.

When I first started publishing audio books in 1999 the big debate back then was when would publishers move from cassette to CD and possibly to the CD's even trendier and sexier cousin the 'MP3 CD' format? We dragged our feet and hedged our bets and most of us missed out on the MP3 CD moment completely. Well, actually the MP3 CD moment never really got off the starting blocks due to the fact that nobody had an MP3 CD player in their car! In fact, it's true to say that such a large proportion of UK cars still only have a cassette player that you can now purchase iPod cassette adaptors.

The digital revolution

Despite this, in 2006 the cassette market finally bit the dust and the CD reigned as the 'new kid on the block' for what seemed like all of ... oh! ... at least ten minutes! Then the 'Digital Revolution' happened, and for some people new to this game it's fair to say that it might feel like it simply arrived overnight and hit the ground running; but actually this is not the case at all.

I can remember meetings with various companies in the early 2000s to discuss the burgeoning download market and what implications this would have for audio book publishers. However, there were a variety of issues that held things up. First, we had no clauses in the majority of our contracts with our authors to cover the sales of the format of 'download'. In fact, in some cases the contracts still mentioned 'vinyl'! Secondly, the quality of sound from MP3 players was at that time markedly inferior. Most of the players sounded just like one of those old transistor radios I'd had as a kid in the '70s, with that crackly, tinny sound so reminiscent of listening to the top twenty countdown every Sunday night. Thirdly, most of them had a storage capacity that was very limited. Finally, the ability to actually download anything was not particularly easy either.

So everyone carried on arguing about the merits of cassette versus CD and completely ignored what was going on in that funny, nerdy world of 'digital', with the idea in the back of our minds that when they improved the sound quality and people actually started buying MP3 players we would 'get back to them on it'.

Everything changed

And then Apple launched the iPod on 23 October 2001 and everything changed. As early as 1997 a company called Audible.com had been launched, and in 2003

Apple and Audible joined as partners enabling Audible to offer audio book content via the lucrative and wide-ranging iTunes website. This proved to be the crucial step that finally saw digital downloads of audio books take off. In 2005, Audible launched Audible Air – software that makes it possible to download audio books over the air – wirelessly and directly to devices such as a smartphone or a PDA (personal digital assistant).

At around this same time other websites such as Spoken Network, Audioville and many others were launching sites and opening up the competition for the market. We watched with interest to see what Amazon had in the pipeline. They announced plans in June 2005 to open their own company to supply download audio book content, but then it all went very quiet for a few years. Then, at the end of January 2008, they announced that they had bought Audible which would continue to be run as an independent company under the leadership of Don Katz.

New technologies

Another potential market beginning to emerge is coming from the direction of the mobile phone industry. Nokia have recently released a piece of beta software called Nokia Audiobooks. The application is essentially a compression tool optimized for voice, dubbed 'AMR-WB' (Adaptive Multi Rate - Wideband – a speech coding standard which provides excellent speech quality due to a wider speech bandwidth of 50–7,000 Hz). The package includes the Nokia Audiobook Manager for compressing the audio on your PC, and the Nokia Audiobook Player for listening to audio books on your handset.

As a standard audio book can have a running time from anywhere between 3 and 20 hours long, the size of the audio file after it has been compressed into MP3 format is still prohibitively large for most handsets. The AMR-WB codec reduces the audio down by about five to ten times more than MP3 does, thus making audio books a real possibility. It is technology such as this which will make listening to an audio book on your phone much more feasible.

New innovations

Mobile phone companies are increasingly approaching audio book publishers for content to offer their customers. Sony launched their Sony Ericsson K750i with a free copy of the *Da Vinci Code* audio book on a 256MB memory stick. They also had a fun competition on their website, which involved solving a series of puzzles against the clock, based on the best-seller by Dan Brown. If you were lucky enough to get through to the final round you had the chance to win a trip to Paris. This is an interesting and innovative piece of cross-promotion between book publisher and a technology company, and is indicative of the way in which we can work together for mutual benefit in this new world of ours.

The state of the audio market

A recent survey by the US Audio Publishing Association says that audio book turnover there is calculated at £923 million retail value. The US has been selling digital downloads of audio books for some years now and their download turnover is reported to be 10 per cent of their total figure.

Here in the UK recent figures compiled by the UK Audio Publishers Association confirm that the total audio book turnover in 2006/07 was worth £72.4 million at retail value, with net download sales estimated at approximately £1.1 million for the same period.

It is too early for audio publishers in the UK to say with confidence where this market is heading, but we have seen strong and speedy growth of download sales for our products since Audible launched its UK site in June 2005 and since the launch of other UK based sites such as Spoken Network and Audioville.

⌐○ Popular audio genres

1. Crime and thriller
2. Popular fiction
3. Classical fiction
4. Radio comedy

Suddenly audio books are hip! No longer are we labelled the poor relations of the book publishing world. The download revolution is enabling us to increase our output of audio products, which can only be of benefit to authors, agents, actors, publishers and, most of all and more importantly, the many audio book fans out there who find it so hard to find a comprehensive range of products in the high street. Instead of only having access to a few hundred titles, suddenly they have access to hundreds of thousands. And I can't resist once again quoting the late, great Harold Macmillan: 'We've never had it so good! Long may it continue!'

Ali Muirden is Audio Publisher at Macmillan Digital Audio and is Chair of the Audiobook Publishing Association.

Reflections on the lives of others

The British love affair with biography shows no sign of weakening.
Barry Turner investigates

A passion for biography is peculiarly British. Our American and European cousins recognize the urge but nowhere else is the addiction quite so strong. Every bookshop and library has shelves stuffed with life stories ranging from the ghosted ephemera of current celebrities to the heavy tomes on the men and women who have changed history. It is a sobering, some might say a chilling, thought that the *Times Literary Supplement* gives more space to new biographies than to new fiction.

The literary obsession

At the core of this literary obsession is a tradition trailing back to the most successful biography of all time – the life of Jesus Christ by four of his disciples. Thousands of sequels have appeared since Matthew, Mark, Luke and John went on record. It is reckoned that in the last two hundred years, Christ has inspired over 60,000 biographies, a rate of production that may even be accelerating. There is a story told of A.N. Wilson, while trawling the London Library for a biographical subject, finding that the most written about personalities were Lord Byron and Jesus Christ. Knowing that popularity breeds popularity, he opted for the Saviour and joined the elect of best-sellers.

The Christian heritage cannot alone explain Britain's love affair with biography. But the Protestant ethic imposed on the individual to pursue his own destiny helps to narrow the inquiry, as too does the still widely accepted educational dictum set by Thomas Carlyle that history is 'nothing but the biographies of great men.'

A British strength

Victoria Glendinning, whose inquiries into private lives include Edith Sitwell, Rebecca West, Anthony Trollope and Leonard Woolf, believes that we tell stories about people from the past to compensate for the absence of folk literature. 'The instinct behind it is to hold the past in memory, not only for its own sake but in order to inspire, warn, structure or otherwise illuminate or manipulate the present.' Like folk literature, the best biographies are powerfully imaginative. To interpret the significance of a life in popular format calls on the skills of a novelist. It is surely no coincidence that the British strength in biography emerges from a rich culture of story telling.

There is a distinct correspondence between Peter Ackroyd's novel *Hawksmoor* and his biography of Charles Dickens. Both take an interpretative leap from the known facts with, in the first case, invented and, in the second, guessed at dialogue and incidents. Both are powerful books leaving the reader with a clearer and more memorable impression of the characters portrayed than any conventional biography.

A difficult genre to master

Many writers and critics vigorously contest the overlap between fact and fiction. Virginia Woolf warned against mixing 'the truth of real life and the truth of fiction' arguing that 'the imagination will not serve two masters simultaneously'. The purpose of biography is to establish truth while fiction, whatever its underlying aim, can fly unhindered into the realms of fantasy.

But where is the literary paragon whose perception is so acute as to fix on the absolute truth about a person? All source material – diaries, letters, memories of friends and enemies – is to some degree suspect. Because nobody tells the truth about themselves, what is said of them by their peers, however close the relationship, is inevitably subjective.

Biographers themselves connive at distortion when they write about characters they admire or have come to admire. To embark on such a life with all the necessary enthusiasm and commitment is to fall in love a little with the subject. It is then that self-censorship can stop the writer short of telling all that he knows. There may be a tacit understanding with the subject's family that some things are best kept hidden. More insidiously and often unconsciously, a reputation is enhanced by ignoring or sidelining rival claims to distinction. Show me an actor's biography which praises rival performances.

The changing shape of the biography

Then again, what we define as biographical truth has shifted dramatically over the generations. Long after Dr Johnson called for robust biographies 'which tell not how any man became great, but how he was made happy; not how he lost the favour of his prince, but how he became discontented with himself', a formula faithfully adopted by his amanuensis, the preferred style of biography was to praise all that was decent and respectable, an example to lesser beings, while ignoring the character blemishes that might detract from the core sermon. On dusty shelves, the weighty volumes commemorating Victorian worthies are still to be found, unread then, unreadable now.

The mould was broken by Lytton Strachey with his iconoclastic study of *Eminent Victorians*. Read today Strachey's disclosures are mild bordering on anodyne. But such was his impact that, with him, biography entered a new age. Honest appraisal, however, still had its limits. Strachey could not introduce his own sexuality into open discussion. Indeed sex, other than its most conventional form, was strictly out of biographical bounds until well into the twentieth century. By then the fashion for psychiatric analysis, pioneered by Freud in his clinical studies, had reshaped the common understanding of character and personality. In search of the inner self of his subject, no biographer could any longer afford to avert his eyes from what went on in the bedroom.

A thoroughly modern biography

A question now is how far the search for truth is used merely as an excuse for titillation. Though a knowledge of sexual inclinations can illuminate a life, truth can be the first casualty when salacious revelations are made the chief selling point of a biography. It is all a matter of degree and relevance. There can be no doubt that the literary reputation of Vita Sackville-West was given a lift by a biography which highlighted her unconventional sex life. And it took Fiona MacCarthy's life of Eric Gill to bring about a serious reappraisal of the artist's work even if, at the same time, he acquired notoriety for having had sex with his daughters and his dog.

But lofty motives are often hard to detect. One way or another, sex sells. In her recently published confession *Shoot the Widow*, the American biographer, Meryle Secrest, who has written nine artistic lives ranging from Kenneth Clark to Salvador Dali, readily concedes that 'prurience titillates, the more the better, leading to bigger sales and better royalties for the writer who is, not to put too fine a point on it, making money from others' misfortunes'.

With refreshing honesty, Meryle Secrest admits to a purely commercial approach to biography. 'Deciding on a subject is mostly a cold-blooded business of weighing the subject against potential markets, timelines, the availability of material and the likelihood of getting the story.' In other words, what readers want, the readers get. Even those not driven by prurience find comfort, as David Ellis (*Literary Lives*) puts it, 'in believing that those who have achieved distinction in a certain field are not merely as humanly fallible as they are, but much more so'.

A story with personality

How far, then, can biography justify its popularity? At the top end of the market heavyweight biographies of cultural celebrities can so alter our perceptions that we end up knowing more about the personalities than the work they created. To choose names at random, the Bloomsbury Group (including Lytton Strachey), Trollope, Hemingway, Orwell, Sylvia Plath, even Dickens, are almost certainly more read about than read.

But that is better than nothing. A reviewer of *Henry James Goes to Paris* comments that 'its biographical element is clearly designed to win a larger audience than a book of criticism on James might expect to find', adding that the author, Peter Brooks, believes that biography is 'one of the few forms that a literary critic can use to reach a large audience'.

This is equally true of historians. Much in our island story is made more palatable by giving it personality. Best of all, from the booksellers' point of view, is when a famous name is attached to another famous name – Roy Jenkins on Gladstone, Asquith and Churchill; William Hague on Pitt and Wilberforce; Douglas Hurd on Peel.

A rather loftier justification is offered by Richard Holmes, chronicler of Shelley, Johnson and Coleridge, who sees biography as shaping self-awareness. 'In studying the lives of others, we slowly learn the truth about ourselves. As Plutarch said, biography is a mirror to the age. We should be able to look into it freely and without fear.'

Start notes for biographers

- For the beginner, launching from the top with a big name biography is unlikely unless (a) you are related or have a close association with the subject, (b) you have access to a cache of letters and papers no one else has managed to find, (c) you can boast the reputation of an expert on the life and times of the subject, (d) you yourself are so well known that any publisher will be glad to enlist your marketing value.
- An email to a front line publisher suggesting a biography of Tony Blair on the basis of meeting him a couple of times on the Sedgefield train will not get you very far.
- For those looking for personalities who might qualify for a biography, local and regional archives can provide rich pickings. Historical figures who are not themselves world beaters but lived in interesting times may be fascinating in their own right and attract a wide readership. Innovative scientists and entrepreneurs and forgotten artists and writers (not necessarily mediocre, just neglected) fall into this category.
- Achievers who make no claim to writing ability are in the market for ghost biographers. This requires the skill of putting yourself in the place of the subject. Since these biographies, mostly of actors and sports personalities, are written in the first person (it is the subject who is supposed to be telling the story) the ghostwriter needs to adopt another voice. It can get spooky.
- A biography does not have to be of a person. The genre now embraces famous buildings; works of art; clubs and societies; towns, cities, villages and rivers.
- Source materials should be read with a cynical eye. Politicians are inclined to write letters and diaries that favour their record. Interviews rarely yield the whole truth. If in doubt, ask the same questions over again and compare answers. Elderly people muddle dates. Celebrities are renowned for their selective memories.

Biography as gossip

This leaves a thick layer of biography that makes no claim to serious intent. The preoccupation with celebrity, fired by news media in search of spacefillers, has spun off into books with the lives of famous people yearning for greater fame on the strength of being famous. In what has been called 'biography as gossip' (think of *Hello!* and *OK!* articles between hard covers) little if anything of consequence is disclosed. Most books on the royals fall into this category along with ghost-written quickies on current sports stars and front cover beauties. None of them do any harm except to the publishers who pay over the odds for potential bestsellers which can, all too quickly, wind up on the remainder shelves.

Mis lit

At the other extreme are the misery memoirs, invariably featuring child abuse. Why anyone should want volumes of human suffering, let alone pay good money for them, is hard to understand. But there it is. Mis lit accounts of horrible childhoods scored eleven places in the 100 best-selling paperbacks last year. One bookshop has an entire section devoted to 'Painful Lives'. Of all types of biography, this is where the overlap between truth and fiction is most apparent. The temptation to overdo the agony for the sake of higher sales must be overwhelming. At least we can assume this from the frequent disputes over the veracity of misery memoirs. You don't have to be a trained psychiatrist to detect the Freudian slips.

Biography will hold its appeal and will for ever be controversial. Skipping over the rubbish there are those who dismiss the study of individual lives because it fails to yield knowledge that enriches the imagination. A.S. Byatt sees biography as essentially backward looking, more an exercise in nostalgia than a force for creativity. The risk, she warns, is that we substitute biography for thought.

Barry Turner has written biographies of actors Richard Burton, John Le Mesurier and Denholm Elliott, of theatrical producer Bernard Delfont and, most challenging of all, of the redoubtable Elaine Blond, the youngest daughter of the Marks of Marks & Spencer.

References

Peter Brooks, *Henry James Goes to Paris*. Princeton University Press, 2007.
David Ellis, *Literary Lives*. Routledge, 2002.
Nigel Hamilton, *Biography: A Brief History*. Harvard University Press, 2007.
Meryle Secrest, *Shoot the Widow*. Knopf, 2007.

The re-engineering of UK Poetry Ltd

Chris Hamilton-Emery

I'd better say from the outset that this article will differ from those in previous editions of *The Writer's Handbook*, commenting as it does on the news at the beginning of 2008, and perhaps differing in other ways from what might normally be expected. First, it doesn't provide advice on how or where to get published; I reasoned that there is already plenty of advice from organizations listed in these pages, on websites and from other pundits, which will help the neophyte writer to find publication, or abandon their desire in favour of accountancy or plumbing or arts administration. Secondly, it's not meant to provide a lengthy overview of the range, diversity or recent achievements of UK publishers (scarily, the list of poetry publishers remains one of the most astonishingly large sections in the *Handbook*). There is no shortage of outlets for writers and, like Moore's law on computing power, the number of writers seeking publication doubles every two years. Finally, this article isn't about the decline of the art in the face of injudicious funding cuts and contraction in the market. I ought to make it clear from the outset that more poetry books are being published than ever before, funding has increased, and more poetry is being sold. Poetry is booming; at least for some. What distinguishes the successful poetry publisher from the unsuccessful is a concentration on selling; no poetry book sells itself.

In this article I will look at publicity and some techniques for marketing writers and writing. Sweeping generalizations will sit beside specific advice, so be prepared. I'll characterize this as a discussion of 'satellites and hubs', which might sound arcane but is actually quite a transparent way of looking at how the Web works in pushing and pulling traffic to points of interest. Then I will survey the substantial changes in the landscape of state-funded poetry and consider its future.

What makes poetry sell?

In a word, *publicity*. Given that the world of poetry is not a meritocracy (if it ever was) and that the best poetry does not necessarily succeed in today's markets (if it ever did), I'd suggest that the largest single driver for sales is publicity. How one acquires publicity is an important matter. Some poets are married to publicists. Some hire them. However one begins building one's profile within the media, an important bit of advice is to be nice to work with: be supportive, adaptive, welcoming, productive, attentive but never noisy. Be a good colleague. It sounds trite, but nice works.

The role of publicity

The role of publicity is to draw the attention of the world of media to the poet, and people like working with people who make life easy, people who get things done, people who enable rather than disable processes, events and products. The range of available media has expanded considerably with the growth of the Web. Make no mistake that public profile, driven by media involvement, is the most secure way to establish long-term writing reputations. What publicity helps to do is confirm in the minds of the book buyer that a writer is a choice, by which I mean that publishers are primarily involved in constructing choices for consumers, and that choices are as much about limiting the range of options available to a book buyer as they are about establishing a unique value offering. Put in other words, publicity can effectively block out the competition.

Literary relationships matter

Perhaps the quickest route to building such a profile is the literary friendship; the literary world is small, it has great longevity (accompanied by a long, long memory). Word gets around fast. As in all walks of life, relationships matter, and *who* you know is as important as what you know in the world of literature. So, line up your ducks and build up those literary friendships. Genuine ones work best. If you want to be read, you should spend as much time on this as you do on reading and writing. Why? Because word of mouth within the literary community is the most effective way to secure contracts, find work and build readers.

Perhaps the easiest way to do this is to choose your mentor (or tutor) from the growing number of creative writing degrees, and find a university place which you can use as a springboard. Though remember that the quality of the teaching is less important than who it is you can gain access to. Choosing to pay a poet for private mentoring is equally effective, and may serve you better in terms of actually improving your writing *and* opening doors on the world of poetry.

'Hubs' and 'satellites' on the Web

As well as concentrating on the human relationships necessary to develop your writing profile, the most powerful ally you have in developing a readership remains the Web. Let's look at two things: pushing and pulling traffic towards you. As a writer you are looking to improve your 'discoverability'. Here is a simple tick list you should have:

- A personal website
- A Wikipedia entry
- One or more blogs (each registered with Technorati)
- A MySpace profile (and make use of the MySpace blogging tools)
- A Facebook profile
- A YouTube account
- Be a contributing member of poetry bulletin boards and LISTSERVs

These seven things really do work. The more time you can spend on your blog and networking, the better. It might sound rather artificial and many will try to tell you that only writing counts; frankly, this isn't true. There are plenty of unread great books. Success in literature is far more complex than writing good works. In our culture it's not unknown for *terribly* written books to be huge commercial successes.

A personal website is a hub: it is largely static and passive in nature. Your job is to make it both content- and feature-rich. The best way to achieve this is to write an exemplary piece of content, something which has considerable equity within a subject. It can be about anything, as long as it's exceptionally good. This piece of content can provide the hub of your site, it may become a main point of entry for visitors to the site, driven by links and indexing within search engines. Search engines discover content based on a hierarchy of information, so make sure that key words are placed within <H1> and <H2> HTML tags. From these access points you may design routes to other parts of your site and you should plan how to navigate people there, but one solid piece of content can attract tens of thousands of visitors. This content *pulls* traffic to you.

> **Top tip**
> • See Jon Reed's article on how to promote yourself online (page 9) for more ideas

The other points in the list above are all 'satellites': their role is to *push* traffic your way by linking to key parts of your website – they are remote places on the Web designed to raise awareness about you and your writing with the sole intention of driving traffic towards you (or to your preferred destination, which may be an Amazon transaction page). You should endeavour to multiply the number of satellites linking to your home page.

Product linking or stealing traffic the easy way

Here's a little insight for publishers now: trust me, this advice is worth its weight in gold. The more money spent on publicity, the more traffic builds on the Web; there is a direct correlation. That traffic swimming around an author or title can be utilized to drive consumers in altogether different directions: watch carefully what your competitors are spending their money on and steal their traffic. Here are three tips:

• Use Google AdWords to build time-bound targeted campaigns which link to your competitor's highlight titles. Use the ads to drive traffic towards your own highlight authors. For example, Dewsbury Publishing are spending £10,000 on the publicity for Gwendoline Frisby's debut *Eye Sauce*. There's a lot of talk about Gwendoline and the chatter is driving interest on the Web and lots of gossip on the blogs and bulletin boards. An AdWord campaign using Gwendoline's name and the title of her book can be used to announce your own highlight title, Deirdre Benn's *Sunlight of my Mind*. The more money Dewsbury spend on Gwendoline, the more traffic will be created, and for a lot less than £10,000 you can filter off that traffic and remind people of your own books. Every time someone searches on Gwendoline, they find out about Deirdre. Better still, link your product to your competitors' current top ten titles.

- Use Amazon to sell proactively. Listmania is a terrific feature in Amazon: simply create lists which feature one of your competitor's lead titles and make the rest of the list your own books. Every time someone searches on the leading title, they're presented with lists that highlight your own books, too. Both these techniques are parasitic in nature, but used carefully and strategically, you can push traffic towards your author; traffic that has been financed by your competitors.

- Many publishers also use polling and reviews to redirect customers to other products. We trust customer reviews more than publisher quotes, so the effective use of commissioned Amazon reviews can steer and steal traffic.

Pressuring the state publishers

Christina Patterson has argued that state funding doesn't guarantee excellence (*Independent*, 10 October 2007). I agree with her and go further in suggesting that innovation is rarely the product of continuous subsidy, but maybe the result of carefully focused investment. Let's dive straight in on the key change in the world of poetry: the change in funding. As you read this, changes in funding poetry businesses have come into effect; few were as drastic as they seemed, some were countered by canvassing support from the literary community, but some have now come into effect and the result is that poetry, whose funding actually increased, faces the psychological change that it must become more competitive, and selling books is now a priority for all state publishers. The delicate shift in relations between funders and the funded is to refocus public sector businesses on sales as the main source of income, at least for those wanting to grow and develop.

The problem with arts funding

The Arts Council, to my mind, should finance businesses that measure up in their race to a sustainable expandable model built upon selling books people really want. If a business isn't interested in selling a product, it's hard to see it as a publisher at all. Except of the vanity kind. In some cases, it may be the taxpayer financing such vanities. My preference is that funding should attach to consumption and not production. For culture is a transaction between writer and reader.

Another problem with arts funding is that it often misses those businesses which are capable of moving poetry forward, but don't qualify for funding, haven't applied for funding, or don't subscribe to the current hegemony around best practice in the arts, as well as those whose writers don't fit the current models or who actually constitute a new model based on different reading demand. So, another part of the matrix of funding practice should be to seek out new private sector businesses who need investment to grow. Salt has certainly benefited from this approach from Arts Council England and I can vouch that it worked.

The new Renaissance

As you will have gathered, I'm in favour of the broad changes in arts funding practice advanced by the Department for Culture, Media and Sport and the Arts Council, and I'm inclined to agree with MP James Purnell's assertion that we are on the brink of a new Renaissance (*Guardian*, 5 January 2008). His comments echo Sir Christopher Frayling's from 2007 (as I write this, the current Chair of the Arts Council), who argued that we are in a golden age for the arts (*Guardian*, 22 July 2006), and I think where poetry is concerned this is certainly true. Despite the threat of funding cuts (which, to date, haven't transpired) and the implications of a business process re-engineering of UK Poetry Ltd, I can't recall a more exciting time for writers and readers, though this is largely due to the huge changes in the technological world, rather than in the world of patronage and literature development.

To be clear, after sustained progressive investment in the arts over the past ten years, much of the poetry world still behaves as if it were moribund, fragile and disenfranchised. But it doesn't need further financial homeopathy, it needs radical surgery, and now it might get it. Poetry must stand up and fight for its central place in our diverse and expansive cultural world. It must be more attractive than fiction, more substantial than Xbox 360, harder hitting than *Murphy's Law*. It must be socially engaged. Sales are no longer lamentable, but no one buys what they can't choose.

Focus on customers

How should the poetry community respond to changes in its public funding? It should welcome it and compete robustly with other art forms and entertainments: poetry is not isolated from contemporary fiction, cinema, theatre, television, radio, music downloads, eating and drinking. For poetry to continue to grow, writers and publishers must compete in the broader world of consumption; this means a full engagement with cultural consumption and for writers to seek a direct relationship with their readers.

To complement this, poetry publishers must turn 180 degrees away from their funders and refocus on their customers, building sales rather than constructing the case for the next grant. We should see a major change in literature, a reconnection between the art and its communities. What has been needed is a more direct response between the poets and their readership, and this is necessitated by sales. If your work is relevant and you actively seek a readership, it will sell; if you do not acquire a readership, it cannot be sold. The renaissance in poetry is this reconnection between poet and reader through book sales. It is the transaction at the heart of a community of readers, and provides the best measure of the life of poetry today. If we want to know more about poetry, we must read it.

Changes in choice

The gateway to poetry is changing, too, and those who have attempted to control consumption, including publishers like me, are facing a world that is increasingly diversifying in its tastes and predilections. The entire nature of poetry is shifting away from a monolithic narrow offering managed by a small range of collaborators to an explosion of new poetries built around small communities or markets. This democratization of poetry brings with it considerable pressures: there are no adequate means of establishing value, no critical framework can prefer one version of the art over another, and there is no centre from which to assess the extremities and absurdities of the art.

An age of abundance and fragmentation

What we are left with is the measure of consumption and cohesive bands of consumers, micromarkets, which may attempt to vocalize their taste over other markets. But, for the writer, we are in an age of abundance and fragmentation. There is no mainstream and no avant-garde. No centre or periphery. But there is plenty of choice. Freed from the artificial constraints of regular funding, some poetry will explode into the new space provided by these changing conditions. Perhaps the biggest architectural contrast will lie between the state-owned poetry publishers and the private sector businesses.

Most publishers spend money to construct choices in the mind of consumers, to provide limits to the range on offer, and to constrain diversity, preferably to the contents of their own publishing programme. Most people want to be told what to buy and choose their advisors carefully, hence the success of the Richard & Judy Book Club and the like. The Web has begun to impact upon such brand building and selection: it is eroding patterns of judgement and they are being replaced by communities of readers sharing their experiences online in social networking sites, bulletin boards and LISTSERVs. Within the next five years, the shape of poetry consumption will shift profoundly and the mechanisms used to control what you buy, what you think is available to you, will change dramatically.

'Risk', 'innovation' and 'excellence'

The Arts Council is refocusing on 'risk', 'innovation' and 'excellence' (the bywords of the state's new arts regime). To understand these words, an economic context is needed to help us to understand that poetry businesses now need to focus on the sole aim of achieving financial independence through an expansion and diversification of the market, and to help businesses to focus on sales and growth.

The reality, it seems, is that our culture is a political tool, if not a weapon, of the state, and patrons are an unavoidable part of our artistic heritage. Art is so often the gorgeous detritus of barons and bishops. Even Milton wrote on demand for the republic. What we need is sound business judgement in the twentieth century, and a focus on independence and commercial achievement in our arts businesses. That should not be a matter of aesthetic value, but an assessment of the vigour and future prospects of a poetry business in serving its customers and building new audiences. It needs to be based on profit.

The poetry problem

There's simply no point to my mind funding stuff no one wants, and if no one wants it we should stop producing it. This is clearly heretical, but looked at more closely it recognizes the fundamental problem of our times: that certain forms of funding cause a fracture between the art we want to buy and the art we don't. Too often we end up without the former and awash in the latter. If something doesn't sell, it's not the reader's fault; it is not a matter of general ignorance. Now poetry *is* for everyone, every last person on the planet, and if we aren't buying it, the problem *might* be examined as follows:

* 'Modern poetry is a load of rubbish and doesn't rhyme or scan' (this is usually expressed by geriatric slubberdegullions who normally write to *The Mail on Sunday* about garbage collecting).
* 'Poetry needs financial help because no one wants to buy it' (publishers are known to express this view, but largely those publishers who never meet booksellers, or buy books, or try to sell them to anyone).
* 'Modern poetry is full of fusty old buggers, most of them blokes, who are like, just totally out of touch with the word on the street' (this is the General Elitist Conspiracy Theory, which can be expanded along racial, gender, religious or sexual grounds).
* 'Modern poetry is for toffs and elitist weirdos with big hair' (okay, it is notable how many poetry editors have Big Hair Syndrome, chiefly at Faber's offices. Allegedly!).

You can expand this list with your own points. Much of the debate in the past seemed to point to the consumer being too dim to appreciate high art. The problem though is that we have a simple disconnection between the producer and the consumer or, more clearly, a fracture between writer and reader. I think funded publishers can all too easily end up seeing the Arts Council as the primary source of revenue, and this may lead to an unsustainable dependency culture which bears no relation to supply and demand of the living art. When money isn't an issue, too often the wrong poets are chosen.

Many splutter at the notion that there is real demand for high culture, but this is simply untrue: people love poetry, even very difficult poetry. It is an art central to being human, and especially to being a child.

Increasing capacity, reducing complexity, financing new writing, encouraging participation, seeking broader social inclusion, encouraging live audiences: all these have been tried and yet the biggest deliverer of poetry to its consumer remains the simple book, sold over the counter of a good shop.

Is UK Poetry Ltd 'fit for purpose'?

To get the blood moving in our veins, I'll write something outrageous here: most poetry publishing is deeply amateur. Most editors have no knowledge of how to run a publishing business, have never worked in publishing, and are focused on their personal view of the art and on simply manufacturing books. The notion that books serve an identifiable need within a defined market (even if it is an avowedly lofty need) seems alien. More time is spent on editing than on selling (sometimes more on pontificating than addressing customers). More time is spent on production than publicity. Is it any wonder poetry sells relatively poorly when so little attempt is made to actually sell it?

The poetry business has no future without sales

If you are a poetry publisher and haven't got a sales director, the next piece of valuable advice in this article is to go and appoint one now. Only sales will save your business from the coming changes in our global poetry markets.

Literature is always being reborn, re-imagined, regenerated, bought and sold, and each generation creates its own poetry out of the creative economy that sustains and directs it. Most literature also dies. Let's consider if we currently get the poetry we deserve. I say this largely because poetry is as much an issue of writers and publishers as it is a matter of the management. The management is currently a cause of some concern, and you could be forgiven for thinking that the management today in Britain is the Arts Council. Or is it perhaps the publisher? It is, in fact, the paying reader.

How many literature businesses make that basic mistake? Perhaps nearly all of them. No poetry business is sustainable without paying readers. If you want to develop literature, your focus has to be on sales development. For writers this means that the route to market lies with businesses that innovate saleable products. From the other angle, what innovative literature businesses need are writers who are capable of discovering or working with audiences.

The poetry business has no future without sales. And sales means consumers, not patrons. This isn't to deny the enormous value arts funding can have, it is merely to recognize that funding which does not lead to sales has failed and as such should be diverted into practices and businesses which sell more effectively. This in turn builds a sustainable economy for successful writers.

Compete or die

It may be unpopular to express the following view, but all art is deeply competitive in nature, and when my own literary business was in its infancy, it seemed anti-competitive that many of my contemporaries in the poetry sector received considerable amounts of permanent funding when my business did not. What mechanism is there that can provide fair, sustainable investment in art? I can't conceive of one. In his robust defence of Marks & Spencer's decaying sales back in Christmas 2007, Stuart Rose remarked to a BBC news anchor that they were not cutting prices, they were finding the right price point for consumers; he went on to clarify that high street retailers competed on value, and that value is price divided by quality. I stared at my 16p half eaten banana and tried to calculate its value to me. I just couldn't find the numeric representation of quality.

The judgement question

Some may conceive that this is about judicious patronage, others that this is about using editorial power to jostle the competition out of the way. My personal conviction was, and still is, that competition was to be enhanced and introduced at every level. I have a deep belief that readers construct literature, and that the world of poetry is as much an economic and commercial world as it is one of aesthetic insight and cultural necessity. In fact it is almost a necessity that the poetry economy be liberalized in order for us to see more clearly how our readers select our writers in defining any cultural and aesthetic moment. We cannot classify books as literature if they are not read, except in a historical and past tense. Is it the role of the academic to make such analyses? No, for literature is not the product of the academy, it is the product of the living culture and its consumption of books.

Poetry may be asynchronous in its impact, in that the readers who create literature may ignore today's talent in favour of discoveries from the near past. Literature is highly contingent. We may feel that the cream rises to the top of the milk, whilst some may quip that so does the effluent. All we can say for sure is that in 2009, the management of literature will be filled with change. In these pages you can discover it for yourself. I would love to join you on that journey, but it takes a lifetime and I'm already on my way; but I hope our paths cross.

Chris Emery's poetry has appeared in The Age, *Jacket, Magma, Poetry London, Poetry Review, Poetry Wales, PN Review, Quid and* The Rialto. *He was anthologized in* New Writing 8 *in 1999. A first full-length poetry collection,* Dr Mephisto, *was published by Arc in 2002 and a new collection of poetry,* Radio Nostalgia, *was published by Arc in 2006. Chris Emery is publishing director of Salt Publishing, an independent literary press based in Cambridge. He was awarded an American Book Award in 2006 and Salt Publishing won the Nielsen Innovation of the Year prize in 2008.*

The worst of times, the best of times

Kim Fletcher tracks the fortunes of the freelance journalist

Changing fortunes

There was a *cri de cœur* on the *Guardian* website the other day, from a journalist responding to a blog item about falling sales in the newspaper industry: 'What gets me ... is the fact that editors and senior executives and top columnists now earn more than ever. It's a bit like Wall Street or the City: it doesn't really matter how badly they perform, they are *valued*. More than that, until they are fired, they are *indispensable*.'

That sounds rather encouraging, doesn't it? If there's money sloshing around the publishing industry, let's have some of it over here. But the contributor hadn't finished: 'As a freelance down the years, I have noticed how fees for pieces have gone down and down. Whereas in the early 1990s I would occasionally be paid as much as £1,000 for a piece, nowadays it's more like £150. But for those at the top, even as they preside over a seemingly disastrous decline, business – personal business, that is – couldn't be better.'

Rising to competition

The writer made things sound bleak for those of us who don't happen to be sitting in a big office suite. Naturally I nodded in agreement. But the next evening I found myself sitting next to a confident freelance journalist at a dinner party. He said business had never been better. It was, he explained, a fantastic time to be a writer. The Internet had opened up opportunities, papers were bigger than ever, his competitors were lazy and unimaginative. If you were good and had something special to offer, you could name your price.

'I tell them my rate. Sometimes they say it is too much. I tell them that's fine, to go and find someone else. And you know what? They come back ten minutes later and say they want the piece.' Was he spinning me a salesman's line? If he was, he was wasting his time, for it is years since I commissioned a story from anyone.

As it happened, I also knew the man who sounded so unhappy on the *Guardian* blog. He wrote under a pen name, but I recognized the writing style of a former colleague. He was always a moaner, even in those happy days when Sunday newspapers gave reporters a week to cover a single story and paid good money to freelances.

A unique industry

Just because he's a whinger doesn't mean he's wrong. When I was sacked as the editor of a Sunday paper and began to fill my Saturdays travelling round football grounds to file match reports for *The Sunday Times*, I was surprised to find the rate for the day was less than I had been paying news reporters to work a shift five years before, in the days when I was a news editor.

It's also true that salaries of many of the men and women at the top of the media bear no relation to the profits of the publications in which they appear. I know several editors whose generous salaries exacerbate the substantial losses sustained by their papers each year. If business leaders were on deals like theirs, they would get a terrible kicking in the City pages. I also know columnists on salaries close to seven figures. They are the ones who are thought to have box appeal, though no one knows whether the readers would disappear if they did.

Then there are the rest of us, starting with the graduates who come out of university with £15,000 worth of debts to compete for jobs on local papers that start at £12,000 a year. Or that the best they can do is work experience that doesn't actually pay.

A unique history

So what are we in? A golden age for freelance writers who can research, write and file work more efficiently and to a wider range of outlets than ever before? Or a disastrous new world that sees us writing for tuppence in competition with amateurs who perform for free?

Anyone starting out now must be amazed to hear how it used to be. As recently as the 1980s there was a sense that, while journalism was not the best paying industry, it encouraged a certain style. Did we really travel everywhere first class? Were we truly encouraged to take out cash advances every week? Were teams of writers and photographers put up in five star hotels for nights on end?

There are rumoured still to be magazines that live on in the old manner, expecting the writers they commission to book expensive hotels and take several weeks in research. My dinner companion, the freelance writer, seemed confident that he was doing business with one or two of them. But they are the exceptions in refusing to downgrade the value of words.

For over the last twenty-five years the publishing industry has abandoned notions of vanity and started to run itself like any other business, seeking ways to increase productivity and to drive down costs. The great defeat of the print unions, which promised to herald a period of prosperity for writers, was merely the prelude to a bigger inquiry into costs, focusing naturally on those who wrote the stories. Advertising departments bring in money. Journalists spend it.

A new business model

Then, as newspapers began to control their costs more efficiently, the Internet arrived. With its belief in providing free material, funded by advertising, the Web was a clear challenge to the business model of the newspaper industry. Where was the sense in providing online for nothing copy that you expected your newspaper readers to pay for. It was no wonder that most publishers were relieved to see the dotcom crash at the turn of the century. With any luck, it would go away.

'Opportunities'

It didn't, of course. Now they are torn between what business euphemistically calls 'challenges' – i.e. threats and problems – and 'opportunities'. The same is true of writers, who now have the opportunity to write much more and have the challenge of getting paid for it. To use another business term that is in vogue, the industry of writing copy has become 'commoditized', which is to say that editors look to buy it on price rather than on the basis of any distinguishing characteristics.

Publishers speak no longer of things called articles, or stories, or features. They talk of 'content', something that you spread across websites, that sounds as if it can be bought by the bucket. Everyone is looking for content – ideally a lot of it, as if it is a kind of inexpensive emulsion paint that will cover the seemingly endless walls of a website.

Once you see words merely as 'content' it is natural to wish to obtain them as cheaply as possible. If you suspect that many readers fail to differentiate between different qualities of 'content', why pay more than you have to? If you are paying freelances, get them to do it as cheaply as possible. If you still employ journalists, make sure you get more work out of them.

The devaluation of the written word?

The effects of cost reduction and productivity drive were discussed recently by Nick Davies in his book *Flat Earth News*, where he argues that most of what appears in newspapers is wholly, mainly or partially constructed from second-hand material, provided by news agencies and by the public relations industry. Davies commissioned researchers at Cardiff University to investigate the work that had gone into 2,000 UK news stories from the *Times, Telegraph, Guardian, Independent* and *Daily Mail*. In the course of their research, they noted the volume of work now expected of journalists. He explained the results in *The Guardian*:

> For each of the 20 years from 1985, they dug out figures for the editorial staffing levels of all the Fleet Street publications and compared them with the amount of space they were filling. They discovered that the average Fleet Street journalist now is filling three times as much space as he or she was in 1985. In other words, as a crude average, they have only one-third of the time that they used to have to do their jobs. Generally, they don't find their own stories, or check their content, because they simply don't have the time.

Mind you, given the remarkably small amount of space many staff journalists were expected to fill in the old days, a threefold increase in productivity does not necessarily mean a two-thirds reduction in the time they have to check and find stories – some of that extra productivity can be found from spending less time in the pub, abandoning those heavy-drinking lunches and no longer sitting around the office moaning.

Or greater scope for creativity?

It is true however that they are now called on to do more than merely file a story. Staff journalists may be expected to write at least two versions – a fast one to go up immediately on the website and a more considered one to run later and in the paper. They may also appear in a video clip or be asked to knock off a blog.

If staff do it, what about contributors? On the face of it, we'd be mad: where is the economic sense in freelances going to a newspaper office to record a piece? So why, when I wrote a fortnightly column on the media for *The Guardian*, did I regularly pedal off to the paper's offices to record a podcast media discussion? It wasn't for the money, because there wasn't any.

Vanity

I guess it was out of vanity. The medium was new. It was interesting. Even if we suspected that the audience was in hundreds rather than thousands, it seemed important to take part. If I didn't, who would appear instead? Why let Andrew Neil pontificate about the media in my place? So I justified free podcast trips to myself on the basis that they built and protected my 'brand', just as I continue to justify cost-ineffective trips to television studios as brand extension rather than ego satisfaction. Besides, I argued, it was good to get away from the home office for a couple of hours, important to be in the swim, useful to meet people. These were rational justifications, but overlooked the fact that *The Guardian* staff journalists were being paid for their time in the studio. Do you think *The Guardian* gets plumbers out on call on the same arrangement?

The reality

We have to face the fact that many newspapers pay so little because they can. There is no shortage of people who want to write. When the Internet is full of bloggers who write for nothing, it becomes harder to maintain the monetary value of comment pieces and to assert the superiority of professional writers. And some of those bloggers write well.

Opportunities

So what's to be done? The first thing is to take advantage of the rising number of opportunities. There is more writing to be done than ever, and if it is hard to obtain the big fees that once obtained, we can make up the money by writing for more people. If the Internet has done a disservice in reducing the value of much of our work, it has also provided the tools that enable us to research, write and file that work more efficiently.

The second thing is to follow the advice of the contented freelance I met at dinner. The thing that does lead to price differentiation – i.e. charging a better rate – is the quality of the product. Newspapers will pay more for something if they cannot find it elsewhere. They will also pay more for a superior service that delivers the kind of articles that go in as written. In an age when everyone can be a writer, the real value is in finding people with new things to say or better ways of saying them.

These are the pieces that papers believe will bring in an audience, the pieces that differentiate them, in a commoditized world, from their rivals. But who sets the value? It is the editors, of course. You are worth only as much as they are prepared to pay you, a value that will rise if they are in competition for your signature. Are these fees capable of objective calculation? Not really, which is why the whole process looks so unfair to the unhappy writer with whom we started.

Kim Fletcher writes and broadcasts and runs a presentation coaching business. He is a former editor of the Independent on Sunday, *was editorial director of The Telegraph Group and is chairman of the National Council for the Training of Journalists.*

Page to screen

Barry Turner on the journey from novel to screenplay

From novelist to screenwriter used to be a natural progression. Maybe progression is the wrong word. There was and is more money to be made from putting words into actors' mouths but the frustration of working with the oddballs of the screen trade can be hard for writers who are used to doing their own thing. Too many of them have wound up with a bottle as their best friend. Whether this has acted as a dreadful warning or whether producers have got wise to the fact that books and movies are two different products needing different skills, it is true that fewer novelists now set their rose-tinted sights on Hollywood. They still want the money but are happiest when their books are optioned by a major studio. They can then get back to doing what they do best while leaving others to translate their work into moving images.

The million dollar deal

But, of course, nothing is ever that simple. For a start, the seven-figure book offers that attract press headlines are misleading. The typical option, giving exclusive rights for one or two years, is hedged about with conditions which let the production company off the hook if it proves too hard or too expensive to attract a marketable director and actors. Less than one in ten options is 'exercised'. For the rest, the author banks an upfront payment, on average ten to twenty thousand dollars, and learns to live with disappointment.

On a more optimistic note, a third of the movies produced in Hollywood are based on books and plays. It is easy to see why. A successful novel will have a fee base attached: a promise, if not a surefire guarantee, that it will translate into profitable cinema. In effect the author provides a cheap means of testing the market. The money paid for the film rights to a novel, even if sizeable in publishing terms, is as nothing compared to the cost of creating a screenplay from scratch.

Knowing the business

For the author, the trick is not to lose this initial advantage. Unless you happen to be a trained lawyer and a financial whiz combined, the first step is to get on the right side of an agent who is immersed in the business. Only a specialist knows who, if anyone, to take seriously from a crowd of chancers. The film industry attracts more con artists than an estate agents' convention. Moreover, the complexity of making a deal barely needs elaborating when film contracts run to fifty pages or more.

Producers come in all guises. The independent with a shared desk may have the backup of a major studio; but then again he may simply be gambling on hitting the jackpot. An offer may come from an actor chasing a meaty role or from a scriptwriter short on ideas of his own. As a general rule, an eagerness to produce is in inverse proportion to the capacity to deliver.

Getting the best deal

Writing in a recent edition of *The Author*, Giles Foden, whose novel *The Last King of Scotland* became a much lauded movie, warned of the contentious issues that can arise when filming is about to start. The buyout fee or purchase price will almost certainly tantalize with the offer of a two or three per cent share of the profits. The production sums involved make this sound like a lottery win. But as Giles Foden points out, 'Hollywood accountants like spinning profits into deficits by constantly introducing costs after the fact'. The two to three per cent may well end up worthless.

The best advice is to try for a purchase price tied to a percentage of the budget. Producers will usually insist that there is a cap on the amount paid, but authors, or their agents, can be equally tough on setting a minimum.

Other Foden tips include an insistence on the film having the title of the book and on the right to an image from the movie for the cover of a new edition. Don't dispose carelessly of associated rights that might turn out to be lucrative: the stage adaptation, for example. And if you expect or hope for anything like visits to the set, tickets and first-class travel to premieres across the Atlantic, make sure they are in the contract.

Surrendering proprietorial claim

Even the success stories have their downside. Authors who sell out to the film industry must know that they have surrendered all proprietorial claims. Any resemblance between an original book and the screen adaptation will be entirely coincidental and there is nothing the author can do about it. Tom Clancy, whose box-office hits include *The Hunt for Red October* and *Patriot Games* compares selling the film rights to pimping your daughter. A more optimistic view is taken by Nick Hornby. 'The way I see it, selling the film rights to a book is a no-lose situation for a writer. Someone gives you money for work you have already done; if the film is bad, you will attract a few more readers, and if it is good (or failing that, successful), chances are that your readership will increase.'

But suppose the ambitions of an author go further than simply selling a project on the best terms. Success brings huge rewards. In Hollywood, there is even a decent living to be made out of failure, the scripts that fail to make it on to screen. The downside is a work ethic that requires writers to submerge their identities. Big studios are liable to engage three or more writers on any single production. Often the rewrites are so extensive as to lose even a hint of the original work.

'Film is a collaborative business'

Faced with demands to amend and adapt, sensitive writers can be forgiven their occasional outbursts. There is the story of Dorothy Parker throwing a typewriter through her office window. It crashed at the feet of a party of tourists on their round of the Paramount Studios. They looked up to see a distraught Miss Parker peering through the broken glass. 'Let us out,' she shrieked. 'We're as sane as you are!' That was in 1938 and though writers are no longer herded together and told to get on with it, nothing else has changed. During his days in Writers' Block, David Mamet commented that 'film is a collaborative business; bend over'.

Movie-writing takes time, which is the other big drawback to the screen trade. While an author might expect to spend a year on a novel, a screenwriter should not be too surprised if his words are still unspoken after a decade of toing and froing. He will be on to other work by then but the frustration of non-fulfilment can be hard.

This article is reproduced with permission from The Screenwriter's Handbook, *edited by Barry Turner and published by Macmillan Publishers Ltd.*

Getting it together

The Writer's Handbook *salutes the first great English dictionary*

We are in for a year of Johnsonian mania. The 300th anniversary of the birth of the Lichfield prodigy – author, poet, journalist, critic and purveyor of memorable one-liners – has festival organizers in overdrive. At *The Writer's Handbook* we will be raising a glass to celebrate the single greatest achievement of the revered doctor, one that merits the thanks of all writers in English.

The story of Dr Johnson's Dictionary, the premiere arbiter of the language for 150 years, has been told many times, starting with Boswell's *Life* but never better than by Henry Hitchings in *Defining the World* (John Murray, 2005), an enlightening and entertaining tribute to a cultural icon.

A momentous undertaking

Sponsored by a consortium of booksellers led by Robert Dodsley, Johnson signed up for his mammoth work on 18 June 1746. The dictionary was to be published in instalments and Johnson was to be paid 1,500 guineas (around £150,000 in today's money), also in instalments. If this sounds like the offer that cannot be refused, Hitchings reminds us of the challenge that Johnson faced. While there were other dictionaries in circulation they were either hard word glossaries, preoccupied with words that were often already out of date or of such slapdash methodology that, in one case, 'cat' was defined as 'a creature well known' and 'to wash' as 'to cleanse by washing'. Nowhere was there an English dictionary that gave meaningful sense to language as it was used.

The French were better equipped with their Académie française, founded in 1635, with the express purpose of creating an authoritative dictionary. But this intellectual powerhouse took fifty-five years to complete the task. Johnson, working as an independent freelance, was under pressure to get the job finished quickly. He gave himself three years but, unsurprisingly, failed to meet his deadline.

The labour involved was prodigious. Johnson's 'plan' for the Dictionary, published a year after he started on his mission, stressed the need to select 'the words and phrases used in the general intercourse of life, or found in the works of those whom we commonly style polite writers'. This called for the collection, copying and collation of some 110,000 quotations, plus those he eventually rejected, in support of 42,773 entries. He also promised to rationalize spellings and to give guidance on correct pronunciation.

Johnson's personal taste for words

Johnson was unusual among lexicographers in exploring scientific texts for words that might usefully be understood by the layman, but his greatest difficulty was in defining common, everyday words. To play the game is to appreciate what Johnson was up against. How many definitions would you give for 'in'? Johnson had sixteen. His explanation of 'to put' runs to more than three pages in the first edition of the Dictionary.

Johnson had no time for indelicate words which offended his own sense of propriety and, doubtless a consideration, the sensitivities of his prospective readership. Though when applauded by a lady on his omission of 'naughty words' Johnson could not resist the put-down. 'I have not daubed my fingers, Madam. I find, however, that you have been looking for them.'

Other words of which Johnson disapproved he dismissed as cant, variously described as 'a corrupt dialect used by beggars and vagabonds' and 'a particular form of speaking peculiar to some certain class or body of men'. What he would make now of the mangled construction of English that comes with the excessive use of technological jargon, defies imagination.

Less endearing is Johnson's distaste for foreign words, or 'barbarisms' as he called them. As Hitchings points out there are no entries in the Dictionary for 'bourgeois', 'unique', 'champagne' or 'cutlet', though all were in common use. While fear of French encroachment was understandable (it was then the only language with claims to world status), Johnson was surely misguided in his efforts to 'fix' for all time a constitution for English that would not allow for borrowings from other languages. That English is now indispensable for international communication may be partly credited to the relative simplicity of its basic grammatical rules but more particularly to its readiness to adopt and adapt to create a rich vocabulary. It is a lesson that the Académie française, with its devotion to linguistic purity, has still to learn.

A book for every household

But Johnson is not to be underrated. In its complete form, the Dictionary made its first appearance in 1755. Hostile critics immediately picked up on examples of slapdash research (the musical entries testify to Johnson's tin ear) and plain errors. But, with necessary revisions, the Dictionary came to be as standard for household libraries as the Bible, a starting point for many great literary adventures. For all his concerns of an unfettered language descending into chaos, Johnson would surely now admit, in this respect at least, he was in error. His disarming excuse would be the one he gave to a dinner guest who asked how he came to one blatantly inaccurate definition: 'Ignorance, Madam, pure ignorance.'

Barry Turner

English as Johnson understood it

A selection from his Dictionary

> It is strange that there should be so little
> reading in the world and so much writing.
> *Johnson, 1 May 1783*

a'ffluence. The act of flowing to any place; concourse. It is almost always used figuratively.

a'leconner. An officer in the city of London, whose business is to inspect the measures of publick houses. Four of them are chosen or rechosen annually by the common-hall of the city; and whatever might be their use formerly, their places are now regarded only as sine-cures for decayed citizens.

amato'rculist. A little insignificant lover; a pretender to affection.

anthropophagi'nian. A ludicrous word, formed by Shakespeare from *anthropophagi*, for the sake of a formidable sound. Go, knock, and call; he'll speak like an *anthropophaginian* unto thee: knock, I say. Shakespeare, *Merry Wives of Windsor*.

belda'm. An old woman; generally a term of contempt, marking the last degree of old age, with all its faults and miseries.

bru'mal. Belonging to the winter.

churme. A confused sound; a noise.

chyle. The white juice formed in the stomach by digestion of the aliment, and afterwards changed into blood.

cliente'le. The condition or office of a client. A word scarcely used.

cli'mate. (1) A space upon the surface of the earth, measured from the equator to the polar circles; in each of which spaces the longest day is half an hour longer than in that nearer to the equator. From the polar circles to the poles climates are measured by the increase of a month. (2) In the common and popular sense, a region, or tract of land, differing from another by the temperature of the air.

co'mmoner. A prostitute.

libra'rian. One who transcribes or copies books.

mo'rtgage. A dead pledge; a thing put into the hands of a creditor.

mo'untebank. A doctor that mounts a bench in the market, and boasts his infallible remedies and cures.

mouse. The smallest of all beasts; a little animal haunting houses and corn fields, destroyed by cats.

to park. To inclose as in a park.

promi'scuous. Mingled; confused; undistinguished.

trapes. An idle slatternly woman.

tre'acle. A medicine made up of many ingredients.

Johnson's only entry under X:

X Is a letter, which, though found in Saxon words, begins no word in the English language.

Clearing permissions

An often difficult issue for non-fiction authors[1]

If you want to include or quote any substantial amount of material written or produced by someone else in your work you will in most cases need to clear permission. Clearing permissions to use material belonging to a third party can be a time consuming and expensive business. So it is always worth asking yourself: 'is this really necessary?' and 'might it be less work to produce an original alternative myself?'. This article discusses what third party material does and does not require permission and provides some guidelines about how to go about clearing it. If you do need to use third party material, you may not need to ask permission if the work is particularly old, and therefore out of copyright, or if the use being made falls under a copyright exception. In order to determine whether permission will be required, it is necessary to understand something about copyright law.

About copyright in the UK

Although the details can be quite complex the essence of copyright law is relatively straightforward. Copyright is an automatic right, belonging to the creator of an original literary, musical, dramatic or artistic work that has involved skill and labour to produce. It also covers sound recordings, films, broadcasts and the typographical arrangement of published works.

Copyright gives its owner the exclusive right to do the following:[2]

- Copy the work
- Issue, rent, lend or communicate the work to the public
- Perform, show or play the work in public
- Make an adaptation of the work or to do any of the above in relation to an adaptation

Copyright subsists in a literary, musical, dramatic or artistic work from the moment it has been created and nowadays lasts (in Europe) for 70 years from the end of the year in which the creator dies.

Copyright may be assigned by the creator to another person or organization. So whilst the initial copyright holder is the creator, copyright, or other rights, may be subsequently assigned or leased to another party. In the case of employees

---

1 This article is an attempt to guide you through a complex issue but is not a substitute for legal advice and neither the author or Macmillan will be responsible for any loss or damage as a result of its use.
2 Copyright, Designs and Patents Act 1988, http://www.ipo.gov.uk/cdpact1988.pdf

creating copyright material in the course of their work, their employer would be the copyright holder, unless other agreements had been made. Any use of copyright material therefore requires the permission of the copyright holder, or the holder of the rights that you wish to make use of.

The following list (though not exhaustive) gives some examples of types of copyright material for which permission needs to be sought for reproduction, unless the use falls under a copyright exception.

Extracts or reproduction in their entirety of:

- Books, newspapers, journals or magazines
- Tables and diagrams
- Screenshots and any material from websites
- Cartoons
- Song lyrics, poetry or drama
- Maps
- Photographs and any other illustrations or artwork
- Logos

Exceptions to copyright

UK copyright law provides for some exceptions to copyright which enable you to make use of copyright material, such as that listed above, without the prior consent of the rights holder. These include, for example, making single copies or taking short extracts of works for certain kinds of educational and library use, and an exception for those with a visual or physical impairment that prevents those persons from being able to access a generally available version of a work. There is also a group of exceptions for 'fair dealing' for the purposes of:

- Non-commercial research or private study
- Criticism and review
- Reporting current events

The Publishers Association's recently published permissions guidelines,[3] explain that:

> 'Fair dealing' is an essential pre-condition of all three activities – the copyright exceptions do not cover *any* kind of, for example, research or private study, but only fair *dealing* for those purposes. What is 'fair dealing' is a subjective test which may vary depending on the facts of each case, and quite possibly on the motives of the person doing the copying. Other factors may include whether the original work is already published, how extensive – and important – the extracts taken from the same work are in relation to the whole work, and in some cases how frequent.

They also advise that:

> Future interpretation of copyright exceptions by UK and other EU courts is likely now to be subject to the Berne Convention '3-step test', which limits exceptions to

3   See http://www.publishers.org.uk/en/home/copyright/copyright_guidelines

1. certain special cases;
2. which do not conflict with normal exploitation of the work, and
3. which do not unreasonably prejudice the legitimate interests of the rightsholder.

The most important of these tests for publishers is number 2, ruling out use which conflicts with normal exploitation by the publisher. For most practical purposes, this means that copyright exceptions should not permit rival commercial use, and judges are likely to bear this in mind in interpreting [fair dealing exceptions].

The Society of Authors[4] suggest that in determining what constitutes fair dealing, the following points might be considered:

- The length and importance of the quotation(s)
- The amount quoted in relation to your commentary
- The extent to which your work competes with or rivals the work quoted
- The extent to which works quoted are saving you work

There is very little case law in this area, and the measure of what constitutes 'fair dealing' is qualitative rather than quantitative. If you intend to use copyright work under these exceptions, it is therefore worth spending some time considering whether or not that use really is 'fair dealing'.

Use of copyright work under the 'fair dealing' exceptions is permitted for any type of copyright work (except a photograph) provided it is accompanied by a sufficient acknowledgement. This would usually mean an acknowledgement to the author and title of the copyright work, and details of where and when it was published.

Moral rights

Alongside copyright, the 1988 Copyright, Designs and Patents Act introduced for the first time into UK law protection of moral rights which have long been recognized in continental Europe. Moral rights are additional rights, distinct from copyright, and include the rights of paternity, integrity and attribution. It is important to be aware of moral rights in addition to copyright when reusing someone else's work.

- Paternity (which needs to be formally asserted by the author) is the right of the author to be identified as the creator of the work whenever it is performed or commercially exploited.
- Integrity is the author's (automatic) right to object to derogatory treatment of his or her work. The definition of 'treatment' according to the Act is an 'addition to, deletion from, or alteration to or adaptation of a copyright work (other than a translation or transcription of a musical work)' and 'derogatory' is defined as a 'distortion or mutilation of the work' or that which 'is in any way prejudicial to the honour or reputation of the author'.

4   http://www.societyofauthors.org/publications/quick_guide_permissions/

- Attribution (also automatic) is the right not to be falsely attributed as the author of a work.

Remember that the owner of the copyright may not be the owner of the moral rights of a particular work.

Finding rights holders

If you wish to use in whole or in part any intellectual property protected by copyright, and the use is not covered by any of the exceptions to copyright, you will need to seek the permission of the rights holder or holders. In the case of published works, the creator or original copyright holder may well have assigned her copyright, or given exclusive publication rights, to the publisher – which is often the best place to start your search. Of course the publisher may have subsequently sold the rights on, or been taken over or subsumed by another publisher, or have reverted the rights they had back to the original creator of the work. It can therefore be a time consuming and complex business just tracking down the appropriate person to seek permission from.

A few resources may prove helpful in tracking down rights holders. The online catalogues of the British Library and Library of Congress can be useful in identifying publishers of written works, and this book contains contact information for publishers – you should direct your enquiry to the rights or permissions departments. It is also worth consulting the WATCH files at http://tyler.hrc. utexas.edu The WATCH (writers, artists and their copyright holders) files are an online database of copyright contacts for writers and artists. For artistic works DACS (Design and Artists Copyright Society) may be a useful starting point (http://www.dacs.org.uk).

Even if you want to reuse material you have authored yourself, check any agreements you have signed relating to its previous publication – you may well have assigned exclusive rights even if you retained copyright, in which case you will have to go back to your publisher and ask permission – although this would usually be granted without charge.

Many publishers will put details of their permissions policy on their website (including their interpretation of fair dealing), and specify what information they require and the form in which you should make your request. Some will even have standard forms to print out or complete online, so it is worth checking website advice before contacting publishers.

Making a permissions application

Having tracked down the rights holder you will then need to write and request permission to reuse the material in question. When writing to request permission you will need to be able to provide the following information:

- Title and author of work from which material is taken
- Relevant page numbers or other reference
- Details of any changes or alterations you plan to make (it's often a good idea to include a copy to show intended changes – be aware that they may not be permitted by the rights holder)

- Information about the context in which the material is being reused – for example publication details – print run, price, author, title, publication date or other relevant details

Allow plenty of time to receive an answer: four to six weeks is generally considered a standard response time.

Most rights holders will reply with a permissions agreement which will detail the rights you have been granted. It is worth checking replies carefully to ensure you have been granted all the rights that you require. In particular rights might be limited by time period, language, territory, publication format, print run or number of people accessing the material. The permissions agreement should also specify the form of acknowledgement to be used. Be aware that the rights holder may deny your request altogether, especially if you have made substantial changes to the original.

Permissions fees

Depending on how much material you have requested permission to reuse, there is likely to be a fee. Fees will vary enormously between rights holders, and will depend on the source material that you wish to reproduce. Song lyrics or poetry are likely to command a higher price than reproduction of a diagram from an academic textbook. The fee will also vary in accordance with the likely financial return of your project. A book with a large print run will usually attract a higher fee than a book with a more limited market. If you are using material from a scientific, technical or medical book, and the original publisher of the material and the publisher of your work are signatories to the STM Permissions Guidelines, you may find you can reuse limited extracts free of charge. For more information and details of the publishers who are signatories, visit the STM website: http://www.stm-assoc.org/stm-permission-guidelines/

If the fee being charged seems exceptionally high, given the amount of material being used and the likely financial return of your project, it is worth trying to negotiate a lower rate with the rights holder. If the fee makes inclusion of an item prohibitive, even after negotiation, you will need to consider an alternative or even dropping the use of the item altogether. For this reason it is imperative that you start the process of permissions clearance as early as possible. Discovering you might have to drop something right at the end of the process might mean rewriting or searching for a replacement and beginning the clearance process over again, causing delays which could seriously jeopardize your project.

Finally, having cleared permission (and paid, where appropriate) to use the material, you can proceed with your project! All that is needed at this point is to ensure that your publisher or whoever is responsible for distributing your work is aware of the third party material you have used and any conditions associated with its use.

Permissions clearance checklist

- Think carefully about whether you need to use any third party material in your project: it may well be time consuming and expensive to do so
- Allow plenty of time for permissions clearance
- Find out whether the material you want to use is still in copyright
- Check whether your use of the material will be covered by a copyright exception
- Make sure the intended use does not infringe any moral rights
- Establish who the rights holders are and check their permissions policies where possible
- Ensure you have detailed information about how you will reuse the material and include this in your application
- Be prepared to chase up your application, maybe several times
- If you receive permission, check that the rights you are being granted are sufficient for your requirements (or your publisher's)
- Be prepared to negotiate fees if they seem unusually high
- If a fee is charged, pay the invoice promptly: the agreement is likely to be invalid unless payment has been made
- Make sure the appropriate acknowledgements are made in your work and ensure that your publisher is aware of any other terms and conditions that apply

More information

- For further information about copyright, visit the website of the UK Intellectual Property Office www.ipo.gov.uk
- For information on US copyright law, visit www.copyright.com
- The Publishers Association's website has an area dedicated to copyright information (including their recently published permissions guidelines), which can be found at http://www.publishers.org.uk/en/home/copyright/copyright_guidelines/
- A sample permissions application letter, and other useful information, is available from *The Writer's Handbook Online*.

Clare Hodder is currently Head of Rights at Palgrave Macmillan. She is Chair of the PA Academic and Professional Division Copyright Committee and also sits on the ALPSP Copyright Committee.

Settling accounts

Ian Spring takes an expert look at the latest Budget and explains how writers can be tax-wise

When he was Chancellor of the Exchequer, Norman Lamont said

Tax should be

- Simple and certain
- Fair and reasonable
- Easy to collect

Unfortunately, over the years, tax law has become ever more complicated and more and more difficult for the 'ordinary' person to understand it. Until recently it was inconceivable that tax changes would be passed which had retrospective effect, but this is no longer the case and is happening. With the recent merger of the Inland Revenue and Customs and Excise, the former have taken on some of the powers and practices of the latter and have become more aggressive in enforcing the law, as they see it, in their never ending endeavour to increase the tax take.

Tax is not simple and a taxpayer cannot be certain that a tax planning action legally taken now will still be legal in years to come. Tax and its ramifications, such as tax credits, are increasingly viewed by the general public as being unfair and unreasonable and, whilst tax collection is regarded by the Revenue as being easier, this is only because taxpayers are now responsible for paying rather than the Revenue being responsible for collecting.

It is against this background that authors have to comply with the law and complete and submit a tax return each year.

Income tax

What is a professional writer for tax purposes?

Writers are professionals while they are writing regularly with the intention of making a profit; or while they are gathering material, researching or otherwise preparing a publication.

A professional freelance writer is taxed under section 5 Income Tax (Trading and Other Income) Act 2005. The taxable income is the amount receivable, either directly or by an agent, on his behalf, less expenses wholly and exclusively laid out for the purpose of the profession. If expenses exceed income, the loss can either be set against other income of the same or preceding years or carried forward and set against future income from writing. If tax has been paid on that other income, a repayment can be obtained, or the sum can be offset against other tax liabilities. Special loss relief can apply in the opening years of the profession. Losses made in the first four years can be set against income of up to three earlier years.

Where a writer receives very occasional payments for isolated articles, it may not be possible to establish that these are profits arising from carrying on a continuing profession. In such circumstances these 'isolated transactions' may be assessed under section 687 Income Tax (Trading and Other Income) Act 2005. Again, expenses may be deducted in arriving at the taxable income but, if expenses exceed income, the loss can only be set against the profits from future isolated transactions or other income assessable under section 687.

Authors are taxed under a system called 'Self Assessment', where the onus is on the individual to declare income and expenses correctly. Each writer therefore has to decide whether profits arise from a professional or occasional activity. The consequences of getting it wrong can be expensive by way of interest, penalties and surcharges on additional tax subsequently found to be due. If in any doubt the writer should seek professional advice.

Income

A writer's income includes fees, advances, royalties, commissions, sale of copyrights, reimbursed expenses, etc. from any source anywhere in the world, whether or not brought to the UK (non-UK resident or domiciled writers should seek professional advice).

Agents

It should be borne in mind that the agent stands in the shoes of the principal. It is not always realized that when the agent receives royalties, fees, advances, etc. on behalf of the author those receipts became the property of the author on the date of their receipt by the agent. This applies for income tax and value added tax purposes.

Expenses

A writer can normally claim the following expenses:

(a) Secretarial, typing, proofreading, research. Where payment for these is made to the author's wife or husband they should be recorded and entered in the spouse's tax return as earned income which is subject to the usual personal allowances. If payments reach relevant levels, PAYE should be operated.

(b) Telephone, faxes, Internet costs, computer software, postage, stationery, printing, equipment maintenance, insurance, dictation tapes, batteries, any equipment or office requisites used for the profession.

(c) Periodicals, books (including presentation copies and reference books) and other publications necessary for the profession; however, amounts received from the sale of books should be deducted.

(d) Hotels, fares, car running expenses (including repairs, petrol, oil, garaging, parking, cleaning, insurance, road fund tax, depreciation), hire of cars or taxis in connection with

(i)   Business discussions with agents, publishers, co-authors, collaborators, researchers, illustrators, etc.

(ii)   Travel at home and abroad to collect background material.

As an alternative to keeping details of full car running costs, a mileage rate can be claimed for business use. This rate depends on the engine size and varies from year to year. This is known as the Fixed Profit Car Scheme and is available to writers whose turnover does not exceed the VAT registration limit, currently £67,000.

(e)   Publishing and advertising expenses, including costs of proof corrections, indexing, photographs, etc.

(f)   Subscriptions to societies and associations, press cutting agencies, libraries, etc. incurred wholly for the purpose of the profession.

(g)   Rent, council tax, water rates, etc., the proportion being determined by the ratio of the number of rooms, used exclusively for the profession, to the total number of rooms in the residence. But see note on capital gains tax below.

(h)   Lighting, heating, cleaning. A carefully calculated figure of the business use of these costs can be claimed as a proportion of the total.

(i)   Agent's commission, accountancy charges and legal charges incurred wholly in the course of the profession including the cost of defending libel actions, damages insofar as they are not covered by insurance, and libel insurance premiums. However, where in a libel case damages are awarded to punish the author for having acted maliciously, the action becomes quasi-criminal and costs and damages may not be allowed.

(j)   TV and video rental (which may be apportioned for private use), and cinema or theatre tickets, if wholly for the purpose of the profession.

(k)   Capital allowances. The Finance Bill 2008 includes proposals that, with effect from 6 April 2008, capital allowances will be as shown below. For businesses whose accounting period spans 6 April 2008, the rate will be a mixture of the old and new rates, apportioned pro rata by time:

(i)   On motor cars the allowance is 20 per cent in the first year and 20 per cent of the reducing balance in each successive year, limited to £3,000 each year. Prior to 6 April 2008, the allowances were 25 per cent. Balances brought forward at 6 April 2008 will be relieved at 20 per cent.

(ii)   For all other business equipment, e.g. TV, radio, hi-fi sets, tape and video recorders, dictaphones, office furniture, photographic equipment, etc., all businesses, including authors, will be able to claim the new Annual Investment Allowance (AIA). The AIA allows relief at 100 per cent on the first £50,000 spent on business equipment each year. This should be more than enough to meet the needs of most authors. If expenditure in a year does exceed £50,000, an allowance of 20 per cent is given against the balance.

For small businesses (which most authors will be), prior to 6 April 2008, there was a first year allowance of 50 per cent of the expenditure on business equipment. After the first year there was an annual writing down allowance of 25 per cent of the reducing balance.

All allowances for both cars and equipment will be reduced to exclude personal (non-professional) use where necessary.

(l) Lease rent. The cost of lease rent of equipment is allowable; also on cars, subject to restrictions for private use, and for expensive cars.

(m) Other expenses incurred wholly and exclusively for professional purposes. (Entertaining expenses are not allowable in any circumstances.)

Note: It is essential to keep detailed records. Diary entries of appointments, notes of fares and receipted bills are much more convincing to the Inland Revenue who are very reluctant to accept estimates.

The Self Assessment regime makes it a legal requirement for proper accounting records to be kept. These records must be sufficient to support the figures declared in the tax return.

In addition to the above, tax relief is available on:

(a) Premiums to a pension scheme. On 6 April 2006 a new pension-scheme tax regime came into effect which replaced all previous rules for occupational, personal pension and retirement annuity schemes. An individual can make, and tax relief is available on, contributions up to the higher of the full amount of relevant income or £3,600, subject to a maximum annual limit. That limit for 2007/08 was £225,000 and for 2008/09 is £235,000.

(b) Gift Aid payments to charities – for any amount.

Capital gains tax

The exemption from capital gains tax which applies to an individual's main residence does not apply to any part of that residence which is used exclusively for business purposes. The appropriate proportion of any increase in value of the residence, since purchase or 31 March 1982 if later, can be taxed when the residence is sold, subject to adjustment for inflation to March 1998 and subsequent period of ownership, at the individual's highest rate of tax.

Writers who own their houses should bear this in mind before claiming expenses for the use of a room for writing purposes. Arguments in favour of making such claims are that they afford some relief now, while capital gains tax in its present form may not stay for ever. Also, where a new house is bought in place of an old one, the gain made on the sale of the first study may be set off against the cost of the study in the new house, thus postponing the tax payment until the final sale. For this relief to apply, each house must have a study and the author must continue his profession throughout. On death there is an exemption of the total capital gains of the estate.

Alternatively, writers can claim that their use is non-exclusive and restrict their claim to the cost of extra lighting, heating and cleaning to avoid any capital gains tax liability.

Can a writer average out his income over a number of years for tax purposes?

Writers are able to average their profits (made wholly or mainly from creative works) over two or more consecutive years. If the profits of the lower year are less than 70 per cent of the profits of the higher year, or the profits of one year (but not both) are nil, then the author will be able to claim to have the profits averaged. Where the profits of the lower year are more than 70 per cent but less than 75 per cent of the profits of the higher year, a pro rata adjustment is made to both years to reduce the difference between them.

It is also possible to average out income within the terms of publishers' contracts, but professional advice should be taken before signature. Where a husband and wife collaborate as writers, advice should be taken as to whether a formal partnership agreement should be made or whether the publishing agreement should be in joint names.

Is a lump sum paid for an outright sale of the copyright, or part of it, exempt from tax?

No. All the money received from the marketing of literary work, by whatever means, is taxable. Some writers, in spite of clear judicial decisions to the contrary, still seem to think that an outright sale of, for instance, the film rights in a book is not subject to tax. The averaging relief described above should be considered.

Remaindering

To avoid remaindering, authors can usually purchase copies of their own books from the publishers. Monies received from sales are subject to income tax but the cost of books sold should be deducted because tax is only payable on the profit made.

Is there any relief where old copyrights are sold?

No. There was relief available until April 2001 but it was then withdrawn. The averaging relief described above should be considered.

Are royalties payable on the publication of a book abroad subject to both foreign tax as well as UK tax?

Where there is a Double Taxation Agreement between the country concerned and the UK, then on the completion of certain formalities no tax is deductible at source by the foreign payer, but such income is taxable in the UK in the ordinary way. Where there is no Double Taxation Agreement, credit will be given against UK tax for overseas tax paid. A complete list of countries with which the UK has conventions for the avoidance of double taxation may be obtained from the Centre for Non-Residents, Inland Revenue, St John's House, Merton Road, Bootle, Merseyside L69 9BB, or a local tax office.

Residence abroad

Writers residing abroad will, of course, be subject to the tax laws ruling in their country of residence, and as a general rule royalty income paid from the UK can be exempted from deduction of UK tax at source, providing the author is carrying on his profession abroad. A writer who is intending to go and live abroad should make early application for future royalties to be paid without deduction of tax to the Centre for Non-Residents (address as above). In certain circumstances writers resident in the Irish Republic are exempt from Irish income tax on their authorship earnings.

Are grants or prizes taxable?

The law is uncertain. Some Arts Council grants are now deemed to be taxable, whereas most prizes and awards are not, though it depends on the conditions in each case. When submitting the Self Assessment annual returns, such items should be excluded but reference made to them in the 'Additional Information' box on the self-employment (or partnership) pages.

What is the item 'Class 4 N.I.C.' which appears on my Self Assessment return?

All taxpayers who are self-employed pay an additional national insurance contribution if their earned income exceeds a figure which varies each year. This contribution is described as Class 4 and is calculated when preparing the return. It is additional to the self-employed Class 2 contribution but confers no additional benefits and is a form of levy. It applies to men aged under 65 and women under 60.

Should an author use a limited company?

The tax regime has made the incorporation of businesses attractive and there can be advantages in so doing even for businesses with relatively low levels of profit, although these advantages have been reduced in recent Budgets. However, there are disadvantages in all cases and there are particular considerations for authors. The advice of an accountant, knowledgeable about the affairs of authors, should be sought if incorporation is contemplated.

Value added tax

Value added tax (VAT) is a tax currently levied at 17.5 per cent on:

(a) The total value of taxable goods and services supplied to consumers
(b) The importation of goods into the UK
(c) Certain services or goods from abroad if a taxable person receives them in the UK for the purpose of their business

Who is taxable?

A writer resident in the UK whose turnover from writing and any other business, craft or art on a self-employed basis is greater than £67,000 annually, before deducting an agent's commission, must register with HM Revenue & Customs as a taxable person. Turnover includes fees, royalties, advances, commissions, sale of copyright, reimbursed expenses, etc.

A business is required to register:

• at the end of any month if the value of taxable supplies in the past 12 months has exceeded the annual threshold; or

- if there are reasonable grounds for believing that the value of taxable supplies in the next 12 months will exceed the annual threshold.

Penalties will be claimed in the case of late registration. A writer whose turnover is below these limits is exempt from the requirements to register for VAT but may apply for voluntary registration, which will be allowed at the discretion of HM Revenue & Customs.

A taxable person collects VAT on outputs (turnover) and deducts VAT paid on inputs (taxable expenses) and, where VAT collected exceeds VAT paid, must remit the difference to HM Revenue & Customs. In the event that input exceeds output, the difference will be refunded by HM Revenue & Customs.

Inputs (expenses)

Taxable at the standard rate if supplier is registered	Taxable at the zero or special rate	Not liable to VAT
Rent of certain commercial premises	Books (zero)	Rent of non-commercial premises
Advertisements in newspapers, magazines, journals and periodicals	Coach, rail and air travel (zero)	Postage
Agent's commission (unless it relates to monies from overseas)	Agent's commission (on monies from overseas)	Services supplied by unregistered persons
Accountant's and solicitor's fees for business matters	Domestic gas and electricity (5%)	Subscriptions to the Society of Authors, PEN, NUJ, etc.
Agency services (typing, copying, etc.)	Insurance	
Word processors, typewriters and stationery		*Outside the scope of VAT*
Artists' materials		PLR (Public Lending Right)
Photographic equipment		Profit shares
Tape recorders and tapes		Investment income
Hotel accommodation		
Taxi fares		
Motorcar expenses		
Telephone		
Theatres and concerts		

Note: This table is not exhaustive.

Outputs (turnover)

A writer's outputs are taxable services supplied to publishers, broadcasting organizations, theatre managements, film companies, educational institutions, etc. A taxable writer must invoice all the persons (either individuals or organizations) in the UK to whom supplies have been made, for fees, royalties or other considerations, plus VAT. An unregistered writer cannot and must not invoice for VAT. A taxable writer is not obliged to collect VAT on royalties or other fees paid by publishers or others overseas. In practice, agents usually collect VAT for the registered author, but these authors must remember to do so if they receive income that does not go through their agent, e.g. payments for public readings.

Remit to Customs

The taxable writer adds up the VAT which has been paid on taxable inputs, deducts it from the VAT received and remits the balance to Customs. Business with HM Customs is conducted through the local VAT offices of HM Revenue & Customs which are listed in local telephone directories, except for VAT returns which are sent direct to the HM Revenue & Customs VAT Central Unit, Alexander House, 21 Victoria Avenue, Southend on Sea, Essex SS99 1AA.

Accounting

A taxable writer is obliged to account to HM Revenue & Customs at quarterly intervals. Returns must be completed and sent to VAT Central Unit by the dates shown on the return. Penalties can be charged if the returns are late.

It is possible to account for the VAT liability under the Cash Accounting Scheme (leaflet 731), whereby the author accounts for the output tax when the invoice is paid or royalties etc. are received. The same applies to the input tax, but as most purchases are probably on a 'cash basis', this will not make a considerable difference to the author's input tax. This scheme is only applicable to those with a taxable turnover of less than £1,350,000 and therefore is available to the majority of authors. The advantage of this scheme is that the author does not have to account for VAT before receiving payments, thereby relieving the author of a cash flow problem.

If turnover is less than or expected to be less than £1,350,000 it is possible to pay VAT by monthly instalments, with a final balance at the end of the year (see leaflet 732). This annual accounting method also means that only one VAT return is submitted.

Flat rate scheme

Small businesses can elect to pay VAT under a flat rate scheme (FRS). This is open to businesses with business income up to £150,000 a year. Under the normal VAT accounting rules, each item of turnover and every claimed expense must be recorded and supported by evidence, e.g. invoices, receipts, etc. Under the FRS, detailed records of sales and purchases do not have to be kept. A record of gross income (including zero rated and excepted income) is maintained and a flat rate percentage is applied to the total. This percentage is then paid over to HM Revenue & Customs. The percentage varies from one profession or business to another, but for authors it is 11 per cent. In the first year of registration this percent is reduced by 1%.

The aim of the scheme is to reduce the amount of time and money spent in complying with VAT regulations, and this is to be welcomed. However, there are disadvantages:

- The detailed records of income and expenses are still going to be required for taxation purposes.
- Invoices on sales are issued in the normal way.
- The percentage is applied to all business income. So 11 per cent VAT will effectively be paid on income from abroad, zero rated under the normal basis, and PLR, otherwise exempt.

For many authors, normal VAT accounting has imposed a good, timely discipline for dealing with accounting and taxation matters.

Registration

A writer will be given a VAT registration number which must be quoted on all VAT correspondence. It is the responsibility of those registered to inform those to whom they make supplies of their registration number. The taxable turnover limit which determines whether a person who is registered for VAT may apply for cancellation of registration is £65,000.

Voluntary registration

A writer whose turnover is below the limits may apply to register. If the writer is paying a relatively large amount of VAT on taxable inputs – agent's commission, accountant's fees, equipment, materials, or agency services, etc. – it may make a significant improvement in the net income to be able to offset the VAT on these inputs. A writer who pays relatively little VAT may find it easier, and no more expensive, to remain unregistered.

Fees and royalties

A taxable writer must notify those to whom he or she makes supplies of the VAT registration number at the first opportunity. One method of accounting for and paying VAT on fees and royalties is the use of multiple stationery for 'self-billing', one copy of the royalty statement being used by the author as the VAT invoice. A second method is for the recipient of taxable outputs to pay fees, including authors' royalties, without VAT. The taxable writer then renders a tax invoice for the VAT element, for which a second payment of this element will be made. This scheme is cumbersome but does involve only taxable authors. Fees and royalties from abroad will count as payments for exported services and will accordingly be zero-rated.

Agents and accountants

A writer is responsible to HM Revenue & Customs for making VAT returns and payments. Neither an agent nor an accountant nor a solicitor can remove the responsibility, although they can be helpful in preparing and keeping VAT returns and accounts. Their professional fees or commission will, except in rare cases where the adviser or agent is himself unregistered, be taxable at the standard rate and will represent some of a writer's taxable inputs.

Income tax

An unregistered writer can claim some of the VAT paid on taxable inputs as a business expense allowable against income tax. However, certain taxable inputs fall into categories which cannot be claimed under the income tax regulations. A taxable writer, who has already claimed VAT on inputs, cannot charge it as a business expense for the purposes of income tax.

Certain services from abroad

A taxable author who resides in the UK and who receives certain services from abroad must account for VAT on those services at the appropriate tax rate on the sum paid for them. Examples of the type of services concerned include the services of lawyers, accountants, consultants and the provision of information and copyright permissions.

Inheritance tax

Inheritance tax was introduced in 1984 to replace capital transfer tax, which had in turn replaced estate duty, the first of the death taxes of recent times. Paradoxically, inheritance tax has reintroduced a number of principles present under the old estate duty.

The general principle now is that all assets owned at death are chargeable to tax (currently 40 per cent) except for the first £312,000 (the nil rate band) of the estate and any assets passed to a surviving spouse, civil partner or a charity. The Finance Bill 2008 includes legislation that the proportion of any unused nil rate band on a person's death can be transferred to the estate of their spouse or civil partner, if he or she dies after 9 October 2007. This means that wills, which contain provisions for the nil rate band to go to a discretionary trust (a common tax planning procedure in recent years), should be reviewed.

Gifts made more than seven years before death are exempt, but those made within this period may be taxed on a sliding scale. No tax is payable at the time of making the gift.

In addition, each individual may currently make gifts of up to £3,000 in any year and these will be considered to be exempt. A further exemption covers any number of annual gifts not exceeding £250 to any one person.

If the £3,000 is not fully utilized in one year, any unused balance can be carried forward to the following year (but no later). Gifts out of income, which do not reduce one's living standards, are also exempt if they are part of normal expenditure.

At death, all assets are valued. They will include any property, investments, life policies, furniture and personal possessions, bank balances and, in the case of authors, the value of copyrights. All, with the sole exception of copyrights, are capable (as assets) of accurate valuation and, if necessary, can be turned into cash. The valuation of copyright is, of course, complicated and frequently gives rise to difficulty. Except where they are bequeathed to the owner's husband or wife, very real problems can be left behind by the author.

Experience has shown that a figure based on two to three years' past royalties may be proposed by the Inland Revenue in their valuation of copyright. However, this may not be reasonable and may require negotiation. If a book is running out of print or if, as in the case of educational books, it may need revision at the next reprint, these factors must be taken into account. In many cases the fact that the author is no longer alive and able to make personal appearances, or provide publicity, or write further works, will result in lower or slower sales. Obviously, this is an area in which help can be given by the publishers, and in particular one needs to know what their future intentions are, what stocks of the books remain, and what likelihood there will be of reprinting.

There is a further relief available to authors who have established that they have been carrying on a business for at least two years prior to death. It has been possible to establish that copyrights are treated as business property and in these circumstances inheritance tax 'business property relief' is available. This relief at present is 100 per cent so that the tax saving can be quite substantial. The Inland Revenue may wish to be assured that the business is continuing and consideration should therefore be given to (i) the appointment, in the author's will, of a literary executor, who should be a qualified business person; or (ii), in certain circumstances, the formation of a partnership between the author and spouse, or other relative, to ensure that it is established that the business is continuing after the author's death.

If the author has sufficient income, consideration should be given to building up a fund to cover future inheritance tax liabilities. One of a number of ways to do this would be to take out a whole life assurance policy which is assigned to the children, or other beneficiaries, the premiums for which are within the annual exemption of £3,000. The capital sum payable on the death of the assured is exempt from inheritance tax.

Tax credits

Working authors may also be eligible to claim Child Tax Credit and Working Tax Credit.

Child tax credit is the main way that families receive money for their younger children and for 16–19 year olds in education. A claim is based on your income and you can claim whether or not you are working. All families (with children) with an income up to £58,175 a year (or up to £66,350 a year if your child is under one) can claim. To qualify, you must be aged 16 or over, usually live in the UK and, if single, (or separated) claim according to your individual circumstances. If you are married or a man and woman living together (or a same sex couple from December 2005) you must claim together based on joint circumstances and income.

Working Tax Credit is a payment to top up earnings of the lower paid (whether employed or self–employed, including those who do not have children). To qualify, you must be aged 16 or over and you or your partner responsible for a child and you must work at least 16 hours a week. If you are part of a couple with children, greater amounts may be claimed if you jointly work at least 30 hours a week, provided one of you works for at least 16 hours. Childless couples cannot add their hours together to qualify for the 30-hour element. If you do not have children and you do not have a disability, you must be aged 25 or over and work at least 30 hours a week. It is also possible that your working tax credit will be increased to assist with the cost of registered or approved childcare.

* * *

Anyone wondering how best to order his affairs for tax purposes should consult an accountant with specialized knowledge in this field. Experience shows that a good accountant is well worth his fee which, incidentally, so far as it relates to professional matters, is an allowable expense.

The information contained in this section has been prepared by Ian Spring of Moore Stephens, Chartered Accountants, who will be pleased to answer questions on tax problems. Please write to Ian Spring, c/o The Writer's Handbook, 34 Ufton Road, London N1 5BX.

UK Publishers

AA Publishing
The Automobile Association, Fanum House, Basingstoke RG21 4EA
℡ 01256 491524   ℻ 01256 491974
AAPublish@TheAA.com
www.theAA.com
Publisher *David Watchus*

Publishes maps, atlases and guidebooks, motoring, travel and leisure; also children's books. About 100 titles a year.

Abacus
▷ Little, Brown Book Group UK

ABC-CLIO
PO Box 1437, Oxford OX4 9AZ
℡ 01865 481403   ℻ 01865 481482
salesuk@abc-clio.com
www.abc-clio.com

Formerly Clio Press Ltd. Publishes academic and general reference works, history and social studies. Markets, outside North America, the Web and reference books of the American parent company.
£ Royalties twice-yearly.

Absolute Press
Scarborough House, 29 James Street West, Bath BA1 2BT
℡ 01225 316013   ℻ 01225 445836
office@absolutepress.co.uk
www.absolutepress.co.uk
Managing Director *Jon Croft*

Founded 1979. Publishes food and wine-related subjects. About 15 titles a year. No unsolicited mss. Synopses and ideas for books welcome.
£ Royalties twice-yearly.

Abson Books London
5 Sidney Square, London E1 2EY
℡ 020 7790 4737   ℻ 020 7790 7346
absonbooks@aol.com
www.absonbooks.co.uk
Publisher *M.J. Ellison*
Manager *Sharon Wright*

Founded 1971 in Bristol. Publishes language and dialect glossaries and curiosities. 2 titles in 2007.

No unsolicited mss; synopses and ideas for books welcome.
£ Royalties twice-yearly.

Academic Press
▷ Elsevier Ltd

Acair Ltd
7 James Street, Stornoway, Isle of Lewis HS1 2QN
℡ 01851 703020   ℻ 01851 703294
info@acairbooks.com
www.acairbooks.com

Specializing in matters pertaining to the Gaidhealtachd, Acair publishes books in Gaelic and English on Scottish history, culture and the Gaelic language. 75% of their children's books are targeted at primary school usage and are published exclusively in Gaelic.
£ Royalties twice-yearly.

Acumen Publishing Limited
15A Lewins Yard, East Street, Chesham HP5 1HQ
℡ 01494 794398   ℻ 01494 784850
steven.gerrard@acumenpublishing.co.uk
www.acumenpublishing.co.uk
Managing Director *Steven Gerrard*

Founded in 1998 as an independent publisher for the higher education market. Publishes academic books on philosophy, classics and history of ideas. No unsolicited mss. Synopses and ideas for books welcome; send written proposal as per Acumen guidelines.
£ Royalties annually.

Addison Wesley
▷ Pearson

Adlard Coles Nautical
▷ A.&C. Black Publishers Ltd

African Books Collective
PO Box 721, Oxford OX1 9EN
www.africanbookscollective.com

Founded 1990. Collectively owned by its 17 founder publishers. Exclusive distribution in North America through Michigan State University Press, UK, Europe and Commonwealth countries outside Africa for 116 African participating publishers. Concentration is on scholarly/academic, literature and children's books. Mainly concerned with the promotion and dissemination of African-published material outside Africa. Supplies African-published books to African libraries and organizations, and publishes resource books on the African publishing industry. TITLES *The African Writers' Handbook*; *African Publishers Networking Directory*; *The Electronic African Bookworm: A Web Navigator*; *Courage and Consequence: Women Publishing in Africa*; *African Scholarly Publishing*.

Age Concern Books

1268 London Road, London SW16 4ER
Ⓣ 020 8765 7200   Ⓕ 020 8765 7211
books@ace.org.uk
www.ageconcern.org.uk/bookshop
Approx. Annual Turnover £400,000+

Publishing arm of Age Concern England. Publishes non-fiction only (no fiction or biographies), including the *Your Rights* series, computing titles for the over 50s, financial handbooks and a series of *Carers Handbooks*; also a wide range of books and training packs for professionals. About 10 titles a year. Synopses and ideas welcome.

Airlife Publishing
▷ The Crowood Press Ltd

Akros Publications

33 Lady Nairn Avenue, Kirkcaldy KY1 2AW
Ⓣ 01592 651522
www.akrospublications.co.uk
Publisher *Duncan Glen*

Founded 1965. Publishes poetry collections, pamphlets and anthologies; literary essays and studies; local histories. Also *Z2O* poetry magazine. No fiction. About 10 titles a year. No unsolicited mss.
Ⓔ Royalties twice-yearly.

Alcemi
▷ Y Lolfa Cyf

Ian Allan Publishing Ltd

Riverdene Business Park, Molesey Road, Hersham KT12 4RG
Ⓣ 01932 266600   Ⓕ 01932 266601
info@ianallanpublishing.co.uk
www.ianallanpublishing.com
Chairman *David Allan*
Managing Director *Tristan Hilderley*

Specialist transport publisher – atlases, maps, railway, aviation, road transport, military, maritime, reference, Freemasonry. Manages distribution and sales for third party publishers. IMPRINTS **Ian Allan Publishing**; **Classic** Aviation titles; **Midland Publishing** (see entry); **OPC**

Railway titles; **Lewis**; **Lewis Masonic**. About 100 titles a year. Send sample chapter and synopsis with s.a.e.

J.A. Allen & Co.

An imprint of Robert Hale Ltd, Clerkenwell House, 45–47 Clerkenwell Green, London EC1R 0HT
Ⓣ 020 7251 2661   Ⓕ 020 7490 4958
allen@halebooks.com
www.halebooks.com
Publisher *Cassandra Campbell*
Approx. Annual Turnover £1 million

Founded 1926 as part of J.A. Allen & Co. (The Horseman's Bookshop) Ltd. Bought by **Robert Hale Ltd** in 1999. Publishes equine and equestrian non-fiction. About 20 titles a year. Mostly commissioned but willing to consider unsolicited mss of technical/instructional material related to all aspects of horses and horsemanship.
Ⓔ Royalties twice-yearly.

Allen Lane
▷ Penguin Group (UK)

Allison & Busby

13 Charlotte Mews, London W1T 4EJ
Ⓣ 020 7580 1080   Ⓕ 020 7580 1180
susie@allisonandbusby.com
www.allisonandbusby.com
Publishing Director *Susie Dunlop*
Approx. Annual Turnover £1 million

Founded 1967. Publishes fiction, historical fiction, crime fiction, true crime, pop culture, romance and fantasy. TITLES *Republic* trilogy by Jack Ludlow; *The Railway Detective* series Edward Marston; *Boomsday* Christopher Buckley; *The Morganville Vampire* series Rachel Caine. About 90 titles a year. No unsolicited mss.

✳*Authors' update* 'A testament to the ability of smaller publishers to survive in the retail jungle' (*Publishing News*), A&B is moving into the fantasy and science fiction market. The crime list is thriving.

Allyn & Bacon
▷ Pearson

Alma Books Ltd

London House, 243–253 Lower Mortlake Road, Richmond TW9 2LL
Ⓣ 020 8948 9550   Ⓕ 020 8948 5599
info@almabooks.com
www.almabooks.com
Chairman *Elisabetta Minervini*
Managing Director *Alessandro Gallenzi*
Approx. Annual Turnover £450,000

Established in 2005 by founders of **Hesperus Press**. Acquired **Calder Publications** in 2007.

Publishes general fiction, biography, history. 20 titles in 2007. Accepts submissions by post. Ⓔ Royalties annually.

❋*Authors' update* Alma is Spanish for 'soul', an apt name say the company founders for a publisher that 'regards a book as an aesthetic artefact rather than a mass produced commodity'.

Alpha Press
▷ Sussex Academic Press

Alton Douglas Books
▷ Brewin Books Ltd

Amber Lane Press Ltd
Cheorl House, Church Street, Charlbury OX7 3PR
☏ 01608 810024   ⊞ 01608 810024
info@amberlanepress.co.uk
www.amberlanepress.co.uk
Chairman *Brian Clark*
Managing Director/Editorial Head *Judith Scott*

Founded 1979 to publish modern play texts. Publishes plays and books on the theatre. TITLES *Whose Life is it Anyway?* Brian Clark; *The Dresser* Ronald Harwood; *Once a Catholic* Mary O'Malley (play texts); *Strindberg and Love* Eivor Martinus. About 4 titles a year. 'Expressly *not* interested in poetry.' No unsolicited mss.
Ⓔ Royalties twice-yearly.

Amsco
▷ Omnibus Press

Andersen Press Ltd
20 Vauxhall Bridge Road, London SW1V 2SA
☏ 020 7840 8703/8701   ⊞ 020 7233 6263
anderseneditorial@randomhouse.co.uk
www.andersenpress.co.uk
Managing Director/Publisher *Klaus Flugge*
Editorial Director *Rona Selby*
Editor, Fiction *Elizabeth Maude*

Founded 1976 by Klaus Flugge and named after Hans Christian Andersen. Publishes children's high-quality picture books and fiction sold in association with Random House Children's Books. Seventy per cent of the books are sold as co-productions abroad. TITLES *Elmer* David McKee; *I Want My Potty* Tony Ross; *little.com* Ralph Steadman; *Cat in the Manger* Michael Foreman; *Preston Pig Books* Colin McNaughton; *Junk* Melvin Burgess; *You* Sandra Glover; *I Love You, Blue Kangaroo* Emma Chichester Clark. Unsolicited mss welcome for picture books; synopsis in the first instance for books for young readers up to age 12. No poetry or short stories.
Ⓔ Royalties twice-yearly.

❋*Authors' update* The secret of Andersen's success is publishing children's books that are loved by adults. That and a wicked sense of fun that has been known to upset the education establishment – and hooray for that.

Chris Andrews Publications
15 Curtis Yard, North Hinksey Lane, Oxford OX2 0LX
☏ 01865 723404   ⊞ 01865 725294
chris.andrews1@btclick.com
www.cap-ox.co.uk
Directors *Chris Andrews, Virginia Andrews*

Founded 1982. Publishes coffee-table, scenic travel guides. Also calendars, diaries, cards and posters in the central area of the UK and the Channel Islands. TITLES *Romance of Oxford*; *Romance of the Cotswolds*; *Romance of the Thames & Chilterns*. Also owns the **Oxford Picture Library** (see entry under *Picture Libraries*). Unsolicited synopses and ideas for travel/guide books in the areas listed above will be considered; phone in the first instance.

The Angels' Share
▷ Neil Wilson Publishing Ltd

Anness Publishing Ltd
Hermes House, 88–89 Blackfriars Road, London SE1 8HA
☏ 020 7401 2077   ⊞ 020 7633 9499
info@anness.com
www.aquamarinebooks.com
www.lorenzbooks.com
www.southwaterbooks.com
Chairman/Managing Director *Paul Anness*
Publisher/Partner *Joanna Lorenz*
Approx. Annual Turnover £15.7 million

Founded 1989. Publishes highly illustrated co-edition titles: general non-fiction – cookery, crafts, interior design, gardening, photography, decorating, lifestyle, health, reference, military, transport and children's. IMPRINTS **Lorenz Books; Aquamarine; Hermes House; Southwater**. About 200 titles a year.

Anova Books
10 Southcombe Street, London W14 0RA
☏ 020 7605 1400   ⊞ 020 7605 1401
info@anovabooks.com
firstinitialsurname@anovabooks.com
www.anovabooks.com porticobooks.co.uk
CEO *Robin Wood*
Managing Director *Polly Powell*
Approx. Annual Turnover £14 million

Anova Books, formerly Chrysalis Books Group, specializes in illustrated non-fiction, publishing under the IMPRINTS **B.T. Batsford; Collins & Brown; Conway; National Trust Books; Paper Tiger; Pavilion; Portico; Robson Books; Salamander**.

IMPRINTS **B.T. Batsford** Associate Publisher *Tina Persaud* Founded in 1843 as a bookseller, and began publishing in 1874. 'A world leader in books on chess, arts and craft.' Publishes non-fiction, architecture, heritage, bridge and chess, film and entertainment, fashion, crafts and hobbies. About 100 titles a year.

Collins & Brown Associate Publisher *Katie Cowan* Publishes a range of lifestyle categories especially in the areas of practical art, photography and needlecrafts.

Conway Associate Publisher *John Lee* Publishes naval history, maritime culture, ship modelling and military history.

National Trust Books Associate Publisher *Tina Persaud* Imprint of Anova Books since 2005. Publishes heritage, gardening, architecture, gift and cookery-based titles. Unsolicited mss not welcome.

Paper Tiger Associate Publisher *Katie Cowan* (Part of Collins & Brown) Publishes science fiction and fantasy art.

Pavilion Associate Publisher *Polly Powell* Publishes illustrated books in biography, cookery, gardening, humour, art, interiors, music, sport and travel. Ideas and synopses for non-fiction titles considered.

Pavilion Children's Books Associate Publisher *Ben Cameron* Publishes innovative, informative and fun books for children of all ages.

Portico Associate Publisher *Tom Bromley* Publishes cutting edge non-fiction which is fresh, funny and forthright.

Robson Books Associate Publisher *Tom Bromley* Founded 1973. Publishes general non-fiction including biography, cookery, gardening, sport and travel. About 60 titles a year. Unsolicited synopses and ideas for books welcome (s.a.e. essential for reply).

Salamander Publisher *Polly Powell* Founded 1973. Publishes colour illustrated books on collecting, cookery, interiors, gardening, music, crafts, military, aviation, sport and transport. Also a wide range of books on American interest subjects. No unsolicited mss but synopses and ideas for the above subjects welcome.

Anthem Press
▷ **Wimbledon Publishing Company**

Antique Collectors' Club
Sandy Lane, Old Martlesham, Woodbridge
IP12 4SD
☎ 01394 389950　🖷 01394 389999
sales@antique-acc.com
www.antiquecollectorsclub.com
Managing Director *Diana Steel*

Founded 1966. Publishes specialist books on art, antiques and collecting, decorative arts, architecture, gardening and fashion. The Price Guide series was introduced in 1968 with the first edition of *Price Guide to Antique Furniture*. Subject areas include furniture, silver and gold, metalwork, jewellery, glass, textiles, art reference, ceramics, horology. TITLES *Roses for Your Garden*; *Paul Nash – John Nash Design* Brian Webb and Peyton Skipwith; *Biba. The Biba Experience* Alwyn Turner; *Understanding Jewellery* David Bennett and Daniela Mascetti. Also publishes subscription-only magazine, *Antique Collecting*, published 10 times a year. Unsolicited synopses and ideas for books and magazine articles welcome; no mss.
💷 Royalties twice-yearly as a rule, but can vary.

Anvil Press Poetry Ltd
Neptune House, 70 Royal Hill, London SE10 8RF
☎ 020 8469 3033　🖷 020 8469 3363
anvil@anvilpresspoetry.com
www.anvilpresspoetry.com
Managing Director *Peter Jay*

Founded 1968. England's oldest independent publishing house dedicated to poetry. Publishes new poets in English and in translation as well as translated classics. AUTHORS include Paul Celan, Carol Ann Duffy, Martina Evans, Michael Hamburger, Dennis O'Driscoll, A.B. Jackson, Stanley Moss and Greta Stoddart. Classic translations include Apollinaire, Baudelaire, Dante, Hikmet, Neruda, Tagore and Verlaine. Preliminary enquiry required for translations. Unsolicited book-length collections of poems are welcome from writers whose work has appeared in poetry magazines or literary journals. Please enclose adequate return postage.

⊛*Authors' update* **With a little help from the Arts Council, Anvil has become one of the foremost publishers of living poets.**

Apex Publishing Ltd
PO Box 7086, Clacton on Sea CO15 5WN
☎ 01255 428500
mail@apexpublishing.co.uk
www.apexpublishing.co.uk
Managing Editor *Susan Kidby*
Publishing & Production Manager *Chris Cowlin*

Founded 2002. Independent publishing company for unknown and established authors. Publishes self-help, non-fiction, crime, history, humour, sport, biography, autobiography, pets, quizzes, religion and travel. 150 titles in print. Welcomes unsolicited mss, synopses and ideas for books. Approach in writing to the address above with a copy of the full ms.
💷 Royalties twice-yearly.

Apollos
▷ Inter-Varsity Press

Apple
▷ Quarto Publishing under UK Packagers

Appletree Press Ltd
The Old Potato Station, 14 Howard Street South,
Belfast BT7 1AP
☎ 028 9024 3074   🖷 028 9024 6756
reception@appletree.ie
www.appletree.ie
Managing Director *John Murphy*

Founded 1974. Publishes gift books, plus general
non-fiction of Irish and Scottish interest. TITLES
In St Patrick's Footsteps; *Ireland's Ancient
Stones*; *Scottish Lighthouses*; *Northern Ireland –
International Football Facts*. No unsolicited mss;
send initial letter or synopsis.
🖹 Royalties twice-yearly in the first year, annually
thereafter.

Aquamarine
▷ Anness Publishing Ltd

Arc Publications Ltd
Nanholme Mill, Shaw Wood Road, Todmorden
OL14 6DA
☎ 01706 812338   🖷 01706 818948
arc.publications@btconnect.com
www.arcpublications.co.uk
Editorial Director *Tony Ward*
Associate Editors *John Kinsella* (international),
Jo Shapcott (UK), *Jean Boase-Beier* (translations),
Angela Jarman (music)

Founded in 1969 to specialize in the publication of
contemporary poetry from new and established
writers both in the UK and abroad. AUTHORS
Philip Salom (Australia), Soleiman Adel (Algeria),
Regina Deriva & Larissa Miller (Russia), Mourid
Barghouti (Palestine), Keki Daruwalla (India),
Mary O'Donnell (Ireland), Brian Henry & Jill
Bialosky (USA), Razmic Davoyan (Armenia),
Victor Rodriguez Nunez (Cuba), Glen Baxter,
Jackie Wills, Paul Stubbs, Liz Almond & Michael
O'Neill (UK).

IMPRINT **Arc Music** specializes in profiles of
contemporary composers (particularly where
none have existed hitherto) and symposia which
take a 'new approach' to well-visited territory.
Commissioned work only. About 25 titles a
year. Unable to accept unsolicited material
for the foreseeable future. 'Please consult
our submissions policy on the website before
submitting any work.'
🖹 Royalties as per contracts.

Arcadia Books
15–16 Nassau Street, London W1W 7AB
☎ 020 7436 9898   🖷 020 7436 9898

info@arcadiabooks.co.uk
www.arcadiabooks.co.uk
Managing Director *Gary Pulsifer*
Associate Publisher *Daniela de Groote*
Managing Editor *Angeline Rothermundt*
Approx. Annual Turnover £500,000

Independent publishing house established in
1996. Specializes in quality translated fiction
from around the world. IMPRINTS **Arcadia**;
BlackAmber; Bliss; EuroCrime. TITLES *L'Oréal
Took My Home: The Secrets of a Theft* Monica
Waitzfelder; *The Hite Report on Women Loving
Women* Shere Hite; *The Autobiography of the
Queen* Emma Tennant; *No Trace* Barry Maitland.
Does not welcome unsolicited material.
🖹 Royalties twice-yearly.

⌖*Authors' update* **One of the lucky ones to
emerge from the recent Arts Council funding
reassessment, Arcadia had its grant restored to
give translated fiction a welcome boost.**

Arcane
▷ Omnibus Press

Architectural Press
▷ Elsevier Ltd

Ardis
▷ Gerald Duckworth & Co. Ltd

Argentum
▷ Aurum Press Ltd

Aris & Phillips
▷ Oxbow Books

Armchair Traveller
▷ Haus Publishing

Arnefold Publishing
PO Box 22, Maidstone ME14 1AH
☎ 01622 759591   🖷 01622 209193
Chairman/Managing Director *I.G. Mann*

Publishes original non-fiction and selected
fiction, and non-fiction reprints. No new fiction.
IMPRINTS **Arnefold; George Mann Books;
Recollections**. 'Will consider non-fiction titles
(which authors find hard to place) on a joint
venture/shared profit basis if they are worthy
of publication or can be made so. Unsolicited
material is welcomed provided it is understood
that if not accompanied by return postage it will
neither be read nor returned; that no new fiction
is wanted and that joint-venture publications may
be suggested. We are not vanity publishers.'
🖹 Royalties annually.

Arris Publishing Ltd
12 Main Street, Adlestrop, Moreton in Marsh
GL56 0YN
☎ 01608 658758   🖷 01608 659345

gcs@arrisbooks.com
www.arrisbooks.com
Joint Managing Directors *Geoffrey Smith,
Michel Moushabeck*
Publishing Director *Victoria Huxley*

Founded in 2001. Publishes politics, history, biography, travel, art and culture. IMPRINTS **Arris Books** *Geoffrey Smith* TITLES *9/11: The New Pearl Harbour; Hitler's British Slaves;* **Chastleton Travel** TITLES *Traveller's Wildlife Guides; A Traveller's History of Cyprus; 100 Best Paintings in London.* About 20 titles a year. No unsolicited mss; synopses and ideas for books welcome. Send introductory letter, synopsis and specimen chapter by post; no e-mail submissions or CDs. No poetry or fiction.
Ⓔ Royalties annually.

Arrow
▷ The Random House Group Ltd

Artech House
46 Gillingham Street, London SW1V 1AH
☏ 020 7596 8750   ℻ 020 7630 0166
splumtree@artechhouse.co.uk
www.artechhouse.com
CEO (USA) *William M. Bazzy*
Commissioning Editor *Dr Simon Plumtree*

Founded 1969. European office of Artech House Inc., Boston. Publishes books for practising professionals in biomedical engineering, bioinformatics, MEMS, nanotechnology, microwaves, radar, antennas and propagation, electro-magnetic analysis, wireless communications, telecommunications, space applications, remote sensing, solid state technology and devices, technology management. 60–70 titles a year. Unsolicited mss and synopses in the specialized areas listed are considered.
Ⓔ Royalties twice-yearly

Ashgate Publishing Ltd
Gower House, Croft Road, Aldershot GU11 3HR
☏ 01252 331551   ℻ 01252 34440 (Ashgate and Gower)/020 7440 7530 (Lund Humphries)
info@ashgatepublishing.com
www.ashgate.com
www.gowerpub.com and
www.lundhumphries.com
Chairman *Nigel Farrow*

Founded 1967. Publishes business and professional titles under the Gower imprint, humanities, social sciences, law and legal studies under the Ashgate imprint, and art and art history under the Lund Humphries imprint. Acquired Lund Humphries, publisher of art books and exhibition catalogues in 1999. DIVISIONS **Ashgate** *Adrian Shanks* Social sciences; *John Smedley* History/Variorum collected studies; *Heidi May*

Music; *Erika Gaffney* Literary studies; *Tom Gray* History; *Sarah Lloyd* Theology and religious studies; *Guy Loft* Aviation studies. **Gower** *Jonathan Norman* Business, management and training; *Alison Kirk* Law and legal studies. **Lund Humphries** *Lucy Myers*. Access the websites for information on submission of material.

Ashgrove Publishing
27 John Street, London WC1N 2BX
☏ 020 7831 5013   ℻ 020 7831 5011
gm073@dial.pipex.com
www.ashgrovepublishing.com
Chairman/Managing Director *Brad Thompson*

Acquired by Hollydata Publishers in 1999, Ashgrove has been publishing for over 20 years. Publishes mind, body, spirit, health, cookery, sports. 3 titles in 2007. No unsolicited mss; approach with letter and outline in the first instance.
Ⓔ Royalties twice-yearly.

Ashwell Publishing Limited
▷ Olympia Publishers

Atlantic Books
26–27 Boswell Street, London WC1N 3JZ
☏ 020 7269 1610   ℻ 020 7430 0916
enquiries@groveatlantic.co.uk
www.groveatlantic.co.uk
Chairman & CEO *Toby Mundy*
Managing Director *Daniel Scott*
Editor-in-Chief *Ravi Mirchandani*

Founded 2000. A subsidiary of **Grove/Atlantic Inc.** (see entry under *US Publishers*), New York. Publishes literary and crime fiction, history, current affairs, biography and memoir, politics, popular science and reference books. 90 titles a year. Strictly no unsolicited material accepted.
Ⓔ Royalties twice-yearly.

✱*Authors' update* For regular book buyers, Atlantic is synonymous with quality non-fiction and an ever strengthening fiction list.

Atlantic Europe Publishing Co. Ltd
Greys Court Farm, Greys Court, Nr Henley on Thames RG9 4PG
☏ 01491 628188   ℻ 01491 628189
writers@atlanticeurope.com
www.AtlanticEurope.com and
www.curriculumVisions.com
Director *Dr B.J. Knapp*

Publishes full-colour, highly illustrated primary school level text books. Not interested in any other material. Main focus is on National Curriculum titles, especially in the fields of religion, science, technology, social history and geography. About 50 titles a year. Unsolicited synopses and ideas for non-fiction curriculum-

based books welcome by e-mail only – does not accept material sent by post.
Ⓔ Fees paid.

Atom
▷ Little, Brown Book Group UK

Aurora Metro
2 Oriel Court, The Green, Twickenham TW2 5AG
☎ 020 8898 4488   🖷 020 8898 0735
info@aurorametro.com
www.aurorametro.com
Managing Director *Cheryl Robson*
Submissions *Aidan Jenkins*

Founded in 1989 to publish new writing. Over 100 authors published including Germaine Greer, Meera Syal, Benjamin Zephaniah, Carole Hayman. International work in translation and fiction, children's, drama, biography, cookery, non-fiction. About 50 titles a year. No unsolicited mss; send synopses and ideas by e-mail first.
Ⓔ Royalties annually.

Aurum Press Ltd
7 Greenland Street, London NW1 0ND
☎ 020 7284 7160   🖷 020 7485 4902
editorial@aurumpress.co.uk
www.aurumpress.co.uk
Managing Director *Bill McCreadie*
Editorial Director *Graham Coster*
Approx. Annual Turnover £4.5 million

Founded 1976. Acquired by the Quarto Group in August 2004. High-quality, illustrated/non-illustrated adult non-fiction in the areas of biography, sport, current affairs, military history, travel, music and film. IMPRINTS **Argentum** Practical photography books; **Jacqui Small** High-quality lifestyle books. About 70 titles a year. No unsolicited mss or e-mailed submissions.
Ⓔ Royalties twice-yearly.

✱*Authors' update* **With a sharp eye on the popular non-fiction market, Aurum combines quality publishing with strong marketing.**

Austin & Macauley Publishers Limited
CGC-33-01, 25 Canada Square, Canary Wharf, London E14 5LQ
☎ 020 7038 8212   🖷 020 7038 8100
editors@austinmacauley.com
www.austinmacauley.com
Chief Editor *Annette Longman*
Executive Editor *David Calvert*

Founded 2005. Publishes fiction and non-fiction, autobiography, biography, cookery, environmental issues, gardening, humour, memoirs, post-1945 fiction, true life experiences, children's, crime, fitness, health, romance, sport, self-help, women's and marital issues (including divorce). Expanding educational list: reference books, business

studies, cultural matters, current affairs, English, mathematics, mythology, politics, human interest, war, music, theatre, social sciences, cinema and TV. About 30 titles a year. Unsolicited synopses and ideas considered. Postal submissions must include return postage.
Ⓔ Royalties twice-yearly.

Authentic Media
9 Holdom Avenue, Bletchley, Milton Keynes MK1 1QR
☎ 01908 364200   🖷 01908 648550
info@authenticmedia.co.uk
www.authenticmedia.co.uk
Publishing Director *Mark Finnie*
Editorial Coordinator *Liz Williams*
Approx. Annual Turnover £2 million

A division of IBS-STL UK Ltd. Publishes Christian books on evangelism, discipleship, biography and mission for Evangelical Alliance and Keswick Ministries. IMPRINT **Paternoster** Founded 1935. Editorial Director *Robin Parry* Publishes academic, religion and learned/church/life-related books and journals. TITLES *Relational Leadership*; *After Christendom*. Over 50 titles a year. Unsolicited mss, synopses and ideas for books welcome. No poetry or fiction.
Ⓔ Royalties twice-yearly.

AuthorHouse UK
500 Avebury Boulevard, Milton Keynes MK9 2BE
☎ 0800 197 4150   🖷 0800 197 4151
info@authorhouse.co.uk
www.authorhouse.co.uk
President & CEO *Kevin Weiss*
Vice President *W. Herbert Senft III*
UK Managing Director *Tim Davies*

Founded 1997. Acquired by Betram Capital in January 2007. Publishes fiction, biography and memoirs, current affairs, history, politics, humour, travel, health, spirituality, sport and academic. Provides a comprehensive range of publishing and promotional services including editorial, design, production, distribution and marketing paid for by the author who retains editorial and creative control.

Authors OnLine
19 The Cinques, Gamlingay, Sandy SG19 3NU
☎ 01767 652005   🖷 01767 652005
theeditor@authorsonline.co.uk
www.authorsonline.co.uk
Submissions: Freephone 0800 107 2423
Managing Director/Editor *Richard Fitt*
Submissions Editor *Mrs Gaynor Johnson*
Approx. Annual Turnover £250,000

Founded 1998. A service for authors wishing to self-publish. Publishes new and reverted rights work in both electronic format via their website and traditional hard-copy mainly using digital print-on-demand technology. 80 titles in 2007. All genres welcome. Submit mss by e-mail or post (disk or CD-ROM) to the Submissions Editor. Ⓔ Payment 60% net, paid quarterly.

Autumn Publishing

Appledram Barns, Birdham Road, Chichester PO20 7EQ
Ⓣ 01243 531660   Ⓕ 01243 538160
autumn@autumnpublishing.co.uk
www.autumnpublishing.co.uk
Managing Director *Michael Herridge*
Editorial Director *Lyn Coutts*

Founded 1976. Division of Bonnier Media Ltd. Publishes baby and toddler books, children's activity, sticker and early learning books. About 200 titles a year. No responsibility accepted for the return of unsolicited mss.

Avon
▷ HarperCollins Publishers Ltd

Award Publications Limited

The Old Riding School, The Welbeck Estate, Worksop S80 3LR
Ⓣ 01909 478170   Ⓕ 01909 484632
info@awardpublications.co.uk
www.awardpublications.co.uk

Founded 1958. Publishes children's books, both fiction and reference. IMPRINT **Horus Editions**. No unsolicited mss, synopses or ideas.

Azure
▷ Society for Promoting Christian Knowledge

Badger Publishing Ltd

15 Wedgwood Gate, Pin Green Industrial Estate, Stevenage SG1 4SU
Ⓣ 01438 356907   Ⓕ 01438 747015
orders@badger-publishing.co.uk
www.badger-publishing.co.uk

Founded 1989. Educational publisher. 48 titles in 2007. No unsolicited mss; synopses and ideas for books considered. Approach by e-mail.
Ⓔ Royalties twice-yearly.

Baillière Tindall
▷ Elsevier Ltd

Duncan Baird Publishers

Castle House, 75–76 Wells Street, London W1T 3QH
Ⓣ 020 7323 2229   Ⓕ 020 7580 5692
info@dbairdpub.co.uk
Managing Director *Duncan Baird*
Editorial Director *Bob Saxton*
Approx. Annual Turnover £7.5 million

Founded in 1992 to publish and package co-editions overseas and went on to launch its own publishing operation in 1998. Publishes illustrated

cultural reference, world religions, health, mind, body and spirit, lifestyle, graphic design. 80 titles in 2007. No unsolicited mss. Synopses and ideas welcome; approach in writing in the first instance with s.a.e. No fiction or UK-only subjects.
Ⓔ Royalties twice-yearly.

Bantam/Bantam Press
▷ Transworld Publishers

Barefoot Books Ltd

124 Walcot Street, Bath BA1 5BG
Ⓣ 01225 322400   Ⓕ 01225 322499
info@barefootbooks.co.uk
www.barefootbooks.com
Publisher *Tessa Strickland*

Founded 1993. Publishes children's picture books, particularly new and traditional stories from a wide range of cultures. No unsolicited mss. See website for submission guidelines.
Ⓔ Royalties twice-yearly.

✱*Authors' update* **Writers of children's books would do well to keep track of Barefoot which, from small beginnings, is building a quality list that must be the envy of bigger publishers.**

Barrington Stoke

18 Walker Street, Edinburgh EH3 7LP
Ⓣ 0131 225 4113   Ⓕ 0131 225 4140
barrington@barringtonstoke.co.uk
www.barringtonstoke.co.uk
Chairman *David Croom*
Managing Director *Sonia Raphael*
Editorial Head *Kate Paice*
Approx. Annual Turnover £900,000

Founded in 1998 to publish books for 'dyslexic, struggling and reluctant' young readers. DIVISIONS **Fiction** for 8–12-year olds, reading at age 8; **Teenage Fiction** for age 12+, reading at age 8; **4u2read.OK** for children aged 8–13, with a reading age of 7; **gr8reads** for age 13+, with a reading age of 8; **fyi** for ages 10–14, fiction with stacks of facts; **Reality Checks** age 10–14, true-life stories; **Reloaded Series**: myths retold for ages 10–14, with a reading age of 8. Also books for teachers and resources to accompany fiction books. Commissioned via literary agents only. *No unsolicited material.*
Ⓔ Royalties twice-yearly.

B.T. Batsford
▷ Anova Books

BBC Active
▷ Pearson

BBC Books
▷ The Random House Group Ltd

BBC Children's Books
▷ Penguin Group (UK)

BBC Worldwide

Media Centre, 201 Wood Lane, London W12 7TQ
℡ 020 8433 2000
www.bbcworldwide.com
CEO *John Smith*
Coo *Sarah Cooper*

The main commercial arm and wholly owned subsidiary of the BBC. Consists of seven core businesses: Global Channels, Global TV Sales, Magazines, Content & Production, Digital Media, Home Entertainment and Global Brands. Third largest consumer magazine publisher in the UK with sales of nearly 100 million copies a year. Titles include *BBC Good Food, BBC Gardeners' World Magazine* and *BBC Top Gear Magazine.* Acquired a 75% stake in **Lonely Planet Publications Ltd** in October 2007.

Beautiful Books Ltd

36–38 Glasshouse Street, London W1B 5DL
℡ 020 7734 4448   🅔 020 3070 0764
office@beautiful-books.co.uk
www.beautiful-books.co.uk
Managing Director *Simon Petherick*
Approx. Annual Turnover £500,000

Founded 2004. Publishes adult fiction and non-fiction. IMPRINTS **Bloody Books** Horror fiction; **Burning House** Contemporary fiction. Unsolicited mss, synopses and ideas welcome. Approach by letter in the first instance with outline, synopsis and initial 30pp plus biographical information. Include s.a.e.

Belair

▷ **Folens Limited**

Berg Publishers

1st Floor, Angel Court, 81 St Clements Street, Oxford OX4 1AW
℡ 01865 245104   🅔 01865 791165
enquiry@bergpublishers.com
www.bergpublishers.com
Managing Director *Kathryn Earle*
Editorial Director *Tristan Palmer*

Publishes scholarly books in the fields of fashion, cultural studies, social sciences, history and humanities. TITLES *Beans,* Ken Albala; *Thinking Through Craft* Glenn Adamson; *Art as Far as the Eye Can See* Paul Virilio; *The Fashion Reader* eds. Linda Welters and Abby Lillethun. IMPRINT **Oswald Wolff Books**. About 65 titles a year plus 11 journals: *Fashion Theory, Home Cultures, Textile: The Journal of Cloth and Culture, Cultural Politics, The Senses and Society* and *Material Religion, Food, Culture and Society, Cultural and Social History, Anthrozoos, Time and Mind, The Journal of Modern Craft.* No unsolicited mss. Synopses and ideas for books welcome.
🅔 Royalties annually.

Berghahn Books

3 Newtec Place, Magdalen Road, Oxford OX4 1RE
℡ 01865 250011   🅔 01865 250056
publisher@berghahnbooks.com
www.berghahnbooks.com
Chairman/Managing Director *Marion Berghahn*
Approx. Annual Turnover £700,000

Founded 1994. Academic publisher of books and journals. TITLES *Imperial Germany, 1871–1918; Children of Palestine; Critical Interventions; A Different Kind of War.* OVERSEAS ASSOCIATE Berghahn Books Inc., New York. About 90 titles a year. No unsolicited mss; will consider synopses and ideas for books. Approach by e-mail in the first instance. No fiction or trade books.
🅔 Royalties annually.

Berlitz Publishing

58 Borough High Stret, London SE1 1XF
℡ 020 7403 0284   🅔 020 7403 0290
berlitz@apaguide.co.uk
www.berlitzpublishing.com
Managing Director *Jeremy Westwood*

Founded 1970. Acquired by the Langenscheidt Publishing Group in February 2002. Publishes travel and language-learning products only: visual travel guides, phrasebooks and language courses. SERIES *Pocket Guides; Berlitz Complete Guide to Cruising and Cruise Ships; Phrase Books; Pocket Dictionaries; Business Phrase Books; Self-teach: Rush Hour Commuter Cassettes; Think & Talk; Berlitz Kids.* No unsolicited mss.

BeWrite Books

▷ **entry under Electronic Publishing and Other Services**

BFI Publishing

Palgrave Macmillan, Porters North, 4 Crinan Street, London N1 9XW
℡ 020 7833 4000   🅔 020 7843 4650
bfipublishing@palgrave.com
www.palgrave.com/bfi

Founded 1980. The publishing imprint of the **British Film Institute**. Publishes academic, schools and general film/television-related books and resources. TITLE *The Cinema Book* Pam Cook. SERIES *BFI Film Classics; BFI TV Classics; BFI Screen Guide; World Directors.* About 30 titles a year. E-mail submissions preferred.
🅔 Royalties annually.

BFP Books

Focus House, 497 Green Lanes, London N13 4BP
℡ 020 8882 3315   🅔 020 8886 3933
mail@thebfp.com
Chief Executive *John Tracy*
Commissioning Editor *Stewart Gibson*

Founded 1982. The publishing arm of the Bureau of Freelance Photographers. Publishes illustrated books on photography, mainly aspects of freelancing and marketing pictures. No unsolicited mss but ideas welcome.

Clive Bingley Books
▷ Facet Publishing

Birlinn Ltd
West Newington House, 10 Newington Road, Edinburgh EH9 1QS
℡ 0131 668 4371   🖷 0131 668 4466
info@birlinn.co.uk
www.birlinn.co.uk
Managing Director *Hugh Andrew*

Founded 1992. Publishes local and military history, adventure, Gaelic, humour, Scottish history, Scottish reference, guidebooks and folklore. IMPRINT **John Donald Publishers** (see entry). Acquired **Polygon** in 2002 (see entry) and Tuckwell Press in 2005 and **Mercat Press** in 2007 (see entry). 170 titles in 2005 across all imprints. No unsolicited mss; synopses and ideas welcome. Ⓔ Royalties paid.

⌗*Authors' update* One of the fastest growing regional publishers.

Bitter Lemon Press
37 Arundel Gardens, London W11 2LW
℡ 020 7727 7927   🖷 020 7460 2164
books@bitterlemonpress.com
www.bitterlemonpress.com
Directors *Laurence Colchester, François von Hurter*

Founded 2003. Specializes in thrillers, *romans noir* and other contemporary fiction from abroad. 8 titles in 2008. No unsolicited material; submissions from agents only. Ⓔ Royalties annually.

Black & White Publishing Ltd
99 Giles Street, Edinburgh EH6 6BZ
℡ 0131 625 4500   🖷 0131 625 4501
mail@blackandwhitepublishing.com
www.blackandwhitepublishing.com
Director *Campbell Brown*

Founded 1990. Publishes general fiction and non-fiction, including memoirs, sport, cookery and humour. IMPRINTS **Itchy Coo** Scots language resources for use in schools. Text only submissions via website, or synopsis and sample chapter by post with s.a.e. or return postage. Ⓔ Royalties twice-yearly.

A.&C. Black Publishers Ltd
38 Soho Square, London W1D 3HB
℡ 020 7758 0200   🖷 020 7758 0222
enquiries@acblack.com
www.acblack.com
Managing Director *Jill Coleman*
Deputy Managing Director *Jonathan Glasspool*

Publishes children's and educational books, including music, for 0–16 year-olds, practical arts, visual arts, ceramics and glass, natural history, ornithology, nautical, reference, business books, dictionaries, fitness, sport, performing arts and writing books. Acquisitions brought the Herbert Press' art, design and general books, Adlard Coles' and Thomas Reed's sailing lists, plus *Reeds Nautical Almanac*, Christopher Helm, Pica Press and T&AD Poyser's natural history and ornithology lists, children's educational publisher Andrew Brodie and Whitakers Almanack into A.&C. Black's stable. Bought by **Bloomsbury Publishing** in May 2000. Acquired **Methuen**'s drama list in 2006. IMPRINTS **Adlard Coles Nautical; Thomas Reed; Christopher Helm; T&AD Poyser; The Herbert Press; Andrew Brodie; Methuen Drama.** TITLES *Who's Who; Writers' & Artists' Yearbook; Children's Yearbook; Whitaker's Almanack; New Mermaids; White Wolves; Methuen Drama Modern Plays.* 496 titles in 2008. Initial enquiry appreciated before submission of mss.
Ⓔ Royalties vary according to contract.

Black Ace Books
PO Box 7547, Perth PH2 1AU
℡ 01821 642822   🖷 01821 642101
www.blackacebooks.com
Managing Director *Hunter Steele*

Founded 1991. Publishes new fiction, Scottish and general; some non-fiction including biography, history, philosophy and psychology. IMPRINTS **Black Ace Books; Black Ace Paperbacks** TITLES *Succeeding at Sex and Scotland, Or the Case of Louis Morel* Hunter Steele; *La Tendresse* Ken Strauss MD; *Count Dracula (The Authorized Version)* Hagen Slawkberg; *Caryddwen's Cauldron* Paul Hilton; *The Sinister Cabaret* John Herdman. No children's, poetry, cookery, DIY, religion. 36 titles in print. 'No submissions at all, please, without first checking our website for details of current requirements and submission guidelines.'
Ⓔ Royalties twice-yearly.

Black Lace
▷ The Random House Group Ltd

Black Spring Press Ltd
Curtain House, 134–146 Curtain Road, London EC2A 3AR
℡ 020 7613 3066   🖷 020 7613 0028
general@blackspringpress.co.uk
Director *Robert Hastings*

Founded 1986. Publishes fiction and non-fiction, literary criticism, biography. TITLES *The Lost Weekend* Charles Jackson; *The Big Brass Ring* Orson Welles; *The Tenant* Roland Topor; *Julian*

Maclaren-Ross Collected Memoirs. About 5 titles a year. No unsolicited mss.
Ⓔ Royalties annually.

Black Swan
▷ Transworld Publishers

BlackAmber
▷ Arcadia Books

Blackstaff Press Ltd
4C Heron Wharf, Sydenham Business Park, Belfast BT3 9LE
Ⓣ 028 9045 5006　　Ⓕ 028 9046 6237
info@blackstaffpress.com
www.blackstaffpress.com
Managing Editor *Patsy Horton*

Founded 1971. Publishes mainly, but not exclusively, Irish interest books, fiction, poetry, history, sport, cookery, politics, illustrated editions, natural history and humour. About 20 titles a year. Unsolicited mss considered, but preliminary submission of synopsis plus short sample of writing preferred. Return postage *must* be enclosed. No submissions by e-mail.
Ⓔ Royalties twice-yearly.

⊞*Authors' update* **Regional and proud of it, this Belfast publisher is noted for bringing on young talent and for 'wonderfully well-presented catalogues and promotional material.'**

Blackstone Publishers
62 Durham Road, London E12 5AX
Ⓣ 020 8991 0605
mail@blackstonepublishers.com
mail@redarrowbooks.com
www.blackstonepublishers.com
www.redarrowbooks.com
Editor *Nina Rogers*
Publisher *John Kurtz*

Founded 2007. Publishes fiction, including romance, women's fiction, mysteries, suspense, science fiction, horror, fantasy, humour, gay, lesbian, erotica; non-fiction, including memoirs, biography, history, war, reference; children's, picture books, young reader and teen fiction.
IMPRINT **Red Arrow Books**. Unsolicited mss, synopses and ideas for books (with s.a.e.) welcome.
Ⓔ Royalties twice-yearly.

Blackwell Publishing
▷ Wiley-Blackwell

John Blake Publishing Ltd
3 Bramber Court, 2 Bramber Road, London W14 9PB
Ⓣ 020 7381 0666　　Ⓕ 020 7381 6868
words@blake.co.uk
Managing Director *John Blake*
Deputy Managing Director *Rosie Ries*

Founded 1991 and expanding rapidly. Bought the assets of Smith Gryphon Ltd in 1997 and acquired Metro Publishing in 2001. Publishes mass-market non-fiction. No fiction, children's, specialist or non-commercial. About 100 titles a year. No unsolicited mss; synopses and ideas welcome. Please enclose s.a.e.
Ⓔ Royalties twice-yearly.

⊞*Authors' update* **With a tabloid journalist's talent for knowing what makes the front page and the front of shop display, John Blake is a first call for celebrity memoirs and celebrity brand titles.**

Blandford Press
▷ Octopus Publishing Group

Bliss
▷ Arcadia Books

Bloodaxe Books Ltd
Highgreen, Tarset NE48 1RP
Ⓣ 01434 240500　　Ⓕ 01434 240505
editor@bloodaxebooks.com
www.bloodaxebooks.com
Managing/Editorial Director *Neil Astley*

Publishes poetry, literature and criticism, and related titles by British, Irish, European, Commonwealth and American writers. Ninety-five per cent of the list is poetry. TITLES include seven major anthologies, *Staying Alive*; *real poems for unreal times*; *Being Alive*, all edited by Neil Astley; *The Bloodaxe Book of 20th Century Poetry* ed. Edna Longley; *Modern Women Poets* ed. Deryn Rees-Jones; editions of Fleur Adcock, David Constantine, Helen Dunmore, Roy Fisher, Jackie Kay, Brendan Kennelly, Mary Oliver, J.H. Prynne, Peter Reading, Ken Smith, Anne Stevenson, C.K. Williams and Benjamin Zephaniah. About 30–40 titles a year. Unsolicited poetry mss welcome; send a sample of no more than 10 poems with s.a.e., 'but if you don't read contemporary poetry, don't bother'. No e-mail submissions of any kind.
Ⓔ Royalties annually.

⊞*Authors' update* **Assisted by regional Arts Council funding, Bloodaxe is one of the liveliest and most innovative of poetry publishers with a list that takes in some of the best of the younger poets.**

Bloody Books
▷ Beautiful Books Ltd

Bloomsbury Publishing Plc
36 Soho Square, London W1D 3QY
Ⓣ 020 7494 2111　　Ⓕ 020 7434 0151
www.bloomsbury.com
Founder and Chief Executive *Nigel Newton*
Executive Director *Richard Charkin*

Publishing Directors *Alexandra Pringle* (Editor-in-Chief), *Liz Calder, Kathy Rooney, Sarah Odedina, Michael Fishwick*
Approx. Annual Turnover £150 million

Founded 1986. Many of its authors have gone on to win prestigious literary prizes, including J.K. Rowling's *Harry Potter and the Philosopher's Stone, Harry Potter and the Chamber of Secrets* and *Harry Potter and the Prisoner of Azkaban* which won the **Nestlé Smarties Book Prize** in 1997, 1998 and 1999 respectively. Margaret Atwood's *The Blind Assassin* won the **Booker Prize** in 2000.

Started Bloomsbury USA in 1998. Acquired **A.&C. Black Publishers Ltd** in May 2000 (see entry), Peter Collin Publishing Ltd in September 2002, Berlin Verlag in 2003 and Walker Publishing Inc in New York in 2004. Publishes literary fiction and non-fiction, including general reference; also audiobooks. AUTHORS include Margaret Atwood, T.C. Boyle, Sophie Dahl, Jeffrey Eugenides, Neil Gaiman, Daniel Goleman, David Guterson, Sheila Hancock, John Irving, Jay McInerney, Tim Pears, Celia Rees, J.K. Rowling, Ben Schott, Will Self, Donna Tartt, Rupert Thomson, Barbara Trapido, Benjamin Zephaniah. No unsolicited submissions.
Ⓔ Royalties twice-yearly.

⌗*Authors' update* It was the publishing coup of the year. No, not a bestselling title but rather a bestselling personality. Richard Charkin was seduced from the top job in Macmillan to become executive director at Bloomsbury. Initially, his job was to lead the cash rich publisher on an acquisition spree but within weeks his role had expanded to take control of Bloomsbury's transatlantic trade business. While he has a foundation of strong selling fiction and non-fiction titles, the challenge for Charkin and founder chairman, Nigel Newton is to prove a commercial life after Harry Potter. Would that the rest of us had such pleasurable problems.

Blue Door
⊳ HarperCollins Publishers Ltd

BMM
⊳ SportsBooks Limited

Boatswain Press
⊳ Kenneth Mason Publications Ltd

Bobcat
⊳ Omnibus Press

Bodleian Library
Communications & Publishing Office, Bodleian Library, Broad Street, Oxford OX1 3BG
☎ 01865 277627   ⓕ 01865 277187

publishing@bodley.ox.ac.uk
www.bodleianbookshop.co.uk
Head of Publishing *Samuel Fanous*

Founded 1602. Publishes trade and academic works relating to the Bodleian Library collections of books and manuscripts. TITLES *A Month at the Front: The Diary of an Unknown Soldier; Queen Elizabeth's Book of Oxford; Laurel for Libby.* 10 titles in 2006. No unsolicited mss. Synopses and ideas for book related to the collections will be considered; approach should be made by post, addressed to the Commissioning Editor.

The Bodley Head/Bodley Head Children's Books
⊳ The Random House Group Ltd

Boltneck Publications Limited
Head Office: Westpoint, 78 Queens Road, Clifton, Bristol BS8 1QX
☎ 0117 985 8709
cfwb@radmorebirch.wanadoo.co.uk
www.boltneck.co.uk
Submissions: King's Cote, Valley Road, Finmere MK18 4AL
Managing Director *David Thomas*
Publishing Director *Clive Birch*

Founded 2004. Linked to Medavia Media Agency. Publishes fiction and popular non-fiction on topical subjects or people of interest. DIVISION **Medavia Publishing** *Clive Birch* TITLES *No Big Deal; A Decent Man.* IMPRINT **Boltneck Business Publications** *David Thomas*. 4 titles in 2006. No unsolicited mss. Initial letter, synopsis, sample chapter, c.v. and s.a.e.
Ⓔ Royalties paid annually.

Bonnier Books (UK)
Appledram Barns, Birdham Road, Chichester PO20 7EQ
☎ 01243 531660   ⓕ 01243 774433
CEO *Des Higgins*

Founded in February 2007. Part of the Swedish media group, Bonnier AB. Publishes illustrated lifestyle and cookery titles. TITLES *Bride and Groom Cookbook; Healthy Heart Cookbook.* 45 titles in 2007.

Book Blocks
⊳ CRW Publishing Ltd

Book Guild Publishing
Pavilion View, 19 New Road, Brighton BN1 1UF
☎ 01273 720900   ⓕ 01273 723122
info@bookguild.co.uk
www.bookguild.co.uk
Chairman *George M. Nissen, CBE*
Managing Director *Carol Biss*

Founded 1982. Publishes fiction, children's and non-fiction, including human interest and biography/memoirs. Expanding mainstream list. Subject areas: **Human Interest** TITLE *Wit and Wisdom of British Prime Ministers* Phil Dampier and Ashley Walton. **Biography** TITLES *On the Trail of Arthur Conan Doyle* Paul Spiring and Brian Pugh; *Hollywood Heat* Steve Rowland. **General Non-Fiction** TITLE *The Beautiful Game is Over* John Samuels. **Fiction** TITLES *Sins of the Flesh* Julian Fane; *I'm Still Alive* Nigel Macbeth. **Children's** TITLE *Wizzy-Woo and His Brand New Friends* Helga Hopkins. About 90 titles a year. Unsolicited mss, ideas and synopses welcome. £ Royalties twice-yearly.

⊛*Authors' update* There are two sides to Book Guild Publishing, not always coexisting in happy harmony. The favoured image is of a conventional publisher with a strong list of general titles. But also on offer is a service where authors cover the costs of publication. Unlike many vanity publishers, Book Guild Publishing clearly sets out its terms of agreement. Still, the outlay can be high relative to sales and first-time authors should not be misled into believing that they have found an easy route to fame and fortune.

Border Lines Biographies
▷ Seren

Boulevard Books & The Babel Guides
71 Lytton Road, Oxford OX4 3NY
☎ 01865 712931
ray.keenoy@gmail.com
www.babelguides.com
Managing Director *Ray Keenoy*

Specializes in contemporary world fiction by young writers in English translation. Existing or forthcoming series of fiction from Brazil, Italy, Latin America, Low Countries, Greece, and elsewhere. The Babel Guides series of popular guides to fiction in translation started in 1995. **Babel Guides to Fiction in Translation** Series Editor *Ray Keenoy* TITLES *Eminent Hungarians*; *Babel Guide to Italian Fiction in Translation*; *Babel Guide to the Fiction of Portugal, Brazil & Africa in Translation*; *Babel Guide to French Fiction in English Translation*; *Babel Guide to Jewish Fiction*; *Babel Guide to Brazilian Fiction*. Suggestions and proposals for translations of contemporary fiction welcome. Also seeking contributors to forthcoming Babel Guides (all literatures).
£ Royalties annually.

Bound Biographies Limited
Heyford Park House, Heyford Park, Bicester OX25 5HD
☎ 01869 232911   🖷 01869 232698
office@boundbiographies.com
www.boundbiographies.com

Managing Director *Michael Oke*
Editorial Head *Dr A.J. Gray*
Approx. Annual Turnover £250,000

Founded in 1992 to assist in the writing and production of low numbers of private life stories. Print-on-demand facilities to provide short runs of paper or hardback books to complement the Bound Biographies leather-bound range. Publishes autobiographies predominantly although novels, poetry and special interest books are considered. TITLES *Red Tails in the Sunset* Bryn Williams; *The Long Straw* Mike Nicholson. 40 titles in 2007. Unsolicited material welcome; approach by post, telephone or e-mail.
£ Royalties annually.

Bounty
▷ Octopus Publishing Group

Bowker (UK) Ltd
1st Floor, Medway House, Cantelupe Road, East Grinstead RH19 3BJ
☎ 01342 310450   🖷 01342 310486
sales@bowker.co.uk
www.bowker.co.uk
Managing Director *Doug McMillan*

Part of the Cambridge Information Group (CIG). Publishes bibliographic references used by publishers, libraries and retailers throughout the world to source new book information. TITLES *Books In Print* and *Global Books in Print*. Unsolicited material will not be read.

Boxtree
▷ Macmillan Publishers Ltd

Marion Boyars Publishers Ltd
24 Lacy Road, London SW15 1NL
☎ 020 8788 9522   🖷 020 8789 8122
rebecca@marionboyars.com
www.marionboyars.co.uk
Editor *Rebecca Gillieron*
Editor, Non-fiction *Kit Maude*

Founded 1975, formerly Calder and Boyars. Publishes biography and autobiography, fiction, literature and criticism, music, philosophy, psychology, sociology and anthropology, theatre and drama, film and cinema, women's studies. AUTHORS include Georges Bataille, Ingmar Bergman, Heinrich Böll, Jean Cocteau, Carlo Gébler, Julian Green, Ivan Illich, Pauline Kael, Ken Kesey, Toby Litt, Kenzaburo Oe, Hubert Selby Jr, Lawrence Potter, Eudora Welty, Judith Williamson, Hong Ying, Elif Shafak and Riverbend. About 30 titles a year. Unsolicited mss not welcome for fiction or poetry; submissions from agents only. Unsolicited synopses and ideas welcome for non-fiction. OVERSEAS ASSOCIATES Marion Boyars Publishers Inc.
£ Royalties annually.

Bradt Travel Guides

23 High Street, Chalfont St Peter SL9 9QE
℡ 01753 893444   ℻ 01753 892333
info@bradtguides.com
www.bradtguides.com
Managing Director *Donald Greig*
Editorial Director *Adrian Phillips*
Approx. Annual Turnover £1 million

Founded in 1974 by Hilary Bradt. Specializes in travel guides to off-beat places and quirky travel and wildlife-related titles. SERIES Country guides and island guides (Azores, Syria, Namibia); wildlife guides (Madagascar, Galapagos, Arctic, Antarctica, Southern Africa); mini guides to cities (Budapest, Dubrovnik, Bratislava, Tallinn) and the 'Eccentric' series (Britain, London, France, America, Oxford, Cambridge, Edinburgh). Worldwide distribution. 50 titles in 2007. No unsolicited mss; synopses and relevant ideas (not travelogues) welcome.
Ⓔ Royalties twice-yearly.

✵*Authors' update* **Judged to be the best of guide publishers by *Wanderlust* magazine, Bradt may well have hit on the antidote to online travel guides. A list of unlikely tourist destinations such as Burkina Faso and Kabul has achieved such success to inspire plans for doubling the number of titles by 2010.**

Nicholas Brealey Publishing

3–5 Spafield Street, London EC1R 4QB
℡ 020 7239 0360   ℻ 020 7239 0370
rights@nicholasbrealey.com
www.nicholasbrealey.com
Managing Director *Nicholas Brealey*

Founded 1992. Independent publishing group focusing on innovative trade/professional books covering business and economics, intelligent self-help, popular psychology and the increasingly active fields of travel writing and crossing cultures.. The group has offices in Boston and includes Intercultural Press. TITLES *It's All Greek to Me!*; *Almost French*; *Your Writing Coach*; *Authentic Happiness*; *50 Psychology Classics*; *The Cult of the Amateur*; *The 80/20 Principle*; *Don't Tell Mum I Work on the Rigs She Thinks I'm a Piano Player in a Whorehouse*. No fiction, poetry or leisure titles. 30 titles a year. No unsolicited mss; synopses and ideas welcome.

✵*Authors' update* **Looks to be succeeding in breaking away from the usual computer-speak business manuals to publish information and literate texts. Lead titles have a distinct trans-Atlantic feel.**

The Breedon Books Publishing Co. Ltd

3 The Parker Centre, Mansfield Rd, Derby DE21 4SZ
℡ 01332 384235   ℻ 01332 292755
steve.caron@breedonpublishing.co.uk
www.breedonbooks.co.uk
Owner/Managing Director *Steve Caron*
Approx. Annual Turnover £1 million

Founded 1983. Publishes football and sport, local history, old photographs, heritage. About 45 titles a year. Unsolicited mss, synopses and ideas welcome if accompanied by s.a.e. No poetry or fiction.

Martin Breese International

19 Hanover Crescent, Brighton BN2 9SB
℡ 01273 687555
MBreese999@aol.com
www.abracadabra.co.uk
Chairman/Managing Director *Martin R. Breese*

Founded in 1975 to produce specialist conjuring books. No unsolicited submissions.

Brewin Books Ltd

Doric House, 56 Alcester Road, Studley B80 7LG
℡ 01527 854228   ℻ 01527 852746
admin@brewinbooks.com
www.brewinbooks.com
Chairman/Managing Director *Alan Brewin*
Company Secretary *Julie Brewin*
Director *Alistair Brewin*

Founded 1976. Publishes books on all aspects of Midland life and history including social, hospital, police, military, transport and family histories as well as biographies. TITLES *Hidden City, Haunted City* Peter Leather; *The Streets of Brum – Part 4* Carl Chinn; *As I Recall: Memories of a Bevin Boy* Derek Hollows. IMPRINTS **Brewin Books; Alton Douglas Books; History Into Print**. About 30 titles a year. Not interested in children's, poetry, short stories or novels. Approach by letter, but do not send full mss. Unsolicited synopses and ideas welcome with s.a.e.
Ⓔ Royalties twice-yearly.

Bright 'I's
▷ Infinite Ideas

Bristol Classical Press
▷ Gerald Duckworth & Co. Ltd

Bristol Phoenix Press
▷ University of Exeter Press

British Academic Press
▷ I.B. Tauris & Co. Ltd

The British Academy

10 Carlton House Terrace, London SW1Y 5AH
℡ 020 7969 5200   ℻ 020 7969 5300

secretary@britac.ac.uk
www.britac.ac.uk
Publications Officer *James Rivington*
Assistant Publications Officer *Janet English*
Publications Assistant *Amrit Bangard*

Founded 1902. The primary body for promoting scholarship in the humanities and social sciences, the Academy publishes many series stemming from its own long-standing research projects, or series of lectures and conference proceedings. Main subjects include history, philosophy and archaeology. SERIES *Auctores Britannici Medii Aevi*; *Early English Church Music*; *Fontes Historiae Africanae*; *Records of Social and Economic History*. About 20 titles a year. Proposals for these series are welcome and are forwarded to the relevant project committees. The British Academy is a registered charity and does not publish for profit.
Ⓔ Royalties only when titles have covered their costs.

The British Computer Society

First Floor, Block D, North Star House, North Star Avenue, Swindon SN2 1FA
☎ 0845 300 4417   ⓕ 01793 480270
publishing@hq.bcs.org.uk
www.bcs.org/books
Chief Executive *David Clarke*
Head of Publishing & Information Products *Elaine Boyes*
Book Publisher *Matthew Flynn*
Approx. Annual Turnover £20 million (Society)

Founded 1957. BCS is the leading professional and learned society in the field of computers and information systems. Publishes books which support the professional, academic and practical needs of the IT community. 6 titles in 2007. Unsolicited material welcome; submissions form and guide on website page: www.bcs.org/books/writer
Ⓔ Royalties annually.

The British Library

96 Euston Road, London NW1 2DB
☎ 020 7412 7535   ⓕ 020 7412 7768
blpublications@bl.uk
www.bl.uk
Head of Publishing *David Way*
Publishing Manager *Catherine Britton*

Founded 1979 as the publishing arm of The British Library to publish works based on the historic collections and related subjects. Publishes bibliographical reference, manuscript studies, illustrated books based on the Library's collections, and book arts. TITLES *1000 Years of English Literature*; *The Books of Henry VIII and His Wives*; *Futurist Typography*. About 50 titles a year. Unsolicited mss, synopses and ideas welcome if related to the history of the book, book arts or bibliography. No fiction or general non-fiction.
Ⓔ Royalties annually.

The British Museum Press

38 Russell Square, London WC1B 3QQ
☎ 020 7323 1234   ⓕ 020 7436 7315
www.britishmuseum.org
Managing Director *Brian Oldman*
Director of Publishing *Rosemary Bradley*

The book publishing division of The British Museum Company Ltd. Founded 1973 as British Museum Publications Ltd; relaunched 1991 as British Museum Press. Publishes books based on the British Museum's collection, including ancient history, archaeology, ethnography, art history, exhibition catalogues, guides, children's books and all official publications of the British Museum. TITLES *The First Emperor: China's Terracotta Army*; *The Parthenon Sculptures in the British Museum*; *500 Things to Know About the Ancient World*; *Indian Art in Detail*. About 45 titles a year. Synopses and ideas for books welcome.
Ⓔ Royalties twice-yearly.

Andrew Brodie
▷ A.&C. Black Publishers Ltd

Brown Skin Books

PO Box 57421, London E5 0ZD
☎ 020 8986 1115
info@brownskinbooks.co.uk
www.brownskinbooks.co.uk
Chairman *Dr John Lake*
Managing Director *Vastiana Belfon*

Founded 2002. Publishes 'quality, intelligent erotic fiction by black women around the world' and a new series of erotic crime thrillers. TITLES *Body and Soul* Jade Williams; *Personal Business* Isabel Baptiste; *Scandalous* and *A Darker Shade of Blue* Angela Campion; *Playthings* Faith Graham; *Online Wildfire* Crystal Humphries; *Strip Poker* and *Beg Me* Lisa Lawrence; *The Singer* Aisha DuQuesne; *Sorcerer* Tamzin Hall. 11 titles in 2007. No unsolicited mss; synopses and sample chapters welcome. Send by e-mail or post. No poetry.
Ⓔ Royalties annually.

Brown, Son & Ferguson, Ltd

4–10 Darnley Street, Glasgow G41 2SD
☎ 0141 429 1234   ⓕ 0141 420 1694
info@skipper.co.uk
www.skipper.co.uk
Chairman/Joint Managing Director *T. Nigel Brown*

Founded 1850. Specializes in nautical textbooks, both technical and non-technical. Also Scottish

one-act/three-act plays. Unsolicited mss, synopses and ideas for books welcome.
Ⓔ Royalties annually.

Bryntirion Press

Bryntirion, Bridgend CF31 4DX
Ⓣ 01656 655886   Ⓕ 01656 665919
office@emw.org.uk
www.emw.org.uk
Publications Manager *Huw Kinsey*
Welsh Language Publications *Gethin Rhys*
Approx. Annual Turnover £40,000

Owned by the Evangelical Movement of Wales. Publishes Christian books in English and Welsh. 50 titles in 2007 No unsolicited mss.
Ⓔ Royalties annually.

Burning House
▷ Beautiful Books Ltd

Burns & Oates
▷ The Continuum International Publishing Group Limited

Business Education Publishers Ltd

Evolve Business Centre, Cygnet Way, Rainton Bridge Business Park, Houghton-le-Springs DH4 5QY
Ⓣ 0191 305 5165
info@bepl.com
www.bepl.com
Managing Director *Mrs A. Murphy*
Approx. Annual Turnover £400,000

Founded 1981. Publishes business education, economics and law for BTEC and GNVQ reading. Currently expanding into further and higher education, computing, IT, business, travel and tourism, occasional papers for institutions and local government administration. Unsolicited mss and synopses welcome.
Ⓔ Royalties annually.

Business Plus
▷ Hachette Livre UK

Buster Books
▷ Michael O'Mara Books Ltd

Butterworth Heinemann
▷ Elsevier Ltd

Cadogan Guides
▷ New Holland Publishers (UK) Ltd

Calder Publications Ltd

51 The Cut, London SE1 8LF
Ⓣ 020 7633 0599   Ⓕ 020 7928 5930
info@calderpublications.com
www.calderpublications.com
Managing Director/Editorial Head *Alessandro Gallenzi*

Acquired in April 2007 by independent publishers **Alma Books**. A publishing company which has grown around the tastes and contacts of John Calder, the iconoclast of the literary establishment. The list has a reputation for controversial and opinion-forming publications; Samuel Beckett is perhaps the most prestigious name. The list includes all of Beckett's prose and poetry. Publishes autobiography, biography, drama, literary fiction, literary criticism, music, opera, poetry, politics, sociology, ENO opera guides. AUTHORS Antonin Artaud, Marguerite Duras, Martin Esslin, Erich Fried, P.J. Kavanagh, Robert Menasse, Robert Pinget, Luigi Pirandello, Alain Robbe-Grillet, Nathalie Sarraute, L.F. Celine, Eva Figes, Claude Simon, Howard Barker (plays). No new material accepted.
Ⓔ Royalties annually.

✱*Authors' update* There was plenty of interest when this idiosyncratic publisher decided to put himself and his valuable backlist up for sale. In the event, the new owners are of the same breed of independent thinkers. Calder will continue to commission new work.

California University Press
▷ University Presses of California, Columbia & Princeton Ltd

Cambridge University Press

The Edinburgh Building, Shaftesbury Road, Cambridge CB2 8RU
Ⓣ 01223 312393   Ⓕ 01223 315052
www.cambridge.org
Chief Executive *Stephen R.R. Bourne*
Managing Director, Publishing, Academic and Professional (books and journals in H&SS and STM) *A.M.C. Brown*
Managing Director, Publishing, Cambridge Learning (ELT and Education) *J.M. Pieterse*

The oldest printer and publisher in the world, now a global organization with a regional structure operating in the UK, Europe, Middle East & Africa, in the Americas and in Asia-Pacific. Has warehousing centres in Cambridge, New York, Melbourne, Madrid, Cape Town, São Paulo, Singapore, New Delhi and Tokyo, with offices and agents in many other countries. Publishing includes major ELT courses; tertiary textbooks, monographs and journals; scientific and medical reference; professional lists in law, management and engineering; educational coursebooks for the National Curriculum; and e-learning materials for schools. Recent ventures in e-pedagogy include the companies Global Grid for Learning and Cambridge-Hitachi. Publishes at all levels from primary school to postgraduate research and professional. Also Bibles, prayer books and over 230 academic journals. Around

25,000 authors in 116 countries and between 1500 and 2000 new titles a year. Synopses and ideas for books are welcome (and preferable to the submission of unsolicited mss). No fiction or poetry.
£ Royalties twice-yearly.

✳*Authors' update* Dedicated to an ever-expanding publication programme, CUP has been an active buyer of overseas publishers with growth concentrated on the Americas and Asia. Academic books are finding more niche markets but the bonanza is in ELT.

Camden Press Ltd

43 Camden Passage, London N1 8EA
℡ 020 7226 4673
Chairman *Bob Borzello*

Founded 1985. Publishes social issues; all books are launched in connection with major national conferences. DIVISION **Publishing for Change** *Bob Borzello* TITLE *Living with the Legacy of Abuse.* IMPRINT **Mindfield** TITLES *Hate Thy Neighbour: The Race Issue*; *Therapy on the Couch*. No unsolicited material. Approach by telephone in the first instance.
£ Royalties annually.

Campbell Books
▷ Macmillan Publishers Ltd

Candle
▷ Lion Hudson plc

Canongate Books Ltd

14 High Street, Edinburgh EH1 1TE
℡ 0131 557 5111   ℻ 0131 557 5211
info@canongate.co.uk
www.canongate.net
Also at: Lower Basement, 151 Chesterton Road, London W10 6ET
Publisher/Managing Director *Jamie Byng*
Publishing Director *Anya Serota*
Editorial Director *Nick Davies*
Approx. Annual Turnover £8.06 million

Founded 1973. In 2002, won both 'Publisher of the Year' at the British Book Awards and The Booker Prize with *Life of Pi* by Yann Martel. Publishes a wide range of literary fiction and non-fiction. Historically, there is a strong Scottish slant to the list but its output is increasingly international, especially now with its joint venture with Grove Atlantic in the US and co-publishing agreement with Text Publishing in Australia. Canongate also has a growing reputation for originating unusual projects (typified by *The Pocket Canons* and more recently by *The Myths* series, launched in 2005). Key AUTHORS include Michel Faber, Yann Martel, Alasdair Gray, John Fante, James Meek and Louise Welsh.
£ Royalties twice-yearly.

✳*Authors' update* Marking another stage in its climb to major publisher status, Canongate is to publish the entire run of Peanuts comic strips to capture a substantial share of the gift market. But the core of the business is still imaginative writing. Heavy investment in new titles has proved to be a profitable strategy.

Canterbury Press
▷ SCM – Canterbury Press Ltd

Capall Bann Publishing

Auton Farm, Milverton TA4 1NE
℡ 01823 401528   ℻ 01823 401529
enquiries@capallbann.co.uk
www.capallbann.co.uk
Chairman *Julia Day*
Editorial Head *Jon Day*

Founded in 1993, Capall Bann now has well over 300 titles in print. Family-owned and run company, 'operated by people with real experience in the topics we publish' which include British traditions, folklore, animals, alternative healing, environmental, Celtic lore, mind, body and spirit. TITLES *Tarot Therapy*; *How to Talk with Fairies*; *Mary Magdalene, Lost Goddess, Lost Gospels*; *Crystals Folklore & Healing*; *Understanding Star Children*. About 40 titles a year. Unsolicited proposals for books welcome. No fiction or poetry.
£ Royalties twice-yearly.

Jonathan Cape/Jonathan Cape Children's Books
▷ The Random House Group Ltd

Capstone Publishing (A Wiley Company)

John Wiley & Sons Ltd, The Atrium, Southern Gate, Chichester PO19 8SQ
℡ 01243 779777
info@wiley-capstone.co.uk
www.wiley.com

Founded 1997. Part of **John Wiley & Sons Inc**. Publishes business, personal development, lifestyle and humour books.

Carcanet Press Ltd

4th Floor, Alliance House, 30 Cross Street, Manchester M2 7AQ
℡ 0161 834 8730   ℻ 0161 832 0084
info@carcanet.co.uk
www.carcanet.co.uk
Owner *Folio Holdings*
Chairman *Kate Gavron*
Managing Director/Editorial Director *Michael Schmidt, OBE, FRSL*

Having started the company as an undergraduate hobby, Robert Gavron bought the company in 1983 and established strong European, Commonwealth and American links. Winner of the **Sunday Times Small Publisher of the Year**

award in 2000, it took over the Oxford Poets list from **Oxford University Press** in 1999, which it now publishes as a distinct imprint. Primarily a poetry publisher but also publishes academic, literary biography, fiction in translation and translations. AUTHORS John Ashbery, Eavan Boland, Donald Davie, Natalia Ginzburg, Robert Graves, Elizabeth Jennings, Hugh MacDiarmid, Edwin Morgan, Sinead Morrissey, Les Murray, Frank O'Hara, Richard Price, Frederic Raphael, C.H. Sisson, Charles Tomlinson, Jane Yeh. About 40 titles a year, including *PN Review* (six issues yearly). Poetry submissions (hard copy only): 6–10 poems with covering letter and return postage. Prospective writers should familiarize themselves with the Carcanet list before submitting work.
£ Royalties annually.

⊞*Authors' update* Now one of the UK's leading literary publishers with a diverse list of modern and classic poetry, Carcanet stays close to the source of literary creativity with its own postgraduate Writing School at Manchester Metropolitan University (see entry under *UK and Irish Writers' Courses*).

Cardiff Academic Press

St Fagans Road, Fairwater, Cardiff CF5 3AE
☏ 029 2056 0333   🖷 029 2055 4909
drakegroup@btinternet.co.uk
www.drakegroup.co.uk
Managing Director *R.G. Drake*

Academic publishers and Drake Audio Visual.

Carlton Publishing Group

20 Mortimer Street, London W1T 3JW
☏ 020 7612 0400   🖷 020 7612 0401
enquiries@carltonbooks.co.uk
www.carltonbooks.co.uk
Managing Director *Jonathan Goodman*
Editorial Director *Piers Murray Hill*

Founded 1992. Carlton Publishing Group is an independent publishing house. It has three main divisions: **Carlton Books** Illustrated leisure and entertainment books aimed at the mass market. Subjects include history, sport, puzzles, health, popular science, children's non-fiction, popular culture, music, fashion and design. **André Deutsch** Autobiography, biography, history, current affairs, narrative non-fiction and the arts. **Prion Books** Humour, nostalgia, drink and classic literature. No unsolicited mss; synopses and ideas welcome. No novels, science fiction, poetry or children's fiction.

Carroll & Brown Publishers Limited

20 Lonsdale Road, London NW6 6RD
☏ 020 7372 0900   🖷 020 7372 0460
mail@carrollandbrown.co.uk
www.carrollandbrown.co.uk
Editorial Director *Amy Carroll*

Publishes practical parenting, health, fitness, recreations, mind, body and spirit. TITLES *Your Pregnancy Bible*; *Practical Wabi Sabi*; *Beading for Beginners*. Synopses and ideas for illustrated books welcome; approach in writing in the first instance, enclosing s.a.e. for reply. No fiction
£ Fees or royalties paid.

Cassell
▷ The Orion Publishing Group Limited

Cassell Illustrated
▷ Octopus Publishing Group

Kyle Cathie Ltd

122 Arlington Road, London NW1 7HP
☏ 020 7692 7215   🖷 020 7692 7260
general.enquiries@kyle-cathie.com
www.kylecathie.co.uk

Founded 1990 to publish and promote 'books we have personal enthusiasm for'. Publishes non-fiction: cookery, food and drink, health and beauty, mind, body and spirit, gardening, homes and interiors, reference and occasional books of classic poetry. TITLES *Fantastico* Gino D'Acampo; *Easy Entertaining* Darina Allen; *The Green Beauty Bible* Sarah Stacey and Josephine Fairley; *Jekka's Complete Herb Book* Jekka McVicar; *Anton's Dance Class* Anton Du Beke. About 25 titles a year. No unsolicited mss. 'Synopses and ideas are considered in the fields in which we publish.'
£ Royalties twice-yearly.

Catholic Truth Society (CTS)

40–46 Harleyford Road, London SE11 5AY
☏ 020 7640 0042   🖷 020 7640 0046
f.martin@cts-online.org.uk
www.cts-online.org.uk
Chairman *Most Rev. Paul Hendricks*
General Secretary *Fergal Martin*

Founded originally in 1868 and re-founded in 1884. Publishes religious books – Roman Catholic; a variety of doctrinal, moral, biographical, devotional and liturgical publications, including a large body of Vatican documents and sources. Unsolicited mss, synopses and ideas welcome if appropriate to their list.
£ Royalties annually.

Catnip Publishing Ltd

14 Greville Street, London EC1N 8SB
☏ 020 7138 3650   🖷 020 7138 3658
martin@catnippublishing.co.uk
www.catnippublishing.co.uk
Managing Director *Robert Snuggs*
Directors *Andrea Reece, Martin West*
Approx. Annual Turnover £300,000

Children's books publisher formed from a merger of Southwood Books and Happy Cat Books in 2005. Concentrating on developing fiction, particularly the Happy Cat First Readers series and fiction for the 8–10 age group. IMPRINTS **Catnip** *Martin West* New books from Joan Lingard, Dominic Barker, Sarah Matthias; **Happy Cat** TITLE *Scaredy Squirrel at the Beach* Melanie Watt; SERIES **Happy Cat First Readers** (5–7 years); **Talking it Through**. 44 titles in 2007. Unsolicited submissions for fiction considered; send outline synopsis, including proposed age range and length, by e-mail. No non-fiction and, currently, only limited interest in picture books. £ Royalties twice-yearly.

Causeway Press
▷ Pearson

CBA Publishing
St Mary's House, 66 Bootham, York YO30 7BZ
☎ 01904 671417   ⓕ 01904 671384
catrinaappleby@britarch.ac.uk
www.britarch.ac.uk
Publications Officer *Catrina Appleby*
British Archaeology Editor *Mike Pitts*
Young Archaeologists' Club Communication Officer *Nicky Milsted*
Approx. Annual Turnover £130,000

Publishing arm of the **Council for British Archaeology**. Publishes academic archaeology reports, practical handbooks, *Internet Archaeology*; *British Archaeology* (bi-monthly magazine), *Young Archaeologist* (Young Archaeologists' Club magazine), monographs, archaeology and education. TITLES *Historic Landscape Analysis: A Handbook*; *Cloth and Clothing in Early Anglo-Saxon England*; *War Art*. Please contact by telephone before submitting mss and proposals.

CBD Research Ltd
Chancery House, 15 Wickham Road, Beckenham BR3 5JS
☎ 020 8650 7745   ⓕ 020 8650 0768
cbd@cbdresearch.com
www.cbdresearch.com
Managing Director *S.P.A. Henderson*
Approx. Annual Turnover £500,000

Founded 1961. Publishes directories and other reference guides to sources of information. No fiction. IMPRINT **Chancery House Press** Non-fiction of an esoteric/specialist nature for 'serious researchers and the dedicated hobbyist' TITLE *Alas, Poor Sherlock: The Imperfections of the World's Greatest Detective (to say nothing of his medical friend)* Joseph Green and Peter Ridgway Watt. About 6 titles a year.
£ Royalties quarterly.

CCV
▷ The Random House Group Ltd

Cengage Learning (EMEA) Ltd
High Holborn House, 50–51 Bedford Row, London WC1R 4LR
☎ 020 7067 2500   ⓕ 020 7067 2600
www.cengage.co.uk
CEO (Worldwide) *Ron Dunn*
CEO/Managing Director (Cengage Learning EMEA) *Tom Davy*
Publishing Director (Cengage Learning EMEA) *John Yates*

Founded 1993. Formerly Thomson Learning, part of Cengage Learning and as such has offices worldwide with the UK office being reported to by Copenhagen (for Europe), Greece (Middle East) and South Africa. Acquired by funds represented by Apax Partners and OMERS Capital Partners in 2007. Publishes higher and vocational education, ELT. TITLES *Management and Cost Accounting* Drury; *Economics* Mankiw and Taylor. Unsolicited material aimed at students is welcome but telephone in the first instance to check out the idea.
£ Royalties vary according to contract.

Centaur Press
▷ Open Gate Press

Century
▷ The Random House Group Ltd

CF4K
▷ Christian Focus Publications

CHA
▷ The Random House Group Ltd

Chambers Harrap Publishers Ltd
7 Hopetoun Crescent, Edinburgh EH7 4AY
☎ 0131 556 5929   ⓕ 0131 556 5313
admin@chambersharrap.co.uk
www.chambersharrap.co.uk
Managing Director *Patrick White*
Editorial Director *Vivian Marr*
Office Manager *Esther Fulton*

Specialists in dictionaries and reference publishing in print and online: English dictionaries and thesauruses for everyone from primary school students to professional writers, including an expanding adult learners' list; general reference dictionaries covering everything from biography to the unexplained; bilingual dictionaries, study aids, and phrase books; factfinders; crossword dictionaries; puzzles and games books; books for word lovers; slang dictionaries; and language and writing guides. A recent acquisition is the Brewer's list of reference books. Part of **Hachette-Livre (UK)**. About 75 titles a year. Send synopsis with accompanying letter rather than completed mss.

✱*Authors' update* Bucking the trend in a declining reference book market, Chambers' distinctive red livery and quirky titles, including a dictionary uniquely noted for a sense of humour, continue to make a strong impact.

Chancery House Press
▷ CBD Research Ltd

Channel 4 Books
▷ Transworld Publishers

Paul Chapman Publishing Ltd
▷ Sage Publications

Chapman Publishing
4 Broughton Place, Edinburgh EH1 3RX
☎ 0131 557 2207
chapman-pub@blueyonder.co.uk
www.chapman-pub.co.uk
Managing Editor *Dr Joy Hendry*

A venture devoted to publishing works by 'the best of Scottish writers, both up-and-coming and established, published in *Chapman*, Scotland's leading literary magazine'. Now publishing a wider range of works though the broad policy stands. Mainly poetry and, more occasionally, drama, short stories and books of contemporary importance in 20th-century Scotland. TITLES *Winter Barley* George Gunn; *Lure* Dilys Rose; *Wild Women Series – Wild Women of a Certain Age* Magi Gibson; *Ye Cannae Win* Janet Paisley. About 2 titles a year. No unsolicited mss.
£ Royalties annually.

Charnwood
▷ F.A. Thorpe Publishing

Chartered Institute of Personnel and Development (CIPD)
151 The Broadway, London SW19 1JQ
☎ 020 8612 6200   ℻ 020 8612 6201
publish@cipd.co.uk
www.cipd.co.uk/bookstore
Publishing Manager *Ruth Lake*
Approx. Annual Turnover £2.3 million

Part of CIPD Enterprises Limited. Publishes on personnel, training and management. A list of around 160 titles including looseleaf and online subscription products, student textbooks and professional information resources. New product proposals welcome.
£ Royalties annually or flat fees paid.

Chastleton Travel
▷ Arris Publishing Ltd

Chatto & Windus
▷ The Random House Group Ltd

Cherrytree Books
▷ Evans Brothers Ltd

Chicken House Publishing
2 Palmer Street, Frome BA11 1DS
☎ 01373 454488   ℻ 01373 454499
chickenhouse@doublecluck.com
Chairman/Managing Director *Barry Cunningham*

Children's publishing house founded in 2000. Acquired by Scholastic Inc. in May 2005. Publishes books 'that are aimed at real children' – fiction, original picture books. Aiming to publish about 25 titles a year. 'We are always on the lookout for new talent.' Unsolicited fiction and picture book material welcome; send letter with synopsis, three sample chapters and s.a.e.
£ Royalties twice-yearly.

✱*Authors' update* Set up as a 'small, creative company', Chicken House is now part of Scholastic. But editorial freedom is preserved 'with no rules about what to publish as long as it's good'.

Child's Play (International) Ltd
Ashworth Road, Bridgemead, Swindon SN5 7YD
☎ 01793 616286   ℻ 01793 512795
office@childs-play.com
www.childs-play.com
Chief Executive *Neil Burden*

Founded in 1972, Child's Play is an independent publisher specializing in learning through play, whole child development, life-skills and values. Publishes books, games and A-V materials. TITLES *Big Hungry Bear; There Was an Old Lady; Our Cat Cuddles; Royston Knapper; Ten Beads Tall; Pocket Pals; Sliders; Roly Poly Books; Animal Lullabies; Monkey's Clever Tale; What's the Time, Mr Wolf?; Sign and Sing-Along; Just Like Me; The Ding Dong Bag; The Flower.* Unsolicited mss welcome. Send s.a.e. for return or response. Expect to wait two months for a reply.
£ Outright or royalty payments are subject to negotiation.

Chimera
▷ Pegasus Elliot Mackenzie Publishers Ltd

Christian Focus Publications
Geanies House, Fearn, Tain IV20 1TW
☎ 01862 871011   ℻ 01862 871699
info@christianfocus.com
www.christianfocus.com
Owner *Balintore Holdings plc*
Chairman *R.W.M. Mackenzie*
Managing Director *William Mackenzie*
Editorial Manager *Willie Mackenzie*
Children's Editor *Catherine Mackenzie*

Approx. Annual Turnover £1.5 million

Founded 1979 to produce children's books for the co-edition market. Now a major producer of Christian books. Publishes adult and children's books, including some fiction for children but not adults. No poetry. Publishes for all English-speaking markets, as well as the UK. IMPRINTS **CF4K** Children's books; **Christian Focus** General books; **Mentor** Study books; **Christian Heritage** Classic reprints. About 90 titles a year. Unsolicited mss, synopses and ideas welcome from Christian writers. See website for submission criteria.
Ⓔ Royalties annually.

Christian Heritage
▷ Christian Focus Publications

Chrome Dreams
▷ entry under Audio Books

Chrysalis Books Group
▷ Anova Books

Churchill Livingstone
▷ Elsevier Ltd

Churchwarden Publications Ltd

PO Box 420, Warminster BA12 9XB
☎ 01985 840189   📠 01985 840243
enquiries@churchwardenbooks.co.uk
www.churchwardenbooks.co.uk
Managing Director *John Stidolph*

Founded 1974. Publishes books and stationery for churchwardens and church administrators including *The Churchwarden's Yearbook*.

Cicerone Press

2 Police Square, Milnthorpe LA7 7PY
☎ 01539 562069   📠 01539 563417
info@cicerone.co.uk
www.cicerone.co.uk
Managing/Editorial Director *Jonathan Williams*

Founded 1969. Guidebook publisher for outdoor enthusiasts. No fiction or poetry. TITLES *Hillwalker's Guide to Mountaineering; Tour of Mont Blanc; Coast to Coast Trail.* SERIES include *Alpine Walking; International Walking; British Long-Distance Trails, Cycling* and *Winter Activities.* About 30 titles a year. No unsolicited mss; synopses and ideas considered.
Ⓔ Royalties twice-yearly.

Cico Books

20–21 Jockey Fields, London WC1R 4BW
☎ 020 7025 2280   📠 020 7025 2281
mail@cicobooks.co.uk
Publisher *Cindy Richards*

Founded in 1999 and acquired by **Ryland, Peters & Small Limited** in 2006. Publishes highly illustrated lifestyle books covering interiors,

crafts, mind, body and spirit and gift. About 30 titles a year. No unsolicited mss but synopses and ideas welcome.
Ⓔ Royalties twice-yearly.

Cisco Press
▷ Pearson

Clairview Books Ltd

Hillside House, The Square, Forest Row RH18 5ES
☎ 0870 486 3526
office@clairviewbooks.com
www.clairviewbooks.com
Managing Director *Mr S. Gulbekian*
Approx. Annual Turnover £100,000

Founded 2000. Publishes non-fiction. General books challenging conventional thinking: mind, body and spirit, current affairs, the arts, health and therapy. TITLES *Imperial America* Gore Vidal; *The Biodynamic Food and Cookbook* Wendy Cook; *My Descent Into Death* Howard Storm. 5 titles in 2007. No unsolicited material; send initial letter of enquiry. No poetry or fiction.
Ⓔ Royalties annually.

T&T Clark
▷ The Continuum International Publishing Group Ltd

James Clarke & Co.

PO Box 60, Cambridge CB1 2NT
☎ 01223 350865   📠 01223 366951
publishing@jamesclarke.co.uk
www.jamesclarke.co.uk
Managing Director *Adrian Brink*

James Clarke and Co Ltd was founded in 1859 in Fleet Street, London, mainly as a magazine publisher. It produced the highly influential religious magazine, Christian World, which by the outbreak of the First World War was selling over 100,000 copies a week, and was the leading nonconformist weekly. It soon began to publish books and by the 1920s it became an exclusively book publishing company, building up a list in religion and in alternative medicine and spirituality. **The Lutterworth Press** (see entry) became associated with James Clarke & Co in 1984. James Clarke & Co's current list concentrates on academic, scholarly and reference works, specializing in theology, history, literature and related subjects. Publications include the Library of Theological Translations and the Library of Ecclesiastical History, and the highly regarded translation of Patrology, edited by Angelo di Berardino. Book proposals to be submitted in writing.

✱*Authors' update* **A leading theological publisher with high intellectual standards.**

Classic
▷ Ian Allan Publishing Ltd

Collector's Library/Collector's Library Editions in Colour
▷ CRW Publishing Ltd

Peter Collin Publishing Ltd
▷ Bloomsbury Publishing Plc

Collins
▷ HarperCollins Publishers Ltd

Collins & Brown
▷ Anova Books

Colourpoint Books
Colourpoint House, Jubilee Business Park, 21 Jubilee Road, Newtownards BT23 4YH
☎ 028 9182 6339   🖷 028 9182 1900
sales@colourpoint.co.uk
www.colourpoint.co.uk
Partners *Wesley Johnston, Malcolm Johnston, Norman Johnston* (transport editor), *Sheila M. Johnston* (commissioning editor)

Founded 1993. Publishes school textbooks and transport (covering the whole of the British Isles), plus books of Irish and general interest. No fiction. About 25 titles a year. Unsolicited material accepted but approach in writing in the first instance; include return postage, please.
💷 Royalties twice-yearly.

Columbia University Press
▷ University Presses of California, Columbia & Princeton Ltd

Communication Ethics
▷ Troubador Publishing Ltd

Compendium Publishing Ltd
43 Frith Street, London W1D 4SA
☎ 020 7287 4570   🖷 020 7494 0583
alan.greene@compendiumpublishing.com
www.compendiumpublishing.com
Managing Director *Alan Greene*
Editorial Director *Simon Forty*
Rights *Christine Delaborde*
Approx. Annual Turnover £4 million

Founded 2000. Publishes and packages for international publishing companies, general illustrated non-fiction: history, reference, hobbies, children's and educational transport and militaria. 70 titles in 2008 No unsolicited mss; synopsis and ideas preferred.
💷 Royalties twice-yearly or fees per project.

Condor
▷ Souvenir Press Ltd

Conran Octopus
▷ Octopus Publishing Group

Constable & Robinson Ltd
3 The Lanchesters, 162 Fulham Palace Road, London W6 9ER
☎ 020 8741 3663   🖷 020 8748 7562
enquiries@constablerobinson.com
www.constablerobinson.com
Chairman *Nick Robinson*
Managing Directors *Jan Chamier, Nova Jayne Heath*
Editorial Director, Crime list *Krystyna Green*
Commissioning Director, Mammoth series *Pete Duncan*
Manager, Overcoming series *Fritha Saunders*

Archibald Constable, Walter Scott's publisher and the originator of the '3 decker novel', published his first titles in Edinburgh in 1795. Constable & Robinson Ltd are now in the third century of publishing under the Constable imprint, still independent, with shareholders who work or have worked in the business running the company. IMPRINTS **Constable** Commissioning Editors *Becky Hardie, Andreas Campomar, Leo Hollis* Current affairs, biography, general and military history, psychology, travel, photography and crime fiction mainly in hardback. **Robinson** Crime fiction in paperback, the Mammoth Book Of series, the Overcoming psychology series, popular health and self-help, military history and true crime. **Right Way** Commissioning Editor *Judith Mitchell.* The very successful Elliot Right Way, publisher of practical author-expert self-help guides, was acquired by Constable & Robinson in 2007. The company will be developing the list which already includes market leaders in many mainstream and specialist areas including cookery and food, puzzles and jokes, hobbies and handicrafts, property, legal and business. 140 titles. Unsolicited sample chapters, synopses and ideas for books welcome. No mss; *no e-mail submissions.* Enclose return postage.

The Continuum International Publishing Group Limited
The Tower Building, 11 York Road, London SE1 7NX
☎ 020 7922 0880   🖷 020 7922 0881
www.continuumbooks.com
CEO *Oliver Gadsby*
Approx. Annual Turnover £10 million

Founded in 1999 by a buy-out of the academic and religious publishing of Cassell and the acquisition of Continuum New York. Publishes academic, religious and general books.

DIVISIONS **Academic Humanities** Publisher *Anna Sandeman* Education, social sciences, literature, film and music, philosophy (including the **Thoemmes** imprint), linguistics, biblical studies and theology printed under the **T&T**

Clark imprint. TITLES *Getting the Buggers to Behave; Teaching in Further Education; What Philosophy Is; The Guerrilla Film Makers Handbook.* **General Trade & Continuum Religion** Publishing Director *Robin Baird-Smith* Publishes under **Continuum Burns & Oates** (RC books) TITLES *Seven Basic Plots; The Home We Build Together.* IMPRINTS **Continuum; T&T Clark Intl; Thoemmes; Burns & Oates; Hambledon Continuum.** 550 titles in 2007. Unsolicited synopses and ideas within the subject areas listed above are welcome; approach in writing in the first instance. OVERSEAS SUBSIDIARIES The Continuum International Publishing Group Inc., New York and Harrisburg. Ⓔ Royalties twice-yearly.

✱*Authors' update* **A major reorganization has put some projects on hold but a return to growth is promised with new history and politics lists leading the way.**

Conway
▷ Anova Books

Thomas Cook Publishing
Thomas Cook Business Park, Coningsby Road, Peterborough PE3 8SB
☎ 01733 416477   🖷 01733 416688
Head of Travel Books *John Sadler*

Part of the Thomas Cook Group Plc, publishing commenced in 1873 with the first issue of *Cook's Continental Timetable*. Publishes guidebooks, maps and timetables. About 130 titles a year. No unsolicited mss; synopses and ideas welcome as long as they are travel-related.
Ⓔ Royalties annually.

Leo Cooper
▷ Pen & Sword Books Ltd

Corgi
▷ Transworld Publishers

Corgi Children's Books
▷ The Random House Group Ltd

Country Publications Ltd
The Water Mill, Broughton Hall, Skipton BD23 3AG
☎ 01756 701381   🖷 01756 701326
editorial@dalesman.co.uk
www.dalesman.co.uk
Book Editor *Mark Whitley*

Publishers of *Countryman, Dalesman, Cumbria* and *Down Your Way* magazines, and regional books covering the North of England. Subjects include walking, guidebooks, history, humour and folklore. About 10 titles a year. Will consider mss on subjects listed above, relating to the North.
Ⓔ Royalties annually.

Countryside Books
2 Highfield Avenue, Newbury RG14 5DS
☎ 01635 43816   🖷 01635 551004
info@countrysidebooks.co.uk
www.countrysidebooks.co.uk
Publisher *Nicholas Battle*

Founded 1976. Publishes local interest paperbacks on regional subjects, generally by English county. Local history, dialects, genealogy, walking, photographic, aviation and military, some transport. Over 400 titles available. About 50 titles a year. Unsolicited mss and synopses welcome but no fiction, poetry, natural history or personal memories. Check the website for suitability before any submission.
Ⓔ Royalties twice-yearly.

Crème de la Crime Ltd
PO Box 523, Chesterfield S40 9AT
☎ 01246 520835   🖷 01246 520835
info@cremedelacrime.com
www.cremedelacrime.com
Managing Director *Lynne Patrick*
Approx. Annual Turnover £60,000

Founded in 2003 to discover and publish crime fiction by new authors. 5 titles scheduled for 2008. Welcomes enquiries and unsolicited submissions, 'especially from talented but unpublished authors who are urged to read the detailed guidelines on the website; or send £2.50 with A4 s.a.e. and two first-class stamps for hard copy'.
Ⓔ Royalties annually.

Cressrelles Publishing Co. Ltd
10 Station Road Industrial Estate, Colwall, Malvern WR13 6RN
☎ 01684 540154   🖷 01684 540154
simonsmith@cressrelles.co.uk
www.cressrelles.co.uk
Managing Director *Leslie Smith*

Publishes a range of local interest books and drama titles. IMPRINTS **J. Garnet Miller** Plays and theatre texts; **Kenyon-Deane** Plays and drama textbooks; **New Playwrights' Network** Plays. About 6–12 new play titles a year. Submissions welcome.

Crimson Publishing
Westminster House, Kew Road, Richmond TW9 2ND
☎ 020 8334 1600   🖷 020 8334 1601
info@crimsonpublishing.co.uk
www.crimsonpublishing.co.uk
Chairman *David Lester*

Founded 2006. Part of Crimson Business Ltd. Non-fiction publisher of books 'to improve the way you live'. Subjects covered: business, travel, working abroad, gap year, careers and education, parenting.

DIVISIONS **Crimson Publishing** *Holly Bennion*; **Trotman Publishing** (see entry). IMPRINTS **Crimson; Crimson Professional; Vacation Work; Trotman; White Ladder**. 20 titles in 2007. Welcomes carefully considered unsolicited mss, synopses and ideas for books in the subject areas listed above. No fiction or books on topics not listed. Approach by e-mail.
Ⓔ Royalties twice-yearly.

Crombie Jardine Publishing Limited

Office 2, 3 Edgar Buildings, George Street, Bath BA1 2FJ
Ⓣ 01225 464445
catriona@crombiejardine.com
www.crombiejardine.com
Sales & Publishing Director *David Crombie*
Publishing Director *Catriona Jardine*

Founded in 2004. Specializes in fun, quirky and topical humour books only. TITLES *The Little Book of Chavs*; *The World's Funniest Proverbs*; *Shag Yourself Slim*; *Make Your Own Sex Toys*; *What Shat That?* 100 titles in 2007. No postal submissions. Send synopsis by e-mail in the first instance.
Ⓔ Flat fee paid upfront; no royalties.

✱*Authors' update* **Having cut their teeth with Michael O'Mara Books, the co-founders of Crombie Jardine are out to grab a share of the young adult humour market.**

Crossway
▷ Inter-Varsity Press

Crown House Publishing

Crown Buildings, Bancyfelin, Carmarthen SA33 5ND
Ⓣ 01267 211345   Ⓕ 01267 211882
books@crownhouse.co.uk
www.crownhouse.co.uk
Managing Director *David Bowman*

Founded 1998. Publishes academic and trade titles in the areas of psychotherapy, education, business training and development, mind, body and spirit. No fiction. 'The aim of the list is to both demystify the latest psychological advances, particularly in the fields of Accelerated Learning, Neuro-Linguistic Programming (NLP) and hypnosis.' 30 titles in 2007. Unsolicited submissions, via the website, are welcome.
Ⓔ Royalties twice-yearly.

The Crowood Press Ltd

The Stable Block, Crowood Lane, Ramsbury, Marlborough SN8 2HR
Ⓣ 01672 520320   Ⓕ 01672 520280
enquiries@crowood.com
www.crowood.com
Chairman *John Dennis*
Managing Director *Ken Hathaway*

Publishes sport and leisure titles, including animal and land husbandry, climbing and walking, maritime, country sports, equestrian, fishing and shooting; also crafts, dogs, gardening, DIY, theatre, natural history, aviation, military history and motoring. IMPRINT **Airlife Publishing** Specialist aviation titles for pilots, historians and enthusiasts. About 70 titles a year. Preliminary letter preferred in all cases.
Ⓔ Royalties annually.

CRW Publishing Ltd

69 Gloucester Crescent, London NW1 7EG
Ⓣ 020 7485 5764   Ⓕ 0870 751 7254
marcus.clapham@crw-publishing.co.uk
www.collectors-library.com
Also at: 6 Turville Barns, Eastleach, Cirencester GL7 3QB
Chairman *Ken Webb*
Editorial Director *Marcus Clapham*
Approx. Annual Turnover £2 million

Founded 2003. Publishes literary classics, gift, children's, boxed sets. IMPRINTS **Collector's Library; Collector's Library Editions in Colour; Book Blocks**. 20 titles in 2007. No unsolicited mss.
Ⓔ Royalties annually.

Benjamin Cummings
▷ Pearson

James Currey Publishers

73 Botley Road, Oxford OX2 0BS
Ⓣ 01865 244111   Ⓕ 01865 246454
editorial@jamescurrey.co.uk
www.jamescurrey.co.uk
Chairman *James Currey*
Managing Director/Editorial Director *Douglas H. Johnson*
Editorial Manager *Lynn Taylor*
Administrator *Vanessa Hinkley*

Founded 1985. A specialist publisher of academic paperback books on Africa and the Third World: history, anthropology, economics, sociology, politics and literary criticism. Approach in writing by post with synopsis if material is 'relevant to our needs'.
Ⓔ Royalties annually.

Custom Publishing
▷ The Orion Publishing Group Limited

D&B Publishing

PO Box 18, Hassocks BN6 9AA
Ⓣ 01273 711443

info@dandbpublishing.com
www.dandbpublishing.com
Joint Managing Directors *Dan Addelman,
Byron Jacobs*

Founded 2002. Publishes games books,
specializing in gambling, primarily. 8 titles
in 2007. Unsolicited mss, synopses and ideas
welcome; approach by e-mail in the first instance.
Ⓔ Royalties annually.

Terence Dalton Ltd

Water Street, Lavenham, Sudbury CO10 9RN
Ⓣ 01787 249291   Ⓕ 01787 248267
www.terencedalton.com
Director/Editorial Head *Elisabeth Whitehair*

Founded 1967. Part of the Lavenham Group Plc, a
family company. Non-fiction only. No unsolicited
mss. Ideas welcome; telephone in the first
instance.
Ⓔ Royalties annually.

Darton, Longman & Todd Ltd

1 Spencer Court, 140–142 Wandsworth High
Street, London SW18 4JJ
Ⓣ 020 8875 0155   Ⓕ 020 8875 0133
tradesales@darton-longman-todd.co.uk
www.dltbooks.com
Editorial Director *Brendan Walsh*
Editor/Rights Officer *Claudine Nightingale*
Approx. Annual Turnover £1 million

A leading independent publisher of books on
spirituality and religion. Unique among UK
religious publishers in being jointly owned
and managed by all its staff members. While
predominantly Christian, DLT publishes books
from different backgrounds and traditions.
TITLES *Jerusalem Bible*; *New Jerusalem Bible*;
NRSV Bible: Catholic Edition; *Return of the
Prodigal Son*; *God of Surprises*; *The Enduring
Melody*; *Hostage in Iraq*. About 45 titles a year.
Information on submissions available on the
website.
Ⓔ Royalties twice-yearly.

David & Charles Publishers

Brunel House, Forde Close, Newton Abbot
TQ12 4PU
Ⓣ 01626 323200   Ⓕ 01626 323317
postmaster@davidandcharles.co.uk
www.davidandcharles.co.uk
Managing Director & Publisher *Sara Domville*
Head of Publishing *Alison Myer*

Founded 1960, owned by F+W Publications.
Publishes illustrated non-fiction for international
markets, specializing in needlecraft, crafts, art
techniques, practical photography, military
history and equestrian. No fiction, poetry
or memoirs. TITLES *Ghosts Caught on Film*;
Sew Pretty Homestyle; *John Howe Fantasy Art

Workshop*; *Keeping Chickens*. About 200 titles a
year. Unsolicited mss will be considered if return
postage is included; synopses and ideas welcome
for the subjects listed above.
Ⓔ Royalties twice-yearly or flat fees.

Christopher Davies Publishers Ltd

PO Box 403, Swansea SA1 4YF
Ⓣ 01792 648825   Ⓕ 01792 648825
chris@cdaviesbookswales.com
Managing Director/Editorial Head *Christopher
T. Davies*
Approx. Annual Turnover £20,000

Founded 1949 to promote and expand Welsh-
language publications. Publishes biography,
cookery, history, sport and literature of Welsh
interest. TITLES *An A–Z of Wales and the Welsh*;
Welsh Birthplaces; *Carwyn: A Personal Memoir*;
Who's Who in Welsh History. About 2 titles a year.
No unsolicited mss. Synopses and ideas for books
welcome.
Ⓔ Royalties twice-yearly.

✱*Authors' update* **A favourite for Celtic readers
and writers.**

Giles de la Mare Publishers Ltd

PO Box 25351, London NW5 1ZT
Ⓣ 020 7485 2533   Ⓕ 020 7485 2534
gilesdelamare@dial.pipex.com
www.gilesdelamare.co.uk
Chairman/Managing Director *Giles de la Mare*
Approx. Annual Turnover £30,000

Founded 1995 and commenced publishing
in April 1996. Publishes mainly non-fiction,
especially art and architecture, biography, history,
music and travel. TITLES *Short Stories, Vols I, II
& III* Walter de la Mare; *Handsworth Revolution*
David Winkley; *The Life of Henry Moore* Roger
Berthoud; *Becoming an Orchestral Musician*
Richard Davis; *Romanesque Churches of France:
A Traveller's Guide* Peter Strafford; *Blindness
and the Visionary* John Coles; *Tricks Journalists
Play* Dennis Barker; *Venice: The Anthology Guide*
Milton Grundy; *Inherit the Truth 1939–1945* Anita
Lasker-Wallfisch; *Calatafimi* Angus Campbell.
Unsolicited mss, synopses and ideas welcome
after initial telephone call.
Ⓔ Royalties twice-yearly.

Debrett's Ltd

18–20 Hill Rise, Richmond TW10 6UA
Ⓣ 020 8939 2250   Ⓕ 020 8939 2251
people@debretts.co.uk
www.debretts.co.uk
Head of Publishing *Liz Wyse*
Chairman *Conrad Free*

Biographical reference plus etiquette, correct
form and other social guides; also diaries. Book
TITLES include triennial *Debrett's Peerage &

Baronetage and annual *People of Today* (as book, CD-ROM and online); *Manners for Men*; *Etiquette for Girls*; *Correct Form*; *Debrett's Wedding Guide*. Book and article proposals welcome.

Dedalus Ltd

Langford Lodge, St Judith's Lane, Sawtry PE28 5XE
℡ 01487 832382
info@dedalusbooks.com
www.dedalusbooks.com
Chairman *Juri Gabriel*
Managing Director *Eric Lane*
Approx. Annual Turnover £200,000

Founded 1983. Publishes contemporary European fiction and classics and original literary fiction. TITLES *The Dedalus Book of the Occult*; *The Arabian Nightmare* Robert Irwin; *Memoirs of a Gnostic Dwarf* David Madsen; *The Double Life of Daniel Glick* Maurice Caldera; *Music in a Foreign Language* Andrew Crumey (winner of the Saltire Best First Book Award in 1994); *The Dedalus Book of Literary Suicides* Gary Lachman; *The Dream Maker* Mikka Haugaard.

DIVISIONS/IMPRINTS **Original Fiction in Paperback**; **Dedalus Europe 1992–2012**; **Literary Concept Books**. Welcomes submissions for original fiction and books suitable for its list but 'most people sending work in have no idea what kind of books Dedalus publishes and merely waste their efforts'. Author guidelines on website. Particularly interested in intellectually clever and unusual fiction. A letter about the author should always accompany any submission. No replies without s.a.e.
Ⓔ Royalties annually.

⌖*Authors' update* **Put at risk by the withdrawal of its Arts Council grant, Dedalus was rescued by a cash injection from Informa publishing group. Thus, virtue and originality are rewarded.**

JM Dent
▷ The Orion Publishing Group Ltd

Despatches
▷ Reportage Press

André Deutsch Ltd
▷ Carlton Publishing Group

Digital Press
▷ Elsevier Ltd

Dinas
▷ Y Lolfa Cyf

Discovered Authors

Roslin Road, London W3 8DH
℡ 0844 800 5215
authors@discoveredauthors.co.uk
(Author Management Dept.)
www.discoveredauthors.co.uk
Managing Director *Graham Miller*

Founded 2006. Part of the Albemarle Group. Independent publisher offering various routes to publication, including self-publishing. Publishes fiction (adult and children's), non-fiction, academic and business. IMPRINTS **Discovered Authors Diamond** Traditional, mainstream publishing TITLES *Laughing Star*; *Haven't We Been Here Before?*; **Discovered Authors Revival** 'Enables out-of-print authors to get their work back into print and available for sale' TITLE *Harold the King*; **Four O Clock Press** Assisted publishing imprint TITLE *Dwelling Place for Dragons*; **Horizon Press** Launched in March 2007. About 200 titles in 2007. Unsolicited material welcome; approach by e-mail or mss can be uploaded via website link.
Ⓔ Royalties twice-yearly.

⌖*Authors' update* **After a troubled year with communications with writers in trouble, the company says it is putting its house in order but it is yet to be seen if Discovered Authors can fulfil its objective of overlapping conventional and self publishing. As ever in these circumstances, first-time authors need to be entirely clear as to who pays for what.**

John Donald Publishers Ltd

West Newington House, 10 Newington Road, Edinburgh EH9 1QS
℡ 0131 668 4371   ℻ 0131 668 4466
info@birlinn.co.uk
www.birlinn.co.uk
Managing Director *Hugh Andrew*

Bought by **Birlinn Ltd** in 1999. Publishes academic and scholarly, archaeology, architecture, textbooks, guidebooks, local, and social history. New books are published as an imprint of Birlinn Ltd. About 30 titles a year.

Donhead Publishing Ltd

Lower Coombe, Donhead St Mary, Shaftesbury SP7 9LY
℡ 01747 828422   ℻ 01747 828522
jillpearce@donhead.com
www.donhead.com
Contact *Jill Pearce*

Founded 1990 to specialize in publishing how-to books for building practitioners; particularly interested in architectural conservation material. Publishes building, architecture and heritage only. TITLES *Windows*; *Stone Conservation, Plastering*; *Preserving Post-War Heritage*; *Stone Cleaning*; *Architecture 1900*; *Encyclopaedia of Architectural*

Terms; *Cleaning Historic Buildings*; *Brickwork*; *Practical Stone Masonry*; *Conservation of Timber Buildings*; *Surveying Historic Buildings*; *Journal of Architectural Conservation* (3 issues yearly). 6 titles a year. Unsolicited mss, synopses and ideas welcome.

Dorling Kindersley Ltd

Part of the Penguin Group, 80 Strand, London WC2R ORL

☎ 020 7010 3000 📠 020 7010 6060

www.dk.com

CEO *Gary June*

Founded 1974. Packager and publisher of illustrated non-fiction: cookery, crafts, gardening, health, travel guides, atlases, natural history and children's information and fiction. Launched a US imprint in 1991 and an Australian imprint in 1997. Acquired Henderson Publishing in 1995 and was purchased by Pearson plc for £311 million in 2000.

DIVISIONS Adult: **Travel/Reference** Publisher *Douglas Amrine*; **General/Lifestyle** Publisher *John Roberts*. Children's: **Reference** Publisher *Miriam Farby*; **PreSchool/Primary** Publisher *Sophie Mitchell*. IMPRINTS **Ladybird**; **Ladybird Audio**; **Funfax**; **Eyewitness Guides**; **Eyewitness Travel Guides**. TITLES *BMA Complete Family Health Encyclopedia*; *RHS A–Z Encyclopedia of Garden Plants*; *Children's Illustrated Encyclopedia*; *The Way Things Work*. Unsolicited synopses/ideas for books welcome.

⊞*Authors' update* **A much slimmed version of the old DK is heavily dependent on 'soft learning' products for education markets. The aim is to make Dorling Kindersley products distinctive against low priced rivals and retailers' own brand product, not to mention the Internet. Most titles are team efforts with writers and illustrators working closely with an in-house editor.**

Doubleday
▷ Transworld Publishers

Doubleday Children's Books
▷ The Random House Group Ltd

Drake Educational Associates

St Fagans Road, Fairwater, Cardiff CF5 3AE

☎ 029 2056 0333 📠 029 2956 0313

info@drakeed.com

www.drakeed.com

Managing Director *R.G. Drake*

Literacy, phonics and language development games and activities. Ideas and scripts in these fields welcome. Also resources for subject areas in the primary school and modern languages at KS2.

Dref Wen

28 Church Road, Whitchurch, Cardiff CF14 2EA

☎ 029 2061 7860 📠 029 2061 0507

gwilym@drefwen.com

Chairman *R. Boore*

Managing Director *G. Boore*

Founded 1970. Publishes Welsh language and bilingual children's books, Welsh and English educational books for Welsh learners.

Ⓔ Royalties annually.

Gerald Duckworth & Co. Ltd

First Floor, 90–93 Cowcross Street, London EC1M 6BF

☎ 020 7490 7300 📠 020 7490 0080

info@duckworth-publishers.co.uk

www.ducknet.co.uk

Managing Director *Peter Mayer*

Editorial Director (Academic) *Deborah Blake*

Approx. Annual Turnover £2 million

Founded 1898 by Gerald Duckworth. Original publishers of Virginia Woolf. Other early authors include Hilaire Belloc, John Galsworthy, D.H. Lawrence and George Orwell. Duckworth is a general trade publisher whose authors include John Bayley, Mary Warnock, Joan Bakewell and J.J. Connolly. In addition to its trade list, Duckworth has a strong academic division. Acquired by Peter Mayer in 2003.

IMPRINTS/DIVISIONS **Bristol Classical Press** Classical texts and modern languages; **Ardis** Russian literature; **Duckworth Academic**; **Duckworth General**. No unsolicited mss; synopses and sample chapters only. Enclose s.a.e. or return postage for response/return.

Ⓔ Royalties twice-yearly.

Dunedin Academic Press Ltd

Hudson House, 8 Albany Street, Edinburgh EH1 3QB

☎ 0131 473 2397 📠 01250 870920

mail@dunedinacademicpress.co.uk

www.dunedinacademicpress.co.uk

Director *Anthony Kinahan*

Founded 2001. Publishes academic and serious general non-fiction. Considerable experience in academic and professional publishing. 14 titles in 2007. Not interested in science (other than earth sciences), poetry, fiction or children's. Check website for active subject areas. No unsolicited mss. Synopses and ideas welcome. Approach first in writing outlining proposal and identifying market. Proposal guidelines are available via the website.

Ⓔ Royalties annually.

Ebury Publishing/Ebury Press
▷ The Random House Group Ltd

Economist Books
▷ Profile Books

Eden
▷ Transworld Publishers

Edinburgh University Press
22 George Square, Edinburgh EH8 9LF
☎ 0131 650 4218   ⊞ 0131 662 0053
www.eup.ed.ac.uk
Chairman *Ivon Asquith*
Chief Executive *Timothy Wright*
Deputy Chief Executive/Head of Book Publishing
Jackie Jones
Senior Commissioning Editors *Nicola Carr,
Sarah Edwards*
Commissioning Editors *Carol Macdonald,
Esmé Watson*

Publishes academic and scholarly books (and
journals): African studies, ancient history and
classics, film and media studies, Islamic studies,
linguistics, literary criticism, philosophy, politics,
Scottish studies and religious studies. About 120
titles a year. E-mail submissions accepted; consult
website for guidelines.
£ Royalties paid annually.

Egmont Press
3rd Floor, Beaumont House, Avonmore Road,
London W14 8TS
☎ 020 7605 6600   ⊞ 020 7605 6601
initial.surname@euk.egmont.com
www.egmont.co.uk
Managing Director *Robert McMenemy*
Director of Press *Cally Poplak*

Founded 1878. Publishes children's books: picture
books, fiction (ages 14–16). IMPRINT **Two Heads**
Consumer-led fiction. AUTHORS include Michael
Morpurgo, Jenny Nimmo, William Nicholson,
Helen Oxenbury, Enid Blyton, Andy Stanton,
Julia Golding, Jan Fearnley and Lydia Monks. 'We
accept unsolicited mss but will not acknowledge
or respond to individual submissions unless
successful.'

Egmont Publishing
239 Kensington High Street, London W8 6SA
☎ 020 7761 3500/7605 6600
⊞ 020 7761 3510/7605 6601
initial.surname@euk.egmont.com
www.egmont.co.uk
Managing Director *Robert McMenemy*
Director of Publishing *David Riley*

Publishes children's books: annuals, activity
books, novelty books, film and TV tie-ins.
Licensed character list includes Thomas the Tank
Engine, Barbie, Mr Men, Miffy, Postman Pat,
Action Man.

✱*Authors' update* The fourth largest children's
book publishing group in the UK, Egmont is
moving into the US market with the aim of
becoming one of America's top ten children's
publishers within five years.

Eland Publishing Ltd
Third Floor, 61 Exmouth Market, London
EC1R 4QL
☎ 020 7833 0762   ⊞ 020 7833 4434
info@travelbooks.co.uk
www.travelbooks.co.uk
Directors *Rose Baring, John Hatt,
Barnaby Rogerson*
Approx. Annual Turnover £250,000

Eland reprints classics of travel literature with a
backlist of 70 titles, including *Naples '44* Norman
Lewis; *Travels with Myself and Another* Martha
Gelhorn; *Portrait of a Turkish Family* Irfan
Orga; *Jigsaw* Sybille Bedford. About 6–12 titles a
year. No unsolicited mss. Postcards and e-mails
welcome.
£ Royalties annually.

Element
▷ HarperCollins Publishers Ltd

11:9
▷ Neil Wilson Publishing Ltd

Edward Elgar Publishing Ltd
Glensanda House, Montpellier Parade,
Cheltenham GL50 1UA
☎ 01242 226934   ⊞ 01242 262111
info@e-elgar.co.uk
www.e-elgar.com
Managing Director *Edward Elgar*

Founded 1986. International publisher in
economics, the environment, public policy,
business and management and law. TITLES
Who's Who in Economics (3rd ed.); *Handbook of
Environmental and Resource Economics*; *Who's
Who in the Management Sciences*. 300 titles in
2007. No unsolicited mss; synopses and ideas in
the subject areas listed above welcome. Approach
by letter or e-mail; no telephone inquiries.

Elliot Right Way
▷ Constable & Robinson Ltd

Elliott & Thompson
27 John Street, London WC1N 2BX
☎ 020 7831 5013   ⊞ 020 7831 5011
www.elliottthompson.com
Publishers *Brad Thompson, Lorne Forsyth*
Approx. Annual Turnover £100,000

Founded 2001. Publishes literary fiction,
biography, belles lettres, reprints of classic male
writers. 11 titles in 2006. New proposals welcome
(send synopsis and sample pages initially) but
s.a.e. essential.

Elm Publications/Training (wholly owned subsidiary of Elm Consulting Ltd)

Seaton House, Kings Ripton, Huntingdon
PE28 2NJ
☎ 01487 773254
elm@elm-training.co.uk
www.elm-training.co.uk
Managing Director *Sheila Ritchie*

Founded 1977. Publishes textbooks, teaching aids, educational resources and educational software in the fields of business and management for adult learners. Books and teaching/training resources are generally commissioned to meet specific business, management and other syllabuses. About 30 titles a year. Ideas are welcome; initial approach by e-mail with outline, or by a brief telephone call.
Ⓔ Royalties annually.

Elsevier Ltd

The Boulevard, Langford Lane, Kidlington, Oxford OX5 1GB
☎ 01865 843000   ℻ 01865 843010
www.elsevier.com
Managing Director *Anna Moon (Oxford)*
CEO, Science & Technology (books & journals) *Herman van Campenhout*

Parent company Reed Elsevier, Amsterdam. Incorporates Pergamon Press and Harcourt Publishers International. Academic and professional reference books, scientific, technical and medical books, journals, CD-ROMs and magazines. IMPRINTS **Academic Press; Architectural Press; Butterworth Heinemann; Digital Press; Elsevier; Elsevier Advanced Technology; Focal Press; Gulf Professional Press; JAI; Made Simple Books; Morgan Kauffman; Newnes; North-Holland.**

DIVISION
Elsevier (Health Sciences)
32 Jamestown Road, London NW1 7BY
☎ 020 7424 4200 ℻ 020 7483 2293
www.elsevier-health.com
CEO Health Sciences (books & journals) *Brian Nairn*, Managing Director (UK & Netherlands) *Mary Ging* Publishes scientific, technical, medical books and journals. IMPRINTS **Baillière Tindall; Churchill Livingstone; Mosby; Pergamon; Saunders.** No unsolicited mss, but synopses and project proposals welcome.
Ⓔ Royalties annually.

Emissary Publishing

PO Box 33, Bicester OX26 4ZZ
☎ 01869 323447   ℻ 01869 322552
Editorial Director *Val Miller*

Founded 1992. Publishes mainly humorous paperback books including the complete set of Peter Pook novels; no poetry or children's. No unsolicited mss or synopses.
Ⓔ Royalties paid according to contract.

Emma Treehouse Ltd

The Studio, Church Street, Nunney BA11 4LW
☎ 01373 836233   ℻ 01373 836299
richard.powell4@virgin.net
sales@emmatreehouse.com
www.emmatreehouse.com
Co-Directors *Richard Powell, David Bailey*
Approx. Annual Turnover £1 million

Founded 1992. Publishes children's pre-school novelty books. No mss. Illustrations, synopses and ideas for books welcome; write in the first instance.
Ⓔ Fees paid; no royalties.

Empiricus Books
▷ Janus Publishing Company Ltd

English Heritage (Publishing)

Kemble Drive, Swindon SN2 2GZ
☎ 01793 414497   ℻ 01793 414769
www.english-heritage.org.uk
Head of Publishing *John Hudson*
Commercial Publishing Manager *Adèle Campbell*
Publishing Manager *Robin Taylor*

English Heritage's publishing programme reflects the organization's core interests in the historic environment of England, covering archaeology, architecture and general history titles. 20 titles a year. 'All titles relate directly to the work of the organization, so we do not accept unsolicited material.'

Enitharmon Press

26B Caversham Road, London NW5 2DU
☎ 020 7482 5967   ℻ 020 7284 1787
books@enitharmon.co.uk
www.enitharmon.co.uk
Director *Stephen Stuart-Smith*

Founded 1967. An independent company with an enterprising editorial policy, Enitharmon has established itself as one of Britain's leading literary presses. Patron of 'the new and the neglected, Enitharmon also prides itself on the success of its collaborations between writers and artists, now published by its associate company', **Enitharmon Editions**. Publishes poetry, literary criticism, fiction, art and photography. TITLES *Out of the Blue* Simon Armitage; *Talking about Aldo* Jim Dine; *Family Values* Maureen Duffy; *Supreme Being* Martha Kapos; *Wavelengths* Michael Longley; *O Vinho* Paula Rego. No unsolicited mss.
Ⓔ Royalties according to contract.

Epworth

c/o mph, 4 John Wesley Road, Werrington, Peterborough PE4 6ZP

☎ 01733 325002   📠 01733 384180

www.mph.org.uk

www.scm-canterburypress.co.uk

Chair *The Revd Michael Townsend*

Imprint of **mph**. Publishes Christian books only: philosophy, theology, biblical studies, pastoralia, social concern and Methodist studies. No fiction, poetry or children's. About 12 titles a year. Unsolicited mss, synopses and ideas welcome; send sample chapter and contents with covering letter and s.a.e.

💷 Royalties annually.

Erotic Review Books

17 Harwood Road, London SW6 4QP

☎ 020 7736 5800   📠 020 7736 6330

eros@eroticprints.org

www.eroticprints.org

Managing Director *Mr J. Maclean*

Publishes books of erotic art and photography and erotic literature. Unsolicited material accepted but approach in writing in the first instance with a summary or first chapter.

EuroCrime

▷ Arcadia Books

Euromonitor International plc

60–61 Britton Street, London EC1M 5UX

☎ 020 7251 8024   📠 020 7608 3149

info@euromonitor.com

www.euromonitor.com

Chairman *R.N. Senior*

Managing Director *T.J. Fenwick*

Approx. Annual Turnover £30 million

Founded 1972. International business information publisher specializing in library and professional reference books, market reports, electronic databases. Publishes business reference, market analysis and information directories only. DIVISIONS **Market Analysis** *A. Irwin*; **Reference Books & Directories** *S. Solomon*. About 2,000 titles a year.

💷 Payment is generally by flat fee.

Evans Brothers Ltd

2A Portman Mansions, Chiltern Street, London W1U 6NR

☎ 020 7487 0920   📠 020 7487 0921

sales@evansbrothers.co.uk

www.evansbooks.co.uk

Managing Director *Stephen Pawley*

International Publishing Director *Brian Jones*

UK Publisher *Su Swallow*

Approx. Annual Turnover £4.5 million

Founded 1908 by Robert and Edward Evans. Publishes UK children's and educational books, and educational books for Africa, the Caribbean and Latin America. IMPRINTS **Cherrytree Books; Zero to Ten**. OVERSEAS ASSOCIATES in Kenya, Cameroon, Sierra Leone; Evans Bros (Nigeria Publishers) Ltd. About 175 titles a year. 'Submissions welcome but cannot be returned. Unable to respond to individual proposals that are not accepted.'

💷 Royalties annually.

Everyman

▷ The Orion Publishing Group Ltd

Everyman Chess

▷ Gloucester Publishers Plc

Everyman's Library

Northburgh House, 10 Northburgh Street, London EC1V 0AT

☎ 020 7566 6350   📠 020 7490 3708

books@everyman.uk.com

Publisher *David Campbell*

Approx. Annual Turnover £3.5 million

Founded 1906. Publishes hardback classics of world literature, pocket poetry anthologies, children's books and travel guides. Publishes no new titles apart from poetry anthologies; only classics (no new authors). AUTHORS include Bulgakov, Bellow, Borges, Heller, Marquez, Nabokov, Naipaul, Orwell, Rushdie, Updike, Waugh and Wodehouse. No unsolicited mss.

💷 Royalties annually.

University of Exeter Press

Reed Hall, Streatham Drive, Exeter EX4 4QR

☎ 01392 263066   📠 01392 263064

uep@exeterpress.co.uk

www.exeterpress.co.uk

Publisher *Simon Baker*

Founded 1956. Publishes academic books: archaeology, classical studies and ancient history, history, maritime studies, English literature (especially medieval), film history, performance studies and books on Exeter and the South West. IMPRINT **Bristol Phoenix Press** Commissioning Editor *John Betts* Classics and ancient history. About 30 titles a year. Proposals welcomed in the above subject areas.

💷 Royalties annually.

Exley Publications Ltd

16 Chalk Hill, Watford WD19 4BG

☎ 01923 474480   📠 01923 800440

editorial@exleypublications.co.uk

www.helenexleygiftbooks.com

Editorial Director *Helen Exley*

Founded 1976. Independent family company. Publishes giftbooks, quotation anthologies and humour. No submissions, please.

Expert Books
> Transworld Publishers

Eyewitness Guides/Eyewitness Travel Guides
> Dorling Kindersley Ltd

Faber & Faber Ltd

3 Queen Square, London WC1N 3AU
☎ 020 7465 0045   ⊞ 020 7465 0034
www.faber.co.uk
Chief Executive *Stephen Page*
Approx. Annual Turnover £14.5 million

Geoffrey Faber founded the company in the 1920s, with T.S. Eliot as an early recruit to the board. The original list was based on contemporary poetry and plays (the distinguished backlist includes Eliot, Auden and MacNeice). Publishes poetry and drama, children's, fiction, film, music, politics, biography. In June 2008 launched **Faber Finds**, a new imprint bringing back classic titles via print on demand. DIVISIONS Fiction *Lee Brackstone, Hannah Griffiths, Angus Cargill* AUTHORS P.D. James, Peter Carey, Rachel Cusk, Giles Foden, Michael Frayn, Jane Harris, Tobias Hill, Kazuo Ishiguro, Barbara Kingsolver, Milan Kundera, Hanif Kureishi, John Lanchester, John McGahern, Rohinton Mistry, Lorrie Moore, Andrew O'Hagan, DBC Pierre, Jane Smiley; **Children's** *Julia Wells* AUTHORS Philip Ardagh, Terry Deary, Ricky Gervais, Russell Stannard, G.P. Taylor; **Film** *Walter Donohue* and **Plays** *Dinah Wood* AUTHORS Samuel Beckett, Alan Bennett, David Hare, Brian Friel, Patrick Marber, Harold Pinter, Tom Stoppard, Woody Allen, John Boorman, Joel and Ethan Coen, John Hodge, Martin Scorsese; **Music** *Belinda Matthews* AUTHORS John Bridcut, Humphrey Burton, Jonathan Carr, Rupert Christiansen, Nicholas Kenyon, Richard Morrison, Susan Tomes, John Tyrrell, Elizabeth Wilson; **Poetry** *Paul Keegan, Matthew Hollis* AUTHORS Simon Armitage, Douglas Dunn, Seamus Heaney, Ted Hughes, Paul Muldoon, Daljit Nagra, Tom Paulin; **Non-fiction** *Neil Belton, Julian Loose* AUTHORS John Carey, Simon Garfield, John Gray, Jan Morris, Francis Spufford, Frances Stonor Saunders, Jenny Uglow. ⓔ Royalties twice-yearly.

✱*Authors' update* One of the few generally recognizable publishing brands, Faber is venturing beyond books to embrace events, seminars and other profitable spinoffs. But the conventional output continues to expand with children's and non-fiction titles performing well. Faber is the prime mover in the successful Independent Alliance of publishers.

Facet Publishing

7 Ridgmount Street, London WC1E 7AE
☎ 020 7255 0590/0505 (text phone)
⊞ 020 7255 0591
info@facetpublishing.co.uk
www.facetpublishing.co.uk
Managing Director *John Woolley*

Publishing arm of **CILIP: The Chartered Institute of Library and Information Professionals** (formerly the Library Association). Publishes library and information science, monographs, reference and bibliography aimed at library and information professionals. IMPRINTS **Library Association Publishing; Clive Bingley Books; Facet Publishing**. Over 200 titles in print, including *The New Walford* and *AACR2*. 25–30 titles a year. Unsolicited mss, synopses and ideas welcome provided material falls firmly within the company's specialist subject areas. ⓔ Royalties annually.

Fernhurst Books
> Wiley Nautical

David Fickling Books
> The Random House Group Ltd

Fig Tree
> Penguin Group (UK)

Findhorn Press Ltd

305A The Park, Findhorn IV36 3TE
☎ 01309 690582   ⊞ 01309 690036
info@findhornpress.com
www.findhornpress.com
Director *Thierry Bogliolo*

Founded 1971. Publishes mind, body and spirit, New Age and healing. 24 titles in 2007. Unsolicited synopses and ideas via e-mail only if they come within Findhorn's subject areas. No children's books, fiction or poetry. 'Please check our website and look at the type of books we publish prior to contacting us.' ⓔ Royalties twice-yearly.

Firefly Publishing
> Helter Skelter Publishing

First & Best in Education Ltd

Hamilton House, Earlstrees Court, Earlstrees Road, Corby NN17 4HH
☎ 01536 399005   ⊞ 01536 399012
editorial@firstandbest.co.uk
www.firstandbest.co.uk
www.shop.firstandbest.co.uk
Publisher *Tony Attwood*
Editor *Anne Cockburn*

Publishers of over 1,000 educational books of all types for all ages of children and for parents and teachers. Also small series of books on

marketing. All books are published as being suitable for photocopying and/or as electronic books. 'Currently looking for books on school management and administration. Also some secondary subjects. No primary subjects, no fiction.' TITLES *Raising Grades Through Study Skills*; *Business Sponsorship of Secondary Schools*; *Policy Documents for Day Nurseries and Nursery Units*. IMPRINT **School Improvement Reports**. Check the website for submission guidelines.
Ⓔ Royalties twice-yearly.

Fitzgerald Publishing

89 Ermine Road, Ladywell, London SE13 7JJ
☎ 020 8690 0597
fitzgeraldbooks@yahoo.co.uk
Managing Editors *Tim Fitzgerald, Michael Fitzgerald*
General Editor *Andrew Smith*

Founded 1974. Specializes in scientific studies of insects and spiders. TITLES *The Tarantula*; *Keeping Spiders and Insects in Captivity*; *Tarantulas of the USA*; *Scorpions of Medical Importance* (books) and *Earth Tigers – Tarantulas of Borneo*; *Desert Tarantulas* (TV/video documentaries). 1–2 titles a year. Unsolicited mss, synopses and ideas for books welcome. Also considers video scripts for video documentaries.

Fitzjames Press
▷ **Motor Racing Publications**

Five Star
▷ **Serpent's Tail**

Floris Books

15 Harrison Gardens, Edinburgh EH11 1SH
☎ 0131 337 2372   📠 0131 347 9919
floris@florisbooks.co.uk
www.florisbooks.co.uk
Managing Director *Christian Maclean*
Editors *Christopher Moore, Gale Winskill*
Approx. Annual Turnover £550,000

Founded 1977. Publishes books related to the Steiner movement, including The Christian Community, as well as arts & crafts, children's (including fiction with a Scottish theme), history, religious, science, social questions and Celtic studies. No unsolicited mss except for the Kelpies children's books (not picture books). Synopses and ideas are welcome for all books except children's.
Ⓔ Royalties annually.

Focal Press
▷ **Elsevier Ltd**

Folens Limited

Waterslade House, Thame Road, Haddenham HP23 5DP
☎ 0870 609 1235   📠 0870 609 1236
folens@folens.com
www.folens.com
Chairman *Dirk Folens*
Managing Director *Adrian Cockell*

Founded 1987. A leading educational publisher. IMPRINTS **Folens**; **Belair**. About 150 titles a year. Unsolicited mss, synopses and ideas for educational books welcome.
Ⓔ Royalties annually.

Fort Publishing Ltd

Old Belmont House, 12 Robsland Avenue, Ayr KA7 2RW
☎ 01292 880693   📠 01292 270134
fortpublishing@aol.com
www.fortpublishing.co.uk
Chairman *Agnes Jane McCarroll*
Managing Director *James McCarroll*
Approx. Annual Turnover £95,000

Founded 1999. Publishes general non-fiction, history, sport, crime and local interest. TITLES *Glasgow Then and Now*; *Great Hull Stories*; *Evil Scotland*; *Ten Days That Shook Rangers*. 40 titles in 2007. No unsolicited mss; send synopses and ideas for books by post or e-mail.
Ⓔ Royalties twice-yearly.

Foulsham Publishers

The Publishing House, Bennetts Close, Slough SL1 5AP
☎ 01753 526769   📠 01753 535003
reception@foulsham.com
www.foulsham.com
Chairman/Managing Director *B.A.R. Belasco*
Approx. Annual Turnover £2.5 million

Founded 1800 and now one of the few remaining independent family companies to survive takeover. Publishes non-fiction on most subjects including lifestyle, travel guides, family reference, cookery, diet, health, DIY, business, self improvement, self development, astrology, dreams, MBS. No fiction. IMPRINT **Quantum** Mind, body and spirit titles. TITLES *Classic 1000 Cocktails*; *Brit Guide to Orlando and Walt Disney World*; *Old Moore's Almanack*; *Raphael's Astrological Ephemeris*. Around 60 titles a year. Unsolicited mss, synopses and ideas welcome – in hard copy.
Ⓔ Royalties twice-yearly.

Fountain Press

Newpro UK Ltd., Old Sawmills Road, Faringdon SN7 7DS
☎ 01367 242411   📠 01367 241124

sales@newprouk.co.uk
www.newprouk.co.uk
Publisher *C.J. Coleman*
Approx. Annual Turnover £800,000

Founded 1923 when it was part of the Rowntree
Trust Social Service. Owned by the British
Electric Traction Group until 1982 when it
was bought out by H.M. Ricketts. Acquired
by Newpro UK Ltd in July 2000. Publishes
mainly photography and natural history. TITLES
Photography and Digital 'Workshop'; *Antique
and Collectable Cameras*; *Camera Manual*
(series). About 2 titles a year. Unsolicited mss and
synopses welcome.
Ⓔ Royalties twice-yearly.

✱*Authors' update* **Highly regarded for
production values, Fountain has the reputation
for involving authors in every stage of the
publishing process.**

Four O Clock Press
▷ Discovered Authors

Fourth Estate
▷ HarperCollins Publishers Ltd

Free Association Books Ltd
PO Box 37664, London NW7 2XU
☎ 020 8906 0396   🖷 020 8906 0006
info@fabooks.com
www.fabooks.com
Managing Director/Publisher *T.E. Brown*

Publishes psychoanalysis and psychotherapy,
psychology, cultural studies, sexuality and
gender, women's studies, applied social sciences.
TITLES *Drug Use and Cultural Context* eds.
Ross Coomber and Nigel South; *Skin Disease*
Ann Maguire; *Caring for the Dying at Home*
Gill Pharoah; *A History of Group Study and
Psychodynamic Organizations* Amy L. Fraher.
Always send a letter in the first instance
accompanied by a book outline. OVERSEAS
ASSOCIATES ISBS, USA; Astam, Australia.
Ⓔ Royalties twice-yearly.

W. H. Freeman
Palgrave Macmillan, Brunel Road, Houndmills,
Basingstoke RG21 6XS
☎ 01256 329242   🖷 01256 330688
www.palgrave.com
President *Elizabeth Widdicombe (New York)*
Editorial Director *Margaret Hewinson*

Following integration into the BFW (Bedford,
Freeman, Worth) College Group, USA, W. H.
Freeman now publishes academic educational
and textbooks in biochemistry, biology and
zoology, chemistry, economics, mathematics
and statistics, natural history, neuroscience,
palaeontology, physics. W. H. Freeman's editorial
office is in New York (Basingstoke is a sales and
marketing office only) but unsolicited mss can go
through Basingstoke. Those which are obviously
unsuitable will be sifted out; the rest will be
forwarded to New York.
Ⓔ Royalties annually.

Samuel French Ltd
52 Fitzroy Street, London W1T 5JR
☎ 020 7387 9373   🖷 020 7387 2161
theatre@samuelfrench-london.co.uk
www.samuelfrench-london.co.uk
Chairman *Leon Embry*
Managing Director *Vivien Goodwin*

Founded 1830 with the object of acquiring acting
rights and publishing plays. Publishes plays only.
About 50 titles a year. Unsolicited mss considered
only after introductory letter addressed to the
Performing Rights Department.
Ⓔ Royalties twice-yearly for books; performing
royalties monthly, subject to a minimum amount.

✱*Authors' update* **Thrives on the amateur
dramatic societies who are forever in need
of play texts. Editorial advisers give serious
attention to new material but a high proportion
of the list is staged before it goes into print.
New writers are advised to try one-act plays,
much in demand by the amateur dramatic
societies but rarely turned out by established
playwrights.**

The Friday Project
▷ HarperCollins Publishers Ltd

Frontline Books
5A Accommodation Road, Golders Green,
London NW11 9ED
☎ 020 8455 5559   🖷 01226 734438
michael@frontline-books.com
www.frontline-books.com
Managing Director *Charles Hewitt*

New military imprint of **Pen & Sword Books Ltd**.
Publishes a wide range of military history topics
and periods, from Ancient Greece and Rome to
the present day. 20 titles in 2007. Send synopsis
and sample chapter giving an outline of the book's
theme and contents, chapter headings and word
extent, if possible. Indication of sources would
be helpful together with number of and type of
illustrations, if relevant, plus personal details
that underline qualifications for writing the book
rather than a formal c.v. Enclose s.a.e. for return
of material. No fiction.
Ⓔ Royalties twice-yearly.

David Fulton Publishers Ltd
2 Park Square, Milton, Abingdon OX14 4RN
☎ 020 7017 6000   🖷 020 8996 3622
info@routledge.co.uk
www.routledge.co.uk
Owner *Taylor & Francis Group*

Senior Publishers *Bruce Roberts* (textbooks), *Monica Lee* (classroom)
Approx. Annual Turnover £2.4 million

Founded 1987. Part of Routledge Education. Publishes textbooks and practical books for students, trainee and working teachers at all levels of the curriculum (foundation stage to post compulsory). Special educational needs books also published in collaboration with the National Association for Special Educational Needs (NASEN); some overlap into other areas of health and social care, therapy, educational psychology, and speech and language therapy. A growing list of early years books for pre-school and nursery staff. About 100 titles a year. No unsolicited mss; synopses and ideas for books welcome.
Ⓔ Royalties annually.

Funfax
▷ **Dorling Kindersley Ltd**

Fusion Press
▷ **Vision**

Gaia Books
▷ **Octopus Publishing Group**

Garnet Publishing Ltd

8 Southern Court, South Street, Reading RG1 4QS
📞 0118 959 7847   📠 0118 959 7356
dan@garnetpublishing.co.uk
www.garnetpublishing.co.uk
Editorial Manager *Dan Nunn*

Founded 1992 and purchased Ithaca Press in the same year. Publishes art, architecture, photography, archive photography, cookery, travel classics, comparative religion, Islamic culture and history, foreign fiction in translation. Core subjects are Middle Eastern but list is rapidly expanding to be more general.

IMPRINTS **Garnet Publishing** TITLES *Simply Lebanese*; *The Art and Architecture of Islamic Cairo*; *What Did We Do To Deserve This?*; **Ithaca Press** Specializes in post-graduate academic works on the Middle East, political science and international relations TITLES *The Making of the Modern Gulf States*; *Reform in the Middle East Oil Monarchies*; *The United States and Persian Gulf Security*. About 20 titles a year. Unsolicited mss not welcome; write with outline and ideas first plus current c.v.
Ⓔ Royalties twice-yearly.

Geddes & Grosset

David Dale House, New Lanark ML11 9DJ
📞 01555 665000   📠 01555 665694
Publishers *Ron Grosset, R. Michael Miller*
Approx. Annual Turnover £2.5 million

Founded 1989. Publisher of children's and reference books. Unsolicited mss, synopses and ideas welcome. No adult fiction.

The Geological Society Publishing House

Unit 7, Brassmill Enterprise Centre, Brassmill Lane, Bath BA1 3JN
📞 01225 445046   📠 01225 442836
sales@geolsoc.org.uk
enquiries@geolsoc.org.uk
www.geolsoc.org.uk
Commissioning Editor *Angharad Hills*

Publishing arm of the Geological Society which was founded in 1807. Publishes postgraduate texts in the earth sciences. 30 titles a year. Unsolicited synopses and ideas welcome.

Stanley Gibbons Publications

7 Parkside, Christchurch Road, Ringwood BH24 3SH
📞 01425 472363   📠 01425 470247
info@stanleygibbons.com
www.stanleygibbons.com
Chief Executive *M. Hall*
Editorial Head *H. Jefferies*

Long-established force in the philatelic world with 150 years in the business. Publishes philatelic reference catalogues and handbooks. Reference works relating to other areas of collecting may be considered. TITLES include *Stanley Gibbons British Commonwealth Stamp Catalogue*; *Stamps of the World*. Foreign catalogues include Japan and Korea, Portugal and Spain, Germany, Middle East, Balkans, China. Monthly publication: *Gibbons Stamp Monthly* (see entry under *Magazines*). www.collectorcafe.com – Internet site covering full range of collectibles – editorial input always considered. About 20 titles a year. Unsolicited mss, synopses and ideas welcome.
Ⓔ Royalties by negotiation.

Gibson Square

47 Lonsdale Square, London N1 1EW
📞 020 7096 1100   📠 020 7993 2214
info@gibsonsquare.com
www.gibsonsquare.com
Chairman *Martin Rynja*
Approx. Annual Turnover £600,000

Founded 2001. Publishes exclusively non-fiction: biography, current affairs, politics, cultural criticism, psychology, history, travel, art history, philosophy. Books must have high publicity profile. IMPRINTS **Gibson Square** TITLES *Londonistan* Melanie Phillips; *House of Bush House of Saud* Craig Unger; *Blowing up Russia* Alexander Litvinenko; *Two Lipsticks & a Lover* Helena Frith Powell; *If I Did It* O.J. Simpson. Welcomes unsolicited mss, synopses and ideas. Send self-addressed, franked, return envelope.

E-mailed submissions will be read but reply sent only if interested. No fiction.
ⓔ Royalties twice-yearly.

⊛*Authors' update* Dedicated to publishing books 'on the cutting edge', Gibson Square is one of the few publishers to embrace serious political issues.

Ginn
▷ Pearson (Heinemann)

GL Assessment
The Chiswick Centre, 414 Chiswick High Road, London W4 5TF
☎ 020 8996 3333   ⓕ 020 8742 8390
www.gl-assessment.co.uk

GL Assessment is the new name for nferNelson, founded 1981. A division of Granada Learning Ltd. Publishes educational and psychological assessments and online tests. Main interest is in educational and clinical and occupational assessment.

Gloucester Publishers Plc
10 Northburgh Street, London EC1V 0AT
☎ 020 7253 7887   ⓕ 020 7490 3708
markbicknell@everyman.uk.com
www.everymanchess.com
Managing Director *Mark Bicknell*
Approx. Annual Turnover £1 million

Publishes exclusively academic and leisure books relating to chess. IMPRINT **Everyman Chess** TITLES *Garry Kasparov: My Great Predecessors, Vols 1–7*; *Play Winning Chess* Yasser Seirawan; *Art of Attack in Chess* Vladimir Vukovich. 30 titles in 2007/8. No unsolicited material.
ⓔ Royalties annually.

GMC Publications Ltd
166 High Street, Lewes BN7 1XU
☎ 01273 477374   ⓕ 01273 402866
pubs@thegmcgroup.com
www.thegmcgroup.com
Joint Managing Directors *J.A.J. Phillips, J.A.B. Phillips*
Managing Editor *Gerrie Purcell*

Founded 1974. Publisher and distributor of craft and leisure books and magazines, covering topics such as woodworking, DIY, architecture, photography, gardening, cookery, art, puzzles and games, reference, humour, TV and film. About 50 illustrated books and 11 magazines a year. Unsolicited mss, synopses and ideas for books welcome. No fiction.
ⓔ Royalties twice-yearly.

Godsfield Press
▷ Octopus Publishing Group

Gollancz
▷ The Orion Publishing Group Limited

Gomer Press/Gwasg Gomer
Llandysul Enterprise Park, Llandysul SA44 4JL
☎ 01559 363090   ⓕ 01559 363758
gwasg@gomer.co.uk
www.gomer.co.uk
www.pontbooks.co.uk
Chairman/Managing Director *J.E. Lewis*
Publishing Director *Mairwen Prys Jones*
Editors (adult books – Welsh) *Bethan Mair, Bryan James*
Editor (adult books – English) *Ceri Wyn Jones*
Editors (children's books – Welsh) *Sioned Lleinau, Helen Evans, Rhiannon Davies*
Pont Books (children's books – English) Editor *Viv Sayer*

Founded 1892. Publishes adult fiction and non-fiction, children's fiction and educational material in English and in Welsh. IMPRINTS **Gomer; Pont Books**. 125 titles a year. Unsolicited mss welcome. Prior enquiry recommended. Mss must have a Welsh dimension.
ⓔ Royalties twice-yearly.

Gower
▷ Ashgate Publishing Ltd

Graham-Cameron Publishing & Illustration
The Studio, 23 Holt Road, Sheringham NR26 8NB
☎ 01263 821333   ⓕ 01263 821334
enquiry@graham-cameron-illustration.com
www.graham-cameron-illustration.com
Also at: 59 Hertford Road, Brighton BN1 7GG
☎ 01273 385890
Editorial Director *Mike Graham-Cameron*
Art Director *Helen Graham-Cameron*

Founded 1984 as a packaging operation. Publishes illustrated information and educational books. TITLES *Up From the Country*; *In All Directions*; *The Holywell Story*; *Flashbacks*; *Let's Look at Dairying*. Has 37 contracted book illustrators concentrating on educational and children's books. *Absolutely no* unsolicited mss.
ⓔ Royalties annually.

Granta Books
Twelve Addison Avenue, London W11 4QR
☎ 020 7605 1360   ⓕ 020 7605 1361
www.granta.com
Managing Director *David Graham*
Publisher *Philip Gwyn Jones*
Editorial Director *Sara Holloway*

Founded 1979. Acquired by Sigrid Rausing in 2006. Publishes general non-fiction and some literary fiction. About 70 titles a year. No unsolicited mss; synopses and sample chapters welcome.

ⓔ Royalties twice-yearly.

✱*Authors' update* **With changes in top management, there are moves to revitalize the Granta list with an increase in the number of titles and the commissioning of more mainstream books.**

W. Green (Scotland)
▷ Sweet & Maxwell Group

Green Books

Foxhole, Dartington, Totnes TQ9 6EB
ⓣ 01803 863260 ⓕ 01803 863843
edit@greenbooks.co.uk
www.greenbooks.co.uk
Chairman *Satish Kumar*
Publisher *John Elford*
Approx. Annual Turnover £500,000

Founded in 1987 with the support of a number of Green organizations. Closely associated with *Resurgence* magazine. Publishes high-quality books on a wide range of Green issues, including economics, politics and the practical application of Green thinking. No fiction or books for children. TITLES *The Transition Handbook* Rob Hopkins; *Timeless Simplicity* John Lane; *Allotment Gardening* Susan Berger. No unsolicited mss. Synopses and ideas welcome but check guidelines on the website in the first instance.
ⓔ Royalties twice-yearly.

Green Print
▷ The Merlin Press Ltd

Gresham Books Ltd

46 Victoria Road, Summertown, Oxford OX2 7QD
ⓣ 01865 513582 ⓕ 01865 512718
info@gresham-books.co.uk
www.gresham-books.co.uk
Managing Director *Paul Lewis*
Approx. Annual Turnover £350,000

A small specialist publishing house. Publishes hymn and service books for schools and churches, school histories, also craft-bound choir and orchestral folders. No unsolicited material but ideas welcome.

Grub Street

4 Rainham Close, London SW11 6SS
ⓣ 020 7924 3966/7738 1008 ⓕ 020 7738 1009
post@grubstreet.co.uk
www.grubstreet.co.uk
Managing Director *John Davies*

Founded 1982. Publishes cookery, food and wine, military and aviation history books. About 30 titles a year. Unsolicited mss and synopses welcome in the above categories but please enclose return postage.
ⓔ Royalties twice-yearly.

Guinness World Records Ltd

3rd Floor, 184–192 Drummond Street, London NW1 3HP
ⓣ 020 7891 4567 ⓕ 020 7891 4501
sales@guinnessworldrecords.com
www.guinnessworldrecords.com
Managing Director *Alistair Richards*

First published in 1955, the annual Guinness World Records book is published in more than 100 countries and 28 languages and is the highest-selling copyright book of all time, with more than 3 million copies sold annually across the globe. In addition to *Guinness World Records* the company publishes *Gamer's Edition*. Acquired by Jim Pattison Group (JPG) in February 2008. Contact from prospective researchers, editors and designers welcome.

✱*Authors' update* **It had to happen. With its sale to the Jim Pattison Group, Guinness World Records has joined the creator of Ripley's Believe It or Not, often thought to have been the original inspiration for Guinness. Some claim that World Records has dumbed down in recent years but you have to hand it to an annual that can sell close to 600,000 copies.**

Gulf Professional Press
▷ Elsevier Ltd

Gwasg Carreg Gwalch

12 Iard yr Orsaf, Llanrwst LL26 0EH
ⓣ 01492 642031 ⓕ 01492 641502
books@carreg-gwalch.co.uk
www.carreg-gwalch.co.uk
Managing Editor *Myrddin ap Dafydd*

Founded in 1980. Publishes Welsh language; English books of Welsh interest – history, folklore, guides and walks. About 90 titles a year. Unsolicited mss, synopses and ideas welcome if of Welsh interest.
ⓔ Royalties paid.

Hachette Children's Books

338 Euston Road, London NW1 3BH
ⓣ 020 7873 6000
www.hachettechildrens.co.uk
www.franklinwatts.co.uk
www.orchardbooks.co.uk
www.hodderchildrens.co.uk
www.waylandbooks.co.uk
Owner *Hachette Book Group Inc.*
Managing Director *Marlene Johnson*
Approx. Annual Turnover £32 million

Formed by the combining of Watts Publishing Group with Hodder Children's Books in April 2005. Publishes children's non-fiction, reference, information, gift, fiction, picture and novelty books and audio books. IMPRINTS **Hodder Children's Books** *Anne McNeil* Fiction, picture books, novelty, general non-fiction and audio. **Orchard Books** Picture book and novelties *Kate Burns*; Fiction *Penny Morris*; **Franklin Watts** *Rachel Cooke* Non-fiction and information books. **Wayland** *Joyce Bentley* Non-fiction and information books. About 1,100 titles a year. Unsolicited material is not considered. Ⓔ Royalties twice-yearly.

Hachette Livre

French publishing conglomerate which embraces **Chambers Harrap Publishers**, **Hachette Livre UK**, **Orion Publishing Group**, **Octopus Publishing Group**, **Little, Brown Book Group**. For further information, see imprint entries and entry under *European Publishers*.

Hachette Livre UK

338 Euston Road, London NW1 3BH
Ⓣ 020 7873 6000   Ⓕ 020 7873 6024
www.hachettelivre.co.uk
Group Chief Executive *Tim Hely Hutchinson*

Hodder Headline was formed in June 1993 through the merger of Headline Book Publishing and Hodder & Stoughton and was renamed Hachette Livre UK in August 2007. The company was acquired by WHSmith plc in 1999 and subsequently by Hachette Livre S.A. in October 2004. Purchased John Murray (Publishers) Ltd in 2002. About 2,000 titles a year.

DIVISIONS

Headline Publishing Group Managing Director *Martin Neild*, Deputy Managing Director and Chief Operating Officer *Kerr MacRae*, Deputy Managing Director *Jane Morpeth* Publishes commercial and literary fiction (hardback and paperback) and popular non-fiction including autobiography, biography, food and wine, history, humour, popular science, sport and TV tie-ins. IMPRINTS **Business Plus**; **Headline**; **Headline Review**; **Little Black Dress**. AUTHORS Lyn Andrews, Emily Barr, Louise Bagshawe, Martina Cole, Janet Evanovich, Victoria Hislop, Wendy Holden, Jonathan Kellerman, Jill Mansell, Maggie O'Farrell, Sheila O'Flanagan, James Patterson, Jed Rubenfeld, Karen Rose, Pamela Stephenson, Penny Vincenzi.

Hodder & Stoughton General CEO *Martin Neild*, Managing Director *Jamie Hodder-Williams*, Deputy Managing Director *Lisa Highton*; **Non-fiction** *Rowena Webb*; **Sceptre** *Carole Welch*; **Fiction** *Carolyn Mays*; **Audio** (see entry under *Audio Books*). Publishes commercial and literary fiction; biography, autobiography, history, self-help, humour, travel and other general interest non-fiction; audio. IMPRINTS **Hodder & Stoughton**; **Hodder Paperbacks**; **Sceptre**; **Mobius**. AUTHORS Melvyn Bragg,

Jeffrey Deaver, Charles Frazier, Elizabeth George, Thomas Keneally, Stephen King, John le Carré, Andrew Miller, David Mitchell, Jodi Picoult, Rosamunde Pilcher, Mary Stewart, Fiona Walker. No unsolicited mss.

Hodder Children's Books (See **Hachette Children's Books**)

Hodder & Stoughton Faith Books Director *Wendy Grisham*, Bibles & Hodder Christian Books *David Moloney*. Publishes TNIV, NIV and NIrV Bibles, Christian books, autobiography, biography, TV-tie-ins, gift/humour books, self-help. IMPRINTS **Hodder & Stoughton**; **Hodder Paperbacks**; **TNIV**; **NIV**; **NIrV**.

Hodder Education Group CEO *Philip Walters* Publishes in the following areas: **Schools** *Lis Tribe*; **Consumer Education** *Katie Roden*; **Further Education/Higher Education Textbooks** *Alexia Chan*; **Health Sciences** *Joanna Koster*. IMPRINTS **Hodder Education**; **Teach Yourself**; **Hodder Arnold**; **Hodder Gibson** (see entry).

Hachette Books Scotland Publisher *Bob McDevitt* Commercial and literary fiction and non-fiction.

Hachette Books Ireland Managing Director and Publisher *Breda Purdue* Commercial and literary fiction and non-fiction (see entry under *Irish Publishers*)

John Murray (Publishers) Ltd (see entry).

✱*Authors' update* **Depending on which set of figures you believe, Hachette commands by far the biggest, the biggest or close to the biggest share of the UK market. For the Hodder imprints, the emphasis is very much on bestsellers with, for example, a concerted effort to dominate the popular end of the crime market with American imports. At the same time the search is on for new talent which can benefit from imaginative and energetic marketing. The losers are midlist authors who do not aspire to mass sales. Downloadable versions of most top titles will be on offer by year end.**

Halban Publishers

22 Golden Square, London W1F 9JW
Ⓣ 020 7437 9300   Ⓕ 020 7437 9512
books@halbanpublishers.com
www.halbanpublishers.com
Directors *Peter Halban, Martine Halban*

Founded 1986. Independent publisher. Publishes biography, autobiography and memoirs, history, literature and Judaica. 8–10 titles a year. No unsolicited material. Approach by letter in first instance. Unsolicited e-mails will be deleted. Ⓔ Royalties twice-yearly for first two years, thereafter annually in December.

Robert Hale Ltd

Clerkenwell House, 45–47 Clerkenwell Green,
London EC1R 0HT

☎ 020 7251 2661   📠 020 7490 4958
enquire@halebooks.com
www.halebooks.com
Chairman/Managing Director *John Hale*

Founded 1936. Family-owned company. Publishes
adult fiction (but not interested in category
romance or science fiction) and non-fiction. No
specialist material (education, law, medical or
scientific). Acquired NAG Press Ltd in 1993 with
its list of horological, gemmological, jewellery
and metalwork titles, and **J.A. Allen & Co.** in
1999 with its extensive list of horse and dog books
– *The Building Blocks of Training* Debby Lush;
Better Jumping Carol Mailer (see entry). TITLES
Non-fiction: *Don't Do It If You Can't Keep It Up:
A Lover's Guide* Gerd de Ley; *Build Your Own
Dream House in France* David Murray; Fiction:
This Rage of Echoes Simon Clark; *Little Marvel
and Other Stories* Wendy Perriam. Over 200 titles
a year. Unsolicited mss, synopses and ideas for
books welcome.
💷 Royalties twice-yearly.

Halsgrove

Halsgrove House, Ryelands Farm Estate, Bagley
Green, Wellington TA21 9PZ

☎ 01823 653777   📠 01823 665294
sales@halsgrove.com
www.halsgrove.com
Publisher *Simon Butler*

Founded in 1990 and now a leading publisher
and distributor of regional books under the
Halsgrove, **Halstar** and **Ryelands** IMPRINTS.
Publishes books of regional interest throughout
the UK: local history (including the *Community
History* series), biography, photography and art,
mainly in hardback. No fiction or poetry. 170
titles in 2008. Unsolicited mss, synopses and
ideas for books of regional interest welcome.
💷 Royalties annually.

Halstar
▷ Halsgrove

Hambledon Continuum
▷ The Continuum International Publishing Grooup
Limited

Hamish Hamilton
▷ Penguin Group (UK)

Hamlyn Octopus
▷ Octopus Publishing Group

Hammersmith Press Ltd

496 Fulham Palace Road, London SW6 6JD
☎ 020 7736 9132   📠 020 7348 7521
gmb@hammersmithpress.co.uk
www.hammersmithpress.co.uk
Managing Director *Georgina Bentliff*

Founded 2004. Publishes health, nutrition, diet,
academic medicine, and 'literary medicine',
i.e. literary works that relate to the practice of
medicine. TITLES *Traditional Herbal Medicines*
Lakshman Karalliedde; *Shadows in Wonderland*
Colin Ludlow. 6 titles in 2007. No unsolicited
mss; synopses and ideas for books welcome;
approach by e-mail. No fiction or children's
material.
💷 Royalties twice-yearly.

Happy Cat
▷ Catnip Publishing Ltd

Harlequin Mills & Boon Limited

Eton House, 18–24 Paradise Road, Richmond
TW9 1SR
☎ 020 8288 2800   📠 020 8288 2898
www.eharlequin.com
www.millsandboon.co.uk
Managing Director *Guy Hallowes*
Editorial Director *Karin Stoecker*
Approx. Annual Turnover £18.97 million

Founded 1908. Owned by the Canadian-based
Torstar Group. Publishes a wide range of women's
fiction including romantic novels.

IMPRINTS **Mills & Boon Modern** *Tessa Shapcott*
(50–55,000 words) Alpha males and attractive
women swept up in intense emotions set against
a backdrop of international locations, luxury
and wealth; passion is guaranteed. **Mills & Boon
Modern Heat** *Bryony Green* (50,000–55,000
words) Delivers a feel-good experience, focusing
on the kind of relationship that women aged 18 to
35 aspire to. Young characters in urban settings
meet, flirt, share experiences, have great sex and
fall in love, finally making a commitment that
will bind them forever. Based around emotional
issues, other concerns – e.g. jobs and friendships
– are also touched upon and resolved in upbeat
way. **Mills & Boon Romance** *Kimberley Young*
(50–55,000 words) Contemporary upbeat and
heroine-focused romance, driven by strongly
emotional conflicts which are believable and
relevant to today's women. Capturing the depth
of emotion and sheer excitement of falling in
love in a variety of international settings. **Mills &
Boon Medical** *Sheila Hodgson* (50–55,000 words)
Modern medical practice provides a unique
background to highly emotional contemporary
romances. **Mills & Boon Historical Romance**
Linda Fildew (80–95,000 words) Richly textured,
emotionally intense historical novels covering
a wide range of settings and periods from
ancient civilizations up to and including the
Second World War. Over 600 titles a year. Other

submissions: also accepting submissions in all general women's fiction genres (contemporary and historical) for longer series and single title imprints. Tip sheets and guidelines are available from the Mills & Boon or eHarlequin websites (see above) or Harlequin Mills & Boon Editorial Dept. (telephone or write sending s.a.e.). Agented and unagented submissions accepted. Send query letter, synopsis and first three chapters in the first instance.

✱*Authors' update* **Light romance still provides a good income for writers who can adapt to the clearly defined M&B formula. Just think, M&B has a network of 12,000 authors worldwide and every title scores six-figure sales. But be warned. M&B gets 2,000 unsolicited manuscripts a year. Latest initiatives include a crime list with five titles appearing every two months. There are plans to make up to 2000 titles available in digital format.**

HarperCollins Publishers Ltd

77–85 Fulham Palace Road, London w6 8jb
☏ 020 8741 7070   🖷 020 8307 4440
www.harpercollins.co.uk
Also at: Westerhill Road, Bishopbriggs, Glasgow G64 2QT ☏ 0141 772 3200 🖷 0141 306 3119
Owner *News Corporation*
CEO/Publisher *Victoria Barnsley*
Managing Director *Amanda Ridout*
Approx. Annual Turnover £173.4 million (UK)

HarperCollins is one of the top book publishers in the UK, with a wider range of books than any other publisher; from cutting-edge contemporary fiction to enduring classics; from school textbooks to celebrity memoirs; and from downloadable dictionaries to prize-winning literature. The wholly-owned division of News Corporation also publishes a wide selection of non-fiction including history, celebrity memoirs, biography, popular science, mind, body and spirit, dictionaries, maps and reference books. HarperCollins is also one of the largest education publishers in the UK. Authors include many award-winning and international bestsellers such as Isabel Allende, Paulo Coelho, Josephine Cox, Michael Crichton, Cathy Kelly, Judith Kerr, Doris Lessing, Frank McCourt, Tony Parsons, Nigel Slater and Diana Wynne Jones. Bestselling licensed properties include Dr Seuss and Noddy. About 1,200 titles a year.

Harper Fiction

Publisher *Lynne Drew* IMPRINTS **HarperCollins** Publishing Directors *Susan Watt, Julia Wisdom*; **Voyager** Fantasy/science fiction Publishing Director *Jane Johnson* General, historical fiction, crime and thrillers and women's fiction.

Harper Non-Fiction

Managing Director/Publisher *Belinda Budge*, Editorial Directors *Viv Bowler* (cookery & lifestyle), *Natalie Jerome* (popular culture), Publishing Director *Jonathan Taylor* Sporting biographies, guides and histories. IMPRINT **HarperThorsons/Element** Publisher *Carole Tonkinson* (mind, body & spirit), *Sally Potter* (media tie-ins).

Avon

Managing Director *Caroline Ridding*, Editorial Director *Maxine Hitchcock* Popular fiction. IMPRINTS **HarperCollins** Publishing Director *David Brawn* (Agatha Christie, J.R.R. Tolkien, C.S. Lewis) Media-related books from film companions to celebrity autobiographies and TV tie-ins.

HarperCollins Children's Books

Managing Director *Mario Santos*, Publisher *Ann-Janine Murtagh*, Brand & Properties Developer *Claire Harding* Quality picture books for under-7s; fiction for age 6 up to young adult; TV and film tie-ins and properties Publishing Directors *Gillie Russell* (fiction), *Sue Buswell* (picture books).

PRESS BOOKS DIVISION

Managing Director *John Bond* IMPRINTS **Fourth Estate** Publishing Director/Publisher *Nick Pearson* Fiction, literary fiction, current affairs, popular science, biography, humour, travel; **HarperPress** General trade non-fiction and fiction Publishing Directors *Arabella Pike* (non-fiction), *Clare Smith* (fiction); **HarperPerennial** Paperback imprint. Publishing Director *Paul Baggaley*; **Blue Door** Publisher *Patrick Janson-Smith* Fiction; **The Friday Project** Managing Director *Clare Christian*, Publisher *Scott Pack*, Managing Editor *Heather Smith* Fiction, non-fiction and children's.

COLLINS REFERENCE AND GEO

Acting Managing Director *Katie Fulford* Guides and handbooks, Times World Atlases, Collins World Atlases, phrase books and manuals on popular reference, art instruction, illustrated, cookery and wine, crafts, DIY, gardening, military, natural history, pet care, pastimes Publishing Directors *Denise Bates* (illustrated reference), *Sheena Barclay* (world atlases), Associate Publisher *Myles Archibald* (natural history), Publishing Director *Helen Gordon* Maps, atlases, street plans and leisure guides. IMPRINTS **Collins**; **Collins Gem**; **Times Books**; **Jane's**.

Collins Language

Managing Director *Robert Scriven* Bilingual and English dictionaries, English dictionaries for foreign learners. Publishing Directors *Helen Newstead* (digital development), *Elaine Higgleton* (editorial).

Collins Education

Managing Director *Nigel Ward*
Books, CD-ROMs and online material for UK
primary and secondary schools and colleges.

✱*Authors' update* **The first publisher to digitize
its content, HarperCollins posts online entire
copies of latest works by leading authors.
Now there are plans to publish all major titles
in print and digital format simultaneously.
HarperCollins has also taken a lead in
consumer research. Such energetic innovation
along with strong sales in supermarkets and
other outlets beyond the traditional high street
make a lot of authors very happy.**

Harrap
▷ Chambers Harrap Publishers Ltd

Harriman House Ltd

3A Penns Road, Petersfield GU32 2EW
☎ 01730 233870   🖷 01730 233880
contact@harriman-house.com
www.harriman-house.com
Managing Director *Myles Hunt*
Head of Rights *Suzanne Anderson*

Founded 1994; commenced publishing in 2001.
Independent publisher of finance, trading,
business, economic and political books, covering
a wide range of subjects from personal finance,
small business and lifestyle through to stock
market investing, trading and professional guides.
30 titles a year. Unsolicited mss, synopses and
ideas for book welcome. Initial approach by
e-mail.
💷 Royalties twice-yearly.

Harvard University Press

Fitzroy House, 11 Chenies Street, London
WC1E 7EY
☎ 020 7306 0603   🖷 020 7306 0604
info@HUP-MITpress.co.uk
www.hup.harvard.edu
Director *William Sisler*
General Manager *Ann Sexsmith*

European office of **Harvard University Press**,
USA. Publishes academic and scholarly works in
history, politics, philosophy, economics, literary
criticism, psychology, sociology, anthropology,
women's studies, biological sciences, classics,
history of science, art, music, film, reference. All
mss go to the American office: 79 Garden Street,
Cambridge, MA 02138 (see entry under *US
Publishers*).

Harvill Secker
▷ The Random House Group Ltd

Haus Publishing

26 Cadogan Court, Draycott Avenue, London
SW3 3BX
☎ 020 7584 6738   🖷 020 7584 9501
haus@hauspublishing.com
www.hauspublishing.co.uk
Managing Director *Barbara Schwepcke*

DIVISIONS/IMPRINTS **Life&Times** Non-academic
biographies of well-known and lesser-known
personalities TITLES *Tito*; *Simone de Beauvoir*;
Prime Ministers of the 20th Century; *Patrice
Lumumba*; *Mussolini*. **Armchair Traveller**
Literary travel series TITLES *Mumbai to Mecca*;
Venice for Lovers; *The Liquid Continent: A
Mediterranean Trilogy*. **HausBooks** Hardback
biographies TITLES *Ellen Terry*; *Rommel*;
Diaghilev. 26 titles in 2006. Unsolicited mss,
synopses and ideas within the subject areas above
are welcome; approach by e-mail or post only. No
children's, fiction or poetry.
💷 Royalties twice-yearly.

Haynes Publishing

Books Division, Sparkford, Near Yeovil BA22 7JJ
☎ 01963 440635   🖷 01963 440823
bookseditorial@haynes.co.uk
www.haynes.co.uk
Chairman *John H. Haynes*, OBE
Approx. Annual Turnover £29.2 million

Founded 1960. Leading publisher of car and
motorcycle manuals, with operations in the UK,
USA, Australia and Sweden. Alongside the long-
established manuals written by in-house technical
authors, the UK-based company also publishes
a wide range of books by external authors
on all areas of transport – cars, motorcycles,
motorsport, aviation, military, maritime, railways,
cycling and caravans. In addition, manuals in
many new subject areas are being developed,
with strong ranges being built in computing,
family health and home DIY. Book proposals and
submissions are welcome in all the described
subject areas, as are any suggestions for manuals
on new topics. Contact *Mark Hughes*, Editorial
Director, Books Division.
💷 Royalties twice-yearly.

✱*Authors' update* **Having sold Sutton
Publishing, Haynes is back to its core
programme of publishing 'practical and
hands-on' books along with tie-in products
such as T-shirts and toys. A range of illustrated
children's books is promised for the autumn.**

Headline/Headline Review
▷ Hachette Livre UK

Heinemann
▷ Pearson (Heinemann)

William Heinemann
> The Random House Group Ltd

Helicon Publishing
RM plc, New Mill House, 183 Milton Park, Abingdon OX14 4SE
☏ 01235 823816
helicon@rm.com
www.helicon.co.uk
Business Manager *Caroline Dodds*

A division of RM plc. Helicon is a general reference database publisher for the UK, US and Australian markets. Licenses quality reference content for online, print, and CD-ROM publications. TITLES include the flagship single-volume *Hutchinson 2005 Encyclopedia* published under licence from **Hodder Headline**. CD-ROM TITLES include *The Hutchinson Encyclopedia*; *The Hutchinson Science Reference Suite*; *The Hutchinson History Reference Suite*; *The Hutchinson Music Reference Suite*.

Christopher Helm
> A.&C. Black Publishers Ltd

Helm Information Ltd
Crowham Manor, Main Road, Westfield, Hastings TN35 4SR
☏ 01424 882422   ☑ 01424 882817
amandahelm@helm-information.co.uk
www.helm-information.co.uk
Director *Amanda Helm*

Founded 1990. Publishes academic books for students and university libraries. SERIES *The Critical Assessments of Writers in English* (collected criticism); *Literary Sources & Documents* (primary source material on various themes/events/cultural/aesthetic movements ranging from the American Civil War to the Gothic Revival); *The Dickens Companions*; and *Icons*, a series which exposes the processes by which a figure, historical or fictional, achieves iconic status, e.g. Faust, Robin Hood. Will consider ideas and proposals provided they are relevant to the series listed above.
Ⓔ Royalties annually.

Helter Skelter Publishing
18A Radbourne Road, London SW12 0DZ
☏ 020 8673 6320
info@helterskelterpublishing.com
www.helterskelterpublishing.com
Art Director *Graeme Milton*

Founded 1995. Publishes music and film books only. IMPRINTS **Helter Skelter Publishing**; **Firefly Publishing**. 15 titles a year. Unsolicited mss, synopses and ideas welcome.

Henderson Publishing
> Dorling Kindersley Ltd

Ian Henry Publications Ltd
20 Park Drive, Romford RM1 4LH
☏ 01708 749119   ☑ 01708 736213
info@ian-henry.com
www.ian-henry.com
Managing Director *Ian Wilkes*

Founded 1976. Publishes local history, humour, transport history and Sherlockian pastiches. TITLES *Sketches of Upminster*; *Romford – Upminster Branch Line*. 1–2 titles a year. No unsolicited mss. Synopses and ideas for books welcome.
Ⓔ Royalties twice-yearly.

The Herbert Press
> A.&C. Black Publishers Ltd

Hermes House
> Anness Publishing Ltd

Nick Hern Books
The Glasshouse, 49a Goldhawk Road, London W12 8QP
☏ 020 8749 4953   ☑ 020 8735 0250
info@nickhernbooks.demon.co.uk
www.nickhernbooks.co.uk
Chairman/Managing Director *Nick Hern*
Submissions Editor *Matt Applewhite*

Founded 1988. Fully independent since 1992. Publishes books on theatre and film: from practical manuals to plays and screenplays. About 60 titles a year. No unsolicited playscripts. Synopses, ideas and proposals for other theatre material welcome. Not interested in material unrelated to the theatre or cinema.

University of Hertfordshire Press
University of Hertfordshire, College Lane, Hatfield AL10 9AB
☏ 01707 281354   ☑ 01707 284666
UHPress@herts.ac.uk
www.herts.ac.uk/UHPress
Contact *Jane Housham*

Founded 1992. Publishes academic books on Gypsies, literature, regional and local history, education. IMPRINTS **University of Hertfordshire Press**; **Hertfordshire Publications**. TITLES *Here to Stay: The Gypsies and Travellers of Britain* Colin Clark and Margaret Greenfields; *Selling Shakespeare to Hollywood: The marketing of filmed Shakespeare adaptations from 1989 into the new millennium* Emma French; *Teachers' Legal Rights and Responsibilities* Jon Berry. 15 titles a year. New publishing proposals are welcome; proposal form available on the website.

Hesperus Press Limited
4 Rickett Street, London SW6 1RU
☏ 020 7610 3331   ☑ 020 7610 3217
www.hesperuspress.com

Managing Editor *Katherine Venn*

Founded 2001. Publishes lesser-known works in both original English and English translation by classic authors. TITLES *Hyde Park Gate News* Virginia Woolf; *Sarrasine* Honoré de Balzac; *No Man's Land* Graham Greene; *Hadji Murat* Lev Tolstoy; *Aller Retour New York* Henry Miller; *The Watsons* Jane Austen. *Modern Classics* series launched in 2005; *Brief Lives* series of newly commissioned short biographies of celebrated literary figures launched in 2008; contemporary European fiction in translation since 2005. 46 titles in 2008. Does not accept unsolicited mss. £ Royalties annually.

Authors' update Much praised for matching book design with quality content.

History Into Print
> Brewin Books Ltd

The History Press Ltd

The Mill, Brimscombe, Stroud GL5 2QG
℡ 01453 883300   ℻ 01453 883233
sales@thehistorypress.co.uk
www.thehistorypress.co.uk
Chairman *Andy Nash*
CEO *Tony Morris*
Editorial Head *Laura Perehinec*
Approx. Annual Turnover £15 million

Founded in 2007 having acquired the assets of NPI Media. The UK's largest local and specialist history publisher. IMPRINTS include **The History Press; Jarrold; Nonsuch; Pathfinder; Phillimore; Pitkin; Spellmount** (see entry); **Stadia; Sutton; Tempus.** 950 titles in 2007. Unsolicited submissions welcome; approach by e-mail in the first instance. No fiction or children's books. OVERSEAS SUBSIDIARIES The History Press Inc, Sutton Verlag, Editions Alan Sutton. £ Royalties twice-yearly.

Authors' update After a chaotic 2007 when advances and royalties were notable for their absence, NPI Media was sold to The History Press. An overhaul of accounting procedures and staff cutbacks promised a brighter future for aggrieved authors.

HMSO
> TSO

Hobsons Plc

Challenger House, 42 Adler Street, London E1 1EE
℡ 020 7958 5000   ℻ 020 7958 5001
www.hobsons.uk.com
Chairman *Martin Morgan*
Group Managing Director *Christopher Letcher*

Founded 1974. Part of the Daily Mail & General Trust. Publishes course and career guides, under exclusive licence and royalty agreements for CRAC (Careers Research and Advisory Centre), directories and specialist titles for students and young professionals. TITLES *GET 2009; The Hobsons Global MBA Guide 2009.*

Hodder & Stoughton
> Hachette Livre UK

Hodder Arnold
> Hachette Livre UK

Hodder Children's Books
> Hachette Children's Books

Hodder Gibson

2a Christie Street, Paisley PA1 1NB
℡ 0141 848 1609   ℻ 0141 889 6315
hoddergibson@hodder.co.uk
www.hoddergibson.co.uk
Managing Director *John Mitchell*

Part of the **Hachette Livre** UK group. Publishes educational textbooks and revision guides specifically for Scotland and mainly for secondary education. About 25 titles a year. Synopses/ideas preferred to unsolicited mss. £ Royalties annually.

Hodder Headline Ltd
> Hachette Livre UK

Honeyglen Publishing Ltd

78 Kenton Court, Kensington High Street, London W14 8NW
℡ 020 7602 2876   ℻ 020 7602 2876
Directors *N.S. Poderegin, J. Poderegin*

Founded 1983. A small publishing house whose output is 'extremely limited'. Publishes history, philosophy of history, biography and selective fiction. No children's or science fiction. TITLES *The Soul of China; The Soul of India; Woman and Power in History; Lost World – Tibet; A Child of the Century* all by Amaury de Riencourt; *With Duncan Grant in South Turkey* Paul Roche; *Vladimir, The Russian Viking* Vladimir Volkoff; *The Dawning* Milka Bajic-Poderegin; *Quicksand* Louise Hide. Unsolicited mss welcome.

Honno Welsh Women's Press

c/o Canolfan Merched y Wawr, Vulcan Street, Aberystwyth SY23 1JH
℡ 01970 623150   ℻ 01970 623150
post@honno.co.uk
www.honno.co.uk
Editor *Caroline Oakley*

Founded in 1986 by a group of women who wanted to create more opportunities for women in publishing. A co-operative operation which publishes mainly fiction and autobiography. 'Contact us for current calls for short story submissions.' 7 titles a year. Welcomes mss and ideas for books from women who are Welsh or

have a significant Welsh connection only Send as hard copy, not by e-mail.
ⓔ Royalties annually.

Horizon Press
▷ Discovered Authors

Horus Editions
▷ Award Publications Limited

How To Books Ltd
Spring Hill House, Spring Hill Road, Begbroke, Oxford OX5 1RX
☎ 01865 375794   🖷 01865 379162
info@howtobooks.co.uk
www.howtobooks.co.uk
Managing Director *Giles Lewis*
Editorial Director *Nikki Read*

An independent publishing house, founded in 1991. Publishes non-fiction, self-help reference books. How To titles are practical, accessible books that enable their readers to achieve their goals in life and work. How To authors must have first-hand experience of the subject about which they are writing. Subjects covered include management, leisure learning, career choices and career development, living and working abroad, small business and self employment, study skills & student guides, creative writing and property. IMPRINTS **Transita** (see entry); **Springhill**. 250 titles a year. Submit outline followed by sample chapter.

Human Horizons
▷ Souvenir Press Ltd

John Hunt Publishing Ltd
▷ O Books/John Hunt Publishing Ltd

Hurst Publishers, Ltd
41 Great Russell Street, London WC1B 3PL
☎ 020 7255 2201
hurst@atlas.co.uk
www.hurstpub.co.uk
Chairman/Managing Director/Editorial Head
Michael Dwyer

Founded 1967. An independent company publishing contemporary history, politics, religion (not theology) and anthropology. About 25 titles a year. No unsolicited mss. Synopses and ideas welcome.
ⓔ Royalties annually.

Hutchinson/Hutchinson Children's Books
▷ The Random House Group Ltd

Hymns Ancient & Modern Ltd
St Mary's Works, St Mary's Plain, Norwich NR3 3BH
☎ 01603 612914   🖷 01603 624483
admin@scm-canterburypress.co.uk
www.scm-canterburypress.co.uk

Publishing Director, SCM Press and Canterbury Press *Christine Smith*
Publisher, RMEP *Valerie Bingham*
Approx. Annual Turnover £5 million

Publishing subsidiary **SCM-Canterbury Press Ltd** (see entry) controls **SCM Press**, **Canterbury Press** and **RMEP** IMPRINTS which cover hymn books, liturgical material, academic and general religious books, and multi-faith religious and social education resources for pupils and teachers. Media subsidiary **G.J. Palmer & Sons Ltd** includes *Church Times* (see entry under *Magazines*); *The Sign* and *Home Words* (monthly) and *Crucible* (quarterly). About 100 titles a year. Ideas for new titles welcome but no unsolicited mss.
ⓔ Royalties annually.

Icon Books Ltd
Editorial, Production, Finance: Old Dairy, Brook Road, Thriplow, Cambridge SG8 7RG
☎ 01763 208008   🖷 01763 208080
info@iconbooks.co.uk
www.iconbooks.co.uk
Marketing, Sales, Press, Rights: Omnibus Business Centre, 39–41 North Road, London N7 9DP
☎ 020 7697 9695
Managing Director *Peter Pugh*
Publishing Director *Simon Flynn*
Editorial Director *Duncan Heath*

Founded 1992. SERIES *Introducing* Graphic introductions to key figures and ideas in the history of science, philosophy, psychology, religion and the arts. TITLE *Introducing Quantum Theory*. Publishes 'provocative and intelligent' non-fiction in science, politics and philosophy. TITLE *Why do People Hate America?* IMPRINT **Wizard Books** Children's non-fiction and game books including SERIES *Fighting Fantasy* (adventure gamebooks). Submit synopsis only. OVERSEAS ASSOCIATE Totem Books, USA, distributed by National Book Network.
ⓔ Royalties twice yearly.

Imprint Academic
PO Box 200, Exeter EX5 5YX
☎ 01392 851550   🖷 01392 851187
keith@imprint.co.uk
www.imprint-academic.com
Publisher *Keith Sutherland*

Founded 1980. Publishes books and journals in politics, philosophy and psychology for both academic and general readers. Book series include *St. Andrews Studies in Philosophy and Public Affairs* and *Societas: Essays in political and cultural criticism*. Unsolicited mss, synopses and ideas welcome with return postage only.

The In Pinn
> Neil Wilson Publishing Ltd

Incomes Data Services
> Sweet & Maxwell Group

Independent Music Press
PO Box 69, Church Stretton SY6 6WZ
℡ 01694 720049   🖷 01694 720049
info@impbooks.com
www.impbooks.com
Managing Director *Martin Roach*

Founded 1992. Publishes music biography and youth culture. No jazz or classical. TITLES include biographies of My Chemical Romance, Green Day, Muse, The Killers, The Cure, Dave Grohl, The Streets, Oasis, Prodigy as well as subculture classics such as *Scooter Boys* and Ian Hunter's *Diary of a Rock 'n' Roll Star*. 8 titles a year. Approach by e-mail, enclosing biography and synopsis only.
£ Royalties twice-yearly.

Independent Voices
> Souvenir Press Ltd

Infinite Ideas
36 St Giles, Oxford OX1 3LD
℡ 01865 514888   🖷 01865 514777
info@infideas.com
www.infideas.com
Joint Managing Directors *David Grant, Richard Burton*
Approx. Annual Turnover £1.5 million

Founded 2004. Publishes inspirational self-help. SERIES 52 **Brilliant Ideas** Subject areas cover health and relationships; careers, finance and personal development; sports, hobbies and games; leisure and lifestyle; art, literature and music. No fiction, illustrated or children's books. IMPRINT **Bright 'I's** Business books, launched in 2006. 40 titles a year. No unsolicited mss. Synopses and ideas welcome; initial contact by e-mail or letter.
£ Flat fees paid.

Informa Law
Informa House, 30–32 Mortimer Street, London
W1W 7RE
℡ 020 7017 4600
Managing Director *Fotini Liontou*
Publishing Director *Andrew Cooney*

Part of LLP Professional Publishing, a trading division of Informa Publishing Group Ltd. Publishes a range of maritime law, commercial law, insurance law, banking law, intellectual property law publications including newsletters, law reports, books, journals and magazines aimed at senior management and professional practices.

Inspire
c/o mph, 4 John Wesley Road, Werrington, Peterborough PE4 6ZP
℡ 01733 384188   🖷 01733 384180
chiefexecutive@mph.org.uk
www.mph.org.uk

Imprint of **mph**. Publishes accessible Christian books for the wider ecumenical market. Spirituality, social concern, worship resources, mission and evangelism, discipleship, prayer. No fiction or children's. Up to 12 titles a year. Synopses and ideas welcome; send two sample chapters and proposal with covering letter and s.a.e. for return of mss.
£ Royalties annually.

Inter-Varsity Press
IVP Book Centre, Norton Street, Nottingham
NG7 3HR
℡ 0115 978 1054   🖷 0115 942 2694
ivp@ivpbooks.com
www.ivpbooks.com
Chairman *Ralph Evershed*
Chief Executive *Brian Wilson*

Founded mid-1930s as the publishing arm of Universities and Colleges Christian Fellowship, it has expanded to wider Christian markets worldwide. Publishes Christian belief and lifestyle, reference and bible commentaries. No secular material or anything which fails to empathize with orthodox Protestant Christianity. IMPRINTS **IVP**; **Apollos**; **Crossway** TITLES *The Bible Speaks Today* series and New Testament CD-ROM; *Pierced for Our Transgressions* Jeffery, Ovey and Sach; *The Living Church* John Stott. About 50 titles a year. Synopses and ideas welcome.
£ Royalties twice-yearly.

Isis Publishing
7 Centremead, Osney Mead, Oxford OX2 0ES
℡ 01865 250333   🖷 01865 790358
sales@isis-publishing.co.uk
www.isis-publishing.co.uk

Part of the Ulverscroft Group Ltd. Publishes large-print books – fiction and non-fiction – and unabridged audio books. Together with **Soundings** (see entry under *Audio Books*) produces around 4,000 titles on audio tape and CD. AUTHORS Lee Child, Martina Cole, Terry Pratchett, Susan Sallis. No unsolicited mss as Isis undertakes no original publishing.
£ Royalties twice-yearly.

Itchy Coo
> Black & White Publishing Ltd

Ithaca Press
> Garnet Publishing Ltd

IVP
> Inter-Varsity Press

J. Garnet Miller
> Cressrelles Publishing Co. Ltd

Jacqui Small
> Aurum Press Ltd

JAI
> Elsevier Ltd

Jane's Information Group

Sentinel House, 163 Brighton Road, Coulsdon CR5 2YH

☎ 020 8700 3700

info.uk@janes.com

www.janes.com

Managing Director *Scott Key*

Chief Content Officer *Ian Kay*

Founded in 1898 by Fred T. Jane with the publication of *All the World's Fighting Ships*. Acquired by IHS Inc. in 2007. Recent focus has been on growth opportunities in its core business and in enhancing the performance of initiatives like Jane's information available online and on in-depth intelligence centres. Publishes Web journals, magazines and yearbooks on defence, aerospace, security and transport topics, with details of equipment and systems; plus directories. Full consulting arm offers bespoke offerings to clients. Also *Jane's Defence Weekly* (see entry under *Magazines*).

DIVISIONS **Magazines** TITLES *Jane's Defence Weekly*; *Jane's International Defence Review*; *Jane's Airport Review*; *Jane's Navy International*; *Jane's Islamic Affairs Analyst*; *Jane's Missiles and Rockets*. **Publishing for Defence, Aerospace, Transport** TITLES *Defence, Aerospace Yearbooks*. **Transport** TITLE *Transportation Yearbooks*. **Security** *Sean Howe* TITLES *Jane's Intelligence Review*; *Foreign Report*; *Jane's Sentinel* (regional security assessment); *Jane's Police Review*. CD-ROM and electronic development and publication. Over 100 titles a year. Unsolicited mss, synopses and ideas for reference/yearbooks welcome.

Janus Publishing Company Ltd

105–107 Gloucester Place, London W1U 6BY

☎ 020 7486 6633   ⓕ 020 7486 6090

publisher@januspublishing.co.uk

www.januspublishing.co.uk

Managing Director *Jeannie Leung*

Publishes fiction, human interest, memoirs, philosophy, mind, body and spirit, religion and theology, social questions, popular science, history, spiritualism and the paranormal, poetry and young adults. IMPRINTS **Janus Books** Subsidy publishing; **Empiricus Books** Non-subsidy publishing. TITLES *The Anarchists in the Spanish Civil War*; *Nature of the Self*; *Politics and Human Nature*; *Napoleon 1813*; *King David*; *Chameleon Candidate*; *Love Emporium*; *Digby*; *The Naked Emperor*. 'Two of our authors won the European Literary Award in 2004.' About 400 titles in print. Unsolicited mss welcome. Agents in the USA, Europe and Asia.

ⓔ Royalties twice-yearly.

✱*Authors' update* **Authors may be asked to cover their own productions costs but Janus has moved into conventional publishing with its Empiricus imprint.**

Jarrold
> The History Press Ltd

Michael Joseph
> Penguin Group (UK)

JR Books Ltd

10 Greenland Street, London NW1 0ND

☎ 020 7284 7163   ⓕ 020 7485 4902

jeremyr@jrbooks.com

www.jrbooks.com

Managing Director *Jeremy Robson*

Founded October 2006. JR Books publishes a wide range of non-fiction subjects, including biography, politics, music, humour, history, sport and self-help. 19 titles, spring 2008. Unsolicited synopses and ideas (with s.a.e.) welcome. Approach by mail. No fiction, children's or academic books.

ⓔ Royalties twice-yearly.

Kahn & Averill

9 Harrington Road, London SW7 3ES

☎ 020 8743 3278   ⓕ 020 8743 3278

kahn@averill23.freeserve.co.uk

Managing Director *Mr M. Kahn*

Founded 1967 to publish children's titles but now specializes in music titles. A small independent publishing house. No unsolicited mss; synopses and ideas for books considered.

ⓔ Royalties twice-yearly.

Kenilworth Press Ltd (An imprint of Quiller Publishing Ltd)

Wykey House, Wykey, Shrewsbury SY4 1JA

☎ 01939 261616   ⓕ 01939 260991

info@quillerbooks.com

www.kenilworthpress.co.uk

Managing Director *Andrew Johnston*

Founded in 1989 and acquired by **Quiller Publishing Ltd** in 2005. Publishes equestrian books only. 10 titles in 2007. No unsolicited mss; synopses and ideas for books welcome. Send by post or e-mail.

ⓔ Royalties twice-yearly.

Kenyon-Deane
▷ Cressrelles Publishing Co. Ltd

Laurence King Publishing Ltd
4th Floor, 361–373 City Road, London EC1V 1LR
☎ 020 7841 6900   📠 020 7841 6910
enquiries@laurenceking.co.uk
www.laurenceking.co.uk
Chairman *Nick Perren*
Managing Director *Laurence King*
Commissioning Editor, Architecture
Philip Cooper
Commissioning Editor, Design *Jo Lightfoot*
Approx. Annual Turnover £6 million

Publishes illustrated books in the fields of graphic design, fashion and textiles, architecture, interiors, product design and art. TITLES *A World History of Art*; *The Design Encyclopedia*; *How to Be a Graphic Designer Without Losing Your Soul*; *1000 New Designs and Where to Find Them*; *100 Years of Fashion Illustration* and the *Portfolio* series for students of art and design. About 60 titles a year. Unsolicited material welcome; send synopsis by post or e-mail (commissioning@laurenceking.co.uk).

Kingfisher
▷ Macmillan Publishers Ltd

Jessica Kingsley Publishers Ltd
116 Pentonville Road, London N1 9JB
☎ 020 7833 2307   📠 020 7837 2917
post@jkp.com
www.jkp.com
Also at: 400 Market Street, Suite 400, Philadelphia, PA 19106, USA
Managing Director and Publisher *Jessica Kingsley*
Senior Acquisitions Editor *Stephen Jones*

Founded 1987. Independent, international publisher of books for professionals, academics and the general reader on autism, disability, special education, arts therapies, child psychology, mental health, practical theology and social work. IMPRINT **Singing Dragon** Tai Chi, martial arts and Qigong. 150 titles a year. 'We are actively publishing and commissioning in all these areas. We welcome suggestions for books and proposals from prospective authors. Proposals should consist of an outline of the book, a contents list, assessment of the market and author's c.v., and should be addressed to *Jessica Kingsley*. Complete manuscript should not be sent.' OVERSEAS SUBSIDIARY in Philadelphia, USA.
💷 Royalties twice-yearly.

Kluwer Law International
250 Waterloo Road, London SE1 8RD
☎ 020 7981 0656   📠 020 7981 0587
simon.bellamy@kluwerlaw.com
www.kluwerlaw.com
Also at: Zuidpoolsingel 2a, 2408 ZE Alphen aan den Rijn, The Netherlands
Publisher *Simon Bellamy*

Founded 1995. Parent company: Wolters Kluwer Group. Publishes international law. About 200 titles a year, including online, CD-ROM, loose-leaf journals and monographs. Unsolicited synopses and ideas for books on law at an international level welcome.
💷 Royalties annually.

Kogan Page Ltd
120 Pentonville Road, London N1 9JN
☎ 020 7278 0433   📠 020 7837 3768/6348
kpinfo@kogan-page.co.uk
www.kogan-page.co.uk
Chairman *Philip Kogan*
Managing Director *Helen Kogan*
Approx. Annual Turnover £5.4 million

Founded 1967 by Philip Kogan to publish *The Industrial Training Yearbook*. Publishes business and management reference books and monographs, careers, marketing, personal finance, personnel, small business, training and industrial relations, transport. Further expansion is planned, particularly in the finance and online development areas, yearbooks and directories. Has initiated a number of electronic publishing projects and provision of EP content. About 130 titles a year.
💷 Royalties twice-yearly.

⊞*Authors' update* **Having celebrated 'forty years of independent publishing', Kogan Page chairman Philip Kogan tells us the secret of survival for a small fish in an overcrowded pool. It is to have sound and preferably original ideas and to be able to assess markets but, above all, to 'be good at housework – proper accounting and financial control'. Himself having thrived on these principles, Kogan Page authors can feel in safe hands. Latest results have benefited from strong sales in the US.**

Ladybird
▷ Dorling Kindersley Ltd

Landmark Publishing Ltd
Ashbourne Hall, Cokayne Avenue, Ashbourne DE6 1EJ
☎ 01335 347349   📠 01335 347303
landmark@clara.net
www.landmarkpublishing.co.uk
Chairman *Mr R. Cork*
Managing Director *Mr C.L.M. Porter*
Approx. Annual Turnover £450,000

Founded in 1996. Publishes itinerary-based travel guides, regional, industrial countryside and local

history. About 120 titles in 2007. No unsolicited mss; telephone in the first instance.
Ⓔ Royalties annually.

Lawrence & Wishart Ltd

99A Wallis Road, London E9 5LN
☎ 020 8533 2506   🖷 020 8533 7369
lw@lwbooks.co.uk
www.lwbooks.co.uk
Managing Director/Editor *Sally Davison*

Founded 1936. An independent publisher with a substantial backlist. Publishes current affairs, cultural politics, economics, history, politics and education. TITLES *After Blair; Labour Legends Russian Gold; After Iraq; Making Sense of New Labour.* 10 titles a year.
Ⓔ Royalties annually, unless by arrangement.

The Learning Institute

No. 1 Overbrook Business Centre, Blackford, Wedmore BS28 4PA
☎ 01934 713563   🖷 01934 713492
courses@inst.org
www.inst.org
Managing Director *Kit Sadgrove*

Founded 1994 to publish home-study courses in vocational subjects such as garden design, writing and computing. Publishes subjects that show the reader how to work from home, gain a new skill or enter a new career. Interests include home working and self employment, especially in twenty-first century jobs. TITLES *Diploma in Interior Design; Become a Garden Designer.* Author's guidelines sent on receipt of s.a.e. No unsolicited mss; send synopses and ideas only.
Ⓔ Royalties quarterly.

Legend Press Limited

13a Northwold Road, London N16 7HL
☎ 020 7249 6901
info@legendpress.co.uk
submissions@legendpress.co.uk
www.legendpress.co.uk
Managing Director *Tom Chalmers*
Publishing Executive *Emma Howard*

Publishes contemporary, diverse and cutting-edge fiction. Is 'committed to providing writers with an advantageous platform from which to develop and succeed. We focus strongly on reaching new audiences and on working with other publishers to drive the independent sector and promote new and different work within the mainstream market.' Acquired PaperBooks Ltd in April 2008. 10 titles in 2008; doubling list year-on-year. Send three chapters and synopsis to the submissions e-mail address above.
Ⓔ Royalties twice-yearly.

✱*Authors' update* Credited with the youngest CEO in the business (Tom Chalmers is just 28),

Legend is driven by enthusiasm for 'fresh new fiction from authors who are fun and inspiring to work with'.

Dewi Lewis Publishing

8 Broomfield Road, Heaton Moor, Stockport SK4 4ND
☎ 0161 442 9450   🖷 0161 442 9450
mail@dewilewispublishing.com
www.dewilewispublishing.com
Owners *Dewi Lewis, Caroline Warhurst*
Approx. Annual Turnover £280,000

Founded 1994. Publishes photography, visual arts and fiction. TITLES *Industry of Souls* Martin Booth (Booker Prize shortlist, 1998); *Wolfy and the Strudelbakers* Zvi Jagendorf (Booker Prize longlist, 2001; Sagittarius Prize, 2002); *Common Sense* Martin Parr; *New York 1954–5* William Klein. IMPRINT **Dewi Lewis Media** Publishes non-fiction, biography, sports and celebrity books. TITLES *David Beckham: Made in Manchester* Eamonn and James Clarke; *500 Flowers* Roger Camp. 16 titles in 2006. Not currently accepting new fiction submissions. For all submissions it is essential to check the website first.
Ⓔ Royalties annually.

Lewis/Lewis Masonic
▷ Ian Allan Publishing Ltd

Library Association Publishing
▷ Facet Publishing

Library of Wales
▷ Parthian

Life&Times
▷ Haus Publishing

Frances Lincoln Ltd

4 Torriano Mews, Torriano Avenue, London NW5 2RZ
☎ 020 7284 4009   🖷 020 7485 0490
firstname
and
initial
of
surname@frances-lincoln.com
www.franceslincoln.com
Managing Director *John Nicoll*
Approx. Annual Turnover £7 million

Founded 1977. Publishes highly illustrated non-fiction: gardening, art and interiors, architecture, parenting, walking and climbing, children's picture and information books; also stationery. DIVISIONS **Adult Non-fiction** *Jo Christian* TITLES *Chatsworth* Duchess of Devonshire; *Grow Your Own Vegetables* Joy Larkcom; *Pictorial Guides to the Lakeland Fells* A. Wainwright; **Children's General Fiction and Non-fiction** *Janetta*

Otter-Barry TITLE *The Wanderings of Odysseus* Rosemary Sutcliffe, illus. Alan Lee; **Stationery** *Anna Sanderson* TITLES *RHS Diary and Address Book*; *British Library Diary*. About 200 titles a year. Synopses and ideas for books considered. Ⓔ Royalties twice-yearly.

Linden Press
▷ Open Gate Press

Linford Romance/Linford Mystery/Linford Western
▷ F.A. Thorpe Publishing

Lion Hudson plc
Wilkinson House, Jordan Hill Road, Oxford OX2 8DR
☏ 01865 302750   🖷 01865 302757
info@lionhudson.com
www.lionhudson.com
Managing Director *Paul Clifford*
Approx. Annual Turnover £8.5 million

Founded 1971. A Christian book publisher, strong on illustrated books for a popular international readership, with rights sold in over 150 languages worldwide. Publishes a diverse list with Christian viewpoint the common denominator. All ages, from board books for children to multi-contributor adult reference, educational, paperbacks and colour co-editions and gift books. IMPRINTS **Lion** and **Lion Children's** *Kate Leech*; **Candle** *Carol Jones*; **Monarch** *Paul Sweeney* (see **Monarch Books**). About 175 titles a year. Unsolicited mss accepted provided they have a positive Christian viewpoint intended for a Christian or wide general and international readership. Ⓔ Royalties twice-yearly.

Little Black Dress
▷ Hachette Livre UK

Little Books Ltd
73 Campden Hill Towers, 112 Notting Hill Gate, London W11 3QW
☏ 020 7792 7929
www.littlebooks.net
Managing Director *Max Hamilton-Little*

Founded 2001. Publishes general trade books, history, biography, health, natural history. IMPRINTS **Max**; **Max Press**. About 20–25 books a year. No unsolicited material. Ⓔ Royalties annually.

Little, Brown Book Group UK
100 Victoria Embankment, London EC4Y 0DY
☏ 020 7911 8000   🖷 020 7911 8100
uk@littlebrown.co.uk
www.littlebrown.co.uk
Owner *Hachette Book Group*
CEO/Publisher *Ursula Mackenzie*

Approx. Annual Turnover £60 million

Founded 1988 as Little, Brown & Co. (UK) and became Time Warner Book Group UK in 2002. Purchased by Hachette in March 2006 and renamed Little, Brown Book Group UK. Began by importing its US parent company's titles and in 1990 launched its own illustrated non-fiction list. Two years later the company took over former Macdonald & Co. In 2007 the company bought Piatkus Books. Publishes hardback and paperback fiction, literary fiction, crime, science fiction and fantasy; and general non-fiction including true crime, biography and autobiography, cinema, history, humour, popular science, travel, reference, sport. IMPRINTS **Little, Brown** *Ursula Mackenzie, Richard Beswick, Tim Whiting* Hardback fiction and general non-fiction; **Abacus** *Richard Beswick, Tim Whiting, Jenny Parrott* Literary fiction and non-fiction paperbacks; **Atom** *Tim Holman, Darren Nash* Young adult/teen paperbacks; **Hachette Digital** *Sarah Shrubb* (see entry under *Audio Books*); **Orbit** *Tim Holman, Darren Nash* Science fiction and fantasy; **Sphere** *Antonia Hodgson, David Shelley, Joanne Dickinson, Hilary Hale, Louise Davies, Adam Strange* Mass-market fiction and non-fiction hardbacks and paperbacks; **Virago Press** *Lennie Goodings* (see entry); **Piatkus Books** *Gill Bailey, Helen Stanton, Emma Dunford* (see entry). Approach in writing in the first instance. 320 titles in 2007. No unsolicited mss. Ⓔ Royalties twice-yearly.

✱*Authors' update* **Though now part of the Hachette group, Little Brown continues to be run as an independent entity publishing a wide range of commercial titles, mostly popular by virtue of quality.**

Little Tiger Press
An imprint of Magi Publications, 1 The Coda Centre, 189 Munster Road, London SW6 6AW
☏ 020 7385 6333   🖷 020 7385 7333
info@littletiger.co.uk
www.littletigerpress.com
Publisher *Monty Bhatia*
Associate Publisher *Jude Evans*
Commissioning Editor *Stephanie Stansbie*
Editor *Jo Collins*
Submissions Editor *Laura Roberts*

Little Tiger Press imprint publishes children's picture and novelty books for ages 0–7. No texts over 1,000 words. About 30 titles a year. Unsolicited mss, synopses and new ideas welcome. See website for submission guidelines. Ⓔ Royalties annually.

✱*Authors' update* **A thriving small publisher with imaginative output.**

Liverpool University Press

4 Cambridge Street, Liverpool L69 7ZU
☎ 0151 794 2233   🖷 0151 794 2235
robblo@liv.ac.uk
www.liverpool-unipress.co.uk
Managing Director/Editorial Head *Robin Bloxsidge*

LUP's primary activity is the publication of academic and scholarly books and journals but it also has a limited number of trade titles. Its principal focus is on the arts and social sciences, in which it is active in a variety of disciplines.
TITLES *Liverpool 800: Culture, Character and History; Irish, Catholic and Scouse: The History of the Liverpool Irish 1800–1939; Writing Liverpool; Liverpool and Transatlantic Slavery; So Spirited a Town; Public Sculptures of South London; sk-interfaces; The Culture of Capital; Cityscape: Ben Johnson's Liverpool; William Roscoe: Commerce and Culture; A Gallery to Play to: The Story of the Mersey Poets; The Cultural Values of Europe.*
30–40 titles a year.
🖭 Royalties annually.

Livewire Books for Teenagers
▷ The Women's Press

Lonely Planet Publications Ltd

2nd Floor, 186 City Road, London EC1V 2NT
☎ 020 7106 2100   🖷 020 7106 2101
go@lonelyplanet.co.uk
www.lonelyplanet.com
Owner *Lonely Planet (Australia)*
Editorial Head *Kat Browning*

'For over 30 years, Lonely Planet's on-the-ground research and no-holds-barred opinion by our team of expert travel writers has inspired and guided independent travellers to explore the world around them.' With over 500 titles in print, publications include travel guidebooks, downloadable digital guides, phrasebooks, travel literature, pictorial books and How To guides. No unsolicited mss; synopses and ideas welcome. Author guidelines available at www.lonelyplanet.com/help/guide

Lorenz Books
▷ Anness Publishing Ltd

Luath Press Ltd

543/2 Castlehill, The Royal Mile, Edinburgh EH1 2ND
☎ 0131 225 4326   🖷 0131 225 4324
gavin.macdougall@luath.co.uk
www.luath.co.uk
Director *G.H. MacDougall*

Founded 1981. Publishes mainly books with a Scottish connection. Current list includes fiction, poetry, guidebooks, walking and outdoor, history, folklore, politics and global issues, cartoons, biography, food and drink, environment, music and dance, sport. SERIES *On the Trail Of; The Quest For*. About 30–40 titles a year. Over 200 titles in print. Unsolicited mss, synopses and ideas welcome; 'committed to publishing well-written books worth reading'.
🖭 Royalties paid.

Lucky Duck Publishing
▷ Sage Publications

Lulu.com

www.lulu.com

Founded 2002. A self publishing, print-on-demand service for individuals and companies. Produces a variety of digital content including books, movies, brochures, portfolios, music, video, software, calendars, posters and artwork with no set-up fee and no requirement to buy copies. Authors supply a digital file of their book and a print-ready version of the file is created by the company. Lulu handles sales and pays the royalty specified by the client with a small mark-up as commission. Offers two retail distribution services for a fee. Full details available on the website.

✱***Authors' update*** **Authors who find it hard to get into print now have recourse to an option that is not vanity publishing, at least not in the strict sense of the word. Launched in the US in 2002, Lulu is a website designed to carry anything that writers can throw at it, and for free. If browsers show an interest in a particular title, the book can be produced to order and shipped direct. To stay in business and, by all accounts, thrive Lulu takes a twenty per cent share of any royalties to which an author is entitled. Publishing services such as editing, proofing and marketing must be contracted separately. Anyone joining the game should not expect too much – as in all publishing luck plays a big part – but as an economic proposition Lulu has the edge on conventional vanity and self publishing operations. The company's list is expanding by around 90,000 titles a year with 15,000 new registrations a week. But John Morrison, a Lulu author, warns 'this is not a solution for those who want to sell their books in commercial quantities into Waterstone's'.**

Lund Humphries
▷ Ashgate Publishing Ltd

The Lutterworth Press

PO Box 60, Cambridge CB1 2NT
☎ 01223 350865   🖷 01223 366951
publishing@jamesclarke.co.uk
www.lutterworth.com

Managing Director *Adrian Brink*

The Lutterworth Press was founded as the Religious Tract Society in Georgian London, with its headquarters just off Fleet Street, in order to provide improving literature for young people and adults. Since then it has published many tens of thousands of titles, ranging from small pamphlets for children to erudite academic works. Became well known to generations of British children because of its publication of the *Boy's Own Paper* and the *Girl's Own Paper*. Since 1984 it has been an imprint of **James Clarke & Co** (see entry). The Lutterworth current list concentrates in several areas, and some of the work, although aimed at a general readership, is at quite a high level. TITLES include: *King Saul. The True Story of the First Messiah* Adam Green; *The Guide to Norfolk Churches* D.P. Mortlock and C.V. Roberts; *Missing the Point. The rise of High Modernity and the decline of everything else* John Elsom. Approach in writing with ideas in the first instance.
Ⓔ Royalties annually.

Macdonald & Co.
▷ Little, Brown Book Group UK

McGraw-Hill Education
McGraw-Hill House, Shoppenhangers Road, Maidenhead SL6 2QL
Ⓣ 01628 502500 Ⓕ 01628 770224
www.mcgraw-hill.co.uk
General Manager *Shona Mullen*

Owned by US parent company, founded in 1888. Began publishing in Maidenhead in 1965, having had an office in the UK since 1899. Publishes business, economics, finance, accounting, computing science, social sciences and engineering for the academic, student and professional markets. Acquired **Open University Press** in 2002 (see entry). Around 200 titles a year. See website for author guidelines. Unsolicited proposals via website only.
Ⓔ Royalties twice-yearly.

MacLehose Press
▷ Quercus Publishing

Macmillan Publishers Ltd
The Macmillan Building, 4 Crinan Street, London N1 9XW
Ⓣ 020 7833 4000 Ⓕ 020 7843 4640
www.macmillan.com
Owner *Verlagsgruppe Georg von Holtzbrinck*
Chief Executive *Annette Thomas*
Approx. Annual Turnover £350 million (Book Publishing Group)

Founded 1843. Macmillan is one of the largest publishing houses in Britain, publishing approximately 1,400 titles a year. In 1995, Verlagsgruppe Georg von Holtzbrinck, a major German publisher, acquired a majority stake in the Macmillan Group and in 1999 purchased the remaining shares. In 1996, Macmillan bought Boxtree, the successful media tie-in publisher; in 1997, it purchased the Heinemann English language teaching list from Reed Elsevier, and it acquired Kingfisher, the popular children's book publisher, in 2007. The educational publishing division was strengthened by the acquisitions of the Mexican list, Ediciones Castillo and the Argentine company Puerto de Palos. No unsolicited material, except for the **Macmillan New Writing** list.

DIVISIONS

Palgrave Macmillan Brunel Road, Houndmills, Basingstoke, Hampshire RG21 6XS
Ⓣ 01256 329242 Ⓕ 01256 328339 Managing Director *Dominic Knight*; **College** *Margaret Hewison*; **Scholarly & Reference** *Sam Burridge*; **Journals** *David Bull*. Publishes textbooks, monographs and journals in academic and professional subjects. Publications in both print and electronic format.

Macmillan Education 4 Between Towns Road, Oxford OX4 3PP Ⓣ 01865 405700 Ⓕ 01865 405701 info@macmillan.com www.macmillaneducation. com Chief Executive Officer *Julian Drinkall*, Publishing Directors *Sue Bale* (dictionaries), *Alison Hubert* (Africa, Caribbean, Midde East and East Asia), *Julie Kniveton* (Latin America), *Angela Lilley* (ELT), *Kate Melliss* (Europe) Publishes a wide range of ELT titles and educational materials for the international education market from Oxford and through 30 subsidiaries worldwide.

Pan Macmillan 20 New Wharf Road, London N1 9RR Ⓣ 020 7014 6000 Ⓕ 020 7014 6001 www.panmacmillan.com Managing Director *Anthony Forbes Watson* Publishes under **Macmillan, Pan, Picador, Macmillan New Writing, Sidgwick & Jackson, Boxtree, Macmillan Children's Books, Kingfisher, Macmillan Digital Audio, Campbell Books, Young Picador, Rodale**. No unsolicited mss. Visit the writers' area on www.panmacmillan.com for useful articles and information on books that may be helpful in getting your book published, or visit the Macmillan New Writing website (see below) for details of the MNW programme, which does accept unsolicited mss of first novels.

IMPRINTS

Macmillan (founded 1843) Fiction: Publishing Director *Maria Rejt*, Editorial Director *Imogen Taylor* Publishes hardback commercial fiction including genre fiction, romantic, crime and thrillers. IMPRINT **Tor** (founded 2003) Publishes science fiction, fantasy and thrillers. Non-Fiction: Publisher *Richard Milner*, Editorial

Director *Georgina Morley* Publishes serious and general non-fiction: autobiography, biography, economics, history, philosophy, politics and world affairs, psychology, popular science, trade reference titles.

Pan (founded 1947) Paperback imprint for Pan Macmillan.

Picador (founded 1972) Publishing Director *Maria Rejt* Publishes literary international fiction, non-fiction and poetry.

Sidgwick & Jackson (founded 1908) Editorial Director *Ingrid Connell* Publishes popular non-fiction in hardback and trade paperback with strong personality or marketable identity, from celebrity and show business to music and sport. Also military history list.

Boxtree (founded 1986) Editorial Director *Jon Butler* Publishes brand and media tie-in titles, including TV, film, music and Internet, plus entertainment licences, pop culture, humour and event-related books. TITLES *Dilbert; James Bond; Purple Ronnie; Wallace & Gromit; The Onion.*

Macmillan Children's Books (New Wharf Road address) Managing Director *Emma Hopkin*, Publishing Director *Rebecca McNally*; Fiction: Editorial Director *Sarah Dudman*; Non-Fiction and Poetry: Editorial Director *Gaby Morgan*; Picture Books and Gift Books: Editorial Director *Suzanne Carnell*. **Kingfisher** Editorial Director *Melissa Fairley*; **Campbell Books** Editorial Director *Sarah Fabiny*; **Young Picador** Publisher *Melissa Fairley* Publishes fiction, non-fiction and poetry in paperback and hardback.

Macmillan New Writing (at New Wharf Road address) Publishing Director *Maria Rejt*, Commissioning Editor *Will Atkins* Founded in 2006 as a way of finding talented new writers who might otherwise go undiscovered. MNW publishes full-length novels from authors who have not previously published a novel. All genres considered and all submissions assessed, but mss must be complete. No advance, but the author pays nothing and receives a royalty of 20% on net sales. Submissions must be sent by e-mail only, via the MNW website (www. macmillannewwriting.com).
Ⓔ Royalties annually or twice-yearly depending on contract.

⊞*Authors' update* **With the surprise departure of CEO Richard Charkin to fresh pastures at Bloomsbury and the elevation of Annette Thomas, formerly head of the science publishing division, other management changes followed in quick succession. Given Thomas's track record as a digital innovator, (Nature.com is now one of the biggest science platforms) an acceleration of online activity can**

be expected. **Among major initiatives** *The New Palgrave Dictionary of Economics* **has a double life, print and online, while Pan Macmillan has launched a blog to promote discussion about the future of books, writing and publishing. More – our own** *Writer's Handbook* **website (www.thewritershandbook.com) is now up and running. Children's books have been given a boost by the acquisition of Kingfisher while for quality adult books, Picador is limiting its hardback publishing to expensive limited editions; all else will be paperback. Macmillan New Writing is attracting rising stars. The deal for first-time authors is a fixed contract with no advance but with a twenty per cent royalty. A new title appears each month. Crime and children's lists are thriving.**

Made Simple Books
▷ Elsevier Ltd

Mainstream Publishing Co. (Edinburgh) Ltd
7 Albany Street, Edinburgh EH1 3UG
☎ 0131 557 2959   🖷 0131 556 8720
bill.campbell@mainstreampublishing.com
www.mainstreampublishing.com
Directors *Bill Campbell, Peter MacKenzie*
Approx. Annual Turnover £3.5 million

Publishes art, autobiography/biography, current affairs, health, sport, history, illustrated and fine editions, photography, politics and world affairs, popular paperbacks. TITLES *Soldier Five* Mike Coburn; *The Real Nureyev* Carolyn Soutar; *Woodward's England* Mick Collins. Over 80 titles a year. Ideas for books considered, but they should be preceded by a letter, synopsis and s.a.e. or return postage.
Ⓔ Royalties twice-yearly.

Management Books 2000 Ltd
Forge House, Limes Road, Kemble, Cirencester GL7 6AD
☎ 01285 771441   🖷 01285 771055
info@mb2000.com
www.mb2000.com
Publisher *Nicholas Dale-Harris*
Approx. Annual Turnover £500,000

Founded 1993 to develop a range of books for executives and managers working in the modern world of business. 'Essentially, the books are working books for working managers, practical and effective.' Publishes business, management, self-development and allied topics as well as sponsored titles. Launched the *In Ninety Minutes* series of compact guide books for managers in 2004, offering advice, ideas and practical help across a range of highly relevant business topics in an hour and a half of study. New ideas for this series are welcome. About 24 titles a year.

Unsolicited mss, synopses and ideas for books welcome.

Manchester University Press

Oxford Road, Manchester M13 9NR
☎ 0161 275 2310   🖷 0161 274 3346
mup@manchester.ac.uk
www.manchesteruniversitypress.co.uk
Publisher/Chief Executive *David Rodgers*
Head of Editorial *Matthew Frost*
Approx. Annual Turnover £2 million

Founded 1904. MUP is Britain's third largest university press, with a list marketed and sold worldwide. Remit consists of occasional trade publications but mainly A-level and undergraduate textbooks and research monographs. Publishes in the areas of literature, TV, film, theatre and media, history and history of art, design, politics, economics, sociology and international law. DIVISIONS **Humanities** *Matthew Frost*; **History/Art History** *Emma Brennan*; **Politics and Law** *Tony Mason*. About 140 titles a year. Unsolicited mss welcome.
£ Royalties annually.

George Mann Books
▷ Arnefold Publishing

Manson Publishing Ltd

73 Corringham Road, London NW11 7DL
☎ 020 8905 5150   🖷 020 8201 9233
manson@mansonpublishing.com
www.mansonpublishing.com
Chairman/Managing Director *Michael Manson*

Founded 1992. Publishes highly illustrated books for study and reference. Subject areas covered include medicine, veterinary medicine, earth science, plant science, agriculture and microbiology. About 10 titles a year. No unsolicited mss; synopses and ideas will be considered.
£ Royalties twice-yearly.

Marshall Cavendish Ltd

5th Floor, 32–38 Saffron Hill, London EC1N 8FH
☎ 020 7421 8120
mcelt@marshallcavendish.co.uk
www.marshallcavendish.co.uk
www.mcelt.com
Managing Director *Chris Jenner*

Founded 1956. Owned by Times Publishing, Singapore, Marshall Cavendish is a major publisher of books. Publishes non-fiction and educational books. DIVISIONS **Trade** *Chris Jenner* TITLE *Football Handbook*; **Education** *Catherine Whitaker* TITLE *Just Right*. About 100 titles a year. Unsolicited mss, synopses and ideas for books welcome; initial approach by e-mail.
£ Royalties annually.

Martin Books
▷ Simon & Schuster UK Ltd

Kenneth Mason Publications Ltd

The Book Barn, Westbourne, Emsworth PO10 8RS
☎ 01243 377977   🖷 01243 379136
info@kennethmason.co.uk
www.kennethmason.co.uk
Chairman *Kenneth Mason*
Managing Director *Piers Mason*

Founded 1958. Publishes diet, health, fitness, nutrition and nautical. No fiction. IMPRINTS **Boatswain Press; Research Disclosure**. Initial approach by letter with synopsis only.
£ Royalties twice-yearly in first year, annually thereafter.

Matador
▷ Troubador Publishing Ltd

Max/Max Press
▷ Little Books Ltd

Kevin Mayhew Publishers

Buxhall, Stowmarket IP14 3BW
☎ 01449 737978   🖷 01449 737834
info@kevinmayhewltd.com
www.kevinmayhew.com
Chairman/Commissioning Editor *Kevin Mayhew*

Founded in 1976. One of the leading sacred music and Christian book publishers in the UK. Publishes religious titles – liturgy, sacramental, devotional, also children's books and school resources. See website for submission guidelines.
£ Royalties annually.

Meadowside Children's Books

185 Fleet Street, London EC4A 2HS
☎ 020 7400 1092   🖷 020 7400 1037
info@meadowsidebooks.com
www.meadowsidebooks.com
Owner *D.C. Thomson*
Publisher *Simon Rosenheim*
Editor *Lucy Cuthew*
Art Director *Sarah Wilson*
Rights Manager *Ellie Wharton*

Founded in September 2003. Publishes distinctive picture books, alongside a range of novelty and character based junior fiction titles. Current growth in young adult fiction list. Over 100 titles a year. Unsolicited mss, synopses and ideas accepted. Picture book mss: no more than 1,000 words. Fiction mss: first three chapters with synopsis. Send to the editorial department by post or e-mail. Only successful submissions will be answered.

Medavia Publishing
▷ Boltneck Publications Limited

Melrose Books

St Thomas Place, Ely CB7 4GG
☎ 01353 646608   ⊞ 01353 646602
info@melrosebooks.co.uk
www.melrosebooks.com
Chairman *Richard A. Kay*
Managing Director *Nicholas S. Law*
Approx. Annual Turnover £100,000

Independent subsidy publisher established by
Melrose Press in 2004. To-date has published
general and children's fiction, literature, fantasy,
biography, travel, religion and science reference.
Welcomes submissions by e-mail or post
Ⓔ Royalties 65%.

Melrose Press Ltd

St Thomas Place, Ely CB7 4GG
☎ 01353 646600   ⊞ 01353 646601
tradesales@melrosepress.co.uk
www.melrosepress.co.uk
Chairman *Richard A. Kay*
Managing Director *Nicholas S. Law*
Approx. Annual Turnover £2 million

Founded 1960. Took on its present name in 1969.
Publishes biographical who's who reference only
(not including *Who's Who*, which is published by
A.&C. Black). About 10 titles a year.

Mentor
▷ Christian Focus Publications

Mercat Press

West Newington House, 10 Newington Road,
Edinburgh EH9 1QS
☎ 0131 668 4371   ⊞ 0131 668 4466
info@birlinn.co.uk
www.mercat.birlinn.co.uk
Managing Editors *Seán Costello, Tom Johnstone*

Founded 1971. Imprint of **Birlinn Ltd**. Publishes
fiction and non-fiction, mainly of Scottish
interest. Subject areas include biography,
photography and walking guides. TITLES
Edinburgh: A New Perspective Jason Baxter; *The
Worms of Euston Square* William Sutton; *Bruar's
Rest* Jess Smith; *Mackerel at Midnight* Ethel G.
Hofman; *West Highland Way, Official Guide* Bob
Aitken and Roger Smith; *25 Walks* SERIES; *River
of Memory* Joe Pieri; *Scottish Cookery* Catherine
Brown. 20 titles in 2007. Unsolicited synopses,
preferably with sample chapters, are welcome. No
new poetry.
Ⓔ Royalties annually.

The Merlin Press Ltd

Suite 4, 96 Monnow Street, Monmouth NP25 3EQ
☎ 01600 775663   ⊞ 01600 775663
info@merlinpress.co.uk
www.merlinpress.co.uk
Managing Director *Anthony W. Zurbrugg*

Director *Adrian Howe*

Founded 1956. Publishes in the area of history,
philosophy and politics. No fiction. IMPRINTS
Merlin Press; Green Print. TITLES *Socialist
Register* (annual); the Chartist Studies SERIES;
Positive Education. About 10 titles a year.
Ⓔ Royalties annually.

Merrell Publishers Ltd

Head office: 81 Southwark Street, London SE1 0HX
☎ 020 7928 8880   ⊞ 020 7928 1199
mail@merrellpublishers.com
www.merrellpublishers.com
US office: 740 Broadway, Suite 1202, New York,
NY 10003 ☎ 001 212 529 2270 ⊞ 001 212 529 2273
info@merrellpublishersusa.com
Managing Director *Hugh Merrell*
Editorial Director *Julian Honer*
US Director *Joan Brookbank*
Approx. Annual Turnover £2.8 million

Founded 1993. Publishes art, architecture, design,
gardening, photography, cars and motorcycles.
48 titles in 2007. Unsolicited synopses and ideas
for books welcome. Send c.v. and synopsis giving
details of the book's target markets and funding of
illustrations.
Ⓔ Royalties annually.

Methodist Publishing House
▷ mph

Methuen Drama
▷ A.&C. Black Publishers Ltd

Methuen Publishing Ltd

8 Artillery Row, London SW1P 1RZ
☎ 020 7798 1600   ⊞ 020 7828 1244
sales@methuen.co.uk
www.methuen.co.uk
Managing Director *Peter Tummons*

Founded 1889. Methuen was owned by Reed
International until it was bought by Random
House in 1997. Purchased by a management buy-
out team in 1998. Acquired **Politico's Publishing**
in 2003 (see entry). Publishes fiction and
non-fiction; travel, sport, humour. No unsolicited
mss; synopses and ideas welcome. Prefers to be
approached via agents or a letter of inquiry. No
first novels, cookery books, personal memoirs.
Ⓔ Royalties twice-yearly.

Metro Publishing
▷ John Blake Publishing Ltd

Michelin Maps & Guides

Hannay House, 39 Clarendon Road, Watford
WD17 1JA
☎ 01923 205240/205254 (sales)   ⊞ 01923 205241
www.Viamichelin.co.uk
Sales Manager *Ian Murray*

Founded 1900 as a travel publisher. Publishes travel guides, maps and atlases.

Midland Publishing

An imprint of Ian Allan Publishing Ltd, 4 Watling Drive, Hinckley LE10 3EY
☎ 01455 25549   🖷 01455 255495
midlandbooks@compuserve.com
Publisher *Nick Grant*

Imprint of **Ian Allan Publishing Ltd**. Publishes aviation books. No wartime memoirs. No unsolicited mss; synopses and ideas welcome.
Ⓔ Royalties twice-yearly.

Miller's
▷ Octopus Publishing Group

Millivres Prowler Limited

Unit M, Spectrum House, 32–34 Gordon House Road, London NW5 1LP
☎ 020 7424 7400   🖷 020 7424 7401
info@millevres.co.uk
www.millivres.co.uk

Publishes various magazines including *Gay Times*, *Diva Magazine*, *AXM* monthly magazine and *The Pink Paper*, the national free gay newspaper.

✱*Authors' update* **No longer interested in books; the focus is entirely on journals.**

Mills & Boon
▷ Harlequin Mills & Boon Ltd

Milo Books Limited

The Old Weighbridge, Station Road, Wrea Green, Preston PR4 2PH
☎ 01772 672900   🖷 01772 687727
info@milobooks.com
www.milobooks.com
Managing Director *Peter Walsh*

Founded 1997. Publishes non-fiction: true crime and sport. 12 titles a year. Unsolicited mss, synopses and ideas for books. 'Authors should note our specialist areas. We cannot promise to return all material submitted.' Approach in writing in the first instance. 'If we are interested in pursuing an idea, we will call to talk it through in more depth.' No fiction or poetry.
Ⓔ Royalties twice-yearly.

Mindfield
▷ Camden Press Ltd

Mitchell Beazley
▷ Octopus Publishing Group

Mobius
▷ Hachette Livre UK

Monarch Books

Lion Hudson plc, Wilkinson House, Jordan Hill Road, Oxford OX2 8DR
☎ 01865 302750   🖷 01865 302757
monarch@lionhudson.com
Editorial Director *Tony Collins*

An imprint of **Lion Hudson plc**. Publishes an independent list of Christian books across a wide range of concerns. IMPRINT **Monarch** Upmarket paperback list with Christian basis and strong social concern agenda including psychology, future studies, politics, mission, theology, leadership, fiction and spirituality. About 35 titles a year. Unsolicited mss, synopses and ideas welcome.

Morgan Kauffman
▷ Elsevier Ltd

Mosby
▷ Elsevier Ltd

Motor Racing Publications

PO Box 1318, Croydon CR9 5YP
☎ 020 8654 2711   🖷 020 8407 0339
john@mrpbooks.co.uk
www.mrpbooks.co.uk
Chairman/Editorial Head *John Blunsden*

Founded soon after the end of World War II to concentrate on motor-racing titles. Fairly dormant in the mid 1960s but was reactivated in 1968 by a new shareholding structure. John Blunsden later acquired a majority share and major expansion followed in the 1970s. Publishes motor-sport history, classic and performance car collection and restoration, race track and off-road driving and related subjects. IMPRINTS **Fitzjames Press; Motor Racing Publications**. About 2–4 titles a year. No unsolicited mss. Send synopses and ideas in specified subject areas in the first instance.
Ⓔ Royalties twice-yearly.

mph

4 John Wesley Road, Werrington, Peterborough PE4 6ZP
☎ 01733 325002   🖷 01733 384180
www.mph.org.uk
Chair *Eric Jarvis*
Chief Executive *Neil Joubert*
Approx. Annual Turnover £2 million

Founded 1800. Owned by the Methodist Church. Publishes a wide range of books, magazines and resources which are sold in the UK and overseas. Also *Epworth Review* quarterly magazine. IMPRINTS **Epworth** (see entry); **Inspire** (see entry); **Methodist Publishing House**. About 30 titles a year. Unsolicited mss, synopses and ideas

welcome; send sample chapter and contents with covering letter and s.a.e. for return of mss.
Ⓔ Royalties annually.

Murdoch Books UK Ltd

Erico House, 6th Floor North, 93–99 Upper Richmond Road, London SW15 2TG
Ⓣ 020 8785 5995　Ⓕ 020 8785 5985
info@murdochbooks.co.uk
www.murdochbooks.co.uk
Publisher *Kay Scarlett (Australia)*

Owned by Australian publisher Murdoch Books Pty Ltd. Publishes full-colour non-fiction: homes and interiors, gardening, cookery, craft, DIY, history, travel and narrative non-fiction. About 80 titles a year. Contact inquiry@murdochbooks.com.au in the first instance.

John Murray (Publishers) Ltd

338 Euston Road, London NW1 3BH
Ⓣ 020 7873 6000　Ⓕ 020 7873 6446
firstname.lastname@johnmurrays.co.uk
www.hachettelivre.co.uk
Chief Executive *Martin Neild*
Managing Director *Roland Philipps*

Founded 1768. Part of **Hachette Livre UK**. Publishes general trade books. **Non-fiction** *Eleanor Birne*; **Fiction** *Heather Barrett, Kate Parkin*. 50 titles a year. No unsolicited material; send preliminary letter.
Ⓔ Royalties twice-yearly.

❋*Authors' update* **Shifting towards mainstream books since its acquisition by Hodder Headline but still interested in serious non-fiction, John Murray has justifiably retained its reputation for quality books.**

Myriad Editions

59 Lansdowne Place, Brighton BN3 1FL
Ⓣ 01273 720000
info@myriadeditions.com
www.myriadeditions.com
Chairman *Robert J. Benewick*
Managing Director *Candida Lacey*

Myriad Editions is an independent publisher, founded in 1993. In addition to its award-winning State of the World atlases it has expanded its publishing programme to include original fiction, innovative non-fiction and documentary comic books. TITLES *365 Ways to Change the World* Michael Norton; *The Atlas of Climate Change* (winner Planeta Environment Book of the Year). 5–10 titles in 2007. No unsolicited mss. Synopses and ideas for books in the specified subject areas only. Approach by e-mail in the first instance.
Ⓔ Royalties annually.

Myrmidon Books

Rotterdam House, 116 Quayside, Newcastle upon Tyne NE1 3DY
Ⓣ 0191 206 4005　Ⓕ 0191 206 4001
submissions@myrmidonbooks.com
www.myrmidonbooks.com
Chairman/Managing Director *Ed Handyside*
Literary Editor *Anne Westgarth*
Approx. Annual Turnover £200,000

Founded 2006. Publishes adult trade fiction. 3 titles in 2007. Full mss from literary agents only; first three chapters from unrepresented writers. No short stories, novellas or works under 65,000 words, children's books, graphic novels, non-fiction. Submit hard copy only after checking requirements on the website.
Ⓔ Royalties quarterly.

NAG Press Ltd
▷ **Robert Hale Ltd**

The National Archives

Kew TW9 4DU
Ⓣ 020 8392 5289　Ⓕ 020 8487 1974
catherine.bradley@nationalarchives.gov.uk
www.nationalarchives.gov.uk
Publisher *Catherine Bradley*

Publishes non-fiction books relating to the thousand years of historical records held in The National Archives. Specialist areas: general history, family history, military history and intelligence history. TITLES *The Cecils. Power and privilege behind the throne* David Loades; *British Intelligence. Secrets, spies and sources* Stephen Twigge, Edward Hampshire and Graham Macklin; *New Lives for Old. The story of Britain's child migrants* Roger Kershaw and Janet Sacks; *Will's Will. The last wishes of William Shakespeare* Simon Trussler. SERIES **Crime Archive** TITLES *Dr Crippen*; *Ruth Ellis*; *Mrs Maybrick*. Also publishes *Ancestors*, a full-colour family history magazine. 20 titles in 2006.

National Trust Books
▷ **Anova Books**

Nautical Data Ltd

The Book Barn, Westbourne, Emsworth PO10 8RS
Ⓣ 01243 389352　Ⓕ 01243 379136
info@nauticaldata.com
www.nauticaldata.com
Managing Director *Piers Mason*

Founded 1999. Publishes pilots and nautical reference. No unsolicited mss; synopses and ideas welcome. No fiction or non-nautical themes.
Ⓔ Royalties twice-yearly.

NCVO Publications

Regent's Wharf, 8 All Saints Street, London
N1 9RL
① 020 7713 6161/HelpDesk: 0800 2798 798
Ⓕ 020 7713 6300
ncvo@ncvo-vol.org.uk
www.ncvo-vol.org.uk/publications
Publications and Promotions Coordinator *Daniel Fluskey*
Approx. Annual Turnover £140,000

Founded 1928. Publishing imprint of the National
Council for Voluntary Organisations, embracing
former Bedford Square Press titles and NCVO's
many other publications. The list reflects
NCVO's role as the representative body for the
voluntary sector. Publishes directories, good
practice information on management and trustee
development, finance and employment titles and
information of primary interest to the voluntary
sector. TITLES *The Voluntary Agencies Directory*;
The Good Trustee Guide; *The Good Campaigns
Guide*; *The Good Financial Management
Guide*; *The Good Employment Guide*; *The Good
Management Guide*; *The Good Membership
Guide*; *The UK Voluntary Sector Almanac*. No
unsolicited mss as all projects are commissioned
in-house.

Thomas Nelson & Sons Ltd
▷ Nelson Thornes Limited

Nelson Thornes Limited

Delta Place, 27 Bath Road, Cheltenham GL53 7TH
① 01242 267100   Ⓕ 01242 221914
info@nelsonthornes.com
www.nelsonthornes.com
Managing Director *Linden Harris*

Now part of Infinitas Learning, Nelson Thornes
was formed in 2000 by the merger of Thomas
Nelson and Stanley Thornes. Educational
publisher (AQA endorsed) of printed and
electronic product, from pre-school to higher
education. Unsolicited mss, synopses and ideas
for books welcome if appropriate to specialized
lists.
Ⓔ Royalties annually.

New Beacon Books Ltd

76 Stroud Green Road, London N4 3EN
① 020 7272 4889   Ⓕ 020 7281 4662
newbeaconbooks@btconnect.com
www.newbeaconbooks.co.uk
Managing Director *Sarah White*
Approx. Annual Turnover £120,000

Founded 1966. Publishes fiction, history, politics,
poetry and language, all concerning black people.
No unsolicited material.
Ⓔ Royalties annually.

New Holland Publishers (UK) Ltd

Garfield House, 86–88 Edgware Road, London
W2 2EA
① 020 7724 7773   Ⓕ 020 7258 1293 (editorial)
postmaster@nhpub.co.uk
www.newhollandpublishers.com
Managing Director *Steve Connolly*
Publishing Director *Rosemary Wilkinson*
Approx. Annual Turnover £8.5 million

With its headquarters in London, New Holland
Publishers (UK) Ltd is the International
Publishing Division of Johnnic Communications,
one of Africa's leading publishing groups, with
offices in Australia and New Zealand. Acquired
Cadogan Guides in April 2007. Publishes
non-fiction, practical and inspirational books
across a range of categories, including food
and drink, DIY, practical art, crafts, interiors,
health and fitness, gardening, pets, humour
and gift, reference, history, sports and outdoor
pursuits, natural history, general travel books
and Globetrotter travel guides. TITLES *Men and
Sheds*; *The Allotment Handbook*; *The Big Book of
Handmade Cards*; *The BTO Garden Bird Book*;
The Specialist Series; *Dugouts*; *Diving with Giants*;
Party Knits; *How to Decorate with Wallpaper*;
Meena Pathak Celebrates Indian Cooking; *Abbeys
and Priories of England*. No unsolicited mss;
synopses and ideas welcome.

✱*Authors' update* Having taken over the highly
regarded Cadogan Guides, New Holland
is clearly confident of holding off the free
travel websites. But for writers this is a highly
competitive market.

New Playwrights' Network
▷ Cressrelles Publishing Co. Ltd

Newnes
▷ Elsevier Ltd

Nexus
▷ The Random House Group Ltd

nferNelson
▷ GL Assessment

Nia
▷ The X Press

Nielsen BookData

3rd Floor, Midas House, 62 Goldsworth Road,
Woking GU21 6LQ
① 0870 777 8710   Ⓕ 0870 777 8711
info.bookdata@nielsen.com
www.nielsenbookdata.co.uk
Commercial Director *Ann Betts*
Senior Manager, Publishing Services *Peter
Mathews*

BookData creates and maintains a unique database of timely, accurate and content rich records for English-language books and other published media, collected from publishers in over 70 countries. The enriched data includes: price, availability, table of contents (where applicable), subject classifications, market rights, publisher and distributor details, cover/jacket images and book descriptions. The data is disseminated worldwide to booksellers and libraries in a variety of formats including data feed, CD-ROMs and online services.

Nightingale Books
▷ Pegasus Elliot Mackenzie Publishers Ltd

NMS Enterprises Limited – Publishing
National Museums Scotland, Chambers Street, Edinburgh EH1 1JF
☎ 0131 247 4026   ⌷ 0131 247 4012
publishing@nms.ac.uk
www.nms.ac.uk/books
Director *Lesley A. Taylor*
Approx. Annual Turnover £123,000

Publishes non-fiction related to the National Museums of Scotland collections: academic and general; children's; archaeology, history, decorative arts worldwide, history of science, technology, natural history and geology, poetry. TITLES *Silver: Made in Scotland*; *Tartan: The Highland Habit*; *Beyond the Palace Walls: Islamic Art from the State Hermitage Museum*; *Nicholas and Alexandra: The Last Tsar and Tsarina*; *Minerals of Scotland*; *Audubon in Edinburgh*; *Commando Country*; *Weights and Measures in Scotland: a European Perspective* (2005 Saltire Society/National Library of Scotland Research Book of the Year). No unsolicited mss; only interested in synopses and ideas for books which are genuinely related to NMS collections and to Scotland in general.
Ⓔ Royalties twice-yearly.

Nonsuch
▷ The History Press Ltd

North-Holland
▷ Elsevier Ltd

Northcote House Publishers Ltd
Horndon House, Horndon, Tavistock PL19 9NQ
☎ 01822 810066   ⌷ 01822 810034
northcote.house@virgin.net
www.northcotehouse.co.uk
Managing Director *Brian Hulme*

Founded 1985. Publishes a series of literary critical studies, in association with the British Council, *Writers and their Work*; education management, literary criticism, educational dance and drama. 30 titles in 2007. 'Well-thought-out proposals,

including contents and sample chapter(s), with strong marketing arguments welcome.'
Ⓔ Royalties annually.

W.W. Norton & Company Ltd
Castle House, 75–76 Wells Street, London W1T 3QT
☎ 020 7323 1579   ⌷ 020 7436 4553
office@wwnorton.co.uk
www.wwnorton.co.uk
Managing Director *Alan Cameron*

Subsidiary of the US parent company founded in 1923 (see entry under *US Publishers*). Publishes non-fiction and academic books.

Nottingham University Press
Manor Farm, Church Lane, Thrumpton NG11 0AX
☎ 0115 983 1011   ⌷ 0115 983 1003
orders@nup.com
www.nup.com

Initially concentrated on agricultural and food sciences titles but has now branched into new areas including engineering, lifesciences, medicine and law. TITLES *Porcine Meat Inspection*; *Farriery: The Whole Horse Concept*; *Nutrition-based Health*; *Paradigms in Pig Science*; *Biotechnology and Genetic Engineering Review*.
Ⓔ Royalties twice-yearly.

NWP
▷ Neil Wilson Publishing Ltd

O Books/John Hunt Publishing Ltd
The Bothy, Deershot Lodge, Park Lane, Ropley SO24 0BE
☎ 01962 773768   ⌷ 01962 773769
office1@o-books.net
www.o-books.net
Approx. Annual Turnover £1.5 million

Publishes children's and world religions as well as mind, body and spirit titles. IMPRINTS **John Hunt Publishing**; **O Books**. About 50 titles a year. Unsolicited material welcome.

Oak
▷ Omnibus Press

Oberon Books
521 Caledonian Road, London N7 9RH
☎ 020 7607 3637   ⌷ 020 7607 3629
info@oberonbooks.com
www.oberonbooks.com
Publishing Director *James Hogan*
Managing Director *Charles D. Glanville*
Editor *Dan Steward*

A leading theatre publisher, Oberon publishes play texts (usually in conjunction with a production), and books on theatre and dance. Specializes in contemporary plays and translations of European classics. IMPRINTS

Oberon Modern Plays; Oberon Classics.
Publishes for the National Theatre, Royal Opera
House, Royal Court, English Touring Theatre,
LAMDA, Gate Theatre, Bush Theatre, West
Yorkshire Playhouse, Chichester Festival Theatre
and many other London and regional theatre
companies. Publishes over 300 writers and
translators including Rodney Ackland, Tariq Ali,
Howard Barker, John Barton, Ranjit Bolt, Pam
Gems, Tanika Gupta, Sir Peter Hall, Bernard
Kops, Adrian Mitchell, Sir John Mortimer,
Adriano Shaplin, Laura Wade. An extensive
classics list includes performance (edited)
versions of Shakespeare including *King Lear*; *A
Midsummer Night's Dream*; *Romeo and Juliet*
and *Twelfth Night*, as well as new translations of
plays by Chekhov, Molière, Strindberg and others.
About 80 titles a year. Send manuscripts, with
information about forthcoming productions to
Will Hammond (will@oberonbooks.com). 'We
do not provide readers' reports or feedback about
work submitted.'

⊛*Authors' update* A first call for theatre books,
Oberon has confirmed its lead in a niche
market by sponsoring the Sheridan Morley
Prize for theatrical biography (see entry under
Prizes).

Octagon Press Ltd

78 York Road, London W1H 1DP
Ⓣ 020 7193 6456   Ⓕ 020 7117 3955
admin@octagonnpress.com
www.octagonpress.com
Director *Saira Shah*
Approx. Annual Turnover £50,000

Founded 1960. Publishes travel, biography,
literature, folklore, psychology, philosophy with
the focus on East-West studies. 4–5 titles a year.
Ⓔ Royalties annually.

Octopus Publishing Group

2–4 Heron Quays, London E14 4JP
Ⓣ 020 7531 8400   Ⓕ 020 7531 8650
www.octopus-publishing.co.uk
Chief Executive *Alison Goff*
Approx. Annual Turnover £45 million (Group)

Formed following a management buyout of Reed
Consumer Books from Reed Elsevier plc in 1998.
Bought by French publishers Hachette Livre in
2001 and acquired Cassell Illustrated and the lists
of Ward Lock and Blandford Press; also, Gaia
Books acquired in 2004.

Conran Octopus Ⓕ 020 7531 8627 info-co@
conran-octopus.co.uk www.conran-octopus.
co.uk Publishing Director *Lorraine Dickey* Quality
illustrated lifestyle books, particularly interiors,
design, cookery, gardening and crafts.

Hamlyn Octopus Ⓕ 020 7531 8562 info-ho@
hamlyn.co.uk www.hamlyn.co.uk Publisher
Jane Birch Practical books for home and family,
cookery, gardening, craft, sport, health, parenting
and pet care.

Mitchell Beazley/Miller's Ⓕ 020 7537 0773
info-mb@mitchell-beazley.co.uk www.mitchell-
beazley.co.uk Publisher *David Lamb* Quality
illustrated reference books, particularly food and
wine, gardening, interior design and architecture,
antiques, general reference.

Philip's Ⓕ 020 7531 8460 george.philip@philips-
maps.co.uk www.philips-maps.co.uk Managing
Director *To be appointed* World atlases, globes,
astronomy, road atlases, encyclopaedias, thematic
reference.

Bounty Ⓕ 020 7531 8607 bountybooksinfo-bp@
bountybooks.co.uk Publishing and International
Sales Director *Polly Manguel* Bargain and
promotional books.

Cassell Illustrated Ⓕ 020 731 8624 Publishing
Director *Mathew Clayton*

Godsfield Press Ⓕ 020 7531 8562 enquiries@
godsfieldpress.com Publisher *Jane Birch* Mind,
body and spirit.

Gaia Books Publisher *Jane Birch* Alternative
health and natural/eco titles.

Spruce Ⓕ 020 7531 8604 ljiljana.baird@octopus-
publishing.co.uk www.octopusbooks.co.uk
Publisher *Ljiljana Baird* Gift, cookery, biography,
popular culture, photography, mind body spirit,
inspiration, parenting.
Ⓔ Royalties twice-yearly/annually, according to
contract in all divisions.

⊛*Authors' update* The emphasis is on
mainstream trade books at the popular end of
the market. The latest initiative is to launch
in the US with content repackaged for an
American audience.

Old Street Publishing Ltd

28–32 Bowling Green Lane, London EC1R 0BJ
Ⓣ 020 7837 1600   Ⓕ 020 7900 6563
info@oldstreetpublishing.co.uk
www.oldstreetpublishing.co.uk
Chairman *David Reynolds*
Managing Director *Ben Yarde-Buller*

Founded in 2006 by Ben Yarde-Buller and David
Reynolds, co-founder of Bloomsbury Publishing.
Publishes eclectic fiction including literary fiction,
crime/thriller and translation. General non-
fiction and humour 'of the better sort'. No poetry,
technical or children's books. 25 titles in 2007.
No unsolicited mss. Synopses and ideas welcome;
approach by e-mail.
Ⓔ Royalties twice-yearly.

Olympia Publishers

60 Cannon Street, London EC4N 6NP
☎ 020 7618 6424   ⊞ 020 7002 1100
editors@olympiapublishers.com
www.olympiapublishers.com
Chief Editor *G. Bartlett*
Managing Editor *M.B. Anderson*

Founded 2004. Publishes fiction, memoirs, human interest, chick lit, military, naval, children's and non-fiction. General interest titles: crime, history, humour, mind, body and soul, self help, science fiction. Specializes in books for international markets by Far Eastern authors and Europeans with relevant background experience. TITLES *The Sunzi Series of Educational Books for South Africa*; *Far Side of Freedom*; *Diplomatic Dog of Barbados*. 40-50 titles a year. See website for submission guidelines. IMPRINTS **Olympia Publishers** Subsidy publishers; **Ashwell Publishing Limited** Non-subsidy publishers TITLES *Business Studies Evaluated*. 40–50 titles a year. See website for submission guidelines. Unsolicited synopses, poetry and short stories also considered. Postal submissions must be accompanied by s.a.e.
Ⓔ Royalties twice-yearly.

⊛*Authors' update* Note that this is a two-in-one operation. Submissions turned down by Ashwell for conventional publication are likely to be passed on to Olympia where the author will be asked to contribute to production costs.

Michael O'Mara Books Ltd

16 Lion Yard, Tremadoc Road, London SW4 7NQ
☎ 020 7720 8643   ⊞ 020 7627 8953
firstname.lastname@mombooks.com
www.mombooks.com
www.mombooks.com/busterbooks
Chairman *Michael O'Mara*
Managing Director *Lesley O'Mara*
Editorial Directors *Toby Buchan, Lindsay Davies, Louise Dixon*
Publishing Director (Buster Books) *Philippa Wingate*
Approx. Annual Turnover £7 million

Founded 1985. Independent publisher. Publishes general non-fiction, humour, anthologies, miscellanies, royalty and celebrity biographies. TITLES *I Before E* Judy Parkinson; *The Mums' Book* and *The Dads' Book*; *The Timewaster Letters* Robin Cooper; *Almost a Celebrity* James Whale; *The Book of Senior Moments* Shelley Klein. IMPRINT **Buster Books** Children's TITLES *The Boys' Book* and *The Girls' Book*; *Do You Doodle?*; *Beautiful Doodles*; board books, picture books, novelty books, humour and quirky non-fiction. 49 titles in 2007. Unsolicited mss, synopses and ideas for books welcome.

Ⓔ Royalties twice-yearly.

⊛*Authors' update* A big jump in sales to the US secured the 2008 International Achievement prize from the Independent Publishers Guild. Most of the creative work for bestselling humour titles is done in-house but strong ideas for the gift market may well attract interest.

Omnibus Press

Music Sales Ltd, 14/15 Berners Street, London W1T 3LJ
☎ 020 7612 7400   ⊞ 020 7612 7545
chris.charlesworth@musicsales.co.uk
www.omnibuspress.com
Owner *Robert Wise*
Editorial Head *Chris Charlesworth*

Founded 1971. Independent publisher of music books, rock and pop biographies, song sheets, educational tutors, cassettes, videos and software. A Division of the Music Sales Group of Companies. IMPRINTS **Amsco**; **Arcane**; **Bobcat**; **Oak**; **Omnibus**; **Vision On**; **Wise Publications**. About 25 titles a year. Unsolicited mss, synopses and ideas for music-related titles welcome.

Oneworld Publications

185 Banbury Road, Oxford OX2 7AR
☎ 01865 310597   ⊞ 01865 310598
info@oneworld-publications.com
www.oneworld-publications.com
Editorial Director *Juliet Mabey*

Founded 1986. Independent publisher of intelligent adult non-fiction for both the trade and academic markets, across a range of subjects including current affairs, philosophy, history, politics, popular science, psychology and world religions. SERIES include **Oneworld Beginners' Guides**; **Coping With** (self-help) and **Makers of the Muslim World**. TITLES *Why Don't Spiders Stick to their Webs?*; *Body Shopping*; *25 Big Ideas: The Science that Changed our World*; *Can a Robot be Human?* 60 titles a year. No unsolicited mss; synopses and ideas welcome, but should be accompanied by s.a.e. for return of material and/or notification of receipt. No autobiographies, fiction, poetry or children's.
Ⓔ Royalties annually.

Onlywomen Press Ltd

40 St Lawrence Terrace, London W10 5ST
☎ 020 8354 0796
onlywomenpress@btconnect.com
www.onlywomenpress.com
Editorial Director *Lilian Mohin*

Founded 1974. Publishes feminist lesbian literature: fiction, poetry, literary criticism and children's books. Unsolicited mss and proposals welcome.

OPC
▷ Ian Allan Publishing Ltd

Open Gate Press incorporating Centaur Press
51 Achilles Road, London NW6 1DZ
☎ 020 7431 4391   📠 020 7431 5129
books@opengatepress.co.uk
www.opengatepress.co.uk
Managing Directors *Jeannie Cohen,
Elisabeth Petersdorff*

Founded in 1989 to provide a forum for
psychoanalytic social and cultural studies.
Publishes psychoanalysis, philosophy,
social sciences, politics, literature, religion,
environment. SERIES *Psychoanalysis and Society.*
IMPRINTS **Open Gate Press; Centaur Press;
Linden Press**. No unsolicited mss.
£ Royalties annually.

Open University Press
McGraw-Hill House, Shoppenhangers Road,
Maidenhead SL6 2QL
☎ 01628 502500   📠 01628 770224
enquiries@openup.co.uk
www.openup.co.uk

Founded 1977 as an imprint independent of the
Open University's course materials. Acquired
by **McGraw-Hill Education** in 2002. Publishes
academic and professional books in the fields
of education, sociology, health, psychology,
psychotherapy and counselling, higher education
and study skills materials. No economics or
anthropology. Not interested in anything outside
the social sciences. About 100 titles a year. No
unsolicited mss; enquiries/proposals only.
£ Royalties annually.

Orbit
▷ Little, Brown Book Group UK

Orchard Books
▷ Hachette Children's Books

The Orion Publishing Group Limited
Orion House, 5 Upper St Martin's Lane, London
WC2H 9EA
☎ 020 7240 3444   📠 020 7240 4822
www.orionbooks.co.uk/pub/index.htm
Chairman *Arnaud Nourry*
Chief Executive *Peter Roche*
Deputy Chief Executive *Malcolm Edwards*
Approx. Annual Turnover £70 million

Founded 1992. Incorporates Weidenfeld &
Nicolson, JM Dent, Gollancz and Cassell.

TRADE DIVISION

Trade Managing Director *Lisa Milton* IMPRINTS
Orion Fiction Publishing Director *Jon Wood*
Hardcover fiction; **Orion Non-Fiction** Publishing
Director *Ian Marshall* Hardcover non-fiction;

Orion Children's Publisher *Fiona Kennedy*
Children's fiction/non-fiction; **Gollancz** Editorial
Directors *Simon Spanton, Jo Fletcher* Science
fiction and fantasy. **Weidenfeld Non-Fiction**
Publisher *Alan Samson* General non-fiction,
history and military; **Weidenfeld Fiction**
Publisher *Kirsty Dunseath* History; **Weidenfeld
Illustrated** Editor-in-Chief *Michael Dover*;
Custom Publishing Director *Elizabeth Leath.*

PAPERBACK DIVISION Managing Director *Susan
Lamb* IMPRINTS **Orion; Phoenix; Everyman**.

✱*Authors' update* **Confirming anecdotal
evidence of a squeeze on midlist non-fiction
authors, Weidenfeld has sliced back its output
from seven to four new titles a month. What has
gone, says George Weidenfeld, is the 'middle of
the road journalistic popularization'.**

Osprey Publishing Ltd
Midland House, West Way, Botley, Oxford
OX2 0PH
☎ 01865 727022   📠 01865 727017/727019
info@ospreypublishing.com
www.ospreypublishing.com
Chairman *Philip Sturrock*
Managing Director *Rebecca Smart*
Editorial Director *Anita Baker*

Publishes illustrated history of war and
warfare and military aviation, also general
military history. Founded 1969, Osprey became
independent from **Reed Elsevier** in February
1998 and launched Osprey Publishing Inc. as a
wholly owned subsidiary in North America in
2004. Acquired **Shire Publications Ltd** in 2007
(see entry). SERIES include *Men-at-Arms; Fortress;
Campaign; Essential Histories; Warrior; Elite; New
Vanguard; Osprey Modelling; Duel; Aircraft of the
Aces; Aviation Elite Units; Combat Aircraft.* About
125 titles a year including 20 non-series titles
a year. No unsolicited mss; ideas and synopses
welcome.
£ Royalties twice-yearly.

Peter Owen Publishers
73 Kenway Road, London SW5 0RE
☎ 020 7373 5628/7370 6093   📠 020 7373 6760
aowen@peterowen.com
www.peterowen.com
Chairman *Peter Owen*
Editorial Director *Antonia Owen*

Founded 1951. Publishes biography, general
non-fiction, English-language literary fiction and
translations, history, literary criticism, the arts
and entertainment. 'No genre or children's fiction;
the company is taking on no first novels or short
stories at present.' AUTHORS Jane Bowles, Paul
Bowles, Yuri Druzhnikov, Shusaku Endo, Philip
Freund, Anna Kavan, Jean Giono, Anaïs Nin,
Jeremy Reed, Peter Vansittart. About 25 titles

a year. Unsolicited synopses welcome for non-fiction material. Please call or e-mail in advance to submit fiction. Mss should be preceded by a descriptive letter and synopsis with s.a.e. OVERSEAS ASSOCIATES worldwide.
Ⓔ Royalties twice-yearly.

⊕*Authors' update* Peter Owen has been described as 'a publisher of the old and idiosyncratic school'. He has seven Nobel prizewinners on his list to prove it.

Oxbow Books

10 Hythe Bridge Street, Oxford OX1 2EW
Ⓣ 01865 241249   Ⓕ 01865 794449
editorial@oxbowbooks.com
www.oxbowbooks.com
Editorial *Clare Litt*

Founded 1983. Publishes academic archaeology, Egyptology, ancient and medieval history. IMPRINTS **Aris & Phillips** Greek, Latin and Hispanic texts and translations; **Windgather Press** Landscape archaeology and history. About 60 titles a year.

Oxford University Press

Great Clarendon Street, Oxford OX2 6DP
Ⓣ 01865 556767   Ⓕ 01865 556646
enquiry@oup.com
www.oup.com
Chief Executive *Henry Reece*
Approx. Annual Turnover £453 million

A department of Oxford University, OUP started as the university's printing business and developed into a major publishing operation in the 19th century. Publishes academic works in all formats (print and online): dictionaries, lexical and non-lexical reference, scholarly journals, student texts, schoolbooks, ELT materials, music, bibles, paperbacks, and children's books. Around 6,000 titles a year.

DIVISIONS **Academic** *T.M. Barton* Academic and higher education titles in major disciplines, dictionaries, non-lexical reference and trade books. OUP welcomes first-class academic material in the form of proposals or accepted theses. **Journals** *M. Richardson* Academic and research journals. **Education** *K. Harris* National Curriculum courses and support materials as well as children's literature. **ELT** *P.R.C. Marshall* ELT courses and dictionaries for all levels.

OVERSEAS BRANCHES/SUBSIDIARIES Sister company in USA with branches or subsidiaries in Argentina, Australia, Brazil, Canada, China, East Africa, India, Japan, Malaysia, Mexico, Pakistan, Southern Africa, Spain and Turkey. Offices in France, Greece, Italy, Poland, Taiwan, Thailand.
Ⓔ Royalties twice-yearly.

⊕*Authors' update* The biggest publisher to benefit from charitable status, OUP is under attack from rivals over its tax exempt status. English and bilingual dictionaries are the core activity with OUP capturing over half the English and a third of the bilingual market.

Palgrave Macmillan
▷ Macmillan Publishers Ltd

G.J. Palmer & Sons Ltd
▷ Hymns Ancient & Modern Ltd

Pan
▷ Macmillan Publishers Ltd

Paper Tiger
▷ Anova Books

PaperBooks Ltd
▷ Legend Press Limited

Parthian

The Old Surgery, Napier Street, Aberteifi (Cardigan) SA43 1ED
Ⓣ 01239 612059
parthianbooks@yahoo.co.uk
www.parthianbooks.co.uk
Also: **Library of Wales** Trinity College, Carmarthen SA31 3EP
Ⓣ 01267 676633 Ⓕ 01239 612059
www.libraryofwales.org
Publisher *Richard Davies*
Fiction/Poetry Editor *Lucy Llewellyn*
Editor, Library of Wales *Professor Dai Smith*

Founded 1993. Publishes contemporary Welsh writing in English including novels, poetry and drama, also translations from Welsh. A keen interest in short stories. Its list includes many young writers. TITLES *Oh Dad! A Search for Robert Mitchum* Lloyd Robson; *Fresh Apples* Rachel Trezise (winner of the EDS Dylan Thomas Award); *The Long Dry* Cynan Jones (winner of the Betty Trask Award); *Martha, Jack and Shanco* Caryl Lewis. New non-fiction list and increasing interest in translation with new titles planned from Catalan and Basque. Also publishes **Library of Wales** classic series including Raymond Williams' *Border Country*, ed. Prof. Dai Smith. 12–15 titles a year. Synopsis with first three sample chapters welcome. 'An idea of what we do is always an advantage.'
Ⓔ Royalties paid annually.

Paternoster
▷ Authentic Media

Pathfinder
▷ The History Press Ltd

Pavilion/Pavilion Children's Books
▷ Anova Books

Payne-Gallway
> Pearson (Heinemann)

Pearson

UK Head office: Edinburgh Gate, Harlow
CM20 2JE
☎ 01279 623623
www.pearsoned.com

The world's largest educational publisher was created by the merger of Addison Wesley Longman, Financial Times Management and **Simon & Schuster**'s educational list. Acquired Heinemann (see **Pearson Heinemann**) in May 2007. Publishes across a wide range of curriculum subjects from primary students to professional practitioners. IMPRINTS **Addison Wesley; Allyn & Bacon; BBC Active; Benjamin Cummings; Causeway Press; Cisco Press; Penguin Longman; Prentice Hall; Prentice Hall Business; QUE Publishing; SAM Publishing; Wharton; York Notes**. OVERSEAS ASSOCIATES worldwide. Unsolicited mss should be addressed to the appropriate department: **ELT** *Alissa Heilbron*; **Schools** *Carol Hayward*; **University & Professional Level** *Louise Smith*.
Ⓔ Royalties twice-yearly.

Pearson (Heinemann)

Halley Court, Jordan Hill, Oxford OX2 8EJ
☎ 01865 888000 (customer services)/311366
🖷 01865 314641
enquiries@pearson.com
www.heinemann.co.uk

Heinemann, now part of **Pearson**, produces educational products under the following IMPRINTS: **Heinemann, Rigby, Ginn, Payne-Gallway** and **Raintree**. Pearson provides a range of published resources, teachers' support, and pupil and student material in all core subjects for all ages. For submissions, write to 'Market Planning' at the address above.
Ⓔ Royalties twice-yearly/annually, according to contract.

Pegasus Elliot Mackenzie Publishers Ltd

Sheraton House, Castle Park, Cambridge CB3 0AX
☎ 01223 370012   🖷 01223 370040
editors@pegasuspublishers.com
www.pegasuspublishers.com
Senior Editor *D.W. Stern*
Editor *R. Sabir*

Publishes fiction and non-fiction, general interest, biography, autobiography, children's, history, humour, fantasy, self-help, educational, science fiction, poetry, travel, war, memoirs, crime and eroticism, also Internet and e-books. IMPRINTS **Vanguard Press; Nightingale Books; Chimera** TITLES *Count the Petals of the Moon Daisy* Martin Kirby; *Good Monday Morning* Dr Funké Baffour;

Modern Day Hero Nick Pettitt, GM, QGM; *Missing In Medway* Alex Mabon, *Deep Siege* Charles Ingram; *The Prince's Thorn* Mary Mackie. Unsolicited mss, synopses and ideas considered if accompanied by return postage.
Ⓔ Royalties twice-yearly.

✳*Authors' update* Liable to ask authors to contribute to production costs.

Pen & Sword Books Ltd

47 Church Street, Barnsley S70 2AS
☎ 01226 734222   🖷 01226 734438
editorial@pen-and-sword.co.uk
www.pen-and-sword.co.uk
Commissioning Editor, Aviation *Peter Coles*
Commissioning Editor, Military & Maritime *Henry Wilson*
Commissioning Editor, History, Local History & Family History *Rupert Harding*
Commissioning Editor, Ancient History *Philip Sidnell*

One of the leading military history publishers in the UK. Publishes non-fiction only, specializing in naval and aviation history, WW1, WW2, Napoleonic, autobiography and biography. Also publishes *Battleground* series for battlefield tourists. IMPRINTS **Leo Cooper; Frontline Books** (see entry); **Pen & Sword Aviation; Pen & Sword Family History; Pen & Sword Maritime; Pen & Sword Military; Remember When** (see entry); **Seaforth Publishing** (see entry); **Wharncliffe Books** (see entry). About 250 titles a year. Unsolicited synopses and ideas welcome; no unsolicited mss.
Ⓔ Royalties twice-yearly.

Penguin Group (UK)

A Pearson Company, 80 Strand, London WC2R 0RL
☎ 020 7010 3000   🖷 020 7010 6060
www.penguin.co.uk
Group Chairman & Chief Executive *John Makinson*
CEO Penguin UK *Peter Field*
Managing Director: Penguin *Helen Fraser*
Approx. Annual Turnover £121 million

Owned by Pearson plc. The world's best known book brand and for more than 70 years a leading publisher whose adult and children's lists include fiction, non-fiction, poetry, drama, classics, reference and special interest areas. Reprints and new work.

DIVISIONS

Penguin General Books Managing Director *Tom Weldon* Adult fiction and non-fiction is published in hardback under **Michael Joseph, Viking, Fig Tree** and **Hamish Hamilton** imprints. Paperbacks come under the **Penguin** imprint. IMPRINTS

Viking Publisher *Venetia Butterfield* Publishing Director *Tony Lacey*; **Hamish Hamilton** Publishing Director *Simon Prosser*; **Michael Joseph** Managing Director *Louise Moore* Does not accept unsolicited mss; **Fig Tree** *Juliet Annan*.

Penguin Press Managing Director *Stefan McGrath* Publishing Directors *Stuart Proffitt*, *Simon Winder* Academic adult non-fiction, reference, specialist and classics. IMPRINTS **Allen Lane; Penguin Classics; Penguin Reference** Publishing Director *Adam Freudenheim* No unsolicited mss.

Dorling Kindersley Ltd (see entry).

Frederick Warne Managing Director *Stephanie Barton* Classic children's publishing and merchandising including *Beatrix Potter™*; *Flower Fairies*; *Orlando*. **Ventura** Publisher *Stephanie Barton* Producer and packager of *Spot* titles by Eric Hill.

Ladybird (see **Dorling Kindersley Ltd**).

BBC Children's Books *Stephanie Barton* Children's imprint launched in 2004.

Puffin Managing Director *Francesca Dow* (poetry and picture books) Publishing Director *Sarah Hughes* (fiction), Editorial Directors *Louise Bolongaro* (picture books), *Kate Hayler* (characters) Leading children's paperback list, publishing in virtually all fields including fiction, non-fiction, poetry, picture books, media-related titles. No unsolicited mss; synopses and ideas welcome.

Penguin Audiobooks (see entry under *Audio Books*).

ePenguin Penguin Digital Publisher *Jeremy Ettinghausen* E-books list launched in 2001. OVERSEAS ASSOCIATES worldwide. Ⓔ Royalties twice-yearly.

⊞*Authors' update* One of the few publishers to boast a strong brand image, Penguin is in third place behind Random House and Hachette in terms of UK market share. An ambitious digitization programme is under way with 50% of the backlist scheduled to be digitized by year end. Puffin has led an 'outstanding' year for children's publishing. In contrast to some rivals, Penguin Press is still interested in serious non-fiction.

Penguin Longman
▷ **Pearson**

Pennant Books Ltd
PO Box 5675, London W1A 3FB
☎ 020 7287 1841
info@pennantbooks.com
www.pennantbooks.com
Managing Director *Cass Pennant*

Approx. Annual Turnover £250,000

Independent publisher, founded in 2005. Publishes non-fiction, biography, sport, true crime and fan culture. No fiction or children's books. 16 titles in 2007. No unsolicited mss; send synopses and ideas for books by post.
Ⓔ Royalties twice-yearly.

Pergamon
▷ **Elsevier Ltd**

Persephone Books
59 Lamb's Conduit Street, London WC1N 3NB
☎ 020 7242 9292   ⓕ 020 7242 9272
sales@persephonebooks.co.uk
www.persephonebooks.co.uk
Managing Director *Nicola Beauman*

Founded 1999. Publishes reprint fiction and non-fiction, mostly 'by women, for women and about women'. TITLES *Miss Pettigrew Lives for a Day* Winifred Watson; *Kitchen Essays* Agnes Jekyll; *The Far Cry* Emma Smith; *The Priory* Dorothy Whipple. 4 titles a year. No unsolicited material.
Ⓔ Royalties twice-yearly.

Phaidon Press Limited
Regent's Wharf, All Saints Street, London N1 9PA
☎ 020 7843 1000   ⓕ 020 7843 1010
initialandname@phaidon.com
www.phaidon.com
Group Chairman & Publisher *Richard Schlagman*
Chairman *Andrew Price*
Editorial Directors *Amanda Renshaw, Emilia Terragni*
Approx. Annual Turnover £22.2 million

Publishes quality books on the visual arts and culture, including fine art, art history, architecture, design, photography, decorative arts, music and performing arts, cookery and children's books. Also produces videos. DIVISIONS (with Editorial Heads) **Architecture and Design** *Emilia Terragni*; **Contemporary Art** Craig Garrett; **Photography** *Denise Wolff*; **Art & Academic** *David Anfam*; **Cooking/Food** *Emilia Terragni*; **Children's** *Amanda Renshaw*. About 100 titles a year. Unsolicited mss welcome but 'only a small amount of unsolicited material gets published'.
Ⓔ Royalties twice-yearly.

⊞*Authors' update* Claiming, with some justification, to have the best art and design list of any publisher. With French and German subsidiaries, Phaidon's European scope gives it added appeal here and in the US for titles that might not otherwise enter the English language market. Commended for tight management and strong marketing.

Philip's
▷ **Octopus Publishing Group**

Phillimore
> The History Press Ltd

Phoenix
> The Orion Publishing Group Limited

Piatkus Books

100 Victoria Embankment, London EC4Y 0DY
☎ 020 7911 8030   ℻ 020 8911 8100
info@littlebrown.co.uk
www.piatkus.co.uk *and* www.littlebrown.co.uk
Owner *Hachette Book Group*
CEO/Publisher *Ursula Mackenzie*
COO *David Kent*
Approx. Annual Turnover £10 million

Founded 1979 by Judy Piatkus, the company was sold to Hachette in autumn 2007. Piatkus non-fiction specializes in lifestyle, health, self-help, mind, body and spirit, popular psychology, biography and business titles. Piatkus fiction includes new novels from many of the world's bestselling international authors. DIVISIONS **Non-fiction** Editorial Director *Gill Bailey* TITLES *The Holford Low GL Diet Made Easy* Patrick Holford; *Women Who Think Too Much* Susan Nolan-Hoeksema; *Mind Body Bible* Mark Atkinson. **Fiction** Senior Editor *Emma Dunford* TITLES *High Noon* Nora Roberts; *It's in His Kiss* Julia Quinn; *Creation in Death* J.D. Robb. About 175 titles a year (70 of which are fiction). Piatkus is expanding its range of books and welcomes synopses together with first three chapters.
🄴 Royalties twice-yearly.

✳*Authors' update* Selling out to Little Brown does not seem to have dented the Piatkus enthusiasm for bright ideas. Publisher-author relations are said to be excellent.

Pica Press
> A.&C. Black Publishers Ltd

Picador
> Macmillan Publishers Ltd

Piccadilly Press

5 Castle Road, London NW1 8PR
☎ 020 7267 4492   ℻ 020 7267 4493
books@piccadillypress.co.uk
www.piccadillypress.co.uk
Publisher/Managing Director *Brenda Gardner*
Approx. Annual Turnover £1.4 million

Founded 1983. Independent publisher of children's and parental books. 30 titles in 2007. Welcomes approaches from authors 'but we would like them to know the sort of books we do. It is frustrating to get inappropriate material. They should check in their local libraries, bookshops or look at our website. We will send a catalogue (please enclose s.a.e.).' No adult or cartoon-type material.
🄴 Royalties twice-yearly.

Pictorial Presentations
> Souvenir Press Ltd

Pimlico
> The Random House Group Ltd

Pinwheel – A Division of Alligator Books Ltd

Gadd House, Arcadia Avenue, London N3 2JU
☎ 020 8371 6622   ℻ 020 8371 6633
Publishing Director *Linda Cole*

Children's non-fiction and novelty titles. Illustrated non-fiction for children from 3–12 years. Cloth and novelty books for children from 0–5 years. Unsolicited mss will not be returned.

Pitkin
> The History Press

Pluto Press Ltd

345 Archway Road, London N6 5AA
☎ 020 8348 2724   ℻ 020 8348 9133
pluto@plutobooks.com
www.plutobooks.com
Chair *Roger Van Zwanenberg*
Managing Director/Publisher *Anne Beech*

Founded 1970. Has developed a reputation for innovatory publishing in the field of non-fiction. Publishes academic and scholarly books across a range of subjects including cultural studies, politics and world affairs and more general titles on key aspects of currrent affairs. About 70 titles a year. Prospective authors are encouraged to consult the website for guidelines on submitting proposals.

Pocket Books
> Simon & Schuster UK Limited

Pocket Mountains Ltd

6 Church Wynd, Bo'ness EH51 0AN
℻ 01506 500405
info@pocketmountains.com
www.pocketmountains.com
Managing Directors *Robbie Porteous, April Simmons*
Approx. Annual Turnover £150,000

Founded in 2003 to publish a series of guidebooks to the Scottish hills and now publishing other guides covering Europe: mountaineering, walking, cycling and wildlife books. 19 titles in 2007. Unsolicited mss, synopses and ideas welcome; approach by e-mail or letter. No fiction, children's or cookery.
🄴 Royalties quarterly.

Poisoned Pen Press UK Ltd

6 Rodney Place, Clifton, Bristol BS8 4HY
☎ 0117 907 5949
info@poisonedpenpressuk.com
www.poisonedpenpressuk.com
Managing Director *Jennifer Muller*

Editorial Head *Barbara Peters*

UK subsidiary of Poisoned Pen Press US, founded in 1996. Publishes adult crime fiction and mystery books. Unsolicited mss welcome; electronic submissions only to: editor@poisonedpenpress.com
€ Royalties annually.

The Policy Press

Fourth Floor, Beacon House, Queen's Road, Bristol BS8 1QU
☎ 0117 331 4054 📠 0117 331 4093
tpp-info@bristol.ac.uk
www.policypress.org.uk
Managing Director *Alison Shaw*

Specialist social science publisher 'committed to publishing work that will impact on research, learning, policy and practice'.

Politico's Publishing

8 Artillery Row, London SW1P 1RZ
☎ 020 7798 1600 📠 020 7828 1244
sales@methuen.co.uk
www.politicospublishing.co.uk
Chairman/Managing Director *Peter Tummons*
Publishing Consultant *Alan Gordon Walker*

Founded 1998. Acquired by **Methuen Publishing Ltd** in April 2003. Publishes political books. 25 titles in 2007. Unsolicited mss, synopses and ideas welcome; telephone in the first instance.
€ Royalties twice-yearly.

Polity Press

65 Bridge Street, Cambridge CB2 1UR
☎ 01223 324315 📠 01223 461385
www.polity.co.uk

Founded 1984. Publishes archaeology and anthropology, criminology, economics, feminism, general interest, history, human geography, literature, media and cultural studies, medicine and society, philosophy, politics, psychology, religion and theology, social and political theory, sociology. Unsolicited synopses and ideas for books welcome. 'Please do not send mss.'
€ Royalties annually.

Polygon

West Newington House, 10 Newington Road, Edinburgh EH9 1QS
☎ 0131 668 4371 📠 0131 668 4466
info@birlinn.co.uk
www.birlinn.co.uk
Publishing Director *Neville Moir*
Managing Editor *Alison Rae*

Bought by **Birlinn Ltd** in 2002. Publishes literary fiction, crime fiction and poetry. About 40 titles a year. 'We do not accept unsolicited work from authors.'
€ Royalties twice-yearly.

Pont Books
▷ Gomer Press

Pop Universal
▷ Souvenir Press Ltd

Portico
▷ Anova Books

Portland Press Ltd

3rd Floor, Eagle House, 16 Procter Street, London WC1V 6NX
☎ 020 7280 4110 📠 020 7280 4169
editorial@portlandpress.com
www.portlandpress.com
Chairman *Professor John Clark*
Managing Director *Rhonda Oliver*
Managing Editor *Pauline Starley*
Approx. Annual Turnover £2.5 million

Founded 1990 to expand the publishing activities of the Biochemical Society (1911). Publishes biochemisty and medicine for graduate, postgraduate and research students. Expanding the list to include schools and general readership. TITLES *Drugs and Ergogenic Aids to Improve Sport Performance*; *The Cell Biology of Inositol Lipids and Phosphates*; *Fundamentals of Enzyme Kinetics*. 3 titles in 2007. Unsolicited mss, synopses and ideas welcome. No fiction.
€ Royalties twice-yearly.

Portobello Books

Twelve Addison Avenue, London W11 4QR
☎ 020 7605 1380 📠 020 7605 1361
mail@portobellobooks.com
www.portobellobooks.com
Owners *Sigrid Rausing, Eric Abraham, Philip Gwyn Jones*
Managing Director *David Graham*
Senior Editor *Laura Barber*
Associate Publisher *Tasja Dorkofikis*
Rights Director *Angela Rose*
Approx. Annual Turnover £500,000

Founded in 2005 by Philip Gwyn Jones, formerly publishing director of HarperCollins' literary imprint Flamingo. Owners acquired **Granta** in 2005. Publishes international literature and activist non-fiction. About 25 original titles a year. No unsolicited mss; synopses and ideas for books welcome. Approach by post to the submissions editor in the first instance. No science fiction, romance or genre fiction of any kind, no gardening, cookery, mind, body and spirit or reference.

T&AD Poyser
▷ A.&C. Black Publishers Ltd

Preface Publishing
▷ The Random House Group Ltd

Prentice Hall/Prentice Hall Business
▷ Pearson

Prestel Publishing Limited
4 Bloomsbury Place, London WC1A 2QA
☎ 020 7323 5004   🖷 020 7636 8004
sales@prestel-uk.co.uk
www.prestel.com
UK Editor *Philippa Hurd*
Editor-in-Chief (Germany) *Curt Holtz*

Founded 1924. Publishes art, architecture, photography, design, fashion, children's and general illustrated books. No fiction. Unsolicited mss, synopses and ideas welcome. Approach by post or e-mail.

Princeton University Press
▷ University Presses of California, Columbia & Princeton Ltd

Prion Books
▷ Carlton Publishing Group

Profile Books
3a Exmouth House, Pine Street, Exmouth Market, London EC1R 0JH
☎ 020 7841 6300   🖷 020 7833 3969
info@profilebooks.com
www.profilebooks.com
Publisher & Managing Director *Andrew Franklin*
Editorial Director *Stephen Brough*
Approx. Annual Turnover £9.09 million

Founded 1996. Publishes general non-fiction: history, biography, current affairs, popular science, politics, business management. Also publishers of *The Economist* books. IMPRINT **Serpent's Tail** (see entry). No unsolicited mss; phone or send preliminary letter.

✱*Authors' update* **An object lesson on how to succeed in publishing. From small beginnings not much more than a decade ago, Profile has built up an impressive non-fiction list, matched now by Serpent's Tail's innovative fiction.**

ProQuest
The Quorum, Barnwell Road, Cambridge CB5 8SW
☎ 01223 215512   🖷 01223 215513
www.proquest.co.uk
Vice President, Technology/General Manager *John Taylor*

Part of Cambridge Information Group. Provides access to and navigation of more than 125 billion digital pages of the world's scholarship, from arts and literature to science, technology and medicine. ProQuest's content is available to researchers through libraries of all types, and includes the world's largest digital newspaper archive, periodical databases comprising the output of more than 9,000 titles and covering more than 500 years.

Psychology Press (An imprint of Taylor & Francis Group, an Informa plc business)
27 Church Road, Hove BN3 2FA
☎ 020 7017 6000   🖷 020 7017 6336
www.psypress.co.uk
Managing Director *Mike Forster*
Deputy Managing Director *Rohays Perry*

Publishes psychology textbooks, monographs, professional books, tests and numerous journals which are available in both printed and online formats. Send detailed proposal plus c.v. and sample chapters. Proposal requirements and submission procedures available on the website. No poetry, fiction, travel or astrology.
£ Royalties according to contract.

Publishing House
Trinity Place, Barnstaple EX32 9HG
☎ 01271 328892   🖷 01271 328768
info@vernoncoleman.com
www.vernoncoleman.com
Managing Director *Vernon Coleman*
Publishing Director *Sue Ward*
Approx. Annual Turnover £750,000

Founded 1989. Self-publisher of fiction, health, humour, animals, politics. Over 90 books published. TITLES *Mrs Caldicot's Cabbage War*; *England our England*; *Bodypower*; *Bilbury Chronicles*; *Second Innings*; *It's Never Too Late*; *Alice's Diary*; *Food for Thought*; *Gordon is a Moron*, all by Vernon Coleman. No submissions.

Puffin
▷ Penguin Group (UK)

Pushkin Press Ltd
12 Chester Terrace, London NW1 4ND
☎ 020 7730 0750   🖷 020 7730 1341
books@pushkinpress.com
www.pushkinpress.com
Chairman *Melissa Ulfane*

Publishes novels and essays in translation drawn from the best of classic and contemporary European literature.

QED Publishing
226 City Road, London EC1V 2TT
☎ 020 7812 8600   🖷 020 7253 4370
stevee@quarto.com
www.qed-publishing.co.uk
www.qeb-publishing.com
Owner *Quarto Publishing plc*
Chairman, Quarto Group *Laurence Orbach*
Director of Co-edition Publishing, Quarto Group *Piers Spence*
Publisher *Steve Evans*
Creative Director *Zeta Davies*

Approx. Annual Turnover £2.6 million

Founded in 2003 by the Quarto Group. Publishes children's: education, reference, literacy, numeracy, fiction, geography, history, picture books, non-fiction. TITLES *QED Start Series*; *QED Learn Computing*; *QED Learn Art*; *QED You and Your Pet*; *QED Maths Club*; *QED Machines at Work*; *QED Down on the Farm*. 74 titles in 2008. No unsolicited material; initial approach by e-mail. No adult books. OVERSEAS SUBSIDIARY QEB Publishing Inc., USA.
Ⓔ Fees paid.

Quadrille Publishing Ltd

Alhambra House, 27–31 Charing Cross Road, London WC2H 0LS
Ⓣ 020 7839 7117   Ⓕ 020 7839 7118
info@quadrille.co.uk
www.quadrille.co.uk
Chairman *Sir David Cooksey*
Managing Director *Alison Cathie*
Editorial Director *Jane O'Shea*
Approx. Annual Turnover £8.09 million

Founded in 1994 with a view to producing a small list of top-quality illustrated books. Publishes non-fiction, including cookery, gardening, interior design and decoration, craft, health. TITLES *Life's Too F***ing Short* Janet Street Porter; *Maze, The Cookbook* Jason Atherton; *Urban Garden Handbook* Joe Swift; *Men's Knits* Erika Knight. 40 titles in 2007. Synopses and ideas for books welcome. No fiction or children's books.
Ⓔ Royalties twice-yearly.

Quantum
▷ Foulsham Publishers

Quartet Books

27 Goodge Street, London W1T 2LD
Ⓣ 020 7636 3992   Ⓕ 020 7637 1866
quartetbooks@easynet.co.uk
Chairman *Naim Attallah*

Founded 1972. Independent publisher. Publishes contemporary literary fiction including translations, popular culture, biography, music, history, politics and some photographic books. Unsolicited sample chapters with return postage welcome; no poetry, romance or science fiction. Submissions by disk or e-mail are not accepted.
Ⓔ Royalties twice-yearly.

✱*Authors' update* A recent relaunch promises 'fresh and exciting new material' with up to 20 titles a year.

Quarto Publishing Plc
▷ entry under UK Packagers

QUE Publishing
▷ Pearson

Quercus Publishing

21 Bloomsbury Square, London WC1A 2NS
Ⓣ 020 7291 7200   Ⓕ 020 7291 7201
firstname.surname@quercusbooks.co.uk
www.quercusbooks.co.uk
Chairman *Anthony Cheetham*
Chief Executive *Mark Smith*
Approx. Annual Turnover £8.6 million

Founded in 2005 as a commercial publisher of quality fiction and non-fiction. DIVISIONS **Trade Publishing** Editor-in-Chief *Jon Riley* TITLE *The Tenderness of Wolves*; **Children's Publishing** Editorial Head *Roisin Heycock* TITLE *The Magic Thief*; **Contract Publishing** Managing Director *Wayne Davies* TITLE *Speeches That Changed the World*; **MacLehose Press** Publisher *Christopher MacLehose* TITLE *The Girl With the Dragon Tattoo*; **Quercus Audiobooks** *Laura Palmer* TITLE *Measuring the World*. 100 titles in 2007. No unsolicited material.
Ⓔ Royalties annually.

✱*Authors' update* Company policy is to focus on new and underdeveloped writers – 'refugees from big groups who weren't high enough up the pecking order to get into the sunshine of the big market spends'. The crime list has proved the greatest success with several previously middle-ranking authors benefiting from strong marketing to emerge as bestsellers.

Quiller Press (An imprint of Quiller Publishing Ltd)

Wykey House, Wykey, Shrewsbury SY4 1JA
Ⓣ 01939 261616   Ⓕ 01939 261606
info@quillerbooks.com
www.countrybooksdirect.com
Managing Director *Andrew Johnston*

Specializes in sponsored books and publications sold through non-book trade channels as well as bookshops. Publishes architecture, biography, business and industry, collecting, cookery, DIY, gardening, guidebooks, humour, reference, sports, travel, wine and spirits. IMPRINTS **Kenilworth Press** (see entry); **The Sportsman's Press** (see entry); **Swan Hill Press** (see entry). About 10 titles a year. Unsolicited mss only if the author sees some potential for sponsorship or guaranteed sales.
Ⓔ Royalties twice-yearly.

Quince Books Limited

209 Hackney Road, London E2 8JL
Ⓣ 0774 295 0013
info@quincebooks.com
www.quincebooks.com
Editor *Sara Hibbard*

Founded 2005. Publishes commercial ficton. About 15 books a year. Welcomes submissions; send synopsis and three chapters by e-mail. £ Royalties annually.

Radcliffe Press
▷ I.B. Tauris & Co. Ltd

Radcliffe Publishing Ltd
18 Marcham Road, Abingdon OX14 1AA
☎ 01235 528820   📠 01235 528830
contact.us@radcliffemed.com
www.radcliffe-oxford.com
Managing Director *Andrew Bax*
Editorial Director *Gillian Nineham*
Approx. Annual Turnover £2 million

Founded 1987. Medical publishing house which began by specializing in books for general practice and health service management. Publishes clinical, management, health policy books, training materials, journals and CD-ROMs. 95 titles in 2007. Unsolicited mss, synopses and ideas welcome. No non-medical books.
£ Royalties twice-yearly.

Raintree
▷ Pearson (Heinemann)

The Random House Group Ltd
Random House, 20 Vauxhall Bridge Road, London SW1V 2SA
☎ 020 7840 8400   📠 020 7233 6058
enquiries@randomhouse.co.uk
www.randomhouse.co.uk
Chief Executive/Chairman *Gail Rebuck*
Deputy CEO *Ian Hudson*
Group Managing Director *Peter Bowron*
Approx. Annual Turnover £271 million (UK)

The Random House Group is the UK's leading trade publisher comprising over 40 diverse imprints in five separate substantially autonomous companies: CHA, CCV & Random House Enterprises, Ebury Publishing, Transworld (see entry) and Random House Children's Books.

CCV

Managing Director *Richard Cable*, Publisher *Dan Franklin*

IMPRINTS **Jonathan Cape** ☎ 020 7840 8576 📠 020 7233 6117 Publisher *Dan Franklin* Biography and memoirs, current affairs, fiction, history, photography, poetry, politics and travel. **Yellow Jersey Press** ☎ 020 7840 8621 📠 020 7223 6117 Editorial Director *Tristan Jones* Narrative sports books. **Harvill Secker** ☎ 020 7840 8893 📠 020 7233 6117 Publishing Director *Geoff Mulligan* Principally literary fiction with some non-fiction. Literature in translation, English literature, quality thrillers. **Chatto**

& Windus ☎ 020 7840 8745 📠 020 7233 6117 Publishing Director *Alison Samuel* Memoirs, current affairs, essays, literary fiction, history, poetry, politics, philosophy and translations. **Pimlico** ☎ 020 7840 8608 📠 020 7233 6117 Publishing Director *Rachel Cugnoni* Quality non-fiction paperbacks, specializing in history. **Vintage** ☎ 020 7840 8608 📠 020 7233 6117 Publishing Director *Rachel Cugnoni* Quality paperback fiction and non-fiction. Vintage (founded in 1990) has been described as one of the 'greatest literary success stories in recent British publishing.' **The Bodley Head** ☎ 020 7840 8553 📠 020 7233 6117 Publishing Director *Will Sulkin* History, science, current affairs, politics, biography, popular culture. **Square Peg** ☎ 020 7840 8621 📠 020 7233 6117 Editorial Director *Rosemary Davidson* Eclectic and commercial non-fiction, humour, novelties and gift books. Unsolicited mss with s.a.e.

RANDOM HOUSE ENTERPRISES

IMPRINTS **Preface Publishing** 1 Queen Anne's Gate, London SW1H 9BT ☎ 020 7840 8827 Publisher *Trevor Dolby*, Publishing Director *Rosie de Courcy* Commercial fiction and non-fiction. Unsolicited mss accepted at submissionspreface@ randomhouse.co.uk

Virgin Books (Vauxhall Bridge Road address) ☎ 020 7840 8400 📠 020 7840 8383 Chairman *Richard Cable*, Editorial *Carolyn Thorne, Edward Faulkner, Louisa Joyner* General non-fiction: biography and autobiography, memoir, popular reference, health and lifestyle, sport, travel, film and TV, music, humour, true crime. IMPRINTS **Black Lace** Erotic fiction by women for women; **Nexus** *Adam Nevill* Erotic fiction.

CHA

Managing Director *Susan Sandon*

IMPRINTS **Century** ☎ 020 7840 8569 📠 020 7233 6127 Publishing Director *Mark Booth* General fiction and non-fiction including commercial fiction, autobiography, biography, history and self-help. **William Heinemann** ☎ 020 7840 8400 📠 020 7233 6127 Publishing Director *Jason Arthur* Fiction and general non-fiction: crime, thrillers, literary fiction, translations, history, biography, science. No unsolicited mss. **Hutchinson** ☎ 020 7840 8400 📠 020 7233 7870 Publishing Director *Caroline Gascoigne* Fiction: literary and women's fiction, crime, thrillers. Non-fiction: current affairs, biography, memoirs, history, politics, travel, adventure. **Random House Books** (incorporating **Random House Business Books**) ☎ 020 7840 8451 📠 020 7233 6127 Publisher *Nigel Wilcockson* Non-fiction: social and cultural history, current affairs, popular culture, reference, business and economics. **Arrow** ⓣ 020 7840 8557 ⓕ 020 7840 6127 Publishing Director *Kate Elton* Mass-market paperback fiction and non-fiction. **Random House Audio Books** ⓣ 020 7840 8519 Commissioning Editor *Zoe Howes* (see entry under *Audio Books*)

EBURY PUBLISHING

(Vauxhall Bridge Road address) ⓣ 020 7840 8400 ⓕ 020 7840 8406

Managing Director *Fiona MacIntyre*, Publisher *Jake Lingwood*

IMPRINTS **Ebury Press** Senior Publishing Director (non-illustrated & fiction) *Hannah MacDonald*, Publishing Director (illustrated, including cookery, family reference, history, culture and gift) *Carey Smith*, Editorial Director (fiction) *Gillian Green*, Publishing Director (popular non-fiction) *Andrew Goodfellow*. **Vermilion** Publishing Director *Clare Hulton* Personal development, health, diet, relationships and parenting. **Rider** Publishing Director *Judith Kendra* Psychology, biography, current affairs, travel, history, philosophy, personal development, spirituality, the paranormal. **Time Out Guides** Senior Publishing Brand Manager *Luthfa Begum* Travel and walking guides, film, reference and lifestyle published in a unique partnership with *Time Out* Group. **BBC Books** Senior Publishing Director *Shirley Patton*.

RANDOM HOUSE CHILDREN'S BOOKS DIVISION

61–63 Uxbridge Road, London W5 5SA
ⓣ 020 8579 2652 ⓕ 020 8231 6737

Managing Director *Philippa Dickinson*, Publisher (fiction) *Annie Eaton*, Editorial Directors (fiction) *Charlie Sheppard, Kelly Hurst*, Publisher (Colour and Custom Publishing) *Fiona Macmillan*, Editorial Director (picture books) *Helen Mackenzie-Smith*, Senior Commissioning Editor (picture books) *Natascha Biebow* Picture books, fiction, poetry, non-fiction and audio cassettes under IMPRINTS **Bodley Head Children's Books**; **Corgi Children's Books**; **Doubleday Children's Books**; **Hutchinson Children's Books**; **Jonathan Cape Children's Books**; **Red Fox Children's Books**; **Tamarind Books** and **David Fickling Books** (31 Beaumont Street, Oxford OX1 2NP ⓣ 01865 339000 ⓕ 01865 339009 email dfickling@ randomhouse.co.uk www.davidficklingbooks. co.uk) Publisher *David Fickling*, Senior Editor *Bella Pearson* Children's fiction and picture books.
ⓔ Royalties twice-yearly for the most part.

✱*Authors' update* In a tight race with Hachette for biggest share of the UK market, Random is in hot putsuit of new readers via the Net. Currently 500 titles can be browsed online. Soon it will be 5,000. The Random House

widget can be added to blogs and websites and allows for secure purchase. This is marketing with a capital M and a huge encouragement to authors who may feel jittery in the age of publishing uncertainty. On the conventional front, Bodley Head has relaunched with a quality non-fiction list and, new on the street, Preface has been set up to capture readable books, fiction and non-fiction, that might otherwise escape the notice of the corporate scouts; newcomers should take heart.

Ransom Publishing Ltd

51 Southgate Street, Winchester SO23 9EH
ⓣ 01962 862307   ⓕ 05601 148881
rebecca@ransom.co.uk
www.ransom.co.uk
Marketing Director *Rebecca Pash*

Ransom's book programme is designed to develop reading skills (and an enthusiasm for reading) in everyone from young children through to adults. Many of Ransom's books are 'high interest age, low reading age'; both fact and fiction. The list includes well known authors such as David Orme, whose Boffin Boy series has proved hugely popular with reluctant readers. Another success is Dark Man. Aimed at adults with a very low reading age, the series won the 2006 Educational Resources Award. Interested in hearing from authors writing for reluctant and struggling readers. Initial enquiries by e-mail with details of the submission and three sample chapters. 'A response cannot be guaranteed but we try to respond to all submissions within a month.' Materials returned if supplied with s.a.e.

Reader's Digest Association Ltd

11 Westferry Circus, Canary Wharf, London E14 4HE
ⓣ 020 7715 8000   ⓕ 020 7715 8181
gbeditorial@readersdigest.co.uk
www.readersdigest.co.uk
Managing Director *Andrew Lynam-Smith*
Editorial Director, Books *Julian Browne*

Editorial office in the USA (see entry under *US Publishers*). Publishes health, gardening, natural history, cookery, history, DIY, computer, travel and word books. About 30 titles a year through trade and direct marketing channels.

Reaktion Books

33 Great Sutton Street, London EC1V 0DX
ⓣ 020 7253 1071   ⓕ 020 7253 1208
info@reaktionbooks.co.uk
www.reaktionbooks.co.uk
Managing Director *Michael R. Leaman*

Founded 1985. Publishes art history, architecture, animal studies, Asian studies, cultural studies, design, film, geography, history, natural history,

photography and travel literature (*not* travel guides or journals). TITLE *Boxing: A Cultural History* Kasia Boddy; SERIES **Critical Lives** TITLES *Walter Benjamin* Esther Leslie; *Georges Bataille* Stuart Kendall; **Animal** TITLES *Swan* Peter Young; *Cow* Hannah Velten; **Exposures** TITLES *Photography and Spirit* John Harvey; *Photography and Cinema* David Campany. About 45 titles a year.
£ Royalties twice-yearly.

Reardon Publishing

PO Box 919, Cheltenham GL50 9AN
☎ 01242 231800
reardon@bigfoot.com
www.reardon.co.uk
www.cotswoldbookshop.com
www.antarcticbookshop.com
Managing Editor *Nicholas Reardon*

Founded in the mid-1970s. Family-run publishing house specializing in local interest and tourism in the Cotswold area and books on Antarctica. Member of the **Outdoor Writers & Photographers Guild**. Publishes walking and driving guides. TITLES *The Cotswold Way*; *The Cotswold Way Map*; *Cotswold Walkabout*; *Cotswold Driveabout*; *The Donnington Way*; *The Haunted Cotswolds*. Also distributes for other publishers such as Ordnance Survey. 10 titles a year. Unsolicited mss, synopses and ideas welcome with return postage only.
£ Royalties twice-yearly.

Recollections
▷ Arnefold Publishing

Red Arrow Books
▷ Blackstone Publishers

Red Fox Children's Books
▷ The Random House Group Ltd

Red Squirrel Publishing

Suite 235, 77 Beak Street, London W1F 9DB
☎ 0845 123 3803
info@redsquirrelbooks.com
www.redsquirrelbooks.com
Managing Director *Henry Dillon*
Approx. Annual Turnover £300,000

Red Squirrel Publishing specializes in citizenship test study guides. Its range of Life in the UK Test study guides has helped thousands of people achieve their goal of becoming British since the first edition was published in January 2006. Now in its third edition, the original definitive guide is expressly designed to help people pass the Life in the UK Test. 2 titles in 2007. Welcomes postal submissions. No fiction, poetry or children's books.
£ Royalties annually.

Thomas Reed
▷ A.&C. Black Publishers Ltd

William Reed Directories

Broadfield Park, Crawley RH11 9RT
☎ 01293 613400   📠 01293 610322
directories@william-reed.co.uk
www.therightinfo.co.uk
Associate Publisher *Clare Dackombe*

William Reed Directories, a division of William Reed Business Media, was established in 1990. Its portfolio includes 13 titles covering the food, drink, non-food, catering, hospitality, retail and export industries, including European versions. The titles are produced as directories, market research reports, exhibition catalogues and electronic publishing.

Reed Elsevier Group plc

1–3 Strand, London WC2N 5JR
☎ 020 7930 7077
www.reedelsevier.com
Also at: 125 Park Avenue, 23rd Floor, New York, NY 10017, USA ☎ 001 212 309 5498
📠 001 212 309 5480
And: Raderweg 29, 1043 NX Amsterdam, The Netherlands
☎ 00 31 20 485 2222 📠 00 31 20 618 0325
Chief Executive Officer UK *Sir Crispin Davis*
Approx. Annual Turnover £5.4 million

Reed Elsevier Group plc is a world-leading publisher and information provider. It is owned equally by its two parent companies, Reed Elsevier Plc and Reed Elsevier NV. No unsolicited material.

Regency House Publishing Limited

Niall House, 24–26 Boulton Road, Stevenage SG1 4QX
☎ 01438 314488   📠 01438 311303
regency-house@btconnect.com
www.regencyhousepublishing.com
Managing Director *Nicolette Trodd*

Founded 1991. Publisher and packager of mass-market non-fiction. No fiction. No unsolicited material.

Remember When

47 Church Street, Barnsley S70 2AS
☎ 01226 734222   📠 01226 734438
enquiries@pen-and-sword.co.uk
www.pen-and-sword.co.uk
Commissioning Editor *Fiona Shoop*

An imprint of **Pen & Sword Books Ltd**, Remember When covers the nostalgia and antiques collecting range of titles. No unsolicited mss; synopses and idea welcome.
£ Royalties twice-yearly.

Reportage Press

26 Richmond Way, London W1Z 8LY
① 07971 461935
info@reportagepress.com
www.reportagepress.com
Chair *Rosie Whitehouse*
Managing Director *Charlotte Eagar*

Founded in May 2007 by journalists Rosie
Whitehouse and Charlotte Eagar to publish
books on foreign affairs and books set in foreign
countries. **Non-Fiction** *Rosie Whitehouse* TITLE
To the End of Hell Denise Affonço; **Fiction**
Charlotte Eagar TITLE *Under the Sun* Justin Kerr
Smiley. IMPRINT **Despatches** *Rosie Whitehouse*
TITLE *Behind the Spanish Barricades* John
Langdon-Davies. 6 titles in 2007. Unsolicited mss,
synopses and ideas welcome; send synopses by
e-mail and mss by post. No academic books or
books without a foreign theme.

Research Disclosure
▷ Kenneth Mason Publications Ltd

Richmond House Publishing Company Ltd

70–76 Bell Street, Marylebone, London NW1 6SP
① 020 7224 9666   ⑤ 020 7224 9688
sales@rhpco.co.uk
www.rhpco.co.uk
Managing Directors *Gloria Gordon, Spencer Block*

Publishes directories for the theatre and
entertainment industries. TITLES *British Theatre
Directory*; *Artistes and Agents*; *London Seating
Plan Guide*. Synopses and ideas welcome.

Rider
▷ The Random House Group Ltd

Rigby
▷ Pearson (Heinemann)

Right Way
▷ Constable & Robinson Ltd

Rising Stars UK Ltd

22 Grafton Street, London W1S 4EX
① 020 7495 6793   ⑤ 020 7495 6796
info@risingstars-uk.com
www.risingstars-uk.com
Chairman *David Fulton*
Managing Director *Andrea Carr* (andreacarr@
risingstars-uk.com)
Publisher *Camilla Erskine* (camillaerskine@
risingstars-uk.com)
Approx. Annual Turnover £4 million

Founded 2002. Educational publisher of books,
readers and software for schools. Winner of
the ERA Supplier of the Year award for 2007.
Unsolicited mss, synopses and ideas welcome;
approach by e-mail.

✱*Authors' update* A 'lively publishing
programme' and a 'clear understanding of its
schools market' persuaded the Independent
Publishers Guild to name Rising Stars as 2008
Education Publisher of the Year.

RMEP
▷ SCM – Canterbury Press Ltd

Robinson
▷ Constable & Robinson Ltd

Robson Books
▷ Anova Books

Rodale
▷ Macmillan Publishers Ltd

RotoVision

Sheridan House, 114 Western Road, Hove
BN3 1DD
① 01273 727268   ⑤ 01273 727269
sales@rotovision.com
www.rotovision.com
www.myspace.com/rotovision
Managing Director *Piers Spence*
Publisher *April Sankey*

Founded 1971. International publishers of graphic
design, photography, digital design, advertising,
film, product design and packaging design books.
TITLES include *World's Top Photographers: Nudes*;
What Is Graphic Design For?; *Designs of the
Times*; *First Steps in Digital Design*; *500 Digital
Photography Hints, Tips and Techniques*; *Print
and Production Techniques for Brochures and
Catalogues*. 35 titles a year. No unsolicited mss;
written synopses and ideas welcome. No phone
calls, please. No academic or fiction.
€ Flat fee paid.

Roundhouse Group

Maritime House, Basin Road North, Hove
BN41 1WR
① 01273 704962   ⑤ 01273 704963
sales@roundhousegroup.co.uk
www.roundhousegroup.co.uk
President *Geoff Cowen*
Editorial Head *Alan Goodworth*
Director, Roundabout (Children's Book Division)
Bill Gills

Founded 1991. Represents and distributes a broad
range of non-fiction publishers from the UK,
Europe, USA, Canada, Australia, and elsewhere.
No unsolicited mss.
€ Royalties twice-yearly.

Routledge (An imprint of Taylor & Francis Group, an Informa plc business)

2 Park Square, Milton Park, Abingdon OX14 4RN
① 020 7017 6000   ⑤ 020 7017 6699
www.routledge.com

CEO *Roger Horton*
Managing Director *Jeremy North*
Publishing Directors *Claire L'Enfant, Alan Jarvis*

Routledge is an international academic imprint publishing books primarily for university students, researchers, academics and professionals. Most Routledge authors are academics with specialist knowledge of their subject. Publishes everything from core text books to research monographs and journals in the following subject areas: humanities, including anthropology, archaeology, classical studies, communications and media studies, cultural studies, history, language and literature, music, philosophy, religious studies, theatre studies and sports studies; social sciences, including geography, politics, military and strategic studies, sociology and urban studies; business, management, law and economics; education; Asian and Middle Eastern studies. Over 2,000 titles a year. Send a detailed proposal plus c.v. and sample chapters. Proposal requirements and submission procedures can be found on the website. No poetry, fiction, travel or astrology.
£ Royalties annually or twice-yearly, according to contract.

RYA (Royal Yachting Association)

RYA House, Ensign Way, Hamble, Southampton SO31 4YA
☎ 023 8060 4100   ⊞ 023 8060 4299
phil.williamsellis@rya.org.uk
www.rya.org.uk
Chief Executive *Rod Carr*
Publishing Editor/Publications Manager *Phil Williams-Ellis*
Approx. Annual Turnover £12million+

The Royal Yachting Association, the UK governing body representing sailing, windsurfing, motorboating, powerboat racing both at sea and on inland waters, was founded in 1875. Publishes sports instruction books relating to specific training courses, technical, legal and boating advice, all written by experts in their field. TITLES *RYA Weather Handbook*; *RYA Navigation Exercises*; *RYA Powerboat Handbook*; *RYA Go Sailing!* 10 titles a year. No unsolicited mss; synopses and ideas for books welcome. Approach by e-mail or post in the first instance.
£ Royalties twice-yearly.

Ryelands
▷ Halsgrove

Ryland Peters & Small Limited

20–21 Jockey's Fields, London WC1R 5BW
☎ 020 7025 2200   ⊞ 020 7025 2201
info@rps.co.uk
www.rylandpeters.com
Managing Director *David Peters*

Publishing Director *Alison Starling*
Art Director *Leslie Harrington*

Founded 1996. Publishes highly illustrated lifestyle books aimed at an international market, covering home and garden, food and drink, babies and kids and gift. No fiction. Acquired **Cico Books** in 2006. No unsolicited mss; synopses and ideas welcome.
£ Royalties twice-yearly.

Sage Publications

1 Oliver's Yard, 55 City Road, London EC1Y 1SP
☎ 020 7324 8500   ⊞ 020 7324 8600
info@sagepub.co.uk
www.sagepub.co.uk
Managing Director *Stephen Barr*
Editorial Director *Ziyad Marar*

Founded 1971. Publishes academic books and journals in humanities and the social sciences, education and science, technology and medicine. Bought academic and professional books publisher Paul Chapman Publishing Ltd in 1998 and Lucky Duck Publishing in 2004.
£ Royalties twice-yearly.

St Pauls Publishing

187 Battersea Bridge Road, London SW11 3AS
☎ 020 7978 4300   ⊞ 020 7978 4370
editions@stpauls.org.uk
www.stpauls.ie
Publisher *Fr. Celso Dodilano*

Publishing division of the Society of St Paul. Began publishing in 1914 but activities were fairly limited until around 1948. Publishes religious material mainly: theology, scripture, catechetics, prayer books and biography. About 20 titles a year. Unsolicited mss, synopses and ideas welcome with s.a.e.

Salamander
▷ Anova Books

SAM Publishing
▷ Pearson

Sandstone Press Ltd

PO Box 5725, One High Street, Dingwall IV15 9WJ
☎ 01349 862583   ⊞ 01349 862583
info@sandstonepress.com
www.sandstonepress.com
Managing Director *Robert Davidson*

Founded 2002. Publishes education and non-fiction. 8 titles in 2007. No unsolicited mss; synopses and ideas for books welcome. Initial approach by e-mail, please.

Saqi Books

26 Westbourne Grove, London W2 5RH
☎ 020 7221 9347   ⊞ 020 7229 7492

enquiries@saqibooks.com
www.saqibooks.com
Chairman/Managing Director *André Gaspard*
Commissioning Editor *Lara Frankena*

Founded 1981, initially as a specialist publisher of books on the Middle East and Arab world but now includes Central Asia, South Asia and European fiction. Publishes non-fiction – academic and illustrated. About 20 titles a year. Welcomes unsolicited material; approach by post or e-mail.
Ⓔ Royalties annually.

⊛*Authors' update* A blunt warning from the managing editor that unsolicited manuscripts are not returned. Saqi is not alone in this. Authors should retain copies of everything they send out.

Saunders
▷ Elsevier Ltd

Savitri Books Ltd
▷ entry under UK Packagers

SB Publications
14 Bishopstone Road, Seaford BN25 2UB
Ⓣ 01323 893498   Ⓕ 01323 893860
sbpublications@tiscali.co.uk
www.sbpublications.co.uk
Owner *Mrs Lindsay Woods*

Founded 1987. Specializes in local history, including themes illustrated by old picture postcards and photographs; also travel, guides (town, walking) and railways. TITLES *Sussex As She Wus Spoke*; *The Neat and Nippy Guide to Brighton*; *Pre-Raphaelite Trail in Sussex*. Also provides marketing and distribution services for local authors. 12 titles a year.
Ⓔ Royalties annually.

Sceptre
▷ Hachette Livre UK

Scholastic Ltd
Euston House, 24 Eversholt Street, London NW1 1DB
Ⓣ 020 7756 7756   Ⓕ 020 7756 7795
www.scholastic.co.uk
Chairman *M.R. Robinson*
Group Managing Director *Kate Wilson*
Approx. Annual Turnover £32.3 million

Founded 1964. Owned by US parent company, Scholastic Inc. who acquired **Chicken House Publishing** in May 2005. Publishes children's fiction and non-fiction and education for primary schools.

DIVISIONS

Scholastic Children's Books Managing Director *Elaine McQuade* (Euston House address)

IMPRINT **Scholastic** TITLES *Horrible Histories* Terry Deary, illus. Martin Brown; *His Dark Materials Trilogy* Philip Pullman; *Mortal Engines* Philip Reeve.

Educational Publishing (Villiers House, Clarendon Avenue, Leamington Spa CV32 5PR Ⓣ 01926 887799) Managing Director *Denise Cripps* Professional books and classroom materials for primary teachers, plus magazines such as *Child Education Plus*, *Junior Education Plus*, *Junior Focus*, *Nursery Education*, *Literacy Time Plus*.

Scholastic Book Clubs Managing Director *To be appointed* Windrush Park, Witney, Oxford OX29 0YT Ⓣ 01993 893456 Ⓕ 01993 776813 'SBC is the UK's number one Schools Book Club, providing the best books at the best prices and supports teachers in the process.'

School Book Fairs (Dolomite Avenue, Coventry Business Park, Coventry CV5 6UE) Managing Director *To be appointed.* The Book Fair Division sells directly to children, parents and teachers in schools through 25,000 week-long book events held in schools throughout the UK.
Ⓔ Royalties twice-yearly.

⊛*Authors' update* After a difficult year, Scholastic is back on track with strong growth in children's books and diversification into picture books. On the education side, teachers can now buy direct online via the Scholastic website.

SCM – Canterbury Press Ltd
13–17 Long Lane, London EC1A 9PN
Ⓣ 020 7776 7540   Ⓕ 020 7776 7556
admin@scm-canterburypress.co.uk
www.scm-canterburypress.co.uk
Publishing Director *Christine Smith*
Commissioning Editors *Natalie K. Watson (SCM)*, *Valerie Bingham (RMEP)*
Approx. Annual Turnover £2.6 million

Part of **Hymns Ancient & Modern Ltd** (see entry). Has three publishing imprints: **SCM Press** Text and reference books for the study of theology, philosophy of religion, ethics and religious studies; **Canterbury Press** Religious titles for the general market, liturgy, spirituality and church resources. **RMEP** Classroom resources for religious education in schools. AUTHORS include Rowan Williams, Ronald Blythe, Helen Prejean. 120 titles a year. Refer to website for guidelines on submitting proposals.
Ⓔ Royalties annually.

⊛*Authors' update* Leading publisher of religious ideas with well-deserved reputation for fresh thinking. At SCM, 'questioning theology is the norm'.

Scottish Cultural Press/Scottish Children's Press

Unit 6, Newbattle Abbey Business Park, Newbattle Road, Dalkeith EH22 3LJ

📞 0131 660 6366   📠 0870 285 4846

info@scottishbooks.com

www.scottishbooks.com

Directors *Avril Gray, Brian Pugh*

Founded 1992. Publishes Scottish interest titles, including cultural literature, poetry, archaeology, local history. DIVISION **S.C.P. Children's Ltd** (trading as **Scottish Children's Press**) Children's fiction and non-fiction. Unsolicited mss, synopses and ideas accepted provided return postage is included, but *always* telephone before sending material, please. 'Mss sent without advance telephone call and return postage will be destroyed.'

💷 Royalties paid.

Seafarer Books Ltd

102 Redwald Road, Rendlesham, Woodbridge IP12 2TE

📞 01394 420789   📠 01394 461314

info@seafarerbooks.com

www.seafarerbooks.com

Managing Director *Patricia M. Eve*

Founded 1968. Publishes sailing titles and stories of the sea in a greatly expanded list reflecting a broader understanding of sailing narrative, maritime history and the crafts, arts and letters of the sea. AUTHORS Erskine Childers, Maldwin Drummond, Tony Groom, Tristan Lovering, Ian Tew, Peter Villiers, Ewen Southby-Tailyour. About 6 titles a year. No unsolicited mss; preliminary letter essential before making any type of submission.

💷 Royalties twice-yearly.

Seaforth Publishing (an imprint of Pen & Sword Books)

47 Church Street, Barnsley S70 2AS

📞 01226 734222   📠 01226 734438

info@seaforthpublishing.com

www.seaforthpublishing.com

Managing Director *Charles Hewitt*

Founded 2007. Specialist maritime imprint of **Pen & Sword Books Ltd**. Publishes maritime history and general maritime books. TITLES *The Wooden Walls*; *Iron & Steel Navies*; *Sail & Traditional Craft*; *The Merchant Marine*; *Ship Modelling*. Unsolicited submissions welcome; send synopsis and sample chapter in printed format (double line spacing, single side). No fiction.

💷 Royalties twice-yearly.

SeaNeverDry Publishing

PO Box 400, Twickenham TW1 9FB

📞 020 8892 2855

orders@seaneverdrypub.com

www.seaneverdrypub.com

Chairman *B. de Vaatt*

Managing Director *D.R. Abbott*

Approx. Annual Turnover £300,000

Founded 2003. Publishes general fiction, erotica and humour. DIVISIONS **General Fiction** *S. Hytte* TITLES *Dropping in on Idi*; *The Book*; *Empire*; **Humour** *D.R. Abbott* TITLES *Fake Honesty*; *His/Her Words of Love*. Synopses and ideas for books welcome; approach by e-mail acceptable. No academic texts or science fiction.

💷 Royalties quarterly.

✳*Authors' update* **Not to be confused with vanity publishing, SeaNeverDry Publishing offers advanced technology to produce a low cost, low volume print run to test the market. If the book takes off, well, good luck; the author gets a higher than average royalty of 15%. But there may be a long stretch between hope and realization.**

Search Press Ltd

Wellwood, North Farm Road, Tunbridge Wells TN2 3DR

📞 01892 510850   📠 01892 515903

searchpress@searchpress.com

www.searchpress.com

Managing Director *Martin de la Bédoyère*

Commissioning Editor *Rosalind Dace*

Founded 1970. Publishes full-colour art, craft, needlecrafts – papermaking and papercrafts, painting on silk, art techniques and embroidery. No unsolicited mss; synopsis with sample chapter welcome.

💷 Royalties annually.

✳*Authors' update* **Widely regarded as the major independent in arts and crafts publishing, Search Press is steadily increasing its list with fifty new titles promised for this year.**

Seren

57 Nolton Street, Bridgend CF31 3AE

📞 01656 663018

general@seren-books.com

www.seren-books.com

Chairman *Richard Houdmon*

Publisher *Mick Felton*

Approx. Annual Turnover £150,000

Founded 1981 as a specialist poetry publisher but has now moved into general literary publishing with an emphasis on Wales. Publishes poetry, fiction, literary criticism, drama, biography, art, history and translations of fiction.

DIVISIONS **Poetry** *Amy Wack* AUTHORS Owen Sheers, Pascale Petit, Sheenagh Pugh, Kathryn Gray, Tiffany Atkinson, John Haynes. **Fiction** *Penny Thomas* AUTHORS Leonora Brito, Richard

Collins, Lloyd Jones, Emyr Humphreys. **Art, Literary Criticism**, History, **Translations** *Mick Felton* IMPRINT **Border Lines Biographies** TITLES *Bruce Chatwin; Dennis Potter; Mary Webb; Wilfred Owen; Raymond Williams.* About 25 titles a year. Unsolicited mss, synopses and ideas for books accepted.
Ⓔ Royalties twice-yearly or annually.

Serpent's Tail

3a Exmouth House, Pine Street, London EC1R 0JH
Ⓣ 020 7841 6300   Ⓕ 020 7833 3969
info@serpentstail.com
www.serpentstail.com
Director *Pete Ayrton*

Founded 1986. Since 2007, an imprint of **Profile Books**. Publishes fiction and non-fiction in paperback; literary and mainstream work, and work in translation. IMPRINT **Five Star**. No unsolicited mss.

Severn House Publishers

9–15 High Street, Sutton SM1 1DF
Ⓣ 020 8770 3930   Ⓕ 020 8770 3850
info@severnhouse.com
www.severnhouse.com
Chairman *Edwin Buckhalter*
Editorial *Amanda Stewart*

Founded 1974. A leader in library fiction publishing. Publishes hardback fiction: romance, science fiction, horror, fantasy, crime. Also large print and trade paperback editions of their fiction titles. About 140 titles a year. No unsolicited material. Synopses/proposals preferred through *bona fide* literary agents only. OVERSEAS ASSOCIATE Severn House Publishers Inc., New York.
Ⓔ Royalties twice-yearly.

✱*Authors' update* **Having established a strong hardback list, Severn House has now entered the paperback market with its crime and romance titles.**

Sheldon Press
▷ Society for Promoting Christian Knowledge

Sheldrake Press

188 Cavendish Road, London SW12 0DA
Ⓣ 020 8675 1767   Ⓕ 020 8675 7736
enquiries@sheldrakepress.co.uk
www.sheldrakepress.co.uk
Publisher *Simon Rigge*

Founded in 1979 as a book packager and commenced publishing under its own imprint in 1991. Publishes illustrated non-fiction: history, travel, style, cookery and stationery. TITLES *The Victorian House Book; The Shorter Mrs Beeton; The Power of Steam; The Railway Heritage of Britain; Wild Britain; Wild France; Wild Spain;*

Wild Italy; Wild Ireland; Amsterdam: Portrait of a City and Kate Greenaway stationery books. 1 title in 2007. Synopses and ideas for books welcome, but not interested in fiction.

Shepheard-Walwyn (Publishers) Ltd

15 Alder Road, London SW14 8ER
Ⓣ 020 8241 5927
books@shepheard-walwyn.co.uk
www.shepheard-walwyn.co.uk
Managing Director *Anthony Werner*
Approx. Annual Turnover £100,000

Founded 1972. 'We regard books as food for heart and mind and want to offer a wholesome diet of original ideas and fresh approaches to old subjects.' Publishes general non-fiction in four main areas: gift books in calligraphy and/or illustrated; biography, ethical economics, perennial philosophy. About 6–7 titles a year. Synopses and ideas for books welcome.
Ⓔ Royalties twice-yearly.

The Shetland Times Ltd

Gremista, Lerwick ZE1 0PX
Ⓣ 01595 693622   Ⓕ 01595 694637
publishing@shetland-times.co.uk
www.shetland-books.co.uk
Managing Director *Brian Johnston*
Publications Manager *Charlotte Black*

Founded 1872 as publishers of the local newspaper. Book publishing followed thereafter plus publication of monthly magazine, *Shetland Life*. Publishes anything with Shetland connections – local and natural history, music, crafts, maritime. Prefers material with a Shetland theme/connection. 8 titles in 2007.
Ⓔ Royalties annually.

Shire Publications Ltd

Midland House, West Way, Botley, Oxford OX2 0PH
Ⓣ 01865 727022   Ⓕ 01865 727017/727019
shire@shirebooks.co.uk
www.shirebooks.co.uk
Managing Director *Rebecca Smart*
Publisher *Nicholas Wright*

Founded 1962. Publishes original non-fiction paperbacks. Acquired by **Osprey Publishing Ltd** in 2007. 32 titles in 2008. No unsolicited material; send introductory letter with detailed outline of idea.
Ⓔ Royalties annually.

✱*Authors' update* **You don't have to live in the country to write books for Shire but it helps. With titles like** *Church Fonts, Haunted Inns* **and** *Discovering Preserved Railways* **there is a distinct rural feel to the list. Another way of putting it, to quote owner John Rotheroe, Shire specializes in 'small books on all manner of obscure** subjects', clearly a successful formula since the publisher was recently snapped up by Osprey which is promising further expansion of the list.

Short Books

3A Exmouth House, Pine Street, Exmouth Market, London EC1R 0JH
☎ 020 7833 9429   🖷 020 7833 9500
emily@shortbooks.co.uk
www.shortbooks.co.uk
Editorial Directors *Rebecca Nicolson, Aurea Carpenter*

Founded 2001. Publishes general adult and children's non-fiction and fiction; biography, history and politics, humour and children's narrative history. No unsolicited mss.

❋*Authors' update* The latest move is into fiction starting with three new titles a year.

Sidgwick & Jackson
▷ Macmillan Publishers Ltd

Sigma Press

5 Alton Road, Wilmslow SK9 5DY
☎ 01625 531035   🖷 01625 531035
info@sigmapress.co.uk
www.sigmapress.co.uk
Owners *Diana Beech, Graham Beech*
Chairman/Managing Director *Graham Beech*

Founded in 1980 as a publisher of technical books, Sigma Press now publishes mainly in the leisure area. Publishes outdoor, local heritage, adventure and biography. DIVISION **Sigma Leisure** TITLES *The Bluebird Years: Donald Campbell and the pursuit of speed* (biography); *In Search of Swallows & Amazons* (biography); *Walks in Ancient Wales*; *All-Terrain Pushchair Walks*; *Discovering Manchester*. About 10 titles a year. No unsolicited mss; synopses and ideas welcome. No poetry or novels required.
🖅 Royalties twice-yearly.

Simon & Schuster UK Limited

Africa House, 64–78 Kingsway, London WC2B 6AH
☎ 020 7316 1900   🖷 020 7316 0331
www.simonsays.co.uk
CEO/Managing Director *Ian S. Chapman*
Publishing Director *Suzanne Baboneau*
Publishers, Non-fiction *Mike Jones, Angela Herlihy*
Publishers, Fiction *Suzanne Baboneau, Kate Lyall-Grant*
Pocket Paperback Publisher *Julie Wright*
Approx. Annual Turnover £31.2 million

Founded 1986. Sister company of the leading American company. Publishes hardback (**Simon & Schuster**) and paperback (**Pocket Books**) commercial and literary fiction, media tie-ins and non-fiction, including autobiography, biography, sport, business, history, popular science, politics, health, self-help, mind, body and spirit, humour, travel and other general interest non-fiction.

DIVISIONS **Martin Books** Director *Janet Copleston* Specializes in bespoke publishing and branded editions, and the Simon & Schuster UK cookery list. **Simon & Schuster Children's Books** Publishing Director *Ingrid Selberg*, Editor (fiction) *Venetia Gosling*, Editor (picture books) *Emma Blackburn* Picture books, novelty and fiction titles. No unsolicited mss.
🖅 Royalties twice-yearly.

❋*Authors' update* Helped by one mega seller (*The Secret*) and a thriving children's division, Simon & Schuster is well placed for further expansion. A row in the US over digital rights has led to a welcome agreement with authors to allow them to claim back control over their work after sales dip below a minimum threshold.

Singing Dragon
▷ Jessica Kingsley Publishers Ltd

Smith Gryphon Ltd
▷ John Blake Publishing Ltd

Colin Smythe Ltd

38 Mill Lane, Gerrards Cross SL9 8BA
☎ 01753 886000   🖷 01753 886469
cs@colinsmythe.co.uk
www.colinsmythe.co.uk
Managing Director *Colin Smythe*
Approx. Annual Turnover £3.45 million

Founded 1966. Publishes Anglo-Irish literature, drama; criticism and Irish history. Also acts as agent for Terry Pratchett. About 5 titles a year. No unsolicited mss.
🖅 Royalties annually/twice-yearly.

Snowbooks Ltd

120 Pentonville Road, London N1 9JN
☎ 020 7837 6482   🖷 020 7837 6348
manuscripts@snowbooks.com
www.snowbooks.com
Managing Director *Emma Barnes*

Founded 2003. Independent publisher of mainly fiction and some non-fiction. 'Keen on promoting open, friendly, productive relationships with authors and retailers.' TITLES *Adept* Robert Finn; *Needle in the Blood* Sarah Bower; *The Idle Thoughts of an Idle Fellow* J.K. Jerome; *Boxing Fitness* Ian Oliver; *Monster Trilogy* David Wellington; *City Cycling* Richard Ballantine; *Plotting for Beginners* Sue Hepworth and Jane Linfoot; *The Other Eden* and *Sand Daughter* Sarah Bryant. 15–30 titles a year. Unsolicited mss 'very welcome', by e-mail only. Read the submission guidelines in the authors' section of the website first: www.snowbooks.com/authors.html

Ⓔ Royalties are payable on *net* receipts.

✱*Authors' update* A welcome indication that ideals and good business sense can go together. There are only two conditions for taking on a book: an editor must like it and believe it will sell. Snowbooks has already chalked up a 'Nibbie' award as Small Publisher of the Year and in 2008 was named Trade Publisher of the Year by the Independent Publishers Guild.

Society for Promoting Christian Knowledge (SPCK)

36 Causton Street, London SW1P 4ST
Ⓣ 020 7592 3900   Ⓕ 020 7592 3939
www.spck.org.uk
www.sheldonpress.co.uk
Chief Executive Officer *Simon Kingston*

Founded 1698, SPCK is the third oldest publisher in the country. IMPRINTS **SPCK** Publishing Director *Joanna Moriarty* Theology, academic, liturgy, prayer, spirituality, biblical studies, educational resources, mission, pastoral care, gospel and culture, worldwide. **Sheldon Press** Editor *Fiona Marshall* Popular medicine, health, self-help, psychology. **Azure** Senior Editor *Alison Barr* General spirituality.
Ⓔ Royalties annually.

✱*Authors' update* Theology with a liberal agenda which may well be the reason for the widely publicized split between publishing and the more orthodox bookshop management.

Sophia Books
▷ Rudolf Steiner Press

Southwater
▷ Anness Publishing Ltd

Southwood Books
▷ Catnip Publishing Ltd

Souvenir Press Ltd

43 Great Russell Street, London WC1B 3PD
Ⓣ 020 7580 9307/7637 5711   Ⓕ 020 7580 5064
souvenirpress@ukonline.co.uk
Chairman/Managing Director *Ernest Hecht*

Independent publishing house. Founded 1951. Publishes academic and scholarly, animal care and breeding, antiques and collecting, archaeology, autobiography and biography, business and industry, children's, cookery, crafts and hobbies, crime, educational, fiction, gardening, health and beauty, history and antiquarian, humour, illustrated and fine editions, magic and the occult, medical, military, music, natural history, philosophy, poetry, psychology, religious, sociology, sports, theatre and women's studies. Souvenir's Human Horizons series for

the disabled and their carers is one of the most pre-eminent in its field and recently celebrated 33 years of publishing for the disabled.

IMPRINTS/SERIES Condor; Human Horizons; Independent Voices; Pictorial Presentations; Pop Universal; The Story-Tellers. TITLES *Beckett before Beckett* Brigitte Le Juez; *Desperadoes* Ron Hansen; *Trans Fats: The Time Bomb in Your Food* Maggie Stanfield; *Fishing's Best Short Stories* ed. Paul D. Staudohar; *Knickers: An Intimate Appraisal* Rosemary Hawthorne; *Guerilla Warfare* Che Guevara; *Beyond Stammering* David McGuire. About 55 titles a year. Unsolicited mss considered but initial letter of enquiry and outline always required in the first instance.
Ⓔ Royalties twice-yearly.

✱*Authors' update* One of the last of the great independents, Souvenir covers every subject which happens to appeal to Ernest Hecht. But authors should not expect sentimentality. Commercial potential is the first priority.

SPCK
▷ Society for Promoting Christian Knowledge

Special Interest Model Books Ltd

PO Box 327, Poole BH15 2RG
Ⓣ 01202 649930   Ⓕ 01202 649950
chrlloyd@globalnet.co.uk
www.specialinterestmodelbooks.co.uk
Contact *Chris Lloyd*
Commissioning Editor *To be appointed*

Publishes aviation, model engineering, leisure and hobbies, modelling, electronics, wine and beer making. Send synopses rather than completed mss.
Ⓔ Royalties twice-yearly.

Speechmark Publishing Ltd

70 Alston Drive, Bradwell Abbey, Milton Keynes MK13 9HG
Ⓣ 01908 326944   Ⓕ 01908 326960
info@speechmark.net
www.speechmark.net
Managing Director *Catherine McAllister*
Publishing Manager *Sue Christelow*

Speechmark publishes practical books, games and *ColorCards*® for health, education and special needs practitioners and students. Unsolicited mss, synopses and ideas welcome. Approach by letter or e-mail.
Ⓔ Royalties twice-yearly.

Spellmount

Cirencester Road, Chalford GL6 8PE
Ⓣ 01285 760000   Ⓕ 01285 760076
info@thehistorypress.co.uk
www.thehistorypress.com
Publisher *Shaun Barrington*

Approx. Annual Turnover £500,000

Founded in 1983 and now an imprint of **The History Press Ltd**. Publishes history and military history. About 75 titles a year. Synopses and ideas for books in these specialist fields welcome. Enclose return postage.

Sphere
▷ Little, Brown Book Group UK

SportsBooks Limited

PO Box 422, Cheltenham GL50 2YN
☎ 01242 256755   🖷 01242 254694
randall@sportsbooks.ltd.uk
www.sportsbooks.ltd.uk
Chairman & Managing Director *Randall Northam*
Approx. Annual Turnover £150,000

Founded 1995. Publishes sports books including biographies, statistical and practical. TITLES include *Memories of George Best*; *Ultimate Goals*; *The Irish Uprising*; *Stan Greenberg's Olympic Almanack*; *George Lohmann*; *Growing up with Subbuteo*; *Fitba Gallimaufry*. IMPRINT **BMM**. 15 titles in 2008. No unsolicited mss. Synopses and ideas welcome. Approach by letter or e-mail.

The Sportsman's Press (An imprint of Quiller Publishing Ltd)

Wykey House, Wykey, Shrewsbury SY4 1JA
☎ 01939 261616   🖷 01939 261606
info@quillerbooks.com
www.countrybooksdirect.com
Managing Director *Andrew Johnston*

Specializes in books on all country subjects and general sports including fishing, fencing, shooting, equestrian, cookery, gunmaking and wildlife art. About 5 titles a year.
Ⓔ Royalties twice-yearly.

Springer-Verlag London Ltd

Ashbourne House, The Guildway, Old Portsmouth Road, Guildford GU3 1LP
☎ 01483 734666   🖷 01483 734411
www.springer.com
General Manager *Beverley Ford*
Approx. Annual Turnover £5 million

The UK subsidiary of Springer Science & Business Media. Publishes science, technical and medical books and journals. Specializes in computing, engineering, medicine, mathematics, chemistry, biosciences. All UK published books are sold through Springer's German and US companies as well as in the UK. Globally: 4,000 books and 1,250 journals; UK: 150 titles plus 27 journals. Not interested in social sciences, fiction or school books but academic and professional science mss or synopses welcome.
Ⓔ Royalties annually.

Springhill
▷ How To Books Ltd

Spruce
▷ Octopus Publishing Group

Square Peg
▷ The Random House Group Ltd

Stadia
▷ The History Press Ltd

Stainer & Bell Ltd

PO Box 110, 23 Gruneisen Road, London N3 1DZ
☎ 020 8343 3303   🖷 020 8343 3024
post@stainer.co.uk
www.stainer.co.uk
Managing Directors *Carol Y. Wakefield, Keith M. Wakefield*
Publishing Director *Nicholas Williams*
Approx. Annual Turnover £793,000

Founded 1907 to publish sheet music. Publishes music and religious subjects related to hymnody. Unsolicited synopses/ideas for books welcome. Send letter enclosing brief précis.
Ⓔ Royalties annually.

The Stationery Office
▷ TSO

Rudolf Steiner Press

Hillside House, The Square, Forest Row RH18 5ES
☎ 01342 824433   🖷 01342 826437
office@rudolfsteinerpress.com
www.rudolfsteinerpress.com
Chairman *Mr P. Martyn*
Manager *Mr S. Gulbekian*
Approx. Annual Turnover £170,000

Founded in 1925 to publish the work of Rudolf Steiner and related materials. Publishes non-fiction: spirituality and philosophy. IMPRINT **Sophia Books** *S. Gulbekian* TITLES *From Stress to Serenity*; *Homemaking as a Social Art*. 15 titles in 2007. No unsolicited material.
Ⓔ Royalties annually.

The Story-Tellers
▷ Souvenir Press Ltd

Straightline Publishing Ltd

29 Main Street, Bothwell G71 8RD
☎ 01698 853000   🖷 01698 854208
Director *Patrick Bellew*
Editor *James Hunt*

Founded 1989. Publishes magazines and directories – trade and technical – books of local interest. TITLES *Wired In*; *enterprising*. No unsolicited material.
Ⓔ Royalties annually.

Studymates Limited

PO Box 225, Abergele, Conwy LL18 9AY
☎ 01745 832863   🖷 01745 826606
info@studymates.co.uk
www.studymates.co.uk
Managing Editor *Graham Lawler, MA*

Authors need to be practising or recently-retired teachers/lecturers, or published authors for the writers' guides. Ideas in writing first; unsolicited mss will not be considered. New list of business books planned under the IMPRINT **Studymates Professional**, including ABER –personal development & self-help books, but again Studymates will not accept unsolicited mss. 'Please approach in writing in the first instance and outline how your idea will complement the current list.'

Summersdale Publishers Ltd

46 West Street, Chichester PO19 1RP
☎ 01243 771107   🖷 01243 786300
submissions@summersdale.com
www.summersdale.com
Owners/Directors *Stewart Ferris,
Alastair Williams*
Editorial Director *Jennifer Barclay*
Approx. Annual Turnover £2 million

Founded 1990. Publishes travel literature, self-help, humour and gift books, general non fiction. TITLES *Finding Normal; How to Get Published; A Year in the Scheisse; Essential Questions to Ask When Buying a Property in Spain; The Cook, the Rat and the Heretic: Living in the Shadow of Rennes-le-Chateau.* 75 titles a year. No unsolicited mss; synopsis and ideas welcome by e-mail.
Ⓔ Royalties paid.

Sussex Academic Press

PO Box 139, Eastbourne BN24 9BP
☎ 01323 479220   🖷 01323 478185
edit@sussex-academic.co.uk
www.sussex-academic.co.uk
Managing Director *Anthony Grahame*
Approx. Annual Turnover £310,000

Founded 1994. Academic publisher. DIVISION/ IMPRINT **Sussex Academic** TITLES *Chinese Religions: Beliefs & Practices; Picture Imperfect: Photography and Eugenics, 1879-1940.* **Alpha Press** TITLE *Watercraft on World Coins, Volume I: Europe 1800-2006.* 45 titles in 2007. No unsolicited material; send letter of inquiry in the first instance.
Ⓔ Royalties annually.

Sutton
▷ The History Press Ltd

Swan Hill Press (An imprint of Quiller Publishing Ltd)

Wykey House, Wykey, Shrewsbury SY4 1JA
☎ 01939 261616   🖷 01939 261606
info@quillerbooks.com
www.countrybooksdirect.com
Managing Director *Andrew Johnston*

Specializes in practical books on all country and field sports activities. Publishes books on fishing, shooting, gundog training, falconry, equestrian, deer, cookery, wildlife art. About 30 titles a year. Unsolicited mss must include s.a.e.
Ⓔ Royalties twice-yearly.

Sweet & Maxwell Group

100 Avenue Road, London NW3 3PF
☎ 020 7393 7000   🖷 020 7393 7010
sweetandmaxwellcustomerservices@thomson.com
www.sweetandmaxwell.thomson.com
Managing Director *Peter Lake*

Founded 1799. Part of the Thomson Corporation. Publishes materials in all media; looseleaf works, journals, law reports, CD-ROMs and online. The legal and professional list is varied and contains academic titles as well as treatises and reference works in the legal and related professional fields.

IMPRINTS **Sweet & Maxwell; W. Green (Scotland); Incomes Data Services; Thomson Round Hall (Ireland).** Over 1,100 products, including 180 looseleafs, 80 periodicals, more than 40 digital products and online information services and 200 new titles each year. Ideas welcome. Writers with legal/professional projects in mind are advised to contact the Legal Business Unit at the earliest possible stage in order to lay the groundwork for best design, production and marketing of a project.
Ⓔ Royalties and fees according to contract.

Tamarind Books
▷ The Random House Group Ltd

Tango Books Ltd

PO Box 32595, London W4 5YD
☎ 020 8996 9970   🖷 020 8996 9977
sales@tangobooks.co.uk
www.tangobooks.co.uk
Directors *David Fielder, Sheri Safran*
Approx. Annual Turnover £1.1 million

Founded 1981. Children's books publisher with international co-edition potential: pop-ups, three-dimensional, novelty, picture and board books; 1,000 words maximum. About 20 titles a year. Preliminary letter or e-mail with sample material. Material posted must include s.a.e. for return.
Ⓔ Primarily, fees paid.

Taschen UK

Fifth Floor, 1 Heathcock Court, 415 Strand, London WC2R 0NS
☎ 020 7845 8585   ⊞ 020 7836 3696
www.taschen.com

UK office of the German photographic, art and architecture publisher. Editorial office in Cologne (see entry under *European Publishers*).

I.B.Tauris & Co. Ltd

6 Salem Road, London W2 4BU
☎ 020 7243 1225   ⊞ 020 7243 1226
mail@ibtauris.com
www.ibtauris.com
Chairman/Publisher *Iradj Bagherzade*
Managing Director *Jonathan McDonnell*

Founded 1984. Independent publisher. Publishes general non-fiction and academic in the fields of international relations, religion, current affairs, history, politics, cultural, media and film studies, Middle East studies. Distributes **Philip Wilson Publishers** and The Federal Trust worldwide.
IMPRINTS **Tauris Parke Books** Illustrated books on architecture, travel, design and culture.
Tauris Parke Paperbacks Trade titles, including history, travel and biography. **British Academic Press** and **Tauris Academic Studies** Academic monographs. **Radcliffe Press** Colonial history and biography. Unsolicited synopses and book proposals welcome.
£ Royalties twice-yearly.

⊞*Authors' update* Imprint Radcliffe Press may require authors to contribute towards costs of publication.

Taylor & Francis (An imprint of Taylor & Francis Group, an Informa plc business)

2 Park Square, Milton Park, Abingdon OX14 4RN
☎ 020 7017 6000   ⊞ 020 7017 6699
www.tandf.co.uk
Managing Director (UK) *Jeremy North*
President (US) *Emmett Dages*
Vice President (US) *John Lavender*

Specializes in scientific and technical books for university students and academics in the following subjects: built environment, architecture, planning and civil engineering; biology, including genetics, plant architecture, molecular biology and biotechnology. About 600 titles a year. Send detailed proposal plus c.v. and sample chapters. Proposal requirements and submission procedures can be found on the website. No poetry, fiction, travel or astrology.
£ Royalties annually.

⊞*Authors' update* One of those companies that barely registers with authors until they look at some of the famous imprints that are gathered under the corporate umbrella. Very much

into higher education and reference. Further expansion is predicted, particularly in the US market.

Teach Yourself
▷ Hachette Livre UK

Telegram

26 Westbourne Grove, London W2 5RH
☎ 020 7229 2911   ⊞ 020 7229 7492
info@telegrambooks.com
www.telegrambooks.com
Chairman/Managing Director *André Gaspard*
Commissioning Editor *Rebecca O'Connor*

Independent publisher of literary fiction from around the world. 15 titles a year. Welcomes unsolicited material; approach by post or e-mail.
£ Royalties annually.

Telegraph Books

111 Buckingham Palace Road, London SW1W 0DT
☎ 020 7931 2887   ⊞ 020 7931 2929
www.books.telegraph.co.uk
Owner *Telegraph Media Group Ltd*
Publisher *Morven Knowles*
Approx. Annual Turnover £3 million

Concentrates on Telegraph branded books in association/collaboration with other publishers. Publishes general non-fiction: reference, personal finance, health, gardening, cookery, heritage, guides, humour, puzzles and games. Mainly interested in books if a Telegraph link exists. About 50 titles a year. No unsolicited material.
£ Royalties twice-yearly.

Templar Publishing

The Granary, North Street, Dorking RH4 1DN
☎ 01306 876361   ⊞ 01306 889097
editorial@templarco.co.uk
www.templarco.co.uk
Managing Director/Editorial Head *Amanda Wood*
Approx. Annual Turnover £18 million

Founded 1981. A division of The Templar Company plc. Publisher and packager of quality novelty and gift books, picture books, children's illustrated non-fiction and fiction for the 8 to 12 age range. 100 titles a year. Synopses and ideas for books welcome. 'We are particularly interested in picture book mss and ideas for new novelty concepts.'
£ Royalties by arrangement.

⊞*Authors' update* Winner of the Independent Publisher of the Year award, Templar has been praised for its 'Ology' series of books and their accompanying website, said to be 'polished and professional.'

Temple Lodge Publishing Ltd

Hillside House, The Square, Forest Row RH18 5ES
☎ 01342 824000
office@templelodge.com
www.templelodge.com
Chairman *Mr R. Pauli*
Chief Editor *Mr S. Gulbekian*
Approx. Annual Turnover £65,000

Founded in 1990 to develop the work of Rudolf
Steiner (see also **Rudolf Steiner Press**). Publishes
non-fiction: mind, body and spirit, current
affairs, health and therapy. 12 titles in 2007. No
unsolicited material; send initial letter of enquiry.
No poetry or fiction.
€ Royalties annually.

Tempus
▷ The History Press Ltd

Thames and Hudson Ltd

181A High Holborn, London WC1V 7QX
☎ 020 7845 5000   🖷 020 7845 5050
mail@thameshudson.co.uk
www.thamesandhudson.com
Chairman *Thomas Neurath*
Managing Director *Jamie Camplin*
Approx. Annual Turnover £25.7 million

Publishes art, archaeology, architecture and
design, biography, fashion, garden and landscape
design, graphics, history, illustrated and fine
editions, mythology, photography, popular
culture, style, travel and topography. SERIES
World of Art; *Hip Hotels*; *StyleCity*; *New Horizons*;
Family LifeStyle; *Earth From the Air*; *The Way
We Live*; *Photofile*. TITLES *Art Since 1900*; *Mirror
of the World: A New History of Art*; *The Artist's
Yearbook*; *Magnum Magnum*; *The Great LIFE
Photographers*; *Hockney's Pictures*; *Sensation*;
The Shock of the New; *Germaine Greer's The Boy*;
Street Sketchbook; *The Seductive Shoe*; *Fashion
Illustration Next*; *Cartier*; *Factory Records*; *XS
Green*; *The Seventy Wonders of China*; *Ancient
Rome on Five Denarii a Day*; *Freemasonry*;
Reuters – Sport in the 21st Century. 200 titles a
year. Send preliminary letter and outline before
mss.
€ Royalties twice-yearly.

Third Millennium Publishing

Third Millennium Information Ltd, 2–5 Benjamin
Street, London EC1M 5QL
☎ 020 7336 0144   🖷 020 7608 1188
info@tmiltd.com
www.tmiltd.com
Development Director *Dr Joel Burden*
Approx. Annual Turnover £1.1 million

Founded 1999 as a highly illustrated book
publisher in the international museum, heritage
and art gallery markets. Also some 'coffee-table'

titles in association with various universities,
Oxbridge colleges, independent schools and
military organizations. 10–15 titles annually.
Unsolicited material welcome; approach in
writing with c.v., book synopsis, target market
and any supporting information. No fiction.
€ Royalties annually.

Thoemmes
▷ The Continuum International Publishing Group
Limited

Thomson Learning
▷ Cengage Learning (EMEA) Ltd

Thomson Round Hall (Ireland)
▷ Sweet & Maxwell Group

Stanley Thornes (Publishers) Ltd
▷ Nelson Thornes Ltd

F.A. Thorpe Publishing

The Green, Bradgate Road, Anstey LE7 7FU
☎ 0116 236 4325   🖷 0116 234 0205
Group Chief Executive *Robert Thirlby*
Approx. Annual Turnover £6 million

Founded in 1964 to supply large print books
to libraries. Part of the Ulverscroft Group Ltd.
Publishes fiction and non-fiction large print
books. No educational, gardening or books that
would not be suitable for large print. IMPRINTS
Charnwood; **Ulverscroft**; **Linford Romance**;
Linford Mystery; **Linford Western**. No
unsolicited material.

Thorsons
▷ HarperCollins Publishers Ltd

Time Out Guides
▷ The Random House Group Ltd

Times Books
▷ HarperCollins Publishers Ltd

Timewell Press

10 Porchester Terrace, London W2 3TL
☎ 0870 760 5250   🖷 0870 760 5250
info@timewellpress.com
www.timewellpress.com
Chairman *Gerard Noel*

Founded 1997. Publishes literary international
fiction and non-fiction. AUTHORS include
Anthony Blond, Francis Fulford, Leslie Grantham,
Ralph Lownie, C.S. Nicholls, M-J. Lancaster,
Jonathan Gathorne-Hardy (J.R. Ackerley Prize,
2005), Charles Campion (Gourmand World
Cookbook Awards, 2004), Anthony Powell,
George Orwell, D.J. Taylor. 6 titles in 2007. No
unsolicited mss; send synopses and ideas with
letter and s.a.e.
€ Royalties twice-yearly.

Titan Books

144 Southwark Street, London SE1 0UP
☎ 020 7620 0200   ⓕ 020 7803 1990
editorial@titanemail.com
www.titanbooks.com
Managing Director *Nick Landau*
Editorial Director *Katy Wild*

Founded 1981. Now a leader in the publication of graphic novels and film and television tie-ins. Publishes comic books/graphic novels, film and television titles. IMPRINT **Titan Books** TITLES *Batman*; *Bones: The Official Companion*; *The Simpsons*; *Sweeney Todd: The Demon Barber of Fleet Street*; *Star Wars*; *Superman*; *The Winston Effect: The Art and History of Stan Winston Studio*; *Transformers*; *24*; *The Official Companion Season 6*; *Wallace & Gromit*. About 200 titles a year. No unsolicited fiction or children's books. Ideas for film and TV titles considered; send synopsis/outline with sample chapter. No e-mail submissions. Author guidelines available.
ⓔ Royalties twice-yearly.

Tor
▷ Macmillan Publishers Ltd

Transita Ltd

Spring Hill House, Spring Hill Road, Begbroke, Oxford OX5 1RX
☎ 01865 375794   ⓕ 01865 379162
info@transita.co.uk
www.transita.co.uk
Managing Director *Giles Lewis*
Editorial Director *Nikki Read*

Imprint of **How To Books Ltd**. Specializes in contemporary fiction for women over 45. Transita are not accepting submissions at present; check the website for updates.

Transworld Publishers, A division of the Random House Group Ltd

61–63 Uxbridge Road, London W5 5SA
☎ 020 8579 2652   ⓕ 020 8579 5479
info@transworld-publishers.co.uk
www.booksattransworld.co.uk
Managing Director *Larry Finlay*
Publisher *Bill Scott-Kerr*
Senior Publishing Director *Francesca Liversidge*
Approx. Annual Turnover £85 million

Founded 1950. A subsidiary of **Random House, Inc.**, New York, which in turn is a wholly-owned subsidiary of **Bertelsmann AG**, Germany. Publishes general fiction and non-fiction, gardening, sports and leisure. IMPRINTS **Bantam** *Francesca Liversidge*; **Bantam Press** *Sally Gaminara*; **Corgi** & **Black Swan** *Linda Evans*; **Doubleday** *Marianne Velmans*; **Channel 4 Books** *Doug Young*; **Eden** *Susanna Wadeson*; **Expert Books** *Susanna Wadeson*; **Transworld Ireland**

Eoin McHugh. AUTHORS Monica Ali, Kate Atkinson, Charlotte Bingham, Dan Brown, Bill Bryson, Lee Child, Jilly Cooper, Richard Dawkins, Ben Elton, Frederick Forsyth, David Gemmell, Tess Gerritsen, Robert Goddard, Joanne Harris, Stephen Hawking, D.G. Hessayon, John Irving, Sophie Kinsella, Anne McCaffrey, Paul McKenna, Andy McNab, John O'Farrell, Terry Pratchett, Patricia Scanlan, Gerald Seymour, Danielle Steel, Joanna Trollope, Robert Winston. No unsolicited mss. OVERSEAS ASSOCIATES Random House Australia Pty Ltd; Random House New Zealand; Random House (Pty) Ltd (South Africa).
ⓔ Royalties twice-yearly.

Travel Publishing Ltd

2nd Floor Offices, Frosbisher House, 64–66 Ebrington Street, Plymouth PL4 4AQ
☎ 01752 276660   ⓕ 01752 276699
info@travelpublishing.co.uk
www.travelpublishing.co.uk
Directors *Peter Robinson, Chris Day*

Founded in 1997 by two former directors of Reed Elsevier plc. Publishes travel guides covering places of interest, accommodation, food, drink and specialist shops in Britain and Ireland. SERIES *Hidden Places*; *Hidden Inns*; *Country Pubs*; *Country Living Rural Guides* (in conjunction with *Country Living* magazine); *Off the Motorway*. Over 40 titles in print. Welcomes unsolicited material; send letter in the first instance.
ⓔ Royalties twice-yearly.

Trentham Books Ltd

Westview House, 734 London Road, Stoke-on-Trent ST4 5NP
☎ 01782 745567/844699   ⓕ 01782 745553
tb@trentham-books.co.uk
www.trentham-books.co.uk
Directors *Dr Gillian Klein, Barbara Wiggins*
Approx. Annual Turnover £1 million

Publishes education (nursery, school to higher), social sciences, intercultural studies, gender studies and law for professional readers *not* for children and parents. Also academic and professional journals. No fiction, biography or poetry. Over 30 titles a year. Unsolicited mss, synopses and ideas welcome if relevant to their interests. Material only returned if adequate s.a.e. sent.
ⓔ Royalties annually.

Trident Press Ltd

Empire House, 175 Piccadilly, London W1J 9TB
☎ 020 7491 8770   ⓕ 020 7491 8664
admin@tridentpress.com
www.tridentpress.com
Managing Director *Peter Vine*
Approx. Annual Turnover £550,000

Founded 1997. Publishes TV tie-ins, natural history, travel, geography, underwater/marine life, history, archaeology and culture. DIVISIONS **Fiction/General Publishing** *Paula Vine*; **Natural History** *Peter Vine*. TITLES *Red Sea Sharks*; *The Elysium Testament*; *BBC Wildlife Specials*; *UAE in Focus*. No unsolicited mss; synopses and ideas welcome, particularly TV tie-ins. Approach in writing or *brief* communications by e-mail, fax or telephone.
£ Royalties annually.

Trotman

Westminster House, Kew Road, Richmond TW9 2ND
☎ 020 8334 1600   ⊞ 020 8334 1601
www.trotman.co.uk
Commercial Director *Tom Lee*
Approx. Annual Turnover £3 million

Division of **Crimson Publishing**. Publishes general careers books, higher education guides, teaching support material, employment and training resources. TITLES *Degree Course Offers*; *The Student Book*; *Which Degree?*; *Trotman Green Guides*; *Students' Money Matters*; *Careers Uncovered*; *Real Life Guides*; *Real Life Issues*; *Careers 2009*. About 60 titles a year. Unsolicited material welcome.
£ Royalties twice-yearly.

Troubador Publishing Ltd

9 De Montfort Mews, Leicester LE1 7FW
☎ 0116 255 9311   ⊞ 0116 244 9323
books@troubador.co.uk
www.troubador.co.uk
Chairman *Jane Rowland*
Managing Director *Jeremy Thompson*
Approx. Annual Turnover £500,000

Founded 1998. Publishes fiction, children's, academic, non-fiction, translation and poetry. Offers self-publishing agreement for fiction or non-fiction under its Matador imprint. DIVISIONS **Academic** *Jane Rowland*; **Fiction/Non-Fiction** *Jeremy Thompson*. IMPRINTS **Matador**; **Troubador Italian Studies**; **Communication Ethics** TITLES *Energy Beyond Oil*; *Grazia Deledda: A Biography*; *The Ethics of Teaching Practice*. 120 titles in 2007. Unsolicited mss welcome, by e-mail only. No synopses or ideas for books.
£ Royalties quarterly.

TSO (The Stationery Office)

St Crispins, Duke Street, Norwich NR3 1PD
☎ 01603 622211
www.tso.co.uk
Chief Executive *Richard Dell*
Approx. Annual Turnover £250 million

Formerly HMSO, which was founded in 1786. Became part of the private sector in October 1996. Publisher of material sponsored by Parliament, government departments and other official bodies. Also commercial publishing in the following broad categories: business and professional, environment, transport, education and law. 11,000 new titles each year with 50,000 titles in print.

Tuckwell Press
▷ Birlinn Ltd

Twenty First Century Publishers Ltd

Braunton Barn, Kiln Lane, Isfield TN22 5UE
☎ 01892 522802
TFCP@btinternet.com
www.twentyfirstcenturypublishers.com
Chairman *Fred Piechoczek*

Founded 2002. Publishes general fiction, financial thrillers and crime. Welcomes submissions by e-mail: manuscripts@connectfree.co.uk (contact *Fred Piechoczek*). Send brief synopsis in the body of the e-mail with extracts from the book in a file attachment (two to three chapters from anywhere in the book). No non-fiction or children's.
£ Royalties twice-yearly.

20/20
▷ The X Press

Two Heads
▷ Egmont Press

Two Ravens Press Ltd

Green Willow Croft, Rhiroy, Lochbroom, Ullapool IV23 2SF
☎ 01854 655307
info@tworavenspress.com
www.tworavenspress.com
Managing Director *Sharon Blackie*

Founded 2006. Publishes literary fiction, poetry, selected literary non-fiction. No commercial non-literary fiction, genre fiction (unless it's literary), anything connected with celebrities, children's books, local history, gift books. **Fiction** *Sharon Blackie* TITLES *One True Void* Dexter Petley; *The Falconer* Alice Thompson; **Poetry** *David Knowles* TITLES *The Atlantic Forest* George Gunn; *The Zig Zag Woman* Maggie Sawkins; **Non-Fiction** *Sharon Blackie* TITLE *Blazing Paddles, Dances with Waves* Brian Wilson. 12 titles in 2007. Unsolicited material welcome subject to initial e-mail enquiry with synopsis and short biog in body of e-mail. 'We do not accept unsolicited mss by mail and do not have time to respond to telephone enquiries (all information about how to submit is on our website).'
£ Royalties twice-yearly.

⊞*Authors' update* One to watch. A well received first year fiction list has attracted plaudits from the trade press. It looks as if Scotland is to be blessed with another strong independent publisher.

Ulverscroft
▷ F.A. Thorpe Publishing

University Presses of California, Columbia & Princeton Ltd

1 Oldlands Way, Bognor Regis PO22 9SA
☎ 01243 842165   🖷 01243 842167
lois@upccp.demon.co.uk

Publishes academic titles only. US-based editorial offices. Enquiries only. Over 200 titles a year.

Usborne Publishing Ltd

83–85 Saffron Hill, London EC1N 8RT
☎ 020 7430 2800   🖷 020 7430 1562
mail@usborne.co.uk
www.usborne.com
Managing Director *Peter Usborne*
Publishing Director *Jenny Tyler*
Approx. Annual Turnover £28 million

Founded 1973. Publishes non-fiction, fiction, art and activity books, puzzle books and music for children, young adults and pre-school. Some titles for parents. Up to 250 titles a year. Non-fiction books are written in-house to a specific format and therefore unsolicited mss are not normally welcome. Ideas which may be developed in-house are sometimes considered. Fiction for children will be considered. Keen to hear from new illustrators and designers.
🖳 Royalties twice-yearly.

⊞*Authors' update* From specializing in children's reference books, Usborne has moved on to 'all round' publishing with, for example, the recent appearance of a popular History of Britain. But the main growth is in fiction and baby books,

Vacation Work
▷ Crimson Publishing

Vanguard Press
▷ Pegasus Elliot Mackenzie Publishers Ltd

Ventura
▷ Penguin Group (UK)

Vermilion
▷ The Random House Group Ltd

Verso

6 Meard Street, London W1F 0EG
☎ 020 7437 3546   🖷 020 7734 0059
enquiries@verso.co.uk
www.versobooks.com
Managing Director *Giles O'Bryen*

Approx. Annual Turnover £2 million

Formerly New Left Books which grew out of the *New Left Review*. Publishes politics, history, sociology, economics, philosophy, cultural studies, feminism. TITLES *The Occupation* Patrick Cockburn; *Planet of Slums* Mike Davis; *Polemics* Alain Badiou; *The Soviet Century* Moshe Lewin; *NHS plc: The Privatisation of our Health Care* Allyson M. Pollock; *Street-Fighting Years* Tariq Ali; *Planet of Slums* Mike Davis. No unsolicited mss; synopses and ideas for books welcome. OVERSEAS OFFICE in New York.
🖳 Royalties annually.

⊞*Authors' update* Dubbed by the *Bookseller* as 'one of the most successful small independent publishers'. The publishing programme is set to expand with the backlist supported by print on demand.

Viking
▷ Penguin Group (UK)

Vintage
▷ The Random House Group Ltd

Virago Press

Little, Brown Book Group UK, 100 Victoria Embankment, London EC4Y 0DY
☎ 020 7911 8000   🖷 020 7911 8100
virago.press@littlebrown.co.uk
www.virago.co.uk
Publisher *Lennie Goodings*
Editor, Virago Modern Classics *Donna Coonan*
Editor, Virago *Elise Dillsworth*

An imprint of **Little, Brown Book Group UK**. Founded in 1973 by Carmen Callil, Virago publishes women's literature, both fiction and non-fiction. IMPRINT **Virago Modern Classics** 19th and 20th-century fiction reprints by writers such as Daphne du Maurier, Angela Carter and Edith Wharton. Virago AUTHORS include Margaret Atwood, Maya Angelou, Jennifer Belle, Waris Dirie, Sarah Dunant, Germaine Greer, Michèle Roberts, Gillian Slovo, Talitha Stevenson, Natasha Walter, Sarah Waters. 50 titles a year. 'We do not currently accept any unsolicited submissions.'
🖳 Royalties twice-yearly.

⊞*Authors' update* Still the first name in women's publishing. A team of young editors favours writers with something new to say.

Virgin Books
▷ The Random House Group Ltd

Vision

101 Southwark Street, London SE1 0JF
☎ 020 7928 5599   🖷 020 7928 8822
info@visionpaperbacks.co.uk
www.visionpaperbacks.co.uk

Managing Director *Sheena Dewan*

Founded 1996. Non-fiction publisher. IMPRINTS **Vision; Vision Paperbacks; Fusion Press**. 18 titles in 2007. Unsolicited mss, synopses and ideas for books welcome. Initial approach by e-mail, without attachments. No fiction or poetry.
Ⓔ Royalties twice-yearly.

Vision On
▷ Omnibus Press

The Vital Spark
▷ Neil Wilson Publishing Ltd

Voyager
▷ HarperCollins Publishers Ltd

University of Wales Press
10 Columbus Walk, Brigantine Place, Cardiff CF10 4UP
☎ 029 2049 6899   🖷 029 2049 6108
post@uwp.co.uk
www.uwp.co.uk
Director *Ashley Drake*
Commissioning Editor *Sarah Lewis*
Assistant Commissioning Editor *Ennis Akpinar*
Approx. Annual Turnover £1 million

Founded 1922. Primarily a publisher of academic and scholarly books in English and Welsh. Subject areas include: history, political philosophy, Welsh and Celtic Studies, Literary Studies, European Studies and Medieval Studies. SERIES include *Political Philosophy Now*; *Religion and Culture in the Middle Ages*; *Iberian and Latin American Studies*; *Gender Studies in Wales*; *Gothic Literary Studies*; *Kantian Studies*; *Histories of Europe*; *French and Francophone Studies*; *European Crime Fiction*. 85 titles a year. Unsolicited mss considered but see website for further guidance.
Ⓔ Royalties annually.

Walker Books Ltd
87 Vauxhall Walk, London SE11 5HJ
☎ 020 7793 0909   🖷 020 7587 1123
editorial@walker.co.uk
www.walkerbooks.co.uk
Managing Director *Helen McAleer*
Publisher *Jane Winterbotham*
Editors *Gill Evans* (fiction), *Deirdre McDermott* (picture books), *Caroline Royds* (early learning, non-fiction & gift books), *Denise Johnstone-Burt* (picture books, board & novelty), *Loraine Taylor* (new media, backlist and books plus)
Approx. Annual Turnover £47.7 million (Group)

Founded 1979. Publishes illustrated children's books, children's fiction and non-fiction. TITLES *Maisy* Lucy Cousins; *Where's Wally?* Martin Handford; *We're Going on a Bear Hunt* Michael Rosen and Helen Oxenbury; *Can't You Sleep,*

Little Bear? Martin Waddell and Barbara Firth; *Guess How Much I Love You* Sam McBratney and Anita Jeram; *Alex Rider* series by Anthony Horowitz. About 300 titles a year.
Ⓔ Royalties twice-yearly.

✱*Authors' update* Dedicated to 'groundbreaking books'. Noted for close working relationships with authors.

Wallflower Press
6 Market Place, London W1W 8AF
☎ 020 7436 9494
info@wallflowerpress.co.uk
www.wallflowerpress.co.uk
Editorial Director *Yoram Allon*
Chief Editor *Del Cullen*
Approx. Annual Turnover £300,000

Founded 1999. Publishes trade and academic books devoted to cinema and the moving image. SERIES *Short Cuts* Introductory undergraduate books; *Director's Cuts* Studies on significant international filmmakers; *24 Frames* Anthologies focusing on national and regional cinemas. Over 35 titles a year. Unsolicited mss, synopses and proposals welcome. No fiction or academic material not related to the moving image.
Ⓔ Royalties annually.

Ward Lock
▷ Octopus Publishing Group

Ward Lock Educational Co. Ltd
BIC Ling Kee House, 1 Christopher Road, East Grinstead RH19 3BT
☎ 01342 318980   🖷 01342 410980
wle@lingkee.com
www.wardlockeducational.com
Owner *Ling Kee (UK) Ltd*

Founded 1952. Publishes educational books (primary, middle, secondary, teaching manuals) for all subjects, specializing in maths, science, geography, reading and English and currently focusing on Key Stages 1 and 2.

Frederick Warne
▷ Penguin Group (UK)

Franklin Watts
▷ Hachette Children's Books

Wayland
▷ Hachette Children's Books

Weidenfeld & Nicolson
▷ The Orion Publishing Group Ltd

Wharncliffe Books
47 Church Street, Barnsley S70 2AS
☎ 01226 734222   🖷 01226 734438
enquiries@wharncliffebooks.co.uk
www.wharncliffebooks.co.uk

Commissioning Editor *Rupert Harding*

An imprint of **Pen & Sword Books Ltd**. Wharncliffe is the book and magazine publishing arm of an old-established, independently owned newspaper publishing and printing house. Publishes local history throughout the UK, focusing on nostalgia and old photographs. SERIES *Foul Deeds*; *Local History Companions*; *Pals*; *Aspects*. No unsolicited mss; synopses and ideas welcome.
£ Royalties twice-yearly.

Wharton
▷ Pearson

Which? Books

2 Marylebone Road, London NW1 4DF
☎ 020 7770 7000   🖷 020 7770 7660
rebecca.leach@which.co.uk
www.which.co.uk
Editorial Director *Helen Parker*

Founded 1957. Publishing arm of Which? Publishes non-fiction: information and reference on personal finance, property, divorce, consumer law. All titles offer direct value to the consumer. IMPRINT **Which? Books** TITLES *The Good Food Guide*; *Wills and Probate*; *Be Your Own Financial Adviser*. 10–12 titles a year. No unsolicited mss; send synopses and ideas only.
£ Royalties, if applicable, twice-yearly.

White Ladder
▷ Crimson Publishing

Whittet Books Ltd

South House, Yatesbury Manor, Yatesbury SN11 8YE
☎ 01672 539004   🖷 01672 555555
annabel@whittet.dircon.co.uk
www.whittetbooks.com
Owner *A. Whittet & Co.*
Managing Director *Annabel Whittet*

Publishes reference books on natural history, pets, poultry, horses, domestic livestock, horticulture, rural interest. 4 titles in 2006. Synopses and ideas for books on the subjects listed are welcome. Enclose s.a.e.
£ Royalties twice-yearly.

Wild Goose Publications

Iona Community, 4th Floor, The Savoy Centre, 140 Sauchiehall Street, Glasgow G2 3DH
☎ 0141 332 6292   🖷 0141 332 1090
admin@ionabooks.com
www.ionabooks.com
Editorial Head *Sandra Kramer*

The publishing house of the Iona Community, established in the Celtic Christian tradition of St Columba, publishes books and CDs on holistic spirituality, social justice, political and peace issues, healing, innovative approaches to worship, song and material for meditation and reflection.

Wiley Europe Ltd

The Atrium, Southern Gate, Chichester PO19 8SQ
☎ 01243 779777   🖷 01243 775878
europe@wiley.co.uk
www.wiley.com
Senior Vice President Europe & International Development *Stephen Smith*

Founded 1807. US parent company. Publishes professional, reference trade and text books, scientific, technical and biomedical.

Wiley Nautical

John Wiley and Sons Ltd., The Atrium, Southern Gate, Arundel BN19 8SQ
☎ 01243 779777
nautical@wiley.co.uk
www.wileynautical.com
Commissioning Editor *David Palmer*

Books for people who love watersports. Publishes over 100 watersports and sailing books, including 'Cruising Companions' and the series of books previously published under the Fernhurst Books imprint. Synopses and ideas welcome.

Wiley-Blackwell

9600 Garsington Road, Oxford OX4 2DQ
☎ 01865 776868   🖷 01865 714591
www.blackwellpublishing.com
Senior Vice President, Wiley-Blackwell *Eric Swanson*

Wiley-Blackwell was formed from the acquisition of Blackwell Publishing (Holdings) Ltd by John Wiley & Sons, Inc. Blackwell's publishing programme has merged with Wiley's global scientific, technical and medical business. The merged operation is now the largest of the three of John Wiley & Sons, Inc.; its other businesses are Professional/Trade and Higher Education.

✳*Authors' update* Wiley's acquisition of Blackwell Publishing has prompted concern among librarians that it will reduce choice but the two lists are largely complimentary with Blackwell strong in social sciences and humanities while Wiley has the lead in life and physical sciences. Academic authors are attracted by the American connection. Online publishing is set to increase.

Neil Wilson Publishing Ltd

Suite Ex 8 The Pentagon Centre, 44 Washington Street, Glasgow G3 8AZ
☎ 0141 221 1117   🖷 0141 221 5363
info@nwp.co.uk
www.nwp.co.uk
Managing Director/Editorial Director *Neil Wilson*
Approx. Annual Turnover £110,000

Founded 1992. Publishes Scottish interest and history, biography, humour and hillwalking, whisky; also Scottish cookery and travel. IMPRINTS **The In Pinn** Outdoor pursuits; **The Angels' Share** Whisky, drink and food-related subjects; **The Vital Spark** Humour; **NWP** History, biography, reference, true crime; **11:9** Scottish fiction. About 6 titles a year. Unsolicited mss, synopses and ideas welcome. No politics, academic or technical.
€ Royalties twice-yearly.

Philip Wilson Publishers Ltd

109 Drysdale Street, The Timber Yard, London N1 6ND
☎ 020 7033 9900   ▤ 020 7033 9922
pwilson@philip-wilson.co.uk
www.philip-wilson.co.uk
Chairman *Philip Wilson*

Founded 1976. Publishes art, art history, antiques and collectibles. About 14 titles a year.

Wimbledon Publishing Company

75–76 Blackfriars Road, London SE1 8HA
☎ 020 7401 4200   ▤ 020 7401 4201
info@wpcpress.com
www.wpcpress.com
Managing Director *Kamaljit Sood*
General Manager *Renu Sood*

Founded in 1992 as a publisher of school texts, going on to launch Anthem Press, an academic and trade imprint focusing on world history, politics, economics, international affairs, literature and culture. IMPRINTS **Anthem Press** *Tej Sood*; **WPC Education** *K. Sood*. Welcomes mss, synopses and ideas for books.
€ Royalties annually.

Windgather Press
▷ Oxbow Books

Wingedchariot Press

7 Court Royal, Eridge Road, Tunbridge Wells TN4 8HT
info@wingedchariot.com
www.wingedchariot.com
Editor *Ann Arscott*

Founded in 2005 'to bring the best of children's books in translation to the UK for the first time'. About 4 titles a year. Will consider books in their original language with illustrations. E-mail synopsis and scan of drawings. No English language books.

Wise Publications
▷ Omnibus Press

WIT Press

Ashurst Lodge, Ashurst, Southampton SO40 7AA
☎ 023 8029 3223   ▤ 023 8029 2853
marketing@witpress.com
www.witpress.com
Owner *Computational Mechanics International Ltd, Southampton*
Chairman/Managing Director/Editorial Head *Professor C.A. Brebbia*

Founded in 1980 as Computational Mechanics Publications to publish engineering analysis titles. Changed to WIT Press to reflect the increased range of publications. Publishes scientific and technical, mainly at postgraduate level and above, including architecture, environmental engineering, bioengineering. 50 titles in 2007. Unsolicited mss, synopses and ideas welcome; approach by post or e-mail. No non-scientific or technical material or lower level (school and college-level texts). OVERSEAS SUBSIDIARY Computational Mechanics, Inc., Billerica, USA.
€ Royalties annually.

Wizard Books
▷ Icon Books Ltd

Oswald Wolff Books
▷ Berg Publishers

The Women's Press

27 Goodge Street, London W1T 2LD
☎ 020 7636 3992   ▤ 020 7637 1866
www.the-womens-press.com
Chairman *Naim Attallah*
Approx. Annual Turnover £1 million

Part of the Namara Group. First title published in 1978. Publishes women only: quality fiction and non-fiction. Fiction usually has a female protagonist and a woman-centred theme. International writers and subject matter encouraged. Non-fiction: books for and about women generally; gender politics, race politics, disability, feminist theory, health and psychology, literary criticism. IMPRINTS **Women's Press Classics; Livewire Books for Teenagers** Fiction and non-fiction series for young adults. No mss without previous letter, synopsis and sample material.
€ Royalties twice-yearly.

Woodhead Publishing Ltd

Abington Hall, Abington, Cambridge CB21 6AH
☎ 01223 891358   ▤ 01223 893694
wp@woodhead-publishing.com
www.woodheadpublishing.com
Chairman *Richard Dawes*
Managing Director *Martin Woodhead*
Approx. Annual Turnover £1.8 million

Founded 1989. Publishes engineering, materials technology, textile technology, finance and investment, food technology, environmental science. DIVISION **Woodhead Publishing** *Martin Woodhead.* About 60 titles a year. Unsolicited material welcome.
Ⓔ Royalties annually.

Wordsworth Editions Ltd

8B East Street, Ware SG12 9HJ
Ⓣ 01920 465167 Ⓕ 01920 462267
enquiries@wordsworth-editions.com
www.wordsworth-editions.co.uk
Directors *E.G. Trayler, Derek Wright*
Approx. Annual Turnover £2 million

Founded 1987. Publishes classics of English and world literature, reference books, poetry, children's classics, mystery and the supernatural and special editions. About 75 titles a year. No unsolicited mss.

WPC Education
▷ Wimbledon Publishing Company

The X Press

PO Box 25694, London N17 6FP
vibes@xpress.co.uk
www.xpress.co.uk
Editorial Director *Dotun Adebayo*
Publisher *Steve Pope*

Launched in 1992 with the cult bestseller *Yardie*, The X Press publishes black-interest fiction. Also, general fiction and children's fiction. IMPRINTS **The X Press; Nia; 20/20.** About 26 titles a year. Send mss rather than synopses or ideas (enclose s.a.e.). No poetry.

⌘*Authors' update* **The X Press has had a tough year but has fought back with energetic marketing and looks set to survive albeit on a more modest level.**

Y Lolfa Cyf

Talybont, Ceredigion SY24 5AP
Ⓣ 01970 832304 Ⓕ 01970 832782
ylolfa@ylolfa.com
www.ylolfa.com/
Managing Director *Garmon Gruffudd*
General Editor *Lefi Gruffudd*
English Editor *Gwen Davies*
Approx. Annual Turnover £1 million

Founded 1967. Small company which publishes mainly in Welsh; has its own four-colour printing and binding facilities. Publishes Welsh language publications; Celtic language tutors; English language books for the Welsh and Celtic tourist trade. Expanding slowly. TITLES *My Kingdom of Books* Richard Booth; *The Welsh Learner's Dictionary* Heini Gruffudd; *The Fight for Welsh Freedom* Gwynfor Evans; *Celtic Vision* John

Merion Morris. IMPRINTS **Dinas** Part-author-subsidized imprint for non-mainstream books of Welsh interest in English and Welsh. **Alcemi** English language fiction. About 50 titles a year. Write first with synopses or ideas.
Ⓔ Royalties twice-yearly.

Yale University Press (London)

47 Bedford Square, London WC1B 3DP
Ⓣ 020 7079 4900 Ⓕ 020 7079 4901
sales@yaleup.co.uk
www.yalebooks.co.uk

Founded 1961. Owned by US parent company. Publishes art history, decorative arts, history, religion, music, politics, current affairs, biography and history of science. About 350 titles (worldwide) a year. Unsolicited mss and synopses welcome if within specialized subject areas.
Ⓔ Royalties annually.

⌘*Authors' update* **A publisher with a talent for turning out serious books which find a general market.**

Yellow Jersey Press
▷ The Random House Group Ltd

York Notes
▷ Pearson

Young Picador
▷ Macmillan Publishers Ltd

Zambezi Publishing Ltd

PO Box 221, Plymouth PL2 2YJ
Ⓣ 01752 367300 Ⓕ 01752 350453
info@zampub.com
www.zampub.com
Chair *Sasha Fenton*
Managing Director *Jan Budkowski*

Founded 1999. Publishes non-fiction: mind, body and spirit, self-help, finance and business, including the 'Simply' series. About 12 titles a year. Send biography, synopsis and sample chapter by mail. Brief e-mail communication acceptable but *no* attachments, please.
Ⓔ Royalties twice-yearly.

Zed Books Ltd

7 Cynthia Street, London N1 9JF
Ⓣ 020 7837 4014 Ⓕ 020 7833 3960
sales@zedbooks.net
www.zedbooks.co.uk
Approx. Annual Turnover £1.4 million

Founded 1976. Publishes international and Third World affairs, development studies, gender studies, environmental studies, human rights and specific area studies. No fiction, children's or poetry. DIVISIONS **Development & Environment** *Tamsine O'Riordan, Ellen McKinlay*; **Gender Studies.** TITLES *Rogue State: A Guide to the*

World's Only Superpower William Blum; *Staying Alive* Vandana Shiva; *The Autobiography of Nawal* Nawal El Saadawi. About 60 titles a year. No unsolicited mss; synopses and ideas welcome. £ Royalties annually.

Zero to Ten
▷ Evans Brothers Ltd

Irish Publishers

International Reply Coupons (IRCs) For return postage, IRCs are required (*not* UK postage stamps). These are available from post offices: letters, 60 pence; mss according to weight. To calculate postage from the Republic of Ireland to the UK go to www.anpost.ie/ AnPost and click on 'Calculate the Postage'.

An Gúm

27 Sr. Fhreidric Thuaidh, Baile Átha Cliath 1
℡ 00 353 1 889 2800   ℻ 00 353 1 873 1140
angum@forasnagaeilge.ie
www.gaeilge.ie
Senior Editor *Seosamh Ó Murchú*
Editors *Máire Nic Mheanman*

Founded 1926. Formerly the Irish language publications branch of the Department of Education and Science. Has now become part of the North/South Language Body, Foras na Gaeilge. Publishes educational, children's, young adult, music, lexicography and general. Little adult fiction or poetry. About 50 titles a year. Unsolicited mss, synopses and ideas for books welcome. Also welcomes reading copies of first and second level school textbooks with a view to translating them into the Irish language.
£ Royalties annually.

Anvil Books

45 Palmerston Road, Dublin 6
℡ 00 353 1 497 3628
Managing Director *Rena Dardis*

Founded 1964 with emphasis on Irish history and biography. IMPRINT **The Children's Press**. Publishes Irish history, biography (particularly 1916–22), folklore and children's fiction. No adult fiction, poetry, fantasy, short stories or full-colour books for children. About 4 titles a year. 'Because of promotional requirements, only books by Irish-based authors considered and only books of Irish interest.' Send synopsis only with IRCs (no UK stamps); unsolicited mss not returned.
£ Royalties annually.

Ashfield Press

30 Linden Grove, Blackrock, Co. Dublin
℡ 00 353 1 288 9808
susanwaine@ashfieldpress.com
www.ashfieldpress.com
Directors *Susan Waine, John Davey, Gerry O'Connor*

Publishes Irish-interest non-fiction books. TITLES *Lady Icarus; The Weather is a Good Storyteller; Secret Sights II*. Unsolicited mss and synopses welcome. ASSOCIATE COMPANY Ashfield Press Publishing Services.
£ Royalties twice-yearly.

Atrium
▷ Cork University Press

Attic Press
▷ Cork University Press

Blackhall Publishing Ltd

33 Carysfort Avenue, Blackrock, Co. Dublin
℡ 00 353 1 278 5090   ℻ 00 353 1 278 4446
info@blackhallpublishing.com
www.blackhallpublishing.com
Managing Director *Gerard O'Connor*
Commissioning Editor *Elizabeth Brennan*
Editor/Sales & Marketing *Eileen O'Brien*

Publishes business, management, law, social studies and life issues. Main subject areas include accounting, finance, management, HRM, social studies and law books aimed at both students and professionals in the industry. Also provides legal publishing services including Statute Law Revision, Law Reporting, Legislative Drafting and Subscription Services. 20 titles a year. Unsolicited mss and synopses welcome.
£ Royalties annually.

Bradshaw Books

Tigh Filí, Cork Arts Theatre, Carroll's Quay, Cork
℡ 00 353 21 450 9274
info@tighfili.com
www.tighfili.com
Managing Director *Maire Bradshaw*
Project Manager *Leslie Ryan*

Founded 1985. Publishes poetry, women's issues, children's books. Organizers of the annual Cork Literary Review poetry manuscript competition. SERIES *Cork Literary Review* and *Eurochild Anthology of Poetry and Art.*
Ⓔ Royalties not generally paid.

Brandon/Mount Eagle Publications
Dingle, Co. Kerry
Ⓣ 00 353 66 915 1463   Ⓕ 00 353 66 915 1234
www.brandonbooks.com
Publisher *Steve MacDonogh*
Approx. Annual Turnover €500,000

Founded in 1997. Publishes Irish fiction, biography, memoirs and other non-fiction. About 15 titles a year. Not seeking unsolicited mss.

Brookside
▷ New Island

Edmund Burke Publisher
Cloonagashel, 27 Priory Drive, Blackrock, Co. Dublin
Ⓣ 00 353 1 288 2159   Ⓕ 00 353 1 283 4080
deburca@indigo.ie
www.deburcararebooks.com
Managing Director *Eamonn De Búrca*
Publications Managers *Regina De Búrca-McAuley, William De Búrca*
Approx. Annual Turnover €320,000

Small family-run business publishing historical and topographical and fine limited-edition books relating to Ireland. TITLES *The Great Book of Irish Genealogies*, 5 vols.; *The Annals of the Four Masters*, 7 vols; *Flowers of Mayo* (illus. Wendy Walsh); *O'Curry's Manners and Customs of the Ancient Irish*; *Joyce's Irish Names of Places.* Unsolicited mss welcome. No synopses or ideas.
Ⓔ Royalties annually.

The Children's Press
▷ Anvil Books

Church of Ireland Publishing
Church of Ireland House, Church Avenue, Rathmines, Dublin 6
Ⓣ 00 353 1 492 3979   Ⓕ 00 353 1 492 4770
susan.hood@rcbdub.org
www.cip.ireland.anglican.org
Owner *Representative Church Body of the Church of Ireland*
Chairman *Dr Kenneth Milne*
Publications Officer *Dr Susan Hood*

Founded 2004. Publishes official publications of the Church of Ireland: theological, doctrinal, historical and administrative, and facilitates publication for other church-related bodies. 6 titles in 2007. No unsolicited material.

Cló Iar-Chonnachta
Indreabhán, Connemara, Galway
Ⓣ 00 353 91 593307   Ⓕ 00 353 91 593362
cic@iol.ie
www.cic.ie
General Manager *Deirdre Ní Thuathail*
Approx. Annual Turnover €500,000

Founded 1985. Publishes fiction, poetry, plays, teenage fiction and children's, mostly in Irish, including translations. Also publishes cassettes of writers reading from their own works. TITLES *Seosamh Ó hÉanaí: Nár fhágha mé bás choíche* Liam Mac Con Iomaire; *Mise an Fear Ceoil: Séamus Ennis Dialann Taistil 1942-1946* Ríonach Uí Ógáin; *Rí an Fhocail & Máirtín Ó Cadhain sa gCnocán Glas (DVD)*; *Lá an Phaoraigh* Seán Óg and Aoife de Paor; *On a Rock in the Middle of the Ocean* Lillis Ó Laoire. 12 titles in 2007.
Ⓔ Royalties annually.

The Collins Press
West Link Park, Doughcloyne, Wilton, Cork
Ⓣ 00 353 21 434 7717
enquiries@collinspress.ie
www.collinspress.ie
Managing Director *Con Collins*
Approx. Annual Turnover €750,000

Founded 1989. Publishes general non-fiction, Irish interest and children's non-fiction books. 25 titles in 2007. No unsolicited mss; synopses and ideas welcome; approach by e-mail. See website for submission guidelines. No academic, technical, professional, fiction, poetry, literary criticism or short stories.

The Columba Press
55A Spruce Avenue, Stillorgan Industrial Park, Blackrock, Co. Dublin
Ⓣ 00 353 1 294 2556   Ⓕ 00 353 1 294 2564
sean@columba.ie
(editorial)
info@columba.ie
(general)
www.columba.ie
Chairman *Neil Kluepfel*
Managing Director *Seán O'Boyle*
Approx. Annual Turnover €1.02 million

Founded 1985. Small company committed to growth. Publishes religious and counselling titles. TITLES *Already Within* Donal O'Leary; *Celtic Prayers and Reflections* Jenny Child. IMPRINT **Currach Press** (see entry). 40 titles in 2007. Backlist of 250 titles. Unsolicited ideas and synopses preferred rather than full mss.
Ⓔ Royalties twice-yearly.

Cork University Press

Youngline Industrial Estate, Pouladuff Road, Togher, Cork
℡ 00 353 21 490 2980 📠 00 353 21 431 5329
corkuniversitypress@ucc.ie
www.corkuniversitypress.com
Owner *University College Cork*
Publications Director *Mike Collins*
Editor *Sophie Watson*

Founded 1925. Relaunched in 1992, the Press publishes academic and some trade titles. TITLES *Jack's World*; *Lyn's Escape*; *The Creators*; *After Bloody Sunday*; *Sport and Society in Victorian Ireland*. Also a biannual journal, *The Irish Review*, an interdisciplinary cultural review. No fiction. IMPRINTS **Attic Press**; **Atrium**. 15 titles in 2008. Unsolicited Irish interest related synopses and ideas welcome for textbooks, academic monographs, trade non-fiction, illustrated histories and journals.
Ⓔ Royalties annually.

Currach Press

55A Spruce Avenue, Stillorgan Industrial Park, Blackrock, Co. Dublin
℡ 00 353 1 294 2556 📠 00 353 1 294 2564
jo@currach.ie
www.currach.ie
Publisher *Jo O'Donoghue*

Founded 2002. An imprint of **The Columba Press**. Publishes non-fiction titles of Irish interest, including politics, history, biography, sport, music and criticism. TITLES *A Memoir* Terry de Valera; *The Harcourt Street Line* Brian Mac Aongusa; *Short Hands, Long Pockets* Eddie Hobbs. Unsolicited ideas welcome; synopses rather than full mss preferred.

A.&A. Farmar Ltd

78 Ranelagh Village, Dublin 6
℡ 00 353 1 496 3625 📠 00 353 1 497 0107
afarmar@iol.ie
www.aafarmar.ie
Managing Director *Anna Farmar*
Production Director *Tony Farmar*

Founded 1992. Publishes Irish social history, business, food and wine. About 10 titles a year. No unsolicited mss. Synopses and ideas welcome; approach by letter or e-mail in the first instance.

Flyleaf Press

4 Spencer Villas, Glenageary, Co. Dublin
℡ 00 353 1 284 5906
books@flyleaf.ie
www.flyleaf.ie
Managing Director *Dr James Ryan*
Sales & Administration *Brian Smith*

Founded 1981. Concentrates on family history/genealogy. No fiction. TITLES *Irish Records*; *Longford and its People*; *Tracing Your Kerry Ancestors*; *Tracing Your Limerick Ancestors*. Unsolicited mss, synopses and ideas for books welcome.
Ⓔ Royalties twice-yearly.

Four Courts Press Ltd

7 Malpas Street, Dublin 8
℡ 00 353 1 453 4668 📠 00 353 1 453 4672
info@fourcourtspress.ie
www.fourcourtspress.ie
Chairman/Managing Director *Michael Adams*
Director *Martin Healy*

Founded 1972. Publishes mainly scholarly books in the humanities. About 65 titles a year. Synopses and ideas for books welcome.
Ⓔ Royalties annually.

The Gallery Press

Loughcrew, Oldcastle, Co. Meath
℡ 00 353 49 854 1779 📠 00 353 49 854 1779
gallery@indigo.ie
www.gallerypress.com
Managing Director *Peter Fallon*

Founded 1970. Publishes Irish poetry and drama. 12 titles in 2007. Currently, only interested in work from Irish poets and dramatists (plays must have had professional production). Unsolicited mss welcome but prefers to send submission notes to potential writers in the first instance.
Ⓔ Royalties paid.

Gill & Macmillan

10 Hume Avenue, Park West, Dublin 12
℡ 00 353 1 500 9500 📠 00 353 1 500 9599
www.gillmacmillan.ie
Chairman *M.H. Gill*
Managing Director *Dermot O'Dwyer*
Approx. Annual Turnover €12 million

Founded 1968 when M.H. Gill & Son Ltd and Macmillan Ltd formed a jointly owned publishing company. Publishes biography/autobiography, history, current affairs, literary criticism (all mainly of Irish interest), guidebooks and cookery. Also educational textbooks for secondary and tertiary levels. Contacts: *Anthony Murray* (educational), *Marion O'Brien* (college), *Fergal Tobin* (general). About 80 titles a year. Unsolicited synopses and ideas welcome.
Ⓔ Royalties subject to contract.

Hachette Books Ireland

8 Castlecourt Centre, Castleknock, Dublin 15
℡ 00 353 1 824 6288 📠 00 353 1 824 6289
info@hbgi.ie
www.hbgi.ie
Managing Director, Publishing *Breda Purdue*
Publisher *Ciara Considine*
Senior Commissioning Editor *Claire Rourke*
Editorial Manager *Ciara Doorley*

Founded 2002. Irish division of **Hachette Livre UK**. Publishes general fiction and non-fiction: memoirs, biography, sport, politics, current affairs, humour, food and drink. 29 titles in 2008. Consult the website for submission guidelines. Ⓔ Royalties twice-yearly.

Institute of Public Administration
57–61 Lansdowne Road, Dublin 4
☎ 00 353 1 240 3600   🖷 00 353 1 269 8644
sales@ipa.ie
www.ipa.ie
Director-General *John Cullen*
Publisher *Eileen Kelly*
Approx. Annual Turnover €1.3 million

Founded in 1957 by a group of public servants, the Institute of Public Administration is the Irish public sector management development agency. The publishing arm of the organization is one of its major activities. Publishes academic and professional books and periodicals: history, law, politics, economics and Irish public administration for students and practitioners. TITLES *Leadership, Structures and Accountability in the Public Service – Priorities for The Next Phase of Reform*; *Ireland 2022: Towards One Hundred Years of Self-Government*; *Beyond Educational Disadvantage*; *Comparing Public Administrations*; *Advisers or Advocates?*; *The Impact of State Agencies on Social Policy, An Investigation of the Measurement of Poverty in Ireland*; *Foundations of an Ever Closer Union: An Irish Perspective on Fifty Years Since the Treaty of Rome*. About 10 titles a year. No unsolicited mss; synopses and ideas welcome. No fiction or children's publishing.
Ⓔ Royalties annually.

Irish Academic Press Ltd
44 Northumberland Road, Ballsbridge, Dublin 4
☎ 00 353 1 668 8244   🖷 00 353 1 660 1610
info@iap.ie
www.iap.ie
Chairman *Stewart Cass (London)*
Dublin Office *Rachel Milotte*

Founded 1974. Publishes academic monographs and humanities. Unsolicited mss, synopses and ideas welcome.
Ⓔ Royalties annually.

The Liffey Press Ltd
Ashbrook House, 10 Main Street, Raheny, Dublin 5
☎ 00 353 1 851 1458   🖷 00 353 1 851 1459
info@theliffeypress.com
www.theliffeypress.com
Managing Director/Publisher *David Givens*

Founded 2001. Publishes general interest, Irish-focused non-fiction. 20 titles in 2007. Synopses, ideas and proposals with sample chapters welcome; approach by letter or e-mail in the first instance. No fiction, children's or books not of interest to Irish readers.
Ⓔ Royalties twice-yearly.

The Lilliput Press
62–63 Sitric Road, Arbour Hill, Dublin 7
☎ 00 353 1 671 1647   🖷 00 353 1 671 1233
info@lilliputpress.ie
www.lilliputpress.ie
Chair *Kathy Gilfillan*
Managing Director *Antony Farrell*
Approx. Annual Turnover €400,000

Founded 1984. Publishes non-fiction: literature, history, autobiography and biography, ecology, essays; criticism; fiction and poetry. TITLES *What the Curlew Said* John Moriaty; *The Companion* Lorcan Roche; *Ireland's Other Poetry: Anonymous to Zorimus* Wyse, Jackson and McDonnell; *Watching the Door* Kevin Myers; *The Dublin Edition of Ulysses* James Joyce. About 18 titles a year. Unsolicited mss, synopses and ideas welcome. No children's or sport titles.
Ⓔ Royalties annually.

Marino Books
▷ Mercier Press Ltd

Maverick House Publishers
Office 19, Dunboyne Business Park, Dunboyne, Co. Meath
☎ 00 353 1 825 5717   🖷 00 353 1 686 5036
info@maverickhouse.com
www.maverickhouse.com
Managing Director *Jean Harrington*

Founded 2001. Publishes biography and autobiography, current affairs, non-fiction, politics, sport. TITLES *Welcome to Hell: One Man's Fight for Life Inside the Bangkok Hilton* Colin Martin; *The General and I: The Untold Story of Martin Cahill's Hotdog Wars* Wolfgang Eulitz; *Miss Bangkok* Bua Boonmee; *Hell in Barbados* Terry Donaldson. Unsolicited material welcome; send by post. No fiction, children's books or poetry.
Ⓔ Royalties twice-yearly.

Mercier Press Ltd
Douglas Village, Cork
☎ 00 353 21 489 9858   🖷 00 353 21 489 9887
info@mercierpress.ie
www.mercierpress.ie
Chairman *John Spillaner*
Managing Director *Clodagh Feehan*

Founded 1944. One of Ireland's largest publishers with a list of approx 250 Irish interest titles.
IMPRINTS **Mercier Press**; **Marino Books** Editorial Director *Mary Feehan* Children's, politics, history, folklore, biography, mind, body and spirit, current

affairs, women's interest. TITLES *The Course of Irish History*; all of John B. Keane's works; *Beyond Prozac*; *It's a Long Way from Penny Apples*; *Ireland's Master Storyteller*. Unsolicited synopses and ideas welcome.
Ⓔ Royalties annually.

Merlin Publishing
Newmarket Hall, Cork Street, Dublin 8
Ⓣ 00 353 1 453 5866   Ⓕ 00 353 1 453 5930
publishing@merlin.ie
www.merlinwolfhound.com
Managing Director *Chenile Keogh*
Editorial Director *Aoife Barrett*

Founded 2000. Member of **Clé**. Publishes true crime, film, music, art, biography, general non-fiction, history and gift books. IMPRINT **Wolfhound Press** Founded 1974. Non-fiction. TITLES *Famine*; *Eyewitness Bloody Sunday*; *Father Browne's Titanic Album*. About 15 titles a year. Two sample chapters of unsolicited mss (with synopses and s.a.e.) welcome by e-mail (aoife@merlin.ie). See Merlin website for submission guidelines and proposal form.

Mount Eagle
▷ Brandon/Mount Eagle Publicaions

National Library of Ireland
Kildare Street, Dublin 2
Ⓣ 00 353 1 603 0200   Ⓕ 00 353 1 676 6690
info@nli.ie
www.nli.ie

Founded 1877. Publishes books and booklets based on the library's collections; folders of historical documents; academic and specialist books; reproduction folders and CD-ROMs. TITLES *Ulysses Unbound: A Reader's Companion to James Joyce's Ulysses* Terence Killeen; *Cooper's Ireland: Drawings and Notes from an Eighteenth Century Gentleman* Peter Harbison; *Into the Light: An Illustrated Guide to the Photographic Collections in the National Library of Ireland* Sarah Rouse; *WB Yeats: Works and Days* James Quin, Eilis Ni Dhuibhne, Ciara McDonnell (book to accompany Yeats exhibition at the National Library – exhibition running until January 2009); *Librarians, Poets and Scholars: A festschrift for Donall O Luanaigh* ed. Felix M. Larkin. 1 title in 2007.

New Island
2 Brookside, Dundrum Road, Dundrum, Dublin 14
Ⓣ 00 353 1 298 9937/298 3411
Ⓕ 00 353 1 298 7912
inka.hagen@newisland.ie
www.newisland.ie
Managing Director/Editorial Head *Edwin Higel*
Editorial Manager *Deirdre Nolan*

Founded 1990. Publishes fiction, Irish non-fiction, poetry and drama. Branching out into popular fiction and memoirs. IMPRINT **Brookside**. About 25 titles a year. Unsolicited mss, synopses and ideas welcome. Send three chapters and synopsis by post.
Ⓔ Royalties twice-yearly

The O'Brien Press Ltd
12 Terenure Road East, Rathgar, Dublin 6
Ⓣ 00 353 1 492 3333   Ⓕ 00 353 1 492 2777
books@obrien.ie
www.obrien.ie
Managing Director *Ivan O'Brien*
Publisher *Michael O'Brien*

Founded 1974. Publishes business, true crime, biography, music, travel, sport, Celtic subjects, food and drink, history, humour, politics, reference. Children's publishing – mainly fiction for every age from tiny tots to teenage. Illustrated fiction SERIES *Panda Cubs* (3 years+); *Pandas* (5 years+); *Flyers* (6 years+); *Red Flag* (8 years+). Novels (10 years+): contemporary, historical, fantasy. Some non-fiction: mainly historical, and art and craft, resource books for teachers. No poetry, adult fiction or academic. Unsolicited mss (sample chapters only), synopses and ideas for books welcome. No e-mail submissions. Submissions will not be returned.
Ⓔ Royalties annually.

Oak Tree Press
19 Rutland Street, Cork
Ⓣ 00 353 21 431 3855   Ⓕ 00 353 21 431 3496
info@oaktreepress.com
www.oaktreepress.com
Owners *Brian O'Kane, Rita O'Kane*
Managing Director *Brian O'Kane*

Founded 1991. Specialist publisher of business and professional books with a focus on small business start-up and development. 5 titles in 2006. Unsolicited mss and synopses welcome; send to the managing director, at the address above.
Ⓔ Royalties annually.

On Stream Publications Ltd
Currabaha, Cloghroe, Co. Cork
Ⓣ 00 353 21 438 5798
info@onstream.ie
www.onstream.ie
Chairman/Managing Director *Roz Crowley*

Founded 1992. Formerly Forum Publications. Publishes academic, cookery, wine, general health and fitness, local history, railways, photography and practical guides. TITLES *The Health Squad Guide to Health and Fitness*; *A Kingdom of Wine: A Celebration of Ireland's Winegeese*; *At Home in Renvyle*; *Stop Howling at the Moon*; *101 Bedtime*

Stories for Managers. About 3 titles a year. Synopses and ideas welcome. No children's books. £ Royalties annually.

Penguin Ireland

25 St Stephen's Green, Dublin 2
℡ 00 353 1 661 7695   ℻ 00 353 1 661 7696
info@penguin.ie
www.penguin.ie
Managing Director *Michael McLoughlin*

Founded 2002. Part of the **Penguin Group UK**. DIVISIONS **Commercial fiction/non-fiction** Editorial Director *Patricia Deevy* TITLES *In My Sister's Shoes* Sinéad Moriaty; *This Champagne Mojito is the Last Thing I Own* Ross O'Carroll Kelly; *Secret Diary of a Demented Housewife* Niamh Greene; *Under My Skin* Alison Jameson. **Literary fiction/non-fiction** Editor *Brendan Barrington* TITLES *Connemara: Listening to the Wind* Tim Robinson; *Notes from a Turkish Whorehouse* Philip Ó Ceallaigh; *Keys to the Kingdom* Jack O'Connor; *In Your Face* Lia Mills; *A Book of Uncommon Prayer* Theo Dorgan. 16 titles in 2007. Unsolicited mss, synopses and ideas for books welcome; send by post.
£ Royalties twice-yearly.

Poolbeg Press Ltd

123 Grange Hill, Baldoyle, Dublin 13
℡ 00 353 1 832 1477   ℻ 00 353 1 832 1430
poolbeg@poolbeg.com
www.poolbeg.com
Managing Director *Kieran Devlin*
Publisher *Paula Campbell*
Non-Fiction Editor *Brian Langan*

Founded 1976 to publish the Irish short story and has since diversified to include all areas of fiction (literary and popular), children's fiction and non-fiction, and adult non-fiction: history, biography and topics of public interest. AUTHORS discovered and first published by Poolbeg include Maeve Binchy, Marian Keyes, Sheila O'Flanagan, Cathy Kelly and Patricia Scanlan. 'Our slogan is Poolbeg. com – The *Irish* for Bestsellers!' IMPRINTS **Poolbeg** (paperback and hardback); **Poolbeg For Children**. About 40 titles a year. Unsolicited mss, synopses and ideas welcome (mss preferred). No drama.
£ Royalties twice-yearly.

Royal Dublin Society

Science Section, Ballsbridge, Dublin 4
℡ 00 353 1 240 7217   ℻ 00 353 1 660 4014
science@rds.ie
www.rds.ie/science
Development Executive, Science & Technology *Dr Claire Mulhall*

Founded 1731 for the promotion of agriculture, science and the arts, and throughout its history has published books and journals towards this end. Publishes conference proceedings and books on the history of Irish science. TITLES *Agricultural Development for the 21st Century*; *The Right Trees in the Right Places*; *Agriculture & the Environment*; *Water of Life*; *Science, Blueprint for a National Irish Science Centre*; *Science Education in Crisis*; *Science in the Service of the Fishing Industry*; occasional papers in *Irish Science & Technology* series. SERIES Science and Irish Culture TITLES *Science and Irish Culture: Why the History of Science Matters*; *Science and Ireland – Value for Society*; *It's Part of What We Are*.
£ Royalties not generally paid.

Royal Irish Academy

19 Dawson Street, Dublin 2
℡ 00 353 1 676 2570   ℻ 00 353 1 676 2346
publications@ria.ie
www.ria.ie/publications
Executive Secretary *Patrick Buckley*
Managing Editor *Ruth Hegarty*

Founded in 1785, the Academy has been publishing since 1787. Core publications are journals but more books published in last 15 years. Publishes academic, Irish interest and Irish language. About 7 titles a year. Welcomes mss, synopses and ideas of an academic standard, but currently encourages publications which will popularize some aspect of the sciences and humanities. Address correspondence to the managing editor of publications.

Tír Eolas

Newtownlynch, Doorus, Kinvara, Co. Galway
℡ 00 353 91 637452   ℻ 00 353 91 637452
info@tireolas.com
www.tireolas.com
Publisher/Managing Director *Anne Korff*

Founded 1987. Publishes books and guides on ecology, archaeology, folklore and culture. TITLES *The Book of the Burren*; *The Shores of Connemara*; *Not a Word of a Lie*; *The Book of Aran*; *Kinvara, A Seaport Town on Galway Bay*; *A Burren Journal*; *Alive, Alive-O, The Shellfish and Shellfisheries of Ireland*; *The Burren Wall*; *Corrib Country, Guide & Map*; *Shannon Valley, Guide & Map*. Unsolicited mss, synopses and ideas for books welcome. No specialist scientific and technical, fiction, plays, school textbooks or philosophy.
£ Royalties annually.

University College Dublin Press

Newman House, 86 St Stephen's Green, Dublin 2
℡ 00 353 1 477 9812/9813   ℻ 00 353 1 477 9821
ucdpress@ucd.ie
www.ucdpress.ie
Executive Editor *Barbara Mennell*

Founded 1995. Academic publisher. 16 titles in 2007. Unsolicited mss, synopses and ideas welcome by post. Approach in writing. No non-academic, journals, poetry, novels or books based on academic conferences.
Ⓔ Royalties annually.

Veritas Publications

7–8 Lower Abbey Street, Dublin 1
Ⓣ 00 353 1 878 8177   Ⓕ 00 353 1 878 6507
publications@veritas.ie
www.veritas.ie
Director *Maura Hyland*

Managing Editor *Ruth Kennedy*

Founded 1969 to supply religious textbooks to schools and later introduced a wide-ranging general list. Part of the Catholic Communications Institute. Publishes books on religious, ethical, moral, societal and social issues. 35 titles a year. Unsolicited mss, synopses and ideas for books welcome.
Ⓔ Royalties annually.

Wolfhound Press
▷ Merlin Publishing

European Publishers

Austria

Springer-Verlag GmbH
Sachsenplatz 4–6, A–1201 Vienna
T 00 43 1 3302415   F 00 43 1 3302426
springer@springer.at
www.springer.at

Founded 1924. Austria's largest scientific
publisher. Publishes academic and textbooks,
journals and translations: architecture, art,
cell biology, chemistry, computer science, law,
medicine, neurosurgery, neurology, nursing,
psychiatry, psychotherapy.

Verlag Carl Ueberreuter GmbH
Alser Strasse 24, A–1090 Vienna
T 00 43 1 404440
www.ueberreuter.de

Founded 1548. Austria's largest privately-owned
publishing house. Publishes children's and young
adult titles; adult non-fiction includes health and
nutrition, history, culture and politics.

Paul Zsolnay Verlag GmbH
Prinz Eugen Strasse 30, A–1040 Vienna
T 00 43 1 505 7661-0   F 00 43 1 505 7661-10
info@zsolnay.at
www.zsolnay.at

Founded 1923. Publishes biography, fiction,
general non-fiction, history, poetry.

Belgium

Brepols Publishers NV
Begijnhof 67, B–2300 Turnhout
T 00 32 14 44 80 20   F 00 32 14 42 89 19
info@brepols.net
www.brepols.net

Founded 1796. Independent, academic humanities
publisher of books, monographs, journals, CD-
ROMs and online. Bibliography, encyclopedias,
reference, history, archaeology, language,
literature, art, art history, music, philosophy,
architecture.

Uitgeverij Lannoo NV
Kasteelstraat 97, B–8700 Tielt
T 00 32 51 42 42 11   F 00 32 51 40 11 52
lannoo@lannoo.be
www.lannoo.be

Founded 1909. Publishes general non-fiction,
art, architecture and interior design, cookery,
biography, children's, gardening, health and
nutrition, history, management, self-help,
religion, travel, young adult fiction, photography,
psychology.

Standaard Uitgeverij
Mechelsesteenweg 203, B–2018 Antwerp
T 00 32 3 285 72 00   F 00 32 3 285 72 99
info@standaarduitgeverij.be
www.standaarduitgeverij.be

Founded 1919. Publishes education, fiction,
poetry, humour, maps, children's and young adult.

Universitaire Pers Leuven
Minderbroedersstraat 4, B-3000 Leuven
T 00 32 16 32 53 45   F 00 32 16 32 53 52
info@upers.kuleuven.be
www.lup.be

Founded 1971. Academic publisher of
anthropolgy, archaeology, architecture, art, arts,
economics, geology, history, law, philosophy,
political science, psychology, sociology, science
and technology.

Bulgaria

Kibea Publishing Company
Komplex Lulin–3, Block 329, Entrance 7, App.
130, BG–1336 Sofia
T 00 359 2 925 0152   F 00 359 2 925 0748
office@kibea.net
www.kibea.net

Founded 1991. Publishes fiction and illustrated
books; encyclopedias, children's, anthologies, art,
history, health and lifestyle guides, alternative
and popular medicine, psychology and
psychoanalysis, philosophy, religion.

Lettera

62 Rhodope Street, POBox 802, BG–4000 Polvdiv
① 00 359 3 262 8027   ⓕ 00 359 3 262 8027
office@lettera.bg
www.lettera.bg

Founded 1991. Educational publisher of textbooks
and school aids, dictionaries, languages, audio
and video; reference and contemporary fiction.

Naouka I Izkoustvo

11 Slaveikov Square, BG–1000 Sofia
① 00 359 2 987 4790
nauk_izk@sigma-bg.com

Founded 1948. Publishes dictionaries, grammar
books, language textbooks, translations,
psychology, philosophy, essays.

Czech Republic

Atlantis Ltd

PS 374 Ceská 15, CZ–602 00 Brno
① 00 420 542 213 221   ⓕ 00 420 542 213 221
atlantis-brno@volny.cz
www.volny.cz/atlantis

Publishes fiction, language and linguistics, literary
studies, poetry, religion, social science.

Brána a.s.

Jankovcova 18/938, CZ–170 37 Prague 7
① 00 420 220 191 313   ⓕ 00 420 220 191 313
info@brana-knihy.cz
www.brana-knihy.cz

Publishes fiction, non-fiction, art, architecture,
encyclopedias, language and linguistics, literary
studies, medicine, health, social science, sport,
tourism.

Dokorán Ltd

Zborovská 40, CZ–150 00 Prague 5
① 00 420 257 320 803   ⓕ 00 420 257 320 805
dokoran@dokoran.cz
www.dokoran.cz

Founded 2001. Publishes fiction, non-fiction,
translations, natural sciences (physics,
mathematics, nature), arts, history, social science.

Olympia Publishing Co. Inc.

Klimentská 1, CZ–110 15 Prague 1
① 00 420 224 810 146   ⓕ 00 420 222 312 137
olympia@mbox.vol.cz
www.iolympia.cz

Publishes fiction, non-fiction, children's,
entertainment, leisure, popular science, science
fiction, sport, textbooks, tourism.

Denmark

Forlaget Apostrof ApS

Postboks 2580, DK–2100 Copenhagen OE
① 00 45 3920 8420   ⓕ 00 45 3920 8453
apostrof@apostrof.dk
www.apostrof.dk

Founded 1980. Publishes psychology and
psychiatry.

Blackwell Munksgaard

1 Rosenørns Allé, DK–1970 Frederiksberg C
① 00 45 7733 3333   ⓕ 00 45 7733 3377
info@msk.blackwellpublishing.com
www.blackwellmunksgaard.com

Founded 1917. Part of Wiley-Blackwell. Publishes
academic books and journals for higher
education, research and professional markets.

Borgens Forlag A/S

Valbygardsvej 33, DK–2500 Valby
① 00 45 3615 3615   ⓕ 00 45 3615 3616
post@borgen.dk
www.borgen.dk

Founded 1948. Publishes fiction and general
non-fiction: biography, children's and young
adult, history, thrillers, humour, professional,
psychology.

Forlaget Forum

Postbox 2252, DK–1019 Copenhagen K
① 00 45 3341 1800   ⓕ 00 45 3341 1801
gb@gb-forlagene.dk
www.forlagetforum.dk

Founded 1940. Part of Gyldendalske
Boghandel-Nordisk Forlag A/S. Publishes
fiction, bibliography, history, humour, mysteries,
children's and young adults.

Gyldendal

Klareboderne 3, DK–1001 Copenhagen K
① 00 45 3375 5555
gyldendal@gyldendal.dk
www.gyldendal.dk

Founded 1770. Publishes fiction, children's
and young adult, directories, art, biography,
education, history, how-to, medicine, music,
poetry, philosophy, psychiatry, reference, general
and social sciences, sociology, textbooks.

Høst & Søn Publishers Ltd

Købmagergade 62, DK–1150 Copenhagen K
① 00 45 3341 1800   ⓕ 00 45 3341 1801
gb@gb-forlagene.dk
www.hoest.dk

Founded 1836. Publishes fiction, biography,
children's, crafts, environmental studies, history.

Lindhardt og Ringhof Forlag A/S

Vognmagergade 11, DK–1148 Copenhagen K
℡ 00 45 3369 5000   🖷 00 45 3369 5001
info@lindhardtogringhof.dk
www.lrforlag.dk

Founded 1971. Merged with Aschehoug Dansk
Forlag A/S in 2008. Publishes fiction and general
non-fiction.

Nyt Nordisk Forlag Arnold Busck A/S

Landemærket 11, 5. sal, DK–1119 Copenhagen K
℡ 00 45 3373 3575   🖷 00 45 3314 0115
nnf@nytnordiskforlag.dk
www.nytnordiskforlag.dk

Founded 1896. Publishes fiction, architecture, art,
academic, cookery, design, education, gardening,
health, natural history, philosophy, reference,
religion, leisure, medicine, nursing, psychology,
textbooks, travel.

Det Schønbergske Forlag

Landemærket 11, 5. sal, DK–1119 Copenhagen K
℡ 00 45 3373 3585   🖷 00 45 3373 3586
schoenberg@nytnordiskforlag.dk
www.nytnordiskforlag.dk

Founded 1857. Part of **Nyt Nordisk Forlag
Arnold Busck A/S**. Publishes directories, fiction,
non-fiction, reference, textbooks.

Tiderne Skifter Forlag A/S

Læderstræde 5, 1. sal, DK–1201 Copenhagen K
℡ 00 45 3318 6390   🖷 00 45 3318 6391
tiderneskifter@tiderneskifter.dk
www.tiderneskifter.dk

Founded 1973. Publishes fiction, literature
and literary criticism, essays, poetry, drama,
translations, history, art, politics, psychoanalysis,
sexual politics, ethnicity, photography.

Finland

Gummerus Oy

PO Box 749, SF–00101 Helsinki
℡ 00 358 9 584301   🖷 00 358 9 5843 0200
www.gummerus.fi

Founded 1872. Publishes fiction and general
non-fiction.

Karisto Oy

PO Box 102, SF–13101 Hämeenlinna
℡ 00 358 3 63151   🖷 00 358 3 616 1565
kustannusliike@karisto.fi
www.karisto.fi

Founded 1900. Publishes fiction (including
crime, thrillers and fantasy), general non-fiction
(including health, history, family and childcare,
self improvement, pets, fishing), juvenile and
young adult.

Otava Publishing Co. Ltd

Uudenmaankatu 8–12, SF–00120 Helsinki
℡ 00 358 9 19961
otava@otava.fi
www.otava.fi

Founded 1890. Publishes fiction, translations,
general non-fiction, children's and young adult,
multimedia, education, textbooks.

Tammi Publishers

PO Box 410, SF–00101 Helsinki
℡ 00 358 9 6937 621   🖷 00 358 9 6937 6266
tammi@tammi.fi
www.tammi.fi

Founded 1943. Finland's third largest publisher.
Publishes fiction, general non-fiction, children's
and young adult, education, translations. Part of
the Bonnier Group.

WSOY (Werner Söderström Osakeyhtiö)

PO Box 222, SF–00121 Helsinki
℡ 00 358 9 616 81   🖷 00 358 9 6168 3560
firstname.lastname@wsoy.fi
www.wsoy.fi

Founded 1878. Publishes fiction, translation,
general non-fiction, education, dictionaries,
encyclopedias, textbooks, children's and young
adult.

France

Éditions Arthaud

Flammarion Groupe, 87 quai Panhard-et-
Levassor, F–75647 Paris Cedex 13
℡ 00 33 1 40 51 31 00
contact@arthaud.fr
www.arthaud.fr

Founded 1890. Imprint of **Éditions Flammarion**.
Publishes illustrated books, photography, travel.

Éditions Belfond

12 avenue d'Italie, F–75627 Paris Cedex 13
℡ 00 33 1 44 16 05 00
www.belfond.fr

Founded 1963. Publishes fiction and general
non-fiction.

Éditions Bordas

89 blvd Auguste-Blanqui, F–75013 Paris
℡ 00 33 1 44 39 54 45
www.editions-bordas.com

Founded 1946. Publishes education and
general non-fiction: dictionaries, directories,
encyclopedias, reference.

Éditions Calmann-Lévy

31 rue de Fleurus, F–75006 Paris
☎ 00 33 1 49 54 36 00   🅵 00 33 1 45 44 86 32
www.editions-calmann-levy.com

Founded 1836. Part of **Hachette Livre**. Publishes fiction, science fiction, fantasy, biography, history, humour, economics, philosophy, psychology, psychiatry, social sciences, sociology, sport.

Éditions Denoël

9 rue du Cherche-Midi, F–75278 Paris Cedex 06
☎ 00 33 1 44 39 73 73   🅵 00 33 1 44 39 73 90
denoel@denoel.fr
www.denoel.fr

Founded 1932. Part of Gallimard Groupe. Publishes fiction, science fiction, fantasy, thrillers, art, art history, biography, directories, government, history, philosophy, political science, psychology, psychiatry, reference.

Librairie Arthème Fayard

13 due du Montparnasse, F-75006
☎ 00 33 1 45 49 82 00
www.fayard.fr

Founded 1857. Part of **Hachette Livre**. Publishes fiction, biography, directories, maps, atlases, history, dance, music, philosophy, religion, reference, sociology, general science.

Éditions Flammarion

87 quai Panhard-et-Levassor, F–75647 Paris Cedex 13
☎ 00 33 1 40 51 31 00
www.editions.flammarion.com

Imprint of Groupe Flammarion. Publishes general fiction, science fiction, thrillers, essays, adventure, literature, literary criticism; non-fiction: art, architecture, biography, children's, cookery, crafts, design, gardening, history, lifestyle, natural history, plants, popular science, reference, medicine, nursing, wine.

Éditions Gallimard

5 rue Sébastien-Bottin, F–75328 Paris Cedex 07
☎ 00 33 1 49 54 42 00   🅵 00 33 1 45 44 94 03
www.gallimard.fr

Founded 1911. Publishes fiction, poetry, art, biography, history, music, philosophy, children's and young adult.

Éditions Grasset & Fasquelle

61 rue des Saints-Pères, F–75006 Paris
☎ 00 33 1 44 39 22 00   🅵 00 33 1 42 22 64 18
www.edition-grasset.fr

Founded 1907. Part of **Hachette Livre**. Publishes fiction and general non-fiction, biography, essays, literature, literary criticism, thrillers, translations, philosophy.

Hachette Livre

43 quai de Grenelle, F–75905 Paris Cedex 15
☎ 00 33 1 43 92 30 00   🅵 00 33 1 43 92 30 30
www.hachette.com

Founded 1826. With the acquisition of Time Warner Book Group, Hachette, already the owner of Hodder Headline and Orion, is now the biggest publisher in the British market and the third largest publisher in the world behind Pearson and Bertelsmann. Publishes fiction and general non-fiction, bibliographies, bilingual, children's, directories, education, encyclopedias, general engineering, government, language and linguistics, reference, general science, textbooks.

Éditions Robert Laffont

24 avenue Marceau, F–75381 Paris Cedex 08
☎ 00 33 1 53 67 14 00
www.laffont.fr

Founded 1987. Publishes fiction and non-fiction.

Éditions Larousse

21 rue du Montparnasse, F–75283 Paris Cedex 06
☎ 00 33 1 44 39 44 00
www.larousse.fr

Founded 1852. Publishes animals, art, bilingual, children's and young adult, childcare, cinema, cookery, dictionaries, directories, drama, encyclopedias, food, games, gardening, health, home interest, history, medicine, natural history, nursing, music, dance, self-help, psychology, reference, general science, language arts, linguistics, literature, religion, sport, technology, textbooks.

Éditions Jean-Claude Lattès

17 rue Jacob, F–75006 Paris
☎ 00 33 1 44 41 74 00   🅵 00 33 1 43 25 30 47
www.editions-jclattes.fr

Founded 1968. Part of **Hachette Livre**. Publishes fiction and general non-fiction.

Éditions Magnard

20 rue Berbier-du-Mets, F–75647 Paris Cedex 13
☎ 00 33 1 44 08 85 85   🅵 00 33 1 44 08 49 79
contact@magnard.fr
www.magnard.fr

Founded 1933. Part of Groupe Albin-Michel. Publishes education, foreign language and bilingual books, juvenile and young adults, textbooks.

Michelin Éditions des Voyages

46 ave de Breteuil, F–75007 Paris
☎ 00 33 1 45 66 11 70

Founded 1900. Publishes travel: maps and atlases.

Les Éditions de Minuit SA

7 rue Bernard-Palissy, F–75006 Paris
℡ 00 33 1 44 39 39 20   🖷 00 33 1 45 44 82 36
www.leseditionsdeminuit.fr

Founded 1942. Publishes fiction, essays, literature, literary criticism, philosophy, social sciences, sociology.

Les Éditions Nathan

9 rue Méchain, F–75686 Paris
℡ 00 33 1 45 87 50 00   🖷 00 33 1 45 87 53 43
www.nathan.fr

Founded 1881. Publishes fiction and non-fiction for children and young adults. Adult reference, leisure and natural history; education, technical and training.

Presses de la Cité

12 ave d'Italie, F–75627 Paris Cedex 13
℡ 00 33 1 44 16 05 00   🖷 00 33 1 44 16 05 01
pressesdelacite@placedesediteurs.com
www.pressesdelacite.com

Founded 1947. Imprint of **Editions Belfond**. Publishes fiction and general non-fiction, science fiction, fantasy, history, mysteries, romance, biography.

Presses Universitaires de France (PUF)

6 ave Reille, F–75685 Paris Cedex 14
℡ 00 33 1 58 10 31 00
www.puf.com

Founded 1921. Academic publisher of scientific books, dictionaries, economics, encyclopedias, geography, history, law, linguistics, literature, philosophy, psychology, psychiatry, political science, reference, sociology, textbooks.

Éditions du Seuil

27 rue Jacob, F–75261 Paris Cedex 06
℡ 00 33 1 40 46 50 50
www.seuil.com

Founded 1935. Part of the Groupe Martinière. Publishes fiction and non-fiction, translations, literature, literary criticism, essays, poetry, biography, classics, illustrated, history.

Les Editions de la Table Ronde

14 rue Séguier, F–75006 Paris
℡ 00 33 1 40 46 70 70   🖷 00 33 1 40 46 71 01
editionslatableronde@editionslatableronde.fr
www.editionslatableronde.fr

Founded 1944. Part of Gallimard Groupe. Publishes fiction and general non-fiction, biography, history, psychology, psychiatry, religion, sport.

Librairie Vuibert

20 rue Berbier-du-Mets, F–75647 Paris Cedex 13
℡ 00 33 1 44 08 49 00   🖷 00 33 1 44 08 49 69
www.vuibert.com

Founded 1877. Part of Groupe Albin-Michel. Publishes textbooks.

Germany

Verlag C.H. Beck (OHG)

Wilhelmstrasse 9, D–80801 Munich
℡ 00 49 89 38 189-0   🖷 00 49 89 38 189-398
www.beck.de

Founded 1763. Publishes general non-fiction, anthropology, archaeology, art, CD-ROMs, dance, dictionaries, directories, economics, essays, encyclopedias, history, illustrated, journals, language arts, law, linguistics, literature, literary criticism, music, philosophy, professional, reference, social sciences, sociology, textbooks, theology.

C. Bertelsmann

Neumarkter Strasse 28, D–81673 Munich
℡ 00 49 89 4136-0

Founded 1835. Imprint of the **Random House Group**. Publishes fiction and general non-fiction, art, autobiography, biography, government and politics.

Carlsen Verlag GmbH

Postfach 50 03 80, D–22703 Hamburg
℡ 00 49 40 39804-0   🖷 00 49 40 39804-390
info@carlsen.de
www.carlsen.de

Founded 1953. Publishes children's and comic books.

Deutscher Taschenbuch Verlag GmbH & Co. KG

Friedrichstrasse 1a, D–80801 Munich
℡ 00 49 89 38167-0   🖷 00 49 89 34 64 28
verlagdtv.de
www.dtv.de

Founded 1961. Publishes fiction and general non-fiction; biography, business, child care and development, children's, dictionaries, directories, economics, education, encyclopedias, fantasy, government, health and nutrition, history, poetry, psychiatry, psychology, philosophy, politics, reference, religion, literature, literary criticism, essays, thrillers, translations, travel, young adult.

S. Fischer Verlag GmbH

Hedderichstrasse 114, D–60596 Frankfurt am Main
℡ 00 49 69 6062-0   🖷 00 49 69 6062-214
www.fischerverlage.de

Founded 1886. Part of the **Holtzbrinck Group**. Publishes fiction, general non-fiction, biography, business, children's and young adult, history, literature, natural history, politics, psychology, reference.

Carl Hanser Verlag GmbH & Co. KG

Postfach 86 04 20, D–81631 Munich

📞 00 49 89 998 30 0   📠 00 49 89 98 48 09

info@hanser.de

www.hanser.de/verlag

Founded in 1928. Publishes international and German contemporary literature; children's and juveniles; specialist books on engineering and technology, natural science, computers and computer science, economics and management, CD-ROMs, journals, academic, textbooks, translations.

Heyne Verlag

Bayerstrasse 71–73, D–80335 Munich

📞 00 49 89 4136-0

www.heyne.de

Founded 1934. Part of **Random House Group**. Publishes fiction, mystery, romance, humour, science fiction, fantasy, biography, cookery, film, history, how-to, occult, psychology, psychiatry.

Hoffmann und Campe Verlag GmbH

Harvestehuder Weg 42, D–20149 Hamburg

📞 00 49 40 44188-0   📠 00 49 40 44188-202

email@hoca.de

www.hoca.de

Founded 1781. Publishes fiction and general non-fiction; art, audio, biography, dance, history, illustrated, journals, music, poetry, philosophy, psychology, psychiatry, general science, social sciences.

Verlagsgruppe Georg von Holtzbrinck GmbH

Gänsheidestrasse 26, D–70184 Stuttgart

📞 00 49 711 2150-0   📠 00 49 711 2150-269

info@holtzbrinck.com

www.holtzbrinck.com

Founded 1948. One of the world's largest publishing groups with 12 book publishing houses and 40 imprints. Also publishes the newspapers, *Handelsblatt* and *Die Zeit*.

Hüthig GmbH & Co. KG

Postfach 10 28 69, D–69018 Heidelberg

📞 00 49 6221 489-0   📠 00 49 6221 489-279

www.huethig.de

Founded 1925. Germany's fourth largest professional publisher: CD-ROMs, multimedia, journals, academic, textbooks, translations.

Ernst Klett Verlag GmbH

Rotebühlstrasse 77, D–70178 Stuttgart

📞 00 49 711 6672 1333

www.klett.de

Founded 1897. Educational publisher: dictionaries, encyclopedias, textbooks.

Verlagsgruppe Lübbe GmbH & Co. KG

Scheidtbachstrasse 23–31, D–51469 Bergisch Gladbach

📞 00 49 2202 121-0   📠 00 49 2202 121-928

www.luebbe.de

Founded 1963. Independent publisher of fiction and general non-fiction, audio, art, biography, history, how-to.

Rowohlt Verlag GmbH

Hamburgerstr 17, D–21465 Reinbeck

📞 00 49 40 72 72-0   📠 00 49 40 72 72-319

info@rowohlt.de

www.rowohlt.de

Founded 1908. Publishes general non-fiction.

Springer GmbH

Tiergartenstrasse 17, D–69121 Heidelberg

📞 00 49 6221 487-0

www.springer.de

Founded 1842. Part of Springer Science+Business Media Deutschland GmbH. Leading specialist publisher of science, technology and medical books, journals, CD-ROMs, databases. Subjects include astronomy, engineering, computer science, economics, geography, law, linguistics, mathematics, philosophy, psychology, architecture, construction and transport – with eighty per cent of publications in English.

Suhrkamp Verlag

Postfach 101945, D–60019 Frankfurt am Main

📞 00 49 69 75601-0   📠 00 49 69 75601-314

www.suhrkamp.de

Founded 1950. Publishes fiction, poetry, biography, cinema, film, theatre, philosophy, psychology, psychiatry, general science.

Taschen GmbH

Hohenzollernring 53, D–50672 Cologne

📞 00 49 221 201 80-0   📠 00 49 221 25 49 19

contact@taschen.com

www.taschen.com

Founded 1980. Publishes photography, art, erotica, architecture and interior design.

Ullstein Buchverlage GmbH

Friedrichstrasse 126, D–10117 Berlin

📞 00 49 30 23456-300   📠 00 49 30 23456-303

info@ullstein-buchverlage.de

www.ullsteinbuchverlage.de

Founded 1903. Publishes general fiction and non-fiction; literature, mystery, biography, business, economics, health, gift books, memoirs, politics.

WEKA Holding GmbH & Co. KG

Postfach 13 31, D–86438 Kissing

⊤ 00 49 8233 23-0   ⨏ 00 49 8233 23-195

info@weka-holding.de

www.weka-holding.de

Founded 1973. Germany's largest professional publisher.

Hungary

Akadémiai Kiadó

Prielle Kornélia u. 19/D, H–1516 Budapest

⊤ 00 36 1 464 8282   ⨏ 00 36 1 464 8251

info@akkrt.hu

www.akkrt.hu

Hungary's oldest publishing house. Founded 1828. Publishes scholarly reference, economics, education, dictionaries, languages, journals, scientific.

Jelenkor Kiadó Szolgáltató Kft.

Munkácsy Mihály u. 30/A, H–7621 Pécs

⊤ 00 36 72 314 782   ⨏ 00 36 72 532 047

editors@jelenkor.com

www.jelenkor.com

Founded 1993. Independent publishing house. Publishes fiction, philosophy, poetry.

Kiss József Kónyvkiadó, Kereskedel mi és Reklám Kft.

Hamzsabégi út 31, H–1114 Budapest

⊤ 00 36 209 9140   ⨏ 00 36 466 0703

konyvhet@alexero.hu

www.konyv7.hu

Publishes fiction, crime, film, history, politics, theatre, translations.

Kráter Association and Publishing House

Búzavirág Street, H–2013 Pomáz

⊤ 00 36 26 328 491   ⨏ 00 36 26 328 491

info@krater.hu

www.krater.hu

Publishes poetry and literature.

Móra Könyvkiadö Zrt. (Móra Publishing House)

Váci út 19, H–1134 Budapest

⊤ 00 36 1 320 4740   ⨏ 00 36 1 320 5382

mora@mora.hu

www.mora.hu

Founded 1950. Publishes books for children of all age groups – fiction and school books, popular science, poems, bath books and teenage horror. Recently extended range to include adult fiction and non-fiction.

Italy

Adelphi Edizioni SpA

Via S. Giovanni sul Muro 14, I–20121 Milan

⊤ 00 39 2 725731   ⨏ 00 39 2 89010337

info@adelphi.it

www.adelphi.it

Founded 1962. Publishes literature, literary criticism, anthropology, art, autobiography, archaeology, architecture, biography, biology, cinema, economics, history, linguistics, mathematics, medicine, music, philosophy, photography, politics, psychiatry, religion, general science, sociology, theatre, translations.

Bompiani

Via Mecenate 91, I–20138 Milan

⊤ 00 39 2 5095 2876   ⨏ 00 39 2 5065 2788

www.rcslibri.it

Founded 1929. Part of RCS Libri publishing group. Publishes fiction and general non-fiction, autobiography, biography, art, illustrated, memoirs, philosophy.

Bulzoni Editore SRL

Via dei Liburni 14, I–00185 Rome

⊤ 00 39 6 4455207   ⨏ 00 39 6 4450355

bulzoni@bulzoni.it

www.bulzoni.it

Founded 1969. Publisher of college textbooks.

Cappelli Editore

Via Farini 14, I–40124 Bologna

⊤ 00 39 51 239060   ⨏ 00 39 51 239286

info@cappellieditore.com

www.cappellieditore.com

Founded 1851. Publishes reference, textbooks, juvenile and young adult.

Garzanti Libri SpA

Via Gasparotto 1, I–20124 Milan

⊤ 00 39 2 0062 3201

info@garzantilibri.it

www.garzantilibri.it

Founded 1861. Publishes fiction and non-fiction, literature, literary criticism, encyclopedias, essays, biography, cookery, crime, dictionaries, history, memoirs, philosophy, poetry, reference, textbooks.

Giunti Editoriale SpA

Via Bolognese 165, I–50139 Florence

⊤ 00 39 55 5062 1   ⨏ 00 39 55 5062 398

www.giunti.it

Founded 1841. Publishes fiction and non-fiction, literature, essays, art, archaeology, alternative medicine, children's, cookery, crafts, dictionaries, education, health and beauty, history, illustrated, journals, language and linguistics, leisure, multimedia, music, psychology, reference, science, textbooks, tourism. Italian publishers of National Geographical Society books.

Casa Editrice Longanesi SpA

Via Gherardini 10, I–20145 Milan
T 00 39 2 34597620   F 00 39 2 34597212
info@longanesi.it
www.longanesi.it

Founded 1946. Mass market paperback publisher of fiction, adventure, archaeology, art, biography, essays, fantasy, history, journalism, maritime, philosophy, popular science, thrillers.

Arnoldo Mondadori Editore SpA

Via Mondadori 1, I–20090 Segrate (Milan)
T 00 39 2 75421   F 00 39 2 75422302
www.mondadori.it

Founded 1907. Italy's largest publisher. Publishes fiction and non-fiction, mystery, romance, art, biography, children's and young adult, directories, history, how-to, journals, medicine, music, poetry, philosophy, psychology, reference, religion, general science, education, textbooks.

Società editrice il Mulino

Strada Maggiore 37, I–40125 Bologna
T 00 39 51 256011   F 00 39 51 256034
info@mulino.it
www.mulino.it

Founded 1954. Publishes economics, government, history, law, language arts, linguistics, literature, literary criticism, philosophy, political science, psychology, psychiatry, religion, social sciences, sociology, textbooks, journals.

Gruppo Ugo Mursia Editore SpA

Via Melchiorre Gioia 45, I–20124 Milan
T 00 39 2 6737 8500   F 00 39 2 6737 8605
info@mursia.com
www.mursia.com

Founded 1941. Publishes fiction, art, biography, directories, education, history, maritime, philosophy, reference, religion, sport, general science, social sciences, textbooks, juvenile and young adult.

Rizzoli Editore

Via Mecenate 91, I–20138 Milan
T 00 39 2 50951
www.rcslibri.it

Founded 1945. Part of RCS Libri publishing group. Publishes fiction and non-fiction, juvenile and young adult.

Società Editrice Internazionale – SEI

Corso Regina Margherita 176, I–10152 Turin
T 00 39 11 5227 1   F 00 39 11 5211 320
www.seieditrice.com

Founded 1908. Publishes children's, dictionaries, education, encyclopedias, textbooks.

Sonzogno

Via Mecenate 91, I–20138 Milan
T 00 39 2 50951   F 00 39 2 5065361
www.rcslibri.it

Founded 1818. Part of RCS Libri publishing group. Publishes fiction, mysteries, and general non-fiction.

Sperling e Kupfer Editori SpA

Via Marco D'Aviano 2, I–20131 Milan
T 00 39 2 28523 1   F 00 39 2 28523 277
info@sperling.it
www.sperling.it

Founded 1899. Publishes fiction and general non-fiction; young adult, biography, health, history, science, sport, thrillers.

Sugarco Edizioni Srl

Via don Gnocchi 4, I–20148 Milan
T 00 39 2 407 8370   F 00 39 2 407 8493
info@sugarcoedizioni.it
www.sugarcoedizioni.it

Founded 1956. Publishes fiction, biography, history, how-to, philosophy.

Todariana Editrice

Via Gardone 29, I–29139 Milan
T 00 39 2 5681 2953   F 00 39 2 5521 3405
toeura@tin.it
www.todariana-eurapress.it

Founded 1967. Publishes fiction, poetry, science fiction, fantasy, literature, literary criticism, language arts, linguistics, psychology, psychiatry, social sciences, sociology, travel.

The Netherlands

A.W. Bruna Uitgevers BV

Postbus 40203, NL–3504 AA Utrecht
T 00 31 30 247 0411   F 00 31 30 241 0018
info@awwbruna.nl
www.awbruna.nl

Founded 1868. Publishes fiction and general non-fiction; mysteries, thrillers, computer science, CD-ROMs, history, philosophy, psychology, psychiatry, general science, social sciences, sociology.

Uitgeverij BZZTÔH BV

Laan van Meerdervoort 10, NL–2517 AJ The Hague
T 00 31 70 363 2934   F 00 31 70 363 1932
info@bzztoh.nl
www.bzztoh.nl

Founded 1970. Publishes fiction, mysteries, romance, literature, literary criticism, foreign language, bilingual, general non-fiction, astrology, biography, cookery, finance, self-help, health and nutrition, lifestyle, occult, sport, thrillers, translations, travel.

Uitgeversmaatschappij J. H. Kok BV

Postbus Box 5018, NL–8260 GA Kampen
℡ 00 31 38 339 2555
www.kok.nl

Founded 1894. Publishes fiction, history, religion, directories, encyclopedias, reference, psychology, general science, social sciences, sociology, textbooks, juvenile and young adult.

Uitgeverij J.M. Meulenhoff BV

Postbus 100, NL–1000 AC Amsterdam
℡ 00 31 20 553 3500   🖷 00 31 20 625 1135
info@meulenhoff.nl
www.meulenhoff.nl

Founded 1895. Publishes international co-productions, fiction and general non-fiction. Specializes in Dutch and translated literature.

Pearson Education Benelux

Postbus 75598, NL–1070 AN Amsterdam
℡ 00 31 20 575 5800   🖷 00 31 20 664 5334
www.pearsoneducation.nl

Founded 1942. Publishes education, business, computer science, directories, economics, management, reference, textbooks, technology.

Uitgeverij Het Spectrum BV

Postbus 2073, NL–3500 GB Utrecht
℡ 00 31 30 265 0650   🖷 00 31 30 262 0850
het@spetrum.nl
www.spectrum.nl

Founded 1935. Publishes general non-fiction, children's, dictionaries, encyclopedias, health, illustrated, language, management, parenting, reference, science, sport, travel.

Springer Science+Business Media BV

Van Godewijckstraat 30, NL–3311 GX Dordrecht
℡ 00 31 78 657 6000   🖷 00 31 78 657 6555
www.springer.com

Publishes ebooks, CD-ROMs, databases and journals; biomedicine, computer science, economics, engineering, humanities, life sciences, medicine, social sciences, mathematics, physics.

Uitgeverij Unieboek BV

Postbus 97, NL–3990 DB Houten
℡ 00 31 30 799 8300   🖷 00 31 30 799 8398
info@unieboek.nl
www.unieboek.nl

Founded 1891. Publishes fiction: mysteries, romance, thrillers; general non-fiction, business, lifestyle, travel, reference, juvenile and young adults.

Norway

H. Aschehoug & Co (W. Nygaard)

Postboks 363 Sentrum, N–0102 Oslo
℡ 00 47 22 400 400   🖷 00 47 22 206 395
epost@aschehoug.no
www.aschehoug.no

Founded 1872. Publishes fiction and general non-fiction, children's, reference, popular science, hobbies, education and textbooks, translations.

Cappelen Damm AS

Postboks 350 Sentrum, N–0101 Oslo
℡ 00 47 21 616 500   🖷 00 47 21 365 040
www.cappelen.no

Founded 1829. Owned by the Egmont Group. Publishes fiction (including foreign fiction), general non-fiction, maps and atlases, juvenile and young adult, dictionaries, directories, encyclopedias, reference, religion, textbooks.

Gyldendal Norsk Forlag

Postboks 6860, St Olavs Plass, N–0130 Oslo
℡ 00 47 22 034 100   🖷 00 47 22 034 105
gnf@gyldendal.no
www.gyldendal.no

Founded 1925. Publishes fiction, science fiction, children's, dictionaries, directories, encyclopedias, journals, reference, textbooks.

Tiden Norsk Forlag

Postboks 6704, St Olavs Plass, N–0130 Oslo
℡ 00 47 23 327 660   🖷 00 47 23 327 697
tiden@tiden.no
www.tiden.no

Founded 1933. Part of **Gyldendal Norsk Forlag** group. Publishes Norwegian and translated fiction and general non-fiction.

Poland

Ameet Sp. zo.o.

Przybyszewskiego 176/178, PL–93–120 Lodz
℡ 00 48 42 676 27 78   🖷 00 48 42 676 28 19
ameet@ameet.com.pl
www.ameet.pl

Children's publisher.

Nasza Ksiegarnia Publishing House

Sarabandy 24c, PL–02–868 Warsaw
℡ 00 48 22 643 93 89   🖷 00 48 22 643 70 28
www.nk.com.pl

Founded 1921. Publishes books for children and young adults .

Polish Scientific Publishers PWN
ul. Miodowa 10, PL–00–251 Warsaw
℡ 00 48 22 69 54 180   📠 00 48 22 69 54 288
www.pwn.pl

Part of the PWN Group. Leading publisher of scientific, educational, professional and reference material.

Publicat SA
ul. Chlebowa 24, PL–61-003 Poznan
℡ 00 48 61 652 92 52   📠 00 48 61 652 92 00
office@publicat.pl
www.publicat.pl

Publishes fiction and non-fiction; biography, children's, crime, education, geography, history, religion.

Portugal

Bertrand Editora Lda
Rua Anchieta 15, P–1249–060 Lisbon
℡ 00 351 213 476 122
info@bertrand.pt
www.bertrand.pt

Founded 1727. Publishes bilingual, dictionaries, encyclopedias, juvenile and young adult.

Editorial Caminho SARL
Avenida Almirante Gago Coutinho 121, P–17100–029 Lisbon
℡ 00 351 21 842 9830   📠 00 351 21 842 9849
www.editorial-caminho.pt

Founded 1975. Publishes fiction and non-fiction, children's and young adult; art, biography, economics, history, linguistics, music, philosophy, photography, psychology, political science, religion.

Civilização Editora
Rua Alberto Aires de Gouveia 27, P–4050–023 Porto
℡ 00 351 22 605 0917   📠 00 351 22 605 0999
info@civilizacao.pt
www.civilizacao.pt

Founded 1921. Publishes fiction, art, biography, cookery, dictionaries, economics, encyclopedias, history, political science, religion, sociology, social sciences, theatre, children's and young adult.

Publicações Europa-América Lda
Rua Francisco Lyon de Castro 2, Apartado 8, P–2725–354 Mem Martins
℡ 00 351 21 926 7700   📠 00 351 21 926 7771
www.europa-america.pt

Founded 1945. Publishes fiction and non-fiction, children's, directories, reference, textbooks.

Gradiva–Publicações S.A.
Rua Almeida e Sousa, 21 R/C Esq, P–1399–041 Lisbon
℡ 00 351 21 397 4067   📠 00 351 21 395 3471
geral@gradiva.mail.pt
www.gradiva.pt

Founded 1981. Publishes academic, fiction, literature, literary criticism, children's, juvenile and young adult, directories, illustrated, reference, translations.

Livros Horizonte Lda
Rua das Chagas 17–1 Dto, P–1200–106 Lisbon
℡ 00 351 21 346 6917
www.livroshorizonte.pt

Founded 1953. Publishes children's and young adult, education, textbooks.

Editorial Verbo SA
Av António Augusto de Aguiar 148, P–1069–019 Lisbon
℡ 00 351 21 380 1100   📠 00 351 21 386 5397
www.editorialverbo.pt

Founded 1958. Publishes dictionaries, education, encyclopedias, history, general science, juvenile and young adult.

Spain

Alianza Editorial SA
Juan Ignacio Luca de Tena 15, E–28027 Madrid
℡ 00 34 91 393 8890   📠 00 34 91 320 4408
www.alianzaeditorial.es

Founded 1966. Part of **Grupo Anaya**. Publishes fiction, poetry, art, history, mathematics, philosophy, government, political science, social sciences, sociology, general science.

Groupo Anaya
Juan Ignacio Luca de Tena 15, E–28027 Madrid
℡ 00 349 1 393 8600   📠 00 349 1 742 6631
www.anaya.es

Founded 1959. Publishes education, reference, literature, multimedia, children's and young adult.

EDHASA (Editora y Distribuidora Hispano – Americana SA)
Av Diagonal 519–521, 2 piso, E–08029 Barcelona
℡ 00 34 93 494 9720   📠 00 34 93 419 4584
info@edhasa.es
www.edhasa.es

Founded 1946. Publishes fiction, literature, literary criticism, fantasy, essays, biography, history.

Editorial Espasa-Calpe SA

Carreterade Irún Km 12, 200, E–28049 Madrid
℡ 00 34 91 358 9689   🖷 00 34 91 358 9364
www.espasa.com

Founded 1926. Part of Groupo Planeta. Publishes
fiction, children's, biography, essays, education,
literature, CD-ROMs, dictionaries, encyclopedias,
reference, maps and atlases.

Grijalbo

Travessera de Gracia 47–49, E–08021 Barcelona
℡ 00 34 93 366 0300   🖷 00 34 93 366 0449
www.grijalbo.com

Founded 1962. An imprint of **Random House
Mondadori**. Publishes fiction (historical
and thrillers) and non-fiction (self-help and
inspirational).

Ediciónes Hiperión SL

Calle de Salustiano Olózaga 14, E–28001 Madrid
℡ 00 34 91 557 6015   🖷 00 34 91 435 8690
info@hiperion.com
www.hiperion.com

Founded 1976. Publishes literature, essays, poetry,
bilingual, foreign language, children's, religion,
translations.

Grupo Editorial Luis Vives

Xaudaró 25, E–28034 Madrid
℡ 00 34 91 334 4883   🖷 00 34 91 334 4882
dediciones@edelvives.es
www.edelvives.es

Founded 1890. Educational publisher.

Editorial Molino SL

Pérez Galdós 36, E–08012 Barcelona
℡ 00 34 93 226 0625
www.rba.es/holding

Founded 1933. Part of Grupo RBA. Publishes
children's and young adult books.

Pearson Educacion SA

Ribera del Loira 28, E–28042 Madrid
℡ 00 34 91 382 8300   🖷 00 34 91 382 8329
pearson.educacion@pearson.com
www.pearsoneducacion.com

Founded 1942. Educational publishers.

Editorial Planeta SA

Calle Còrsega 273–279, E–08008 Barcelona
℡ 00 34 93 228 5800
www.planeta.es

Founded 1952. Part of Groupo Planeta SA.
Publishes fiction and general non-fiction.

Random House Mondadori

Travessera de Gracia 47–49, E–08021 Barcelona
℡ 00 349 3 366 0300   🖷 00 349 3 366 0449
www.randomhousemondadori.es

Publishes fiction and non-fiction; children's and
young adult, art, bilingual dictionaries, biography,
economics, essays, guides, illustrated, literature,
memoirs, poetry, reference, science, self-help,
sociology, technology, thrillers,

Editorial Seix Barral SA

Avda Diagonal 662–664, 7°, E–08034 Barcelona
℡ 00 349 3 496 7003   🖷 00 349 3 496 7004
editorial@seix-barral.es
www.seix-barral.es

Founded 1911. Part of Grupo Planeta SA. Foreign
language publisher of literature.

Tusquets Editores

Cesare Cantù 8, E–08023 Barcelona
℡ 00 349 3 253 0400
www.tusquets-editores.com

Founded 1968. Publishes fiction, art,
autobiography, biography, cinema, cookery, food,
history, essays, literature, memoirs, philosophy,
poetry, social sciences, theatre.

Sweden

Bonnier Books

PO Box 3159, SE–103 63 Stockholm
℡ 00 46 8 696 8000
info@bok.bonnier.se
www.bok.bonnier.se

Founded 2002. Part of the Bonnier AB group.
Publishes fiction, non-fiction and children's
books.

Albert Bonniers Förlag

PO Box 3159, SE-103 63 Stockholm
℡ 00 46 8 696 8620
info@abforlag.bonnier.se
www.albertbonniersforlag.com

Founded 1837. Part of Bonnier Books. Publishes
fiction and general non-fiction.

Bra Böcker AB

PO Box 892, SE-20180 Malmö
℡ 00 46 40 665 4693
www.bbb.se

Founded 1965. Publishes fiction and non-fiction.

Brombergs Bokförlag AB

PO Box 12886, SE-112 98 Stockholm
℡ 00 46 8 562 620 80   🖷 00 46 8 562 620 85
info@brombergs.se
www.brombergs.se

Founded 1975. Publishes literary fiction and
general non-fiction.

Bokförlaget Forum AB

PO Box 70321, SE-107 23 Stockholm
ⓉⓉ 00 46 8 696 8440   Ⓕ 00 46 8 696 8367
www.forum.se

Founded 1944. Publishes fiction (including foreign) and general non-fiction.

Bokförlaget Natur och Kultur

PO Box 27 323, SE-102 54 Stockholm
Ⓣ 00 46 8 453 8600   Ⓕ 00 46 8 453 8790
info@nok.se
www.nok.se

Founded 1922. Publishes fiction and general non-fiction; education and textbooks.

Norstedts Förlag

Box 2052, SE-103 12 Stockholm
Ⓣ 00 46 8 769 88 50   Ⓕ 00 46 8 769 88 64
info@norstedts.se
www.norstedts.se

Founded 1823. Sweden's oldest publishing house. Publishes fiction and general non-fiction.

Rabén och Sjögren Bokförlag

PO Box 2052, SE-103 12 Stockholm
Ⓣ 00 46 8 769 8800   Ⓕ 00 46 8 769 8813
info@raben.se
www.raben.se

Founded 1941. Part of Norstedts Förlagsgrupp. Publishes children's and young adult books.

B. Wählströms Bokförlag AB

PO Box 6630, SE-113 84 Stockholm
Ⓣ 00 46 8 728 2300   Ⓕ 00 46 8 618 9761
info@wahlstroms.se
www.wahlstroms.se

Founded 1911. Part of Forma Publishing Group AB. Publishes fiction and general non-fiction, juvenile and young adult.

Switzerland

Arche Verlag AG

Niederdorfstr 90, CH-8001 Zurich
Ⓣ 00 41 1 252 24 10   Ⓕ 00 41 1 261 11 15
info@arche-verlag.com
www.arche-verlag.com

Founded 1944. Publishes literature and literary criticism, essays, biography, fiction, poetry, music, dance, travel.

Diogenes Verlag AG

Sprecherstr 8, CH-8032 Zurich
Ⓣ 00 41 1 254 85 11   Ⓕ 00 41 1 252 84 07
info@diogenes.ch
www.diogenes.ch

Founded 1952. Publishes fiction, essays, literature, literary criticism, art, children's.

Neptun-Verlag

Erlenstrasse 2, CH-8280 Kreuzlingen
Ⓣ 00 41 71 677 96 55   Ⓕ 00 41 72 677 96 50
neptun@bluewin.ch
www.neptunart.ch

Founded 1946. Publishes non-fiction and children's books.

Orell Füssli Verlag

Dietzingerstr 3, CH-8036 Zurich
Ⓣ 00 41 1 466 77 11   Ⓕ 00 41 1 466 74 12
www.ofv.ch

Founded 1519. Publishes accountancy, architecture, art, art history, business, leisure, photography, politics, psychology.

Payot Libraire

Case Postale 6730, CH-1002 Lausanne
Ⓣ 00 41 21 341 32 52   Ⓕ 00 41 21 341 32 17
www.payot-libraire.ch

Founded 1875. Publishes fiction and general non-fiction; arts and entertainment, audio, business, children's, education, language, lifestyle, professional, reference, travel.

Sauerländer Verlage AG

Ausserfeldstr. 9, CH-5036 Oberentfelden
Ⓣ 00 41 62 836 86 86   Ⓕ 00 41 62 836 86 20
verlag@sauerlaender.ch
www.sauerlaender.ch

Founded 1807. Publishes education and general non-fiction.

US Publishers

Anyone corresponding with the big American publishers should know that they are wary of any package or letter that looks in any way suspicious. This, of course, begs the question what is or what is not suspicious which is as difficult to answer as what makes or does not make a good novel. The best policy is an initial enquiry by e-mail.

International Reply Coupons (IRCs)

For return postage, send IRCs, available from post offices. Letters, 60 pence; mss according to weight. For current postal rates to the UK, go to the United States Postal Service website at www.usps.com and click on 'Calculate Postage'.

ABC–CLIO

PO Box 1911, Santa Barbara CA 93116–1911
☎ 001 805 968 1911   🖷 001 805 685 9685
www.abc-clio.com
President/CEO *Ronald J. Boehm*

Founded 1953. Publishes academic and reference, focusing on history and social studies; books and multimedia. UK SUBSIDIARY **ABC-Clio Ltd**, Oxford. No unsolicited mss; synopses and ideas welcome.

Abingdon Press

PO Box 801, Nashville TN 37202–0801
☎ 001 615 749 6290   🖷 001 615 749 6056
www.abingdonpress.com

Founded 1789. Publishes non-fiction: religious (Protestant), reference, professional and academic texts. IMPRINTS **Dimensions for Living**; **Kingswod Books**. Over 100 titles a year.

Harry N. Abrams, Inc.

115 West 18th Street, New York NY 10011
☎ 001 212 206 7715   🖷 001 212 229 0312
www.hnabooks.com
VP/Editor-in-Chief *Eric Himmel*

Founded 1950. Publishes illustrated books: art, natural history, photography. No fiction. About 200 titles a year.

Academy Chicago Publishers

363 West Erie Street, 7E, Chicago IL 60610
☎ 001 312 751 7300   🖷 001 312 751 7306
info@academychicago.com
www.academychicago.com
President & Senior Editor *Anita Miller*

Founded 1975. Publishes mainstream fiction; non-fiction: art, history, mysteries. No romance, children's, young adult, religious, sexist or avant-garde. See website for submission guidelines.

Ace Books
▷ **Penguin Group (USA) Inc.**

University of Alabama Press

Box 870380, Tuscaloosa AL 35487–0380
☎ 001 205 348 5180   🖷 001 205 348 9201
www.uapress.ua.edu
Director *Daniel J.J. Ross*

Founded 1945. Publishes American history, Latin American history, American religious history; African-American and Native American studies, Judaic studies, archaeology, anthropology, enthnohistory and American letters. About 70 titles a year. Submissions are not invited in poetry, fiction or drama.

Aladdin Paperbacks
▷ **Simon & Schuster Children's Publishing**

University of Alaska Press

PO Box 756240, University of Alaska, Fairbanks
AK 99775–6240
☎ 001 907 474 5831   🖷 001 907 474 5502
fypress@uaf.edu
www.uaf.edu/uapress

Publishes scholarly works about Alaska and the North Pacific rim, with a special emphasis on circumpolar regions, history, anthropology, natural history. 12 titles a year. Queries and proposals welcome (see 'For Authors' page on the website for proposal guidelines).

Alpha Books
▷ **Penguin Group (USA) Inc.**

AMACOM Books

American Management Association, 1601 Broadway, New York NY 10019

☎ 001 212 586 8100
www.amacombooks.org
Editor-in-Chief *Adrienne Hickey*

Founded 1972. Owned by American Management Association. Publishes business books only, including general management, business communications, sales and marketing, finance, computers and information systems, human resource management and training, career/personal growth skills. About 80 titles a year.

Amistad
▷ HarperCollins Publishers, Inc.

Julie Andrews Collection
▷ HarperCollins Publishers, Inc.

Anvil
▷ Krieger Publishing Co.

Shaye Areheart Books
▷ Crown Publishing Group

University of Arizona Press
355 S. Euclid Avenue, Suite. 103, Tucson AZ 85719
☎ 001 520 621 1441　🖷 001 520 621 8899
uapress@uapress.arizona.edu
www.uapress.arizona.edu
Director *Christine Szuter*

Founded 1959. Publishes academic non-fiction, particularly with a regional/cultural link, plus Native-American and Hispanic literature. About 55 titles a year.

University of Arkansas Press
McIlroy House, 105 N. McIlroy Avenue, Fayetteville AR 72701
☎ 001 479 575 3246　🖷 001 479 575 6044
www.uapress.com
Director *Lawrence J. Malley*

Founded 1980. Publishes scholarly monographs, poetry and general trade books. Particularly interested in scholarly works in history, politics and literary criticism. About 20 titles a year.

Atheneum Books for Young Readers
▷ Simon & Schuster Children's Publishing

Atlantic Monthly Press
▷ Grove/Atlantic Inc.

Atria Books
▷ Simon & Schuster Adult Publishing Group

Avalon Travel
▷ Perseus Books Group

Avery
375 Hudson Street, New York NY 10014
☎ 001 212 366 2000　🖷 001 212 366 2643
us.penguingroup.com
Publisher *Megan Newman*

Founded 1976. Imprint of **Penguin Group (USA) Inc.** Publishes alternative and complementary health care, nutrition, disease prevention, diet, supplements, functional foods, mind/body therapies, holistic healing, narrative medical and science non-fiction. About 30 titles a year. No unsolicited mss; synopses and ideas welcome if accompanied by s.a.e./IRCs.
🖭 Royalties twice-yearly.

Avon
▷ HarperCollins Publishers, Inc.

Back Bay Books
▷ Hachette Book Group USA

Ballantine Books
▷ Random House Publishing Group

Banner Books
▷ University Press of Mississippi

Bantam Dell Publishing Group
1745 Broadway, New York NY 10019
☎ 001 212 782 9000
www.randomhouse.com/bantamdell
President & Publisher *Irwyn Applebaum*

Founded 1945. The largest mass market paperback publisher in the USA. A division of the **Random House Publishing Group.** Publishes general commercial fiction and non-fiction, young readers and children's: mysteries, science fiction and fantasy, romance, health and nutrition, self-help.

DIVISIONS/IMPRINTS

Bantam Books; Delacorte Press; Dell; Delta; Dial Press; Spectra.

Barron's Educational Series, Inc.
250 Wireless Boulevard, Hauppauge NY 11788
☎ 001 631 434 3311　🖷 001 631 434 3723
barrons@barronseduc.com
www.barronseduc.com
Chairman/CEO *Manuel H. Barron*
President *Ellen Sibley*

Founded 1941. Publishes adult non-fiction, children's fiction and non-fiction, test preparation materials and language materials, cookbooks, pets, hobbies, sport, photography, health, business, law, computers, art and painting. No adult fiction. Over 300 titles a year. Unsolicited mss, synopses and ideas for books welcome.
🖭 Royalties twice-yearly.

Basic Books/Basic Civitas
▷ Perseus Books Group

Beacon Press
25 Beacon Street, Boston MA 02108
☎ 001 617 742 2110
www.beacon.org

Director *Helene Atwan*

Founded 1854. Publishes general non-fiction. About 60 titles a year. Does not accept unsolicited mss. For further information, refer to the website.

Berkley
▷ Penguin Group (USA) Inc.

Bison Books
▷ University of Nebraska Press

Blackbirch Press
▷ Thomson Gale

Blackwell Publishing
2121 State Avenue, Ames IA 50014–8300
☎ 001 515 292 0140   ℻ 001 515 292 3348
acquisitions@blackwellprofessional.com
www.blackwellprofessional.com
www.blackwellpublishing.com
Vice President, Professional Publishing *Antonia Seymour*

Publishes reference books and textbooks on plant science, applied chemistry, food science, human dentistry, aquaculture, agriculture, animal science and veterinary medicine. No fiction, trade books or poetry. 60 titles a year. Submissions welcome if within the subject areas listed above.
Ⓔ Royalties annually

Boyds Mills Press
815 Church Street, Honesdale PA 18431
☎ 001 570 253 1164
contact@boydsmillspress.com
www.boydsmillspress.com
Publisher *Stephen Roxburgh*

A subsidiary of Highlights for Children, Inc. Founded 1990 as a publisher of children's picture and activity books. Publishes children's and young adult's fiction, non-fiction and poetry under five IMPRINTS: **Boyds Mills Press; Calkins Creek; Front Street; Lemniscaat; Wordsong**. About 80 titles a year. Unsolicited material welcome; submission guidelines available on the website.

Brassey's, Inc.
▷ Potomac Books, Inc.

Broadway
▷ Random House Publishing Group

Bulfinch Press
▷ Hachette Book Group USA

Business Plus
▷ Hachette Book Group USA

Caedmon
▷ HarperCollins Publishers, Inc.

University of California Press
2120 Berkeley Way, Berkeley CA 94704
☎ 001 510 642 4247   ℻ 001 510 643 7127
askucp@ucpress.edu
www.ucpress.edu
Director *Lynne Withey*

Founded 1893. Publishes scholarly and scientific non-fiction; some fiction in translation. About 200 books and 50 journals annually. Preliminary letter with outline preferred.

Calkins Creek
▷ Boyds Mills Press

Canongate US
▷ Grove/Atlantic Inc.

Carolrhoda Books, a division of Lerner Publishing Group, Inc.
241 First Avenue N, Minneapolis MN 55401
☎ 001 612 332 3344   ℻ 001 612 332 7615
www.lernerbooks.com

Founded 1969. A division of the **Lerner Publishing Group, Inc**, Carolrhoda publishes hardcover originals. Averages 8–10 picture books each year for ages 3–8; 6 fiction titles for ages 7–18; 2–3 non-fiction titles for various ages. No longer accepting unsolicited submissions.

Center Street
▷ Hachette Book Group USA

Charlesbridge Publishing
85 Main Street, Watertown MA 02472
☎ 001 617 926 0329   ℻ 001 617 926 5775
books@charlesbridge.com
www.charlesbridge.com
President/Publisher *Brent Farmer*

Publishes both picture books and transitional 'bridge books' (books ranging from early readers to middle-grade chapter books); fiction; and non-fiction with a focus on nature, science, social studies and multicultural topic.

The University of Chicago Press
1427 East 60th Street, Chicago IL 60637
☎ 001 773 702 7700   ℻ 001 773 702 9756
www.press.uchicago.edu

Founded 1891. Publishes academic non-fiction only.

Children's Press
▷ Scholastic Library Publishing

Chronicle Books LLC
680 Second Street, San Francisco CA 94107
☎ 001 415 537 4200   ℻ 001 415 537 4460
frontdesk@chroniclebooks.com
www.chroniclebooks.com
President/Publisher *Jack Jensen*

Founded 1967. Publishes illustrated and non-illustrated adult trade and children's books as well as stationery and gift items. About 200 titles a year. Query or submit outline/synopsis and sample chapters and artwork.

Arthur H. Clarke Company
▷ University of Oklahoma Press

Clarkson Potter
▷ Crown Publishing Group

Collins
▷ HarperCollins Publishers, Inc.

Columbia University Press
61 West 62nd Street, New York NY 10023
☎ 001 212 459 0600   ℻ 001 212 459 3678
www.columbia.edu/cu/cup
President & Director *James D. Jordan*
Associate Director *Jennifer Crewe*

Founded 1893. Publishes scholarly and general interest non-fiction, reference, translations of Asian literature. No fiction or poetry. Welcomes unsolicited material if the subject fits their programme (see website). Approach by regular mail only.

Joanna Cotler Books
▷ HarperCollins Publishers, Inc.

Crocodile Books, USA
▷ Interlink Publishing Group, Inc.

Crown Publishing Group
1745 Broadway, New York NY 10019
☎ 001 212 782 9000
www.randomhouse.com/crown
VP & Publisher *Tina Constable*

Founded 1933. Division of the **Random House Publishing Group**. Publishes popular trade fiction and non-fiction.

IMPRINTS **Clarkson Potter** *Lauren Shakely* Illustrated books: cookery, gardening, style, decorating and design; **Crown Business** Business books; **Crown** *Steve Ross* General fiction and non-fiction; **Harmony Books** *Shaye Areheart* New Age, spirituality, religion, some fiction; **Shaye Areheart Books** *Shaye Areheart* Fiction; **Three Rivers Press** *Philip Patrick* Non-fiction paperbacks. Mss submissions by agents only.

Currency
▷ Doubleday Broadway Publishing Group

DaCapo
▷ Perseus Books Group

DAW Books, Inc.
375 Hudson Street, 3rd Floor, New York NY 10014–3658
☎ 001 212 366 2096   ℻ 001 212 366 2090
daw@penguingroup.com
www.dawbooks.com

Publishers *Elizabeth R. Wollheim, Sheila E. Gilbert*
Associate Editor *Peter Stampfel*

Founded 1971 by Donald and Elsie Wollheim as the first mass-market publisher devoted to science fiction and fantasy. An imprint of **Penguin Group (USA) Inc.** Publishes science fiction/fantasy, and some horror. No short stories, anthology ideas or non-fiction. About 42 titles a year. Unsolicited mss, synopses and ideas for books welcome.
£ Royalties twice-yearly.

Del Rey
▷ Random House Publishing Group

Delacorte Press
▷ Bantam Dell Publishing Group

Dell
▷ Bantam Dell Publishing Group

Delta Books
▷ Bantam Dell Publishing Group

Dial Books for Young Readers
345 Hudson Street, New York NY 10014–3657
☎ 001 212 366 2000
Queries *Submissions Coordinator*

Founded 1961. A division of Penguin Books for Young Readers. Publishes children's books, including picture books, beginning readers, fiction and non-fiction for middle grade and young adults.

IMPRINTS

Hardcover only: **Dial Books for Young Readers**; **Dial Easy-to-Read**. 50 titles a year. Accepts unsolicited picture book mss and up to ten pages for longer works with query letter. 'Do not send self-addressed envelope with submissions. Dial will respond within four months if interested in a manuscript.'
£ Royalties twice-yearly.

Dial Press
▷ Bantam Dell Publishing Group

Dimensions for Living
▷ Abingdon Press

Doubleday Broadway Publishing Group
1745 Broadway, New York NY 10019
☎ 001 212 782 9000
www.randomhouse.com
President & Publisher *Stephen Rubin*

The Doubleday Broadway Publishing Group, a division of the **Random House Publishing Group**, was formed by a merger of Doubleday and Broadway Books in 1998. Publishes fiction and non-fiction. DIVISION/IMPRINTS **Currency**; **Nan A. Talese Books**. No unsolicited material.

Downtown Press
▷ Simon & Schuster Adult Publishing Group

Thomas Dunne Books
▷ St Martin's Press LLC

Dutton/Dutton's Children's Books
▷ Penguin Group (USA) Inc.

East Gate Books
▷ M.E. Sharpe, Inc.

Ecco
▷ HarperCollins Publishers, Inc.

Eerdmans Publishing Company

2140 Oak Industrial Drive, NE, Grand Rapids
MI 49505
☎ 001 616 459 4591   🖷 001 616 459 6540
info@eerdmans.com
www.eerdmans.com
President *William B. Eerdmans Jr*

Founded in 1911 as a theological and reference
publisher. Gradually began publishing in other
genres with authors like C.S. Lewis, Dorothy
Sayers and Malcolm Muggeridge on its lists.
Publishes religious: theology, biblical studies,
ethical and social concern, social criticism and
children's, religious history, religion and literature.

DIVISIONS **Children's** *Shannon White* Picture
books, novels and biographies for the general
trade market; **Other** *Jon Pott*. About 120 titles
a year. Unsolicited mss, synopses and ideas
welcome.
💷 Royalties twice-yearly.

Eos
▷ HarperCollins Publishers, Inc.

Faber & Faber, Inc.

19 Union Square West, New York NY 10003
☎ 001 212 741 6900   🖷 001 212 633 9385
www.fsgbooks.com/faberandfaber.htm
Senior Editor *Denise Oswald*

Founded 1976. An affiliate of **Farrar, Straus &
Giroux, Inc.** Publishes primarily non-fiction
books for adults with a focus on arts and
entertainment; film, theatre, music, popular
culture and literary and cultural criticism.
Unsolicited mss accepted; please query first and
include IRCs with letter.

FaithWords
▷ Hachette Book Group USA

FalconGuides
▷ The Globe Pequot Press

Farrar, Straus & Giroux, Inc.

19 Union Square West, New York NY 10003
☎ 001 212 741 6900
www.fsgbooks.com

Founded 1946. Publishes general and literary
fiction, non-fiction, poetry and children's books.
No unsolicited material.

Firebird
▷ Penguin Group (USA) Inc.

Fireside
▷ Simon & Schuster Adult Publishing Group

First Avenue Editions
▷ Lerner Publishing Group

Five Star
▷ Thomson Gale

5-Spot
▷ Hachette Book Group USA

Forever
▷ Hachette Book Group USA

Free Press
▷ Simon & Schuster Adult Publishing Group

Samuel French, Inc.

45 West 25th Street, New York NY 10010–2751
☎ 001 212 206 8990   🖷 001 212 206 1429
info@samuelfrench.com
www.samuelfrench.com
Managing Editor *Roxane Heinze-Bradshaw*

Founded 1830. Publishes plays in paperback:
London, Broadway and off-Broadway hits,
light comedies, mysteries, one-act plays and
plays for young audiences. About 50–60 titles
a year. Unsolicited mss welcome. No synopses.
OVERSEAS ASSOCIATES in London, Toronto,
Sydney and Johannesburg.
💷 Royalties annually (books); twice-yearly
(amateur productions); monthly (professional
productions).

Front Street
▷ Boyds Mills Press

Gale
▷ Thomson Gale

Laura Geringer Books
▷ HarperCollins Publishers, Inc.

The Globe Pequot Press

PO Box 480, Guilford CT 06437
☎ 001 203 458 4500   🖷 001 203 458 4601
info@gobepequot.com
www.globepequot.com

Founded 1947. Publishes regional and
international travel, lifestyle, current events,
how-to and outdoor recreation. IMPRINTS include
**Skirt! Books; Lyons; FalconGuides; Insiders'
Guides**. Unsolicited mss, synopses, and ideas
welcome; growth categories include books for

women and lifestyle. See website for submission guidelines.

Gotham
▷ Penguin Group (USA) Inc.

Graham & Whiteside
▷ Thomson Gale

Grand Central Publishing
▷ Hachette Book Group USA

Graphic Universe
▷ Lerner Publishing Group

Green Willow Books
▷ HarperCollins Publishers, Inc.

Greenhaven Press
▷ Thomson Gale

Greenwood-Heinemann
▷ Houghton Mifflin Harcourt Publishing Company

Griffin
▷ St Martin's Press LLC

Grolier
▷ Scholastic Library Publishing

Grosset & Dunlap
▷ Penguin Group (USA) Inc.

Grove/Atlantic Inc.
841 Broadway, 4th Floor, New York NY 10003
① 001 212 614 7850   ② 001 212 614 7886
www.groveatlantic.com
President/Publisher *Morgan Entrekin*
Managing Editor *Michael Hornburg*

Founded 1952. Publishes general fiction and non-fiction. IMPRINTS **Atlantic Monthly Press; Canongate US; Grove Press**. Mss submissions by agents only.

Hachette Book Group USA
237 Park Avenue, New York NY 10017
① 001 212 364 1100
www.hachettebookgroupusa.com
CEO & Chairman *David Young*

Founded in 2006 following the acquisition of the Time Warner Book Group by Hachette Livre. Publishes fiction and non-fiction, children's and young adult, audio books.

DIVISIONS **Grand Central Publishing; FaithWords; Center Street; Little, Brown and Company; Little, Brown Books for Young Readers; Orbit; Yen Press; Hachette Book Group Digital Media**.

IMPRINTS **Back Bay Books; Bulfinch Press; Business Plus; 5-Spot; Forever; Hachette Audio; LB Kids; Springboard Press; Poppy; Twelve; Vision; Wellness Central**. About 660 titles a year. No unsolicited mss.

G.K. Hall
▷ Thomson Gale

Harcourt
▷ Houghton Mifflin Harcourt Publishing Company

Harmony Books
▷ Crown Publishing Group

HarperCollins Publishers, Inc.
10 East 53rd Street, New York NY 10022
① 001 212 207 7000
www.harpercollins.com
President/Publisher US General Books *Michael Morrison*

Founded 1817. Subsidiary of News Corporation. Publishes general and literary fiction, general non-fiction, business, children's, reference and religious books. About 1,700 titles a year.

GENERAL BOOKS GROUP IMPRINTS

Amistad; Avon; Avon A; Avon Inspire; Avon Red; Caedmon; Collins; Collins Design; Ecco; Eos; Harper; Harper Paperbacks; Harper Perennial; Harper Perennial Modern Classics; HarperAudio; HarperEntertainment; HarperCollins; HarperCollins e-Books; HarperLuxe; Harper One; Rayo; William Morrow; Morrow Cookbooks.

CHILDREN'S IMPRINTS

Amistad; Eos; Greenwillow Books; HarperCollins Children's Audio; HarperCollins Children's Books; HarperFestival; HarperEntertainment; HarperTeen; HarperTrophy; Joanna Cotler Books; Julie Andrews Collection; Katherine Tegen Books; Laura Geringer Books; Rayo. No unsolicited material.

Harvard University Press
79 Garden Street, Cambridge MA 02138
① 001 401 531 2800   ② 001 401 531 2001
contact_HUP@harvard.edu
www.hup.harvard.edu
Editor-in-Chief *Michael Fisher*

Founded 1913. Publishes scholarly non-fiction only; humanities, social sciences, sciences.

University of Hawai'i Press
2840 Kolowalu Street, Honolulu HI 96822–1888
① 001 808 956 8255   ② 001 808 988 6052
uhpbooks@hawaii.edu
www.uhpress.hawaii.edu
Director *William H. Hamilton*
Executive Editor *Patricia Crosby*

Founded 1947. Publishes scholarly books pertaining to East Asia, Southeast Asia, Hawaii and the Pacific. Unsolicited synopses and ideas welcome; approach by mail. No poetry, children's books or any topics other than Asia and the Pacific.

Ⓔ Royalties twice-yearly.

Hiddenspring Books
▷ Paulist Press

Hill Street Press LLC

191 E. Broad Street, Suite 216, Athens GA 30601–2848

☏ 001 706 613 7200   Ⓕ 001 706 613 7204

info@hillstreetpress.com

hillstreetpress.com

VP/Editor-in-Chief *Judy Long*

Founded 1998. Publishes books on the American South – fiction and non-fiction. No poetry, erotica, romance, children's or young adults. About 20 titles a year. Unsolicited material welcome; approach in writing in the first instance.

Hippocrene Books, Inc.

171 Madison Avenue, New York NY 10016

☏ 001 212 685 4371

info@hippocrenebooks.com

www.hippocrenebooks.com

President/Editorial Director *George Blagowidow*

Founded 1970. Publishes general non-fiction and reference books. Particularly strong on foreign language dictionaries, language studies and international cookbooks. No fiction. Send brief summary, table of contents and one chapter for appraisal. S.a.e. essential for response. For manuscript return include sufficient postage cover (IRCs).

Holiday House, Inc.

425 Madison Avenue, New York NY 10017

☏ 001 212 688 0085   Ⓕ 001 212 421 6134

www.holidayhouse.com

Vice President/Editor-in-Chief *Mary Cash*

Publishes children's general fiction and non-fiction (pre-school to secondary). About 50 titles a year. Send query letters only. IRCs for reply must be included. Submission guidelines available on the website.

Henry Holt & Company, Inc.

175 Fifth Avenue, New York NY 10010

☏ 001 646 307 5095   Ⓕ 001 212 633 0748

www.henryholt.com

President/Publisher *John Sterling*

Editor-in-Chief, Adult Trade *Jennifer Barth*

Founded in 1866, Henry Holt is one of the oldest publishers in the United States. Part of Holtzbrinck Publishing Holdings. Publishes fiction, by both American and international authors, biographies, children's, health, science, history, politics, ecology and psychology.

DIVISIONS/IMPRINTS **Adult Trade; Books for Young Readers; Metropolitan Books; Owl Books; Times Books**. About 250 titles a year. No unsolicited mss.

Holt McDougal
▷ Houghton Mifflin Harcourt Publishing Company

Houghton Mifflin Harcourt Publishing Company

222 Berkeley Street, Boston MA 02116–3764

☏ 001 617 351 5000   Ⓕ 001 617 351 1125

www.hmco.com

Contact *Submissions Editor*

Founded 1832. Acquired Harcourt in 2008. Publishes literary fiction and general non-fiction, including autobiography, biography and history. Also school textbooks; children's fiction and non-fiction.

DIVISIONS/SUBSIDIARY COMPANIES **Houghton Mifflin Harcourt School Division; Houghton Mifflin Harcourt Trade & Reference Division; Houghton Mifflin Harcourt Learning Technology; Houghton Mifflin Harcourt International Publishers; Houghton Mifflin Harcourt Supplemental Publishers; Greenwood-Heinemann; Holt McDougal; The Riverside Publishing Co**. About 100 titles a year. Unsolicited adult mss no longer accepted. Send synopses, outline and sample chapters for children's non-fiction; complete mss for children's fiction. Do not include s.a.e. 'You will not hear from us regarding the status of your submission unless we are interested, in which case you can expect to hear back within 16 weeks. We cannot track your manuscript and regret that we cannot respond personally to each submission, but we do consider each and every submission we receive. Mss we are not interested in will be recycled.'

Howard Books
▷ Simon & Schuster Adult Publishing Group

HPBooks
▷ Penguin Group (USA) Inc.

Hudson Street Press
▷ Penguin Group (USA) Inc.

Humanity Books
▷ Prometheus Books

University of Illinois Press

1325 South Oak Street, Champaign IL 61820–6903

☏ 001 217 333 0950   Ⓕ 001 217 244 8082

uipress@uillinois.edu

www.press.uillinois.edu

Director *Willis G. Regier*

Associate Director and Editor-in-Chief *Joan Catapano*

Founded 1918. Publishes non-fiction, scholarly and general, with special interest in Americana, women's studies, African–American studies, film, religion, American music and regional books. About 140–150 titles a year.

Indiana University Press
601 North Morton Street, Bloomington
IN 47404–3797
☏ 001 812 855 8817   🖷 001 812 855 8507
iupress@indiana.edu
www.indiana.edu/~iupress
Director *Janet Rabinowitch*

Founded 1950. Publishes scholarly non-fiction in the following subject areas: African studies, anthropology, Asian studies, African-American studies, bioethics, environment and ecology, film, folklore, history, Jewish studies, Middle East studies, military history, music, paleontology, philanthropy, philosophy, politics, religion, Russian and East European studies, women's studies. About 140 books and 30 journals a year. Query in writing in first instance.

Insiders' Guides
▷ The Globe Pequot Press

Interlink Publishing Group, Inc.
46 Crosby Street, Northampton MA 01060
☏ 001 413 582 7054   🖷 001 413 582 7057
info@interlinkbooks.com
www.interlinkbooks.com
Owner/Publisher *Michel Moushabeck*
Editor *Pam Thompson*

Founded 1987. Publishes international fiction, travel, politics, cookbooks. Specializes in Middle East titles and ethnicity. IMPRINTS **Crocodile Books, USA** Editorial Head *Ruth Lane Moushabeck* Children's books; **Olive Branch Press** Editorial Head *Phyllis Bennis* Political books. About 50-60 titles a year. See submission guidelines on the website before making an approach.
🖳 Royalties annually.

University of Iowa Press
100 Kuhl House, 119 West Park Road, Iowa City
IA 52242–1000
☏ 001 319 335 2000   🖷 001 319 335 2055
uipress@uiowa.edu
www.uiowapress.org
Director *Holly Carver*

Founded 1969 as a small scholarly press publishing about five books a year. Now publishing about 35 annually in a variety of scholarly fields, plus local interest, short stories, creative non-fiction and poetry anthologies. No unsolicited mss; query first. Unsolicited ideas and synopses welcome.
🖳 Royalties annually.

Jove
▷ Penguin Group (USA) Inc.

University Press of Kansas
2502 Westbrooke Circle, Lawrence
KS 66045-4444
☏ 001 785 864 4154   🖷 001 785 864 4586
upress@ku.edu
www.kansaspress.ku.edu
Director *Fred M. Woodward*

Founded 1946. Became the publishing arm for all six state universities in Kansas in 1976. Publishes scholarly books in American history, legal studies, presidential studies, American studies, political philosophy, political science, military history and environmental history. About 55 titles a year. Proposals welcome.
🖳 Royalties annually.

Kar-Ben Publishing
▷ Lerner Publishing Group

The Kent State University Press
307 Lowry Hall, PO Box 5190, Kent OH 44242
☏ 001 330 672 7913   🖷 001 330 672 3104
ksupress@kent.edu
www.kentstateuniversitypress.com
Director *Will Underwood*
Assistant Director/Editor-in-Chief *Joanna Hildebrand Craig*

Founded 1965. Publishes scholarly works in history, literary studies and general non-fiction with an emphasis on American history and literature. 30–35 titles a year. Queries welcome; no mss.

The University Press of Kentucky
663 South Limestone Street, Lexington
KY 40508–4008
☏ 001 859 257 8434   🖷 001 859 323 1873
www.kentuckypress.com
Director *Stephen M. Wrinn*

Founded 1943. Publishes scholarly and general interest books in the humanities and social sciences. 58 titles in 2006. No unsolicited mss; send synopses and ideas for books by post or e-mail. No fiction, drama, poetry, translations, memoirs or children's books.
🖳 Royalties annually.

KidHaven Press
▷ Thomson Gale

Kingswood Books
▷ Abingdon Press

Krieger Publishing Co.
PO Box 9542, Melbourne FL 32902–9542
☏ 001 321 724 9542   🖷 001 321 951 3671
info@krieger-publishing.com
www.krieger-publishing.com

CEO *Robert E. Krieger*
President *Donald E. Krieger*
Vice-President *Maxine D. Krieger*

Founded 1970. Publishes science, education, ecology, humanities, history, mathematics, chemistry, space science, technology and engineering.

IMPRINTS/SERIES **Anvil; Exploring Community History; Open Forum; Orbit; Professional Practices; Public History.** Unsolicited mss welcome. Not interested in synopses/ideas or trade type titles.

Large Print Press
▷ Thomson Gale

Lark
▷ Sterling Publishing Co. Inc.

Latino Voices
▷ Northwestern University Press

LB Kids
▷ Hachette Book Group USA

Lemniscaat
▷ Boyds Mills Press

Lerner Publishing Group
241 First Avenue N., Minneapolis MN 55401
☎ 001 612 332 3344   🖷 001 612 332 7615
info@lernerbooks.com
www.lernerbooks.com
Chairman *Harry J. Lerner*
President & Publisher *Adam M. Lerner*

Founded 1959. Publishes children's non-fiction, fiction and curriculum material for all grade levels.

DIVISIONS/IMPRINTS **Lerner Publications; Carolrhoda Books** (see entry); **LernerClassroom; First Avenue Editions; Millbrook Press; Twenty-First Century Books; Kar-Ben Publishing; Graphic Universe.** No puzzle, song or alphabet books, text books, workbooks, or plays. No longer accepting unsolicited submissions.

Little Simon
▷ Simon & Schuster Children's Publishing

Little, Brown and Company/Little, Brown Books for Young Readers
▷ Hachette Book Group USA

Llewellyn Publications
2143 Wooddale Drive, Woodbury MN 55125
☎ 001 651 291 1970   🖷 001 612 291 1908
www.llewellyn.com
President & Publisher *Carl L. Weschcke*

Division of Llewellyn Worldwide Ltd. Founded 1901. Publishes self-help and how-to: astrology, alternative health, tantra, Fortean studies, tarot, yoga, Santeria, dream studies, metaphysics, magic, witchcraft, herbalism, shamanism, organic gardening, women's spirituality, graphology, palmistry, parapsychology. Also fiction with an authentic magical or metaphysical theme. About 100 titles a year. Unsolicited mss welcome; proposals preferred. IRCs essential in all cases. Submission guidelines available on the website.

Louisiana State University Press
3990 West Lakeshore Drive, Baton Rouge LA 70808
☎ 001 225 578 6295   🖷 001 225 578 6461
lsupress@lsu.edu
www.lsu.edu/lsupress
Director *MaryKatherine Callaway*

Publishes fiction and non-fiction: Southern and American history, Southern and American literary criticism, regional trade books, biography, poetry, Atlantic World studies, media studies, environmental studies, geography, political science and music (jazz). About 85 titles a year. See website for submission guidelines.

Lucent Books
▷ Thomson Gale

Lyons
▷ The Globe Pequot Press

Margaret K. McElderry Books
▷ Simon & Schuster Children's Publishing

McFarland & Company, Inc., Publishers
PO Box 611, Jefferson NC 28640
☎ 001 336 246 4460   🖷 001 336 246 5018
info@mcfarlandpub.com
www.mcfarlandpub.com
Executive Vice President *Rhonda Herman*
Editorial Director *Steve Wilson*
Acquisitions Editor *Gary Mitchem*

Founded 1979. A library reference and upper-end speciality market press publishing scholarly books in many fields: international studies, performing arts, popular culture, sports, automotive history, women's studies, music and fine arts, chess, history and librarianship. Specializes in general reference. Especially strong in cinema studies. No fiction, poetry, children's, New Age or inspirational/devotional works. About 320 titles a year. No unsolicited mss; send query letter first. Synopses and ideas welcome; submissions by mail preferred.
Ⓔ Royalties annually.

The McGraw-Hill Companies, Inc.
1221 Avenue of the Americas, New York NY 10020–1095
☎ 001 212 904 2000
www.mcgraw-hill.com
Chairman/President/CEO *Harold W. McGraw*

Founded 1888. Parent of **McGraw-Hill Education** which has offices in the UK (see entry under *UK Publishers*). Publishes a wide range of educational, professional, business, science, engineering and computing books.

Macmillan Reference USA
▷ Thomson Gale

University of Massachusetts Press
PO Box 429, Amherst MA 01004
☎ 001 413 545 2217   🖷 001 413 545 1226
info@umpress.umass.edu
www.umass.edu/umpress
Director *Bruce Wilcox*
Senior Editor *Clark Dougan*

Founded 1963. Publishes scholarly, general interest, African-American, ethnic, women's and gender studies, cultural criticism, economics, fiction, literary criticism, poetry, philosophy, biography, history. About 40 titles a year. Unsolicited mss considered but query letter preferred in the first instance. Synopses and ideas welcome.

Metropolitan Books
▷ Henry Holt & Company, Inc.

The University of Michigan Press
839 Greene Street, Ann Arbor MI 48104–3209
☎ 001 734 764 4388   🖷 001 734 615 1540
www.press.umich.edu

Founded 1930. Publishes scholarly and trade non-fiction, fiction, textbooks, translation, literary criticism, new media studies, theatre, music, economics, political science, history, classics, law studies, gender studies, English as a second language.

Milet Publishing LLC
333 North Michigan Avenue, Suite 530, Chicago IL 60601
info@milet.com
www.milet.com
Contact *Editorial Director*

Founded 1995. Publishes children's picture books in English and dual language, and language learning books. Welcomes synopses and ideas for books. Queries may be made by e-mail but all submissions – proposal, outline or synopsis with sample text and/or artwork – by post only. 'Please review submission guidelines and existing titles on our website before submitting. We like bold, original stories and artwork, universal and/or multicultural.'

Millbrook Press
▷ Lerner Publishing Group

University of Minnesota Press
111 Third Avenue South, Suite 290, Minneapolis MN 55401

☎ 001 612 627 1970   🖷 001 612 627 1980
www.upress.umn.edu
Executive Editor *Richard Morrison*

Founded 1927. Publishes academic books for scholars and selected general interest titles: American studies, anthropology, art and aesthetics, cultural theory, film and media studies, gay and lesbian studies, geography, literary theory, political and social theory, race and ethnic studies, sociology and urban studies. No original fiction or poetry. About 110 titles a year. No unsolicited mss. Welcomes synopses and proposals sent by mail (*no e-mails*). See website for submission details.

Minotaur
▷ St Martin's Press LLC

University Press of Mississippi
3825 Ridgewood Road, Jackson MS 39211
☎ 001 601 432 6205   🖷 001 601 432 6217
press@ihl.state.ms.us
www.upress.state.ms.us
Director *Seetha Srinivasan*
Assistant Director/Editor-in-Chief *Craig Gill*

Founded 1970. Non-profit book publisher partially supported by the eight state universities. Publishes scholarly and trade titles in literature, history, American culture, Southern culture, African-American studies, women's studies, popular culture, folklife, ethnic, performance, art, architecture, photography and other liberal arts. IMPRINTS **Banner Books; Muscadine Books**. Represented worldwide. UK REPRESENTATIVE: **Roundhouse Group**. About 60 titles a year. Send letter of enquiry, prospectus, table of contents and sample chapter prior to submission of full mss.
🅔 Royalties annually.

University of Missouri Press
2910 LeMone Boulevard, Columbia MO 65201
☎ 001 573 882 7641   🖷 001 573 884 4498
upress@umsystem.edu
www.system.missouri.edu/upress
Director *Beverly Jarrett*

Founded 1958. Publishes academic: history, intellectual history, journalism, literary criticism and related humanities disciplines, and occasional volumes of short stories. About 70 titles a year. Best approach is by letter. Send synopses for academic work together with author c.v.

Modern Library
▷ Random House Publishing Group

William Morrow
▷ HarperCollins Publishers, Inc.

Muscadine Books
▷ University Press of Mississippi

NAL
▷ Penguin Group (USA) Inc.

University of Nebraska Press
1111 Lincoln Mall, Lincoln NE 68588–0630
☎ 001 402 472 3581   🖷 001 402 472 6214
www.nebraskapress.unl.edu
Interim Director *Ladette Randolph*

Founded 1941. Publishes scholarly, Native
American studies, history of the American
West, literary and cultural studies, music,
Jewish studies, military history, sports history,
environmental history. IMPRINT **Bison Books**.
No unsolicited mss; welcomes synopses and ideas
for books. Send enquiry letter in the first instance
with description of project and sample material.
No original fiction, children's books or poetry.
Ⓔ Royalties annually.

University of Nevada Press
Morrill Hall Mail Stop 0166, Reno
NV 89557–0076–0166
☎ 001 775 784 6573
www.nvbooks.nevada.edu
Director *Joanne O'Hare*

Founded 1961. Publishes scholarly and popular
books; serious fiction, Native American studies,
natural history, Western Americana, Basque
studies and regional studies. Unsolicited material
welcome if it fits in with areas published, or offers
a 'new and exciting' direction.

University Press of New England
One Court Street, Suite 250, Lebanon NH 03766
☎ 001 603 448 1533   🖷 001 603 448 7006
university.press@dartmouth.edu
www.upne.com
Director *Michael Burton*
Editor-in-Chief *Phyllis Deutsch*

Founded 1970. A scholarly books publisher
sponsored by six institutions of higher education
in the region: Brandeis, Dartmouth, University of
Vermont, Tufts, the University of New Hampshire
and Northeastern. Publishes general and scholarly
non-fiction. OVERSEAS ASSOCIATES Canada:
University of British Columbia; UK, Europe,
Middle East: Eurospan University Press Group;
Australia, New Zealand, Asia & the Pacific:
East-West Export Books. About 80 titles a year.
Unsolicited material welcome.
Ⓔ Royalties annually.

University of New Mexico Press
1312 Basehart Road SE, Albuquerque
NM 87106–4363
☎ 001 505 277 2346   🖷 001 505 272 7141
unmpress@unm.edu
www.unmpress.com
Director *Luther Wilson*

Editor-in-Chief *William 'Clark' Whitehorn*

Founded 1929. Publishes scholarly, regional
books, biography and fiction. No how-to,
humour, self-help or technical.

New Seeds
▷ Shambhala Publications, Inc.

Newman Press
▷ Paulist Press

University of North Texas Press
PO Box 311336, Denton TX 76203–1336
☎ 001 940 565 2142   🖷 001 940 565 4590
rchrisman@unt.edu
www.unt.edu/untpress
Director *Ronald Chrisman*
Managing Editor *Karen DeVinney*

Founded 1987. Publishes regional interest,
contemporary, social issues, Texas history,
military history, music, women's issues,
multicultural. Publishes annually the winners of
the Vassar Miller Poetry Prize and the Katherine
Anne Porter Prize in Short Fiction. About 15 titles
a year. Synopses and ideas welcome. Do not send
full ms until requested.
Ⓔ Royalties annually.

Northwestern University Press
629 Noyes Street, Evanston IL 60208
☎ 001 847 491 2046   🖷 001 847 491 8150
nupress@northwestern.edu
www.nupress.northwestern.edu
Chairman *Peter Hayes*
Managing Director *Donna Shear*

Publishes general and academic books:
philosophy, theatre and performance studies,
Slavic studies, Latino fiction and biography,
contemporary fiction, poetry. No children's, art
books. IMPRINTS **Latino Voices, TriQuarterly**
Fiction and poetry Editorial Head *Henry
Carrigan*. 60 titles in 2007. Unsolicited mss,
synopses and ideas welcome; approach by mail.
No unsolicited mss in fiction or poetry, please.
Ⓔ Royalties annually.

W.W. Norton & Company, Inc.
500 Fifth Avenue, New York NY 10110
☎ 001 212 354 5500   🖷 001 212 869 0856
www.wwnorton.com
VP/Editor-in-Chief *Starling R. Lawrence*

Founded 1923. Publishes fiction and non-fiction,
college textbooks and professional books. About
400 titles a year. Submission guidelines available
on the website.

Ohio University Press
19 Circle Drive, The Ridges, Athens
OH 45701–2979
☎ 001 740 593 1155   🖷 001 740 593 4536

www.ohioswallow.com
Director *David Sanders*

Founded 1964. Publishes academic, regional and general trade books. IMPRINT **Swallow Press**. Synopses and ideas for books welcome. Send detailed synopsis, sample chapters and c.v. with covering letter. See website for author guidelines. No children's, how-to or genre fiction.
Ⓔ Royalties annually.

University of Oklahoma Press

2800 Venture Drive, Norman OK 73069–8216
Ⓣ 001 405 325 2000   Ⓕ 001 405 325 4000
www.oupress.com
Director *B. Byron Price*
Publisher *John Drayton*

Founded 1928. Publishes general scholarly non-fiction only: American Indian studies, American West, classical studies, anthropology, natural history and political science. IMPRINT **Arthur H. Clarke Company** Publisher *Robert Clark* About 100 titles a year. See website for submissions policy.

Olive Branch Press
▷ Interlink Publishing Group, Inc.

One World
▷ Random House Publishing Group

Open Forum
▷ Krieger Publishing Co.

Orbit
▷ Krieger Publishing Co.

Orbit
▷ Hachette Book Group USA

Owl Books
▷ Henry Holt & Company, Inc.

Palgrave
▷ St Martin's Press llc

Paragon House

1925 Oakcrest Avenue, Suite 7, St Paul MN 55113–2619
Ⓣ 001 651 644 3087   Ⓕ 001 651 644 0997
paragon@paragonhouse.com
www.paragonhouse.com
Executive Director *Dr Gordon L. Anderson*

Founded 1982. Publishes non-fiction: reference and academic. Subjects include history, religion, philosophy, spirituality, Jewish interest, political science, international relations, psychology.
Ⓔ Royalties twice-yearly.

Paulist Press

997 Macarthur Road, Mahwah NJ 07430
Ⓣ 001 201 825 7300   Ⓕ 001 201 825 8345

pmcmahon@paulistpress.com
www.paulistpress.com
Managing Director *Fr. Lawrence Boadt, CSP*

Founded in 1858 and owned by Paulist Fathers. Publishes religion, spirituality, theology, pastoral resources, ecumenical and interfaith works. No fiction, memoirs or poetry. Planning to increase children's (spirituality) and catechetical list of titles. IMPRINTS **Hiddenspring Books** *Paul McMahon*; **Stimulus Books** *Fr. Lawrence Boadt, CSP*; **Newman Press** *Fr. Lawrence Boadt, CSP*. About 85 titles a year. Submissions welcome for the Paulist Press imprint only. Initial approach by e-mail but accepts hard copy submissions only.
Ⓔ Royalties twice-yearly.

Pearson Scott Foresman

1900 E Lake Avenue, Glenview IL 60025
Ⓣ 001 847 729 3000
www.scottforesman.com
President *Paul McFall*

Founded 1896. Part of Pearson Education. Publishes elementary education materials. No unsolicited material.

Pelican Publishing Company

1000 Burmaster Street, Gretna LA 70053
Ⓣ 001 504 368 1175
www.pelicanpub.com
Editor-in-Chief *Nina Kooij*

Publishes general non-fiction: popular history, cookbooks, travel, art, business, biography, children's, architecture, collectibles guides and motivational. About 75 titles a year. Initial enquiries required for all submissions.

Penguin Group (USA) Inc.

375 Hudson Street, New York NY 10014
Ⓣ 001 212 366 2000   Ⓕ 001 212 366 2666
www.penguingroup.com
CEO *David Shanks*
President *Susan Petersen Kennedy*

The second-largest trade book publisher in the world. Publishes fiction and non-fiction in hardback and paperback; adult and children's.

ADULT DIVISION IMPRINTS Hardcover: **Avery** (see entry); **Dutton; G.P. Putnam's Sons; Gotham; Hudson Street Press; Jeremy P. Tarcher/Putnam; The Penguin Press; Portfolio; Riverhead Books; Sentinel; Viking**; Trade Paperback: **Ace Books; Alpha Books; Berkley; HPBooks; NAL; Perigee; Penguin Books; Plume; Riverhead; Roc**. Mass Market Paperback: **Ace; Berkley; DAW Books** (see entry); **Jove**.

YOUNG READERS DIVISION: Hardcover: **Dial Books for Young Readers** (see entry); **Dutton Children's Books; Frederick Warne; G.P. Putnam's Sons** (see entry); **Philomel**

Books; **Viking Children's Books**. Paperback: **Firebird**; **Puffin Books**; **Razorbill**; **Speak**. Mass Merchandise: **Grosset & Dunlap**; **Price Stern Sloan**. No unsolicited mss.
Ⓔ Royalties twice-yearly.

University of Pennsylvania Press
3905 Spruce Street, Philadelphia PA 19104–4112
Ⓣ 001 215 898 6261 Ⓕ 001 215 898 0404
www.pennpress.org
Director *Eric Halpern*
Editor-in-Chief *Peter Agree*

Founded 1890. Publishes serious non-fiction: scholarly, reference, professional, textbooks and semi-popular trade. No original fiction or poetry. Over 100 titles a year. No unsolicited mss but synopses and ideas for books welcome.
Ⓔ Royalties vary according to sales prospects and other relevant factors.

Perennial
▷ HarperCollins Publishers, Inc.

Perigee
▷ Penguin Group (USA) Inc.

Perseus Books Group
387 Park Avenue S., 12th Floor, New York NY 10016
Ⓣ 001 212 340 8100 Ⓕ 001 212 340 8115
www.perseusbooksgroup.com
President/CEO *David Steinberger*

Founded 1997. IMPRINTS **Avalon Travel** Guide books; **Basic Books** Psychology, science, politics, sociology, current affairs, history; **Basic Civitas** African and African-American studies; **DaCapo** Art, biography, film, history, humour, music, photography, sport; **PublicAffairs** (see entry); **Running Press** General non-fiction, children's, illustrated books; **Seal Press** 'Books by women for women'; **Vanguard Press** Fiction and non-fiction; **Westview** Undergraduate and graduate textbooks. No unsolicited mss.

Philomel Books
▷ Penguin Group (USA) Inc.

Picador USA
▷ St Martin's Press LLC

Players Press
PO Box 1132, Studio City CA 91614-0132
Ⓣ 001 818 789 4980
President/CEO *Robert Gordon*

Founded 1965 as a publisher of plays; now publishes across the entire range of performing arts: plays, musicals, theatre, film, cinema, television, costume, puppetry, plus technical theatre and cinema material. No unsolicited mss; synopses/ideas welcome. Send query letter s.a.e. for response.

Ⓔ Royalties twice-yearly.

Plume
▷ Penguin Group (USA) Inc.

Pocket Books/Pocket Star
▷ Simon & Schuster Adult Publishing Group

Poppy
▷ Hachette Book Group USA

Portfolio
▷ Penguin Group (USA) Inc.

Potomac Books, Inc.
22841 Quicksilver Drive, Dulles VA 20166
Ⓣ 001 703 661 1548 Ⓕ 001 703 661 1547
www.potomacbooksinc.com
Publisher *Sam Dorrance*
Senior Editors *Hilary Claggett, Elizabeth Demers*
Sports Editor *Kevin Cuddihy*

Formerly Brassey's, Inc., founded 1984 (acquired in 1999 by Books International of Dulles, Virginia). Publishes non-fiction titles on topics of history (especially military and diplomatic history), world and US affairs, US foreign policy, defence, intelligence, biography and sport. About 80 titles a year. No unsolicited mss; query letters/synopses welcome.
Ⓔ Royalties annually.

Presidio Press
▷ Random House Publishing Group

Price Stern Sloan
▷ Penguin Group (USA) Inc.

Primary Source Media
▷ Thomson Gale

Princeton University Press
41 William Street, Princeton NJ 08540–5237
Ⓣ 001 609 258 4900 Ⓕ 001 609 258 6305
www.pup.princeton.edu
Editor-in-Chief/Executive Editor *Brigitta van Rheinberg*

Founded 1905. Publishes academic and trade books. About 200 titles a year. No unsolicited mss. Synopses and ideas considered. Send letter to editor-in-chief in the first instance. No e-mail submissions. No fiction. OVERSEAS ASSOCIATE Princeton University Press, 3 Market Place, Woodstock OX20 1SY, UK.
Ⓔ Royalties annually.

Professional Practices
▷ Krieger Publishing Co.

Prometheus Books
59 John Glenn Drive, Amherst NY 14228–2197
Ⓣ 001 716 691 0133 Ⓕ 001 716 691 0137
marketing@prometheusbooks.com
www.prometheusbooks.com

www.pyrsf.com
Chairman *Paul Kurtz*
Editor-in-Chief *Steven L. Mitchell*

Founded 1969. Publishes books and journals: educational, scientific professional, library, popular science, science fiction and fantasy, philosophy and young readers. IMPRINTS **Humanity Books; Pyr™**. About 100 titles a year. Unsolicited mss, synopses and ideas for books welcome.
Ⓔ Royalties twice-yearly.

Public History
▷ Krieger Publishing Co.

PublicAffairs
250 West 57th Street, Suite 1321, New York
NY 10107
Ⓣ 001 212 397 6666   Ⓕ 001 212 397 4777
publicaffairs@perseusbooks.com
www.publicaffairsbooks.com
Publisher *Susan Weinberg*

Founded 1997. Part of the **Perseus Books Group**. Publishes current affairs, biography, history and journalism.

Puffin Books
▷ Penguin Group (USA) Inc.

G.P. Putnam's Sons (Children's)
345 Hudson Street, New York NY 10014
Ⓣ 001 212 366 2000
www.penguingroup.com
President/Publisher *Nancy Paulsen*
Associate Editorial Director *Susan Kochan*

A children's book imprint of Penguin Young Readers Group, a member of **Penguin Group (USA) Inc.** Publishes picture books, middle-grade fiction and young adult fiction.

Pyr™
▷ Prometheus Books

The Random House Publishing Group
1745 Broadway, New York NY 10019
Ⓣ 001 212 782 9000
www.randomhouse.com
President/Publisher *Gina Centrello*

Publishes fiction, non-fiction, science fiction and graphic novels in hardcover, trade paperback and mass market.

IMPRINTS

Ballantine Books; Broadway; Del Rey; Modern Library; One World; Presidio Press; Random House; Random House Trade Paperbacks; Villard. 675 titles in 2007. No unsolicited mss.

Rayo
▷ HarperCollins Publishers, Inc.

Razorbill
▷ Penguin Group (USA) Inc.

Reader's Digest Association Inc
Reader's Digest Road, Pleasantville
NY 10570–7000
Ⓣ 001 914 238 1000
www.rd.com
President/CEO *Mary Berner*

Publishes home maintenance reference, cookery, DIY, health, gardening, children's books; videos and magazines.

Riverhead Books
▷ Penguin Group (USA) Inc.

The Riverside Publishing Co.
▷ Houghton Mifflin Harcourt Publishing Company

Roc
▷ Penguin Group (USA) Inc.

The Rosen Publishing Group, Inc.
29 East 21st Street, New York NY 10010
Ⓣ 001 212 777 3017   Ⓕ 001 212 777 0277
www.rosenpublishing.com
President *Roger Rosen*

Founded 1950. Publishes non-fiction books (supplementary to the curriculum, reference and self-help) for a young adult audience. Reading levels are years 7–12, 4–6 (books for teens with literacy problems), and 5–9. Subjects include conflict resolution, character building, health, safety, drug abuse prevention, self-help and multicultural titles. For all imprints, write with outline and sample chapters.

Running Press
▷ Perseus Books Group

Rutgers University Press
100 Joyce Kilmer Avenue, Piscataway
NJ 08854–8099
Ⓣ 001 732 445 7762   Ⓕ 001 732 445 7039
rutgerspress.rutgers.edu
Director *Marlie Wasserman*
Associate Director/Editor-in-Chief *Leslie Mitchner*

Founded 1936. Publishes scholarly books, regional, social sciences and humanities. About 80 titles a year. Unsolicited mss, synopses and ideas for books welcome. No original fiction or poetry.

S&S Libros eñ Espanol
▷ Simon & Schuster Adult Publishing Group

St James Press
▷ Thomson Gale

St Martin's Press LLC

175 Fifth Avenue, New York NY 10010
☎ 001 646 307 5151   🖷 001 212 420 9314
inquiries@stmartins.com
www.stmartins.com
CEO (Macmillan) *John Sargent*
President/Publisher (Trade Division) *Sally Richardson*

Founded 1952. A division of **Macmillan** in the US, owned by German publishing group Holtzbrinck, GmbH, St Martin's Press made its name and fortune by importing raw talent from the UK to the States and has continued to buy heavily in the UK. Publishes general fiction, especially mysteries and crime; and adult non-fiction: history, self-help, political science, travel, biography, scholarly, popular reference, college textbooks.

IMPRINTS

Picador USA; **Griffin** (trade paperbacks); **St Martin's Paperbacks** (mass market); **Thomas Dunne Books**; **Minotaur**; **Palgrave**; **Truman Talley Books**. All submissions via legitimate literary agents only.

Scarecrow Press Inc.

4501 Forbes Boulevard, Suite 200, Lanham MD 20706
☎ 001 301 459 3366   🖷 001 301 429 5748
www.scarecrowpress.com
President *Jed Lyons*
Publisher/ Editorial Director *Edward Kurdyla*

Founded 1950 as a short-run publisher of library reference books. Acquired by **University Press of America, Inc.** in 1995 which is now part of Rowman and Littlefield Publishing Group. Publishes reference, scholarly and monographs (all levels) for libraries. Reference books in all areas except sciences, specializing in the performing arts, music, cinema and library science. About 160 titles a year. Unsolicited mss welcome but material will not be returned unless requested and accompanied by return postage. Unsolicited synopses and ideas for books welcome.

Schirmer Reference
▷ Thomson Gale

Scholarly Resources, Inc.
▷ Thomson Gale

Scholastic Library Publishing

90 Sherman Turnpike, Danbury CT 06816
☎ 001 800 621 1115
www.scholastic.com/aboutscholastic/divisions/slp.htm
President *Greg Worrell*

Publishes print and online juvenile non-fiction, encyclopedias, speciality reference sets and picture books.

DIVISIONS/IMPRINTS **Children's Press**; **Grolier**; **Grolier Online**; **Franklin Watts**. About 500 titles a year.

Charles Scribner's Sons
▷ Thomson Gale

Scribner/Scribner Classics
▷ Simon & Schuster Adult Publishing Group

Sentinel
▷ Penguin Group (USA) Inc.

Seven Stories Press

140 Watts Street, New York NY 10013
☎ 001 212 226 8760   🖷 001 212 226 1411
www.sevenstories.com
Publisher *Daniel Simon*
Subsidiary Rights *Anna Lui*

Founded 1995. Publishes, literature, translations, politics, media studies, popular culture, health. About 50 titles a year. No unsolicited mss; synopses and ideas from agents welcome or send query letter. No e-mail submissions. No children's or business books. OVERSEAS SUBSIDIARIES Turnaround Publishing Services, UK and EU; Palgrave Macmillan, Australia; Macmillian Publishers NZ Ltd., New Zealand; Liberty Books (Pte.) Ltd., Pakistan; Pen International Pte. Ltd., Singapore; Stephan Phillips (Pty.) Ltd., Southern Africa; Publishers Group Canada, Canada.

Shambhala Publications, Inc.

PO Box 308, Boston MA 02117
www.shambhala.com

Publishes contemplative religion, Buddhism, Christianity, philosophy, self-help, health, creativity and fiction. IMPRINTS **Trumpeter**; **New Seeds**; **Weatherhill**. 50 titles in 2007. Welcomes submissions – by post only; no e-mails. No original poetry.

M.E. Sharpe, Inc.

80 Business Park Drive, Armonk NY 10504
☎ 001 914 273 1800   🖷 001 914 273 2106
info@mesharpe.com
www.mesharpe.com
President/CEO *M.E. Sharpe*
Snr VP/COO *Vincent Fuentes*

Founded 1958. Publishes books and journals in the social sciences and humanities, both original works and translations in Asian and East European studies. IMPRINTS **East Gate Books**; **Sharpe Focus**; **Sharpe Reference** *Patricia Kolb*. 80 titles in 2006. No unsolicited material.
Ⓔ Royalties annually.

Simon & Schuster Adult Publishing Group (Division of Simon & Schuster, Inc)

1230 Avenue of the Americas, New York NY 10020
☎ 001 212 698 7000   ⓕ 001 212 698 7007
www.simonsays.com
President & CEO *Carolyn K. Reidy*

Publishes fiction and non-fiction.

DIVISIONS

Atria Books VP & Executive Editorial Director
Emily Bestler; **Free Press** VP & Editorial Director
Dominick Anfuso; **Touchstone-Fireside** VP &
Editor-in-Chief *Trish Todd*; **Howard Books**
EVP & Publisher *John Howard*; **Pocket Books**
VP & Editorial Director *Maggie Crawford*;
Scribner VP & Editor-in-Chief *Nan Graham*;
Simon &Schuster VP & Editorial Director
Alice Mayhew; **Simon Spotlight Entertainment**
Editorial Director *Tricia Boczkowski*.

IMPRINTS

**Atria Books; Downtown Press; Fireside; Free
Press; Howard Books; Pocket Books; Pocket
Star; Scribner; Scribner Classics; S&S Libros eñ
Espanol; Simon & Schuster; Simon Spotlight
Entertainment; Strebor Books; Threshold
Editions; Touchstone; Washington Square
Press**. No unsolicited mss.
ⓔ Royalties twice-yearly.

Simon & Schuster Children's Publishing

1230 Avenue of the Americas, New York NY 10020
☎ 001 212 698 7200
www.simonsayskids.com
President/Publisher *Rick Richter*

A division of the Simon & Schuster Consumer
Group. Publishes pre-school to young adult,
picture books, hardcover and paperback fiction,
non-fiction, trade, library and mass-market titles.

IMPRINTS

Aladdin Paperbacks Picture books, paperback
fiction and non-fiction reprints and originals,
and limited series for ages pre-school to young
adult; **Atheneum Books for Young Readers**
Picture books, hardcover fiction and non-fiction
books across all genres for ages three to young
adult; **Little Simon** Mass-market novelty books
(pop-ups, board books, colouring & activity) and
merchandise (book and audio cassette) for ages
birth through eight; **Margaret K. McElderry
Books** Picture books, hardcover fiction and
non-fiction trade books for children ages three to
young adult; **Simon & Schuster Books for Young
Readers** Picture books, hardcover fiction and
non-fiction for children ages three to young adult;
Simon Pulse Paperbacks for teenagers; **Simon
Spotlight** Devoted exclusively to children's media
tie-ins and licensed properties. About 480 titles a
year. No unsolicited material.

Simon Pulse/Simon Spotlight
▷ Simon & Schuster Children's Publishing

Skirt! Books
▷ The Globe Pequot Press

Sleeping Bear Press
▷ Thomson Gale

Southern Illinois University Press

1915 University Press Drive, SIUC Mail Code
6806, Carbondale IL 62901
☎ 001 618 453 2281   ⓕ 001 618 453 1221
www.siu.edu/~siupress
Director *Lain Adkins*

Founded 1956. Publishes scholarly and general
interest non-fiction books and educational
materials. 50 titles a year.

Speak
▷ Penguin Group (USA) Inc.

Spectra
▷ Bantam Dell Publishing Group

Springboard Press
▷ Hachette Book Group USA

Stackpole Books

5067 Ritter Road, Mechanicsburg PA 17055
☎ 001 717 796 0411   ⓕ 001 717 796 0412
www.stackpolebooks.com
President *M. David Detweiler*
Vice President/Publisher *Judith Schnell*

Founded 1933. Publishes outdoor sports, fishing,
hunting, nature, crafts, Pennsylvania and regional,
military reference, history. About 130 titles a year.
ⓔ Royalties twice-yearly.

Stanford University Press

1450 Page Mill Road, Palo Alto CA 94304–1124
☎ 001 650 723 9434   ⓕ 001 650 725 3457
info@www.sup.org
www.sup.org
Director *Dr Geoffrey R.H. Burn*

Founded 1925. Publishes non-fiction: scholarly
works in all areas of the humanities, social
sciences, history and literature, also professional
lists in business, economics and law. About 150
titles a year. No unsolicited mss; query in writing
first.

State University of New York Press

194 Washington Avenue, Suite 305, Albany
NY 12210–2384
☎ 001 518 472 5000   ⓕ 001 518 472 5038
info@sunypress.edu
www.sunypress.edu
Editor-in-Chief *Jane Bunker*

Founded 1966. Part of the Research Foundation
of SUNY, the Press is one of the largest university

presses in the USA. Publishes scholarly and trade books in the humanities, social sciences and books of regional interest on New York State. Strengths in philosophy, religious studies, education and Asian studies. 183 titles in 2007. Unsolicited submissions welcome. Send material to *Jane Bunker*, Editor-in-Chief.
Ⓔ Royalties annually.

Sterling Publishing Co. Inc.
387 Park Avenue South, New York NY 10016
Ⓣ 001 212 532 7160
www.sterlingpub.com
VP/Publisher *Andrew Martin*
President/CEO *Charles Nurnberg*

Founded 1949. A subsidiary of Barnes & Noble, Inc. Publishes illustrated non-fiction: reference and how-to books on arts and crafts, home improvement, history, photography, children's, woodworking, pets, hobbies, gardening, games and puzzles, general non-fiction. IMPRINT **Lark** Crafts, children's, photography. Submission guidelines available.

Stimulus Books
▷ Paulist Press

Strebor Books
▷ Simon & Schuster Adult Publishing Group

Swallow Press
▷ Ohio University Press

Syracuse University Press
621 Skytop Road, Suite 110, Syracuse
NY 13244–5290
Ⓣ 001 315 443 5534   Ⓕ 001 315 443 5545
www.SyracuseUniversityPress.syr.edu
Director *Alice R. Pfeiffer*

Founded 1943. Publishes scholarly books in the following areas: Middle East studies, Middle East literature in translation, Irish studies, Jewish studies, sports, gender and globalization, Iroquois studies, women and religion, medieval studies, religion and politics, television and popular culture, geography and regional books. Distributes for the Moshe Dayan Center, Jusoor. About 50 titles a year. No unsolicited mss. Send query letter with IRCs.
Ⓔ Royalties annually.

The TAFT Group
▷ Thomson Gale

Nan A. Talese
▷ Doubleday Broadway Publishing Group

Jeremy P. Tarcher
▷ Penguin Group (USA) Inc.

Katherine Tegan Books
▷ HarperCollins Publishers, Inc.

Temple University Press
1601 N. Broad Street, 083–42, Philadelphia
PA 19122–6099
Ⓣ 001 215 204 8787   Ⓕ 001 215 204 4719
www.temple.edu/tempress
Editor-in-Chief *Janet M. Francendese*

Founded 1969. Publishes scholarly books. Authors generally academics. About 60 titles a year. Letter or e-mail of inquiry with brief outline. Include fax/e-mail address.

University of Tennessee Press
110 Conference Center, 600 Henley Street, Knoxville TN 37996–4108
Ⓣ 001 865 974 3321   Ⓕ 001 865 974 3724
www.utpress.org
Acquisitions Editor *Scott Danforth*

Founded 1940. Publishes scholarly and regional non-fiction (American Studies only). Submission guidelines are posted on the website.

University of Texas Press
PO Box 7819, Austin TX 78722
Ⓣ 001 512 471 7233/471 4278 (editorial)
Ⓕ 001 512 232 7178
utpress@uts.cc.utexas.edu
www.utexas.edu/utpress/
www.utexaspress.com
Director *Joanna Hitchcock*
Assistant Director/Editor-in-Chief *Theresa J. May*

Publishes scholarly and regional non-fiction: anthropology, architecture, classics, environmental studies, humanities, social sciences, geography, language studies, literary modernism; Latin American/Latino/Mexican American/Middle Eastern/Native American studies, natural history and ornithology, regional books (Texas and the southwest). About 100 titles a year and 11 journals. Unsolicited material welcome in above subject areas only. Author guidelines available on the website.
Ⓔ Royalties annually.

Thomson Gale
27500 Drake Road, Farmington Hills MI 48331
Ⓣ 001 248 699 4253   Ⓕ 001 248 699 8070
www.gale.com
President *Gordon Macomber*

Part of Cengage Learning, Thomson Gale is a world leader in information and educational publishing. Addresses all types of information needs, from homework help to health questions and business profiles, in a variety of formats: books, Web-based material and microfilm.
BRANDS/IMPRINTS **Blackbirch Press; Five Star; Graham & Whiteside; Greenhaven Press; G.K. Hall; KidHaven Press; Large Print Press; Lucent Books; Macmillan Reference USA; Primary Source Media; Schirmer Reference; Scholarly**

Resources, Inc.; Charles Scribner's Sons;
Sleeping Bear Press; St James Press; The TAFT
Group; Thorndike Press; Twayne Publishers;
UXL; Walker Large Print; Wheeler Publishing.

Thorndike Press
▷ Thomson Gale

Three Rivers Press
▷ Crown Publishing Group

Threshold Editions
▷ Simon & Schuster Adult Publishing Group

Times Books
▷ Henry Holt & Company, Inc.

Touchstone
▷ Simon & Schuster Adult Publishing Group

Transaction Publishers Ltd
c/o Rutgers – The State University of New Jersey,
35 Berrue Circle, Piscataway NJ 08854–8042
☎ 001 732 445 2280   ⊟ 001 732 445 3138
ihorowitz@transactionpub.com
www.transactionpub.com
Also at: 390 Campus Drive, Somerset, NJ 08873
Chairman *I.L. Horowitz*
President *Mary E. Curtis*
Vice President *Scott B. Bramson*

Founded 1962. Independent publisher of
academic social scientific books, periodicals
and serials. Publisher of record in international
social research. 170 titles in 2006. All unsolicited
material should be submitted in hard copy and
full draft. E-mail submissions are not accepted.
'Decision-making is rapid.'
Ⓔ Royalties average 10% hardcover originals; 7½%
paperback.

TriQuarterly
▷ Northwestern University Press

Truman Talley Books
▷ St Martin's Press LLC

Trumpeter
▷ Shambhala Publications, Inc.

Twayne Publishers
▷ Thomson Gale

Twelve
▷ Hachette Book Group USA

Twenty-First Century Books
▷ Lerner Publishing Group

Tyndale House Publishers, Inc.
351 Executive Drive, Carol Stream IL 60188
☎ 001 630 668 8303
www.tyndale.com
President *Mark D. Taylor*

Founded 1962. Books cover a wide range of
categories from non-fiction to gift books,
theology, doctrine, Bibles, fiction, children's and
youth. Also produces video material, calendars
and audio books for the same market. No
poetry. Non-denominational religious publisher
of around 300 titles a year for the evangelical
Christian market. No unsolicited mss. Synopses
and ideas considered. Writer's guidelines available
on the website.

University Press of America, Inc.
4501 Forbes Boulevard, Suite 200, Lanham
MD 20706
☎ 001 301 459 3366
www.univpress.com
VP/Director *Judith L. Rothman*

Founded 1975. An imprint of Rowman &
Littlefield Publishing Group. Publishes scholarly
monographs, college and graduate level
textbooks. No children's, elementary or high
school. About 300 titles a year. Submit outline or
request proposal questionnaire.

UXL
▷ Thomson Gale

Vanguard Press
▷ Perseus Books Group

Viking/Viking Children's Books
▷ Penguin Group (USA) Inc.

Villard
▷ Random House Publishing Group

University of Virginia Press
PO Box 400318, Charlottesville VA 22904–4318
☎ 001 434 924 3468
vapress@virginia.edu
www.upress.virginia.edu
Director *Penelope J. Kaiserlian*

Founded 1963. Publishes academic books in
humanities and social science with concentrations
in American history, African-American studies,
architecture, Victorian literature, Caribbean
literature and ecocriticism. 50–60 titles a year.
No unsolicited mss; will consider synopses
and ideas for books if in their specific areas of
concentration. Approach by mail for full proposal;
e-mail for short inquiry. No fiction, children's,
poetry or academic books outside interests
specified above.
Ⓔ Royalties annually.

Vision
▷ Hachette Book Group USA

Walch Education
40 Walch Drive, Portland ME 04103
☎ 001 207 772 2846   ⊟ 001 207 772 3105
www.walch.com

President *Al Noyes*

Editor-in-Chief *Erica Varney*

Founded 1927. Publishes supplementary educational materials for middle and secondary schools across a wide range of subjects, including English/language arts, literacy, special needs, mathematics, social studies, science and school-to-career. Always interested in ideas from secondary school teachers who develop materials in the classroom. Proposal letters, synopses and ideas welcome.

Walker & Co.

175 Fifth Avenue, 8th Floor, New York NY 10010

☎ 001 646 438 6075   ℻ 001 212 727 0984

firstname.lastname@bloomsburyusa.com

www.walkerbooks.com (trade non-fiction)

www.walkeryoungreaders.com (BFYR)

Contact *Submissions Editor*

Founded 1959. A division of **Bloomsbury Publishing**. Publishes children's fiction and non-fiction and adult non-fiction. Please contact the following editors in advance before sending any material to be sure of their interest, then follow up as instructed: **Trade Non-fiction** *George Gibson* Permission and documentation must be available with mss. Submit prospectus first, with sample chapters and marketing analysis. **Books for Young Readers** *Emily Easton* Fiction and non-fiction for all ages. Query before sending non-fiction proposals. Especially interested in picture books – fiction and non-fiction, historical and contemporary fiction for middle grades and young adults. BFYR will consider unsolicited submissions. See website for guidelines.

Walker Large Print
▷ Thomson Gale

Frederick Warne
▷ Penguin Group (USA) Inc.

Washington Square Press
▷ Simon & Schuster Adult Publishing Group

Washington State University Press

PO Box 645910, Washington State University, Pullman WA 99164–5910

☎ 001 509 335 3518   ℻ 001 509 335 8568

wsupress@wsu.edu

wsupress.wsu.edu

Editor-in-Chief *Glen Lindeman*

Founded 1928. Publishes hardcover originals, trade paperbacks and reprints. Publishes mainly on the history, prehistory, culture, politics and natural history of the Northwest United States (Washington, Idaho, Oregon, Montana, Alaska and British Columbia). Subjects include Native American, maritime, railroad, biography, cooking/food history, nature/environment, politics. 4–6 titles a year. Submission guidelines are available on the website.

Franklin Watts
▷ Scholastic Library Publishing

Weatherhill
▷ Shambhala Publications, Inc.

Wellness Central
▷ Hachette Book Group USA

Westview
▷ Perseus Books Group

Wheeler Publishing
▷ Thomson Gale

John Wiley & Sons, Inc.

111 River Street, Hoboken NJ 07030–5774

☎ 001 201 748 6000   ℻ 001 201 748 6088

info@wiley.com

www.wiley.com

President/CEO *William J. Pesce*

Founded 1807. Wiley's core publishing programme includes scientific, technical, medical and scholarly journals, encyclopedias, books and online products and services; professional and consumer books, and subscription and information services in all media; educational materials for undergraduate, graduate students and life-long learners. Submission guidelines available on the website under 'Resources for Authors'.

The University of Wisconsin Press

1930 Monroe Street, 3rd Floor, Madison WI 53711–2059

☎ 001 608 263 1110   ℻ 001 608 263 1120

uwiscpress@uwpress.wisc.edu

www.wisc.edu/wisconsinpress

Senior Acquisitions Editor *Raphael Kadushin*

Founded 1936. Publishes scholarly, general interest non-fiction and regional books about Wisconsin and the mid-west. About 50 titles a year. No unsolicited mss. Detailed author guidelines available on the website.

Wordsong
▷ Boyds Mills Press

Yen Press
▷ Hachette Book Group USA

Zondervan

5300 Patterson Avenue SE, Grand Rapids MI 49530

☎ 001 616 698 6900

www.zondervan.com

President/CEO *Bruce E. Ryskamp*

Founded 1931. Subsidiary of **HarperCollins Publishers, Inc.** Publishes Protestant religion, Bibles, books, audio & video, computer software, calendars and speciality items.

Includes FREE online access to **www.thewritershandbook.com**

Other International Publishers

Australia

ACER Press

19 Prospect Hill Road, Camberwell, Victoria 3124
T 00 61 3 9277 5555   F 00 61 3 9277 5500
www.acer.edu.au

Founded 1930. The publishing arm of the
Australian Council for Educational Research.
Publishes education, human resources,
psychology, parent education and mental health,
occupational therapy.

Allen & Unwin Pty Ltd

PO Box 8500, St Leonards, Sydney, NSW 1590
T 00 61 2 8425 0100   F 00 61 2 9906 2218
info@allenandunwin.com
www.allenandunwin.com

Founded 1976. Independent publisher of fiction,
general non-fiction, academic (social sciences,
health) and children's. IMPRINTS **Arena** Fiction
(adventure, crime, fantasy, thrillers, chick
lit) and non-fiction (biography, DIY, health,
humour, self-help, travel); **Jacana Books** Natural
history; **Inspired Living** Mind, body and spirit,
psychology, spirituality. 250 titles in 2008.

Arena
⊳ Allen & Unwin Pty Ltd

Cliffs Notes
⊳ John Wiley & Sons Australia Ltd

Currency Press Pty Ltd

PO Box 2287, Strawberry Hills, NSW 2012
T 00 61 2 9319 5877   F 00 61 2 9319 3649
enquiries@currency.com.au
www.currency.com.au

Founded 1971. Performing arts publisher –
handbooks, biography, crical studies, directories
and reference – drama, theatre, music, dance, film
and video.

For Dummies
⊳ John Wiley & Sons Australia Ltd

Frommer's
⊳ John Wiley & Sons Australia Ltd

Hachette Livre Australia

Level 17, 207 Kent Street, Sydney, NSW 2000
T 00 61 2 8248 0800   F 00 61 2 8248 0810
www.hha.com.au

Founded 1958. Part of the **Hachette Livre
Publishing Group**. Publishes popular fiction,
literature, general non-fiction, autobiography,
biography, current affairs, health, history, lifestyle,
memoirs, self-help, sport, travel.

Margaret Hamilton Books
⊳ Scholastic Australia Pty Limited

HarperCollins Publishers (Australia) Pty Ltd

PO Box 321, Pymble, NSW 2073
T 00 61 2 9952 5000   F 00 61 2 9952 5555
www.harpercollins.com.au

Founded 1989. Part of the News Corporation.
Publishes fiction and non-fiction, children's and
illustrated books.

Inspired Living
⊳ Allen & Unwin Pty Ltd

Jacana Books
⊳ Allen & Unwin Pty Ltd

Jacaranda
⊳ John Wiley & Sons Australia Ltd

Kangaroo Press
⊳ Simon & Schuster (Australia) Pty Ltd

Macmillan Publishers Australia Pty Ltd

1–3/15–19 Claremont Street, South Yarra, Victoria
3141
T 00 61 3 9825 1000   F 00 61 3 9825 1010
www.macmillan.com.au

Founded 1904. Part of **Macmillan Publishers**,
UK. Publishes fiction and non-fiction, academic,
children's, reference, education, professional,
scientific, technical, medical. DIVISIONS
Macmillan Education Primary, secondary
and ELT; **Palgrave Macmillan** Academic and
reference; **Pan Macmillan** (see entry); **Macquarie
Dictionary Publishers** Reference (print and
online);

Macquarie Dictionary Publishers
▷ Macmillan Publishers Australia Pty Ltd

McGraw-Hill Australia & New Zealand Pty Ltd
Locked Bag 2233, Business Centre, North Ryde, NSW 1670
℡ 00 61 2 9900 1800
www.mcgraw-hill.com.au

Founded 1964 (Australia), 1974 (NZ). Owned by **The McGraw-Hill Companies, Inc**. Publishes textbooks and educational material for primary, secondary and higher education. Also professional (medical, general and reference).

Melbourne University Publishing Ltd
187 Grattan Street, Carlton, Victoria 3053
℡ 00 61 3 9342 0300 🖷 00 61 3 9342 0399
mup-info@unimelb.edu.au
www.mup.unimelb.edu.au

Founded 1922. Academic publishers; general non-fiction, archaeology, architecture, art, biography, business, economics, education, finance, history, language, linguistics, law, literature, maritime, military history, natural history, philosophy, politics, psychology, psychiatry, reference, religion, science and technology, sociology, sport, travel, zoology. IMPRINTS **Melbourne University Press; Miegunyah Press**. About 70 titles a year.

Miegunyah Press
▷ Melbourne University Publishing Ltd

Omnibus Books
▷ Scholastic Australia Pty Limited

Oxford University Press (Australia, New Zealand & the Pacific)
GPO Box 2784, Melbourne, Victoria 3001
℡ 00 61 3 9934 9123 🖷 00 61 3 9934 9100
www.oup.com.au

Founded 1908. Owned by **Oxford University Press**, UK. Publishes for the college, school and trade markets.

Palgrave Macmillan
▷ Macmillan Publishers Australia Pty Ltd

Pan Macmillan Australia Pty Ltd
Level 25, 1 Market Street, Sydney, NSW 2000
℡ 00 61 2 9285 9100 🖷 00 61 2 9285 9190
www.panmacmillan.com.au

Founded 1983. Part of **Macmillan Publishers**, UK. Publishes fiction, crime fiction, thrillers, literature; general non-fiction, biography, children's, history, self-help, travel.

Pearson Education Australia Pty Ltd
Locked Bag 507, Frenchs Forest, NSW 1640
℡ 00 61 3 9454 2200 🖷 00 61 3 9453 0089
www.pearson.com.au

Part of Pearson Plc. Australia's largest educational publisher, formed in 1998 from a merger of Addison Wesley Longman Australia and Prentice Hall Australia.

Penguin Group (Australia)
PO Box 701, Hawthorn, Victoria 3124
℡ 00 61 3 9811 2400 🖷 00 61 3 9811 2620
www.penguin.com.au

Founded 1946. Publishes adult and children's general non-fiction and fiction.

University of Queensland Press
PO Box 6042, St Lucia, Queensland 4067
℡ 00 61 7 3365 2127
uqp@uqp.uq.edu.au
www.uqp.uq.edu.au

Founded 1948. Publishes academic, textbooks, children's and young adult, general non-fiction and fiction, literature, literary criticism, short stories, essays, poetry, art, architecture, biography, current affairs, history, indigenous studies, memoirs, military history, reference.

Random House Australia
Level 3, 100 Pacific Highway, North Sydney, NSW 2060
℡ 00 61 2 9954 9966 🖷 00 61 2 9954 4562
random@randomhouse.com.au
www.randomhouse.com.au

Part of the **Random House Group**. Publishes popular and literary fiction, non-fiction, children's and illustrated.

Scholastic Australia Pty Limited
PO Box 579, Gosford, NSW 2250
℡ 00 61 2 4328 3555
www.scholastic.com.au

Founded 1968. Part of **Scholastic Inc**. Educational and children's publisher of books and multimedia. IMPRINTS **Scholastic Press; Omnibus Books; Margaret Hamilton Books**. About 100 titles a year.

Science Press
Bag 7023, Marrickville, NSW 1475
℡ 00 61 2 9516 1122 🖷 00 61 2 9550 1915
www.sciencepress.com.au

Founded 1948. Educational publisher.

Simon & Schuster (Australia) Pty Ltd
PO Box 33, Pymble, NSW 2073
℡ 00 61 2 9983 6600 🖷 00 61 2 9988 4293

Founded 1987. Part of **Simon & Schuster Inc.**, USA. Publishes fiction and general non-fiction, children's books. IMPRINTS **Simon & Schuster Australia** Biography, health, parenting, relationships; **Kangaroo Press** Crafts, gardening, military history.

UWA Press

(M419) University of Western Australia, 35 Stirling Highway, Crawley, WA 6009
☎ 00 61 8 6488 3670   🖷 00 61 8 6488 1027
admin@uwapress.uwa.edu.au
www.uwapress.uwa.edu.au

Founded 1935. Publishes academic, general non-fiction, essays, literature, literary criticism, history, natural history, children's, maritime history, reference, literary studies.

John Wiley & Sons Australia Ltd

PO Box 1226, Milton, Queensland 4064
☎ 00 61 7 3859 9755   🖷 00 61 7 3859 9715
brisbane@johnwiley.com.au
www.johnwiley.com.au

Founded 1954. Owned by **John Wiley & Sons Inc.**, USA. Publishes general non-fiction and education books, dictionaries, encyclopedias, maps and atlases, reference, textbooks. IMPRINTS **Wiley; Jacaranda; WrightBooks; For Dummies; Frommer's; Cliffs Notes**.

WrightBooks
▷ John Wiley & Sons Australia Ltd

Canada

Anchor Canada
▷ Random House of Canada Ltd

Bond Street
▷ Random House of Canada Ltd

Canadian Scholars' Press, Inc

180 Bloor Street West, Suite 801, Toronto, Ontario M5S 2V6
☎ 001 416 929 2774
info@cspi.org
www.cspi.org

Founded 1986. Independent publisher of academic and trade books. IMPRINTS **Women's Press; Kellom Books**.

Doubleday Canada
▷ Random House of Canada Ltd

Emblem Editions
▷ McClelland & Stewart Ltd

H.B. Fenn & Company Ltd

34 Nixon Road, Bolton, Ontario L7E 1W2
☎ 001 905 951 6600   🖷 001 905 951 6601
www.hbfenn.com

Founded 1977. Publishes adult and young adult sports books.

Firefly Books Ltd

66 Leek Crescent, Richmond Hill, Ontario L4B 1H1
☎ 001 416 499 8412   🖷 001 416 499 8313
www.fireflybooks.com

Founded 1977. Publishes children's and young adult non-fiction. No unsolicited submissions.

Fitzhenry & Whiteside Limited

195 Allstate Parkway, Markham, Ontario L3R 4T8
☎ 001 905 477 9700
godwit@fitzhenry.ca
www.fitzhenry.ca

Founded 1966. Publishes reference, biography, history, poetry, photography, sport, children's and young adult books. About 200 titles a year.

Douglas Gibson Books
▷ McClelland & Stewart Ltd

Harlequin Enterprises Ltd

225 Duncan Mill Road, Don Mills, Ontario M3B 3K9
☎ 001 416 445 5860
www.eharlequin.com

Founded 1949. Canada's second largest publishing house. Publishes romantic and women's fiction.

HarperCollins Canada Ltd

2 Bloor Street E, 20th Floor, Toronto, Ontario M4W 1A8
☎ 001 416 975 9334   🖷 001 416 975 5223
www.harpercanada.com

Founded 1989. Publishes commerical and literary fiction, non-fiction, children's, reference, cookery and religious books.

Kellom Books
▷ Canadian Scholars' Press, Inc

Knopf Canada
▷ Random House of Canada Ltd

LexisNexis Canada Ltd

Suite 200, 181 University Avenue, Toronto, Ontario M5H 3M7
☎ 001 416 862 7656   🖷 001 416 862 8073
www.lexisnexis.ca

Founded 1912. Publishes law books, CD-ROMs, journals, newsletters, law reports.

McClelland & Stewart Ltd

75 Sherbourne Street, Fifth Floor, Toronto, Ontario M5A 2P9
☎ 001 416 598 1114   🖷 001 416 598 7764
mail@mcclelland.com
www.mcclelland.com

Founded 1906. Publishes fiction and general non-fiction, poetry. IMPRINTS **Douglas Gibson Books;**

Emblem Editions; Tundra Books (see entry). About 100 titles a year.

McGraw-Hill Ryerson Ltd

300 Water Street, Whitby, Ontario L1N 9B6
☏ 001 905 430 5000
www.mcgrawhill.ca

Founded 1944. Subsidiary of **The McGraw-Hill Companies, Inc.** Publishes education and professional.

Nelson Education

1120 Birchmount Road, Scarborough, Ontario M1K 5G4
☏ 001 416 752 9448   ℻ 001 416 752 8101
inquire@nelson.com
www.nelson.com

Part of the Thomson Corporation. Publishes directories, education, professional, reference, textbooks.

New Star Books Ltd

107–3477 Commercial Street, Vancouver, British Columbia V5N 4E8
☏ 001 604 738 9429   ℻ 001 604 738 9332
info@newstarbooks.com
www.newstarbooks.com

Founded 1974. Publishes fiction and non-fiction, literature, poetry, history, sociology, social sciences.

Oxford University Press, Canada

70 Wynford Drive, Don Mills, Ontario M3C 1J9
☏ 001 416 441 2941   ℻ 001 416 444 0427
www.oup.com/ca

Founded 1904. Owned by **Oxford University Press**, UK. Publishes for college, school and trade markets.

Pearson Canada

26 Prince Andrew Place, Toronto, Ontario M3C 2T8
☏ 001 416 447 5101   ℻ 001 416 443 0948
www.pearsoncanada.ca/index.html

Founded 1966. A division of Pearson Plc and Canada's largest publisher. Publishes fiction, non-fiction and educational material in English and French.

Penguin Group (Canada)

90 Eglinton Avenue East, Suite 700, Toronto, Ontario M4P 2Y3
☏ 001 416 925 2249   ℻ 001 416 925 0068
info@penguin.ca
www.penguin.ca

Founded 1974. Owned by Pearson Plc. Publishes fiction and non-fiction books and audio cassettes.

Random House of Canada Ltd

One Toronto Street, Unit 300, Toronto, Ontario M5C 2V6
☏ 001 416 364 4449   ℻ 001 416 364 6863
www.randomhouse.ca

Founded 1944. Publishes fiction and non-fiction and children's. IMPRINTS **Anchor Canada** Fiction and non-fiction; **Bond Street** Fiction and non-fiction; **Doubleday Canada** Fiction (literary and commercial), business, history, memoir, journalism young adult books; **Knopf Canada** Fiction and non-fiction; **Seal Books** Primarily reprints of fiction and some non-fiction; **Vintage Canada** Paperback fiction and non-fiction.

Scholastic Canada Ltd

175 Hillmount Road, Markham, Ontario L6C 1Z7
☏ 001 905 887 7323   ℻ 001 905 887 3639
www.scholastic.ca

Founded 1957. Publishes (in English and French) children's books and educational material.

Seal Books
▷ Random House of Canada Ltd

University of Toronto Press, Inc.

10 St Mary Street, Suite 700, Toronto, Ontario M4Y 2W8
☏ 001 416 978 2239
info@utpress.utoronto.ca
www.utpress.utoronto.ca

Founded 1901. Publishes scholarly books and journals. About 150 titles a year.

Tundra Books

75 Sherbourne Street, 5th Floor, Toronto, Ontario M5A 2P9
☏ 001 416 598 4786
tundra@mcclelland.com
www.tundrabooks.com

Founded 1967. Division of **McClelland & Stewart Ltd**. Publishes (in English and French) children's and young adult books.

Vintage Canada
▷ Random House of Canada Ltd

John Wiley & Sons Canada Ltd

5353 Dundas Street W., Suite 400, Toronto, Ontario M9B 6H8
www.wiley.ca

Founded 1968. Subsidiary of **John Wiley & Sons Inc.**, USA. Publishes professional, reference and textbooks.

Women's Press
▷ Canadian Scholars' Press, Inc

India

Affiliated East West Press Pvt Ltd

G–1/16 Ansari Road, Darya Ganj, New Delhi 110 002

☎ 00 91 11 2327 9113
affiliate@vsnl.com
www.aewpress.com

Founded 1962. Publishes textbooks, CD-ROMs, electronic books.

Caring

▷ Vision Books Pvt. Ltd

S. Chand & Co Ltd

7361 Ram Nagar, Qutub Road, New Delhi 110 055

☎ 00 91 11 2367 2080   ⓕ 00 91 11 2367 7446
info@schandgroup.com
www.schandgroup.com

Founded 1917. Publishes computer science, dictionaries, encyclopedias, engineering, reference, textbooks and educational CDs.

HarperCollins Publishers India Ltd

A 53, Sector 57, Noida, UP

☎ 00 91 120 404 4800
contact@harpercollins.co.in
www.harpercollins.co.in

A joint venture between the India Today Group and **HarperCollins Publishers**. Publishes fiction and non-fiction; autobiography, astrology, biography, children's, cinema, cookery, crime, current affairs, dictionaries, fantasy, gardening, graphic novels, home interest, literature, memoirs, mind, body and spirit, poetry, politics, popular science, philosophy, psychology, puzzles, reference, religion, science fiction, short stories, sport, thrillers, travel.

Hind Pocket Books Pvt. Ltd

18–19 Dilshad Garden, G.T. Road, Shahadara, Delhi 110 095

☎ 00 91 11 229 7792   ⓕ 00 9111 228 2332

Founded 1957. Publishes fiction and general non-fiction; literature, health and self-help.

Jaico Publishing House

127 M.G. Road, Mumbai 400 023

☎ 00 91 22 267 6702   ⓕ 00 91 22 265 6412
jaicobom@bom5.vsnl.net.in
www.jaicobooks.com/home.asp

Founded 1946. Publishes computer science, engineering, finance, health and nutrition, information technology, law, reference, literature, self-help, philosophy, religion, history, government, political science, sociology, textbooks.

LexisNexis India

C–33 Inner Circle, Connaught Place, New Delhi 110 001

☎ 00 91 11 4355 9999   ⓕ 00 91 11 4151 3388
help.in@lexisnexis.co.in
www.lexisnexis.co.in

Part of Reed Elsevier. Publishes law, taxation, government and business books in print form, online and CD-ROM.

Macmillan India Ltd

315–316 Raheja Chambers, 12 Museum Road, Bangalore 560 011

☎ 00 91 80 2558 7878
bgledit@macmillan.co.in
www.macmillanindia.com

Founded 1893. Part of **Macmillan Publishers**, UK. Publishes education, non-fiction, reference, dictionaries, encyclopedias.

McGraw-Hill Education (India)

B-4, Sector 63, Distt. Gautam Budh Nagar, Noida, UP 201 301

☎ 00 91 9512 0438 3400
www.tatamcgrawhill.com

Founded 1970. Educational publisher.

Munshiram Manoharlal Publishers Pvt Ltd

PO Box 5715, 54 Rani Jhansi Road, New Delhi 110 055

☎ 00 91 11 2367 1668   ⓕ 00 91 11 2361 2745
info@mrmlonline.com
www.mrmlbooks.com

Founded 1952. Publishes academic, dictionaries, anthropology, art and art history, autobiography, architecture, archaeology, biography, environment, religion, philosophy, geography, history, humanities, language arts and linguistics, law, literature, medicine, numismatics, music, dance, theatre, Asian studies, sociology, travel.

National Publishing House

2/35 Ansari Road, New Delhi 110 002

☎ 00 91 11 2327 5267
info@nationalpublishinghouse.com
www.nationalpublishinghouse.com

Founded 1945. Publishes non-fiction, children's, history, literature, science, textbooks.

Orient Paperbacks

5 A/8 Ansari Road, First Floor, Darya Ganj, Delhi 110 002

☎ 00 91 11 2327 8877   ⓕ 00 91 11 2327 8879
mail@orientpaperbacks.com
www.orientpaperbacks.com

Founded 1975. Division of **Vision Books Pvt. Ltd**. Publishes (in English and Hindi) fiction and general non-fiction; humour, health and nutrition, literature, New Age, puzzles, reference, religion, self-help.

Oxford University Press India
1st Floor, YMCA Library Building, 1 Jai Singh Road, New Delhi 110 001
℡ 00 91 11 4360 0300 🅕 00 91 11 2336 0897
admin.in@oup.com
www.oup.co.in

Founded 1912. Owned by **Oxford University Press**, UK. Publishes academic, education, dictionaries. Reference subjects include arts, architecture, biography, natural history, memoirs.

Rajpal
▷ **Vision Books Pvt. Ltd**

Vision Books Pvt. Ltd
24, Feroze Gandhi Road, Lajpat Nagar III, New Delhi 110 024
℡ 00 91 11 2983 6470 🅕 00 91 11 2983 6490
editor@visionbooksindia.com
www.visionbooksindia.com

Founded 1975. Specializes in books on business and management, finance and taxation, careers, current affairs, religion. IMPRINTS **Vision Books**; **Orient Paperbacks** (see entry); **Caring**; **Rajpal**.

New Zealand

Auckland University Press
University of Auckland, Private Bag 92019, Auckland
℡ 00 64 9 373 7528 🅕 00 64 9 373 7465
aup@auckland.ac.nz
www.auckland.ac.nz/aup

Founded 1966. Publishes academic, biography, government, political science, history, social sciences, sociology, poetry, literature, literary criticism, Maori studies.

Canterbury University Press
University of Canterbury, Private Bag 4800, Christchurch
℡ 00 64 3 364 2914 🅕 00 64 3 364 2044
mail@cup.canterbury.ac.nz
www.cup.canterbury.ac.nz

Founded 1964. Academic publishers; general non-fiction; history, marine biology, natural history.

The Caxton Press
113 Victoria Street, Christchurch
℡ 00 64 3 353 0734 🅕 00 64 3 365 7840
www.caxton.co.nz
Founded 1935. Publishes general non-fiction.

Hachette Livre New Zealand Ltd
4 Whetu Place, Mairangi Bay, Auckland
℡ 00 64 9 478 1000 🅕 00 64 9 478 1010
www.hha.com.au

Founded 1971. Part of the **Hachette Livre Publishing Group**. Publishes fiction and general non-fiction; biography, business, children's, cookery, humour, sport.

HarperCollins Publishers (New Zealand) Ltd
PO Box 1, Auckland
℡ 00 64 9 443 9400 🅕 00 64 9 443 9403
editors@harpercollins.co.nz
www.harpercollins.co.nz

Founded 1888. Publishes fiction, literature, children's, business, cookery, gardening, reference, religion.

Huia (NZ) Ltd
PO Box 17-335, 39 Pipitea Street, Wellington, Aotearoa
℡ 00 64 4 473 9262 🅕 00 64 4 473 9265
info@huia.co.nz
www.huia.co.nz

Founded 1991. Independet publisher of Maori cultural history and language, children's books in Maori and English, fiction.

LexisNexis New Zealand
PO Box 472, Wellington
℡ 00 64 4 385 1479 🅕 00 64 4 385 1598
www.lexisnexis.co.nz

Founded 1914. Part of the LexisNexis Group. Publishes law, professional and textbooks in book and electronic formats.

Macmillan Publishers New Zealand
6 Ride Way, Albany, Auckland
℡ 00 64 9 414 0350 🅕 00 64 9 414 0351
www.macmillan.co.nz

Education publishers – general, school and academic books; fiction and non-fiction for all ages.

McGraw-Hill New Zealand Pty Ltd
▷ **McGraw-Hill Australia & New Zealand Pty Ltd**

Oxford University Press New Zealand
▷ **Oxford University Press (Australia)**

Pearson Education New Zealand Ltd
46 Hillside Road, Auckland 10
℡ 00 64 9 444 4968 🅕 00 64 9 444 4957
www.pearson.co.nz

Founded in 1998 from the merger of Addison Wesley Longman New Zealand and Prentice Hall New Zealand. Educational publishers.

Penguin Books (NZ) Ltd
Private Bag 102 902, North Shore Mail Centre, Auckland 0745
℡ 00 64 9 442 7400   🖷 00 64 9 442 7401
publishing@penguin.co.nz
www.penguin.co.nz

Founded 1978. Owned by **Penguin UK**. Adult and children's fiction and non-fiction.

Random House New Zealand
Private Bag 102950, North Shore Mail Centre, Auckland
℡ 00 64 9 444 7197   🖷 00 64 9 444 7524
editor@randomhouse.co.nz
www.randomhouse.co.nz

Publishes fiction and non-fiction (cooking, gardening, art, natural history) and children's.

Raupo Publishing (NZ) Ltd
Private Bag 102950, Birkenhead, Auckland
℡ 00 64 9 441 2960   🖷 00 64 9 480 4999
info@raupopublishing.co.nz
www.raupopublishing.co.nz

Founded 1988. Publishes fiction and general non-fiction, biography, children's, cookery, health, history, natural history, regional interests, reference, self help, travel.

Victoria University Press
PO Box 600, Wellington
℡ 00 64 4 463 6580   🖷 00 64 4 463 6581
victoria-press@vuw.ac.nz
www.vuw.ac.nz/vup

Founded 1979. Academic publishers; new fiction, biography, poetry, literature, essays, New Zealand history, Maori topics. About 25 titles a year.

Bridget Williams Books Ltd
PO Box 5482, Wellington
℡ 00 64 4 473 8128
info@bwb.co.nz
www.bwb.co.nz

Founded 1990. Independent publisher of New Zealand history, Maori experience, contemporary issues and women's studies.

South Africa

Heinemann International Southern Africa
PO Box 781940, Sandton 2146
℡ 00 27 11 322 8600   🖷 00 27 11 322 8716
www.heinemann.co.za

Founded 1986. Educational publisher.

University of Kwa-Zulu–Natal Press
Private Bag X01, Scottsville 3209
℡ 00 27 33 260 5226   🖷 00 27 33 260 5801
books@ukzn.ac.za
www.ukznpress.co.za

Founded 1947. Publishes academic and general books; children's, African literature, poetry, economics, military history, natural sciences, social sciences.

LexisNexis Butterworths (Pty) Ltd South Africa
PO Box 792, Durban 4000
℡ 00 27 31 268 3111   🖷 00 27 31 268 3109
www.lexisnexis.co.za

Publishes professional, accountancy, finance, business, law, taxation and economics books in print form, online and CD-ROM.

Macmillan South Africa
PO Box 32484, Braamfontein, Gauteng 2017, Johannesburg
℡ 00 27 11 484 0916   🖷 00 27 11 484 3129
info@macmillan.co.za
www.macmillan.co.za

Founded 1972. Part of **Macmillan Publishers Ltd**, UK. Educational publisher.

Maskew Miller Longman
PO Box 396, Cape Town 8000
℡ 00 27 21 532 6000   🖷 00 27 21 531 8103
www.mml.co.za

Founded 1893. Owned jointly by **Pearson Education** and Caxton Publishers and Printers Ltd. Publishes education, ELT and teacher support books.

Oshun
▷ Struik Publishers (Pty) Ltd

Oxford University Press Southern Africa (Pty) Ltd
PO Box 12119, N1 City, Cape Town 7463
℡ 00 27 21 596 2300   🖷 00 27 21 596 1234
oxford.za@oup.com
www.oup.com/za

Founded 1915. Parent company: **Oxford University Press**, UK. Publishes academic, educational and general books.

Shuter & Shooter (Pty) Ltd
21c Cascades Crescent, Pietermaritzburg, KwaZulu-Natal 3201
℡ 00 27 33 347 6100   🖷 00 27 33 346 6120
www.shuter.co.za

Founded 1925. Publishes educational and general books.

Struik Publishers (Pty) Ltd

PO Box 1144, Cape Town 8000

Ⓣ 00 27 21 462 4360   Ⓕ 00 27 21 462 4377

info@struik.co.za

www.struik.co.za

Founded 1962. Publishes general non-fiction, illustrated, art and culture, lifestyle, natural history, travel. IMPRINTS **Oshun** Fiction, biography, health, humour, parenting, reference, self help, South African interest; **Two Dogs** Men's interests; **Zebra** Fiction and non-fiction.

Two Dogs
▷ Struik Publishers (Pty) Ltd

Wits University Press

PO Wits, Johannesburg 2050

Ⓣ 00 27 11 484 5907   Ⓕ 00 27 11 484 5971

witspress@wup.wits.ac.za

witspress.wits.ac.za

Founded 1922. Publishes academic, biography and memoirs, economics, heritage, history, business, politics, popular science, women's writing.

Zebra
▷ Struik Publishers (Pty) Ltd

Poetry Presses

Agenda Editions
The Wheelwrights, Fletching Street, Mayfield
TN20 6TL
☎ 01435 873703
editor@agendapoetry.co.uk
www.agendapoetry.co.uk
Editor *Patricia McCarthy*

Separate collections of poetry. Very few
published in a year and not many unsolicited.
Seek publication in the magazine first – see entry
under *Poetry Magazines*.

Akros Publications
33 Lady Nairn Avenue, Kirkcaldy KY1 2AW
☎ 01592 651522
www.akrospublications.co.uk
Contact *Duncan Glen*

Publisher of Scottish poetry and literary criticism.
See also *ZED20* magazine and entry under *UK
Publishers*.

Anvil Press Poetry Ltd
Neptune House, 70 Royal Hill, London SE10 8RF
☎ 020 8469 3033   ☒ 020 8469 3363
anvil@anvilpresspoetry.com
www.anvilpresspoetry.com
Contact *Peter Jay*

Contemporary British poetry and poetry in
translation. See entry under *UK Publishers*.

Arc Publications Ltd.
Nanholme Mill, Shaw Wood Road, Todmorden
OL14 6DA
☎ 01706 812338   ☒ 01706 818948
arc.publications@btconnect.com
www.arcpublications.co.uk
Editorial Director *Tony Ward*

Contemporary poetry from new and established
writers using English as their first language
from the UK and abroad. See entry under *UK
Publishers*.

Arrowhead Press
70 Clifton Road, Darlington DL1 5DX
editor@arrowheadpress.co.uk
www.arrowheadpress.co.uk
Contact *Roger Collett*

Quality books and pamphlets of contemporary
poetry.

Atlantean Publishing
38 Pierrot Steps, 71 Kursaal Way, Southend on Sea
SS1 2UY
atlanteanpublishing@hotmail.com
www.geocities.com/dj_tyrer/atlantean_pub.html
Contact *David John Tyrer*

Single author broadsheets and poetry/prose
booklets. See also **Awen, Monomyth, Bard,
Garbaj** and **The Supplement** magazines.

Barque Press
70A Cranwich Road, London N16 5JD
☎ 020 7502 0906
www.barquepress.com
Contact *Andrea Brady*

Poetry.

BB Books
Spring Bank, Longsight Road, Copster Green,
Blackburn BB1 9EU
☎ 01254 249128
Contact *Dave Cunliffe*

Post-Beat poetics and counterculture theoretic.
Iconoclastic rants and anarchic psycho-cultural
tracts. See also **Global Tapestry Journal**.

Between the Lines
▷ entry under Small Presses

Beyond the Cloister Publications
14 Lewes Crescent, Brighton BN2 1FH
☎ 01273 687053
www.beyondthecloister.com
Contact *Hugh Hellicar*

Anthologies of poetry and slim volumes to launch
poets.

Blinking Eye Publishing
PO Box 175, Hexham NE46 9AW
☎ 01434 600345
judy@blinking-eye.co.uk
www.blinking-eye.co.uk
Editor *Judy Walker*

Promotes the work of writers over 50 – through poetry and short story competitions and publication of the winners' collections.

Bloodaxe Books Ltd

Highgreen, Tarset NE48 1RP
℡ 01434 240500   🖷 01434 240505
editor@bloodaxebooks.com
www.bloodaxebooks.com
Contact *Neil Astley*

Britain's leading publisher of new poetry. No submissions by e-mail attachments. See entry under *UK Publishers*.

Blue Butterfly Publishers

13 Irvine Way, Inverurie AB51 4ZR
℡ 01467 625986   🖷 01467 625986
blue7butterfly@which.net
madill.prodigynet.co.uk/bbp
Contact *Betty Madill*

Christian poetry. 'Poetry to be submitted through our poetry competitions only.' Entry forms from address above or from the website.

Bluechrome Publishing

PO Box 109, Portishead, Bristol BS20 7ZJ
anthony@bluechrome.co.uk
www.bluechrome.co.uk
Contact *Anthony Delgrado*

Founded 2002. Poetry, literary and experimental fiction. Submission details available on the website.

Bradshaw Books

Tigh Fili, Cork Arts Theatre, Carroll's Quay, Cork, Republic of Ireland
℡ 00 353 21 450 9274
admin@tighfili.com
www.tighfili.com
Contact *Maire Bradshaw*

See entry under *Irish Publishers*

Bridge Pamphlets

PO Box 309, Aylsham, Norwich NR11 6LN
mail@therialto.co.uk
www.therialto.co.uk
Contact *Michael Mackmin*

An imprint of **Rialto Publications**.

The Brodie Press

www.brodiepress.co.uk

See entry under *Small Presses*.

Bullseye Publications

5 Camptoun, North Berwick EH39 5BA
℡ 01620 880311
alancharlesgay@aol.com
Contact *Alan Gay*

Calder Wood Press

1 Beachmont Court, Dunbar EH42 1YF
℡ 01368 864953
colin.will@zen.co.uk
www.calderwoodpress.co.uk
Contact *Colin Will*

Founded 1997. Publishes pamphlets and short-run publications. No unsolicited mss.

Carcanet Press Ltd

Major poetry publisher. See entry under *UK Publishers*; also publishes *PN Review* magazine.

Chapman Publishing

See entry under *UK Publishers*.

Cinnamon Press

Ty Meirion, Glan yr afon, Tanygrisiau, Blaenau Ffestiniog LL41 3SU
℡ 01766 832112
jan@cinnamonpress.com
www.cinnamonpress.com
Editor *Dr Jan Fortune-Wood*

Poetry, fiction, some non-fiction. Growing list from Wales, UK & international. Regular literary competitions. See also **Envoi** magazine.

The Collective Press

c/o Penlanlas Farm, Llantilio Pertholey, Y-fenni NP7 7HN
℡ 01873 856350   🖷 01873 858357
jj@jojowales.co.uk
www.welshwriters.com
Contact *John Jones & Frank Olding*

Non-profit promoter and publishers of contemporary poetry.

Day Dream Press

39 Exmouth Street, Swindon SN1 3PU
℡ 01793 523927
Contact *Kevin Bailey*

Small poetry collections. Now associated with **Bluechrome Press**. See also *HQ Poetry Magazine (Haiku Quarterly)*.

The Dedalus Press

13 Moyclare Road, Baldoyle, Dublin 13, Republic of Ireland
℡ 00 353 1 839 2034   🖷 0870 127 2089
office@dedaluspress.com
www.dedaluspress.com
Publisher *Pat Boran*

Contemporary Irish poetry and poetry from around the world in English translation.

Diehard Publishers

91–93 Main Street, Callander FK17 8BQ
℡ 0131 229 7252
Contact *Ian W. King*

Poetry and drama, mainly Scottish. See also **Poetry Scotland** magazine.

Dionysia Press

127 Milton Road West, 7 Duddingston House Courtyard, Edinburgh EH15 1JG
Contact *Denise Smith*

Collections of poetry, short stories, novels and translations. See also **Understanding** magazine.

DogHouse

PO Box 312, Tralee, Co. Kerry, Republic of Ireland
℡ 00 353 667 137547   🖷 00 353 667 137547
info@doghousebooks.ie
www.doghousebooks.ie
Contact *Noel King*

Poetry and short story collections by individuals (Irish born only). About 4–6 titles a year.

Dreadful Night Press

82 Kelvin Court, Glasgow G12 0AQ
℡ 0141 339 7988
dreadfulnight1@aol.com

Dream Catcher

Jasmine Cottage, 4 Church Street, Market Rasen LN8 3ET
℡ 01673 844325
paulsuther@hotmail.com
www.dreamcatchermagazine.co.uk
Contact *Paul Sutherland*

Short fiction, poetry, reviews, artwork, biographies. See also *Dream Catcher* magazine.

Driftwood Publications

Apartment 608, Beetham Plaza, 124 The Strand, Liverpool L2 0XJ
℡ 0151 236 7550   🖷 0151 524 0216
janet.speedy@tesco.net
Contact *Brian Wake*

New work by new and established poets more suited to the page than the stage.

Egg Box Publishing

25 Brian Avenue, Norwich NR1 2PH
mail@eggboxpublishing.com
www.eggboxpublishing.com
Contact *Nathan Hamilton & Gordon Smith*

New poetry and fiction; translations also considered. Submissions by e-mail only.

Enitharmon Press

26B Caversham Road, London NW5 2DU
℡ 020 7482 5967   🖷 020 7284 1787
books@enitharmon.co.uk
www.enitharmon.co.uk
Contact *Stephen Stuart-Smith*

Poetry and literary criticism. See entry under *UK Publishers*.

Essence Press

8 Craiglea Drive, Edinburgh EH10 5PA
jaj@essencepress.co.uk
www.essencepress.co.uk
Contact *Julie Johnstone*

Handbound editions of poetry, poetry postcards. Interested in concrete poetry and nature. See also **Island** magazine.

Etruscan Books

Stowe Lane, Exbourne EX20 3RY
atetruscan@aol.com
www.e-truscan.co.uk
Contact *Nicholas Johnson*

Modernist, sound, visual poetry, Gaelic, lyric poetry, US/UK poets.

Fal Publications

PO Box 74, Truro TR1 1XS
℡ 01208 851205
info@falpublications.co.uk
www.falpublications.co.uk
Contact *Victoria Field*

Poetry and literature from Cornwall.

Feather Books

PO Box 438, Shrewsbury SY3 0WN
℡ 01743 872177   🖷 01743 872177
john@waddysweb.freeuk.com
www.waddysweb.freeuk.com
Contact *Rev. J. Waddington-Feather*
General Manager *Paul Evans*

Christian poetry, drama, hymns, novels. See entry under *Small Presses*; also **The Poetry Church** magazine and *Feather's Miscellany*.

Five Leaves Publications

PO Box 8786, Nottingham NG1 9AW
℡ 0115 969 3597
info@fiveleaves.co.uk
www.fiveleaves.co.uk
Contact *Ross Bradshaw*

Fiction, poetry, Jewish interest, social history. See entry under *Small Presses*.

Flambard Press

Stable Cottage, East Fourstones, Hexham NE47 5DX
℡ 01434 674360   🖷 01434 674178
flambardpress@btinternet.com
www.flambardpress.co.uk
Contact *Peter Lewis & Will Mackie*

Concentrates on poetry and literary fiction. Consult the website for submission information.

Flarestack Publishing

8 Abbot's Way, Pilton BA4 4BN
℡ 01749 890019

cannula.dementia@virgin.net
www.flarestack.co.uk
Contact *Charles Johnson*

Poetry. See also **Obsessed With Pipework** magazine.

Forward Press

Remus House, Coltsfoot Drive, Woodston, Peterborough PE2 9JX
℡ 01733 890099   ℻ 01733 313524
info@forwardpress.co.uk
www.forwardpress.co.uk
Contact *The Editorial Team*

General poetry and short fiction anthologies.

The Frogmore Press

42 Morehall Avenue, Folkestone CT19 4EF
℡ 07751 251689
www.frogmorepress.co.uk
Managing Editor *Jeremy Page*

A forum for contemporary poetry, prose and artwork. See also **The Frogmore Papers** magazine.

The Gallery Press

Loughcrew, Oldcastle, Co. Meath, Republic of Ireland
℡ 00 353 49 854 1779   ℻ 00 353 49 854 1779
gallery@indigo.ie
www.gallerypress.com
Contact *Peter Fallon*

Poems, plays and prose by contemporary Irish writers.

The Goldsmith Press

Newbridge, Co. Kildare, Republic of Ireland
℡ 00 353 45 433613   ℻ 00 353 45 434648
viv1@iol.ie
Contact *Vivienne Abbott*

Gomer Press/Gwasg Gomer

Llandysul Enterprise Park, Llandysul SA44 4JL
℡ 01559 363090   ℻ 01559 363758
www.gomer.co.uk
English Books for Adults *Ceri Wyn Jones*
(ceri@gomer.co.uk)
English Books for Children *Viv Sayer*
(viv@gomer.co.uk)
Publishing Director *Mairwen Prys Jones*
(mairwen@gomer.co.uk)

Welsh interest. See entry under *UK Publishers.*

Green Arrow Publishing

6 Green Bank, Stacksteads, Bacup OL13 8LQ
℡ 01706 870357/07967 315270 (mobile)
mail@johndench.demon.co.uk
Contact *John Dench*

Collaborative publishing scheme and other services for writers. See also **Scriptor** magazine.

HappenStance Press

21 Hatton Green, Glenrothes KY7 4SD
nell@happenstancepress.com
www.happenstancepress.com
Contact *Helena Nelson*

Poetry publishing; mainly chapbooks and usually first collections. Submission guidelines on the website. See also **Sphinx** magazine.

Harpercroft

24 Castle Street, Crail KY10 3SH
℡ 01333 451744
jarvie522@btinternet.com
Contact *Gordon Jarvie*

Poetry, local interest.

Headland Publications

Ty Coch, Galltegfa, Ruthin LL15 2AR
℡ 0151 625 9128   ℻ 0151 625 9128
headlandpublications@hotmail.co.uk
www.headlandpublications.co.uk
Contact *Gladys Mary Coles*

Fine editions of poetry; anthologies.

Hearing Eye

c/o 99 Torriano Avenue, London NW5 2RX
hearing_eye@torriano.org
www.torriano.org/hearing_eye/
Contact *John Rety*

Poetry/literature publishers based in Kentish Town.

Hilltop Press

4 Nowell Place, Almondbury, Huddersfield HD5 8PD
booksmusicfilmstv.com/HilltopPress.htm
Contact *Steve Sneyd*

Specialist publisher of science fiction and dark fantasy poetry, and background material.

Hippopotamus Press

22 Whitewell Road, Frome BA11 4EL
℡ 01373 466653   ℻ 01373 466653
rjhippopress@aol.com
Contact *Roland John & Mansell Pargitter*

First collections of verse from those with a track record in the magazines.

Honno Welsh Women's Press

c/o Canolfan Merched Y Wawr, Vulcan Street, Aberystwyth SY23 1JH
℡ 01970 623150   ℻ 01970 623150
post@honno.co.uk
www.honno.co.uk
Editor *Caroline Oakley*

The Welsh women's press – novels, short stories, poetry and autobiographical anthologies. Must have a Welsh connection. See entry under *UK Publishers.*

I*D Books

Connah's Quay Library, Wepre Drive, Connah's Quay, Deeside

☎ 01244 830485

Poetry, short fiction, local history – mainly self-publishing by associated writers' group.

Iron Press

5 Marden Terrace, Cullercoats, North Shields NE30 4PD

☎ 0191 253 1901   ☏ 0191 253 1901

ironpress@blueyonder.co.uk

www.ironpress.co.uk

Contact *Peter Mortimer*

Poetry and fiction.

Katabasis

10 St Martin's Close, London NW1 0HR

☎ 020 7485 3830

katabasis@katabasis.co.uk

www.katabasis.co.uk

Contact *Dinah Livingstone*

Down-to-earth and Utopian poetry and prose from home an abroad – English and Latin American. No unsolicited mss.

Kates Hill Press

126 Watsons Green Road, Kates Hill, Dudley DY2 7LG

kateshillpress@blueyonder.co.uk

www.kateshillpress.co.uk

Fiction and social history from the West Midlands or with a West Midlands theme. Some poetry.

Kettillonia

24 South Street, Newtyle PH12 8UQ

☎ 01828 650615

james@kettillonia.co.uk

www.kettillonia.co.uk

Editor *James Robertson*

Occasional publisher of pamphlets. No unsolicited manuscripts.

The King's England Press

Cambertown House, Commercial Road, Goldthorpe, Rotherham S63 9BL

☎ 01484 663790   ☏ 01484 663790

sales@kingsengland.com

www.kingsengland.com

www.pottypoets.com

Contact *Steve Rudd*

History, folklore, children's poetry.

KT Publications

16 Fane Close, Stamford PE9 1HG

☎ 01780 754193

Editor *Kevin Troop*

Always looking for new material of the highest standard. See also **The Third Half** magazine.

Leaf Books

Gti Suite, Valleys Innovation Centre, Navigation Park, Abercynon CF45 4SN

☎ 029 20810726

contact@leafbooks.co.uk

www.leafbooks.co.uk

Poetry and Micro-Fiction competition anthologies: getting new writers into print.

Leafe Press

www.leafepress.com

Contact *Alan Baker*

Pamphlets of contemporary poetry with occasional full-length collections.

Malfunction Press (1969)

Rose Cottage, 3 Tram Lane, Buckley CH7 3JB

rosecot@presford.freeserve.co.uk

Editor *Peter E. Presford*

Mainly dedicated to science fiction, fantasy, light horror.

Mare's Nest

41 Addison Gardens, London W14 0DP

☎ 020 7603 3969

www.maresnest.co.uk

Contact *Pamela Clunies-Ross*

Icelandic literature.

Mariscat Press

10 Bell Place, Edinburgh EH3 5HT

☎ 0131 343 1070

hamish.whyte@btinternet.com

Contact *Hamish Whyte & Diana Hendry*

Currently publishing poetry pamphlets only.

Masque Publishing

PO Box 3257, 84 Dinsdale Gardens, Littlehampton BN16 9AF

masque_pub@tiscali.co.uk

myweb.tiscali.co.uk/masquepublishing

Contact *Lisa Stewart*

For self-publishers. See also **Decanto** magazine.

Mews Press

English Department, Owen 11, Sheffield Hallam University, Howard Street, Sheffield S1 1WB

☎ 0114 225 4163

s.l.earnshaw@shu.ac.uk

extra.shu.ac.uk/mews-press

Contact *Dr Steven Earnshaw*

Work by writers associated with Sheffield Hallam University's MA Writing, including *Matter* and the *Ictus*.

Mudfog Press

Arts Development, The Stables, Stewart Park, The Grove, Marton, Middlesborough TS7 8AR

www.mudfog.co.uk

Community press publishing poetry and short fiction only from writers in the Tees Valley and adjoining rural areas.

New Hope International

20 Werneth Avenue, Gee Cross, Hyde SK14 5NL
www.geraldengland.co.uk
Contact *Gerald England*

Poetry booklet publisher. No longer considering unsolicited mss.

New Island

2 Brookside, Dundrum Road, Dublin 14, Republic of Ireland
www.newisland.ie
Publisher *Edwin Higel*
Editorial Manager *Deirdre Nolan*

See entry under *Irish Publishers*.

nthposition

38 Allcroft Road, London NW5 4NE
☎ 020 7485 5002
val@nthposition.com
www.nthposition.com
Editor *Val Stevenson*
Poetry Editor *Todd Swift* (toddswift@clara.co.uk)

Eclectic and award-winning mix of politics and opinion, travel writing, fiction and poetry, reviews and interviews and some high weirdness.

Object Permanence

First Floor, 16 Ruskin Terrace, Glasgow G12 8DY
robinpurves@yahoo.co.uk
www.objectpermanence.co.uk
Contact *Robin Purves*

The Old Stile Press

Catchmays Court, Llandogo, Nr Monmouth NP25 4TN
☎ 01291 689226
oldstile@dircon.co.uk
www.oldstilepress.com
Contact *Frances & Nicolas McDowall*

Fine, hand-printed books with text and images. No unsolicited mss.

The Once Orange Badge Poetry Press

PO Box 184, South Ockenden RM15 5WT
☎ 01708 852827
orangebadge@poetry.fsworld.co.uk
Contact *D. Martyn Heath*

A4 and A5 pamphlets – ideal for first collections. E-mail submissions preferred. See also **The Once Orange Badge Poetry Supplement** magazine.

The One Time Press

Model Farm, Linstead Magna, Halesworth IP19 0DT
☎ 01986 785422
www.onetimepress.com

Contact *Peter Wells*

Poetry of the forties and occasionally some contemporary work in limited editions. Illustrated.

Original Plus

17 High Street, Maryport CA15 6BQ
☎ 01900 812194
smithsssj@aol.com
members.aol.com/smithsssj/index.html
Contact *Sam Smith*

See also **The Journal** magazine.

Palores Publishing

11a Penryn Street, Redruth TR15 2SP
☎ 01209 218209
les.merton@tesco.net
Editor *Les Merton*

Non-profit-making publisher helping writers and poets to self publish.

Parthian

The Old Surgery, Napier Street, Cardigan SA43 1ED
☎ 01239 612059
parthianbooks@yahoo.co.uk
www.parthianbooks.co.uk

New Welsh writing. See entry under *UK Publishers*.

Partners

289 Elmwood Avenue, Feltham TW13 7QB
partners_writing_group@hotmail.com
Contact *Ian Deal*

Competitions and poetry pamphlets. See also **A Bard Hair Day**, **Imagenation** and **Poet Tree** magazines.

Peepal Tree Press Ltd

17 King's Avenue, Leeds LS6 1QS
☎ 0113 245 1703
contact@peepaltreepress.com
www.peepaltreepress.com
Contact *Jeremy Poynting*

Peepal Tree Press is the home of challenging and inspiring literature from the Caribbean and Black Britain. Online bookstore.

Penniless Press

100 Waterloo Road, Ashton, Preston PR2 1EP
☎ 01772 736421
www.pennilesspress.co.uk
Contact *Alan Dent*

Novels, short fiction, poetry, plays.

Peterloo Poets

The Old Chapel, Sand Lane, Calstock PL18 9QX
☎ 01822 833473

publisher@peterloopoets.com
www.peterloopoets.com
Publishing Director *Harry Chambers*

Pigasus Press

13 Hazely Combe, Arreton, Isle of Wight PO30 3AJ
mail@pigasuspress.co.uk
www.pigasuspress.co.uk
Contact *Tony Lee*

Publisher of science fiction magazine and genre poetry.

Pikestaff Press

Ellon House, Harpford, Sidmouth EX10 0NH
📞 01395 568941
Contact *Robert Roberts*

Contemporary poetry belonging to the English tradition.

Pipers' Ash Ltd

Pipers Ash, Church Road, Christian Malford, Chippenham SN15 4BW
📞 01249 720563   📠 0870 056 8916
pipersash@supamasu.co.uk
www.supamasu.co.uk
Contact *Mr A. Tyson*

Collections of 60 poems arranged in their best order, grouped on common themes, with an appropriate title. See entry under *Small Presses*.

Poems in the Waiting Room

PO Box 488, Richmond TW9 4SW
pitwr@blueyonder.co.uk
www.pitwr.pwp.blueyonder.co.uk
Editor *Michael Lee*

Arts in Health charity supplying quarterly batches of poetry cards for NHS patients nationwide.

Poetry Now

Remus House, Coltsfoot Drive, Woodston, Peterborough PE2 8JX
📞 01733 8998101   📠 01733 313524
poetrynow@forwardpress.co.uk
www.forwardpress.co.uk
Editorial Manager *Michelle Afford*

Lively, personal and contemporary. See also **Forward Press**.

Poetry Powerhouse Press

▷ Performance Poetry Society under Organizations of Interest to Poets

Poetry Press Ltd

26 Park Grove, Edgware HA8 7SJ
📞 020 8958 6499
poetrypress@yahoo.co.uk
www.judyk.co.uk
Contact *Judy Karbritz*

Anthologies – rhyming poetry welcome.

Poetry Salzburg

Dept. of English, University of Salzburg, Akademiestr. 24, A–5020 Salzburg, Austria
📞 00 43 662 8044 4424   📠 00 43 662 8044 167
editor@poetrysalzburg.com
www.poetrysalzburg.com
Contact *Wolfgang Görtschacher*

Publishes books of poetry, poetry in translation, literary criticism. See also **Poetry Salzburg Review** magazine.

Poetry Wednesbury

25 Griffiths Road, West Bromwich B71 2EH
📞 07950 591455
ppatch66@hotmail.com
www.poetrywednesbury.co.uk
Contact *Geoff Stevens*

Occasional CDs, DVDs, leaflets and booklets. See also **The Firing Squad** and **Purple Patch** magazines. Public meetings for readings, workshops, etc. monthly at The Old Post Office, Holyhead Road, Wednesbury.

Poets Anonymous

70 Aveling Close, Purley CR8 4DW
poets@poetsanon.org.uk
www.poetsanon.org.uk
Contact *Peter L. Evans*

Anthologies and collections of predominantly south London poets. See also **Poetic Licence** magazine.

Precious Pearl Press

41 Grantham Road, Manor Park, London E12 5LZ
Contact *P.G.P. Thompson*

Romantic, lyrical, spiritual, inspirational and mystical poetry; a traditionalist poetry press. See also **Rubies in the Darkness** magazine.

PS Avalon

Box 1865, Glastonbury BA6 8YR
info@psavalon.com
www.psavalon.com
Contact *Will Parfitt*

Quality books of psychospiritual poetry and for personal and spiritual development.

Puppet State Press

1 West Colinton House, 40 Woodhall Road, Edinburgh EH13 0DU
📞 0131 441 9693
richard@puppetstate.com
www.puppetstate.com
Contact *Richard Medrington*

QQ Press

York House, 15 Argyle Terrace, Rothesay, Isle of Bute PA20 0BD
Contact *Alan Carter*

Collections of poetry plus poetry anthologies. Enquiries should be marked 'Collections' and sent with an s.a.e. or two IRCs if from abroad. See also **Quantum Leap** magazine.

Rack Press

The Rack, Kinnerton, Presteigne LD8 2PF
☎ 01547 560411
rackpress@nicholasmurray.co.uk
www.nicholasmurray.co.uk/RackPress.html
Blog: www.rackpress.blogspot.com
Contact *Nicholas Murray*

Welsh poetry pamphlet imprint with an international vision.

Ragged Raven Press

1 Lodge Farm, Snitterfield, Stratford upon Avon CV37 0LR
☎ 01789 730320
raggedravenpress@aol.com
www.raggedraven.co.uk
Contact *Bob Mee & Janet Murch*

Poetry. See also **Iota** magazine.

Raunchland Publications

26 Alder Grove, Dunfermline KY11 8RP
raunchland@hotmail.com
www.raunchland.co.uk
Contact *John Mingay*

Limited edition poetry/graphics booklets and online publications.

Reality Street Editions

63 All Saints Street, Hastings TN34 3BN
info@realitystreet.co.uk
www.realitystreet.co.uk
Contact *Ken Edwards*

New writing from Britain, Europe and America. No unsolicited mss.

Red Candle Press

1 Chatsworth Court, Outram Road, Southsea PO5 1RA
www.members.tripod.com/redcandlepress
Contact *M.L. McCarthy*

Founded 1970. Traditionalist poetry press. See also **Candelabrum Poetry** magazine.

Rialto Publications

PO Box 309, Aylsham, Norwich NR11 6LN
mail@therialto.co.uk
www.therialto.co.uk
Contact *Michael Mackmin*

'We want to publish first collections by poets of promise.' See also **The Rialto** magazine and **Bridge Pamphlets** press.

Rive Gauche Publishing

69 Lower Redland Road, Bristol BS6 6SP
☎ 0117 974 5106

pat@patvtwest.co.uk
twinset1969@hotmail.com
www.patvtwest.co.uk
Contact *PVT West*

Poetry by women writing and/or performing in Bristol.

Route Publishing

PO Box 167, Pontefract WF8 4WW
☎ 01977 797695
info@route-online.com
www.route-online.com
Contact *Ian Daley & Isabel Galan*

Contemporary fiction and performance poetry. Online magazine publishes downloadable books.

Salt Publishing

PO Box 937, Great Wilbraham, Cambridge CB1 5JX
☎ 01223 882220   ⊕ 01223 882260
sales@saltpublishing.com
www.saltpublishing.com
Contact *Chris Hamilton-Emery*

International poetry and poetics. 2008 winner of the Nielsen Innovation of the Year prize.

Second Light Publications

9 Greendale Close, London SE22 8TG
dilyswood@tiscali.co.uk
www.secondlightlive.co.uk
www.poetrypf.co.uk/secondlight.html
www.esch.dircon.co.uk/second/second.htm
Contact *Dilys Wood*
Administrator *Anne Stewart*
(editor@poetrypf.co.uk)

Publishes high-calibre women's poetry, mainly anthologies, occasional collections (by invitation).

Seren

57 Nolton Street, Bridgend CF31 3AE
☎ 01656 663018
seren@seren-books.com
www.seren-books.com
Contact *Mick Felton*

Poetry, fiction, lit crit, biography, essays. See entry under *UK Publishers*; also **Poetry Wales** magazine.

Shearsman Books

58 Velwell Road, Exeter EX4 4LD
☎ 01392 434511   ⊕ 01392 434511
editor@shearsman.com
www.shearsman.com
Contact *Tony Frazer*

Publishes poetry almost exclusively. Contact editor before submitting or consult the submissions page at the Shearsman website for the most up-to-date instructions. See also **Shearsman** magazine.

Shoestring Press

19 Devonshire Avenue, Beeston, Nottingham
NG9 1BS
☎ 0115 925 1827   ℻ 0115 925 1827
info@shoestringpress.co.uk
www.shoestring-press.com
Contact *John Lucas*

Poetry.

Sixties Press

89 Connaught Road, Sutton SM1 3PJ
☎ 020 8286 0419
Contact *Barry Tebb*

Small press concentrating on poetry and novellas.
See also **Leeds Poetry Quarterly** and **Literature
and Psychoanalysis** magazines.

Smokestack Books

PO Box 408, Middlesbrough TS5 6WA
☎ 01642 813997
info@smokestack-books.co.uk
Contact *Andy Croft*

Champions poets who are unconventional,
unfashionable, radical or left-field.

Spectacular Diseases

83b London Road, Peterborough PE2 9BS
Contact *Paul Green*

Innovative poetry and some prose. Translations
of both.

Stinging Fly Press

PO Box 6016, Dublin 8, Republic of Ireland
stingingfly@gmail.com
www.stingingfly.org

New Irish and international writing. See also **The
Stinging Fly** magazine.

Talking Pen

12 Derby Crescent, Moorside, Consett DH8 8DZ
☎ 01207 505724
Contact *Steve Urwin*

See also **Moodswing** magazine.

Templar Poetry

PO Box 7082, Bakewell DE45 9AF
☎ 01629 582500
info@templarpoetry.co.uk
www.templarpoetry.co.uk
Contact *Alexander McMillen*

Publishes new poets and poetry via an annual
open pamphlet and collection competition,
and launches new titles at the Derwent Poetry
Festival in Derbyshire in October. Poets are also
occasionally commissioned to submit work for
publication. 'Templar Poetry is committed to
the professional publication and development of
excellent contemporary poetry.'

Vane Women Press

31 Garthorne Avenue, Darlington DL3 9XL
☎ 01325 468464
marg.rule@ntl.com
www.vanewomen.co.uk
Contact *Margaret Rule*

Work by women of the North East.

Waterloo Press

126 Furze Croft, Hove BN3 1PF
☎ 01273 202876
www.waterloopress.com
Contact *Simon Jenner*

Poetry and periodical publisher; please phone for
submissions. See also **Eratica** magazine.

Waywiser Press

The Cottage, 14 Lyncroft Gardens, Ewell KT17 1UR
☎ 020 8393 7055   ℻ 020 8393 7055
waywiserpress@aol.com
www.waywiser-press.com
Managing Editor *Philip Hoy*

Independent literary press. See entry under *Small
Presses.*

Wendy Webb Books

9 Walnut Close, Taverham, Norwich NR8 6YN
tipsforwriters@yahoo.co.uk
Contact *Wendy Webb*

Poetry books with rules of new and traditional
forms; Arthur legend, Norfolk, Pantoums.
See also **Norfolk Poets and Writers – Tips
Newsletter** magazine.

West House Books

40 Crescent Road, Nether Edge, Sheffield S7 1HN
☎ 0114 258 6035
alan@nethedge.demon.co.uk
www.westhousebooks.co.uk
Contact *Alan Halsey*

Contemporary poetry, poets' prose and related
work.

Wild Women Press

Flat 10, The Common, Windermere LA23 1JH
wildwomenpress.blogspot.com

Not-for-profit small press run by founders
Victoria Bennett and Adam Clarke. 'We are
unable to accept or reply to, or return, unsolicited
manuscripts so please don't send them. Any
opportunity will be posted on our blog.'

Worple Press

See entry under *Small Presses.*

Poetry Magazines

Acumen

6 The Mount, Higher Furzeham, Brixham
TQ5 8QY
☎ 01803 851098
pwoxley@aol.com
www.acumen-poetry.co.uk
Contact *Patricia Oxley*

Good poetry, intelligent articles and wide-ranging reviews.

Aesthetica

PO Box 371, York YO23 1WL
☎ 01904 527560
info@aestheticamagazine.com
www.aestheticamagazine.com
Managing Editor *Cherie Federico*
Poetry Editor *Bruce Corrie*
Fiction Editor *David Martin*

Contemporary glossy bi-monthly: literature, art, music, film, theatre, news, reviews, interviews and features. Distribution throughout the UK in WH Smith and Borders.

Agenda Poetry Magazine

The Wheelwrights, Fletching Street, Mayfield
TN20 6TL
☎ 01435 873703
editor@agendapoetry.co.uk
www.agendapoetry.co.uk
Editor *Patricia McCarthy*

Poems, essays, reviews. Visit the website for further information and supplements to Agenda. Submissions: up to five poems, with s.a.e. and e-mail address. Young poets (and young artists) and essayists (age 15 to late thirties) also encouraged in the journal and in the online Broadsheets. Subscriptions: (one vol. = 4 issues = one subscription) £28 individual (£22 OAPs/students); £35 libraries/institutions. See **Agenda Editions** press.

Aireings

submissions
@aireings.co.uk
www.aireings.co.uk/
Contact *Lesley Quayle & Linda Marshall*

Online poetry mag: poetry, prose, reviews, b&w artwork.

Aquarius

Flat 4, Room B, 116 Sutherland Avenue, London
W9 2QP
☎ 020 7289 4338
www.geocities.com/eddielinden
Contact *Eddie S. Linden*

Literary magazine; prose and poetry. No emailed submissions.

Areopagus

48 Cornwood Road, Plympton, Plymouth PL7 1AL
editor@areopagus.org.uk
www.areopagus.org.uk
Contact *Julian Barritt*

A Christian-based arena for creative writers. Subscription: £12 p.a.

Awen

38 Pierrot Steps, 71 Kursaal Way, Southend on Sea
SS1 2UY
atlanteanpublishing@hotmail.com
www.geocities.com/dj_tyrer/awen.html
Contact *David John Tyrer*

Poetry and vignette-length fiction of any style/genre. See also **Bard, Monomyth, Garbaj** and **The Supplement** magazines and **Atlantean Publishing** press.

Bard

38 Pierrot Steps, 71 Kursaal Way, Southend on Sea
SS1 2UY
atlanteanpublishing@hotmail.com
www.geocities.com/dj_tyrer/bard.html
Contact *David John Tyrer*

Short poetry. See also **Awen, Garbaj, Monomyth** and **The Supplement** magazines and **Atlantean Publishing** press.

A Bard Hair Day

289 Elmwood Avenue, Feltham TW13 7QB
Contact *Ian Deal*

Contemporary poetry magazine. See also **Partners** press, **Imagenation** and **Poet Tree** magazines.

Blithe Spirit

12 Eliot Vale, Blackheath, London SE3 0UW
☎ 020 8318 4677
www.britishhaikusociety.org
Contact *Graham High*

Journal of the British Haiku Society. Haiku and related forms.

Brittle Star

PO Box 56108, London E17 0AY
☎ 020 8802 1507
magazine@brittlestar.org.uk
www.brittlestar.org.uk
Contact *Louise Hooper*

Poetry, short stories and articles on contemporary poetry.

Candelabrum Poetry Magazine

1 Chatsworth Court, Outram Road, Southsea PO5 1RA
www.members.tripod.com/redcandlepress
Contact *M.L. McCarthy*

Formalist poetry magazine for people who like poetry rhythmic and shapely. See also **Red Candle Press**.

Cannon's Mouth

22 Margaret Grove, Harborne, Birmingham B17 9JH
☎ 0121 426 6413
greg@cannonpoets.co.uk
www.cannonpoets.co.uk
Contact *Greg Cox*

New poetry, criticism, articles, information.

Carillon

19 Godric Drive, Brinsworth, Rotherham S60 5NA
editor@carillonmag.org.uk
www.carillonmag.org.uk
Contact *Graham Rippon*

Eclectic poetry and prose.

Chanticleer Magazine

6/1 Jamaica Mews, Edinburgh EH3 6HN
chanticleer@blueyonder.co.uk
Contact *Richad Livermore*

Poetry and ideas – not necessarily in that order.

Chapman

4 Broughton Place, Edinburgh EH1 3RX
☎ 0131 557 2207
chapman-pub@blueyonder.co.uk
www.chapman-pub.co.uk
Contact *Joy Hendry*

The best in Scottish and international writing, well-established writers and the up-and-coming. See entry under *Magazines*.

Current Accounts

16–18 Mill Lane, Horwich, Bolton BL6 6AT
☎ 01204 962110
bswscribe@aol.com
hometown.ao.co.uk/bswscribe/myhomepage/writing.html
Contact *Rod Riesco*

Poetry, short fiction, articles. Magazine of the Bank Street Writers' Group. Submissions also accepted from non-members (by e-mail or by post with s.a.e.).

Dandelion Arts Magazine

24 Frosty Hollow, East Hunsbury NN4 0SY
☎ 01604 701730   🖷 01604 701730
Editor/Publisher *Jacqueline Gonzalez-Marina*

Biannual, international publication. No religious or political material. Essential to become a subscriber when seeking publication: UK, £17 p.a.; Europe, £30; RoW, US$90. See also **The Student Magazine**.

Decanto

PO Box 3257, 84 Dinsdale Gardens, Littlehampton BN16 9AF
masque_pub@tiscali.co.uk
myweb.tiscali.co.uk/masquepublishing
Contact *Lisa Stewart*

Non-conformist poetry magazine; any style considered, not just contemporary. See also **Masque Publishing** press.

Dream Catcher

Jasmine Cottage, 4 Church Street, Market Rasen LN8 3ET
☎ 01673 844325
paulsuther@hotmail.com
www.dreamcatchermagazine.co.uk
www.inpressbooks.co.uk
Editor *Paul Sutherland*
Business Manager *Gary McKeone*
(gary.mckeone@googlemail.com)

Short fiction, poetry, reviews, artwork, biographies. See also **Dream Catcher** press.

Earth Love

PO Box 11219, Paisley PA1 2WH
www.earthlove.741.com
Contact *Tracy Patrick*

Poetry magazine for the environment; proceeds to conservation charities.

Eastern Rainbow

17 Farrow Road, Whaplode Drove, Spalding PE12 0TS
p_rance@yahoo.co.uk
uk.geocities.com/p_rance/pandf.htm
Contact *Paul Rance*

Focuses on 20th century culture via poetry, prose and art. See also **Peace and Freedom** magazine.

The Engine

3 Ardgreenan Drive, Belfast BT4 3FQ
☎ 028 9065 9866   🖷 028 9032 2767
clitophon@yahoo.com
www.theengine.net
Contact *Paul Murphy*

Journal of art, poetry, reviews and short stories.

Envoi

Ty Meirion, Glan yr afon, Tanygrisiau, Blaenau Ffestiniog LL41 3SU
☎ 01766 832112
Contact *Jan Fortune-Wood*

Poetry, sequences, features, reviews, letters pages, competitions. Celebrating 50th anniversary and 150th issue in 2008. See also **Cinnamon Press**.

Equinox

134b Joy Lane, Whitstable CT5 4ES
☎ 01227 282718
dordi.barbara@neuf.fr
www.poetrymagazines.org.uk/equinox
Contact *Barbara Dordi*

Contemporary poetry invited. No e-mail submissions.

Eratica

126 Furze Croft, Hove BN3 1PF
☎ 01273 202876
drjenner@ntlworld.com
Contact *Simon Jenner*

Biannual journal with colour plates – focus on poetry, strong on music and art. See also **Waterloo Press**.

Feather's Miscellany

▷ **Feather Books under Small Presses**

The Firing Squad

barzdo@tiscali.co.uk
www.purplepatchpoetry.co.uk
Contact *Alex Barzdo*

Protest poetry website. E-mail submissions only. See also **Purple Patch** magazine and **Poetry Wednesbury** press.

First Offense

Syringa, The Street, Stodmarsh, Canterbury CT3 4BA
☎ 01227 721249
tim@firstoffense.co.uk
www.firstoffense.co.uk
Contact *Tim Fletcher*

Magazine for contemporary poetry; not traditional but is received by most ground-breaking poets.

Flaming Arrows

Sligo-Leitrim Arts, V E C, Riverside, Sligo, Republic of Ireland
☎ 00 353 71 914 7304   🖷 00 353 71 914 3093
leojregan@yahoo.ie
Editor *Leo Regan*

Poetry of the spirit: mystical, contemplative, metaphysical, grounded in the senses. Worldwide. Written submissions by post only with e-mail address for reply.

The French Literary Review

134b Joy Lane, Whitstable CT5 4ES
☎ 01227 282718
dordi.barbara@neuf.fr
Contact *Barbara Dordi*
Also at: 7 rue de la Chapelle, 11240 Alaigne, Aude, France

A5 illustrated magazine with poems, stories and articles with a French connection. 'Looking for lively, contemporary poems.'

The Frogmore Papers

21 Mildmay Road, Lewes BN7 1PJ
☎ 07751 251689
J.N.Page@sussex.ac.uk
www.frogmorepress.co.uk
Editor *Jeremy Page*

Founded 1983. Biannual. Poetry, prose and artwork. See also **The Frogmore Press**.

Garbaj

38 Pierrot Steps, 71 Kursaal Way, Southend on Sea SS1 2UY
atlanteanpublishing@hotmail.com
www.geocities.com/dj_tyrer/garbaj.html
Contact *D.J. Tyrer*

Humourous/non-pc poetry, vignette-length fiction, fake news, etc. See also **Awen**, **Bard**, **Monomyth** and **The Supplement** magazines and **Atlantean Publishing** press.

Global Tapestry Journal

Spring Bank, Longsight Road, Copster Green, Blackburn BB1 9EU
☎ 01254 249128
Contact *Dave Cunliffe*

Global Bohemia, post-Beat and counterculture orientation. See also **BB Books** press.

Haiku Scotland

2 Elizabeth Gardens, Stoneyburn EH47 8PB
haiku.scotland@btinternet.com
Contact *Frazer Henderson*

Haiku, senryu, aphorism, epigram, reviews and articles.

Handshake

5 Cross Farm, Station Road North, Fearnhead,
Warrington WA2 0QG
Contact *John Francis Haines*

Newsletter of **The Eight Hand Gang**, an
association of UK sci-fi poets.

Harlequin

PO Box 23392, Edinburgh EH8 7YZ
harlequinmagazine@yahoo.com
www.harlequinmagazine.com
Contact *Jim Sinclair*

High quality poetry and artwork of intense
beauty, mysticism and wisdom.

HQ Poetry Magazine (Haiku Quarterly)

39 Exmouth Street, Swindon SN1 3PU
☎ 01793 523927
Contact *Kevin Bailey*

General poetry mag with slight bias towards
imagistic/haikuesque work

Imagenation

289 Elmwood Avenue, Feltham TW13 7QB
Contact *Ian Deal*

Poetry and artwork magazine. See also **Partners**
press, **A Bard Hair Day** and **Poet Tree**
magazines.

Inclement

White Rose House, 8 Newmarket Road, Fordham,
Ely CB7 5LL
inclement_poetry_magazine@hotmail.com
Contact *Michelle Foster*

All forms and styles of poetry. Responses within
a month.

The Interpreter's House

19 The Paddox, Oxford OX2 7PN
www.interpretershouse.org.uk
Contact *Merryn Williams*

Poems and stories up to 2,500 words; new
and established writers. Check website for
subscription details.

Iota

1 Lodge Farm, Snitterfield, Stratford upon Avon
CV37 0LR
☎ 01789 730320
iotapoetry@aol.com
www.iotapoetry.co.uk
Contact *Bob Mee & Janet Murch*

Poetry, reviews; long and short poems welcome.
No epics. See also **Ragged Raven Press**.

Irish Pages

The Linen Hall Library, 17 Donegall Square
North, Belfast BT1 5GB
☎ 028 9043 4800
irishpages@yahoo.co.uk
www.irishpages.org
Contact *Chris Agee*

'Ireland's premier literary journal.' Outstanding
writing from Ireland and overseas: poetry, short
fiction, essays, literary journalism, nature-writing,
etc. See entry under *Magazines*.

Island

8 Craiglea Drive, Edinburgh EH10 5PA
jaj@essencepress.co.uk
www.essencepress.co.uk
Contact *Julie Johnstone*

A distinctive space for writing inspired by nature
and exploring our place within the natural world.
See also **Essence Press**.

The David Jones Journal

The David Jones Society, 22 Gower Road, Sketty,
Swansea SA2 9BY
☎ 01792 206144   ☒ 01792 470385
anne.price-owen@sihe.ac.uk
www.sihe.ac.uk/davidjones
Contact *Anne Price-Owen*

Articles, poetry, information, reviews and
inspired works.

The Journal

17 High Street, Maryport CA15 6BQ
☎ 01900 812194
smithsssj@aol.com
members.aol.com/smithsssj/index.html
Contact *Sam Smith*

Poems in translation alongside poetry written in
English. See also **Original Plus** press.

Krax

63 Dixon Lane, Wortley, Leeds LS12 4RR
Contact *Andy Robson*

Light-hearted, contemporary poetry, short fiction
and graphics.

Leeds Poetry Quarterly

89 Connaught Road, Sutton SM1 3PJ
☎ 020 8286 0419
Contact *Barry Tebb*

New poems, reviews, articles on literary and
psychoanalytic matters. See also **Sixties Press** and
Literature and Psychoanalysis magazine.

Linkway

The Shieling, The Links, Burry Port SA16 0HU
☎ 01554 834486   ☒ 01554 834486
fay@faydavies.orangehome.co.uk
Contact *Fay C. Davies*

A publication for writers and friends; a general
interest magazine for the whole family. No crude
language.

Literature and Psychoanalysis
89 Connaught Road, Sutton SM1 3PJ
☎ 020 8286 0419
Contact *Barry Tebb*

New poems, reviews, articles on literary and
psychoanalytical matters. See also **Sixties Press**
and **Leeds Poetry Quarterly** magazine.

The London Magazine
▷ entry under Magazines

Magma
43 Keslake Road, London NW6 6DH
magmapoetry@ntlworld.com
Submissions to:
contributions@magmapoetry.com
www.magmapoetry.com
Contact *David Boll*

New poetry plus poetry reviews and interviews.

Markings
▷ entry under Magazines

Modern Poetry in Translation
The Queen's College, Oxford OX1 4AW
☎ 01865 244701
administrator@mptmagazine.com
www.mptmagazine.com
Contact *David & Helen Constantine*

Publishes and promotes poetry in English
translation.

Monkey Kettle
monkeykettle@hotmail.com
www.monkeykettle.co.uk
Editor *Matthew Taylor*

Biannual poetry, prose, photos, articles in Milton
Keynes and further afield.

Monomyth
38 Pierrot Steps, 71 Kursaal Way, Southend on Sea
SS1 2UY
atlanteanpublishing@hotmail.com
www.geocities.com/dj_tyrer/monomyth.html
Contact *David John Tyrer*

Poetry, prose and articles; all genres, styles and
lengths considered. New writers welcome. See
also **Awen**, **Bard**, **Garbaj** and **The Supplement**
magazines and **Atlantean Publishing** press.

Moodswing
12 Derby Crescent, Moorside, Consett DH8 8DZ
☎ 01207 505724
Contact *Steve Urwin*

Pocket broadsheet – light/dark – psychologically
charged. See also **Talking Pen** press

Mslexia
PO Box 656, Newcastle upon Tyne NE99 1PZ
☎ 0191 261 6656   ☏ 0191 261 6636

postbag@mslexia.co.uk
www.mslexia.co.uk
Editor *Daneet Steffens*

National quarterly magazine for women who
write. Advice, inspiration, news, reviews,
interviews, etc. Contributor guidelines on the
website. See entry under *Magazines*.

Neon Highway
37 Grinshill Close, Liverpool L8 8LD
poetshideout@yahoo.com
www.neonhighway.co.uk
Editors *Alice Lenkiewicz, Jane Marsh,*
Dee McMahon, Matt Fallaize

Innovative and experimental poetry/arts
magazine.

Never Bury Poetry
Bracken Clock, Troutbeck Close, Hawkshaw,
Bury BL8 4LJ
☎ 01204 884080
n.b.poetry@zen.co.uk
Contact *Jean Tarry*

Founded 1989. Quarterly. International
reputation. Each issue has a different theme.

New Welsh Review
PO Box 170, Aberystwyth SY23 1WZ
☎ 01970 628410
admin@newwelshreview.com
www.newwelshreview.com
Editor *Francesca Rhydderch*

Vibrant literary quarterly magazine which
showcases the best new writing from Wales. Also
includes features and reviews. See entry under
Magazines.

Norfolk Poets and Writers – Tips Newsletter
9 Walnut Close, Taverham, Norwich NR8 6YN
tipsforwriters@yahoo.co.uk
wendywebb.mysite.orange.co.uk
Editor *Wendy Webb*

Tips for Writers; six magazines per year;
competitions, themes and forms. See also **Wendy**
Webb Books press.

The North
The Poetry Business, Bank Street Arts, 32–40
Bank Street, Sheffield S1 2DS
edit@poetrybusiness.co.uk
www.poetrybusiness.co.uk
Contact *Peter Sansom & Ann Sansom*

Contemporary poetry and articles, extensive
reviews.

Obsessed With Pipework
8 Abbot's Way, Pilton BA4 4BN
☎ 01749 890019

cannula.dementia@virgin.net
www.flarestack.co.uk
Contact *Charles Johnson*

Quarterly magazine of new poetry 'to surprise and delight'. See also **Flarestack Publishing** press.

The Once Orange Badge Poetry Supplement

PO Box 184, South Ockendon RM15 5WT
☎ 01708 852827
onceorangebadge@poetry.fsworld.co.uk
Contact *D. Martyn Heath*

A4 poetry supplement for everyone whose life has been touched by disability in some way. See also **The Once Orange Badge Poetry Press**.

Open Wide Magazine

40 Wingfield Road, Lakenheath IP27 9HR
☎ 07790 962317
contact@openwidemagazine.co.uk
www.openwidemagazine.co.uk
Contact *James Quinton*

'The UK's best arts magazine', publishing short fiction, punchy poetry, reviews and interviews plus OWM Extra, a new online supplement.

Other Poetry

29 Western Hill, Durham DH1 4RL
☎ 0191 386 4058
www.otherpoetry.com
Managing Editor *Michael Standen*

Founded 1979. Thrice-yearly, 60–70 poems per issue. Process of selection involves all four editors. Token payment. Submit up to four poems (with name on each sheet) plus s.a.e.

Parameter

PO Box 220, Wythenshawe, Manchester M23 0WE
editor@parametermagazine.org
www.parametermagazine.org

Partners Annual Poetry Competition

289 Elmwood Avenue, Feltham TW13 7QB
partners_writing_group@hotmail.com
Contact *Ian Deal*

Showcases poems entered into the Partners annual open poetry competitions.

Peace and Freedom

17 Farrow Road, Whaplode Drove, Spalding PE12 0TS
☎ 01406 330242
p_rance@yahoo.co.uk
uk.geocities.com/p_rance/pandf.htm
Contact *Paul Rance*

Poetry, prose, art mag; humanitarian, environmental, animal welfare. See also **Eastern Rainbow** magazine.

Peer Poetry International

26 (wh) Arlington House, Bath Street, Bath BA1 1QN
☎ 01225 445298
peerpoetry@msn.com
Contact *Paul Amphlett*

Biannual; 60 poets per issue. Concentrating on e-books, with particular emphasis on haiku.

The Penniless Press

100 Waterloo Road, Ashton, Preston PR2 1EP
☎ 01772 736421
www.pennilesspress.co.uk
Contact *Alan Dent*

Quarterly for the poor pocket and the rich mind. Poetry, fiction, essays, reviews. See also **Penniless Press Publications**.

Pennine Ink Magazine

The Gallery, Mid-Pennine Arts, Yorke Street, Burnley BB11 1HD
☎ 01282 432992
sheridansdandl@yahoo.co.uk
Contact *Laura Sheridan*

Quality poetry reflecting traditional and modern trends.

Pennine Platform

Frizingley Hall, Frizinghall Road, Bradford BD9 4LD
☎ 01274 541015
nicholas.bielby@virgin.net
www.pennineplatform.co.uk
Contact *Nicholas Bielby*

Biannual poetry magazine; eclectic and serious-minded. Considers hard copy submissions only.

Planet

PO Box 44, Aberystwyth SY23 3ZZ
☎ 01970 611255   ⊞ 01970 611197
planet.enquiries@planetmagazine.org.uk
www.planetmagazine.org.uk
Editor *Helle Michelsen*

The Welsh Internationalist: current affairs, arts, environment.

PN Review

▷ Carcanet Press Ltd under UK Publishers

Poet Tree

289 Elmwood Avenue, Feltham TW13 7QB
Contact *Ian Deal*

Contemporary poetry magazine. See also **Partners** press, **A Bard Hair Day** and **Imagenation** magazines.

Poetic Hours

43 Willow Road, Carlton, Nottingham NG4 3BH
erranpublishing@hotmail.com
www.poetichours.homestead.com
Editor *Nick Clark*

Non-profit supporter of Third World charities.
See also **Erran Publishing**.

Poetic Licence

70 Aveling Close, Purley CR8 4DW
poets@poetsanon.org.uk
www.poetsanon.org.uk
Contact *Peter L. Evans*

Original unpublished poems and drawings. See
also **Poets Anonymous** press.

The Poetry Church

Feather Books, PO Box 438, Shrewsbury SY3 0WN
☎ 01743 872177   🖷 01743 872177
john@waddysweb.freeuk.com
www.waddysweb.freeuk.com
Contact *Rev. J. Waddington-Feather*

Quarterly magazine of Christian poetry and
prayers. See **Feather Books** under *Small Presses*.

Poetry Combination Module

P.E. F Productions, 196 High Road, London
N22 8HH
page84direct@yahoo.co.uk
www.geocities.com/andyfloydplease
Contact *Mr P.E. F*

Available late spring, early autumn and mid-
winter. Printable from the website. Poetry,
anti-poetry, aphorism and artwork magazine by
P.E. F and guest artists. Submissions welcome,
preferably by e-mail.

Poetry Cornwall/Bardhonyeth Kernow

11a Penryn Street, Redruth TR15 2SP
☎ 01209 218209
poetrycornwall@yahoo.com
www.poetrycornwall.freeservers.com
Editor *Les Merton*

Publishes poets from around the world, including
'Meet the Editors', poetry in its original language
with English translation and poetry in Kernewek
and Cornish dialect. Three issues a year.
Submission guidelines on the website.

Poetry Express

Studio 11, Bickerton House, 25–27 Bickerton
Road, London N19 5JT
☎ 020 7281 4654
simon@survivorspoetry.org.uk
www.survivorspoetry.com
Contact *Dr Simon Jenner & Others*

Biannual newsletter from **Survivors' Poetry** (see
entry under *Organizations of Interest to Poets*).

Poetry Ireland News

2 Proud's Lane, Dublin 2, Republic of Ireland
☎ 00 353 1 478 9974   🖷 00 353 1 478 0205
publications@poetryireland.ie
www.poetryireland.ie
Contact *Paul Lenehan*

Bi-monthly newsletter. See also **Poetry Ireland
Review** magazine.

Poetry Ireland Review/Eigse Eireann

2 Proud's Lane, Dublin 2, Republic of Ireland
☎ 00 353 1 478 9974   🖷 00 353 1 478 0205
poetry@iol.ie
www.poetryireland.ie
Contact *Eiléan Ní Chuilleanáin*

Quarterly journal of poetry and reviews. See also
Poetry Ireland News magazine.

Poetry London

81 Lambeth Walk, London SE11 6DX
☎ 020 7735 8880   🖷 020 7735 8880
editors@poetrylondon.co.uk
www.poetrylondon.co.uk
Poetry Editor *Maurice Riordan*
Listings *Gyonghi Vegh*
Reviews *Scott Verner*

Published three times a year, *Poetry London*
includes poetry by new and established writers,
reviews of recent collections and anthologies,
articles on issues relating to poetry, and an
encyclopædic listings section of virtually
everything to do with poetry in the capital and
elsewhere in the UK.

Poetry Nottingham

11 Orkney Close, Stenson Fields, Derby DE24 3LW
adrian.buckner@btopenworld.com
Editor *Adrian Buckner*

Poetry, articles, reviews, published thrice-yearly.

Poetry Reader

Scottish Poetry Library, 5 Crichton's Close,
Canongate, Edinburgh EH8 8DT
☎ 0131 557 2876   🖷 0131 557 8393
inquiries@spl.org.uk
www.spl.org.uk
www.readingroom.spl.org.uk
Contact *Robyn Marsack*

Newsletter of the **Scottish Poetry Library**.

Poetry Review

Poetry Society, 22 Betterton Street, London
WC2H 9BU
☎ 020 7420 9883   🖷 020 7240 4818
poetryreview@poetrysociety.org.uk
www.poetrysoc.org.uk
Editor *Fiona Sampson*

The senior review of contemporary poetry.
Quarterly.

Poetry Salzburg Review

Dept. of English, University of Salzburg, Akademiestr. 24, A–5020 Salzburg, Austria
☎ 00 43 662 8044 4424   ⓕ 00 43 662 8044 167
editor@poetrysalzburg.com
www.poetrysalzburg.com
Contact *Wolfgang Görtschacher*
Editorial board: *David Miller* (99 Mitre Road, London SE1 8PT); *William Bedford* (The Croft, 3 Manor Farm Court, Purton Stoke SN5 4LA); *Robert Dassanowsky* (Dept. of Languages & Cultures, University of Colorado, 1420 Austin Bluffs, Colorado CO 80933, USA)

Poetry magazine, formerly *The Poet's Voice*. Published at the University of Salzburg. Publishes new poetry, translations, interviews, review-essays and artwork.

Poetry Scotland

91–93 Main Street, Callander FK17 8BQ
www.poetryscotland.co.uk
Contact *Sally Evans*

All-poetry broadsheet with Scottish emphasis. See also **Diehard Publishers** press.

Poetry Wales

57 Nolton Street, Bridgend CF31 3BN
☎ 01656 663018   ⓕ 01656 649226
poetrywales@seren-books.com
www.poetrywales.co.uk
Editor *Zöe Skoulding*

An international magazine with a reputation for fine writing and criticism. See also **Seren Books** under *UK Publishers*.

Premonitions

13 Hazely Combe, Arreton, Isle of Wight PO30 3AJ
☎ 01983 865668
mail@pigasus.press.co.uk
www.pigasuspress.co.uk
Contact *Tony Lee*

Magazine of science fiction and horror stories, with genre poetry and artwork. See also **Pigasus Press**.

Presence

12 Grovehall Avenue, Leeds LS11 7EX
martin.lucas2@btinternet.com
haiku-presence.mysite.orange.co.uk
Contact *Martin Lucas*

Haiku, senryu, tanka, renku and related poetry in English.

Pulsar Poetry Magazine

34 Lineacre Close, Grange Park, Swindon SN5 6DA
☎ 01793 875941
pulsar.ed@btopenworld.com
www.pulsarpoetry.com

Editor *David Pike*

Hard-hitting/inspirational poetry with a message and meaning. Biannual.

Purple Patch

25 Griffiths Road, West Bromwich B71 2EH
ppatch66@hotmail.com
www.purplepatchpoetry.co.uk
Contact *Geoff Stevens*

Poetry mag founded 1976. Includes reviews and gossip column. See also **The Firing Squad** magazine and **Poetry Wednesbury** press.

Quantum Leap

York House, 15 Argyle Terrace, Rothesay, Isle of Bute PA20 0BD
Contact *Alan Carter*

User-friendly magazine – encourages new writers; all types of poetry – pays. Mark envelope 'Guidelines' and enclose s.a.e. or two IRCs if from abroad. See also **QQ Press**.

The Quiet Feather

25 Sidney Street, Oxford OX4 3AG
☎ 01865 727927
editors@thequietfeather.co.uk
www.thequietfeather.co.uk
Contact *Dominic Hall & Taissa Csaky*

Lively and eclectic quarterly: short stories, poetry, travel, essays, cartoons, photographs, line drawings; all welcome.

The Radiator

Flat 10, 21 Greenheys Road, Liverpool L8 0SX
Contact *Scott Thurston*

Publishes commissioned essays on poetry by contemporary poets.

Rainbow Poetry News

74 Marina, St Leonards-on-Sea TN38 0BJ
☎ 01424 444072
beyondcloister@hotmail.co.uk
Contact *Hugh C. Hellicar*

Quarterly with poems, articles and news of poetry recitals in London and the South East but with overseas readership.

Read the Music

20 Wharfedale Street, Wednesbury WS10 9AG
☎ 07970 441110
mooncrow@tiscali.co.uk
www.poetrywednesbury.co.uk
Contact *Brendan Hawthorne*

Original poetry based on music and its influences.

The Reater
Wrecking Ball Press, 24 Cavendish Square, Hull
HU3 1SS
editor@wreckingballpress.com
www.wreckingballpress.com
Contact *Shane Rhodes*

No flowers, just blunt, chiselled poetry.

The Recusant
www.therecusant.moonfruit.com
Editor *Alan Morrison*

Non-conformist poetry, prose, polemic, reviews
and articles. Seeks to provide a home for
contemporary writing that goes against the grain,
in subject more than style. Socially-inclined work
with a left-field slant particularly welcome. Visit
the website for submission guidelines.

Red Poets – Y Beirdd Coch
26 Andrew's Close, Heolgerrig, Merthyr Tydfil
CF48 1SS
01685 376726
marc.jones@phonecoop.coop
Contact *Mike Jenkins & Marc Jones*

A magazine of left-wing poetry from Wales and
beyond: socialist, republican and anarchist.

Reflections
PO Box 178, Sunderland SR1 1DU
reflections1@fastmail.fm

Poetry by and for men and women of goodwill.

The Rialto
PO Box 309, Aylsham, Norwich NR11 6LN
mail@therialto.co.uk
www.therialto.co.uk
Contact *Michael Mackmin*

Excellent poetry in a clear environment. Buy
online at www.impressbooks.co.uk See also
Rialto Publications press.

Roundyhouse
3 Crown Street, Port Talbot SA13 1BG
srjones@alunbooks.co.uk
Contact *Sally R. Jones*

Poems and articles on poetry. Reviews. Eclectic
approach.

Rubies In the Darkness
41 Grantham Road, Manor Park, London E12 5LZ
Contact *P.G.P. Thompson*

Romantic, lyrical, spiritual, inspirational,
traditional and mystical poetry. See also **Precious
Pearl Press**.

Sable
PO Box 33504, London E9 7YE
info@sablelitmag.org
www.sablelitmag.org

Contact *Kadija Sesay*

Literary magazine for writers of African,
Caribbean and Asian descent. Poetry, prose,
memoirs, travel, etc.

Scar Tissue
Pigasus Press, 13 Hazley Combe, Arreton, Isle of
Wight PO30 3AJ
Contact *Tony Lee*

SF genre poetry, short fiction, cartoons and
artwork.

Scribbler!
Remus House, Coltsfoot Drive, Woodston,
Peterborough PE2 9JX
01733 890066   01733 313524
youngwriters@forwardpress.co.uk
www.youngwriters.co.uk
Editor *Donna Samworth*

Thrice-yearly magazine by 7–11 year-olds: poetry,
stories, guest authors, workshops, features.

Scriptor
6 Green Bank, Stacksteads, Bacup OL13 8LQ
01706 870357/07967 315270 (mobile)
mail@johndench.demon.co.uk
Contact *John Dench*

Poetry, short stories, essays from the UK;
published biennially. See also **Green Arrow
Publishing** press.

The Seventh Quarry – Swansea Poetry
Magazine
Dan-Y-Bryn, 74 Cwm Level Road, Brynhyfryd,
Swansea SA5 9DY
www.myspace.com/peterthabitjones
Contact *Peter Thabit Jones*
Consulting Editor, America *Vince Clemente*

Quality poetry from Swansea and beyond.
Submission details available on the website.

Shearsman
58 Velwell Road, Exeter EX4 4LD
01392 434511   01392 434511
editor@shearsman.com
www.shearsman.com
Contact *Tony Frazer*

Mainly poetry, some prose, some reviews. Poetry
in the modernist tradition. No fiction. See also
Shearsman Books press.

The ShOp: A Magazine Of Poetry
Skeagh, Schull, Co. Cork, Republic of Ireland
Theshop@theshop-poetry-magazine.ie
(not for submissions)
www.theshop-poetry-magazine.ie
Contact *John Wakeman & Hilary Wakeman*

International but with emphasis on Irish poetry.

Smiths Knoll

Goldings, Goldings Lane, Leiston IP16 4EB
Contact *Joanna Cutts & Michael Laskey*

Poems worth re-reading; two-week turnaround for submissions.

Smoke

The Windows Project, 96 Bold Street, Liverpool L1 4HY
① 0151 709 3688
www.windowsproject.co.uk
Contact *Dave Ward*

Poetry, graphics, short prose. Biannual; 24pp.

South

PO Box 3744, Cookham, Maidenhead SL6 9UY
Contact *Chrissie Williams*

Poetry magazine from the southern counties of England that welcomes poets from across the world. Poems judged anonymously.

Southword

Frank O'Connor House, 84 Douglas Street, Cork, Republic of Ireland
www.munsterlit.ie
Contact *Rosemary Canavan*

Biannual, 120 pages of poetry, fiction, visual art and book reviews.

Sphinx

21 Hatton Green, Glenrothes KY7 4SD
nell@happenstancepress.com
www.happenstancepress.com
Contact *Helena Nelson*

A feature & review-based magazine focusing on poetry publishing: who, how and why. No unsolicited submissions. See also **Happenstance Press**.

Springboard

Corrimbla, Ballina, Co. Mayo, Republic of Ireland
bobgroom@eircom.net
Contact *Robert Groom*

Short fiction, poetry and articles.

The Stinging Fly

PO Box 6016, Dublin 8, Republic of Ireland
stingingfly@gmail.com
www.stingingfly.org
Contact *Declan Meade*

New Irish and international writing. Poetry, short fiction, author interviews, essays and book reviews. See also **Stinging Fly Press**. Submissions are considered from January to March each year.

The Student Magazine

24 Frosty Hollow, East Hunsbury NN4 0SY
① 01604 701730   ⓕ 01604 701730
Editor & Publisher *Jacqueline Gonzalez-Marina*

Poetry, articles, interviews, art and illustration. Biannual magazine. No religious or political material. 'Essential to become a subscriber when seeking publication': UK, £17 p.a.; Europe, £30; RoW, US$90. See also **Dandelion Arts Magazine**.

The Supplement

38 Pierrot Steps, 71 Kursaal Way, Southend on Sea SS1 2UY
atlanteanpublishing@hotmail.com
www.geocities.com/dj_tyrer/mms.html
Contact *D.J. Tyrer*

Non-fiction: articles, news, reviews, letters, etc. See also **Awen**, **Bard**, **Garbaj** and **Monomyth** magazines and **Atlantean Publishing** press.

Taliesin

Academi, Mount Stuart House, Mount Stuart Square, Cardiff CF10 5FQ
① 029 2047 2266   ⓕ 029 2049 2930
taliesin@academi.org
www.academi.org
Contact *Manon Rhys & Christine James*

Wales' leading Welsh language literary journal, published three times a year.

Tears in the Fence

38 Hod View, Stourpaine, Nr Blandford Forum DT11 8TN
① 01258 456803   ⓕ 01258 454026
david@davidcaddy.wanadoo.co.uk
www.myspace.com/tearsinthefence
Contact *David Caddy*

A magazine looking for the unusual, perceptive, risk-taking, lived and visionary literature.

10th Muse

33 Hartington Road, Southampton SO14 0EW
a.jordan@surfree.co.uk
Contact *Andrew Jordan*

Publishes poetry and reviews as well as short prose (usually no more than 2,000 words) and graphics. E-mail for information only; no submissions by e-mail.

The Third Half

16 Fane Close, Stamford PE9 1HG
① 01780 754193
Editor *Kevin Troop*

Searches for good poetry and fiction. See also **KT Publications** press.

Time Haiku

Basho-an, 105 King's Head Hill, London E4 7JG
① 020 8529 6478

facey@aol.com
Contact *Erica Facey*
Editor *Doreen King*

Founded 1994. Promotes haiku, tanka and haiku-related forms. Aims to create accessibility to all people interested in this form of poetry as well as established poets. A biannual journal and newsletter are available to subscribers.

Understanding

127 Milton Road West, 7 Duddingston House Courtyard, Edinburgh EH15 1JG
Contact *Denise Smith*

Original poetry, short stories, parts of plays, reviews and articles. See also **Dionysia Press**.

Wasafiri

The Open University in London, 1–11 Hawley Crescent, London NW1 8NP
☏ 020 7556 6110   🖷 020 7556 6187
wasafiri@open.ac.uk
www.wasafiri.org
Editor *Sushiela Nasta*

Deputy Editor *Sharmilla Beezmohun*
Assistant Editor *Nisha Jones*
Editorial Manager *Teresa Palmiero*
Reviews Editor *Mark Stein*

Literary journal of international and diasporic writing.

Wordsmith

Remus House, Coltsfoot Drive, Woodston, Peterborough PE2 9JX
☏ 01733 890066   🖷 01733 313524
youngwriters@forwardpress.co.uk
www.youngwriters.co.uk
Editor *Claire Tupholme*

Poetry, stories, guest authors, features in a quarterly magazine for 11–18-year-olds. See also **Scribbler!** magazine.

Zed20

33 Lady Nairn Avenue, Kirkcaldy KY1 2AW
☏ 01592 651522
Contact *Duncan Glen*

See also **Akros Publications** press.

Organizations of Interest to Poets

Key advice

A survey of some of the societies, groups and other bodies which may be of interest to practising poets. Organizations not listed should send details to the Editor for inclusion in future editions.

Academi (Yr Academi Gymreig)

3rd Floor, Mount Stuart House, Mount Stuart Square, Cardiff Bay CF10 5FQ
☎ 029 2047 2266   🖷 029 2049 2930
post@academi.org
www.academi.org
Glyn Jones Centre: Wales Millennium Centre, Cardiff
North West Wales Office: Ty Newydd, Llanystumdwy, Cricieth, Gwynedd LL52 0LW
West Wales Office: Dylan Thomas Centre, Somerset Place, Swansea SA1 1RR
Chief Executive *Peter Finch*

The writers' organization of Wales with special responsibility for literary activity, writers' residencies, writers on tour, festivals, writers' groups, readings, tours, exchanges and other development work. Academi awards financial bursaries annually, runs a criticism service and organizes the **Welsh Book of the Year Awards**. Yr Academi Gymreig/The Welsh Academy operates the Arts Council of Wales franchise for Wales-wide literature development. It has offices in Cardiff and fieldworkers elsewhere in Wales. The Academi sponsors a range of annual contests including the John Tripp Award For Spoken Poetry and the prestigious **Academi Cardiff International Poetry Competition**. Publishes *A470*, a bi-monthly literary information magazine. See also entry under *Professional Associations and Societies.*

Apples & Snakes

The Albany, Deptford, London SE8 4AG
☎ 0845 521 3460   🖷 0845 521 3461
info@applesandsnakes.org
www.myspace.com/applesandsnakespoetry
www.applesandsnakes.org
Director *Geraldine Collinge*

Works nationwide to promote popular, high-quality and cross-cultural poetry; programmes live events for new and established poets including open mic events; operates a Poetry in Education scheme where poets run workshops and perform in schools, prisons and community settings; coordinates the professional development of poets, with opportunities or training mentoring, national touring and residencies.

The Arvon Foundation
▷ entry under UK and Irish Writers' Courses

The British Haiku Society

38 Wayside Avenue, Hornchurch RM12 4LL
☎ 01772 251827
www.britishhaikusociety.org
General Secretary *Doreen King*

Formed in 1990. Promotes the appreciation and writing of haiku and related forms; welcomes overseas members. Provides tutorials, workshops, readings, critical comment and information. Specialist advisers are available. Runs a haiku library and administers a haiku contest. The journal, *Blithe Spirit*, is produced quarterly. The Society has active local groups and issues a regular newsletter. Current membership details can be obtained from the website.

Camelford Poetry Workshops

The Indian King, Garmoe Cottage, 2 Trefrew Road, Camelford PL32 9TP
☎ 01840 212161
Director *Helen Wood*

A weekly half-day workshop where regular participants write together, bring work in progress for criticism and share experience of publication. They also organize a programme of visiting poets who run a writing workshop one Saturday every other month followed by a public reading. The alternate months see a Saturday Novel Writing Surgery led by Karen Hayes.

Contemporary Poetics Research Centre

School of English and Humanities, Birkbeck
College, Malet Street, London WC1E 7HX
w.rowe@bbk.ac.uk
c.watts@bbk.ac.uk
www.bbk.ac.uk/eh/research/
contemppoeticscentre
Contacts *Will Rowe & Carol Watts*

Dedicated to fostering research, performance
and practice in all modes and formats of
contemporary poetry. It runs a Creative Reading
Workshop, holds readings, performances,
talks and debates and organizes conferences,
the latest of which was E Poetry London 2005.
It collaborates closely with Royal Holloway
University of London and the Centre for Cultural
Poetics at the University of Southampton, e.g.
on the British Electronic Poetry Centre at www.
soton.ac.uk/~bepc Poets associated with the
Centre include Ulli Freer, Caroline Bergvall and
Sean Bonney. The Centre publishes the web
journal, www.pores.bbk.ac.uk

The Eight Hand Gang

5 Cross Farm, Station Road North, Fearnhead,
Warrington WA2 0QG
Secretary *John F. Haines*

An association of SF poets. Publishes *Handshake*,
a single-sheet newsletter of SF poetry and
information available free in exchange for an s.a.e.

57 Productions

57 Effingham Road, Lee Green, London SE12 8NT
☎ 020 8463 0866   🖷 020 8463 0866
info@57productions.com
www.57productions.com
Contact *Paul Beasley*

Specializes in the promotion of poetry and
its production through an agency service, a
programme of events and a series of audio
publications. Services are available to event
promoters, festivals, education institutions and
the media. Poets represented include Jean 'Binta'
Breeze, Adrian Mitchell, Patience Agbabi and
many others. 57 Productions' series of audio
cassettes, CDs and Poetry in Performance
compilations offer access to some of the most
exciting poets working in Britain today – check
their Poetry Jukebox on their excellent website.

The Football Poets

4 The Retreat, Butterow, Stroud GL5 2LS
☎ 01453 757376
editors@footballpoets.org
ctm@crispinthomas.orangehome.co.uk
www.footballpoets.org
Editor & Performance *Crispin Thomas*
Co-Editors *Simon Williams, Peter Goulding*

The Football Poets exist to promote and
encourage the writing, reading and performing
of football poetry. Formed in 1995, they perform
extensively and also provide comprehensive
football poetry workshops in schools, football
clubs, prisons and communities. Their website is
a fast, free and entertaining mix of literature and
soccer poetry from around the world. Anyone
may contribute. The site has recently been
archived by the British Library.

Performance Poetry Society

PO Box 11178, Birmingham B11 4WP
☎ 0121 242 6644
performancepoetry@yahoo.com
Branch contact & rehearsal room:
The Old Meeting House, behind 20/22
Wolverhampton Street, Dudley
Contact *Sandra Dennis*
Dudley Branch Contact *Jim MacCool* (☎ 01384
258191)

PPS was founded in 1999 by members of the
Birmingham Midland Institute to serve the
interests of performance poets and poetry in
performance throughout the UK and Ireland.
Initiated October as National Poetry Month in
2000. Organizes an extensive national tour each
autumn, visiting small halls, colleges, schools and
prisons. The Society has an in-house publishing
company, Poetry Powerhouse Press.

The Poetry Archive

PO Box 286, Stroud GL6 1AL
☎ 01453 832090   🖷 01453 836450
info@poetryarchive.org
www.poetryarchive.org
Directors *Richard Carrington, Andrew Motion*

A permanent online archive of audio recordings
of work by poets who write in English. New
60-minute recordings are made for the Archive
by a wide-ranging list of contemporary poets;
extracts from those recordings are available
on the website launched in 2005 at www.
poetryarchive.org and the full-length recordings
are available on CD for purchase by mail order.
The site also contains historic recordings by now-
dead poets, a separate archive of recordings for
young children, and a wealth of educational and
intepretative information. A registered charity,
the Archive intends to ensure that all significant
poets are recorded for posterity and that their
recordings are widely valued and enjoyed.

The Poetry Book Society

Fourth Floor, 2 Tavistock Place, London
WC1H 9RA
☎ 020 7833 9247   🖷 020 7833 5990
info@poetrybooks.co.uk
www.poetrybooks.co.uk
www.poetrybookshoponline.com

www.childrenspoetrybookshelf.co.uk
Director *Chris Holifield*
Editorial/Marketing Officer *Pinda Bryars*
Sales/Membership Officer *James Knapton*

For readers, writers, students and teachers of poetry. Founded in 1953 by T.S. Eliot and funded by the Arts Council, the PBS is a unique membership organization and book club providing up-to-date and comprehensive information about poetry from publishers in the UK and Ireland. Members receive the quarterly *Poet Selectors' Choice* and the quarterly *PBS Bulletin* packed with articles by poets, poems, news, listings and access to discounts of at least 25% off featured titles. Subscriptions start at £12. The PBS website (www.poetrybooks.co.uk) has a recruitment area and a closed section for members. PBS also offers a range of over 30,000 poetry titles at www.poetrybookshoponline.com and it also sells the Poetry Archive CDs and has a new section called SoundBlast for performance poets' CDs. During 2005 the PBS relaunched the Children's Poetry Bookshelf (www.childrenspoetrybookshelf.co.uk) range of books and new membership schemes for parents, grandparents and libraries, adding an annual competition for child poets a year later. Also runs the annual **T.S. Eliot Prize** for the best collection of new poetry, which has a Shadowing Scheme for schools. In 2004 PBS ran the Next Generation Poets promotion.

The Poetry Business

Bank Street Arts, 32–40 Bank Street, Sheffield S1 2DS
edit@poetrybusiness.co.uk
www.poetrybusiness.co.uk
Directors *Peter Sansom, Ann Sansom*

Founded in 1986, the Business publishes *The North* magazine and books and pamphlets under the Smith/Doorstop imprint. It runs an annual competition and organizes writing days and a Writing School. Send an s.a.e. for full details.

Poetry Can

12 Great George Street, Bristol BS1 5RH
Ⓣ 0117 933 0900
admin@poetrycan.co.uk
www.poetrycan.co.uk
Director *Colin Brown*
Administrator *Peter Hunter*

Based in Bristol and founded in 1995, Poetry Can is a poetry development agency working across Bristol and the South West Region. It organizes events such as the annual Bristol Poetry Festival, an education programme and provides information and advice concerning all aspects of poetry to individuals, agencies and organizations. It also hosts the Literature South West website.

Poetry Ireland/Eigse Eireann

2 Proud's Lane, Dublin 2, Republic of Ireland
Ⓣ 00 353 1 478 9974   Ⓕ 00 353 1 478 0205
poetry@iol.ie
www.poetryireland.ie
Education Ⓣ 00 353 671 4216
Writers in Schools scheme Ⓣ 00 353 1 475 8601
Director *Joseph Woods*

Poetry Ireland is the national organization for poetry in Ireland, with its four core activities being readings, publications, education and an information and resource service. Organizes readings by Irish and international poets countrywide. Through its website, telephone, post and public enquiries, Poetry Ireland operates as a clearing house for everything pertaining to poetry in Ireland. Operates the Writers in Schools scheme. Publishes *Poetry Ireland News*, a bi-monthly newsletter containing information on events, competitions and opportunities. *Poetry Ireland Review* appears quarterly and is the journal of record for poetry in Ireland; current editor: *Peter Sirr*.

The Poetry Library

Royal Festival Hall, Level 5, London SE1 8XX
Ⓣ 020 7921 0943/0664   Ⓕ 020 7921 0700
info@poetrylibrary.org.uk
www.poetrylibrary.org.uk
www.poetrymagazines.org.uk
Librarians *Chris McCabe, Miriam Valencia*

Founded by the Arts Council in 1953. A collection of 100,000 books, pamphlets, CDs, magazines and videos of modern poetry since 1912, from Georgian to Rap, representing all English-speaking countries and including translations into English by contemporary poets. Two copies of each title are held, one for loan and one for reference. A wide range of poetry magazines and ephemera from all over the world are kept along with cassettes, CDs and videos for consultation, with many available for loan. There is a children's poetry section with teacher's resource collection.

An information service compiles lists of poetry magazines, competitions, publishers, groups and workshops which are available from the Library on receipt of a large s.a.e., or direct from the website. The main website also has a noticeboard for lost quotations through which it tries to identify lines or fragments of poetry which have been sent in by other readers.

Poetry pf

20 Clovelly Way, Orpington BR6 0WD
Ⓣ 01689 811394
editor@poetrypf.co.uk
www.poetrypf.co.uk
Contact *Anne Stewart*

Poet showcase site. Creates and manages searchable and contactable internet presence for poets. Publishes Poem Cards for public sale. Poetry-related print and design services and project work.

The Poetry School

81 Lambeth Road, London SE11 6DX
☎ 0845 223 5274   🖷 020 8223 0439
programme@poetryschool.com
www.poetryschool.com
Contact *The Administrator*

Funded by Arts Council England, London, The Poetry School offers a wide range of courses and workshops on writing and reading poetry and is open to anyone regardless of experience or formal qualifications. Tutors include Mimi Khalvati, Linda Chase, Graham Fawcett, Pascale Petit, Myra Schneider and Penelope Shuttle. The School provides a forum for practitioners to share experiences, develop skills and extend appreciation of the traditional and innovative aspects of their art. There are Special Events with visiting poets such as Marilyn Hacker, Jorie Graham, Molly Peacock, Galway Kinnell, Paul Muldoon, Alfred Corn and Sharon Olds. The Poetry School is a Registered Charity No. 1069314.

The Poetry Society

22 Betterton Street, London WC2H 9BX
☎ 020 7240 9881   🖷 020 7240 4818
info@poetrysociety.org.uk
www.poetrysociety.org.uk
Director *Jules Mann*

Founded in 1909 which ought to make it venerable, the Society exists to help poets and poetry thrive in Britain. In the past decade it has undergone a renaissance, reaching out from its Covent Garden base to promote the national health of poetry in a range of imaginative ways. Membership costs £35 for individuals. *Poetry News* membership is £15. Current activities include:

* Quarterly magazine of new verse, views and criticism, *Poetry Review*.
* Quarterly newsletter, *Poetry News*, with lively relevant articles for members.
* Promotions, events and cooperation with Britain's many literature festivals, poetry venues and poetry publishers.
* Competitions and awards, including the annual **National Poetry Competition** with a £5,000 first prize.
* A manuscript diagnosis service, *The Poetry Prescription*, which gives detailed reports on submissions. Reduced rates for members.

* Provides information and advice, publishes books, posters and resources for schools and libraries. Education membership costs £50 (secondary) or £30 (primary) which includes free poetry anthologies, lesson plans and a subscription to Poems on the Underground. Publications include *The Poetry Book for Primary Schools* and *Jumpstart Poetry for the Secondary School*, colourful poetry posters for Keystages 1, 2, 3 and 4. Many of Britain's most popular poets – including Michael Rosen, Roger McGough and Jackie Kay – contribute, offering advice and inspiration.

*The Society's online poetry classroom is at www.poetryclass.net

* The Poetry Café serving snacks and drinks to members, friends and guests, part of The Poetry Place, a venue for many poetry activities – readings, workshops and poetry launches. This space is available for bookings (☎ 020 7420 9887).

Poetry Landmarks of Britain, a free resource on the Society's website, packed with regional poetry places, publishers, venues and regular events.

The Poetry Trust

The Cut, 9 New Cut, Halesworth IP9 8BY
☎ 01986 835950   🖷 01986 874524
info@thepoetrytrust.org
www.thepoetrytrust.org
Director *Naomi Jaffa*

One of the UK's flagship poetry organizations, delivering a year-round programme of live events, creative education opportunities, courses, prizes and publications. Major events include the international **Aldeburgh Poetry Festival** held annually in early November (6th–8th in 2009) and the **Jerwood Aldeburgh First Collection Prize**. Registered Charity No. 1102893.

Point Editions

Ithaca, Apdo. 125, E–03590 Altea, Spain
☎ 00 34 96 584 2350   🖷 00 34 96 584 2350
elpoeta@point-editions.com
www.point-editions.com
Also at: Rekkemsestraat 167, B–8510 Marke, Belgium
Director *Germain Droogenbroodt*

Founded as POetry INTernational in 1984, Point is based in Spain and Belgium. A multilingual publisher of contemporary verse from *established* poets, the organization has brought out more than 80 titles in at least eight languages, including English. Point Editions runs the original work alongside a verse translation into Dutch made in cooperation with the poet. The organization's website features the world's best known and unknown poets in English, Spanish and Dutch. Point also co-organizes several international poetry festivals.

The Railway Prism

☎ 01202 624334
myabbott@tiscali.co.uk
myweb.tiscali.co.uk/railwayprism
Contact *Horace Jay*

The Ralway Prism website is a focal point for lovers of railway poetry. Preferably unpublished submissions welcome about railways old and new, reflecting industrial and social history, childhood nostalgia and modern rail travel both above and below ground. 'We're after atmosphere rather than dry facts.'

Regional Arts Councils

See **Arts Councils and Regional Offices**, pages 572–3.

Scottish Pamphlet Poetry

No 4 Cottage, High Swinton, Masham HG4 4JH
hazel@scottish-pamphlet-poetry.biz
www.scottish-pamphlet-poetry.com
Administrator *Hazel Cameron*

A publisher collective of more than 50 mainly Scottish-based presses promoting poetry published as pamphlets. A pamphlet is defined as not less than six pages but no more than 30 and with a first edition of less than 300 copies. The organization has an excellent website, publishes a video, runs book fairs and events, and administers the annual Callum Macdonald Memorial Award created to recognize publishing skill and effort in the pamphlet form.

Scottish Poetry Library

5 Crichton's Close, Canongate, Edinburgh EH8 8DT
☎ 0131 557 2876   🖷 0131 557 8393
inquiries@spl.org.uk
www.spl.org.uk
Director *Robyn Marsack*
Librarian *Julie Johnstone*

A comprehensive reference and lending collection of work by Scottish poets in Gaelic, Scots and English, plus the work of British, European and international poets. Stock includes books, audio, videos, news cuttings and magazines. Borrowing is free to all. Services include a postal lending scheme, for which there is a small fee; enquiries; schools workshops throughout Scotland on application; exhibitions; bibliographies; publications; general information in the field of poetry. Also available is an online catalogue and index to Scottish poetry periodicals. There is a Friends' scheme costing £25 annually. Friends

receive a newsletter and other benefits and support the library.

Second Light

9 Greendale Close, London SE22 8TG
dilyswood@tiscali.co.uk
www.secondlightlive.co.uk
www.poetrypf.co.uk/secondlight.html
www.esch.dircon.co.uk/second/second.htm
Director *Dilys Wood*
Administrator *Anne Stewart* (editor@poetrypf.co.uk)

A network of over 350 women poets, major names and less well-known. With its publishing arm, Second Light Publications, the network aims to develop and promote women's poetry. Issues a twice-yearly newsletter with poetry, articles and reviews; runs an annual poetry competition; holds residential workshops, readings; has published three anthologies of women's poetry.

Spiel Unlimited

20 Coxwell Street, Cirencester GL7 2BH
☎ 01285 640470/07814 830031
spiel@arbury.freeserve.co.uk
info@spiel.wanadoo.co.uk
Directors *Marcus Moore, Sara-Jane Arbury*

'Spoken word, written word, anywhere, everywhere' with two writers who put a positive charge in live literature by organizing quirky and original events such as Slam!Fests, Spontaneity Days and Living Room Poetry performances, hosting UK poetry slams and running workshops for schools and adults. Also produce *SPIEL*, a monthly e-mail newsletter. Specialists in breathing new life into literature.

Survivors' Poetry

Studio 11, Bickerton House, 25–27 Bickerton Road, London N19 5JT
☎ 020 7281 4654   🖷 020 7281 7894
info@survivorspoetry.org.uk
www.survivorspoetry.com
National Outreach:
royb@survivorspoetry.org.uk
National Mentoring Scheme & *Poetry Express*:
simon@survivorspoetry.org.uk
Contacts *Dr. Simon Jenner, Roy Birch, Roy Holland, Blanche Donnery*

A unique national literature organization promoting poetry by survivors of mental distress through workshops, readings and performances to audiences all over the UK. It was founded in 1991 by four poets with first-hand experience of the mental health system. Survivors' community outreach work provides training and performance workshops and publishing projects. For the organization's quarterly magazine, see *Poetry Express*. Survivors work with those who have survived mental distress and those who empathize with their experience.

Ty Newydd Writers' Centre

Llanystumdwy, Cricieth LL52 0LW
℡ 01766 522811   ℻ 01766 523095
post@tynewydd.org
www.tynewydd.org
Executive Director *Sally Baker*

Run by the Taliesin Trust, an independent, Arvon-style residential writers centre established in the one-time home of Lloyd George in North Wales. The programme (in both Welsh and English) has a regular poetry content. (See also *UK and Irish Writers' Courses.*) Fees start at £215 for weekends and £420 for week-long courses. Bursaries available. Among the many tutors to-date have been: Gillian Clarke, Owen Sheers, Robert Minhinnick, Liz Lochhead, Michael Longley, Ian Macmillan and Jo Shapcott. Send for the centre's descriptive leaflets and programme of courses.

The Western Writers' Centre/Ionad Scriobhneoirí Chaitlin Maud

34 Nuns Island, Galway, Republic of Ireland
℡ 00 353 91 533594
westernwriters@eircom.net
www.twwc.ie

Director *Fred Johnston*
Administrator *Marvelle Maguire*

Founded 2001.The only writers' centre in the West Shannon region, named after the late Caitlin Maud, prominent Irish-language activist, poet and singer. Courses, organized readings and workshops are held in the Galway City and County area. Small library of literary periodicals. News and reviews of books in the 'Kiosque!' section of the website. Publishes new short stories and poems in Gaelic and English. Accepts original work for online publication in Gaelic, English and French. No payment for publication.

writers inc

14 Somerset Gardens, London SE13 7SY
℡ 020 8305 8844   ℻ 020 9469 2147
admin@writersinc-london.org.uk
www.writersinc-london.org.uk
Administrator *John Hole*

Founded by poets Sue Hubbard and Mario Petrucci, writers inc provides a forum for writers to develop their skills. Offers writing workshops on Sunday afternoons at the Barbican Library in the City of London, regular residential writing weekends at the Abbey, Sutton Courtnay near Oxford and week-long writing holidays in Andalucia. Runs the annual Writers-of-the-Year Competition.

Small Presses

Aard Press

c/o Aardverx, 31 Mountearl Gardens, London
SW16 2NL
Managing Editors *Dawn Redwood, D. Jarvis*

Founded 1971. Publishes artists' bookworks, experimental/visual poetry and art theory, topographics, ephemera, international mail-art and performance art documentation. Very small editions. No unsolicited material or proposals.
£ Royalties not paid. No sale-or-return deals.

Accent Press

The Old School, Upper High Street, Bedlinog
CF46 6SA
℡ 01443 710930   ℻ 0870 130 6934
info@accentpress.co.uk
www.accentpress.co.uk
Managing Editor *Hazel Cushion*

Founded 2002. Publishes general fiction, erotic fiction, health, humour and biography. 24 titles a year. Visit the website for submission guidelines.
£ Royalties twice-yearly.

Allardyce, Barnett, Publishers

14 Mount Street, Lewes BN7 1HL
℡ 01273 479393
www.abar.net
Publisher *Fiona Allardyce*
Managing Editor *Anthony Barnett*

Founded 1981. Publishes art, literature and music. IMPRINT **Allardyce Book**. About 3 titles a year. Unsolicited mss and synopses cannot be considered.

Alloway Publishing Limited

54–58 Mill Square, Catrine KA5 6RD
℡ 01290 551122   ℻ 01290 551122
alloway@stenlake.co.uk
Managing Editor *Richard Stenlake*

Founded 1982. Acquired by **Stenlake Publishing Limited** in 2008. Publishes local history and crafts. No unsolicited mss; e-mail in the first instance with outline proposal.
£ Royalties annually.

Anglo-Saxon Books

25 Brocks Road, Swaffham PE37 7XG
℡ 0845 430 4200
www.asbooks.co.uk
Managing Editor *Tony Linsell*

Founded 1990 to promote a greater awareness of and interest in early English history, language and culture. Publishes Anglo-Saxon history, culture, language. About 5 titles a year. Please phone before sending mss.
£ Royalties at standard rate.

M.&M. Baldwin

24 High Street, Cleobury Mortimer, Kidderminster DY14 8BY
℡ 01299 270110
mb@mbaldwin.free-online.co.uk
Managing Editor *Dr Mark Baldwin*

Founded 1978. Publishes local interest/history, WW2 codebreaking and inland waterways books. Up to 5 titles a year. Unsolicited mss, synopses and ideas for books welcome (not general fiction).
£ Royalties paid.

Barny Books

The Cottage, Hough on the Hill, near Grantham
NG32 2BB
℡ 01400 250246   ℻ 01400 251737
www.barnybooks.biz
Managing Director/Editorial Head *Molly Burkett*
Business Manager *Jayne Thompson*

Founded with the aim of encouraging new writers and illustrators. Publishes a wide variety of books including adult fiction and non-fiction. Offers schools' projects where students help to produce books. About 12–15 titles a year. Too small a concern to have the staff/resources to deal with unsolicited mss. Writers with strong ideas should approach Molly Burkett by letter in the first instance. Also runs a readership and advisory service for new writers (£20 fee for short stories or illustrations; £50 for full-length stories).
£ Division of profits 50/50.

BB Books

▷ entry under Poetry Presses

Between the Lines

The Cottage, 14 Lyncroft Gardens, Ewell KT17 1UR
☎ 020 8393 7055   🖷 020 8393 7055
btluk@aol.com
www.interviews-with-poets.com
Editorial Board *Peter Dale, Philip Hoy, J.D. McClatchy*
Editorial Assistant *Ryan Roberts*

Founded 1998. Publishes in book form extended interviews with leading contemporary poets. Thirteen volumes currently in print, featuring Peter Dale, John Ashbery, Charles Simic, Ian Hamilton, Donald Justice, Seamus Heaney, Richard Wilbur, Thom Gunn, Donald Hall, Anthony Hecht, Anthony Thwaite, Michael Hamburger and W.D. Snodgrass. Each volume includes a career sketch, a comprehensive bibliography and a representative selection of quotations from the poets' critics and reviewers. More recent volumes include photographs as well as previously uncollected poems. A second series, featuring slightly younger poets, was launched in autumn 2006. The first volumes in that series feature Dick Davis, Rachel Hadas and Timothy Steele.
£ Royalties not paid.

Bluemoose Books Limited

25 Sackville Street, Hebden Bridge HX7 7DJ
☎ 01422 842731
dufk@aol.com
www.bluemoosebooks.com
Managing Editor *Hetha Duffy*
Publisher *Kevin Duffy*

Founded 2006. Publishes fiction, non-fiction and children's books. 2 titles in 2006. No unsolicited mss; send first three chapters and synopsis either by post (include s.a.e.) or via e-mail.
£ Royalties annually.

The Book Castle

12 Church Street, Dunstable LU5 4RU
☎ 01582 605670   🖷 01582 662431
bc@book-castle.co.uk
www.book-castle.co.uk
Managing Editor *Paul Bowes*
Assistant Editor *Sally Siddons*

Founded 1986. Publishes non-fiction of local interest (Bedfordshire, Hertfordshire, Buckinghamshire, Oxfordshire, the Chilterns). Over 120 titles in print. 12 titles a year. Unsolicited mss, synopses and ideas for books welcome.
£ Royalties paid.

Brilliant Publications

Unit 10, Sparrow Hall Farm, Edlesborough, Dunstable LU6 2ES
☎ 01525 222292   🖷 01525 222720
info@brilliantpublications.co.uk
www.brilliantpublications.co.uk
Publisher *Priscilla Hannaford*

Founded 1993. Publishes practical resource books for teachers, parents and others working with 0–13-year-olds. About 10–15 titles a year. Submit synopsis and sample pages in the first instance. Most of the books have black and white insides and many have pupil sheets that may be photocopied. Potential authors are strongly advised to look at the format of existing books before submitting synopses. 'We do not publish children's picture books.'
£ Royalties twice-yearly.

The Brodie Press

thebrodiepress@hotmail.com
www.brodiepress.co.uk
Project Editor *Penny Price*

Founded 2002. Publishes poetry and anthologies. SERIES *The Brodie Poets*, launched in 2003, including Julie-ann Rowell, Poetry Book Society Pamphlet Choice, Winter 2003. Send synopses and proposals by e-mail in the first instance.

Charlewood Press

7 Weavers Place, Chandlers Ford SO53 1TU
☎ 023 8026 1192
gponting@clara.net
www.home.clara.net/gponting/index-page11.html
Managing Editors *Gerald Ponting, Anthony Light*

Founded 1987. Publishes local history books and walks booklets on Hampshire and adjacent counties, especially around the Fordingbridge area. Mss in field of Hampshire local history (only) considered.

Chrysalis Press

7 Lower Ladyes Hills, Kenilworth CV8 2GN
☎ 01926 855223
editor@margaretbuckley.com
www.margaretbuckley.com
Managing Editor *Brian Boyd*

Founded 1994. Publishes fiction, literary criticism and biography. No unsolicited mss.
£ Royalties paid.

CNP Publications
▷ Lyfrow Trelyspen

Contact Publishing Ltd

Unit 346, 176 Finchley Road, London NW3 6BT
☎ 020 7193 1782
info@contact-publishing.co.uk
www.contact-publishing.co.uk
Managing Director *Anne Kontoyannis*

Founded 2003. Publishes general fiction, non-fiction, self help and New Age. No children's, political, sports or westerns. 'Interested in seeing

mss by new authors with original angles and new ideas.' See website for author guidelines. Send synopsis, three sample chapters and outline in the first instance.
£ Royalties twice-yearly.

The Cosmic Elk

68 Elsham Crescent, Lincoln LN6 3YS
℡ 01522 820922
post@cosmicelk.co.uk
www.cosmicelk.co.uk
www.cosmicelk.com
Contact *Heather Hobden*

Founded 1988. Publishes easily updated print-on-demand books on science, history and the history of science. Websites designed and maintained, with or without associated printed books. 'Please check the CosmicElk website first where you will find the information you need and contact us by e-mail through the website.'

Crescent Moon Publishing and Joe's Press

PO Box 393, Maidstone ME14 5XU
℡ 01622 729593
cresmopub@yahoo.co.uk
www.crescentmoon.org.uk
Managing Editor *Jeremy Robinson*

Founded 1988 to publish critical studies of figures such as D.H. Lawrence, Thomas Hardy, André Gide, Walt Disney, Rilke, Leonardo da Vinci, Mark Rothko, C.P. Cavafy and Hélène Cixous. Publishes literature, criticism, media, art, feminism, painting, poetry, travel, guidebooks, cinema and some fiction. Literary magazine, *Passion*, launched February 1994 (quarterly). *Pagan America*, twice-yearly anthology of American poetry. About 15–20 titles per year. Unsolicited synopses and ideas welcome but approach in writing first and send an s.a.e. Do not send whole mss.
£ Royalties negotiable.

Crime Express
▷ Five Leaves Publications

Crossbridge Books

Tree Shadow, Berrow Green, Martley WR6 6PL
℡ 01886 821128
crossbridgebooks@btinternet.com
www.crossbridgebooks.com
Managing Director *Eileen Mohr*

Founded 1995. Publishes Christian books for adults and children. IMPRINT **Mohr Books**. No new submissions for the foreseeable future.

Day Books

Orchard Piece, Crawborough, Charlbury OX7 3TX
℡ 01608 811196   ℻ 01608 811196
diaries@day-books.com
www.day-books.com

Managing Editor *James Sanderson*

Founded in 1997 to publish a series of great diaries from around the world. Unsolicited mss, synopses and ideas welcome. Include postage if return of material is required.

The Dragonby Press

30 King Street, Winterton DN15 9TP
℡ 01724 737254
rich@rahwilliams.orangehome.co.uk
Managing Editor *Richard Williams*

Founded 1987 to publish affordable bibliography for reader, collector and dealer. About 3 titles a year. Unsolicited mss, synopses and ideas welcome for bibliographical projects only.
£ Royalties paid.

Dramatic Lines

PO Box 201, Twickenham TW2 5RQ
℡ 020 8296 9502   ℻ 020 8296 9503
mail@dramaticlinespublishers.co.uk
www.dramaticlines.co.uk
Managing Editor *John Nicholas*

Founded to promote drama for all. Publications with a wide variety of theatrical applications including classroom use and school assemblies, drama examinations, auditions, festivals, theatre group performance, musicals and handbooks. Unsolicited drama-related mss, proposals and synopses welcome; enclose s.a.e.
£ Royalties paid.

Edgewell Publishing

5A Front Street, Prudhoe NE42 5HJ
℡ 01661 835330   ℻ 01661 835330
keithminton@btconnect.com
www.edgewell-publishing.co.uk
Editor *Keith Minton*
Sub Editor *Deborah Moran*

Quarterly short story and poetry magazine (*New Lively Tales*) to encourage writing from new and established writers, also children. Writers from the North East especially welcome. Please send by e-mail, where possible.
£ 'No royalties; writers maintain copyright.'

Educational Heretics Press

113 Arundel Drive, Bramcote, Nottingham NG9 3FQ
℡ 0115 925 7261   ℻ 0115 925 7261
www.edheretics.gn.apc.org
Directors *Janet & Roland Meighan*

Non-profit venture which aims to question the dogmas of schooling in particular and education in general, and establish the logistics of the next personalized learning system. No unsolicited material. Enquiries only.
£ Royalties paid on recent contracts.

Eilish Press

4 Collegiate Crescent, Broomhall Park, Sheffield
S10 2BA
☎ 07973 353964
eilishpress@hotmail.co.uk
eilishpress.tripod.com
Head of Marketing *Suzi Kapadia*

Specializes in academic work and popular non-
fiction in the areas of women's studies, human
rights and anti-racism. Also produces children's
literature with humanitarian themes. No adult
fiction. Unsolicited mss cannot be accepted at this
time. However, ideas for children's books may be
sent by e-mail.
ⓔ Royalties annually.

Enable Enterprises

PO Box 1974, Coventry CV6 4FD
☎ 0800 358 8484   ☒ 0800 358 8484
writers@enableenterprises.com
www.enableenterprises.com
Chief Executive *Simon Stevens*

Enable Enterprises provides a wide range of
accessibilty and disability services including
publications on relevant issues. Welcomes
unsolicited material related to accessibility and
disability issues.

Fand Music Press

Glenelg, 10 Avon Close, Petersfield GU31 4LG
☎ 01730 267341   ☒ 01730 267341
paul@fandmusic.com
www.fandmusic.com
Managing Editor *Peter Thompson*

Founded in 1989 as a sheet music publisher, Fand
Music Press has expanded its range to include CD
recordings and books on music. Also publishes
poetry and short stories. 10 titles a year. No
unsolicited mss. Write with ideas in the first
instance.

Feather Books

PO Box 438, Shrewsbury SY3 0WN
☎ 01743 872177   ☒ 01743 872177
john@waddysweb.freeuk.com
www.waddysweb.freeuk.com
Managing Director *Rev. John Waddington-Feather*
Music Director *David Grundy*
Drama Director/Recordings Manager *Tony Reavil*

Founded 1980 to publish writers' group work. All
material has a strong Christian ethos. Publishes
poetry (mainly, but not exclusively, religious);
Christian mystery novels (the Revd. D.I. Blake
Hartley series); Christian children's *Quill
Hedgehog* novels; seasonal poetry collections, *The
Poetry Church*, a quarterly magazine and *Feather's
Miscellany*. Produces poetry, drama and hymn
CD/cassettes; also the anthology 'Pilgrimages' ed.
Professor Walter Nash ('a milestone in Christian

poetry publishing'). About 20 titles a year. All
correspondence to include s.a.e., please.

Fidra Books

219 Bruntsfield Place, Edinburgh EH10 4DH
☎ 0131 447 1917
vanessa@fidrabooks.com
www.fidrabooks.com
Managing Editor *Vanessa Robertson*

Founded 2005. Publishes children's fiction,
specializing in reprints of sought-after 20th
century children's fiction. Will consider original
work if it fits with existing list. No unsolicited
material; initial approach by letter or e-mail.
ⓔ Royalties quarterly.

Five Leaves Publications

PO Box 8786, Nottingham NG1 9AW
☎ 0115 969 3597
info@fiveleaves.co.uk
www.fiveleaves.co.uk
Contact *Ross Bradshaw*

Founded 1995 (taking over the publishing
programme of Mushroom Bookshop). Publishes
fiction, poetry, social history, politics and Jewish
interest. Also publishes for the Anglo-Catalan
Society. IMPRINT **Crime Express** Crime fiction
novellas. Titles normally commissioned. About 15
titles a year.
ⓔ Royalties paid.

Frontier Publishing

Windetts, Kirstead NR15 1EG
☎ 01508 558174
frontier.pub@macunlimited.net
www.frontierpublishing.co.uk
Managing Editor *John Black*

Founded 1983. Publishes travel, photography,
sculptural history and literature. 2–3 titles a year.
No unsolicited mss; synopses and ideas welcome.
ⓔ Royalties paid.

Galactic Central Publications

Imladris, 25A Copgrove Road, Leeds LS8 2SP
gcp@philsp.com
www.philsp.com
Managing Editor *Phil Stephensen-Payne*

Founded 1982 in the US. Publishes science
fiction bibliographies. All new publications
originate in the UK. No new titles planned at
present. Unsolicited mss, synopses and ideas (for
bibliographies) welcome.

Galore Park Publishing Ltd

19/21 Sayer's Lane, Tenterden TN30 6BW
☎ 01580 764242   ☒ 01580 764242
info@galorepark.co.uk
www.galorepark.co.uk
Managing Director *Nicholas Oulton*

Founded 1999. Publishes school textbooks and children's books. 124 titles in 2007. No unsolicited mss. Synopses and ideas welcome; send by e-mail. Ⓔ Royalties twice-yearly.

Glosa Education Organisation

PO Box 18, Richmond TW9 2GE
www.glosa.org
Managing Editor *Wendy Ashby*

Founded 1981. Publishes textbooks, dictionaries and translations for the teaching, speaking and promotion of Glosa (an international, auxiliary language); also a newsletter, *Plu Glosa Nota* and journal, *PGN*. Unsolicited mss and ideas for Glosa books welcome.

Godstow Press

60 Godstow Road, Wolvercote, Oxford OX2 8NY
Ⓣ 01865 556215
info@godstowpress.co.uk
www.godstowpress.co.uk
Managing Editors *Linda Smith, David Smith*

Founded in 2003 for the publication of creative work with a philosophical/spiritual content. 2–3 titles a year. Considers works self-published by the author for inclusion in the catalogue. The press works on direct sales rather than through usual trade outlets. Synopses and ideas welcome but no unsolicited mss.

Graffeg

2 Radnor Court, 256 Cowbridge Road East, Cardiff CF5 1GZ
Ⓣ 029 2037 7312   Ⓕ 029 2039 8101
info@graffeg.com
www.graffeg.com
Managing Editor *Peter Gill*

Founded 2002. Publishes travel, food and drink, Wales, landscapes, photography. 8 titles in 2007; list of 17 titles. Unsolicited material welcome by post. Ⓔ Royalties twice-yearly.

Grant Books

The Coach House, New Road, Cutnall Green, Droitwich WR9 0PQ
Ⓣ 01299 851588   Ⓕ 01299 851446
golf@grantbooks.co.uk
www.grantbooks.co.uk
Managing Editor *H.R.J. Grant*

Founded 1978. Publishes golf-related titles only: course architecture, history, biography, etc., but not instructional, humour or fiction material. About 3 titles a year. Mss, synopses and ideas welcome. Ⓔ Royalties paid.

Great Northern Publishing

PO Box 202, Scarborough YO11 3GE
Ⓣ 01723 581329   Ⓕ 01723 581329
books@greatnorthernpublishing.co.uk
www.greatnorthernpublishing.co.uk
Production Manager/Senior Editor *Mark Marsay*

Small, independent, award-winning, family-owned company founded in 1999. Publishers of books, magazines and journals. Also provides full book, magazine and print production services to individuals, businesses, charities, museums and other small publishers. Publishes non-fiction (mainly military), adult erotic material and, occasionally, fiction. No romance, religious, political, feminist, New Age, medical or children's books. Publishers of the bi-monthly magazines, *The Great War* and *Jade* (see entries under *Magazines*). Mail order and Internet-based bookshop stocking selected titles alongside its own. Website provides submission guidelines and current publishing requirements. No unsolicited phone calls or mss; send letter or text-only e-mail in the first instance. Ⓔ Royalties twice-yearly.

Griffin Publishing Associates

61 Grove Avenue, London N10 2AL
Ⓣ 020 8444 0815/0778 842 5262
griffin.publishing.associates@googlemail.com
Managing Editor *Ali Ismail*

Founded 2007. Publishes all kinds of fiction (plays, poetry and prose) with an emphasis on science fiction, fantasy and horror. Unsolicited mss, synopses and ideas for books welcome. Approach by post (return postage must be included) or telephone. Ⓔ Royalties annually.

GSSE

11 Malford Grove, Gilwern, Abergavenny NP7 0RN
Ⓣ 01873 830872
GSSE@zoo.co.uk
www.gsse.org.uk
Owner/Manager *David P. Bosworth*

Publishes books and booklets describing classroom practice (at all levels of education and training). Ideas welcome – particularly from practising teachers, lecturers and trainers describing how they use technology in their teaching. Ⓔ Royalties by arrangement.

Haunted Library

Flat 1, 36 Hamilton Street, Hoole, Chester CH2 3JQ
Ⓣ 01244 313685
pardos@globalnet.co.uk
www.users.globalnet.co.uk/~pardos/GS.html
Managing Editor *Rosemary Pardoe*

Founded 1979. Publishes the *Ghosts and Scholars M.R. James Newsletter* twice a year, featuring articles, news and reviews (no fiction). Ⓔ Royalties not paid.

Headpress

Suite 306, The Colourworks, 2a Abbot Street,
London E8 3DP
℡ 0845 330 1844   ℻ 020 7249 6395
office@headpress.com
www.headpress.com
Managing Editor *David Kerekes*

Founded 1991. Publishes *Headpress* journal and
books devoted to film, music, popular culture,
the strange and esoteric. No fiction or poetry.
Unsolicited material welcome; send letter and
outline in the first instance.
Ⓔ Royalties/flat fees vary.

Heart of Albion Press

2 Cross Hill Close, Wymeswold, Loughborough
LE12 6UJ
℡ 01509 880725
albion@indigogroup.co.uk
www.hoap.co.uk
Managing Editor *R.N. Trubshaw*

Founded 1990 to publish local history. Publishes
folklore, local history and mythology. About 8
titles a year. Synopses relating to these subjects
welcome.
Ⓔ Royalties negotiable.

Ignotus Press

BCM–Writer, London WC1N 3XX
℡ 0840 230 2980
ignotuspress@eircom.net
www.ignotuspress.com
Managing Editor *Suzanne Ruthven*

Founded 1994. Publishes mind, body and spirit
books and fiction. 5 titles in 2007. No unsolicited
mss; send outline by e-mail or post after reading
submission guidelines on the website.
Ⓔ Royalties twice-yearly.

Immanion Press

8 Rowley Grove, Stafford ST17 9BJ
℡ 01785 613299
editorial@immanion-press.com
www.immanion-press.com
Managing Editor *Storm Constantine*

Founded 2003. Publishes genre fiction (science
fiction, fantasy, horror), slipstream fiction;
esoteric non-fiction. IMPRINT **Megalithica Books**
Commissioning Editor *Taylor Ellwood* (info@
immanion-press.com) Non-fiction, specifically
cutting edge esoteric work. Submissions by
e-mail or post (enclose s.a.e.). Website gives full
submission details.
Ⓔ Royalties paid.

Infinity Junction

PO Box 64, Neston DO CH64 0WB
infin-info@infinityjunction.com
www.infinityjunction.com

Managing Editor *Neil Gee*

Established originally as a self-publishing,
self-help organization in 1991, Infinity Junction
became a publisher in 2001. Currently offers a
variety of services to authors, including advice,
free web display space and full commercial
publication. Authors are strongly advised to read
the detailed information on the website before
making contact. E-mail communication preferred.
'We cannot undertake to read unsolicited whole
mss.'

Inner Sanctum Publications

75 Greenleaf Gardens, Polegate BN26 6PQ
℡ 01323 484058
www.innersanctumpublications.com
Managing Editor *Mary Hession*

Founded in 1999 to publish spiritual books. Initial
approach via e-mail (address available on the
website) or by post, including s.a.e.

Iolo

38 Chaucer Road, Bedford MK40 2AJ
℡ 01234 301718/07909 934866   ℻ 01234 301718
dedwydd1@ntlworld.com
www.dedwyddjones.co.uk
Managing Director *Dedwydd Jones*

Publishes Welsh theatre-related material and
campaigns for a Welsh National Theatre. Ideas on
Welsh themes welcome; approach in writing.

Ivy Publications

72 Hyperion House, Somers Road, London
SW2 1HZ
℡ 020 8671 6872
Proprietor *Ian Bruton-Simmonds*

Founded 1989. Publishes educational, science,
fiction, philosophy, children's, travel, literary
criticism, history, film scripts. No unsolicited
mss; send two pages, one from the beginning and
one from the body of the book, together with
synopsis (one paragraph) and s.a.e. No cookery,
gardening or science fiction.
Ⓔ Royalties annually.

The Jupiter Press

The Coach House, Gatacre, Claverley, Nr
Wolverhampton WV5 7AW
℡ 01746 710694   ℻ 01746 710158
gordonthomasdrury@btinternet.com
Managing Editor *Gordon Thomas Drury*

Founded 1995. Looking for niche market and
information publications, also quiz and game
subjects. Interested in holistic, clairvoyance and
esoteric subjects. Synopses and ideas for books
welcome. Send idea and sample chapter (include
contact telephone number and return postage).
Unsolicited mss welcome 'but may request a

reader's fee'. Currently seeking books and ideas for SERIES entitled *How To Make Money At ....*
Ⓔ Royalties paid.

Kittiwake

3 Glantwymyn Village Workshops, Glantwymyn, Nr Machynlleth SY20 8LY
☎ 01650 511314   🖷 01650 511314
perrographics@btconnect.com
www.kittiwake-books.com
Managing Editor *David Perrott*

Founded 1986. Publishes guidebooks only, with an emphasis on careful design/production. Specialist research, writing, cartographic and electronic publishing services available. Unsolicited mss, synopses and ideas for guidebooks welcome.
Ⓔ Royalties paid.

The Lindsey Press

Unitarian Headquarters, 1–6 Essex Street, Strand, London WC2R 3HY
☎ 020 7240 2384   🖷 020 7240 3089
ga@unitarian.org.uk
Convenor *Kate Taylor*

Established at the end of the 18th century as a vehicle for disseminating liberal religion. Adopted the name of The Lindsey Press at the beginning of the 20th century (after Theophilus Lindsey, the great Unitarian Theologian). Publishes books reflecting liberal religious thought or Unitarian denominational history. Also worship material – hymn books, collections of prayers, etc. 1 title in 2007. No unsolicited mss; synopses and ideas welcome.
Ⓔ Royalties not paid.

The Linen Press

75c (13) South Oswald Road, Edinburgh EH9 2HH
lynn@linenpressbooks.co.uk
www.linenpressbooks.co.uk
Contact *Lynn Michell*

Founded 2006. 'Run by women, for women. We are looking for beautifully crafted writing which speaks to readers who possess a wealth of life experience and wisdom.' Publishes literary fiction and memoir as well as writing which does not fit comfortably in either genre. No poetry, children's books, romantic fiction or science fiction. Happy to receive submissions from emerging as well as established writers as long as they fulfil the criteria and ethos of the Linen Press. Please read the submission guidelines before sending in a proposal.

Lyfrow Trelyspen

The Roseland Institute, Gorran, St Austell PL26 6NT
☎ 01726 843501   🖷 01726 843501
trelispen@care4free.net
www.theroselandinstitute.co.uk

Managing Editor *Dr James Whetter*

Founded 1975. Publishes works on Cornish history, biography, essays, etc. Also **CNP Publications** which publishes the quarterly journal *The Cornish Banner/An Baner Kernewek*. 1–2 titles a year. Unsolicited mss, synopses and ideas welcome.
Ⓔ Royalties not paid.

The Maia Press

82 Forest Road, London E8 3BH
☎ 020 7683 8141   🖷 020 7683 8141
maggie@maiapress.com
www.maiapress.com
Contacts *Maggie Hamand, Jane Havell*

Founded 2002. Publishes fiction by known and new writers. 6 titles a year. No unsolicited mss, write with c.v. and synopsis only. S.a.e. essential.
Ⓔ Royalties twice-yearly.

Maypole Editions

18 The Lindens, Shifnal TF11 8AB
Managing Editor *Barry Taylor*

Founded 1987. Publishes fiction, drama, filmscripts and poetry. 3 titles in 2007. No submissions until 2010 at which time an s.a.e. will be required. 'No reply means no'.

Megalithica Books
▷ Immanion Press

Mercia Cinema Society

29 Blackbrook Court, Durham Road, Loughborough LE11 5UA
☎ 01509 218393
Mervyn.Gould@virgin.net
www.merciacinema.org
Managing Editor *Kate Taylor*
Series Editor *Mervyn Gould*

Founded 1980 to foster research into the history of picture houses. Publishes books and booklets on the subject, including cinema circuits and chains. Books are often tied in with specific geographical areas. Quarterly journal: *Mercia Bioscope*, editor *Paul Smith* (Paul@Smith2004. fs.net.co.uk). 63 titles to date. Unsolicited mss (preferably on disk); synopses and ideas.
Ⓔ Royalties paid. Six free copies to the author.

Meridian Books

40 Hadzor Road, Oldbury B68 9LA
☎ 0121 429 4397
meridian.books@btopenworld.com
Managing Editor *Peter Groves*

Founded 1985 as a small home-based enterprise following the acquisition of titles from Tetradon Publications Ltd. Publishes walking and regional guides. 4–5 titles a year. Unsolicited mss,

synopses and ideas welcome if relevant. Send s.a.e. if mss to be returned.
£ Royalties paid.

Neil Miller Publications

c/o Ormonde House, 49 Ormonde Road, Hythe CT21 6DW
contact@neilmillersbooks.co.uk
www.neilmillersbooks.co.uk
Also at: 4 Rue D'Equiire Bergueneuse, Pas de Calais, France 62134
Managing Editor *Neil Miller*

Founded 1994. Publishes novels and collections of short stories on any subject, fact or fiction: memoirs, autobiography, mystery, horror, war, comedy, science fiction. Interested in working with new and published writers. Evaluation and critique service available. 'Send your novel or selection of short stories for consideration plus our reader's fee of £9.75. For this you receive a critique plus two of our latest authors' books and have the opportunity of being published by us.' Enclose s.a.e. for return of work. No unrequested tales.

Millers Dale Publications

7 Weavers Place, Chandlers Ford, Eastleigh SO53 1TU
☎ 023 8026 1192
gponting@clara.net
www.home.clara.net/gponting/index-page10.html
Managing Editor *Gerald Ponting*

Founded 1990. Publishes books on local history related to central Hampshire. Also books related to slide presentations by Gerald Ponting. Ideas for local history books on Hampshire (only this topic) considered.

Mirage Publishing

PO Box 161, Gateshead NE8 1WY
☎ 0870 720 9498
sales@miragepublishing.com
www.miragepublishing.com
Managing Director *Sharon Anderson*

Founded 1998. Publishes mind, body and spirit, autobiography, biography. 9 titles in 2007. No unsolicited mss; send synopses and ideas by mail or e-mail.
£ Royalties twice-yearly.

Mohr Books
▷ Crossbridge Books

Need2Know

Remus House, Coltsfoot Drive, Woodston PE2 9JX
☎ 01733 898103   ☏ 01733 313524
sales@n2kbooks.com
www.n2kbooks.com

Founded 1995 'to fill a gap in the market for self-help books'. Need2Know is an imprint of **Forward Press** (see entry under *Poetry Presses*). Publishes contemporary health and lifestyle issues. No unsolicited mss. Call in the first instance.
£ Advance plus 15% royalties.

Jane Nissen Books

Swan House, Chiswick Mall, London W4 2PS
☎ 020 8994 8203   ☏ 020 8742 8198
jane@nissen.demon.co.uk
www.janenissenbooks.co.uk
Publisher *Jane Nissen*

Founded 2000. Publishes reprints of children's fiction only. Winner of the 2007 **Eleanor Farjeon Award**. 3 titles in 2007; list of 30 titles. Does not consider submitted material as all titles published are reprints of classics or fogotten children's books.

The Nostalgia Collection

Silver Link Publishing Ltd, The Trundle, Ringstead Road, Great Addington, Kettering NN14 4BW
☎ 01536 330588   ☏ 01536 330588
sales@nostalgiacollection.com
www.nostalgiacollection.com
Managing Editor *Will Adams*

Founded 1985. Small independent company specializing in illustrated post-war nostalgia titles including railways, trams, ships and other transport subjects, towns and cities, villages and rural life, rivers and inland waterways, and industrial heritage under the **Silver Link** and **Past and Present** imprints.
£ Fees/royalties paid.

Nyala Publishing

4 Christian Fields, London SW16 3JZ
☎ 020 8764 6292/0115 981 9418
☏ 020 8764 6292/0115 981 9418
nyala.publishing@geo-group.co.uk
www.geo-group.co.uk
Editorial Director *J.F.J. Douglas*

Founded 1996. Publishing arm of Geo Group. Publishes biography, travel and general non-fiction. Also offers a wide range of printing and publishing services. Reading, editing and proofreading. No unsolicited mss; synopses and ideas considered.
£ Royalties annually.

Orpheus Publishing House

4 Dunsborough Park, Ripley Green, Ripley, Guildford GU23 6AL
☎ 01483 225777   ☏ 01483 225776
orpheuspubl.ho@btinternet.com
Managing Editor *J.S. Gordon*

Founded 1996. Publishes 'well-researched and properly argued' books in the fields of occult science, esotericism and comparative philosophy/

religion. 'Keen to encourage good (but sensible) new authors.' In the first instance, send maximum three-page synopsis with s.a.e.
Ⓔ Royalties by agreement.

Packard Publishing Limited
Forum House, Stirling Road, Chichester PO19 7DN
Ⓣ 01243 537977   Ⓕ 01243 537977
info@packardpublishing.co.uk
www.packardpublishing.com
Chairman/Managing Director *Michael Packard*

Founded in 1977 to distribute overseas publishers' lists in biology, biochemistry and ecology. Publishes academic & professional: school/university interface, postgraduate – mainly in land management and applied ecology; agriculture, forestry, nature conservation, rural studies, landscape architecture and garden design; some languages (French, Arabic). SERIES Instructional books on garden and landscape design, and monographs or critical biographies in garden and landscape design and history. 32 titles to date. No unsolicited mss. Synopses and ideas in relevant areas welcome. Telephone first.
Ⓔ Royalties twice-yearly in first year, then annually.

Panacea Press Limited
86 North Gate, Prince Albert Road, London NW8 7EJ
Ⓣ 020 7722 8464   Ⓕ 020 7586 8187
ebrecher@panaceapress.net
www.panaceapress.net
Managing Editor *Erwin Brecher, PhD*

Founded as a self-publisher but now open for non-fiction from other authors. Material of academic value considered provided it commands a wide general market. No unsolicited mss; synopses and ideas welcome. No telephone calls. Approach by e-mail, fax or letter.
Ⓔ Royalties annually.

Parapress
The Basement, 9 Frant Road, Tunbridge Wells TN2 5SD
Ⓣ 01892 512118   Ⓕ 01892 512118
office@parapress.myzen.co.uk
www.parapress.co.uk
Managing Editor *Elizabeth Imlay*

Founded 1993. Publishes animals, autobiography, biography, history, literary criticism, military and naval, music, self-help. Some self-publishing. About 3 titles a year.

Past and Present
▷ The Nostalgia Collection

Paupers' Press
37 Quayside Close, Turney's Quay, Trent Bridge,

Nottingham NG2 3BP
Ⓣ 0115 986 3334   Ⓕ 0115 986 3334
books@pauperspress.com
www.pauperspress.com
Managing Editor *Colin Stanley*

Founded 1983. Publishes extended essays in booklet form (about 15,000 words) on literary criticism and philosophy. 'Sometimes we stray from these criteria and produce full-length books, but only to accommodate an exceptional ms.' Limited hardback editions of bestselling titles. Centre for Colin Wilson studies. Critical essays on his work encouraged. About 6 titles a year. No unsolicited mss but synopses and ideas for books welcome.
Ⓔ Royalties paid.

Peepal Tree Press Ltd
▷ entry under Poetry Presses

Pen Press Publishing Ltd
25 Eastern Place, Brighton BN2 1GJ
Ⓣ 0845 108 0530   Ⓕ 01273 261434
info@penpress.co.uk
www.penpress.co.uk
Managing Director *Lynn Ashman*

Founded 1996. Publishing across a wide range of categories, Pen Press helps new authors to self-publish. Distribution, promotion and marketing included in self-publishing deal. Publisher, not author, pays for reprints and shares in sales revenue. 700 titles to date. Write, phone or e-mail for submission form and full details. Return form with full ms (digital or hard copy).
Ⓔ Royalties 45%, payable twice-yearly.

Perfect Publishers Ltd
23 Maitland Avenue, Cambridge CB4 1TA
Ⓣ 01223 424422   Ⓕ 01223 424414
editor@perfectpublishers.co.uk
www.perfectpublishers.co.uk
Managing Editor *S.N. Rahman*

A print on demand book publishing company, founded 2005. Offers self-publishing services. Publishes books of all genres. Submissions by e-mail or letter.
Ⓔ 'We pay 100% royalties twice-yearly.'

Pipers' Ash Ltd
'Pipers' Ash', Church Road, Christian Malford, Chippenham SN15 4BW
Ⓣ 01249 720563   Ⓕ 0870 0568916
pipersash@supamasu.co.uk
www.supamasu.co.uk
Managing Editor *Mr A. Tyson*

Founded 1976. The company's publishing activities include individual collections of contemporary short stories, science fiction short stories, poetry, plays, short novels, local histories,

children's fiction, philosophy, biographies, translations and general non-fiction. 12 titles a year. Synopses and ideas welcome; 'new authors with potential will be actively encouraged'. Offices in New Zealand and Australia.
Ⓔ Royalties annually.

Playwrights Publishing Co.

70 Nottingham Road, Burton Joyce, Nottingham NG14 5AL
☎ 0115 931 3356
playwrightspublishingco@yahoo.com
geocities.com/playwrightspublishingco
Managing Editors *Liz Breeze, Tony Breeze*

Founded 1990. Publishes one-act and full-length plays. Unsolicited scripts welcome. No synopses or ideas. Reading fees: £15 one act; £30 full length (waived if evidence of professional performance or if writer is unwaged). S.a.e. required.
Ⓔ Royalties paid.

Pomegranate Press

Dolphin House, 51 St Nicholas Lane, Lewes BN7 2JZ
☎ 01273 470100   ℻ 01273 470100
pomegranatepress@aol.com
www.pomegranate-press.co.uk
Managing Editor *David Arscott*

Founded in 1992 by writer/broadcaster David Arscott, who also administers the **Sussex Book Club**. Specializes in books about Sussex and self publishing. IMPRINT **Pomegranate Practicals** How-to books.
Ⓔ Royalties twice-yearly.

David Porteous Editions

PO Box 5, Chudleigh, Newton Abbot TQ13 0YZ
☎ 01626 853310   ℻ 01626 853663
editorial@davidporteous.com
www.davidporteous.com
Publisher *David Porteous*

Founded 1992 to produce colour illustrated books on hobbies and leisure for the UK and international markets. Publishes crafts, hobbies, art techniques and needlecrafts. No poetry or fiction. 3–4 titles a year. Unsolicited mss, synopses and ideas welcome if return postage included.
Ⓔ Royalties twice-yearly.

Praxis Books

Crossways Cottage, Walterstone HR2 0DX
☎ 01873 890695
rebeccatope@btinternet.com
www.rebeccatope.com
Proprietor *Rebecca Smith*

Founded 1992. Publishes reissues of the works of Sabine Baring-Gould, memoirs, diaries and local interest. 21 titles to date. Unsolicited mss accepted with s.a.e. No fiction. Editing and advisory service available. Funding negotiable. 'I am most likely to accept work with a clearly identifiable market.'

QueenSpark Books

Room 211, 10–11 Pavilion Parade, Brighton BN2 1RA
☎ 01273 571710
info@queensparkbooks.org.uk
www.queensparkbooks.org.uk

A community writing and publishing group. Since the early 1970s QueenSpark Books has published 90 titles, mainly featuring the lives of local people. Writing workshops and groups held on a regular basis. No unsolicited mss, please.

Radikal Phase Publishing House Ltd

Willow Court, Cordy Lane, Underwood NG16 5FD
☎ 01773 764288   ℻ 01773 764282
sales@phaseprint.com
www.phaseprint.com
Managing Director *Kevin Marks*

Founded 2001. Publishes radical revelation and technical electrical books. IMPRINTS **William Ernest** *Kevin Marks*; **Radikal Phase** *Philip Gardiner*. Welcomes unsolicited material; approach in writing in the first instance.
Ⓔ Royalties twice-yearly.

Robinswood Press

30 South Avenue, Stourbridge DY8 3XY
☎ 01384 397475   ℻ 01384 440443
info@robinswoodpress.com
www.robinswoodpress.com
Managing Editor *Christopher J. Marshall*

Founded 1985. Publishes children's, educational, teaching resources, SEN. Also collaborative publishing, e.g., with Camphill Foundation. About 30–50 titles a year. Unsolicited mss, synopses and ideas welcome. See website.
Ⓔ Royalties paid.

Rockpool Children's Books Ltd

15 North Street, Marton CV23 9RJ
☎ 01926 633114
stuart@rockpoolchildrensbooks.com
www.rockpoolchildrensbooks.com
Managing Director *Stuart Trotter*

Established 2006. Publishes children's picture books, novelty, board books. 16 titles published to date. Unsolicited mss, synopses and ideas for books welcome; send by e-mail but ensure all illustration samples are low-res files (PDF or JPEG). No short stories, novels or non-fiction.

St James Publishing

Earsby Street, London W14 8SH
☎ 020 7348 1799   ℻ 020 7348 1795
stjamespublishing@stjamesschools.co.uk
Managing Editors *Linda Smith, David Smith*

Founded in 1995 to provide teaching materials with spiritual substance for St James Independent Schools: 'Good Books for Fine Minds'. Also many workbooks and textbooks for the study of English and mathematics. No unsolicited mss, synopses or ideas.

Serif

47 Strahan Road, London E3 5DA
☎ 020 8981 3990   ᖴ 020 8981 3990
stephen@serif.books.co.uk
www.serifbooks.co.uk
Managing Editor *Stephen Hayward*

Founded 1993. Publishes cookery, Irish and African studies, travel writing and modern history; no fiction. Ideas and synopses welcome; no unsolicited mss.
£ Royalties paid.

Silver Link

▷ The Nostalgia Collection

Spacelink Books

115 Hollybush Lane, Hampton TW12 2QY
☎ 020 8979 3148
www.spacelink.fsworld.co.uk
Managing Director *Lionel Beer*

Founded 1967. Named after a UFO magazine published in the 1960/70s. Publishes non-fiction titles connected with UFOs, Fortean phenomena and paranormal events. Publishers of *TEMS News* for the Travel and Earth Mysteries Society and *News Brief* for the MWB Railway Society. Distributors of a wide range of related titles and magazines. No unsolicited mss; send synopses and ideas.
£ Royalties/fees according to contract.

Stenlake Publishing Limited

54–58 Mill Square, Catrine KA5 6RD
☎ 01290 552233
enquiries@stenlake.co.uk
www.stenlake.co.uk

Publishes illustrated local history, railways, shipping, aviation and industrial. Acquired **Alloway Publishing Limited** in February 2008 (see entry). 30 titles in 2007. Unsolicited mss, synopses and ideas welcome if accompanied by s.a.e. or sent by e-mail. Freelance writers with experience in above fields also sought for specific commissions.
£ Royalties or fixed fee paid.

Stone Flower Limited

PO Box 1513, Ilford IG1 3QU
renalomen@hotmail.com
Managing Editor *L.G. Norman*

Founded 1989. Publishes humour and general fiction. No submissions. All material commissioned or generated in-house.

Superscript

404 Robin Square, Newtown SY16 1HP
☎ 01588 650452
drjbford@yahoo.co.uk
www.dubsolution.org
Editor *Max Fuller*
Chair *Paul Binding*
Secretary *Ray Pahl*

Founded 2002. Publishes literary, philosophical and political fiction, humanities and social sciences. 13 titles in 2007. Initial enquiries by letter, telephone or e-mail to the editor.
£ Royalties vary with each contract.

Sylph Editions

5 St Swithun's Terrace, Lewes BN7 1UJ
☎ 01273 471706   ᖴ 01273 808247
or@sylpheditions.com
www.sylpheditions.com
Managing Editor *Ornan Rotem*

Founded 2006. Publishes limited-edition art and photography books, monographs, theoretical essays and different forms of experimental writing. 5 titles in 2007. Ideas welcome; approach by e-mail.
£ Royalties twice-yearly.

Tarquin Publications

99 Hatfield Road, St Albans AL1 4JL
☎ 0870 143 2568   ᖴ 0845 456 6385
sales@tarquinbooks.com
www.tarquinbooks.com
Managing Editor *Andrew Griffin*

Founded 1970 as a hobby which gradually grew and now publishes mathematical, cut-out models, teaching and pop-up books. Other topics covered if they involve some kind of paper cutting or pop-up scenes. About 5 titles a year. No unsolicited mss; letter with 1–2 page synopses welcome.
£ Royalties paid.

Tartarus Press

Coverley House, Carlton-in-Coverdale, Leyburn DL8 4AY
☎ 01969 640399   ᖴ 01969 640399
tartarus@pavilion.co.uk
www.tartaruspress.com
Proprietor *Raymond Russell*
Editor *Rosalie Parker*

Founded 1987. Publishes fiction, short stories, reprinted classic supernatural fiction and reference books. About 10 titles a year. 'Please do not send submissions. We cater to a small, collectable market and commission the fiction we publish.'

Tharpa Publications

Conishead Priory, Ulverston LA12 9QQ
☎ 01229 588599   ᖴ 01229 483919

info.uk@tharpa.com
www.tharpa.com
Managing Editor *Neil Elliott*

Founded 1980. Publishes non-fiction, religion, Buddhism and meditation. 2 titles in 2007. No unsolicited material.

Tindal Street Press Ltd

217 The Custard Factory, Gibb Street, Birmingham B9 4AA
☎ 0121 773 8157
alan@tindalstreet.co.uk
www.tindalstreet.co.uk
Publishing Director *Alan Mahar*

Founded in 1998 to publish contemporary original fiction from the English regions. Publishes original fiction only – novels and short story collections. No local history, memoirs or poetry. Published Man Booker 2003 shortlisted *Astonishing Splashes of Colour* and Costa First Novel 2007 *What Was Lost*. 6 titles in 2008. Approach with a letter, synopsis and three chapters, with s.a.e. if return required.
£ Royalties paid.

Tlön Books Publishing Ltd

64 Arthurdon Road, London SE4 1JU
☎ 020 8690 0642   ⊞ 020 8690 0642
mark.reid@tlon.co.uk
www.tlon.co.uk
Managing Editor *Jason Shelley*

Founded 2002. Publishes fiction, art, photography, poetry and children's books. 8 titles in 2006. Unsolicited mss, synopses and ideas welcome; approach by e-mail.
£ Royalties twice-yearly.

Wakefield Historical Publications

19 Pinder's Grove, Wakefield WF1 4AH
☎ 01924 372748
kate@airtime.co.uk
Managing Editor *Kate Taylor*

Founded 1977 by the Wakefield Historical Society to publish well-researched, scholarly works of regional (namely West Riding) historical significance. 1–2 titles a year. Unsolicited mss, synopses and ideas for books welcome.
£ Royalties not paid.

Watling Street Publishing Ltd

33 Hatherop, Nr Cirencester GL7 3NA
☎ 01285 750212
chris.mclaren@saltwaypublishing.co.uk
Managing Director *Chris McLaren*
Editorial Head *Christine Kidney*

Founded 2001. Publishes non-fiction titles on London and its history; children's and adult. No unsolicited mss. Ideas and synopses welcome. Approach in writing or by e-mail with a one-page proposal. 'Not interested in anything not related to London.'

Waywiser Press

The Cottage, 14 Lyncroft Gardens, Ewell KT17 1UR
☎ 020 8393 7055   ⊞ 020 8393 7055
waywiserpress@aol.com
www.waywiser-press.com
Managing Editor *Philip Hoy*
Editorial Advisers *Joseph Harrison, Clive Watkins, Greg Williamson*

Founded 2002. An independent literary press, publishing poetry, fiction and other kinds of literary work by new as well as established writers; recent publications include a compendium volume of seven interviews with American poets, published previously as separate volumes. The press also runs a literary award, the **Anthony Hecht Poetry Prize** (see entry under *Prizes*). Submission details available on the website.
£ Royalties paid.

Whitchurch Books Ltd

67 Merthyr Road, Whitchurch, Cardiff CF14 1DD
☎ 029 2052 1956
whitchurchbooks@btconnect.com
Managing Director *Gale Canvin*

Founded 1994. Publishes local interest, particularly local history. Welcomes unsolicited mss, synopses and ideas on relevant subjects. Initial approach by phone or in writing.
£ Royalties twice-yearly.

Whittles Publishing

Dunbeath Mains Cottages, Dunbeath KW6 6EY
☎ 01593 731333   ⊞ 01593 731400
info@whittlespublishing.com
www.whittlespublishing.com
Publisher *Dr Keith Whittles*

Publisher in geomatics, civil and structural engineering, geotechnics, manufacturing and materials technology and fuel and energy science. Also publishes non-technical books within the following areas: maritime, pharology, military history, landscape and nature writing. Unsolicited mss, synopses and ideas welcome on appropriate themes.
£ Royalties annually.

William Ernest
▷ Radikal Phase Publishing House Ltd

Willow Bank Publishers Ltd

E–Space North, 181 Wisbech Road, Littleport, Ely CB6 1RA
☎ 01353 865405
editorial@willowbankpublishers.co.uk
www.willowbankpublishers.co.uk
Managing Editor *Christopher Sims*

Founded 2005. Publishes all genres, fiction and non-fiction. Unsolicited mss, synopses and ideas for books welcome; send via post in the first instance.
£ Royalties annually.

Witan Books

Cherry Tree House, 8 Nelson Crescent, Cotes Heath, via Stafford ST21 6ST
T 01782 791673
witan@mail.com
Director & Managing Editor *Jeff Kent*

Founded in 1980 for self-publishing and commenced publishing other writers in 1991. Publishes general books, including biography, education, environment, geography, history, politics, popular music and sport. 1 or 2 titles a year. Unsolicited mss, synopses and ideas welcome (include s.a.e.).
£ Royalties paid.

Worple Press

PO Box 328, Tonbridge TN9 1WR
T 01732 368958

theworpleco@aol.com
www.theworplepress.co.uk
Managing Editors *Peter Carpenter, Amanda Knight*

Founded 1997. Independent publisher specializing in poetry, art and alternative titles. Write or phone for catalogue and flyers. 4 titles a year. No unsolicited mss.
£ Royalties paid.

Zymurgy Publishing

Hoults Estate, Walker Road, Newcastle upon Tyne NE6 2HL
T 0191 276 2425   F 0191 276 2425
martin.ellis@ablibris.com
zymurgypublishing.com
Chairman *Martin Ellis*

Founded 2000. Publishes adult non-fiction, ranging from full colour illustrated hardbacks to mass market paperbacks. 4 titles in 2007. Synopses and ideas for books welcome; initial contact by telephone or e-mail. No unsolicited mss.
£ Royalties twice-yearly.

Audio Books

BBC Audiobooks Ltd

St James House, The Square, Lower Bristol Road, Bath BA2 3BH

☎ 01225 800000   ℻ 01225 310771

bbcaudiobooks@bbc.co.uk

www.bbcworldwide.com

Managing Director *Paul Dempsey*

Publishing Director *Jan Paterson*

BBC Audiobooks Ltd was established in 2003 with the integration of BBC Radio Collection, Cover to Cover and Chivers Audio Books. Since then it has become a leading trade and library publisher in the UK, publishing a wide range of entertainment on a variety of formats including CD, cassette, MP3-CD and downloads, podcasts, etc.

DIVISIONS

Trade Titles range from original books and full-cast radio dramas through to abridged readings and full, unabridged recordings. TITLES *War and Peace*; *The Odyssey*; *Captain Corelli's Mandolin*; the complete Sherlock Holmes canon. IMPRINTS **BBC Audio – Children's** From pre-school nursery rhymes to modern classics. TITLES *The Chronicles of Narnia*; *His Dark Materials* trilogy. **BBC Audio – Radio Collection** Comedy classics from TV and radio. TITLES *The Goons*; *Hancock's Half Hour*; *Fawlty Towers*; *The News Quiz*; *Just a Minute*. **BBC Audio** Contemporary comedy. TITLES *The Mighty Boosh*; *Have I Got News For You*; *Little Britain*. **BBC Audio – Lifestyle** A range of titles to help the listener improve or learn. TITLES include the *Glenn Harrold's Ultimate Guide to ...* series. **BBC Audio – Radio 4** TITLES *This Sceptred Isle*; *Letter from America*; *Churchill Remembered*. **Science Fiction & Fantasy** TITLES *The Hitchhiker's Guide to the Galaxy*; *Doctor Who*; *Journey into Space*; *The Lord of the Rings*.

Library IMPRINTS **Chivers Audio Books**; **Chivers Children's Audio Books** Founded 1980. Publishes over 500 titles annually. Complete and unabridged books for adults and children are sold to libraries and to the public by direct mail through **The Audiobook Collection**. CDs and cassettes covering bestselling fiction and popular non-fiction.

Bloomsbury Publishing

▷ entry under UK Publishers

Chivers Audio Books/Chivers Children's Audio Books

▷ BBC Audiobooks Ltd

Chrome Dreams

12 Seaforth Avenue, New Malden KT3 6JP

☎ 020 8715 9781   ℻ 020 8241 1426

mail@chromedreams.co.uk

www.chromedreams.co.uk

Managing Director *Rob Johnstone*

A small record company and publisher founded 1998 to produce audio-biographies of current rock and pop artists and legendary performers on CD and, more recently, books on the same subjects. Ideas for biographies welcome.

Corgi Audio

Transworld Publishers, 61–63 Uxbridge Road, London W5 5SA

☎ 020 8579 2652

www.booksattransworldpublishers.co.uk

Managing Director *Larry Finlay*

Publisher *Bill Scott-Kerr*

A division of the **Random House Group Ltd**. Publishes fiction, autobiography, humour and travel writing on audio. TITLES include Terry Pratchett's *Discworld* series; Bill Bryson's *The Life and Times of the Thunderbolt Kid* and *A Short History of Nearly Everything*; Sophie Kinsella's *Shopaholic* series and *The Lollipop Shoes* by Joanne Harris.

CSA Word

6a Archway Mews, 241a Putney Bridge Road, London SW15 2PE

☎ 020 8871 0220   ℻ 020 8877 0712

info@csaword.co.uk

www.csaword.co.uk

Managing Director *Clive Stanhope*

Audio Director *Victoria Williams*

Editorial and Production Assistant *Rebecca Fenton*

Founded 1989. Publishes fiction, non-fiction, children's, short stories, poetry, travel, biographies, classics. Over 150 titles to-date on CD and available as downloads. Tends to favour quality/classic/nostalgic/timeless literature. Titles include *Carry on Jeeves* P.G. Wodehouse; *Room with a View* E.M. Forster; *Brideshead Revisited* Evelyn Waugh; *William's Happy Days* Richmal Crompton; *Lady Chatterley's Lover* D.H. Lawrence; *Bunter Does His Best* Frank Richards; *The Prince* Nicolo Machiavelli; *Seven Pillars of Wisdom* T.E. Lawrence. Also produces programmes for the BBC and Channel 4 – ideas welcome – see entry under *Film, TV and Radio Producers*. A list of 175 titles.

CYP

The Fairway, Bush Fair, Harlow CM18 6LY
℡ 01279 444707   ℻ 01279 445570
enquiries@cyp.co.uk
www.kidsmusic.co.uk
Operations Director *Mike Kitson*

Founded 1978. Publishes music DVDs and audiobooks for young children; educational, entertainment. TV music and soundtrack production.

57 Productions
▷ entry under Organizations of Interest to Poets

Hachette Digital

100 Victoria Embankment, London EC4Y 0DY
℡ 020 7911 8044   ℻ 020 7911 8100
sarah.shrubb@littlebrown.co.uk
www.littlebrown.co.uk
Publisher *Ursula Mackenzie*
Commissioning Editor *Sarah Shrubb*

Launched in 2003 with titles from bestselling authors Alexander McCall Smith and Mitch Albom. Publishes fiction, humour, poetry and non-fiction. TITLES include new fiction from authors such as Patricia Cornwell and Sarah Waters, non-fiction such as Sharon Osbourne's bestselling autobiography *Extreme*, and unabridged classics such as Patricia Highsmith's *The Talented Mr Ripley* and Joseph Heller's *Catch 22*. A list of 141 titles.

HarperCollins AudioBooks

77–85 Fulham Palace Road, London W6 8JB
℡ 020 8741 7070
www.harpercollins.co.uk
Director of Audio and E-books *David Roth-Ey*
Senior Editor *Nicola Townsend*

The HarperCollins audio list was launched in 1990. Publishes a wide range including popular and classic fiction, non-fiction, children's,

Shakespeare, poetry and self help (Thorsons' imprint) on cassette, CD and digital download. AUTHORS Agatha Christie, Ian McEwan, C.S. Lewis, Roald Dahl, Dr Seuss, Lemony Snicket, J.R.R. Tolkien, Bernard Cornwell, Enid Blyton, Nick Butterworth, Judith Kerr.

Hodder & Stoughton Audiobooks

338 Euston Road, London NW1 3BH
℡ 020 7873 6000   ℻ 020 7873 6194
rupert.lancaster@hodder.co.uk
www.hodder.co.uk
Publisher *Rupert Lancaster*

Launched in 1994 with the aim of publishing outstanding authors from within the Hodder group and commissioning independent audio titles. Publishes fiction and non-fiction. AUTHORS include Stephen King, John LeCarré, Alan Titchmarsh, Elizabeth George, Joanne Harris, Haruki Murakami, Jodi Picoult, Pam Ayres, David Mitchell, Linda Smith, John Humphries, Al Murray, Dickie Bird, Peter Robinson and Charles Frazier.

Ladybird Audio
▷ Dorling Kindersley Ltd under UK Publishers

Laughing Stock Productions

81 Charlotte Street, London W1T 4PP
℡ 020 7637 7943   ℻ 020 7436 1666
Managing Director *Michael O'Brien*

Founded 1991. Issues a wide range of comedy cassettes/CDs from family humour to alternative comedy. TITLES *Red Dwarf*; *Shirley Valentine* (read by Willy Russell); *Rory Bremner*; *Peter Cook Anthology*; *Sean Hughes*; *John Bird and John Fortune*; *Eddie Izzard*. 12–16 titles a year.

Macmillan Digital Audio

20 New Wharf Road, London N1 9RR
℡ 020 7014 6040   ℻ 020 7014 6141
a.muirden@macmillan.co.uk
www.panmacmillan.co.uk
Owner *Macmillan Publishers Ltd*
Audio Publisher *Alison Muirden*

Founded 1995. Publishes adult fiction, non-fiction and autobiography, focusing mainly on lead book titles and releasing audio simultaneously with hard or paperback publication. Also publishes children's audio titles. Won Audio Publisher of the Year at the 2003 Spoken Word Awards and has won many other Spoken Word Awards for individual audio titles in previous years About 40–60 titles a year. Submissions not accepted.

Naxos AudioBooks

40a High Street, Welwyn AL6 9EQ
℡ 01438 717808   ℻ 01438 717809
naxos_audiobooks@compuserve.com
www.naxosaudiobooks.com

Owner *HNH International, Hong Kong/ Nicolas Soames*
Managing Director *Nicolas Soames*

Founded 1994. Part of Naxos, the classical budget CD company. Publisher of the Year in the 2001 Spoken Word Awards. Publishes classic and modern fiction, non-fiction, children's and junior classics, drama and poetry. TITLES *Ulysses* Joyce; *King Lear* Shakespeare; *History of the Musical* Fawkes; *Just So Stories* Kipling. A list of 400 titles.

Oakhill Publishing Ltd

PO Box 3855, Bath BA1 3WW
☎ 01225 874355   🖷 01225 874442
info@oakhillpublishing.com
Managing Director *Julian Batson*

Founded in 2005 by Julian Batson, formerly with Chivers Press (now trading as BBC Audiobooks) to offer a wider selection of titles for the library market. Produces complete and unabridged audiobooks of popular fiction and non-fiction for adults and children. TITLES *The Gathering* Anne Enright; *The Visible World* Mark Slouka; *The Sixth Wife* Suzannah Dunn; *Dirty Bertie* David Roberts and Alan Macdonald; *Free Lance* Paul Stewart and Chris Riddell; *My Swordhand is Singing* Marcus Sedgwick. 99 titles in 2007. Suggestions for titles from literary agents only.

Orion Audio Books (Division of the Orion Publishing Group Ltd)

Orion House, 5 Upper St Martin's Lane, London WC2H 9EA
☎ 020 7520 4425   🖷 020 7240 4822
pandora.white@orionbooks.co.uk
www.orionbooks.co.uk
Publisher *Pandora White*

Orion Audio has released over 500 titles since it was founded in 1998. It publishes fiction, non-fiction, humour, autobiographies, children's, poetry, science, crime and thrillers. AUTHORS include Maeve Binchy, Ian Rankin, Robert Crais, Michael Connelly, Francesca Simon (*Horrid Henry* series), Harlan Coben, Dan Brown, Sally Gardner, Michelle Paver, Meg Cabot and Erica James. In 2006, Orion expanded into the digital download market via www.audible.co.uk

Penguin Audiobooks

80 Strand, London WC2R 0RL
☎ 020 7010 3000   🖷 020 7010 6060
audio@penguin.co.uk
www.penguin.co.uk
Head of Audio Publishing *Jeremy Ettinghausen*

Launched in November 1993 and has rapidly expanded since then to reflect the diversity of Penguin Books' list. Publishes mostly fiction, both classical and contemporary, non-fiction, autobiography and an increasing range of digital audiobooks as well as children's titles under the **Puffin Audiobooks** imprint. Contemporary AUTHORS include: Zadie Smith, Eoin Colfer, John Mortimer, Roald Dahl, Nicci French, Gervase Phinn, Charlie Higson, Marina Lewycka, Niall Fergusson, Claire Tomalin. About 30 titles a year.

Puffin Audiobooks
▷ Penguin Audiobooks

Quercus Audiobooks
▷ Quercus Publishing under UK Publishers

Random House Audio Books

20 Vauxhall Bridge Road, London SW1V 2SA
☎ 020 7840 8519   🖷 020 7931 7672
Owner *The Random House Group Ltd.*
Commissioning Editor *Zoe Howes*

The audiobooks division of Random House started early in 1991. Acquired the Reed Audio list in 1997. Publishes fiction and non-fiction. AUTHORS include Monica Ali, Louis de Bernières, Sebastian Faulks, John Grisham, Mark Haddon, Robert Harris, Thomas Harris, Tony Hawks, Andy McNab, Ruth Rendell, Kathy Reichs, Chris Ryan.

Rickshaw Productions Limited

Suite 125, 99 Warwick Street, Leamington Spa CV32 4RB
☎ 0780 3553214   🖷 01926 402490
info@thelisteningzone.com
www.thelisteningzone.com
Commissioning Editor *Ms L.J. Fairgrieve*

Founded in 1998 as audiobook publishers, Rickshaw Productions Limited are now concentrating solely on their audiobook/music downloading website (www.thelisteningzone. com). Apart from their mainstream suppliers, the website also includes work by established and non-established writers and musicians, together with features such as *BenchTalk* and the *Speaker Zone*. Uniquely, their audiobook and music titles are acquired from publishers and individuals from around the world. Available for download are full-length audiobooks, short stories, poetry, non-fiction, self-help, lectures, interviews, new music titles, read/performed in Dutch, English, French, German, Italian and Spanish, with more languages being added at a later date. Enquiries by e-mail or telephone. Submissions by snail-mail *only*; salacious material will not be accepted. Two sample chapters, one-page synopsis and s.a.e. to cover return by Recorded Delivery.

L.M. Roberts Limited

5 Townwalk, Chesterfield S41 7TZ
☎ 01246 204222
enquiries@lmroberts.com
www.lmroberts.com

Managing Director *Lee Michael Roberts*

Founded 2007. Specializes in audiobooks at budget prices. An expanding selection of children's, drama and general fiction (Austen, Shakespeare, Conan Doyle, Conrad, Wilde), enhanced with specially selected music and sound effects. Planned expansion of distribution in both the UK and worldwide. Welcomes ideas for audiobooks.

Shakespeare Appreciated
▷ SmartPass Ltd

Simon & Schuster Audio

Africa House, 64–78 Kingsway, London
WC2B 6AH

☎ 020 7316 1900   ⨏ 020 7316 0332
info@simonandschuster.co.uk
www.simonsays.co.uk
Audio Manager *Rhedd Lewis*

Simon & Schuster Audio began by distributing their American parent company's audio products. Moved on to repackaging products specifically for the UK market and in 1994 became more firmly established in this market with a huge rise in turnover. Publishes adult fiction, self help and business titles. TITLES *Above Suspicion* Lynda la Plante; *Rosie* Alan Titchmarsh; *The 7 Habits of Highly Effective People* Stephen R. Covey; *Chronicles* Bob Dylan; *Angels and Demons* and *Deception Point* Dan Brown; *Kate Remembered* A. Scott Berg; *The Kite Runner* and *A Thousand Splendid Suns* Khaled Hosseini; *The Secret* Rhonda Byrne.

SmartPass Ltd

15 Park Road, Rottingdean, Brighton BN2 7HL
☎ 01273 300742
info@smartpass.co.uk
www.smartpass.co.uk
www.shakespeareappreciated.com
www.spaudiobooks.com

Managing Director *Phil Viner*
Creative Director *Jools Viner*

Founded 1999. SmartPass produces audio education study resources, full-cast drama with commentary analysis and study strategies for English Literature set texts – novels, plays and poetry. Available in student and teacher formats, including audio linked study materials. TITLES include *A Kestrel for a Knave*; *Animal Farm*; *Pride and Prejudice*; *Great Expectations*; *The Mayor of Casterbridge*; *Shakespeare: the works*; *War Poetry*; *An Inspector Calls*; *Twelfth Night*; *Henry V*; *Romeo and Juliet*. IMPRINTS **Shakespeare Appreciated** Full-cast unabridged dramas with commentary explaining who's who and what's going on plus historical insights and background information; created for adults rather than students, for enjoyment rather than education. TITLES *Twelfth Night*; *Othello*; *King Lear*. **SPAudiobooks** Full-cast unabridged dramas of classic and cult texts TITLES *The Antipope* Robert Rankin; *Macbeth*; *Henry V*; *Romeo and Juliet*. Welcomes ideas from authors who are teachers and from agents who represent the authors of studied texts.

Soundings Audio Books

Isis House, Kings Drive, Whitley Bay NE26 2JT
☎ 0191 253 4155   ⨏ 0191 251 0662
www.isis-publishing.co.uk

Founded in 1982. Part of the Ulverscroft Group Ltd. Together with **Isis Publishing** publishes fiction and non-fiction; crime, romance. AUTHORS include Lyn Andrews, Rita Bradshaw, Lee Child, Catherine Cookson, Alexander Fullerton, Joanne Harris, Anna Jacobs, Robert Ludlum, Patrick O'Brian, Pamela Oldfield, Susan Sallis, Judith Saxton, Mary Jane Staples, Sally Worboyes. About 190 titles a year. Many also available on CD.

SPAudiobooks
▷ SmartPass Ltd

UK Packagers

Aladdin Books Ltd

2/3 Fitroy Mews, London W1T 6DF

☏ 020 7383 2084   🖷 020 7388 6391

sales@aladdinbooks.co.uk

www.aladdinbooks.co.uk

Managing Director *Charles Nicholas*

Founded in 1979 as a packaging company but with joint publishing ventures in the UK and USA. Commissions children's fully illustrated, non-fiction reference books. IMPRINTS **Aladdin Books** *Bibby Whittaker* Children's reference; **Nicholas Enterprises** *Charles Nicholas* Adult non-fiction; **The Learning Factory** *Charles Nicholas* Early learning concepts 0–4 years. TITLES *World Issues*; *Science Readers*; *The Atlas of Animals*. About 40 titles a year. Will consider synopses and ideas for children's non-fiction with international sales potential only. No fiction.
🖸 Fees usually paid instead of royalties.

The Albion Press Ltd

Spring Hill, Idbury OX7 6RU

☏ 01993 831094

Chairman/Managing Director *Emma Bradford*

Founded 1984. Commissions illustrated trade titles, particularly children's. TITLES *The Great Circle: A History of the First Nations* Neil Philip; *Fairy Tales of Hans Christian Andersen* Isabelle Brent. About 2 titles a year. Unsolicited synopses and ideas for books not welcome.
🖸 Royalties paid; fees paid for introductions and partial contributions.

Amber Books Ltd

Bradleys Close, 74–77 White Lion Street, London N1 9PF

☏ 020 7520 7600   🖷 020 7520 7606/7

enquiries@amberbooks.co.uk

www.amberbooks.co.uk

Managing Director *Stasz Gnych*

Rights Director *Sara Ballard*

Publishing Manager *Charles Catton*

Founded 1989. Commissions history, encyclopedias, military, aviation, transport, sport, combat, survival and fitness, cookery, lifestyle, naval history, crime, childrens, schools and library sets and general reference. No fiction, poetry or biography. 80 titles in 2007. No unsolicited material outside subject areas listed above.
🖸 Fees paid.

BCS Publishing Ltd

2nd Floor, Temple Court, 109 Oxford Road, Cowley OX4 2ER

☏ 01865 770099   🖷 01865 770050

bcs-publishing@dsl.pipex.com

Managing Director *Steve McCurdy*

Approx. Annual Turnover £150,000

Commissions general interest non-fiction for the international co-edition market.

Bender Richardson White

PO Box 266, Uxbridge UB9 5BD

☏ 01895 832444   🖷 01895 835213

brw@brw.co.uk

Partners *Lionel Bender, Kim Richardson, Ben White*

Founded 1990 to produce illustrated non-fiction for children, adults and family reference for publishers in the UK and USA. 60–70 titles a year. Unsolicited material not welcome.
🖸 Fees paid.

Book House

▷ **Salariya Book Company Ltd**

Breslich & Foss Ltd

Unit 2A, Union Court, 20–22 Union Road, Clapham, London SW4 6JP

☏ 020 7819 3990   🖷 020 7819 3998

sales@breslichfoss.com

Directors *Paula Breslich, K.B. Dunning*

Approx. Annual Turnover £650,000

Packagers of adult non-fiction titles, including interior design, crafts, gardening, health and children's non-fiction and picturebooks. Unsolicited mss welcome but synopses preferred. Include s.a.e. with all submissions.
🖸 Royalties paid twice-yearly.

Brown Wells and Jacobs Ltd

Forresters Hall, 25–27 Westow Street, London SE19 3RY

☏ 020 8771 5115   🖷 020 8771 9994

graham@bwj-ltd.com
www.bwj.org
Managing Director *Graham Brown*

Founded 1979. Commissions non-fiction, novelty, pre-school and first readers, natural history and science. About 40 titles a year. Unsolicited synopses and ideas for books welcome.
Ⓔ Fees paid.

Cameron & Hollis

PO Box 1, Moffat DG10 9SU
Ⓣ 01683 220808   Ⓕ 01683 220012
info@cameronbooks.co.uk
www.cameronbooks.co.uk
Directors *Ian A. Cameron, Jill Hollis*
Approx. Annual Turnover £400,000

Commissions contemporary art, design, collectors' reference, decorative arts, architecture, children's non-fiction and film (serious critical works only). About 6 titles a year.
Ⓔ Payment varies with each contract.

Compendium Publishing Ltd
▷ entry under UK Publishers

Diagram Visual Information Ltd

34 Elaine Grove, London NW5 4QH
Ⓣ 020 7485 5941   Ⓕ 020 7485 5941
brucerobertson@diagramgroup.com
Managing Director *Bruce Robertson*

Founded 1967. Producer of library, school, academic and trade reference books. About 10 titles a year.
Ⓔ Fees paid; no payment for sample material/submissions for consideration.

Eddison Sadd Editions

St Chad's House, 148 King's Cross Road, London WC1X 9DH
Ⓣ 020 7837 1968   Ⓕ 020 7837 2025
info@eddisonsadd.com
www.eddisonsadd.com
Managing Director *Nick Eddison*
Editorial Director *Ian Jackson*
Approx. Annual Turnover £2.5 million

Founded 1982. Produces a wide range of popular illustrated non-fiction with books and interactive kits published in 30 countries. Many titles suit gift markets. Incorporates Bookinabox Ltd. Ideas and synopses, rather than mss, are welcome but titles must have international appeal.
Ⓔ Royalties paid twice-yearly; fees paid when appropriate.

Elwin Street Productions

144 Liverpool Road, London N1 1LA
Ⓣ 020 7700 6785   Ⓕ 020 7700 6031
info@elwinstreet.com
www.elwinstreet.com
Managing Director *Silvia Langford*

Editorial Manager *Erin King*
Approx. Annual Turnover £1 million

Founded 2003. Produces illustrated non-fiction in quirky humour, gift, reference and practical life including self-help and parenting. 22 titles in 2007. Synopses and ideas for books welcome; approach by post only with s.a.e. No cookery or gardening. All books must have an international angle and not of UK interest only.
Ⓔ Fees paid.

Erskine Press

The White House, Sandfield Lane, Eccles, Norwich NR16 2PB
Ⓣ 01953 887277   Ⓕ 01953 888361
erskpres@aol.com
www.erskine-press.com
Chief Executive *Crispin de Boos*
Approx. Annual Turnover £65,000

Specialist publisher of books on Antarctic exploration – facsimiles, diaries, previously unpublished works and first English translations of expeditions of the late 19th and early 20th centuries. Publisher of general interest biographies with special reference to Norfolk. Recent publications include a number of Second World War related subjects. Also produces scholarly reprints and limited edition publications for academic/business organizations ranging from period print reproductions to facsimiles of rare and important books no longer available. 6 titles in 2007. No unsolicited mss. Ideas welcome.
Ⓔ Royalties paid twice-yearly.

Expert Publications Ltd

Sloe House, Halstead CO9 1PA
Ⓣ 01787 474744   Ⓕ 01787 474700
expert@lineone.net
Chairman *Dr. D.G. Hessayon*

Founded 1993. Produces the *Expert* series of books by Dr. D.G. Hessayon. Currently 24 titles in the series, including *The Flower Expert; The Evergreen Expert; The Vegetable & Herb Expert; The Flowering Shrub Expert; The Container Expert; The Orchid Expert*. No unsolicited material.

Haldane Mason Ltd

PO Box 34196, London NW10 3YB
Ⓣ 020 8459 2131   Ⓕ 020 8728 1216
info@haldanemason.com
www.haldanemason.com
Editorial Director *Sydney Francis*
Art Director *Ron Samuel*

Founded 1995. Commissions mainly children's illustrated non-fiction. Children's books are published under the **Red Kite Books** imprint. Adult list consists mainly of mind, body and spirit plus alternative health books under the

Neal's Yard Remedies banner; children's age range 0–15. Unsolicited synopses and ideas welcome but approach by phone or e-mail first. No adult fiction.
€ Fees paid.

The Ilex Press Limited

The Old Candlemakers, West Street, Lewes BN7 2NZ
☎ 01273 487440   ⊞ 01273 487441
surname@ilex-press.com
www.ilex-press.com
Managing Director *Stephen Paul*
Rights Director *Roly Allen*

Founded 1999. Sister company of **The Ivy Press Limited**. Commissions titles on digital art, design and photography as well as on all aspects of website design and graphics software. Also publishing reference works on popular culture, comics, etc. No fiction. Unsolicited synopses and ideas welcome; send a *brief* idea outline (3 or 4 pages) and a letter.
€ Fees paid.

The Ivy Press Limited

The Old Candlemakers, West Street, Lewes BN7 2NZ
☎ 01273 487440   ⊞ 01273 487441
surname@ivy-group.co.uk
www.ivy-press.com
Managing Director *Stephen Paul*
Publisher *Jason Hook*

Founded 1996. Sister company of **The Ilex Press Limited**. Commissions illustrated non-fiction books covering subjects such as art, lifestyle, popular culture, health, self-help and humour. No fiction. Unsolicited synopses and ideas welcome; send a *brief* idea outline (3 or 4 pages) and a letter.
€ Fees paid.

The Learning Factory
▷ **Aladdin Books Ltd**

Lennard Associates Ltd

Windmill Cottage, Mackerye End, Harpenden AL5 5DR
☎ 01582 715866   ⊞ 01582 715866
stephenson@lennardqap.co.uk
Chairman/Managing Director *Adrian Stephenson*

Founded 1979. Now operating as a packager/production company only with no trade distribution. 2 titles in 2007. No unsolicited mss.
€ Both fees and royalties by arrangement.

Lexus Ltd

60 Brook Street, Glasgow G40 2AB
☎ 0141 556 0440   ⊞ 0141 556 2202
peterterrell@lexusforlanguages.co.uk
www.lexusforlanguages.co.uk
Managing/Editorial Director *P.M. Terrell*

Founded 1980. Compiles bilingual reference, language and phrase books. TITLES Rough Guide phrasebooks; Langenscheidt dictionaries; *HarperCollins English-Chinese*; *Oxford Italian Pocket Dictionary*. Own series of *Travelmates* published in 2004; launched a new beginner's course for learning Chinese, *The Chinese Classroom*, in 2007. About 5 titles a year. Unsolicited material considered although books are mostly commissioned. Freelance contributors employed for a wide range of languages.
€ Generally flat fee.

Lionheart Books

10 Chelmsford Square, London NW10 3AR
☎ 020 8459 0453   ⊞ 020 8451 3681
lionheart.brw@btinternet.com
Senior Partner *Lionel Bender*
Partner *Madeleine Samuel*

A design/editorial packaging team. Titles are primarily commissioned from publishers. Highly illustrated non-fiction for children aged 8–14, mostly natural history, history and general science. See also **Bender Richardson White**. About 20 titles a year.
€ Generally flat fee.

M&M Publishing Services

33 Warner Road, Ware SG12 9JL
☎ 01920 466003
mikemoran@moran01.wanadoo,co.uk
www.photography-london.co.uk
Proprietors *Mike Moran, Maggie Copeland*

Project managers, production and editorial, packagers and publishers. TITLES *MM Publisher Database*; *MM Printer Database* (available in UK, European and international editions).

Market House Books Ltd

Suite B, Elsinore House, 43 Buckingham Street, Aylesbury HP20 2NQ
☎ 01296 484911   ⊞ 01296 437073
books@mhbref.com
www.markethousebooks.com
Directors *John Daintith, Peter Sapsed, Anne Kerr*

Founded 1970. Formerly Laurence Urdang Associates. Compiles dictionaries, encyclopedias and reference. TITLES *Collins English Dictionary*; Facts on File dictionaries; *Grolier Bibliographical Encyclopedia of Scientists* (10 vols); *Larousse Thematica* (6 vols); *The Macmillan Encyclopedia*; Macmillan Dictionaries; Oxford Dictionaries of: *Business, Science, Medicine*, etc.; *Penguin Dictionary of Electronics*; *Penguin Rhyming Dictionary, etc.* About 15 titles a year. Unsolicited material not welcome as most books are compiled in-house.
€ Fees paid.

Monkey Puzzle Media Ltd

Gissing's Farm, Fressingfield IP21 5SH

☎ 01379 588044   📠 01379 588055

info@monkeypuzzlemedia.com

Chairman/Managing Director *Roger Goddard-Coote*

Commissioning and List Development *Paul Mason*

Founded 1998. Packager of adult and children's non-fiction for trade, school, library and mass markets. No fiction or textbooks. About 80 titles a year. Synopses and ideas welcome. Include s.a.e. for return.

💷 Fees paid; no royalties.

Nicholas Enterprises

▷ Aladdin Books Ltd

Orpheus Books Limited

6 Church Green, Witney OX28 4AW

☎ 01993 774949   📠 01993 700330

info@orpheusbooks.com

www.orpheusbooks.com

Directors *Nicholas Harris, Sarah Hartley*

Founded 1993. Commissions children's non-fiction. 20 titles in 2007. No unsolicited material.

💷 Fees paid.

Playne Books Limited

Park Court Barn, Trefin, Haverfordwest SA62 5AU

☎ 01348 837073   📠 01348 837063

playne.books@virgin.net

Editorial Director *Gill Davies*

Design & Production *David Playne*

Founded 1987. Commissions early learning titles for young children – fun ideas with an educational slant and novelty books. Also highly illustrated and practical books on any subject. Synopses and ideas by prior arrangement only. Vanity publications welcomed.

💷 Royalties paid on payment from publishers. Fees sometimes paid instead of royalties.

Mathew Price Ltd

The Old Glove Factory, Bristol Road, Sherborne DT9 4HP

☎ 01935 816010   📠 01935 816310

mathewp@mathewprice.com

www.mathewprice.com

Chairman/Managing Director *Mathew Price*

Approx. Annual Turnover £500,000

Founded 1983. Commissions full-colour novelty and picture books and fiction for young children plus children's non-fiction for all ages. TITLES *Magnificent Mazes; Creepies; Tractor Factory; Join-In Stories for the Very Young; Treasure Hunt.* Mss should be double-spaced, typed on one side of paper. Enclose s.a.e. with submission and keep a copy of everything that is sent.

💷 Fees sometimes paid instead of royalties.

Quarto Publishing plc

The Old Brewery, 6 Blundell Street, London N7 9BH

☎ 020 7700 6700   📠 020 7700 4191

www.quarto.com

Chairman & CEO *Laurence Orbach*

Director of Co-edition Publishing *Piers Spence*

Founded 1976. Britain's largest book packager. Acquired Marshall Editions in 2002. Commissions illustrated non-fiction, including painting, graphic design, how-to, lifestyle, visual arts, history, cookery, gardening, crafts. Publishes under the Apple imprint. Unsolicited synopses/ideas for books welcome.

💷 Flat fees paid.

Red Kite Books

▷ Haldane Mason Ltd

Regency House Publishing Limited

▷ entry under UK Publishers

Salariya Book Company Ltd

25 Marlborough Place, Brighton BN1 1UB

☎ 01273 603306   📠 01273 693857

salariya@salariya.com

www.salariya.com

www.book-house.co.uk

www.scribblersbook.com

Managing Director *David Salariya*

Editor *Steven Hayes*

Art Director *Carolyn Franklin*

Founded 1989. Books for children: history, art, music, science, architecture, education, nature, environment, fantasy, folktales, multicultural, fiction, young readers and picture books. IMPRINTS **Book House; Scribblers** Highly illustrated books in all subjects for babies and children, from pre-school board books to teenage graphic novels. For fiction, submit complete c.v., ms for picture books; outline/synopsis and one sample story or chapter for collections, novels or graphic novels. Response in four months if s.a.e. included. Illustrations: no original artwork. Send c.v., promotion sheet to be kept on file. Samples returned if s.a.e. included.

💷 Payment by arrangement.

Savitri Books Ltd

25 Lisle Lane, Ely CB7 4AS

☎ 01353 654327   📠 01353 654327

munni@savitribooks.demon.co.uk

Managing Director *Mrinalini S. Srivastava*

Approx. Annual Turnover £200,000

Founded 1983 and since 1998, Savitri Books has also become a publisher in its own right (textile crafts). Keen to work 'very closely with authors/illustrators and try to establish long-term relationships with them, doing more books with the same team of people'. Commissions illustrated non-fiction: biography, history, travel. About 7 titles a year. Unsolicited synopses and ideas for books 'very welcome'.

Scribblers
▷ Salariya Book Company Ltd

Stonecastle Graphics Ltd

Highlands Lodge, Chartway Street, Sutton Valence ME17 3HZ
☎ 01622 844414
paul@stonecastle-graphics.com
sue@stonecastle-graphics.com
www.stonecastle-graphics.com
Partner *Paul Turner*
Editorial Head *Sue Pressley*
Approx. Annual Turnover £300,000

Founded 1976. Formed additional design/ packaging partnership, **Touchstone**, in 1983 (see entry). Commissions illustrated non-fiction general books: motoring, health, sport, leisure, lifestyle and children's non-fiction. TITLES *Spirit of the Jungle*; *Skyscrapers*; *Snakes*; *Baby Animals*; *Learn to Play the Piano*; *Learn to Dance*. 20 titles in 2007. Unsolicited synopses and ideas for books welcome
₤ Fees paid.

Templar Publishing
▷ entry under UK Publishers

Toucan Books Ltd

Third Floor, 89 Charterhouse Street, London EC1M 6HR
☎ 020 7250 3388   ₣ 020 7250 3123
Managing Director *Ellen Dupont*

Founded 1985. Specializes in international co-editions and fee-based editorial, design and production services. Commissions illustrated non-fiction for children and adults. No fiction or non-illustrated titles. About 15 titles a year.
₤ Royalties paid twice-yearly; fees paid in addition to or instead of royalties.

Touchstone Books Ltd

Highlands Lodge, Chartway Street, Sutton Valence ME17 3HZ
☎ 01622 844414   ₣ 01622 844414

paul@touchstone-books.net
sue@touchstone-books.net
www.touchstone-books.net
Directors *Paul Turner, Sue Pressley, Nick Wigley, Trevor Legate*

Founded in 2006, Touchstone Books Ltd is an independent publisher of lavishly-illustrated books on motoring, art and general-interest subjects. TITLES *100 Years of Brooklands: The Birthplace of British Motorsport and Aviation*; *Fighting Force: The 90th Anniversary of the Royal Air Force* (officially endorsed by the RAF); *Harley-Davidson: Lifestyle*. 6 titles in 2007. Unsolicited synopses and ideas for books welcome.
₤ Fees paid.

David West Children's Books

7 Princeton Court, 55 Felsham Road, London SW15 1AZ
☎ 020 8780 3836   ₣ 020 8780 9313
dww@btinternet.com
www.davidwestchildrensbooks.com

Founded 1992. Commissions children's illustrated reference books. No fiction or adult books. 140 titles in 2006. Unsolicited ideas and synopses welcome; approach in writing in the first instance
₤ Fees/royalties paid annually.

Working Partners Ltd

Stanley House, St Chad's Place, London WC1X 9HH
☎ 020 7841 3939   ₣ 020 7841 3940
enquiries@workingpartnersltd.co.uk
www.workingpartnersltd.co.uk
Chairman *Ben Baglio*
Managing Director *Chris Snowdon*
Contact *James Noble*

Specializes in quality mass-market fiction for leading children's publishers including Bloomsbury, HarperCollins, Hodder, Macmillan, Orchard, OUP, Random House and Scholastic. First chapter books to young adult. No picture books. **Working Partners Two** handles adult fiction across all popular genres on a similar basis. Welcomes approaches from interested writers.
₤ Both fees and royalties by arrangement.

Electronic Publishing and Other Services

www.ABCtales.com

Unit 3N, Leroy House, 436 Essex Road, London
N1 3QP
☎ 020 7682 1865   🖷 020 7682 1867
tcook@abctales.com
www.abctales.com
Owner *Burgeon Creative Ideas Ltd*
Editor *Tony Cook*

Founded by A. John Bird, MBE, co-founder of
the *Big Issue* magazine, Tony Cook and Gordon
Roddick. ABCtales is a free website dedicated to
publishing and developing new writing. Content
is predominantly short stories, autobiography and
poetry. Anyone can upload creative writing to the
website for free.

Arbour E-Books

2 Clonmel Road, Stirchley, Birmingham B30 2BU
☎ 0121 459 5659
mike@arbour-ebooks.com
www.arbour-ebooks.com
Owner/Editor *Mike Hill*

Electronic publisher. 'A small fee is charged for
electronic submission checking by a specialist.
If accepted, the e-book is produced, published
and sold for a negotiated percentage of sales.'
Short stories, serials or poems of social comedy/
drama preferred. A negotiable fee is applicable to
commercial artists and companies for catalogues
and marketing e-books published. Some
website work may be undertaken for all clients.
Submissions must be Microsoft Word compatible,
English and to a high standard of spelling and
casual grammar. Paper submissions not accepted.

Authors OnLine

▶ entry under UK Publishers

BeWrite Books

32 Bryn Road South, Wigan WN4 8QR
contact@bewrite.net
www.bewrite.net
Managing Director *Cait Myers*
Editorial Director *Neil Marr*

Unsolicited mss and synopses welcome (using
the online form on the website www.bewrite.
net/bookshop/submission_requirements.htm).

'All offers promptly acknowledged and draft
manuscript-to-publication time shorter than most
other publishing houses.' About 20 titles a year.
£ Royalties quarterly (no advance).

Books 4 Publishing

Drovers House, The Auction Yard, Craven Arms
SY7 9BZ
☎ 0870 777 3339   🖷 01547 530845
editor@books4publishing.com
www.books4publishing.com
Owner *Corvedale Media Ltd*
Managing Director *Mark Oliver*

Founded 2000. Specializes in showcasing
unpublished books and helping new authors gain
recognition for their work by displaying synopses
and up to 5,000 words on the Books 4 Publishing
website and promoting titles to publishers and
literary agents. Mss, synopses and ideas welcome;
initial enquiries by e-mail, post or submission
form on website.

Crime Scene Scotland

crimescenescotland@yahoo.co.uk
www.crimescenescotland.com/
Editor *Russel D. McLean*

News, reviews and 'a celebration of the darker
aspects of crime fiction'. Interviews and
commentary on the changing world of the genre.
Comment, reviews or articles should be e-mailed
for the attention of Russel E. McLean with
'CRIME SCENE SCOTLAND' in the header.

e-book-downloads-uploads.com

63 Lulworth Avenue, Preston PR2 2BE
☎ 07722 209009
editor@e-book-downloads-uploads.com
info@e-book-downloads-uploads.com
www.e-book-downloads-uploads.com
Contact *H. Pritchard*

A new e-book publisher, established in 2006,
'willing to give opportunities to writers who by
no fault of their own cannot get their work to
market'. Will consider all submissions and all
categories apart from poetry. Approach by e-mail
with short synopsis attached.
£ Royalties twice-yearly.

Fern HousE-Publishing

19 High Street, Haddenham, Ely CB6 3XA
☎ 01353 740222
epub@fernhouse.com
www.fernhouse.com
Managing Editor *Rodney Dale*

Founded 1995. Has published mainly non-fiction (13 titles in print) but now entering the field of e-publishing. Open to submissions – 'but only of the highest quality'.

50connect – www.50connect.co.uk

Claremont House, 70–72 Alma Road, Windsor SL4 3EZ
☎ 01753 850606   🖷 0800 634 2250
admin@50connect.com
www.50connect.co.uk
Editor *Rachael Hannan*
(rhannan@50connect.com)

Launched in 2000, and awarded 'esuperbrand status' in 2006, www.50connect.co.uk is a leading editorial-led lifestyle website for today's over-45s. Covers finance, health, travel, overseas retirement, living abroad, retirement, food, wine, theatre, music, gardening, pets and much more. Also strong community element including forums and a chat room. Submissions welcome but please contact via e-mail in the first instance with some previous examples of work published.

Fledgling Press

7 Lennox Street, Edinburgh EH4 1QB
☎ 0131 343 2367
info@fledglingpress.co.uk
www.fledglingpress.co.uk
Director *Zander Wedderburn*

Founded 2000. Internet publisher, aiming to be a launching pad for new authors. Special interest in authentic writing about the human condition, including autobiography, diaries, poetry and fictionalized variations on these. Monthly online competition (www.canyouwrite.com) with small prizes for short pieces. Also free books and reports in the areas of shiftwork and working time. Send mss and other details by e-mail from the website or by post. Links into short-run book production. 20 titles to date.

Justis Publishing Ltd

Grand Union House, 20 Kentish Town Road, London NW1 9NR
☎ 020 7267 8989   🖷 020 7267 1133
enquiries@justis.com
www.justis.com

Founded 1986. Electronic publisher of UK and European legal and offical information on CD-ROM, online and the Internet.

Lulu
▷ entry under UK Publishers

New Authors Showcase
▷ Barrie James Literary Agency under UK Literary Agents

Notes from the Underground
▷ entry under Magazines

Online Originals

Priory Cottage, Wordsworth Place, London NW5 4HG
☎ 020 7267 4244
editor@onlineoriginals.com
www.onlineoriginals.com
Managing Director *David Gettman*
Technical Director *Neill Sanders*

Publishes book-length works on the Internet and as print on demand or short run library editions. Acquires global electronic rights in literary fiction, intellectual non-fiction, drama, and youth fiction (ages 8–16). No poetry, fantasy, how-to, self-help, picture books, cookery, hobbies, crafts or local interest. TITLES *Being and Becoming* Christopher Macann (4 vols.); *The Jealous God* Martin Boyland; *Lard and Speck* Andrew Cogan; *Quintet* Frederick Forsyth; *The Angels of Russia* Patricia Leroy. 4 titles in 2007. Unsolicited mss, synopses and proposals for books welcome. *All* authors must have Internet access. Unique peer-review, automated submissions system, accessed via the website address above. No submissions on paper or disk.
💷 Royalties paid annually (50% royalties on e-book price of £6).

WritersServices
▷ entry under Useful Websites

UK Literary Agents

* = Members of the **Association of Authors' Agents**

Sheila Ableman Literary Agency*

48–56 Bayham Place, London NW1 0EU
☎ 020 7388 7222
sheila@sheilaableman.co.uk
Contact *Sheila Ableman*

Founded 1999. Handles non-fiction including history, science, biography and autobiography. Specializes in celebrity ghostwriting and TV tie-ins. No poetry, children's, gardening or sport. COMMISSION Home 15%; US & Translation 20%. Unsolicited mss welcome. Approach in writing with publishing history, c.v., synopsis, three chapters and s.a.e. for return. No reading fee.

The Agency (London) Ltd*

24 Pottery Lane, London W11 4LZ
☎ 020 7727 1346   ℻ 020 7727 9037
info@theagency.co.uk
Contacts *Stephen Durbridge, Leah Schmidt, Norman North, Julia Kreitman, Bethan Evans, Hilary Delamere, Katie Haines, Faye Webber, Nick Quinn, Fay Davies, Ian Benson, Jago Irwin*

Founded 1995. Deals with writers and rights for TV, film, theatre, radio scripts and children's fiction. Also handles directors. Only existing clients for adult fiction or non-fiction. COMMISSION Home 10%; US by arrangement. Send letter with s.a.e. No unsolicited mss. No reading fee.

Aitken Alexander Associates Ltd*

18–21 Cavaye Place, London SW10 9PT
☎ 020 7373 8672   ℻ 020 7373 6002
reception@aitkenalexander.co.uk
www.aitkenalexander.co.uk
Contacts *Gillon Aitken, Clare Alexander, Lesley Thorne* (also film/TV)
Associated Agents *Anthony Sheil, Mary Pachnos*

Founded 1977. Handles fiction and non-fiction. No plays, scripts or children's fiction unless by existing clients. CLIENTS include Caroline Alexander, Pat Barker, Nicholas Blincoe, Gordon Burn, Jung Chang, John Cornwell, Sarah Dunant, Susan Elderkin, Sebastian Faulks, Helen Fielding, Jane Goldman, Germaine Greer, Mark Haddon, Susan Howatch, Liz Jensen, John Keegan, V.S. Naipaul, Jonathan Raban, Piers Paul Read, Michèle Roberts, Nicholas Shakespeare, Gillian Slovo, Matt Thorne, Colin Thubron, Salley Vickers, A.N. Wilson, Robert Wilson. COMMISSION Home 10%; US & Translation 20%; Film/TV 10%. Send preliminary letter, with half-page synopsis and first 30pp of sample material, and adequate return postage, in the first instance. No reading fee.

The Ampersand Agency Ltd*

Ryman's Cottages, Little Tew OX7 4JJ
☎ 01608 683677   ℻ 01608 683449
info@theampersandagency.co.uk
www.theampersandagency.co.uk
Contacts *Peter Buckman, Anne-Marie Doulton*
Consultants *Peter Janson-Smith, Patrick Neale*

Founded 2003. Handles literary and commercial fiction and non-fiction; contemporary and historical novels, crime, thrillers, biography, women's fiction, history, memoirs. No scripts unless by existing clients. No poetry, science fiction or fantasy. CLIENTS Helen Black, S.J. Bolton, Georgette Heyer estate, Druin Burch, Martin Conway, Vanessa Curtis, Will Davis, Cora Harrison, Michael Hutchinson, Beryl Kingston, Miriam Morrison, Philip Purser, Ivo Stourton, Vikas Swarup, Nick van Bloss, Michael Walters. COMMISSION Home 12½–15%; US 15–20%; Translation 20%. Translation rights handled by **The Buckman Agency.** Unsolicited mss, synopses and ideas welcome, as are e-mail enquiries, but send sample chapters and synopsis by post (s.a.e. required if material is to be returned). No reading fee.

Darley Anderson Literary, TV & Film Agency*

Estelle House, 11 Eustace Road, London SW6 1JB
☎ 020 7385 6652   ℻ 020 7386 5571
enquiries@darleyanderson.com
www.darleyanderson.com
www.darleyandersonchildrens.com
Contacts *Darley Anderson* (thrillers), *Zoe King* (non-fiction), *Julia Churchill* (children's books), *Camilla Bolton* (crime and mystery), *Ella Andrews* (women's fiction), *Emma White* (Head of Rights), *Madeleine Buston* (Rights Manager), *Rosi Bridge* (Finance)

Founded 1988. Run by an ex-publisher with a knack for spotting talent and making great deals – many for six and seven figure advances. Handles commercial fiction and non-fiction, children's fiction and non-fiction; also selected scripts for film and TV. No academic books or poetry. Special fiction interests: all types of thrillers and crime (American/hard boiled/cosy/historical); women's fiction (sagas, chick-lit, contemporary, love stories, 'tear jerkers', women in jeopardy) and all types of American and Irish novels. Non-fiction interests: celebrity autobiographies, biographies, sports books, 'true life' women in jeopardy, revelatory history and science, popular psychology, self improvement, diet, beauty, health, finance, fashion, animals, humour/cartoon, gardening, cookery, inspirational and religious. CLIENTS include Anne Baker, Alex Barclay, Constance Briscoe, Paul Carson, Cathy Cassidy, Lee Child, Lisa Clark, Martina Cole, John Connolly, Margaret Dickinson, Clare Dowling, Tana French, Milly Johnson, Danny King, Adrienne Kress, Patrick Lennon, Lesley Pearse, Lynda Page, Adrian Plass, Carmen Reid, Rebecca Shaw, Graeme Sims, Ahmet Zappa. COMMISSION Home 15%; US & Translation 20%; Film/TV/Radio 20%. OVERSEAS ASSOCIATES APA Talent and Literary Agency (LA/Hollywood); Liza Dawson Literary Agency (New York); and leading foreign agents throughout the world. Send letter, synopsis and first three chapters; plus s.a.e. for return. No reading fee.

Anubis Literary Agency

7 Birdhaven Close, Lighthorne, Warwick CV35 0BE

Ⓣ 01926 642588   Ⓕ 01926 642588
Contact *Steve Calcutt*

Founded 1994. Handles genre fiction in the following categories: science fiction, fantasy and horror. No other material considered. COMMISSION Home 15%; US & Translation 20%. Works with **The Marsh Agency** on translation rights. In the first instance send 50 pages with a one-page synopsis (s.a.e. essential). No telephone calls. No reading fee.

Artellus Limited

30 Dorset House, Gloucester Place, London NW1 5AD

Ⓣ 020 7935 6972   Ⓕ 020 7487 5957
www.artellusltd.co.uk
Chairman *Gabriele Pantucci*
Director *Leslie Gardner*
Associates *Darryl Samaraweera, Eleanor Day*

Founded 1986. Full-length and short mss. Handles crime, science fiction, historical, contemporary and literary fiction; non-fiction: art history, current affairs, biography, general history,

science. Works directly in the USA and with agencies internationally. COMMISSION Home 10%; Overseas 20%. Sample chapters in the first instance. Selective reader's service available for a fee. Return postage essential.

Author Literary Agents

53 Talbot Road, London N6 4QX
Ⓣ 020 8341 0442/0776 7022659 (mobile)
Ⓕ 020 8341 0442
agile@authors.co.uk
Contact *John Havergal*

Founded 1997. 'We put to leading publishers and producers strong new novels and graphic media ideas.' COMMISSION (VAT extra) Book, Screen & Internet production rights – Home 15%; Overseas & Translation 25%. Send s.a.e. with first chapter, scene or section writing sample, plus half-to-one page outline. Please include graphics samples, if applicable. No reading fee.

The Bell Lomax Moreton Agency

James House, 1 Babmaes Street, London SW1Y 6HF

Ⓣ 020 7930 4447   Ⓕ 020 7925 0118
agency@bell-lomax.co.uk
Executives *Eddie Bell, Pat Lomax, Paul Moreton, June Bell*

Established 2002. Handles quality fiction and non-fiction, biography, children's, business and sport. No unsolicited mss without preliminary letter. No scripts. No reading fee.

Lorella Belli Literary Agency (LBLA)*

54 Hartford House, 35 Tavistock Crescent, Notting Hill, London W11 1AY
Ⓣ 020 7727 8547   Ⓕ 0870 787 4194
info@lorellabelliagency.com
www.lorellabelliagency.com
Contact *Lorella Belli*

Founded 2002. Handles full-length fiction (from literary to genre) and general non-fiction. Particularly interested in first novelists, journalists, multi-cultural and international writing, books on or about Italy. No children's, fantasy, science fiction, short stories, poetry, plays, screenplays or academic books. CLIENTS include Michael Bess, Sean Bidder, Zöe Brân, Annalisa Coppolaro-Nowell, Emily Giffin, Nisha Minhas, Alanna Mitchell, Angela Murrills, Jennifer Ouellette, Robet Ray, Sheldon Rusch, Grace Saunders, Dave Singleton, Rupert Steiner, Diana Winston. COMMISSION Home 15%; US, Translation & Dramatic Rights 20%. Works in conjunction with leading associate agencies in most countries and a film/TV agency in London. Also represents American, Canadian, Australian and European agencies in the UK. Welcomes approaches from new authors. Send full proposal plus two chapters for non-fiction, and short synopsis plus first three chapters for fiction. S.a.e. essential. No reading fee. Revision suggested where appropriate.

Blake Friedmann Literary Agency Ltd*

122 Arlington Road, London NW1 7HP
☎ 020 7284 0408   🖷 020 7284 0442
firstname@blakefriedmann.co.uk
www.blakefriedmann.co.uk
Contacts *Carole Blake* (books), *Julian Friedmann* (film/TV), *Conrad Williams* (original scripts/radio), *Isobel Dixon* (books), *Oli Munson* (books)

Founded 1977. Handles all kinds of fiction from genre to literary; a varied range of specialist and general non-fiction, plus scripts for TV, radio and film. No poetry, science fiction or short stories (unless from existing clients). Special interests: commercial women's fiction, intelligent thrillers, literary fiction, upmarket non-fiction. CLIENTS include Gilbert Adair, Jane Asher, Edward Carey, Elizabeth Chadwick, Anne de Courcy, Anna Davis, Barbara Erskine, Ann Granger, Ken Hom, Billy Hopkins, Peter James, Deon Meyer, Lawrence Norfolk, Gregory Norminton, Joseph O'Connor, Sheila O'Flanagan, Siân Rees, Michael Ridpath, Craig Russell, Tess Stimson, Michael White.

COMMISSION Books: Home 15%; US & Translation 20%. Radio/TV/Film: 15%. OVERSEAS ASSOCIATES 24 worldwide. Unsolicited mss welcome but initial letter with synopsis and first two chapters preferred. Letters should contain as much information as possible on previous writing experience, aims for the future, etc. No reading fee.

Luigi Bonomi Associates Limited (LBA)*

91 Great Russell Street, London WC1B 3PS
☎ 020 7637 1234   🖷 020 7637 2111
info@bonomiassociates.co.uk
www.bonomiassociates.co.uk
Contacts *Luigi Bonomi, Amanda Preston, Molly Stirling*

Handles commercial and literary fiction, thrillers, crime and women's fiction; non-fiction: history, science, parenting, lifestyle, diet, health; teen and young adult fiction. CLIENTS include James Barrington, Chris Beardshaw, Gennaro Contaldo, Nick Foulkes, David Gibbins, Richard Hammond, Jane Hill, John Humphrys, Graham Joyce, Simon Kernick, James May, Niki Monaghan, Mike Morley, Gillian McKeith, Richard Madeley & Judy Finnigan, Sue Palmer, William Petre, Melanie Phillips, Jem Poster, John Rickards, Prof. Bryan Sykes, Mitch Symons, Alan Titchmarsh, Sally Worboyes, Sir Terry Wogan. COMMISSION Home 15%; US & Translation 20-25%. 'We are very

keen to find new authors and help them develop their careers.' No scripts, poetry, science fiction/fantasy, children's storybooks. *Send synopsis and first chapter only.* 'We do *not* reply by e-mail so return postage/s.a.e. essential for reply.' No reading fee.

BookBlast Ltd

PO Box 20184, London W10 5AU
☎ 020 8968 3089
www.bookblast.com
Contact *Address material to the Company*

Handles fiction and non-fiction. Memoirs, travel, popular culture, multicultural writing only. Reading very selectively. No new authors taken on except by recommendation. Editorial advice given to own authors. Initiates in-house projects. Also offers translation consultancy. Film, TV and radio rights sold in works by existing clients. COMMISSION Home 12%; US & Translation 20%; Film, TV & Radio 20%. No reading fee.

Bookseeker Agency

PO Box 7535, Perth PH2 1AF
☎ 01738 620688
bookseeker@blueyonder.co.uk
Contact *Paul Thompson*

Founded 2005. Handles poetry and any creative writing except non-fiction; also some illustration and visual art. COMMISSION negotiable 'but not exceeding 10%'. Approach in writing, giving a personal introduction, a summary of work so far and any current projects. No mss or sample chapters; synopsis or half-a-dozen poems acceptable. No reading fee.

Alan Brodie Representation Ltd

6th Floor, Fairgate House, 78 New Oxford Street, London WC1A 1HB
☎ 020 7079 7990   🖷 020 7079 7999
info@alanbrodie.com
www.alanbrodie.com
Contacts *Alan Brodie, Sarah McNair, Lisa Foster*

Founded 1989. Handles theatre, film and TV scripts. No books. COMMISSION Home 10%; Overseas 15%. Preliminary letter plus professional recommendation and c.v. essential. No reading fee but s.a.e. required.

Jenny Brown Associates*

33 Argyle Place, Edinburgh EH9 1JT
☎ 0131 229 5334
info@jennybrownassociates.com
www.jennybrownassociates.com
Contacts *Jenny Brown, Mark Stanton*
Children's *Lucy Juckes*
Crime *Allan Guthrie*

Founded 2002. Handles literary fiction, crime writing, writing for children; non-fiction:

biography, history, sport, music, popular culture. No poetry, science fiction, fantasy or academic. CLIENTS include Lin Anderson, Jeff Connor, Jennie Erdal, Alex Gray, Laura Hird, Roger Hutchinson, Sara Maitland, Laura Marney, Jonathan Rendall, Suhayl Saadi, Paul Torday. COMMISSION Home 12½%; US & Translation 20%. No unsolicited mss. Approach in writing with letter, synopsis and first two chapters. Include c.v. and s.a.e.

Felicity Bryan*

2A North Parade, Banbury Road, Oxford OX2 6LX
☎ 01865 513816   📠 01865 310055
agency@felicitybryan.com
www.felicitybryan.com
Agents *Felicity Bryan, Catherine Clarke, Sally Holloway, Caroline Wood*

Founded 1988. Handles fiction of various types and non-fiction with emphasis on history, biography, science and current affairs. No scripts for TV, radio or theatre. No crafts, picture books, how-to, science fiction or light romance. CLIENTS include Carlos Acosta, David Almond, Karen Armstrong, Simon Blackburn, Artemis Cooper, Isla Dewar, John Dickie, Sally Gardner, A.C. Grayling, Tim Harford, Sadie Jones, Diarmaid MacCulloch, John Man, Martin Meredith, James Naughtie, Linda Newbery, John Julius Norwich, Iain Pears, Robin Pilcher, Rosamunde Pilcher, Matt Ridley, Meg Rosoff, Miriam Stoppard, Roy Strong, Adrian Tinniswood, Colin Tudge, Eleanor Updale, Lucy Worsley and the estates of Robertson Davies and Humphrey Carpenter. COMMISSION Home 15%; US & Translation 20%. OVERSEAS ASSOCIATES **Andrew Nurnberg**, Europe; several agencies in US. Please approach by letter. See website for submission guidelines. No e-mail submissions. No reading fee.

The Buckman Agency

Ryman's Cottages, Little Tew OX7 4JJ
☎ 01608 683677   📠 01608 683449
r.buckman@talk21.com
j.buckman@talk21.com
Partners *Rosie Buckman, Jessica Buckman*

Founded in the early 1970s, the agency specializes in foreign rights and represents leading authors and agents from the UK and US. COMMISSON 20% (including sub-agent's commission). No unsolicited mss.

Brie Burkeman*

14 Neville Court, Abbey Road, London NW8 9DD
☎ 0870 199 5002   📠 0870 199 1029
brie.burkeman@mail.com
Contact *Brie Burkeman, Isabel White*

Founded 2000. Handles commercial and literary full-length fiction and non-fiction. Film and

theatre scripts. No academic, text, poetry, short stories, musicals or short films. Also associated with Serafina Clarke Ltd and independent film/TV consultant to literary agents. CLIENTS include Richard Askwith, Alexandra Carew, Alastair Chisholm, Kitty Ferguson, Joanne Harris, Shaun Hutson, Robin Norwood, David Savage, Steven Sivell, Gerald Wilson. COMMISSION Home 15%; Overseas 20%. Unsolicited e-mail attachments will be deleted without opening. No reading fee but return postage essential.

Juliet Burton Literary Agency*

2 Clifton Avenue, London W12 9DR
☎ 020 8762 0148   📠 020 8743 8765
juliet.burton@btinternet.com
Contact *Juliet Burton*

Founded 1999. Handles fiction and non-fiction. Special interests crime and women's fiction. No plays, film scripts, articles, poetry or academic material. COMMISSION Home 15%; US & Translation 20%. Approach in writing in the first instance; send synopsis and two sample chapters with s.a.e. No e-mail submissions. No unsolicited mss. No reading fee.

Campbell Thomson & McLaughlin Ltd*

50 Albemarle Street, London W1S 4BD
☎ 020 7493 4361   📠 020 7495 8961
submissions@ctmcl.co.uk
Contact *Charlotte Bruton*

Founded 1931. Handles fiction and general non-fiction, excluding children's. No plays, film/TV scripts, articles, short stories or poetry. Translation rights handled by **The Marsh Agency**. No unsolicited mss or synopses. Preliminary enquiry essential, by letter or e-mail. No reading fee.

Capel & Land Ltd*

29 Wardour Street, London W1D 6PS
☎ 020 7734 2414   📠 020 7734 8101
rosie@capelland.co.uk
www.capelland.com
Contact *Georgina Capel*

Handles fiction and non-fiction. Also film and TV. CLIENTS Kunal Basu, John Gimlette, Andrew Greig, Dr Tristram Hunt, Andrew Roberts, Simon Sebag Montefiore, Stella Rimington, Diana Souhami, Louis Theroux, Fay Weldon. COMMISSION Home, US & Translation 15%. Send sample chapters and synopsis with covering letter and s.a.e. (if return required) in the first instance. No reading fee.

Casarotto Ramsay and Associates Ltd

Waverley House, 7–12 Noel Street, London W1F 8GQ
☎ 020 7287 4450   📠 020 7287 9128

agents@casarotto.co.uk
www.casarotto.uk.com
Film/TV/Radio *Jenne Casarotto, Charlotte Kelly, Jodi Shields, Elinor Burns, Miriam James, Rachel Holroyd, Mark Casarotto, Abby Singer*
Stage *Tom Erhardt, Mel Kenyon*

Handles scripts for TV, theatre, film and radio. CLIENTS include: **Theatre** Alan Ayckbourn, Caryl Churchill, Christopher Hampton, David Hare, Sarah Kane estate, Mark Ravenhill. **Film** Laura Jones, Neil Jordan, Nick Hornby, Shane Meadows, Purvis & Wade, Lynne Ramsay. **TV** Howard Brenton, Amy Jenkins, Susan Nickson, Jessica Hynes. COMMISSION Home 10%. OVERSEAS ASSOCIATES worldwide. No unsolicited material without preliminary letter.

Celia Catchpole

56 Gilpin Avenue, London SW14 8QY
☎ 020 8255 4835
www.celiacatchpole.co.uk
Contact *Celia Catchpole*

Founded 1996. Handles children's books – artists and writers. No TV, film, radio or theatre scripts. No poetry. COMMISSION Home 10% (writers) 15% (artists); US & Translation 20%. Works with associate agents abroad. Will consider complete picture books and the first two chapters of longer scripts. S.a.e. essential.

Chapman & Vincent*

7 Dilke Street, London SW3 4JE
☎ 020 7352 5582 📠 020 7352 5582
chapmanvincent@hotmail.co.uk
Directors *Jennifer Chapman, Gilly Vincent*

A small agency handling non-fiction work – often highly illustrated – and whose clients come mainly from personal recommendation. The agency is not actively seeking clients but is happy to consider really original work. CLIENTS include George Carter, Leslie Geddes-Brown, Lucinda Lambton, Rowley Leigh and Eve Pollard. COMMISSION Home 15%; US & Europe 20%. Works with The Elaine Markson Agency. Please do not telephone or submit by fax. Submissions: two sample chapters and s.a.e. Will consider e-mail submissions without attachments. No reading fee.

Mic Cheetham Associates

50 Albemarle Street, London W1S 4BD
☎ 020 7495 2002 📠 020 7495 8961
info@miccheetham.com
www.miccheetham.com
Contacts *Mic Cheetham, Simon Kavanagh*

Established 1994. Handles general and literary fiction, crime and science fiction, and some specific non-fiction. No film/TV scripts apart from existing clients. No children's, illustrated books or poetry. CLIENTS include Iain Banks, Simon Beckett, Carol Birch, Anita Burgh, Laurie Graham, M. John Harrison, Toby Litt, Ken MacLeod, China Miéville, Antony Sher, Janette Turner Hospital. COMMISSION Home 15%; US & Translation 20%. No unsolicited mss. Approach in writing with publishing history, first two chapters and return postage. No reading fee.

Judith Chilcote Agency*

8 Wentworth Mansions, Keats Grove, London NW3 2RL
☎ 020 7794 3717
judybks@aol.com
Contact *Judith Chilcote*

Founded 1990. Handles commercial fiction, TV tie-ins, health and nutrition and celebrity autobiography and current affairs. COMMISSION Home 15%; Overseas 20–25%. *No* academic, science fiction, children's, short stories, film scripts or poetry. *No approaches by e-mail.* Send letter with c.v., synopsis, three chapters and s.a.e. for return. No reading fee.

Teresa Chris Literary Agency Ltd*

43 Musard Road, London W6 8NR
☎ 020 7386 0633
Contacts *Teresa Chris, Charles Brudenell-Bruce*

Founded 1989. Handles crime, general, women's, commercial and literary fiction, and non-fiction: history, biography, health, cookery, lifestyle, gardening, etc. Specializes in crime fiction and commercial women's fiction. No scripts. Film and TV rights handled by co-agent. No poetry, short stories, fantasy, science fiction or horror. CLIENTS include Stephen Booth, Martin Davies, Tamara McKinley, Marguerite Patten, Eileen Ramsay, Debby Holt. COMMISSION Home 10%; US & Translation 20%. OVERSEAS ASSOCIATES Patty Moosbrugger Literary Agency, USA; representatives in all other countries. Unsolicited mss welcome. Send query letter with first two chapters plus two-page synopsis (s.a.e. *essential*) in first instance. No reading fee.

Mary Clemmey Literary Agency*

6 Dunollie Road, London NW5 2XP
☎ 020 7267 1290 📠 020 7813 9757
mcwords@googlemail.com
Contact *Mary Clemmey*

Founded 1992. Handles fiction and non-fiction – high-quality work with an international market. No science fiction, fantasy or children's books. TV, film, radio and theatre scripts from existing clients only. US clients: Frederick Hill Bonnie Nadell Inc., Lynn C. Franklin Associates Ltd, The Miller Agency, Roslyn Targ, Weingel-Fidel Agency Inc, Betsy Amster Literary. COMMISSION Home 15%; US & Translation 20%. OVERSEAS

ASSOCIATE Elaine Markson Literary Agency, New York. No unsolicited mss and no e-mail submissions. Approach by letter only in the first instance giving a description of the work (include s.a.e.). No reading fee.

Jonathan Clowes Ltd*

10 Iron Bridge House, Bridge Approach, London NW1 8BD

☎ 020 7722 7674   🖷 020 7722 7677

Contacts *Ann Evans, Lisa Thompson*

Founded 1960. Pronounced 'clewes'. Now one of the biggest fish in the pond and not really for the untried unless they are true high-flyers. Fiction and non-fiction, plus scripts. No textbooks or children's. Special interests: situation comedy, film and television rights. CLIENTS include David Bellamy, Len Deighton, Elizabeth Jane Howard, Doris Lessing, David Nobbs, Gillian White and the estate of Kingsley Amis. COMMISSION Home & US 15%; Translation 19%. OVERSEAS ASSOCIATES **Andrew Nurnberg Associates**; Sane Töregard Agency. No unsolicited mss; authors come by recommendation or by successful follow-ups to preliminary letters.

Elspeth Cochrane Personal Management

16 Trinity Close, The Pavement, London SW4 0JD

☎ 020 7622 3566

elspethcochrane@talktalk.net

Contact *Elspeth Cochrane*

Founded 1960. Handles fiction, non-fiction, biographies, screenplays and plays. No children's fiction. CLIENTS include Royce Ryton, Robert Tanitch. COMMISSION 12½%. No unsolicited mss. Phone before submitting work.

Rosica Colin Ltd

1 Clareville Grove Mews, London SW7 5AH

☎ 020 7370 1080   🖷 020 7244 6441

Contact *Joanna Marston*

Founded 1949. Handles all full-length mss, plus theatre, film, television and sound broadcasting but few new writers being accepted. COMMISSION Home 10%; US & Translation 20%. Preliminary letter with return postage essential; writers should outline their writing credits and whether their mss have previously been submitted elsewhere. May take 3–4 months to consider full mss; synopsis preferred in the first instance. No reading fee.

Conville & Walsh Limited*

2 Ganton Street, London W1F 7QL

☎ 020 7287 3030   🖷 020 7287 4545

info@convilleandwalsh.com

www.convilleandwalsh.com

Directors *Clare Conville, Patrick Walsh, Peter Tallack* (book rights), *Alan Oliver* (finance)

Foreign Rights *Jake Smith-Bosanquet*

Book Development *Susan Armstrong, Robert Dinsdale*

Established in 2000 by Clare Conville (ex-A.P. Watt) and Patrick Walsh (ex-Christopher Little Literary Agency). Handle all genres of fiction, non-fiction and children's worldwide but no poetry, screenplays or short stories. 'Agency taste is generally upmarket to mass-market: literary and commercial fiction.' Fiction CLIENTS range from the Booker winner DBC Pierre to John Llewellyn Rhys prize-winner Sarah Hall and the Orwell Prize winner Delia Jarrett-Macauley. Commercial novelists include Isabel Wolff, Nick Harkaway, Matt Dunn and Michael Cordy. Non-fiction CLIENTS Tom Holland, Helen Castor, Simon Singh, Richard Wiseman, Ian Stewart, Ruth Padel, Misha Glenny, Michael Bywater. Also handles the estate of artist Francis Bacon. Prestigious children's and young adult list ranges from John Burningham to Steve Voake and the estate of Astrid Lindgren. COMMISSION Home 15%; US & Translation 20%. Send first three chapters, one-page synopsis, covering letter and s.a.e. No reading fee.

Jane Conway-Gordon Ltd*

1 Old Compton Street, London W1D 5JA

☎ 020 7494 0148   🖷 020 7287 9264

jconway_gordon@dsl.pipex.com

Contact *Jane Conway-Gordon*

Founded 1982. Handles fiction and general non-fiction. No poetry, science fiction, children's or short stories. COMMISSION Home 15%; US & Translation 20%. OVERSEAS ASSOCIATES Lyons Literary LLC, New York; plus agencies throughout Europe and Japan. Unsolicited mss welcome; preliminary letter and return postage essential. No reading fee.

Rupert Crew Ltd*

1A King's Mews, London WC1N 2JA

☎ 020 7242 8586   🖷 020 7831 7914

info@rupertcrew.co.uk (correspondence only)

www.rupertcrew.co.uk

Contacts *Doreen Montgomery, Caroline Montgomery*

Founded 1927. International representation, handling volume and subsidiary rights in fiction and non-fiction properties. No screenplays, plays or poetry, journalism or short stories, science fiction or fantasy. COMMISSION Home 15%; Elsewhere 20%. Preliminary letter and return postage essential. No reading fee.

Curtis Brown Group Ltd*

Haymarket House, 28/29 Haymarket, London SW1Y 4SP

☎ 020 7393 4400   🖷 020 7393 4401

cb@curtisbrown.co.uk
www.curtisbrown.co.uk
CEO *Jonathan Lloyd*
Director of Operations *Ben Hall*
Directors *Jacquie Drewe, Jonny Geller,
Nick Marston, Sarah Spear*
Books *Jonny Geller* (MD, Book Division),
*Jonathan Lloyd, Camilla Hornby, Jonathan Pegg,
Vivienne Schuster, Elizabeth Sheinkman,
Janice Swanson, Gordon Wise, Karolina Sutton*
Joint Head of Foreign Rights *Betsy Robbins,
Kate Cooper*
Foreign Rights Agents *Carol Jackson,
Katie McGowan, Elizabeth Iveson* (on behalf
of ICM books), *Daisy Meyrick* (on behalf of
ICM books)
Film/TV/Theatre *Nick Marston* (MD,
Media Division), *Tally Garner, Ben Hall,
Joe Phillips, Ben Hall, Amanda Davis*
Actors *Grace Clissold, Maxine Hoffman,
Sarah MacCormick, Sarah Spear, Kate Staddon,
Claire Stannard, Lucy Johnson, Mary Fitzgerald*
Presenters *Jacquie Drewe*

Founded 1899. Agents for the negotiation in all markets of novels, general non-fiction, children's books and associated rights (including multimedia) as well as play, film, theatre, TV and radio scripts. OVERSEAS ASSOCIATES in Australia and the US. Send outline for non-fiction and short synopsis for fiction with two/three sample chapters and autobiographical note. No reading fee. Return postage essential. No submissions by e-mail. See further submission guidelines on website. Also represents playwrights, film and TV writers and directors, theatre directors and designers, TV and radio presenters and actors.

Judy Daish Associates Ltd
2 St Charles Place, London W10 6EG
☎ 020 8964 8811   ☎ 020 8964 8966
Contacts *Judy Daish, Tracey Elliston,
Howard Gooding*

Founded 1978. Theatrical literary agent. Handles scripts for film, TV, theatre and radio. No books. Preliminary letter essential. No unsolicited mss.

Caroline Davidson Literary Agency*
5 Queen Anne's Gardens, London W4 1TU
☎ 020 8995 5768   ☎ 020 8994 2770
cdla@ukgateway.net
www.cdla.co.uk
Contact *Caroline Davidson*

Founded 1988. Handles a wide range of original fiction and non-fiction. Non-fiction includes archaeology, architecture, art, biography, climate change, current affairs, design, gardening, health, history, medicine, natural history, reference, science. Finished, polished first novels positively welcomed. No thrillers, crime, fantasy, science

fiction. Short stories, children's, plays, poetry not considered. CLIENTS Peter Barham, Nigel Barlow, Andrew Beatty, Andrew Dalby, Emma Donoghue, Chris Greenhalgh, Ed Husain, Tom Jaine, Simon Unwin. COMMISSION Home & Commonwealth, US, Translation 12½%; 20% if sub-agents are involved in foreign sales or films. Send an initial letter giving details of the project, plus c.v. and s.a.e. Non-fiction writers should include a detailed proposal with chapter synopsis. Fiction writers should submit the first 50 and last ten pages of their novel. No response to submissions without return postage and s.a.e. or to those sent by fax or e-mail.

Merric Davidson Literary Agency
▷ MBA Literary Agents Ltd

Felix de Wolfe
Kingsway House, 103 Kingsway, London
WC2B 6QX
☎ 020 7242 5066   ☎ 020 7242 8119
info@felixdewolfe.com
Contact *Felix de Wolfe*

Founded 1938. Handles quality fiction only, and scripts. No non-fiction or children's. CLIENTS include Jan Butlin, Jeff Dowson, Brian Glover, Sheila Goff, Aileen Gonsalves, John Kershaw, Ray Kilby, Bill MacIlwraith, Angus Mackay, Gerard McLarnon, Malcolm Taylor, David Thompson, Paul Todd, Dolores Walshe. COMMISSION Home 12½%; US 20%. No unsolicited mss. No reading fee.

The Dench Arnold Agency
10 Newburgh Street, London W1F 7RN
☎ 020 7437 4551   ☎ 020 7439 1355
contact@dencharnold.co.uk
www.dencharnold.co.uk
Contacts *Elizabeth Dench, Michelle Arnold*

Founded 1972. Handles scripts for TV and film. CLIENTS include Peter Chelsom. COMMISSION Home 10–15%. OVERSEAS ASSOCIATES include William Morris/Sanford Gross and C.A.A., Los Angeles. Unsolicited mss will be read, but a letter with sample of work and c.v. (plus s.a.e.) is required.

Dorian Literary Agency (DLA)*
Upper Thornehill, 27 Church Road, St Marychurch, Torquay TQ1 4QY
☎ 01803 312095
Contact *Dorothy Lumley*

Founded 1986. Handles popular genre fiction for adults: romance, historicals, sagas; crime and thrillers; science fiction, fantasy, horror. No short stories, poetry, autobiography, film/TV scripts or plays. CLIENTS include Gillian Bradshaw, Gary Gibson, Kate Hardy, Stephen Jones, Brian Lumley, Amy Myers, Rosemary Rowe, Lyndon

Stacey. COMMISSION Home 12½% for new clients; US 15%; Translation 20–25%. Works with agents in most countries for translation. Introductory letter with outline and 1–3 chapters (with return postage/s.a.e.) only, please. No enquiries or submissions by telephone, fax or e-mail. No reading fee.

Toby Eady Associates Ltd

9 Orme Court, London W2 4RL
☎ 020 7792 0092   🖷 020 7792 0879
toby@tobyeady.demon.co.uk
www.tobyeadyassociates.co.uk
Contacts *Toby Eady, Laetitia Rutherford*

Handles fiction, and non-fiction. Special interests: China, Middle East, Africa, India. CLIENTS include Mark Burnell, Michaela Clarke, Bernard Cornwell, John Carey, Yasmin Crowther, Rana Dasgupta, Fadia Faqir, Sophie Gee, Samson Kambalu, Richard Lloyd Parry, Julia Lovell, Francesca Marciano, Lawrence Potter, Deborah Scroggins, Rachel Seiffert, Samia Serageldin, John Stubbs, Diane Wei Liang, Robert Winder, Fan Wu Xinran Xue. COMMISSION Home 15%; Elsewhere 20%. OVERSEAS ASSOCIATES USA: ICM; France: La Nouvelle Agence; Holland: Jan Michael; Italy, Spain, Germany, Portugal, Scandinavia: Deal Direct; China, Taiwan: The Bardon Chinese Media Agency; Japan: The English Agency; Korea: The Eric Yang Agency; Eastern Europe/Russia: Prava I Prevodi; Turkey: Akcali Copyright; Czech Republic: Kristin Olson; Hungary: Katai & Bolza; Greece: JLM. Approach by personal recommendation. No film/TV scripts or poetry.

Eddison Pearson Ltd*

West Hill House, 6 Swains Lane, London N6 6QS
☎ 020 7700 7763   🖷 020 7700 7866
info@eddisonpearson.com
Contact *Clare Pearson*

Founded 1996. Small, personally-run agency. Handles children's books, fiction and non-fiction, poetry. CLIENTS include Sue Heap, Robert Muchamore, Valerie Bloom. COMMISSION Home 10%; US & Translation 15–20%. Enquiries and submissions by e-mail only. Please e-mail for up-to-date submission guidelines by return. No reading fee.

Edwards Fuglewicz*

49 Great Ormond Street, London WC1N 3HZ
☎ 020 7405 6725   🖷 020 7405 6726
Contacts *Ros Edwards, Helenka Fuglewicz, Julia Forrest*

Founded 1996. Handles literary and commercial fiction (not children's, science fiction, horror or fantasy); non-fiction: biography, history, popular culture. COMMISSION Home 15%; US &

Translation 20%. No unsolicited mss or e-mail submissions.

Faith Evans Associates*

27 Park Avenue North, London N8 7RU
☎ 020 8340 9920   🖷 020 8340 9410
faithevanslituk@hotmail.com

Founded 1987. Small agency. CLIENTS include Melissa Benn, Eleanor Bron, Caroline Conran, Alicia Foster, Midge Gillies, Ed Glinert, Vesna Goldsworthy, Cate Haste, Jim Kelly, Helena Kennedy, Seumas Milne, Tom Paulin, Sheila Rowbotham, the estate of Lorna Sage, Rebecca Stott, Harriet Walter, Francesca Weisman, Elizabeth Wilson. COMMISSION Home 15%; US & Translation 20%. OVERSEAS ASSOCIATES worldwide. List full; no submissions, please.

The Feldstein Agency

123–125 Main Street, 2nd Floor, Bangor BT20 4AE
☎ 028 9147 2823
paul@thefeldsteinagency.co.uk
www.thefeldsteinagency.co.uk
Contacts *Paul Feldstein, Susan Feldstein*

Founded 2007. Handles commercial fiction, literary fiction and criticism, romantic fiction, science fiction, thrillers, short stories, essays, adventure, autobiography/biography, business, cookery, crime, current affairs, football, golf, guidebooks, heritage, history, humanities, humour, Celtic and Irish interest, journalism, leisure, lifestyle, local history, media, memoirs, military, music, mysteries, philosophy, politics, sociology, travel, war, wine, women's interest and world affairs. CLIENTS Mary O'Sullivan, Ray Strobel, Adrian White. COMMISSION Home 15%; US & Translation 20%. Unsolicited mss, sample chapters and synopses welcome; approach by e-mail. No children's books. No reading fee.

Laurence Fitch Ltd

Mezzanine, Quadrant House, 80–82 Regent Street, London W1B 5AU
☎ 020 7734 9911
information@laurencefitch.com
www.laurencefitch.com
Contact *Brendan Davis*

Founded 1952, incorporating the London Play Company (1922) and in association with Film Rights Ltd (1932). Handles children's and horror books, scripts for theatre, film, TV and radio only. CLIENTS include Carlo Ardito, Hindi Brooks, John Chapman & Ray Cooney, Jeremy Lloyd, Dave Freeman, John Graham, Robin Hawdon, Glyn Robbins, Lawrence Roman, Gene Stone, the estate of Dodie Smith, Edward Taylor. COMMISSION UK 10%; Overseas 15%. OVERSEAS ASSOCIATES worldwide. No unsolicited mss. Send synopsis with sample scene(s)/first chapters in the first instance. No reading fee.

Jill Foster Ltd

9 Barb Mews, Brook Green, London W6 7PA
☎ 020 7602 1263   🖷 020 7602 9336
www.jflagency.com
Contacts *Jill Foster, Alison Finch,*
Simon Williamson, Dominic Lord, Gary Wild

Founded 1976. Handles scripts for TV, radio, film and theatre. No fiction, short stories or poetry. CLIENTS include Ian Brown, Phil Ford, Nev Fountain and Tom Jamieson, Rob Gittins, David Lane, Jim Pullin and Fraser Steele, Pete Sinclair, Paul Smith, Peter Tilbury, Roger Williams, Susan Wilkins. COMMISSION Home 12½%; Books, US & Translation 15%. No unsolicited mss; approach by letter in the first instance. No approaches by e-mail. No reading fee.

Fox & Howard Literary Agency*

4 Bramerton Street, London SW3 5JX
☎ 020 7352 8691   🖷 020 7352 8691
Contacts *Chelsey Fox, Charlotte Howard*

Founded 1992. A small agency, specializing in non-fiction, that prides itself on 'working closely with its authors'. Handles biography, history and popular culture, reference, business, mind, body and spirit, health and personal development, popular psychology. COMMISSION Home 15%; US & Translation 20%. No unsolicited mss; send letter and synopsis with s.a.e. for response. No reading fee.

Fraser Ross Associates

6 Wellington Place, Edinburgh EH6 7EQ
☎ 0131 657 4412/0131 553 2759
kjross@tiscali.co.uk
lindsey.fraser@tiscali.co.uk
www.fraserross.co.uk
Contacts *Lindsey Fraser, Kathryn Ross*

Founded 2002. Handles children's books, adult literary and mainstream fiction. No poetry, short stories, adult fantasy and science fiction, academic, scripts. CLIENTS Thomas Bloor, Sally J. Collins, Chris Fisher, Vivian French, Janey Louise Jones, Tanya Landman, Iain MacIntosh, Jack McLean, Michaela Morgan, Dugald Steer, Linda Strachan. Send preliminary letter, synopsis, first three chapters or equivalent, c.v. and s.a.e. No reading fee.

Futerman, Rose & Associates*

91 St Leonards Road, London SW14 7BL
☎ 020 8255 7755   🖷 020 8286 4860
enquiries@futermanrose.co.uk
www.futermanrose.co.uk
Contact *Guy Rose*

Founded 1984. Handles fiction, biography, show business, current affairs, teen fiction and scripts for TV and film. No children's, science fiction or fantasy. CLIENTS include Larry Barker, Christian Piers Betley, David Bret, Tom Conti, Iain Duncan Smith, Sir Martin Ewans, Yvette Fielding, Susan George, Paul Hendy, Russell Warren Howe, Keith R. Lindsay, Eric MacInnes, Paul Marx, Valerie Mendes, Max Morgan-Witts, Ciarán O'Keeffe, Lynsey de Paul, Erin Pizzey, John Repsch, Liz Rettig, Peter Sallis, Pat Silver-Lasky, Paul Stinchcombe, Gordon Thomas, Bill Tidy, Dr Mark White, Toyah Willcox, Simon Woodham, Allen Zeleski. No unsolicited mss. Send preliminary letter with brief resumé, synopsis, first 20pp and s.a.e.

Jüri Gabriel

35 Camberwell Grove, London SE5 8JA
☎ 020 7703 6186
Contact *Jüri Gabriel*

Handles quality fiction and non-fiction and (almost exclusively for existing clients) film, TV and radio rights. Jüri Gabriel worked in television, wrote books for 20 years and is chairman of Dedalus publishers. No short stories, articles, verse or books for children. CLIENTS include Maurice Caldera, Diana Constance, Miriam Dunne, Matt Fox, Paul Genney, Pat Gray, Mikka Haugaard, Robert Irwin, John Lucas, David Madsen, Richard Mankiewicz, Karina Mellinger, David Miller, Andy Oakes, John Outram, Philip Roberts, Dr Stefan Szymanski, Frances Treanor, Dr Terence White, Dr Robert Youngson. COMMISSION Home 10%; US & Translation 20%. Unsolicited mss ('two-page synopsis and three sample chapters in first instance, please') welcome if accompanied by return postage and letter giving sufficient information about author's writing experience, aims, etc.

Eric Glass Ltd

25 Ladbroke Crescent, London W11 1PS
☎ 020 7229 9500   🖷 020 7229 6220
eglassltd@aol.com
Contacts *Janet Glass, Sissi Liechtenstein*

Founded 1934. Handles fiction, non-fiction and scripts for publication or production in all media. No poetry, short stories or children's works. CLIENTS include Herbert Appleman, Pierre Chesnot, Charles Dyer, Henry Fleet, Tudor Gates, Philip Goulding, Pauline Macaulay, Brendan Murray and the estates of Rodney Ackland, Marc Camoletti, Jean Cocteau, Warwick Deeping, William Douglas Home, Philip King, Robin Maugham, Alan Melville, Beverley Nichols, Jack Popplewell, Jean-Paul Sartre. COMMISSION Home 15%; US & Translation 20%. OVERSEAS ASSOCIATES in the US, Australia, Czech Republic,

France, Germany, Greece, Holland, Italy, Japan, Poland, Scandinavia, Slovakia, South Africa, Spain. No unsolicited mss. Return postage required. No reading fee.

David Godwin Associates

55 Monmouth Street, London WC2H 9DG
☎ 020 7240 9992   📠 020 7395 6110
assistant@davidgodwinassociates.co.uk
www.davidgodwinassociates.co.uk
Contact *David Godwin*

Founded 1996. Handles literary and general fiction, non-fiction, biography. No scripts, science fiction or children's. No reading fee. COMMISSION Home 15%; Overseas 15%. 'Please contact the office before making a submission.'

Graham Maw Christie Literary Agency

19 Thornhill Crescent, London N1 1BJ
☎ 020 7812 9937
enquiries@grahammawchristie.com
www.grahammawchristie.com
Contacts *Jane Graham Maw, Jennifer Christie*

Founded 2005. Handles general non-fiction. No children's or poetry.

Annette Green Authors' Agency*

1 East Cliff Road, Tunbridge Wells TN4 9AD
☎ 01892 558262   📠 01892 558262
annette@annettegreenagency.co.uk
www.annettegreenagency.co.uk
Contact *Address material to the Agency*

Founded 1998. Handles literary and general fiction and non-fiction, popular culture and current affairs, science, music, film, history, biography, older children's and teenage fiction. No dramatic scripts or poetry. CLIENTS include Andrew Baker, Bill Broady, Meg Cabot, Simon Conway, Justin Hill, Mary Hogan, Anvar Khan, Maria McCann, Adam Macqueen, Ian Marchant, Professor Charles Pasternak, Kirsty Scott, Bernadette Strachan, Elizabeth Woodcraft. COMMISSION Home 15%; US & Translation 20%. Letter, synopsis, sample chapters and s.a.e. essential. No reading fee.

Christine Green Authors' Agent*

6 Whitehorse Mews, Westminster Bridge Road, London SE1 7QD
☎ 020 7401 8844   📠 020 7401 8860
info@christinegreen.co.uk
www.christinegreen.co.uk
Contact *Christine Green*

Founded 1984. Handles fiction (general and literary) and general non-fiction. No scripts, poetry or children's. COMMISSION Home 10%; US & Translation 20%. Initial letter, synopsis and first three chapters preferred. No reading fee but return postage essential.

Louise Greenberg Books Ltd*

The End House, Church Crescent, London N3 1BG
☎ 020 8349 1179   📠 020 8343 4559
louisegreenberg@msn.com
Contact *Louise Greenberg*

Founded 1997. Handles full-length literary fiction and serious non-fiction only. COMMISSION Home 15%; US & Translation 20%. DRAMATIC ASSOCIATE **Micheline Steinberg Associates**; CHILDREN'S ASSOCIATE **Sarah Manson Literary Agent**. *No telephone approaches.* No reading fee. S.a.e. essential.

Greene & Heaton Ltd*

37 Goldhawk Road, London W12 8QQ
☎ 020 8749 0315   📠 020 8749 0318
www.greeneheaton.co.uk
Contacts *Carol Heaton, Judith Murray, Antony Topping, Linda Davis, Will Francis*

A medium-sized agency with a broad range of clients. Handles all types of fiction and non-fiction. No original scripts for theatre, film or TV. CLIENTS include Mark Barrowcliffe, Bill Bryson, Jan Dalley, Marcus du Sautoy, Hugh Fearnley-Whittingstall, Michael Frayn, P.D. James, William Leith, Mary Morrissy, C.J. Sansom, William Shawcross, Sarah Waters. COMMISSION Home 15%; US & Translation 20%. OVERSEAS ASSOCIATES worldwide. No reply to unsolicited submissions without s.a.e. and/or return postage.

Gregory & Company Authors' Agents* (formerly Gregory & Radice)

3 Barb Mews, London W6 7PA
☎ 020 7610 4676   📠 020 7610 4686
info@gregoryandcompany.co.uk (general enquiries)
www.gregoryandcompany.co.uk
Contact *Jane Gregory*
Editorial Manager *Stephanie Glencross*
Rights Manager *Claire Morris*
Rights Executive *Jemma McDonagh*

Founded 1987. Handles all kinds of fiction and general non-fiction. Special interest fiction – literary, commercial, women's fiction, crime, suspense and thrillers. 'We are particularly interested in books which will also sell to publishers abroad.' No original plays, film or TV scripts (only published books are sold to film and TV). No science fiction, fantasy, poetry, academic or children's books. No reading fee. Editorial advice given to own authors. COMMISSION Home 15%; US, Translation, Radio/TV/Film 20%. Is well represented throughout Europe, Asia and US. No unsolicited mss; send a preliminary letter with c.v., synopsis, first three chapters and future writing plans (plus return postage). Short submissions by fax or e-mail (maryjones@

gregoryandcompany.co.uk) in the first instance; maximum five pages.

David Grossman Literary Agency Ltd

118b Holland Park Avenue, London W11 4UA

☎ 020 7221 2770   🖷 020 7221 1445

Contact *Submissions Dept.*

Founded 1976. Handles full-length fiction and general non-fiction – good writing of all kinds and anything healthily controversial. No verse or technical books for students. No original screenplays or teleplays (only works existing in volume form are sold for performance rights). Generally works with published writers of fiction only but 'truly original, well-written novels from beginners' will be considered. COMMISSION Rates vary for different markets. OVERSEAS ASSOCIATES throughout Europe, Asia, Brazil and the US. Best approach by preliminary letter giving full description of the work and, in the case of fiction, with the first 50 pages. 'Please ensure material is printed in at least 1½ spacing on one side of paper only; that all pages are numbered consecutively throughout and in minimum 11pt typeface.' All material must be accompanied by return postage. No approaches or submissions by fax or e-mail. No unsolicited mss. No reading fee.

Gunnmedia Ltd

50 Albemarle Street, London W1S 4BD

☎ 020 7351 5141

ali@gunnmedia.co.uk

www.gunnmedia.co.uk

Contacts *Ali Gunn, Mariam Keen, Sarah McFadden*

Head of Foreign Rights *Diana McKay*

Founded 2005. Handles commercial fiction & non-fiction. No scripts. CLIENTS include Jenny Colgan, Martina Reilly, Sarah Webb, Brian Freeman, Mil Millington, Alison Penton Harper. COMMISSION Home 15%; US & Translation 20%. ASSOCIATES Gelfman Schneider, LJK LITERARY, Fletcher Parry. Unsolicited mss, sample chapters and synopses welcome; approach in writing. No reading fee.

The Rod Hall Agency Limited

6th Floor, Fairgate House, 78 New Oxford Street, London W1A 1HB

☎ 020 7079 7987   🖷 0845 638 4094

office@rodhallagency.com

www.rodhallagency.com

Contact *Charlotte Knight*

Founded 1997. Handles drama for film, TV and theatre and writers-directors. Does not represent writers of episodes for TV series where the format is provided but represents originators of series. CLIENTS include Simon Beaufoy (*The Full Monty*), Jeremy Brock (*Mrs Brown*), Liz Lochhead (*Perfect Days*), Martin McDonagh (*The Pillowman*), Simon Nye (*Men Behaving Badly*). COMMISSION Home 10%; US & Translation 15%. No reading fee.

The Hanbury Agency*

28 Moreton Street, London SW1V 2PE

☎ 020 7630 6768

maggie@hanburyagency.com

www.hanburyagency.com

Contacts *Margaret Hanbury, Genevieve Carden*

Personally run agency representing quality fiction and non-fiction. No plays, scripts, poetry, children's books, fantasy, horror. CLIENTS include George Alagiah, J.G. Ballard, Simon Callow, Jane Glover, Bernard Hare, Judith Lennox, Katie Price. COMMISSION Home 15%; Overseas 20%. No unsolicited approaches via e-mail.

Roger Hancock Ltd

4 Water Lane, London NW1 8NZ

☎ 020 7267 4418   🖷 020 7267 0705

Contact *Material should be addressed to the Submissions Department*

Founded 1960. Special interests: comedy drama and light entertainment. COMMISSION Home 10%; Overseas 15%. Unsolicited mss not welcome. Initial phone call required.

Antony Harwood Limited*

103 Walton Street, Oxford OX2 6EB

☎ 01865 559615   🖷 01865 310660

mail@antonyharwood.com

www.antonyharwood.com

Contacts *Antony Harwood, James Macdonald Lockhart*

Founded 2000. Handles fiction and non-fiction. CLIENTS Amanda Craig, Peter F. Hamilton, Alan Hollinghurst, A.L. Kennedy, Douglas Kennedy, Dorothy Koomson, Chris Manby, George Monbiot, Garth Nix, Tim Parks. COMMISSION Home 15%; US & Translation 20%. Send letter and synopsis with return postage in the first instance. No reading fee.

A.M. Heath & Co. Ltd*

6 Warwick Court, London WC1R 5DJ

☎ 020 7242 2811   🖷 020 7242 2711

www.amheath.com

Contacts *Bill Hamilton, Sara Fisher, Sarah Molloy, Victoria Hobbs, Euan Thorneycroft*

Founded 1919. Handles fiction, general non-fiction and children's. No dramatic scripts, poetry, picture books or short stories. CLIENTS include Christopher Andrew, Bella Bathurst, Anita Brookner, Geoff Dyer, Katie Fforde, Lesley Glaister, Graham Hancock, Conn Iggulden, Hilary Mantel, Hilary Norman, Susan Price, John Sutherland, Adam Thorpe, Barbara Trapido.

COMMISSION Home 10–15%; US & Translation 20%; Film & TV 15%. OVERSEAS ASSOCIATES in the US, Europe, South America, Japan and the Far East. Preliminary letter and synopsis essential. No reading fee.

Rupert Heath Literary Agency

50 Albemarle Street, London W1S 4BD
℡ 020 7788 7807   🖷 020 7691 9331
emailagency@rupertheath.com
Contact *Rupert Heath*

Founded 2000. Handles literary and general fiction and non-fiction, including history, biography and autobiography, current affairs, popular science, the arts and popular culture. No scripts or poetry. COMMISSION Home 15%; US & Translation 20%. OVERSEAS ASSOCIATES worldwide. Approach with e-mail initially ('telling us a bit about yourself and the book you wish to submit'). No reading fee.

Henser Literary Agency

174 Pennant Road, Llanelli SA14 8HN
℡ 01554 753520   🖷 01554 753520
Contact *Steve Henser*

Founded 2002. Handles fiction: mystery and general; science fiction and fantasy; translation between English and Japanese; non-fiction book on Japan. Also TV, film, radio and theatre scripts. No horror or sadism. No reading fee. COMMISSION Home 15%; US 20%. No unsolicited material. Not looking for additional clients at present. No reading fee.

hhb agency ltd*

122 Arlington Road, London N1 7HP
℡ 020 7485 0044
heather@hhbagency.com
elly@hhbagency.com
www.hhbagency.com
Contacts *Heather Holden-Brown, Elly James*

Founded in 2005 by Heather Holden-Brown, a publishing editor for 20 years with Waterstone's, Harrap, BBC Books and Headline. Handles non-fiction: history and politics, contemporary autobiography/biography, popular culture, entertainment and TV, business, family memoir, food and cookery a speciality. No scripts. COMMISSION Home 15%. No unsolicited mss; please contact by e-mail or telephone before sending any material. No reading fee.

David Higham Associates Ltd*

5–8 Lower John Street, Golden Square, London W1F 9HA
℡ 020 7434 5900   🖷 020 7437 1072
dha@davidhigham.co.uk
www.davidhigham.co.uk
Books *Anthony Goff, Bruce Hunter, Veronique Baxter, Lizzy Kremer,*

Caroline Walsh (children's), *Andrew Gordon, Georgia Glover, Alice Williams*
Scripts *Nicky Lund, Georgina Ruffhead*

Founded 1935. Handles fiction, general non-fiction (biography, history, current affairs, etc.) and children's books. Also scripts. CLIENTS include John le Carré, J.M. Coetzee, Stephen Fry, Jane Green, James Herbert, Alexander McCall Smith, Lynne Truss, Jacqueline Wilson. COMMISSION Home 15%; US & Translation 20%; Scripts 10%. Preliminary letter with synopsis essential in first instance. No e-mail submissions. No reading fee. See website for further information.

Vanessa Holt Ltd*

59 Crescent Road, Leigh-on-Sea SS9 2PF
℡ 01702 473787   🖷 01702 473787
Contact *Vanessa Holt*

Founded 1989. Handles general fiction, non-fiction and non-illustrated children's books. No scripts, poetry, academic or technical. Specializes in crime fiction, commercial and literary fiction, and particularly interested in books with potential for sales abroad and/or to TV. COMMISSION Home 15%; US & Translation 20%; Radio/TV/Film 15%. Represented in all foreign markets. Approach by letter only and no unsolicited mss.

Kate Hordern Literary Agency

18 Mortimer Road, Clifton, Bristol BS8 4EY
℡ 0117 923 9368
katehordern@blueyonder.co.uk
Contact *Kate Hordern*

Founded 1999. Handles quality fiction and general non-fiction. CLIENTS Richard Bassett, Jeff Dawson, Leigh Eduardo, Kylie Fitzpatrick, Duncan Hewitt, J.T. Lees, Will Randall. COMMISSION Home 15%; US & Translation 20%. OVERSEAS ASSOCIATES VVV Agency, France; Carmen Balcells Agency, Spain; Synopsis Agency, Russia and various agencies in Asia. Approach in writing (or by e-mail) in the first instance with details of project. New clients taken on very selectively. Synopsis required for fiction; proposal/chapter breakdown for non-fiction. Sample chapters on request only. S.a.e. essential. No reading fee.

Valerie Hoskins Associates Limited

20 Charlotte Street, London W1T 2NA
℡ 020 7637 4490   🖷 020 7637 4493
vha@vhassociates.co.uk
Contacts *Valerie Hoskins, Rebecca Watson*

Founded 1983. Handles scripts for film, TV and radio. Special interests feature films, animation and TV. COMMISSION Home 12½%; US 20% (maximum). No unsolicited scripts; preliminary letter of introduction essential. No reading fee.

Independent Talent Group Limited*

Oxford House, 76 Oxford Street, London W1D 1BS

☎ 020 7636 6565   🖷 020 7323 0101

Contacts *Sue Rodgers, Jessica Sykes, Cathy King, Greg Hunt, Hugo Young, Michael McCoy, Duncan Heath, Paul Lyon-Maris*

Founded 1973. Specializes in scripts for film, theatre, TV and radio. No reading fee.

Indepublishing CIA – Consultancy for Independent Authors

77 Acre Road, Kingston upon Thames KT2 6ES

☎ 07768 864364

joanna@indepublishing.com

www.indepublishing.com

Contact *Joanna Anthony*

Founded 2005. Handles women's fiction, crime fiction, contemporary fiction, travelogues and biographies. No TV, film, radio or theatre scripts. COMMISSION Home 10%. Send e-mail with synopsis and publishing plans in the first instance. £75 reading fee 'only upon invitation'.

The Inspira Group

5 Bradley Road, Enfield EN3 6ES

☎ 020 8292 5163   🖷 0870 139 3057

darin@theinspiragroup.com

www.theinspiragroup.com

Managing Director *Darin Jewell*

Founded 2001. Handles children's books, crime fiction, lifestyle/relationships, science fiction/fantasy, humour, non-fiction. No scripts. CLIENTS Michael G.R. Tolkien, Simon Brown, Mike Haskins, Clive Whichelow, Simon Hall. COMMISSION Home & US 15%. Unsolicited mss and synopses welcome; approach by e-mail or telephone. No reading fee.

Intercontinental Literary Agency*

Centric House, 390–391 Strand, London WC2R 0LT

☎ 020 7379 6611   🖷 020 7240 4724

ila@ila-agency.co.uk

www.ila-agency.co.uk

Contacts *Nicki Kennedy, Sam Edenborough, Tessa Girvan, Katherine West*

Founded 1965. Represents translation rights only on behalf of Lucas Alexander Whitley; Luigi Bonomi Associates; Mulcahy & Viney; Mary Cunnane Agency (NSW); Wade & Doherty; John Beaton Writers' Agent; Jane Conway-Gordon; Short Books; Free Agents; Bell Lomax Moreton; The Agency (Hilary Delamere clients only); Stimola Literary Studio (New York); Elyse Cheney Literary Associates (New York); Barer Literary (New York); Park Literary Group (New York); Harold Matson Company (New York); The Steinberg Agency (New York); The Turnbull Agency (John Irving); and selected authors on behalf of various other agencies. No unsolicited submissions accepted.

International Scripts

1A Kidbrooke Park Road, London SE3 0LR

☎ 020 8319 8666   🖷 020 8319 0801

internationalscripts@btinternet.com

Contacts *Pat Hornsey, Jill Lawson*

Founded 1979 by Bob Tanner. Handles most types of books (non-fiction and fiction) and scripts for most media. No poetry, articles or short stories. CLIENTS include Jane Adams, Zita Adamson, Ashleigh Bingham, Frances Burke, Simon Clark, Ann Cliff, Nicholas Clough, David Stuart Davies, June Gadsby, Robert A. Heinlein, Anna Jacobs, Margaret James, Anne Jones, Richard Laymon, Trevor Lummis, Chris Marr, K.T. McCaffrey, Margaret Muir, Nick Oldham, Chris Pascoe, Mary Ryan, Ruth Searle, Anne Whitfield, Janet Woods. COMMISSION Home 15%; US & Translation 20%. OVERSEAS ASSOCIATES include Ralph Vicinanza, USA; Thomas Schlück, Germany. Preliminary letter, one-page synopsis, wordage, plus s.a.e. required. No unsolicited mss by post or e-mail accepted.

Barrie James Literary Agency

Rivendell, Kingsgate Close, Torquay TQ2 8QA

☎ 01803 326617

mail@newauthors.org.uk

www.newauthors.org.uk

Contact *Barrie E. James*

Founded 1997. Includes New Authors Showcase an internet site for new writers and poets to display their work to publishers and others. *No unsolicited mss.* First contact: s.a.e. or e-mail.

Janklow & Nesbit (UK) Ltd*

33 Drayson Mews, London W8 4LY

☎ 020 7376 2733   🖷 020 7376 2915

queries@janklow.co.uk

www.janklowandnesbit.co.uk

Agents *Tif Loehnis, Claire Paterson, Jenny McVeigh*

Head of Rights *Rebecca Folland*

Founded 2000. Handles fiction and non-fiction; commercial and literary. Send full outline (non-fiction), synopsis and three sample chapters (fiction) plus informative covering letter and return postage. No reading fee. No poetry, plays or film/TV scripts. US rights handled by **Janklow & Nesbit Associates** in New York (see entry under *US Literary Agents*).

Johnson & Alcock*

Clerkenwell House, 45–47 Clerkenwell Green, London EC1R 0HT

☎ 020 7251 0125   🖷 020 7251 2172

info@johnsonandalcock.co.uk

www.johnsonandalcock.co.uk

Contacts *Andrew Hewson, Michael Alcock, Anna Power, Ed Wilson*

Founded 1956. Handles literary and commercial fiction, children's 9+ fiction; general non-fiction including current affairs, biography, memoirs, history, culture, design, lifestyle and health, film and music, graphic novels. No poetry, screenplays, technical or academic material. COMMISSION Home 15%; US & Translation 20%. For fiction, send first three chapters, full synopsis and covering letter. For non-fiction, send covering letter with details of writing and other relevant experience, plus full synopsis. No response to e-mail submissions. No reading fee but return postage essential. Will give editorial advice on work with exceptional promise.

Jane Judd Literary Agency*

18 Belitha Villas, London N1 1PD

☎ 020 7607 0273   🖷 020 7607 0623

Contact *Jane Judd*

Founded 1986. Handles general fiction and non-fiction: women's fiction, crime, thrillers, literary fiction, humour, biography, investigative journalism, health, women's interests and travel. 'Looking for good contemporary women's fiction but not Mills & Boon-type; also historical fiction and non-fiction.' No scripts, academic, gardening, short stories or DIY. CLIENTS include Andy Dougan, Cliff Goodwin, Jill Mansell, Jonathon Porritt, Rosie Rushton, Manda Scott, David Winner. COMMISSION Home 10%; US & Translation 20%. Approach with letter, including synopsis, first chapter and s.a.e. Initial telephone call helpful in the case of non-fiction.

Michelle Kass Associates*

85 Charing Cross Road, London WC2H 0AA

☎ 020 7439 1624   🖷 020 7734 3394

office@michellekass.co.uk

Contacts *Michelle Kass, Andrew Mills*

Founded 1991. Handles literary fiction, film and television primarily. COMMISSION Home 10%; US & Translation 15–20%. No mss accepted without preliminary phone call. No e-mail submissions. No reading fee.

Frances Kelly*

111 Clifton Road, Kingston upon Thames KT2 6PL

☎ 020 8549 7830   🖷 020 8547 0051

Contact *Frances Kelly*

Founded 1978. Handles non-fiction, including illustrated: biography, history, art, self-help, food & wine, complementary medicine and therapies, finance and business books; and academic non-fiction in all disciplines. No scripts except for existing clients. COMMISSION Home 10%; US & Translation 20%. No unsolicited mss. Approach by letter with brief description of work or synopsis, together with c.v. and return postage.

Knight Features

20 Crescent Grove, London SW4 7AH

☎ 020 7622 1467   🖷 020 7622 1522

peter@knightfeatures.co.uk

Contacts *Peter Knight, Samantha Ferris, Gaby Martin, Andrew Knight*

Founded 1985. Handles motor sports, cartoon books, puzzles, business, history, factual and biographical material. No poetry, science fiction or cookery. CLIENTS include Frank Dickens, Gray Jolliffe, Angus McGill, Chris Maslanka, Barbara Minto. COMMISSION dependent upon authors and territories. OVERSEAS ASSOCIATES United Media, US; Auspac Media, Australia; Puzzle Co., New Zealand. No unsolicited mss and no e-mail submissions. Send letter accompanied by c.v. and s.a.e. with synopsis of proposed work.

LAW*

14 Vernon Street, London W14 0RJ

☎ 020 7471 7900   🖷 020 7471 7910

Contacts *Mark Lucas, Julian Alexander, Araminta Whitley, Alice Saunders, Peta Nightingale, Philippa Milnes-Smith* (children's), *Aisha Mobin* (children's)

Founded 1996. Handles full-length commercial and literary fiction, non-fiction and children's books. No fantasy (except children's), plays, poetry or textbooks. Film and TV scripts handled for established clients only. COMMISSION Home 15%; US & Translation 20%. OVERSEAS ASSOCIATES worldwide. Unsolicited mss considered; send brief covering letter, short synopsis and two sample chapters. S.a.e. essential. No e-mailed or disk submissions.

LBA

▷ Luigi Bonomi Associates Limited

Barbara Levy Literary Agency*

64 Greenhill, Hampstead High Street, London NW3 5TZ

☎ 020 7435 9046   🖷 020 7431 2063

Contacts *Barbara Levy, John Selby*

Founded 1986. Handles general fiction, non-fiction, and film and TV rights. COMMISSION Home 15%; US 20%; Translation by arrangement, in conjunction with **The Buckman Agency**. US ASSOCIATE Marshall Rights. No unsolicited mss. No reading fee.

Limelight Management*

33 Newman Street, London W1T 1PY

☎ 020 7637 2529   🖷 020 7637 2538

limelight.management@virgin.net

www.limelightmanagement.com

Contacts *Fiona Lindsay, Mary Bekhait*

Founded 1991. Handles literary fiction and quality commercial fiction. Film and TV scripts handled for established clients only. Full-length mss. Also specialists in general non-fiction in the areas of cookery, gardening, antiques, interior design, wine, art and crafts and health. No academic text, poetry, short stories, plays, musicals or short films. COMMISSION Home 15%; US & Translation 20%. Unsolicited mss welcome; send preliminary letter (s.a.e. essential). No reading fee.

The Christopher Little Literary Agency*

Eel Brook Studios, 125 Moore Park Road, London SW6 4PS
☎ 020 7736 4455   ℻ 020 7736 4490
info@christopherlittle.net
firstname@christopherlittle.net
www.christopherlittle.net
Contacts *Christopher Little*

Founded 1979. Handles commercial and literary full-length fiction and non-fiction. No poetry, plays, science fiction, fantasy, textbooks, illustrated children's or short stories. Film scripts for established clients only. AUTHORS include Steve Barlow and Steve Skidmore, Paul Bajoria, Andrew Butcher, Janet Gleeson, Carol Hughes, Alastair MacNeill, Robert Mawson, Haydn Middleton, A.J. Quinnell, Christopher Matthew, Robert Radcliffe, J.K. Rowling, Darren Shan, Wladyslaw Szpilman, John Watson, Pip Vaughan-Hughes, Gorillaz, Christopher Hale, Peter Howells, Gen. Sir Mike Jackson, Lauren Liebenberg, Shiromi Pinto, Dr Nicholas Reeves, Shayne Ward, Angela Woolfe, Anne Zouroudi. COMMISSION Home 15%; US, Canada, Translation, Audio & Motion Picture 20%.

London Independent Books

26 Chalcot Crescent, London NW1 8YD
☎ 020 7706 0486   ℻ 020 7724 3122
Proprietor *Carolyn Whitaker*

Founded 1971. A self-styled 'small and idiosyncratic' agency. Handles fiction and non-fiction reflecting the tastes of the proprietor. All subjects considered (except computer books and young children's), providing the treatment is strong and saleable. Scripts handled only if by existing clients. Special interests: boats, travelogues, commercial fiction, science fiction and fantasy, teenage fiction. COMMISSION Home 15%; US & Translation 20%. No unsolicited mss; letter, synopsis and first two chapters with return postage the best approach. No reading fee.

The Andrew Lownie Literary Agency*

36 Great Smith Street, London SW1P 3BU
☎ 020 7222 7574   ℻ 020 7222 7576
lownie@globalnet.co.uk
www.andrewlownie.co.uk
Contact *Andrew Lownie*

Founded 1988. Specializes in non-fiction, especially history, biography, current affairs, military history, UFOs, reference and packaging celebrities and journalists for the book market. No poetry, short stories or science fiction. Formerly a journalist, publisher and himself the author of 12 non-fiction books, Andrew Lownie's CLIENTS include Richard Aldrich, Juliet Barker, the Joyce Cary estate, Tom Devine, Duncan Falconer, Jonathan Fryer, Laurence Gardner, Cathy Glass, Timothy Good, David Hasselhoff, Robert Holden, Lawrence James, Robert Jobson, Damien Lewis, Julian Maclaren-Ross estate, Patrick MacNee, Norma Major, Sir John Mills, Tom Pocock, Nick Pope, Martin Pugh, Desmond Seward, David Stafford. COMMISSION Worldwide 15%. Preferred approach by e-mail in format suggested on website.

Lucas Alexander Whitley
▷ LAW

Lucy Luck Associates (In association with Aitken Alexander Associates Ltd)

18–21 Cavaye Place, London SW10 9PT
☎ 020 7373 8672
lucy@lucyluck.com
www.lucyluck.com
Contact *Lucy Luck*

Founded 2006. Handles quality fiction and non-fiction. No TV, film, radio or theatre scripts; no illustrated or children's books. CLIENTS include Jon Hotten, Doug Johnstone, Ewan Morrison, Philip Ó Ceallaigh, Catherine O'Flynn, Rebbecca Ray, Adam Thorpe. COMMISSION Home 15%; US & Translation 20%. Send sample chapters by post. No reading fee.

Lutyens and Rubinstein*

231 Westbourne Park Road, London W11 1EB
☎ 020 7792 4855   ℻ 020 7792 4833
Partners *Sarah Lutyens, Felicity Rubinstein*
Submissions *Susannah Godman*

Founded 1993. Handles adult fiction and non-fiction books. No TV, film, radio or theatre scripts. COMMISSION Home 15%; US & Translation 20%. Unsolicited mss accepted; send introductory letter, c.v., two chapters and return postage for all material submitted. No reading fee.

Luxton Harris Ltd*

2 Deanery Street, London W1K 1AU
☎ 020 7629 8325
rebecca.winfield@btopenworld.com
Contacts *Rebecca Winfield, Jonathan Harris, David Luxton*

Founded 2003. Specializes in sport and general non-fiction; biography, popular culture, history. No children's books, science fiction, fantasy, horror or poetry. COMMISSION Home 15%; US &

Translation 20%. No unsolicited mss. Please send a letter of enquiry or e-mail in the first instance.

Duncan McAra

28 Beresford Gardens, Edinburgh EH5 3ES
☎ 0131 552 1558   🖷 0131 552 1558
duncanmcara@mac.com
Contact *Duncan McAra*

Founded 1988. Handles fiction (literary fiction) and non-fiction, including art, architecture, archaeology, biography, military, travel and books of Scottish interest. COMMISSION Home 10%; Overseas 20%. Preliminary letter, synopsis and sample chapter (including return postage) essential. No reading fee.

Bill McLean Personal Management

23B Deodar Road, London SW15 2NP
☎ 020 8789 8191
Contact *Bill McLean*

Founded 1972. Handles scripts for all media. No books. CLIENTS include Dwynwen Berry, Graham Carlisle, Phil Clark, Pat Cumper, Jane Galletly, Patrick Jones, Tony Jordan, Bill Lyons, John Maynard, Michael McStay, Les Miller, Ian Rowlands, Jeffrey Segal, Richard Shannon, Ronnie Smith, Barry Thomas, Garry Tyler, Frank Vickery, Laura Watson, Mark Wheatley. COMMISSION Home 10%. No unsolicited mss. Phone call or introductory letter essential. No reading fee.

Eunice McMullen Ltd

Low Ibbotsholme Cottage, Off Bridge Lane, Troutbeck Bridge, Windermere LA23 1HU
☎ 01539 448551
eunicemcmullen@totalise.co.uk
www.eunicemcmullen.co.uk
Contact *Eunice McMullen*

Founded 1992. Handles all types of children's fiction. Especially interested in 9plus and has 'an excellent' list of picture book authors and illustrators. CLIENTS include Wayne Anderson, Jon Berkeley, Sam Childs, Caroline Jayne Church, Jason Cockcroft, Ross Collins, Charles Fuge, Angela McAllister, David Melling, Angie Sage, Gillian Shields, David Wood. COMMISSION Home 10%; US 15%; Translation 20%. *No unsolicited scripts.* Telephone enquiries only.

Andrew Mann Ltd*

1 Old Compton Street, London W1D 5JA
☎ 020 7734 4751   🖷 020 7287 9264
info@andrewmann.co.uk
www.andrewmann.co.uk
Contacts *Anne Dewe, Tina Betts, Louise Burns*

Founded 1975. Handles fiction, general non-fiction, children's and film, TV, theatre, radio scripts. COMMISSION Home 15%; US & Translation 20%. OVERSEAS ASSOCIATES various.

Unsolicited mss accepted only with preliminary letter, synopsis, first three chapters; s.a.e. essential. E-mail submissions for synopses only, but 'we do not open attachments.' No reading fee.

Sarah Manson Literary Agent

6 Totnes Walk, London N2 0AD
☎ 020 8442 0396
info@sarahmanson.com
www.sarahmanson.com
Contact *Sarah Manson*

Founded 2002. Handles quality fiction for children and young adults. COMMISSION Home 10%; Overseas & Translation 20%. Please consult website for submission guidelines. No reading fee.

Marjacq Scripts Ltd

34 Devonshire Place, London W1G 6JW
☎ 020 7935 9499   🖷 020 7935 9115
subs@marjacq.com
www.marjacq.com
Contact *Philip Patterson (books), Luke Speed (film/TV)*

Handles fiction and non-fiction, literary and commercial, graphic novels and comic books as well as film, TV, screenwriters and directors. No poetry. COMMISSION Home 10%; Overseas 20%. New work welcome; send brief letter, synopsis and approx. first 50 pages plus s.a.e. No reading fee.

The Marsh Agency Ltd*

50 Albemarle Street, London W1S 4BD
☎ 020 7493 4361   🖷 020 7495 8961
enquiries@marsh-agency.co.uk
www.marsh-agency.co.uk
Managing Director *Paul Marsh*
Contacts *Geraldine Cooke, Jessica Woollard, Caroline Hardman*
Foreign Rights *Camilla Ferrier*

Founded in 1994 as international rights specialist for British, American and Canadian agencies. Expanded to act as agents handling fiction and non-fiction. Specializes in authors with international potential. See website for further information on individual agents' areas of interest. COMMISSION Home 15%; Elsewhere 20%; Film & TV 15%. See also **Paterson Marsh Ltd**. No plays, scripts or poetry. Unsolicited mss considered. Electronic submissions to go through the website.

MBA Literary Agents Ltd*

62 Grafton Way, London W1T 5DW
☎ 020 7387 2076   🖷 020 7387 2042
firstname@mbalit.co.uk
www.mbalit.co.uk
Contacts *Diana Tyler, John Richard Parker, Meg Davis, Laura Longrigg, David Riding, Sophie Gorell Barnes, Susan Smith, Jean Kitson*

Founded 1971. Handles fiction and non-fiction, TV, film, radio and theatre scripts. Works in conjunction with agents in most countries. Also UK representative for Donald Maass Agency, Frances Collin Literary Agency, Martha Millard Literary Agency and the JABberwocky Agency. In November 2006, the Merric Davidson Literary Agency became part of MBA. COMMISSION Home 15%; Overseas 20%; Theatre/TV/Radio 10%; Film 10–20%. See website for submission details.

McKernan Literary Agency & Consultancy
5 Gayfield Square, Edinburgh EH1 3NW
☎ 0131 557 1771
maggie@mckernanagency.co.uk
www.mckernanagency.co.uk
Contact *Maggie McKernan*

Founded 2005. Works in conjunction with **Capel & Land Ltd**. Handles fiction, both literary and commercial; also non-fiction. COMMISSION Home 15%; US & Translation 15%. Send s.a.e. for return of ms. No reading fee.

Christy Moore Ltd
▷ Sheil Land Associates Ltd

William Morris Agency (UK) Ltd*
Centre Tower, 103 Oxford Street, London WC1A 1DD
☎ 020 7534 6800   📠 020 7534 6900
www.wma.com
Books *Eugenie Furniss* (Head of Book Division), *Rowan Lawton, Cathryn Summerhayes, Raffaella De Angelis* (translation rights)
TV *Holly Pye* (Head of TV), *Sophie Laurimore, Isabella Zoltowski*
Film *Lucinda Prain*

Worldwide talent and literary agency with offices in New York, Beverly Hills, Nashville, Miami and Shanghai. Handles film and TV scripts, TV formats; fiction and general non-fiction. COMMISSION Film & TV 10%; UK Books 15%; USA Books & Translation 20%. Accepts e-mail submissions only. Send synopsis and three sample chapters (50 pages or fewer) to ldnsubmissions@wma.com No reading fee.

Michael Motley Ltd
The Old Vicarage, Tredington, Tewkesbury GL20 7BP
☎ 01684 276390   📠 01684 297355
Contact *Michael Motley*

Founded 1973. Handles only full-length mss (adult: 60,000+; children's – all ages – and humour: length variable). No short stories or journalism. No science fiction, horror, poetry or original dramatic material. COMMISSION Home 10%; US 15%; Translation 20%. New clients by referral only: no unsolicited material considered. No reading fee.

Judith Murdoch Literary Agency*
19 Chalcot Square, London NW1 8YA
☎ 020 7722 4197
Contact *Judith Murdoch*

Founded 1993. Handles full-length fiction only, especially accessible literary and popular women's fiction. No science fiction/fantasy, children's, poetry or short stories. CLIENTS include Anne Bennett, Tony Black, Alison Bond, Meg Hutchinson, Lola Jaye, Lisa Jewell, Pamela Jooste, Jessie Keane, Catherine King, Eve Makis. Translation rights handled by **The Marsh Agency**. COMMISSION Home 15%; US & Translation 20%. No unsolicited mss; approach in writing only enclosing first two chapters and brief synopsis. Submissions by e-mail cannot be considered. Return postage/s.a.e. essential. Editorial advice given. No reading fee.

The Narrow Road Company
182 Brighton Road, Coulsdon CR5 2NF
☎ 020 8763 9895   📠 020 8763 2558
richardireson@narrowroad.co.uk
Contact *Richard Ireson*

Founded 1986. Part of the Narrow Road Group. Theatrical literary agency. Handles scripts for TV, theatre, film and radio. No novels or poetry. CLIENTS include Ron Aldridge, Geoff Aymer, Steve Gooch, Joe Graham, Renny Krupinski, Simon McAllum, Tim Sanders. No unsolicited mss or e-mail attachments; approach by letter with c.v. only. Interested in writers with some experience in theatre and radio

The Maggie Noach Literary Agency*
Unit 4, 246 Acklam Road, London W10 5YG
☎ 07506 717726
info@mnla.co.uk
www.mnla.co.uk
Contact *Address material to the Company.*

Pronounced 'no-ack'. Handles fiction, general non-fiction and children's. COMMISSION Home 15%; US & Translation 20%. No unsolicited mss; submissions by arrangement only.

Andrew Nurnberg Associates Ltd*
Clerkenwell House, 45–47 Clerkenwell Green, London EC1R 0QX
☎ 020 7417 8800   📠 020 7417 8812
www.andrewnurnberg.com
Directors *Andrew Nurnberg, Sarah Nundy, D. Roger Seaton, Vicky Mark*

Agent for UK and international authors, including children's writers. Sells rights internationally on behalf of UK/US agencies and publishers. Branches in Moscow, Budapest, Prague, Sofia, Warsaw, Riga, Beijing and Taipei. COMMISSION Home 15%; US & Translation 20%. See website for full information and submission guidelines.

Alexandra Nye

'Craigower', 6 Kinnoull Avenue, Dunblane
FK15 9JG
☎ 01786 825114
Contact *Alexandra Nye*

Founded 1991. Handles fiction and topical
non-fiction. Special interests: literary fiction
and history. CLIENTS include Dr Tom Gallagher,
Harry Mehta, Robin Jenkins. COMMISSION
Home 10%; US 20%; Translation 15%. Unsolicited
mss welcome (s.a.e. essential for return). No
phone calls. Preliminary approach by letter, with
synopsis, preferred. Reading fee for supply of
detailed report.

Deborah Owen Ltd*

78 Narrow Street, Limehouse, London E14 8BP
☎ 020 7987 5119/5441   🄕 020 7538 4004
do@deborahowen.co.uk
Contact *Deborah Owen*

Founded 1971. Small agency specializing in only
representing three authors direct around the
world. *No new authors.* CLIENTS Amos Oz, David
Owen and Delia Smith. COMMISSION Home 15%;
US & Translation 15%.

Paterson Marsh Ltd*

50 Albemarle Street, London W1S 4BD
☎ 020 7297 4312   🄕 020 7495 8961
www.patersonmarsh.co.uk
Contact *Stephanie Ebdon*

World rights representatives of authors
and publishers primarily in the UK and
US, specializing in books for mental health
professionals. CLIENTS include the estates of
Sigmund and Anna Freud, Donald Winnicott,
Michael Balint and Wilfred Bion. Also interested
in serious and popular non-fiction. Visit website
for further information and submissions.
COMMISSION 20% (including sub-agent's
commission). See also **The Marsh Agency Ltd**.
No fiction, scripts, poetry, children's, articles or
short stories. Unsolicited mss considered. Send
outline and sample chapter.

John Pawsey

60 High Street, Tarring, Worthing BN14 7NR
☎ 01903 205167   🄕 01903 205167
Contact *John Pawsey*

Founded 1981. Handles non-fiction only:
biography, current affairs, popular culture, sport.
Special interests: sport and biography. No fiction,
children's, drama scripts, poetry, short stories,
journalism, academic or submissions from
outside the UK. CLIENTS include David Rayvern
Allen, William Fotheringham, Don Hale, Patricia
Hall, Roy Hudd OBE, Gary Imlach, Anne Mustoe,
Gavin Nesham, Matt Rendell. COMMISSION
Home 12½%; US & Translation 19–25%. OVERSEAS

ASSOCIATES in the US, Japan, South America and
throughout Europe. Preliminary letter with s.a.e.
essential (no e-mail submissions). No reading fee.

Maggie Pearlstine Associates*

31 Ashley Gardens, Ambrosden Avenue, London
SW1P 1QE
☎ 020 7828 4212   🄕 020 7834 5546
maggie@pearlstine.co.uk
Contact *Maggie Pearlstine*

Founded 1989. Small agency representing a select
few authors. CLIENTS include Matthew Baylis,
Roy Hattersley, Charles Kennedy, Quentin Letts,
Claire Macdonald, Prof. Raj Persaud, Prof. Lesley
Regan, Henrietta Spencer-Churchill, Prof. Robert
Winston. No new authors. Translation rights
handled by **Aitken Alexander Associates Ltd**.

Peters Fraser & Dunlop Group Ltd
▷ PFD

PFD*

Drury House, 34–43 Russell Street, London
WC2B 5HA
☎ 020 7344 1000   🄕 020 7836 9539/7836 9541
postmaster@pfd.co.uk
www.pfd.co.uk
Chief Executive Officer *Caroline Michel*
Managing Director *Lesley Davey*
Books *Caroline Michel, Michael Sissons,
Annabel Merullo, Marcella Edwards*
Foreign Rights *Louisa Pritchard*
Film, TV, Theatre, Presenters, Public
Speakers *Gemma Hirst, Jessica Cooper,
Alexandra Henderson, Michelle Archer*
Children's Books *Suzy Jenvey*
Creative Director *Sue Douglas*
PFD New York *Erin Edmison*

PFD represents authors of fiction and non-fiction,
children's writers, screenwriters, playwrights,
documentary makers, technicians, presenters and
public speakers throughout the world. PFD has
85 years of international experience in all media.
Outline sample chapters and author biographies
should be addressed to the books department.
Material should be submitted on an exclusive
basis; or in any event disclose if material is being
submitted to other agencies or publishers. Return
postage essential. No reading fee. Response to
e-mail submissions cannot be guaranteed. See
website for submission guidelines.

Pollinger Limited*

9 Staple Inn, London WC1V 7QH
☎ 020 7404 0342   🄕 020 7242 5737
info@pollingerltd.com
www.pollingerltd.com
Managing Director *Lesley Pollinger*
Agents *Lesley Pollinger, Joanna Devereux,
Tim Bates, Ruth Needham*

Permissions *Electronic form available on website*
Rights Manager *rightsmanager@pollingerltd.com*
Consultants *Leigh Pollinger, Joan Deitch*

CLIENTS include Michael Coleman, Catherine Fisher, Philip Gross, Kelly McKain, Jeremy Poolman, Nicholas Rhea and Michal Snunit. Also the estates of H.E. Bates, Erskine Caldwell, Rachel Carson, D.H. Lawrence, John Masters, Carson McCullers, Alan Moorehead, W. Heath Robinson, Eric Frank Russell, Clifford D. Simak and other notables. COMMISSION Home 15%; Translation 20%. Overseas, theatrical and media associates. Submissions policy outlined on the website.

Shelley Power Literary Agency Ltd*

13 rue du Pré Saint Gervais, 75019 Paris, France
℡ 00 33 1 42 38 36 49   🖷 00 33 1 40 40 70 08
shelley.power@wanadoo.fr
Contact *Shelley Power*

Founded 1976. Shelley Power works between London and Paris. This is an English agency with London-based administration/accounts office and the editorial office in Paris. Handles general commercial fiction, quality fiction, business books, self-help, true crime, investigative exposés, film and entertainment. No scripts, short stories, children's or poetry. COMMISSION Home 10%; US & Translation 19% (new clients: Home 12½%; US & Translation 20%). Preliminary letter with brief outline of project (plus return postage as from UK or France) essential. 'We do not consider submissions by e-mail.' No reading fee.

PVA Management Limited

Hallow Park, Worcester WR2 6PG
℡ 01905 640663   🖷 01905 641842
maggie@pva.co.uk
www.pva.co.uk
Managing Director *Paul Vaughan*

Founded 1978. Handles non-fiction only. COMMISSION 15%. Send synopsis and sample chapters together with return postage.

Redhammer Management Ltd*

186 Bickenhall Mansions, Bickenhall Street, London W1U 6BX
℡ 020 7486 3465   🖷 020 7000 1249
info@redhammer.info
www.redhammer.info
Vice President *Peter Cox*

'Provides in-depth management for a small number of highly talented authors.' Willing to take on unpublished authors who are professional in their approach and who have major international potential, ideally for, book, film and/or tv. CLIENTS Martin Bell OBE, Garry Bushell, Brian Clegg, Audrey Eyton, Maria Harris, Lucy Johnson, Senator Orrin Hatch, Amanda Lees, Hon. Nicholas Monson, Prof. Ann Oakley,

Michelle Paver, Kellie Santin, Carolyn Soutar, Carole Stone, Donald Trelford, David Yelland. Submissions considered only if the guidelines given on the website have been followed. Do not send unsolicited mss by post. No radio or theatre scripts. No reading fee.

Robinson Literary Agency Ltd*

Block A511, The Jam Factory, 27 Green Walk, London SE1 4TT
℡ 020 7096 1460
info@rlabooks.co.uk
www.rlabooks.co.uk
Managing Director *Peter Robinson*
Agent *Sam Copeland*

Founded 2005. Handles fiction, general non-fiction, popular culture. Also, TV documentaries and presenters. CLIENTS Joanne Harris, Steve Jones, Ian Rankin, David Starkey. COMMISSION Home 15%; US & Translation 20%. Unsolicited material welcome. No reading fee.

Rogers, Coleridge & White Ltd*

20 Powis Mews, London W11 1JN
℡ 020 7221 3717   🖷 020 7229 9084
Chairman *Deborah Rogers*
Managing Director *Peter Straus*
Directors *Gill Coleridge, Pat White, David Miller, Zoe Waldie, Laurence Laluyaux, Stephen Edwards*
Other Agents *Catherine Pellegrino, Hannah Westland*

Founded 1967. Handles full-length mss fiction, non-fiction and children's books. COMMISSION Home 15%; US & Translation 20%. No submissions by fax or e-mail.

Uli Rushby-Smith Literary Agency

72 Plimsoll Road, London N4 2EE
℡ 020 7354 2718   🖷 020 7354 2718
Contact *Uli Rushby-Smith*

Founded 1993. Handles fiction and non-fiction, commercial and literary, both adult and children's. Film and TV rights handled in conjunction with a sub-agent. No plays or poetry. COMMISSION Home 15%; US & Translation 20%. Represents UK rights for Columbia University Press. Approach with an outline, two or three sample chapters and explanatory letter in the first instance (s.a.e. essential). No disks. No reading fee.

Rosemary Sandberg Ltd

6 Bayley Street, London WC1B 3HB
℡ 020 7304 4110
rosemary@sandberg.demon.co.uk
Contact *Rosemary Sandberg*

Founded 1991. In association with **Ed Victor Ltd**. Specializes in children's writers and illustrators. COMMISSION 10%. No unsolicited mss as client list is currently full.

The Sayle Literary Agency*

1 Petersfield, Cambridge CB1 1BB
☎ 01223 303035 ℻ 01223 301638
www.sayleliteraryagency.com
Contact *Rachel Calder*

Handles literary, crime and general fiction, current affairs, social issues, travel, biography, history, general non-fiction. No plays, poetry, children's, textbooks, technical, legal or medical books. COMMISSION Home 15%; US & Translation 20%. OVERSEAS ASSOCIATES Dunow & Carlson & Lerner Literary Agency; Darhansoff, Verrill Feldman; Anne Edelstein Literary Agency; New England Publishing Associates, USA; translation rights handled by **The Marsh Agency**; film rights by **Sayle Screen Ltd**. No unsolicited mss. Preliminary letter essential, including a brief biographical note and a synopsis plus two or three sample chapters. Return postage essential. No reading fee.

Sayle Screen Ltd

11 Jubilee Place, London SW3 3TD
☎ 020 7823 3883 ℻ 020 7823 3363
info@saylescreen.com
www.saylescreen.com
Agents *Jane Villiers, Matthew Bates, Toby Moorcroft*

Specializes in writers and directors for film and television. Also deals with theatre and radio. Works in association with **The Sayle Literary Agency**, **Greene & Heaton Ltd**, **Robinson Literary Agency Ltd** representing film and TV rights in novels and non-fiction. CLIENTS include Andrea Arnold, Shelagh Delaney, Marc Evans, Margaret Forster, John Forte, Rob Green, Mark Haddon, Christopher Monger, Paul Morrison, Gitta Sereny, Sue Townsend. No unsolicited material without preliminary letter. No e-mail submissions.

The Sharland Organisation Ltd

The Manor House, Manor Street, Raunds NN9 6JW
☎ 01933 626600
tso@btconnect.com
www.sharlandorganisation.co.uk
Contacts *Mike Sharland, Alice Sharland*

Founded 1988. Specializes in national and international film and TV negotiations. Also negotiates multimedia, interactive TV deals and computer game contracts. Handles scripts for film, TV, radio and theatre; also non-fiction. Markets books for film and handles stage, radio, film and TV rights for authors. No scientific, technical or poetry. COMMISSION Home 15%; US & Translation 20%. OVERSEAS ASSOCIATES various. No unsolicited mss. Preliminary enquiry by letter or phone essential.

Sheil Land Associates Ltd*
(incorporating Richard Scott Simon Ltd 1971 and Christy Moore Ltd 1912)

52 Doughty Street, London WC1N 2LS
☎ 020 7405 9351 ℻ 020 7831 2127
info@sheilland.co.uk
Agents, UK & US *Sonia Land, Vivien Green, Ian Drury*
Film/Theatrical/TV *Sophie Janson, Lucy Fawcett*
Foreign *Gaia Banks, Sophie LeGrand*

Founded 1962. Handles full-length, literary and commercial fiction and non-fiction, including: politics, history, military history, gardening, science, thrillers, crime, romance, drama, science fiction, fantasy, biography, business, travel, cookery, memoirs, humour, UK and foreign estates. Also theatre, film, radio and TV scripts. CLIENTS include Peter Ackroyd, Benedict Allen, Charles Allen, Pam Ayres, Melvyn Bragg, Steven Carroll, David Cohen, Anna del Conte, Judy Corbalis, Elizabeth Corley, Seamus Deane, Greg Dyke, Rosie Goodwin, Chris Ewan, Jean Goodhind, Robert Green, Susan Hill, Richard Holmes, HRH The Prince of Wales, Ian Johnstone, Irene Karafilly, Richard Mabey, Graham Rice, Robert Rigby, Steve Rider, Martin Riley, Diane Setterfield, Tom Sharpe, Martin Stephen, Jeffrey Tayler, Andrew Taylor, Rose Tremain, Barry Unsworth, Kevin Wells, Prof. Stanley Wells, Neil White, John Wilsher, Paul Wilson and the estates of Catherine Cookson, Patrick O'Brian, Penelope Mortimer, Jean Rhys and F.A. Wesley. COMMISSION Home 15%; US & Translation 20%. OVERSEAS ASSOCIATES Georges Borchardt, Inc. (Richard Scott Simon). US film and TV representation: CAA, APA, and others. Welcomes approaches from new clients either to start or to develop their careers. Preliminary letter with s.a.e. essential. No reading fee.

Caroline Sheldon Literary Agency Ltd*

70–75 Cowcross Street, London EC1M 6EJ
☎ 01983 760205/020 7336 6550
carolinesheldon@carolinesheldon.co.uk
pennyholroyde@carolinesheldon.co.uk
www.carolinesheldon.co.uk
Also at: Thorley Manor Farm, Thorley, Yarmouth PO41 0SJ
Contacts *Caroline Sheldon, Penny Holroyde*

Founded 1985. Handles fiction and children's books. Special interests: all fiction for women, sagas, suspense, contemporary, chick-lit, historical fiction, fantasy and comic novels. Non-fiction: true life stories, memoirs, humour, quirky, animal interest. Children's books: fiction for all age groups, contemporary, comic, fantasy and all major genres including fiction series. Picture book stories. Non-fiction. Children's illustration: all quality illustrations. COMMISSION Home 15%; US & Translation 20%; Film & TV 15%. Send submissions by post or e-mail to the Thorley address above. 'Include introductory information about yourself and ambitions and the first three chapters of your book.' If sending by e-mail, type 'Submission' in subject line. Send large s.a.e. for postal submissions. No reading fee. 'We do not represent TV or film scripts except for current clients. Editorial advice given on work of exceptional promise.'

Dorie Simmonds Agency*

Riverbank House, 1 Putney Bridge Approach, London SW6 3JD
☎ 020 7736 0002   ⊞ 020 7736 0010
dhsimmonds@aol.com
Contact *Dorie Simmonds*

Handles a wide range of subjects including commercial fiction and non-fiction, children's books and associated rights. Specialities include contemporary personalities and historical biographies, commercial women's fiction and children's books. COMMISSION Home & US 15%; Translation 20%. Outline required for non-fiction; a short synopsis for fiction with 2–3 sample chapters, and a c.v. with writing experience/publishing history. No reading fee. Return postage essential.

Jeffrey Simmons

15 Penn House, Mallory Street, London NW8 8SX
☎ 020 7224 8917
jasimmons@btconnect.com
Contact *Jeffrey Simmons*

Founded 1978. Handles biography and autobiography, cinema and theatre, fiction (both quality and commercial), history, law and crime, politics and world affairs, parapsychology and sport (but not exclusively). No science fiction/fantasy, children's books, cookery, crafts, hobbies or gardening. Film scripts handled only if by book-writing clients. Special interest in personality books of all sorts and fiction from young writers (i.e. under 40) with a future. COMMISSION Home 10–15%; US & Foreign 15%. Writers become clients by personal introduction or by letter, enclosing a synopsis if possible, a brief biography, a note of any previously published books, plus a list of any publishers and agents who have already seen the mss.

Richard Scott Simon Ltd
▷ Sheil Land Associates Ltd

Sinclair-Stevenson

3 South Terrace, London SW7 2TB
☎ 020 7581 2550   ⊞ 020 7581 2550
Contact *Christopher Sinclair-Stevenson*

Founded 1995. Handles biography, current affairs, travel, history, fiction, the arts. No scripts, children's, academic, science fiction/fantasy. CLIENTS include Jennifer Johnston, J.D.F. Jones, Ross King, Christopher Lee, Andrew Sinclair and the estates of Alec Guinness, John Cowper Powys and John Galsworthy. COMMISSION Home 10%; US 15%; Translation 20%. OVERSEAS ASSOCIATE T.C. Wallace Ltd, New York. Translation rights handled by **David Higham Associates**. Send synopsis with s.a.e. in the first instance. No reading fee.

Robert Smith Literary Agency Ltd*

12 Bridge Wharf, 156 Caledonian Road, London N1 9UU
☎ 020 7278 2444   ⊞ 020 7833 5680
robertsmith.literaryagency@virgin.net
Contact *Robert Smith*

Founded 1997. Handles non-fiction; biography, memoirs, current affairs, entertainment, true crime, cookery and lifestyle. No scripts, fiction, poetry, academic or children's books. CLIENTS Kate Adie (serializations), Martin Allen, Amanda Barrie (serializations), Kevin Booth, Jamie Cameron Stewart, Judy Cook, Stewart Evans, Neil and Christine Hamilton, James Haspiel, Nikola T. James, Muriel Jakubait, Lois Jenkins, Siobhan Kennedy-McGuinness, Roberta Kray, Jean MacColl, Ann Ming, Michelle Morgan, Theo Paphitis, Mike Reid, Frances Reilly, Keith Skinner, Jayne Sterne. COMMISSION Home 15%; US & Translation 20%. No unsolicited mss. Send a letter and synopsis in the first instance. No reading fee.

The Standen Literary Agency

53 Hardwicke Road, London N13 4SL
☎ 020 8889 1167   ⊞ 020 8889 1167
info@standenliteraryagency.com
www.standenliteraryagency.com
Contact *Mrs Yasmin Standen*

Founded 2004. Handles fiction, both adult and children's. 'We are interested in first time writers.' No TV/radio scripts. CLIENTS include Jonathan Yeatman Biggs, Zara Kane, Zoe Marriott, Andrew Murray. COMMISSION Home 15%; Overseas 20%. Follow the submissions procedure on the website. 'With regard to non-fiction, please contact us in the first instance.' No reading fee.

Elaine Steel

110 Gloucester Avenue, London NW1 8HX
☎ 020 7483 2681
ecmsteel@aol.com
Contact *Elaine Steel*

Founded 1986. Handles scripts, screenplays and books. No technical or academic. CLIENTS include Les Blair, Anna Campion, Michael Eaton,

Pearse Elliott, Gwyneth Hughes, Brian Keenan, Troy Kennedy Martin, James Lovelock, Rob Ritchie, Albie Sachs, Ben Steiner. COMMISSION Home 10%; US & Translation 20%. Initial phone call preferred.

Abner Stein*

10 Roland Gardens, London SW7 3PH
℡ 020 7373 0456   ℻ 020 7370 6316
Contact *Arabella Stein*

Founded 1971. Mainly represents US agents and authors but handles some full-length fiction and general non-fiction. No scientific, technical, etc. No scripts. COMMISSION Home 10%; US & Translation 20%.

Micheline Steinberg Associates

104 Great Portland Street, London W1W 6PE
℡ 020 7631 1310
info@steinplays.com
www.steinplays.com
Contacts *Micheline Steinberg, Helen MacAuley*

Founded 1988. Specializes in plays for stage, TV, film, animation and radio. Represents writers for film and TV rights in fiction and non-fiction on behalf of book agents. COMMISSION Home 10%; Overseas 15–20%. Works in association with agents in the USA and overseas. Return postage essential.

Shirley Stewart Literary Agency*

3rd Floor, 4A Nelson Road, London SE10 9JB
℡ 020 8293 3000
Director *Shirley Stewart*

Founded 1993. Handles fiction and non-fiction. No scripts, children's, science fiction, fantasy or poetry. COMMISSION Home 10–15%; US & Translation 20%. OVERSEAS ASSOCIATE Curtis Brown Ltd, New York; Carolyn Swayze Agency, Canada. Will consider unsolicited material; send letter with two or three sample chapters in the first instance. S.a.e. essential. Submissions by fax or on disk not accepted. No reading fee.

Sarah Such Literary Agency

81 Arabella Drive, London SW15 5LL
℡ 020 8876 4228   ℻ 020 8878 8705
sarah@sarahsuch.com
www.sarahsuch.com
Director *Sarah Such*

Founded 2006. Former publisher with 18 years' experience. Handles high-quality literary and commercial non-fiction and fiction including children's books. 'Always looking for original work and new talented writers.' COMMISSION Home 15%; US & Translation 20%. OVERSEAS ASSOCIATES worldwide. Works by recommendation but welcomes unsolicited synopses and sample chapter submissions. TV/

film scripts for established clients only. No radio or theatre scripts, poetry, fantasy, self-help or short stories. Contact by e-mail with brief synopsis, author biog and first two chapters (as Word attachment). S.a.e. essential if postal submission. No unsolicited mss and no telephone submissions. No reading fee.

Sunflower Literary Agency

BP 14, Lauzerte 82110 FRANCE
submission@sunflowerliteraryagency.com
www.sunflowerliteraryagency.com
Senior Editor *Paul Muller*
Correspondence Secretary *David Sherriff*

Founded 2003. Handles full-length fiction, especially thrillers (techno a plus); literary fiction, especially political and/or social satire (controversy a plus); erotica, in any of the above genres. No poetry, screenplays, non-fiction, short fiction, 'who-dunnits' or autobiography. COMMISSION Home 15% (first book), 10% (later books); USA & Translation: additional 5% to base rate. 'We *only* consider work from *unpublished* authors (or major changes of genre)'. Electronic submissions only: first contact via the website then e-mail only. Please do not write to the above address in the first instance. All requirements listed on the web page must be fulfilled before e-mail contact is made. No fees charged for first reading, see website for full terms.

The Susijn Agency Ltd

3rd Floor, 64 Great Titchfield Street, London W1W 7QH
℡ 020 7580 6341   ℻ 020 7580 8626
info@thesusijnagency.com
www.thesusijnagency.com
Contacts *Laura Susijn, Nicola Barr*

Founded April 1998. Specializes in selling rights worldwide in literary fiction and non-fiction. Also represents non-English language authors and publishers for UK, US and translation rights worldwide. COMMISSION Home 15%; US & Translation 15–20%. Send preliminary letter, synopsis and first two chapters by post. No reading fee.

The Tennyson Agency

10 Cleveland Avenue, Wimbledon Chase, London SW20 9EW
℡ 020 8543 5939
submissions@tenagy.co.uk
www.tenagy.co.uk
Contacts *Christopher Oxford, Adam Sheldon*

Founded 2001. Specializes in theatre, radio, television and film scripts. Related material considered on an ad hoc basis. No short stories, children's, poetry, travel, military/historical, academic, fantasy, science fiction or sport.

CLIENTS include Vivienne Allen, Tony Bagley, Kristina Bedford, Alastair Cording, Caroline Coxon, Iain Grant, Jonathan Holloway, Philip Hurd-Wood, Joanna Leigh, Steve MacGregor, Antony Mann, Ken Ross, John Ryan, Walter Saunders, Graeme Scarf, Diane Speakman, Diana Ward and the estate of Julian Howell. COMMISSION Home 12½–15%; Overseas 20%. No unsolicited material; send introductory letter with author's résumé and proposal/outline of work. No reading fee.

Lavinia Trevor Literary Agency*

29 Addison Place, London W11 4RJ
📞 020 7603 5254   📠 0870 129 0838
www.laviniatrevor.co.uk
Contact *Lavinia Trevor*

Founded 1993. Handles general fiction (literary and commercial) and non-fiction, including popular science. No fantasy, science-fiction, poetry, academic, technical or children's books. No TV, film, radio, theatre scripts. COMMISSION rate by agreement with author. 'Sorry – no unsolicited submissions.'

Jane Turnbull*

Mailing address: Barn Cottage, Veryan, Truro TR2 5QA
📞 01872 501317
jane.turnbull@btinternet.com
www.janeturnbull.co.uk
London office: 58 Elgin Crescent, London W11 2JJ
📞 020 7727 9409
Contact *Jane Turnbull*

Founded 1986. Handles fiction and non-fiction but specializes in biography, history, current affairs, self-help and humour. Translation rights handled by **Aitken Alexander Associates Ltd**. COMMISSION Home 10%; US & Foreign 20%. No unsolicited mss. Approach with letter in the first instance. No reading fee.

TVmyworld

14 Dean Street, London W1D 3RS
📞 020 7437 4188
business@tvmyworld.com
www.TVmyworld.com
Contacts *Mark Maco, Malcolm Rasala*

Specializes in books, television, movies, internet tv, games, second life. Has a production arm making motion pictures, television, commercials and internet television Send first 30 pages only initially.

United Agents Limited*

12–26 Lexington Street, London W1F 0LE
www.unitedagents.co.uk
Contacts *Simon Trewin, Caroline Dawnay, Jim Gill, Sarah Ballard, Rosemary Scoular,*
Rosemary Canter, Anna Webber, Jane Wills, Jessica Craig

Founded 2007. Handles fiction, non-fiction, biography, children's. Also TV, film, radio and theatre scripts. Unsolicited material welcome; see submissions policy on the website. No reading fee.

Ed Victor Ltd*

6 Bayley Street, Bedford Square, London WC1B 3HE
📞 020 7304 4100   📠 020 7304 4111
mary@edvictor.com
Contacts *Ed Victor, Maggie Phillips, Sophie Hicks, Grainne Fox, Charlie Campbell*
Foreign Rights Manager *Morag O'Brien*

Founded 1976. Handles a broad range of material but leans towards the more commercial ends of the fiction and non-fiction spectrums. Excellent children's book list. No poetry, scripts or academic. Takes on very few new writers and does not accept unsolicited submissions. After trying his hand at book publishing and literary magazines, Ed Victor, an ebullient American, found his true vocation. Strong opinions, very pushy and works hard for those whose intelligence he respects. Loves nothing more than a good title auction. CLIENTS include John Banville, Herbie Brennan, Eoin Colfer, Frederick Forsyth, A.A. Gill, Josephine Hart, Jack Higgins, Nigella Lawson, Kathy Lette, Allan Mallinson, Andrew Marr, Danny Scheinmann, Janet Street-Porter and the estates of Douglas Adams, Raymond Chandler, Dame Iris Murdoch, Sir Stephen Spender and Irving Wallace. COMMISSION Home & US 15%; Translation 20%. No unsolicited mss.

Wade & Doherty Literary Agency Ltd

33 Cormorant Lodge, Thomas More Street, London E1W 1AU
📞 020 7488 4171   📠 020 7488 4172
rw@rwla.com
bd@rwla.com
www.rwla.com
Contacts *Robin Wade, Broo Doherty*

Founded 2001. Handles general fiction and non-fiction including children's books. Specializes in military history and crime books. No scripts, poetry, plays or short stories. CLIENTS include Louise Cooper, Adam Guillain, Georgina Harding, Caroline Kington, Helen Oyeyemi, Lance Price, Ewen Southby-Tailyour. COMMISSION Home 10%; Overseas & Translation 20%. 'Fees negotiable if a contract has already been offered.' Send detailed synopsis and first 10,000 words by e-mail with a brief biography. No reading fee.

Cecily Ware Literary Agents

19C John Spencer Square, London N1 2LZ
☎ 020 7359 3787   🖷 020 7226 9828
info@cecilyware.com
Contacts *Cecily Ware, Gilly Schuster,*
Warren Sherman

Founded 1972. Primarily a film and TV script agency representing work in all areas: drama, children's, series/serials, adaptations, comedies, etc. COMMISSION Home 10%; US 10–20% by arrangement. No unsolicited mss or phone calls. Approach in writing only. No reading fee.

Watson, Little Ltd*

48–56 Bayham Place, London NW1 0EU
☎ 020 7388 7529   🖷 020 7388 8501
office@watsonlittle.com
Contacts *Mandy Little, James Wills*

Handles fiction, commercial women's fiction, crime and literary fiction. Non-fiction special interests include history, science, popular psychology, self-help and general leisure books. Also children's fiction and non-fiction. No short stories, poetry, TV, play or film scripts. Not interested in purely academic writers. COMMISSION Home 15%; US & Translation 20%. OVERSEAS ASSOCIATE The Marsh Agency; FILM & TV ASSOCIATES The Sharland Organisation Ltd; MBA Literary Agents Ltd; USA Howard Morhaim (adult); The Chudney Agency (children). Informative preliminary letter and synopsis with return postage essential.

A.P. Watt Ltd*

20 John Street, London WC1N 2DR
☎ 020 7405 6774   🖷 020 7831 2154
apw@apwatt.co.uk
www.apwatt.co.uk
Directors *Caradoc King, Linda Shaughnessy,*
Derek Johns, Georgia Garrett,
Natasha Fairweather, Sheila Crowley

Founded 1875. The oldest-established literary agency in the world. Handles full-length typescripts, including children's books, screenplays for film and TV. No poetry, academic or specialist works. CLIENTS include Monica Ali, Trezza Azzopardi, David Baddiel, Sebastian Barry, Quentin Blake, Melvin Burgess, Marika Cobbold, Michael Cox, Helen Dunmore, Nicholas Evans, Giles Foden, Esther Freud, Janice Galloway, Martin Gilbert, Nadine Gordimer, Linda Grant, Reginald Hill, Michael Holroyd, Michael Ignatieff, Mick Jackson, Philip Kerr, Dick King-Smith, India Knight, John Lanchester, Alison Lurie, Jan Morris, Andrew O'Hagan, Susie Orbach, Tony Parsons, Caryl Phillips, Philip Pullman, James Robertson, Jancis Robinson, Jon Ronson, Elaine Showalter, Zadie Smith, Graham Swift, Fiona Walker and the estates of Graves and Maugham. COMMISSION

Home 15%; US & Translation 20%. No unsolicited mss accepted.

Josef Weinberger Plays

12–14 Mortimer Street, London W1T 3JJ
☎ 020 7580 2827   🖷 020 7436 9616
general.info@jwmail.co.uk
www.josef-weinberger.com
Contact *Michael Callahan*

Josef Weinberger is both agent and publisher of scripts for the theatre. CLIENTS include Ray Cooney, John Godber, Peter Gordon, Debbie Isitt, Arthur Miller, Sam Shepard, John Steinbeck. OVERSEAS REPRESENTATIVES in the USA, Canada, Australia, New Zealand, India, South Africa and Zimbabwe. No unsolicited mss; introductory letter essential. No reading fee.

John Welch, Literary Consultant & Agent

Mill Cottage, Calf Lane, Chipping Camden GL55 6JQ
☎ 01386 840237   🖷 01386 840568
johnwelch@cyphus.co.uk
Contact *John Welch*

Founded 1992. Handles military aviation and naval history, and history in general. No fiction, poetry, children's books or scripts for radio, TV, film or theatre. Already has a full hand of authors so no new authors being considered. CLIENTS include Alexander Baron, Michael Calvert, Patrick Delaforce, David Holbrook, Timothy Jenkins, Sybil Marshall, Ewart Oakeshott, Peter Reid, Norman Scarfe, Anthony Trew, Peter Trew, David Wragg. COMMISSION Home 10%.

Eve White Literary Agent*

1a High Street, Kintbury RG17 9TJ
☎ 01488 657656
eve@evewhite.co.uk
www.evewhite.co.uk
Contact *Eve White*

Founded 2003. Handles full-length adult and children's fiction and non-fiction. No poetry, short stories or textbooks. CLIENTS Susannah Corbett, Carolyn Ching, Jimmy Docherty, Rae Earl, Shanta Everington, David Flavell, Marguerite Hann Syme, Abie Longstaff, Vijay Medtia, Gillian Rogerson, Tabitha Suzuma, Andy Stanton. COMMISSION Home 15%; US & Translation 20%. Please see website for up-to-date submission requirements. No initial approach by e-mail or telephone. No reading fee.

Dinah Wiener Ltd*

12 Cornwall Grove, Chiswick, London W4 2LB
☎ 020 8994 6011   🖷 020 8994 6044
Contact *Dinah Wiener*

Founded 1985. Handles fiction and general non-fiction: auto/biography, popular science, cookery.

No scripts, children's or poetry. CLIENTS include Valerie-Anne Baglietto, Malcolm Billings, Guy Burt, David Deutsch, Wendy K. Harris, Jenny Hobbs, Mark Jeffery, Michael Lockwood, Daniel Snowman, Rachel Trethewey, Marcia Willett. COMMISSION Home 15%; US & Translation 20%. Approach with preliminary letter in first instance, giving full but brief c.v. of past work and future plans. Mss submitted must include s.a.e. and be typed in double-spacing.

The Wylie Agency (UK) Ltd

17 Bedford Square, London WC1B 3JA
☏ 020 7908 5900   ☏ 020 7908 5901
mail@wylieagency.co.uk

Handles fiction and non-fiction. No scripts or children's books. COMMISSION Home 10%; US 15%; Translation 20%. The Wylie Agency does not accept unsolicited submissions. Enquire by letter or e-mail before submitting. Any submission must be accompanied by return postage/s.a.e.

Agency consultants

Agent Research & Evaluation, Inc. (AR&E)

425 North 20th Street, Philadelphia, PA 19130, USA
☏ 001 215 563 1867   ☏ 001 215 563 6797
info@agentresearch.com
www.agentresearch.uk.net
Contact *Bill Martin*

US consultancy founded in 1996. Provides authors, editors and other professionals with data on clients and sales made by literary agents in the USA, UK and Canada, i.e. who sells what to whom for how much. Information is culled from the trade and general press, and collection has been continuous since 1980. Individual reports on specific agents and various forms of manuscript-agent match-up services are available. See website for pricing or send s.a.e. Publishes *Talking Agents*, the AR&E newsletter. The website offers a free service of agent verification.

UK Literary Scouts

Louise Allen-Jones Literary Scouts

5c Old Town, London SW4 0JT
☎ 020 7720 2453   🖷 020 7627 3510
Contacts *Louise Allen-Jones* (louise@louiseallenjones.com), *Lucy Abrahams* (lucy@louiseallenjones.com), *Sally Page* (sally@louiseallenjones.com)

Scouts for Ullstein Buchverlage (Ullstein hc/ppbk, List, Claassen, Schroeder, Econ, Propyläen), Germany; AW Bruna (Bruna, VIP, Signatuur) and Meulenhoff Boekerij (De Boekerij, Forum, Mynx, Arena, Meulenhoff), The Netherlands; Editoria Record, Brazil; Sperling & Kupfer, Italy; Keter Books, Israel; Oceanida, Greece; The English Agency, Japan; BBC TV Drama, UK.

Badcock & Rozycki Literary Scouts

1 Old Compton Street, London W1D 5JA
☎ 020 7734 7997   🖷 020 7734 6886
Contacts *June Badcock* (june@litscouts.co.uk), *Barbara Rozycki* (barbara@litscouts.co.uk), *Claire Holt* (claire@litscouts.co.uk), *Hollie Paterson* (hollie@litscouts.co.uk)

Scouts for Wilhelm Heyne Verlag and Diana Verlag, Germany; Unieboek BV and Het Spectrum, The Netherlands; RCS Libri Group, Italy; Editions Jean-Claude Lattès, France; Ediciones Salamandra, Spain; Ellinika Grammata, Greece; Werner Söderström Osakeyhtiö, Finland; Forum, Sweden; Damm division of Cappelen Damm A/S, Norway; Lindhardt & Ringhof Forlag, Denmark; Verold division of Bjartur-Verold, Iceland; Rupa & Co Publishers, India; Blueprint Pictures, UK.

Natasha Farrant

77 Starfield Road, London W12 9SN
☎ 020 8746 1857
natasha.farrant@btinternet.com
Contact *Natasha Farrant*

Scouts for children's fiction on behalf of Hachette Jeunesse, France; Edizione Piemme, Italy; Carlsen Verlag, Germany; Jane Southern Literary Scout; Laika Entertainment (animation).

Anne Louise Fisher

29 D'Arblay Street, London W1F 8EP
☎ 020 7494 4609   🖷 020 7494 4611
annelouise@alfisher.co.uk
Contacts *Anne Louise Fisher, Catherine Eccles*

Scouts for Doubleday/Broadway Books/Nan Talese/Spiegel & Grau, US; Doubleday/Bond Street Books, Canada; Librarie Plon, Univers Poche/Pocket Jeunesse, France; C. Bertelsmann, Knaus, Blanvalet, DVA/Siedler/Pantheon, C. Bertelsmann Jugenbuch, Germany; Arnoldo Mondadori Editore, Oscar and Mondadori Ragazzi, Italy; Albert Bonniers Forlag, Sweden; Otava, Finland; Gyldendal Norsk Forlag, Norway; Gyldendal, Denmark; Mouria, Holland; Random House Mondadori and Montena, Spain; Editora Nova Fronteira, Brazil, Patakis Publications, Greece; Heyday Films, UK.

Folly Marland, Literary Scout

6 Elmcroft Street, London E5 0SQ
☎ 020 8986 0111   🖷 020 8986 0111
fmarland@pobox.com
Contact *Folly Marland*

Scouts for Scherz Verlag, Krüger Verlag, Germany; Livani, Greece; FMG: Truth & Dare, Pimento, The Netherlands; Village Books, Japan; Kowalski, Italy.

Rosalind Ramsay Limited

Third Floor, 23 Monmouth Street, London WC2H 9DD
☎ 020 7836 0054
ros@rosalindramsay.com
www.rosalindramsay.com
Contacts *Rosalind Ramsay*

Assistants *Ben Fowler, Carla Burnett*
Office Manager *Lindsey Clarke*

Scouts for Ambo Anthos and Uitgeverij Artemis, The Netherlands; Kadokawa Shoten, Japan; Kiepenheuer and Witsch, Germany; Droemer Knaur, Germany; Kinneret Zmora Bitan Dvir, Israel; Norstedts Forlagsgrupp, Sweden; Il Saggiatore, Edizioni EL, Italy; Santillana Ediciones General, Spain; Editora Objetiva, Brazil; Channel 4 and FilmFour, UK.

Heather Schiller

heather.schiller@virgin.net
Contact *Heather Schiller*

Scouts for Prometheus, The Netherlands; Cappelen, Norway; Piper, Germany; Bompiani, Italy; Urano, Spain.

Petra Sluka

3 Dry Bank Road, Tonbridge TN10 3BS
℡ 01732 353694
petrasluka@petrasluka.plus.com
Contact *Petra Sluka*

Scouts for Verlagsgruppe Lübbe, Gemany; De Fontein/De Kern, Holland; Ediciones B in Spain; Bra Böcker, Sweden; Japan Uni Agency, Inc., Japan; Harlenic Hallas, Greece.

Jane Southern

11 Russell Avenue, Bedford MK40 3TE
℡ 01234 400147
jane@janesouthern.com
Contact *Jane Southern*

UK Scout for Der Club, Germany; The House of Books, The Netherlands; Belfond, Presses de la Cité and France Loisirs, France; Newton Compton, Italy; Damm Forlag, Sweden; Planeta Group, Spain; Dom Quixote, Portugal; Minoas, Greece; Owls Agency, Japan.

Sylvie Zannier-Betts

114 Springfield Road, Brighton BN1 6DE
℡ 01273 557370   🖷 01273 557370
szannier@lineone.net
Contact *Sylvie Zannier-Betts*

Scouts for Uitgeverij De Geus, The Netherlands; S. Fischer Verlag, Germany; Gummerus Publishers, Finland; Alfabeta Bokfoerlag, Sweden; Borgen Forlag, Denmark. Exclusive agent in the UK for S. Fischer Verlag, Germany.

PR Consultants

Claire Bowles Publicity
Ashley Court, Ashley, Market Harborough
LE16 8HF
☎ 01858 565800   🖷 01858 565811
www.clairebowlespr.co.uk
Contact *Claire Bowles*

Represents non-fiction: The Good Pub Guide; Oz
Clarke. CLIENTS include Workman (US); Artisan
(US); Ebury Press; BBC Books; Time Out; David
& Charles; Readers Digest; Anova Books

Maria Boyle Communications Limited
36 Shalstone Road, Mortlake, London SW14 7HR
☎ 020 8876 8444
maria@mbcomms.co.uk
www.mbcomms.co.uk
Director *Maria Boyle*

Handles PR projects for some of the UK's
leading publishers. Named as one of the top
three independent PR consultants in the UK by
PR Week. Works for a range of consumer and
corporate clients. 'One area of specialism includes
developing and delivering media campaigns
that jump off the books pages and feature in the
main body of newspapers and magazines, on
national high profile TV and radio programmes
and online.' CLIENTS include Penguin Classics;
Pearson Booktime; DK Rough Guides;
HarperCollins and a number of high-profile
bestselling authors.

Cameron Publicity and Marketing
35 Edward Road, Farnham GU9 8NP
☎ 07903 951957
ben@cameronpm.co.uk
www.cameronpm.co.uk
Contact *Ben Cameron*

Represents publishers and authors. CLIENTS
include Anova Books; Anthem Press; Hachette
Children's Books (Hodder and Orchard); Pavilion
Books; Smith TCI Publicity.

Colbert Macalister PR
7 Killieser Avenue, London SW2 5NU
☎ 020 8671 6615
ailsa@colmacpr.co.uk
www.colbertmacalister.co.uk

Contacts *Ailsa Macalister, Diana Colbert*

Represents publishers' lists and authors. CLIENTS
include John Blake Publishing; Random House;
Faber & Faber; Hay House; Bonnie; Robert Hale;
Absolute Press.

Colman Getty
28 Windmill Street, London W1T 2JJ
☎ 020 7631 2666   🖷 020 7631 2699
info@colmangetty.co.uk
www.colmangetty.co.uk

Handles prizes, book campaigns, authors,
publishers' lists, professional associations.
CLIENTS include The Man Booker Prize for
Fiction; World Book Day; National Poetry Day;
Icon Books; Frankfurt Book Fair.

Day Five Ltd
1 Archibald Road, London N7 0AN
☎ 020 7619 0098
eobr@blueyonder.co.uk
Contact *Emma O'Bryen*

Represents publishers' lists and individual book
projects. CLIENTS include Frances Lincoln
Publishers; Macmillan; Natural History Museum.

FMcM Associates
3rd Floor, Colonial Buildings, 59–61 Hatton
Garden, London EC1N 8LS
☎ 020 7405 7422   🖷 020 7405 7424
publicity@fmcm.co.uk
www.fmcm.co.uk
Managing Director *Fiona McMorrough*

Founded 1998. Represents authors, publishers'
lists, literary festivals, industry associations.
Winner of two Publishers Publicity Circle
Awards, and the inaugural Hospital Award
for Creative Contribution to Book Publishing.
Specializes in publishing, literature, the arts and
events. Clients include; Marian Keyes; Maggie
O'Farrell; Tracy Chevalier; Salley Vickers;
Elizabeth Noble; Rosie Thomas; Lionel Shriver;
Lisa Jardine; Ronan Bennett; Michael Wood; Griff
Rhys Jones; Althorp Literary Festival; Canongate;
Cape; Penguin; Little, Brown; Headline;
HarperCollins; Fourth Estate; Guardian Books;

Preface Publishing; Brandon Books; Independent Alliance.

Cathy Frazer PR

Old Hayle Barn, Hayle Farm, Marle Place Road, Horsmonden TN12 8DZ

☎ 01892 724156   🖷 01892 724155

cathyfrazer@hotmail.com

Contact *Cathy Frazer*

Handles illustrated non-reference, consumer education, children's books; subjects include wine, food, antiques, sport, languages, health, parenting. Individual authors also represented. Over 20 years of publishing experience. CLIENTS include Hodder Education; Hodder Children's Books; Virgin; Mitchell Beazley; Millers; DBP; Kingfisher; Carol Vorderman; Uri Geller; Viscount Linley; Sir Patrick Moore; Carol Smillie; Lorraine Kelly; Peter Allis; Jonty Hearnden; Eric Knowles; Judith Miller; Marguerite Patten; Sarah Kennedy.

Giant Rooster PR

9 Hereford Road, London W2 4AB

☎ 020 7792 4701

gina@giantroosterpr.co.uk

Contact *Gina Rozner*

Established 1997. Handles authors, publishers' lists and prizes. CLIENTS include Weidenfeld & Nicolson; Simon & Schuster; Poetry Book Society (T.S. Eliot Prize); Welsh Books Council; Constable & Robinson.

Gotch Solutions

Flat 1, 10 Holmdene Avenue, London SE24 9LF

☎ 020 7733 0882

gotchsolutions@btinternet.com

Contact *Corinne Gotch*

General PR, particularly children's.

Katie Hambly PR

6 Lockington Avenue, Hartley, Plymouth PL3 5QP

☎ 01752 778211   🖷 01752 778211

katie.hambly1@btinternet.com

Contact *Katie Hambly*

Handles authors and currently working for Weidenfeld & Nicolson, part of the Orion Publishing Group.

Idea Generation

11 Chance Street, London E2 7JB

☎ 020 7749 6850

hector@ideageneration.co.uk

www.ideageneration.co.uk

Managing Director *Hector Proud*

Arts and entertainment PR consultancy, working across the arts, photography, publishing, museums, film, TV, media and lifestyle sectors. Specializes in PR for illustrated, arts

and photographic books. CLIENTS include Gloria Books; Genesis Publications; Vision On Publishing; Dazed Publishing; Waddell Publishing.

Andrea Marks Public Relations

146 Edgwarebury Lane, Edgware HA8 8NE

☎ 020 8958 4398   🖷 020 8905 3727

info@andreamarks.co.uk

www.andreamarks.co.uk

Contact *Andrea Marks*

Corporate PR for publishers, booksellers, publishers' lists (not individual titles), prizes and awards, events, projects and initiatives, professional associations.

MGA

190 Shaftesbury Avenue, London WC2H 8JL

☎ 020 7836 4774   🖷 020 7836 4775

mga@mga-pr.com

www.mga-pr.com

Managing Director *Beth Macdougall*

Director *Bethan Jones*

Formerly Macdougall Gabriel Associates. Handles publishers' lists and authors. CLIENTS include (publishers' lists) English Heritage; National Archives; Souvenir Press; Chaucer Press; Mercury Junior; Asia Ink and Psychology News Press. Selected books for Penguin, HarperCollins, Michael Joseph, Weidenfeld & Nicolson, Oberon Books, Constable & Robinson and Alastair Sawday. Authors include Brian Aldiss, Bowvayne and Dinah Lampitt/Deryn Lake.

Louise Page PR

☎ 020 8741 5663   🖷 020 8741 5663

louise@lpagepr.co.uk

Contact *Louise Page*

PR for fiction and non-fiction authors. CLIENTS include Maeve Binchy; Martina Cole; Wendy Holden.

Platypus PR

2 Tidy Street, Brighton BN1 4EL

☎ 01273 692215

info@platypuspr.com

www.platypuspr.com

Contact *Jeff Scott*

Represents publishers, authors, including self-published authors. CLIENTS include Cambridge University Press; Simon & Schuster; Dorling Kindersley; Pearson Education; Continuum; Clairview Books; Gibson Square; various independent publishers. Books handled have won the Sunday Times Political Book of the Year 2005, Economist Best Book 2006: Politics & Current Affairs and Financial Times Best Business Book of 2007.

Nicky Potter

181 Alexandra Park Road, London N22 7UL
☎ 020 8889 9735
nicpot@dircon.co.uk
Contact *Nicky Potter*

Represents publishers' lists, prizes and professional organizations. CLIENTS include Booktrust (for the Children's Laureate); School Library Association (SLA School Librarian of the Year Award); Marsh Award for Children's Literature in Translation; Frances Lincoln; Barrington Stoke.

Publishing Services

9 Curwen Road, London W12 9AF
☎ 020 8222 6800   🖷 020 8222 6799
susanne@publishing-services.co.uk
www.publishing-services.co.uk

Handles publishers and authors. Third-party associates include book distributor Central Books; sales agency Signature Book Services. Publishers: mainly niche publishers such as educational/reference publisher Lucas Publications and history specialist Peter Lang. Authors: the emphasis is on media-friendly authors capable of selling a minimum of 5,000 copies. Areas of interest include politics/current affairs; English language/reference; history, biography, autobiography; quirky idiosyncratic books; country/green. No children's books, poetry, plays or screenplays. In the spring of 2008 held a series of seven Society of Authors seminars.

Sonia Pugh PR

☎ 01375 891063
sonia.pugh@ntlworld.com
Contact *Sonia Pugh*

Independent PR consultancy which handles illustrated non-fiction books in all areas including launches, etc.

Sally Randall PR

Forty Acre Oast, Woodchurch, Ashford TN26 3PW
☎ 01233 860670
sallyrandall@reynoldsm.f2s.com
Contact *Sally Randall*

Handles general fiction and non-fiction, independent publishers' lists, individual authors/titles, arts. CLIENTS include Meet the Author; Galore Park; Simon & Schuster; Michael O'Mara.

Claire Sawford PR

Studio One, 7 Chalcot Road, London NW1 8LH
☎ 020 7722 4114
cs@cspr.uk.net
www.cspr.uk.net
Contact *Claire Sawford*

Handles authors, publishers, prizes and professional associations. Specializes in non-fiction: art, architecture, fashion, design, biography. CLIENTS include V&A Publications; RIBA Bookshops; Royal Academy of Arts; National Portrait Gallery; Gill Hicks.

Irish Literary Agents

The Book Bureau Literary Agency
7 Duncairn Avenue, Bray, Co. Wicklow
℡ 00 353 1 276 4996   ℻ 00 353 1 276 4834
thebookbureau@oceanfree.net
Contact *Ger Nichol*

Handles general and literary fiction. Special interest in women's fiction, crime and thrillers and some non-fiction. COMMISSION Home 10%; Overseas 20%. Works with foreign associates. Will suggest revision. Send preliminary letter, synopsis and first three chapters (single line spacing preferred); return postage essential (IRCs only from UK and abroad). No reading fee.

Font International Literary Agency
Hollyville House, Hollybrook Road, Clontarf, Dublin 3
℡ 00 353 1 853 2356
info@fontlitagency.com
Contacts *Aine McCarthy, Ita O'Driscoll*

Founded 2003. Handles book-length adult fiction and non-fiction from previously published writers (no children's, drama, sci-fi, erotic, technical or poetry). COMMISSION 15–20%; Translation 20–25%. ASSOCIATE **The Marsh Agency**, London for translation rights. No unsolicited mss. 'As an initial contact, please query our interest in your property either by post only.' Include details of writing and other media experience. S.a.e. required. 'We regret that we cannot discuss queries over the phone.' No reading fee.

Marianne Gunn O'Connor Literary Agency
Morrison Chambers, Suite 17, 32 Nassau Street, Dublin 2
mgoclitagency@eircom.net
Contact *Marianne Gunn O'Connor*

Founded 1996. Handles commercial and literary fiction, non-fiction: biography, health and children's fiction. CLIENTS include Patrick McCabe, Cecelia Ahern, Claudia Carroll, Morag Prunty, Gisele Scanlon (Goddess Guide), Chris Binchy, Anita Notaro, Noelle Harrison, Julie Dam, Mike McCormack, Paddy McMahon, Thrity Engineer, John Lynch, Helen Falconer, Brinda Charry, Peter Murphy, Louise Douglas, Carla Gylnn, Caroline McFarlane-Watts. COMMISSION UK 15%; Overseas 20%; Film & TV 20%. Translation rights handled by Vicki Satlow Literary Agency, Milan. No unsolicited mss; send preliminary enquiry letter plus half-page synopsis per e-mail.

The Lisa Richards Agency
108 Upper Leeson Street, Dublin 4
℡ 00 353 1 637 5000   ℻ 00 353 1 667 1256
faith@lisarichards.ie
www.lisarichards.ie
Contact *Faith O'Grady*

Founded 1998. Handles fiction and general non-fiction. CLIENTS include Charlie Bird, Helena Close, June Considine, Matt Cooper, Denise Deegan, Gary Duggan, Christine Dwyer Hickey, Karen Gillece, Tara Heavey, Paul Howard (Ross O'Carroll-Kelly), Arlene Hunt, Roisin Ingle, Alison Jameson, George Lee, Declan Lynch, Roisin Meaney, Pauline McLynn, Anna McPartlin, Damien Owens, Trisha Rainsford, Kevin Rafter, Ray Scannell, Eirin Thompson. COMMISSION Home 10%; UK 15%; US & Translation 20%; Film & TV 15%. OVERSEAS ASSOCIATE **The Marsh Agency** for translation rights. Approach with proposal and sample chapter for non-fiction, and 3–4 chapters and synopsis for fiction (s.a.e. essential). No reading fee.

Jonathan Williams Literary Agency
Rosney Mews, Upper Glenageary Road, Glenageary, Co. Dublin
℡ 00 353 1 280 3482   ℻ 00 353 1 280 3482
Contact *Jonathan Williams*

Founded 1980. Handles general trade books: fiction, auto/biography, travel, politics, history, music, literature and criticism, gardening, cookery, sport and leisure, humour, reference, social questions, photography. Some poetry. No plays, science fiction, children's books, mind, body and spirit, computer books, theology, multimedia, motoring, aviation. COMMISSION Home 10%; US & Translation 20%. OVERSEAS ASSOCIATES Piergiorgio Nicolazzini Literary Agency, Italy; Linda Kohn, International Literature Bureau, Holland and Germany; Antonia Kerrigan Literary Agency, Spain; Tuttle-Mori Agency Inc., Japan. No reading fee 'unless the author wants a very fast opinion'. Initial approach by phone or letter.

US Literary Agents

* = Members of the **Association of Authors' Representatives, Inc.**

Dominick Abel Literary Agency, Inc.*
146 West 82nd Street, Suite 1B, New York
NY 10024
℡ 001 212 877 0710   ℻ 001 212 595 3133
Contact *Dominick Abel*

Established 1975. No scripts, children's books or
poetry. COMMISSION Home 15%; Translation 20%.
No unsolicited material. Send query letter with
s.a.e. in the first instance. No reading fee.

Miriam Altshuler Literary Agency*
53 Old Post Road North, Red Hook NY 12571
℡ 001 845 758 9408
www.miriamaltshulerliteraryagency.com/

Founded 1994. Handles literary and commercial
fiction; general non-fiction, narrative non-fiction,
memoirs, psychology, biography, young adult
fiction. No romance, science fiction, mysteries,
poetry, westerns, fantasy, how-to, techno thrillers
or self-help. COMMISSION Home 15%; Foreign &
Translation 20%. Send query letter and synopsis
in the first instance with return postage or s.a.e.
(no e-mail or fax queries). No reading fee.

The Axelrod Agency*
55 Main Street, PO Box 357, Chatham NY 12037
℡ 001 518 392 2100
steve@axelrodagency.com
Contact *Steven Axelrod*

Founded 1983. Handles commercial women's
fiction, romantic fiction and mysteries. No
non-fiction or scripts. COMMISSION Home 15%;
Translation 20%. Welcomes submissions by post
(include s.a.e.) or e-mail. No reading fee.

Malaga Baldi Literary Agency
233 West 99th Street, Suite 19C, New York
NY 10025
℡ 001 212 222 3213
baldibooks@gmail.com
Contact *Malaga Baldi*

Founded 1986. Handles quality fiction and non-
fiction. No scripts. No westerns, men's adventure,
science fiction/fantasy, romance, how-to, young
adult or children's. COMMISSION 15%. OVERSEAS
ASSOCIATES **Abner Stein**, UK; Owl Agency;
Eliane Benisti, France; Marsh Agency. Writers of

fiction should send query letter describing the
novel plus IRCs. For non-fiction, approach in
writing with a proposal, table of contents and two
sample chapters. No reading fee.

The Balkin Agency, Inc.*
PO Box 222, Amherst MA 01004
℡ 001 413 548 9835   ℻ 001 413 548 9836
rick62838@crocker.com
Contact *Richard Balkin*

Founded 1973. Handles adult non-fiction only.
COMMISSION Home 15%; Foreign 20%. No reading
fee for outlines and synopses.

Loretta Barrett Books, Inc.*
101 Fifth Avenue, New York NY 10003
℡ 001 212 242 3420
www.lorettabarrettbooks.com
Contact *Loretta Barrett*

Founded 1990. Handles all non-fiction and fiction
genres except children's, poetry, science fiction/
fantasy and historical romance. Specializes in
women's fiction, history, spirituality. No scripts.
COMMISSION Home 15%; Foreign 20%. No e-mail
submissions. Send query letter with biography
and return postage only in the first instance.

Meredith Bernstein Literary Agency, Inc.*
2095 Broadway, Suite 505, New York NY 10023
℡ 001 212 799 1007   ℻ 001 212 799 1145
Contacts *Meredith Bernstein*

Founded 1981. Handles commercial and literary
fiction, mysteries and non-fiction (women's
issues, biography, memoirs, health, current
affairs, crafts). COMMISSION Home & Dramatic
15%; Translation 20%. OVERSEAS ASSOCIATES
Abner Stein, UK; Lennart Sane, Holland,
Scandinavia and Spanish language; Thomas
Schluck, Germany; Bardon Chinese Media
Agency; William Miller, Japan; Frederique
Porretta, France; Agenzia Letteraria, Italy.

Bleecker Street Associates, Inc.*
532 LaGuardia Place, #617, New York NY 10012
℡ 001 212 677 4492   ℻ 001 212 388 0001

Founded 1984. Handles fiction: women's, mystery,
suspense, literary; non-fiction: history, women's

interests, parenting, health, relationships, psychology, sports, sociology, current events, biography, spirituality, New Age, business. No poetry, children's, westerns, science fiction, professional, academic. COMMISSION Home 15%; Foreign 25%. Send query letter in the first instance. Will only respond if envelope and return postage enclosed. No phone calls, faxes or e-mails. No reading fee.

Georges Borchardt, Inc.*

136 East 57th Street, New York NY 10022
☎ 001 212 753 5785   ⓕ 001 212 838 6518

Founded 1967. Works mostly with established/published authors. Specializes in fiction, biography, and general non-fiction of unusual interest. COMMISSION Home, UK & Dramatic 15%; Translation 20%. UK ASSOCIATE **Sheil Land Associates Ltd (Richard Scott Simon)**, London. Unsolicited mss not read.

Barbara Braun Associates, Inc.*

151 West 19th Street, 4th Floor, New York NY 10011
☎ 001 212 604 9023   ⓕ 001 212 604 9041
barbara@barbarabraunagency.com
www/barbarabraunagency.com
Contacts *Barbara Braun, John Baker*

Founded 1995. Handles literary and mainstream, women's and historical fiction, also serious non-fiction, including psychology and biography. Also young adult and mystery books. Specializes in art history, archaeology, architecture and cultural history. No scripts, poetry, science fiction. COMMISSION Home 15%; Foreign 20%. OVERSEAS ASSOCIATE Chandler Crawford Literary Agency. No unsolicited mss; send query letter in the first instance. See website for submission guidelines. No reading fee.

Browne & Miller Literary Associates*

410 S. Michigan Avenue, Suite 460, Chicago IL 60605
☎ 001 312 922 3063   ⓕ 001 312 922 1905
mail@browneandmiller.com
www.browneandmiller.com
President *Danielle Egan-Miller*

Founded 1971. Formerly known as Multimedia Product Development. Handles commercial and literary fiction, and practical non-fiction with wide appeal. No scripts, juvenile, science fiction or poetry. COMMISSION Home 15%; Foreign & Translation 20%. OVERSEAS ASSOCIATES in Europe, Latin America, Japan and Asia. In the first instance, send query letter with return postage. No unsolicited material. No reading fee.

Pema Browne Ltd, Literary Agents

11 Tena Place, Valley Cottage NY 10989
ppbltd@optonline.net
www.pemabrowneltd.com
Contact *Pema Browne*

Founded 1966. ('Pema rhymes with Emma.') Handles mass-market mainstream and hardcover fiction: romance, business, children's picture books and young adult; non-fiction: how-to and reference. COMMISSION Home 20%; Translation 20%; Overseas authors 20%. No unsolicited mss; send query letter with IRCs. No fax or e-mail queries. Also handles illustrators' work. No longer accepting screenplays. 'We are accepting very few new clients at this time.'

Sheree Bykofsky Associates, Inc.*

PO Box 706, Brigantine NJ 08203
☎ 001 212 244 4144
www.shereebee.com
Contact *submitbee@aol.com*

Founded 1991. Handles adult fiction and non-fiction. No scripts. No children's, young adult, horror, science fiction, romance, westerns, occult or supernatural. COMMISSION Home 15%; UK (including sub-agent's fee) 20%. No unsolicited mss. Send query letter first with brief synopsis or outline and writing sample (1–3 pp) for fiction. IRCs essential for reply or return of material. 'Please do not send material via methods that require signature, such as FedEx, etc.' No phone calls. See website for submission guidelines. No reading fee.

Maria Carvainis Agency, Inc.*

Rockefeller Center, 1270 Avenue of the Americas, Suite 2320, New York NY 10020
☎ 001 212 245 6365   ⓕ 001 212 245 7196
mca@mariacarvainisagency.com
President *Maria Carvainis*

Founded 1977. Handles fiction: literary and mainstream, contemporary women's, mystery, suspense, historical, young adult novels; non-fiction: business, women's issues, memoirs, health, biography, medicine. No film scripts unless from writers with established credits. No science fiction. COMMISSION Domestic & Dramatic 15%; Translation 20%. No faxed or e-mailed queries. No unsolicited mss; they will be returned unread. Queries only, with IRCs for response. No reading fee.

Castiglia Literary Agency*

1155 Camino del mar, Suite 510, Del Mar CA 92014
☎ 001 858 755 8761   ⓕ 001 858 755 7063
www.castigliaagency.com
Contacts *Julie Castiglia, Winifred Golden, Sally Van Haitsma, Deborah Ritchken*

Founded 1993. Handles fiction: literary, mainstream, ethnic; non-fiction: narrative, biography, business, science, health, parenting, memoirs, psychology, women's and contemporary issues. Specializes in science, biography and literary fiction. No scripts, horror or fantasy. COMMISSION Home 15%; Foreign & Translation 25%. No unsolicited material; send query letter only in the first instance. No reading fee.

Jane Chelius Literary Agency, Inc.*

548 Second Street, Brooklyn, New York NY 11215
℡ 001 718 499 0236   ℻ 001 718 832 7335
queries@janechelius.com
www.janechelius.com
Contact *Jane Chelius*

Founded 1995. Handles popular and literary fiction; narrative and how-to non-fiction. No children's, young adult, screenplays, scripts and poetry. COMMISSION Home 15% Foreign 20%. Submission guidelines available on the website.

Linda Chester & Associates*

Rockefeller Center, 630 Fifth Avenue, Suite 2036, New York NY 10111
℡ 001 212 218 3350
www.lindachester.com
Contact *Linda Chester*

Founded 1978. Handles literary and commercial fiction and non-fiction in all subjects. No scripts, children's or textbooks. COMMISSION Home & Dramatic 15%; Translation 25%. No unsolicited mss or queries. No reading fee for solicited material.

William Clark Associates*

154 Christopher Street, Suite 3C, New York NY 10014
℡ 001 212 675 2784
query@wmclark.com
www.wmclark.com
Contact *William Clark*

Founded 1997. Handles non-fiction, mainstream literary fiction and some young adult books. No scripts, horror, science fiction, fantasy, diet or mystery. COMMISSION Home 15%; Foreign & Translation 20%. OVERSEAS ASSOCIATES **Ed Victor Ltd**; **Andrew Nurnberg Associates Ltd** (translation rights). Synopses should be part of query letter; sample chapters, complete mss on request only. Contact via e-mail only with concise description of work, synopsis/outline, biographical information and publishing history, if any. No reading fee.

Frances Collin Literary Agent*

PO Box 33, Wayne PA 19087–0033
℡ 001 610 254 0555   ℻ 001 610 254 5029
www.francescollin.com
Contact *Frances Collin*

Founded 1948. Successor to Marie Rodell. Handles general fiction and non-fiction. No scripts. OVERSEAS ASSOCIATES worldwide. No unsolicited mss. Send query letter only, with IRCs for reply, for the attention of *Sarah Yake*. No fax or telephone queries, please. No reading fee. Rarely accepts non-professional writers or writers not represented in the UK.

Don Congdon Associates, Inc.*

156 Fifth Avenue, Suite 625, New York NY 10010–7002
℡ 001 212 645 1229   ℻ 001 212 727 2688
Contacts *Don Congdon, Michael Congdon, Susan Ramer, Cristina Concepcion*

Founded 1983. Handles fiction and non-fiction. No academic, technical, romantic fiction or scripts. COMMISSION Home 15%; UK & Translation 19%. OVERSEAS ASSOCIATES worldwide. No unsolicited mss. Query letter with return postage in the first instance. No reading fee.

Curtis Brown Ltd*

10 Astor Place, New York NY 10003
℡ 001 212 473 5400
Book Rights *Laura Blake Peterson, Katherine Fausset, Peter L. Ginsberg, Emilie Jacobson, Ginger Knowlton, Maureen Walters, Mitchell Waters, Elizabeth Harding, Ginger Clark*
Film & TV Rights *Timothy Knowlton, Holly Frederick*
Translation *Dave Barbor*

Founded 1914. Handles general fiction and non-fiction. Also some scripts for film, TV and theatre. OVERSEAS ASSOCIATES Representatives in all major foreign countries. No unsolicited mss; queries only, with IRCs for reply. No reading fee.

Sandra Dijkstra Literary Agency*

1155 Camino Del Mar, PMB 515, Del Mar CA 92014
℡ 001 858 755 3115   ℻ 001 858 794 2822
Contact *Taryn Fagerness*

Founded 1981. Handles quality and commercial non-fiction and fiction, including some genre fiction. No scripts. No westerns, science fiction or poetry. Specializes in quality fiction including women's and multicultural fiction, mystery/ thrillers, children's literature, narrative non-fiction, psychology, self-help, science, health, business, memoirs, biography, current affairs and history. 'Dedicated to promoting new and original voices and ideas.' COMMISSION Home 15%; Translation 20%. OVERSEAS ASSOCIATES **Abner Stein**, UK; Agence Hoffman, Germany; Licht & Burr, Scandinavia; Luigi Bernabo, Italy; Sandra Bruna, Spain/Portugal; Caroline Van Gelderen, Netherlands; La Nouvelle Agence,

France; The English Agency, Japan; Bardon-Chinese Media Agency, China/Taiwan; Prava I Prevodi, Eastern Europe; Tuttle Mori, Thailand; Synopsis, Russia/Baltic States; Maxima Creative Agency, Indonesia; Graal, Poland. For fiction send brief synopsis (one page) and first 50 pages; for non-fiction send proposal with overview, chapter outline, author biog, 1–2 sample chapters and profile of competition. All submissions should be accompanied by IRCs. No reading fee.

Dunham Literary Inc.*

156 Fifth Avenue, Suite 625, New York NY 10010
☎ 001 212 929 0994   🖷 001 212 929 0904
www.dunhamlit.com
Contacts *Jennie Dunham, Blair Hewes*

Founded 2000. Handles literary fiction, non-fiction and children's (from picture books through young adult). No scripts, romance, westerns, horror, science-fiction/fantasy or poetry. COMMISSION Home 15%; Foreign & Translation 20%. OVERSEAS ASSOCIATE UK & Europe: **A.M. Heath & Co. Ltd**. The Rhoda Weyr Agency is now a division of Dunham Literary Inc. No unsolicited material. Send query letter (with return postage) giving information on the author and ms. No e-mail or faxed queries; see the website for further contact information. No reading fee.

Dystel & Goderich Literary Management*

One Union Square West, Suite 904, New York NY 10003
☎ 001 212 627 9100   🖷 001 212 627 9313
www.dystel.com
Contacts *Jane Dystel, Miriam Goderich, Stacey Glick, Michael Bourret, Jim McCarthy, Lauren E. Abramo, Adina Kahn, Chasya Milgrom*

Founded 1994. Handles non-fiction and fiction. Specializes in politics, history, biography, cookbooks, current affairs, celebrities, commercial and literary fiction. No reading fee.

Educational Design Services, LLC

5750 Bou Avenue, Ste. 1508, N. Bethesda MD 20852
☎ 001 301 881 8611
blinder@educationaldesignservices.com
www.educationaldesignservices.com
President *Bertram Linder*

Founded 1979. Specializes in texts and professional development materials for the education and school market. COMMISSION Home 15%; Foreign 25%. 'E-submissions' welcome. IRCs must accompany mail submissions.

Ethan Ellenberg Literary Agency*

548 Broadway, Suite 5E, New York NY 10012
☎ 001 212 431 4554   🖷 001 212 941 4652
agent@ethanellenberg.com
www.ethanellenberg.com
Contact *Ethan Ellenberg*

Founded 1984. Handles fiction: commercial, genre, literary and children's; non-fiction: history, biography, science, health, cooking, current affairs. Specializes in commercial fiction, thrillers, suspense and romance. No scripts, poetry or short stories. COMMISSION Home 15%; Translation 20%. Prefers submissions by mail with return postage. For fiction send synopsis and first 50 pages; for non-fiction send proposal and sample chapters, if available. No reading fee. For e-mail submissions send query letter only; no attachments.

Jeanne Fredericks Literary Agency, Inc.*

221 Benedict Hill Road, New Canaan CT 06840
☎ 001 203 972 3011   🖷 001 203 972 3011
jeanne.fredericks@gmail.com
www.jeannefredericks.com
Contact *Jeanne Fredericks*

Founded 1997. Handles quality adult non-fiction, usually of a practical and popular nature by authorities in their fields. Specializes in health, gardening, science/nature/environment, business, self-help, reference. No fiction, juvenile, poetry, essays, politics, academic or textbooks. COMMISSION Home 15%; Foreign 25% (with co-agent) or 20% (direct). No unsolicited material; send query by e-mail (no attachments) or letter with return postage in the first instance. No reading fee.

Robert A. Freedman Dramatic Agency, Inc.*

Suite 2310, 1501 Broadway, New York NY 10036
☎ 001 212 840 5760
President *Robert A. Freedman*
Senior Vice-President *Selma Luttinger*
Vice-President *Marta Praeger*

Founded 1928 as Brandt & Brandt Dramatic Department, Inc. Took its present name in 1984. Works mostly with established authors. Send letter of enquiry first with s.a.e. Specializes in plays, film and TV scripts. COMMISSION Dramatic 10%. Unsolicited mss not read.

Gelfman Schneider Literary Agents, Inc.*

250 West 57th Street, Suite 2122, New York NY 10107
☎ 001 212 245 1993   🖷 001 212 245 8678
Contacts *Deborah Schneider, Jane Gelfman*

Founded 1919 (London), 1980 (New York). Formerly John Farquharson Ltd. Works mostly with established/published authors. Specializes in general trade fiction and non-fiction. No poetry, short stories or screenplays. COMMISSION Home 15%; Dramatic 15%; Foreign 20%. OVERSEAS ASSOCIATE **Curtis Brown Group Ltd**, UK. No reading fee for outlines. Submissions must be accompanied by IRCs. No e-mail queries please.

Sanford J. Greenburger Associates, Inc.*

55 Fifth Avenue, 15th Floor, New York NY 10003
☎ 001 212 206 5600   ℻ 001 212 463 8718
www.greenburger.com
Contacts *Heide Lange, Faith Hamlin, Daniel Mandel, Matt Bialer, Jeremy Katz, Tricia Davey*

Handles fiction and non-fiction. No unsolicited mss. First approach with query letter, sample chapter and synopsis. No reading fee.

The Charlotte Gusay Literary Agency

10532 Blythe Avenue, Los Angeles CA 90064
☎ 001 310 559 0831   ℻ 001 310 559 2639
gusay1@ca.rr.com (queries only)
www.gusay.com
Owner/President *Charlotte Gusay*

Founded 1989. Handles fiction, non-fiction, young adult/teen and screenplays (feature scripts/plays to film). No short stories, poetry. Specializes in novels, narrative non-fiction, gardening and travel. COMMISSION Home 15%; Foreign & Translation 25%. No unsolicited mss; send one-page query letter and s.a.e. in the first instance. No reading fee.

Joy Harris Literary Agency, Inc.*

156 Fifth Avenue, Suite 617, New York NY 10010
☎ 001 212 924 6269   ℻ 001 212 924 6609
Contact *Joy Harris*

Handles adult non-fiction and fiction. COMMISSION Home 15%; Foreign 20%. Query letter in the first instance. No reading fee.

John Hawkins & Associates, Inc.*

71 West 23rd Street, Suite 1600, New York NY 10010
☎ 001 212 807 7040   ℻ 001 212 807 9555
www.jhalit.com
Contacts *John Hawkins, William Reiss, Anne Hawkins, Warren Frazier, Moses Cardona*

Founded 1893. Handles film and TV rights. COMMISSION Apply for rates. No unsolicited mss; send queries with 1–3-page outline and one-page c.v. IRCs necessary for response. No reading fee.

The Jeff Herman Agency, LLC

PO Box 1522, Stockbridge MA 01262
☎ 001 413 298 0077   ℻ 001 413 298 8188
jeff@jeffherman.com
www.jeffherman.com
Contact *Jeffrey H. Herman*

Handles all areas of non-fiction, textbooks and reference, business, spiritual and psychology. No scripts. COMMISSION Home 15%; Translation 10%. No unsolicited mss. Query letter with IRCs in the first instance. No reading fee. Jeff Herman publishes a useful reference guide to the book trade called *Jeff Herman's Guide to Book Editors, Publishers & Literary Agents* (Three Dog Press).

The Barbara Hogenson Agency, Inc.*

165 West End Avenue, Suite 19–C, New York NY 10023
☎ 001 212 874 8084   ℻ 001 212 362 3011
Owner *Barbara Hogenson*
Contracts Manager *Nicole Verity*

Founded 1994. Handles non-fiction, literary fiction and plays. No science fiction, romance or children's books. Specializes in theatre-related books and literary fiction. COMMISSION Home 15%; Foreign & Translation 20%. OVERSEAS ASSOCIATES **David Higham Associates Ltd**, UK; Lora Fountain, France; Andrew Nurnberg, Eastern Europe; Japan Uni, Japan. No unsolicited mss; query letter only. Recommendation from current clients preferred or reference to a client's work. No reading fee.

Janklow & Nesbit Associates

445 Park Avenue, New York NY 10022–2606
☎ 001 212 421 1700   ℻ 001 212 980 3671
postmaster@janklow.com
Partners *Morton L. Janklow, Lynn Nesbit*
Senior Vice-President *Anne Sibbald*
Agents *Tina Bennett, Luke Janklow, Richard Morris, Eric Simonoff*

Founded 1989. Handles fiction and non-fiction; commercial and literary. See also **Janklow & Nesbit (UK) Ltd** under *UK Literary Agents*. No unsolicited mss.

JCA Literary Agency, Inc.

174 Sullivan Street, New York NY 10012
www.jcalit.com/
Contact *Tom Cushman*

Founded 1978. Handles general fiction and non-fiction. No scripts, poetry, science fiction/fantasy or children's books. COMMISSION Home 15%; Foreign 20%. OVERSEAS ASSOCIATE **Vanessa Holt Ltd**, UK. No unsolicited mss. No reading fee.

Kirchoff/Wohlberg, Inc.*

866 United Nations Plaza, Suite 525, New York NY 10017
☎ 001 212 644 2020   ℻ 001 212 223 4387
www.kirchoffwohlberg.com
Authors' Representative *Liza Pulitzer-Voges*

Founded 1930. Handles books for children and young adults. No adult material. No scripts for TV, radio, film or theatre. Send letter of enquiry with synopsis or outline and writing sample plus IRCs for reply or return. No reading fee.

Harvey Klinger, Inc.*

300 West 55th Street, Suite 11V, New York NY 10019
queries@harveyklinger.com
www.harveyklinger.com
Contact *Harvey Klinger*

Founded 1977. Handles mainstream fiction and non-fiction. Specializes in commercial and literary fiction, psychology, health and science. No scripts, poetry, computer or children's books. COMMISSION Home 15%; Foreign 25%. OVERSEAS ASSOCIATES in all principal countries. Welcomes unsolicited material; send by e-mail or post (no faxes). No reading fee.

Linda Konner Literary Agency*

10 West 15 Street, Suite 1918, New York NY 10011
☎ 001 212 691 3419
Contact *Linda Konner*

Founded 1996. Handles non-fiction only, specializing in health, self-help, diet/fitness, pop psychology, relationships, parenting, personal finance. Also some pop culture/celebrities. COMMISSION Home 15%; Foreign 25%. Books must be written by or with established experts in their field. No scripts, fiction, poetry or children's books. No unsolicited material; send one-page query with return postage. No reading fee.

Peter Lampack Agency, Inc.

551 Fifth Avenue, Suite 1613, New York NY 10176
☎ 001 212 687 9106
alampack@verizon.net
Contact *Andrew Lampack*

Founded in 1977. Handles commercial fiction: male action and adventure, contemporary relationships, mysteries and suspense, literary fiction; also non-fiction from recognized experts in a given field, plus biographies. Handles theatrical, motion picture, and TV rights from book properties. No original scripts or screenplays, series or episodic material. COMMISSION Home & Dramatic 15%; Translation & UK 20%. Best approach by letter in first instance. No reply without s.a.e. 'We will respond within three weeks and invite the submission of manuscripts which we would like to examine.' No reading fee. No unsolicited mss.

Michael Larsen/Elizabeth Pomada Literary Agency*

1029 Jones Street, San Francisco CA 94109
☎ 001 415 673 0939
larsenpoma@aol.com
www.larsenpomada.com
Contact (non-fiction) *Michael Larsen*
Contact (fiction) *Elizabeth Pomada*

Founded 1972. Handles adult fiction and non-fiction – literary and commercial. No scripts, poetry, children's, science fiction. COMMISSION Home 15%; Foreign 20–30%. ASSOCIATE Laurie McLean (represents science fiction and fantasy and young adult books. OVERSEAS ASSOCIATES **David Grossman Literary Agency Ltd**, British rights; Chandler Crawford (foreign rights). Fiction: send first ten pages with 2-page synopsis and s.a.e.; non-fiction: e-mail first ten pages with 2-page synopsis in the body of the e-mail (no attachments). Consult website for guidelines.

The Ned Leavitt Agency*

70 Wooster Street, Suite 4F, New York NY 10012
www.nedleavittagency.com
President *Ned Leavitt*
Agent *Britta Steiner Alexander*

Ned Leavitt specializes in creativity, spirituality, health and literary fiction; Britta Alexander specializes in smart non-fiction for 20 and 30-somethings (dating, relationships, business) and commercial fiction. No screenplays or genre fiction. COMMISSION Home 15%. Submissions by recommendation only; see guidelines on the website. No reading fee.

Lescher & Lescher Ltd*

346 East 84th Street, New York NY 10028
☎ 001 212 529 1790   🖷 001 212 529 2716
Contacts *Robert Lescher, Susan Lescher*

Founded 1964. Handles a broad range of serious non-fiction including current affairs, history, biography, memoirs, politics, law, contemporary issues, popular culture, food and wine, literary and commercial fiction including mysteries and thrillers; some children's books. Specializes in wine books and cookbooks. No poetry, science fiction, New Age, spiritual, romance. COMMISSION Home 15%; Foreign 20%. No unsolicited manuscripts – please query first. No reading fee.

Sterling Lord Literistic, Inc.

65 Bleecker Street, New York NY 10012
☎ 001 212 780 6050   🖷 001 212 780 6095
info@sll.com
www.sll.com
Contacts *Philippa Brophy, Chris Calhoun, Laurie Liss*

Founded 1979. Handles all genres, fiction and non-fiction. COMMISSION Home 15%; UK & Translation 20%. No unsolicited mss. Prefers letter outlining all non-fiction. No reading fee.

Lowenstein-Yost Associates Inc.*

121 West 27th Street, Suite 601, New York NY 10001
☎ 001 212 206 1630   🖷 001 212 727 0280
www.lowensteinyost.com
Agents *Barbara Lowenstein, Nancy Yost, Zoe Fishman, Natanya Wheeler*

Founded 1976. Handles fiction: upmarket, commercial and multicultural, thrillers, mysteries, women's fiction of all kinds including historicals, paranormals and romantic suspense, fantasy and young adult. Non-fiction: narrative, business, health, spirituality, psychology, relationships, politics, history, memoirs, natural science, women's issues, social issues, pop culture, parenting, personal finance. COMMISSION Home 15%; Foreign & Translation 20%. OVERSEAS ASSOCIATES in all major countries. Fiction: send query letter with short synopsis, first chapter and s.a.e. Include any prior literary credits (previously published titles and reviews, writing courses, awards/grants); non-fiction: send query letter, project overview, list of credentials, media appearances, previous titles, reviews and s.a.e. No reading fee.

The Margret McBride Literary Agency*

7744 Fay Avenue, Suite 201, La Jolla CA 92037
☎ 001 858 454 1550   ℻ 001 858 454 2156
staff@mcbridelit.com
www.mcbrideliterary.com
Submissions Manager *Michael Daley*
Assistant to Margret McBride *Faye Atchison*
Vice President/Associate Agent *Donna DeGutis*

Founded 1981. Handles non-fiction mostly, plus business and some fiction. No scripts, science fiction, fantasy, romance or children's books. Specializes in business and management. COMMISSION Home 15%; Overseas 15–25%; Translation 25%. No unsolicited mss. Send query letter with brief synopsis and s.a.e. only. Check the submission guidelines on the website before making contact.

Carol Mann Agency*

55 Fifth Avenue, New York NY 10003
☎ 001 212 206 5635   ℻ 001 212 675 4809
www.carolmannagency.com
Contacts *Carol Mann, Emily Nurkin, Kristy Mayer*

Founded 1977. Handles literary and commercial fiction and narrative non-fiction. No scripts or genre fiction. COMMISSION Home 15%; Foreign 20%. No unsolicited material; send query letter with return postage. No reading fee.

Manus & Associates Literary Agency, Inc.*

425 Sherman Avenue, Suite 200, Palo Alto CA 94306
ManusLit@ManusLit.com
www.ManusLit.com
Also at: 445 Park Avenue, New York, NY 10022
Contacts (California) *Jillian Manus, Jandy Nelson, Stephanie Lee, Penny Nelson, Dena Fischer*
Contact (New York) *Janet Manus*

Handles commercial and literary fiction, also young adult and middle grade; non-fiction, including true crime, self-help, memoirs, history, pop culture and popular science. *No scripts, science fiction/fantasy, westerns, romance, horror, poetry or children's books.* COMMISSION Home 15%; Foreign 25%. No unsolicited mss. Fiction: send query letter and first 30 pages; non-fiction: query letter and proposal. Include return postage. All submissions must be made to the California office. No reading fee.

Mews Books Ltd

c/o Sidney B. Kramer, 20 Bluewater Hill, Westport CT 06880
☎ 001 203 227 1836   ℻ 001 203 227 1144
mewsbooks@aol.com (initial contact only; submission by regular mail)
Contacts *Sidney B. Kramer, Fran Pollak*

Founded 1970. Handles adult fiction and non-fiction, children's, pre-school and young adult. No scripts, short stories or novellas (unless by established authors). Specializes in cookery, medical, health and nutrition, scientific non-fiction, children's and young adult. COMMISSION Home 15%; Film & Translation 20%. Unsolicited material welcome. Presentation must be professional and should include brief summary of plot/characters ('avoid a reviewer's point of view when describing plot'), one or two sample chapters, personal credentials and targeted market, all suitable for forwarding to a publisher. Send by regular mail, not e-mail. No reading fee. Requests exclusivity while reading and information if material has been circulated. Charges for photocopying, postage expenses, telephone calls and other direct costs. Principal agent is an attorney and former publisher (a founder of Bantam Books, Corgi Books, London). Offers consultation service through which writers can get advice on a contract or on publishing problems.

Doris S. Michaels Literary Agency Inc.*

1841 Broadway, Suite 903, New York NY 10023
☎ 001 212 265 9474   ℻ 001 212 265 9480
query@dsmagency.com
www.dsmagency.com
Contact *Doris Michaels*

Handles literary fiction that has a commercial appeal and strong screen potential and women's fiction; non-fiction: current affairs, biography and memoirs, self-help, humour, history, health, classical music, sports, women's issues, social sciences and pop culture. No action/adventure, suspense, science fiction, romance, New Age, religion/spirituality, gift books, art books, fantasy, thrillers, mysteries, westerns, occult and supernatural, horror, historical fiction, poetry, children's literature, humour or travel books. COMMISSION Home 15% Send query letter via e-mail with a one-page synopsis and include a short paragraph detailing credentials. No reading fee. .

Howard Morhaim Literary Agency*

30 Pierrepont Street, Brooklyn NY 11201
📞 001 718 222 8400 📠 001 718 222 5056
www.morhaimliterary.com
Contact *Howard Morhaim*

Founded 1979. Handles general adult and young adult fiction and non-fiction. No scripts poetry or religious. COMMISSION Home 15%; UK & Translation 20%. OVERSEAS ASSOCIATES worldwide. Send query letter with synopsis and sample chapters for fiction; query letter with outline or proposal for non-fiction. Include return postage. No unsolicited mss. No reading fee.

Henry Morrison, Inc.

PO Box 235, Bedford Hills NY 10507–0235
📞 001 914 666 3500 📠 001 914 241 7846
hmorrison1@aol.com
Contact *Henry Morrison*

Founded 1965. Handles general fiction, crime and science fiction, and non-fiction. No scripts unless by established writers. COMMISSION Home 15%; UK & Translation 25%. Unsolicited material welcome but send query letter with outline (1–5 pp) in the first instance. No reading fee.

The Jean V. Naggar Literary Agency*

216 East 75th Street, Suite 1-E, New York City NY 10021
mglick@jvnla.com
Contacts *Jean Naggar, Alice Tasman, Mollie Glick, Jennifer Weltz, Jessica Regel*

Founded 1978. Handles strong mainstream fiction, literary fiction, memoirs, biography, sophisticated self-help, popular science and psychology. No scripts. COMMISSION Home 15%; Foreign 20%. No unsolicited material; send query letter *only* with return postage initially. No reading fee.

B.K. Nelson Literary Agency

84 Woodland Road, Pleasantville NY 10570
📞 001 914 741 1322 📠 001 914 741 1324
bknelson4@cs.com
www.bknelson.com
Also at: 1565 Paseo Vida, Palm Springs, CA 92264
📞 001 760 778 8800 📠 001 914 778 0034
President *Bonita K. Nelson*
Vice President *Leonard 'Chip' Ashbach*
Editorial Director *John W. Benson*

Founded 1979. Specializes in novels, business, self-help, how-to, political, autobiography, celebrity biography. Major motion picture and TV documentary success. COMMISSION 20%. Lecture

Bureau for Authors founded 1994; Foreign Rights Catalogue established 1995; BK Nelson Infomercial Marketing Co. 1996, primarily for authors and endorsements, and BKNelson, Inc. for motion picture production in 1998. Signatory to Writers Guild of America, West (WGAW). No unsolicited mss. Letter of inquiry. Reading fee charged.

Richard Parks Agency*

PO Box 693, Salem NY 12865
📞 001 518 854 9466 📠 001 518 854 9466
rp@richardparksagency.com
Contact *Richard Parks*

Founded 1989. Handles general trade fiction and non-fiction: literary novels, mysteries and thrillers, commercial fiction, science fiction, biography, pop culture, psychology, self-help, parenting, medical, cooking, gardening, history, etc. No scripts. No technical or academic. COMMISSION Home 15%; UK & Translation 20%. OVERSEAS ASSOCIATES **The Marsh Agency**; **Barbara Levy Literary Agency**. No unsolicited mss. Fiction read by referral only. Non-fiction query with s.a.e. Faxed or e-mail queries will *not* be considered. No reading fee.

Alison J. Picard Literary Agent

PO Box 2000, Cotuit MA 02635
📞 001 508 477 7192
📠 001 508 477 7192 (notify before faxing)
ajpicard@aol.com
Contact *Alison Picard*

Founded 1985. Handles mainstream and literary fiction, contemporary and historical romance, children's and young adult, mysteries and thrillers; plus non-fiction. No short stories or poetry. Rarely any science fiction and fantasy. Particularly interested in expanding non-fiction titles. COMMISSION 15%. OVERSEAS ASSOCIATE **John Pawsey**, UK. Approach with written query. No reading fee.

Pinder Lane & Garon-Brooke Associates Ltd*

159 West 53rd Street, Suite 14–C, New York NY 10019
📞 001 212 489 0880
Owner Agents *Dick Duane, Robert Thixton*

Founded 1951. Fiction and non-fiction. No category romance, westerns or mysteries. COMMISSION Home 15%; Dramatic 10–15%; Foreign 30%. OVERSEAS ASSOCIATES **Abner Stein**, UK; Translation: Rights Unlimited. No unsolicited mss. First approach by query letter. No reading fee.

PMA Literary & Film Management, Inc.

PO Box 1817, Old Chelsea Station, New York NY 10113
📞 001 212 929 1222 📠 001 212 206 0238

queries@pmalitfilm.com
www.pmalitfilm.com
President *Peter Miller*

Founded 1976. Commercial fiction and non-fiction. Specializes in books with motion picture and television potential, and in true crime. No poetry, pornography, non-commercial or academic. COMMISSION Home 15%; Dramatic 10–15%; Foreign 20–25%. No unsolicited mss. Approach by letter with one-page synopsis.

The Aaron M. Priest Literary Agency*

708 Third Avenue, 23rd Floor, New York NY 10017
℡ 001 212 818 0344   ℻ 001 212 573 9417
www.aaronpriest.com

Founded 1974. Handles literary and commercial fiction. No scripts, children's fantasy or sci-fi. Submissions by e-mail only; refer to the website for full details. No reading fee.

Susan Ann Protter Literary Agent*

110 West 40th Street, Suite 1408, New York NY 10018
℡ 001 212 840 0480   ℻ 001 212 840 1132
sapla@aol.com
www.susanannprotter.com
Contact *Susan Ann Protter*

Founded 1971. Handles general fiction, mysteries, thrillers, science fiction and fantasy; non-fiction: history, general reference, biography, science, health, current affairs and parenting. No romance, poetry, westerns, religious, children's or sport manuals. No scripts. COMMISSION Home & Dramatic 15%. OVERSEAS ASSOCIATES **Abner Stein**, UK; agents in all major markets. First approach with letter, including IRCs. No reading fee.

Quicksilver Books, Literary Agents

508 Central Park Avenue, Suite 5101, Scarsdale NY 10583
℡ 001 914 722 4664   ℻ 001 914 722 4664
quickbooks@optonline.net
President *Bob Silverstein*

Founded 1973. Handles literary fiction and mainstream commercial fiction: blockbuster, suspense, thriller, contemporary, mystery and historical; and general non-fiction, including self-help, psychology, holistic healing, ecology, environmental, biography, fact crime, New Age, health, nutrition, cookery, enlightened wisdom and spirituality. No scripts, science fiction and fantasy, pornography, children's or romance. COMMISSION Home & Dramatic 15%; Translation 20%. UK material being submitted must have universal appeal for the US market. Unsolicited material welcome but must be accompanied by IRCs for response, together with biographical details, covering letter, etc. No reading fee.

Raines & Raines*

103 Kenyon Road, Medusa NY 12120
℡ 001 518 239 8311   ℻ 001 518 239 6029
Contacts *Theron Raines, Joan Raines, Keith Korman*

Founded 1961. Handles general non-fiction. No scripts. COMMISSION Home 15%; Foreign & Translation 20%. No unsolicited material; send one-page letter in the first instance. No reading fee.

Reece Halsey North Literary Agency*

98 Main Street, #704, Tiburon CA 94920
℡ 001 415 789 9191   ℻ 001 415 789 9177
info@reecehalseynorth.com (Phil Lang)
www.reecehalseynorth.com
Contacts *Kimberley Cameron, Elizabeth Evans*

The Reece Halsey Agency was founded 1957. Aldous Huxley, Upton Sinclair and William Faulkner have been among their clients. Represents literary and mainstream fiction, non-fiction. No scripts, poetry or children's books. COMMISSION Home 15%; Foreign 20%. Send query letter with first 10–50 pages together with return postage. Will accept e-mail submissions from abroad. No reading fee.

Helen Rees Literary Agency*

376 North Street, Boston MA 02113–2103
℡ 001 617 227 9014   ℻ 001 617 227 8762
reesagency@reesagency.com
Contact *Helen Rees*
Associates *Ann Collette, Lorin Rees*

Founded 1982. Specializes in books on health and business; also handles biography, autobiography and history; quality fiction. No scholarly or technical books. No scripts, science fiction, children's, poetry, photography, short stories, cookery. COMMISSION Home 15%; Foreign 20%. No e-mail queries or attachments. Send query letter with IRCs. No reading fee.

Ann Rittenberg Literary Agency, Inc.*

30 Bond Street, New York NY 10012
℡ 001 212 684 6936   ℻ 001 212 684 6929
www.rittlit.com
Contacts *Ann Rittenberg, Penn Whaling*

Founded 1991. Handles literary fiction, serious narrative non-fiction, biography/memoirs, cultural history, upmarket women's fiction and thrillers. No scripts, romance, science fiction, self-help, inspirational and nothing at genre level. COMMISSION Home 15%; Foreign 20%. Send query letter, sample chapters and synopses by post; no e-mails. Enclose s.a.e. No reading fee.

B.J. Robbins Literary Agency*

5130 Bellaire Avenue, North Hollywood CA 91607
℡ 001 818 760 6602

robbinsliterary@aol.com
Contact *B.J. Robbins*

Founded 1992. Handles literary fiction, narrative and general non-fiction. No scripts, genre fiction, romance, horror, science fiction or children's books. COMMISSION Home 15%; Foreign & Translation 20%. OVERSEAS ASSOCIATES **Abner Stein** and **The Marsh Agency**, UK. Send covering letter with first three chapters or e-mail query in the first instance. No reading fee.

Linda Roghaar Literary Agency, LLC*

133 High Point Drive, Amherst MA 01002
℡ 001 413 256 1921
info@LindaRoghaar.com
www.lindaroghaar.com
Contact *Linda L. Roghaar*

Founded 1997. Handles fiction and specializes in non-fiction titles. No romance, science fiction or horror. COMMISSION Home 15%; Foreign & Translation rate varies. Send query by e-mail or letter with return postage (for fiction, include the first five pages). No reading fee.

The Rosenberg Group*

23 Lincoln Avenue, Marblehead MA 01945
℡ 001 781 990 1341   ℻ 001 781 990 1344
www.rosenberggroup.com
Contact *Barbara Collins Rosenberg*

Founded 1998. Handles non-fiction (please see website for areas of interest), fiction, specializing in romance (single title and category) and women's; trade non-fiction with a special interest in sports; also handles college textbooks for the first and second year courses. No scripts, science fiction, true crime, inspirational fiction, children's and young adult. COMMISSION Home 15%; Foreign 25%. No unsolicited material; send query letter by post only. No reading fee.

Russell & Volkening, Inc.*

50 West 29th Street, #7E, New York NY 10001
℡ 001 212 684 6050   ℻ 001 212 889 3026
Contacts *Tim Seldes, Jesseca Salky* (adult), *Carrie Hannigan* (children's), *Rosanna Bruno* (mystery & permissions)

Founded 1940. Handles literary fiction and non-fiction; science, history, current affairs, politics, mystery, thriller, true crime, children's. Specializes in politics and current affairs. No scripts, science fiction/fantasy or very prescriptive non-fiction. COMMISSION Home 15%; Foreign 20%. No unsolicited mss; sample chapters or synopsis with query and covering letter in the first instance. No reading fee.

The Sagalyn Literary Agency*

4922 Fairmont Avenue, Suite 200, Bethesda MD 20814
℡ 001 301 718 6440   ℻ 001 301 718 6444
info@sagalyn.com
www.sagalyn.com

Founded 1980. Handles mostly upmarket nonfiction with some fiction. No screenplays, romance, science fiction/fantasy, children's literature. No unsolicited material. See website for submissions procedure. No reading fee.

Victoria Sanders & Associates LLC*

241 Avenue of the Americas, Suite 11H, New York NY 10014
℡ 001 212 633 8811   ℻ 001 212 633 0525
queriesvsa@hotmail.com
www.victoriasanders.com
Contacts *Victoria Sanders, Diane Dickensheid*

Founded 1992. Handles general trade fiction and non-fiction, plus ancillary film and television rights. COMMISSION Home & Dramatic 15%; Translation 20%. Please send all queries via e-mail.

Jack Scagnetti Talent & Literary Agency

5118 Vineland Avenue, Suite 106, North Hollywood CA 91601
℡ 001 818 762 3871
www.jackscagnetti.com
Contact *Jack Scagnetti*

Founded 1974. Works mostly with established/ published authors. Handles non-fiction, fiction, film and TV scripts. No reading fee. COMMISSION Home & Dramatic 10% (scripts), 15% (books); Foreign 15%.

Schiavone Literary Agency, Inc.

Corporate Offices: 236 Trails End, West Palm Beach FL 33413–2135
℡ 001 561 966 9294   ℻ 001 561 966 9294
profschia@aol.com
www.publishersmarketplace.com/members/ profschia
New York branch office: 3671 Hudson Manor Terrace, Suite 11H, Bronx, NY 10463
℡/℻ 001 718 548 5332
CEO *James Schiavone* (Florida office)
President *Jennifer Duvall* (New York; jendu77@ aol.com)

Founded 1996. Handles fiction and non-fiction (all genres). Specializes in biography, autobiography, celebrity memoirs. No poetry. COMMISSION Home 15%; Foreign & Translation 20%. OVERSEAS ASSOCIATES in Europe and Asia. No unsolicited mss; send query letter only with s.a.e. and IRCs. No response without s.a.e. For fastest response, e-mail queries of one page (no attachments) are acceptable and encouraged. No reading fee.

Susan Schulman, A Literary Agency*
454 West 44th Street, New York NY 10036
☎ 001 212 713 1633   🖷 001 212 581 8830
schulman@aol.com
Submissions Editor (books) *Emily Uhry*
Submissions Editor (plays) *Linda Kiss*
Rights & Permissions Editor *Eleanora Tevís*

Founded 1980. Specializes in non-fiction of all types but particularly in health and psychology-based self-help for men, women and families. Other interests include business, memoirs, the social sciences, biography, language and international law. Fiction interests include contemporary fiction, including women's, mysteries, historical and thrillers with a cutting edge. 'Always looking for something original and fresh.' Represents properties for film and theatre, and works with agents in appropriate territories for translation rights. COMMISSION Home & Dramatic 15%; Translation 20%. No unsolicited mss. Query first, including outline and three sample chapters with IRCs. No reading fee.

Scovil Chichak Galen Literary Agency, Inc.*
276 Fifth Avenue, Suite 708, New York NY 10001
☎ 001 212 679 8686   🖷 001 212 679 6710
info@scglit.com
www.scglit.com
Contacts *Russell Galen, Anna M. Ghosh, Jack Scovil*

Founded 1993. Handles all categories of books. No scripts. COMMISSION Home 15%; Foreign 20%. No unsolicited material; send query by e-mail or letter. No reading fee.

Scribblers House® LLC Literary Agency
PO Box 1007 Cooper Station, New York NY 10276–1007
query@scribblershouse.net
www.scribblershouse.net
Agents *Stedman Mays, Garrett Gambino*

Founded 2003 by Stedman Mays, founding member of Clausen, Mays & Tahan Literary Agency. Handles health, medical, diet, nutrition, the brain, psychology, self-help, how-to, business, personal finance, memoirs, biography, history, politics, writing books, language, relationships, sex, pop culture, spirituality, gender issues, parenting. No fiction, young adult or screenplays. COMMISSION 15%. Send e-mail query letter with brief description of the book, what's fresh and new about it, and a brief author biog including credentials for writing it. Consult the website for more submission information.

Rosalie Siegel, International Literary Agency, Inc.*
1 Abey Drive, Pennington NJ 08534
☎ 001 609 737 1007   🖷 001 609 737 3708
Contact *Rosalie Siegel*

Founded 1978. A one-woman, highly selective agency that takes on only a limited number of new projects. Handles fiction and non-fiction (especially narrative). Specializes in French history and literature; Europe in general; American history, social history. No science fiction, photography, illustrated art books or children's. COMMISSION Home 15%; Foreign 20%. OVERSEAS ASSOCIATE **Louise Greenberg**, UK; plus associates worldwide. No unsolicited material.

Michael Snell Literary Agency
PO Box 1206, Truro MA 02666–1206
☎ 001 508 349 3718
President *Michael Snell*
Vice President *Patricia Smith*

Founded 1980. Adult non-fiction, especially psychology, health, parenting, science, business and women's issues. Specializes in business and pet and animal books (professional and reference to popular trade how-to); general how-to and self-help on all topics, from diet and exercise to parenting, relationships, health, sex, psychology, personal finance, and dogs, cats and horses, plus literary and suspense fiction. COMMISSION Home 15%. No unsolicited mss. Send outline and sample chapter with return postage for reply. No reading fee for outlines. Brochure available on how to write a book proposal. Model proposal also for sale directly to prospective clients. Author of *From Book Idea to Bestseller*, published by Prima. Rewriting, developmental editing, collaborating and ghostwriting services available on a fee basis. Send IRCs.

Spectrum Literary Agency*
320 Central Park West, Suite 1-D, New York NY 10025
☎ 001 212 362 4323   🖷 001 212 362 4562
www.spectrumliteraryagency.com
President and Agent *Eleanor Wood*
Agent *Lucienne Diver*

Founded 1976. Handles science fiction, fantasy, mystery, suspense and romance. No scripts. Not interested in self-help, New Age, religious fiction/non-fiction, children's books, poetry, short stories, gift books or memoirs. COMMISSION Home 15%; Translation 20%. Send query letter in first instance with s.a.e. Faxed or electronic submissions are not accepted. No reading fees.

Philip G. Spitzer Literary Agency, Inc.*
50 Talmage Farm Lane, East Hampton NY 11937
☎ 001 631 329 3650
Contact *Philip Spitzer*

Founded 1969. Works mostly with established/published authors. Specializes in general non-fiction and fiction – thrillers. COMMISSION Home & Dramatic 15%; Foreign 20%. No reading fee for outlines.

Stimola Literary Studio, LLC*

306 Chase Court, Edgewater NJ 07020
℡ 001 201 945 9353   🖷 001 201 945 9353
LtryStudio@aol.com
www.stimolaliterarystudio.com
Contact *Rosemary B. Stimola*

Founded 1997. Handles children's books – pre-school through young adult – fiction and non fiction. Specializes in picture books, middle/young adult novels. No adult fiction. COMMISSION Home 15%; Foreign 20%. No unsolicited material; send query e-mail (no attachments). No reading fee.

Barbara W. Stuhlmann Author's Representative

PO Box 276, Becket MA 01223–0276
℡ 001 413 623 5170
Contact *Barbara Ward Stuhlmann*

Founded 1954. COMMISSION Home 10%; Foreign 15%; Translation 20%. Query first with IRCs, including sample chapters and synopsis of project. No reading fee. 'No new clients at this time.'

Patricia Teal Literary Agency*

2036 Vista del Rosa, Fullerton CA 92831
℡ 001 714 738 8333   🖷 001 714 738 8333
Contact *Patricial Teal*

Founded 1978. Handles women's fiction: series and single-title works; commercial and popular non-fiction. Specializes in romantic fiction. No scripts, science fiction, fantasy, horror, academic texts. COMMISSION Home 15%; Foreign 20%. No unsolicited material. Send query letter with s.a.e. in the first instance. No reading fee.

S©ott Treimel NY*

434 Lafayette Street, New York NY 10003
℡ 001 212 505 8353   🖷 001 212 505 0664

Founded 1995. Handles children's books only, from concept/board books to teen fiction. No picture book texts or filmscripts. COMMISSION Home 15%; Foreign 20%. No unsolicited material. Query first. 'Prefer writers and illustrators recommended by professional authors or editors.' No reading fee.

2M Communications Ltd*

121 West 27th Street, Suite 601, New York NY 10001
℡ 001 212 741 1509   🖷 001 212 691 4460
morel@bookhaven.com
www.2mcommunications.com
Contact *Madeleine Morel*

Founded 1982. Specializes in representing ghostwriters. No unsolicited mss; send letter with sample pages and IRCs. No reading fee.

Wales Literary Agency, Inc.*

PO Box 9428, Seattle WA 98109–0428
℡ 001 206 284 7114
waleslit@waleslit.com
www.waleslit.com
Contacts *Elizabeth Wales, Neal Swain*

Founded 1988. Handles quality fiction and non-fiction. No westerns, romance, self help, science fiction or horror. Special interest in 'Pacific Rim', West Coast and Pacific Northwest stories. COMMISSION Home 15%; Dramatic & Translation/Foreign 20%. No unsolicited mss; send query letter. No e-mail queries longer than one page and no attachments. No reading fee.

John A. Ware Literary Agency

392 Central Park West, New York NY 10025
℡ 001 212 866 4733   🖷 001 212 866 4734
Contact *John Ware*

Founded 1978. Specializes in non-fiction: biography, history, current affairs, investigative journalism, science, nature, inside looks at phenomena, medicine and psychology (academic credentials required). No personal memoirs. Also handles literary fiction, mysteries/thrillers, sport, oral history, Americana and folklore. COMMISSION Home & Dramatic 15%; Foreign 20%. Unsolicited mss not read. Send query letter first with IRCs to cover return postage. No reading fee.

The Wendy Weil Agency, Inc.*

232 Madison Avenue, Suite 1300, New York NY 10016
℡ 001 212 685 0030   🖷 001 212 685 0765
info@wendyweil.com
www.wendyweil.com
Agents *Wendy Weil, Emily Forland*
Associate/Assistant *Emma Patterson*

Founded 1987. Handles fiction (commercial and literary), non-fiction (journalism, memoirs, etc.). No scripts, science fiction, romance, visual books, self-help, cookery. OVERSEAS ASSOCIATES **David Higham Associates**; Paul & Peter Fritz Agency. Send query letter with synopsis. No reading fee.

Cherry Weiner Literary Agency

28 Kipling Way, Manalapan NJ 07726
℡ 001 732 446 2096   🖷 001 732 792 0506
Contact *Cherry Weiner*

Founded 1977. Handles all types of fiction: science fiction and fantasy, mainstream, romance, mystery, westerns and some non-fiction. COMMISSION 15%. No submissions except through referral. No reading fee.

Rhoda Weyr Agency
▷ Dunham Literary Agency

Writers House, LLC.*
21 West 26th Street, New York NY 10010
℡ 001 212 685 2400   🖷 001 212 685 1781
www.writershouse.com
Contacts *Albert Zuckerman, Amy Berkower, Merrilee Heifetz, Susan Cohen, Susan Ginsburg, Robin Rue, Simon Lipskar, Steven Malk, Jodi Reamer, Dan Lazar, Rebecca Sherman, Ken Wright, Dan Conaway, Michele Rubin*

Founded 1973. Handles all types of fiction, including children's and young adult, plus narrative non-fiction: history, biography, popular science, pop and rock culture as well as how-to, business and finance, and New Age. Specializes in popular and literary fiction, women's novels, thrillers and children's. No scripts. No professional or scholarly. COMMISSION Home & Dramatic 15%; Foreign 20%. Albert Zuckerman is author of *Writing the Blockbuster Novel*, published by Little, Brown & Co. and Warner Paperbacks. For consideration of unsolicited mss, send one-page letter of enquiry, 'explaining why your book is wonderful, briefly what it's about and outlining your writing background'. No reading fee.

Susan Zeckendorf Associates, Inc.*
171 West 57th Street, Suite 11B, New York NY 10019
℡ 001 212 245 2928
Contact *Susan Zeckendorf*

Founded 1979. Handles non-fiction of all kinds: self help, social history, biography; commercial fiction. No scripts, romance, science fiction or children's books. COMMISSION Home 15%; Foreign & Translation 20%. ASSOCIATES Rosemarie Buckman (translation); **Abner Stein**, UK. No unsolicited material; send query letter with s.a.e. in the first instance. No reading fee.

Agency Consultants

Agent Research & Evaluation, Inc. (AR&E)
▷ entry under Agency Consultants (p.273)

Book Clubs

David Arscott's Sussex Book Club

Dolphin House, 51 St Nicholas Lane, Lewes
BN7 2JZ
℡ 01273 470100   🅕 01273 470100
sussexbooks@aol.com

Founded January 1998. Specializes in books
about the county of Sussex. Represents all the
major publishers of Sussex books and offers a
wide range of titles. Free membership without
obligation to buy.

Artists' Choice

PO Box 3, Huntingdon PE28 0QX
℡ 01832 710201   🅕 01832 710488
www.artists-choice.co.uk

Specializes in books for the amateur artist at all
levels of ability.

Baker Books

Manfield Park, Cranleigh GU6 8NU
℡ 01483 267888   🅕 01483 267409
bakerbooks@dial.pipex.com
www.bakerbooks.co.uk

Book clubs for schools: Funfare for ages 3–8,
Bookzone, ages 8–13 and My Word, ages 13–16.
Four issues per year operated in the UK and
overseas.

BCA (Book Club Associates)

Hargreaves Road, Groundwell, Swindon SN25 5BG
℡ 01793 723547
www.bca.co.uk

BCA, a wholly-owned subsidiary of Bertelsmann
AG, is Britain's largest book club organization.
Clubs include: Books For Children (BFC),
BooksDirect, Fantasy & SF, History Guild,
Mango, Military & Aviation Book Society, World
of Mystery & Thriller, Railway Book Club, The
Softback Preview (TSP), WorldBooks.

Bibliophile Books

Unit 5, Datapoint, 6 South Crescent, London
E16 4TL
℡ 020 7474 2474   🅕 020 7474 8589
orders@bibliophilebooks.com
www.bibliophilebooks.com

New books covering a wide range of subjects at
discount prices. Write, phone, fax or e-mail for
free catalogue of 3,000 titles issued 10 times a year.

Book Club Associates

▷ BCA

Cygnus Books

PO Box 15, Llandeilo SA19 6YX
℡ 01550 777701   🅕 01550 777569
enquiries@cygnus-books.co.uk
www.cygnus-books.co.uk

'Bookseller offering books for your next step in
spirituality and complementary health care.' Over
1,000 hand-picked titles. Publishes *The Cygnus
Review* magazine which features 30–40 reviews
on new mind, body and spirit titles each month.

The Folio Society

44 Eagle Street, London WC1R 4FS
℡ 020 7400 4222   🅕 020 7400 4242
enquiries@foliosoc.co.uk
www.foliosociety.com

Fine editions of classic fiction, history and
memoirs; also some children's classics.

Letterbox Library

71–73 Allen Road, London N16 8RY
℡ 020 7503 4801   🅕 020 7503 4800
info@letterboxlibrary.com
www.letterboxlibrary.com

Children's book cooperative. Hard and softcover
specialist in multicultural and inclusive books for
children from one to teenage.

Poetry Book Society

See entry under *Organizations of Interest to Poets*

Readers' Union Ltd

Brunel House, Forde Close, Newton Abbot
TQ12 4PU
℡ 01626 323200   🅕 01626 323318
www.readersunion.co.uk

Book clubs with specific interests: Anglers' Book
Club, The Craft Club, Craftsman's Book Club,
Country Sports Book Club, Equestrian Society,
The Gardeners Society, Needlecrafts with Cross
Stitch, Painting for Pleasure, Photographers' Book
Club, Puzzles Plus.

Scholastic Book Clubs

See **Scholastic Ltd** under *UK Publishers*

Magazines

Abraxas Unbound

57 Eastbourne Road, St Austell PL25 4SU
☎ 01726 64975   ᖴ 01726 64975
lordcrashingbore@btinternet.com
www.abrax7.stormloader.com
Owner *Paul Newman*
Editors *Paul Newman, Pamela Smith-Rawnsley*

Founded 1991. Large Crown Quarto, biannual paperback of over 250 pages, available both from address above and Lulu Publishing (www.lulu.com/lord-crashingbore). Combines leading articles by Colin Wilson and other scholars and critics with short stories, poems, reviews, translations and features on philosophy, existentialism and ideas. Favours challenging essays on the new and radical slants on the old (e.g. Colin Wilson on Edmund Husserl). Contributors include D.M. Thomas, Beryl Bainbridge, John Shand, Matthew Coniam, Gary Allen, Roger Morris and A.R. Lamb.
FICTION Flexible but prefers short stories of 1,000–2,000 words. 'Think of a writer like Wolfgang Borchert who wrote briefly, vividly and fiercely.'
POETRY Open to all styles, rhymed and irregular but prefers a creative, original use of language rather than prosy verse.
€ 'Flattery' and complimentary copy.

Acclaim
▷ The New Writer

Accountancy

145 London Road, Kingston Upon Thames KT2 6SR
☎ 020 8247 1389   ᖴ 020 8247 1424
accountancynews@cch.co.uk
www.accountancymagazine.com
Owner *Wolters Kluwer UK*
Editor *Chris Quick*
Circulation 119,249

Founded 1889. MONTHLY. Written ideas welcome.
FEATURES *Lesley Bolton* Accounting, tax, business-related articles of high technical content aimed at professional/managerial readers. Maximum 2,000 words. € Payment by arrangement.

Accountancy Age

32–34 Broadwick Street, London W1A 2HG
☎ 020 7316 9000   ᖴ 020 7316 9250
accountancy_age@incisivemedia.com
www.accountancyage.com
Owner *Incisive Media*
Editor-in-Chief *Damian Wild*
Editor *Gavin Hinks* (☎ 020 7316 9608)
Circulation 60,368

Founded 1969. WEEKLY. Unsolicited mss welcome. Ideas may be suggested in writing provided they are clearly thought out.
FEATURES *Kevin Reed* Topics right across the accountancy, business and financial world. Max. 1,400 words. € Payment negotiable.

ACE Tennis Magazine

Advantage Publishing (UK) Ltd, First Floor, Barry House, 20–22 Worple Road, London SW19 4DH
☎ 020 8947 0100   ᖴ 020 8947 0117
paul.morgan@acemag.co.uk
Owner *Tennis GB*
Editor *Paul Morgan*
Circulation 47,766

Founded 1996. MONTHLY specialist tennis magazine. News and features. No unsolicited mss; send feature synopses by e-mail in the first instance. No tournament reports.

Acoustic

Oyster House Media Ltd, Oyster House, Hunter's Lodge, Kentisbeare EX15 2DY
☎ 01884 266100   ᖴ 01884 266101
editor@acousticmagazine.com
www.acousticmagazine.com
Owner *Oyster House Media Ltd*
Editor *Russell Welton*
Circulation 25,000

Launched 2004. BI-MONTHLY. Reviews, news, features and interviews related to acoustic guitars, acoustic instruments and acoustic instrument manufacturing. Approach by e-mail.

Acumen
▷ under Poetry Magazines

Aeroplane

IPC Media Ltd., Blue Fin Building, 110 Southwark Street, London SE1 0SU
① 020 3148 4100
aeroplane_monthly@ipcmedia.com
www.aeroplanemonthly.com
Owner *IPC Country & Leisure Media Ltd*
Editor *Michael Oakey*
Circulation 35,704

Founded 1973. MONTHLY. Historic aviation and aircraft preservation from its beginnings to the 1960s. No poetry. Will consider news items and features written with authoritative knowledge of the subject, illustrated with good quality photographs.
NEWS *Tony Harmsworth* Max. 500 words.
FEATURES *Michael Oakey* Max. 3,000 words.
Approach by letter or e-mail in the first instance.
Ⓔ £60 per 1,000 words; £10–40 per picture used.

AIR International

PO Box 100, Stamford PE9 1XQ
① 01780 755131 Ⓕ 01780 757261
malcolm.english@keypublishing.com
www.airinternational.com
Owner *Key Publishing Ltd*
Editor *Malcolm English*
Circulation 12,041

Founded 1971. MONTHLY. Civil and military aircraft magazine. Unsolicited mss welcome but initial approach by phone or in writing preferred.

AirForces Monthly

PO Box 100, Stamford PE9 1XQ
① 01780 755131 Ⓕ 01780 757261
edafm@keypublishing.com
Owner *Key Publishing Ltd*
Editor *Alan Warnes*
Circulation 20,439

Founded 1988. MONTHLY. Modern military aviation magazine. Unsolicited contributions welcome but initial approach by phone or in writing preferred.

All About Soap

Hachette Filipacchi (UK) Ltd, 64 North Row, London W1K 7LL
① 020 7150 7000 Ⓕ 020 7150 7681
allaboutsoap@hf-uk.com
www.allaboutsoap.co.uk
Editor *Johnathon Hughes*
Circulation 100,048

Founded 1999. FORTNIGHTLY. Soap opera storylines, television gossip and celebrity features. No unsolicited contributions; make initial contact by phone or e-mail.

Amateur Gardening

Westover House, West Quay Road, Poole BH15 1JG
① 01202 440840 Ⓕ 01202 440860
Owner *IPC Media (A Time Warner Company)*
Editor *Tim Rumball*
Circulation 43,884

Founded 1884. WEEKLY. Topical and practical gardening articles. News items are compiled and edited in-house, generally. Approach by e-mail in the first instance.
Ⓔ Payment negotiable.

Amateur Photographer

IPC Media Ltd., Blue Fin Building, 110 Southwark Street, London SE1 0SU
① 020 3148 4138
amateurphotographer@ipcmedia.com
Owner *IPC Media (A Time Warner Company)*
Editor *Damien Demolder*
Circulation 24,567

Founded 1884. WEEKLY. For the competent amateur with a technical interest. Freelancers are used but writers should be aware that there is ordinarily no use for words without pictures.

Amateur Stage

Hampden House, 2 Weymouth Street, London W1W 5BT
① 020 7636 4343 Ⓕ 020 7636 2323
cvtheatre@aol.com
Owner *Platform Publications Ltd*
Editor *Mark Thorburn*

Topics of interest include amateur premières, technical developments within the amateur forum and items relating to landmarks or anniversaries in the history of amateur societies. Approach in writing only (include s.a.e. for return of mss). Ⓔ No payment.

Ancestors

▷ **The National Archives under UK Publishers**

Angler's Mail

IPC Media Ltd., Blue Fin Building, 110 Southwark Street, London SE1 0SU
① 020 3148 4159
anglersmail@ipcmedia.com
www.anglersmail.com
Owner *IPC Media (A Time Warner Company)*
Editor *Tim Knight*
Circulation 32,400

Founded 1965. WEEKLY. Angling news and matches. Interested in pictures, stories, tip-offs and features. Approach the news desk by telephone. Ⓔ £10–40 per 200 words; pictures, £20–40.

Animal Action

Wilberforce Way, Southwater, Horsham RH13 9RS
☎ 0300 123 0392   🖷 0300 123 0392
publications@rspca.org.uk
www.rspca.org.uk
Owner *RSPCA*
Editor *Sarah Evans*
Circulation 50,000

BI-MONTHLY RSPCA youth membership
magazine. Articles (pet care, etc.) are written
in-house. Good-quality animal photographs
welcome.

Animals and You

D.C. Thomson & Co. Ltd, 2 Albert Square,
Dundee DD1 9QJ
☎ 01382 223131   🖷 01382 322214
animalsandyou@dcthomson.co.uk
Owner *D.C. Thomson & Co. Ltd*
Editor *Margaret Monaghan*
Circulation 38,441

Founded 1998. THREE-WEEKLY magazine aimed
at girls, aged 7–11 years who love animals.
Contains cute 'n' cuddly pin-ups, pet tips and
advice, animal charity information, photo stories,
puzzles, readers' pictures and drawings. Most
work done in-house but will consider short
features, photographic or with good illustrations.
Approach by e-mail.

Antique Collecting

▷ Antique Collectors' Club under UK Publishers

Antiquesnews incorporating Antiques & Art Independent

PO Box 483, Winchester SO23 3BL
☎ 07000 268478
antiquesnews@hotmail.com
www.antiquesnews.co.uk
Owners *A.J. & M.K. Keniston*
Publisher/Editor *Tony Keniston*

Founded 1997. WEEKLY. Up-to-date information
online for the British antiques and art trade.
News, photographs, gossip and controversial
views on all aspects of the fine art and antiques
world welcome. Articles on antiques and fine arts
themselves are not featured. Approach in writing
with ideas.

Apex Advertiser

PO Box 7086, Clacton-on-Sea CO15 5WN
☎ 01255 428500
mail@apexpublishing.co.uk
www.apexadvertiser.co.uk
Owner *Apex Publishing Ltd*
Editor *Chris Cowlin*

Circulation 12,000

Founded 2002. MONTHLY advertising magazine
including guides and information, features,
listings and reviews. Approach by post, only.

Apollo

22 Old Queen Street, London SW1H 9HP
☎ 020 7961 0107
editorial@apollomag.com
www.apollo-magazine.com
Owner *The Spectator (1828) Ltd*
Chief Executive *Andrew Neil*
Editor *Michael Hall*

Founded 1925. MONTHLY. Specialist articles on art
and antiques, collectors and collecting, exhibition
and book reviews, information on dealers and
auction houses. Unsolicited mss welcome but
please make initial contact by e-mail. Interested
in new research on fine and decorative arts of all
eras, and architecture.

The Architects' Journal

Greater London House, Hampstead Road,
London NW1 7EJ
☎ 020 7728 4574   🖷 020 7728 4666
firstname.surname@emap.com
www.architectsjournal.co.uk
Owner *Emap Construct*
Editor *Kieran Long*
Circulation 11,694

WEEKLY trade magazine dealing with all aspects
of the industry. No unsolicited mss. Approach in
writing with ideas.

Architectural Design

John Wiley & Sons, 3rd Floor, International
House, Ealing Broadway Centre, London W5 5DB
☎ 020 8326 3800   🖷 020 8326 3801
Owner *John Wiley & Sons Ltd*
Editor *Helen Castle*
Development Editor *Mariangela Palazzi-Williams*
Circulation 5,000

Founded 1930. BI-MONTHLY. Sold as a book and
journal as well as online, *AD* charts theoretical
and topical developments in architecture. Format
consists of a minimum of 128pp, the first part
dedicated to a theme compiled by a specially
commissioned guest-editor; the back section
(AD+) carries series and more current one-off
articles. Unsolicited mss not welcome generally,
though journalistic contributions will be
considered for the back section.

The Architectural Review

Greater London House, Hampstead Road,
London NW1 7EJ
☎ 020 7728 4589
www.arplus.com

Editor *Paul Finch*
Circulation 18,180

MONTHLY international professional magazine dealing with architecture and all aspects of design. No unsolicited mss. Approach in writing with ideas.

Arena

Mappin House, 4 Winsley Street, London W1W 8HF
📞 020 7182 8000   📠 020 7182 8529
hollie.moat@bauerconsumer.co.uk
Owner *Bauer Consumer Media*
Acting Editor *Mat Smith*
Circulation 25,232

MONTHLY style and general interest magazine for men. Intelligent articles and profiles.
FEATURES Fashion, lifestyle, film, television, politics, business, music, media, design, art and architecture. Ideas for articles will be considered; send by e-mail.

The Armourer Magazine

PO Box 161, Congleton CW12 3WJ
📞 01260 278044   📠 01260 278044
editor@armourer.co.uk
www.armourer.co.uk
Owner *Beaumont Publishing Ltd*
Editor *Irene Moore*
Circulation 15,000

Founded 1994. BI-MONTHLY. Militaria, military antiques, military history, events and auctions, with an emphasis on the First and Second World Wars. Unsolicited military-related contributions are welcome but 'we do not have an editorial budget'. Approach by e-mail in the first instance.

Art Monthly

4th Floor, 28 Charing Cross Road, London WC2H 0DB
📞 020 7240 0389   📠 020 7497 0726
info@artmonthly.co.uk
www.artmonthly.co.uk
Owner *Britannia Art Publications Ltd*
Editor *Patricia Bickers*
Circulation 6,000

Founded 1976. TEN ISSUES YEARLY. News and features of relevance to those interested in modern and contemporary visual art. Unsolicited mss welcome. Contributions should be addressed to the deputy editor, accompanied by s.a.e.
FEATURES Always commissioned. Interviews and articles of up to 2,000 words on art theory, individual artists, contemporary art history and issues affecting the arts (e.g. funding and arts education). Exhibition and book reviews: 750–1,000 words.NEWS Brief reports (250–300 words) on art issues.
💷 Payment negotiable.

The Art Newspaper

70 South Lambeth Road, London SW8 1RL
📞 020 7735 3331   📠 020 7735 3332
contact@theartnewspaper.com
www.theartnewspaper.com
Owner *Umberto Allemandi & Co. Publishing*
Editor *Cristina Ruiz*
Circulation 22,000

Founded 1990. MONTHLY. Tabloid format with hard news on the international art market, news, museums, exhibitions, archaeology, conservation, books and current debate topics. Length 250–2,000 words. No unsolicited mss. Approach with ideas in writing. Commissions only.

Art Review

1 Sekforde Street, London EC1R 0BE
📞 020 7107 2760   📠 020 7107 2761
office@artreview.com
www.artreview.com
Owner *Dennis Hotz*
Editor *Mark Rappolt*
Deputy Editor *Skye Sherwin*
Circulation 35,000

Founded 1949. MONTHLY magazine of international contemporary art. Features, profiles, previews and reviews of artists and exhibitions worldwide. 'We have a very international focus so it's important that the artist is exhibiting internationally and has good gallery representation.'
NEWS All items written in-house.
FEATURES *Mark Rappolt* Max. 1,500 words. E-mail pitch and c.v. at least three months in advance of related show/publication.
💷 £400–500.

The Artist

Caxton House, 63–65 High Street, Tenterden TN30 6BD
📞 0158076 3673   📠 0158076 5411
www.theartistmagazine.co.uk
Owner/Editor *Sally Bulgin*
Circulation 23,000

Founded 1931. MONTHLY. Art journalists, artists, art tutors and writers with a good knowledge of art materials are invited to write to the editor with ideas for practical and informative features about art, techniques, techniques and artists.

Athletics Weekly

83 Park Road, Peterborough PE1 2TN
📞 01733 898440   📠 01733 898441
results@athletics-weekly.co.uk
www.athleticsweekly.com
Owner *Descartes Publishing*
Editor *Jason Henderson*
Results Editor *Paul Halford*

Circulation 14,000

Founded 1945. WEEKLY. Covers track and field, road, fell, cross-country, race walking. Results, fixtures, interviews, technical coaching articles and news surrounding build up to 2012 Olympics. NEWS *Jon Mulkeen* Max. 400 words. FEATURES *Jason Henderson* Max. 2,000 words. Approach in writing. Ⓔ Payment negotiable.

The Author

84 Drayton Gardens, London SW10 9SB
Ⓣ 020 7373 6642
theauthor@societyofauthors.org
Owner *The Society of Authors*
Editor *Andrew Rosenheim*
Manager *Kate Pool*
Circulation 9,000

Founded 1890. QUARTERLY journal of the **Society of Authors**. Most articles are commissioned.

Auto Express

30 Cleveland Street, London W1T 4JD
Ⓣ 020 7907 6000   Ⓕ 020 7907 6234
editorial@autoexpress.co.uk
www.autoexpress.co.uk
Owner *Dennis Publishing*
Editor *David Johns*
Managing Editor *Graham Hope*
News and Features Editor *Mat Watson*
Circulation 81,786

Founded 1988. WEEKLY consumer motoring title with news, drives, tests, investigations, etc. NEWS News stories and tip-offs welcome. Fillers 100 words max.; leads 300 words. FEATURES Ideas welcome. No fully-written articles – features will be commissioned if appropriate. Max. 2,000 words.

Autocar

Haymarket Consumer Media, Teddington Studios, Broom Road, Teddington TW11 9BE
Ⓣ 020 8267 5630   Ⓕ 020 8267 5759
autocar@haymarket.com
www.autocar.co.uk
Owner *Haymarket Consumer Media,*
Editor *Chas Hallett*
News Editor *Dan Stevens*
Circulation 66,505

Founded 1895. WEEKLY. All news stories, features, interviews, scoops, ideas, tip-offs and photographs welcome. Ⓔ Payment negotiable.

Aviation News

HPC, Drury Lane, St Leonards on Sea TN38 9BJ
Ⓣ 01424 205536   Ⓕ 01424 443693
editor@aviation-news.co.uk
www.aviation-news.co.uk
Owner *Hastings Printing Company*

Editor *David Baker*
Circulation 21,000

Founded 1939. MONTHLY review of aviation for those interested in military, commercial and business aircraft and equipment – old and new. Will consider articles on military and civil aircraft, airports, air forces, current and historical subjects. No fiction. FEATURES 'New writers always welcome and if the copy is not good enough it is returned with guidance attached.' Max. 5,000 words. NEWS Items are always considered from new sources. Max. 300 words. Ⓔ Payment negotiable.

Balance

Diabetes UK, 10 Parkway, London NW1 7AA
Ⓣ 020 7424 1000   Ⓕ 020 7424 1001
balance@diabetes.org.uk
Owner *Diabetes UK*
Editor *Martin Cullen*
Circulation 250,000

Founded 1935. BI-MONTHLY. Unsolicited mss are not accepted. Writers may submit a brief proposal in writing. Only topics relevant to diabetes will be considered. NEWS Short pieces about activities relating to diabetes and the lifestyle of people with diabetes. Max. 150 words. FEATURES Medical, diet and lifestyle features written by people with diabetes or with an interest and expert knowledge in the field. General features are mostly based on experience or personal observation. Max. 1,500 words. Ⓔ NUJ rates.

The Banker

Number One Southwark Bridge, London SE1 9HL
Ⓣ 020 7873 3000
stephen.timewell@ft.com
www.thebanker.com
Owner *FT Business*
Editor-in-Chief *Stephen Timewell*
Circulation 28,060

Founded 1926. MONTHLY. News and features on banking, finance and capital markets worldwide and technology.

BBC Gardeners' World Magazine

Media City, 201 Wood Lane, London W12 7TQ
Ⓣ 020 8433 2000   Ⓕ 020 8433 3986
adam.pasco@bbc.co.uk
www.gardenersworld.com
Owner *BBC Worldwide Publishing Ltd*
Editor *Adam Pasco*
Deputy Editor *Lucy Hall*
Commissioning Editor *Kevin Smith*

Circulation 238,673

Founded 1991. MONTHLY. Gardening advice, ideas and inspiration. No unsolicited mss. Approach by phone or in writing with ideas – interested in features about exceptional small gardens. Also interested in any exciting new gardens showing good design and planting ideas as well as new plants and the stories behind them. Including sections on Grow & Eat, problem solving, wildlife gardening, shopping and buyers' guides, plus practical projects.

BBC Good Food

Media Centre, 201 Wood Lane, London W12 7TQ
☎ 020 8433 2000   ⓕ 020 8433 3931
goodfood@bbc.com
www.bbcgoodfood.com
Owner *BBC Worldwide Publishing Ltd*
Editor *Gillian Carter*
Circulation 365,978

Founded 1989. MONTHLY recipe magazine featuring celebrity chefs. No unsolicited mss.

BBC History Magazine

Tower House, Fairfax Street, Bristol BS1 3BN
☎ 0117 927 9009
historymagazine@bbcmagazinesbristol.com
www.bbcmagazinesbristol.com
Owner *Bristol Magazines Ltd*
Editor *Dave Musgrove*
Circulation 58,395

Founded 2000. MONTHLY. General news and features on British and international history, with books and CD reviews, listings of history events, TV and radio history programmes and regular features for those interested in history and current affairs. Winner, British Society of Magazine Editors 'Editor of the Year Award', 2002, for general interest and current affairs titles. Will consider submissions from academic or otherwise expert historians/archaeologists as well as, occasionally, from historically literate journalists who include expert analysis and historiography with a well-told narrative. Ideas for regular features are welcome. Also publishes cartoons, quizzes and crosswords. 'We cannot guarantee to acknowledge all unsolicited mss.'
FEATURES should be pegged to anniversaries or forthcoming books/TV programmes, current affairs topics, etc. 750–3,000 words.
NEWS 400–500 words. Send short letter or e-mail with synopsis, giving appropriate sources, pegs for publication dates, etc.
ⓔ Payment negotiable.

BBC Music Magazine

9th Floor, Tower House, Fairfax Street, Bristol BS1 3BN
☎ 0117 927 9009   ⓕ 0117 934 9008

music@bbcmagazinesbristol.com
www.bbcmusicmagazine.com
Owner *BBC Worldwide Publishing Ltd*
Editor *Oliver Condy*
Circulation 47,104

Founded 1992. MONTHLY. All areas of classical music. Not interested in unsolicited material. Approach with ideas by e-mail or fax.

BBC Top Gear Magazine

Media Centre, 201 Wood Lane, London W12 7TQ
☎ 020 8433 2000
www.topgear.com
Owner *BBC Magazines*
Editorial Director *Michael Harvey*
Circulation 200,286

Founded 1993. MONTHLY general interest motoring magazine. No unsolicited material as most features are commissioned.

BBC Wildlife Magazine

14th Floor, Tower House, Fairfax Street, Bristol BS1 3BN
☎ 0117 927 9009   ⓕ 0117 927 9008
wildlifemagazine@bbcmagazinesbristol.com
www.bbcwildlifemagazine.com
Owner *Bristol Magazines Ltd*
Editor *Sophie Stafford*
Features Editor *Fergus Collins*
News & Travel Editor *James Fair*
Circulation 47,114

Founded 1963. Formerly *Wildlife*; became *BBC Wildlife Magazine* in 1983. THIRTEEN ISSUES A YEAR. Welcomes idea pitches but not unsolicited mss. Read guidelines on website to improve chances of success.
FEATURES Most features commissioned from excellent writers with expert knowledge of wildlife or conservation subjects. Max. 2,000 words.
NEWS Most news stories commissioned from known freelancers. Max. 400 words.
ⓔ Features, £200–450; news, £80–120.

Bee World
▶ Journal of Apicultural Research

Bella

H. Bauer Publishing, Academic House, 24–28 Oval Road, London NW1 7DT
☎ 020 7241 8000   ⓕ 020 7241 8056
Owner *H. Bauer Publishing*
Editor *Julia Davis*
Circulation 221,661

Founded 1987. WEEKLY. Women's magazine specializing in real-life, human interest stories.

FEATURES *Unity Wroe* Contributions welcome for some sections of the magazine.

Best

33 Broadwick Street, London W1F 0DQ
☎ 020 7339 4000   ℻ 020 7339 4580
best@acp-natmag.co.uk
Owner *National Magazine Company/ACP*
Editor *Michelle Hather*
Circulation 330,770

Founded 1987. WEEKLY women's magazine.
Multiple features, news, short stories on all topics
of interest to women. Important for would-be
contributors to study the magazine's style which
differs from many other women's weeklies.
Approach in writing with s.a.e.
FEATURES *Helen Garston, Charlotte Seligman*
Max. 1,500 words. No unsolicited mss.
FICTION *Pat Richardson* Short story slot:
900–1,200 words. Send s.a.e. for guidelines.
€ Payment negotiable.

Best of British

Church Lane Publishing Ltd, Bank Chambers, 27a
Market Place, Market Deeping PE6 8EA
☎ 01778 342814
mail@british.fsbusiness.co.uk
www.bestofbritishmag.co.uk
Owner *Church Lane Publishing Ltd*
Editor-in-Chief *Ian Beacham*
Assistant Editor *Linne Matthews*

Founded 1994. MONTHLY magazine celebrating all
things British, both past and present. Emphasis
on nostalgia – memories from the 1940s, 1950s
and 1960s. Study of the magazine is advised in the
first instance. All preliminary approaches should
be made in writing or by e-mail.

The Big Issue and The Big Issue South West

1–5 Wandsworth Road, London SW8 2LN
☎ 020 7526 3200   ℻ 020 7526 3201
helena.drakakis@bigissue.com
www.bigissue.com
Editor-in-Chief *A. John Bird*
Editor *Charles Howgego*
Associate Editor *Helena Drakakis*
Circulation 87,750; 15,097 (South West); 174,770
(Group)

Founded 1991. WEEKLY. An award-winning
campaigning and street-wise general interest
magazine sold in London, the Midlands, the
North East and South of England. Separate
regional editions sold in Manchester, Scotland,
Wales and the South West.
NEWS *Helena Drakakis* Hard-hitting exclusive
stories with emphasis on social injustice aimed at
national leaders.
ARTS *Charles Howgego* Interested in interviews
and analysis ideas. Reviews written in-house.
Send synopses to arts editor.
FEATURES Interviews, campaigns, comment,
opinion and social issues reflecting a varied

and informed audience. Balance includes social
issues but mixed with arts and cultural features.
Freelance writers used each week commissioned
from a variety of contributors. Best approach in
the first instance is to e-mail or post synopses to
Editor *Charles Howgego* with examples of work.
Max. 1,500 words.

The Big Issue Cymru

55 Charles Street, Cardiff CF10 2GD
☎ 029 2025 5670   ℻ 029 2025 5672
editor@bigissuecymru.co.uk
www.bigissuecymru.co.uk
Editor *Rachell Howells*
Circulation 10,330

Founded 1993. WEEKLY. Hard-hitting social and
news features plus arts and entertainment.
FEATURES/NEWS *Rachel Howells* Welcomes
contributions but must be relevant to a Welsh
readership. Approach by e-mail. Lead in of four
to six weeks. Max. 1,000 words (features); 350
(news). No poetry or fiction.

The Big Issue in Scotland

2nd Floor, 43 Bath Street, Glasgow G2 1HW
☎ 0141 352 7260   ℻ 0141 333 9049
editorial@bigissuescotland.com
www.bigissuescotland.com
Editor *Paul McNamee*
Circulation 31,401

WEEKLY. Coverage of social issues, arts and
culture, sport, interviews; news and hard-hitting
journalism.

The Big Issue in the North

10 Swan Street, Manchester M4 5JN
☎ 0161 831 5550   ℻ 0161 831 5577
kevin.gopal@bigissuenorth.co.uk
www.thebiglifecompany.com
Owner *The Big Life Company*
Editor *Kevin Gopal*
Circulation 30,192

Launched 1992. WEEKLY. Entertainment,
celebrity interviews, political and environmental
issues and issues from the social sector as well
as homelessness. Welcomes news stories not
covered by the national press; approach by e-mail.

The Big Issue South West

Founded 1994. WEEKLY. Hard-hitting, socially
aware news and features, local arts and
entertainment. Welcomes news features with a
regional relevance. No poetry or fiction.
Editorial dealt with by London office – see entry
for *The Big Issue*.

Bike

Bauer Automotive Ltd, Media House, Lynch
Wood, Peterborough PE2 6EA
☎ 01733 468099   ℻ 01733 468196

bike@bauerconsumer.co.uk
www.bikemagazine.co.uk
Owner *Bauer Automotive Ltd*
Editor *Steve Rose*
Circulation 81,579

Founded 1971. MONTHLY. Broad-based motorcycle magazine. Approach by e-mail.

Bird Life Magazine

The RSPB, UK Headquarters, The Lodge, Sandy SG19 2DL
℡ 01767 680551   ℻ 01767 683262
derek.niemann@rspb.org.uk
Owner *Royal Society for the Protection of Birds*
Editor *Derek Niemann*
Circulation 90,000
Founded 1965. BI-MONTHLY. Bird, wildlife and nature conservation for 8–12-year-olds (RSPB Wildlife Explorers members). No unsolicited mss. No 'captive/animal welfare' articles.

FEATURES Unsolicited material rarely used.
NEWS News releases welcome. Approach in writing in the first instance.

Bird Watching

Media House, Lynchwood, Peterborough PE2 6EA
℡ 01733 468000   ℻ 01733 468387
kevin.wilmot@emap.com
www.birdwatching.co.uk
Editor *Kevin Wilmot*
Circulation 19,763

Founded 1986. MONTHLY magazine for birdwatchers at all levels of experience. Offers practical advice of where and when to see birds, how to identify them, product reviews (particularly binoculars, telescopes and digiscoping equipment), extensive bird sightings, 'Go Birding' walks (10 each issue), news and general articles.
FEATURES Interested in authoratative articles on bird behaviour and bird watching/sites in the UK. Please send synopsis, not the whole piece. Max. 1,000 words.
NEWS UK news pre-eminent, preferably with pictures. Max. 350 words.
£ Fees negotiable.

Birds

The RSPB, UK Headquarters, The Lodge, Sandy SG19 2DL
℡ 01767 680551   ℻ 01767 683262
rob.hume@rspb.org.uk
www.rspb.org.uk
Owner *Royal Society for the Protection of Birds*
Editor *R.A. Hume*
Circulation 608,975

QUARTERLY magazine which covers not only wild birds but also other wildlife and related conservation topics. No interest in features on

pet birds or 'rescued' sick/injured/orphaned ones. Content refers mostly to RSPB work so opportunities for freelance work on other subjects are limited but some freelance submissions are used in most issues. Phone or e-mail to discuss.

Birdwatch

B403A The Chocolate Factory, 5 Clarendon Road, London N22 6XJ
℡ 020 8881 0550   ℻ 020 8881 0990
editorial@birdwatch.co.uk
www.birdwatch.co.uk
Owner *Solo Publishing*
Editor *Dominic Mitchell*
Circulation 14,000

Founded 1992. MONTHLY magazine featuring illustrated articles on all aspects of birds and birdwatching, especially in Britain. No unsolicited mss. Approach in writing with synopsis of 100 words max. Annual **Birdwatch Bird Book of the Year** award (see entry under *Prizes*).
FEATURES *Dominic Mitchell* Unusual angles/personal accounts, if well-written. Articles of an educative or practical nature suited to the readership. Max. 2,000 words.
FICTION *Dominic Mitchell* Very little opportunity although occasional short story published. Max. 1,500 words.
NEWS *Dominic Mitchell* Very rarely use external material.
£ £50 per 1,000 words.

Bizarre

30 Cleveland Street, London W1T 4JD
℡ 020 7907 6000   ℻ 020 7907 6516
bizarre@dennis.co.uk
www.bizarremag.com
Owner *Dennis Publishing*
Editor *David McComb*
Circulation 50,132

Founded 1997. MONTHLY magazine featuring amazing stories and images from around the world. No fiction, poetry, illustrations, short snippets.
FEATURES *Kate Hodges* No unsolicited mss. Send synopsis by post or e-mail.
£ Payment negotiable.

Black Beauty & Hair

2nd Floor, Culvert House, Culvert Road, Battersea, London SW11 5HD
℡ 020 7720 2108   ℻ 020 7498 3023
info@blackbeautyandhair.com
www.blackbeautyandhair.com
Owner *Hawker Consumer Publications Ltd*
Publisher *Pat Petker* (℡ ext. 207)
Editor *Irene Shelley* (℡ ext. 215)
Circulation 19,697
BI-MONTHLY with one annual special: *The Hairstyle Book* in October; and a *Bridal*

Supplement in the April/May issue. Black hair and beauty magazine with emphasis on authoritative articles relating to hair, beauty, fashion, health and lifestyle. Unsolicited contributions welcome. FEATURES Beauty and fashion pieces welcome from writers with a sound knowledge of the Afro-Caribbean beauty scene plus bridal features. Minimum 1,000 words.
£ £100 per 1,000 words.

Black Static

TTA Press, 5 Martins Lane, Witcham, Ely CB6 2LB
info@ttapress.demon.co.uk
www.ttapress.com
Owner *TTA Press*
Editor *Andy Cox*

Formerly *The Third Alternative*, founded 1993. BIMONTHLY colour horror magazine: short fiction, news and gossip, film and television, manga and anime books. Publishes talented newcomers alongside famous authors. Unsolicited mss welcome if accompanied by s.a.e. or e-mail address for overseas submissions (max. 8,000 words). Fiction submissions as hard copy only; feature suggestions can be e-mailed. Potential contributors are advised to study the magazine. Contracts are exchanged and payment made upon acceptance. Winner of British Fantasy Awards, International Horror Guild Award.

Bliss Magazine

Coach and Horses Passage, The Pantiles, Tunbridge Wells TN2 5UJ
℡ 01892 500106   ℻ 01892 545666
bliss@panini.co.uk
www.mybliss.co.uk
Owner *Panini UK Ltd*
Editor *Leslie Sinoway*
Circulation 151,729

Founded 1995. MONTHLY teenage lifestyle magazine for girls. No unsolicited mss; 'e-mail the Features Editor, *Angeli Milburn* (amilburn@panini.co.uk) with an idea and then follow up with a call.'
NEWS Worldwide teenage news. Max. 200 words.
FEATURES Real life teenage stories with subjects willing to be photographed. Reports on teenage issues. Max. 1,500 words.

Blueprint

6–14 Underwood Street, London N1 7JQ
℡ 020 7549 2542
vrichardson@wilmington.co.uk
www.blueprintmagazine.co.uk
Owner *Wilmington Media*
Editor *Vicky Richardson*
Deputy Editor *Tim Abrahams*
Senior Staff Writer *Peter Kelly*

Circulation 6,693

Founded 1983. MONTHLY. 'Crossing the boundaries of design and architecture (from graphics and product design to interiors and urban development), Blueprint examines their impact on society in general.' Features, news and reviews. Welcomes ideas and pitches for articles on topics relevant to the magazine; no completed features. Approach by e-mail.

The Book Collector

PO Box 12426, London W11 3GW
℡ 020 7792 3492   ℻ 020 7792 3492
editor@thebookcollector.co.uk
(editorial)
www.thebookcollector.co.uk
Owner *The Collector Ltd*
Editor *Nicolas J. Barker*

Founded 1950. QUARTERLY magazine on bibliography and the history of books, book-collecting, libraries and the book trade.

Book World Magazine

2 Caversham Street, London SW3 4AH
℡ 020 7351 4995   ℻ 020 7351 4995
leonard.holdsworth@btinternet.com
Owner *Christchurch Publishers Ltd*
Editor *James Hughes*
Circulation 5,500

Founded 1980. MONTHLY news and reviews for serious book collectors, librarians, antiquarian and other booksellers. No unsolicited mss. Interested in material relevant to literature, art and book collecting. Send letter in the first instance.

The Bookseller

5th Floor, Endeavour House, 189 Shaftesbury Avenue, London WC2H 8TJ
℡ 020 7420 6006   ℻ 020 7420 6103
www.theBookseller.com
Owner *Nielsen Business Media*
Editor-in-Chief *Neill Denny*

Trade journal of the book trade, publishing, retail and libraries – the essential guide to the book business since 1858. Trade news and features, including special features, company news, publishing trends, bestseller data, etc. Unsolicited mss rarely used as most writing is either done in-house or commissioned from experts within the trade. Approach in writing first.

Boxing Monthly

40 Morpeth Road, London E9 7LD
℡ 020 8986 4141   ℻ 020 8986 4145
mail@boxing-monthly.demon.co.uk
www.boxing-monthly.co.uk
Owner *Topwave Ltd*
Editor *Glyn Leach*

Circulation 25,000

Founded 1989. MONTHLY. International coverage of professional boxing; previews, reports and interviews. Unsolicited material welcome. Interested in small hall shows and grass-roots knowledge. No big fight reports. Approach in writing in the first instance.

Boyz

18 Brewer Street, London W1F 0SH
☎ 020 7025 6100   🖷 020 7025 6109
davidb@boyz.co.uk
Managing Editor *David Bridle*
Circulation 55,000

Founded 1991. WEEKLY entertainment and features magazine aimed at a gay readership covering clubs, fashion, TV, films, music, theatre, celebrities and the London gay scene in general. Unsolicited mss welcome.

Brand Literary Magazine

Creative Writing Programme, EPS, King William Court, University of Greenwich, Maritime Campus, Park Row, Greenwich, London SE10 9LS
☎ 020 8331 8952
info@brandliterarymagazine.co.uk
www.brandliterarymagazine.co.uk
Editor *Nina Rapi*
Circulation 1,000

Biannual literary magazine, launched in 2007, focusing on the short form; left of field work and international writing. Welcomes stories, plays, poems and creative non-fiction; no genre or historical material. Approach by post; see submission guidelines on the website.
FICTION *Nina Rapi* 'Looking for writing that takes risks and has a strong voice.' Max. 2,500 words.
POETRY *Cherry Smyth* Max. 40 lines.
PLAYS *Nina Rapi, Harry Derbyshire* (Advisor) Bold, short plays and performance text. Max. ten minutes performance time.
💷 £30.

Brides

Vogue House, Hanover Square, London W1S 1JU
☎ 020 7499 9080   🖷 020 7152 3369
www.bridesmagazine.co.uk
Owner *Condé Nast Publications Ltd*
Editor *Deborah Joseph*
Circulation 68,259

Founded 1955. BI-MONTHLY. Much of the magazine is produced in-house, but a good, relevant feature on cakes, jewellery, music, flowers, etc. is always welcome. Max. 1,000 words. Prospective contributors should make initial contact by e-mail.

British Birds

4 Harlequin Gardens, St Leonards-on-Sea TN37 7PF
☎ 01424 755155   🖷 01424 755155
editor@britishbirds.co.uk
Editor *Dr R. Riddington*
Circulation 6,000

Founded 1907. MONTHLY ornithological journal. Features main papers on topics such as behaviour, distribution, ecology, identification and taxonomy; annual *Report on Rare Birds in Great Britain* and *Rare Breeding Birds in the UK*; sponsored competition for Bird Photograph of the Year. Unsolicited mss welcome from ornithologists.
FEATURES Well-researched, original material relating to Western Palearctic birds welcome.
NEWS *Adrian Pitches* (adrian.pitches@ blueyonder.co.uk) Items ranging from conservation to humour. Max. 200 words.
💷 Payment for photographs, drawings, paintings and main papers.

British Chess Magazine

The Chess Shop, 44 Baker Street, London W1U 7RT
☎ 020 7486 8222   🖷 020 7486 3355
bcmchess@compuserve.com
www.bcmchess.co.uk
Director/Editor *John Saunders*

Founded 1881. MONTHLY. Emphasis on tournaments, the history of chess and chess-related literature. Approach in writing with ideas. Unsolicited mss not welcome unless from qualified chess experts and players.

British Journalism Review

Sage Publications Ltd, 1 Oliver's Yard, 55 City Road, London EC1Y 1SP
☎ 020 7324 8500   🖷 020 7324 8756
bill.hagerty@btinternet.com
www.bjr.org.uk
www.bjr.sagepub.com
Owner *BJR Publishing Ltd*
Editor *Bill Hagerty*

Founded 1988. QUARTERLY. Media magazine providing a forum for the news media. Freelance contributions welcome: non-academic articles on all aspects of the news media only. No academic papers. Approach by e-mail. Max. 2,000 words.

British Medical Journal

BMA House, Tavistock Square, London WC1H 9JR
☎ 020 7387 4499   🖷 020 7383 6418
editor@bmj.com
www.bmj.com
Owner *British Medical Association*
Editor *Dr Fiona Goodlee*

Circulation 122,982

One of the world's leading general medical journals.

British Philatelic Bulletin

35–50 Rathbone Place, London W1T 1HQ

☎ 020 7441 4744

www.royalmail.com/stamps

Owner *Royal Mail*

Editor *John Holman*

Circulation 15,000

Founded 1963. MONTHLY bulletin giving details of forthcoming British stamps, features on older stamps and postal history, and book reviews. Welcomes photographs of interesting, unusual or historic letter boxes.

FEATURES Articles on all aspects of British philately. Max. 1,500 words.

NEWS Reports on exhibitions and philatelic events. Max. 500 words. Approach in writing in the first instance.

Ⓔ £75 per 1,000 words.

British Railway Modelling

The Maltings, West Street, Bourne PE10 9PH

☎ 01778 391027   ⓕ 01778 425437

johne@warnersgroup.co.uk

www.brmodelling.co.uk

Owner *Warners Group Publications Plc*

Managing Editor *David Brown*

Editor *John Emerson*

Assistant Editor *Tony Wright*

Circulation 25,000

Founded 1993. MONTHLY. A general magazine for the practising modeller. Ideas are welcome. Interested in features on quality models, from individual items to complete layouts. Approach in writing first.

FEATURES Articles on practical elements of the hobby, e.g. locomotive construction, kit conversions, etc. Layout features and articles on individual items which represent high standards of the railway modelling art. Max. length 6,000 words (single feature).

NEWS News and reviews containing the model railway trade, new products, etc. Max. length 1,000 words.

Broadcast

Greater London House, Hampstead Road, London NW1 7EJ

☎ 020 7728 5542   ⓕ 020 7728 5555

Owner *Emap Communications*

Editor *Lisa Campbell*

Circulation 10,585

Founded 1959. WEEKLY. Opportunities for freelance contributions. Write to the relevant editor in the first instance.

FEATURES *Emily Booth* Any broadcasting issue.

Max. 1,500 words.

NEWS *Chris Curtis* Broadcasting news. Max. 350 words.

Ⓔ £270 per 1,000 words.

Build It

1 Canada Square, Canary Wharf, London E14 5AP

☎ 020 7772 8452   ⓕ 020 7772 8599

buildit@oceanmedia.co.uk

www.buildit-online.co.uk

Owner *Ocean Media*

Acting Editor *Duncan Hayes*

Circulation 16,613

Founded 1990. MONTHLY magazine covering self-build, conversion and renovation. Unsolicited material welcome on self-build case studies as well as articles on technical construction, architecture and design and dealing with builders. Max. length 2,500 words. Approach by post or e-mail.

Building Magazine

3rd Floor, Ludgate House, 245 Blackfriars Road, London SE1 9UY

☎ 020 7560 4141   ⓕ 020 7560 4080

dchevin@cmpi.biz

www.building.co.uk

Owner *CMP Information Ltd*

Editor *Denise Chevin*

Circulation 24,198

WEEKLY magazine for the whole construction industry, from architects and engineers to contractors and house builders. News and features. No unsolicited contributions as all material is commissioned, although legal items welcome.

The Burlington Magazine

14–16 Duke's Road, London WC1H 9SZ

☎ 020 7388 1228   ⓕ 020 7388 1230

editorial@burlington.org.uk

www.burlington.org.uk

Owner *The Burlington Magazine Publications Ltd*

Managing Director *Kate Trevelyan*

Editor *Richard Shone*

Deputy Editor *Bart Cornelis*

Associate Editor *Jane Martineau*

Contributing Editor *John-Paul Stonard*

Founded 1903. MONTHLY. Unsolicited contributions welcome on the subject of art history provided they are previously unpublished. All preliminary approaches should be made in writing.

EXHIBITION REVIEWS Usually commissioned, but occasionally unsolicited reviews are published if appropriate. Max. 1,000 words.

ARTICLES Max. 4,500 words. Ⓔ Max. £150.

SHORTER NOTICES Max. 2,000 words. Ⓔ Max. £80.

Buses

PO Box 14644, Leven KY9 1WX
① 01333 340637   ⑤ 01333 340608
buseseditor@btconnect.com
www.busesmag.com
Owner *Ian Allan Publishing*
Editor *Alan Millar*
Circulation 13,000

Founded 1949. MONTHLY magazine of the bus and
coach industry; vehicles, companies and people.
Specialist articles welcome; no general articles
with no specific bus content. E-mail enquiry
welcome.

Business Brief

PO Box 582, Five Oaks, St Saviour JE4 8XQ
① 01534 611600   ⑤ 01534 611610
mspeditorial@msppublishing.com
Owner *MSP Publishing*
Editor *Peter Body*
Circulation 6,500 (Jersey & Guernsey)

Founded 1989. MONTHLY magazine covering
business developments in the Channel Islands
and how they affect the local market. Styles itself
as the magazine for business people rather than
just a magazine about business. Interested in
business-orientated articles only – 800 words
max. Approach the editor by e-mail in the first
instance with telephone follow-up.
Ⓔ Payment negotiable.

Business Traveller

2nd Floor, Cardinal House, Albemarle Street,
London W1S 4TE
① 020 7647 6330   ⑤ 020 7647 6331
editorial@businesstraveller.com
www.businesstraveller.com
Owner *Panacea Publications*
Editorial Director *Tom Otley*
Circulation 500,000 (worldwide)

MONTHLY. Consumer publication. Opportunities
exist for freelance writers. Would-be contributors
are strongly advised to study the magazine or the
website first. Approach in writing with ideas.
Ⓔ Payment varies.

Buzz Extra

IBRA, 16 North Road, Cardiff CF10 3DY
① 029 2037 2409   ⑤ 05601 135640
mail@ibra.org.uk
www.ibra.org.uk
Owner *International Bee Research Association*
Editor *Richard Jones*

Informative, full-colour newsletter of the
International Bee Research Association.
QUARTERLY (March, June, September and
December).

The Cambridgeshire Journal Magazine

Cambridge Newspapers Ltd, Winship Road,
Milton CB24 6PP
① 01223 434409   ⑤ 01223 434415
Owner *Cambridge Newspapers Ltd*
Editor *Debbie Tweedie*
Circulation 11,000

Founded 1990. MONTHLY magazines covering
items of local interest – history, people,
conservation, business, places, food and wine,
fashion, homes and gardens.
FEATURES 750–1,500 words max., plus pictures.
Approach the editor with ideas in the first
instance.

Campaign

174 Hammersmith Road, London W6 7JP
① 020 8267 4683   ⑤ 020 8267 4927
campaign@haymarket.com
www.brandrepublic.com/campaign
Owner *Haymarket Publishing Ltd*
Editor *Claire Beale*
Circulation 10,112

Founded 1968. WEEKLY. Lively magazine serving
the advertising and related industries. Freelance
contributors are best advised to write in the first
instance.
FEATURES Ideas welcome.
NEWS Relevant news stories of up to 320 words.
Ⓔ Payment negotiable.

Camping and Caravanning

Greenfields House, Westwood Way, Coventry
CV4 8JH
① 0845 130 7631   ⑤ 024 7647 5413
www.campingandcaravanningclub.co.uk
Owner *The Camping and Caravanning Club*
Features Editor *Sue Taylor*
Circulation 209,042

Founded 1901. MONTHLY. Interested in journalists
with camping and caravanning knowledge. Write
with ideas for features in the first instance.
FEATURES Outdoor pieces in general, plus items
on specific regions of Britain. Max. 1,200 words.
Illustrations to support text essential.

Canals and Rivers

PO Box 618, Norwich NR7 0QT
① 01603 708930
chris@themag.fsnet.co.uk
www.canalsandrivers.com
Owner *A.E. Morgan Publications Ltd*
Editor *Chris Cattrall*
Circulation 14,000

Covers all aspects of waterways, narrow boats
and cruisers. Contributions welcome. Make initial
approach by post or e-mail.
FEATURES Waterways, narrow boats and motor
cruisers, cruising reports, practical advice,

etc. Unusual ideas and personal comments are particularly welcome. Max. 2,000 words. Articles should be supplied in PC Windows format disk or e-mailed with JPEG images.

NEWS Items of up to 200 words welcome on the Inland Waterways System, plus photographs if possible.

Ⓔ Features, around £60 per page; news, £20.

Candis

Newhall Lane, Hoylake CH47 4BQ
Ⓣ 0844 545 8102   Ⓕ 0844 545 8103
debbie@candis.co.uk
www.candis.co.uk
Owner *Newhall Publications*
Editor *Debbie Atwell*
Circulation 300,147

MONTHLY. General features on health and family life – no fashion or beauty items or first-person accounts of illness. No unsolicited contributions as most features are individually commissioned but fiction will be considered (max. 2,000 words). Send by e-mail (fiction@candis.co.uk).

Car Mechanics

Kelsey Publishing Group, Cudham Tithe Barn, Berry's Hill, Cudham TN16 3AG
Ⓣ 01959 541444/01733 771292 (editor)
cm.mag@kelsey.co.uk
Owner *Kelsey Publishing Group*
Editor *Peter Simpson*
Circulation 35,000

MONTHLY. Practical guide to maintenance and repair of post-1978 cars for DIY and the motor trade. Unsolicited mss, with good-quality colour prints or transparencies, sent 'at sender's risk'. Initial approach by letter or phone strongly recommended, 'but please study a recent copy for style and content first'. No non-technical material, 'travelogues', cartoons or fiction.
FEATURES Good, entertaining and well-researched technical material welcome, especially anything presenting complex matters clearly and simply. PAYMENT negotiable 'but generous for the right material. We normally require words and pictures and can rarely use words alone.'

Caravan Magazine

IPC Media Ltd, Leon House, 233 High Street, Croydon CR9 1HZ
Ⓣ 020 8726 8000   Ⓕ 020 8726 8299
caravan@ipcmedia.com
www.caravanmagazine.co.uk
Owner *IPC Media (A Time Warner Company)*
Editor *Victoria Heath*
Circulation 15,774

Founded 1933. MONTHLY. Unsolicited mss welcome. Approach in writing with ideas. All correspondence should go direct to the editor.

FEATURES Touring with strong caravan bias, technical/DIY features and how-to section. Max. 1,500 words.
Ⓔ Payment by arrangement.

Cat World

Ancient Lights, 19 River Road, Arundel BN18 9EY
Ⓣ 01903 884988   Ⓕ 01903 885514
laura@catworld.co.uk
www.catworld.co.uk
Owner *Ashdown Publishing Ltd*
Editor *Laura Quiggen*

Founded 1981. MONTHLY. Unsolicited mss welcome but initial approach in writing preferred. No poems or fiction.
FEATURES Lively, first-hand experience features on every aspect of the cat. Breeding features and veterinary articles by acknowledged experts only. Preferred length 750 or 1,700 words. Accompanying pictures should be good quality and sharp. Submissions by e-mail (MS Word attachment) if possible, or by disk with accompanying hard copy and s.a.e. for return.

Caterer and Hotelkeeper

Quadrant House, The Quadrant, Sutton SM2 5AS
Ⓣ 020 8652 3500   Ⓕ 020 8652 8213
caterernews@rbi.co.uk
www.caterersearch.com
Owner *Reed Business Information*
Editor *Mark Lewis*
Circulation 18,807

Founded 1878. WEEKLY journal of the catering industry: news, case studies, products. Contributions welcome; approach by e-mail.

The Catholic Herald

Lamb's Passage, Bunhill Row, London EC1Y 8TQ
Ⓣ 020 7448 3603   Ⓕ 020 7256 9728
editorial@catholicherald.co.uk
www.catholicherald.co.uk
Editor *Luke Coppen*
Features Editor *Ed West*
Literary Editor *Stav Sherez*
Circulation 22,000

WEEKLY. Interested mainly in straight Catholic issues but also in general humanitarian matters, social policies, the Third World, the arts and books. Ⓔ Payment by arrangement.

Celebs on Sunday
▷ Sunday Mirror under National Newspapers

Chapman

4 Broughton Place, Edinburgh EH1 3RX
Ⓣ 0131 557 2207
chapman-pub@blueyonder.co.uk
www.chapman-pub.co.uk
Owner/Editor *Dr Joy Hendry*

Circulation 2,000

Founded 1970. Scotland's quality literary magazine. Features poetry, short works of fiction, criticism, reviews and articles on theatre, politics, language and the arts. Gives priority to Scottish writers but also has work from UK and international authors. Unsolicited material welcome if accompanied by s.a.e. Approach in writing unless discussion is needed. Priority is given to full-time writers.

FEATURES Topics of literary interest, especially Scottish literature, theatre, culture or politics. Max. 5,000 words.

FICTION Short stories, occasionally novel extracts if self-contained. Max. 6,000 words.

SPECIAL PAGES Poetry, both UK and non-UK in translation (mainly, but not necessarily, European).

Ⓔ Payment by negotiation.

Chat

IPC Media Ltd., Blue Fin Building, 110 Southwark Street, London SE1 0SU

Ⓣ 020 3148 5000

www.ipcmedia.com/pubs/chat.htm

Owner *IPC Media (A TimeWarner Company)*

Editor *Gilly Sinclair*

Circulation 519,413

Founded 1985. WEEKLY general interest women's magazine. Unsolicited mss considered; approach in writing with ideas. Not interested in contributors 'who have never bothered to read *Chat* and therefore don't know what type of magazine it is. *Chat* does not publish fiction.'

FEATURES *Ingrid Millar* Human interest and humour. Max. 1,000 words.

Ⓔ Up to £1,500 maximum.

Chemist & Druggist

Riverbank House, Angel Lane, Tonbridge TN9 1SE

Ⓣ 01732 377487   Ⓕ 01732 367065

chemdrug@cmpmedica.com

www.chemistanddruggist.co.uk

Owner *CMP Medica*

Editor *Gary Paragpuri*

Circulation 13,675

Founded 1859. WEEKLY news magazine for the UK community pharmacy.

FEATURES practice, politics, professional matters, clinical articles and business advice. Contact by phone or e-mail to discuss features ideas.

NEWS Contact News Editor with news articles relating to local pharmacy matters, local pharmaceutical industry events and pharmacists in the news. Max. 200 words.

Ⓔ Payment standard rates.

Choice

First Floor, 2 King Street, Peterborough PE1 1LT

Ⓣ 01733 555123   Ⓕ 01733 427500

editorial@choicemag.co.uk

www.choicemag.co.uk

Owner *Choice Publishing Ltd*

Editor *Norman Wright*

Circulation 85,000

Monthly full-colour, lively and informative magazine for people aged 50 plus which helps them get the most out of their lives, time and money after full-time work.

FEATURES Real-life stories, hobbies, interesting (older) people, British heritage and countryside, involving activities for active bodies and minds, health, relationships, book/entertainment reviews. Unsolicited mss read (s.a.e. for return of material); write with ideas and copies of cuttings if new contributor. No phone calls, please.

RIGHTS/MONEY All items affecting the magazine's readership are written by experts. Areas of interest include pensions, state benefits, health, finance, property, legal.

Ⓔ Payment by arrangement.

Church Music Quarterly

19 The Close, Salisbury SP1 2EB

Ⓣ 01722 424848   Ⓕ 01722 424849

cmq@rscm.com

www.rscm.com

Owner *Royal School of Church Music*

Guest Editor *Julian Elloway*

Circulation 14,000

QUARTERLY. Contributions welcome. Phone in the first instance.

FEATURES Articles on church music or related subjects considered. Max. 2,100 words. Ⓔ £60 per page.

Church of England Newspaper

Central House, 142 Central Street, London EC1V 8AR

Ⓣ 020 7417 5800   Ⓕ 020 7216 6410

cen@churchnewspaper.com

www.churchnewspaper.com

Owner *Religious Intelligence*

Editor *Colin Blakely*

Circulation 9,500

Founded 1828. WEEKLY. Almost all material is commissioned but unsolicited mss are considered.

FEATURES Preliminary enquiry essential. Max. 1,200 words.

NEWS Items must be sent promptly to the e-mail address above and should have a church/Christian relevance. Max. 200–400 words.

Ⓔ Payment negotiable.

Church Times

13–17 Long Lane, Smithfield, London EC1A 9PN

Ⓣ 020 7776 1060   Ⓕ 020 7776 1086

news@churchtimes.co.uk
features@churchtimes.co.uk
www.churchtimes.co.uk
Owner *Hymns Ancient & Modern*
Editor *Paul Handley*
Circulation 29,500

Founded 1863. WEEKLY. Unsolicited mss considered.
FEATURES *Christine Miles* Articles and pictures (any format) on religious topics. Max. 1,500 words.
NEWS *Helen Saxbee* Occasional reports (commissions only) and up-to-date photographs.
€ Features, £120 per 1,000 words; news, by arrangement.

Classic & Sports Car

Teddington Studios, Broom Road, Teddington
TW11 9BE
☎ 020 8267 5399   🖷 020 8267 5318
letters.c&sc@haymarket.com
www.classicandsportscar.com
Owner *Haymarket Consumer Media*
Editor *James Elliott*
Circulation 81,000

Founded 1982. MONTHLY. Best-selling classic car magazine covering all aspects of buying, selling, owning and maintaining historic vehicles. Will consider news stories and features involving classic cars but 'be patient because hundreds of unsolicited material received monthly'. Initial contact by e-mail (james.elliott@haymarket.com).

Classic Bike

Bauer Automotive, Media House, Lynchwood, Peterborough Business Park, Peterborough
PE2 6EA
☎ 01733 468000   🖷 01733 468092
classic.bike@bauerconsumer.co.uk
www.classicbike.co.uk
Owner *Bauer Automotive*
Editor *Hugo Wilson*
Features Editor *Jim Moore*
Circulation 41,285

Founded 1978. MONTHLY. Mainly pre-1990 classic motorcycles. Freelancers must contact the editor before making submissions. Approach in writing or by e-mail.
NEWS Genuine news with good illustrations, if possible, suitable for a global audience. Max. 400 words.
FEATURES British motorcycle industry inside stories, technical features 'that can be understood by all', German, Spanish and French machine features, people. Restoration features, riding features; 'good reads'. Max. 2,000 words.
SPECIAL PAGES How-to features, oddball machines, stunning pictures, features with a fresh slant.
€ Payment negotiable.

Classic Boat

IPC Media Ltd, Leon House, 233 High Street, Croydon CR9 1HZ
☎ 020 8726 8000   🖷 020 8726 8195
cb@ipcmedia.com
www.classicboat.co.uk
Owner *IPC Media (A Time Warner Company)*
Editor *Dan Houston*
Circulation 12,505

Founded 1987. MONTHLY. Traditional boats and classic yachts, old and new; maritime history. Unsolicited mss, particularly if supported by good photos, are welcome. Sail and power boat pieces considered. Approach in writing with ideas. Interested in well-researched stories on all nautical matters. News reports welcome. Contributor's notes available online.
FEATURES Boatbuilding, boat history and design, events, yachts and working boats. Material must be well-informed and supported where possible by good-quality or historic photos. Max. 3,000 words. Classic Boat is defined by excellence of design and construction – the boat need not be old and wooden!
NEWS New boats, restorations, events, boatbuilders, etc. Max. 500 words.
€ Features, £75–100 per published page; news, according to merit.

Classic Cars

Media House, Peterborough Business Park, Lynchwood, Peterborough PE2 6EA
☎ 01733 468582   🖷 01733 468888
classic.cars@emap.com
www.classiccarsmagazine.co.uk
Owner *EMAP Ltd*
Editor *Phil Bell*
Circulation 40,162

Founded 1973. TWELVE ISSUES YEARLY. International classic car magazine containing entertaining and informative articles about classic cars, events and associated personalities. Contributions welcome. € Payment negotiable.

Classic Land Rover World
▷ Land Rover World

Classic Stitches

D.C. Thomson & Co. Ltd, 80 Kingsway East, Dundee DD4 8SL
☎ 01382 223131   🖷 01382 452491
editorial@classicstitches.com
www.classicstitches.com
Owner *D.C. Thomson & Co. Ltd*
Editor *Mrs Bea Neilson*
Circulation 10,000

Founded 1994. BI-MONTHLY magazine with exclusive designs to stitch, textile-based features, designer profiles, book reviews and 'what's new shopping' information. Will consider features covering textiles – historical and modern, designer profiles and original designs in context with the style of the magazine (no items on knitting or crochet). Approach by e-mail or letter with short synopsis to avoid clashes. For designs, send sketches, swatches and brief description of the idea.

FEATURES *Bea Neilson* Three features per issue with accompanying photography. Max. 2,000 words.

NEWS *Liz O'Rourke* Max. 300 words.

SPECIAL PAGES *Liz O'Rourke* Details only.

WEBSITE *Allison Hay* Sub-editor. Short, craft-orientated features. Max 500 words.

Classical Guitar

1 & 2 Vance Court, Trans Britannia Enterprise Park, Blaydon on Tyne NE21 5NH
℡ 0191 414 9000   ℻ 0191 414 9001
classicalguitar@ashleymark.co.uk
www.classicalguitarmagazine.com
Owner *Ashley Mark Publishing Co.*
Editor *Maurice Summerfield*

Founded 1982. MONTHLY.

FEATURES *Oliver McGhie* Usually written by staff writers. Max. 1,500 words.

NEWS *Thérèse Wassily Saba* Small paragraphs and festival concert reports welcome.

REVIEWS *Tim Panting* Concert reviews of up to 250 words are usually written by staff reviewers.
£ Features, by arrangement; no payment for news.

Classical Music

241 Shaftesbury Avenue, London WC2H 8TF
℡ 020 7333 1742   ℻ 020 7333 1769
classical.music@rhinegold.co.uk
www.rhinegold.co.uk
Owner *Rhinegold Publishing Ltd*
Editor *Keith Clarke*

Founded 1976. FORTNIGHTLY. A specialist magazine using precisely targeted news and feature articles aimed at the music business. Most material is commissioned but professionally written unsolicited mss are occasionally published. Freelance contributors may approach in writing with an idea but should familiarize themselves beforehand with the style and market of the magazine. £ Payment negotiable.

Climber

Warners Group Publications plc, West Street, Bourne PE10 9PH
℡ 01778 391000
climberscomments@warnersgroup.co.uk
www.climber.co.uk
Owner *Warners Group Publications plc*
Editor *Kate Burke*

Founded 1962. MONTHLY. Features articles and news on climbing and mountaineering in the UK and abroad. No unsolicited mss.

Closer

Endeavour House, 189 Shaftesbury Avenue, London WC2H 8JG
℡ 020 7437 9011   ℻ 020 7859 8600
closer@bauerconsumer.co.uk
www.closeronline.co.uk
Owner *Bauer Consumer Media*
Editor *Lisa Burrow*
Circulation 548,594

Launched 2002. WEEKLY celebrity magazine with a 'tongue-in-cheek' approach. No unsolicited material; approach by e-mail in the first instance.

Club International

2 Archer Street, London W1D 7AW
℡ 020 7292 8000   ℻ 020 7734 5030
mattb@paulraymond.com
Editor *Matt Berry*
Circulation 80,000

Founded 1972. MONTHLY. Features and short humorous items aimed at young male readership aged 18–30.

FEATURES Max. 1,000 words.

SHORTS 200–750 words.
£ Payment negotiable.

Coach and Bus Week

Rouncy Media Ltd, 3 The Office Village, Forder Way, Cygnet Park, Hampton, Peterborough PE7 8GX
℡ 01733 293240   ℻ 0845 280 2927
jacqui.grobler@rouncymedia.co.uk
www.CBWonline.com
Owner *Rouncy Media*
Editorial Director *Andrew Sutcliffe*
Circulation 4,000

WEEKLY magazine, aimed at coach and bus operators. Interested in coach and bus industry-related items only: business, financial and legal. Contact by letter, e-mail, telephone or via website.

Coin News

Token Publishing Ltd, Orchard House, Duchy Road, Heathpark, Honiton EX14 1YD
℡ 01404 46972   ℻ 01404 44788
info@tokenpublishing.com
www.tokenpublishing.com
Owners *J.W. Mussell, Carol Hartman*
Editor *J.W. Mussell*
Circulation 10,000

Founded 1964. MONTHLY. Contributions welcome. Approach by phone in the first instance.

FEATURES Opportunity exists for well-informed authors 'who know the subject and do their homework'. Max. 2,000 words.
£ Payment by arrangement.

Company

National Magazine House, 72 Broadwick Street,
London W1F 9EP
☏ 020 7439 5000   🖷 020 7312 3797
company.mail@natmags.co.uk
www.company.co.uk
Owner *National Magazine Co. Ltd*
Editor *Victoria White*
Circulation 250,075

MONTHLY. Glossy women's magazine appealing
to the independent and intelligent young woman.
A good market for freelancers: 'We look for great
newsy features relevant to young British women'.
Keen to encourage bright, new, young talent,
but uncommissioned material is rarely accepted.
Feature outlines are the only sensible approach in
the first instance. Max. 1,500–2,000 words. Send
to *Emma Justice*, Features Editor. 💷 £250 per
1,000 words.

Computer Arts

30 Monmouth Street, Bath BA1 2BW
☏ 01225 442244   🖷 01225 732295
ca.mail@futurenet.co.uk
www.computerarts.co.uk
Owner *Future Publishing*
Publisher *Liz Taylor*
Editor *Paul Newman*
Circulation 21,452

Founded 1995. MONTHLY. The world of computer
arts and digital creative design. 3D, web design,
Photoshop, digital video. No unsolicited mss.
Interested in tutorials, profiles, tips, software and
hardware reviews. Approach by post or e-mail.

Computer Weekly

Quadrant House, The Quadrant, Sutton SM2 5AS
☏ 020 8652 3122   🖷 020 8652 8979
computer.weekly@rbi.co.uk
www.computerweekly.com
Owner *Reed Business Information*
Editor *Brian McKenna*
Circulation 143,000

Founded 1966. Freelance contributions welcome.
FEATURES Always looking for good new writers
with specialized industry knowledge. Max. 1,800
words.
NEWS *Bill Goodwin* Some openings for regional
or foreign news items. Max. 300 words.
💷 Payment negotiable.

Computeractive

Incisive Media Ltd, 32–34 Broadwick Street,
London W1A 2HG
☏ 020 7316 9000   🖷 020 7316 9520
news@computeractive.co.uk
www.computeractive.co.uk
Owner *Incisive Media Ltd*
Editor *Paul Allen*

Circulation 200,307
Launched 1998. FORTNIGHTLY. Consumer and
computer technology magazine. No unsolicited
material.

Computing

Incisive Media Ltd, 32–34 Broadwick Street,
London W1A 2HG
☏ 020 7316 9000   🖷 020 7316 9160
feedback@computing.co.uk
www.computing.co.uk
Owner *Incisive Media Ltd*
Editor *Bryan Glick*
Circulation 115,000

Founded 1973. WEEKLY newspaper for IT
professionals. Unsolicited articles welcome.
Please enclose s.a.e. for return.
NEWS *Emma Nash*
💷 Up to £300 per 1,000 words.

Condé Nast Traveller

Vogue House, Hanover Square, London W1S 1JU
☏ 020 7499 9080   🖷 020 7493 3758
editorcntraveller@condenast.co.uk
www.cntraveller.com
Owner *Condé Nast Publications*
Editor *Sarah Miller*
Circulation 85,106

Founded 1997. MONTHLY travel magazine.
Proposals rather than completed mss preferred.
Approach in writing in the first instance. No
unsolicited photographs. 'The magazine has a no
freebie policy and no writing can be accepted on
the basis of a press or paid-for trip.'

Construction News

Greater London House, Hampstead Road,
London NW1 7EJ
☏ 020 7728 4629   🖷 020 7728 4400
cneditorial@emap.com
www.cnplus.co.uk
Owner *Emap*
Editor *Nick Edwards*
Features Editor *Emma Crates*
News Editor *David Rogers*
Circulation 21,203

WEEKLY. Since 1871, *Construction News* has
provided news to the industry. Readership ranges
from construction site to the boardroom, from
building and specialist construction through to
civil engineering. Initial approach by e-mail.

Coop Traveller

River Publishing, Victory House, 14 Leicester
Place, London WC2H 7BZ
☏ 020 7413 9308   🖷 020 7306 0304
cooptraveller@riverltd.co.uk
www.riverltd.co.uk
Owner *River Publishing*

Editor *Heather Farmbrough*
Circulation 220,000

Founded 2001. THREE ISSUES YEARLY. Magazine for Co-op Travel customers. Travel articles and travel-related news for the independent traveller. Approach by telephone, fax, letter or e-mail in the first instance.

FEATURES Ideas for exciting travel features, including destinations, accommodation, activities etc. Max. 1,000 words.

NEWS Stories that reveal fascinating facts for holidaymakers, including interesting products, new destinations and developments in the travel market. Max. 300 words.
Ⓔ Payment negotiable.

Cosmopolitan

National Magazine House, 72 Broadwick Street, London W1F 9EP
☎ 020 7439 5000   🖷 020 7439 5016
cosmo.mail@natmags.co.uk
www.cosmopolitan.co.uk
Owner *National Magazine Co. Ltd*
Editor *Louise Court*
Deputy Editor *Helen Daly*
Circulation 460,276

MONTHLY. Designed to appeal to the mid-twenties, modern-minded female. Popular mix of articles, with emphasis on relationships and careers, and hard news. No fiction. Will rarely use unsolicited mss but always on the look-out for 'new writers with original and relevant ideas and a strong voice'. All would-be writers should be familiar with the magazine. E-mail for submission guidelines.

Counselling at Work

Association for Counselling at Work, BACP, BACP House, 15 St John's Business Park, Lutterworth LE17 4HB
☎ 01455 883300   🖷 01455 550243
acw@bacp.co.uk
www.counsellingatwork.org.uk
Owner *British Association for Counselling and Psychotherapy*
Editor *Rick Hughes*
Circulation 1,600

Founded 1993. QUARTERLY official journal of the Association for Counselling at Work, a Division of BACP. Looking for well-researched articles (500–2,400 words) about *any* aspect of workplace counselling. Mss from those employed as counsellors or in welfare posts are particularly welcome. Photographs accepted. No fiction or poetry. Send A4 s.a.e. for writer's guidelines and sample copy of the journal. Ⓔ No payment.

Country Homes and Interiors

IPC Media Ltd., Blue Fin Building, 110 Southwark Street, London SE1 0SU
☎ 020 3148 5000
countryhomes@ipcmedia.com
Owner *IPC Media (A Time Warner Company)*
Editor *Rhoda Parry*
Circulation 86,603

Founded 1986. MONTHLY. The best approach for prospective contributors is with an idea in writing as unsolicited mss are not welcome.

FEATURES *Jean Carr* Monthly personality interviews of interest to an intelligent, affluent readership (women and men), aged 25–44. Max. 1,200 words.

HOUSES *Vivienne Ayers* Country-style homes with excellent design ideas. Length 1,000 words.
Ⓔ Payment negotiable.

Country Life

IPC Media Ltd., Blue Fin Building, 110 Southwark Street, London SE1 0SU
☎ 020 3148 4444
www.countrylife.co.uk
Owner *IPC Media (A Time Warner Company)*
Editor-in-Chief *Mark Hedges*
Circulation 40,821

Established 1897. WEEKLY. *Country Life* features articles which relate to architecture, countryside, wildlife, rural events, sports, arts, exhibitions, current events, property and news articles of interest to town and country dwellers. Strong informed material rather than amateur enthusiasm. 'Uncommissioned material is rarely accepted. We regret we cannot be liable for the safe custody or return of any solicited or unsolicited materials.'

Country Living

National Magazine House, 72 Broadwick Street, London W1F 9EP
☎ 020 7439 5000   🖷 020 7439 5093/5077
www.countryliving.co.uk
Owner *National Magazine Co. Ltd*
Editor *Susy Smith*
Circulation 195,159

Magazine aimed at both country dwellers and town dwellers who love the countryside. Covers people, conservation, wildlife, houses (gardens and interiors) and rural businesses. No unsolicited mss

Country Smallholding

Archant Publishing, Fair Oak Close, Exeter Airport Business Park, Clysts Honiton, Nr Exeter EX5 2UL
☎ 01392 888475   🖷 01392 888550
editorial.csh@archant.co.uk
www.countrysmallholding.com

Owner *Archant Publishing*
Editor *Diane Cowgill*
Circulation 20,896

Founded 1975. MONTHLY magazine for small
farmers, smallholders, practical landowners and
for anyone interested in self sufficiency at all
levels both in town and in the country. Articles
welcome on organic growing, keeping poultry,
livestock and other animals, crafts, cookery,
herbs, building and energy. Articles should
be detailed and practical, based on first-hand
knowledge and experience. Approach in writing
or by e-mail.

Country Walking

Media House, Lynchwood, Peterborough
Business Park, Peterborough PE2 6EA
☎ 01733 468000
country.walking@emap.com
www.countrywalking.co.uk
www.livefortheoutdoors.com
Owner *Emap Plc*
Editor *Jonathan Manning*
Circulation 44,549

Founded 1987. MONTHLY magazine focused on
walking in the finest landscapes of the UK and
Europe. News, gear tests and profiles of celebrity
walkers also feature. Special monthly pull-out
section of detailed walking routes. Original
high quality photography and ideas for features
considered, although very few unsolicited mss
accepted. Approach by letter or e-mail.
SPECIAL PAGES Routes section of magazine,
including outline Ordnance Survey map
and step-by-step directions. Accurately and
recently researched walk and fact file including
photographs and points of interest en route.
Please contact *Jane Ward* for guidelines
(unsolicited submissions generally not accepted
for this section).
£ Payment not negotiable.

The Countryman

The Water Mill, Broughton Hall, Skipton
BD23 3AG
☎ 01756 701381   ᖴ 01756 701326
editorial@thecountryman.co.uk
www.thecountryman.co.uk
Owner *Country Publications Ltd*
Editor *Paul Jackson*
Circulation 19,817

Founded 1927. Monthly. Countryside feature ideas
welcome by e-mail, phone or in writing. Articles
supplied with top quality illustrations (digital,
colour transparencies, archive b&w prints and
line drawings) are far more likely to be used. Max.
article length 1,200 words.

The Countryman's Weekly (incorporating Gamekeeper and Sporting Dog)

Unit 2, Lynher House, 3 Bush Park, Estover,
Plymouth PL6 7RG
☎ 01752 762990   ᖴ 01752 771715
editorial@countrymanswweekly.com
Owner *Diamond Publishing Ltd*
Editor *David Venner*

Founded 1895. WEEKLY. Unsolicited material
welcome.
FEATURES on any country sports topic. Max.
1,000 words.
£ Payment rates available on request.

Countryside Alliance Update

The Old Town Hall 367 Kennington Road,
London SE11 4PT
☎ 020 7840 9200   ᖴ 020 7793 8484
www.countryside-alliance.org.uk
Owner *Countryside Alliance*
Editor *Georgina Kester*
Circulation 70,000

Founded 1996. QUARTERLY membership
magazine on country sports and conservation
issues. No unsolicited mss.

The Cricketer International
▷ The Wisden Cricketer

Crimewave

TTA Press, 5 Martins Lane, Witcham, Ely CB6 2LB
info@ttapress.demon.co.uk
www.ttapress.com
Owner *TTA Press*
Editor *Andy Cox*
Circulation 8,000

Founded 1998. BIANNUAL large paperback format
magazine of cutting edge crime and mystery
fiction. 'The UK's only magazine specializing in
crime short stories, publishing the very best from
across the spectrum.' Every issue contains stories
by authors who are household names in the crime
fiction world but room is found for lesser known
and unknown writers. Submissions welcome
(not via e-mail) with appropriate return postage.
Potential contributors are advised to study the
magazine. Contracts exchanged upon acceptance.
£ Payment on publication.

Crucible
▷ Hymns Ancient & Modern Ltd under UK Publishers

Cumbria

The Water Mill, Broughton Hall, Skipton
BD23 3AG
☎ 01535 637983
editorial@cumbriamagazine.co.uk
www.dalesman.co.uk
Owner *Country Publications Ltd*
Editor *Terry Fletcher*

Circulation 12,505

Founded 1951. MONTHLY. County magazine of strong regional and countryside interest, focusing on the Lake District. Unsolicited mss welcome. Max. 1,500 words. Approach in writing, by phone or e-mail with feature ideas. £ Payment negotiable.

Cycle Sport

IPC Media Ltd, Leon House, 233 High Street, Croydon CR9 1HZ

☎ 020 8726 8463   🖷 020 8726 8499

cyclesport@ipcmedia.com

cycling@ipcmedia.com

www.cyclesport.co.uk

Owner *IPC Media (A Time Warner Company)*

Managing Editor *Robert Garbutt*

Deputy Editor *Edward Pickering*

Circulation 18,165

Founded 1993. MONTHLY magazine dedicated to professional cycle racing. Unsolicited ideas for features welcome.

Cycling Weekly

IPC Media Ltd, Leon House, 233 High Street, Croydon CR9 1HZ

☎ 020 8726 8462   🖷 020 8726 8499

cycling@ipcmedia.com

www.cyclingweekly.co.uk

Owner *IPC Media (A Time Warner Company)*

Publishing Director *Keith Foster*

Editor *Robert Garbutt*

Circulation 27,609

Founded 1891. WEEKLY. All aspects of cycle sport covered. Unsolicited mss and ideas for features welcome. Approach in writing with ideas. Fiction rarely used. 'It is important that you are familiar with the format so read the magazine first.'
FEATURES Cycle racing, coaching, technical material and related areas. Max. 2,000 words. Most work commissioned but interested in seeing new work.
NEWS Short news pieces, local news, etc. Max. 300 words.
£ Features, around £60–120 per 1,000 words (quality permitting); news, £15 per story.

Dalesman

The Water Mill, Broughton Hall, Skipton BD23 3AG

☎ 01756 701381   🖷 01756 701326

editorial@dalesman.co.uk

www.dalesman.co.uk

Owner *Country Publications Ltd*

Editor *Paul Jackson*

Circulation 37,687

Founded 1939. Now the biggest-selling regional publication of its kind in the country. MONTHLY magazine with articles of specific Yorkshire

interest. Unsolicited mss welcome; receives approximately ten per day. Initial approach in writing, by phone or e-mail. Max. 1,500 words. £ Payment negotiable.

Dance Theatre Journal

Laban, Creekside, London SE8 3DZ

☎ 020 8691 8600   🖷 020 8691 8400

dtj@laban.org

www.myspace.com/dancetheatrejournal

Owner *Laban*

Editor *Martin Hargreaves*

Circulation 2,000

Founded 1982. QUARTERLY. Interested in features on every aspect of the contemporary dance scene, particularly issues such as the funding policy for dance, critical assessments of choreographers' work and the latest developments in the various schools of contemporary dance. Unsolicited mss welcome. Length 1,000–3,000 words

Dance Today

Dancing Times Ltd, Clerkenwell House, 45–47 Clerkenwell Green, London EC1R 0EB

☎ 020 7250 3006   🖷 020 7253 6679

dancetoday@dancing-times.co.uk

www.dance-today.co.uk

Owner *The Dancing Times Ltd*

Editor *Katie Gregory*

Editorial Assistants *Alison Kirkman, Jon Gray*

Circulation 3,500

Founded in 1956 as the *Ballroom Dancing Times* and relaunched as *Dance Today* in 2001. MONTHLY magazine for anyone interested in social and competitive dance. Styles include ballroom and Latin American, salsa, tango, flamenco, tea dancing and sequence. Related topics covered include health and fitness and dance fashion. Contributions welcome; send c.v. with sample work by post or e-mail to the editor.

The Dancing Times

The Dancing Times Ltd, Clerkenwell House, 45–47 Clerkenwell Green, London EC1R 0EB

☎ 020 7250 3006   🖷 020 7253 6679

editorial@dancing-times.co.uk

www.dancing-times.co.uk

Owner *The Dancing Times Ltd*

Editors *Mary Clarke, Jonathan Gray*

Founded 1910. MONTHLY. Freelance suggestions welcome from specialist dance writers and photographers only. Approach in writing.

The Dandy

D.C. Thomson & Co. Ltd, Albert Square, Dundee DD1 8QJ

☎ 01382 223131   🖷 01382 322214

editor@dandy.co.uk

www.dandy.com

Owner *D.C. Thomson & Co. Ltd*

Editor *Craig Graham*

Founded 1937. WEEKLY comic. Original, humorous comic strips for 7–12-year-olds, featuring established characters such as Desperate Dan as well as new ones like Jak. Generally, unsolicited contributions are welcome but 'please read the comic before sending any material. Try a one-off script for an established, current *Dandy* character. Send by e-mail or post but never send your only copy.' Include s.a.e. for reply. No non-script material such as puzzles, poetry, etc.

Darts World

81 Selwood Road, Croydon CR0 7JW
℡ 020 8650 6580   🖷 020 8654 4343
dartsworld@blueyonder.co.uk
www.dartsworld.com
Owner *World Magazines Ltd*
Editor *A.J. Wood*
Circulation 12,263

FEATURES Single articles or series on technique and instruction. Max. 1,200 words.
FICTION Short stories with darts theme. Max. 1,000 words.
NEWS Tournament reports and general or personality news required. Max. 800 words.
£ Payment negotiable.

Day by Day

Woolacombe House, 141 Woolacombe Road, London SE3 8QP
℡ 020 8856 6249
Owner *Loverseed Press*
Editor *Patrick Richards*
Circulation 25,000

Founded 1963. MONTHLY. News commentary and digest of national and international affairs, with reviews of the arts (books, plays, art exhibitions, films, opera, musicals) and county cricket and Test reports among regular slots. Unsolicited mss welcome (s.a.e. essential). Approach in writing with ideas. Contributors are advised to study the magazine in the first instance. (Specimen copy £1.50) UK subscription £15; Europe £19; RoW £23.
NEWS *Ronald Mallone* Interested in themes connected with non-violence and social justice only. Max. 600 words.
FEATURES No scope for freelance contributions here.
POEMS *Michael Gibson* Short poems in line with editorial principles considered. Max. 20 lines.
£ Payment negotiable.

Dazed & Confused

112–116 Old Street, London EC1V 9BG
℡ 020 7549 6816   🖷 020 7336 0966
rod@dazedgroup.com
www.dazeddigital.com
Owner *Waddell Ltd*

Editor *Rod Stanley*
Circulation 80,000

Founded 1992. MONTHLY. Cutting edge fashion, music, film, art, interviews and features. No unsolicited material. Approach in writing with ideas in the first instance.

Decanter

Blue Fin Building, 110 Southwark Street, London SE1 0SU
℡ 020 3148 4488   🖷 020 3148 8524
editor@decanter.com
www.decanter.com
Editor *Guy Woodward*
Circulation 40,000

FOUNDED 1975. Glossy wines magazine. Feature ideas welcome; send to editor by post, fax or e-mail. No fiction.
NEWS/FEATURES All items and articles should concern wines, food and related subjects.
£ £250 per 1,000 words.

Delicious

Seven Publishing Group, 20 Upper Ground, London SE1 9PD
℡ 020 7775 7757   🖷 020 7775 7705
info@deliciousmagazine.co.uk
www.deliciousmagazine.co.uk
Owner *Seven Publishing Group*
Editor *Matthew Drennan*
Circulation 104,274

Launched 2003. MONTHLY publication for people who love food and cooking. Recipes – from the simple to the sophisticated – plus features on food and where it comes from. Welcomes freelance contributions; approach by e-mail.

Derbyshire Life and Countryside

61 Friar Gate, Derby DE1 1DJ
℡ 01332 227850   🖷 01332 227860
www.derbyshirelife.co.u
Owner *Archant Life*
Editor *Joy Hales*
Circulation 15,793

Founded 1931. MONTHLY county magazine for Derbyshire. Unsolicited mss and photographs of Derbyshire welcome, but written approach with ideas preferred.

Descent

Wild Places Publishing, PO Box 100, Abergavenny NP7 9WY
℡ 01873 737707
descent@wildplaces.co.uk
www.caving.uk.com
Owner *Wild Places Publishing*
Editor *Chris Howes*
Assistant Editor *Judith Calford*

Founded 1969. BI-MONTHLY magazine for cavers and mine enthusiasts. Submissions welcome from freelance contributors who can write accurately and knowledgeably and in the style of existing content on any aspect of caves, mines or underground structures.

FEATURES General interest articles of under 1,000 words welcome, as well as short foreign news reports, especially if supported by photographs/illustrations. Suitable topics include exploration (particularly British, both historical and modern), expeditions, equipment, techniques and regional British news. Max. 2,000 words.
£ Payment on publication according to page area filled.

Desire

211 Old Street, London EC1V 9NR
☎ 020 7608 6383   📠 020 7608 6380
editorial@trojanpublishing.co.uk
www.desire.co.uk
Owner *Trojan Publishing*
Editor *Ian Jackson*

Founded 1994. SIX ISSUES YEARLY. Britain's only erotic magazine for both women and men, celebrating sex and sensuality with a mix of features, reviews, interviews, photography and erotic fantasy. For a sample copy of the magazine plus contributors' guidelines and rates, please enclose four loose first class stamps.

Destination France

151 Station Street, Burton on Trent DE14 1BG
☎ 01283 742950   📠 01283 742957
alison.bell@wwonline.co.uk
Owner *Waterways World*
Editor *Carmen Konopka*
Circulation 12,000

A stylish, contemporary, QUARTERLY magazine aimed at Francophiles. Covers food, wine, art, architecture, books and lifestyle as well as travel within France. Features include sports, general tourism, property and regional information. Photographs to accompany articles. Approach in writing, preferably by e-mail, in the first instance.
£ Payment negotiable.

Devon Life

Archant House, Babbage Road, Totnes TQ9 5JA
☎ 01803 860910   📠 01803 860926
devonlife@archant.co.uk
www.devonlife.co.uk
Owner *Archant*
Editor *Jan Barwick*
Circulation 16,217

Originally launched in 1963; relaunched 1996. MONTHLY. Articles on all aspects of Devon – places of interest, personalities, events, arts,

history and food. Contributions welcome; send by e-mail to the editor.
FEATURES Words/picture package preferred on topics listed above. Max. 1,000–2,000 words.
FICTION Occasional short stories with a distinctly Devon flavour. Max 1,600 words.
£ Payment by arrangement.

Director

116 Pall Mall, London SW1Y 5ED
☎ 020 7766 8950   📠 020 7766 8840
director-ed@iod.com
Group Editor *Richard Cree*
Circulation 57,431

1991 Business Magazine of the Year. Published by Director Publications Ltd for members of the Institute of Directors and subscribers. Wide range of features from business and business profiles and management thinking to employment and financial issues. Also book reviews. Regular contributors used. Send letter with synopsis/published samples rather than unsolicited mss.
£ £350 per 1,000 words.

Disability Now

6 Market Road, London N7 9PW
☎ 020 7619 7323   📠 020 7619 7331
editor@disabilitynow.org.uk
www.disabilitynow.org.uk
Owner *SCOPE*
Editor *Ian Macrae*
Circulation 20,735

Founded 1984. Leading MONTHLY magazine for disabled people in the UK – people with all disabilities, as well as their families, carers and relevant professionals. Freelance contributions welcome. No fiction. Approach in writing or by e-mail.
FEATURES Covering new initiatives and services, personal experiences and general issues of interest to a wide national readership. Max. 900 words. Disabled contributors welcome.
NEWS Max. 300 words.
SPECIAL PAGES Possible openings for cartoonists.
£ Payment by arrangement.

Diva, lesbian life and style

Spectrum House, 32–34 Gordon House Road, London NW5 1LP
☎ 020 7424 7400   📠 020 7424 7401
edit@divamag.co.uk
www.divamag.co.uk
Owner *Millivres-Prowler Group*
Editor *Jane Czyzselska*

Founded 1994. MONTHLY glossy magazine featuring lesbian news and culture including fashion, lifestyle and satire. Welcomes news, features and photographs. No poetry. Contact the news editor at the e-mail address above with news items and feature ideas and photo/fashion photographs.

Dog World

Somerfield House, Wotton Road, Ashford TN23 6LW

☎ 01233 621877 📠 01233 645669
www.dogworld.co.uk
Owner *DW Media Holdings Ltd*
Editor *Stuart Baillie*
Circulation 17,828

Founded 1902. WEEKLY newspaper for people who are seriously interested in pedigree dogs. Unsolicited mss occasionally considered but initial approach in writing preferred.
FEATURES Well-researched historical items or items of unusual interest concerning dogs. Max. 2,000 words. Photographs of unusual 'doggy' situations occasionally of interest.
NEWS Freelance reports welcome on court cases and local government issues involving dogs.
💷 Features, up to £75; photos, £15.

Dogs Monthly

ABM Publishing Ltd, 61 Great Whyte, Ramsey, Huntingdon PE26 1HJ
☎ 0845 094 8958
www.dogsmonthly.co.uk
Owner *ABM Publishing Ltd*
Editor *Caroline Davis*
Sub Editor *Hannah Roche*
Circulation 11,000

Established in 1983 and relaunched at Crufts in 2008. MONTHLY magazine containing expert, practical advice on caring for and training pet dogs by 'the best of the canine world's writers, vets and trainers'.

Dorset Life – The Dorset Magazine

7 The Leanne, Sandford Lane, Wareham BH20 4DY
☎ 01929 551264 📠 01929 552099
editor@dorsetlife.co.uk
www.dorsetlife.co.uk
Owner/Editor *John Newth*
Circulation 10,000

Founded 1968. MONTHLY magazine about Dorset, both present and past. Unsolicited contributions welcome if specifically about Dorset.

Eastern Eye

Unit 2, 65 Whitechapel Road, London E1 1DU
☎ 020 7650 2000 📠 020 7650 2001
editorial@easterneyeuk.co.uk
www.ethnicmedia.co.uk
Owner *Ethnic Media Group*
Editor *Hamant Verma*

Circulation 20,844

Founded 1989. WEEKLY community paper for the Asian community in Britain. Interested in relevant general, local and international issues. Approach in writing with ideas for submission.

Easy Living

68 Old Bond Street, London W1S 4PH
☎ 020 7499 9080 📠 020 7399 2625
easylivingeditorial@condenast.co.uk
www.easylivingmagazine.com
Owner *Condé Nast*
Editor *Susie Forbes*
Features Director *Lisa Markwell*
Health and Beauty Director *Catherine Turner*
Fashion Director *Liz Thody*
Decorating Editor *Sophie Martell*
Food Editor *David Herbert*
Circulation 196,150

Launched 2005. MONTHLY magazine aimed at women aged 30–50-plus. Focuses on aspects of fashion, home life, beauty and health, food, relationships; stylish and glossy with real-life, practical and useful items.

The Ecologist

Unit 102, Lana House Studios, 116-118 Commercial Street, London E1 6NF
☎ 020 7422 8100 📠 020 7422 8101
editorial@theecologist.org
www.theecologist.org
Owner *Ecosystems Ltd*
Managing Editor *Pat Thomas*
Director *Zac Goldsmith*
Circulation 30,000

Founded 1970. TEN ISSUES YEARLY. Unsolicited mss welcome but best approach is a brief proposal to the editor by e-mail (address above), outlining experience and background and summarizing suggested article.
FEATURES Radical approach to political, economic, social and environmental issues, with an emphasis on rethinking the basic assumptions that underpin modern society. Articles of between 500 and 3,000 words.
💷 Payment negotiable.

The Economist

25 St James's Street, London SW1A 1HG
☎ 020 7830 7000 📠 020 7839 2968
www.economist.com
Owner *Pearson/individual shareholders*
Editor *John Micklethwait*
Circulation 181,374 (UK)

Founded 1843. WEEKLY. Worldwide circulation. Approaches should be made in writing to the editor. No unsolicited mss.

The Edge

65 Guinness Buildings, Hammersmith, London
W6 8BD
☎ 0845 456 9337
www.theedge.abelgratis.co.uk
Editor *David Clark*

QUARTERLY. Reviews and features: film (indie, arts), books, popular culture.
FICTION 2,000+ words, almost any 'type' of story but see site or magazine for examples/tone. Sample issue £5 post free (cheques payable to 'The Edge'). Guidelines on website or for s.a.e.
Ⓔ Payment 'negotiable after we express interest'.

Edinburgh Review

22A Buccleugh Place, Edinburgh EH8 9LN
☎ 0131 651 1415
edinburgh.review@ed.ac.uk
www.edinburghreview.org.uk
Owner *Dept. of English Literature, University of Edinburgh*
Editor *Brian McCabe*

Founded 1802. THREE ISSUES YEARLY. Cultural and literary magazine. Articles on Scottish and international culture, history, art, literature and politics. Each issue is broadly geographically themed; recent issues include the Carribean, Poland and Northern Ireland. Unsolicited contributions welcome, particularly articles, fiction and poetry; send by post or e-mail. 'Writers offering unsolicited essays should contact us by e-mail in the first instance.'

Electrical Times

Purple Media Solutions Ltd., 2nd Floor, 207–215 High Street, Orpington BR6 0PL
☎ 01689 899170　🖷 01689 899171
ben.cronin@purplems.com
Owner *Purple Media Solutions Ltd*
Editor *Ben Cronin*
Circulation 10,516

Founded 1891. MONTHLY. Aimed at electrical contractors, designers and installers. Unsolicited mss welcome but initial approach preferred.

Elle

64 North Row, London W1K 7LL
☎ 020 7150 7000　🖷 020 7150 7670
Owner *Hachette Filipacchi*
Editor *Lorraine Candy*
Features Director *Kerry Potter*
Circulation 203,435

Founded 1985. MONTHLY fashion glossy. Prospective contributors should approach the relevant editor in writing in the first instance, including cuttings. FEATURES Max. 2,000 words. PREVIEW Short articles on hot trends, events, fashion and beauty. Max. 500 words.
Ⓔ Payment negotiable.

Elle Decoration

64 North Row, London W1K 7LL
☎ 020 7150 7000　🖷 020 7150 7671
Owner *Hachette Filipacchi*
Editor *Michelle Ogundehin*
Features Editor *Sophie Davies*
News Editor *Amy Bradford*
Circulation 64,418

Launched 1989. MONTHLY style magazine for the home – 'beautiful homes, great designs, inspiration and ideas'. No unsolicited contributions; approach by e-mail to the appropriate department.

Embroidery

PO Box 42B, East Molesey KT8 9BB
☎ 01260 273891
jo.editor@btinternet.com
www.embroiderersguild.com
Owner *Embroiderers' Guild*
Editor *Joanne Hall*
Circulation 6,000

Founded 1933. BI-MONTHLY. Features articles on historical and foreign embroidery, and contemporary artists' work with illustrations. Covers all forms of textile art, historical and contemporary. Also reviews. Unsolicited material welcome. Max. 1,000 words.
Ⓔ Payment negotiable.

Empire

4th Floor, Mappin House, 4 Winsley Street, London W1W 8HF
☎ 020 7182 8781　🖷 020 7182 8703
empire@emap.com
www.empireonline.com
Owner *Emap East*
Editor *Mark Dinning*
Circulation 181,511

Founded 1989. Launched at the Cannes Film Festival. MONTHLY guide to the movies which aims to cover the world of films in a 'comprehensive, adult, intelligent and witty package'. Although most of *Empire* is devoted to films and the people behind them, it also looks at the developments and technology behind television and DVDs plus music, multimedia and books. Wide selection of in-depth features and stories on all the main releases of the month, and reviews of over 100 films and videos. Contributions welcome but approach in writing first.
FEATURES Behind-the-scenes features on films, humorous and factual features.
Ⓔ Payment by agreement.

The Engineer

50 Poland Street, London W1F 7AX
☎ 020 7970 4000　🖷 020 7970 4123

andrew.lee@centaur.co.uk
www.theengineer.co.uk
Owner *Centaur Media*
Editor *Andrew Lee*
Circulation 31,367

FOUNDED 1856. FORTNIGHTLY news magazine for technology and innovation.
FEATURES Most outside contributions are commissioned but good ideas are always welcome. Max. 2,000 words.
NEWS Scope for specialist regional freelancers, and for tip-offs. Max. 500 words.
TECHNOLOGY Technology news from specialists. Max. 500 words.
£ Payment by arrangement.

The English Garden

Archant Specialist, Jubilee House, 2 Jubilee Place, London SW3 3TQ
☏ 020 7751 4800   F 020 7751 4848
theenglishgarden@archant.co.uk
www.theenglishgarden.co.uk
Owner *Archant Specialist*
Editor *Janine Wookey*
Editor-at-Large *Jackie Bennett* (☏ 01366 328606)
Circulation 34,814 (UK), 64,892 (Worldwide)

Founded 1996. MONTHLY. Features on beautiful gardens with practical ideas on design and planting. No unsolicited mss.
FEATURES *Jackie Bennett* Max. 800 words. Approach in writing in the first instance; send synopsis of 150 words with strong design and planting ideas, feature proposals, or sets of photographs of interesting gardens.

ES
▷ Evening Standard under Regional Newspapers

Esquire

National Magazine House, 72 Broadwick Street, London W1F 9EP
☏ 020 7439 5000   F 020 7439 5675
Owner *National Magazine Co. Ltd*
Editor *Jeremy Langmead*
Circulation 59,800

Founded 1991. MONTHLY. Quality men's general interest magazine. No unsolicited mss or short stories.

Essentials

IPC Media Ltd., Blue Fin Building, 110 Southwark Street, London SE1 0SU
☏ 020 3148 7211
Owner *IPC Media (A Time Warner Company)*
Editor *Julie Barton-Breck*
Circulation 94,122

Founded 1988. MONTHLY women's interest magazine. Unsolicited mss (not originals) welcome if accompanied by s.a.e. Initial approach

in writing preferred. Prospective contributors should study the magazine thoroughly before submitting anything. No fiction.
FEATURES Max. 2,000 words (double-spaced on A4).
£ Payment negotiable.

Essex Life

427 High Road, Woodford Green IG8 0XE
☏ 020 8504 0455
www.essexlifemag.co.uk
Owner *Archant*
Editor *Julian Read*
Circulation 17,000

Founded 1952. MONTHLY county magazine 'celebrating all that is best about Essex'. No unsolicited contributions. Send c.v. and covering letter in the first instance.

Eve

Griffin House, 161 Hammersmith Road, London W6 7JP
☏ 020 8267 8223   F 020 8267 8222
evrp@servicehelpline.co.uk
www.evemagazine.co.uk
Owner *Haymarket Publishing*
Editor *Nic McCarthy*
Deputy Editor *Rachael Ashley*
Features Director *Emma Elms*
Features Editor *Linda Gray*
Circulation 168,270

Founded 2000. MONTHLY. Wide-ranging general interest – aimed at the intelligent 30+ woman. No unsolicited material. Send introductory letter, *recent* writings and outlines for ideas aimed at a specific section. It is essential that would-be contributors familiarize themselves with the magazine.

Eventing
▷ Horse and Hound

Evergreen

PO Box 52, Cheltenham GL50 1YQ
☏ 01242 537900   F 01242 537901
Editor *S. Garnett*
Circulation 48,000

Founded 1985. QUARTERLY magazine featuring articles and poems about Britain. Unsolicited contributions welcome.
FEATURES Britain's natural beauty, towns and villages, nostalgia, wildlife, traditions, odd customs, legends, folklore, crafts, etc. Length 250–2,000 words.
£ £15 per 1,000 words; poems £4.

Eye: The international review of graphic design

174 Hammersmith Road, London W6 7JP
☏ 020 8267 8046   F 020 8264 4900

john.walters@haynet.com
www.eyemagazine.com
Owner *Haymarket Brand Media*
Editor *John L. Walters*
Circulation 9,000

Founded 1990. QUARTERLY. 'The world's most beautiful and authoritative magazine about graphic design and visual culture.' Unsolicited contributions not generally welcomed.
FEATURES Specialist articles about graphic design and designers. Design history by experts. Max. 3,000 words.
REVIEWS Max. 200–1,000 words.
Ⓔ Payment negotiable.

Fabulous
▷ **News of the World under National Newspapers**

Fairgame Magazine
The Baltic Business Centre, Gateshead NE8 3DA
☎ 0191 442 4003   🖷 0191 442 4002
jen@fgmag.com
www.fgmag.com
Editor *Jennifer O'Neill*
Circulation 15,000

Founded 2003. BI-MONTHLY magazine for women football players. Contributions welcome.
FEATURES *Jennifer O'Neill* International reports, player and team profiles, interviews, diet, health and fitness, tactics, training advice, play improvement, fund-raising. Max. 1,500 words.
NEWS *Wilf Frith* (wilf@fgmag.com) Match reports, team news, transfers, injuries, results and fixtures. Max. 600 words.

Families First
Mary Sumner House, 24 Tufton Street, London SW1P 3RB
☎ 020 7222 5533   🖷 020 7222 1591
Owner *Mothers' Union Publishing*
Editor *Catherine Butcher*
Circulation 46,798

Founded 1976. Formerly *Home & Family*. QUARTERLY. Unsolicited mss considered. No fiction or poetry. Features on family life, social issues, marriage, Christian faith, etc. Max. 1,000 words. Ⓔ Payment 'modest'.

Family Tree Magazine
61 Great Whyte, Ramsey, Huntingdon PE26 1HJ
☎ 01487 814050   🖷 01487 711361
sue.f@family-tree.co.uk
www.family-tree.co.uk
Owner *ABM Publishing Ltd*
Editor *Helen Tovey*
Circulation 40,000

Founded 1984. THIRTEEN ISSUES YEARLY. News and features on matters of genealogy. Not interested in own family histories. Approach

in writing with ideas. All material should be addressed to *Peter Watson*.
FEATURES Any genealogically related subject. Max. 2,000 words. No puzzles or fictional articles.
Ⓔ News and features, payment negotiable.

Fancy Fowl
The Publishing House, Station Road, Framlingham IP13 9EE
☎ 01728 622030   🖷 01728 622031
fancyfowl@todaymagazines.co.uk
www.fancyfowl.com
Owner *Today Magazines Ltd*
Editor *Kevin Davis*
Circulation 3,000

Founded 1979. MONTHLY. Devoted entirely to poultry, waterfowl, turkeys, geese, pea fowl, etc. – management, breeding, rearing and exhibition. Outside contributions of knowledgeable, technical poultry-related articles and news welcome. Max. 1,000 words. Approach by letter.
Ⓔ Payment negotiable.

Farmers Weekly
Quadrant House, Sutton SM2 5AS
☎ 020 8652 4911   🖷 020 8652 4005
farmers.weekly@rbi.co.uk
www.fwi.co.uk
Owner *Reed Business Information*
Editor *Jane King*
Circulation 70,315

WEEKLY. PPA Awards winner – 2006 Business Magazine of the Year. For practising farmers and those in the ancillary industries. Unsolicited mss considered.
FEATURES A wide range of material relating to running a farm business: specific sections on arable and livestock farming, farm life, practical and general interest, machinery and business.
NEWS General farming news.
Ⓔ Payment negotiable.

Fast Car
Future Publishing Ltd, 30 Monmouth Street, Bath BA1 2BW
☎ 01225 442244
www.fastcar.co.uk
Owner *Future Publishing Ltd*
Editor *Steve Chalmers*
Circulation 51,414

Founded 1987. THIRTEEN ISSUES YEARLY. Lad's magazine about performance tuning and modifying cars. Covers all aspects of this youth culture including the latest street styles and music. Features cars and their owners, product tests and in-car entertainment. Also includes a free reader ads section.
NEWS Any item in line with the above.

FEATURES Innovative ideas in line with the above and in the *Fast Car* writing style. Generally two to four pages in length. No Kit-car features, race reports or road test reports of standard cars. Copy should be as concise as possible. Ⓔ Payment negotiable.

FHM

Mappin House, 4 Winsley Street, London W1W 8HF

Ⓣ 020 7436 1515   Ⓕ 020 7182 8021
general@fhm.com
www.fhm.com
Owner *Bauer London Lifestyle*
Editor-in-Chief *Anthony Noguera*
Circulation 315,149

Founded in 1986 as a free fashion magazine, FHM evolved to become more male oriented. Best-selling men's magazine covering all areas of men's lifestyle. Published MONTHLY with editions worldwide. No unsolicited mss. Synopses and ideas welcome by e-mail. Ⓔ Payment negotiable.

The Field

IPC Media Ltd., Blue Fin Building, 110 Southwark Street, London SE1 0SU

Ⓣ 020 3148 4772   Ⓕ 020 3148 8179
field_secretary@ipcmedia.com
www.thefield.co.uk
Owner *IPC Media (A Time Warner Company)*
Editor-in-Chief *Jonathan Young*
Circulation 32,366

Founded 1853. MONTHLY magazine for those who are serious about the British countryside and its pleasures. Unsolicited mss welcome but initial approach should be made in writing or by phone. FEATURES Exceptional work on any subject concerning the countryside. Most work tends to be commissioned. Ⓔ Payment varies.

50 Connect – www.50connect.co.uk
▷ entry under Electronic Publishing and Other Services

Film Review

Visual Imagination Ltd, 9 Blades Court, Deodar Road, London SW15 2NU

Ⓣ 020 8875 1520   Ⓕ 020 8875 1588
filmreview@visimag.com
www.visimag.com
Owner *Visual Imagination Ltd*
Editor *Nikki Baughan*
Circulation 50,000

FOUR-WEEKLY. Reviews, profiles, interviews and special reports on films. Unsolicited material considered. PAYMENT negotiable.

Fishing News

Haines House, 21 John Street, London WC1N 2BP
Ⓣ 020 7017 4531 (editor)   Ⓕ 020 7017 4536

tim.oliver@fishingnews.co.uk
www.fishingnews.co.uk
Owner *IntraFish Media*
Editor *Tim Oliver*
Circulation 8,320

Founded 1913. WEEKLY. All aspects of the commercial fishing industry in the UK and Ireland. No unsolicited mss; telephone inquiry in the first instance. Max. 600 words for news and 1,500 words for features. Ⓔ £100 per 1,000 words.

The Fix

info@ttapress.demon.co.uk
www.thefix-online.com
Owner *TTA Press*

Founded 1994. Free online review of short fiction (magazines, anthologies, collections, webzines, podcasts etc), plus related interviews and regular columns. Interested reviewers and feature writers should visit the website and follow the contact information.

Flash: The International Short-Short Story Magazine

Department of English, University of Chester, Parkgate Road, Chester CH1 4BJ

Ⓣ 01244 392715
flash.magazine@chester.ac.uk
www.chester.ac.uk/flash.magazine
Editors *Dr Peter Blair, Dr Ashley Chantler*

Launched 2008. BIANNUAL magazine of short stories of up to 360 words (title included). Articles and reviews considered. Full details on the website.

Flight International

Quadrant House, The Quadrant, Sutton SM2 5AS
Ⓣ 020 8652 3842   Ⓕ 020 8652 3840
flight.international@flightglobal.com
www.flightglobal.com
Owner *Reed Business Information*
Editor *Murdo Morrison*
Circulation 52,222

Founded 1909. WEEKLY. International trade magazine for the aerospace industry, including civil, military and space. Unsolicited mss considered. Commissions preferred – phone with ideas and follow up with letter. E-mail submissions encouraged.
FEATURES *Murdo Morrison* Technically informed articles and pieces on specific geographical areas with international appeal. Analytical, in-depth coverage required, preferably supported by interviews. Max. 1,800 words.
NEWS *Dan Thisdell* Opportunities exist for news pieces from particular geographical areas on specific technical developments. Max. 350 words. Ⓔ NUJ rates.

Flora International

The Fishing Lodge Studio, 77 Bulbridge Road, Wilton, Salisbury SP2 OLE

☎ 01722 743207   ☒ 01722 743207

floramag@aol.com

Publisher/Editor *Maureen Foster*

Circulation 16,000

Founded 1974. BI-MONTHLY magazine for flower arrangers and florists. Unsolicited mss welcome. Approach in writing with ideas. Not interested in general gardening articles.

FEATURES Fully illustrated: colour photographs, transparencies or CD. Flower arranging, flower arrangers' gardens and flower crafts. Floristry items written with practical knowledge and well illustrated are particularly welcome. Average 1,000 words.

Ⓔ £60 per 1,000 words plus additional payment for suitable photographs, by arrangement.

FlyPast

PO Box 100, Stamford PE9 1XQ

☎ 01780 755131   ☒ 01780 757261

flypast@keypublishing.com

www.flypast.com

Owner *Key Publishing Ltd*

Editor *Ken Ellis*

Circulation 39,620

Founded 1981. MONTHLY. Historic aviation and aviation heritage, mainly military, First and Second World War period 1914–1970. Unsolicited mss welcome.

Focus

Bristol Magazines Ltd, Tower House, Fairfax Street, Bristol BS1 3BN

☎ 0117 927 9009 (switchboard)   ☒ 0117 934 9008

focus@bbcmagazinesbristol.com

www.bbcfocusmagazine.com

Owner *Bristol Magazines Ltd, a subsidiary of BBC Worldwide*

Editor *Jheni Osman*

Features *IanTaylor*

Circulation 65,301

Founded 1992. FOUR-WEEKLY. Popular science and discovery. Welcomes *relevant* summaries of original feature ideas by e-mail.

Fortean Times: The Journal of Strange Phenomena

Dennis Publishing, 30 Cleveland Street, London W1T 4JD

☎ 020 7907 6235   ☒ 020 7907 6139

david_sutton@dennis.co.uk

www.forteantimes.com

Editor *David Sutton*

Circulation 21,329

Founded 1973. THIRTEEN ISSUES YEARLY. Accounts of strange phenomena and experiences, curiosities, mysteries, prodigies and portents. Unsolicited mss welcome. Approach in writing with ideas. No fiction, poetry, rehashes or politics.

FEATURES Well-researched and referenced material on current or historical mysteries, or first-hand accounts of oddities. Max. 4,000 words, preferably with good relevant photos/illustrations.

NEWS Concise copy with full source references essential.

Ⓔ Payment negotiable.

Foundation: the international review of science fiction

Journal of the **Science Fiction Foundation** (see entry under *Professional Associations and Societies*).

FQ Magazine

Seymour House, South Street, Bromley BR1 1RH

☎ 020 8460 6060   ☒ 020 8460 6050

andy@touchline.com

www.fqmagazine.co.uk

Owner *3D Media Ltd*

Editor *Andy Tongue*

Launched 2003. BI-MONTHLY. Lifestyle magazine for young fathers. Interviews, reviews, finance, kids' toys, child issues, parenting tips and general fatherhood advice. Ideas welcome; contact the editor by e-mail.

France Magazine

Archant House, Oriel Road, Cheltenham GL50 1BB

☎ 01242 216050   ☒ 01242 216094

editorial@francemag.com

www.francemag.com

Owner *Archant Life France*

Editor *Carolyn Boyd*

Circulation 37,082 (group, including US edition); 19,351 (UK edition)

FOUNDED 1989. MONTHLY magazine containing all things of interest to Francophiles – in English. Approach by e-mail in the first instance.

Freelance Market News

Sevendale House, 7 Dale Street, Manchester M1 1JB

☎ 0161 228 2362   ☒ 0161 228 3533

fmn@writersbureau.com

www.freelancemarketnews.com

Editor *Angela Cox*

MONTHLY. News and information on the freelance writers' market, both inland and overseas. Includes market information, competitions, seminars, courses, overseas openings, etc. Short articles (700 words max.). Unsolicited contributions welcome.

Ⓔ Payment by negotiation.

The Freelance
> National Union of Journalists under Professional Associations and Societies

Front Magazine
Flip Media, 19 Waterside, 44-48 Wharf Road, London N1 7UX
☎ 020 7689 3916
front@frontmag.co.uk
Owner *Flip Media*
Editor *Joe Barnes*
Circulation 85,000

Founded 1998. MONTHLY men's interest magazine. Includes features on sport (including extreme sport), music, crime, war, humour, sex, alternative culture, real life stories and irreverent celebrity interviews. Welcomes contributions. Approach by e-mail.

Garden Answers
Bushfield House, Orton Centre, Peterborough PE2 5UW
☎ 01733 237111
Owner *Bauer Consumer Media*
Editor *Geoff Stebbings*
Circulation 26,880

Founded 1982. MONTHLY. 'It is unlikely that unsolicited manuscripts will be used, as articles are usually commissioned and must be in the magazine style.' Prospective contributors should approach the editor in writing. Interested in hearing from gardening writers on any subject, whether flowers, fruit, vegetables, houseplants or greenhouse gardening.

Garden News
Bushfield House, Orton Centre, Peterborough PE32 9QW
☎ 01733 237111   ⓕ 01733 465779
Owner *Emap Active Publications Ltd*
Editor *Neil Pope*
Circulation 36,020

Founded 1958. Britain's biggest-selling, full-colour gardening WEEKLY. News and advice on growing flowers, fruit and vegetables, plus colourful features on all aspects of gardening especially for the committed gardener. News and features welcome, especially if accompanied by top-quality photos or illustrations. Contact the editor before submitting any material.

The Garden, Journal of the Royal Horticultural Society
RHS Publications, 4th Floor, Churchgate New Road, Peterborough PE1 1TT
☎ 01733 294666   ⓕ 01733 341633
thegarden@rhs.org.uk
www.rhs.org.uk
Owner *The Royal Horticultural Society*

Editor *Ian Hodgson*
Circulation 352,323

Founded 1866. MONTHLY journal of the Royal Horticultural Society. Covers all aspects of the art, science and practice of horticulture and garden making. 'Articles must have depth and substance.' Approach by letter or e-mail with a synopsis in the first instance. Max. 2,500 words.

Gardens Illustrated
Bristol Magazines Ltd, Tower House, Fairfax Street, Bristol BS1 3BN
☎ 0117 314 8774   ⓕ 0117 933 8032
gardens@bbcmagazinesbristol.com
www.gardensillustrated.com
Owner *Bristol Magazines Ltd*
Editor *Juliet Roberts*
Circulation 28,259

Founded 1993. MONTHLY. 'Britain's most distinguished garden magazine' with a world-wide readership. The focus is on garden design, with a strong international flavour. Unsolicited mss are rarely used and it is best that prospective contributors approach the editor with ideas in writing, supported by photographs.

Gay Times
> GT

Gibbons Stamp Monthly
Stanley Gibbons, 7 Parkside, Christchurch Road, Ringwood BH24 3SH
☎ 01425 472363   ⓕ 01425 470247
hjefferies@stanleygibbons.co.uk
www.gibbonsstampmonthly.com
Owner *Stanley Gibbons Ltd*
Editor *Hugh Jefferies*
Circulation 22,000

Founded 1890. MONTHLY. News and features. Unsolicited mss welcome. Make initial approach in writing by telephone or e-mail to avoid disappointment.
FEATURES *Hugh Jefferies* Unsolicited material of specialized nature and general stamp features welcome. Max. 3,000 words but longer pieces can be serialized.
NEWS *John Moody* Any philatelic news item. Max. 500 words.
£ Features, £40–50 per 1,000 words; news, no payment.

Girl Talk/Girl Talk Extra
Media Centre, 201 Wood Lane, London W12 7TQ
☎ 020 8433 2758
samantha.robinson@bbc.co.uk
www.bbcgirltalk.com
Owner *BBC Worldwide*
Editor *Sam Robinson*

Circulation 86,995 (Girl Talk) 40,847 (Girl Talk Extra)

Founded 1995. FORTNIGHTLY/MONTHLY. Lifestyle magazine for girls aged 8–12. Friendship and belonging, celebrity. No unsolicited submissions; send c.v. by e-mail.

Glamour

6–8 Old Bond Street, London W1S 4PH
☎ 020 7499 9080 📠 020 7491 2551
letters@glamourmagazine.co.uk
www.glamour.com
Owner *Condé Nast*
Editor *Jo Elvin*
Circulation 550,066

Founded 2001. Handbag-size glossy women's magazine – fashion, beauty and celebrities. No unsolicited mss; send ideas for features in synopsis form to the Features Editor, *Corrie Jackson*.

Gold Dust Magazine

55 Elmdale Road, London E17 6PN
☎ 020 8765 4665
davidgardiner@worldonline.co.uk
www.golddustmagazine.co.uk
Owner *Omma Velada*
Editors *David Gardiner, Claire Tyne*
Circulation 300+

Launched 2004. Biannual magazine of short stories, flash fiction, short plays, short film scripts and extracts from novels. Articles about writing, book and film reviews. Unsolicited submissions welcome; approach by e-mail.
FICTION *David Gardiner* 'We are always looking for quality work from new or experienced writers. Please read all our submission guidelines before submitting work.' Max. 3,000 words. 💷 £20 best prose; £6 all others.
POETRY *Claire Tyne* Max. 50 lines. 💷 £20 best poem; £6 all others.
FEATURES *David Gardiner* Book reviews. Max. 2,000 words.

Golf Monthly

IPC Media Ltd., Blue Fin Building, 110 Southwark Street, London SE1 0SU
☎ 020 3148 4530 📠 020 3148 8130
golfmonthly@ipcmedia.com
www.golf-monthly.co.uk
Owner *IPC Media (A Time Warner Company)*
Editor *Michael Harris*
Assistant Editor *John Thynne*
Deputy Editor *Neil Tappin*
Senior Staff Writer *Luke Norman*
Circulation 83,002

Founded 1911. MONTHLY. Player profiles, golf instruction, general golf features and columns. Not interested in instruction material from outside contributors. Approach in writing with ideas.
FEATURES Max. 1,500–2,000 words.
💷 Payment by arrangement.

Golf World

Media House, Lynchwood, Peterborough Business Park, Peterborough PE2 6EA
☎ 01733 468000 📠 01733 468671
chris.jones@emap.com
www. todaysgolfer.co.uk/golfworld
Owner *Emap Sport*
Editor *Chris Jones*
Editorial Assistant *Linda Manigan*
Circulation 30,034

Founded 1962. MONTHLY. No unsolicited mss. Approach in writing with ideas.

Good Holiday Magazine

27A High Street, Esher KT10 9RL
☎ 01372 468140 📠 01372 470765
edit@goodholidayideas.com
www.goodskiguide.com
Editor *John Hill*
Circulation 100,000

Founded 1985. QUARTERLY aimed at better-off holiday-makers rather than travellers. Worldwide destinations including Europe and domestic. No unsolicited material but approach in writing or by e-mail with ideas and/or synopsis.
💷 Payment negotiable.

Good Homes Magazine

Media Centre, 201 Wood Lane, London W12 7TQ
☎ 020 8433 2391 📠 020 8433 2691
deargoodhomes@bbc.co.uk
www.bbcgoodhomes.com
Owner *BBC Worldwide Publishing Ltd*
Editor *Bernie Herlihy*
Circulation 127,202

Founded 1998. MONTHLY. Decorating, interiors, readers' homes, shopping, makeovers, property and consumer information. No fiction.
FEATURES *Emily Peck*
PROPERTY & READERS' HOMES *Rachel Watson*
Property features from specialists. Approach by letter with cuttings.

Good Housekeeping

National Magazine House, 72 Broadwick Street, London W1F 9EP
☎ 020 7439 5000 📠 020 7439 5616
firstname.lastname@natmags.co.uk
www.allaboutyou.com/goodhousekeeping
Owner *National Magazine Co. Ltd*
Editor *Louise Chunn*

Circulation 464,041

Founded 1922. MONTHLY glossy. No unsolicited mss. Write with ideas in the first instance to the appropriate editor.
FEATURES *Lucy Moore* Most work is commissioned but original ideas are always welcome. No ideas are discussed on the telephone. Send short synopsis, plus relevant cuttings, showing previous examples of work published. No unsolicited mss.
HEALTH *Julie Powell*. Submission guidelines as for features; no unsolicited mss.

Good Motoring

Station Road, Forest Row RH18 5EN
☎ 01342 825676 ☎ 01342 824847
editor@motoringassist.com
www.roadsafety.org.uk
Owner *Gem Motoring Assist*
Editor *James Luckhurst*
Circulation 52,000

Founded 1932. QUARTERLY motoring, road safety, travel and general features magazine. 1,500 words max. Prospective contributors should approach in writing only.

Good News

9 Spinney Close, West Bridgford, Nottingham NG2 6HH
☎ 0115 923 3424
goonewseditor@ntlworld.com
www.goodnews-paper.org.uk
Owner *Good News Fellowship*
Editor *Andrew Halloway*
Circulation 50,000

Founded 2001. MONTHLY evangelistic newspaper which welcomes contributions. No fiction. Send for sample copy of writers' guidelines in the first instance. No poetry or children's stories.
NEWS Items of up to 500 words (preferably with pictures) 'showing God at work, and human interest photo stories. "Churchy" items not wanted. Testimonies of how people have come to personal faith in Jesus Christ and the difference it has made are always welcome. They do not need to be dramatic!'
WOMEN'S PAGE Relevant items of interest welcome.
Ⓔ Payment negotiable.

The Good Ski Guide

27A High Street, Esher KT10 9RL
☎ 01372 468140 ☎ 01372 470765
edit@goodholidayideas.com
www.goodskiguide.com
Owner *Good Holiday Group*
Editors *John Hill, Nick Dalton*

Circulation 250,000

Founded 1976. FOUR ISSUES YEARLY. Unsolicited mss welcome from writers with a knowledge of skiing and ski resorts. Prospective contributors are best advised to make initial contact in writing as ideas and work need to be seen before any discussion can take place. Ⓔ Payment negotiable.

GQ

Vogue House, Hanover Square, London W1S 1JU
☎ 020 7499 9080 ☎ 020 7495 1679
www.gq.com
Owner *Condé Nast Publications Ltd*
Editor *Dylan Jones*
Circulation 129,520

Founded 1988. MONTHLY. Men's style magazine. No unsolicited material. Write or fax with an idea in the first instance.

Granta

12 Addison Avenue, London W11 4QR
☎ 020 7605 1360 ☎ 020 7605 1361
editorial@granta.com
www.granta.com
Editor *Alex Clark*

QUARTERLY magazine of new writing, including fiction, memoirs, reportage and photography published in paperback book form. Highbrow, diverse and contemporary, often with a thematic approach. Unsolicited mss (including fiction) considered. A lot of material is commissioned. Vital to read the magazine first to appreciate its very particular fusion of cultural and political interests. No reviews or news articles. Access the website for submission guidelines.
Ⓔ Payment negotiable.

Grazia

Bauer Consumer Media, Endeavour House, 189 Shaftesbury Avenue, London WC2H 8JG
☎ 020 7437 9011 ☎ 020 7520 6589
www.graziamagazine.co.uk
Owner *Bauer Consumer Media*
Editor *Jane Bruton*
Circulation 227,083

Launched February 2005. Britain's first WEEKLY glossy aimed at women aged 25 to 45. Features articles on fashion, celebrity, news, beauty and lifestyle.

The Great Outdoors
▷ TGO

The Great War (1914–1918)

PO Box 202, Scarborough YO11 3GE
☎ 01723 581329 ☎ 01723 581329
books@greatnorthernpublishing.co.uk
www.greatnorthernpublishing.co.uk
Owner *Great Northern Publishing*
Editor *Mark Marsay*

Founded 2001. BI-MONTHLY subscription only, non-academic magazine published in A5 format. 'The little magazine dedicated to the Great War (1914–19) and to those who perished and those who returned.' Articles, personal stories and accounts of those who served (men and women of all nationalities) and their families: diaries, anecdotes, letters, postcards, poetry, unit histories, events and memorials, etc. Absolutely no fiction or long academic works expounding historian's personal views. New material welcome but contact editor prior to sending. See website for submission guidelines and editorial content. 'Open door policy: all welcome regardless of ability to write to high standard as all work is carefully edited. No subject or topic excluded.' Sample copy £5.

Grow Your Own

25 Phoenix Court, Hawkins Road, Colchester CO2 8JY
☎ 01206 505979   ⨏ 01206 505945
craig.drever@aceville.co.uk
www.growfruitandveg.co.uk
Owner *Helen Tudor*
Editor *Craig Drever*

Launched 2005. MONTHLY magazine aimed at aspiring self-sufficients giving information on the best ways to grow (and cook) seasonal produce. Contributions are welcome on anything relevant to fruit, vegetables and herbs. Approach by e-mail.
FEATURES & NEWS *Craig Drever* (e-mail address above).

GT (Gay Times)

Unit M, Spectrum House, 32–34 Gordon House Road, London NW5 1LP
☎ 020 7424 7400   ⨏ 020 7424 7401
edit@gaytimes.co.uk
www.gaytimes.co.uk
Owner *Millivres-Prowler Group*
Editor *Joseph Galliano*
Circulation 68,000

Covers all aspects of gay life, plus general interest likely to appeal to the gay community, art reviews, fashion, style and news. Regular freelance writers used. ⓔ Payment negotiable.

Guardian Weekend
▷ The Guardian under National Newspapers

Guiding magazine

17–19 Buckingham Palace Road, London SW1W 0PT
☎ 020 7834 6242   ⨏ 020 7828 8317
guiding@girlguiding.org.uk
www.girlguiding.org.uk
Owner *Girlguiding UK*
Editor *To be appointed*

Circulation 79,000

Founded 1914. MONTHLY. Unsolicited mss welcome provided topics relate to the Movement and/or women's role in society. Ideas in writing appreciated in first instance. No nostalgic, 'when I was a Guide', pieces, please.
ACTIVITY IDEAS Interesting, contemporary ideas and instructions for activities for girls aged 5 to 18+ to do during unit meetings – crafts, games (indoor/outdoor), etc.
FEATURES Topics relevant to today's women. 500 words.
NEWS Items likely to be of interest to members. Max. 100–150 words.
ⓔ Payment negotiable.

H&E Naturist

Burlington Court, Carlisle Street, Goole DN14 5EG
☎ 01405 760298/766769   ⨏ 01405 763815
editor@henaturist.co.uk
www.healthandefficiency.co.uk
Owner *New Freedom Publications Ltd*
Editor *Sara Backhouse*
Circulation 20,000

Founded 1898. MONTHLY naturist magazine.
FEATURES Will consider short features on social nudism, longer features on nudist holidays and nudist philosophy. 90% of every issue is by freelance contributors. 800–1,500 words.
NEWS 'We are always on the lookout for national and international nudist news stories.' 250–500 words. No soft porn, 'sexy' stories or sleazy photographs. Approach by post, e-mail or telephone.

Hair

IPC Media Ltd., Blue Fin Building, 110 Southwark Street, London SE1 0SU
☎ 020 3148 7274
Owner *IPC Media (A Time Warner Company)*
Editor *Louise White*
Circulation 80,319

Founded 1977. TWELVE ISSUES YEARLY hair and beauty magazine. No unsolicited mss, but always interested in good photographs. Approach with ideas in writing.
FEATURES Fashion pieces on hair trends and styling advice.
ⓔ Payment negotiable.

Hairflair & Beauty

Haversham Publications Ltd, Freebournes House, Freebournes Road, Witham CM8 3US
☎ 01376 534540   ⨏ 01376 534546
Owner *Haversham Publications Ltd*
Editor *Ruth Page*
Circulation 27,628

Founded 1982. BI-MONTHLY. Original and interesting hair and beauty-related features written in a young, lively style to appeal to a readership aged 16–35 years. Freelancers are not used.

Harper's Bazaar

National Magazine House, 72 Broadwick Street, London W1F 9EP
☎ 020 7439 5000 📠 020 7439 5506
www.harpersbazaar.co.uk
Owner *National Magazine Co. Ltd*
Editor *Lucy Yeomans*
Circulation 109,033

MONTHLY. Up-market glossy combining the stylish and the streetwise. Approach in writing (not by phone) with ideas.
FEATURES *Francesca Martin* Ideas only in the first instance.
NEWS Snippets welcome if very original.
💷 Payment negotiable.

Health & Fitness Magazine

2 Balcombe Street, London NW1 6NW
☎ 020 7042 4000
mary.comber@futurenet.co.uk
www.healthandfitnessonline.co.uk
Owner *Future Publishing Ltd*
Editor *Mary Comber*
Circulation 31,346

Founded 1983. MONTHLY. Target reader: active, health-conscious women aged 25–40.
FEATURES news and articles on nutrition, exercise, healthy eating, holistic health and well-being. Will consider ideas; approach in writing in the first instance.

Healthy

River Publishing, 14 Leicester Place, London WC2H 7BZ
☎ 020 7413 9359 📠 020 7306 0314
hberesford@riverltd.co.uk
www.healthy-magazine.co.uk
Owner *River Publishing*
Editor *Heather Beresford*
Circulation 188,171

NINE ISSUES YEARLY. Health magazine with authoritative and accessible information and advice. Will consider holistic health and lifestyle features; approach by e-mail.

Heat

Endeavour House, 189 Shaftesbury Avenue, London WC2H 8JG
☎ 020 7437 9011 📠 020 7859 8670
heat@bauerconsumer.co.uk
www.heatworld.co.uk
Owner *Bauer Consumer Media*
Acting Editor *Julian Linley*

Circulation 533,034

Founded January 1999. WEEKLY entertainment magazine dealing with TV, film and radio information, fashion and features, with an emphasis on celebrity interviews and news. Targets 18 to 40-year-old readership, male and female. Articles written both in-house and by trusted freelancers. No unsolicited mss.

Hello!

Wellington House, 69–71 Upper Ground, London SE1 9PQ
☎ 020 7667 8700 📠 020 7667 8716
www.hellomagazine.com
Owner *Hola! (Spain)*
Editor *Kay Goddard*
Commissioning Editor *Linda Newman*
Circulation 405,615

WEEKLY. Owned by a Madrid-based publishing family, *Hello!* has grown faster than any other British magazine since its launch here in 1988. The magazine has editorial offices both in Madrid and London. Major colour features plus regular news pages. Although much of the material is provided by regulars, good proposals do stand a chance. Approach the commissioning editor with ideas in the first instance. No unsolicited mss.
FEATURES Interested in celebrity-based features, with a newsy angle, and exclusive interviews from generally unapproachable personalities.
💷 Payment by arrangement.

Hi-Fi News

IPC Media Ltd, Leon House, 233 High Street, Croydon CR9 1HZ
☎ 020 8726 8311 📠 020 8726 8397
hi-finews@ipcmedia.com
www.hifinews.co.uk
Owner *IPC Media (A Time Warner Company)*
Editor *Paul Miller*
Circulation 11,353

Founded 1956. MONTHLY. Write in the first instance with suggestions based on knowledge of the magazine's style and subject. All articles must be written from an informed technical or enthusiast viewpoint. 💷 Payment negotiable, according to technical content.

High Life

85 Strand, London WC2R 0DW
☎ 020 7550 8000 📠 020 7534 8253
high.life@cedarcom.co.uk
www.cedarcom.co.uk
Owner *Cedar Communications*
Editor *Kerry Smith*
Circulation 200,298

Founded 1973. MONTHLY glossy. British Airways in-flight magazine. Almost all the content is commissioned. No unsolicited mss. Few opportunities for freelancers.

History Today

20 Old Compton Street, London W1D 4TW
☎ 020 7534 8000
p.furtado@historytoday.com
www.historytoday.com
Owner *History Today Trust for the Advancement of Education*
Editor *Peter Furtado*
Circulation 26,191

Founded 1951. MONTHLY. General history and archaeology worldwide, history behind the headlines. Serious submissions only; no 'jokey' material. Approach by post only with s.a.e.

Home & Family
▷ Families First

Homes & Antiques

9th Floor, Tower House, Fairfax Street, Bristol BS1 3BN
☎ 0117 927 9009
www.homesandantiques.com
Owner *BBC Magazines Bristol*
Editor *Angela Linforth*
Circulation 100,972

Founded 1993. MONTHLY traditional home interest magazine with a strong bias towards antiques and collectibles. 'There are plenty of opportunities for freelancers and *Homes & Antiques* is always looking for interesting people and stories to feature in the fields of heritage, collecting, crafts and home design.' No fiction or health and beauty. Approach with ideas by phone or in writing.
FEATURES *Natasha Goodfellow* Pieces commissioned on recce shots and cuttings. Guidelines available on request. Send cuttings of relevant work published. Max. 1,500 words. £ Payment negotiable.

Homes & Gardens

IPC Media Ltd., Blue Fin Building, 110 Southwark Street, London SE1 0SU
☎ 020 3148 5000   📠 020 3148 8165
www.homesand gardens.com
Owner *IPC Media (A Time Warner Company)*
Editor *Deborah Barker*
Circulation 140,185

Founded 1919. MONTHLY. Almost all published articles are specially commissioned. No fiction or poetry. Best to approach in writing with an idea, enclosing snapshots if appropriate.

Horse

IPC Media Ltd., Blue Fin Building, 110 Southwark Street, London SE1 0SU
☎ 020 3148 4609

karen_spinner@ipcmedia.com
www.horsemagazine.com
Owner *IPC Media (A Time Warner Company)*
Acting Editor *Karen Spinner*
Circulation 20,062

Founded 1997. MONTHLY magazine aimed at serious leisure riders and keen competitors. Each month the magazine includes easy-to-follow training tips from some of the world's top riders, alongside horsecare features and the latest news and gossip from the horseworld. Send feature ideas with a short synopsis to the editor.

Horse and Hound

IPC Media Ltd., 9th Floor, Blue Fin Building, 110 Southwark Street, London SE1 0SU
☎ 020 3148 4562
jenny_sims@ipcmedia.com
www.horseandhound.co.uk
Owner *IPC Media (A Time Warner Company)*
Editor *Lucy Higginson*
Circulation 65,631

Founded 1884. WEEKLY. The oldest equestrian magazine on the market. Contains regular veterinary advice and instructional articles, as well as authoritative news and comment on fox hunting, international and national showjumping, horse trials, dressage, driving and endurance riding. Also weekly racing and point-to-points, breeding reports and articles. Regular books and art reviews, and humorous articles and cartoons are frequently published. Plenty of opportunities for freelancers. Unsolicited contributions welcome.
Also publishes a sister monthly publication, *Eventing*, which covers the sport of horse trials comprehensively; Editor *Julie Harding*.
£ NUJ rates.

Horse and Rider

Headley House, Headley Road, Grayshott GU26 6TU
☎ 01428 601020   📠 01428 601030
djm@djmurphy.co.uk
www.horseandrideruk.com
Owner *D.J. Murphy (Publishers) Ltd*
Editor *Nicky Moffatt*
Deputy Editor *Jane Gazzard*
Circulation 46,000

Founded 1949. MONTHLY. Adult readership, largely horse-owning. News and instructional features, which make up the bulk of the magazine, are almost all written in-house or commissioned. New contributors and unsolicited articles are occasionally used. Approach the editor in writing with ideas.

Hotline

The River Group, Victory House, Leicester Square, London WC2H 7BZ
☎ 020 7413 9339
mhigham@riverltd.co.uk
Editor-in-Chief *Mark Higham*
Circulation 124,740

Founded 1997. QUARTERLY on-board magazine for Virgin trains. UK celebrity and travel-based features. Ideas and outlines welcome by post or e-mail.

House & Garden

Vogue House, Hanover Square, London W1S 1JU
☎ 020 7499 9080   ☏ 020 7629 2907
www.houseandgarden.co.uk
Owner *The Condé Nast Publications Ltd*
Editor *Susan Crewe*
Circulation 143,089

Founded 1947. MONTHLY. Most feature material is produced in-house but occasional specialist features are commissioned from qualified freelancers, mainly for the interiors, wine and food sections and travel.
FEATURES *Hatta Byng* Suggestions for features, preferably in the form of brief outlines of proposed subjects, will be considered.

House Beautiful

National Magazine House, 72 Broadwick Street, London W1F 9EP
☎ 020 7439 5000   ☏ 020 7439 5141
Owner *National Magazine Co. Ltd*
Editor *Julia Goodwin*
Circulation 175,019

Founded 1989 and relaunched in November 2003. MONTHLY. Lively magazine offering sound, practical information and plenty of inspiration for those who want to make the most of where they live. Over 100 pages of easy-reading editorial. Regular features about decoration, DIY, food, gardening and occasionally property, plus topical features with a newsy slant to fit the 'Hot Topics' slot. Approach in writing with synopses in the first instance including a brief outline of areas of specialism and examples of previously published articles in a similar vein.

How to Spend It
▷ Financial Times under National Newspapers

i-D Magazine

124 Tabernacle Street, London EC2A 4SA
☎ 020 7490 9710   ☏ 020 7490 9737
editor@i-dmagazine.co.uk
www.i-dmagazine.com
Owner *Levelprint*
Editor *Ben Reardon*

Circulation 74,267

Founded 1980. MONTHLY lifestyle magazine for both sexes with a fashion bias. International. Very hip. Does not accept unsolicited contributions but welcomes new ideas from the fields of fashion, music, clubs, art, film, technology, books, sport, etc. No fiction or poetry. 'We are always looking for freelance non-fiction writers with new or unusual ideas.' A different theme each issue – past themes have included Green politics, taste, films, sex, love and loud dance music – means it is advisable to discuss feature ideas in the first instance.

Ideal Home

IPC Media Ltd., Blue Fin Building, 110 Southwark Street, London SE1 0SU
☎ 020 3148 7357
Owner *IPC Media (A Time Warner Company)*
Editorial Director *Isobel McKenzie-Price*
Circulation 214,483

Founded 1920. MONTHLY glossy. Unsolicited feature ideas are welcome only if appropriate to the magazine. Prospective contributors wishing to submit ideas should do so in writing to the editor. No fiction.
FEATURES Furnishing and decoration of houses, kitchens or bathrooms; interior design, soft furnishings, furniture and home improvements, lifestyle, consumer, gardening, property, food and readers' homes. Length to be discussed with editor.
£ Payment negotiable.

Image Magazine

Upper Mounts, Northampton NN1 3HR
☎ 01604 467000   ☏ 01604 467190
image@northantsnews.co.uk
www.northantsnews.co.uk
Owner *Northamptonshire Newspapers Ltd*
Editor *Ruth Supple*
Circulation 27,466

Founded 1905. MONTHLY general interest regional magazine. No unsolicited mss.
FEATURES Local issues, personalities, businesses, etc., of Northamptonshire, Bedfordshire, Buckinghamshire interest. Max. 500 words.
NEWS No hard news as such, just monthly diary column.
OTHER Regulars on motoring, fashion, beauty, lifestyle and travel. Max. 500 words.
£ Payment for features negotiable.

In Britain

Jubilee House, 2 Jubilee Place, London SW3 3TQ
☎ 020 7751 4800   ☏ 020 7751 4848
inbritain@archant.co.uk
Editor *Andrea Spain*

Circulation 27,882

Founded in the 1930s. BI-MONTHLY. Travel magazine of 'VisitBritain'. Not much opportunity for unsolicited work – approach (by e-mail) with ideas and samples. Words and picture packages preferred (good quality transparencies or digital images by CD or e-mail).

Independent Magazine
▷ The Independent under National Newspapers

Inspire
CPO, Garcia Estate, Canterbury Road, Worthing BN13 1BW
☎ 01903 264556
russbravo@cpo.org.uk
www.inspiremagazine.org.uk
Owner *Christian Publishing & Outreach Ltd*
Editor *Russ Bravo*
Circulation 75,000

MONTHLY. Upbeat, good news Christian magazine featuring human interest stories, growing churches and community transformation. Limited freelance articles used. Contributor's guidelines available.

InStyle
IPC Media Ltd., Blue Fin Building, 110 Southwark Street, London SE1 0SU
☎ 020 3148 5000   🖷 020 3148 8166
firstname_lastname@instyleuk.com
Owner *IPC Media (A Time Warner Company)*
Senior Editor *Kate O'Donnell*
Editor *Trish Halpin*
Deputy Editor *Charlotte Moore*
Fashion & Jewellery Editor *Sophie Hedley*
Features Editor *Kate Bussman*
Fashion Director *Fiona Rubie*
Circulation 179,558

Launched March 2001. MONTHLY. UK edition of US fashion, beauty, celebrity and lifestyle magazine. No unsolicited material.

Interzone: Science Fiction & Fantasy
TTA Press, 5 Martins Lane, Witcham, Ely CB6 2LB
info@ttapress.demon.co.uk
www.ttapress.com
Owner *TTA Press*
Editor *Andy Cox*
Circulation 15,000

Founded 1982. BI-MONTHLY magazine of science fiction and fantasy. Unsolicited mss are welcome 'from writers who have a knowledge of the magazine and its contents'. S.a.e. essential for return.
FICTION 2,000–6,000 words.
FEATURES Book/film reviews, interviews with writers and occasional short articles. Length by arrangement.

£ Fiction, £30 per 1,000 words; features, negotiable.

Investors Chronicle
Number One, Southwark Bridge, London SE1 9HL
☎ 020 7775 6582   🖷 020 7382 8105
www.investorschronicle.co.uk
Owner *Pearson*
Editor *Oliver Ralph*
Deputy Editor *Rosie Carr*
Circulation 33,514

Founded 1860. WEEKLY. Opportunities for freelance contributors in the survey section only. All approaches should be made in writing. Over forty surveys are published each year on a wide variety of subjects, generally with a financial, business or investment emphasis. Copies of survey list and synopses of individual surveys are obtainable from the surveys editor.
£ Payment negotiable.

The Irish Book Review
Mary Immaculate College, Dublin Road, Limerick, Republic of Ireland
eugene.obrien@micul.ie
www.irishbookreview.com
Editor *Eugene O'Brien*

Founded 2005. QUARTERLY. Reviews books of Irish interest and/or by Irish authors. Also publishes features, extracts, articles relating to Irish books, publishing and the book trade. 'Would like to hear from potential reviewers/writers.'

Irish Pages
The Linen Hall Library, 17 Donegall Square North, Belfast BT1 5GB
☎ 028 9043 4800
irishpages@yahoo.co.uk
www.irishpages.org
Editor *Chris Agee*
Circulation 2,800

Founded 2002. BIANNUAL non-partisan, non-sectarian journal publishing writing from Ireland and abroad. 'The most important cultural journal in Ireland at the present moment' (Jonathan Allison, Director of the Yeats Summer School). FEATURES Poetry, short fiction, essays, non-fiction, memoirs, nature-writing, translated work, literary journalism and other autobiographical, historical and scientific writing of literary distinction. Irish language and Ulster Scots writing are published in the original, with English translations. Equal editorial attention is given to established, emergent and new writers. Send submissions to the editor by post.

IRRV Valuer
IRRV, 41 Doughty Street, London WC1N 2LF
☎ 01843 290919   🖷 01843 290919

jcroberts54@hotmail.com
www.irrv.org.uk
Owner *Institute of Revenues, Rating and Valuation*
Managing Editor *John Roberts*
Circulation 2,000

Founded 2001. FOUR issues yearly for property valuation professionals. Well-qualified commentary, analysis and news covering issues affecting valuers, including reforms, legislation and technology. Small amount of appropriate lifestyle coverage. No unsolicited material; approach by telephone, fax, letter or e-mail in the first instance.
FEATURES Ideas for stories must be well sourced and informed.
Ⓔ Payment negotiable.

Jade – The International Erotic Art and Literature Magazine

PO Box 202, Scarborough YO11 3GE
Ⓣ 01723 581329   Ⓕ 01723 581329
books@greatnorthernpublishing.co.uk
www.greatnorthernpublishing.co.uk
Owner *Great Northern Publishing*
Editor *Mark Marsay*

Founded 2002. MONTHLY adult subscription-only magazine. Uncensored, high-res PDF download format since April 2008. Official magazine of The Guild of Erotic Artists. Features new and established international photographers, artists, sculptors and writers in all erotic genres from around the world. Some editorial articles and features within the genre. No advice columns or non-erotica related submissions. Contributors (photographers, artists, sculptors and writers – articles and fiction) should consult the website for submission guidelines, style and current requirements; contact the editor prior to sending material. Sample download copy £5 (contains strong adult content).

Jane's Defence Weekly

Sentinel House, 163 Brighton Road, Coulsdon CR5 2YH
Ⓣ 020 8700 3700   Ⓕ 020 8763 1007
jdw@janes.com
www.janes.com
Owner *Jane's Information Group*
Editor *Peter Felstead*
Circulation 28,100

Founded 1984. WEEKLY. No unsolicited mss. Approach in writing with ideas in the first instance.
FEATURES Current defence topics of worldwide interest. No history pieces. Most features are commissioned. Max. 4,000 words.

Jazz Journal International

3 & 3A Forest Road, Loughton IG10 1DR
Ⓣ 020 8532 0456/0678   Ⓕ 020 8532 0440
Owner *Jazz Journal Ltd*
Editor *Janet Cook*
Circulation 8,000+

Founded 1948. MONTHLY. A specialized jazz magazine, for record collectors, principally using expert contributors whose work is known to the editor. Unsolicited mss not welcome, with the exception of news material (for which no payment is made). It is not a gig guide, nor a free reference source for students.

Jersey Now

PO Box 582, Five Oaks, St Saviour JE4 8XQ
Ⓣ 01534 611743   Ⓕ 01534 611610
eperchard@msppublishing.com
Owner *MSP Publishing*
Managing Editor *Peter Body*
Editor *Elisabeth Perchard*
Circulation 26,000

Founded 1987. QUARTERLY lifestyle magazine for Jersey covering homes, gardens, the arts, Jersey heritage, motoring, boating, fashion, beauty, food and drink, health, travel and technology. Upmarket glossy aimed at an informed and discerning readership. Interested in Jersey-orientated articles only; 1,200 words max. Approach the editor initially. Ⓔ Payment negotiable.

Jewish Chronicle

25 Furnival Street, London EC4A 1JT
Ⓣ 020 7415 1500   Ⓕ 020 7405 9040
editorial@thejc.com
www.thejc.com
Owner *Kessler Foundation*
Editor *David Rowan*
Circulation 32,623

WEEKLY. Unsolicited mss welcome if 'the specific interests of our readership are borne in mind by writers'. Approach in writing, except for urgent current news items. No fiction. Max. 1,500 words for all material.
FEATURES *Alan Montague*
NEWS EDITOR *Jenni Frazer*
FOREIGN NEWS *Daniella Peled*
SUPPLEMENTS *Angela Kiverstein*
Ⓔ Payment negotiable.

Jewish Quarterly

Haskell House, 152 West End Lane, London NW6 1SD
Ⓣ 020 7443 5155 (editorial)/020 8343 4675 (admin)
editor@jewishquarterly.org
and
adminjewishquarterly.org
www.jewishquarterly.org

Publisher *Jewish Literary Trust*
Editor *Rachel Lasserson*

Founded 1953. QUARTERLY illustrated magazine featuring Jewish literature and fiction, politics, art, music, film, poetry, history, dance, community, autobiography, Hebrew, Yiddish, Israel and the Middle East, Judaism, interviews, Zionism, philosophy and holocaust studies. Features a major books and arts section. Unsolicited mss welcome but letter or e-mail preferred in the first instance.

Jewish Telegraph Group of Newspapers

Telegraph House, 11 Park Hill, Bury Old Road, Prestwich, Manchester M25 0HH
☎ 0161 740 9321   🖷 0161 740 9325
manchester@jewishtelegraph.com
www.jewishtelegraph.com
Owner *Jewish Telegraph Ltd*
Editor *Paul Harris* (pharris@jewishtelegraph.com
☎ 0161 741 2633 🖷 0161 740 5555)
Circulation 16,000

Founded 1950. WEEKLY publication with local, national and international news and features. (Separate editions published for Manchester, Leeds, Liverpool and Glasgow.) Unsolicited features on Jewish humour and history welcome.

Journal of Apicultural Research (incorporating Bee World)

IBRA, 16 North Road, Cardiff CF10 3DY
☎ 029 2037 2409   🖷 05601 135640
mail@ibra.org.uk
www.ibra.org.uk
Owner *International Bee Research Association*
Editor *Norman Carreck*
Circulation 1,700

Bee World, founded 1919; Journal of Apicultural Research, 1962. QUARTERLY. High-quality factual journal, including peer-reviewed articles, with international readership. Features on apicultural science and technology. Unsolicited mss welcome but authors should write to the editor for guidelines before submitting material.

Junior Education PLUS

Scholastic Ltd, Villiers House, Clarendon Avenue, Leamington Spa CV32 5PR
☎ 01926 887799   🖷 01926 883331
vpaley@scholastic.co.uk
www.scholastic.co.uk/magazines
Owner *Scholastic Ltd*
Editor *Michelle Guy*
Assistant Editor *Kirstin McCreadie*
Circulation 21,121

'UK's leading magazine, in print and online', aimed at teachers of children aged 7–11 years. Practical articles from teachers about education for this age group are welcome. Max. 900 words. Approach in writing with synopsis.

Kerrang!

Mappin House, 4 Winsley Street, London W1W 8HF
☎ 020 7182 8000   🖷 020 7182 8910
kerrang@bauerconsumer.co.uk
www.kerrang.com
Owner *Bauer Consumer Media*
Editor *Paul Brannigan*
Circulation 76,937

Founded 1981. WEEKLY rock, punk and metal magazine. 'Written by fans for fans.' Will consider ideas for features but not actively seeking new contributors unless expert in specialist fields such as black metal and hardcore, emo, etc.

Koi Magazine

Origin Publishing, Tower House, Fairfax Street, Bristol BS1 3BN
☎ 0117 927 9009   🖷 0117 934 9008
koi@originpublishing.co.uk
www.koimag.co.uk
Owner *Origin Publishing*
Editor *Hilary Clapham*
Circulation 15,000

Launched 1999. FOUR WEEKLY. Practical guide to koi-keeping featuring everything you need to know about the hobby, including koi appreciation and pond construction. Contributions welcome; send press releases, articles and photographs by e-mail.
FEATURES *Hilary Clapham* Max. 4,000 words. 🖷 £150–£400.
NEWS *Beckie Rodgers* Max. 200 words.

The Lady

39–40 Bedford Street, London WC2E 9ER
☎ 020 7379 4717   🖷 020 7836 4620
Editor *Arline Usden*
Circulation 28,911

Founded 1885. WEEKLY. Unsolicited mss are accepted provided they are not on the subject of politics or religion, or on topics covered by staff writers or special correspondents, i.e. fashion and beauty, health, cookery, household, gardening, finance and shopping.
FEATURES Well-researched pieces on British and foreign travel, historical subjects or events; interviews and profiles and other general interest topics. Max. 1,200 words for illustrated two-page articles; 900 words for one-page features; 420 words for first-person 'Viewpoint' pieces. All material should be addressed to the editor with s.a.e. enclosed. Photographs supporting features may be supplied as colour transparencies, b&w prints or on disk. Telephone enquiries about features are not encouraged.

Lake District Life

3 Tustin Court, Port Way, Preston PR2 2YQ
☎ 01772 722022   🖷 01772 736496
roger.borrell@lakedistrict-life.co.uk
www.lakedistrict-life.co.uk
Owner *Archant Life*
Editor *Roger Borrell*
Circulation 23,000

Founded 1947. MONTHLY. A celebration of life
in the Lake District, both past and present.
Unsolicited contributions welcome; approach the
editor by e-mail.

Lancashire Life

3 Tustin Court, Port Way, Preston PR2 2YQ
☎ 01772 722022   🖷 01772 736496
roger.borrell@lancashirelife.co.uk
www.lancashirelife.co.uk
Owner *Archant Life*
Editor *Roger Borrell*
Circulation 23,000

Founded 1947. MONTHLY county magazine.
Features and pictures about Lancashire.
Unsolicited contributions welcome; approach the
editor by e-mail.

Land Rover World

IPC Media Ltd, Leon House, 233 High Street,
Croydon CR9 1HZ
☎ 020 8726 8371   🖷 020 8726 8398
landroverworld@ipcmedia.co.uk
www.landroverworld.co.uk
Owner *IPC Media (A Time Warner Company)*
Editor *John Carroll*
Circulation 30,000

Founded 1994. MONTHLY. Incorporates *Practical
Land Rover World* and *Classic Land Rover World*.
Unsolicited material welcome, especially if
supported by high-quality illustrations.
FEATURES All articles with a Land Rover theme
of interest. Potential contributors are strongly
advised to examine previous issues before starting
work.
💷 Payment negotiable.

Lawyer 2B

50 Poland Street, London W1F 7AX
☎ 0207 970 4000
www.lawyer2b.com
Owner *Centaur Media plc*
Editor *Husnara Begum*
Circulation 30,000

Launched in 2001, *Lawyer 2B* is a sister
publication of *The Lawyer*, circulated to graduate
recruitment professionals, law students and
trainees. FIVE ISSUES PER ACADEMIC YEAR.
Regular opportunities for profiles/features
relating to legal education and law students.

The Lawyer

50 Poland Street, London W1F 7AX
☎ 020 7970 4000
editorial@thelawyer.com
www.thelawyer.com
Owner *Centaur Media plc*
Editor *Catrin Griffiths*
City Editor *Margaret Taylor*
Web Editor *Jon Parker*
Special Reports Editor *Tom Phillips*
Associate Editor (New York) *Matt Byrne*
Circulation 31,507 (print); 134,040 (online)

Launched 1987. WEEKLY plus daily news e-mail
alerts and commentaries. Leading news magazine
for City and commercial lawyers. Strong
international coverage and leading-edge research.
FEATURES *Steve Hoare* Technical legal content,
written in an accessible style. No unsolicited
contributions; initial approach by e-mail with
specific idea.

Lincolnshire Life

County House, 9 Checkpoint Court, Sadler Road,
Lincoln LN6 3PW
☎ 01522 527127   🖷 01522 842000
editorial@lincolnshirelife.co.uk
www.lincolnshirelife.co.uk
Publisher *C. Bingham*
Executive Editor *Josie Thurston*
Circulation 10,000

Founded 1961. MONTHLY county magazine
featuring geographically relevant articles on local
culture, history, personalities, etc. Max. 1,500
words. Contributions supported by three or four
good-quality photographs welcome. Approach in
writing. 💷 Payment varies.

The Lincolnshire Poacher

County House, 9 Checkpoint Court, Sadler Road,
Lincoln LN6 3PW
☎ 01522 527127   🖷 01522 842000
editorial@lincolnshirelife.co.uk
www.lincolnshirelife.co.uk
Publisher *C. Bingham*
Executive Editor *Josie Thurston*
Circulation 5,000

QUARTERLY county magazine featuring
geographically relevant but nostalgic articles on
history, culture and personalities of Lincolnshire.
Max. 2,000 words. Contributions supported
by three or four good-quality photographs/
illustrations appreciated. Approach in writing.
💷 Payment varies.

The List

14 High Street, Edinburgh EH1 1TE
☎ 0131 557 8500   🖷 0131 557 8500
claire.prentice@list.co.uk
www.list.co.uk

Owner *The List Ltd*
Publisher *Robin Hodge*
Editor *Claire Prentice*
Circulation 9,686

Founded 1985. FORTNIGHTLY. Events guide covering Glasgow and Edinburgh. Interviews and profiles of people working in film, theatre, music and the arts. Max. 1,200 words. No unsolicited mss. News material tends to be handled in-house.

Literary Review

44 Lexington Street, London W1F 0LW
℡ 020 7437 9392   ℻ 020 7734 1844
editorial@literaryreview.co.uk
www.literaryreview.co.uk
Editor *Nancy Sladek*
Circulation 15,000

Founded 1979. MONTHLY. Publishes book reviews (commissioned), features and articles on literary subjects. Prospective contributors are best advised to contact the editor in writing. Unsolicited mss not welcome. Runs a monthly competition for 'poems which rhyme, scan and make sense'.
£ Payment 'minuscule'.

Living France

Archant House, Oriel Road, Cheltenham GL50 1BB
℡ 01242 216086   ℻ 01242 216094
editorial@livingfrance.com
www.livingfrance.com
Owner *Archant Life*
Editor *Eleanor O'Kane*

Founded 1989. FOUR-WEEKLY. A magazine for people who are considering or actively purchasing a home in France. Editorial consists of travel features as well as practical, friendly information and advice on living in France or becoming an owner of French property. Will consider articles describing different regions of France, interviews with owners of property or new-build and aspects of living in France. Interested in hearing from professional writers. No unsolicited mss; approach in writing or by e-mail with an idea.

Loaded

IPC Media Ltd., Blue Fin Building, 110 Southwark Street, London SE1 0SU
℡ 020 3148 5000   ℻ 020 3148 8107
andy_sherwood@ipcmedia.com
Owner *IPC Media (A Time Warner Company)*
Editor *Martin Daubney*
Features Editor *Andy Sherwood*
Circulation 115,065

Founded 1994. MONTHLY men's lifestyle magazine featuring music, sport, sex, humour, travel, fashion, hard news and popular culture. Will consider material which comes into these categories;

approach the features editor in writing or by e-mail in the first instance. No fiction or poetry.

Logos

5 Beechwood Drive, Marlow SL7 2DH
℡ 01628 483371   ℻ 01628 477577
logos-marlow@dial.pipex.com
www.logos-journal.org
Owner *LOGOS International Educational Publishing Foundation*
Editor (New York) *Charles M. Levine* (Charlev@Nyc.rr.com)
Editor Emeritus *Gordon Graham*
Business Manager *Betty Graham*

Founded 1990. QUARTERLY. Aims to 'deal in depth with issues which unite, divide, excite and concern the world of books', with an international perspective. Each issue contains six to eight articles of between 3,500 and 7,000 words. 'Logos is a non-profit making professional forum, not a scholarly journal.' Suggestions and ideas for contributions are welcome, and should be addressed to the editor. 'Guidelines for Contributors' available. Contributors write from their experience as authors, publishers, booksellers, librarians, etc.
£ Contributors receive off-prints, a copy of the issue in which their article appears and a 50% concession on the subscription rate.

The London Magazine

Editorial office: 70 Wargrave Road, London N15 6UB
℡ 020 8400 5882 (admin. office)
℻ 020 8994 1713
admin@thelondonmagazine.net
www.thelondonmagazine.net
Publisher *Christopher Arkell*
Editor *Sebastian Barker*
Circulation 1,200

Founded originally in 1732 and relaunched in 2002. BI-MONTHLY review of literature and the arts. Publishes poems, short stories, features, memoirs, and book and performance reviews. Not interested in overtly political material. Do *not* send anything by e-mail or fax. No phone-calls. All submissions should be made by post with an s.a.e. (plus IRCs if overseas) to the editorial office.
FEATURES Max. 6,000 words.
FICTION Max. 6,000 words.
POEMS 'Any length within reason'.
£ Payment negotiable with the editor. Standard fees for most items.

London Review of Books

28 Little Russell Street, London WC1A 2HN
℡ 020 7209 1101   ℻ 020 7209 1102
edit@lrb.co.uk
www.lrb.co.uk
Owner *LRB Ltd*

Editor *Mary-Kay Wilmers*
Circulation 45,905

Founded 1979. FORTNIGHTLY. Reviews, essays and articles on political, literary, cultural and scientific subjects. Also poetry. Unsolicited contributions welcome (approximately 50 received each week). No pieces under 2,000 words. Contact the editor in writing. Please include s.a.e.
£ £200 per 1,000 words; poems, £125.

Lothian Life

Larick House, Whitehouse, Tarbert PA29 6XR
☎ 01880 730360/07734 699607
info@lothianlife.co.uk
www.lothianlife.co.uk
Owner *Pages Editorial & Publishing Services*
Editor *Suse Coon*

Online county magazine for people who live, work or have an interest in the Lothians. Features on successful people, businesses or initiatives. Regular articles by experts in the Lifestyle section including Homes & Gardens, Tastebuds, Out & About, The Arts and Health & Fitness. Phone first to discuss content and timing.
£ See website.

Machine Knitting Monthly

PO Box 1479, Maidenhead SL6 8YX
☎ 01628 783080   📠 01628 633250
mail@machineknittingmonthly.net
www.machineknittingmonthly.net
Owner *RPA Publishing Ltd*
Editor *Anne Smith*

Founded 1986. MONTHLY. Unsolicited mss considered 'as long as they are applicable to this specialist publication. We have our own regular contributors each month but we're always willing to look at new ideas from other writers.' Approach in writing in the first instance.

MacUser

30 Cleveland Street, London W1T 4JD
☎ 020 7907 6000
mailbox@macuser.co.uk
www.macuser.co.uk
Owner *Dennis Publishing*
Editor *Nik Rawlinson*
Circulation 14,502

Launched 1985. FORTNIGHTLY computer magazine.

Management Today

174 Hammersmith Road, London W6 7JP
☎ 020 8267 4610
editorial@managementtoday.co.uk
Owner *Haymarket Media*
Managing Director *Nick Simpson*
Editor *Matthew Gwyther*

Circulation 100,011

General business topics and features. Ideas welcome. Send brief synopsis to the features editor.

marie claire

IPC Media Ltd., Blue Fin Building, 110 Southwark Street, London SE1 0SU
☎ 020 3148 7513   📠 020 3148 8120
marieclaire@ipcmedia.com
www.marieclaire.co.uk
Owner *European Magazines Ltd*
Editor *Marie O'Riordan*
Circulation 330,182

Founded 1988. MONTHLY. An intelligent glossy magazine for women, with strong international features and fashion. No unsolicited mss. Approach with ideas in writing or by e-mail (marieclaireideas@ipcmedia.com). No fiction. FEATURES *Miranda McMinn* Detailed proposals for feature ideas should be accompanied by samples of previous work.

Marketing Week

St Giles House, 50 Poland Street, London W1F 7AX
☎ 020 7970 4000   📠 020 7970 6722
mw.editorial@centaur.co.uk
www.marketingweek.co.uk
Owner *Centaur Communications*
Editor *Stuart Smith*
Circulation 40,069

WEEKLY trade magazine of the marketing industry. Features on all aspects of the business, written in a newsy and up-to-the-minute style. Approach with ideas in the first instance.
FEATURES *Daney Parker*
£ Payment negotiable.

Markings

The Bakehouse, 44 High Street, Gatehouse of Fleet DG7 2HP
☎ 01557 814196/175
info@markings.org.uk
www.markings.org.uk
Editors *John Hudson, Chrys Salt*
Circulation 500+

Launched 1995. BIANNUAL Scottish literary magazine with an international readership. Poetry, contemporary art, short stories, criticism and reviews. Occasionally publishes special features on established or emergent literary figures. Unsolicited submissions are invited in any language but must be accompanied by English translation. 'Markings wishes to be challenging, controversial and loyal to honest and humane writing.' Black and white artwork appears in every issue; artists and illustrators are encouraged to

submit work. Electronic submissions preferred but if sending by post enclose an s.a.e.
FEATURES Any subject relating to the arts. Max. 2,500 words.
FICTION Short stories on any subject. Max. 3,000 words.

Match

Media House, Peterborough Business Park, Lynchwood, Peterborough PE2 6EA
☎ 01733 468008   🖷 01733 468724
match.magazine@bauerconsumer.co.uk
www.matchmag.co.uk
Owner *Bauer Consumer Media*
Editor *James Bandy*
Circulation 113,049

Founded 1979. WEEKLY. The UK's biggest-selling football magazine aimed at 8–14-year-olds. All material is generated in-house by a strong news and features team. Work experience placements often given to trainee journalists and students; several staff have been recruited through this route. Approach in writing or by phone.

Maxim

30 Cleveland Street, London W1T 4JD
☎ 020 7907 6000   🖷 020 7907 6439
editorial.maxim@dennis.co.uk
www.maxim.co.uk
Owner *Dennis Publishing*
Editor *Michael Donlevy*
Circulation 78,463

Established 1995. MONTHLY glossy men's lifestyle magazine featuring sex, grooming, girls, fun-stuff, cars and fashion. No fiction or poetry. Approach in writing in the first instance, sending outlines of ideas only together with examples of published work.

Mayfair

2 Archer Street, Piccadilly Circus, London W1D 7AW
☎ 020 7292 8000   🖷 020 7734 5030
mayfair@paulraymond.com
www.paulraymond.com
Owner *Paul Raymond Publications*
Editor *David Spenser*
Circulation 85,000

Founded 1966. THIRTEEN ISSUES YEARLY. Unsolicited material accepted if pertinent to the magazine and if accompanied by suitable illustrative material. 'We will *only* publish work if we can illustrate it.' Interested in features and humour aimed at men aged 18 to 80; 800–1,000 words. Punchy, bite-sized humour, top ten features, etc. 'Must make the editor laugh.' Also considers erotica.

Mayfair Times

Blandel Bridge House, 56 Sloane Square, London SW1W 8AX
☎ 020 7259 1050   🖷 020 7901 9042
mayfair.times@pubbiz.com
www.pubbiz.com
Owner *Publishing Business Ltd*
Editor *Selma Day*
Circulation 17,000

Founded 1985. MONTHLY. Features on Mayfair of interest to residents, local workers, visitors and shoppers.

Medal News

Token Publishing, Orchard House, Duchy Road, Heathpark, Honiton EX14 1YD
☎ 01404 46972   🖷 01404 44788
info@tokenpublishing.com
www.tokenpublishing.com
Owners *J.W. Mussell, Carol Hartman*
Editor *J.W. Mussell*
Circulation 6,000

Founded 1989. TEN ISSUES YEARLY. Unsolicited material welcome but initial approach by phone or in writing preferred.
FEATURES 'Opportunities exist for well-informed authors who know the subject and do their homework. It would help if a digital copy of the contribution is provided in the form of an e-mail or CD. Relevant illustrations welcome.' Max. 2,500 words.
💷 £30 per 1,000 words.

Media Week

Griffin House, 161 Hammersmith Road, London W6 8BS
☎ 020 8267 5000   🖷 020 8267 8020
firstname.surname@haymarket.com
www.mediaweek.co.uk
Owner *Haymarket Publishing Ltd*
Editor *Steve Barrett*
Circulation 11,588

Founded 1986. WEEKLY trade magazine. UK and international coverage on all aspects of commercial media. Approach in writing with ideas. See website for e-mail addresses.

Melody Maker
▷ New Musical Express

Men's Health

National Magazine House, 33 Broadwick Street, London W1F 0DQ
☎ 020 7339 4400   🖷 020 7339 4444
www.menshealth.co.uk
Owner *Natmag-Rodale Publishing*
Editor *Morgan Rees*

Circulation 240,315

Founded 1994. MONTHLY men's healthy lifestyle magazine covering health, fitness, nutrition, stress and sex issues. No unsolicited mss; will consider ideas and synopses tailored to men's health. No fiction or extreme sports. Approach in writing in the first instance.

MiniWorld Magazine

IPC Media Ltd, Leon House, 233 High Street, Croydon CR9 1HZ

☏ 020 8726 8000   🖷 020 8726 8399

miniworld@ipcmedia.com

www.miniworld.co.uk

Owner *IPC Media (A Time Warner Company)*
Editor *Monty Watkins*
Circulation 37,000

Founded 1991. THIRTEEN ISSUES YEARLY. Car magazine devoted to the Mini. Unsolicited material welcome but prospective contributors are advised to contact the editor.
FEATURES Maintenance, tuning, restoration, technical advice, classified, sport, readers' cars and social history of this cult car.
Ⓔ Payment negotiable.

Mixmag

Development Hell Ltd, 90–92 Pentonville Road, London N1 9HS

☏ 020 7078 8400   🖷 020 7833 9900

www.mixmag.net

Owner *Development Hell Ltd*
Editor *Nick DeCosemo*
Circulation 35,817

Originally launched in 1982 and relaunched by new owners, Development Hell Ltd in 2006. Bestselling MONTHLY dance music magazine and 'leading authority on club culture'. News, features, night-life reviews, fashion, club listings and free mix CDs.

Mizz

Panini UK, Panini House, Coach & Horses Passage, Tunbridge Wells TN2 5UJ

☏ 01892 500100   🖷 01892 545666

mizz@panini.co.uk

www.mizz.com

Owner *Panini UK Ltd*
Editor *Karen O'Brien*
Circulation 71,092

Founded 1985. FORTNIGHTLY magazine for the 10–14-year-old girl.

FEATURES 'We have a full features team and thus do not accept freelance features.'

Mobilise

Mobilise Organisation, Ashwellthorpe, Norwich NR16 1EX

☏ 01508 489449   🖷 01508 488173

enquiries@mobilise.info

www.mobilise.info

Owner *Mobilise Organisation*
Editor *Sophie Douglas*
Circulation 30,000

MONTHLY publication of the Mobilise Organisation, which aims to promote and protect the interests and welfare of disabled people and help and encourage them in gaining increased mobility. Various discounts available for members; membership costs £14 p.a. (single), £18 (joint). The magazine includes information for members plus members' letters. Approach in writing with ideas. Unsolicited mss welcome.

Model Collector

IPC Focus Network, Leon House, 233 High Street, Croydon CR9 1HZ

☏ 020 8726 8000   🖷 020 8726 8299

modelcollector@ipcmedia.com

www.modelcollector.com

Owner *AOL Time Warner/IPC Media*
Editor *Lindsey Amrani*
Circulation 10,254

Founded 1987. THIRTEEN ISSUES YEARLY Britain's best selling die cast magazine. From the latest models to classic 1930's Dinkys. Interested in historical articles about particular models or ranges and reviews of the latest products. Photographs welcome. Not interested in radio controlled models or model railways.
FEATURES Freelancers should contact the editor. Unsolicited material may be considered.
NEWS Trade news welcome. Call *Lindsey Armrani* (☏ 020 8726 8238) to discuss details.

Mojo

Mappin House, 4 Winsley Street, London W1W 8HF

☏ 020 7182 8616   🖷 020 7182 8596

mojo@bauerconsumer.co.uk

www.mojo4music.com

Owner *Bauer Consumer Media*
Editor-in-Chief *Phil Alexander*
Circulation 106,218

Founded 1993. MONTHLY magazine containing features, reviews and news stories about all types of music and its influences. Receives about five mss per day. No poetry, think-pieces on dead rock stars or similar fan worship.
FEATURES Amateur writers discouraged except as providers of source material, contacts, etc.
NEWS All verifiable, relevant stories considered.
REVIEWS Write to Reviews Editor, *Jenny Bulley* with relevant specimen material.
Ⓔ Payment negotiable.

Moneywise

1st Floor, Standon House, Mansell Street, London
E1 8AA

☎ 020 7680 3600 📠 020 7702 0710
rachel.lacey@moneywise.co.uk
www.moneywise.co.uk
Owner *Capital Accumulation Services Ltd*
Editor *Rachel Lacey*
Circulation 25,399

Founded 1990. MONTHLY. No unsolicited mss;
ideas welcome. Make initial approach in writing
to the editorial department (c.v. preferred).

More

Endeavour House, 189 Shaftesbury Avenue,
London WC2H 8JG
☎ 020 7208 3165 📠 020 7208 3595
www.moremagazine.co.uk
Owner *Emap élan Publications*
Editor *Lisa Smosarski*
Circulation 200,033

Founded 1988. WEEKLY women's magazine aimed
at the working woman aged 18–26. Features on
sex and relationships plus celebrity news, style
and fashion. Most items are commissioned;
approach features director with idea. Prospective
contributors are strongly advised to study the
magazine's style before submitting anything.

Mother and Baby

Endeavour House, 189 Shaftesbury Avenue,
London WC2H 8JG
☎ 020 7295 5560
mother&baby@emap.com
Owner *Emap Esprit*
Editor *Miranda Levy*
Editorial Assistant *Lucy Quick*
Circulation 65,856

Founded 1956. MONTHLY. Welcomes suggestions
for feature ideas about pregnancy, newborn
basics, practical babycare, baby development and
childcare subjects. Approaches may be made by
telephone, e-mail or in writing to the Features
Editor, *Jean Jollands*.

Motor Boat & Yachting

IPC Media Ltd., Blue Fin Building, 110 Southwark
Street, London SE1 0SU
☎ 020 3148 4651
mby@ipcmedia.com
www.mby.com
Owner *Time Warner*
Editor *Hugo Andreae*
Technical Editor *David Marsh*
Circulation 17,504

Founded 1904. MONTHLY for those interested in
motor boats and motor cruising.
FEATURES *Hugo Andreae* Cruising features
and practical features especially welcome.

Illustrations/photographs (colour) are just as
important as text. Max. 3,000 words.
NEWS *Rob Peake* Factual pieces. Max. 200 words.
💷 Features: from £100 per 1,000 words or by
arrangement; news: up to £50 per item.

Motor Boats Monthly

9th Floor, Blue Fin Building, 110 Southwark
Street, London SE1 0SU
☎ 020 3148 4664 📠 020 3148 8128
carl_richardson@ipcmedia.com
Owner *IPC Media (A TIme Warner Company)*
Editor *Carl Richardson*
Circulation 16,246

Founded 1987. MONTHLY magazine devoted
to motor boating in the UK. Unsolicited
contributions welcome: cruising articles, DIY
and human interest. No powerboat racing or any
sailing-related material. Approach by e-mail.

Motor Caravan Magazine

IPC Media Ltd, Leon House, 233 High Street,
Croydon CR9 1HZ
☎ 020 8726 8000
helen_avery@ipcmedia.com
www.motorcaravanmagazine.co.uk
Owner *IPC Media (A Time Warner Company)*
Editor *Helen Avery*
Circulation 12,000

Founded 1986. MONTHLY magazine with ideas
about where to go in your motor caravan, expert
advice to keep trips fun and practical advice on
vans and kit. Interested in features on touring,
unusual motorhomes, and activities people
pursue while out and about in their van. E-mail
the editor with ideas. 💷 £60 a page.

Motor Cycle News

Media House, Lynchwood, Peterborough
Business Park, Peterborough PE2 6EA
☎ 01733 468000 📠 01733 468028
MCN@bauerconsumer.co.uk
www.motorcylenews.com
Owner *Bauer Consumer Media*
Editor *Marc Potter*
Circulation 128,801

Founded 1955. WEEKLY. Interested in short
news stories and features on motorcyles and
motorcycle racing. Contact relevant desks direct.

Motorcaravan Motorhome Monthly (MMM)

PO Box 88, Tiverton EX16 7ZN
mmmeditor@warnersgroup.co.uk
www.mmmonline.co.uk
Owner *Warners Group Publications Plc*
Editor *Mike Jago*
Circulation 40,408

Founded 1966. MONTHLY. 'There's no money in
motorcaravan journalism but for those wishing

to cut their first teeth ...' Unsolicited mss welcome if relevant, but ideas in writing preferred in first instance.
FEATURES Caravan site reports. Max. 500 words.
TRAVEL Motorcaravanning trips (home and overseas). Max. 2,000 words.
NEWS Short news items for miscellaneous pages. Max. 200 words.
FICTION Must be motorcaravan-related and include artwork/photos if possible. Max. 2,000 words.
SPECIAL PAGES DIY – modifications to motorcaravans. Max. 1,500 words.
OWNER REPORTS Contributions welcome from motorcaravan owners. Contact the editor for requirements. Max. 2,000 words.
£ Payment varies.

Mountain Magic

27A High Street, Esher KT10 9RL
☎ 01372 468140   ℻ 01372 470765
info@goodholidayideas.com
Editor *John Hill*
Circulation 150,000

QUARTERLY magazine for those seeking healthy holidays away from beaches and cities, such as hill walking, trekking, fishing, windsurfing, etc. No backpacking or serious climbing. No unsolicited material but approach in writing or by e-mail with ideas and/or synopsis. £ Payment negotiable.

Mslexia (For Women Who Write)

PO Box 656, Newcastle upon Tyne NE99 1PZ
☎ 0191 261 6656
postbag@mslexia.demon.co.uk
www.mslexia.co.uk
Owner *Mslexia Publications Limited*
Editor *Daneet Steffens*
Circulation 10,700

Founded 1997. QUARTERLY. Articles, advice, reviews, interviews, events for women writers plus new poetry and prose. Will consider fiction, poetry, features and letters but contributors *must* send for guidelines first or see website for details.

Music Week

Ludgate House, 245 Blackfriars Road, London SE1 9UR
☎ 020 7921 5000
ben@musicweek.com
www.musicweek.com
Owner *CMP Information*
Publisher *Ben Hosken*
Editor *Paul Williams*
Circulation 7,960

Britain's only WEEKLY music business magazine. Also produces a news and data website. Annual music industry contacts book – the *Music Week*

Directory. Free Daily news e-mail. No unsolicited mss. Approach in writing with ideas.
FEATURES Analysis of specific music business events and trends.
NEWS Music industry news only.

Musical Opinion

50 Collinswood Drive, St Leonards on Sea TN38 0NX
☎ 01424 715167   ℻ 01424 712214
musicalopinion2@aol.com
www.musicalopinion.com
Owner *Musical Opinion Ltd*
Editor *Denby Richards*
Circulation 8,500

Founded 1877. Glossy, full-colour BI-MONTHLY magazine. Classical music content, with topical features on music, musicians, festivals, etc., and reviews (concerts, festivals, opera, ballet, jazz, CDs, DVDs, videos, books and printed music). International readership. No unsolicited mss; commissions only. Ideas always welcome though; approach by phone, fax or e-mail, giving telephone number. Visit the website for full information.
£ Payment negotiable.

My Weekly

80 Kingsway East, Dundee DD4 8SL
☎ 01382 223131   ℻ 01382 452491
myweekly@dcthomson.co.uk
www.dcthomson.co.uk
Owner *D.C. Thomson & Co. Ltd*
Editor *Sally Hampton*
Deputy Editor *Fiona Brown* (fbrown@dcthomson.co.uk)
Fiction Editor *Liz Smith* (lsmith@dcthomson.co.uk)
Circulation 162,874

WEEKLY title aimed at women of 50-plus. Mix of fiction, true-life stories, health, fashion, food, finance, gardening, travel. Ideas welcome. Approach in writing.
FEATURES Particularly interested in human interest pieces which appeal to the 50-plus age-group.
FICTION Three stories a week, ranging in content from the emotional to the off-beat and unexpected. E-mail for current guidelines.
£ Payment negotiable.

My Weekly Pocket Novels

D.C. Thomson & Co. Ltd, Albert Square, Dundee DD1 9QJ
☎ 01382 575810   ℻ 01382 322214
tsteele@dcthomson.co.uk
Owner *D.C. Thomson & Co. Ltd*
Editor *Tracey Steele*

WEEKLY. Publishes 25,000–30,000-word romantic stories aimed at the adult market. Send first three chapters and synopsis by post.

The National Trust Magazine

Heelis, Kemble Drive, Swindon SN2 2NA
℡ 01793 817716   🖷 01793 817401
magazine@nationaltrust.org.uk
Owner *The National Trust*
Editor *Sue Herdman*
Deputy Editor *Vicky Sartain*
Circulation 1.82 million

Founded 1968. THREE ISSUES YEARLY. Conservation of historic houses, coast and countryside in England, Northern Ireland and Wales. No unsolicited mss. Approach in writing with ideas.

The Naturalist

c/o University of Bradford, Bradford BD7 1DP
℡ 01274 234212   🖷 01274 234231
m.r.d.seaward@bradford.ac.uk
Owner *Yorkshire Naturalists' Union*
Editor *Prof. M.R.D. Seaward*
Circulation 5,000

Founded 1875. QUARTERLY. Natural history, biological and environmental sciences for a professional and amateur readership. Unsolicited mss and b&w illustrations welcome. Particularly interested in material – scientific papers – relating to the north of England. 🖤 No payment.

Nature

The Macmillan Building, 4–6 Crinan Street, London N1 9XW
℡ 020 7833 4000   🖷 020 7843 4596
nature@nature.com
www.nature.com/nature
Owner *Nature Publishing Group*
Editor *Philip Campbell*
Circulation 65,000

Covers all fields of science, with articles and news on science and science policy only. Scope only for freelance writers with specialist knowledge in these areas.

NB

105 Judd Street, London WC1H 9NE
℡ 020 7388 1266
www.rnib.org.uk/nbmagazine
Owner *Royal National Institute of Blind People*
Editor *Ann Lee*
Circulation 1,500

Founded 1917. MONTHLY. Published in print, braille and on audio CD and e-mail. Unsolicited mss accepted. Authoritative items by professionals or volunteers working in the field of eye health and sight loss welcome. Max. 1,500 words.
🖤 Payment negotiable.

New Humanist

1 Gower Street, London WC1E 6HD
℡ 020 7436 1151   🖷 020 7079 3588
editor@newhumanist.org.uk
www.newhumanist.org.uk
Owner *Rationalist Association*
Editor *Caspar Melville*
Circulation 5,000
Founded 1885. BI-MONTHLY. Unsolicited mss welcome. No fiction.

FEATURES Articles with a humanist perspective welcome in the following fields: religion (critical), humanism, human rights, philosophy, current events, literature, history and science. 2,000 words.
BOOK REVIEWS 750–1,000 words, by arrangement with the editor.
🖤 Payment for features is nominal, but negotiable.

New Internationalist

55 Rectory Road, Oxford OX4 1BW
℡ 01865 811400   🖷 01865 793152
ni@newint.org
www.newint.org/
Owner *New Internationalist Trust*
Co-Editors *Vanessa Baird, David Ransom, Adam Ma'anit, Jess Worth*
Circulation 80,000

Radical and broadly leftist in approach, but unaligned. Concerned with world poverty and global issues of peace and politics, feminism and environmentalism, with emphasis on the Third World. Difficult to use unsolicited material as they work to a theme each month and features are commissioned by the editor on that basis. The way in is to send examples of published or unpublished work; writers of interest are taken up.

New Musical Express

IPC Media Ltd., Blue Fin Building, 110 Southwark Street, London SE1 0SU
℡ 020 3148 5000   🖷 020 3148 8107
www.nme.com
Owner *IPC Media (A Time Warner Company)*
Editor *Conor McNicholas*
Circulation 64,033

Britain's best-selling musical WEEKLY. Now incorporates *Melody Maker*. Freelancers used, but always for reviews in the first instance. Specialization in areas of music is a help.
REVIEWS: ALBUMS *Julian Marshall* Send in examples of work, either published or specially written samples.

New Nation

Unit 2, 65 Whitechapel Road, London E1 1DU
℡ 020 7650 2000   🖷 020 7650 2001

lester@newnation.co.uk
www.newnation.co.uk
Owner *Ethnic Media Group*
Editor *Lester Holloway*
Circulation 73,000

Founded 1996. WEEKLY community paper for the black community in Britain. Interested in relevant general, local and international issues. Approach in writing with ideas for submission.

New Scientist

8th Floor, Lacon House, 84 Theobalds Road, London WC1X 8NS
☎ 020 8652 3500   🖷 020 7611 1250 (news)
www.newscientist.com
Owner *Reed Business Information Ltd*
Editor *Jeremy Webb*
Circulation 174,029 (Worldwide)

Founded 1956. WEEKLY. No unsolicited mss. Approach with ideas (one A4-page synopsis) by fax or e-mail.
FEATURES Commissions only, but good ideas welcome. Max. 3,500 words.
NEWS *Shaoni Bhattacharya* Mostly commissions, but ideas for specialist news welcome. Max. 1,000 words.
REVIEWS are commissioned.
OPINION Unsolicited material welcome if of general/humorous interest and related to science. Max. 1,000 words.
🖳 Payment negotiable.

The New Shetlander

Market House, 14 Market Street, Lerwick ZE1 0JP
☎ 01595 743902   🖷 01595 696787
scss@shetland.org
www.shetlandcss.co.uk
Owner *Shetland Council of Social Service*
Editors *Brian Smith, Laureen Johnson*
Circulation 1,400

Founded 1947. QUARTERLY literary magazine containing short stories, essays, poetry, historical articles, literary criticism, political comment, arts and books. The magazine has two editors and an editorial committee who all look at submitted material. Interested in considering short stories, poetry, historical articles with a northern Scottish or Scandinavian flavour, literary pieces and articles on Shetland. As a rough guide, items should be between 1,000 and 2,000 words although longer mss are considered. Initial approach in writing, please.
🖳 Complimentary copy.

New Statesman

3rd Floor, 52 Grosvenor Gardens, London SW1W 0AU
☎ 020 7730 3444   🖷 020 7259 0181

info@newstatesman.co.uk
www.newstatesman.com
Publisher *Spencer Neal*
Editor *Jason Cowley*
Circulation 26,208

WEEKLY magazine, the result of a merger (1988) of *New Statesman* and *New Society.* Coverage of news, book reviews, arts, current affairs, politics and social reportage. Unsolicited contributions with s.a.e. will be considered. No short stories.
Arts Editor *Alice O'Keefe*
Books Editor *Ian Irvine*

New Theatre Quarterly

Oldstairs, Kingsdown, Deal CT14 8ES
☎ 01304 373448
simontrussler@btinternet.com
www.uk.cambridge.org
Publisher *Cambridge University Press*
Editors *Simon Trussler, Maria Shevtsova*

Founded 1985 (originally launched in 1971 as *Theatre Quarterly*). Articles, interviews, documentation and reference material covering all aspects of live theatre. Recommend preliminary e-mail enquiry before sending contributions. No theatre reviews or anecdotal material.

New Welsh Review

PO Box 170, Aberystwyth SY23 1WZ
☎ 01970 628410   🖷 01970 628410
editor@newwelshreview.com
www.newwelshreview.com
Owner *New Welsh Review Ltd*
Editor *Francesca Rhydderch*
Circulation 3,500

Founded 1988. QUARTERLY Welsh literary magazine in the English language. Welcomes material of literary and cultural relevance to Welsh readers and those with an interest in Wales. Approach in writing in the first instance.
FEATURES Max. 3,000 words.
FICTION Max. 3,000 words.
REVIEWS Max. 800 words.
🖳 Average of £150 (features); £75 (fiction); £40 (reviews); £25 per poem.

The New Writer

PO Box 60, Cranbrook TN17 2ZR
☎ 01580 212626   🖷 01580 212041
editor@thenewwriter.com
www.thenewwriter.com
Publisher *Merric Davidson*
Editor *Suzanne Ruthven*

Founded 1996. Published BI-MONTHLY following the merger between *Acclaim* and *Quartos* magazines. Available by subscription only, TNW continues to offer practical 'nuts and bolts' advice on creative writing but with the emphasis on *forward-looking* articles and features on all aspects of the written word that demonstrate the writer's grasp of contemporary writing and current editorial/publishing policies. Plenty of news, views, competitions, reviews; writers' guidelines available with s.a.e. Monthly e-mail News bulletin is included free of charge in the subscription package.

FEATURES Unsolicited mss welcome. Interested in lively, original articles on writing in its broadest sense. Approach with ideas in writing in the first instance. No material is returned unless accompanied by s.a.e.

FICTION Publishes short-listed entries from competitions and subscriber-only submissions.

POETRY Unsolicited poetry welcome. Both short and long unpublished poems, providing they are original and interesting.

£ Features, £20 per 1,000 words; fiction, £10 per story; poetry, £3 per poem.

Full guidelines and details of annual Prose & Poetry Prizes at the website.

New Writing Scotland

Association for Scottish Literary Studies, c/o Department of Scottish Literature, 7 University Gardens, University of Glasgow G12 8QH
☎ 0141 330 5309   📠 0141 330 5309
nws@asls.org.uk
www.asls.org.uk
Contact *Duncan Jones*

ANNUAL anthology of contemporary poetry and prose in English, Gaelic and Scots, produced by the **Association for Scottish Literary Studies** (see entry under *Professional Associations and Societies*). Will consider poetry, drama, short fiction or other creative prose but not full-length plays or novels, though self-contained extracts are acceptable. Contributors should be Scottish by birth or upbringing, or resident in Scotland. Max. length of 3,500 words is suggested. Send no more than one short story and no more than four poems. Submissions should be accompanied by two s.a.e.s (one for receipt, the other for return of mss). Mss, which must be sent by 30 September, should be typed, double-spaced, on one side of the paper only with the sheets secured at top left-hand corner. Provide covering letter with full contact details but do not put name or address on individual work(s). Prose pieces should carry an approximate word count.

£ £20 per printed page.

New!

The Northern & Shell Building, 10 Lower Thames Street, London EC3R 6EN
☎ 0871 520 7016
karmel.doughty@express.co.uk
Owner *Richard Desmond*
Editor *Kirsty Mouatt*

Circulation 464,727

Founded 2003. Celebrity WEEKLY with true life content plus fashion and beauty. Celebrity interviews and true life submissions welcome. Approach by e-mail.

newbooks

4 Froxfield Close, Winchester SO22 6JW
guy@newbooksmag.com
www.newbooksmag.com
Owner/Editor *Guy Pringle*
Circulation 80,000

Founded 2000. BI-MONTHLY magazine for readers and reading groups with extracts from the best new fiction and free copies to be claimed. No unsolicited contributions. Also publishes *tBkmag* for 8–12-year-olds.

The North
▷ under Poetry Magazines

Notes from the Underground

23 Sutherland Square, London SE17 3EQ
☎ 020 7701 2777
info@notesfromtheunderground.co.uk
www.notesfromtheunderground.co.uk
Joint Editors *Christopher Vernon, Tristan Summerscale*

Free creative writing paper, aimed at bridging the gap between niche literary journals and those free newspapers that have a low quality of editorial content. Publishes short stories, cartoons and original illustrations as well as arts-based non-fiction. A print run of 100,000 copies. All the content from the print edition is available online along with web exclusive stories, videos and podcasts. Contributor information available on the website.

Now

IPC Media Ltd., Blue Fin Building, 110 Southwark Street, London SE1 0SU
☎ 020 3148 6373
Owner *IPC Media (A Time Warner Company)*
Editor *Abigail Blackburn*
Circulation 470,290

Founded 1996. WEEKLY magazine of celebrity gossip, news and topical features aimed at the working woman. Unlikely to use freelance contributions due to specialist content – e.g. exclusive showbiz interviews – but ideas will be considered. Approach in writing; no faxes.

Nursery Education

Scholastic Ltd, Villiers House, Clarendon Avenue, Leamington Spa CV32 5PR
☎ 01926 887799   📠 01926 883331
earlyyears@scholastic.co.uk
www.scholastic.co.uk
Owner *Scholastic Ltd*

Publishing Director *Helen Freeman*
Circulation 15,457

Founded 1997. MONTHLY magazine for early years professionals. Welcomes contributions from freelance journalists specializing in education.
FEATURES Early years issues explored. Max. 1,500 words.
NEWS Relevant and timely news for the early years sector. Max. 400 words.
PRACTICAL IDEAS Themed activities that support the Foundation Stage curriculum and Birth-to-Three activites. Max. 800 words. Approach by letter or e-mail.

Nursing Times

Greater London House, Hampstead Road, London NW1 7EJ
☏ 020 7728 3702   ☏ 020 7728 3700
nt@emap.com
www.nursingtimes.net
Owner *Emap Public Sector*
Editor *Rachel Downey*
Circulation 48,388

Some of *Nursing Times*' feature content is from unsolicited contributions sent on spec. Articles on all aspects of nursing and health care, including clinical information, written in a contemporary style, are welcome.

Nuts

IPC Media Ltd., Blue Fin Building, 110 Southwark Street, London SE1 0SU
☏ 020 3148 5000
nutsmagazine@ipcmedia.com
www.nuts.co.uk
Owner *IPC Media (A Time Warner Company)*
Editor *Dominic Smith*
Circulation 270,053

Launched 2004. WEEKLY magazine for men aged 16–45 'who like chat, women, gossip, cars and football'. Welcomes true life stories and items on gadgets. Approach by e-mail in the first instance.

OK! Magazine

The Northern & Shell Building, 10 Lower Thames Street, London EC3R 6EN
☏ 0871 434 1010   ☏ 0871 434 7305
firstname.lastname@express.co.uk
Owner *Northern & Shell Media/Richard Desmond*
Editor *Lisa Byrne*
Circulation 683,451

Founded 1996. WEEKLY celebrity-based magazine. Welcomes interviews and pictures on well known personalities, and ideas for general features. Approach the editor by phone or fax in the first instance.

Old Glory

Mortons Heritage Media, Mortons Way, Horncastle LN9 6JR
☏ 01507 529300   ☏ 01507 529495
editor@oldglory.co.uk
www.oldglory.co.uk
Owner *Mortons Media Ltd*
Editor *Colin Tyson*

Founded 1988. MONTHLY magazine covering all aspects of transport and industrial heritage both in the UK and overseas. Specializes in vintage vehicles and steam preservation. Unsolicited articles are welcome. Approach by letter addressed to the editor.
FEATURES Restoration projects from steam rollers to windmills. Max. 2,000 words.
NEWS Steam and transport rally reports. News concerning industrial and heritage sites and buildings.

The Oldie

65 Newman Street, London W1T 3EG
☏ 020 7436 8801   ☏ 020 7436 8804
theoldie@theoldie.co.uk
www.theoldie.co.uk
Owner *Oldie Publications Ltd*
Editor *Richard Ingrams*
Circulation 25,859

Founded 1992. MONTHLY general interest magazine with a strong humorous slant for the older person. Submissions welcome; enclose s.a.e. No poetry.

Olive

Media Centre, 201 Wood Lane, London W12 7TQ
☏ 020 8433 1389
firstname.surname@bbc.co.uk
www.olivemagazine.co.uk
Owner *BBC Worldwide Publishing Ltd*
Editorial Coordinator *Danielle Theunissen*
Deputy Editor *Lulu Grimes*
Features Editor *Jessica Gunn* (jess.gunn@bbc.co.uk)
Food Editor *Janine Ratcliffe*
Circulation 90,236

Launched 2003. MONTHLY food magazine. Recipes, restaurants and food lovers' travel. Unsolicited pitches are welcome 'but please note that we will reply *only* to those we plan to commission'.

Opera

36 Black Lion Lane, London W6 9BE
☏ 020 8563 8893   ☏ 020 8563 8635
editor@opera.co.uk
www.opera.co.uk
Owner *Opera Magazine Ltd*
Editor *John Allison*

Circulation 11,500

Founded 1950. MONTHLY review of the current opera scene. Almost all articles are commissioned and unsolicited mss are not welcome. All approaches should be made in writing.

Opera Now

241 Shaftesbury Avenue, London WC2H 8TF
℡ 020 7333 1740   🅵 020 7333 1769
opera.now@rhinegold.co.uk
www.rhinegold.co.uk
Publisher *Rhinegold Publishing Ltd*
Editor-in-Chief *Ashutosh Khandekar*
Deputy Editor *Antonia Couling*

Founded 1989. BI-MONTHLY. News, features and reviews aimed at those involved as well as those interested in opera. No unsolicited mss. All work is commissioned. Approach with ideas in writing.

Organic Gardening

Mortons Media Group, Media Centre, Morton Way, Horncastle LN9 6JR
slawson@mortons.co.uk
Editor *Sarah Lawson*
Circulation 20,000

Founded 1988. MONTHLY. Articles and features on all aspects of gardening based on organic methods. Unsolicited material welcome; 800–2,000 words for features and 100–300 for news items. Poetry not published. Prefers 'hands-on' accounts of projects, problems, challenges and how they are dealt with. Approach in writing.
🅴 Payment by arrangement.

OS (Office Secretary) Magazine

15 Grangers Place, Witney OX28 4BS
℡ 01993 775545   🅵 01993 778884
paul.ormond@peeblesmedia.com
www.peeblesmedia.com
Owner *Peebles Media Group Ltd*
Editor *Paul Ormond*
Circulation 24,976

Founded 1986. BI-MONTHLY. Features articles of interest to secretaries and personal assistants aged 25–60. No unsolicited mss.
FEATURES Informative pieces on technology and practices, office and employment-related topics. Length 1,000 words. Approach with ideas by telephone or in writing.
🅴 Payment by negotiation.

Park Home & Holiday Caravan

IPC Media Ltd, Leon House, 233 High Street, Croydon CR9 1HZ
℡ 020 8726 8252   🅵 020 8726 8299
em_bartlett@ipcmedia.com
www.phhc.co.uk
Owner *IPC Inspire*
Editor *Emma Bartlett*

Circulation 15,000

Founded 1960. THIRTEEN ISSUES YEARLY. News of parks, models, legislation, accessories, lifestyle for those living in residential park homes or owning holiday caravans. Welcomes material specifically related to park home and holiday caravan sites; no general touring features. Approach by e-mail.

PC Format

Future Publishing, 30 Monmouth Street, Bath BA1 2BW
℡ 01225 442244   🅵 01225 732295
pcfmail@futurenet.co.uk
www.pcformat.co.uk
Owner *Future Publishing*
Publisher *Stuart Anderton*
Editor *Adam Ifans*
Circulation 23,764

Founded 1991. FOUR-WEEKLY magazine covering performance hardware and gaming. Welcomes feature and interview ideas in the first instance; approach by e-mail only.

Pensions World

2 Addiscombe Road, Croydon CR9 5AF
℡ 020 8212 1946   🅵 020 8212 1920
stephanie.hawthorne@lexisnexis.co.uk
www.pensionsworld.co.uk
Owner *Reed Elsevier*
Editor *Stephanie Hawthorne*
Circulation 7,685

Launched 1972. MONTHLY magazine for pensions professionals. Pensions, law, investment and retirement issues. Specialist contributors only. No consumer pension stories. Approach by e-mail.
FEATURES Max. 1,500 words.
NEWS Max. 400 words.
🅴 Payment negotiable.

People Management

Personnel Publications Ltd., 17–18 Britton Street, London EC1M 5TP
℡ 020 7324 2729   🅵 020 7324 2791
editorial@peoplemanagement.co.uk
www.peoplemanagement.co.uk
Editor *Rima Evans*
Circulation 127,396

FORTNIGHTLY magazine on human resources, industrial relations, employment issues, etc. Welcomes submissions but apply for 'Guidelines for Contributors' in the first instance.
FEATURES *Jane Pickard/Claire Warren*
NEWS *Anna Scott*
LAW AT WORK *Jill Evans*

People's Friend Pocket Novels

D.C. Thomson & Co. Ltd, Albert Square, Dundee DD1 9QJ
℡ 01382 575938   🅵 01382 322214

shheron@dcthomson.co.uk
Owner *D.C. Thomson & Co. Ltd*
Editor *Sheelagh Heron*

Founded 1938. FORTNIGHTLY. Family and romantic stories of 50,000–55,000 words, aimed at the 30+ age group. Unsolicited contributions welcome. Phone, write or e-mail for guidelines; synopsis and first three chapters to be sent initially.

The People's Friend

80 Kingsway East, Dundee DD4 8SL
℡ 01382 223131   🖷 01382 452491
peoplesfriend@dcthomson.co.uk
Owner *D.C. Thomson & Co. Ltd*
Editor *Angela Gilchrist*
Circulation 335,486

The *Friend* is a fiction magazine, with two serials and several short stories each week. Founded in 1869, it has always prided itself on providing 'a good read for all the family'. Stories should be about ordinary, identifiable characters with the kind of problems the average reader can understand and sympathize with. 'We look for the romantic and emotional developments of characters, rather than an over-complicated or contrived plot. We regularly use period serials and, occasionally, mystery/adventure.' Guidelines on request with s.a.e.
SHORT STORIES Can vary in length from 1,000 words or less to 4,000.
SERIALS Serials of 8–12 instalments.
ARTICLES Short fillers welcome.
£ Payment on acceptance.

Period Living

St Giles House, 50 Poland Street, London
W1F 7AX
℡ 020 7970 4433   🖷 020 7970 4438
period.living@centaur.co.uk
Owner *Centaur*
Editor-in-Chief *Michael Holmes*
Editor *Sara Whelan*
Circulation 52,834

Founded 1992. Formed from the merger of *Period Living* and *Traditional Homes*. Covers interior decoration in a period style, period house profiles, traditional crafts, renovation of period properties.
£ Payment varies according to length/type of article.

The Philosopher

6 Craghall Dene Avenue, Newcastle upon Tyne
NE3 1QR
m.c.bavidge@ncl.ac.uk
http://atschool.eduweb.co.uk/cite/staff/
philosopher/philsocindex.htm
Owner *The Philosophical Society*
Editor *Martin Cohen*

Founded 1913. BIANNUAL journal of the Philosophical Society of Great Britain with an international readership made up of members, libraries and specialist booksellers. Wide range of philosophical interests, but leaning towards articles that present philosophical investigation which is relevant to the individual and to society in our modern era. Accessible to the non-specialist. Will consider articles and book reviews. Notes for Contributors available; see website. As well as short philosophical papers, will accept:
NEWS about lectures, conventions, philosophy groups. Ethical issues in the news. Max. 1,000 words.
REVIEWS of philosophy books (max. 600 words); discussion articles of individual philosophers and their published works (max. 2,000 words).
MISCELLANEOUS items, including graphics, of philosophical interest and/or merit.
£ Free copies.

Piano

241 Shaftesbury Avenue, London WC2H 8TF
℡ 020 7333 1724   🖷 020 7333 1769
piano@rhinegold.co.uk
www.rhinegold.co.uk
Owner *Rhinegold Publishing*
Editor *Jeremy Siepmann*
Deputy Editor *Kimon Daltas*
Circulation 11,000

Founded 1993. BI-MONTHLY magazine containing features, profiles, technical information, news, reviews of interest to those with a serious amateur or professional concern with pianos or their playing. No unsolicited material. Approach with ideas in writing only.

Picture Postcard Monthly

15 Debdale Lane, Keyworth, Nottingham
NG12 5HT
℡ 0115 937 4079   🖷 0115 937 6197
reflections@postcardcollecting.co.uk
www.postcardcollecting.co.uk
Owners *Brian & Mary Lund*
Editor *Brian Lund*
Circulation 4,000

Founded 1978. MONTHLY. News, views, clubs, diary of fairs, sales, auctions, and well-researched postcard-related articles. Might be interested in general articles supported by postcards. Unsolicited mss welcome. Approach by phone, e-mail or in writing with ideas.

Pilot

Archant Specialist, The Mill, Bearwalden Business Park, Wendens Ambo, Saffron Walden CB11 4GB
℡ 01799 544200   🖷 01799 544201
nick.bloom@pilotweb.co.uk
www.pilotweb.co.uk

Publisher *Archant Specialist*
Editor *Nick Bloom*
Circulation 19,181

Founded 1966. MONTHLY magazine for private plane pilots. Much of the magazine is written by invitation from regular contributors. Unsolicited mss welcome but ideas in writing preferred. Non-pilot authors unlikley to be considered. Perusal of any issue of the magazine will reveal the type of material bought. See the website for advice to would-be contributors.
NEWS Contributions, preferably with photographs, need to be as short as possible. See *Pilot Notes* and *Old-Timers* in the magazine.
FEATURES Many articles are unsolicited personal experiences/travel accounts from pilots of private planes; good photo coverage is very important. Max. 5,000 words.

Pink Paper

Unit M, Spectrum House, 32–34 Gordon House Road, London NW5 1LP
☎ 020 7424 7400   🖷 020 7424 7401
editorial@pinkpaper.com
Owner *Millivres Prowler Group*
Editor *Tris Reid-Smith*
Circulation 39,758

Founded 1987. FORTNIGHTLY. Only national newspaper for lesbians and gay men covering politics, social issues, health, the arts, celebrity interviews and all areas of concern to lesbian/gay people. Unsolicited mss welcome. Initial approach by post with an idea preferred. Interested in profiles, reviews, in-depth features and short news pieces.
£ Payment by arrangement.

Poetry Ireland Review
> under Poetry Magazines

Poetry Review
> under Poetry Magazines

Poetry Scotland
> under Poetry Magazines

Poetry Wales
> under Poetry Magazines

PONY

D.J. Murphy (Publishers) Ltd, Headley House, Headley Road, Grayshott GU26 6TU
☎ 01428 601020   🖷 01428 601030
djm@djmurphy.co.uk (text only)
www.ponymag.com
Owner *D.J. Murphy (Publishers) Ltd*
Editor *Janet Rising*
Assistant Editor *Penny Rendall*
Circulation 40,274

Founded 1948. Lively MONTHLY aimed at 10–16-year-olds. News, instruction on riding,

stable management, veterinary care, interviews. Approach in writing with an idea.
FEATURES welcome. Max. 900 words.
NEWS Written in-house. Photographs and illustrations (serious/cartoon) welcome.
£ £65 per 1,000 words.

post plus
> **Sunday Post under National Newspapers**

PR Week

174 Hammersmith Road, London W6 7JP
☎ 020 8267 4429   🖷 020 8267 4509
prweek@haymarket.com
www.prweek.com/uk
Owner *Haymarket Business Publications Ltd*
Editor *Daniel Rogers*
Circulation 16,587

Founded 1984. WEEKLY. Contributions accepted from experienced journalists. Approach in writing with an idea.
£ Payment negotiable.

Practical Boat Owner

IPC Media Ltd, Westover House, West Quay Road, Poole BH15 1JG
☎ 01202 440820   🖷 01202 440860
sarah_norbury@ipcmedia.com
pbo@ipcmedia.com
www.pbo.co.uk
Owner *IPC Media (A Time Warner Company)*
Editor *Sarah Norbury*
Circulation 48,202

Britain's biggest selling boating magazine. MONTHLY magazine of practical information for cruising boat owners. Receives about 1,500 mss per year.
FEATURES Technical articles about maintenance, restoration, modifications to cruising boats, power and sail up to 45ft. European and British regional pilotage articles and cruising guides. Approach in writing with synopsis in the first instance.
£ Payment negotiable.

Practical Caravan

Teddington Studios, Broom Road, Teddington TW11 9BE
☎ 020 8267 5629   🖷 020 8267 5725
practical.caravan@haymarket.com
www.practicalcaravan.com
Owner *Haymarket Consumer Media*
Editor *Nigel Donnelly*
Circulation 45,158

Founded 1967. MONTHLY. Contains caravan reviews, travel features, investigations, products, park reviews. Unsolicited mss welcome on travel relevant only to caravanning/touring vans. No motorcaravan or static van stories. Approach with ideas by phone, letter or e-mail.

FEATURES Must refer to caravanning. Written in friendly, chatty manner. Pictures essential. Max. length 2,000 words.
£ Payment negotiable.

Practical Fishkeeping
Bauer Active, Bushfield House, Bretton, Peterborough PE2 5UW
☏ 01733 237111 ⊞ 01733 465246
www.practicalfishkeeping.co.uk
Owner *Bauer Active*
Editor *Karen Youngs*
Circulation 15,446

THIRTEEN ISSUES YEARLY. Practical articles on all aspects of fishkeeping. Unsolicited mss welcome; approach in writing with ideas. Quality photographs of fish always welcome. No fiction or verse.

Practical Land Rover World
▷ Land Rover World

Practical Parenting
Magicalia Publishing Ltd, Berwick House, 8–10 Knoll Rise, Orpington BR6 0EL
☏ 01689 899032
Owner *Magicalia Publishing Ltd*
Editor *Susie Boone*
Circulation 35,616

Founded 1987. MONTHLY. Practical advice on pregnancy, birth, babycare and childcare, 0–4 years. Submit ideas/synopses by e-mail. Interested in feature articles of up to 3,000 words in length, and in readers' experiences/personal viewpoint pieces of between 750–1,000 words. All material must be written for the magazine's specifically targeted audience and in-house style. Submissions to: *Wendy Parker* (wendy.parker@magicalia.com)
£ Payment negotiable.

Practical Photography
Bauer Active, Media House PE2 6EA, Lynchwood, Peterborough PE2 6EA
☏ 01733 468000 ⊞ 01733 465246
practical.photography@bauerconsumer.co.uk
www.photoanswers.co.uk
Owner *Bauer Active*
Editor-in-Chief *Andrew James*
Circulation 62,303

MONTHLY All types of photography, particularly technique-orientated pictures. No unsolicited mss. Preliminary approach may be made by e-mail. Always interested in new, especially unusual, ideas.
FEATURES Anything relevant to the world of photography, but 'not the sort of feature produced by staff writers. Features on technology, digital imaging techniques and humour are three areas worth exploring. Bear in mind that there is a

three-month lead-in time.' Max. 2,000 words.
£ Payment varies.

Practical Wireless
Arrowsmith Court, Station Approach, Broadstone BH18 8PW
☏ 0845 803 1979 ⊞ 01202 659950
rob@pwpublishing.ltd.uk
www.pwpublishing.ltd.uk
Owner *PW Publishing Ltd*
Editor *Rob Mannion (G3XFD/EI5IW)*
Circulation 20,000

Founded 1932. MONTHLY. News and features relating to Amateur Radio, radio construction and radio communications. Author's guidelines available on request and are essential reading (send s.a.e.). Approach by phone or e-mail with ideas in the first instance. Only interested in hearing from people with an interest in and knowledge of Amateur Radio. Copy (on disk) should be supported where possible by artwork, either illustrations, diagrams or photographs.
£ £54–70 per page.

Practical Woodworking
Magicalia Publishing Ltd, Berwick House, 8–10 Knoll Rise, Orpington BR6 0EL
☏ 08444 122262 ⊞ 01689 899266
practicalwoodworking@magicalia.com
www.getwoodworking.com
Owner *Magicalia Publishing Ltd*
Editor *Neil Mead*
Circulation 7,000

Founded 1965. SIX ISSUES YEARLY. Contains articles relating to woodworking: projects, techniques, tips, etc. Unsolicited mss welcome. Approach with ideas in writing, by phone or e-mail.
FEATURES Projects, techniques, etc.
£ Payment by arrangement.

The Practitioner
245 Blackfriars Road, London SE1 9UY
☏ 020 7921 8113
cshort@cmpmedica.com
www.pulseportfolio.com
Owner *CMP Mediac*
Editor *Corinne Short*
Circulation 35,656

Founded 1868. MONTHLY publication that keeps General Practitioners up to date on clinical issues. All articles are independently commissioned.

Prediction
IPC Focus Network, Leon House, 233 High Street, Croydon CR9 1HZ
☏ 020 8726 8000 ⊞ 020 8726 8296
prediction@ipcmedia.com
www.predictionmagazine.co.uk
Owner *IPC Media (A Time Warner Company)*

Editor *Marion Williamson*
Circulation 13,718

Founded 1936. MONTHLY. Covering astrology and mind, body, spirit topics. Unsolicited material in these areas welcome (about 200–300 mss received every year). Writers' guidelines available on request.

ASTROLOGY Pieces should be practical and of general interest.

FEATURES Articles on divination, shamanism, alternative healing, psychics and other supernatural phenomena considered. Please read a recent copy of the magazine before sending unsolicited material.

Pregnancy & Birth

Endeavour House, 189 Shaftebury Avenue, London WC2H 8JG
☎ 020 7295 5563
p.andb@bauerconsumer.co.uk
Owner *Bauer London Lifestyle*
Acting Editor *Eleanor Dalrymple*
Circulation 44,152

MONTHLY magazine covering all aspects of pregnancy from health to fashion. Regularly commissions features from health journalists. Freelancers should approach by e-mail in the first instance.
£ Payment varies.

Pregnancy, Baby & You

Magicalia Publishing Ltd, Berwick House, 8–10 Knoll Rise, Orpington BR6 0EL
☎ 01689 899284
emma.hartfield@magicalia.com
Owner *Magicalia Publishing Ltd*
Editor *Emma Hartfield*
Deputy Editor *Susannah Parker*
Circulation 18,951

THIRTEEN ISSUES YEARLY focusing on issues affecting health and emotional issues during pregnancy. No unsolicited mss; e-mail the editor with ideas only.

Press Gazette

Paulton House, 8 Shepherdess Walk, London N1 7LB
☎ 020 7324 2385   ⓕ 020 7566 5769
pged@pressgazette.co.uk
www.pressgazette.co.uk
Owner *Wilmington Business Information*
Editor *Dominic Ponsford*
Circulation 5,010

WEEKLY magazine for all journalists – in regional and national newspapers, magazines, broadcasting and online – containing news, features and analysis of all areas of journalism, print and broadcasting. Unsolicited mss welcome; interested in profiles of magazines, broadcasting companies and news agencies, personality profiles, technical and current affairs relating to the world of journalism. Also welcomes gossip for diary page. Approach with ideas by phone, e-mail, fax or post.

Pride

Pride House, 55 Battersea Bridge Road, London SW11 3AX
☎ 020 7228 3110   ⓕ 020 7801 6717
info@pridemagazine.com
Owner/Editor *Carl Cushnie Junior*
Circulation 40,000

Founded 1991. MONTHLY lifestyle magazine for black women with features, beauty, arts and fashion. Approach in writing or by e-mail with ideas.

FEATURES Issues pertaining to the black community. 'Ideas and solicited mss are welcomed from new freelancers.' Max. 2,000 words.

FASHION & BEAUTY *Shevelle Rhule* Freelancers used for short features. Max. 1,000 words.

Prima

National Magazine House, 72 Broadwick Street, London W1F 9EP
☎ 020 7439 5000
prima@natmags.co.uk
www.natmags.co.uk
Owner *National Magazine Company*
Editor *Maire Fahey*
Circulation 284,659
Founded 1986. MONTHLY women's magazine.

HEALTH & FEATURES Coordinator *Carrie Buckle* Mostly practical and real life, written by specialists or commissioned from known freelancers. Unsolicited mss not welcome.

Private Eye

6 Carlisle Street, London W1D 3BN
☎ 020 7437 4017   ⓕ 020 7437 0705
strobes@private-eye.co.uk
www.private-eye.co.uk
Owner *Pressdram*
Editor *Ian Hislop*
Circulation 207,566

Founded 1961. FORTNIGHTLY satirical and investigative magazine. Prospective contributors are best advised to approach the editor in writing. News stories and feature ideas are always welcome, as are cartoons.

Prospect

2 Bloomsbury Place, London WC1A 2QA
☎ 020 7255 1344 (editorial)/1281 (publishing)
ⓕ 020 7255 1279
editorial@prospect-magazine.co.uk
publishing@prospect-magazine.co.uk
www.prospect-magazine.co.uk

Owner *Prospect Publishing Limited*
Editor *David Goodhart*
Circulation 26,767

Founded 1995. MONTHLY. Essays, reviews, short fiction and research on current/international affairs and cultural issues. No news features. Unsolicited contributions welcome, although more useful to approach in writing with ideas in the first instance.

Psychic News

The Coach House, Stansted Hall, Stansted CM24 8UD
☎ 01279 817050 ⊞ 01279 817051
pnadverts@btconnect.com
www.psychicnewsbookshop.co.uk
Owner *Psychic Press 1995 Ltd*
Editor *Susan Farrow*
Circulation 40,000

Founded 1932. *Psychic News* is the world's only WEEKLY spiritualist newspaper. It covers subjects such as psychic research, hauntings, ghosts, poltergeists, spiritual healing, survival after death and paranormal gifts. Unsolicited material considered. ⓔ No payment.

Psychologies

64 North Row, London W1K 7LL
☎ 020 7150 7000
psychologies@hf-uk.com
www.psychologies.co.uk
Owner *Hachette Filipacchi UK*
Editor *Maureen Rice*
Features Editor *Rebecca Alexander* (rebecca.alexander@hf-uk.com)
Assistant Editor *Sarah Maber* (sarah.maber@hf-uk.com)
Circulation 140,162

Founded 2005. MONTHLY. Women's lifestyle magazine covering relationships, family and parenting, health, beauty, social trends, travel, personality and behaviour, spirituality and sex. Consult the submissions guidelines on the website before submitting ideas for features.

Publishing News

7 John Street, London WC1N 2ES
☎ 0870 870 2345 ⊞ 0870 870 0385
mailbox@publishingnews.co.uk
www.publishingnews.co.uk
Owner *Publishing News Ltd*
Editor *Liz Thomson*
Deputy Editor *Roger Tagholm*

WEEKLY newspaper of the book trade. Hardback and paperback reviews and extensive listings of new paperbacks and hardbacks. Interviews with leading personalities in the trade, authors, agents, and features on specialist book areas. 'Almost no submissions accepted.'

Q

Mappin House, 4 Winsley Street, London W1W 8HF
☎ 020 7182 8482 ⊞ 020 7182 8547
firstname.surname@q4music.com
www.q4music.com
Owner *Bauer Consumer Media*
Editor *Paul Rees*
Circulation 131,330

Founded 1986. MONTHLY. Glossy aimed at educated popular music enthusiasts of all ages. Few opportunities for freelance writers. Prospective contributors should approach in writing only to the relevant section editor.

Q-News, The Muslim Magazine

PO Box 4295, London W1A 7YH
☎ 0870 224 5908
info@q-news.com
www.q-news.com
Owner *Fuad Nahdi*
Editor *Fareena Alam*
Circulation 40,000

Founded 1992. MONTHLY British Muslim community magazine concerned with shaping the debate about Muslims in Britain. Covers recruitment, news, opinion and reviews.
FEATURES *Abdul-Rehman Malik* 'Writers wishing to focus on areas of interest to our readership will gain access and credibility among the relevant people.' Max. 3,000 words.
OTHER *Fareena Alam* Analysis on news and current affairs – alternative rather than mainstream viewpoint preferred. Max. 2,000 words.
ⓔ No payment.

Quarterly Review

PO Box 36, Mablethorpe LN12 9AB
☎ 01507 339056 ⊞ 01507 339056
editor@quarterly-review.org
www.quarterly-review.org
Owner *Quarterly Review*
Editor *Derek Turner*

Founded 2007. A revival of the classic journal founded by Sir Walter Scott, Robert Southey and George Canning in 1809. QUARTERLY journal of politics, ideas and culture, with an emphasis on regionalism, small-scale economics, ecology and socio-biology. Approach in writing in the first instance. Articles from 1,000 – 8,000 words.
ⓔ No payment.

Quartos
▷ The New Writer

QWF Magazine

234 Brook Street, Unit 2, Waukesha WI 53188, USA
☎ 01788 334302

info@allwriters.org
qwfsubmissionsusa@yahoo.com
(submissions)
www.allwriters.org (click on QWF)

Semi-annual literary magazine published in
January and July. Founded in 1994 in the UK,
QWF is now published through AllWriters'
Workplace & Workshop in the US. Evoking
emotion is the most important characteristic of a
QWF story. There is no room for a dry reporting
of the facts here. The stories should present the
expanse that is every woman's emotional lifetime
and experience. Only considers fiction written
by women that are previously unpublished
and of less than 5,000 words. Submissions by
e-mail only. Submit only during reading periods:
September 1 and November 31 and March 1
and May 31. Submissions received outside of
the reading period will be returned unread. For
further information and detailed guidelines,
please access the website or contact the editor at
the e-mail address above.

RA Magazine (Royal Academy of Arts Magazine)

Royal Academy of Arts, Burlington House,
Piccadilly, London W1J 0BD
☎ 020 7300 5820   🖷 020 7300 5032
ramagazine@royalacademy.org.uk
www.ramagazine.org.uk
Owner *Royal Academy of Arts*
Editor *Sarah Greenberg*
Circulation 100,000

Founded 1983. QUARTERLY visual arts,
architecture and culture magazine distributed
to Friends of the RA, promoting exhibitions
and publishing articles on art shows, books
and news worldwide. 'There are opportunities
for freelancers who can write about art in an
accessible way. Most articles need to be tied in to
forthcoming exhibitions, books and projects.' No
unsolicited contributions. Approach by telephone
or e-mail.

Racecar Engineering

IPC Media Ltd, Leon House, 233 High Street,
Croydon CR9 1HZ
☎ 020 8726 8364   🖷 0208 726 8398
racecar@ipcmedia.com
www.racecar-engineering.co.uk
Owner *IPC Media (A Time Warner Company)*
Editor *Charles Armstrong-Wilson*
Circulation 15,000

Founded 1990. MONTHLY. In-depth features on
motorsport technology plus news and products.
Interested in receiving news and features from
freelancers.
FEATURES Informed insight into current
motorsport technology. 3,000 words max.

NEWS New cars, products or business news
relevant to motorsport. No items on road cars or
racing drivers. 500 words max. Call or e-mail to
discuss proposal.

Racing Post

1 Canada Square, Canary Wharf, London E14 5AP
☎ 020 7293 3000   🖷 020 7293 3758
editor@racingpost.co.uk
www.racingpost.co.uk
Owner *Centricomm Ltd*
Editor *Bruce Millington*
Circulation 62,983

Founded 1986. DAILY horse racing paper with
some general sport. In 1998, following an
agreement between the owners of *The Sporting
Life* and the *Racing Post*, the two papers merged.

Radio Times

Media Centre, 201 Wood Lane, London W12 7TQ
☎ 020 8433 3400   🖷 020 8433 3160
radio.times@bbc.co.uk
www.radiotimes.com
Owner *BBC Worldwide Limited*
Editor *Gill Hudson*
Circulation 1.05 million

WEEKLY. UK's leading broadcast listings
magazine. The majority of material is provided
by freelance and retained writers, but the
topicality of the pieces means close consultation
with editors is essential. Very unlikely to use
unsolicited material. Detailed BBC, ITV, Channel
4, Channel 5 and satellite television and radio
listings are accompanied by feature material
relevant to the week's output.
🖷 Payment by arrangement.

Radio User

Arrowsmith Court, Station Approach, Broadstone
BH18 8PW
☎ 0845 803 1979   🖷 01202 659950
elaine@pwpublishing.ltd.uk
www.pwpublishing.ltd.uk
Owner *PW Publishing Ltd*
Editor *Elaine Richards*
Circulation 19,000

Founded 1937. In 2006, *Shortwave Magazine*
merged with *Radio Active* magazine to become
Radio User. MONTHLY. Specialist electronics and
radio enthusiasts' magazine covering all aspects of
listening to radio signals. Features, news, regular
columns and projects. Will consider contributions
but recommends that contact be made prior to
submission to check publishing plans.

Rail

Brookfield House, Orton Centre, Peterborough
PE2 5UW
☎ 01733 237111   🖷 01733 288163

rail@emap.com
www.rail-magazine.co.uk
Owner *Emap Active Ltd*
Managing Editor *Nigel Harris*
Circulation 21,369

Founded 1981. FORTNIGHTLY magazine dedicated
to modern railway. News and features, and topical
newsworthy events. Unsolicited mss welcome.
Approach by phone with ideas. Not interested in
personal journey reminiscences. No fiction.
FEATURES By arrangement with the editor. All
modern railway British subjects considered. Max.
2,000 words.
NEWS Any news item welcome. Max. 500 words.
£ Features, varies/negotiable; news, up to £100
per 1,000 words.

Rail Express

Foursight Publications Ltd, 20 Park Street, King's
Cliffe, Peterborough PE8 6XN
☎ 01780 470086   ⊞ 01780 470060
editor@railexpress.co.uk
www.railexpress.co.uk
Owner/Editor *P.A. Sutton*
Circulation c. 13,000

Launched 1996. MONTHLY glossy dealing with
contemporary railway operations in the UK. Well
researched and illustrated news (max. 500 words)
and features (max. 4,000 words) that appeal to
the core enthusiast readership are considered.
Contact with brief in the first instance.

Railway Gazette International

NINE Sutton Court Road, Sutton SM1 4SZ
☎ 020 8652 5200   ⊞ 020 8652 5210
editor@railwaygazette.com
www.railwaygazette.com
Owner *DVV Media UK*
Editor *Chris Jackson*
Industry Editor *Andrew Gratham*
Circulation 9,945

Founded 1835. MONTHLY magazine for senior
railway, metro and tramway managers, engineers
and suppliers worldwide. 'No material for railway
enthusiast publications.' E-mail to discuss ideas in
the first instance.

The Railway Magazine

IPC Media Ltd., Blue Fin Building, 110 Southwark
Street, London SE1 0SU
☎ 020 3148 4683   ⊞ 020 3148 8521
railway@ipcmedia.com
www.ipcmedia.com
Owner *IPC Media (A Time Warner Company)*
Editor *Nick Pigott*
Circulation 34,661

Founded 1897. MONTHLY. Articles, photos and
short news stories of a topical nature, covering
modern railways, steam preservation and railway

history, welcome. Max. 2,000 words, with
sketch maps of routes, etc., where appropriate.
Unsolicited mss welcome. No poetry.
£ Payment negotiable.

Random Acts of Writing

Lower Knock-na-Shalavaig, Struy, Beauly IV4 7JU
randomactsofwriting@fsmail.net
www.randomactsofwriting.co.uk
Owners/Editors *Vikki Trelfer, Jennifer Thomson*
Circulation 200

Founded 2005. THREE ISSUES YEARLY. Promotes
the short story. 'Unsolicited short fiction (max.
3,000 words) welcomed from all, regardless of
previous publication history. Will give advice and
feedback.' Approach by e-mail.
£ Complimentary copy of the magazine plus
discounts on further copies.

Reader's Digest

11 Westferry Circus, Canary Wharf, London
E14 4HE
☎ 020 7715 8000   ⊞ 020 7715 8716
www.readersdigest.co.uk
Owner *Reader's Digest Association Ltd*
Editor *Sarah Sands*
Circulation 712,815

Although in theory, a good market for general
interest features of around 2,500 words very few
are ever accepted. However, 'a tiny proportion'
comes from freelance writers, all of which are
specially commissioned. Opportunities exist for
short humorous contributions to regular features
– 'Life', 'Laughter', '@Work'.
£ Up to £100.

The Reader

Reader Office, 19 Abercromby Square, Liverpool
L69 7ZG
☎ 0151 794 2830
readers@liv.co.uk
www.thereader.co.uk
Editor *Philip Davis*
Circulation 1,200

Founded 1997. QUARTERLY. Poetry, short
fiction, literary articles and essays, thought,
reviews, recommendations for individual
readers and reading groups. Contributions
from internationally lauded and new voices.
Welcomes articles/essays about reading and
about issues in modern life, max. 2,000 words.
Recommendations for good reading, max. 1,000
words. Short stories, max. 2,500 words. Aims
to reach serious readers in non-academic but
thoughtful style. Approach in writing.
£ Payment negotiable.

Readers' Review Magazine
▷ The Self Publishing Magazine

Record Collector

Room 101, The Perfume Factory, 140 Wales Farm Road, Acton, London W3 6UG
Ⓣ 020 8752 8080 Ⓕ 0870 732 6060
alan.lewis@metropolis.co.uk
www.recordcollectormag.com
Owner *Metropolis*
Editor *Alan Lewis*

Founded 1979. MONTHLY. Detailed, well-researched articles welcome on any aspect of record collecting or any collectable artist in the field of popular music (1950s to present day), with complete discographies where appropriate. Unsolicited mss welcome. Approach with ideas by phone or e-mail. Ⓔ Payment negotiable.

Red

64 North Row, London W1K 7LL
Ⓣ 020 7150 7000 Ⓕ 020 7150 7684
lindsay.frankel@hf-uk.com
www.redmagazine.co.uk
Owner *Hachette Filipacchi UK Ltd*
Editor *Sam Baker*
Circulation 224,208

Founded 1998. MONTHLY magazine aimed at the 30-something woman. Will consider ideas sent in 'on spec' but tends to rely on regular contributors.

Report

ATL, 7 Northumberland Street, London WC2N 5RD
Ⓣ 020 7930 6441 Ⓕ 020 7782 1618
www.atl.org.uk
Owner *Association of Teachers and Lecturers*
Editor *Alex Tomlin*
Managing Editor *Victoria Poskitt*
Circulation 160,000

Founded 1978. NINE ISSUES YEARLY during academic terms. Contributions welcome. All submissions should go directly to the editor. Articles should be no more than 800 words and must be of practical interest to the classroom teacher and F.E. lecturers.

Retail Week

Emap, Greater London House, Hampstead Road, London NW1 7EJ
Ⓣ 020 7728 5000
editorial@retail-week.com
www.retail-week.com
Owner *Emap*
Editor *Tim Danaher*
Circulation 10,260

Founded 1988. WEEKLY. Leading news publication for the multiple retail sector. No unsolicited contributions. Limited opportunities for both features and news. Approach by e-mail.
FEATURES *Liz Morrell* Max. 1,600 words.
NEWS *James Thompson*

Reveal

National Magazine Co., National Magazine House, 33 Broadwick Street, London W1F 0DQ
Ⓣ 020 7339 4500 Ⓕ 020 7339 4529
michael@acp-natmag.co.uk
www.natmags.co.uk
Owner *National Magazine Co.*
Editor *Michael Butcher*
Circulation 326,057

Launched 2004. WEEKLY glossy lifestyle magazine with celebrities, real-life stories, cookery, beauty and seven-day TV guide. No unsolicited contributions.

Rugby World

IPC Media Ltd., Blue Fin Building, 110 Southwark Street, London SE1 0SU
Ⓣ 020 3148 4708
paul_morgan@ipcmedia.com
www.rugbyworld.com
Owner *IPC Media (A Time Warner Company)*
Editor *Paul Morgan*
Circulation 48,381

Founded 1960. MONTHLY. Features of special rugby interest only. Unsolicited contributions welcome but s.a.e. essential for return of material. Prior approach by phone or in writing preferred.

Runner's World

33 Broadwick Street, London W1F 0DQ
Ⓣ 020 7339 4400 Ⓕ 020 7339 4220
rwedit@natmag-rodale.co.uk
www.runnersworld.co.uk
Owner *Natmag – Rodale Ltd*
Editor *Andy Dixon*
Circulation 88,567

Founded 1979. MONTHLY magazine giving practical advice on all areas of distance running including products and training, travel features, news and cross-training advice. Personal running-related articles, famous people who run or off-beat travel articles are welcome. Approach with ideas in writing in the first instance.

Running Fitness

1st Floor, South Wing, Broadway Court, Broadway, Peterborough PE1 1RP
Ⓣ 01733 353363 Ⓕ 01733 891342
rf.ed@kelsey.co.uk
www.runningfitnessmag.com
Owner *Kelsey Publishing*
Editor *David Castle*
Circulation 26,000

Founded 1985. MONTHLY. Instructional articles on running, fitness, and lifestyle, plus running-related activities and health.
FEATURES Specialist knowledge an advantage. Opportunities are wide, but approach with ideas in first instance.

NEWS Opportunities for people stories, especially if backed up by photographs.

Saga Magazine

Saga Publishing Ltd, The Saga Building, Enbrook Park, Sandgate, Folkstone CT20 3SE
☎ 01303 771523   ℻ 01303 776699
editor@saga.co.uk
www.saga.co.uk
Owner *Saga Publishing Ltd*
Editor *Kate Bravery*
Circulation 651,096

Founded 1984. MONTHLY magazine that sets out to celebrate the role of older people in society, reflecting their achievements, promoting their skills, protecting their interests, and campaigning on their behalf. A warm personal approach, addressing the readership in an up-beat and positive manner. It has a hard core of celebrated commentators/writers (e.g. Keith Waterhouse, Katharine Whitehorn, Sally Brampton) as regular contributors. Articles mostly commissioned or written in-house but exclusive celebrity interviews welcome if appropriate/relevant. Length 1,000–1,200 words (max. 1,600). No short stories, travel, fiction or poems, please.

Sailing Today

Swanwick Marina, Lower Swanwick, Southampton SO31 1ZL
☎ 01489 585225   ℻ 01489 565054
duncan.kent@sailingtoday.co.uk
www.sailingtoday.co.uk
Owner *Edisey Ltd*
Editor *Duncan Kent*

Founded 1997. MONTHLY practical magazine for cruising sailors. *Sailing Today* covers owning and buying a boat, equipment and products for sailing and is about improving readers' skills, boat maintenance and product tests. Most articles are commissioned but will consider practical features and cruising stories with photos. Approach by telephone or in writing in the first instance.

Sainsbury's Magazine

20 Upper Ground, London SE1 9PD
☎ 020 7775 7775   ℻ 020 7775 7705
edit@sainsburysmagazine.co.uk
Owner *Seven Publishing Ltd*
Editor *Sue Robinson*
Consultant Director *Delia Smith*
Circulation 327,637

Founded 1993. MONTHLY featuring a main core of food and cookery with features, health, beauty, travel, home and gardening. No unsolicited mss. Approach in writing with ideas only in the first instance.

The Salisbury Review

33 Canonbury Park South, London N1 2JW
☎ 020 7226 7791   ℻ 020 7354 0383
salisburyreview@tiscali.co.uk
www.salisburyreview.co.uk
Managing Editor *Merrie Cave*
Consulting Editors *Roger Scruton,
Sir Richard Body, Myles Harris, Lord Charles Cecil*
Literary Editor *Ian Crowther*
Circulation 1,700

Founded 1982. QUARTERLY magazine of conservative thought. Editorials and features from a right-wing viewpoint. Unsolicited material welcome. No fiction or poetry.
FEATURES Max. 2,000 words.
REVIEWS Max. 1,000 words.
Ⓔ Small payment.

Scotland in Trust

91 East London Street, Edinburgh EH7 4BQ
☎ 0131 556 2220   ℻ 0131 556 3300
editor@scotlandintrust.co.uk
www.scotlandintrust.co.uk
Owner *National Trust for Scotland*
Editor *Iain Gale*
Contact *Neil Braidwood*
Circulation 179,082

Founded 1983. THREE ISSUES YEARLY. Membership magazine of the National Trust for Scotland. Contains heritage/conservation features relating to the Trust, their properties and work. No unsolicited mss.

Scotland Means Business
▷ Daily Record under National Newspapers

The Scots Magazine

D.C. Thomson & Co., 2 Albert Square, Dundee DD1 9QJ
☎ 01382 223131
mail@scotsmagazine.com
www.scotsmagazine.com
Owner *D.C. Thomson & Co. Ltd*
Editor *John Methven*
Circulation 37,632

Founded 1739. MONTHLY. Covers a wide field of Scottish interests ranging from personalities to wildlife, climbing, reminiscence, history and folklore. Outside contributions welcome; 'staff delighted to discuss in advance by letter or e-mail'.
EVENTS LISTING *Christina Dolan*
PHOTOGRAPHY/MUSIC *Ian Neilson*
BOOKS *Alison Cook*

The Scottish Farmer

Newsquest Magazines, 200 Renfield Street, Glasgow G2 3QB
☎ 0141 302 7727   ℻ 0141 302 7799

ken.fletcher@thescottishfarmer.co.uk
www.thescottishfarmer.co.uk
Owner *Newsquest Magazines*
Editor *Alasdair Fletcher*
Circulation 19,700

Founded 1893. WEEKLY. Farmer's magazine covering most aspects of Scottish agriculture. Unsolicited mss welcome. Approach with ideas in writing, by fax or e-mail.
FEATURES *Ken Fletcher* Technical articles on agriculture or farming units. 1,000–2,000 words.
NEWS *Ken Fletcher* Factual news about farming developments, political, personal and technological. Max. 800 words.
WEEKEND FAMILY PAGES Rural and craft topics.

Scottish Field

Special Publications, Craigcrook Castle, Craigcrook Road, Edinburgh EH4 3PE
☎ 0131 312 4550   🖷 0131 312 4551
editor@scottishfield.co.uk
www.scottishfield.co.uk
Owner *Wyvex Media Ltd*
Editor *Archie Mackenzie*
Circulation 14,106

Founded 1903. MONTHLY. Scotland's quality lifestyle magazine. Unsolicited mss welcome but writers should study the magazine first.
FEATURES Articles of general interest on Scotland and Scots abroad with good photographs or, preferably, e-mailed images at 300 DPI. Approx. 1,000 words.
£ Payment negotiable.

Scottish Home & Country

42 Heriot Row, Edinburgh EH3 6ES
☎ 0131 225 1724   🖷 0131 225 8129
magazine@swri.demon.co.uk
www.swri.org.uk
Owner *Scottish Women's Rural Institutes*
Editor *Liz Ferguson*
Circulation 11,000

Founded 1924. MONTHLY. Scottish or rural-related issues, health, travel, women's issues and general interest. Unsolicited mss welcome. Commissions are rare and tend to go to established contributors only.

Scouting Magazine

Gilwell House, Gilwell Park, Chingford E4 7QW
☎ 020 8433 7100   🖷 020 8433 7103
scouting.magazine@scout.org.uk
www.scouts.org.uk
Owner *The Scout Association*
Editors *Chris James, Hilary Galloway, Matthew Oakes*
Circulation 80,000

BI-MONTHLY magazine for adults connected to or interested in the Scout Movement. Interested in Scouting-related submissions only. Sent to Members of the Association; not available on subscription
£ Payment by negotiation.

Screen

Gilmorehill Centre for Theatre, Film and Television, University of Glasgow, Glasgow G12 8QQ
☎ 0141 330 5035   🖷 0141 330 3515
screen@arts.gla.ac.uk
www.screen.arts.gla.ac.uk
screen.oxfordjournals.org
Publisher *Oxford University Press*
Editorial Office *Caroline Beven, Elizabeth Anderson*
Editors *Annette Kuhn, John Caughie, Simon Frith, Karen Lury, Jackie Stacey, Sarah Street*
Circulation 1,200

QUARTERLY academic journal of film and television studies for a readership ranging from undergraduates to screen studies academics and media professionals. There are no specific qualifications for acceptance of articles. Straightforward film reviews are not normally published. Check the magazine's style and market in the first instance.

Screen International

Greater London House, Hampstead Road, London NW1 7EJ
☎ 020 7728 5605   🖷 020 7728 5555
michael.gubbins@emap.com
www.screendaily.com
Owner *Emap Communications*
Editor *Michael Gubbins*
Circulation 7,352

The voice of the international film industry. Expert freelance writers used in all areas. No unsolicited mss. Approach by e-mail.
NEWS *Wendy Mitchell*
FEATURES *Rachel Nouchi, Chris Evans*
REVIEWS *Finn Halligan*
BOX OFFICE *Diana Lodderhose*
£ Negotiable on NUJ basis.

Screentrade Magazine

Screentrade Media Ltd, PO Box 144, Orpington BR6 6LZ
☎ 01689 833117   🖷 01689 833117
philip@screentrademagazine.co.uk
philip45other@yahoo.co.uk
www.screentrademagazine.co.uk
Owner *Screentrade Media Ltd*
Editor *Philip Turner*
US Representative, Screentrade Media Ltd
Pamala Stanton (pamscreentrade@aol.com);
☎ 001 616 847 0144/616 405 2643

Circulation 3,000+

Founded 2002. QUARTERLY journal for British, European and North American exhibitors and film distributors. Now also serving Russia, China and India.

FEATURES Items on cinema management, technical, cinema building history, nostalgia, showmanship, book reviews, concession supply, ticketing issues, trade articles including seating, screening, projection/career tips and other managerial matters. Also film and producer/director opinions. Some film reviews but no 'film star' interviews. Most contributions are from within the industry. Items on the state of cinema exhibition, film distribution, cinema architecture, interviews with key industry personnel frequently undertaken. 1,500–3,000 words.

NEWS Topical items (if substantiated) welcome. Events coverage (e.g. festivals, premières) from an exhibitor's viewpoint preferred.

ScriptWriter Magazine

2 Elliott Square, London NW3 3SU
℡ 020 7586 4853　🅕 020 7586 4853
info@scriptwritermagazine.com
www.scriptwritermagazine.com
Owner *Scriptease Ltd*
Managing Editor *Jonquil Florentin*
Editor *Julian Friedmann*
Circulation 1,500

Launched 2001. SIX ISSUES PER YEAR. Magazine for professional scriptwriters covering all aspects of the business and craft of writing for the small and large screen. Interested in serious, in-depth analysis; max. 1,500–3,500 words. E-mail with synopsis sample material and c.v.

Sea Breezes

Mannin Media Group Ltd, Media House, Cronkbourne IM4 4SB
℡ 01624 696573　🅕 01624 661655
seabreezes@manninmedia.co.im
www.seabreezes.co.im
Owner *Print Centres*
Editor *Captain A.C. Douglas*
Circulation 17,000

Founded 1919. MONTHLY. Covers virtually everything relating to ships and seamen. Unsolicited mss welcome; they should be thoroughly researched and accompanied by relevant photographs. No fiction, poetry, or anything which 'smacks of the romance of the sea'.
FEATURES Factual tales of ships, seamen and the sea, Royal or Merchant Navy, sail or power, nautical history, shipping company histories, epic voyages, etc. Length 1,000–4,000 words. 'The most readily acceptable work will be that which shows it is clearly the result of first-hand experience or the product of extensive and

accurate research.' 🅔 £14 per page (about 800 words).

The Self Publishing Magazine (incorporating Readers' Review Magazine)

9 De Montfort Mews, Leicester LE1 7FW
℡ 0116 255 9311　🅕 0116 255 9323
selfpublishing@troubador.co.uk
www.selfpublishingmagazine.co.uk
Owner *Troubador Publishing Ltd*
Editor *Jane Rowland*
Circulation 2,000

Launched 2006. THREE ISSUES YEARLY. Articles on self publishing. Submissions welcome on aspects of authors publishing their own books (design, production, editing, print, marketing). Approach by e-mail. Max. 2,000 words.
BOOK REVIEWS Self-published books only. Max. 300 words.

She Magazine

National Magazine House, 72 Broadwick Street, London W1F 9EP
℡ 020 7439 5000　🅕 020 7312 3940
www.allaboutyou.com
Owner *National Magazine Company Ltd*
Editor *Sian Rees*
Circulation 163,689

Glossy MONTHLY for the 35-plus woman, addressing her needs as a modern individual, a parent and a homemaker. Talks to its readers in an intelligent, humorous and informative way.
FEATURES Approach with feature ideas in writing or e-mail to Deputy Editor, *Carmen Bruegmann* (carmen.bruegmann@natmags.co.uk). No unsolicited material.
🅔 Payment negotiable.

Ships Monthly

IPC Inspire, 222 Branston Road, Burton-upon-Trent DE14 3BT
℡ 01283 542721　🅕 01283 546436
shipsmonthly@ipcmedia.com
www.shipsmonthly.com
Owner *IPC Inspire*
Editor *Iain Wakefield*
Deputy Editor *Nicholas Leach*
Circulation 22,000

Founded 1966. MONTHLY A4 format magazine for ship enthusiasts and maritime professionals. News, photographs and illustrated articles on all kinds of ships – mercantile and naval, past and present. No yachting. Welcomes contributions on company histories, classic and modern ships, ports and harbours. Most articles are commissioned; prospective contributors should telephone or e-mail in the first instance.

Shoot

IPC Media Ltd., Blue Fin Building, 110 Southwark Street, London SE1 0SU
℡ 020 3148 4727   🖷 020 3148 8130
shoot@ipcmedia.com
www.shoot.co.uk
Owner *IPC Media (A Time Warner Company)*
Editor *Frank Tennyson*
Circulation 35,830

Founded 1969. WEEKLY football magazine aimed at younger readers. No unsolicited mss. Present ideas for news, features or colour photo-features to the editor by letter or e-mail. Very limited opportunities for freelance work.
💷 Payment negotiable.

Shooting and Conservation

BASC, Marford Mill, Rossett, Wrexham LL12 0HL
℡ 01244 573000   🖷 01244 573001
jeffrey.olstead@basc.org.uk
www.basc.org.uk
Owner *The British Association for Shooting and Conservation (BASC)*
Editor *Jeffrey Olstead*
Circulation 125,000

SIX ISSUES PER YEAR. Good articles and stories on shooting, conservation and related areas may be considered although most material is produced in-house. Max. 1,500 words.
💷 Payment negotiable.

The Shooting Gazette

PO Box 225, Stamford PE9 2HS
℡ 01780 485350   🖷 01780 485390
will_hetherington@ipcmedia.com
Owner *IPC Media (A Time Warner Company)*
Editor *Will Hetherington*
Circulation 17,200

Launched 1989. MONTHLY aimed at those who enjoy driven game shooting in the UK and around the world.
FEATURES *Will Hetherington* Welcomes articles directly related to interesting or unusual game shooting stories; max. 2,000 words. Approach by e-mail. NEWS *Martin Puddifer* Max. 500 words.

Shooting Times & Country Magazine

IPC Media Ltd., Blue Fin Building, 110 Southwark Street, London SE1 0SU
℡ 020 3148 4741
steditorial@ipcmedia.com
www.shootingtimes.co.uk
Owner *IPC Media (A Time Warner Company)*
Editor *Camilla Clark*
Circulation 26,614

Founded 1882. WEEKLY. Covers shooting, fishing and related countryside topics. Unsolicited contributions considered.
💷 Payment negotiable.

Shout Magazine

D.C. Thomson & Co., Albert Square, Dundee DD1 9QJ
℡ 01382 223131   🖷 01382 200880
shout@dcthomson.co.uk
www.shoutmag.co.uk
Owner *D.C. Thomson Publishers*
Editor *Maria T. Welch*
Circulation 86,618

Founded 1993. FORTNIGHTLY. Pop music, quizzes, emotional, beauty, fashion, soap features.

Shout!

PO Box YR46, Leeds LS9 6XG
℡ 0113 248 5700   🖷 0113 295 6097
shout.magazine@ntlworld.com
www.shoutweb.co.uk
Owner/Editor *Mark Michalowski*
Circulation 7,000

Founded 1995. MONTHLY lesbian/gay and bisexual news, views, arts and scene for Yorkshire; lgb health and politics. Interested in reviews of Yorkshire lgb events, happenings, news, analysis – 300 to 1,000 words max. No fiction, fashion or items with no reasonable relevance to Yorkshire and the north.
💷 £40 per 1,000 words.

Showing World

Robin Aldwood Publications Ltd, PO Box 793, Needham IP6 8WN
℡ 01449 722505
info@showingworldonline.co.uk
www.showingworldonline.co.uk
Owner *Robin Aldwood Publications Ltd*
Editor *Sandy Lee-Wooderson*
Circulation 10,000

Founded 1991. SIX ISSUES YEARLY Features every aspect of showing horses and ponies plus natives, miniatures and donkeys. Knowledgeable and how-to-do-it articles welcome. Photos essential.
💷 Payment negotiable.

Shropshire Magazine

Waterloo Road, Telford TS1 5HU
℡ 01952 288822 (advertising)   🖷 01952 288820
Owner *Shropshire Newspapers Ltd*
Editor *Henry Carpenter*
Founded 1950. MONTHLY. Unsolicited mss welcome but ideas in writing preferred.
FEATURES Personalities, topical items, historical (e.g. family) of Shropshire; also general interest: homes, weddings, antiques, etc. Max. 1,000 words.
💷 Payment negotiable 'but modest'.

Sight & Sound

BFI, 21 Stephen Street, London W1T 1LN
℡ 020 7255 1444   🖷 020 7436 2327

s&s@bfi.org.uk
www.bfi.org.uk/sightandsound
Owner *BFI*
Editor *Nick James*
Circulation 20,283

Founded 1932. MONTHLY. Topical and critical articles on international cinema, with regular columns from the USA and Europe. Length 1,000–5,000 words. Relevant photographs appreciated. Also book, film and DVD release reviews. Approach in writing with ideas.
Ⓔ Payment by arrangement.

The Skier and Snowboarder Magazine

Mountain Marketing Ltd., PO Box 386, Sevenoaks TN13 1AQ
☎ 0845 310 8303
frank.baldwin@skierandsnowboarder.co.uk
www.skierandsnowboarder.co.uk
Publisher *Mountain Marketing Ltd*
Editor *Frank Baldwin*
Circulation 30,000

SEASONAL (from July to May). FIVE ISSUES YEARLY. Outside contributions welcome.
FEATURES Various topics covered, including race and resort reports, fashion, equipment update, artificial slope, club news, new products, health and safety. Crisp, tight, informative copy of 800 words or less preferred.
NEWS All aspects of skiing news covered.
Ⓔ Payment negotiable.

Skymag

News Magazines Ltd, 2 Chelsea Manor Gardens, London SW3 5PN
☎ 020 7198 3000   🖷 020 7198 3232
sky@newsmagazines.co.uk
www.sky.com/skymag
Owner *News Magazines*
Editor *Lysanne Currie*
Circulation 7.03 million

MONTHLY lifestyle and entertainment magazine. Contributions of celebrity interviews and real-life stories welcome. Send by e-mail to the deputy editor, *Nick Chalmers* (nick.chalmers@newsmagazines.co.uk).

Slightly Foxed: The Real Reader's Quarterly

67 Dickinson Court, 15 Brewhouse Yard, London EC1V 4JX
☎ 020 7549 2121/2111   🖷 0870 199 1245
all@foxedquarterly.com
www.foxedquarterly.com
Owner *Slightly Foxed Ltd*
Editors *Gail Pirkis, Hazel Wood*
Circulation 5,000

Founded 2004. 'Reviews of books (fiction and non-fiction) that have stood the test of time or books that have been published recently and are of real quality but which have been overlooked by reviewers and bookshops.' Unsolicited contributions of 'lively, personal, idiosyncratic writing of real quality' are welcome but it is recommended that would-be contributors read the magazine first to gauge its approach. Send e-mail with sample work in the first instance. Not interested in anything that is not actually a review of a book or author.

Slim at Home

25 Phoenix Court, Hawkins Road, Colchester CO2 8JY
☎ 01206 505972
emma@aceville.co.uk
www.slimathome.co.uk
Owner *Aceville Publications (2001) Ltd*
Editor *Rachel Callen*
Circulation 110,000

Launched 2007. MONTHLY National women's magazine for dieters who want to lose weight in their own way.
FEATURES *Naomi Abeykoon* Targeted weight-loss articles must be highly researched and of use to people who want to lose weight. No unrelated health features, quick tips or reviews of products and services. Approach by e-mail.
CELEBS *Rachel Callen* Current celebrity weight-loss stories commissioned. Pictures or shoot must be available.

Smallholder

Editorial: Hook House, Wimblington March PE15 0QL
☎ 01354 741538   🖷 01354 741182
liz.wright1@btconnect.com
www.smallholder.co.uk
Owner *Newsquest Plc*
Editor *Liz Wright*
Circulation 20,000

Founded 1982. MONTHLY. Outside contributions welcome. Send for sample magazine and editorial schedule before submitting anything. Follow up with samples of work to the editor so that style can be assessed for suitability. No poetry or humorous, unfocused personal tales; no puzzles.
FEATURES New writers always welcome, but must have high level of technical expertise – 'not textbook stuff'. 'How to do it' articles with photos welcome. Illustrations and photos paid for. Length 750–1,500 words.
NEWS All agricultural and rural news welcome. Length 200–500 words.
Ⓔ Payment negotiable.

SmartLife International

21–23 Phoenix Court, Hawkins Road, Colchester CO2 8JY
☎ 01206 505924   🖷 01206 505929

stuart@smartlifeint.com
www.smartlifeint.com
Owner *Aceville Publications Ltd*
Editor *Stuart Pritchard*
Circulation 19,411

Launched 2000. TEN ISSUES YEARLY. High-end
lifestyle magazine: consumer technology, interior
design, luxury cars, travel, fashion and grooming.
Welcomes proposals by e-mail.
FEATURES *Jake Stow* 'No good feature ideas that
fit the remit are ruled out.' Max. 2,000 words.
NEWS *Callum Fauser* Generally compiled in-
house.

Snooker Scene

Hayley Green Court, 130 Hagley Road, Hayley
Green, Halesowen, Birmingham B63 1DY
☎ 0121 585 9188   ✆ 0121 585 7117
clive.everton@talk21.com
www.snookersceneonline.com
Owner *Everton's News Agency*
Editor *Clive Everton*
Circulation 10,000

Founded 1971. MONTHLY. No unsolicited mss.
Approach in writing with an idea.

Somerset Life

Archant House, Babbage Road, Totnes TQ9 5JA
☎ 01803 860910   ✆ 01803 860926
somersetlifeinfo@archant.co.uk
www.somerset-life.co.uk
Owner *Archant Life*
Editor *Nicki Lukehurst*
Circulation 14,000

Launched 1994. MONTHLY magazine that
celebrates Somerset with coverage of towns,
villages, the county's history, property, antiques,
fashion, interior design, gardens and motoring.
FEATURES *Nicki Lukehurst* Prefers picture/
word packages, particularly on food and drink,
countryside, heritage, health and beauty, people,
business and the arts. 'Always on the lookout
for new voices (particularly if also a competent
photographer) on a regular or one-off freelance
basis.' Max. 1,500 words. Approach by e-mail. £
Fee negotiable.

Spear's Wealth Management Survey

Luxury Publishing Ltd., 5 Jubilee Place, London
SW3 3TD
☎ 020 7591 2900   ✆ 020 7591 2929
info@luxurypublishing.com
www.luxurypublishing.com
Editor *William Cash*
CEO *Alan O'Sullivan*
Circulation 25,000

Launched 2005. QUARTERLY specialist guide for
'high and ultra-high net worth' individuals to all
aspects of wealth management and the super-rich

lifestyle. Subscription is by invitation, only. No
unsolicited contributions.

Speciality Food

21–23 Phoenix Court, Hawkins Road, Colchester
CO2 8JY
☎ 01206 505981   ✆ 01206 505945
nicola.mallett@aceville.co.uk
www.specialityfoodmagazine.co.uk
Owner *Aceville Publications Ltd*
Editor *Nicola Mallett*
Circulation 8,713

Founded 2002. NINE ISSUES PER YEAR. Trade
magazine focusing on premium food and drink
relevant for the fine food retail sector. Unsolicited
contributions not generally welcome.
FEATURES Subjects such as cheese, dairy and
chocolate; also organic food. Approach by e-mail.
Max. 1,800 words. £ Payment negotiable.

The Spectator

22 Old Queen Street, London SW1H 9HP
☎ 020 7961 0020   ✆ 020 7961 0058
editor@spectator.co.uk
www.spectator.co.uk
Owner *The Spectator (1828) Ltd*
Editor *Matthew d'Ancona*
Books *Mark Amory*
Style Editor *Sarah Standing*
Business Editor *Martin Vander Weyer*
Art Editor *Liz Anderson*
Circulation 75,633

Founded 1828. WEEKLY political and literary
magazine. Prospective contributors should
write in the first instance to the relevant editor.
Unsolicited mss welcome, but no 'follow up'
phone calls, please.
£ Payment nominal.

Spectrum
▷ **Scotland on Sunday under National Newspapers**

Staffordshire Life

The Publishing Centre, Derby Street, Stafford
ST16 2DT
☎ 01785 257700   ✆ 01785 253287
editor@staffordshirelife.co.uk
www.staffordshirelife.co.uk
Owner *Staffordshire Newspapers Ltd*
Editor *Louise Elliott*
Circulation 15,500

Founded 1982. MONTHLY. Full-colour county
magazine devoted to Staffordshire, its
surroundings and people. Contributions welcome.
Approach in writing with ideas.
FEATURES Max. 1,200 words.
£ NUJ rates.

Stage

Stage House, 47 Bermondsey Street, London SE1 3XT

☎ 020 7403 1818   📠 020 7357 9287

editor@thestage.co.uk

www.thestage.co.uk

Owner *The Stage Newspaper Ltd*

Editor *Brian Attwood*

Circulation 23,184

Founded 1880. WEEKLY. No unsolicited mss. Prospective contributors should write with ideas in the first instance.

FEATURES Preference for middle-market, tabloid-style articles. 'Puff pieces', PR plugs and extended production notes will not be considered. Max. 800 words. Profiles: 1,200 words.

NEWS News stories from outside London are always welcome. Max. 300 words.

💷 £100 per 1,000 words.

Stamp Lover

National Philatelic Society, 107 Charterhouse Street, London EC1M 6PT

☎ 020 7490 9610

nps@ukphilately.org.uk

www.ukphilately.org.uk/nps

Owner *National Philatelic Society*

Editor *David Alford*

Circulation 800

Founded 1908. SIX ISSUES YEARLY. Magazine of the National Philatelic Society. Welcomes articles about the hobby; stamps, old and new. Approach by e-mail.

NEWS Information about worldwide philately. Max. 1,000 words. 💷 No payment.

Stamp Magazine

IPC Media Ltd, Leon House, 233 High Street, Croydon CR9 1HZ

☎ 020 8726 8243   📠 020 8726 8299

guy_thomas@ipcmedia.com

Owner *IPC Media (A Time Warner Company)*

Editor *Guy Thomas*

Circulation 12,000

Founded 1934. MONTHLY news and features on the world of stamp collecting from the past to the present day. Interested in articles by experts on particular countries or themes such as subject matter illustrated on stamps – dogs, politics, etc. Approach in writing.

NEWS *Julia Lee* News of latest stamp issues or industry news. Max. 500 words.

FEATURES *Guy Thomas* Any features welcome on famous stamps, rarities, postmarks, postal history, postcards, personal collections. Must be illustrated with colour images ('we can arrange for photography of original stamps').

💷 Payment negotiable.

Steam Railway Magazine

Bushfield House, Peterborough PE2 5UW

☎ 01733 288026

steam.railway@emap.com

Owner *Emap Active Limited*

Editor *Danny Hopkins*

Deputy Editor *Richard Foster*

Circulation 32,889

Founded 1979. FOUR-WEEKLY magazine targeted at all steam enthusiasts interested in the modern preservation movement. Unsolicited material welcome. News reports, photographs, steam-age reminiscences. Approach in writing or by e-mail.

Stella Magazine

▶ The Sunday Telegraph under National Newspapers

The Strad

Orpheus Publications, Newsquest Specialist Media Ltd, 2nd Floor, 30 Cannon Street, London EC4M 6YJ

☎ 020 7618 3095   📠 020 7618 3483

thestrad@orpheuspublications.com

www.thestrad.com

Owner *Newsquest Specialist Media Group*

Editor *Ariane Todes*

Circulation 17,500

Founded 1890. MONTHLY for classical string musicians, makers and enthusiasts. Unsolicited mss accepted occasionally 'though acknowledgement/return not guaranteed'.

FEATURES Profiles of string players, teachers, luthiers and musical instruments, also relevant research. Max. 2,000 words.

REVIEWS *Matthew Rye*

💷 £150 per 1,000 words.

Stuff

Teddington Studios, Broom Road, Teddington TW11 9BE

☎ 020 8267 5036

stuff@haymarket.com

www.stuff.tv

Owner *Haymarket Publishing Ltd*

Editor *Fraser MacDonald*

Circulation 96,866

Launched 1998. MONTHLY guide to new technology with information on all the latest gadgets and gear. Includes a 'Buyers' Guide' with information about the top tried and tested items. No unsolicited contributions.

Suffolk and Norfolk Life

The Publishing House, Framlingham IP13 9EE

☎ 01728 622030   📠 01728 622031

Owner *Today Magazines Ltd*

Editor *Kevin Davis*

Circulation 17,000

Founded 1989. MONTHLY. General interest, local stories, historical, personalities, wine, travel, food. Unsolicited mss welcome. Approach by phone or in writing with ideas. Not interested in anything which does not relate specifically to East Anglia.
FEATURES *Kevin Davis* Max. 1,500 words, with photos.
NEWS *Kevin Davis* Max. 1,000 words, with photos.
SPECIAL PAGES *William Locks* Study the magazine for guidelines. Max. 1,500 words.
Ⓔ £45–£70.

Sugar Magazine

64 North Row, London W1K 7LL
Ⓣ 020 7150 7087
www.sugarmagazine.co.uk
Owner *Hachette Filipacchi (UK)*
Editor *Annabel Brog*
Circulation 157,261

Founded 1994. MONTHLY. Everything that might interest the teenage girl. No unsolicited mss. Will consider ideas or contacts for real-life features. No fiction. Approach in writing in the first instance.

SuperBike Magazine

IPC Media Ltd, Leon House, 233 High Street, Croydon CR9 1HZ
Ⓣ 020 8726 8455   Ⓕ 020 8726 8499
kenny_pryde@ipcmedia.com
www.superbike.co.uk
Owner *IPC Media (A Time Warner Company)*
Publishing Director *Keith Foster*
Editor *Kenny Pryde*
Circulation 47,511

Founded 1977. MONTHLY. Dedicated to all that is best and most exciting in the world of motorcycling. Unsolicited mss, synopses and ideas by e-mail are welcome.

Surrey Life

Holmesdale House, 46 Croydon Road, Reigate RH2 0NH
Ⓣ 01737 247188   Ⓕ 01737 246596
editor@surreylife.co.uk
www.surreylife.co.uk
Owner *Archant Life*
Editor *Caroline Harrap*
Circulation 15,000

Launched 2002. MONTHLY county magazine featuring the best property, food and drinks and celebrity interviews. Unsolicited submissions of Surrey-based feature ideas are welcome; send by e-mail.

Sussex Life

Baskerville Place, 28 Teville Road, Worthing BN11 1UG
Ⓣ 01903 703730   Ⓕ 01903 703770
jonathan.keeble@archant.co.uk
www.susssexlife.co.uk
Owner *Archant Life*
Editor *Jonathan Keeble*
Circulation 42,000

Founded 1965. MONTHLY. Sussex and general interest magazine. Regular supplements on education, fashion, homes and gardens. Interested in investigative, journalistic pieces relevant to the area and celebrity profiles. Unsolicited mss, synopses and ideas welcome; approach by e-mail.

Swimming Times Magazine

41 Granby Street, Loughborough LE11 3DU
Ⓣ 01509 632230   Ⓕ 01509 632233
Owner *Amateur Swimming Association*
Editor *P. Hassall*
Circulation 20,000

Founded 1923. MONTHLY about competitive swimming and associated subjects. Unsolicited mss welcome.
FEATURES Technical articles on swimming, water polo, diving or synchronized swimming. Length and payment negotiable.

The Tablet

1 King Street Cloisters, Clifton Walk, London W6 0GY
Ⓣ 020 8748 8484   Ⓕ 020 8748 1550
thetablet@tablet.co.uk
www.thetablet.co.uk
Owner *The Tablet Publishing Co Ltd*
Editor *Catherine Pepinster*
Circulation 22,897

Founded 1840. WEEKLY. Quality international Roman Catholic periodical featuring articles (political, social, cultural, theological or spiritual) of interest to concerned Christian laity and clergy. 'We welcome ideas for articles; please send a brief outline in advance, setting out what you propose to write about and explaining your credentials.' Unsolicited material is not accepted.
Ⓔ Payment negotiable.

Take a Break

Academic House, 24–28 Oval Road, London NW1 7DT
Ⓣ 020 7241 8000
tab.features@bauer.co.uk
Owner *H. Bauer Publishing Ltd*
Editor *John Dale*
Circulation 1 million

Founded 1990. WEEKLY. True-life feature magazine. Approach with ideas in writing.
NEWS/FEATURES Always on the look-out for

good, true-life stories. Max. 1,200 words.
FICTION Sharp, succinct stories which are
well told and often with a twist at the end.
All categories, provided it is relevant to the
magazine's style and market. Max. 1,000 words.
Ⓔ Payment negotiable.

Take It Easy
▷ The People under National Newspapers

Tate Etc
Millbank, London SW1P 4RG
☎ 020 7887 8724   🖷 020 7887 3940
tateetc@tate.org.uk
www.tate.org.uk
Owner *Tate*
Editorial Director *Bice Curiger*
Editor *Simon Grant*
Circulation 100,000

Relaunched May 2004. THREE ISSUES YEARLY.
Visual arts magazine aimed at a broad readership
with articles blending the historic and the
contemporary. Please send material by post.

Tatler
Vogue House, Hanover Square, London W1S 1JU
☎ 020 7499 9080   🖷 020 7493 1641
www.tatler.co.uk
Owner *Condé Nast Publications Ltd*
Editor *Geordie Greig*
Features Director *Vassi Chamberlain*
Features Editor *Ticky Hedley-Dent*
Editor's Assistant *Fiona Kent*
Features Associate *Richard Dennen*
Circulation 90,590

Up-market glossy from the Condé Nast stable.
New writers should send in copies of either
published work or unpublished material; writers
of promise will be taken up. The magazine works
largely on a commission basis: they are unlikely to
publish unsolicited features but will ask writers to
work to specific projects.

The Teacher
Hamilton House, Mabledon Place, London
WC1H 9BD
☎ 020 7380 4708   🖷 020 7383 7230
teacher@nut.org.uk
www.teachers.org.uk
Owner *National Union of Teachers*
Editor *Elyssa Campbell-Barr*
Circulation 320,000

Journal of the National Union of Teachers,
published EIGHT TIMES YEARLY. News, advice,
information and special features on educational
matters relating to classroom teaching. Will
consider 'anything on spec'. Approach by letter or
e-mail.

TGO (The Great Outdoors)
Newsquest (Herald & Times) Magazines Ltd., 200
Renfield Street, Glasgow G2 3QB
☎ 0141 302 7700   🖷 0141 302 7799
cameron.mcneish@tgomagazine.co.uk
www.tgomagazine.co.uk
Owner *Newsquest*
Editor *Cameron McNeish*
Circulation 10,705

Founded 1978. MONTHLY. Deals with walking,
backpacking and wild country topics. Unsolicited
mss are welcome.
FEATURES Well-written and illustrated items on
relevant topics. Max. 2,500 words. Quality high
resolution digital colour images only, please.
NEWS Short topical items (or photographs). Max.
300 words.
Ⓔ Payment negotiable.

That's Life!
3rd Floor, Academic House, 24–28 Oval Road,
London NW1 7DT
☎ 020 7241 8000   🖷 020 7241 8008
firstname.lastname@bauer.co.uk
www.bauer.co.uk
Owner *H. Bauer Publishing Ltd*
Editor *Jo Checkley*
Circulation 459,634

Founded 1995. WEEKLY. True-life stories, puzzles,
health, homes, parenting, cookery and fun.
FEATURES *Tiffany Sherlock* Max. 1,600 words.

The Third Alternative
▷ Black Static

This England
PO Box 52, Cheltenham GL50 1YQ
☎ 01242 537900   🖷 01242 537901
Owner *This England Ltd*
Editor *Roy Faiers*
Circulation 140,000

Founded 1968. QUARTERLY, with a strong
overseas readership. Celebration of England
and all things English: famous people, natural
beauty, towns and villages, history, traditions,
customs and legends, crafts, etc. Generally a
rural basis, with the 'Forgetmenots' section
publishing readers' recollections and nostalgia.
Up to a hundred unsolicited pieces received each
week. Unsolicited mss/ideas welcome. Length
250–2,000 words.
Ⓔ £25 per 1,000 words.

Time
Blue Fin Building, 110 Southwark Street, London
SE1 0SU
☎ 020 3148 3000
edit_office@timemagazine.com
www.time.com

Owner *Time Warner*
International Editor *Michael Elliott*
Circulation 135,496 (UK)

Founded 1923. WEEKLY current affairs and news magazine. There are few opportunities for freelancers on *Time* as almost all the magazine's content is written by staff members from various bureaux around the world. No unsolicited mss.

Time Out

Universal House, 251 Tottenham Court Road, London W1T 7AB
☎ 020 7813 3000   🖷 020 7813 6001
www.timeout.com
Publisher *Mark Elliott*
Editor *Gordon Thomson*
Deputy Editor *Rachel Halliburton*
Circulation 86,951

Founded 1968. WEEKLY magazine of news, arts, entertainment and lifestyle in London plus listings.
FEATURES *Nina Caplan* 'Usually written by staff writers or commissioned, but it's always worth submitting an idea if particularly apt to the magazine.' Word length varies; up to 2,000 max.
CONSUME SECTION Food and drink, shopping, services, travel, design, property, health and fitness.
£ Payment negotiable.

The Times Educational Supplement

26 Red Lion Square, Holborn, London WC1R 4HQ
☎ 020 3194 3000
🖷 020 3194 3200 (editorial)/3194 3202 (newsdesk)
www.tes.co.uk
Owner *Charter House*
Editor *Karen Dempsey*
Circulation 59,758

Founded 1910. WEEKLY. New contributors are welcome and should fax or e-mail ideas on one sheet of A4 for news (newsdesk@tes.co.uk), features (features@tes.co.uk) or reviews. The main newspaper accepts contributions in the following sections:
COMMENT Weekly slot for a well-informed and cogently argued viewpoint. Max. 750 words. (comment@tes.co.uk).
LEADERSHIP Practical issues for school governors and managers. Max. 500 words.
FE FOCUS Weekly pull-out section covering post-16 education and training in colleges, work and the wider community. Aimed at everyone from teachers/lecturers to leaders and opinion formers in lifelong learning. News, features, comment and opinion on all aspects of college life welcome. Length from 350 words (news) to 1,000 max. (features). Contact *Steve Hook* (steve.hook@tes.co.uk).
TES MAGAZINE Weekly magazine for teachers focusing on looking at their lives, inside and outside the classroom, investigating the key issues of the day and highlighting good practice. Max. 800 words (features@tes.co.uk).

The Times Educational Supplement Scotland

Thistle House, 21-23 Thistle Street, Edinburgh EH2 1DF
☎ 0131 624 8332   🖷 0131 467 8019
scoted@tes.co.uk
www.tes.co.uk/scotland
Owner *TSL Education Ltd*
Editor *Neil Munro*
Circulation 7,157

Founded 1965. WEEKLY. Unsolicited mss welcome.
FEATURES Articles on education in Scotland. Max. 1,000 words.
NEWS Items on education in Scotland. Max. 600 words.

The Times Higher Education

26 Red Lion Square, London WC1R 4HQ
☎ 020 3194 3000   🖷 020 3194 3300
editor@tsleducation.com
www.tsleducation.co.uk
Owner *Charterhouse Private Equity*
Editor *Gerard Kelly*
Circulation 21,140

Founded 1971. WEEKLY. Unsolicited items are welcome but most articles and almost all book reviews are commissioned. 'In most cases it is better to write or e-mail, but in the case of news stories it is all right to phone.'
FEATURES *Alan Thomson* Most articles are commissioned from academics in higher education.
NEWS *Phil Baty* Freelance opportunities very occasionally.
SCIENCE *Zoe Corbyn*
BOOKS *Ann Mroz*
FOREIGN *Alan Thomson*
£ Payment by negotiation.

The Times Literary Supplement

Times House, 1 Pennington Street, London E98 1BS
☎ 020 7782 5000   🖷 020 7782 4966
www.the-tls.co.uk
Owner *The Times Literary Supplement Ltd*
Editor *Peter Stothard*
Circulation 33,768

Founded 1902. WEEKLY review of literature, history, philosophy, science and the arts. Contributors should approach in writing and be familiar with the general level of writing in the *TLS*.
LITERARY DISCOVERIES *Alan Jenkins*
POEMS *Mick Imlah*
NEWS Reviews and general articles concerned with literature, publishing and new intellectual

developments anywhere in the world. Length by arrangement.
Ⓔ Payment by arrangement.

Today's Golfer

Media House, Lynchwood, Peterborough Business Park, Peterborough PE2 6EA
☎ 01733 468000   Ⓕ 01733 468671
editorial@todaysgolfer.co.uk
www.todaysgolfer.co.uk
Owner *Emap Active Ltd*
Editor-in-Chief *Andy Calton*
Associate Editor *Simon Penson*
Circulation 104,000

Founded 1988. MONTHLY. Golf instruction, features, player profiles and news. Most features written in-house but unsolicited mss will be considered. Approach in writing with ideas. Not interested in instruction material from outside contributors.
FEATURES/NEWS *Kevin Brown* Opinion, player profiles and general golf-related features.

Top of the Pops Magazine

Media Centre, 201 Wood Lane, London W12 7TQ
☎ 020 8433 2000
www.tcfmagazine.com
Owner *BBC Worldwide Publishing Ltd*
Editor *Peter Hart*
Circulation 105,025

Founded 1995. FOUR-WEEKLY teenage celebrity magazine with a lighthearted and humorous approach. No unsolicited material.

Total Coarse Fishing

DHP Ltd, 2 Stephenson Close, Drayton Fields, Daventry NN11 8RF
☎ 01327 311999   Ⓕ 01327 311190
gareth@dhpub.co.uk
www.tcfmagazine.com
Owner *David Hall*
Editor *Gareth Purnell*
Circulation 35,000

Founded 2006. MONTHLY magazine for anglers who like to keep their fishing varied and target quality coarse fish from exceptional venues. No unsolicited contributions.
FEATURES Overseas freshwater angling features with 'exceptional' photographs. Max. 2,200 words. Approach by e-mail. Ⓔ £200.

Total Film

2 Balcombe Street, London NW1 6NW
☎ 020 7042 4000   Ⓕ 020 7042 4839
totalfilm@futurenet.co.uk
Owner *Future Publishing*
Editor *Nev Pierce*
Reviews *Matthew Leyland* (matthew.leyland@ futurenet.co.uk)

Features *Jamie Graham* (jamie.graham@ futurenet.co.uk), *Andy Lowe* (andy.lowe@ futurenet.co.uk),
News *Jonathan Dean* (jonathan.dean@ futurenet.co.uk)
Circulation 90,642

Founded 1997. MONTHLY reviews-based movie magazine. Interested in ideas for features, not necessarily tied in to specific releases, and humour items. No reviews or interviews with celebrities/directors. Approach by post or e-mail.

Total Flyfisher

DHP Ltd, 2 Stephenson Close, Drayton Fields, Daventry NN11 8RF
☎ 01327 311999   Ⓕ 01327 311190
steve.cullen@dhpub.co.uk
www.totalflyfisher.com
Owner *David Hall*
Editor *Steve Cullen*
Circulation 17,000

Founded 2003. MONTHLY instructional fly fishing magazine. Welcomes contributions including fishery reports and fly tying articles. No foreign features or advanced skills. Approach by e-mail.
FEATURES Stillwater trout tactics, salmon tactics, river trout/grayling tactics, UK saltwater tactics. Max. 2,500 words. Ⓔ Payment (including pictures) negotiable.
NEWS Stories relevant to UK fly fishers.
FICTION Funny fishing stories. Max. 400 words. Ⓔ Payment negotiable.

Total TV Guide

H. Bauer Publishing, Academic House, 24–28 Oval Road, London NW1 8DT
☎ 020 7241 8000   Ⓕ 020 7241 8042
barbara.miller@bauer.co.uk
www.bauer.co.uk
Owner *H. Bauer Publishing*
Editor *Jon Peake*
Circulation 92,004

Founded 2003. WEEKLY TV listings for those with access to multichannel television, covering everything from movies to sport, drama to children's programmes.
FEATURES *Ben Lawrence* Commissions features relating to current TV programmes; chiefly celebrity interviews and set visits. No unsolicited contributions.

Total Vauxhall

Future Publishing, 30 Monmouth Street, Bath BA1 2BW
☎ 01225 442244   Ⓕ 01225 446019
info@totalvauxhall.co.uk
www.totalvauxhall.co.uk
Owner *Future Publishing*
Editor *Barton Brisland*

Circulation 20,000

Founded 2001. MONTHLY independent newsstand magazine aimed at the Vauxhall enthusiast. Covers new, modified, historical/classic, race and rally Vauxhalls of all kinds. Substantial amount of technical content. Also covers the more interesting parts of the GM family, particularly Opel and Holden. Uses a lot of freelance contributors; 'those who hit deadlines and fulfil the brief get regular work and lots of it.' Feature ideas, news items, historical pieces and potential feature cars welcome but must have a Vauxhall/GM tilt. Call the editor directly for an informal discussion on style, approach and angle.
£ Payment 'surprisingly generous'.

Trail

Bretton Court, Bretton, Peterborough PE3 8DZ
☎ 01733 468363
trail@emap.com
www.trailroutes.com
Owner *Emap Active Publishing Ltd*
Editor *Matthew Swaine*
Circulation 39,349

Founded 1990. MONTHLY. Gear reports, where to walk and practical advice for the hill walker and long distance walker. Inspirational reads on people and outdoor/walking issues. Health, fitness and injury prevention for high level walkers and outdoor lovers. 'We always welcome ideas from writers who understand our readers' needs. No travel pieces. We are interested in pieces that allow our readers to get out more, to enjoy their time outdoors and improve their skills. So don't tell us what *you've* done. Tell our readers what they could do, why they should do it and how best to do it. Inspiration and practical advice in equal measure.' Approach by phone or in writing in the first instance.
FEATURES *Simon Ingram* Very limited requirement for overseas articles, 'written to our style'. Ask for guidelines. Max. 2,000 words. Limited requirement for guided walks articles. Specialist writers only. Ask for guidelines. 750–2,000 words (depending on subject).
£ £100 per 1,000 words.

Traveller

45–49 Brompton Road, London SW3 1DE
☎ 020 7589 0500   🖷 020 7581 8476
traveller@wexas.com
www.traveller.org.uk
Owner *Wexas International Ltd*
Editor *Amy Sohanpaul*
Circulation 23,817

Founded 1970. QUARTERLY travel magazine.
FEATURES High quality, personal narratives of remarkable journeys. Articles should be off-beat, adventurous, authentic. For guidelines, see website. Articles may be accompanied by professional quality, original slides or high-res digital photographs. Freelance articles considered. Max. 1,000 words. Initial contact by e-mail.
£ £200 per 1,000 words.

Tribune

9 Arkwright Road, London NW3 6AN
☎ 020 7433 6410   🖷 020 7433 6410
mail@tribunemagazine.co.uk
Owner *Tribune Publications*
Editor *Chris McLaughlin*

Founded 1937. WEEKLY. Independent Labour publication covering parliament, politics, trade unions, public policy and social issues, international affairs and the arts. Welcomes freelance contributions; approach by telephone or e-mail.
FEATURES *George Osgerby* Anything in line with the topics above. Max. 1,000 words.
NEWS *Keith Richmond* Breaking or forthcoming news events/stories. Max. 480 words.
OTHER Interviews, reviews, cartoons.

Trout Fisherman

Bauer Active Ltd, Bushfield House, Orton Centre, Peterborough PE2 5UW
☎ 01733 237111   🖷 01733 465820
russell.hill@bauerconsumer.co.uk
www.gofishing.co.u
Owner *Bauer Active*
Editor *Russell Hill*
Circulation 25,673

Founded 1977. MONTHLY instructive magazine on trout fishing. Most of the articles are commissioned, but unsolicited mss and quality colour images welcome.
FEATURES Max. 2,500 words.
£ Payment varies.

TVTimes

IPC Media Ltd., Blue Fin Building, 110 Southwark Street, London SE1 0SU
☎ 020 3148 5615   🖷 020 3148 8115
Owner *IPC Media (A Time Warner Company)*
Editor *Ian Abbott*
Circulation 358,511

Founded 1955. WEEKLY magazine of listings and features serving the viewers of independent television, BBC, satellite and radio. Freelance contributions by commission only. No unsolicited contributions.

20x20 magazine

36 Osborne Road, London N4 3SD
info@20x20magazine.com
20x20magazine.com
Owner/Editors *Giovanna Paternò, Francesca Ricci*

Circulation 500

Launched 2008. BIANNUAL. 'A square platform for writings, visuals and cross-bred projects.' Includes three sections – Words: in the shape of fiction, essays, poetry; Visions: drawings and photography; The Blender: where words and visions cross path. Contributions welcome 'although they must fit in our "metawords" spirit'. Max. 1,500 words. Approach by e-mail.

Ulster Business

5b Edgewater Business Park, Belfast Harbour Estate, Belfast BT3 9JQ
☎ 028 9078 3200   ℻ 028 9078 3210
davidcullen@greerpublications.com
www.ulsterbusiness.com
Owner *James Greer*
Editor *David Cullen*
Circulation 6,000

Founded 1987. MONTHLY. General business content with coverage of all sectors: ICT, retail, agriculture, commercial property, etc. 'Opportunities exist for well-written local orientated features, particularly those dealing with local firms or the issues they face; local business personality interviews also useful.' Contact the editor by e-mail. Max. 650 words.

Ulster Tatler

39 Boucher Road, Belfast BT12 6UT
☎ 028 9066 3311   ℻ 028 9038 1915
edit@ulstertatler.com
www.ulstertatler.com
Owner/Editor *Richard Sherry*
Circulation 10,870

Founded 1965. MONTHLY. Articles of local interest and social functions appealing to Northern Ireland's ABC1 population. Welcomes unsolicited material; approach by phone or in writing in the first instance.
FEATURES *Noreen Dorman* Max. 1,500 words.
FICTION *Richard Sherry* Max. 3,000 words

Ultimate Advertiser

PO Box 7086, Clacton on Sea CO15 5WN
☎ 01255 428500
mail@apexpublishing.co.uk
www.ultimateadvertiser.co.uk
Owner *Apex Publishing Ltd*
Editor *Chris Cowlin*
Circulation 5,000

Founded 2005. MONTHLY advertising magazine including guides and information, features, listings and reviews. Approach by post, only.

Uncut

IPC Media Ltd., Blue Fin Building, 110 Southwark Street, London SE1 0SU
☎ 020 3148 6985
www.uncut.co.uk
Owner *IPC Media (A Time Warner Company)*
Editor *Allan Jones*
Associate Editor *Michael Bonner*
Music Editor *John Mulvey*
Art Editor *Marc Jones*
Online News Editor *Farah Ishaq*
Circulation 91,028

Launched 1997. MONTHLY music and film magazine with free CD each issue. Welcomes music and film reviews; approach by e-mail.

The Universe

Gabriel Communications Ltd., 4th Floor, Landmark House, Station Road, Cheadle Hulme SK8 7JH
☎ 0161 488 1700   ℻ 0161 488 1701
joseph.kelly@totalcatholic.com
www.totalcatholic.com
Owner *Gabriel Communications Ltd*
Editor *Joseph Kelly*
Circulation 60,000

Occasional use of new writers, but a substantial network of regular contributors already exists. Interested in a very wide range of material: all subjects which might bear on Christian life. Fiction not normally accepted.
Ⓔ Payment negotiable.

The Vegan

Donald Watson House, 21 Hylton Street, Birmingham B18 6HJ
☎ 0121 523 1730
editor@vegansociety.com
www.vegansociety.com
Owner *Vegan Society*
Editor *Rosamund Raha*
Circulation 7,000

Founded 1944. QUARTERLY. Deals with the ecological, ethical and health aspects of veganism. Unsolicited mss welcome. Max. 2,000 words.
Ⓔ Payment negotiable.

Vogue

Vogue House, Hanover Square, London W1S 1JU
☎ 020 7499 9080   ℻ 020 7408 0559
www.vogue.com
Owner *Condé Nast Publications Ltd*
Editor *Alexandra Shulman*
Circulation 220,325

Launched 1916. Condé Nast Magazines tend to use known writers and commission what's needed, rather than using unsolicited mss. Contacts are useful.
FEATURES *Harriet Quick* Upmarket general interest rather than 'women's'. Good proportion of highbrow art and literary articles, as well as travel, gardens, food, home interest and reviews.

The Voice

GV Media Group Ltd., Northern & Shell Tower, 6th Floor, 4 Selsdon Way, London E14 9GL
☎ 020 7510 0340   🖷 020 7510 0341
yourviews@gvmedia.co.uk
www.voice-online.co.uk
Acting Managing Director *George Ruddock*
Editor *Steve Pope*
Circulation 40,000

Founded 1982. Leading WEEKLY newspaper for black Britons and other minority communities. Includes news, features, arts, sport and a comprehensive jobs section. Illustrations: colour and b&w photos. Open to ideas for news and features on sport, business, community events and the arts.

Volkswagen Golf+

IPC Media Ltd, Leon House, 233 High Street, Croydon CR9 1HZ
☎ 020 8726 8000   🖷 020 8726 8296
thegolfmag@ipcmedia.com
www.thegolf.co.uk
Owner *IPC Media (A Time Warner Company)*
Editor *Fabian Cotter*
Circulation 25,000

Founded 1995. MONTHLY. Features Volkswagen Golfs and other watercooled VAG cars. Articles on tuning, styling, performance, technical and potential feature cars. 'Not interested in anything to do with the game of golf.' Submissions should be made directly to the editor. Articles must show excellent specialist knowledge of the subject matter. 💷 Payment by negotiation.

W.I. Life

104 New Kings Road, London SW6 4LY
☎ 020 7731 5777   🖷 020 7736 4061
Owner *National Federation of Women's Institutes*
Editor *Neal Maidment*
Circulation 201,800

Women's Institute membership magazine. First issue February 2007. EIGHT ISSUES YEARLY. Contains articles on a wide range of subjects of interest to women. Strong environmental country slant with crafts and cookery plus gardening. Contributions, photos and illustrations from WI members welcome. 💷 Payment by arrangement.

Walk Magazine

2nd Floor, Camelford House, 87–90 Albert Embankment, London SE1 7TW
☎ 020 7339 8500   🖷 020 7339 8501
ramblers@ramblers.org.uk
www.ramblers.org.uk
www.walkmag.co.uk
Owner *Ramblers' Association*
Editor *Dominic Bates*
Assistant Editor *Denise Noble*

Circulation 105,000

QUARTERLY. Official magazine of the Ramblers' Association, available to members only. Unsolicited mss welcome. S.a.e. required for return.
FEATURES Freelance features are invited on any aspect of walking in Britain. Length up to 1,500 words, preferably with good photographs. No general travel articles.

Wallpaper

Blue Fin Building, 110 Southwark Street, London SE1 0SU
☎ 020 3148 5000   🖷 020 3148 8119
contact@wallpaper.com
www.wallpaper.com
Owner *IPC Media (A Time Warner Company)*
Editor *Tony Chambers*
Circulation 112,847

Launched 1996. MONTHLY magazine with the best of international design, architecture, fashion, travel and lifestyle.
FEATURES *Nick Compton* Ideas welcome but contributors must be familiar with the voice and subject matter of the magazine. No celebrity items. Approach by e-mail.

The War Cry

101 Newington Causeway, London SE1 6BN
☎ 020 7367 4900   🖷 020 7367 4710
warcry@salvationarmy.org.uk
www.salvationarmy.org.uk/warcry
Owner *The Salvation Army*
Editor *Major Nigel Bovey*
Circulation 58,000

Founded 1879. WEEKLY full-colour magazine containing Christian comment on current issues. Unsolicited mss welcome if appropriate to contents. No fiction or poetry. Approach by phone with ideas.
NEWS relating to Christian Church or social issues. Max. length 500 words.
FEATURES Human interest articles aimed at the 'person-in-the-street'. Max. 500 words.
💷 Payment negotiable.

Waterways World

151 Station Street, Burton on Trent DE14 1BG
☎ 01283 742950   🖷 01283 742957
editorial@waterwaysworld.com
www.waterwaysworld.com
Owner *Waterways World Ltd*
Editor *Richard Fairhurst*
Circulation 16,602

Founded 1972. MONTHLY magazine for inland waterway enthusiasts. Unsolicited mss welcome, provided the writer has a good knowledge of the subject. No fiction.
FEATURES *Richard Fairhurst* Articles (preferably

illustrated) are published on all aspects of inland waterways in Britain and abroad but predominantly recreational boating on rivers and canals.

NEWS *Chris Daniels* Max. 500 words.
Ⓔ £70 per published page.

Web User

IPC Media Ltd., Blue Fin Building, 110 Southwark Street, London SE1 0SU
☎ 020 3148 5000   Ⓕ 020 3148 8122
letters@webuser.co.uk
www.webuser.co.uk
Owner *IPC Media (A Time Warner Company)*
Editor *Claire Woffenden*
Circulation 35,279

Founded 2001. FORTNIGHTLY bestseller Internet magazine for all users of the Internet. No unsolicited material; send e-mail or letter of enquiry in the first instance.

Wedding

IPC Media Ltd., Blue Fin Building, 110 Southwark Street, London SE1 0SU
☎ 020 3148 7800
wedding@ipcmedia.com
Owner *IPC Media (A Time Warner Company)*
Editor *Catherine Westwood*
Circulation 38,026

Founded 1985. BI-MONTHLY offering ideas and inspiration for women planning their wedding. Most features are written in-house or commissioned from known freelancers. Unsolicited mss are not welcome, but approaches may be made in writing.

Weekly News

D.C. Thomson & Co. Ltd., Albert Square, Dundee DD1 9QJ
☎ 01382 223131   Ⓕ 01382 201390
weeklynews@dcthomson.co.uk
Owner *D.C. Thomson & Co. Ltd*
Editor *David Burness*
Circulation 70,576

Founded 1855. WEEKLY. Newsy, family-orientated magazine designed to appeal to the busy housewife. Regulars include true-life stories told in the first person, showbiz, royals and television. Two or three general interest fiction stories each week. Usually commissions, but writers of promise will be taken up. Ⓔ Payment negotiable.

Welsh Country Magazine

Aberbanc, Llandysul SA44 5NP
☎ 01559 372010   Ⓕ 01559 371995
editor@welshcountry.co.uk
www.welshcountry.co.uk
Owner *Equine Marketing Limited*
Editor *Kath Rhodes*

Circulation 19,658

Founded 2004. BI-MONTHLY. Outside contributions welcome. Contact the editor with samples of work before submitting full article. The magazine contains a couple of pages in Welsh but is predominantly written in English.

FEATURES Pesonal experiences or viewpoints, historical or modern, but there needs to be a strong Welsh connection in some form.

NEWS All Welsh news welcome; anything at all commercial is not paid for.
Ⓔ Payment negotiable.

What Car?

Teddington Studios, Teddington Lock, Broom Road, Teddington TW11 9BE
☎ 020 8267 5688
editorial@whatcar.com
www.whatcar.com
Owner *Haymarket Motoring Publications Ltd*
Editor *Steve Fowler*
Circulation 107,812

MONTHLY. The car buyer's bible, *What Car?* concentrates on road test comparisons of new cars, news and buying advice on used cars, as well as a strong consumer section. Some scope for freelancers. No unsolicited mss.
Ⓔ Payment negotiable.

What Hi-Fi? Sound & Vision

Teddington Studios, Teddington Lock, Broom Road, Teddington TW11 9BE
☎ 020 8943 5000   Ⓕ 020 8267 5019
whathifi@haymarket.com
www.whathifi.com
Owner *Haymarket Consumer Media Ltd*
Editor *Clare Newsome*
Circulation 67,531

Founded 1976. THIRTEEN ISSUES YEARLY. Features on hi-fi and home cinema. No unsolicited contributions. Prior consultation with the editor essential.

FEATURES General or more specific items on hi-fi and home cinema pertinent to the consumer electronics market.

REVIEWS Specific product reviews. All material is now generated by in-house staff. Freelance writing no longer accepted.

What Investment

Vitesse Media, Octavia House, 50 Banner Street, London EC1Y 8ST
☎ 020 7250 7026   Ⓕ 020 7250 7011
keiron.root@vitessemedia.co.uk
www.whatinvestment.co.uk
Owner *Vitesse Media plc*
Editor *Keiron Root*
Deputy Editor *Jenny Lowe*

Circulation 21,446

Founded 1982. MONTHLY. Features articles on a variety of savings and investment matters. E-mail ideas to the editor.
FEATURES Max. 2,500 words.
£ £200 per 1,000 words.

What Satellite and Digital TV

2 Balcombe Street, London NW1 6NW
☎ 020 7042 4000
wotsat@futurenet.co.uk
www.wotsat.com
Owner *Future Publishing Ltd*
Editor *Alex Lane*
Circulation 40,000

Founded 1986. MONTHLY including news, technical information, reviews, equipment tests, programme background, listings. Contributions welcome – phone first.
FEATURES In-depth guides to popular/cult shows. Technical tutorials.
NEWS Industry and programming. Max. 250 words.

What's New in Building

Ludgate House, 245 Blackfriars Road, London SE1 9UY
☎ 01732 364422   ℻ 020 7921 4134
mpennington@cmpi.biz
Owner *CMP Information*
Editor *Mark Pennington*
Circulation 27,000

MONTHLY. Specialist magazine covering new products for building. Unsolicited mss not generally welcome. The only freelance work available is rewriting press release material. This is offered on a monthly basis of 25–50 items of about 150 words each. £ £5.25 per item.

Which Caravan

Warners Group Publications plc, The Maltings, West Street, Bourne PE10 9PH
☎ 01778 391165   ℻ 01778 425437
sallyp@warnersgroup.co.uk
www.outandaboutlive.co.uk
Editor *Sally Pepper*

Founded 1987. Buyers' guide for first-time and experienced caravanners and enthusiasts providing product information, advice and testing. Opportunities for caravanning, relevant touring and travel material, also monthly reviews of the best tow cars, all with good-quality colour photographs.

Wild Times

The RSPB, UK Headquarters, The Lodge, Sandy SG19 2DL
☎ 01767 680551   ℻ 01767 683262

derek.niemann@rspb.org.uk
Owner *Royal Society for the Protection of Birds*
Editor *Derek Niemann*

Founded 1965. BI-MONTHLY. Bird, wildlife and nature conservation for under-8-year-olds (RSPB Wildlife Explorers members). No unsolicited mss. No 'captive/animal welfare' articles.
FEATURES Unsolicited material rarely used.
NEWS News releases welcome. Approach in writing in the first instance.

Wine & Spirit

William Reed Publishing, Broadfield Park, Crawley RH11 9RT
☎ 01293 613400   ℻ 01293 610317
david.williams@william-reed.co.uk
www.wine-spirit.com
Owner *William Reed Business Media*
Editor *David Williams*
Circulation 11,044

MONTHLY. New owner, William Reed, has merged *Wine Magazine* with its sister title *Wine & Spirit International* to create *Wine & Spirit*. Still a consumer/specialist magazine, covering wine, spirit and beer, plus food and food/wine-related travel. No unsolicited mss. Prospective contributors should approach in writing or by e-mail.

Wingbeat

The RSPB, UK Headquarters, The Lodge, Sandy SG19 2DL
☎ 01767 680551   ℻ 01767 683262
derek.niemann@rspb.org.uk
Owner *Royal Society for the Protection of Birds*
Editor *Derek Niemann*

Founded 1965. QUARTERLY. Bird, wildlife and nature conservation for 13–18-year-olds (RSPB Wildlife Explorers teenage members). No unsolicited mss. No 'captive/animal welfare' articles.
FEATURES Unsolicited material rarely used.
NEWS News releases welcome. Approach in writing in the first instance.

The Wisden Cricketer

46 Loman Street, London SE1 0EH
☎ 020 7921 9195   ℻ 020 7921 9151
twc@wisdengroup.com
www.thewisdencricketer.com
Owner *BSkyB Publications*
Editor *John Stern*
Deputy Editor *Edward Craig*
Features Editor *Paul Coupar*
Staff Writer *Daniel Brigham*
Circulation 34,559

Founded 2003. MONTHLY. Result of a merger between *The Cricketer International* (1921) and *Wisden Cricket Monthly* (1979). Very few uncommissioned articles are used, but would-be contributors are not discouraged. Approach in writing. £ Payment varies.

Woman

IPC Media Ltd., Blue Fin Building, 110 Southwark Street, London SE1 0SU

☎ 020 3148 6491

woman@ipcmedia.com

www.ipcmedia.com

Owner *IPC Media (A Time Warner Company)*

Editor *Jackie Hatton*

Deputy Editor *Abigail Blackburn*

Circulation 371,351

Founded 1937. WEEKLY. Long-running, popular women's magazine which boasts a readership of over 2.5 million. No unsolicited mss. Most work commissioned. Approach with ideas in writing.

FEATURES *Jenny Vereker* Max. 1,250 words.

BOOKS *Claire Snewin*

Woman and Home

IPC Media Ltd., Blue Fin Building, 110 Southwark Street, London SE1 0SU

☎ 020 3148 7836

w&hmail@ipcmedia.com

www.womanandhome.com

Owner *IPC Media (A Time Warner Company)*

Editor *Sue James*

Circulation 336,022

Founded 1926. MONTHLY. No unsolicited mss. Prospective contributors are advised to write with ideas, including photocopies of other published work or details of magazines to which they have contributed. S.a.e. essential for return of material. All freelance work is specially commissioned.

Woman's Own

IPC Media Ltd., Blue Fin Building, 110 Southwark Street, London SE1 0SU

☎ 020 3148 5000   📠 020 3148 8112

Owner *IPC Media (A Time Warner Company)*

Editor *Karen Livermore*

Features Editor *Rachel Woolston*

Circulation 349,164

Founded 1932. WEEKLY. Prospective contributors should contact the features editor *in writing* in the first instance before making a submission.

Woman's Weekly

IPC Media Ltd., Blue Fin Building, 110 Southwark Street, London SE1 0SU

☎ 020 3148 5000

womansweeklypostbag@ipcmedia.com

Owner *IPC Media (A Time Warner Company)*

Editor *Diane Kenwood*

Deputy Editor *Geoffrey Palmer*

Features Editor *Sue Pilkington*

Fiction Editor *Gaynor Davies*

Circulation 36,073

Founded 1911. Mass-market WEEKLY for the mature woman.

FEATURES General features covering anything and everything of interest to women of forty-plus. Can be newsy and/or emotional but should be informative. Could be campaigning, issue-based, first or third person. Words: ranging from 800 to 2,000. Only experienced journalists. Synopses and ideas should be submitted by e-mail to susan_archer@ipcmedia.com

FICTION Short stories, 1,000–2,500 words; serials, 12,000–30,000 words. Guidelines: should be contemporary and engaging. Humour and wit also welcome. Short stories up to 5,000 words considered for *Woman's Weekly Fiction Special* (see entry). E-mails to maureen_street@ipcmedia.com

Woman's Weekly Fiction Special

IPC Media Ltd., Blue Fin Building, 110 Southwark Street, London SE1 0SU

☎ 020 3148 6600

womansweeklypostbag@ipcmedia.com

www.ipcmedia.com

Owner *IPC Media (A Time Warner Company)*

Editor *Gaynor Davies*

Launched 1998. EIGHT ISSUES A YEAR. Welcomes short stories of between 1,000 and 5,000 words. Guidelines are available from the Fiction Department. See also *Woman's Weekly*.

Women & Golf

Gemini Network Media Ltd., Bloxham Mill Business Centre, Bloxham OX15 4FF

☎ 01295 724106   📠 01295 724108

alison.root@womensgolfnetwork.co.uk

Owner *IPC Media (A Time Warner Company)*

Editor *Jane Carter*

Circulation 18,000

Founded 1991. NINE ISSUES YEARLY. Consumer magazine aimed at amateur lady golfers of all ability levels. Features and photography and news welcome; approach by e-mail or post.

The Woodworker

Magicalia Publishing Ltd., Berwick House, 8–10 Knoll Rise, Orpington BR6 0EL

☎ 01689 899216

mike.lawrence@magicalia.com

www.getwoodworking.com

Owner *Magicalia Publishing Ltd.*

Editor *Mike Lawrence*

Founded 1901. MONTHLY. Contributions welcome; approach with ideas in writing.

FEATURES Articles on woodworking with good photo support appreciated. Max. 2,000 words.

£ Payment negotiable.

World Fishing

Mercator Media Ltd., The Old Mill, Lower Quay,
Fareham PO6 0RA
℡ 01329 825335   🖷 01329 825330
cwills@mercatormedia.com
Owner *Mercator Media Ltd*
Editor *Carly Wills*
Circulation 3,376

Founded 1952. MONTHLY. Unsolicited mss
welcome; approach by phone or in writing with
an idea.
NEWS/FEATURES of a technical or commercial
nature relating to the commercial fishing and fish
processing industries worldwide (the magazine is
read in over 100 different countries). Max. 1,500
words.
🖹 Payment by arrangement.

The World of Interiors

Vogue House, Hanover Square, London W1S 1JU
℡ 020 7499 9080   🖷 020 7493 4013
interiors@condenast.co.uk
www.worldofinteriors.co.uk
Owner *Condé Nast Publications Ltd*
Editor *Rupert Thomas*
Circulation 67,021

Founded 1981. MONTHLY. Best approach by fax
or letter with an idea, preferably with reference
snaps or guidebooks.
FEATURES *Rupert Thomas* Most feature material
is commissioned. 'Subjects tend to be found by
us, but we are delighted to receive suggestions
of interiors, archives, little-known museums,
collections, etc. unpublished elsewhere, and are
keen to find new writers.'

World Ski Guide

27a High Street, Esher KT10 9RL
℡ 01372 468140
edit@goodholidayideas.com
www.goodskiguide.com
Owner *Mountain Leisure Ltd*
Editors *John Hill, Christian Berger*
Circulation 40,000

Launched 2002. ANNUAL. Translation of *Good
Ski Guide* into German. See website for editorial
guidelines.
FEATURES *John Hill* Destination features on ski
resorts worldwide. Max. 900 words. 🖹 Fee £200.

World Soccer

IPC Media Ltd., Blue Fin Building, 110 Southwark
Street, London SE1 0SU
℡ 020 3148 4817   🖷 020 3148 8130
world_soccer@ipcmedia.com
www.worldsoccer.com
Owner *IPC Media (A Time Warner Company)*
Editor *Gavin Hamilton*

Circulation 44,020

Founded 1960. MONTHLY. Unsolicited material
welcome but initial approach by e-mail or in
writing. News and features on world soccer.

Writers' Forum incorporating World Wide Writers

PO Box 6337, Bournemouth BH1 9EH
℡ 01202 586848
editorial@writers-forum.com
www.writers-forum.com
Owner *Select Publisher Services*
Editor *Carl Styants*
Publisher *Tim Harris*

Founded 1993. MONTHLY. Magazine covers all
aspects of the craft of writing. Well written articles
welcome. Write to the editor in the first instance.
Writers Forum Short Story Competition Prizes:
£300 (1st), £150 (2nd), £100 (3rd) with winning
entries published in an anthology. Entrance fee:
non-subscribers: £10; subscribers: £7. Winners
published in every issue.
Writers' Forum Poetry Competition First prize
£100; runners up receive a dictionary. Entrance
fee: £5 for one poem or £7 for two. Winners
published in every issue. Annual subscription:
£33 UK; £46 Worldwide. Send s.a.e. with 66p in
stamps for free back issue.

Writers' News/Writing Magazine

Warners Group Publications, Fifth Floor, 31–32
Park Row, Leeds LS1 5JD
℡ 0113 200 2929   🖷 0113 200 2928
www.writersnews.co.uk
Owner *Warners Group Publications plc*
Publisher *Janet Davison*
Editor, Writers' News *Jonathan Telfer*
Editor, Writing Magazine *Hilary Bowman*

Founded 1989. MONTHLY magazines containing
news and advice for writers. *Writers' News* is
available exclusively by subscription; *Writing
Magazine*, a full-colour glossy publication, is
available by subscription or on newsstands. No
poetry or general items on 'how to become a
writer'. Receive 1,000 mss each year. Approach in
writing or by e-mail.
NEWS Exclusive news stories of interest to writers.
Max. 350 words.
FEATURES How-to articles of interest to
professional writers. Max. 1,000 words.

Yachting Monthly

IPC Media Ltd., Blue Fin Building, 110 Southwark
Street, London SE1 0SU
℡ 020 3148 4872   🖷 020 3148 8128
paul_gelder@ipcmedia.com
www.yachtingmonthly.com
Owner *IPC Media (A Time Warner Company)*
Editor *Paul Gelder*

Circulation 34,402

Founded 1906. MONTHLY magazine for yachting and cruising enthusiasts – not racing. Unsolicited mss welcome, but many are received and not used. Prospective contributors should make initial contact in writing.

FEATURES A wide range of features concerned with maritime subjects and cruising under sail; well-researched and innovative material always welcome, especially if accompanied by high resolution digital images. Max. 1,800 words.
Ⓔ £80–£110 per 1,000 words.

Yachting World

IPC Media Ltd., Blue Fin Building, 110 Southwark Street, London SE1 0SU
☎ 020 3148 4846   🖷 020 3148 8127
yachting_world@ipcmedia.com
www.yachtingworld.com
Owner *IPC Media (A Time Warner Company)*
Editor *Andrew Bray*
Circulation 27,430

Founded 1894. MONTHLY with international coverage of yacht racing, cruising and yachting events. Will consider well researched and written sailing stories. Preliminary approaches should be by phone for news stories and in writing for features.
Ⓔ Payment by arrangement.

Yorkshire Women's Life Magazine

PO Box 113, Leeds LS8 2WX
ywlmagenquiries@btinternet.com
www.yorkshirewomenslife.co.uk
Editor/Owner *Dawn Maria France*
Fashion *Sky Taylor*
Magazine PA *Anna Jenkins*
Diary *Giles Smith*
Circulation 15,000

Founded 2001. THREE ISSUES YEARLY. Features of interest to women along with regional, national, international news and lifestyle articles. Past issues covered health, stress management, living with domestic abuse, pampering breaks for city women, coping with a difficult boss, Windrush awards. 'It is important to study the style of the magazine before submitting material. Send A4 s.a.e. with 46p stamp for copy of submission guidelines. Unsolicited mss and new writers actively encouraged; approach in writing in the first instance with s.a.e.' Magazine available on subscription at the website address.

You & Your Wedding

National Magazine Company, 72 Broadwick Street, London W1F 9EP
☎ 020 7439 5000 (editorial)   🖷 020 7439 2985
cathy.howes@natmags.co.uk
www.youandyourwedding.co.uk

Owner *The National Magazine Company Ltd*
Editor *Colette Harris*
Circulation 60,024

Founded 1985. BI-MONTHLY. Anything relating to weddings, setting up home, and honeymoons. No unsolicited mss. Submit ideas by e-mail only. No phone calls.

You – The Mail on Sunday Magazine
▷ The Mail on Sunday under National Newspapers

Young Voices

GV Media Group Ltd, Northern & Shell Tower, 6th Floor, 4 Selsdon Way, London E14 9GL
☎ 020 7510 0340   🖷 020 7510 0341
yourviews@gvmedia.co.uk
www.young-voices.co.uk
News Editor *Steve Pope*

Founded 2003. MONTHLY glossy magazine aimed at 11–19-year-olds. 'Provides a new outlet for today's youth.' Latest news features, showbiz insight and reviews. Also covers current affairs topics that affect readers' lives.

Young Writer

Fifth Floor, 31–32 Park Row, Leeds LS1 5JD
☎ 0113 200 2929   🖷 0113 200 2928
ftelfer@writersnews.co.uk
www.youngwriter.org
Editor *Jonathan Telfer*

Describing itself as 'The Magazine for Children with Something to Say', *Young Writer* is issued three times a year, at the back-to-school times of Sept., Jan. and April. A forum for young people's writing (fiction and non-fiction, prose and poetry) the magazine is an introduction to independent writing for writers aged five to 18. Ⓔ Payment from £20 to £100 for freelance commissioned articles (these can be from adult writers).

Your Cat Magazine

Roebuck House, 33 Broad Street, Stamford PE9 1RB
☎ 01780 766199   🖷 01780 766416
yourcat@bournepublishinggroup.co.uk
www.yourcat.co.uk
Owner *BPG (Stamford) Ltd*
Editor *Sue Parslow*

Founded 1994. MONTHLY magazine giving practical information on the care of cats and kittens, pedigree and non-pedigree, plus a wide range of general interest items on cats. Will consider 'true life' cat stories (max. 900 words) and quality fiction by published novelists. Send synopsis in the first instance. 'No articles written as though by a cat and no poetry.'

Your Dog Magazine

Roebuck House, 33 Broad Street, Stamford PE9 1RB
☎ 01780 766199   🖷 01780 754774

s.wright@bournepublishinggroup.co.uk
www.yourdog.co.uk
Owner *BPG (Stamford) Ltd*
Editor *Sarah Wright*
Circulation 31,475

Founded 1995. MONTHLY. Practical advice for pet dog owners. Will consider practical features and some personal experiences (no highly emotive pieces or fiction). Telephone in the first instance. NEWS Max. 300–400 words. FEATURES Max. 2,500 words; limited opportunities.
🄴 Payment negotiable.

Your Hair

Origin Publishing, 9th Floor, Tower House, Fairfax Street, Bristol BS1 3BN
🅃 0117 927 9009 🄵 0117 314 8310
michelletiernan@originpublishing.co.uk
www.yourhair.co.uk
Editor *Michelle Tiernan*
Circulation 44,552

Launched 2001. MONTHLY magazine for women aged 21-plus who need some style, inspiration, a whole new look or to keep up with catwalk trends. 'We pride ourselves on wearable, achievable solutions.' No unsolicited contributions but welcome product information from PR agencies.

Your Horse

Bushfield House, Orton Centre, Peterborough PE2 5UW
🅃 01733 465644 🄵 01733 288163
www.yourhorse.co.uk
Owner *Emap*
Editor *Justine Thompson*
Circulation 40,006

'The magazine that aims to make owning and riding horses more rewarding.' Most writing produced in-house but well-targeted articles will always be considered.

Yours Magazine

Media House, Peterborough Business Park, Peterborough PE2 6EA
🅃 01733 468000

yours@emap.com
Owner *Emap Esprit*
Editor *Valery McConnell*
Circulation 327,072

Founded 1973. FORTNIGHTLY. Aimed at a readership aged 50 and over. Submission guidelines on request.
FEATURES Unsolicited mss welcome but must enclose s.a.e. Max. 1,400 words.
FICTION One or two short stories used in each issue. Max. 1,500 words. 🄴 Payment negotiable.

Zest

National Magazine House, 72 Broadwick Street, London W1F 9EP
🅃 020 7439 5000 🄵 020 7312 3750
zest.mail@natmags.co.uk
www.zest.co.uk
Owner *National Magazine Company*
Editor *Alison Pylkkänen*
Deputy Editors *Rebecca Frank, Charlotte Bradshaw*
Circulation 96,611

Founded 1994. MONTHLY. Health, beauty, fitness, nutrition and general well-being. No unsolicited mss. Prefers ideas in synopsis form; approach in writing.

ZOO

Bauer Consumer Media, Mappin House, 4 Winsley Street, London W1W 8HF
🅃 020 7182 8355 🄵 020 7182 8300
info@zooweekly.co.uk
www.zootoday.com
Owner *Bauer Publishing*
Editor *Ben Todd*
Circulation 179,006

Launched 2004. WEEKLY men's lifestyle magazine featuring showbiz, sport, humour, sex, fashion and news.

National Newspapers

Daily Express

The Northern & Shell Building, No. 10 Lower Thames Street, London EC3R 6EN
☎ 0871 434 1010
www.express.co.uk
Owner *Northern & Shell Media/Richard Desmond*
Editor *Peter Hill*
Circulation 727,180

Under owner Richard Desmond, publisher of *OK!* magazine, the paper features a large amount of celebrity coverage. The general rule of thumb is to approach in writing with an idea; all departments are prepared to look at an outline without commitment. Ideas welcome but already receives many which are 'too numerous to count'.
News Editor *Geoff Maynard*
Diary (Hickey Column) *Kathryn Spencer*
Features Editor *Fergus Kelly*
City Editor *Stephen Kahn*
Political Editor *Macer Hall*
Sports Editor *Bill Bradshaw*
Planning Editor (News Desk) should be circulated with copies of official reports, press releases, etc., to ensure news desk cover at all times.

Saturday magazine. Editor *Graham Bailey*
[£] Payment negotiable.

Daily Mail

Northcliffe House, 2 Derry Street, London W8 5TT
☎ 020 7938 6000
Owner *Associated Newspapers/Lord Rothermere*
Editor *Paul Dacre*
Circulation 2.3 million

In-house feature writers and regular columnists provide much of the material. Photo-stories and crusading features often appear; it's essential to hit the right note to be a successful Mail writer. Close scrutiny of the paper is strongly advised.

Not a good bet for the unseasoned. Accepts news on savings, building societies, insurance, unit trusts, legal rights and tax.
News Editor *Keith Poole*
City Editor *Alex Brummer*
'Money Mail' Editor *Tony Hazell*
Political Editor *Benedict Brogan*
Education Correspondent *Laura Clark*
Diary Editor *Richard Kay*
Features Editor *Leaf Kalfayan*
Literary Editor *Sandra Parsons*
Head of Sport *Tim Jotischky*

Femail Editor *Lisa Collins*
Weekend Saturday supplement. Editor *Charlotte Kemp*

Daily Mirror

1 Canada Square, Canary Wharf, London E14 5AP
☎ 020 7293 3000 📠 020 7293 3409
www.mirror.co.uk
Owner *Trinity Mirror plc*
Editor *Richard Wallace*
Circulation 1.48 million

No freelance opportunities for the inexperienced, but strong writers who understand what the tabloid market demands are always needed.
Deputy Editor *Conor Hanna*
News Editor *Anthony Harwood*
Features Editor *Carole Watson*
Political Editor *Bob Roberts*
Business Editor *Clinton Manning*
Showbusiness Diary Editor *Clemmie Moody*
Sports Editor *Dean Morse*

Daily Record

One Central Quay, Glasgow G3 8DA
☎ 0141 309 3000 📠 0141 309 3340
reporters@dailyrecord.co.uk
www.dailyrecord.co.uk
Owner *Trinity Mirror plc*
Editor *Bruce Waddell*
Circulation 396,190

Mass-market Scottish tabloid. Freelance material is generally welcome.
News Editor *Andy Lines*
Features Editor *Melanie Harvey*

Political Editor *Magnus Gardham*
Assistant Editor (Sport) *James Traynor*
Magazine Editor *Jayne Savva*

Scotland Means Business Quarterly business magazine, launched 2002. Editor *Magnus Gardham*

Daily Sport

19 Great Ancoats Street, Manchester M60 4BT
☎ 0161 236 4466 📠 0161 236 4535
www.dailysport.co.uk
Owner *Sport Newspapers Ltd*

Editor-in-Chief *Barry McIlheny*
Circulation 95,060

Tabloid catering for young male readership. Unsolicited material welcome; send to News Editor, *Jane Field-Harris*
Sports Editor *Marc Smith*

Lads Mag Monthly glossy magazine. Editor *Mark Harris*.

Daily Star

The Northern & Shell Building, No. 10 Lower Thames Street, London EC3R 6EN
☎ 0871 434 1010
Owner *Northern & Shell Media/Richard Desmond*
Editor *Dawn Neesom*
Circulation 730,244

In competition with *The Sun* for off-the-wall news and features. Freelance opportunities available.
Deputy Editor *Ben Knowles*
Sports Editor *Howard Wheatcroft*

Daily Star Sunday

The Northern & Shell Building, No. 10 Lower Thames Street, London EC3R 6EN
☎ 0871 520 7424 📠 0871 434 2941
michael.booker@dailystar.co.uk
www.dailystarsunday.co.uk
Owner *Northern & Shell Media/Richard Desmond*
Editor *Gareth Morgan*
Circulation 361,138

Launched in September 2002 in direct competition with *News of the World* and *The People*.

Supplement: *Take5* Lifestyle/showbiz magazine.

The Daily Telegraph

111 Buckingham Palace Road, London SW1W 0DT
☎ 020 7931 2000
www.telegraph.co.uk
Owner *Telegraph Media Group*
Editor *William Lewis*
Circulation 871,598

Unsolicited mss not generally welcome – 'all are carefully read and considered, but very few published.' Contenders should approach the paper in writing, making clear their authority for writing on that subject. No fiction.
Arts Editor *Sarah Crompton*
City Editor *Richard Blackden*
Political Editor *Andrew Porter*
Diary Editor *Tim Walker* Always interested in diary pieces.
Education *John Clare*
Environment *Charles Clover*
Features Editor *Liz Hunt* Most material supplied by commission from established contributors. New writers are tried out by arrangement with the features editor. Approach in writing. Maximum 1,500 words.
Literary Editor *Sam Leith*
Group Managing Editor, Sport *Keith Perry* Occasional opportunities for specialized items.
Style Editor *Georgina Cover*
💷 Payment by arrangement.

Daily Telegraph Weekend Saturday supplement. Editor *Jon Stock*

Telegraph Magazine. Editor *Michele Lavery*

Financial Times

1 Southwark Bridge, London SE1 9HL
☎ 020 7873 3000 📠 020 7873 3076
Andy
Lines
www.ft.com
Owner *Pearson*
Editor *Lionel Barber*
Circulation 448,241

Founded 1888. UK and international coverage of business, finance, politics, technology, management, marketing and the arts. All feature ideas must be discussed with the department's editor in advance. Not snowed under with unsolicited contributions – they get less than any other national newspaper. Approach by e-mail with ideas in the first instance.
Comment and Analysis Editor *Brian Groom*
Arts Editor *Jan Dalley*
City Editor *Andrew Hill*
UK News Editor *Sarah Neville*
FT.com News Editor *Andrew Slade*
Books and Arts Editor, *FT Magazine Rosie Blau*
Literary Editor *Jan Dalley*
Education *David Turner*
Environment Correspondent *Fiona Harvey*
Political Editor *George Parker*
Leisure Industries *Roger Blitz*

Weekend FT Editor *Andy Davis*
How to Spend It Monthly magazine. Editor *Gillian de Bono*
#

The Guardian

119 Farringdon Road, London EC1R 3ER
℡ 020 7278 2332   🖷 020 7837 2114
From Nov. 2008: Kings Place, 90 York Way,
London N1 9AG
firstname.secondname@guardian.co.uk
www.guardian.co.uk
Owner *The Scott Trust*
Editor *Alan Rusbridger*
Circulation 351,031

Of all the nationals *The Guardian* probably offers
the greatest opportunities for freelance writers,
if only because it has the greatest number of
specialized pages which use freelance work. Read
specific sections before submitting mss.
Deputy Editor *Paul Johnson* No opportunities
except in those regions where there is presently
no local contact for news stories.
Literary Editor *Claire Armitstead*
City Editor *Julia Finch*
G2 Editor *Katharine Viner*
'Technology' Editor *Charles Arthur* Thursday
supplement.'
Diary Editor *Jon Henley*
Education Editor *Claire Phipps*
Environment *John Vidal*
Features Editor *Katharine Viner*
'Guardian Society' *Patrick Butler* Focuses on all
public service activity.
Media Editor *Matt Wells*
Political Editor *Patrick Wintour*
Sports Editor *Ben Clissitt*
Women's Page *Kira Cochrane* Runs 3 days a week.

The Guardian Weekend Glossy Saturday issue.
Editor *Merope Mills*
The Guide Editor *Malik Meer*

The Herald (Glasgow)

200 Renfield Street, Glasgow G2 3QB
℡ 0141 302 7000   🖷 0141 302 7171
www.theherald.co.uk
Owner *Gannett UK Ltd*
Editor *Charles McGhee*
Circulation 65,747

One of the oldest national newspapers in the
English-speaking world, *The Herald*, which
dropped its 'Glasgow' prefix in February 1992,
was bought by Scottish Television in 1996 and
by Newsquest in 2003. Lively, quality, national
Scottish daily broadsheet. Approach with ideas in
writing or by phone in first instance.
News Editor *Calum Macdonald*
Arts Editor *Keith Bruce*
Business Editor *Ian McConnell*
Diary *Ken Smith*
Education *Andrew Denholm*
Sports Editor *Donald Cowey*

Herald Magazine Editor *Kathleen Morgan*

The Independent

Independent House, 191 Marsh Wall, London
E14 9RS
℡ 020 7005 2000   🖷 020 7005 2999
www.independent.co.uk
Owner *Independent Newspapers*
Editor *Roger Alton*
Circulation 243,398

Founded 1986. Particularly strong on its arts/
media coverage, with a high proportion of
feature material. Theoretically, opportunities for
freelancers are good. However, unsolicited mss
are not welcome; most pieces originate in-house
or from known and trusted outsiders. Ideas
should be submitted in writing.
Executive News Editor *Dan Gledhill*
Features *Adam Leigh*
Arts Editor *David Lister*
Business Editor *Jeremy Warner*
Education *Richard Garner*
Environment *Michael McCarthy*
Literary Editor *Boyd Tonkin*
Political Editor *Andrew Grice*
Sports Editor *Matt Tench*
Travel Editor *Simon Calder*

The Independent Magazine Saturday supplement.
Editor *Laurence Earle*
The Independent Traveller 32-page Saturday
magazine.
The Information Editor *Jo Ellison*

Independent on Sunday

Independent House, 191 Marsh Wall, London
E14 9RS
℡ 020 7005 2000   🖷 020 7005 2999
www.independent.co.uk
Owner *Independent Newspapers*
Editor *John Mullin*
Circulation 225,403

Founded 1986. Regular columnists contribute
most material but feature opportunites exist.
Approach with ideas in first instance.
Executive Editor (News) *Peter Victor*
Arts Editor *Mike Higgins*
Comment Editor *James Hanning*
Literary Editor *Suzi Feay*
Environment *Geoffrey Lean*
Political Editor *Jane Merrick*
Sports Editor *Mark Padgett*
Travel Editor *Kate Simon*

The New Review supplement. Editor *Bill Tuckey*

International Herald Tribune

6 bis, rue des Graviers, 92521 Neuilly Cedex
℡ 0033 1 4143 9322
🖷 0033 1 4143 9429 (editorial)
iht@iht.com
www.iht.com

Circulation 241,000
Executive Editor *Michael Oreskes*
Managing Editor *Alison Smale*
Features Editor *Katherine Knorr*

Published in France, Monday to Saturday, and circulated in Europe, the Middle East, North Africa, the Far East and the USA. General news, business and financial, arts and leisure. Uses regular freelance contributors. Contributor policy can be found on the website at: www.iht.com/contributor.htm

The Mail on Sunday

Northcliffe House, 2 Derry Street, London w8 5ts
℡ 020 7938 6000   ℻ 020 7937 3829
Owner *Associated Newspapers/Lord Rothermere*
Editor *Peter Wright*
Circulation 2.29 million

Sunday paper with a high proportion of newsy features and articles. Experience and judgement required to break into its band of regular feature writers.
News Editor *David Dillon*
Financial Editor *Lisa Buckingham*
Diary Editor *Katie Nicholl*
Features Editor/Women's Page *Sian James*
Books *Marilyn Warnick*
Political Editor *Simon Walters*
Sports Editor *Malcolm Vallerius*
'Live Night & Day' Editor *Gerard Greaves*
'Review' Editor *George Thwaites*

You – The Mail on Sunday Magazine Colour supplement. Many feature articles, supplied entirely by freelance writers. Editor *Sue Peart*, Features Editor *Rosalind Lowe*

Morning Star

William Rust House, 52 Beachy Road, London
E3 2NS
℡ 020 8510 0815   ℻ 020 8986 5694
newsed@peoples-press.com
Owner *Peoples Press Printing Society*
Editor *John Haylett*
Circulation 9,000

Not to be confused with the *Daily Star*, the *Morning Star* is the farthest left national daily. Those with a penchant for a Marxist reading of events and ideas can try their luck, though feature space is as competitive here as in the other nationals.
News Editor *Dan Coysh*
Features & Arts Editor/Political Editor *Richard Bagley*
Foreign Editor *Dave Williams*
Sports Editor *Greg Leedham*

News of the World

1 Virginia Street, London E98 1NW
℡ 020 7782 1000   ℻ 020 7583 9504

www.newsoftheworld.co.uk
Owner *News International plc/Rupert Murdoch*
Editor *Colin Myler*
Circulation 3.19 million

Highest circulation Sunday paper. Freelance contributions welcome. News and features editors welcome tips and ideas.
Deputy Editor *Jane Johnson*
Assistant Editor (News) *Ian Edmondson*
Features Editor *Jules Stenson*
Political Editor *Ian Kirby*
Sports Editor *Paul McArthy*

Fabulous Colour supplement. Editor *Mandy Appleyard* Glossy magazine, launched in February 2008. Celebrity, fashion and real-life features. Unsolicited mss and ideas welcome.

The Observer

3–7 Herbal Hill, London EC1R 5EJ
℡ 020 7278 2332   ℻ 020 7837 7817
From Nov. 2008: Kings Place, 90 York Way, London N1 9AG
firstname.surname@observer.co.uk
www.observer.co.uk
Owner *Guardian Newspapers Ltd*
Editor *John Mulholland*
Circulation 452,009

Founded 1791. Acquired by Guardian Newspapers from Lonrho in May 1993. Occupies the middle ground of Sunday newspaper politics. Unsolicited material is not generally welcome, 'except from distinguished, established writers'. Receives far too many unsolicited offerings already. No news, fiction or special page opportunities. The newspaper runs annual competitions which change from year to year. Details are advertised in the newspaper.
Executive Editor, News *Ian Katz*
Home News Editor *Lucy Rock*
Foreign News Editor *Tracy McVeigh*
Crime Correspondent *Mark Townsend*
Foreign Affairs Editor *Peter Beaumont*
Home Affairs Editor *Jamie Doward*
Health Correspondent *Denis Campbell*
Social Affairs *Amelia Hill*
Arts and Media Correspondent *Vanessa Thorpe*
Political Editor *Gaby Hinsliff*
Whitehall Correspondent *Jo Revill*
Transport Correspondent *Ned Temko*
Arts Editor *Sarah Donaldson*
'Review' Editor *Jane Ferguson*
Comment Editor *Ruaridh Nicoll*
Deputy Business Editor/City Editor *Richard Wachman*
Business Editor *Ruth Sunderland*
Personal Finance Editor *Jill Insley*
Science Editor *Robin McKie*
Education Correspondent *Anushka Asthana*
Environment Editor *Juliette Jowit*

Escape Editor *Joanne O'Connor*
Literary Editor *William Skidelsky*
Travel Editor *Joanne O'Connor*
Sports Editor *Brian Oliver*

The Observer Magazine Glossy arts and lifestyle
supplement. Editor *Allan Jenkins*
The Observer Sport Monthly Magazine
supplement launched in 2000. Editor *Tim Lewis*
The Observer Food Monthly Launched in 2001.
Editor *Nicola Jeal*
The Observer Music Monthly Launched in 2003.
Editor *Caspar Llewellyn Smith*
Observer Woman Launched in 2006. Editor
Nicola Jeal
The Observer Film Magazine Launched in 2008.
Editor *Akin Ojumu*

The People

1 Canada Square, Canary Wharf, London E14 5AP
☎ 020 7293 3000   ⊟ 020 7293 3517
www.people.co.uk
Owner *Trinity Mirror plc*
Editor *Lloyd Embley*
Circulation 658,905

Slightly up-market version of *The News of the
World*. Keen on exposés and big-name gossip.
Interested in ideas for investigative articles. Phone
in the first instance.
News Editor *Lee Harpin*
Features Editor *Chris Bucktin*
Political Editor *Nigel Nelson*
Sports Editor *Lee Horton*

Take It Easy Magazine supplement. Editor *Maria
Coole* Approach by phone with ideas in first
instance.

Scotland on Sunday

Barclay House, 108 Holyrood Road, Edinburgh
EH8 8AS
☎ 0131 620 8620   ⊟ 0131 620 8491
www.scotlandonsunday.com
Owner *Scotsman Publications Ltd*
Editor *Les Snowdon*
Circulation 67,104

Scotland's top-selling quality broadsheet.
Welcomes ideas rather than finished articles.
Assistant Editor *Kenny Farquharson*
News Editor *Peter Laing*
Arts Editor *Fiona Leith*

Spectrum Colour supplement. Features on
personalities, etc. Editor *Clare Trodden*

The Scotsman

Barclay House, 108 Holyrood Road, Edinburgh
EH8 8AS
☎ 0131 620 8620   ⊟ 0131 620 8617
enquiries@scotsman.com
www.scotsman.com
Owner *Johnston Press*

Editor *Mike Gilson*
Circulation 52,015

Scotland's national newspaper. Many unsolicited
mss come in, and stand a good chance of being
read, although a small army of regulars supply
much of the feature material not written in-house.
See website for contact details.
News Editor *Frank O'Donnell*
Business Editor *Peter MacMahon*
Education *Fiona MacLeod*
Features Editor *Jackie Hunter*
Arts Editor *Andrew Eaton*
Book Reviews *David Robinson*
Sports Editor *Donald Walker*

The Sun

1 Virginia Street, London E98 1SN
☎ 020 7782 4000   ⊟ 020 7782 4108
firstname.lastname@the-sun.co.uk
www.the-sun.co.uk
Owner *News International Ltd/Rupert Murdoch*
Editor *Rebekah Wade*
Deputy Editor *Dominic Mohan*
Circulation 3.13 million

Highest circulation daily. Populist outlook;
very keen on gossip, pop stars, TV soap,
scandals and exposés of all kinds. No room for
non-professional feature writers; 'investigative
journalism' of a certain hue is always in demand,
however.
Head of News *Chris Pharo*
Head of Features *Victoria Newton*
Head of Sport *Steve Waring*
Health Editor *Jane Symons*
Fashion Editor *Erica Davies*

Sunday Express

The Northern & Shell Building, Number 10 Lower
Thames Street, London EC3R 6EN
☎ 0871 434 1010   ⊟ 0871 434 7300
Owner *Northern & Shell Media/Richard Desmond*
Editor *Martin Townsend*
Circulation 673,840

The general rule of thumb is to approach in
writing with an idea; all departments are prepared
to look at an outline without commitment.
News Editor *Stephen Rigley*
Features Editor *Giulia Rhodes*
Business Editor *Lawrie Holmes*
Political Editor *Julia Hartley-Brewer*
Sports Editor *Scott Wilson*

S Fashion and lifestyle magazine for women.
Editor *Louise Robinson* No unsolicited mss. All
contributions are commissioned. Ideas in writing
only.
£ Payment negotiable.

Sunday Herald

200 Renfield Street, Glasgow G2 3QB
☎ 0141 302 7800 🖷 0141 302 7963
editor@sundayherald.com
www.sundayherald.com
Owner *Newsquest*
Editor *Richard Walker*
Circulation 47,926

Launched February 1999. Scottish seven-section compact.
Deputy Editor *David Milne*
Head of News *Bill Mackintosh*
Political Editor *Paul Hutcheon*
Sports Editor *Stephen Penman*
Magazine Editor *Susan Flockhart*

Sunday Mail

One Central Quay, Glasgow G3 8DA
☎ 0141 309 3000 🖷 0141 309 3587
www.sundaymail.co.uk
Owner *Trinity Mirror plc*
Editor *Allan Rennie*
Circulation 485,233

Popular Scottish Sunday tabloid.
News Editor *Brendan McGinty*
Sports Editor *George Cheyne*

7Days Weekly supplement. Editor *Liz Cowan*

Sunday Mirror

1 Canada Square, Canary Wharf, London E14 5AP
☎ 020 7293 3000 🖷 020 7293 3939 (news desk)
www.sundaymirror.co.uk
Owner *Trinity Mirror*
Editor *Tina Weaver*
Circulation 1.34 million

In general terms contributions are welcome, though the paper patiently points out it has more time for those who have taken the trouble to study the market. Initial contact in writing preferred, except for live news situations. No fiction.
News Editor *James Saville* The news desk is very much in the market for tip-offs and inside information. Contributors would be expected to work with staff writers on news stories. Approach by telephone or fax in the first instance.
Finance *Melanie Wright*
Features Editor *Nicky Dawson* 'Anyone who has obviously studied the market will be dealt with constructively and courteously.' Cherishes its record as a breeding ground for new talent.
Sports Editor *David Walker*

Celebs on Sunday Colour supplement. Editor *Mel Brodie*

Sunday Post

2 Albert Square, Dundee DD1 9QJ
☎ 01382 575791 🖷 01382 201064

mail@sundaypost.com
www.sundaypost.com
Owner *D.C. Thomson & Co. Ltd*
Editor *David Pollington*
Circulation 398,151

Contributions should be addressed to the editor.

post plus Monthly colour supplement. Editor *Jan Gooderham*

Sunday Sport

19 Great Ancoats Street, Manchester M60 4BT
☎ 0161 236 4466 🖷 0161 236 4535
www.sundaysport.com
Owner *Sport Newspapers Ltd*
Editor *Nick Appleyard*
Circulation 86,205

Founded 1986. Sunday tabloid catering for a particular sector of the male 15–35 readership. As concerned with 'glamour' (for which, read: 'page 3') as with human interest, news, features and sport. Unsolicited mss are welcome; receives about 90 a week. Approach should be made by phone in the case of news and sports items, by letter for features. All material should be addressed to the news editor.
News Editor *Gary Moran* Off-beat news, human interest, preferably with photographs.
Showbiz Editor *Tanya Jones* Regular items.
Sports Editor *Marc Smith* Hard-hitting sports stories on major soccer clubs and their personalities, plus leading clubs/people in other sports. Strong quotations to back up the news angle essential.
£ Payment negotiable and on publication.

The Sunday Telegraph

111 Buckingham Palace Road, London SW1W 0DT
☎ 020 7931 2000 🖷 020 7931 2936
stnews@telegraph.co.uk
www.telegraph.co.uk
Owner *Press Holdings Limited*
Editor *Ian MacGregor*
Circulation 636,719

Right-of-centre quality Sunday paper which, although traditionally formal, has pepped up its image to attract a younger readership. Unsolicited material from untried writers is rarely ever used. Contact with idea and details of track record.
Deputy Editor *Mark Kleinman*
News Editor *Tim Woodward*
'Seven' Editor *Ross Jones*
City Editor *Mark Kleinman*
Political Editor *Patrick Hennessy*
Education Editor *Julie Henry*
Arts Editor *Chris Hastings*
Environment Editor *David Harrison*
Literary Editor *Michael Prodger*
Diary Editor *Tim Walker*

Stella Magazine Colour supplement. Editor *Anna Murphy*

The Sunday Times

1 Pennington Street, London E98 1ST
☎ 020 7782 5000 ☏ 020 7782 5658
www.sunday-times.co.uk
Owner *News International plc/Rupert Murdoch*
Editor *John Witherow*
Circulation 1.2 million

Founded 1820. Tendency to be anti-establishment, with a strong crusading investigative tradition. Approach the relevant editor with an idea in writing. Close scrutiny of the style of each section of the paper is strongly advised before sending mss. No fiction. All fees by negotiation.
News Editor *Charles Hymas* Opportunities are very rare.
News Review Editor *Eleanor Mills* Submissions are always welcome, but the paper commissions its own, uses staff writers or works with literary agents, by and large. The features sections where most opportunities exist are *Style* and *The Culture*.
Culture Editor *Helen Hawkins*
Business Editor *John Waples*
City Editor *Grant Ringshaw*
Education Correspondent *Geraldine Hackett*
Science and Environment Editor *Jonathan Leake*
Literary Editor *Susannah Herbert*

Sports Editor *Alex Butler*
Style Editor *Tiffanie Darke*

Sunday Times Magazine Colour supplement. Editor *Robin Morgan* No unsolicited material. Write with ideas in first instance.

The Times

1 Pennington Street, London E98 1TT
☎ 020 7782 5000 ☏ 020 7488 3242
www.thetimes.co.uk
Owner *News International plc/Rupert Murdoch*
Editor *James Harding*
Circulation 618,160

Generally right (though features can range in tone from diehard to libertarian). *The Times* receives a great many unsolicited offerings. Writers with feature ideas should approach by letter in the first instance. No fiction.
Deputy Editor *Ben Preston*
Deputy Head of News *John Wellman*
City Editor *David Wighton*
Diary Editor *To be appointed*
Arts Editor *Alex O'Connell*
Literary Editor *Erica Wagner*
Political Editor *Phil Webster*
Sports Editor *Tim Hallissey*

The Times Magazine Saturday supplement. Editor *Gill Morgan*
Times 2 Editor *Emma Tucker*, Features Editor *Michael Harvey*

Regional Newspapers

England

Berkshire

Reading Evening Post

8 Tessa Road, Reading RG1 8NS
☎ 0118 918 3000   🖷 0118 959 9363
editorial@reading-epost.co.uk
Owner *Guardian Media Group*
Editor *Andy Murrill*
Circulation 12,160

Unsolicited mss welcome; one or two received every day. Fiction rarely used. Interested in local news features, human interest, well-researched investigations. Special sections include holidays & travel; children's page (Tues); style page (Wed); business (Mon–Fri); motorcycling; food page; gardening; reviews; rock music (Thurs); motoring (Fri). Also magazines, *Food Monthly* and *Business Monthly*.

Cambridgeshire

Cambridge Evening News

Winship Road, Milton, Cambridge CB24 6PP
☎ 01223 434434   🖷 01223 434415
Owner *Cambridge Newspapers Ltd*
Editor *To be appointed*
Circulation 25,720
News Editor *John Deex*
Business Editor *Jenny Chapman*
Sports Editor *Chris Gill*

Evening Telegraph

57 Priestgate, Peterborough PE1 1JW
☎ 01733 555111   🖷 01733 313147 (editorial)
www.peterboroughtoday.co.uk
Owner *East Midlands Newspapers Ltd*
Editor *Mark Edwards*
Circulation 18,053
News Editor *Rose Taylor*
Assistant Editor (Content) *Alex Gordon*
Business Editor *Paul Grinnell*

Cheshire

Chester Chronicle

Chronicle House, Commonhall Street, Chester CH1 2AA
☎ 01244 340151   🖷 01244 606498
eric.langton@cheshirenews.co.uk
www.chesterchronicle.co.uk
Owner *Trinity Mirror Plc*
Editor-in-Chief *Eric Langton*

All unsolicited feature material will be considered. No payment.

Cleveland

Hartlepool Mail

New Clarence House, Wesley Square, Hartlepool TS24 8BX
☎ 01429 239333   🖷 01429 869024
mail.news@northeast-press.co.uk
www.hartlepoolmail.co.uk
www.peterleemail.co.uk
Owner *Johnston Press Plc*
Editor *Joy Yates*
Circulation 18,223
Deputy Editor *Gavin Foster*
News Editor *Paul Watson*
Sports Editor *Roy Kelly*

Cumbria

News & Star

Newspaper House, Dalston Road, Carlisle CA2 5UA
☎ 01228 612600   🖷 01228 612601
Owner *CN Group Ltd*
Editor *Neil Hodgkinson*
Circulation 23,070
Assistant Editor *Andy Nixon*
Associate Editor *Anne Pickles*
Sports Editor *Phil Rostron*
Women's Page *Jane Loughran*

North West Evening Mail

Abbey Road, Barrow in Furness LA14 5QS
☎ 01229 821835   🖷 01229 840164

news@nwemail.co.uk
www.nwemail.co.uk
Owner *Robin Burgess*
Editor *Jonathan Lee*
Circulation 18,520

All editorial material should be addressed to the editor.
Sports Editor *Frank Cassidy*

Derbyshire

Derby Evening Telegraph

Northcliffe House, Meadow Road, Derby DE1 2BH
☎ 01332 291111
newsdesk@derbytelegraph.co.uk
Owner *Northcliffe Media Group Ltd*
Editor *Steve Hall*
Circulation 42,726

Devon

Express & Echo

Heron Road, Sowton, Exeter EX2 7NF
☎ 01392 442211   🖷 01392 442294 (editorial)
echonews@expressandecho.co.uk
www.thisisexeter.co.uk
Owner *South West Media Group*
Editor *Marc Astley*
Circulation 20,767

Weekly supplements: *Business Week*; *Property Echo*; *Motoring*; *Weekend Echo*.
Head of Content *Lynne Turner*
Features Editor *Becky Moran*
Sports Editor *Richard Davies* (🖷 01392 442416)

Herald Express

Harmsworth House, Barton Hill Road, Torquay TQ2 8JN
☎ 01803 676000   🖷 01803 676228 (editorial)
newsdesk@heraldexpress.co.uk
www.thisissouthdevon.co.uk
Owner *Northcliffe Media Group Ltd*
Editor *Andy Phelan*
Circulation 23,987

Drive scene, property guide, *On The Town* (leisure and entertainment guide), Monday sports, special pages, rail trail, Saturday surgery, nature and conservation column, dance scene, shop scene, The Business. Supplements: *Gardening* (weekly); *Visitors Guide* and *Antiques & Collectibles* (fortnightly). Unsolicited mss generally not welcome. All editorial material should be addressed to the editor in writing.

The Herald

17 Brest Road, Derriford Business Park, Derriford, Plymouth PL6 5AA
☎ 01752 765500   🖷 01752 765527

news@eveningherald.co.uk
www.thisisplymouth.co.uk
Owner *South West Media Group Ltd*
Editor *Bill Martin*
Circulation 36,516

All editorial material to be addressed to the editor or the News Editor, *James Garnett*.

Western Morning News

17 Brest Road, Derriford Business Park, Derriford, Plymouth PL6 5AA
☎ 01752 765500   🖷 01752 765535
wmnnewsdesk@westernmorningnews.co.uk
www.thisisdevon.co.uk
www.thisiscornwall.co.uk
www.thisiswesternmorningnews.co.uk
Owner *South West Media Group Ltd*
Editor *Alan Qualtrough*
Circulation 41,154

Unsolicited mss welcome, but must be of topical and local interest and addressed to the Deputy Editor, *Philip Bowern*.
News Editor *Steve Grant*
Features Editor *Su Carroll*
Sports Editor *Mark Stevens*

Dorset

Daily Echo

Richmond Hill, Bournemouth BH2 6HH
☎ 01202 554601   🖷 01202 299543
neal.butterworth@bournemouthecho.co.uk
www.bournemouthecho.co.uk
Owner *Newsquest Media Group Ltd (a Gannett company)*
Editor *Neal Butterworth*
Circulation 32,441

Founded 1900. Has a strong news and features content and invites specialist articles, particularly on unusual subjects, either contemporary or historical, but only with a local angle or flavour. Special review sections each day, including sport, property, entertainment and culture, heritage, motoring, gardening, the environment and the coastline. Also *Saturday Magazine* and monthly *Society* glossy magazine. All editorial material should be addressed to the Editor, *Neal Butterworth*. 🖷 Payment on publication.

Dorset Echo

Fleet House, Hampshire Road, Granby Industrial Estate, Weymouth DT4 9XD
☎ 01305 830930   🖷 01305 830956
www.dorsetecho.co.uk
Owner *Newsquest Media Group Ltd (a Gannett company)*
Editor *David Murdock*
Circulation 18,803
By-gone days, films, arts, showbiz, brides,

motoring, property, weekend leisure and entertainment including computers and gardening and weekend magazine.
News Editor *Paul Thomas*
Sports Editor *Nigel Dean*

Co. Durham

The Northern Echo

Priestgate, Darlington DL1 1NF
☎ 01325 381313   ℻ 01325 380539
newsdesk@nne.co.uk
www.thenorthernecho.co.uk
Owner *Newsquest (North East) Ltd (a Gannett company)*
Editor *Peter Barron*
Circulation 50,256

Founded 1870. Freelance pieces welcome but telephone first to discuss submission. Assistant Editor (News) *Nigel Burton* Interested in reports involving the North East or North Yorkshire. Preferably phoned in.
Features Editor *Lindsay Jennings* Background pieces to topical news stories relevant to the area. Must be arranged with the features editor before submission of any material.
Business Editor *Owen McAteer*
Sports Editor *Nick Loughlin*
Ⓔ Payment and length by arrangement.

Essex

Echo Newspapers

Newspaper House, Chester Hall Lane, Basildon
SS14 3BL
☎ 0844 477 4512   ℻ 0844 477 4286
Owner *Newsquest Media Group (a Gannett company)*
Editor *Martin McNeill*
Circulation 34,844

Relies almost entirely on staff and regular outside contributors, but will very occasionally consider material sent on spec. Approach the editor in writing with ideas. Although the paper is Basildon-based, its largest circulation is in the Southend area.

Evening Gazette (Colchester)

Oriel House, 43–44 North Hill, Colchester
CO1 1TZ
☎ 01206 506000   ℻ 01206 508274
gazette.newsdesk@nqe.com
www.evening-gazette.co.uk
Owner *Newsquest (Essex) Ltd*
Editor *Irene Kettle*
Circulation 22,131

Monday to Friday daily newspaper servicing north and mid-Essex including Colchester,

Harwich, Clacton, Braintree, Witham, Maldon and Chelmsford. Unsolicited mss not generally used. Relies heavily on regular contributors.
Features Editor *Iris Clapp*

Gloucestershire

Gloucestershire Echo

1 Clarence Parade, Cheltenham GL50 3NY
☎ 01242 271900   ℻ 01242 271848
echo.editor@glosmedia.co.uk
www.thisisgloucestershire.co.uk
Owner *Northcliffe Media Group Ltd*
Editor *Kevan Blackadder*
Circulation 21,074

All material, other than news, should be addressed to the editor.
News Editor *Paul Kennedy*

The Citizen

6–8 The Oxebode, Gloucester GL1 1RZ
☎ 01452 420621   ℻ 01452 420664 (editorial)
citizen.news@glosmedia.co.uk
www.thisisgloucestershire.co.uk
Owner *Northcliffe Newspapers Group Ltd*
Editor *Ian Mean*
Circulation 26,259

News or features material should be addressed to News Editor, *Colin O'Toole* or Features Editor, *Tanya Gledhill*.

Hampshire

The News

The News Centre, Hilsea, Portsmouth PO2 9SX
☎ 023 9266 4488   ℻ 023 9267 3363
newsdesk@thenews.co.uk
www.portsmouth.co.uk
Owner *Portsmouth Printing & Publishing Ltd*
Editor *Mark Waldron*
Circulation 52,531

Unsolicited mss not generally accepted. Approach by letter.
Assistant Editor *Graeme Patfield* General subjects of S.E. Hants interest. Maximum 600 words. No fiction.
Sports Editor *Howard Frost* Sports background features. Maximum 600 words.

The Southern Daily Echo

Newspaper House, Test Lane, Redbridge, Southampton SO16 9JX
☎ 023 8042 4777   ℻ 023 8042 4545
newsdesk@dailyecho.co.uk
www.dailyecho.co.u
Owner *Newsquest Plc*
Editor *Ian Murray*

Circulation 39,174

Unsolicited mss 'tolerated'. Approach the editor in writing with strong ideas; staff supply almost all the material.

Kent

Kent Messenger

6 & 7 Middle Row, Maidstone ME14 1TG
☎ 01622 695666   🖷 01622 664988
messengernews@thekmgroup.co.uk
www.kentonline.co.uk
Owner *Kent Messenger Group*
Editor *Bob Bounds*
Circulation 53,987

Very little freelance work is commissioned.

Medway Messenger

Medway House, Ginsbury Close, Sir Thomas Longley Road, Medway City Estate, Strood, Rochester ME2 4DU
☎ 01634 227800   🖷 01634 715256
medwaymessenger@thekmgroup.co.uk
www.kentonline.co.uk
Owner *Kent Messenger Group*
Editor *Bob Bounds*
Circulation 12,335 (Mon)/20,000 (Fri)

Published Mondays and Fridays with the free *Medway Extra* on Wednesday.
Business Editor *Trevor Sturgess*
Community Editor *David Jones*
Sports Editor *Mike Rees*

Lancashire

Lancashire Evening Post

Olivers Place, Eastway, Fulwood, Preston PR2 9ZA
☎ 01772 254841   🖷 01772 880173
www.lep.co.uk
Owner *Johnston Press plc*
Editor *Simon Reynolds*
Circulation 30,467

Unsolicited mss are not generally welcome; many are received and not used. All ideas in writing to the editor.

Lancashire Telegraph

Newspaper House, High Street, Blackburn BB1 1HT
☎ 01254 678678   🖷 01254 680429
lt_editorial@lancashire.newsquest.co.uk
www.lancashiretelegraph.co.uk
Owner *Newsquest Media Group Ltd (a Gannett company)*
Editor *Kevin Young*
Circulation 30,465

News stories and feature material with an East Lancashire flavour (a local angle, or written by local people) welcome. Approach in writing with an idea in the first instance. No fiction.
News Editor *Andrew Turner*
Features Editor *John Anson*
Sports Editor *Paul Plunkett*
Pictures Editor *Neil Johnson*

Oldham Evening Chronicle

PO Box 47, Union Street, Oldham OL1 1EQ
☎ 0161 633 2121   🖷 0161 652 2111
editorial@oldham-chronicle.co.uk
www.oldham-chronicle.co.uk
Owner *Hirst Kidd & Rennie Ltd*
Editor *Jim Williams*
Circulation 20,976

Motoring, food and wine, lifestyle supplement, business page.
News Editor *Mike Attenborough*

The Bolton News

Newspaper House, Churchgate, Bolton BL1 1DE
☎ 01204 522345   🖷 01204 365068
newsdesk@theboltonnews.co.uk
www.theboltonnews.co.uk
Owner *Newsquest Media Group Ltd (a Gannett company)*
Editor *Ian Savage*
Circulation 29,552

Business, children's page, travel, local services, motoring, fashion and cookery.
News Editor *James Higgins*
Features Editor/Women's Page *Andrew Mosley*

The Gazette (Blackpool)

Avroe House, Avroe Crescent, Blackpool FY4 2DP
☎ 01253 400888   🖷 01253 361870
jon.rhodes@blackpoolgazette.co.uk
www.blackpoolgazette.co.uk
Owner *Johnston Press plc*
Managing Director *Darren Russell*
Editor *David Helliwell*
Circulation 29,189

Unsolicited mss welcome in theory. Approach in writing with an idea. Supplements: *Eve* (women, Tues); *All Stars* (football, youth sport, Wed); *Homes* (Thurs); *The Weekend* (entertainment, Fri); *Wheels* (motoring, Fri); *Life!* magazine (entertainment & leisure, Sat); *Jobs Plus* (jobs and careers, Sat).

Leicestershire

Leicester Mercury

St George Street, Leicester LE1 9FQ
☎ 0116 251 2512   🖷 0116 253 0645
newsdesk@leicestermercury.co.uk
www.thisisleicestershire.co.uk
Owner *Northcliffe Media Group Ltd*
Editor *Nick Carter*

Circulation 70,028
News Editor *Mark Charlton*
Features Editor *Alex Dawson*

Lincolnshire

Grimsby Telegraph

80 Cleethorpe Road, Grimsby DN31 3EH
☎ 01472 360360 📠 01472 372257
newsdesk@grimsbytelegraph.co.uk
Owner *Northcliffe Media Group Ltd*
Editor *Michelle Lalor*
Circulation 34,590

Sister paper of the *Scunthorpe Telegraph.*
Unsolicited mss generally welcome. Approach in
writing. No fiction. Weekly supplements: *Business
Telegraph*; *Property Telegraph*; *Motoring.* All
material to be addressed to the News Editor, *Lucy
Wood.* Particularly welcomes hard news stories;
approach in haste by telephone.

Lincolnshire Echo

Brayford Wharf East, Lincoln LN5 7AT
☎ 01522 820000 📠 01522 804493
news@lincolnshireecho.co.uk
Owner *Northcliffe Media Group Ltd*
Editor *Jon Grubb*
Circulation 22,263

News and sport, best buys, holidays, motoring,
dial-a-service, restaurants, leisure, home
improvement, record reviews, gardening corner,
stars.

Scunthorpe Telegraph

4–5 Park Square, Scunthorpe DN15 6JH
☎ 01724 273273 📠 01724 273101
Owner *Northcliffe Media Group Ltd*
Editor *Mel Cook*
Circulation 20,568

All correspondence should go to the Deputy
Editor, *Dave Atkin.*

London

Evening Standard

Northcliffe House, 2 Derry Street, London W8 5EE
☎ 020 7938 6000 📠 020 7937 2648
www.standard.co.uk
Owner *Associated Newspapers/Lord Rothermere*
Editor *Veronica Wadley*
Circulation 288,157

Long-established evening paper, serving
Londoners with both news and feature material.
Genuine opportunities for London-based
features. Produces a weekly colour supplement,
ES Magazine (Fri) and *Homes & Property* (Wed).
Deputy Editor *Andrew Bordiss,*
Executive Editor *Anne McElvoy*

Managing Editor *Doug Wills*
Features Editor *Simon Davies*
News Editor *Hugh Dougherty*
Arts Editor *Fiona Hughes*
Sports Editor *Martin Chilton*
ES Editor *Catherine Ostler*
Homes & Property Editor *Janice Morley*

Manchester

Manchester Evening News

Number 1 Scott Place, Manchester M3 3RN
☎ 0161 832 7200 📠 0161 211 2030
www.manchestereveningnews.co.uk
Owner *Manchester Evening News Ltd*
Editor *Paul Horrocks*
Circulation 81,326

One of the country's major regional dailies.
Initial approach in writing preferred. No fiction.
Personal Finance (Mon); *Health*; *Holidays* (Tues);
Homes & Property (Wed); *Small Business* (Thurs);
Lifestyle (Fri/Sat).
News Editor *Sarah Lester*
Features Editor *Deanna Delamotta* Regional news
features, personality pieces and showbiz profiles
considered. Maximum 1,200 words.
Sports Editor *Peter Spencer*
Women's Page *Helen Tither*
💷 Payment based on house agreement rates.

Merseyside

Liverpool Daily Post

PO Box 48, Old Hall Street, Liverpool L69 3EB
☎ 0151 227 2000 📠 0151 472 2506
www.liverpooldailypost.co.uk
Owner *Trinity Mirror Plc*
Editor *Mark Thomas*
Circulation 15,581

Unsolicited mss welcome. Receives about six
a day. Approach in writing with an idea. No
fiction. Local, national/international news,
current affairs, profiles (with pictures). Maximum
800–1,000 words.
Features Editor *Emma Johnson*
News Editor *Andy Kelly*
Business Editor *Bill Gleeson*
Sports Editor *Richard Williamson*

Liverpool Echo

PO Box 48, Old Hall Street, Liverpool L69 3EB
☎ 0151 227 2000 📠 0151 236 4682
letters@liverpoolecho.co.uk
www.liverpoolecho.co.uk
Owner *Trinity Mirror Merseyside*
Editor *Alastair Machray*

Circulation 106,401

One of the country's major regional dailies. Unsolicited mss welcome; initial approach with ideas in writing preferred.
News Editor *Maria Breslin*
Features Editor *Jane Haase*
Sports Editor *John Thompson*
Women's Editor *Susan Lee*1

Norfolk

Eastern Daily Press

Prospect House, Rouen Road, Norwich NR1 1RE
℡ 01603 628311   🖷 01603 623872
EDP@archant.co.uk
www.EDP24.co.uk
Owner *Archant Regional*
Editor *Peter Franzen, OBE*
Circulation 64,700

Most pieces by commission only. Supplements: *Centro* (daily); rental, motoring, business, property pages, agriculture, employment (all weekly); *Event* full-colour magazine (Fri); *Saturday* full colour magazine; *Sunday* 24-page supplement.
Deputy Editor *Peter Waters*
Assistant Editor (News) *Paul Durrant*
Head of Feature Content *Sarah Hardy*
Head of Sport *Richard Willner*
Assistant Editor (*Saturday*) *Steve Snelling*

Evening News

Prospect House, Rouen Road, Norwich NR1 1RE
℡ 01603 628311   🖷 01603 219060
tim.williams@archant.co.uk
www.eveningnews24.co.uk
Owner *Archant Ltd*
Editor *James Foster*
Circulation 22,914

Includes special pages on local property, motoring, pop, fashion, arts, entertainments and TV, gardening, local music scene, home and family.
Deputy Editor *Tim Williams* (tim.williams@archant.co.uk)
Features Editor *Derek James* (derek.james@archant.co.uk)

Northamptonshire

Chronicle and Echo

Upper Mounts, Northampton NN1 3HR
℡ 01604 467000   🖷 01604 467190
Owner *Northamptonshire Newspapers*
Editor *David Summers*
Circulation 20,070

Unsolicited mss are 'not necessarily unwelcome but opportunities to use them are rare'. Approach

in writing with an idea. No fiction. Supplements: *Sport* (Mon); *Property Today* (Wed); *Jobs Today* (Thurs); *Motors Today* and *The Guide* (Friday); *Weekend Life* (Sat).
News Editor *Richard Edmondson*
Features Editor/Women's Page *Lily Canter*
Sports Editor *Jeremy Casey*

Evening Telegraph

Newspaper House, Ise Park, Rothwell Road, Kettering NN16 8GA
℡ 01536 506100   🖷 01536 506195 (editorial)
et.newsdesk@northantsnews.co.uk
www.northantset.co.uk
Owner *Johnston Press Plc*
Editor *Jeremy Clifford*
Circulation 22,451

Northamptonshire Business Guide (weekly); *Entertainment Guide* and *Jobs* supplements (Thurs); films and eating out (Fri); and *Home & Garden* – monthly lifestyle supplement including gardening, etc.
Deputy Editor *Neil Pickford*
Assistant Editor *Nick Tite*
News Editor *Kristy Ward*
Sports Editor *Jim Lyon*

Nottinghamshire

Evening Post Nottingham

Castle Wharf House, Nottingham NG1 7EU
℡ 0115 948 2000   🖷 0115 964 4032
Owner *Northcliffe Media Group Ltd*
Editor *Malcolm Pheby*
Circulation 57,699

Unsolicited mss occasionally used. Good local interest only. Maximum 800 words. No fiction. Send ideas in writing.
News Editor *Steven Fletcher*
Deputy Editor *Martin Done*
Sports Editor *Dave Parkinson*

Oxfordshire

Oxford Mail

Osney Mead, Oxford OX2 0EJ
℡ 01865 425262   🖷 01865 425554
nqonews@nqo.com
www.oxfordmail.co.uk
Owner *Newsquest (Oxfordshire) Ltd*
Editor *Simon O'Neill*
Circulation 25,100

Unsolicited mss are considered but a great many unsuitable offerings are received. Approach in writing with an idea, rather than by phone. No fiction.
🖻 All fees negotiable.

Shropshire

Shropshire Star

Ketley, Telford TF1 5HU
☎ 01952 242424   🖷 01952 254605
Owner *Shropshire Newspapers Ltd*
Editor *Sarah-Jane Smith*
Circulation 71,513

No unsolicited mss; approach the editor with ideas in writing in the first instance. No news or fiction.
Head of Features *Carl Jones* Limited opportunities; uses mostly in-house or syndicated material. Maximum 1,200 words.
Sports Editor *Dave Ballinger*

Somerset

Evening Post

Temple Way, Bristol BS99 7HD
☎ 0117 934 3000   🖷 0117 934 3575
epnews@bepp.co.uk
www.thisisbristol.co.uk
Owner *Bristol News & Media Ltd (part of Northcliffe Media Group Ltd)*
Editor *Mike Norton*
Circulation 51,287
News Editor *Robin Perkins*
Newsdesk *Jane Westhead*
Features Editor *David Webb*
Sports Editor *Steve Mellen*

The Bath Chronicle

Windsor House, Windsor Bridge Road, Bath BA2 3AU
☎ 01225 322322   🖷 01225 322291
www.thisisbath.com
Owner *Northcliffe Media Group Ltd*
Editor *Sam Holliday* (s.holliday@bathchron.co.uk)

Local news and features especially welcomed (news@bathchron.co.uk).
Deputy Editor *Paul Wiltshire*
Features Editor *Jackie Chappell*
Sports Editor *Julie Riegal*

Western Daily Press

Temple Way, Bristol BS99 7HD
☎ 0117 934 3000   🖷 0117 934 3574
WDEditor or WDNews or WDFeats@bepp.co.uk
www.westerndailypress.co.uk
Owner *Bristol News and Media Ltd*
Editor *Andy Wright*
Circulation 41,639

Sports Editor *Steve Mellen*
Women's Page *Susie Weldon* (Features Production)

Staffordshire

Burton Mail

65–68 High Street, Burton upon Trent DE14 1LE
☎ 01283 512345   🖷 01283 515351
editorial@burtonmail.co.uk
Owner *Staffordshire Newspapers Ltd*
Editor *Paul Hazeldine*
Circulation 14,658

Fashion, health, wildlife, environment, nostalgia, financial/money; consumer; women's world, rock; property; motoring, farming; what's on, leisure *Weekend Supplement* (Sat).
Sports Editor *Rex Page*

The Sentinel

Sentinel House, Etruria, Stoke on Trent ST1 5SS
☎ 01782 602525   🖷 01782 201167
www.thisisthesentinel.co.uk
Owner *Staffordshire Sentinel News & Media Ltd*
Circulation 61,910
Editor-in-Chief *Mike Sassi*

Weekly sports final supplement. All material should be sent to the Assistant Editor, *Martin Tideswell.*

Suffolk

East Anglian Daily Times

Press House, 30 Lower Brook Street, Ipswich IP4 1AN
☎ 01473 230023   🖷 01473 324776
news@eadt.co.uk
www.eadt.co.uk
Owner *Archant Ltd*
Editor *Terry Hunt*
Circulation 34,392

Founded 1874. Unsolicited mss generally not welcome; three or four received a week and almost none are used. Approach in writing in the first instance. No fiction. Supplements: sport (Mon); business (Tues); jobs (Wed); property (Thurs); motoring (Fri); magazine (Sat).
News Editor *Brad Jones* Hard news stories involving East Anglia (Suffolk, Essex particularly) or individuals resident in the area are always of interest.
Features *Julian Ford* Mostly in-house, but will occasionally buy in when the subject is of strong Suffolk/East Anglian interest. Photo features preferred (extra payment). Special advertisement features are regularly run. Some opportunities here. Maximum 1,000 words.
Deputy Editor, Sport *Colin Adwent*
Women's Page *Victoria Hawkins*

Evening Star

30 Lower Brook Street, Ipswich IP4 1AN
℡ 01473 230023 🄵 01473 324850
Owner *Archant Ltd*
Editor *Nigel Pickover*
Circulation 19,707

News Editor *Jess Nicholls*

Sussex

The Argus

Argus House, Crowhurst Road, Hollingbury,
Brighton BN1 8AR
℡ 01273 544544 🄵 01273 505703
editor@theargus.co.uk
www.theargus.co.uk
Owner *Newsquest (Sussex) Ltd*
Editor *Michael Beard*
Circulation 32,874
News Editor *Frankie Taggart*
Sports Editor *Chris Giles*

Teesside

Evening Gazette

Borough Road, Middlesbrough TS1 3AZ
℡ 01642 234242
www.gazettelive.co.uk
Owner *Trinity Mirror plc*
Editor *Darren Thwaites*
Circulation 50,920

Special pages: health, education, family, business,
property, entertainment, leisure, motoring,
recruitment.
News Editor *Jim Horsley*
Business Features/Commercial Features *Sue Scott*
Sports Editor *Phil Tallentire*
Councils *Sandy McKenzie*
Health *Marie Levy*
Crime *Simon Haworth*

Tyne & Wear

Evening Chronicle

Groat Market, Newcastle upon Tyne NE1 1ED
℡ 0191 232 7500 🄵 0191 232 2256
www.chroniclelive.co.uk
www.icnewcastle.co.uk
Owner *Trinity Mirror Plc*
Editor *Paul Robertson*
Circulation 74,977

Receives a lot of unsolicited material, much of
which is not used. Family issues, gardening,
pop, fashion, cooking, consumer, films and
entertainment guide, home improvements,
motoring, property, angling, sport and holidays.
Approach in writing with ideas. Limited

opportunities for features due to full-time feature
staff.
News Editor *James Marley*
Sports Editor *Paul New*
Women's Interests *Liz Lamb*

Shields Gazette

Chapter Row, South Shields NE33 1BL
℡ 0191 427 4800 🄵 0191 456 8270
www.southtynesidetoday.co.uk
Owner *Northeast Press Ltd*
Editor *John Szymanski*
Circulation 18,726

News & Features Editor *Helen Charlton*

Sunday Sun

ncjMedia, Groat Market, Newcastle upon Tyne
NE1 1ED
℡ 0191 201 6299 🄵 0191 201 6180
colin.patterson@ncjmedia.co.uk
www.sundaysun.co.uk flnm.co.uk (football
website)
Owner *Trinity Mirror Plc*
Editor *Colin Patterson*
Circulation 68,033

All material should be addressed to the
appropriate editor (phone to check), or to the
editor.
Sports Editor *Neil Farrington*

Sunderland Echo

Echo House, Pennywell, Sunderland SR4 9ER
℡ 0191 501 5800 🄵 0191 534 5975
rob.lawson@northeast-press.co.uk
Owner *Johnston Press Plc*
Editor *Rob Lawson*
Circulation 42,910

All editorial material to be addressed to the News
Editor, *Peter Jeffrey* (echo.news@northeast-press.
co.uk).

The Journal

Groat Market, Newcastle upon Tyne NE1 1ED
℡ 0191 232 7500/201 6344 (Newsdesk)
🄵 0191 232 2256/201 6044
jnl.newsdesk@ncjmedia.co.uk
www.journallive.co.uk
Owner *Trinity Mirror Plc*
Editor *Brian Aitken*
Circulation 35,476

Daily platforms include farming and business.
Deputy Editor *Peter Montellier*
Sports Editor *Kevin Dinsdale*
Arts & Entertainment Editor *David Whetstone*
Environment Editor *Tony Henderson*
Business Editor *Iain Laing*

Warwickshire

Coventry Telegraph
Corporation Street, Coventry CV1 1FP
☎ 024 7663 3633   🖷 024 7655 0869
news@coventry-telegraph.co.uk
www.IcCoventry.co.uk
Owner *Trinity Mirror Plc*
Editor *Alan Kirby*
Circulation 46,933

Unsolicited mss are read, but few are published. Approach in writing with an idea. No fiction. All unsolicited material should be addressed to the editor. Maximum 600 words for features.
News Managers *Steve Chilton, Steve Williams*
Features Manager/Women's Page *Tara Cain*
Sports Editor *Rob Madill*
£ Payment negotiable.

Leamington Spa Courier
32 Hamilton Terrace, Leamington Spa CV32 4LY
☎ 01926 457755   🖷 01926 339960
editorial@leamingtoncourier.co.uk
www.leamingtoncourier.co.uk
Owner *Johnston Press Publishing*
Editor *Martin Lawson*
Circulation 13,413

One of the Leamington Spa Courier series which also includes the *Warwick Courier* and *Kenilworth Weekly News*. Unsolicited feature articles considered, particularly matter with a local angle. Telephone with idea first.
News Editor *Peter Ormerod*

West Midlands

Birmingham Mail
PO Box 78, Weaman Street, Birmingham B4 6AT
☎ 0121 236 3366   🖷 0121 233 0271 (editorial)
steve.dyson@birminghammail.net
www.birminghammail.net
Owner *Trinity Mirror Plc*
Editor *Steve Dyson*
Circulation 67,231

Freelance contributions are welcome, particularly topics of interest to the West Midlands and feature pieces offering original and lively comment on family issues, health or education. Also news tips and community picture contributions.
Deputy Editor *Carole Cole*
Assistant Editor (Content) *Alf Bennett*
Features *Paul Fulford*
News *Andy Richards*
Sport *Ken Montgomery*
Pictures *Steve Murphy*
Business *Jon Griffin*

Birmingham Post
Weaman Street, Birmingham B4 6AY
☎ 0121 234 5301   🖷 0121 234 5625
Owner *Trinity Mirror Plc*
Editor *Marc Reeves*
Circulation 12,549

One of the country's leading regional newspapers. Freelance contributions are welcome. Topics of interest to the West Midlands and pieces offering lively, original comment are particularly welcome.
News Editor *Mo Ilyas*
Features Editor *Sarah Probert*

Express & Star
Queen Street, Wolverhampton WV1 1ES
☎ 01902 313131   🖷 01902 319721
Owner *Midlands News Association*
Editor *Adrian Faber*
Circulation 138,780
Deputy Editor *Keith Harrison*
Assistant Editor *Marc Drew*
Features Editor *Emma Farmer*
Business Editor *Jim Walsh*
Sports Editor *Tony Reynolds*
Internet News Editor *Tim Walters*
Women's Editor *Maria Cusine*

Sunday Mercury (Birmingham)
Weaman Street, Birmingham B4 6AY
☎ 0121 234 5567   🖷 0121 234 5877
sundaymercury@mrn.co.uk
www.icbirmingham.co.uk
Owner *Trinity Mirror Plc*
Editor *David Brookes*
Circulation 59,339
Deputy Editor (Features) *Paul Cole*
Head of Content (News) *Tony Larner*
Assistant Editor (Sport) *Lee Gibson*
Lifestyle Editor *Zoe Chamberlain*

Wiltshire

Swindon Advertiser
100 Victoria Road, Swindon SN1 3BE
☎ 01793 501806   🖷 01793 501888
newsdesk@swindonadvertiser.co.uk
www.swindonadvertiser.co.uk
Owner *Newsquest (Wiltshire) Ltd*
Editor *Dave King*
Circulation 21,951

Copy and ideas invited. 'All material must be strongly related or relevant to the town of Swindon or the county of Wiltshire.' Little scope for freelance work. Fees vary depending on material.
Deputy Editor *Pauline Leighton*
News Editor *Kevin Burchall*
Sports Editor *Steve Butt*

Worcestershire

Worcester News

Berrow's House, Hylton Road, Worcester WR2 5JX
℡ 01905 748200   ℱ 01905 742277
Owner *Newsquest (Midlands South) Ltd*
Editor *Kevin Ward*
Circulation 17,668

Community (Tues); jobs (Wed); property (Thurs); entertainment and motoring (Fri).
News Editor *Stephanie Preece*
Sports Editor *Paul Ricketts*

Yorkshire

Huddersfield Daily Examiner

Queen Street South, Huddersfield HD1 3DU
℡ 01484 430000   ℱ 01484 437789
Owner *Trinity Mirror Plc*
Editor *Roy Wright*
Circulation 25,337

Home improvement, home heating, weddings, dining out, motoring, fashion, services to trade and industry.
Deputy Editor *Michael O'Connell*
News Editor *Neil Atkinson*
Features Editor *Andrew Flynn*
Sports Editor *Mel Booth*
Women's Page *Hilarie Stelfox*

Hull Daily Mail

Blundell's Corner, Beverley Road, Hull HU3 1XS
℡ 01482 327111   ℱ 01482 315353
news@mailnewsmedia.co.uk
www.thisishullandeastriding.co.uk
Owner *Northcliffe Media Group Ltd*
Editor *John Meehan*
Circulation 56,208
News Editor *Paul Baxter*
Features Editor *Brian Marshall*

Scarborough Evening News

17–23 Aberdeen Walk, Scarborough YO11 1BB
℡ 01723 363636   ℱ 01723 379033
editorial@scarborougheveningnews.co.uk
www.scarboroughtoday.co.uk
Owner *Yorkshire Regional Newspapers Ltd*
Editor *Ed Asquith*
Circulation 13,626

Special pages include property (Mon); jobs (Thurs); motors (Fri).
Deputy Editor *Sue Wilkinson*
News Editor *Steven Hartley*
Sports Editor *Charles Place*
All other material should be addressed to the editor.

Telegraph & Argus (Bradford)

Hall Ings, Bradford BD1 1JR
℡ 01274 729511   ℱ 01274 723634
www.thetelegraphandargus.co.uk
Owner *Newsquest Media Group Ltd (a Gannett company)*
Editor *Perry Austin-Clarke*
Circulation 36,839

No unsolicited mss – approach in writing with samples of work. No fiction.
Head of News *Martin Heminway*
Features Editor *Emma Clayton*
Local features and general interest. Showbiz pieces. 600–1,000 words (maximum 1,500).

The Doncaster Star

Sunny Bar, Doncaster DN1 1NB
℡ 01302 348500   ℱ 01302 348528
doncaster@sheffieldnewspapers.co.uk
www.thestar.co.uk/doncaster
Owner *Sheffield Newspapers Ltd*
Editor *Alan Powell*
Circulation 3,939

All editorial material to be addressed to the News Editor, *David Kessen*.
Sports Writer *Steve Hossack*

The Press

PO Box 29, 76–86 Walmgate, York YO1 9YN
℡ 01904 653051   ℱ 01904 612853
newsdesk@ycp.co.uk
www.yorkpress.co.uk
Owner *Newsquest Media Group (a Gannett company)*
Editor *Kevin Booth*
Circulation 33,045

Unsolicited mss not generally welcome, unless submitted by journalists of proven ability.
Business Press Pages (daily); *Property Press* (Thurs); *Twenty4Seven* – entertainment (Fri).
Assistant Editor *Fran Clee*
Head of Content *Scott Armstrong*
Picture Editor *Martin Oates*
Sports Editor *Stuart Martel*
£ Payment negotiable.

The Star

York Street, Sheffield S1 1PU
℡ 0114 276 7676   ℱ 0114 272 5978
starnews@sheffieldnewspapers.co.uk
www.thestar.co.uk
Owner *Johnston Press*
Editor *Alan Powell*
Circulation 52,300

Unsolicited mss not welcome, unless topical and local.
News Editor *Charles Smith* Contributions only accepted from freelance news reporters if they relate to the area.

Features Editor *Martin Smith* Very rarely requires outside features, unless on specialized local subjects.
Sports Editor *Bob Westerdale*
Women's Page *Jo Davison*
Ⓔ Payment negotiable.

Yorkshire Evening Post

Wellington Street, Leeds LS1 1RF
☏ 0113 243 2701   Ⓕ 0113 238 8536
eped@ypn.co.uk
www.yorkshireeveningpost.co.uk
Owner *Johnston Press*
Editor *Paul Napier*
Circulation 55,732

Evening sister of the Yorkshire Post.
News Editor *Gillian Haworth*
Features Editor *Jayne Dawson*
Sports Editor *Mark Absolom*
Women's Editor *Jayne Dawson*

Yorkshire Post

Wellington Street, Leeds LS1 1RF
☏ 0113 243 2701   Ⓕ 0113 238 8537
yp.editor@ypn.co.uk
www.yorkshireposttoday.co.uk
Owner *Johnston Press*
Editor *Peter Charlton*
Circulation 49,031

A serious-minded, quality regional daily with a generally conservative outlook. Three or four unsolicited mss arrive each day; all will be considered but initial approach in writing preferred. All submissions should be addressed to the editor. No fiction, poetry or family histories.
Deputy Editor *Duncan Hamilton*
Features Editor *Sarah Freeman* Open to suggestions in all fields (though ordinarily commissioned from specialist writers).
Sports Editor *Matt Reeder*

Northern Ireland

Belfast News Letter

Johnson Press, Northern Ireland Division, 2 Esky Drive, Portadown BT63 5YY
☏ 028 3839 3939   Ⓕ 028 3839 3940
Owner *Johnson Press*
Editor *Darwin Templeton*

Weekly supplements: *Farming Life*; *Business News Letter*; *Lifestyle*; *PM*; *Sports Ulster*.
News Editors *Steven Moore, Karen Grimerson, Ben Lowrie*
Features Editor *Geoff Hill*
Sports Editor *Brian Millar*
Business Editor *Adrienne McGill*
Agricultural Editor *David McCoy*
Political Editor *Stephen Dempster*

Belfast Telegraph

Royal Avenue, Belfast BT1 1EB
☏ 028 9026 4000   Ⓕ 028 9055 4506
newseditor@belfasttelegraph.co.uk
www.belfasttelegraph.co.uk
Owner *Independent News & Media (UK)*
Editor *Martin Lindsay*
Circulation 75,602

Weekly business, property and recruitment supplements.
Deputy Editor *Paul Connolly*
News Editor *Ronan Henry*
Features Editor *Gail Walker*
Sports Editor *Steven Beacom*
Business Editor *Nigel Tilson*

The Irish News

113/117 Donegall Street, Belfast BT1 2GE
☏ 028 9032 2226   Ⓕ 028 9033 7505
newsdesk@irishnews.com
www.irishnews.com
Owner *Irish News Ltd*
Editor *Noel Doran*
Circulation 47,790

All material to appropriate editor (phone to check), or to the news desk.
Assistant Editor *Fiona McGarry*
Features Editor *Joanna Braniff*
Sports Editor *Thomas Hawkins*

Sunday Life

124–144 Royal Avenue, Belfast BT1 1EB
☏ 028 9026 4300   Ⓕ 028 9055 4507
m.hill@belfasttelegraph.co.uk
www.sundaylife.co.uk
Owner *Independent News & Media (UK)*
Editor *Jim Flanagan*
Circulation 67,608
Deputy Editor *Martin Hill*
Sports Editor *Jim Gracey*
Features Editor *Audrey Watson*

Scotland

The Courier and Advertiser

80 Kingsway East, Dundee DD4 8SL
☏ 01382 223131   Ⓕ 01382 454590
courier@dcthomson.co.uk
www.thecourier.co.uk
Owner *D.C. Thomson & Co. Ltd*
Editor *Bill Hutcheon*
Circulation 73,485

Circulates in East Central Scotland. Features occasionally accepted on a wide range of subjects, particularly local/Scottish interest, including finance, insurance, agriculture, motoring, modern homes, lifestyle and fitness. Maximum length: 500 words.

News Editor *Michael Alexander*
Features Editor/Women's Page *Catriona McInnes*
Sports Editor *Graeme Dey*

Daily Record (Glasgow)
▷ National Newspapers

Evening Express (Aberdeen)
PO Box 43, Lang Stracht, Mastrick, Aberdeen
AB15 6DF
☎ 01224 690222　🖷 01224 699575
ee.news@ajl.co.uk
Owner *D.C. Thomson & Co. Ltd*
Editor *Damian Bates*
Circulation 53,384

Circulates in Aberdeen and the Grampian region.
Local, national and international news and
pictures, sport. Family platforms include *What's
On, Counter* (consumer news), *Eating Out Guide,
Family Days Out*. Unsolicited mss welcome 'if on
a controlled basis'.
Deputy Editor *Richard Prest* Freelance news
contributors welcome.
🖹 Payment negotiable.

Evening News
Barclay House, 108 Holyrood Road, Edinburgh
EH8 8AS
☎ 0131 620 8620　🖷 0131 620 8696
www.edinburghnews.com
Owner *Scotsman Publications Ltd*
Editor *John McLellan*
Circulation 50,847

Founded 1873. Circulates in Edinburgh, Fife,
Central and Lothian. Coverage includes:
entertainment, gardening, motoring, shopping,
fashion, health and lifestyle, showbusiness.
Occasional platform pieces, features of topical
and/or local interest. Unsolicited feature material
welcome. Approach the appropriate editor in
writing.
News Editor *Alan Young*
Sports Editor *Graham Lindsay*
🖹 NUJ/house rates.

Evening Telegraph
80 Kingsway East, Dundee DD4 8SL
☎ 01382 223131　🖷 01382 454590
general@eveningtelegraph.co.uk
www.eveningtelegraph.co.uk
Owner *D.C. Thomson & Co. Ltd*
Editor *Gordon Wishart*
Circulation 24,349

Circulates in Tayside, Dundee and Fife. All
material should be addressed to the editor.

Evening Times
200 Renfield Street, Glasgow G2 3QB
☎ 0141 302 7000　🖷 0141 302 6677

news@eveningtimes.co.uk
www.eveningtimes.co.uk
Owner *Newsquest*
Editor *Donald Martin*
Circulation 74,466

Circulates in Glasgow and the west of Scotland.
Supplements: *Job Search*; *City Living*; *Drive times*;
Extra times; *Times Out*; *etc* magazine.
Head of News *Yvonne Flynn*
Features Editor *Garry Scott*
Sports Editor *David Stirling*

Greenock Telegraph
2 Crawfurd Street, Greenock PA15 1LH
☎ 01475 726511　🖷 01475 783734
Owner *Clyde & Forth Press Ltd*
Editor *Wendy Metcalfe*
Circulation 17,430

Circulates in Greenock, Port Glasgow, Gourock,
Kilmacolm, Langbank, Inverkip, Wemyss Bay,
Skelmorlie, Largs. Unsolicited mss considered 'if
they relate to the newspaper's general interests'.
No fiction. All material to be addressed to the
editor.

The Herald/Sunday Herald (Glasgow)
▷ National Newspapers

Paisley Daily Express
14 New Street, Paisley PA1 1YA
☎ 0141 887 7911　🖷 0141 887 6254
pde@s-un.co.uk
www.icrenfrewshire.co.uk
Owner *Scottish & Universal Newspapers Ltd*
Editor *Anne Dalrymple*
Circulation 9,528

Circulates in Paisley, Linwood, Renfrew,
Johnstone, Elderslie, Neilston and Barrhead.
Unsolicited mss welcome only if of genuine local
(Renfrewshire) interest. The paper does not
commission work and will consider submitted
material. Maximum 1,000–1,500 words. All
submissions to the news editor.
News Editor *Gavin Penny*
Sports Reporter *Paul Behan*

The Press and Journal
PO Box 43, Lang Stracht, Mastrick, Aberdeen
AB15 6DF
☎ 01224 690222　🖷 01224 663575
Owner *D.C. Thomson & Co. Ltd*
Editor *Derek Tucker*
Circulation 80,177

A well-established daily that circulates in
Aberdeen, Grampians, Highlands, Tayside,
Orkney, Shetland and the Western Isles. Most
material is commissioned but will consider ideas.
News Editor *Andrew Hebden* Wide variety of hard
or off-beat news and features relating especially,

but not exclusively, to the north of Scotland.
Sports Editor *Alex Martin*
Your Life *Sonja Cox* Saturday lifestyle tabloid
pullout. Features food, fashion, travel, books, arts
and lifestyle.
Ⓔ Payment by arrangement.

Scotland on Sunday (Edinburgh)
▷ National Newspapers)

The Scotsman (Edinburgh)
▷ National Newspapers

Sunday Mail (Glasgow)
▷ National Newspapers

Sunday Post (Dundee)
▷ National Newspapers

Wales

Daily Post
PO Box 202, Vale Road, Llandudno Junction,
Conwy LL31 9ZD
☎ 01492 574455   🖷 01492 574433
welshnews@dailypost.co.uk
Owner *Trinity Mirror Plc*
Editor *Rob Irvine*
Circulation 36,432

Senior Assistant Editor *Mark Brittain*
Features Editor *Sarah Bartley*
News Editor *Debbie James*
Acting Sports Editor *Rob Brady*

Evening Leader
Mold Business Park, Wrexham Road, Mold
CH7 1XY
☎ 01352 707707   🖷 01352 752180
news@eveningleader.co.uk
Owner *NWN Media*
Circulation 21,838
Editor-in-Chief *Barrie Jones*

Circulates in Wrexham, Flintshire, Deeside
and Chester. Special pages/features: motoring,
travel, arts, women's, children's, local housing,
information and news for the disabled, music and
entertainment.
Deputy Editor *Martin Wright*
Assistant Editor *Jonathon Barnett*
Sports Editor *Nick Harrison*

South Wales Argus
Cardiff Road, Maesglas, Newport NP20 3QN
☎ 01633 810000   🖷 01633 777202
www.southwalesargus.co.uk
Owner *Newsquest*
Editor *Gerry Keighley*

Circulation 28,457
Circulates in Newport, Gwent and surrounding
areas.

News Editor *Maria Williams*
Sports Editor *Phil Webb*

South Wales Echo
Thomson House, Havelock Street, Cardiff
CF10 1XR
☎ 029 2058 3622   🖷 029 2058 3624
echo.newsdesk@mediawales.co.uk
www.icwales.co.uk
Owner *Trinity Mirror Plc*
Editor *Mike Hill*
Circulation 46,127

Circulates in South and Mid Glamorgan and
Gwent.
Head of News *Nick Machin*
Head of Features *Alison Stokes*
Head of Sport *Delme Parfitt*

South Wales Evening Post
Adelaide Street, Swansea SA1 1QT
☎ 01792 510000   🖷 01792 514697
postbox@swwmedia.co.uk
www.thisisswansea.co.uk
Owner *Northcliffe Media Group Ltd*
Editor *Spencer Feeney*
Circulation 51,329

Circulates throughout south west Wales.
News Editor *Chris Davies*
Features Editor *Peter Slee*
Sports Editor *David Evans*

Wales on Sunday
Thomson House, Havelock Street, Cardiff
CF10 1XR
☎ 029 2058 3733   🖷 029 2058 3725
Owner *Trinity Mirror plc*
Editor *Tim Gordon*
Circulation 41,199

Launched 1989. Tabloid with sports supplement.
Does not welcome unsolicited mss.
Head of Content *Nick Rippington*
Sports Editor *Paul Banadanato*

Western Mail
Thomson House, Havelock Street, Cardiff
CF10 1XR
☎ 029 2058 3583   🖷 029 2058 3652
www.icwales.co.uk
Owner *Trinity Mirror Plc*
Editor *Alan Edmunds*
Circulation 37,576

Circulates in Cardiff, Merthyr Tydfil, Newport,
Swansea and towns and villages throughout
Wales. Mss welcome if of a topical nature,
of Welsh interest. No short stories or travel.
Approach in writing to the features editor. 'Usual
subjects already well covered, e.g. motoring,
travel, books, gardening. We look for the unusual.'
Maximum 1,000 words. Opportunities also on

women's page. Supplements: *Saturday Magazine*; *Education*; *WM*; *Box Office*; *Welsh Homes*; *Business*; *Sport*; *Motoring*; *Country & Farming*.
Deputy Editor *Ceri Gould*
News Editor *Pal Carey*
Sports Editor *Philip Blanche*
Features Editor *Peter Morrell*

Channel Islands

Guernsey Press & Star
Braye Road, Vale, Guernsey GY1 3BW
℡ 01481 240240   ℻ 01481 240235
newsroom@guernsey-press.com
www.guernsey-press.com
Owner *Guiton Group*
Editor *Richard Digard*
Circulation 16,196

Special pages include children's and women's interest, gardening and fashion.

News Editor *Gemma Hockey*
Sports Editor *Rob Batiste*
Features Editor *Diane Digard*

Jersey Evening Post
PO Box 582, Jersey JE4 8XQ
℡ 01534 611611   ℻ 01534 611622
editorial@jerseyeveningpost.com
www.thisisjersey.com
Owner *Claverley Company*
Editor *Chris Bright*
Circulation 21,100

Special pages: gardening, motoring, property, boating, technology, food and drink, personal finance, health, business.
News Editor *Sue Le Ruez*
Features Editor *Carl Walker*
Sports Editor *Ron Felton*

News Agencies

Associated Press Limited

12 Norwich Street, London EC4A 1BP
☎ 020 7353 1515   📠 020 7353 8118 (Newsdesk)

Material is either generated in-house or by regulars. Hires the occasional stringer. No unsolicited mss.

Dow Jones Newswires

10 Fleet Place, London EC4M 7QN
☎ 020 7842 9900   📠 020 7842 9361 (admin)

A real-time financial and business newswire operated by Dow Jones & Co., publishers in the USA of *The Wall Street Journal*. No unsolicited material.

Famous Pictures and Features Agency

▷ entry under Picture Libraries

National News Press and Photo Agency

4–5 Academy Buildings, Fanshaw Street, London N1 6LQ
☎ 020 7684 3000   📠 020 7684 3030
news@nationalnews.co.uk

All press releases are welcome. Most work is ordered or commissioned. Coverage includes courts, tribunals, conferences, general news, etc. – words and pictures – as well as PR.

The Press Association Ltd

292 Vauxhall Bridge Road, London SW1V 1AE
☎ 020 7963 7000   📠 020 7963 7192 (news desk)
copy@pa.press.net
www.pressassociation.co.uk

No unsolicited material. Most items are produced in-house though occasional outsiders may be used. A phone call to discuss specific material may lead somewhere 'but this is rare'.

Reuters

30 South Colonnade, Canary Wharf, London E14 5EP
☎ 020 7250 1122
www.reuters.com

No unsolicited material.

Solo Syndication Ltd

17–18 Hayward's Place, London EC1R 0EQ
☎ 020 7566 0360   📠 020 7566 0388
wgardiner@solosyndication.com
www.solosyndication.com

Founded 1978. Specializes in worldwide newspaper syndication of photos, features and cartoons. Professional contributors only.

South Yorkshire Sport

6 Sharman Walk, Apperknowle, Sheffield S18 4BJ
☎ 01246 414767/07970 284848 (mobile)
📠 01246 414767
nickjohnson@uwclub.net
Contact *Nick Johnson*

Provides written/broadcast coverage of sport in the South Yorkshire area.

Space Press News and Pictures

Bridge House, Blackden Lane, Goostrey CW4 8PZ
☎ 01477 533403
Scoop2001@aol.com
Editor *John Williams* (07970 213528)
Pictures *Emma Williams* (evwphoto@aol.com; 07976 795494)

Founded 1972. Press and picture agency covering Cheshire and the North West, North Midlands, including Knutsford, Macclesfield, Congleton, Crewe and Nantwich, Wilmslow, Alderley Edge, serving national, regional and local press, TV, radio, and digital picture transmission. Copy and pictures produced for in-house publications and PR. Property, countryside and travel writing. A member of the National Association of Press Agencies (NAPA).

US Media Contacts in the UK

ABC News Intercontinental Inc.
3 Queen Caroline Street, London W6 9PE
T 020 8222 5000   F 020 8222 5020
Director of News Coverage, Europe, Middle East & Africa *Marcus Wilford*

Associated Press
12 Norwich Street, London EC4A 1BP
T 020 7353 1515   F 020 7353 8118 (news)
Vice President – Global Business for Europe, Africa & Middle East *Barry Renfrew*

The Baltimore Sun
11 Kensington Court Place, London W8 5BJ
T 020 7460 2200
Bureau Chief *Todd Richissin*

Bloomberg News
City Gate House, 39–45 Finsbury Square, London EC2A 1PQ
T 020 7330 7500   F 020 7330 7797
London Bureau Chief (Print) *Chris Collins*

Business Week
20 Canada Square, Canary Wharf, London E14 5LH
T 020 7176 6060   F 020 7176 6070
Bureau Chief *Stanley Reed*

Canadian Broadcasting Corporation
43/51 Great Titchfield Street, London W1P 8DD
T 020 7412 9200   F 020 7412 9226
London Bureau Manager *Ann Macmillan*

Canadian Television Network
2nd Floor, The Interchange Building, Oval Road, London NW1 7DZ
T 020 7267 5511   F 020 7267 3772
Bureau Chief *Thomas Kennedy*

CBS News
68 Knightsbridge, London SW1X 7LL
T 020 7887 3000   F 020 7887 3092
Deputy Bureau Chief *Laura Dubowski*

Chicago Tribune Press Service
116 Brompton Road, London SW3 1JJ
T 020 7225 0345
Chief European Correspondent *Tom Hundley*

CNBC
10 Fleet Place, Limeburner Lane, London EC4M 7QS
T 020 7653 9451   F 020 7653 9393
Assignments Editor *Jennifer Callegher*

CNN International
▷ entry under Television and Radio

Dow Jones Newswires
10 Fleet Place, Limeburner Lane, London EC4M 7QN
T 020 7842 9300
Editor (Europe, Middle East, Africa) *Jan Boucek*

Fairchild Publications of New York
20 Shorts Gardens, London WC2H 9AU
T 020 7240 0420
Bureau Chief *Samantha Conti*

Forbes Magazine
Malta House, 36/38 Picadilly, London W1J 0DP
T 020 7534 3900
Managing Director & Publisher (Europe, Middle East & Africa) *Bob Crozier*

Fortune
Blue Fin Building, 110 Southwark Street, London SE1 0SU
Europe Editor *Nelson Schwartz*

Fox News Channel
6 Centaurs Business Park, Grant Way, Isleworth TW7 5QD
T 020 7805 7143   F 020 7805 1111
Bureau Chief *Scott Norvell*

The Globe and Mail
16 St Georges Avenue, London N7 0HD
T 020 7697 9820
European Bureau Chief *Doug Saunders*

International Herald Tribune
40 Marsh Wall, London E14 9TP
T 020 7510 5700
London Correspondent *Eric Pfanner*
See entry under *National Newspapers*

Los Angeles Times
Moreau House, 116–118 Brompton Road, London
SW3 1JJ
T 020 7823 7315   F 020 7823 7308
Bureau Chief *Kim Murphy*

Market News International
Ocean House, 10–12 Little Trinity Lane, London
EC4A 2AR
T 020 7634 1655   F 020 7236 7122
ukeditorial@marketnews.com
London Bureau Chief *David Thomas*

National Public Radio
Room G-10 East Wing, Bush House, Strand,
London WC2B 4PH
T 020 7557 1087   F 020 7379 6486
Bureau Chief *Robert Gifford*

NBC News
4th Floor, 3 Shortlands, Hammersmith, London
W6 8HX
T 020 8600 6600   F 020 8600 6601
Bureau Chief *Chris Hampson*

The New York Times
66 Buckingham Gate, London SW1E 6AU
T 020 7799 5050   F 020 7799 2962
Bureau Chief *John Burns*

Newsweek
Academy House, 36 Poland Street, London
W1F 7LU
T 020 7851 9750   F 020 7851 9762
Bureau Chief *Stryker McGuire*

People Magazine
Blue Fin Building, 110 Southwark Street, London
SE1 0SU
T 020 3148 3000
Bureau Chief *Simon Perry*

Reader's Digest Association Ltd
11 Westferry Circus, Canary Wharf, London
E14 4HE
T 020 7715 8000
See entries under *UK Publishers* and *Magazines*

Time
Blue Fin Building, 110 Southwark Street, London
SE1 0SU
T 020 3148 3000
Bureau Chief *Jef McAllister*
See entry under *Magazines*

Voice of America
London News Centre, 167 Fleet Street, London
EC4A 2EA
T 020 7410 0960
Bureau Chief *Michael Drudge*

Wall Street Journal
10 Fleet Place, Limeburner Lane, London
EC4M 7QN
T 020 7842 9200
Deputy Bureau Chief *Emily Nelson*

Washington Post
Flat 3, 55–56 Hampstead High Street, London
NW3 1QH
T 020 7433 8094
Bureau Chiefs *Mary Jordan, Kevin Sullivan*

Television and Radio

For the latest information on broadcast rates for freelancers access the following websites: the National Union of Journalists' 'The Freelance Fees Guide' database link (www. gn.apc.org/media); the Producers Alliance for Cinema and Television (www.pact.co.uk); the Society of Authors (www.societyofauthors.org); and the Writers' Guild (www.writersguild.org. uk).

BBC TV and Radio

BBC Vision, BBC Audio & Music

www.bbc.co.uk
Director-General *Mark Thompson*
Deputy Director-General *Mark Byford*

The operational structure of the BBC consists of: BBC Vision Group which runs three main areas: production, commissioning and services and includes the terrestrial and digital television channels; Drama, Entertainment and Children's; Factual and Learning. The Audio & Music Group – responsible for TV Music Entertainment; In-house Factual, Specialist Factual and Drama Audio production. The Journalism Group covers News; Sport; Global News, Nations and Regions.

BBC Television Centre, Wood Lane, London
W12 7RJ
☎ 020 8743 8000
Broadcasting House, Portland Place, London
W1A 1AA
☎ 020 7580 4468

TV & RADIO COMMISSIONING:
www.bbc.co.uk/commissioning
Director, BBC Vision *Jana Bennett*
Director of Audio & Music Group *To be appointed*
Controller, BBC One *Jay Hunt*
Controller, BBC Two *Roly Keating*
Controller, Daytime *Liam Keelan*
Controller, BBC Three *Danny Cohen*
Controller, BBC Four *Janice Hadlow*
Controller, Radio 1 and 1Xtra *Andy Parfitt*
Radio 1 documentary submission enquiries: *Joe*

Harland, Executive Producer (joe.harland@bbc. co.uk; ☎ 020 7765 3552)
1Xtra commissioning enquiries: *Charmaine Cozier*, Commissioning Editor, Documentaries (charmaine.cozier@bbc.co.uk)
1Xtra commissions weekly documentaries from independent producers and an in-house production team. Welcomes proposals that may be suitable as a co-commission with another network such as Radio 1 or Asian Network.
Controller, Radio 2 & 6 Music *Lesley Douglas*
Radio 2 commissioning enquiries: *Robert Gallacher*, Editor, Planning & Station Sound (robert.gallacher@bbc.co.uk; ☎ 020 7765 4373)
Controller, Radio 3 *Roger Wright*
Radio 3 commissioning enquiries: *David Ireland*, Commissions & Schedules Manager (david. ireland@bbc.co.uk; ☎ 020 7765 4943)
Controller, Radio 4 & BBC 7 *Mark Damazer*
Radio 4 proposals must be submitted through an in-house department or a registered independent production company (lists of departments and companies are available at www.bbc.co.uk/ commissioning/radio/network/radio4.shtml).
Controller, Five Live and Five Live Sports Extra *Adrian van Klaveren*
Radio Five Live & Asian Network commissioning enquiries: *Michael Hill*, Network Manager (michael.hill@bbc.co.uk; ☎ 020 8624 8945)

BBC Drama, Comedy and Children's

BBC *writersroom* champions new writers across all BBC platforms for drama, comedy and children's programmes, running targeted schemes and workshops linked directly to production. It accepts and assesses unsolicited scripts for all departments: film, TV drama, radio drama, TV narrative comedy and radio narrative comedy. The *writersroom* website (www. bbc.co.uk/writersroom) offers a diary of events, opportunities, competitions, interviews with established writers, submission guidelines and free formatting software. BBC *writersroom* also has a Manchester base which focuses on new writing in the north of England. Writers should address hard copies of original, completed scripts to: Development Manager, BBC *writersroom*, Grafton House, 379–381 Euston Road, London

NW1 3AU. Before sending in scripts, please log on to the website, or send an A5 s.a.e. to *writersroom* at the Grafton House address above for the latest guidelines on submitting unsolicited work. See also www.bbc.co.uk/commissioning/structure/public.shtml

COMMISSIONERS

Controller, Fiction *Jane Tranter*
Head of Drama Commissioning *Ben Stephenson*
Head of Drama Commissioning and Drama Wales *Julie Gardner*
Continuing Comedy Commissioning *Lucy Lumsden*

DRAMA

Controller, Drama Production *John Yorke*
Head of Series and Serials, BBC Drama Production *Kate Harwood*
Creative Director, BBC Drama Production *Manda Levin*
Director, BBC Drama Production *Nicolas Brown*
Executive Producer, EastEnders *Diederick Santer*
Creative Director, New Writing *Kate Rowland*
Head of Radio Drama *Alison Hindell*
Executive Producer (Birmingham) & Editor, The Archers *Vanessa Whitburn*
Executive Producer (World Service Drama) *Marion Nancarrow*
Executive Producer (Manchester Radio Drama) *Sue Roberts*
Editor, Silver Street *James Peries*

COMEDY

Head of Comedy *Mark Freeland*
Head of Radio Entertainment *Paul Schlesinger*

CHILDREN'S

Controller, BBC Children's *Richard Deverell*
Creative Director, CBBC *Anne Gilchrist*
Creative Director, CBeebies *Michael Carrington*
Head of CBBC Drama *Jon East*
Head of CBBC Entertainment *Joe Godwin*
Head of News, Factual & Learning *Reem Nouss*
Head of CBeebies Production, Animation & Acquisitions *Kay Benbow*

BBC News

www.bbc.co.uk/news
news.bbc.co.uk

BBC News is the world's largest news-gathering organization with over 2,500 journalists, 45 bureaux worldwide and 15 networks and services across TV, radio and new media.
Director, News *Helen Boaden*
Deputy Director & Head of Programmes, News *Stephen Mitchell*
Head of Newsgathering *Francesca Unsworth*
Head of Newsroom *Peter Horrocks*
Controller, BBC News 24 *Kevin Bakhurst*

TELEVISION

Editor, 10 o'clock News *Craig Oliver*

Editor, Newsnight *Peter Barron*
Editor, Breakfast *Alison Ford*

RADIO

Editor, Today *Ceri Thomas*
Editor, The World at One/World This Weekend/PM/Broadcasting House *Peter Rippon*
Head of News, Five Live *Matt Morris*
Editor, The World Tonight *Alistair Burnett*
Editor, World Service News Programmes & Current Affairs *Liliane Landor*

CEEFAX

Room 7540, BBC Television Centre, Wood Lane, London W12 7RJ
☎ 020 8576 1801
Editor, Cee🄵 *Paul Brannan*

SUBTITLING

Red Bee Media, Broadcast Centre, 201 Wood Lane, London W12 7TP www.redbeemedia.com
Available via Cee🄵 page 888.

BBC Sport

www.bbc.co.uk/sport
Director, Sport *Roger Mosey*
Head of Programmes & Planning *Philip Bernie*
Head of Football *Niall Sloane*
Head of General Sport *Barbara Slater*

Sports news and commentaries across television, online and Radios 1, 4 and 5 Live.

BBC Religion

New Broadcasting House, Oxford Road, Manchester M60 1SJ
☎ 0161 200 2020   🄵 0161 244 3183
Head of Religion and Ethics *Michael Wakelin*

Regular programmes for television include *Songs of Praise*. Radio output includes *Good Morning Sunday*; *Sunday Half Hour*; *Thought for the Day*; *The Daily Service*.

BBC New Talent

www.bbc.co.uk/newtalent

The BBC's search for new talent covers a constantly changing range of outlets that has included radio producers, presenters, young storytellers, filmmakers and comedy writers. Access the website for latest information.

BBC World Service

PO Box 76, Bush House, Strand, London WC2B 4PH
☎ 020 7240 3456   🄵 020 7557 1258
www.bbc.co.uk/worldservice
Director, World Service *Nigel Chapman*
Director, English Networks & News, BBCWS *Gwyneth Williams*

The World Service broadcasts in English and 32 other languages. The English service is round-the-clock, with news and current affairs as the main component. With over 180 million listeners, excluding countries where research is not possible, it reaches a bigger audience than its five closest international broadcasting competitors combined. BBC World Service is increasingly available throughout the world on local FM stations, via satellite and online as well as through short-wave frequencies. Coverage includes world business, politics, people/events/ opinions, development issues, the international scene, developments in science and technology, sport, religion, music and the arts. BBC World Service broadcasting is financed by a grant-in-aid voted by Parliament amounting to £265 million for 2008/2009.

BBC writersroom
▷ BBC Drama, Comedy and Children's

BBC Nations

BBC Northern Ireland
Broadcasting House, Ormeau Avenue, Belfast BT2 8HQ
☏ 028 9033 8000
bbc.co.uk/northernireland
Controller *Peter Johnston*
Head of Broadcasting/Interactive and Learning *Ailsa Orr*
Chief Operation Officer *Mark Taylor*
Head of Production Planning & Development *Stephen Beckett*
Head of News & Current Affairs *Andrew Colman*
Head of Drama *Patrick Spence*
Head of Entertainment, Events & Sport *Mike Edgar*
Head of Radio *Susan Lovell*
Head of Factual TV *Paul McGuigan*
Commissioning Editor Broadcasting *Fergus Keeling*
Editor Learning *Jane Cassidy*
Editor New Media *David Sims*
Senior Producer Irish Language *Antaine O'Donnaile*
Editor, TV News *Angelina Fusco*
Editor, Radio News *Kathleen Carragher*
Editor, News Gathering *Michael Cairns*
Editor TV Current Affairs *Jeremy Adams*
Editor News Online *Seamus Boyd*
Editor – Foyle *Paul McCauley*

Regular television programmes include *BBC Newsline 6.30* and a wide range of documentary, popular factual and entertainment programmes. Radio stations: BBC Radio Foyle and BBC Radio Ulster (see entries). For further details on television and film scripts contact: *Susan Carson*, Programme Development Executive Television, BBC Northern Ireland Drama, Room 3.07, Blackstaff House, Great Victoria Street,

Belfast, BT2 7BB. (☏ 028 9033 8498; tvdrama.ni@ bbc.co.uk). Radio scripts contact: *Anne Simpson*, Manager, Radio Drama at the same address.

BBC Scotland
40 Pacific Quay, Glasgow G51 1DA
☏ 0141 422 6000
www.bbc.co.uk/scotland
Controller *Ken MacQuarrie*
Chief Operating Officer *Bruce Malcolm*
Head of Programme and Services *Donalda MacKinnon, Maggie Cunningham*
Head of Drama, Television *Anne Mensah*
Head of Drama, Radio *Patrick Rayner*
Head of Factual Programmes *Andrea Miller*
Head of Gaelic *Margaret Mary Murray*
Head of Radio *Jeff Zycinski*
Head of News and Current Affairs *Atholl Duncan*
Head of Sport & Commissioning Editor, Television *Ewan Angus*
Head of Children's *Simon Parsons*

Headquarters of BBC Scotland with centres in Aberdeen, Dundee, Edinburgh, Dumfries, Inverness, Orkney, Shetland and Stornaway. Regular and recent programmes include *Reporting Scotland*, *Sportscene* and *River City* on television and *Good Morning Scotland* and *Macaulay and Co.* on radio.

Aberdeen
Broadcasting House, Beechgrove Terrace, Aberdeen AB9 2ZT
☏ 01224 625233

Dundee
Nethergate Centre, 66 Nethergate, Dundee DD1 4ER
☏ 01382 202481

Edinburgh
The Tun, Holyrood Road, Edinburgh EH8 8JF
☏ 0131 557 5677

Dumfries
Elmbank, Lover's Walk, Dumfries DG1 1NZ
☏ 01387 268008

Inverness
7 Culduthel Road, Inverness 1V2 4ADT
☏ 01463 720720
Editor *Ishbel MacLennan*

Orkney
Castle Street, Kirkwall, Orkney KW15 1DF

Shetland
Pitt Lane, Lerwick, Shetland ZE1 0DW
☏ 01595 694747

Stornaway
Rosebank, Church Street, Stornoway, Isle of Lewis PA87 2LS
☏ 01851 705000

Radio Nan Gaidheal
Rosebank, Church Street, Stornoway, Isle of Lewis PA87 2LS
☎ 01851 705000

BBC Wales

Broadcasting House, Llandaff CF5 2YQ
☎ 029 2032 2000   🖷 029 2055 2973
www.bbc.co.uk/wales
Controller *Menna Richards*
Head of Programmes (Welsh Language) *Keith Jones*
Head of Programmes (English Language) *Clare Hudson*
Head of News & Current Affairs *Mark O'Callaghan*
Head of Drama *Julie Gardner*
Commissioning Editor, Independents and BBC2W *Martyn Ingram*
Producer, Pobol y Cwm *Bethan Jones*
Editor, New Media *Iain Tweedale*

Headquarters of BBC Wales, with regional centres in Bangor, Aberystwyth, Carmarthen, Wrexham and Swansea. BBC Wales television produces up to 12 hours of English language programmes a week, 12 hours in Welsh for transmission on S4C and an increasing number of hours on network services, including *Doctor Who*, *Tribe* and *Torchwood*. BBC2W, the digital channel, was launched in 2001. Regular programmes include *Wales Today*; *Newyddion* (Welsh-language daily news); *Week In Week Out*, *X-Ray* and *Pobol y Cwm* (Welsh-language soap) on television and *Good Morning Wales*; *Good Evening Wales*; *Post Cyntaf* and *Post Prynhawn* on Radio Wales and Radio Cymru. Ideas for programmes should be submitted either by post to *Martyn Ingram*, Commissioning Editor, Room 3021 at the address above or by e-mail to commissioning@bbc.co.uk

Bangor Broadcasting House, Meirion Road, Bangor LL57 2BY
☎ 01248 370880 🖷 01248 351443
Head of Centre *Marian Wyn Jones*

BBC Regions

BBC Asian Network

The Mailbox, Birmingham B1 1RF
☎ 0121 567 6000/0116 201 6772 (newsdesk)
asian.network@bbc.co.uk
www.bbc.co.uk/asiannetwork
Managing Editor *Vijay Sharma*

Commenced broadcasting in November 1996. Broadcasts nationwide. Programmes in English, Bengali, Gujerati, Hindi, Punjabi and Urdu.

BBC Birmingham

The Mailbox, Birmingham B1 1RF
☎ 0121 567 6767   🖷 0121 567 6875

birmingham@bbc.co.uk
www.bbc.co.uk/birmingham
Head of Regional and Local Programmes *David Holdsworth*
Output Editor *Dave Hart*

BBC Birmingham's Documentaries & Contemporary Factual Department produces a broad range of programming for BBC ONE and BBC TWO, including *Gardener's World*; *Coast*; *Desi DNA*; *To Buy or Not to Buy*. Network radio includes: Radio 2, Radio 4 and the Asian Network. Regional & Local Programmes: TV: *Midlands Today*; *Inside Out*; *The Politics Show*; Radio: BBC WM (see entry). BBC Birmingham is the headquarters for BBC English Regions. There is a television Drama Village at the University of Birmingham.

BBC Birmingham Drama
Archibald House, 1059 Bristol Road, Selly Oak, Birmingham B29 6LT
☎ 0121 567 7350
Programmes produced include *Dalziel and Pascoe*; *Doctors*; *The Afternoon Play* and *Brief Encounters* (which champions new writers).

BBC East Midlands (Nottingham)
East Midlands Broadcasting Centre, London Road, Nottingham NG2 4UU
☎ 0115 955 0500 🖷 0115 902 1983
www.bbc.co.uk/nottingham
Head of Regional Programming *Aziz Rashid*
Output Editor *Sally Bowman*

BBC East (Norwich)
The Forum, Millennium Plain, Norwich NR2 1AW
☎ 01603 284700 🖷 01603 284399
www.bbc.co.uk/norfolk
Head of Regional Programming *Tim Bishop*
Output Editor *Dave Betts*

BBC Bristol

Broadcasting House, Whiteladies Road, Bristol BS8 2LR
☎ 0117 973 2211
www.bbc.co.uk/bristol
Head of Regional and Local Programmes *Lucio Mesquita*
Managing Editor, BBC Radio Bristol *Tim Pemberton*
Head of Programmes Leisure and Factual Entertainment *Tom Archer*
Head of Natural History Unit *Neil Nightingale*

BBC Bristol is the home of the BBC's Natural History Unit, producing programmes such as *Blue Planet*; *Life of Mammals*; *British Isles, A Natural History*; *Abyss* and *Planet Earth* for BBC ONE and BBC TWO. It also produces natural history programmes for Radio 4 and Radio 5 Live. The Factual department produces a wide range of television programmes, including *DIY SOS*; *Antiques Roadshow*; *Soul Deep* and *Ray Mears* in addition to radio programmes specializing in history, travel, literature and human interest features for Radio 4.

BBC London

35 Marylebone High Street, London W1U 4QA
☏ 020 7224 2424
www.bbc.co.uk/london
Executive Editor *Mike MacFarlane* (Responsible for BBC tri-media: BBC London News, BBC London 94.9FM and the BBC London website)
Planning Editor *Duncan Williamson*

BBC South East (Tunbridge Wells)

The Great Hall, Mount Pleasant Road, Tunbridge Wells TN1 1QQ
☏ 01892 670000
www.bbc.co.uk/kent
Head of Regional and Local Programmes *Mick Rawsthorne* (Responsible for BBC South East [TV], BBC Radio Kent and BBC Southern Counties Radio)
Editor, South East Today *Quentin Smith*
Managing Editor, Southern Counties Radio *Nicci Holiday*
Managing Editor, Radio Kent *Paul Leaper*
Editor, Newsgathering *Michael Gravesande*
Producer, Inside Out *Linda Bell*

BBC West/BBC South/BBC South West

The three regional television stations, BBC West, BBC South and BBC South West produce the lunchtime and nightly news magazine programmes, as well as *Inside Out*, regular 30-minute local current affairs programmes and regional parliamentary programmes, *The Politics Show*. Each of the regions operates a comprehensive local radio service as well as a range of correspondents specializing in subjects such as health, education, business, home affairs and the environment.

BBC West (Bristol)

Broadcasting House, Whiteladies Road, Bristol BS8 2LR
☏ 0117 973 2211
Head of Regional and Local Programmes *Lucio Mesquita* (Responsible for the BBC West region. TV: BBC Points West (regional news programme), Inside Out West (regional current affairs series) and The Politics Show West. Local Radio: BBC Radio Bristol, BBC Somerset, BBC Radio Gloucestershire, BBC Radio Wiltshire, BBC Radio Swindon and the regional BBC Where I Live websites)
Daily TV Output Editor *Stephanie Marshall*
TV Features Editor *Roger Farrant*
Political Editor *Dave Harvey*

BBC South (Southampton)

Broadcasting House, Havelock Road, Southampton SO14 7PU
☏ 023 8022 6201
Head of BBC South *Mike Hapgood* (Responsible for BBC South TV, BBC Radio Berkshire, BBC Radio Oxford, BBC Radio Solent and online content for the region)
Output Editor, BBC South Today *Lee Desty*
Series Producer, Inside Out *Jane French*

BBC South West (Plymouth)

Broadcasting House, Seymour Road, Mannamead, Plymouth PL3 5BD
☏ 01752 229201
Head of BBC South West *John Lilley* (Responsible for the BBC South West region. TV: BBC Spotlight (regional news programme), Spotlight Channel Islands (regional news programme), Inside Out (regional current affairs series) and The Politics Show South West. Local Radio: BBC Radio Devon, BBC Radio Cornwall, BBC Radio Guernsey, BBC Radio Jersey and the regional BBC Where I Live websites)
Output Editor *Simon Read*
Series Producer, Inside Out *Simon Willis*
Political Editor *Chris Rogers*

BBC Yorkshire/BBC North West/BBC North East & Cumbria

The regional centres at Leeds, Manchester and Newcastle make their own programmes on a bi-media approach, each centre having its own head of regional and local programmes.

BBC Yorkshire

Broadcasting Centre, 2 St Peter's Square, Leeds LS9 8AH
☏ 0113 244 1188
Head of Regional and Local Programmes *Helen Thomas*
Editor, News *Tim Smith*
Assistant News Editor, Look North *Denise Wallace*
Political Editor, North of Westminster *Len Tingle*
Weeklies Editor, Close Up North *Ian Cundall*
Producers, Close Up North *Richard Taylor, Paul Greenan*

BBC North West (Manchester)

New Broadcasting House, Oxford Road, Manchester M60 1SJ
☏ 0161 200 2020
Head of Regional and Local Programmes *Tamsin O'Brien*
Editor, Newsgathering *Jim Clarke*
Output Editor *Cerys Griffiths*
Producer, Inside Out *Deborah van Bishop*
Producer, The Politics Show *Michelle Mayman*

BBC North East & Cumbria (Newcastle upon Tyne)
Broadcasting Centre, Barrack Road, Newcastle Upon Tyne NE99 2NE
☏ 0191 232 1313
Head of Regional and Local Programmes *Wendy Pilmer*
News Editor *Andy Cooper*

BBC Local Radio

BBC local radio stations have their own newsroom which supplies local bulletins and national news service. Many have specialist producers. A comprehensive list of programmes for each is unavailable and would soon be out of date. For general information on programming visit the BBC website and select the local station you want or contact the relevant station direct.

BBC Radio Berkshire
PO Box 104.4, Reading RG4 8FH
☏ 0118 946 4200 (news)   🅵 0118 946 4555
www.bbc.co.uk/berkshire
Editor *Marianne Bell*

BBC Radio Bristol
PO Box 194, Bristol BS99 7QT
☏ 0117 974 1111   🅵 0117 923 8323
radio.bristol@bbc.co.uk
www.bbc.co.uk/bristol/local_radio
Managing Editor *Tim Pemberton*

Wide range of feature material used.

BBC Radio Cambridgeshire
104 Hills Road, Cambridge CB2 1LQ
☏ 01223 259696
cambs@bbc.co.uk
www.bbc.co.uk/radiocambridgeshire
Managing Editor *Jason Horton*

Commenced broadcasting in May 1982.

BBC Radio Cornwall
Phoenix Wharf, Truro TR1 1UA
☏ 01872 275421   🅵 01872 240679
cornwall@bbc.co.uk
www.bbc.co.uk/radiocornwall
Editor *Pauline Causey*

On air from 1983 serving Cornwall and the Isles of Scilly. Broadcasts 110 hours of local programmes weekly including news, phone-ins and debate.

BBC Coventry and Warwickshire
Priory Place, Coventry CV1 5SQ
☏ 024 7655 1000
coventry@bbc.co.uk
warwickshire@bbc.co.uk
www.bbc.co.uk/coventrywarwickshire
Editor *David Clargo*

Commenced broadcasting in January 1990 as CWR. News, current affairs, public service information and community involvement, relevant to its broadcast area. Details of any special events are available on the website.

BBC Radio Cumbria
Annetwell Street, Carlisle CA3 8BB
☏ 01228 592444   🅵 01228 511195
radio.cumbria@bbc.co.uk
bbc.co.uk/cumbria
Editor *Nigel Dyson*

Occasional opportunities for plays and short stories are advertised on-air.

BBC Radio Cymru
Broadcasting House, Llandaff, Cardiff CF5 2YQ
☏ 029 20 322018   🅵 029 20 322473
radio.cymru@bbc.co.uk
www.bbc.co.uk/radiocymru
Editor *Siân Gwynedd*

Welsh language station serving Welsh-speaking communities. Rich mix of news, music, sport, features, current affairs, drama, comedy, religion and education. Writing opportunities: DRAMA During 2008–09, up to 18 new radio plays. COMEDY Occasional opportunities during the year (for details of independent commissions see www.bbc.co.uk/wales/commissioning/).

BBC Radio Derby
PO Box 104.5, Derby DE1 3HL
☏ 01332 361111   🅵 01332 290794
radio.derby@bbc.co.uk
www.bbc.co.uk/derby
Editor *Simon Cornes*

News, sport, information and entertainment.

BBC Radio Devon
Broadcasting House, Seymour Road, Mannamead, Plymouth PL3 5BD
☏ 01752 260323
radio.devon@bbc.co.uk
www.bbc.co.uk/devon
Also at: Walnut Gardens, St David's Hill, Exeter EX4 4DH
Managing Editor *Robert Wallace*
Head of News *Sarah Solftley*

On air since 1983.

BBC Essex
PO Box 765, Chelmsford CM2 9XB
☏ 01245 616000   🅵 01245 492983
essex@bbc.co.uk
www.bbc.co.uk/essex
Managing Editor *Gerald Main*

Programmes are a mix of news, interviews, expert contributors, phone-ins, sport and special interest such as gardening.

BBC Radio Foyle

8 Northland Road, Londonderry BT48 7GD
☎ 028 7137 8600   🖷 028 7137 8666
radio.foyle@bbc.co.uk
www.bbc.co.uk/northernireland
Managing Editor *Paul McCauley*
News Producer *Paul McFadden*

Radio Foyle broadcasts about seven hours of
original material a day, seven days a week to
the north west of Northern Ireland. Other
programmes are transmitted simultaneously with
Radio Ulster. The output ranges from news, sport,
and current affairs to live music recordings and
arts reviews.

BBC Radio Gloucestershire

London Road, Gloucester GL1 1SW
☎ 01452 308585   🖷 01452 309491
radio.gloucestershire@bbc.co.uk
www.bbc.co.uk/radiogloucestershire
Managing Editor *Mark Hurrell*
Assistant Editor *Mark Jones*

News and information covering the large variety
of interests and concerns in Gloucestershire.
Leisure, sport and music, plus African Caribbean
and Asian interests. Regular book reviews and
interviews with local authors in the *Steve Kitchen
Show*.

BBC Guernsey

Broadcasting House, Bulwer Avenue, St
Sampson's, Guernsey GY2 4LA
☎ 01481 200600   🖷 01481 200361
radio.guernsey@bbc.co.uk
www.bbc.co.uk/guernsey
Managing Editor *David Martin*

Opened with its sister station, BBC Radio Jersey,
in March 1982. Broadcasts 80 hours of local
programming a week.

BBC Hereford & Worcester

Hylton Road, Worcester WR2 5WW
☎ 01905 748485   🖷 01905 748006
bbchw@bbc.co.uk
worcester@bbc.co.uk
www.bbc.co.uk/herefordworcester
Also at: 43 Broad Street, Hereford HR4 9HH
Managing Editor *James Coghill*

Has an interest in writers/writing with local
connections.

BBC Radio Humberside

Queen's Court, Queen's Gardens, Hull HU1 3RH
☎ 01482 323232
radio.humberside@bbc.co.uk
www.bbc.co.uk/radiohumberside
Managing Editor *Simon Pattern*

On air since 1971. BBC Radio Humberside has
broadcast short pieces by local writers.

BBC Radio Jersey

18–21 Parade Road, St Helier, Jersey JE2 3PL
☎ 01534 837200   🖷 01534 732569
radiojersey@bbc.co.uk
www.bbc.co.uk/jersey
Managing Editor *Denzil Dudley*
Assistant Editor *Matthew Price*

Local news, current affairs and community items.

BBC Radio Kent

The Great Hall, Mount Pleasant Road, Tunbridge
Wells TN1 1QQ
☎ 01892 670000   🖷 01892 549118
radio.kent@bbc.co.uk
www.bbc.co.uk/radiokent
Managing Editor *Paul Leaper*

Occasional commissions are made for local
interest documentaries and other one-off
programmes.

BBC Radio Lancashire

20–26 Darwen Street, Blackburn BB2 2EA
☎ 01254 262411
lancashire@bbc.co.uk
www.bbc.co.uk/radiolancashire
Managing Editor *John Clayton*

Journalism-based radio station, interested in
interviews with local writers. Contact *Alison
Brown*, Daily Programmes Producer, Monday to
Friday (alison.brown.01@bbc.co.uk).

BBC Radio Leeds

2 St Peter's Square, Leeds LS9 8AH
☎ 0113 244 2131   🖷 0113 224 7316
radio.leeds@bbc.co.uk
www.bbc.co.uk/radioleeds
Managing Editor *Phil Squire*

BBC Radio Leeds is the station for West
Yorkshire, serving 1.3 million potential listeners
since 1968. Broadcasts a mix of news, sport, talk,
music and entertainment, 24 hours a day.

BBC Radio Leicester

9 St Nicholas Place, Leicester LE1 5LB
☎ 0116 251 6688   🖷 0116 251 1463
radio.leicesternews@bbc.co.uk
www.bbc.co.uk/leicester
Managing Editor *Kate Squire*

The first local station in Britain. Occasional
interviews with local authors.

BBC Radio Lincolnshire

Newport, Lincoln LN1 3XY
☎ 01522 511411   🖷 01522 511058
radio.lincolnshire@bbc.co.uk
www.bbc.co.uk/radiolincolnshire
Managing Editor *Charlie Partridge*

Unsolicited material considered only if locally relevant. Maximum 1,000 words: straight narrative preferred, ideally with a topical content.

BBC London

PO Box 94.9, Marylebone High Street, London W1A 6FL
☎ 020 7224 2424   🖷 020 7208 9661 (news)
ldn-planning@bbc.co.uk
www.bbc.co.uk/london
Managing Editor *David Robey*

Formerly Greater London Radio (GLR), launched in 1988, BBC London 94.9 broadcasts news, information, travel bulletins, sport and music to Greater London and the Home Counties.

BBC Radio Manchester

PO Box 951, Oxford Road, Manchester M60 1SJ
☎ 0161 200 2000   🖷 0161 236 5804
radiomanchester@bbc.co.uk
www.bbc.co.uk/radiomanchester
Managing Editor *John Ryan*
News Editor *Mark Elliot*

One of the largest of the BBC local radio stations, broadcasting news, current affairs, phone-ins, help, advice and sport.

BBC Radio Merseyside

PO Box 95.8, Liverpool L69 1ZJ
☎ 0151 708 5500   🖷 0151 794 0988
radio.merseyside@bbc.co.uk
www.bbc.co.uk/liverpool
Managing Editor *Phil Roberts*

Launched in 1967. BBC Radio Merseyside's editorial area covers the whole of Merseyside (Liverpool, Bootle, Birkenhead, Southport, St Helens, etc.) much of North Cheshire (Chester, Warrington, Runcorn, Widnes, Ellesmere Port) and West Lancashire (Skelmersdale, Orrell, Burscough, Ormskirk).

BBC Radio Newcastle

Broadcasting Centre, Barrack Road, Newcastle upon Tyne NE99 1RN
☎ 0191 232 4141
radionewcastle.news@bbc.co.uk
www.bbc.co.uk/tyne
www.bbc.co.uk/wear
Managing Editor *Andrew Robson*
Senior Producer (Programmes) *Sarah Miller*

Commenced broadcasting in January 1971. One of the BBC's big city radio stations in England, Radio Newcastle reaches an audience of 233,000.

BBC Radio Norfolk

The Forum, Millennium Plain, Norwich NR2 1BH
☎ 01603 619331   🖷 01603 284488
radionorfolk@bbc.co.uk
www.bbc.co.uk/radionorfolk
Managing Editor *David Clayton*

Good local ideas and material welcome for features and documentaries *if* directly related to Norfolk.

BBC Radio Northampton

Broadcasting House, Abington Street, Northampton NN1 2BH
☎ 01604 239100   🖷 01604 230709
northampton@bbc.co.uk
www.bbc.co.uk/northamptonshire
Managing Editor *Laura Moss*

Books of local interest are regularly featured and authors and poets are interviewed on merit.

BBC Radio Nottingham

London Road, Nottingham NG2 4UU
☎ 0115 955 0500   🖷 0115 902 1983
radio.nottingham@bbc.co.uk
www.bbc.co.uk/radionottingham
Editor *Sophie Stewart*

Rarely broadcasts scripted pieces of any kind but interviews with authors form a regular part of the station's output.

BBC Radio Orkney

Castle Street, Kirkwall KW15 1DF
☎ 01856 873939   🖷 01856 872908
radio.orkney@bbc.co.uk
Senior Producer *John Fergusson*

Regular programmes include *Around Orkney* (news magazine programme), *Bruck* (swapshop and general features) and *Orky-Ology* (archaeology magazine). As a BBC Community station, all Radio Orkney's programme material is generated from within the Orkney Islands.

BBC Oxford

269 Banbury Road, Oxford OX2 7DW
☎ 08459 311444   🖷 08459 311555
oxford@bbc.co.uk
www.bbc.co.uk/oxford/local_radio
Executive Editor *Steve Taschini*

Restored to its original name in 2000 having been merged with BBC Radio Berkshire in 1995 to create BBC Thames Valley. The station occasionally carries interviews with local authors.

BBC Radio Scotland (Dumfries)

Elmbank, Lover's Walk, Dumfries DG1 1NZ
☎ 01387 268008   🖷 01387 252568
dumfries@bbc.co.uk
Senior Producer *Willie Johnston*

Previously Radio Solway. The station mainly outputs news bulletins (four daily) although it has become more of a production centre with programmes being made for Radio Scotland as well as BBC Radio 2 and 5 Live. Freelancers of a high standard, familiar with Radio Scotland, should contact the producer.

BBC Radio Scotland (Selkirk)

Unit 1, Ettrick Riverside, Dunsdale Road, Selkirk
TD7 5EB

📞 01750 724567 📠 01750 724555
selkirk.news@bbc.co.uk
Senior Broadcaster *Cameron Buttle*

Formerly BBC Radio Tweed. Local news bulletins only.

BBC Radio Sheffield

54 Shoreham Street, Sheffield S1 4RS
📞 0114 273 1177 📠 0114 267 5454
radio.sheffield@bbc.co.uk
www.bbc.co.uk/southyorkshire
Managing Editor *Gary Keown*
Programme Editor *Mike Woodcock*

Writer interviews, writing-related topics and readings on the Rony Robinson show at 11.00 am.

BBC Radio Shetland

Pitt Lane, Lerwick ZE1 0DW
📞 01595 694747 📠 01595 694307
radio.shetland@bbc.co.uk
Senior Producer *Caroline Moyes*

Regular programmes include *Good Evening Shetland*. An occasional books programme highlights the activities of local writers and writers' groups.

BBC Radio Shropshire

2–4 Boscobel Drive, Shrewsbury SY1 3TT
📞 01743 248484 📠 01743 237018
radio.shropshire@bbc.co.uk
bbc.co.uk/shropshire
Managing Editor *Tim Beech*

On air since 1985. Unsolicited literary material rarely used, and then only if locally relevant.

BBC Radio Solent

Broadcasting House, Havelock Road, Southampton SO14 7PU
📞 023 8063 2811
radio.solent.news@bbc.co.uk
www.bbc.co.uk/radiosolent
Managing Editor *Mia Costello*

Broadcasting since 1970.

BBC Somerset

Broadcasting House, Park Street, Taunton
TA1 4DA
📞 01823 323956 📠 01823 332539
somerset@bbc.co.uk
www.bbc.co.uk/somerset
Managing Editor *Simon Clifford*

Informal, speech-based programming with strong news and current affairs output and regular local interest features, including local writing. Poetry and short stories on the breakfast-time *Adam Thomas Programme* and Jo Phillips' mid-morning programme; also arts features on Elise Rayner's *Somerset's Drive* at 5.00 pm.

BBC Southern Counties Radio

Broadcasting House, 40–42 Queens Road, Brighton BN1 3XB
📞 01273 320400
southerncounties.radio@bbc.co.uk
www.bbc.co.uk/southerncounties
Also at: Broadcasting Centre, Guildford GU2 7AP
📞 01483 306306
Managing Editor *Nicci Holliday*

Regular programmes include the breakfast shows: *Breakfast Live in Surrey with Fred Maden* and *Breakfast Live in Sussex with Neil Pringle*.

BBC Radio Stoke

Cheapside, Hanley, Stoke on Trent ST1 1JJ
📞 01782 208080 📠 01782 289115
radio.stoke@bbc.co.uk
www.bbc.co.uk/stoke
Managing Editor *Sue Owen*

On air since 1968, one of the first eight 'experimental' BBC stations. Emphasis on news, current affairs and local topics. Music represents one fifth of total output. Unsolicited material of local interest is welcome – send to News Editor, *James O'Hara*.

BBC Radio Suffolk

Broadcasting House, St Matthew's Street, Ipswich
IP1 3EP
📞 01473 250000 📠 01473 210887
radiosuffolk@bbc.co.uk
www.bbc.co.uk/suffolk
Managing Editor *Peter Cook*

Strongly locally speech-based, dealing with news, current affairs, community issues, the arts, agriculture, commerce, travel, sport and leisure. Programmes sometimes carry interviews with writers.

BBC Radio Swindon

BBC Broadcasting House, 56–58 Prospect Place, Swindon SN1 3RW
📞 01793 513626
radio.swindon@bbc.co.uk
www.bbc.co.uk/england/radiowiltshire
Managing Editor *Tony Worgan*

Local news and current affairs. Local authors are interviewed on Mark Seaman's *Weekend Show*.

BBC Tees

Broadcasting House, Newport Road, Middlesbrough TS1 5DG
📞 01642 225211 📠 01642 211356
tees@bbc.co.uk
www.bbc.co.uk/tees/local_radio
Managing Editor *Matthew Barraclough*
Assistant Editor *Ben Thomas*

Material used is local to Teesside, Co. Durham and North Yorkshire. News, current affairs and features about life in the north east.

BBC Three Counties Radio

1 Hastings Street, Luton LU1 5XL
℡ 01582 637400   🖷 01582 401467
3CR@bbc.co.uk
www.bbc.co.uk/threecounties/local_radio
Managing Editor *Mark Norman*

Covers Bedfordshire, Hertfordshire and Buckinghamshire. Encourages freelance contributions from the community across a wide range of radio output, including interview and feature material. Interested in local history topics (five minutes maximum).

BBC Radio Ulster

Broadcasting House, Ormeau Avenue, Belfast BT2 8HQ
℡ 028 9033 8000
Head of Radio Ulster *Susan Lovell*

Programmes broadcast from 6.30 am to 1.00 am weekdays and from 6.55 am to 1.00 am at weekends. Programmes include: *Good Morning Ulster*; *The Stephen Nolan Show*; *Gerry Anderson*; *Talk Back*; *Evening Extra*; *On Your Behalf*; *Your Place and Mine*; *Sunday Sequence* and *Saturday Magazine*. Comedy, documentary, specialist music and community programming are also included.

BBC Radio Wales

Broadcasting House, Llandaff, Cardiff CF5 2YQ
℡ 029 2032 2000   🖷 029 2032 2674
radio.wales@bbc.co.uk
www.bbc.co.uk/radiowales
Editor *Sally Collins*
Editor, Radio Wales News *Geoff Williams*

Broadcasts news on the hour and half hour throughout weekday daytime programmes; hourly bulletins in the evenings and at weekends. Programmes include *Good Morning Wales* and *The Richard Evans Programme*.

BBC Radio Wiltshire

BBC Broadcasting House, Prospect Place, Swindon SN1 3RW
℡ 01793 513626
radio.wiltshire@bbc.co.uk
www.bbc.co.uk/england/radiowiltshire
Editor *Tony Worgon*

Sister company of BBC Radio Swindon. Regular programmes include Mark Seaman's *Weekend Show* (reviews and author interviews).

BBC WM

The Mailbox, Birmingham B1 1RF
℡ 08453 009956   🖷 0121 567 6025

bbcwm@bbc.co.uk
www.bbc.co.uk/radiowm
Managing Editor *Keith Beech*

Commenced broadcasting in November 1970 as BBC Birmingham and has won four gold Sony awards in recent years. Speech-based station broadcasting to the West Midlands, South Staffordshire and North Worcestershire.

BBC Radio York

20 Bootham Row, York YO30 7BR
℡ 01904 641351   🖷 01904 610937
northyorkshire.news@bbc.co.uk
www.bbc.co.uk/northyorkshire
Editor *Sarah Drummond*

Books of local interest are sometimes previewed.

Independent Television

Channel 4

124 Horseferry Road, London SW1P 2TX
℡ 020 7396 4444   🖷 020 7306 8356
www.channel4.com
Chief Executive *Andy Duncan*
Director of Television and Content *Kevin Lygo*
Head of Channel 4, Commissioning *Julian Bellamy*
Head of Features *Sue Murphy*
Head of Specialist Factual *Ralph Lee*
Head of News & Current Affairs *Dorothy Byrne*
Head of Education and Managing Editor, Commissioning *Janey Walker*
Director of Acquisitions *Jeff Ford*
Head of More4 *Peter Dale*
Head of E4 *Angela Jain*
Director of Radio *Bob Shennan*
Controller, Film and Drama *Tessa Ross*
Commissioning Editor, Comedy and Head of Comedy Films *Caroline Leddy*
Head of Comedy and Entertainment *Andrew Newman*
Head of Sport *Andrew Thompson*
Controller of Broadcasting *Rosemary Newell*
Head of Schedules and T4 *Julie Oldroyd*
Commissioning Editor, Daytime *Adam MacDonald*
Head of Documentaries *Hamish Mykura*

Channel 4 started broadcasting as a national channel in November 1982. It enjoys unique status as the world's only major public service broadcaster funded entirely by its own commercial activities. All programmes are either commissioned from production companies or acquired and are broadcast across the whole of the UK (the analogue service is not broadcast in Wales where it is replaced by S4C). Its Film4 channel, launched in 1998, features modern independent cinema. A second digital entertainment channel, E4, was launched in January 2001, focusing on young audiences. More4, launched in October 2005, features news, current affairs, documentaries and drama targeted at older audiences. All of Channel 4's digital channels are available on Freeview, Cable and Satellite on a free to view basis.

Channel Television

Television Centre, La Pouquelaye, St Helier, Jersey JE1 3ZD

☎ 01534 816816   🖷 01534 816817

www.channelonline.tv

Also at: Television House, Bulwer Avenue, St Sampsons, Guernsey GY2 4LA ☎ 01481 241888 🖷 01481 241878

Managing Director, Broadcast *Karen Rankine*
Managing Director, Commercial *Mike Elsey*
Director of Resources & Transmission
Kevin Banner

Channel Television is the independent television broadcaster to the Channel Islands, serving 150,000 residents, most of whom live on the main islands, Jersey, Guernsey, Alderney and Sark. The station has a weekly reach of more than 83% with local programmes (in the region of five and a half hours each week) at the heart of the ITV service to the islands.

Five

22 Long Acre, London WC2E 9LY

☎ 020 7550 5555

www.five.tv

Interim CEO *Mark White*
Designate CEO *Dawn Airey*
Director of Programmes *Ben Gale*
Controller of Children's Programmes *Nick Wilson*
Senior Programme Controller (News & Current Affairs) *Chris Shaw*

Channel 5 Broadcasting Ltd won the franchise for Britain's third commercial terrestrial television station in 1995 and came on air at the end of March 1997. Regular programmes include *The Wright Stuff* (weekday morning chat show), *The Trisha Goddard Show* (talk show), *CSI: NY* (drama) and *House* (drama).

GMTV

Head Office: The London Television Centre, Upper Ground, London SE1 9TT

☎ 020 7827 7000   🖷 020 7827 7001

www.gm.tv

Chairman *Clive Jones*
Acting Chief Operating Officer *Clive Crouch*
Director of Programmes *Peter McHugh*
Editor *Martin Frizell*
Deputy Editor *Malcolm Douglas*
Operations Director *Di Holmes*
Head of Entertainment *Corinne Bishop*
Deputy Head of Entertainment *Amy Vosburgh*

Winner of the national breakfast television franchise. Jointly owned by ITV plc and Disney. GMTV took over from TV-am on 1 January 1993, with live programming from 6.00 am to 9.25 am. Regular news bulletins, current affairs, topical features, showbiz and lifestyle. Also competitions, travel and weather reports. The week's news and political issues are reviewed on Sundays, followed by children's programming. Launched its digital service, GMTV2, in January 1999, with daily broadcasts from 6.00 am to 9.25 am. GMTV2 moved from ITV2 to ITV4 in March 2008 and is now known as GMTV Digital. Children's programming with some simulcast with GMTV1.

ITN (Independent Television News Ltd)

200 Gray's Inn Road, London WC1X 8XZ

☎ 020 7833 3000   🖷 020 7430 4868

contact@itn.co.uk

www.itn.co.uk

Chief Executive *Mark Wood*
Editor-in-Chief, ITV News *David Mannion*
Editor, Channel 4 News *Jim Gray*

Provider of the main national and international news for ITV and Channel 4 and radio news for IRN. Programmes on ITV: *Lunchtime News*; *London Today*; *Evening News*; *London Tonight*; *News at Ten*, plus regular news summaries and three programmes a day at weekends. Programmes on Channel 4 include the in-depth news analysis programmes *Channel 4 News at Noon*, *Channel 4 News* and *More 4 News*.

ITV plc

200 Gray's Inn Road, London WC1X 8HF

☎ 020 7843 8000

www.itvplc.com

www.itv.com

Executive Chairman *Michael Grade*
Chief Operating Officer *John Cresswell*
Director of Television *Peter Fincham*

See listings below for ITV Anglia, ITV Border, ITV Central, ITV Granada, ITV London, ITV Meridian, ITV Tyne Tees, ITV Wales, ITV West, ITV Westcountry and ITV Yorkshire.

ITV Network

ITV1 comprises 15 regional channels with ITV plc holding the English and Welsh regional ITV licences (Anglia, Border, Central, Granada, London, Meridian, Thames Valley, Tyne Tees, Wales, West, Westcountry and Yorkshire) as well as the digital channels, ITV2, ITV3, ITV4, CiTV and ITVPlay. The remaining ITV licences belong to: SMG plc (Grampian and Scottish), Ulster Television plc (UTV) and Channel Television. Granada is the production arm of ITV plc. For commissioning information go to: www.itv.com/AboutITV/Commissioning-Production/

ITV Anglia Television

Anglia House, Norwich NR1 3JG
℡ 01603 615151
www.itvlocal.com/anglia
Managing Director & Controller of News & Programmes *Neil Thompson*

Broadcasting to the east of England, Anglia Television is a major producer of programmes for the ITV network. Its network factual department currently boasts the largest portfolio of North American documentary commissions for a UK-based production company.

ITV Border Television

Television Centre, Durranhill, Carlisle CA1 3NT
℡ 01228 525101   ℻ 01228 541384
www.itvlocal.com/border
Managing Director *Douglas Merrall*

Border's region covers three different cultures – English, Scottish and Manx. Programming concentrates on documentaries rather than drama. Most scripts are supplied in-house but occasionally there are commissions. Apart from notes, writers should not submit written work until their ideas have been fully discussed.

ITV Central

Gas Street, Birmingham B1 2JT
℡ 0844 881 4000/0808 100 7888 (newsdesk)
central@itvlocal.com
www.itvlocal.com/central
Managing Director *Ian Squires*

Formed in 2004 by the merger of Carlton and Granada, ITV Central is the largest of the UK commercial television networks outside of London.

ITV Granada

Quay Street, Manchester M60 9EA
℡ 0161 832 7211   ℻ 0161 827 2180
granada@itvlocal.com
www.itvlocal.com/central
Director of Production *Claire Poyser*
Controller of Drama *Keiron Roberts*
Controller of Documentaries, History and Science *Bill Jones*

The longest continuous ITC licence holder, broadcasting for over 50 years. Programmes include *Coronation Street* and *Disappearing World*. See also ITV **Productions** under *Film, TV and Radio Producers*.

ITV London

London Television Centre, Upper Ground, London SE1 9LT
℡ 020 7261 8163
newsdesk@itvlondon.com
www.itvlocal.com/london
Managing Director *Christy Swords*

ITV London, formed in February 2004 by the merger of Carlton and LWT, broadcasts to the Greater London region, extending into the counties on its border.

ITV Meridian

Solent Business Park, Whiteley PO15 7PA
℡ 0844 881 2000
www.www.itvlocal.com/meridian
Managing Director *Mark Southgate*

Serves viewers across the South and South-East region.

ITV Thames Valley

Solent Business Park, Whiteley PO15 7PA
℡ 0844 881 2000
Head of News *Robin Britton*

Formed in December 2004 from the two former ITV sub-regions, Central South and Meridian West, ITV Thames Valley is ITV's newest regional news service. It covers an area stretching from Banbury in the north to Winchester in the south, from Swindon in the west to Bracknell in the east.

ITV Tyne Tees

Television House, The Watermark, Gateshead NE11 9SZ
℡ 0844 88 15000   ℻ 0191 404 8710
tynetees@itvlocal.com
www.itvlocal.com/tynetees
Managing Director/Controller of Programmes *Graeme Thompson*
Head of News & Sport *Graham Marples*
Executive Producer & Head of Development, Granada Factual North *Mark Robinson*
Regional Affairs Manager *Brenda Mitchell*

Programming covers politics, news and current affairs, regional documentaries and sport. Regular programmes include *North East Tonight with Jonathan Morrell and Philippa Tomson* and *Around the House* (politics).

ITV Wales

The Television Centre, Culverhouse Cross, Cardiff CF5 6XJ
℡ 0844 881 0100   ℻ 029 2059 7183
wales@itvlocal.com
www.itvlocal.com/wales
Managing Director *Elis Owen*

Fomerly HTV Wales. Programmes include the consumer affairs magazine, *The Ferret* and current affairs series, *Wales this Week*. The company also makes Welsh language programming for S4C, including the series voted the most popular with viewers on the Welsh channel – *Cefn Gwlad* and the current affairs series, *Y Byd ar Bedwar*.

ITV West

470 Bath Road, Bristol BS4 3HG
℡ 0844 881 2345/2308 (newsdesk)

Ⓕ 0844 881 2346
itvwestnews@itv.com (newsdesk)
www.itvlocal.com/west
Director of Programmes *Jane McCloskey*

Formerly HTV. Broadcasts to over two million viewers in the West of England.

ITV West Country

Langage Science Park, Western Wood Way,
Plymouth PL7 5BQ
Ⓣ 0844 881 4900/4800 (newsdesk)
Ⓕ 0844 881 4901
news@westcountry.co.uk (newsdesk)
www.itvlocal.com/westcountry
Director of Programmes *Jane McCloskey*

Broadcasts to Cornwall, Devon, Somerset and west Dorset. Produces regional news, current affairs and features programmes.

ITV Yorkshire

The Television Centre, Leeds LS3 1JS
Ⓣ 0113 243 8283   Ⓕ 0113 244 5107
www.itvlocal.com/yorkshire
Managing Director *David Croft*
Controller of Drama, Leeds *Keith Richardson*
Controller of Comedy Drama and Drama
Features *David Reynolds*

Drama series, comedy drama, single drama, adaptations and long-running series like *Emmerdale* and *Heartbeat*. Always looking for strong writing in these areas, but prefers to find it through an agent. Documentary/current affairs material tends to be supplied by producers; opportunities in these areas are rare but adaptations of published work as a documentary subject are considered. In theory, opportunity exists within series, episode material but the best approach is through a good agent.

S4C

Parc Ty Glas, Llanishen, Cardiff CF14 5DU
Ⓣ 029 2074 7444   Ⓕ 029 2075 4444
s4c@s4c.co.uk
www.s4c.co.uk
Chief Executive *Iona Jones*
Director of Commissioning *Rhian Gibson*

The Welsh 4th Channel, established by the Broadcasting Act 1980, is responsible for a schedule of Welsh and English programmes on the Fourth Channel in Wales. Known as S4C, the analogue service is made up of about 34 hours per week of Welsh language programmes and more than 85 hours of English language output from Channel 4. S4C digidol broadcasts in Welsh exclusively for 80 hours per week. Ten hours a week of the Welsh programmes are provided by the BBC; the remainder are commissioned from ITV1 Wales and independent producers. Drama, comedy, sport, music and documentary are all

part of S4C's programming. Commissioning guidelines can be viewed on www.s4c.co.uk/production

stv

Pacific Quay, Glasgow G51 1PQ
Ⓣ 0141 300 3000   Ⓕ 0141 300 3030
www.stv.tv
Also at: Television Centre, Craigshaw Business Park, West Tullos, Aberdeen AB12 3QH
Ⓣ 01224 848848 Ⓕ 01224 848800
Managing Director, Broadcasting, SMG
Television *Bobby Hain*
Head of News & Current Affairs (Scotland)
Gordon Macmillan
Executive Producer *Henry Eagles*

stv is the ITV (Channel 3) licence holder for the north and central regions of Scotland. The company produces a range of television programmes, covering current affairs, sport, entertainment, documentary and drama, as well as flagship news programmes *Scotland Today* and *North Tonight*.

Teletext Ltd

Building 10, Chiswick Park, 566 Chiswick High Road, London W4 5TS
Ⓣ 0870 731 3000   Ⓕ 0870 731 3001
www.teletext.co.uk
Managing Director *Dr Mike Stewart*
Editor-in-Chief *John Sage*

Broadcasts TV text services on analogue TV on ITV, Channel 4 and Five. Also broadcasts digital text services on Freeview and satellite on Channel 4 and, since 2005, on ITV. In addition, Teletext runs a holidays website and mobile services.

UTV

Ormeau Road, Belfast BT7 1EB
Ⓣ 028 9032 8122   Ⓕ 028 9024 6695
info@u.tv
www.u.tv
Head of Television *Michael Wilson*
Head of News, Current Affairs and Sport
Rob Morrison

Regular programmes on news and current affairs, sport and entertainment.

Cable and Satellite Television

British Sky Broadcasting Ltd (BSkyB)

6 Centaurs Business Park, Grant Way, Isleworth TW7 5QD

☎ 020 7705 3000   🖷 020 7705 3030

www.sky.com

Chief Executive *Jeremy Darroch*
Managing Director, Sky Networks *Sophie Turner-Laing*
Chief Operating Officer *Mike Darcey*
Head of Sky News *John Ryley*
Managing Director, Sky Sports *Vic Wakeling*

Launched in 1989, British Sky Broadcasting gives over 21 million viewers (in more than 8.2 million households) access to movies, news, entertainment and sports channels, and interactive services on Sky digital. Launched in October 1998, Sky digital has more than 500 channels and offers a range of innovative interactive services. 2006 saw the introduction of high definition television (HDTV). Sky Broadband was launched later that year.

WHOLLY-OWNED SKY CHANNELS

Sky Movies/Sky Cinema
Largest TV movie service outside the US with 10 Sky Movie channels and 2 Sky Movie HD channels. Sky Movies show over 90% of the top 100 grossing box-office films of the previous year including classic films, from Westerns to Film Noir, Chaplin to Chevy Chase, as well as a World Cinema strand that includes UK premières of new and old movies.

Sky News
Award-winning 24-hours news service with hourly bulletins and expert comment.

Sky One/Sky Two/Sky Three
One of the most frequently watched non-terrestrial channel.

Sky Sports 1/Sky Sports 2/Sky Sports 3/Sky Sports News/Sky Sports Extra/Sky Sports HD1 & HD2
Around 40,000 hours of sport are broadcast every year across the five Sky Sports channels. Sky Sports 1, 2 and 3 are devoted to live events, support programmes and in-depth sports coverage seven days a week. Sky Sports News provides sports news and the latest results and information 24 hours a day. Sky Sports Extra carries additional sports programming including the award-winning live interactive coverage.

Sky Travel/Sky Travel Extra
Magazine shows, documentaries and teleshopping.

Sky Arts
Wholly-owned Sky channel showcasing the best of the world's arts programmes, 18 hours a day, seven days a week.

JOINT VENTURES

National Geographic; Nickelodeon; Nick Jr; The History Channel; Paramount Comedy Channel; MUTV; Adventure One; The Biography Channel; Attheraces; Chelsea Digital Media.

CNBC Europe

10 Fleet Place, London EC4M 7QS

☎ 020 7653 9300   🖷 020 7653 5956

www.cnbc.com

CEO/President *Mick Buckley*

A service of NBC Universal. 24-hour business and financial news service. Programmes include *Capital Connections*; *Worldwide Exchange*; *Power Lunch Europe*; *European Closing Bell*; *Squawk Box Europe*.

CNN International

Turner House, 16 Great Marlborough Street, London W1P 1DF

☎ 020 7693 1000   🖷 020 7693 08921552

edition.cnn.com

Managing Editor, CNN International Europe, Middle East & Africa *Nick Wrenn*

CNN, the leading global 24-hour news network, is available to one billion people worldwide via the 26 CNN branded TV, Internet, radio and mobile services produced by CNN News Group, a Time Warner company. CNN has major production centres in Atlanta, New York, Los Angeles, London, Hong Kong and Mexico City. The London bureau, the largest outside the USA, is CNN's European headquarters and produces over 50 hours of programming per week. Live business and news programmes, including *Business International*; *World News* and *World Business Today*.

MTV Networks Europe

180 Oxford Street, London W1D 1DS

☎ 020 7284 7777   🖷 020 7284 7788

www.mtvnc.com

Vice Chairman *Bill Roedy*

Established 1987. Europe's 24-hour music and youth entertainment channel, available on cable, via satellite and digitally. Transmitted from London in English across Europe.

Travel Channel

64 Newman Street, London W1T 3EF

☎ 020 7636 5401   🖷 020 7636 6424

www.travelchannel.co.uk

Launched in February 1994. Broadcasts in 14 languages to over 117 countries across Europe, Africa and the Middle East. Programmes and information on the world of travel; destinations reports, lifestyle programmes plus food and drink, sport and leisure pursuits.

National Commercial Radio

Classic FM

30 Leicester Square, London WC2H 7LA
☎ 020 7343 9000   🖷 020 7766 6100
www.classicfm.com
Managing Director *Darren Henley*

Classic FM, Britain's largest national commercial radio station, started broadcasting in September 1992. Plays accessible classical music 24 hours a day and broadcasts news, weather, travel, business information, political/celebrity/general interest talks, features and interviews. Winner of the 'Station of the Year' Sony Award in 1993, 2000 and 2007.

Digital One

30 Leicester Square, London WC2H 7LA
☎ 020 7288 4600   🖷 020 7288 4601
info@digitalone.co.uk
www.ukdigitalradio.com

The UK's national commercial digital radio multiplex operator. Backed by GCap Media and Arqiva, Digital One began broadcasting on 15 November 1999. Digital One broadcast stations include **Classic FM**, **Virgin Radio**, **talkSPORT** and digital-only stations Planet Rock, Core and Capital Life as well as an electronic programme guide and the BT Movio television channels.

talkSPORT

PO Box 1089, London SE1 8WQ
☎ 020 7959 7800
kaytownsend@talksport.co.uk
www.talksport.net
Programme Director *Moz Dee*

Commenced broadcasting in February 1995 as Talk Radio UK. Re-launched January 2000 as TalkSport, the UK's first sports radio station. Acquired by Ulster Television in 2005. Broadcasts 24 hours a day. News items can be e-mailed via the website.

Virgin Radio

1 Golden Square, London W1F 9DJ
☎ 020 7434 1215   🖷 020 7434 1197
www.virginradio.co.uk
Programme Director *David Lloyd*

Music-based station launched in 1973, bought by Chris Evans' Ginger Media Group in December 1997 and acquired by the Scottish Media Group in March 2000.

Independent Local Radio

Capital Radio London

30 Leicester Square, London WC2H 7LA
☎ 020 7766 6000   🖷 020 7766 6012
www.capitalradio.com
Programme Director *Paul Jackson*

Commenced broadcasting in October 1973 as the country's second commercial radio station (the first being LBC, launched a week earlier). Europe's largest commercial radio station.

Central FM Ltd

201–203 High Street, Falkirk FK1 1DU
☎ 01324 611164   🖷 01324 611168
email@centralfm.co.uk
www.centralfm.co.uk
Programme Controller *Gary Muircroft*

Broadcasts music, sport and local news to central Scotland, 24 hours a day.

Clyde 1/Clyde 2

Clydebank Business Park, Clydebank G81 2RX
☎ 0141 565 2200   🖷 0141 565 2265
www.clyde1.com or
www.clyde2.com
Owner *Bauer Radio*
Station Director *Tracey McNellan*

Programmes usually originate in-house or by commission. All documentary material is made in-house. Good local news items always considered.

Cool FM

▷ Downtown Radio

Downtown Radio/Cool FM

Newtownards, Co. Down BT23 4ES
☎ 028 9181 5555   🖷 028 9181 5252
programmes@downtown.co.uk
www.downtown.co.uk
Managing Director *Mark Mahaffy*

Downtown Radio first ran a highly successful short story competition in 1988, attracting over 400 stories. The competition is now an annual event and writers living within the station's transmission area are asked to submit material during the winter and early spring. The competition is promoted in association with Eason Shops. For further information, write to *Florence Ambrose* at the station.

Forth One/Forth 2

Forth House, Forth Street, Edinburgh EH1 3LE
☎ 0131 556 9255   🖷 0131 557 4645
moira.miller@radioforth.com
www.forth.com
Programme Director *Luke McCullough*
Programme Producer, Forth 2 *Moira Millar*

News Editor *Paul Robertson*
Book Reviewer *Lesley Fraser-Taylor*

News stories welcome from freelancers. Music-based programming.

Isle of Wight Radio

Dodnor Park, Newport PO30 5XE
☎ 01983 822557   🖷 01983 822109
www.iwradio.co.uk
Station Manager *Dan Bates*
Programme Controller *Simon Osborne*

Part of The Local Radio Company, Isle of Wight Radio is the island's only radio station broadcasting local news, music and general entertainment including phone-ins and interview based shows. Music, television, film, popular culture are the main areas of interest. *John Hannam Meets* – long-running interview programme – on Sundays.

kmfm

Head Office: Express House, 34–36 North Street, Ashford TN24 8JR
☎ 01233 895825
www.kmfm.co.uk
Group Programme Controller *Steve Fountain*

A wide range of music programming plus news, views and local interest. Part of the Kent Messenger Group.

LBC Radio Ltd

The Chrysalis Building, Bramley Road, London W10 6SP
☎ 020 7314 7300
www.lbc.co.uk
Editorial Director *Jonathan Richards*

LBC 97.3 FM is a talk-based station broadcasting 24 hours a day, providing entertainment, interviews, celebrity guests, music, chat shows, local interest, news and sport. LBC News 1152 AM, the sister station of LBC 97.3 FM, provides 24-hour rolling news.

NorthSound Radio

Abbotswell Road, Aberdeen AB12 3AJ
☎ 01224 337000
www.northsound1.com
www.northsound2.com
Managing Director *Ken Massie*
Programme Director *Chris Thomson*

Features and music programmes 24 hours a day including, mid-morning (9.00 am – midday), *Northsound 2* feature programme.

Premier Radio

22 Chapter Street, London SW1P 4NP
☎ 020 7316 1300   🖷 020 7233 6706
premier@premier.org.uk
www.premier.org.uk

Chief Executive Officer *Peter Kerridge*

Broadcasts programmes that reflect the beliefs and values of the Christian faith, 24 hours a day on 1305, 1413, 1332 MW, Sky digital 0123, Virgin Media 968. Programmes include *Off the Shelf*, book reviews and author interviews, Sundays at 4.00 pm.

Radio XL 1296 AM

KMS House, Bradford Street, Birmingham B12 0JD
☎ 0121 753 5353   🖷 0121 753 3111
Managing Director *Arun Bajaj*

Asian broadcasting for the West Midlands, 24 hours a day. Broadcasts *Love Express* featuring love stories and poems. Writers should send material for the attention of *Satinder Dhaiwal*.

Sabras Radio

Radio House, 63 Melton Road, Leicester LE4 6PN
☎ 0116 261 0666   🖷 0116 266 7776
www.sabrasradio.com
Programme Controller *Don Kotak*

Programmes for the Asian community, broadcasting 24 hours a day.

Spectrum Radio

4 Ingate Place, London SW8 3NS
☎ 020 7627 4433   🖷 020 7627 3409
enquiries@spectrumradio.net
www.spectrumradio.net
Managing Director *Toby Aldrych*
General Manager *John Ogden*

Programmes for a broad spectrum of ethnic groups in London.

Spire FM

City Hall Studios, Malthouse Lane, Salisbury SP2 7QQ
☎ 01722 416644   🖷 01722 416688
www.spirefm.co.uk
Station Director *Karen Bosley*

Music-based programming. Broadcasts to south Wiltshire and west Hampshire.

Sunrise Radio (Yorkshire)

55 Leeds Road, Bradford BD1 5AF
☎ 01274 735043   🖷 01274 728534
www.sunriseradio.fm
Programme Controller, Chief Executive & Chairman *Usha Parmar*

Programmes for the Asian community in West Yorkshire.

Swansea Sound 1170 MW/96.4 FM The Wave

Victoria Road, Gowerton, Swansea SA4 3AB
☎ 01792 511170 (MW)/ 511964 (FM)
🖷 01792 511171 (MW)/511965 (FM)

www.swanseasound.co.uk
Station Director *Carrie Mosley*
Programme Controller *Steve Barnes*
News Editor *Emma Thomas*

Music-based programming on FM while Swansea Sound is interested in a wide variety of material, though news items must be of local relevance. An explanatory letter, in the first instance, is advisable.

Talk 107

9 South Gyle Crescent, Edinburgh Park, Edinburgh EH12 9EB
🕾 0131 316 3107

news@talk107.co.uk
www.talk107.co.uk
Acting Station Director *Matt Allitt*
News Editor *Gwen Lawrie*

Talk 107 is the first commercial speech radio station to be launched in the UK outside London. Commenced broadcasting in February 2006. Serves Edinburgh, Fife and the Lothians in Scotland. News, sport and debate.

96.4 FM The Wave

▷ Swansea Sound 1170 MW

European Television Companies

Austria

ORF(Österreichisher Rundfunk)1/ORF2
Würzburggasse 30, A–1136 Vienna
☎ 00 43 1 87 878-0
www.orf.at

Belgium

RTBF (Radio-Télévision Belge de la Communauté Française) (French language TV & radio)
Boulevard Auguste Reyers 52, B–1044 Brussels
☎ 00 32 2 737 2111
www.rtbf.be

Vlaamse Televisie Maatschappij (VTM) (cable)
Medialaan 1, B–1800 Vilvoorde
☎ 00 32 2 255 3211
www.vtm.be

VRT (Vlaamse Radio- en Televisieomroep) (Dutch language TV)
Auguste Reyerslaan 52, B–1043 Brussels
☎ 00 32 2 741 3111
www.vrt.be

Bosnia and Herzegovina

PBS BiH (Public Broadcasting Service of Bosnia and Herzegovina)
Bulevar Mese Selimovica 12, 71000 Sarajevo
☎ 00 387 33 45 123/461 101
www.pbsbih.ba

Bulgaria

BNT
29 San Stefano str., BG–1504 Sofia
☎ 00 359 2 944 4999
www.bnt.bg

Croatia

HRT (Hrvatska Televizija)
Prisavlje 3, CR–10000 Zagreb
☎ 00 385 1 634 3683
www.hrt.hr

Czech Republic

CT (Ceska Televize)
Kavci Hory, CZ–140 70 Prague 4
☎ 00 420 2 6113 7000
www.czech-tv.cz

Denmark

DR (Danmarks Radio)
Emil Holms Kanal 20, DK–0999 Copenhagen C
☎ 00 45 35 20 3040
www.dr.dk

SBS TV A/S (cable & satellite)
Mileparken 20A, DK–2740 Skovlunde
☎ 00 45 70 10 1010
www.sbstv.dk

TV–2 Danmark
Rugaardsvej 25, DK–5100 Odense C
☎ 00 45 65 91 9191
www.tv2.dk

TV3 (Viasat) (cable & satellite)
Wildersgade 8, DK–1408 Copenhagen K
☎ 00 45 77 30 5500
www.viasat.dk

Estonia

ETV (Eesti Televisioon)
Faehlmanni 12, EE–15029 Tallinn
☎ 00 372 628 4100
www.etv.ee

Finland

MTV3 Finland
Ilmalantori 2, SF–00033 Helsinki
℡ 00 358 10 300300
www.mtv3.fi

YLE FST (Swedish language TV)
Yleisradio Oy PO Box 62, SF–00024 Helsinki
℡ 00 358 9 14801
www.yle.fi/fst

YLE TV1
Yleisradio Oy PO Box 90, SF–00024 Helsinki
℡ 00 358 9 14801
www.yle.fi/tv1

YLE TV2
Yleisradio PO Box 196, SF–33101 Tampere
℡ 00 358 3 3456 111
www.yle.fi/tv2

France

Arte France (cable & satellite)
8 rue Marceau, F–92785 Issy-les-Moulineaux
Cedex 09
℡ 00 33 1 55 00 77 77
www.artefrance.fr

Canal+ (pay TV)
25 rue Leblanc, F–75906 Paris Cedex 15
℡ 00 33 1 44 25 10 10
www.canalplus.fr

France 2/France 3
7 Esplanade Henri de France, F–75907 Paris
Cedex 15
℡ 00 33 1 56 22 42 42 (F2)/30 30 (F3)
www.france2.fr
www.france3.fr

France 5
10–16 rue Horace Vernet, F–92136 Issy-les-
Moulineaux
℡ 00 33 1 56 22 91 91
www.france5.fr

Planète (privately-owned documentary channel)
48 quai du Pont du Jour, F–92659 Boulogne
℡ 00 33 1 71 10 10 20
www.planete.com

Voyage (cable & satellite)
241 Boulevard Péreire, F–75838 Paris
℡ 00 33 1 058 05 58 05
www.voyage.fr

Germany

ARD – Das Erste
Arnulfstrasse 42, D–80335 Munich
℡ 00 49 89 5900 3344
www.daserste.de

Arte Germany (cable & satellite)
Postfach 10 02 13, D–76483 Baden-Baden
℡ 00 49 7221 93690
www.arte.de

DW-TV (Deutsche Welle) (international TV service)
Voltastrasse 6, D–13355 Berlin
℡ 00 49 30 4646 0
www.dw-world.de

ZDF (Zweites Deutsches Fernsehen)
ZDF–Strasse 1, D–55127 Mainz
℡ 00 49 61 31/701
www.zdf.de

Greece

ERT S.A. (Elliniki Radiophonia Tileorassi)
Mesogeion 432, PO Box 19, GR–15342 Athens
℡ 00 30 2 10 606 6000
www.ert.gr

Hungary

DUNA TV (Duna Televízió)
Mészáros utca 48, H–1016 Budapest
℡ 00 36 1 489 1200
www.dunatv.hu

MTV 1/MTV 2 (Magyar Televízió)
Szabadság tér 17, H–1810 Budapest
℡ 00 36 1 353 3200
www.mtv.hu

Republic of Ireland

Radio Telefís Éireann (RTÉ One – RTÉ Two)
Donnybrook, Dublin 4
℡ 00 353 1 208 3111
www.rte.ie

TG4 (Teilefís na Gaelige) (Irish language TV)
Baile na hAbhann, Co. na Gaillimhe
℡ 00 353 91 505050
www.tg4.ie

Italy

RAI (RadioTelevisione Italiana)
Viale Mazzini 14, I–00195 Rome
℡ 00 39 06 3878-1
www.rai.it

RTI SpA (Reti Televisive Italiane)
Viale Europa 44, I–20093 Cologno Monzese
Milano
℡ 00 39 02 2514-1
www.rete4.com

Latvia

LTV1 & LTV7
Zakusalas Embankment 3, LV–1509 Riga
℡ 00 371 7 200 314
www.ltv.lv

Lithuania

LRT (Lietuvos radijas ir televizija)
S. Konarskio 49, LT–2600 Vilnius
℡ 00 370 5 236 3209
www.lrt.lt

The Netherlands

AVRO (Algemene Omroep Vereniging)
Postbus 2, NL–1200 JA Hilversum
℡ 00 31 35 671 79 11
www.central.avro.nl/index.asp

EO (Evangelische Omroep) (Evangelical TV)
Postbus 21000, NL–1202 BB Hilversum
℡ 00 31 35 647 4747
www.eo.nl

IKON (Interkerkelijke Omroep Nederland) (church-based TV)
Postbus 10009, NL–1201 DA Hilversum
℡ 00 31 35 672 72 72
www.omroep.nl/ikon

KRO (Katholieke Radio Omroep) (Catholic TV)
PO Box 23000, NL–1202 EA Hilversum
℡ 00 31 35 671 3911
www.kro.nl

NCRV (Nederlandse Christelijke Radio Vereniging)
Postbus 25000, NL–1202 HB Hilversum
℡ 00 31 35 671 99 11
info.omroep.nl/ncrv

NOS (Nederlandse Omroep Stichting) – Nederland 1, 2 & 3
Postbus 26600, NL–1202 JT Hilversum
℡ 00 31 35 677 92 22
www.nos.nl

NPS (Nederlandse Programma Stichting)
Postbus 29000, NL–1202 MA Hilversum
℡ 00 31 35 677 9333
www.omroep.nl/nps

VPRO
Postbus 11, NL–1200 JC Hilversum
℡ 00 31 35 671 2639
www.vpro.nl

Norway

NRK (Norsk Rikskringkasting)
Bjørnstjerne Bjørnsonsplass 1, N–0340 Oslo
℡ 00 47 23 04 7000
www.nrk.no

TV 2 Norge
Postboks 7222, Nøstegatan 72, N–5020 Bergen
℡ 00 47 55 90 8070
www.tv2.no

Poland

TVP (Telewizja Polska S.A.)
ul. J.P. Woronicza 17, PL–00 999 Warsaw
℡ 00 48 22 547 8000
www.tvp.com.pl

Portugal

RTP 1 & RTP 2 (Radiotelevisão Portuguesa)
Avenida Marechal, Gomes da Costa 37,
P–1849–030 Lisbon
℡ 00 351 21 794 7000
www.rtp.pt

SIC (Sociedade Independente de Comunicação) (cable TV)
Estrada da Outurela 119, P–2799-526 Carnaxide
℡ 00 351 21 417 9406
www.six.pt

TVI (Televisão Independente)
Rua Mário Castelhano 40, Queluz de Baixo,
P–2734–502 Barcarena
℡ 00 351 21 434 7500
www.tvi.pt

Romania

TVR (The Romanian Television Corporation)
Calea Dorobantilor 191, RO–71281 Bucharest 1
℡ 00 40 21 230 6290
www.tvr.ro

Serbia & Montenegro

RTS
Takovska 10, 11000 Beograda
℡ 00 381 11 321 2000
www.rts.co.yu

Slovenia

TV Slovenija
Kolodvorska ulica 2, SL–1550 Ljubljana
☏ 00 386 1 475 211
www.rtvslo.si

Spain

Canal Sur Televisión (Andalucian TV)
Ctra. San Juan de Aznalfarache-Tomares,
E–41920 San Juan de Aznalfarache
☏ 00 34 95 505 4600
www.canalsur.es

Canal+ Espana (pay TV)
Avenida de los Artesanos 6, E–28760 Tres Cantos
Madrid
☏ 00 34 91 736 7940
www.plus.es

ETB-1/ETB-2 (Euskal Telebista S.A.) (Basque TV)
E–48215 Iurreta/Bizkaia
☏ 00 34 94 603 1000
www.eitb.com

TVC (Televisió de Catalunya S.A.) (Catalan TV)
Carrer de la TV 3, E–08970 Saint Joan Despi
☏ 00 34 93 499 9333
www.tvcatalunya.com

TVE-1/ TVE-2 (Television Española)
Prado del Rey, E–28023 Madrid
☏ 00 349 1 346 4000
www.rtve.es

TVV (TVV-Televisio Valenciana) (Valencian TV)
Poligon Acc. Ademuz s/n, E–46100 Burjassot
☏ 00 34 96 318 3000
www.rtvv.es

Sweden

SVT (Sveriges Television)
Oxenstierngatan 26–34, S–105 10 Stockholm
☏ 00 46 8 784 0000
www.svt.se

TV4
Tegeluddsvägen 3–5, S–115 79 Stockholm
☏ 00 46 8 459 4000
www.tv4.se

Switzerland

RTR (Radio e Televisiun Rumantscha) (Romansch language TV)
Via da Masans 2, Plazza dal teater, CH–7002
Cuira
☏ 00 41 81 255 7575
www.rtr.ch

RTSI (Radiotelevisione svizzera di lingua Italiana) (Italian language TV)
Casella postale, CH–6903 Lugano
☏ 00 41 91 803 51 11
www.rtsi.ch

SF DRS (Schweizer Fernsehen) (German language TV)
Fernsehstrasse 1–4, CH–8052 Zurich
☏ 00 41 1 305 66 11
www.sfdrs.tv

SRG SSR idée suisse (Swiss Broadcasting Corp.)
Belpstrasse 48, CH-3000 Berne14
☏ 00 41 31 350 91 11
www.srg-ssr.ch

TSR (Télévision Suisse Romande) (French language TV)
Quai Ernest Ansermet 20, CH–1211 Geneva 8
☏ 00 41 22 708 20 20
www.tsr.ch

Film, TV and Radio Producers

Aardman

Gas Ferry Road, Bristol BS1 6UN
☎ 0117 984 8485   🖷 0117 984 8486
www.aardman.com

Founded 1976. Award-winning animation studio producing feature films and television programmes. Won Oscars for *Creature Comforts* and *Wallace and Gromit*. Other OUTPUT includes *Angry Kid*; *Chicken Run*; *Flushed Away*; *Shaun The Sheep*; *Chop Socky Chooks*; *Purple and Brown*. No unsolicited submissions.

Above The Title

Level 2, 10/11 St Georges Mews, London NW1 8XE
☎ 020 7916 1984   🖷 020 7722 5706
mail@abovethetitle.com
www.abovethetitle.com
Contacts *David Morley, Helen Chattwell*

Producer of radio drama, comedy and factual programmes. OUTPUT includes *Clive Anderson's Chat Room* and *Till the End of the Day – The Kinks Story* (Radio 2); *Unreliable Evidence*; *The Glittering Prizes*; *The Hitchhiker's Guide to the Galaxy*; *Fame & Fortune* and *Dirk Gently's Holistic Detective Agency* (Radio 4); *Features Like Mine* and *Iain Burnside* (Radio 3). Unsolicited mss and ideas welcome with s.a.e. See website for details of how to approach in the first instance by e-mail. 'We encourage and support new writing in every way we can.'

Absolutely Productions Ltd

Unit 19, 77 Beak Street, London W1F 9DB
☎ 020 7644 5575
www.absolutely.biz

Founded in 1988 by a group of writer/performer/ producers including Morwenna Banks, Jack Docherty, Moray Hunter, Pete Baikie, John Sparkes and Gordon Kennedy. Long-established producer for TV, has recently moved into feature films. OUTPUT *Absolutely*; *Welcome to Strathmuir* (a new comedy pilot for BBC Scotland) and *Baggage* (Radio 4).

Abstract Images

117 Willoughby House, Barbican, London EC2Y 8BL
☎ 020 7638 5123
productions@abstract-images.co.uk
Contact *Howard Ross*

Television documentary and drama programming. Also theatre productions. OUTPUT includes *Balm in Gilead*; *Road* and *Bent* (all dramas); *God: For & Against* (documentary); *This Is a Man* (drama/doc). New writers should send synopsis in the first instance.

Acacia Productions Ltd

80 Weston Park, London N8 9TB
☎ 020 8341 9392
acacia@dial.pipex.com
www.acaciaproductions.co.uk
Contact *J. Edward Milner*

Producer of award-winning television and video documentaries; also news reports, corporates and programmes for educational charities. Undertakes basic video training and has a video manual in preparation. OUTPUT includes documentary series entitled *Last Plant Standing*; *A Farm in Uganda*; *Montserrat: Under the Volcano*; *Spirit of Trees* (8 programmes); *Vietnam: After the Fire*; *Macroeconomics – the Decision-makers*; *Greening of Thailand*; *A Future for Forests*. No unsolicited mss.

Acrobat Television

107 Wellington Road North, Stockport SK4 2LP
☎ 0161 477 9090   🖷 0161 477 9191
info@acrobat-tv.co.uk
www.acrobat-tv.co.uk
Contact *David Hill*

All script genres for broadcast and coporate television, including training and promotional scripts, comedy and drama-based material. OUTPUT includes *Make a Stand* (Jack Dee, Gina Bellman and John Thompson for Video Arts); *The Customer View* (Roy Barraclough for Air Products); *Serious About Waves* series (Peter Hart for the Royal Yachting Association); *Fat Face Night* series (Extreme). No unsolicited mss.

Actaeon Films Ltd

50 Gracefield Gardens, London SW16 2ST
☎ 020 8769 3339   ⓕ 0870 134 7980
info@actaeonfilms.com
www.actaeonfilms.com
Producer *Daniel Cormack*
Head of Development *Becky Connell*

Founded 2004. Specializes in short and feature length theatrical motion pictures. OUTPUT includes *Amelia and Michael* (short drama starring Anthony Head). In development: *Golden Apples* (feature-length drama); *The Dead Letters* and *My Brother's Keeper* (both feature-length psychological thrillers). Actaeon Films is a member of the **New Producers Alliance**, **The Script Factory** and the London Filming Partnership. 'We actively encourage new writers and new writing. However, we can only give feedback on scripts with strong potential for developments (all mss are read by a professional reader within four weeks of submission). Scripts returned only if accompanied by s.a.e.'

All Out Productions

50 Copperas Street, Manchester M4 1HS
☎ 0161 834 9955   ⓕ 0161 834 6978
mail@allout.co.uk
www.allout.co.uk
Contact *Richard McIlroy*

Producer of documentaries, features and current affairs programmes for radio. OUTPUT includes *Five Live Report* (BBC Five Live – weekly news documentary); *Lamacq Live* (BBC Radio One – music and social affairs). Ideas welcome but not mss. Approach by e-mail.

Alomo Productions
▷ FremantleMedia Group Ltd

Anglo/Fortunato Films Ltd

170 Popes Lane, London W5 4NJ
☎ 020 8932 7676   ⓕ 020 8932 7491
anglofortunato@aol.com
Contact *Luciano Celentino*

Film, television and video producer/director of action comedy and psych-thriller drama. No unsolicited mss.

Arlington Productions Limited

Cippenham Court, Cippenham Lane, Cippenham, Nr Slough SL1 5AU
☎ 01753 516767   ⓕ 01753 691785

Television producer. Specializes in popular international drama, with *occasional* forays into other areas. 'We have an enviable reputation for encouraging new writers but only accept unsolicited submissions via agents.'

Art & Training Films Ltd

PO Box 3459, Stratford upon Avon CV37 6ZJ
☎ 01789 294910
andrew.haynes@atf.org.uk
www.atf.org.uk
Contact *Andrew Haynes*

Producer of documentaries, drama, commecials and corporate films and video. Submissions via agents considered.

The Ashford Entertainment Corporation Ltd

20 The Chase, Coulsdon CR5 2EG
☎ 020 8668 9609   ⓕ 087 0116 4142
info@ashford-entertainment.co.uk
www.ashford-entertainment.co.uk
Managing Director *Frazer Ashford*

Founded in 1996 by award-winning producer Frazer Ashford to produce character-led documentaries for TV. Examples of work include *Serial Killers*, *Great Little Trains* (winner of RTS Best Regional Documentary) and *Streetlife* (winner of bronze award at the Flagstaff Film Festival). Happy to look at any submissions falling within the remit of character-led documentaries. Send hard copy with an s.a.e.

Avalon

4a Exmoor Street, London W10 6BD
☎ 020 7598 7280
mikea@avalonuk.com
www.avalonuk.com/tv

Founded in 1993, Avalon is a BAFTA-winning production company supplying TV programming to all the British terrestrial channels and most leading satellite channels. Best known for light entertainment programmes such as documentary *Frank Skinner on Frank Skinner* (ITV1) or comedy *Fantasy Football* (BBC2). Now broadening output to include drama, factual and formatted shows. Script, treatment or ideas are welcomed via e-mail.

Baby Cow Productions

77 Oxford Street, London W1D 2ES
☎ 020 7399 1267   ⓕ 020 7399 1262
script@babycow.co.uk
www.babycow.co.uk

Founded in 1989 by Steve Coogan and Henry Normal. Specializes in cutting-edge comedy for television. Has divisions for animation, radio and films. OUTPUT *Marion and Geoff*; *Human Remains*; *The Sketch Show*; *24 Hour Party People*; *I'm Alan Patridge*; *Nighty Night*; *The Mighty Boosh*; *Dating Ray Fenwick*. Dedicated development department always on the lookout for new talent. In the first instance, send a synopsis or treatment and a ten-page extract, plus a DVD or video if the idea is filmed, and a covering letter. If the idea is of interest, Baby Cow will contact the writer. Include all contact details, including e-mail.

Beckmann International

Milntown Lodge, Lezayre Road, Ramsey IM8 2TG
☎ 01624 816585   🖷 01624 816589
sales@beckmanngroup.co.uk
www.beckmanngroup.co.uk
Contacts *Jo White, Stuart Semark*

Isle of Man-based company. Video and television documentary distributor. OUTPUT *Practical Guide to Europe* (travel series); *Maestro* (12-part series on classical composers); *Above and Beyond – The Story of British Aviation* (a celebration of aviation technology).

Big Heart Media

Flat 4, 6 Pear Tree Court, London EC1R 0DW
☎ 020 7608 0352   🖷 020 7250 1138
info@bigheartmedia.com
www.bigheartmedia.com
Contacts *Colin Izod, Beth Newell*

Producer of drama and documetaries for television and video. Ideas/outlines welcome by e-mail, but not unsolicited mss. Keen to enourage new writing.

Blackwatch Productions Limited

2/1, 104 Marlboro Avenue, Glasgow G11 7LE
☎ 0141 339 9996
info@blackwatchtv.com
Company Director *Nicola Black*

Film, television, video producer of drama and documentary programmes. OUTPUT includes *The Paranormal Peter Sellers*; *Snorting Coke with the BBC*; *When Freddie Mercury Met Kenny Everett*; *Designer Vaginas*; *Bonebreakers*; *Luv Bytes*; and *Can We Carry On, Girls?* for Ch4. Also coordinates *Mesh*, animation scheme. Does not welcome unsolicited mss.

Bona Broadcasting Limited

2nd Floor, 9 Gayfield Square, Edinburgh EH1 3NT
☎ 0131 558 1696   🖷 0131 558 1694
enquiries@bonabroadcasting.com
www.bonabroadcasting.com
Contact *Turan Ali*

Producer of award winning-drama and documentary programmes for BBC radio, TV and film projects. No unsolicited mss but send a one-paragraph summary by e-mail in the first instance. Runs radio and TV drama training courses in the UK and internationally.

Brighter Pictures
▷ **Endemol UK Productions**

Buccaneer Films

5 Rainbow Court, Oxhey WD19 4RP
☎ 01923 254000

Contact *Michael Gosling*

Corporate video production and still photography specialists in education and sport. No unsolicited mss.

Cactus TV

373 Kennington Road, London SE11 4PS
☎ 020 7091 4900   🖷 020 7091 4901
touch.us@cactustv.co.uk
cactustv.co.uk

Founded 1994. Specializes in broad-based entertainment, features and chat shows. Produces around 300 hours of broadcast material per year. OUTPUT includes *Richard & Judy*; *Sir Cliff Richard – The Hits I Missed*; *Songs of Bond*; *The British Soap Awards*.

Calon TV

3 Mount Stuart Square, Butetown, Cardiff CF10 5EE
☎ 029 2048 8400   🖷 029 2048 5962
enquiries@calon.tv
www.calon.tv
Head of Development *Andrew Offiler*

Founded 2005. Animated series, mainly for children. OUTPUT includes *Meeow*; *Hilltop Hospital*; *The Hurricanes*; *Tales of the Toothfairies*; *Billy the Cat*; *The Blobs*, as well as the feature films, *Under Milk Wood* and *The Princess and the Goblin*. Write with ideas and sample script in the first instance.

Carnival (Films & Theatre) Ltd

47 Marylebone Lane, London W1U 2NT
☎ 020 7317 1370   🖷 020 7317 1380
info@carnivalfilms.co.uk
www.carnivalfilms.co.uk
Managing Director *Gareth Neame*

Film, TV and theatre producer. OUTPUT FILM: *The Mill on the Floss* (BBC); *Firelight* (Hollywood Pictures/Wind Dancer Productions); *Up on the Roof* (Rank/Granada); *Shadowlands* (Savoy/Spelling); *The Infiltrator* (Home Box Office); *Under Suspicion* (Columbia/Rank/LWT). TELEVISION: *Rosemary & Thyme* (ITV/Granada); *The Grid* (BBC/TNT); *As If* (Ch4/Columbia); *Lucy Sullivan is Getting Married* (ITV); *The Tenth Kingdom* (Sky/NBC); *Agatha Christie's Poirot* (ITV/LWT/A&E); *Every Woman Knows a Secret* and *Oktober* (both for ITV Network Centre); *Hotel Babylon*; *Blott on the Landscape*; *Crime Traveller* and *Bugs 1–4* (all for BBC); *Anna Lee* and *All Or Nothing At All* (LWT); *Head Over Heels* (Carlton); *Jeeves & Wooster* I–IV (Granada); *The Fragile Heart*; *Traffik* and *Porterhouse Blue* (all for Ch4). THEATRE: *How Was it for You?*; *What a Performance*; *Juno & the Paycock*; *Murder is Easy*; *Misery*; *The Ghost Train*; *Map of the Heart*; *Shadowlands*; *Up on the Roof*. No unsolicited mss

Cartwn Cymru

12 Queen's Road, Mumbles, Swansea SA3 4AN
☎ 07771 640400
production@cartwn-cymru.com
Producer *Naomi Jones*

Animation production company. OUTPUT *Toucan 'Tecs* (YTV/S4C); *Funnybones* and *Turandot: Operavox* (both for S4C/BBC); *Testament: The Bible in Animation* (BBC2/S4C); *The Miracle Maker* (S4C/BBC/British Screen/Icon Entertainment International); *Faeries* (HIT Entertainment plc for CITV); *Otherworld* (animated feature film for S4C Films, British Screen, Arts Council of Wales).

Celador Films

39 Long Acre, London WC2E 9LG
☎ 020 7845 6988   🖷 020 7845 6977
www.celador.co.uk
Head of Development *Ivana McKinnon*

Producer of feature films. OUTPUT includes *Dirty Pretty Things*; *Separate Lives*; *Slumdog Millionaire* and Neil Marshall's second feature, *The Descent*. Submissions from established literary and film agents only or talent already known to the company, or from third party producers seeking possible co-production.

Celador Productions

39 Long Acre, London WC2E 9LG
☎ 020 7240 8101   🖷 020 7845 6976
info@celador.co.uk
www.celadorproductions.com
Director of Production *Heather Hampson*
Head of Factual Entertainment *Murray Boland*
Head of Entertainment *Ruth Wrigley*
Head of Radio *Liz Anstee*

Producer of a range of TV and radio programmes, from situation comedy to quizzes, game shows, popular factual entertainment and international co-productions. OUTPUT *You Are What You Eat*; *Perfect Strangers*; *Popcorn*; *It's Been a Bad Week*; *Commercial Breakdown*; *How to Dump Your Mates* and *Three Off the Tee*. 'We are interested in radio scripts but do not accept unsolicited proposals for comedy or entertainment formats. As a relatively small company our script-reading resources are limited.'

Celtic Films Entertainment Ltd

Lodge House, 69 Beaufort Street, London SW3 5AH
☎ 020 7351 0909   🖷 020 7351 4139
info@celticfilms.co.uk
www.celticfilms.co.uk
Contact *Steven Russell*

Founded in 1986 producing high-end TV drama and mainstream feature films. OUTPUT includes fifteen *Sharpe* television films, *Girl from Rio* and *Stefan Kiszko: A Life for a Life*. No unsolicited submissions.

Chameleon Television Ltd

Great Minster House, Lister Hill, Horsforth, Leeds LS18 5DL
☎ 0113 205 0040   🖷 0113 281 9454
allen@chameleontv.com
Contacts *Allen Jewhurst*, *Julia Kirby-Smith*

Film and television drama and documentary producer. OUTPUT includes *Edge of the City*; 'Dispatches' – *Channel 4 News*; *The Family Who Vanished*; *Killing for Honour*; *College Girls*; *Ken Dodd in the Dock*; *Unholy War* (all for Ch4); *Diary of a Mother on the Edge*; *Divorces From Hell*; *Shipman*; *Love to Shop*; *The Marchioness* (all for ITV); *Ted & Sylvia – Love, Loss*; *Hamas Bombers* (BBC); *Liverpool Poets* (Ch5). Scripts not welcome unless via agents but new writing is encouraged.

Channel Television Ltd

Television Centre, St Helier, Jersey JE1 3ZD
☎ 01534 816873   🖷 01534 816889
david@channeltv.co.uk
www.channeltvco.uk
Senior Producer *David Evans*

Producer of TV commercials and corporate material: information, promotional, sales, training and events coverage. CD and DVD production; promotional videos for all types of businesses throughout Europe. No unsolicited mss; new writing/scripts commissioned as required. Interested in hearing from local writers resident in the Channel Islands.

Cheetah Television
▷ Endemol UK Productions

The Children's Film & Television Foundation Ltd

c/o Simon George, Head of Finance, Ealing Studios, Ealing Green, London W5 5EP
☎ 07887 573479
info@cftf.org
www.cftf.org.uk
Chief Executive *Anna Home*

Involved in the development and co-production of films for children and the family, both for the theatric market and for TV.

Cleveland Productions

5 Rainbow Court, Oxhey, Near Watford WD19 4RP
☎ 01923 254000
Contact *Michael Gosling*

Communications in sound and vision A/V production and still photography specialists in education and sport. No unsolicited mss.

COI

Hercules Road, London SE1 7DU
☎ 020 7928 2345
www.coi.gov.uk

Government advertising and marketing communications, and public information films.

Collingwood O'Hare Entertainment Ltd

10–14 Crown Street, London W3 8SB
☎ 020 8993 3666   📠 020 8993 9595
info@crownstreet.co.uk
www.collingwoodohare.com
Producer *Christopher O'Hare*
Head of Development *Helen Stroud*

Film and TV; specializes in children's animation. OUTPUT *The Secret Show* (BBC/BBC Worldwide); *RARG* (award-winning animated film); *Yoko! Jakamoko! Toto!* (CITV); *Daisy-Head Mayzie* (Dr Seuss animated series for Turner Network and Hanna-Barbera); *Gordon the Garden Gnome* (CBBC); *Eddy and the Bear* and *The King's Beard* (both for CITV). Unsolicited mss not welcome 'as a general rule as we do not have the capacity to process the sheer weight of submissions this creates. We therefore tend to review material from individuals recommended to us through personal contact with agents or other industry professionals. We like to encourage new writing and have worked with new writers but our ability to do so is limited by our capacity for development. We can usually only consider taking on one project each year, as development/finance takes several years to put in place.'

The Comedy Unit

6th Floor 53 Bothwell Street, Glasgow G2 6TS
☎ 0141 220 6400
scripts@comedyunit.co.uk
www.comedyunit.co.uk
Managing Directors *April Chamberlain, Colin Gilbert*

Producers of comedy entertainment for children's TV, radio, video and film. OUTPUT includes *Still Game*; *Karen Dunbar Show*; *Offside*; *Chewin' the Fat*; *Only An Excuse*; *Yo! Diary* and *Watson's Wind Up*. Unsolicited mss welcome by post or e-mail.

Company Pictures

Suffolk House, 1–8 Whitfield Place, London W1T 5JU
☎ 020 7380 3900   📠 020 7380 1166
enquiries@companypictures.co.uk
www.companypictures.co.uk
Managing Directors *George Faber, Charles Pattinson*

One of the UK's largest independent film and television drama production companies. Established in 1998. Won Best Independent Production Company at the 2005 Broadcast Awards and the European Producers of the Year Award at the 2004 Monte Carlo Awards. OUTPUT includes *Shameless* (Ch4); *Life and Death of Peter Sellers* (HBO, two Golden Globe Awards, 15 Emmies); *Elizabeth I*, starring Helen Mirren and Jeremy Irons (Ch4/HBO, three Golden Globe Awards, 9 Emmies). Proposals accepted via agents only.

Cosgrove Hall Films

8 Albany Road, Chorlton–cum–Hardy M21 0AW
☎ 0161 882 2500   📠 0161 882 2555
animation@cosgrovehall.com
www.chf.co.uk

Producer of animated children's television programmes, founded in 1976. OUTPUT includes *Andy Pandy*; *Bill and Ben*; *Count Duckula*; *DangerMouse*; *Fifi and the Flowertots*; *Roald Dahl's BFG*; *The Jungle Book* and *The Wind in the Willows*.

The Creative Partnership

13 Bateman Street, London W1D 3AF
☎ 020 7439 7762
sarah.fforde@thecreativepartnership.co.uk
www.thecreativepartnership.co.uk
Contact *Sarah Fforde*

'Europe's largest "one-stop shop" for advertising and marketing campaigns for the film and television industries.' Clients include most major and independent film companies. No scripts. 'We train new writers in-house, and find them from submitted c.v.s. All applicants must have previous commercial writing experience.'

Cricket Ltd

Medius House, 63–69 New Oxford Street, London WC1A 1EA
☎ 020 7845 0300   📠 020 7845 0303
team@cricket-ltd.com
www.cricket-ltd.com
Head of Recruitment *Mary McDonnell*

Film and video, live events and conferences, print and design. 'Communications solutions for business clients wishing to influence targeted external and internal audiences.'

CSA Word

6A Archway Mews, 241A Putney Bridge Road, London SW15 2PE
☎ 020 8871 0220   📠 020 8877 0712
victoria@csaword.co.uk
www.csaword.co.uk
Contacts *Victoria Williams, Clive Stanhope*

Producer of drama, documentaries and readings for radio. OUTPUT *Alfie Elkins & His Little Life* (drama for BBC World Service); *The Hungry Years* (reading for BBC R4); *A Stable Relationship* (feature for BBC R4); *Bob Dylan's Chronicles 1* (reading for BBC R2); *It's a Girl*; *Chat Snaps & Videotape – Two* (documentaries for BBC World Service); *The Glenn Miller Story, Berlin...Soundz Decadent* and *Original Soundtrack Recordings* (documentaries for BBC R2), Unsolicited drama ideas welcome by post or e-mail. 'We encourage new drama ideas.'

Cutting Edge Productions Ltd

27 Erpingham Road, London SW15 1BE
℡ 020 8780 1476   ℻ 020 8780 0102
juliannorridge@btconnect.com
Contact *Julian Norridge*

Corporate and documentary video and television. OUTPUT includes US series on evangelicalism, 'Dispatches' on US tobacco and government videos. No unsolicited mss; 'we commission all our writing to order, but are open to ideas.'

Diverse Production Limited

6 Gorleston Street, London W14 8XS
℡ 020 7603 4567   ℻ 020 7603 2148
www.diverse.tv

Independent production company specializing in popular prime-time formats, strong documentaries, specialist factual, historical, cultural, religious, arts, music and factual entertainment. Recent OUTPUT includes *Ballet Changed My Life: Ballet Hoo!*; *Mission Africa*; *Codex*; *Bear Grylls' Man vs. Wild*; *Tribal Wife*; *Musicality*; *Operatunity*; *Who Wrote the Bible*; *Who You Callin' Nigger?*; *In Search of Tony Blair*; *Britain AD*; *Shock Treatment*; *Beyond Boundaries*; *Escape to the Legion*.

DoubleBand Films

3 Crescent Gardens, Belfast BT7 1NS
℡ 028 9024 3331   ℻ 028 9023 6980
info@doublebandfilms.com
www.doublebandfilms.com
Contacts *Michael Hewitt, Dermot Lavery*

Established 1988. Specializes in documentaries and some drama. Recent productions include *Seven Days that Shook the World* and *War in Mind* (both for Ch4); *Christine's Children* (BBC Northern Ireland; nominated for both the RTS and Celtic Film Festival); *D-Day: Triumph and Tragedy* (BBC NI). No unsolicited scripts.

Drake A-V Video Ltd

89 St Fagans Road, Fairwater, Cardiff CF5 3AE
℡ 029 2056 0333
www.drakeav.com
Contact *Helen Stewart*

Specialists in corporate and educational films. No unsolicited mss.

Charles Dunstan Communications Ltd

42 Wolseley Gardens, London W4 3LS
℡ 020 8994 2328   ℻ 020 8994 2328
Contact *Charles Dunstan*

Producer of film, video and TV for documentary and corporate material. OUTPUT *Renewable Energy* for broadcast worldwide in 'Inside Britain' series; *The Far Reaches* travel series; *The Electric Environment*. No unsolicited scripts.

Electric Airwaves Ltd (Ladbroke Radio)

Essel House, 29 Foley Street, London W1W 7JW
℡ 020 7323 2770   ℻ 020 7079 2080
neil@electricairwaves.com
www.electricairwaves.com
Contacts *Neil Gardner, Richard Bannerman*

Producer of radio drama, documentary, corporate, music, music documentaries and readings. OUTPUT includes *The Woman in White*; *The Darling Buds of May*; *The True History of British Pop*; *The World on a String*; *The Colour of Music* (all for BBC Radio 2); *Don Carlos*; *In the Company of Men* (BBC Radio 3 drama); *Sitting in Limbo* (BBC World Service drama); *Your Vote Counts* (Electoral Commision audio CDs). Unsolicited mss and ideas welcome; send letter in the first instance. 'We are willing to help develop for possible submission to the BBC.'

Endemol UK Productions

Shepherds Building, Central Charecroft Way, London W14 0EE
℡ 0870 333 1700   ℻ 0870 333 1800
info@endemoluk.com
www.endemoluk.com

Major independent producer of TV and digital media content, responsible for over 5,000 hours of programming per year. Incorporates several production brands including Brighter Pictures (factual entertainment shows, reality programming, live events and popular documentaries), Cheetah Television (factual), Initial (prime time entertainment, children's and teens, events and arts shows), Zeppotron (comedy) and Showrunner (drama). OUTPUT includes *Big Brother*; *Deal or No Deal*; *Comic Relief Does Fame Academy*; *Golden Balls*; *Lucy: Teen Transsexual*; *Restoration Village*.

Farnham Film Company Ltd

34 Burnt Hill Road, Lower Bourne, Farnham GU10 3LZ
℡ 01252 710313   ℻ 01252 725855
www.farnfilm.com
Contact *Ian Lewis*

Television and film: intelligent full-length film and children's drama. Unsolicited mss usually welcome but prefers a letter or e-mail to be sent in the first instance. Check website for current requirements.

Fast Films

Christmas House, 213 Chester Road, Castle Bromwich, Solihull B36 0ET
☎ 0121 749 7147/4144
gavinprime@mac.com
Contact *Gavin Prime*

Film and television: comedy, entertainment and animation. No unsolicited mss.

Festival Film and Television Ltd

Festival House, Tranquil Passage, Blackheath, London SE3 0BJ
☎ 020 8297 9999   ℻ 020 8297 1155
info@festivalfilm.com
www.festivalfilm.com
Managing Director *Ray Marshall*
Producer *Matt Marshall*

Television drama and feature films. Best known for its Catherine Cookson Dramas, which became one of ITV's long-running brands. Now developing both TV drama and features. Only accept material submitted through an agent. Will not consider unsolicited material.

Fiction Factory

14 Greenwich Church Street, London SE10 9BJ
☎ 020 8853 5100   ℻ 020 8293 3001
radio@fictionfactory.co.uk
www.fictionfactory.co.uk
Creative Director *John Taylor*

Production company specializing in intelligent entertainments. Recent OUTPUT for BBC Radio includes Marcel Proust's *A La Recherche du Temps Perdu*; *Scam* John Arden; *The Coup* Annie Caulfield; *A Dance to the Music of Time* Anthony Powell. Ideas considered if sent by e-mail. Mss only from agents or writers with a professional track record in the broadcast media.

Film and General Productions Ltd

4 Bradbrook House, Studio Place, London SW1X 8EL
☎ 020 7235 4495   ℻ 020 7245 9853
cparsons@filmgen.co.uk
Contacts *Clive Parsons, Davina Belling*

Film and television drama. Feature films include *True Blue*; *Tea with Mussolini* and *I Am David*. Also *Seesaw* (ITV drama), *The Greatest Store in the World* (family drama, BBC) and *The Queen's Nose* (children's series, BBC). Interested in considering new writing but subject to prior telephone conversation.

Firehouse Productions

42 Glasshouse Street, London W1B 5DW
☎ 020 7439 2220
postie@firehouse.biz
www.firehouse.biz
Contacts *Peter Granger, Mike McLeod*

Corporate films and websites, commercials, DRTV and live production events. OUTPUT includes work for De Beers; Prudential; government and various agencies.

The First Film Company Ltd

3 Bourlet Close, London W1W 7BQ
☎ 020 7436 9490   ℻ 020 7637 1290
info@firstfilmcompany.com
Producer *Roger Randall-Cutler*

Founded 1984. Cinema screenplays. All submissions should be made through an agent.

First Writes Theatre Company

Lime Kiln Cottage, High Starlings, Banham, Norwich NR16 2BS
☎ 01953 888525   ℻ 01953 888974
ellen@first-writes.co.uk
www.first-writes.co.uk
Contact *Ellen Dryden*

Producer of numerous afternoon, Friday and Saturday plays, classic serials and comedy narrative series for BBC Radio 3, Radio 4 and World Service.

Flannel

21 Berwick Street, London W1F 0PZ
☎ 020 7287 9277   ℻ 020 7287 7785
mail@flannel.net
Contact *Kate Haldane*

Producer of drama, documentaries and comedy for television and radio. OUTPUT includes *Woman's Hour* (BBC R4); *The Hendersons' Christmas Party* (five-part Christmas drama, BBC R4). No unsolicited mss. 'Keen to encourage new writing, but must come via an agent. Particularly interested in 45–50 minute dramas for radio. Not in a position to produce plays for stage, but very happy to consider adaptations. Welcomes comedy with some track record.'

Flashback Television Limited

58 Farringdon Road, London EC1R 3BP
☎ 020 7253 8768   ℻ 020 7253 8765
mailbox@flashbacktv.co.uk
www.flashbacktelevision.com
Managing Director *Taylor Downing*
Creative Director *David Edgar*

Founded in 1982, award-winning producer of drama, factual entertainment and documentaries for television. Recent credits include *Beau Brummell*. No unsolicited submissions.

Focus Films Ltd

The Rotunda Studios, Rear of 116–118 Finchley Road, London NW3 5HT

☎ 020 7435 9004   🖷 020 7431 3562

focus@focusfilms.co.uk

www.focusfilms.co.uk

Managing Director *David Pupkewitz*

Contact *Raimund Berens*

Founded 1982. Independent feature film development and production company. OUTPUT *The Book of Eve* (Canadian drama); *The Bone Snatcher* (Horror, UK/Can/SA); *Julia's Ghost* (German co-production); *The 51st State* (feature film); *Secret Society* (comedy drama feature film); *Crimetime* (feature thriller); *Diary of a Sane Man*; *Othello*. Projects in development include *Heaven and Earth*; *Chemical Wedding*; *Surviving Evil*. No unsolicited scripts.

Mark Forstater Productions Ltd

11 Keslake Road, London NW6 6DJ

☎ 020 8933 4375   🖷 020 8933 4375

Contact *Mark Forstater*

Active in the selection, development and production of material for film and TV. OUTPUT *Monty Python and the Holy Grail*; *The Odd Job*; *The Grass is Singing*; *Xtro*; *Forbidden*; *Separation*; *The Fantasist*; *Shalom Joan Collins*; *The Silent Touch*; *Grushko*; *The Wolves of Willoughby Chase*; *Between the Devil and the Deep Blue Sea*; *Doing Rude Things*. No unsolicited scripts.

FremantleMedia Group Ltd

1 Stephen Street, London W1T 1AL

☎ 020 7691 6000   🖷 020 7691 6100

www.fremantlemedia.com

CEO *Tony Cohen*

CEO, talkbackTHAMES *Lorraine Heggessey*

FremantleMedia, formerly known as Pearson Television, is the production arm of the RTL Group, Europe's largest TV and radio company. Acquired Thames Television in 1993 (producer of *The Bill*) and Grundy Worldwide (*Neighbours*) in 1995. Further acquisitions were Witzend Productions (*Lovejoy*) and Alomo Productions in 1996 and TalkBack Productions in 2000. FremantleMedia produces more than 260 programmes in over 40 countries and territories a year. *No unsolicited submissions, please.*

Gaia Communications

20 Pevensey Road, Eastbourne BN21 3HP

☎ 01323 734809/727183   🖷 01323 734809

production@gaiacommunications.co.uk

www.gaiacommunications.co.uk

Producer *Robert Armstrong*

Script Editor *Loni Webb*

Established 1987. Video and TV corporate and documentary. OUTPUT *Discovering* (south east regional tourist and local knowledge series); *Holistic* (therapies and general information); local interest audiobooks.

Noel Gay Television

Shepperton Studios, Studios Road, Shepperton TW17 0QD

☎ 01932 592569   🖷 01932 592172

charles.armitage@virgin.net

CEO *Charles Armitage*

OUTPUT *The Fear* (BBC Choice); *Second Chance* and *I-Camcorder* (both for Ch4); *Hububb* Series 1–5 (BBC); *Frank Stubbs Promotes* and *10%ers* Series 2 (both for Carlton/ITV); *Call Up the Stars* (BBC1); *Smeg Outs* (BBC video); *Red Dwarf*; *Dave Allen* (ITV); *Windrush* (BBC2). Joint ventures and companies include a partnership with Odyssey, a leading Indian commercials, film and TV producer, and the Noel Gay Motion Picture Company, whose credits include *Virtual Sexuality*; *Trainspotting* (with Ch4 and Figment Films); *Killer Tongue*; *Dog Soldiers*; *Fast Sofa* and *Pasty Faces*. Associate NGTV companies are Grant Naylor Productions and Pepper Productions. NGTV is willing to accept unsolicited material from writers but 1–2-page treatments only. No scripts, please.

Ginger Productions

▷ SMG Productions & Ginger Productions

Goldcrest Films International Ltd

65–66 Dean Street, London W1D 4PL

☎ 020 7437 8696   🖷 020 7437 4448

info@goldcrestfilms.com

www.goldcrestfilms.com

Chairman *John Quested*

Contact *Laura Murphy*

Since it was established in 1977, Goldcrest Films has become a leading independent film production company winning many prizes at international festivals including 19 Academy Awards and 28 Baftas. Finances, produces and distributes films and television programmes. OUTPUT includes *Chariots of Fire*; *Gandhi*; *The Killing Fields*; *A Room With a View*; *Local Hero*; *The Mission*; *To End All Wars*. 'We are currently seeking film projects to invest equity in through our Finishing Fund in return for International Sales Rights.' Scripts via agents only.

The Good Film Company Ltd

The Studio, 5–6 Eton Garages, Lambolle Place, London NW3 4PE

☎ 020 7794 6222   🖷 020 7794 4651

yanina@goodfilms.co.uk

www.goodfilms.co.uk

Contact *Yanina Barry*

Commercials and pop videos. Clients include Suzuki, Burberry, Totally London, Hugo Boss, Cadbury's. *No* unsolicited mss.

Granite Productions

Easter Davoch, Tarland, Aboyne AB34 4US
☎ 01339 880175
Contact *Simon Welfare*

Producer of television documentary programmes such as *Nicholas & Alexandra*; *Victoria & Albert* and *Arthur C. Clarke's Mysterious Universe*.

Grant Naylor Productions

▷ **Noel Gay Television**

Green Umbrella Ltd

59 Cotham Hill, Cotham, Bristol BS6 6JR
☎ 0117 906 4336   🖷 0117 923 7003
info@green-umbrella.co.uk
www.green-umbrella.co.uk

Film producers specializing in science and natural history documentaries. OUTPUT includes episodes for *The Natural World*, *Wildlife on One* and original series such as *Living Europe* and *Triumph of Life*. Unsolicited treatments relating to natural history and science subjects are welcome.

Greenwich Village Productions

14 Greenwich Church Street, London SE10 9BJ
☎ 020 8853 5100   🖷 020 8293 3001
tv@fictionfactory.co.uk
www.fictionfactory.co.uk
Contact *John Taylor*

Features, arts, corporate and educational projects, web-movies and new-media productions. Recent OUTPUT includes *Athletes* and *Secret Roads*. Mss only via agents or from writers with a professional track record in the chosen medium.

H2 Business Communications

Shepperton Studios, Shepperton TW17 0QD
☎ 01932 593717   🖷 01932 593718
mail@h2bc.co.uk
www.h2bc.co.uk
Contact *Julie Knight*

Conferences, videos, awards presentations and speaker training.

Hammerwood Film Productions

info@filmangel.co.uk
www.filmangel.co.uk

Film, video and TV drama. OUTPUT *Iceni* (film; co-production with Pan-European Film Productions and Boudicca Film Productions Ltd); *Boudicca – A Celtic Tragedy* (TV series). In pre-production: *The Black Egg* (witchcraft in 17th century England); *The Ghosthunter*; *Iceni* (documentary of the rebellion of AD61); *No Case to Answer* (legal series). 'Authors are

recommended to access www.filmangel.co.uk' (see *Useful Websites*).

Hartswood Films Ltd

Twickenham Studios, The Barons, St Margarets TW1 2AW
☎ 020 8607 8736   🖷 020 8607 8744
films.tv@hartswoodfilms.co.uk
Contact *Elaine Cameron*

Film and TV production for drama, comedy and documentary. OUTPUT *Jekyll*; *Coupling*; *Men Behaving Badly*; *Border Cafe*; *Carrie & Barry*; *Fear, Stress and Anger* (all for BBC); *After Thomas* (ITV drama).

Hat Trick Productions Ltd

10 Livonia Street, London W1F 8AF
☎ 020 7434 2451   🖷 020 7287 9791
info@hattrick.com
www.hattrick.com
Managing Director *Jimmy Mulville*

Television programmes.

Healthcare Productions Limited

The Great Barn, Godmersham Park, Canterbury CT4 7DT
☎ 01227 738279
penny@healthcareproductions.co.uk
www.healthcareproductions.co.uk
Contact *Penny Wilson*

Established in 1989, producing health-based dramas and documentaries for TV and the DVD market. Has won numerous awards from the BMA. Happy to look at new material, which should be submitted by e-mail.

Heritage Theatre Ltd

Unit 1, 8 Clanricarde Gardens, London W2 4NA
☎ 020 7243 2750   🖷 020 7792 8584
rm@heritagetheatre.com
www.heritagetheatre.com
Contact *Robert Marshall*

Video recordings of successful stage plays for distribution on DVD and broadcast. 'We cannot deal with scripts of unproduced plays.'

Holmes Associates

The Studio, 37 Redington Road, London NW3 7QY
☎ 020 7813 4333
holmesassociates@blueyonder.co.uk
Contact *Andrew Holmes*

Prolific originator, producer and packager of documentary, drama and music television and films. OUTPUT has included *Ashes and Sand* (Film 4); *Chunky Monkey* (J&V Films); *Prometheus* (Ch4 'Film 4'); *The Shadow of Hiroshima* (Ch4 'Witness'); *The House of Bernarda Alba* (Ch4/ WNET/Amaya); *Piece of Cake* (LWT); *The Cormorant* (BBC/Screen 2); *John Gielgud Looks*

Back; *Rock Steady*; *Well Being*; *Signals*; *Ideal Home?* (all Ch4); *Timeline* (with MPT, TVE Spain & TRT Turkey); *Seven Canticles of St Francis* (BBC2). Submissions only accepted by e-mail in synopsis form.

Hourglass Productions

27 Prince's Road, London SW19 8RA
☏ 020 8540 8786
productions@hourglass.co.uk
www.hourglass.co.uk
Partners *Martin Chilcott, Jacqueline Chilcott*

Film and video; documentary and drama. OUTPUT BAFTA nominated scientific television documentaries and educational programming. Also current affairs, health and social issues.

Icon Films

1–2 Fitzroy Terrace, Bristol BS6 6TF
info@iconfilms.co.uk
www.iconfilms.co.uk
Contact *Harry Marshall*

Film and TV documentaries. OUTPUT *Nick Baker's Weird Creatures* (five/Animal Planet/ Granada International); *Tom Harrisson – The Barefoot Anthropologist* (BBC). Specializes in factual documentaries. Open-minded to new documentary proposals.

Imari Entertainment Ltd

PO Box 158, Beaconsfield HP9 1AY
☏ 01494 677147   📠 01494 677147
info@imarientertainment.com
Contact *Jonathan Fowke*

TV and video producer, covering all areas of drama, documentary and corporate productions.

Isis Productions

387b King Street, Hammersmith, London W6 9NJ
☏ 020 8748 3042   📠 020 8748 7634
isis@isis-productions.com
www.isis-productions.com
Directors *Nick de Grunwald, Jamie Rugge-Price*

Formed in 1991, Isis Productions focuses on the production of music and documentary programmes. OUTPUT *Imagine – Yusuf Islam* (BBC); *Pet Shop Boys – A Life in Pop*; *Rufus Wainwright*; *Brian Ferry – The Dylan Sessions*; *Ray Davies – The World from My Window*; *James Brown – Soul Survivor* (Ch4); *Bernie Taupin* (ITV 'South Bank Show'); *Iron Maiden, Judas Priest* (Five 'Rock Classics'); Films on *Deep Purple, Metallica, Def Leppard, Lou Reed, Elton John, Elvis Presley, Sex Pistols* (ITV 'Classic Albums 3'); *Simply Red, Nirvana, Cream, Pink Floyd, Motorhead* ('Classic Albums 4'); *England's Other Elizabeth – Elizabeth Taylor* (BBC 'Omnibus').

Isolde Films

28 Twyford Avenue, London W3 9QB
☏ 020 8896 2860
isolde@btinternet.com
www.tonypalmer.org
Contact *Michela Antonello*

Film and TV documentaries. OUTPUT *Wagner*; *Margot*; *Menuhin*; *Maria Callas*; *Testimony*; *In From the Cold*; *Pushkin*; *England, My England* (by John Osborne). Unsolicited material is read, but please send a written outline first.

ITV Productions

The London Television Centre, Upper Ground, South Bank, London SE1 9LT
☏ 020 7620 1620   📠 020 7261 3041
www.itv.com
Controller of ITV Productions (Drama), London *Kate Bartlett*
Director *John Whiston*

ITV Productions is the largest commercial TV production company in the UK. Produces original programmes, co-productions and TV movies for ITV channels and other broadcasters, both in the UK and abroad. OUTPUT includes *Hornblower*; *Poirot*; *Miss Marple*; *Touching Evil*; *Where the Heart Is*; *The Last Detective*; *Jericho*.

JAM Pictures and Jane Walmsley Productions

8 Hanover Street, London W1S 1YE
☏ 020 7290 2676   📠 020 7256 6818
producers@jampix.com
Contacts *Jane Walmsley, Michael Braham*

JAM Pictures was founded in 1996 to produce drama for film, TV and stage. Projects include: *Hillary's Choice* (TV film, A&E Network); *Son of Pocahontas* (TV film, ABC); *Rudy: the Rudy Giuliani Story* (TV film, USA Network); *One More Kiss* (feature, directed by Vadim Jean); *Bad Blood* (UK theatre tour). Jane Walmsley Productions, formed in 1985 by TV producer, writer and broadcaster, Jane Walmsley, has completed award-winning documentaries and features such as *Hot House People* (Ch4). No unsolicited mss. 'Letters can be sent to us, asking if we wish to see mss; we are very interested in quality material, from published or produced writers only, please'.

Justice Entertainment Ltd

PO Box 4377, London W1A 7SX
☏ 020 7287 2355   📠 020 7287 2354
info@timwestwood.com
www.timwestwood.com

TV and radio production company. No unsolicited material.

Keo Films.com Ltd

101 St John Street, London EC1M 4AS
☏ 020 7490 3580   🖷 020 7490 8419
keo@keofilms.com
www.keofilms.com
Contact *Katherine Perry*

Television documentaries and factual
entertainment. OUTPUT includes programmes for
all the main broadcasters including *River Cottage
Gone Fishing*; *River Cottage*; *Meet the Natives* (all
for Ch4); *Return to Lullingstone Castle*; *Blizzard:
Race to the Pole* (BBC2). No unsolicited mss.

Kingfisher Television Productions Ltd

Martindale House, The Green, Ruddington,
Nottingham NG11 6HH
☏ 0115 945 6581   🖷 0115 921 7750
Contact *Tony Francis*

Broadcast television production.

Kudos Film and Televison Ltd

12–14 Amwell Street, London EC1R 1UQ
☏ 020 7812 3270   🖷 020 7812 3271
info@kudosfilmandtv.com
www.kudosfilmandtv.com
Joint Managing Directors *Jane Featherstone,
Stephen Garrett*
Commercial Director *Daniel Isaacs*
Director of Drama *Simon Crawford Collins*

Television dramas include *Life on Mars*; *Spooks*;
Hustle; *Tsunami – The Aftermath*; *The Amazing
Mrs Pritchard*; *Wide Sargasso Sea*; *Scars*; *MI High*;
Child of Mine; *Comfortably Numb*; *Pleasureland*;
The Magician's House and *Psychos*. Feature films
such as *Among Giants* and *Pure*. No unsolicited
mss.

Lagan Pictures Ltd

21 Tullaghbrow, Tullaghgarley, Ballymena
BT42 2LY
☏ 028 2563 9479/07798 528797   🖷 028 2563 9479
laganpictures@btinternet.com
Producer/Director *Stephen Butcher*

Film, video and TV: drama, documentary and
corporate. OUTPUT *A Force Under Fire* (Ulster
TV). In development: *Into the Bright Light of Day*
(drama-doc); *The £10 Float* (feature film); *The
Centre* (drama series). 'We are always interested
in hearing from writers originating from or based
in Northern Ireland or anyone with, preferably
unstereotypical, projects relevant to Northern
Ireland. We do not have the resources to deal
with unsolicited mss, so please write with a brief
treatment/synopsis in the first instance.'

Landseer Productions Ltd

140 Royal College Street, London NW1 0TA
☏ 020 7485 7333

ken@landseerproductions.com
www.landseerfilms.com
Directors *Derek Bailey, Ken Howard*

Film and video production: documentary,
drama, music and arts. OUTPUT *Swinger* (BBC2/
Arts Council); *Auld Lang Syne* and *Retying the
Knot – The Incredible String Band* (both for BBC
Scotland); *Benjamin Zander* ('The Works', BBC2);
*Zeffirelli, Johnnie Ray, Petula Clark, Bing Crosby,
Maxim Vengerov* (all for 'South Bank Show',
LWT); *Death of a Legend – Frank Sinatra* ('South
Bank Show' special); *Routes of Rock* (Carlton); *See
You in Court* (BBC); *Nureyev Unzipped, Gounod's
Faust, The Judas Tree, Ballet Boyz, 4Dance* and
Bourne to Dance (all for Ch4), *Proms in the Park*
(Belfast).

Lilyville Screen Entertainment Ltd

7 Lilyville Road, London SW6 5DP
☏ 020 7471 8989
tony.cash@btclick.com
Contact *Tony Cash*

Drama and documentaries for TV. OUTPUT
Poetry in Motion (series for Ch4); 'South Bank
Show': *Ben Elton* and *Vanessa Redgrave*; *Musique
Enquête* (drama-based French language series,
Ch4); *Sex and Religion* (ITV); *Landscape and
Memory* (arts documentary series for the BBC);
Jonathan Miller's production of the *St Matthew
Passion* for the BBC; major documentary on the
BeeGees for the 'South Bank Show'. Scripts with
an obvious application to TV may be considered.
Interested in new writing for documentary
programmes.

Lime Pictures Limited

Campus Manor, Childwall Abbey Road,
Childwall, Liverpool L16 0JP
☏ 0151 722 9122   🖷 0151 722 1969
www.merseyfilm.biz
Creative Director *Tony Wood*

Originally known as Mersey Television until
All3Media's acquisition in 2005 and a change
of name to Lime Pictures in 2006. Creators of
television dramas: *Brookside*; *Hollyoaks*; *The
Courtroom* (Ch4); *Grange Hill*; *Apparitions*
(BBC).

Loftus Audio Ltd

2a Aldine Street, London W12 8AN
☏ 020 8740 4666
ask@loftusaudiio.co.uk
www.loftusaudiio.co.uk
Contact *Matt Thompson* (radio drama) ☏ 01620
893876

Producer of factual radio programmes,
documentaries, readings (mostly non-fiction) and
drama, as well as audio books and audio guides
for museums and galleries. OUTPUT includes

numerous titles for Radio 4's 'Book of the Week'. 'We are happy to look at brief emailed synopses of radio plays suitable for broadcast in the available drama slots on BBC Radio 3 or 4.'

London Scientific Films Ltd

Dassels House, Dassels, Braughing, Ware SG11 2RW
☏ 01763 289905
lsf@londonscientificfilms.co.uk
www.londonscientificfilms.co.uk
Contact *Mike Cockburn*

Film and video documentary and corporate programming. No unsolicited mss.

Lucida Productions

5 Alleyn Crescent, London SE21 8BN
☏ 020 8761 4344
pj.lucida@tiscali.co.uk
Contact *Paul Joyce*

Television and cinema: arts, adventure, current affairs, documentary, drama and music. OUTPUT has included *Motion and Emotion: The Films of Wim Wenders*; *Dirk Bogarde – By Myself*; *Sam Peckinpah – Man of Iron*; *Kris Kristofferson – Pilgrim*; *Wild One: Marlon Brando*; *Stanley Kubrick: 'The Invisible Man'*; *2001: the Making of a Myth* (Ch4); *Mantrap – Straw Dogs, the final cut* (with Dustin Hoffman). Restoration of the Director's Cut of *The Devils* plus the documentary *Hell on Earth* with Ken Russell and Vanessa Redgrave. Currently in development for documentary projects.

Marchmont Films Ltd

24 Three Cups Yard, Sandland Street, London WC1R 4PZ
office@marchmontfilms.com
www.marchmontfilms.com
Development Executives *Beverley Hills, Daniel Hayes, Andrew Cussens*

Producer of short films and feature projects. OUTPUT includes *Out In the Cold*; *The Green Wave*; *Punch*; *Shag & Go*. Welcomes new writers. See website for current submission criteria.

Jane Marshall Productions

The Coach-House, Westhill Road, Blackdown, Leamington Spa CV32 6RA
☏ 01926 831680
jane@janemarshallproductions.co.uk
Contact *Jane Marshall*

Producer of readings of published work both fiction and non-fiction for BBC Radio. Published work only.

Maverick Television

Progress Works, Heath Mill Lane, Birmingham B9 4AL
☏ 0121 771 1812   ☏ 0121 771 1550
mail@mavericktv.co.uk
www.mavericktv.co.uk
Also at: 40 Church Way, London NW1 1LW
☏ 020 7383 2727
Head of Production *Allan Tott*

Established in 1994, Maverick has a strong reputation for popular factual programming as well as drama. It is now one of network television's most prolific independent suppliers. OUTPUT includes *How to Look Good Naked*; *10 Years Younger*; *Who'll Age Worst?*; *Bollywood Star*; *Fat Chance*; *Born Too Soon*; *VeeTV*; *Trade Secrets*; *Embarrassing Illnesses*; *10 Things You Didn't Know About ...*; *How To Live Longer*; *The Property Chain*; *Male, 33, Seeks Puberty*; *Extreme Engineering*; *Picture This: Accidental Hero*; *Up Your Street*; *The Property Chain*; *Motherless Daughters*; *Highland Bollywood: Black Bag*; *Health Alert: My Teenage Menopause*; *Long Haul*; *Learning to Love the Grey*.

Maya Vision International Ltd

6 Kinghorn Street, London EC1A 7HW
☏ 020 7796 4842   ☏ 020 7796 4580
www.mayavisionint.com
Contact *Tamsin Ranger*

Film and TV: drama and documentary. OUTPUT *Saddam's Killing Fields* (for 'Viewpoint', Central TV); *3 Steps to Heaven* and *A Bit of Scarlet* (feature films for BFI/Ch4); *A Place in the Sun* and *North of Vortex* (dramas for Ch4/Arts Council); *The Real History Show* (Ch4); *In Search of Myths and Heroes*; *In Search of Shakespeare*; *In the Footsteps of Alexander the Great*; *Conquistadors* (BBC documentaries); *Hitler's Search for the Holy Grail*; *Once Upon a Time in Iran* (Ch4 documentaries). Absolutely no unsolicited material; commissions only.

MBP TV

Saucelands Barn, Coolham, Horsham RH13 8QG
☏ 01403 741620   ☏ 01403 741647
info@mbptv.com
www.mbptv.com
Contact *Phil Jennings*

Maker of film and video specializing in programmes covering equestrianism and the countryside. No unsolicited scripts, but always looking for new writers who are fully acquainted with the subject.

Melendez Films

Julia House, 44 Newman Street, London W1T 1QD
☏ 020 7323 5273   ☏ 020 7323 5373
Contact *Steven Melendez*

Independent production company specializing in 2D animation. Also involved in production and film design for clients in England, Spain, Sweden, India and the US, plus website design and 3D animation on the Web. Clients include book publishers, TV companies and advertisers. Winner of international awards for films, particularly of classic books, stories and comic characters. 'We will look at unsolicited projects in outline or synopsis form only. Enclose s.a.e.'

Mendoza Film Productions

3–5 Barrett Street, London W1U 1AY
☎ 020 7935 4674   ⨏ 020 7935 4417
office@mendozafilms.com
www.mendozafilms.com
Contacts *Wynn Wheldon, Debby Mendoza*

Commercials, title sequences (e.g. Alan Bleasdale's *G.B.H.*); party political broadcasts. Currently in pre-production on a feature-length comedy film. Involved with the Screenwriters' Workshop. Unsolicited mss welcome but 'comedies only, please'. Material will not be returned without s.a.e.

Mersey Television
▷ Lime Pictures Limited

Moonstone Films Ltd

☎ 020 7870 7180   ⨏ 0870 005 6839
info@moonstonefilms.co.uk
www.moonstonefilms.co.uk
Contact *Tony Stark*

Television: current affairs, science and history documentaries. OUTPUT *Arafat's Authority* and *Arafat Investigated*, both for BBC 'Correspondent'. Plus various Ch4 News commissions. Unsolicited mss welcome.

Neon

Studio Two, 19 Marine Crescent, Kinning Park, Glasgow G51 1HD
☎ 0141 429 6366   ⨏ 0141 429 6377
stephy@go2neon.com
www.go2neon.com
Contact *Stephanie Pordage*

Television and radio: drama and documentary producers. OUTPUT includes *Brand New Country*; *Asian Overground*; *Peeking Past the Gates of Skibo*. Supports and encourages new writing 'at every opportunity'. Welcomes unsolicited material but telephone in the first instance.

Number 9 Films

Linton House, 24 Wells Street, London W1T 3PH
☎ 020 7323 4060   ⨏ 020 7323 0456
info@number9films.co.uk
Contacts *Stephen Woolley, Elizabeth Karlsen*

Leading feature film producer. OUTPUT includes *Breakfast on Pluto*; *Stoned*; *Mrs Harris*. Forthcoming productions: *And When Did You Last See Your Father?*; *How to Lose Friends and Alienate People*; *Edith and the Lonely Doll*. No unsolicited material.

Odyssey Productions Ltd

72 Tay Street, Newport-on-Tay DD6 8AP
☎ 01382 542070   ⨏ 01382 542070
billykay@sol.co.uk
www.billykay.co.uk
Contact *Billy Kay*

Producer of radio documentaries. OUTPUT includes *Scotland's Black History*; *Gentle Shepherds* (oral history); *Street Kids* (Scottish missionaries working with street kids in Brazil). Ideas for radio documentaries welcome; send letter with one page outlining the idea and programme content.

Omnivision

Pinewood Studios, Iver Heath SL0 0NH
☎ 01753 656329   ⨏ 01753 631145
info@omnivision.co.uk
www.omnivision.co.uk
Contacts *Christopher Morris, Nick Long*

TV and video producers of documentary, corporate, news and sport programming. Also equipment and facilities hire. Interested in ideas; approach by letter or e-mail.

Orlando Television Productions Ltd

Up-the-Steps, Little Tew, Chipping Norton OX7 4JB
☎ 01608 683218   ⨏ 01608 683364
info@orlandomedia.co.uk
www.orlandomedia.co.uk
Contact *Mike Tomlinson*

Producer of TV documentaries and digital multimedia content, with science, health and information technology subjects as a specialization. Approaches by established writers/ journalists to discuss proposals for collaboration are welcome.

Orpheus Productions

6 Amyand Park Gardens, Twickenham TW1 3HS
☎ 020 8892 3172   ⨏ 020 8892 4821
richard-taylor@blueyonder.co.uk
Contact *Richard Taylor*

Television documentaries and corporate work. OUTPUT has included programmes for the BBC, ITV and Ch4 as well as documentaries for the United Nations, the Shell Film Unit and Video Arts. Unsolicited scripts are welcomed with caution. 'Our preference is for the more classically structured documentary that, while being hard-hitting, explores the subtleties and the paradox of an issue – and is not presented by unqualified celebrities.'

Ovation

Upstairs at the Gatehouse, Highgate, London N6 4BD
☎ 020 8340 4256   ⨏ 020 8340 3466

Contact *John Plews*

Corporate video and conference scripts. Unsolicited mss not welcome. 'We talk to new writers from time to time.' Ovation also runs the fringe theatre, 'Upstairs at the Gatehouse'.

Paper Moon Productions

Wychwood House, Burchetts Green Lane, Littlewick Green, Nr. Maidenhead SL6 3QW

☎ 01628 829819   🖷 01628 829819
david@paper-moon.co.uk
Contact *David Haggas*

Broadcast documentaries and corporate communications. Recent OUTPUT includes *Bilbo & Beyond*, an affectionate glimpse into the life and work of the dedicated philologist and fantasy writer J.R.R. Tolkien.

Parallax East Ltd

Victoria Chambers, St Runwald Street, Colchester CO1 1HF

☎ 01206 574909   🖷 01206 575311
assistant@parallaxindependent.co.uk
Contact *Sally Hibbin*

Feature films/television drama. OUTPUT *A Very British Coup*; *Riff-Raff*; *Bad Behaviour*; *Raining Stones*; *Ladybird, Ladybird*; *i.d.*; *Land and Freedom*; *The Englishman Who Went up a Hill But Came Down a Mountain*; *Bliss*; *Jump the Gun*; *Carla's Song*; *The Governess*; *My Name Is Joe*; *Stand and Deliver*; *Dockers*; *Hold Back the Night*; *Bread and Roses*; *Princesa*; *The Navigators*; *Sweet Sixteen*; *Innocence*; *The Intended*; *Blind Flight*; *Yasmin*; *Almost Adult*.

Passion Pictures

3rd Floor, 33–34 Rathbone Place, London W1T 1JN

☎ 020 7323 9933   🖷 020 7323 9030
info@passion-pictures.com
www.passion-pictures.com
Managing Director *Andrew Ruhemann*

Documentary and drama includes: *One Day in September* (Academy Award-winner for Best Feature Documentary, 2000); also commercials and music videos. Unsolicited mss welcome.

Pathé Pictures

14–17 Kenthouse, Market Place, London W1W 8AR

☎ 020 7323 5151   🖷 020 7631 3568
Head of Creative Affairs *Colleen Woodcock*

Produces 4–6 theatrical feature films each year. 'We are pleased to consider all material that has representation from an agent or production company.'

Pearson Television
▷ FremantleMedia Group Ltd

Pelicula Films

59 Holland Street, Glasgow G2 4NJ
☎ 0141 287 9522
Contact *Mike Alexander*

Television producer. Maker of drama documentaries and music programmes for the BBC and Ch4. OUTPUT *As an Eilean (From the Island)*; *The Trans-Atlantic Sessions 1 & 2*; *Nanci Griffith, Other Voices 2*; *Follow the Moonstone*.

Pepper Productions
▷ Noel Gay Television

Photoplay Productions Ltd

21 Princess Road, London NW1 8JR
☎ 020 7722 2500   🖷 020 7722 6662
info@photoplay.co.uk
www.photoplay.co.uk
Contact *Patrick Stanbury*

Documentaries for film, television and video plus restoration of silent films and their theatrical presentation. OUTPUT includes *The Cat and the Canary*; *Orphans of the Storm*; *Cecil B. DeMille: American Epic* and the 'Channel 4 Silents' series of silent film restoration, including *The Wedding March* and *The Iron Mask*. Recently completed *Garbo* and *I'm King Kong!* No unsolicited mss; 'we tend to create and write all our own programmes'.

Picardy Media & Communication

Dalintober Hall, 40 Dalintober Street, Glasgow G5 8NW
☎ 0141 420 0909   🖷 0141 429 3723
jr@picardy.co.uk
www.picardy.co.uk
Head of Production *John Rocchiccioli*

Produces screen-based content for education, training, sales and marketing, HR and induction. Operates in a variety of areas including health, education, social policy, etc. 'Aims to inform through entertainment.' Unsolicited mss welcome; 'keen to encourage new writing'.

Picture Palace Films Ltd

13 Egbert Street, London NW1 8LJ
☎ 020 7586 8763   🖷 020 7586 9048
info@picturepalace.com
www.picturepalace.com
Producer & Chief Executive *Malcolm Craddock*

Leading independent producer of film and TV drama. OUTPUT *Rebel Heart* (BBC1); *Sharpe's Challenge*; *Extremely Dangerous*; *A Life for A Life* and *Frances Tuesday* (all for ITV); *Sharpe's Rifles* (14 films for Carlton TV); *Little Napoleons* (comedy drama, Ch4); *The Orchid House* (drama serial, Ch4); *Tandoori Nights*; *4 Minutes*; *When Love Dies* (all for Ch4); *Ping Pong* (feature film); *Acid House* (Picture Palace North). Material will only be considered if submitted through an agent.

Planet24 Pictures Ltd

Octagon House, Fir Road, Stockport SK7 2NP
℡ 0870 765 8780
planet24picture@aol.com
www.planet24pictures.co.uk
Contact *Mercedes de Dunewíc*

Producer of television and video documentaries plus community/infomercials. OUTPUT includes *Jack the Ripper: The Conspiracies*; *Truth to Tell*; *Pitmen and Politics: The County Built on Coal*. Mss, plot outlines and screenplays considered but e-mail for information first. 'Welcome the input of new writers with radical points of view.'

Plantagenet Films Limited

Ard-Daraich Studio B, Ardgour, Nr Fort William PH33 7AB
℡ 01855 841384   ℻ 01855 841384
norriemaclaren@btinternet.com
Contact *Norrie Maclaren*

Film and television: documentary and drama programming such as *Dig* (gardening series for Ch4); various 'Dispatches' for Ch4 and 'Omnibus' for BBC. Keen to encourage and promote new writing.

Portobello Pictures

12 Addison Avenue, London W11 4QR
℡ 020 7605 1396   ℻ 020 7605 1391
mail@portobellopictures.com
www.portobellopictures.com
Producer *Eric Abraham*
Associate Producer *Kate McCullagh*

Oscar-winning film, television and theatre production company.

Pozzitive Television

Paramount House, 162–170 Wardour Street, London W1F 8AB
℡ 020 7734 3258   ℻ 020 7437 3130
david@pozzitive.co.uk
Contact *David Tyler*

Producer of comedy and entertainment for television and radio. OUTPUT *Dinner Ladies*; *Coogan's Run*; *The 99p Challenge*; *The Comic Side of 7 Days*; *Armando Iannucci's Charm Offensive*. Unsolicited mss of TV and radio comedy welcome. 'No screenplays or stage plays or novels, please. Send hard copy of full sample script. We read everything submitted this way. Sorry, we don't return scripts unless you send an s.a.e.'

Promenade Enterprises Limited

6 Russell Grove, London SW9 6HS
℡ 020 7582 9354
info@promenadeproductions.com
www.promenadeproductions.com
Contact *Nicholas Newton*

Producer of drama for radio and theatre predominantly. Supports new writing but only accepts unsolicited mss via agents or producers.

Redweather

Easton Business Centre, Felix Road, Bristol BS5 0HE
℡ 0117 941 5854   ℻ 0117 941 5851
production@redweather.co.uk
www.redweather.co.uk

Broadcast documentaries on arts and disability, corporate video and CD-ROM, Water Aid, British Oxygen, etc.

Renaissance Films

34–35 Berwick Street, London W1F 8RP
℡ 020 7287 5190   ℻ 020 7287 5191
info@renaissance-films.com

Film production company whose credits include *Henry V*; *Peter's Friends*; *Much Ado About Nothing*; *The Luzhin Defence*; *The Reckoning*; *Inbreeds*; *We Don't Live Here Anymore* and *Candy*.

Richmond Films & Television Ltd

PO Box 33154, London NW3 4AZ
℡ 020 7722 6464   ℻ 020 7722 6232
mail@richmondfilms.com
Contact *Development Executive*

Film and TV: drama and comedy. OUTPUT *Press Gang*; *The Lodge*; *The Office*; *Wavelength*; *Privates*; *in2minds*. No unsolicited scripts.

RS Productions

191 Trewhitt Road, Newcastle upon Tyne NE6 5DY
℡ 07710 064632 (mobile)/00 49 89 5111 5895
enquiries@rsproductions.co.uk
www.rsproductions.co.uk
Contact *Mark Lavender*

Feature films and television: drama series/serials and singles. TV documentaries and series. Working with established and new talent.

Sands Films

119 Rotherhithe Street, London SE16 4NF
℡ 020 7231 2209   ℻ 020 7231 2119
info@sandsfilms.co.uk
www.sandsfilms.co.uk
Contacts *Christine Edzard, Olivier Stockman*

Film and TV drama. OUTPUT *Little Dorrit*; *The Fool*; *As You Like It*; *A Dangerous Man*; *The Long Day Closes*; *A Passage to India*; *Topsy Turvy*; *Nicholas Nickleby*; *The Gangs of New York*; *The Children's Midsummer Night's Dream*. No unsolicited scripts.

Scala Productions Ltd

2nd Floor, 37 Foley Street, London W1W 7TN
℡ 020 7637 5720   ℻ 020 7637 5734

scalaprods@aol.com
Contacts *Ian Prior, Nik Powell*

Production company set up by ex-Palace Productions Nik Powell and Stephen Woolley, who have an impressive list of credits including *Company of Wolves*; *Absolute Beginners*; *Mona Lisa*; *Scandal*; *The Crying Game*; *Backbeat*; *Neon Bible*; *24:7*; *Little Voice*; *Divorcing Jack*; *The Last September*; *Wild About Harry*; *Last Orders*; *A Christmas Carol – The Movie*; *Black and White*; *Leo*; *The Night We Called It A Day*; *Ladies in Lavender*; *Stoned*. In development: *The Coat*; *Du Cane's Boys*; *Celluloid Dream*; *Johnny Bollywood*; *The Rough*; *Black Cockatoo*; *Nightcab*; *White Turks*.

Scope Productions Ltd

180 West Regent Street, Glasgow G2 4RW
☎ 0141 221 4312
laurakingwell@scopeproductions.co.uk
www.scopeproductions.co.uk
Corporate *Laura Kingwell*

Corporate film and video and multimedia communications for clients across all sectors.

Screen First Ltd

The Studios, Funnells Farm, Down Street, Nutley TN22 3LG
☎ 01825 712034
Producer/Director *Paul Madden*

Founded 1985. Focus on animation, arts, children's programmes, documentary and drama. OUTPUT includes *Ivor the Invisible*. No unsolicited scripts.

Screen Ventures Ltd

49 Goodge Street, London W1T 1TE
☎ 020 7580 7448   ☎ 020 7631 1265
info@screenventures.com
www.screenventures.com
Contacts *Christopher Mould, Michael Evans*

Film and TV sales and production: documentary, music videos and drama. OUTPUT *Life and Limb* (documentary, Discovery Health Channel); *Pavement Aristocrats* (SABC); *Woodstock Diary*; *Vanessa Redgrave* and *Genet* (both for LWT 'South Bank Show'); *Mojo Working*; *Burma: Dying for Democracy* (Ch4); *Dani Dares* (Ch4 series on strong women); *Pagad* (Ch4 news report).

Screenhouse Productions Ltd

Chapel Allerton House, 114 Harrogate Road, Leeds LS7 4NY
☎ 0113 266 8881   ☎ 0113 266 8882
paul.bader@screenhouse.co.uk
www.screenhouse.co.uk
Contact *Paul Bader*

Specializes in science TV, documentary, stunts and events, including outside broadcasts. Science prop. and demo workshop supplying working models to, e.g., BBC History *What the Past Did For Us*; ITV, Discovery and museums. OUTPUT includes *Stardate* (BBC2 astronomy series); *Zapped* (Discovery/US/Canada); *The Man Who Invented the Aeroplane* (UKTV/BBC North); *Science Shack*, *Local Heroes*, BBC2, presented by Adam Hart-Davis. 'More likely to consider written up proposals.'

September Films Ltd

Glen House, 22 Glenthorne Road, London W6 0NG
☎ 020 8563 9393   ☎ 020 8741 7214
september@septemberfilms.com
www.septemberfilms.com
Director of Production *Elaine Day*

Factual entertainment and documentary specialists. Feature film OUTPUT includes *Breathtaking*; *House of America*; *Solomon & Gaenor*. No unsolicited submissions, please.

Shell Like

81 Whitfield Street, London W1T 4HG
☎ 020 7255 5204   ☎ 020 7255 5255
enquiries@shelllike.com
www.shelllike.com
Contact *Anna Pollard*

Produces radio commercials. Unsolicited mss and ideas welcome; send by e-mail.

Showrunner
▷ Endemol UK Productions

Sianco Cyf

Pen-y-graig, Llanfaglan, Caernarfon LL54 5RF
☎ 01286 676100/07831 726111 (mobile)
☎ 01286 677616
sian@sianco.tv
Contact *Siân Teifi*

Children's, youth and education programmes, children's drama, people-based documentaries for adults. 'Please note, *we do not accept any unsolicited scripts*.'

Silent Sound Films Ltd

Cambridge Court, Cambridge Road, Frinton on Sea CO13 9HN
☎ 01255 676381   ☎ 01255 676381
thj@silentsoundfilms.co.uk
www.silentsoundfilms.co.uk
www.londonfoodfilmfiesta.co.uk
Contact *Timothy Foster*

Active in European film co-production with mainstream connections in the USA. Special interest in developing stage and film musicals, art house fiction and documentaries on the arts. Initial enquiry via e-mail.

Skyline Productions

10 Scotland Street, Edinburgh EH3 6PS
📞 0131 557 4580   📠 0131 556 4377
leslie@skyline.uk.com
www.skyline.uk.com
Producer/Writer *Leslie Hills*

Produces film and television drama and documentary.

SMG Productions & Ginger Productions

SMG: Pacific Quay, Glasgow G51 1PQ
📞 0141 300 3000
www.smgproductions.tv
Also: Ginger Productions, 3 Waterhouse Square, 138–142 Holborn London EC1N 2NY
📞 020 7882 1020 www.ginger.tv
Managing Director (SMG Productions & Ginger Productions) *Elizabeth Partyka*
Head of Drama *Eric Coulter*
Head of Business Development *Helen Alexander*

SMG Productions, which incorporates Ginger Productions, makes programmes for the national television networks, including ITV, Ch4 and Sky. Specializes in drama, factual entertainment and children's programming. OUTPUT includes *Taggart*; *Rebus*; *Our Daughter Holly*; *Club Reps*.

Somethin Else

20-26 Brunswick Place, London N1 6DZ
📞 020 7250 5500   📠 020 7250 0937
info@somethinelse.com
www.somethinelse.com
Contact *Jez Nelson*

Producer of television, video and radio documentaries, DVD and interactive content. Ideas for TV shows welcome; send letter in the first instance.

Specific Films

25 Rathbone Street, London W1T 1NQ
📞 020 7580 7476   📠 020 7636 6866
info@specificfilms.com
Contact *Michael Hamlyn*

Founded 1991. OUTPUT includes *Mr Reliable* (feature film co-produced by PolyGram and the AFFC); *The Adventures of Priscilla, Queen of the Desert*, co-produced with Latent Image (Australia) and financed by PolyGram and AFFC; *U2 Rattle and Hum*, full-length feature – part concert film/part cinema verité documentary; *Paws* (executive producer); *The Last Seduction 2* (Polygram); and numerous pop promos for major international artists.

Spice Factory (UK) Ltd

14 Regent Hill, Brighton BN1 3ED
📞 01273 739182   📠 01273 749122
info@spicefactory.co.uk
www.spicefactory.co.uk

Founded 1995. Film producers. OUTPUT *Plots With a View* (Christopher Walken, Brenda Blethyn, Alfred Molina, Lee Evans); *Bollywood Queen* (Preeya Kallidas, James McAvoy, Ian McShane); *The Bridge of San Luis Rey* (Robert De Niro, Kathy Bates, Harvey Keitel); *A Different Loyalty* (Sharon Stone, Rupert Everett); *Head in the Clouds* (Charlize Theron, Penelope Cruz); *The Merchant of Venice* (Al Pacino, Jeremy Irons, Joseph Fiennes, Lynn Collins). No unsolicited material accepted.

'Spoken' Image Ltd

8 Hewitt Street, Manchester M15 4GB
📞 0161 236 7522   📠 0161 236 0020
info@spoken-image.com
www.spoken-image.com
Contact *Geoff Allman*

Film, video and TV production for documentary and corporate material. Specializes in high-quality brochures and reports, CD-ROMs, exhibitions, conferences, film and video production for broadcast, industry and commerce.

Square Dog Radio LLP

Kilmagadwood Cottage, Scotlandwell, Kinross KY13 9HY
📞 05600 472247
mike@squaredogradio.co.uk
www.squaredogradio.co.uk
Contact *Mike Hally*
Producer *Mark Whitaker*

Producer of radio documentaries, features and short story readings. OUTPUT includes *Northern Creative Writing Groups* (returning series, for Radio 4). No unsolicited material.

Tony Staveacre Productions

Channel View, Blagdon BS40 7TP
📞 01761 462161   📠 01761 462161
newstaving@btinternet.com
Contact *Tony Staveacre*

Producer of dramas and documentaries as well as music, arts and comedy progammes. Recent OUTPUT *The Wodehouse Notebooks*; *The Liberation of Daphne, Speaking from the Belly*; *Standing Up for Liverpool* (Radio 4); *The Very Thought of You* and *Jigsy* (theatre); *Tango Maestro* (BBC4); *The Old Boys Band* (BBC1); *Mendip Voices* (CD). No unsolicited mss.

Stirling Film & TV Productions Limited

137 University Street, Belfast BT7 1HP
📞 028 9033 3848   📠 028 9043 8644
anne@stirlingtelevision.co.uk
www.stirlingtelevision.co.uk
Contact *Anne Stirling*

Producer of broadcast and corporate programming – documentary, sport, entertainment and lifestyle programmes.

Straight Forward Film & Television Productions Ltd

Building 2, Lesley Office Park, 393 Hollywood Road, Belfast BT4 2LS
☎ 028 9065 1010   🖷 028 9065 1012
enquiries@straightforwardltd.co.uk
www.straightforwardltd.co.uk
Contacts *John Nicholson, Ian Kennedy*

Northern Ireland-based production company specializing in documentary, feature and lifestyle series for both regional and network transmission. OUTPUT includes *We Shall Overcome* (winner of Best Documentary at 1999 Celtic Television Festival for BBC); *Conquering the Normans* (Ch4 Learning – history of Normans in Ireland); *Gift of the Gab* (Ch4 Learning – contemporary Irish writing); *Sportsweek* (BBC Radio Ulster); *On Eagle's Wing* (full stage musical/TV material; story of the Scots/Irish in America); *Fire School*; *Mission Employable*; *Sweet Child of Mine*; *School Challenge*, 3rd series; *World Indoor Bowls* (all for BBC NI); *Awash With Colour* (series, BBC Daytime).

Sunset + Vine Productions Ltd

Elsinore House, 77 Fulham Palace Road, London W6 8JA
☎ 020 7478 7300   🖷 020 7478 7403
www.sunsetvine.co.uk

Sports, children's and music programmes for television. No unsolicited mss. 'We hire freelancers only upon receipt of a commission.'

Table Top Productions

1 The Orchard, Chiswick, London W4 1JZ
☎ 020 8742 0507   🖷 020 8742 0507
alvin@tabletopproductions.com
Contact *Alvin Rakoff*

TV and film. OUTPUT *Paradise Postponed*; *The Adventures of Don Quixote*; *A Voyage Round My Father*; *The First Olympics 1896*; *Dirty Tricks*; *A Dance to the Music of Time*; *Too Marvellous for Words*. Also Dancetime Ltd. No unsolicited mss.

talkbackTHAMES

20–21 Newman Street, London W1T 1PG
☎ 020 7861 8000   🖷 020 7861 8001
reception@talkbackthames.tv
www.talkbackthames.tv
Chief Executive Officer *Lorraine Heggessey*
Chief Operating Officer *Sara Geater*

talkbackTHAMES Productions is a **FremantleMedia** company. OUTPUT includes: comedy, comedy drama, drama, entertainment, documentary and lifestyle programmes. OUTPUT *The Apprentice*; *Green Wing*; *The Bill*; *Britain's Got Talent*; *The X Factor*; *Never Mind the Buzzcocks*; *Grand Designs*; *Property Ladder*.

Talkingheads Production Ltd

2–4 Noel Street, London W1F 8GB
☎ 020 7292 7575   🖷 020 7292 7576
johnsachs@talkingheadsproductions.com
www.talkingheadsproductions.com
Contact *John Sachs*

Feature films. OUTPUT includes *The Merchant of Venice* starring Al Pacino. Will consider scripts; contact by e-mail. 'Somewhere out there is the new Tom Stoppard.'

Tandem TV & Film Ltd

Suite 206, Charleston House, Hemel Hempstead HP1 3AA
☎ 01442 261576   🖷 01442 219250
info@tandemtv.com
www.tandemtv.com
Production Director *Barbara Page*
Creative Director *Terry Page*
Production Controller *Jevan Green*

Internal and external communications, documentaries, drama-documentaries, public relations, sales, marketing and training programmes for, amongst others, the construction, civil engineering, transport, local government and charitable sectors. Welcomes unsolicited mss.

Taylor Made Broadcasts Ltd

3B Cromwell Park, Chipping Norton OX7 5SR
☎ 01608 646444
post@tmtv.co.uk
Contact *Trevor Taylor*

Producer of *Gardeners' Question Time* (BBC Radio 4). No unsolicited mss.

Telemagination Ltd

Royalty House, 72–74 Dean Street, London W1D 3SG
☎ 020 7434 1551   🖷 020 7434 3344
mail@tmation.co.uk
www.telemagination.co.uk
Contact *Beth Parker*

Animation production company. CREDITS include *The Animals of Farthing Wood*; *Noah's Island*; *The Last Polar Bears*; *Little Ghosts*; *Pongwiffy*; *Heidi*; *The Cramp Twins ll*; *Pettson and Findus*; *Rudi and Trudi*. In production: *Littlest Pet Shop*.

Tern Television Productions Ltd

73 Crown Street, Aberdeen AB11 6EX
☎ 01224 211123   🖷 01224 211199
aberdeen@terntv.com
www.terntv.com
Also at: 4th Floor, 114 Union Street, Glasgow G1 3QQ ☎ 0141 204 1717 glasgow@terntv.com

And: 1st Floor, Cotton Court, 38–42 Waring Street, Belfast BT1 2ED ☎ 028 9032 6061
belfast@terntv.com
Contacts *David Strachan, Gwyneth Hardy* (Aberdeen), *Harry Bell* (Glasgow)

Broadcast, television and corporate video productions. Specializes in factual entertainment. Currently developing drama.

Testimony Films

12 Great George Street, Bristol BS1 5RS
☎ 0117 925 8589   ℻ 0117 925 7608
steve.humphries@testimonyfilms.com
Contact *Steve Humphries*

TV documentary producer. Specializes in social history exploring Britain's past using living memory. OUTPUT includes *Hooked: History of Addictions; Married Love* (both Ch4 series); *A Secret World of Sex* (BBC series); *The 50s & 60s in Living Colour; Some Liked It Hot* (both ITV series). Welcomes ideas from those working on life stories and oral history.

Tiger Aspect Productions

7 Soho Street, London W1D 3DQ
☎ 020 7434 6700   ℻ 020 7434 1798
general@tigeraspect.co.uk
www.tigeraspect.co.uk
Head of Entertainment *Clive Tulloh*
Head of Comedy *Sophie Clarke-Jervoise*

Part of IMG Media. Television producer for comedy, drama, documentary and entertainment. OUTPUT *Ross Kemp on Gangs; Omagh; Births, Marriages & Deaths; Kid in the Corner; Country House; Gimme Gimme Gimme; Harry Enfield & Chums; Howard Goodalls' Big Bangs; Playing the Field I, II & III; Streetmate I & II; Let Them Eat Cake; The Vicar of Dibley.* Sister companies: Tiger Aspect Pictures (feature films; OUTPUT includes *Billy Elliot; The League of Gentlemen's Apocalyse*) and **Tigress Productions Ltd** (see entry). Only considers material submitted via an agent or from writers with a known track record.

Tigress Productions Ltd

2 St Paul's Road, Bristol BS8 1LT
☎ 0117 933 5600   ℻ 0117 933 560066
general@tigressproductions.co.uk
www.tigressproductions.co.uk
Managing Director *Andrew Jackson*
Executive Producer *Kath Moore*

Sister company of **Tiger Aspect Productions**. Specializes in wildlife, science adventure and documentary projects. Output includes *Jaguar Adventure with Nigel Marven; Ben's Zoo; Everest: Beyond the Limit.*

Touch Productions Ltd

18 Queen Square, Bath BA1 2HN
☎ 01225 484666   ℻ 01225 483620
erica@touchproductions.co.uk
www.touchproductions.co.uk
Contacts *Erica Wolfe-Murray, Malcolm Brinkworth*

Over the last 20 years, Touch has made a wide range of programmes including award-winning investigations, popular documentaries, medical and science films, revelatory history productions as well as observational, social, religious and arts programmes. Current commissions include *The Human Footprint*, a Ch4 documentary special and various series and documentaries for the BBC, National Geographic, TLC and Animal Planet. Other projects include *Transplanting Memories?; The Boy Who Couldn't Stop Running; Parish in the Sun; Revival* and *Angela's Dying Wish.*

Transatlantic Films Production and Distribution Company

Cabalva Studios, Whitney-on-Wye HR3 6EX
☎ 01497 831428   ℻ 01497 831677
revel@transatlanticfilms.com
www.transatlanticfilms.com
Executive Producer *Revel Guest*

Producer of TV documentaries. OUTPUT *Belzoni* (Ch4 Schools); *Science of Sleep and Dreams; Science of Love* and *Extreme Body Parts* (all for Discovery Health); *Legends of the Living Dead* (Discovery Travel/S4C International); *2025* (Discovery Digital); *How Animals Tell the Time* (Discovery); *Trailblazers* (Travel Channel). No unsolicited scripts. Interested in new writers to write 'the book of the series', e.g. for *Greek Fire* and *History's Turning Points*, but not usually drama script writers.

TV Choice Ltd

PO Box 597, Bromley BR2 0YB
☎ 020 8464 7402   ℻ 020 8464 7845
tvchoiceuk@aol.com
www.tvchoice.uk.com
Contact *Norman Thomas*

Produces a range of educational videos for schools and colleges on subjects such as history, geography, business studies and economics. No unsolicited mss; send proposals only.

Twentieth Century Fox Film Limited

Twentieth Century House, 31–32 Soho Square, London W1D 3AP
☎ 020 7437 7766
www.fox.co.uk

London office of the American giant. Does not accept unsolicited material.

Twenty Twenty Television

20 Kentish Town Road, London NW1 9NX
☎ 020 7284 2020   🖷 020 7284 1810
pch@twentytwenty.tv
www.twentytwenty.tv
Managing Director *Peter Casely-Hayford*
Co-Founder and Consultant Executive Producer
Claudia Milne
Creative Director/Executive Producer
Jamie Isaacs
Head of Factual Programming/Executive
Producer *Tim Carter*

Founded 1982. Large independent television
production company. Concentrates on
documentaries, lifestyle programmes, popular
drama and living history series. OUTPUT includes
The Choir (BBC2); *Bad Lads Army* (ITV1); *How
to Divorce Without Screwing Up Your Children*
(Ch4). Both *Brat Camp* and *That'll Teach Em*
have now aired a third season. Drama enquiries
and scripts should be mailed to *Jamie Isaacs*
(jamieisaacs@twentytwenty.tv).

Twofour Broadcast

5th Floor, 6–7 St Cross Street, London EC1N 8UA
☎ 020 7438 1800   🖷 020 7438 1850
enq@twofour.co.uk
www.twofour.co.uk
Managing Director *Melanie Leach*
Creative Director *Stuart Murphy*

Television production company founded in 1987.
Specializes in factual entertainment, drama and
comedy. Unsolicited material or ideas accepted
via e-mail.

Tyburn Film Productions Limited

Cippenham Court, Cippenham Lane, Cippenham,
Nr Slough SL1 5AU
☎ 01753 516767   🖷 01753 691785

Feature films. Subsidiary of **Arlington
Productions Limited**. No unsolicited
submissions.

The UK Film and TV Production Company Plc

3 Colville Place, London W1T 2BH
☎ 020 7255 1650
Head of Development *Henrietta Fudakowski*

In 2006, *Tsotsi*, UKFTV's first wholly-produced
film won the Oscar for best foreign language film.
Other titles include *Bugs! in 3D*. Telephone ahead
of sending a submission to see if script might be
suitable for the company.

Vera Productions

66–68 Margaret Street, London W1W 8SR
☎ 020 7436 6116   🖷 020 7436 6117
Contact *Phoebe Wallace*

Produces television comedy such as *Bremner, Bird
and Fortune*.

Video Enterprises

12 Barbers Wood Road, High Wycombe HP12 4EP
☎ 01494 534144/07831 875216 (mobile)
🖷 01494 534144
videoenterprises@ntlworld.com
www.videoenterprises.co.uk
Contact *Maurice R. Fleisher*

Video and TV, mainly corporate: business and
industrial training, promotional material and
conferences. No unsolicited material 'but always
ready to try out good new writers'.

VIP Broadcasting

8 Bunbury Way, Epsom KT17 4JP
☎ 01372 721196   🖷 01372 726697
mail@vipbroadcasting.co.uk
Contact *Chris Vezey*

Produces a wide range of radio programmes,
particularly interviews, documentaries, music
programmes and live concerts. Won award for
'Best Sound' at New York Festival 2000. Approach
with idea by e-mail in the first instance; no
unsolicited mss.

Wall to Wall

8–9 Spring Place, London NW5 3ER
☎ 020 7485 7424   🖷 020 7267 5292
mail@walltowall.co.uk
www.walltowall.co.uk
Chief Executive *Alex Graham*

Factual and drama programming. OUTPUT
includes *Who Do You Think You Are?*; *New Tricks*;
A Rather English Marriage; *Glasgow Kiss*; *Sex,
Chips & Rock 'n' Roll*; *The 1940s House*; *Body
Story*; *Neanderthal*; *The Mafia*; *Not Forgotten*;
H.G. Wells.

Jane Walmsley Productions
▷ JAM Pictures

Walsh Bros. Limited

4 Trafalgar Grove, London SE10 9TB
☎ 020 8858 6870
john@walshbros.co.uk
www.walshbros.co.uk
Producer/Director *John Walsh*
Producer/Head of Finance *David Walsh, ACA*
Producer/Head of Development *Maura Walsh*

BAFTA-nominated producers of television, film
drama and documentaries. OUTPUT *Monarch*
(feature film); BAFTA-winning *Don't Make Me
Angry* (Ch 4); Grierson-nominated *Headhunting
the Homeless*; *Trex* (factual series on teenagers at
work in China, Mexico, Vancouver and Alaska);
Trex2 (follow-up series covering Romania,
India, Iceland and Louisiana); *Boyz & Girlz*,
(Derbyshire dairy farm documentary series);
Cowboyz & Cowgirlz (US sequel to hit series
of Brit teens working on a ranch in Montana).

Also arts documentaries: *The Comedy Store*; *Ray Harryhausen* (the work of Hollywood special effects legend). Drama: *The Sleeper*; *The Sceptic and the Psychic*; *A State of Mind*. Initial enquiries to: development@walshbros.co.uk

Paul Weiland Film Company

14 Newburgh Street, London W1F 7RT
☎ 020 7287 6900   🖷 020 7434 0146
info@weilands.co.uk
www.weilands.co.uk

Television commercials and pop promos.

Whistledown Productions Ltd

8A Ayres Street, London SE1 1ES
☎ 020 7407 8001
davidprest@whistledown.net
www.whistledown.net
Managing Director *David Prest*

Producer of Sony Award-winning Landmark Series for BBC Radio 4. Features and documentaries on a wide range of social and historical subjects, contemporary issues and popular culture. OUTPUT includes *Questions Questions* and *The Reunion* (Radio 4); music-based documentaries for Radio 2. 'We welcome contributions and ideas, but phone or e-mail first.'

Wise Buddah Creative Ltd

74 Great Titchfield Street, London W1W 7QP
☎ 020 7307 1600   🖷 020 7307 1601
info@wisebuddah.com
www.wisebuddah.com
Contacts *Chris North, Paul Sam Leese*

Radio production company: documentaries and commercials. Also studio facilities, sound-to-picture/sound design, talent management. No unsolicited material.

Witzend Productions
▷ **FremantleMedia Group Ltd**

Working Title Films Ltd

Oxford House, 76 Oxford Street, London W1D 1BS
☎ 020 7307 3000   🖷 020 7307 3001/2/3

Co-Chairmen *Tim Bevan, Eric Fellner*
Head of Development *Natascha Wharton*
Head of Film *Debra Hayward*
Development Executives *Amelia Granger, Rachael Prior*

Feature films OUTPUT *Elizabeth: The Golden Age*; *Atonement*; *Hot Fuzz*; *Mr Bean's Vacation*; *United 93*; *Smokin' Aces*; *Sixty Six*; *Catch a Fire*; *Gone*; *Nanny McPhee*; *Pride & Prejudice*; *The Interpreter*; *Wimbledon*; *Bridget Jones 2*; *Edge of Reason*; *Shaun of the Dead*; *Love Actually*; *Thunderbirds*; *Ned Kelly*; *Johnny English*; *Bridget Jones's Diary*; *Captain Corelli's Mandolin*; *Ali G Indahouse*; *Billy Elliot*; *Notting Hill*; *Elizabeth*; *Fargo*; *Dead Man Walking*; *French Kiss*; *Four Weddings and a Funeral*; *The Hudsucker Proxy*; *The Tall Guy*; *Wish You Were Here*; *My Beautiful Laundrette*. No unsolicited mss at present.

Wortman Productions UK

48 Chiswick Staithe, London W4 3TP
☎ 020 8994 8886/07976 805976 (mobile)
nevillewortman@beeb.net
Producer *Neville Wortman*

Co-producers with Polestar Pictures Ltd. Feature film and TV production for drama, documentary, entertainment and corporate. OUTPUT 'Lost Ships' series: *White Gold of the Dragon Sea* (Discovery Channel, US); *Neffertiti*; *Resurrected*; *Who Killed Julius Caesar* (Discovery Channel, US/Ch 5, UK); 'Days That Shook the World' series: *Conspiracy to Kill*; *Hitler Bomb Plot*; *Assassinate De Gaulle* (BBC); 'Surviving Disaster' series: *Munich Air Crash*; *Eruption at Mount St Helen*; *Murder in Paradise* (Lion TV/Ch4) *Pevkovsky* ('Nuclear Spies' BBC2 series). Open to new writing, preferably through agents; single page outline, some pages of dialogue; s.a.e. for reply.

Zeppotron
▷ **Endemol UK Productions**

Theatre Producers

For latest minimum rates for writers of plays for subsidized repertory theatres (not Scotland) access the following websites: Theatrical Management Association at www.tmauk.org; and the Writers' Guild at www.writersguild.org.uk

Actors Touring Company
▷ ATC

Almeida Theatre Company

The Almeida Theatre, Almeida Street, London N1 1TA

☎ 020 7288 4900   ℻ 020 7288 4901

www.almeida.co.uk

Artistic Director *Michael Attenborough*

Founded 1980. The Almeida is a full-time producing theatre, presenting a year-round theatre programme in which new light is shed on an eclectic mix of new plays, classic revivals, adaptations and new versions of international work. Previous productions: *Festen*; *The Goat, or Who is Sylvia?*; *The Lady From the Sea*; *The Late Henry Moss*; *Enemies*. No unsolicited mss.

Alternative Theatre Company Ltd

Bush Theatre, Shepherds Bush Green, London W12 8QD

☎ 020 7602 3703   ℻ 020 7602 7614

info@bushtheatre.co.uk

www.bushtheatre.co.uk

Artistic Director *Josie Rourke*

Founded 1972. Trading as The Bush Theatre. Produces up to nine new plays a year, including up to three visiting companies also producing new work: 'we are a writer's theatre'. Previous productions include: *When You Cure Me* Jack Thorne; *Pumpgirl* Abbie Spallen; *Blackbird* Adam Rapp; *The Wexford Trilogy* Billy Roche; *Love and Understanding* Joe Penhall; *This Limetree Bower* Conor McPherson; *Discopigs* Enda Walsh; *The Pitchfork Disney* Philip Ridley; *Caravan* Helen Blakeman; *Beautiful Thing* Jonathan Harvey; *Killer Joe* Tracy Letts; *Shang-a-Lang* Catherine Johnson; *Howie the Rookie* Mark O'Rowe; *The*

Glee Club Richard Cameron; *Adrenalin ... Heart* Georgia Fitch. Scripts are read by a team of associates, then discussed with the management, a process which takes about four months. The theatre offers a small number of commissions. Writers should send scripts (full-length plays only) with small s.a.e. for acknowledgement and large s.a.e. for return of script.

ATC

15–16 Nassau Street, London W1W 7AB

☎ 020 7580 7723   ℻ 020 7580 7724

info@atc-online.com

www.atc-online.com

Artistic Director *Bijan Sheibani*

Collaborates with writers on new work, adaptations and/or translations.

Birmingham Repertory Theatre

Centenary Square, Broad Street, Birmingham B1 2EP

☎ 0121 245 2000   ℻ 0121 245 2100

www.birmingham-rep.co.uk

Contact *Literary Department*

The Birmingham Repertory Theatre aims to provide a platform for the best work from new writers from both within and beyond the West Midlands region. The Rep is committed to the production of new work which reflects both the diversity of contemporary experience and of approaches to writing for the stage. The commissioning of new plays takes place across the full range of the theatre's activities: in the Main House, The Door (which is a dedicated new writing space) and on tour to community venues in the region. The theatre runs a programme of writers' attachments every year in addition to its commissioning policy and maintains close links with *Script* (the regional writers' training agency) and the MPhil in Playwriting Studies at the University of Birmingham. For more information contact the Literary Department.

Black Theatre Co-op
▷ Nitro

Bootleg Theatre Company

23 Burgess Green, Bishopdown, Salisbury SP1 3EL
℡ 01722 421476
colin281@btinternet.com
Contact *Colin Burden*

Founded 1984. Tries to encompass as wide an audience as possible and has a tendency towards plays with socially relevant themes. A good bet for new writing since unsolicited mss are very welcome. 'Our policy is to produce new and/or rarely seen plays and anything received is given the most serious consideration.' Actively seeks to obtain grants to commission new writers for the company. New productions to include: *Asking for It*; *17th Valentine*; *King Squealer*.

Borderline Theatre Company

Darlington New Church, North Harbour Street, Ayr KA8 8AA
℡ 01292 281010   ℻ 01292 618685
enquiries@borderlinetheatre.co.uk
www.borderlinetheatre.co.uk
Producer *Eddie Jackson*

Founded 1974. Borderline is one of Scotland's leading touring companies. Tours to main-house theatres and small venues throughout Scotland. Productions include the world premières of Liz Lochhead's work and *The Wall* by D.C. Jackson. Previous writers have included Dario Fo and John Byrne. Borderline is also committed to commissioning and touring new plays for young people. Please contact the company before submitting synopsis or script.

Bristol Old Vic Theatre Company (Old Vic, Studio & Basement)

King Street, Bristol BS1 4ED
℡ 0117 949 3993   ℻ 0117 949 3996
admin@bristol-old-vic.co.uk
www.bristololdvic.co.uk

Bristol Old Vic actively seeks new writers for development and commission. 'We do not offer a formal reading service but are keen to build relationships with local writers in particular.' NB The theatre is closed until 2009 for a £7 million refurbishment; no submissions during this period.

Bush Theatre
▷ Alternative Theatre Company Ltd

Carnival (Films & Theatre) Ltd
▷ entry under Film, TV and Radio Producers

Citizens Theatre

Gorbals, Glasgow G5 9DS
℡ 0141 429 5561   ℻ 0141 429 7374
info@citz.co.uk
www.citz.co.uk
Artistic Directors *Jeremy Raison, Guy Hollands*
Administrative Director *Anna Stapleton*

No formal new play policy. The theatre has a play reader but opportunities to do new work are limited.

Clwyd Theatr Cymru

Mold, Flintshire CH7 1YA
℡ 01352 756331   ℻ 01352 701558
drama@celtic.co.uk
www.clwyd-theatr-cymru.co.uk
Literary Manager *William James* (william.james@clwyd-theatr-cymru.co.uk)

Clwyd Theatr Cymru produces plays performed by a core ensemble in Mold and tours them throughout Wales (in English and Welsh). Productions are a mix of classics, revivals and contemporary drama. Recent new writing includes: *Two Princes* Meredydd Barker; *Memory* Jonathan Lichtenstein; *Stone City Blue* Ed Thomas; *The Rabbit* Meredydd Barker; *The Journey of Mary Kelly* Siân Evans; *And Now What?* Tim Baker and Sarah Argent; *The Way It Was, Flights of Fancy; Pocketful of Memories; The Ballad of Megan Morgan; Flora's War/Rhyfel Flora; Word for Word/Gair am Air* and *The Secret/Y Gyfrinach*: all Tim Baker. Plays by Welsh writers or on Welsh themes will be considered.

Michael Codron Plays Ltd

Aldwych Theatre Offices, Aldwych, London WC2B 4DF
℡ 020 7240 8291   ℻ 020 7240 8467

Michael Codron Plays Ltd manages the Aldwych Theatre in London's West End. The plays it produces don't necessarily go into the Aldwych but always tend to be big-time West End fare. Previous productions: *Bedroom Farce*; *Blue Orange*; *Copenhagen*; *The Invention of Love*; *Hapgood*; *Uncle Vanya*; *Rise and Fall of Little Voice*; *Arcadia*; *Dead Funny*; *My Brilliant Divorce*; *Dinner*; *Democracy*; *Glorious!*. No particular rule of thumb on subject matter or treatment. The acid test is whether 'something appeals to Michael'. Straight plays rather than musicals.

Colchester Mercury Theatre Limited

Balkerne Gate, Colchester CO1 1PT
℡ 01206 577006   ℻ 01206 769607
info@mercurytheatre.co.uk
www.mercurytheatre.co.uk
Chief Executive & Artistic Director *Dee Evans*

Producing theatre with a wide-ranging audience. New writing encouraged. The theatre has a free playwright's group for adults with a serious commitment to writing plays.

The Coliseum, Oldham

Fairbottom Street, Oldham OL1 3SW
℡ 0161 624 1731   ℻ 0161 624 5318
mail@coliseum.org.uk
www.coliseum.org.uk

Chief Executive/Artistic Director *Kevin Shaw*

The artistic policy of the theatre is to present a high quality and diverse theatre programme with the ambition to commission a new play each year. The Coliseum employs a reader to read all submitted scripts; a written report will be sent to the writer within approximately four months. An s.a.e. should be enclosed for return of the script along with an s.a.e. for acknowledgement of receipt, if required. All submitted scripts must be clearly typed, using one side of A4 for each page, and must contain a page of casting requirements.

Concordance

Finborough Theatre, 118 Finborough Road, London SW10 9ED

Ⓣ 020 7244 7439   Ⓕ 020 7835 1853

admin@concordance.org.uk

www.concordance.org.uk

Artistic Director *Neil McPherson*

Founded in 1981, Concordance is the resident company based at the **Finborough Theatre** (see entry). Presents world premières of new writing and revivals of neglected work with a special commitment to music theatre as well as integrating music into its work and productions featuring the writing of non-theatrical artists – poets, artists, novelists, etc. – presenting their work in a theatrical setting. We accept unsolicited scripts through the Finborough Theatre's submission department. Details at www.finboroughtheatre.co.uk

Contact Theatre Company

Oxford Road, Manchester M15 6JA

Ⓣ 0161 274 3434   Ⓕ 0161 274 0640

info@contact-theatre.org.uk

www.contact-theatre.org

Artistic Director *John E. McGrath*

The RAW Theatre Department at Contact looks at new writing and new work. The title RAW: Rhythm and Words in Theatre emphasizes the fact that the new writing department will concentrate on all forms of writing for theatre – including lyrical writing, and experiments with other artists as well as plays – and is particularly interested in materials that relate to the lives and culture of young people. 'We guarantee to read the first ten pages of whatever you send. From there your writing could be entered into a developmental programme for new writers, BBC Writers Room, or you can access our RAW writing workshop programme.'

Crucible Theatre

55 Norfolk Street, Sheffield S1 1DA

Ⓣ 0114 249 5999   Ⓕ 0114 249 6003

Executive Producer *Mark Feakins*

Literary Associate *Matthew Byam Shaw*

'Although we are interested in all new writing, most of the new work we present will be the result of commissions or prolonged script development with writers in whom we have expressed an interest.'

Derby Playhouse

Closed following the withdrawal of Arts Council funding.

Druid Theatre Company

Flood Street, Galway, Republic of Ireland

Ⓣ 00 353 91 568660   Ⓕ 00 353 91 563109

info@druidtheatre.com

www.druidtheatre.com

Artistic Director *Garry Hynes*

New Writing Manager *Thomas Conway*

Managing Director *Declan Gibbons*

Founded in 1975 and based in Galway, 'Druid has always worked to reinvigorate perceptions of classic dramatic works and to engage with new dramatic works of a challenging, innovative and daring kind'. Druid consistently commissions, develops, and produces new plays by a wide range of emerging and established writers both from Ireland and abroad.

The Dukes

Moor Lane, Lancaster LA1 1QE

Ⓣ 01524 598505   Ⓕ 01524 598519

info@dukes-lancaster.org

www.dukes-lancaster.org

Chief Executive/Artistic Director *Joe Sumsion*

Founded 1971. The only producing house in Lancashire. Wide target market for cinema and theatre. Plays in a 313-seater end-on auditorium and in a 198-seater in-the-round studio (currently undergoing refurbishment). Host for **Litfest** – Lancaster's annual festival of literature. In the summer months open-air promenade performances are held in Williamson Park. DT3 – The Education Centre is the Dukes' newly refurbished space dedicated to work by and for young people.

Dundee Repertory Theatre

Tay Square, Dundee DD1 1PB

Ⓣ 01382 227684   Ⓕ 01382 228609

hwatson@dundeereptheatre.co.uk

www.dundeereptheatre.co.uk

Artistic Director *James Brining*

Founded 1939. Plays to a varied audience. Translations and adaptations of classics, and new local plays. Most new work is commissioned. Interested in contemporary plays in translation and in new Scottish writing. No scripts except by prior arrangement.

Eastern Angles Theatre Company

Sir John Mills Theatre, Gatacre Road, Ipswich
IP1 2LQ
☎ 01473 218202 ℻ 01473 384999
admin@easternangles.co.uk
www.easternangles.co.uk
Artistic Director *Ivan Cutting*
General Manager *Jill Streatfeild*

Founded 1982. Plays to a rural audience for the most part. New work only: some commissioned, some devised by the company, some researched documentaries. Unsolicited mss welcome but scripts or writers need to have some connection with East Anglia. 'We are always keen to develop and produce new writing, especially that which is germane to a rural area.'

Edinburgh Royal Lyceum Theatre

▷ Royal Lyceum Theatre Company

English Touring Theatre

25 Short Street, London SE1 8LJ
☎ 020 7450 1990 ℻ 020 7450 1991
admin@ett.org.uk
www.ett.org.uk
Director *Rachel Tackley*

Founded 1993. National touring company visiting middle-scale receiving houses and arts centres throughout the UK. Mostly mainstream. Largely classical programme, but with increasing interest to tour one modern English play per year. Strong commitment to education work. No unsolicited mss.

Exeter Northcott

Stocker Road, Exeter EX4 4QB
☎ 01392 223999 ℻ 01392 223996
www.exeternorthcott.co.uk

Founded 1967. The Exeter Northcott is the South-west's principal subsidized producing theatre, situated on the University of Exeter campus. The Theatre is continually looking to broaden the base of its audience profile. Aims to develop, promote and produce quality new writing which reflects the life of the region and addresses the audience it serves. Generally works on a commission basis but occasionally options existing new work. The literary policy is currently under review, please contact the Theatre for more details.

Finborough Theatre

118 Finborough Road, London SW10 9ED
☎ 020 7244 7439 ℻ 020 7835 1853
admin@finboroughtheatre.co.uk
www.finboroughtheatre.co.uk
Artistic Director *Neil McPherson*

Founded 1980. 'One of London's leading new writing venues' (*Time Out*). Presents new British writing, revivals of neglected work from the 19th and 20th centuries, music theatre and UK premières of foreign work, particularly from Ireland, the United States and Canada. The theatre is available for hire and the fee is sometimes negotiable to encourage interesting work. Premièred work by Chris Lee, Anthony Neilson, Naomi Wallace, Tony Marchant, Mark Ravenhill, Laura Wade, James Graham. Three times winner of the Pearson Award. Unsolicited scripts are accepted, but please read carefully the details on submissions policy on the website before sending any scripts.

Gate Theatre Company Ltd

11 Pembridge Road, London W11 3HQ
☎ 020 7229 5387 ℻ 020 7221 6055
gate@gatetheatre.co.uk
www.gatetheatre.co.uk
Artistic Directors *Natalie Abrahami, Carrie Cracknell*

Founded 1979. Plays to a mixed, London-wide audience, depending on production. Aims to produce British premières of plays which originate from abroad and translations of neglected classics. Recent productions include *The Sexual Neuroses of Our Parents* by Lukas Bärfuss, translated by Neil Blackadder; *Car Cemetery* by Fernando Arrabal, translated by Barbara Wright; *I am Falling*, devised in collaboration with choreographer Anna Williams and dramturg Jenny Worton; *Press*, co-production created and performed by Pierre Rigal and his company dernière minute; *The Internationalist*, by US writer Anne Washburn. Positively encourages writers from abroad to send in scripts or translations. Most unsolicited scripts are read but plays by writers from the UK will not be accepted. Please address submissions to the Artistic Directors. Always enclose s.a.e. if play needs returning.

Graeae Theatre Company

LVS Resource Centre, 356 Holloway Road, London N7 6PA
☎ 020 7700 2455 ℻ 020 7609 7324
info@graeae.org
www.graeae.org
Minicom 020 7700 8184
CEO/Artistic Director *Jenny Sealey*

Europe's premier theatre company of disabled people, the company tours nationally and internationally with innovative theatre productions highlighting both historical and contemporary disabled experience. Graeae also runs educational programmes available to schools, youth clubs, students and disabled adults nationally and provides vocational training in theatre arts (including playwriting). Unsolicited scripts, from disabled writers, welcome. New work is commissioned.

Hampstead Theatre

Eton Avenue, Swiss Cottage, London NW3 3EU
☎ 020 7449 4200   🖷 020 7449 4201
literary@hampsteadtheatre.com
www.hampsteadtheatre.com
Contact *Katy Silverton*

Hampstead Theatre celebrates its 50th birthday in 2009. Based in a state-of-the art building, the theatre is an intimate space with a flexible stage and an auditorium capable of expanding to seat 325. The artistic policy continues to be the production of British and international new plays and the development of important young writers. 'We are looking for writers who recognize the power of theatre and who have a story to tell. All plays are read and discussed. We give feedback to all writers with potential.' Writers produced in the last five years at Hampstead include Gregory Burke, Dennis Kelly, Nell Leyshon, Sharman Macdonald, Tamsin Oglesby, Debbie Tucker Green and Roy Williams. In earlier years, breakthrough plays by Harold Pinter, David Hare, Mike Leigh, Pam Gems and Stephen Jeffreys were produced.

Harrogate Theatre

Oxford Street, Harrogate HG1 1QF
☎ 01423 502710   🖷 01423 563205
firstname.surname@harrogatetheatre.co.uk
www.harrogatetheatre.co.uk
Artistic Director *To be appointed*

Produces four to five productions a year on the main stage, one of which may be a new play. Annual main stage Youth Theatre production may be commissioned. (Policy on new writing unknown at the time of going to press as a new artistic director had not been appointed.)

Headlong Theatre

Chertsey Chambers, 12 Mercer Street, London WC2H 9QD
☎ 020 7438 9940   🖷 020 7438 9941
info@headlongtheatre.co.uk
www.headlongtheatre.co.uk
Artistic Director *Rupert Goold*

A middle-scale touring company producing established and new plays. On average, one new play or new adaptation a year. Due to forthcoming projects the company is not considering unsolicited scripts at present.

Heritage Theatre Ltd
▷ entry under Film, TV and Radio Producers

The Hiss & Boo Co. Ltd

Nyes Hill, Wineham Lane, Bolney RH17 5SD
☎ 01444 881707
email@hissboo.co.uk
www.hissboo.co.uk
Artistic Director *Ian Liston*

Founded 1976. Produces mainly traditional pantomime, revue, variety and occasional plays for a family audience. No unsolicited scripts; send letter and synopsis in the first instance. No run-of-the mill comedies, historical drama or anything without humour; interested in revue-type shows and unusual thrillers.

Horsecross Arts Ltd

185 High Street, Perth PH1 5UW
☎ 01738 472700   🖷 01738 624576
info@horsecross.co.uk
www.horsecross.co.uk
Creative Director *Ian Grieve*
General Manager *Paul Hackett*

Founded 1935. Combination of one to four-weekly repertoire of plays, incoming tours and studio productions. Unsolicited mss are read when time permits, but the timetable for return of scripts is lengthy. New plays staged by the company are usually commissioned under the SAC scheme.

Hull Truck Theatre Company

Spring Street, Hull HU2 8RW
☎ 01482 224800   🖷 01482 581182
admin@hulltruck.co.uk
www.hulltruck.co.uk
Executive Director *Joanne Gower*
Artistic Directors *John Godber, Gareth Tudor-Price*
General Manager *Nigel Penn*
Associate Director *Nick Lane*

John Godber, of *Teechers, Bouncers* and *Up 'n' Under* fame (the artistic director of this high-profile Northern company since 1984), has very much dominated the scene in the past with his own successful plays. The emphasis is still on new writing but Godber's works continue to be toured extensively. Recent premières/commissions include *On a Shout; Blue Cross Xmas; Sold; Wilde Boyz; Upon on Roof; Christmas Crackers; Crown Prince; My Favourite Summer; Ladies Down Under; Sully*. Most new plays are commissioned. 'New scripts should be addressed to the Literary Development Manager who will attempt as quick a response as possible. Bear in mind the artistic policy of Hull Truck, which is accessibility and popularity.' In general, not interested in musicals or in plays with casts of more than seven.

Stephen Joseph Theatre

Westborough, Scarborough YO11 1JW
☎ 01723 370540   🖷 01723 360506
www.sjt.uk.com
Artistic Director *Alan Ayckbourn*

A two-auditoria complex housing a 165-seat end stage theatre/cinema (the McCarthy) and a 400-seat theatre-in-the-round (the Round). Positive policy on new work. For obvious reasons, Alan Ayckbourn's work features quite strongly but a new writing programme ensures plays from other sources are actively encouraged. Also runs a lunchtime season of one-act plays each summer. Writers are advised however that the SJT is very unlikely to produce an unsolicited script – synopses are preferred. Recent commissions and past productions include: *Soap* Sarah Woods; *Fields of Gold* Alex Jones; *For Starters* Nick Warburton; *Bedtime Stories* Lesley Bruce; *Making Waves* Stephen Clark; *Larkin with Women* Ben Brown; *Amaretti Angels* Sarah Phelps; *Something Blue* Gill Adams; *Clockwatching* and *A Listening Heaven* Torben Betts; *The Star Throwers* Paul Lucas; *Safari Party* Tim Firth; *Drowning on Dry Land*; *Private Fears in Public Places* and *My Sister Sadie* Alan Ayckbourn. 'Writers are welcome to send details of rehearsed readings and productions as an alternative means of introducing the theatre to their work.' Submit to the Literary Department enclosing an s.a.e. for return of mss.

Bill Kenwright Ltd

BKL House, 1 Venice Walk, London W2 1RR
℡ 020 7446 6200   🇫 020 7446 6222
Contact *Bill Kenwright*

Presents both revivals and new shows for West End and touring theatres. Although new work tends to be by established playwrights, this does not preclude or prejudice new plays from new writers. The company has an in-house dramaturg. Scripts should be addressed to Bill Kenwright with a covering letter and s.a.e. 'We have enormous amounts of scripts sent to us although we very rarely produce unsolicited work. Scripts are read systematically. Please do not phone; the return of your script or contact with you will take place in time.'

Komedia

44–47 Gardner Street, North Laine, Brighton BN1 1UN
℡ 01273 647101   🇫 01273 647102
admin@komedia.co.uk
www.komedia.co.uk
Contact *David Lavender*

Founded in 1994, Komedia promotes, produces and presents new work. Mss of new plays welcome.

Library Theatre Company

St Peter's Square, Manchester M2 5PD
℡ 0161 234 1913   🇫 0161 228 6481
ltcadmin@manchester.gov.uk
www.librarytheatre.com

Artistic Director *Chris Honer*

Produces new and contemporary work, as well as the classics. No unsolicited mss. Send outline of the nature of the script first. Encourages new writing through the commissioning of new plays and through a programme of rehearsed readings to help writers' development.

Live Theatre Company

Broad Chare, Quayside, Newcastle upon Tyne NE1 3DF
℡ 0191 261 2694   🇫 0191 232 2224
info@live.org.uk
www.live.org.uk
Artistic Director *Max Roberts*
Chief Executive *Jim Beirne*

Following a £5.5 million redevelopment, Live Theatre is home to a 200-seat cabaret style theatre, a new studio theatre, renovated rehearsal rooms and a series of dedicated writer's rooms. 'The new space combined with the theatre's commitment to new writing has positioned Live Theatre at the forefront of Newcastle's emergence as one of the country's most exciting artistic and cultural regional capitals.' Productions include *The Pitmen Painters*; *Cooking With Elvis* and *NE1* by Lee Hall; *A Nightingale Sang* by C.P. Taylor; *Laughter When We're Dead* by Sean O'Brien; *Falling Together* by Tom Hadaway; *Bones* by Peter Straughan

London Bubble Theatre Company

5 Elephant Lane, London SE16 4JD
℡ 020 7237 4434   🇫 020 7231 2366
admin@londonbubble.org.uk
www.londonbubble.org.uk
Artistic Director *Jonathan Petherbridge*

London Bubble has existed within its community of southeast London for 35 years offering everyone access to enjoying theatre whether as an audience member or theatre-maker. Works with individuals and communities that may otherwise not have a chance to experience theatre, and provides access into the arts for young professionals. Produces two shows a year, a summer tour which brings theatre to London parks and unusual spaces and a Christmas show at a SE London venue. Also has an extensive participatory programme, offering the chance to work collaboratively with theatre professionals to create and stage a piece of theatre, and runs innovative arts education projects in schools locally and further afield.

Lyric Hammersmith

Lyric Square, King Street, London W6 0QL
℡ 0871 221 1722   🇫 020 8741 5965
enquiries@lyric.co.uk
www.lyric.co.uk

Executive Director *Jessica Hepburn*

The main theatre stages an eclectic programme of new and revived classics. Interested in developing projects with writers, translators and adaptors. The Lyric does not accept unsolicited scripts. Its 110-seat studio focuses on work for children, young people and families.

mac

Cannon Hill Park, Birmingham B12 9QH
www.macarts.co.uk

Closed in April 2008 for an 18-month, £13.6 million rebuild and refurbishment.

Manchester Library Theatre
▷ Library Theatre Company

New Vic Theatre

Etruria Road, Newcastle under Lyme ST5 0JG
☏ 01782 717954   ⊕ 01782 712885
admin@newvictheatre.org.uk
www.newvictheatre.org.uk
Artistic Director *Theresa Heskins*

The New Vic is a purpose-built theatre-in-the-round. Produces ten in-house plays each year and is active within the education sector and community. New plays produced are the result of specific commissions. Send synopses and the first ten pages of dialogue, *not* unsolicited scripts. 'We cannot guarantee that unsolicited scripts will be read; they will be returned on receipt of an s.a.e.'

Newpalm Productions

26 Cavendish Avenue, London N3 3QN
☏ 020 8349 0802   ⊕ 020 8346 8257
Contact *Lionel Chilcott*

Very rarely produces new plays (*As Is* by William M. Hoffman, which came from Broadway to the Half Moon Theatre, was an exception to this). National tours and West End productions such as *Peter Pan (The Musical)*; *Noises Off*; *Seven Brides for Seven Brothers* and *Rebecca*, at regional repertory theatres, are more typical examples of Newpalm's work. Both plays and musicals are, however, welcome; synopses are preferable to scripts.

Nimax Theatres Limited

1 Lumley Court, off 402 Strand, London
WC2R 0NB
☏ 0845 434 9290
www.nimaxtheatres.com
CEO *Nica Burns*

West End theatre managers of the Apollo, Duchess, Garrick, Lyric and Vaudeville theatres. Unsolicited scripts are not considered.

Nitro

6 Brewery Road, London N7 9NH
☏ 020 7609 1331   ⊕ 020 7609 1221
info@nitro.co.uk
www.nitro.co.uk
Artistic Director *Felix Cross*

Founded 1979. Formerly Black Theatre Co-op. A music theatre company, 'producing dynamic new work that explores stories by black people in Britain and elsewhere. Nitro productions present popular and epic themes, embracing the vibrant spectrum of black music and dance, appealing to diverse audiences of all ages.'

Norwich Puppet Theatre

St James, Whitefriars, Norwich NR3 1TN
☏ 01603 615564   ⊕ 01603 617578
info@puppettheatre.co.uk
www.puppettheatre.co.uk
Artistic Director *Luis Boy*
General Manager *Ian Woods*

Young/family-centred audience (aged 3–12) but developing shows for adult audiences interested in puppetry. All year round programme plus tours to schools and arts venues. Most productions are based on traditional stories but unsolicited mss welcome if relevant.

Nottingham Playhouse

Nottingham Playhouse Trust, Wellington Circus, Nottingham NG1 5AF
☏ 0115 947 4361   ⊕ 0115 947 5759
www.nottinghamplayhouse.co.uk
Artistic Director *Giles Croft*

Aims to make innovation popular, and present the best of world theatre, working closely with the communities of Nottingham and Nottinghamshire. Unsolicited mss will be read. It normally takes about six months, however, and 'we have never yet produced an unsolicited script. All our plays have to achieve a minimum of 60 per cent audiences in a 732-seat theatre. We have no studio.'

Nottingham Playhouse Roundabout Theatre in Education

Wellington Circus, Nottingham NG1 5AF
☏ 0115 947 4361
andrewb@nottinghamplayhouse.co.uk
www.nottinghamplayhouse.co.uk
Contact *Andrew Breakwell*

Founded 1973. Theatre-in-Education company of the **Nottingham Playhouse**. Plays to a young audience aged 5–18 years of age. 'We are committed to the encouragement of new writing and commission at least two new plays for young people each year. With other major producers in the East Midlands we share the resources of the *Theatre Writing Partnership* which is based at the Playhouse.' See website for philosophy and play details. Please make contact before submitting scripts.

N.T.C. Touring Theatre Company

The Playhouse, Bondgate Without, Alnwick
NE66 1PQ
☏ 01665 602586   🖷 01665 605837
admin@ntc-touringtheatre.co.uk
www.ntc-touringtheatre.co.uk
Contact *Gillian Hambleton*
General Manager *Anna Flood*

Founded 1978. Northumberland Theatre
Company. An Arts Council England revenue
funded organization. Predominantly rural, small-
scale touring company, playing to village halls
and community centres throughout the Northern
region, the Scottish Borders and countrywide.
Productions range from established classics to
new work and popular comedies, but must be
appropriate to their audience. Unsolicited scripts
welcome but are unlikely to be produced. All
scripts are read and returned with constructive
criticism within six months. Writers whose
style is of interest may then be commissioned.
The company encourages new writing and
commissions when possible. Financial constraints
restrict casting to a *maximum* of six.

Nuffield Theatre

University Road, Southampton SO17 1TR
☏ 023 8031 5500   🖷 023 8031 5511
alison.thurley@nuffieldtheatre.co.uk
www.nuffieldtheatre.co.uk
Artistic Director *Patrick Sandford*
Script Executive *John Burgess*

Well known as a good bet for new playwrights,
the Nuffield gets an awful lot of scripts. Produces
two new main stage plays every season. Previous
productions: *Exchange* by Yuri Trifonov
(transl. Michael Frayn) which transferred to
the Vaudeville Theatre; *The Floating Light Bulb*
Woody Allen (British première); *Nelson*, a new
play by Pam Gems; *The Dramatic Attitudes of
Miss Fanny Kemble* Claire Luckham; *The Winter
Wife* Claire Tomalin and *Tchaikovsky and the
Queen of Spades* John Clifford. Open-minded
about subject and style, producing musicals as
well as straight plays. Also opportunities for some
small-scale fringe work. Scripts preferred to
synopses in the case of writers new to theatre. All
will, eventually, be read 'but please be patient. We
do not have a large team of paid readers. We read
everything ourselves.'

Octagon Theatre Trust Ltd

Howell Croft South, Bolton BL1 1SB
☏ 01204 529407   🖷 01204 556502
info@octagonbolton.co.uk
www.octagonbolton.co.uk
Executive Director *John Blackmore*

Founded in 1967 and celebrating its 40th
anniversary from September 2007 to July 2008,
the award-winning Octagon Theatre stages at
least eight main auditorium home-produced
shows a year, and hosts UK touring companies
such as the National Theatre, John Godber's Hull
Truck Theatre Company, Peshkar Productions
and Alan Ayckbourn's Stephen Joseph Theatre.
Also hosts the work of partner companies as
part of its commitment to creative partnerships.
The theatre boasts a thriving and constantly
developing participatory department, activ8,
which operates a highly successful Youth Theatre
as well as initiating and facilitating exciting
education and outreach programmes. The theatre
is keen to encourage new writing but, due to the
lack of a dedicated literary department, is unable
to accept and process new and unsolicited scripts;
the theatre uses **North West Playwrights** as a
reading service instead.

Orange Tree Theatre

1 Clarence Street, Richmond TW9 2SA
☏ 020 8940 0141   🖷 020 8332 0369
admin@orangetreetheatre.co.uk
www.orangetreetheatre.co.uk
Artistic Director *Sam Walters*
Executive Director *Gillian Thorpe*

The Orange Tree, a theatre-in-the-round venue,
presents a broad cross section of work. Past
productions have included plays by Rodney
Ackland, John Galsworthy and a new translation
of Lorca as well as new plays by Oliver Ford
Davies, David Lewis, Ben Brown and Kenneth
Jupp. The theatre no longer considers unsolicited
mss; writers who may wish to approach the
theatre are asked to write first.

Out of Joint

7 Thane Works, Thane Villas, London N7 7NU
☏ 020 7609 0207   🖷 020 7609 0203
ojo@outofjoint.co.uk
www.outofjoint.co.uk
Director *Max Stafford-Clark*
Producer *Graham Cowley*
Literary Associate *Alex Yates*

Founded 1993. Award-winning theatre company
with new writing central to its policy. Produces
new plays which reflect society and its concerns,
placing an emphasis on education activity to
attract young audiences. Welcomes unsolicited
mss. Productions include: *Blue Heart* Caryl
Churchill; *Our Lady of Sligo*, *The Steward of
Christendom* and *Hinterland* Sebastian Barry;
Shopping and Fucking and *Some Explicit
Polaroids* Mark Ravenhill; *A State Affair* and
Talking to Terrorists Robin Soans; *Sliding with
Suzanne* Judy Upton; *The Positive Hour* and *A
Laughing Matter* April De Angelis; *Duck* and *O go
my Man* Stella Feehily; *The Permanent Way* David
Hare; *The Overwhelming* J.T. Rogers; *Feelgood* and

King of Hearts Alistair Beaton; *Testing the Echo* David Edgar.

Paines Plough

4th Floor, 43 Aldwych, London WC2B 4DN
☎ 020 7240 4533   🖷 020 7240 4534
office@painesplough.com
www.painesplough.com
Artistic Director *Roxana Silbert*
Literary Manager *Pippa Ellis*

Founded 1975. Award-winning company commissioning and producing new plays by British and international playwrights. Tours 2–4 plays a year nationally and worldwide for small and middle-scale. Also runs a range of projects focused on identifying and launching emerging playwrights (see website for details). 'We welcome unsolicited scripts and will respond to all submissions. We are looking for original plays that engage with the contemporary world and are written in a distinctive voice.' Recent productions: *Product* Mark Ravenhill; *Long Time Dead* Rona Munro; *After the End* Dennis Kelly; *If Destroyed True* Douglas Maxwell; *Pyrenees* David Greig; *Mercury Fur* Philip Ridley; *The Small Things* Enda Walsh.

Perth Repertory Theatre Ltd
▷ Horsecross Arts Ltd

Plymouth Theatre Royal
▷ Theatre Royal

Polka Theatre

240 The Broadway, Wimbledon SW19 1SB
☎ 020 8543 4888 (box office)   🖷 020 8545 8365
admin@polkatheatre.com
www.polkatheatre.com
Artistic Director *Jon Lloyd*
Executive Director *Stephen Midlane*
Associate Director, New Writing
Richard Shannon

Founded in 1968 and moved into its Wimbledon base in 1979. Leading children's theatre committed to commissioning and producing new plays. Programmes are planned two years ahead and at least three new plays are commissioned each year. 'Many of our scripts are commissioned from established writers. We are, however, keen to develop work from writers new to children's and young people's theatre. We run a new writing programme which includes master classes and workshops. Potential new writers' work is read and discussed on a regular basis; thus we constantly add to our pool of interesting and interested writers. This department is headed by *Richard Shannon*, Associate Director for New Writing.'

Queen's Theatre, Hornchurch

Billet Lane, Hornchurch RM11 1QT
☎ 01708 462362   🖷 01708 462363
info@queens-theatre.co.uk
www.queens-theatre.co.uk
Artistic Director *Bob Carlton*

The Queen's Theatre is a 503-seat theatre producing eight main house productions per year, including pantomime. Aims for distinctive and accessible performances in an identifiable house style focused upon actor/musician shows but, in addition, embraces straight plays, classics and comedies. 'New play/musical submissions are welcome but should be submitted in treatment and not script form.' Also runs a writers' social group which meets on the first Monday of each month, and writers' groups for adults (contact the Education Manager for information). These groups have close links with the Queen's Community Company (of amateur actors as well as the main house company) which workshops and showcases the groups' work. The theatre also runs a biannual New Writing Award scheme which culminates in a festival called Writenow!.

The Really Useful Group Ltd

22 Tower Street, London WC2H 9TW
☎ 020 7240 0880   🖷 020 7240 1204
www.reallyuseful.com

Commercial/West End, national and international theatre producer/co-producer/licensor whose output has included *Joseph and the Amazing Technicolor Dreamcoat*; *Jesus Christ Superstar*; *Cats*; *Evita*; *Song & Dance*; *Daisy Pulls It Off*; *Lend Me a Tenor*; *Starlight Express*; *The Phantom of the Opera*; *Aspects of Love*; *Sunset Boulevard*; *By Jeeves*; *Whistle Down the Wind*; *The Beautiful Game*; *Tell Me On a Sunday*; *Bombay Dreams*; *The Woman in White*; *The Sound of Music*.

Red Ladder Theatre Company

3 St Peter's Buildings, York Street, Leeds LS9 8AJ
☎ 0113 245 5311   🖷 0113 245 5351
rod@redladder.co.uk
www.redladder.co.uk
Artistic Director *Rod Dixon*

Founded 1968. National company touring 1–2 shows a year with a strong commitment to new work and new writers. Aimed at an audience of young people aged between 13–25 years who have little or no access to theatre. Performances held in youth clubs and theatre venues. Recent productions: *Kaahini* Maya Chowdhry; *Door: This Life Was Given to Me* Madani Younis; *Where's Vietnam?* Alice Nutter; *Forgotten Things* Emma Adams. The company is particularly keen to enter into a dialogue with writers with regard to creating new work for young people. E-mail the artistic director for more information at the address above.

Red Shift Theatre Company

jane@redshifttheatreco.co.uk
www.rcdshifttheatreco.co.uk
Artistic Director *Jonathan Holloway*

Founded in 1982, formerly a small-scale touring company, Red Shift has recently stepped up to middle scale 'commercial' production. Interested in new plays on subjects and themes with broad audience appeal. Willing to read treatments but full scripts should be sent by invitation only.

Ridiculusmus

c/o BAC, Lavender Hill, London SW11 5TN
enquiries@ridiculusmus.com
www.ridiculusmus.com
Artistic Directors *Jon Haynes, David Woods*

Founded 1992. Touring company which plays to a wide range of audiences. Productions have included adaptations of *The Importance of Being Earnest*; *Three Men In a Boat*; *The Third Policeman*; *At Swim Two Birds* and original work: *The Exhibitionists*; *Yes, Yes, Yes*; *Say Nothing* and *Ideas Men*. Most recent work: *Nice time, tough time*. Unsolicited scripts not welcome.

Royal Court Theatre

Sloane Square, London SW1W 8AS
☏ 020 7565 5050
🖷 020 7565 5002 (Literary office)
www.royalcourttheatre.com
Literary Manager *Graham Whybrow*

The Royal Court is a leading international theatre producing up to 17 new plays each year in its 400-seat proscenium theatre and 80-seat studio. In 1956 its first director George Devine set out to find 'hard-hitting, uncompromising writers whose plays are stimulating, provocative and exciting'. This artistic policy helped transform post-war British theatre, with new plays by writers such as John Osborne, Arnold Wesker, John Arden, Samuel Beckett, Edward Bond and David Storey, through to Caryl Churchill, Jim Cartwright, Kevin Elyot and Timberlake Wertenbaker. Since 1994 it has produced a new generation of playwrights such as Joe Penhall, Rebecca Prichard, Sarah Kane, Jez Butterworth, Martin McDonagh, Mark Ravenhill, Ayub Khan-Din, Conor McPherson, Roy Williams and many other first-time writers. The Royal Court has programmes for young writers and international writers, and it is always searching for new plays and new playwrights.

Royal Exchange Theatre Company

St Ann's Square, Manchester M2 7DH
☏ 0161 615 6709   🖷 0161 832 0881
jo.combes@royalexchange.co.uk
www.royalexchange.co.uk

Artistic Director *Sarah Frankcom*
Associate Director (Literary) *Jo Combes*

Founded 1976. The Royal Exchange has developed a new writing policy which it finds is attracting a younger audience to the theatre. The company has recently produced new plays by Nick Leather, Chloe Moss, Simon Stephens, Shelagh Stephenson, Brad Fraser, Jim Cartwright and Owen McCafferty. Currently has Gurpreet Bhatti, Linda Marshall Griffiths, Owen McCafferty, Abi Morgan, Sharif Samad, Simon Stephens and Roy Williams on commission. 'We accept unsolicited work from writers based in the UK, and as we will often include a full script report, we limit our reading to one script per writer per year. If you would like your script returned at the end of the process, please include the appropriate postage. Please note that we do not accept musicals or electronic copies of scripts and our current turnaround time is four months. Send scripts to *Jo Combes*.'

Royal Lyceum Theatre Company

Grindlay Street, Edinburgh EH3 9AX
☏ 0131 248 4800   🖷 0131 228 3955
www.lyceum.org.uk
Artistic Director *Mark Thomson*
Administration Manager *Ruth Butterworth*

Founded 1965. Repertory theatre which plays to a mixed urban Scottish audience. Produces classic, contemporary and new plays. Would like to stage more new plays, especially Scottish. No full-time literary staff to provide reports on submitted scripts.

Royal National Theatre

South Bank, London SE1 9PX
☏ 020 7452 3333   🖷 020 7452 3344
www.nationaltheatre.org.uk
Associate Director (Scripts) *Sebastian Born*

The majority of the National's new plays come about as a result of direct commission or from existing contacts with playwrights. There is no quota for new work, though many of the plays presented have been the work of living playwrights especially in the Cottesloe Theatre. Writers new to the theatre would need to be of exceptional talent to be successful with a script here, however the Royal National Theatre Studio helps a limited number of playwrights, through readings, workshops and discussions. Unsolicited scripts considered from the UK but not by e-mail (send s.a.e).

Royal Shakespeare Company

1 Earlham Street, London WC2H 9LL
☏ 020 7845 0515
Daniel.Usztan@rsc.org.uk
www.rsc.org.uk

Company Dramaturg *Jeanie O'Hare*

The RSC is a classical theatre company based in Stratford-upon-Avon. Produces a core repertoire of Shakespeare alongside modern classics and new plays, and the work of Shakespeare's contemporaries. Also commissions adaptations of favourite novels and stories for their Christmas show. Produces a season in London each year of plays first produced in Stratford. 'The RSC works with contemporary writers, encouraging them to write epic plays. The Literary Department seeks out the playwrights we wish to commission. We are unable to read unsolicited work, but we do monitor the work of emerging playwrights in production nationally and internationally.' The Company is currently undergoing a four-year rebuilding programme transforming the Royal Shakespeare Theatre and the entire Waterside complex of studios, rehearsal rooms, actors' cottages and workshops. The theatre is being re-modelled to create a thrust stage within a 'one room' theatre. Waterside is being re-modelled to create state-of-the-art facilities for artists.

7:84 Theatre Company Scotland

Film City Glasgow, 4 Summertown Road, Glasgow G51 2LY
℡ 0141 445 7245
admin@784theatre.com
www.784theatre.com
Artistic Director *Lorenzo Mele*

Founded 1973. One of Scotland's foremost touring theatre companies committed to producing work that addresses current social, cultural and political issues. Recent productions include commissions by Scottish playwrights such as Stephen Greenhorn, Rona Munro and Martin McArdie (*Reasons to be Cheerful*); Christopher Dean (*Boiling a Frog*); Peter Arnott (*A Little Rain*); Stephen Greenhorn (*Dissent*); David Greig (*Caledonia Dreaming*) and the Scottish premières of Tony Kushner's *Angels in America* and Athol Fugard's *Valley Song*. 'The company is committed to a new writing policy that encourages and develops writers at every level of experience, to get new voices and strong messages on to the stage.' New writing development has always been central to 7:84's core activity and has included Summer Schools and rehearsed readings. The company continues to be committed to this work and its development.

Shared Experience

13 Riverside House, 27/29 Vauxhall Grove, London W1F 7SJ
℡ 020 7587 1596   ℻ 020 7735 0374
admin@sharedexperience.org.uk
www.sharedexperience.org.uk
Joint Artistic Directors *Nancy Meckler, Polly Teale*

Founded 1975. Varied audience depending on venue, since this is a touring company. Recent productions have included: *Anna Karenina*, *Mill on the Floss*, and *War and Peace* (1996; all adapt. by Helen Edmundson); *Jane Eyre* (adapt. Polly Teale); Euripides' *Orestes* and *War and Peace* (2008; both adapt. by Helen Edmundson); *Kindertransport* Diana Samuels; *A Doll's House* (transl. Michael Meyer); *The Magic Toyshop* (adapt. by Bryony Lavery); *The Clearing* and *Gone to Earth* Helen Edmundson; *A Passage to India* (adapt. by Martin Sherman); *After Mrs Rochester and Brönte* Polly Teale. No unsolicited mss. Primarily not a new writing company. New plays always commissioned.

Sherman Cymru

Senghennydd Road, Cardiff CF24 4YE
℡ 029 2064 6901   ℻ 029 2064 6902
www.shermancymru.co.uk
Director *Chris Ricketts*

Originally founded in 1973, April 2007 saw a merger between The Sherman Theatre Company and Sgript Cymru to form a new company, Sherman Cymru, with the aim of becoming a powerhouse for drama, contemporary theatre and new writing in Wales and beyond. With a strong commitment to producing new work by emerging and experienced writers, Sherman Cymru now has a Literary Department which provides a reading service, advice and support, mentorship and script development for writers at every level. The company offers bursaries, full commissions and various one-off events and schemes to promote and nurture excellence in new writing for Welsh and Wales-based writers.

Show of Strength

74 Chessel Street, Bedminster, Bristol BS3 3DN
℡ 0117 902 0235   ℻ 0117 902 0196
info@showofstrength.org.uk
www.showofstrength.org.uk
Creative Producer *Sheila Hannon*

Founded 1986. Plays to an informal, younger than average audience. Aims to stage at least one new play each season with a preference for work from Bristol and the South West. Will read unsolicited scripts but a lack of funding means they are unable to provide written reports. Output: *The Wills Girls* Amanda Whittington; *Lags* Ron Hutchinson; *So Long Life* and *Nicholodeon* Peter Nichols. Also, rehearsed readings of new work.

Soho Theatre Company

21 Dean Street, London W1D 3NE
℡ 020 7287 5060   ℻ 020 7287 5061
writers@sohotheatre.com
www.sohotheatre.com
Artistic Director *Lisa Goldman*

The company has an extensive development programme for writers from all levels of experience and various disciplines consisting of a free script-reading service, Open Access workshops, Masterclasses, talks and events. Soho Connect holds free 'Taster Workshops' and longer courses for promising playwrights aged 15–25. The company produces and co-produces year round. Recent productions include: *Leaves of Glass* by Philip Ridley, *The Christ of Coldharbour Lane* by Oladipo Agboluaje, *Baghdad Wedding* by Abdulrazzak (nominated for the Evening Standard's Most Promising Playwright), *God in Ruins* Anthony Neilson (co-production with RSC), *Pure Gold* Michael Bhim (co-production with Talawa) and *Joe Guy* by Roy Williams (co-production with Tiata Fahodzi). Runs the **Verity Bargate Award**, a biennial competition (see entry under *Prizes*).

Sphinx Theatre Company

25 Short Street, London SE1 8LJ
☎ 020 7401 9993
info@sphinxtheatre.co.uk
www.sphinxtheatre.co.uk
Artistic Director *Sue Parrish*

'Sphinx is a feminist theatre company that places women at the centre of its artistic endeavour.' Since 1973 the company has toured groundbreaking productions to small and mid-scale venues. Synopses and ideas are welcome by e-mail or post with s.a.e.

Talawa Theatre Company Ltd

53–55 East Road, London N1 6AH
☎ 020 7251 6644   ℻ 020 7251 5969
hq@talawa.com
www.talawa.com

Founded 1985. 'Aims to provide high quality productions that reflect the significant creative role that black theatre plays within the national and international arena and also to enlarge theatre audiences from the black community.' Previous productions include *Pure Gold*; *High Heel Parrotfish*; *Blues for Mister Charlie*; *The Key Game*. Seeks to provide a platform for new work from up and coming black British writers. Send a copy of the script by post or e-mail. Runs a black script development project and black writers' group. Talawa is funded by Arts Council England, London.

Theatre Absolute

57–61 Corporation Street, Coventry CV1 1GQ
☎ 024 7625 7380
info@theatreabsolute.co.uk
www.theatreabsolute.co.uk
Artistic Director/Writer *Chris O'Connell*
Producer *Julia Negus*
Founded 1992. An independent theatre company which commissions, produces and tours new plays based on a strong narrative text and aimed at audiences aged 15 and upwards. Productions include: *Car*; *Raw, Kid* (Street Trilogy); *cloud:burst*; *Hang Lenny Pope*.

Theatre Royal Plymouth & Drum Theatre Plymouth

Royal Parade, Plymouth PL1 2TR
☎ 01752 230347   ℻ 01752 230499
david.prescott@theatreroyal.com
www.theatreroyal.com
Artistic Director *Simon Stokes*
Artistic Associate *David Prescott*

Stages small, middle and large-scale drama and music theatre. Commissions and produces new plays. The theatre no longer accepts unsolicited playscripts but will consider plays through known channels, i.e. theatre practitioners, regional and national scriptwriters' groups and agents.

Theatre Royal Stratford East

Gerry Raffles Square, London E15 1BN
☎ 020 8279 1104   ℻ 020 8534 8381
writers@stratfordeast.com
www.stratfordeast.com
Contact *New Writing Manager*

Situated in the heart of the East End, the theatre caters for a very mixed audience, in terms of both culture and age range. Produces plays and musicals, youth theatre and local community plays/events, all of which is new work. Special interest in plays and musicals reflecting the culturally diverse communities of London and the UK. New initiatives in developing contemporary British musicals. No longer accepts unsolicited scripts but instead asks for a) synopsis of script, b) sample ten pages of writing, c) brief writer's biography. 'From this information we will decide whether or not to ask for full-length script. We only respond to projects of interest. If no reply within three months assume that the proposal is not right for TRSE.'

Theatre Royal Windsor

Windsor SL4 1PS
☎ 01753 863444   ℻ 01753 831673
info@theatreroyalwindsor.co.uk
www.theatreroyalwindsor.co.uk
Executive Producer *Bill Kenwright*
Theatre Director *Angela Edwards*

Plays to a middle-class, West End-type audience. Produces thirteen plays a year and 'would be disappointed to do fewer than two new plays in a year; always hope to do half a dozen'. Modern classics, thrillers, comedy and farce. Only interested in scripts along these lines.

Theatre Workshop Edinburgh

34 Hamilton Place, Edinburgh EH3 5AX
☎ 0131 225 7942   🖷 0131 220 0112
www.theatre-workshop.com
Artistic Director *Robert Rae*

First ever professional producing theatre to fully include disabled actors in all its productions. Plays to a young, broad-based audience with many pieces targeted towards particular groups or communities. Output has included *D.A.R.E.*; *Threepenny Opera*; *Black Sun Over Genoa* and Beckett's *Endgame*. Particularly interested in issues-based work for young people and minority groups. Frequently engages writers for collaborative work and devised projects. Commissions a significant amount of new writing for a wide range of contexts, from large-scale community plays to small-scale professional productions. Favours writers based in Scotland, producing material relevant to a contemporary Scottish audience.

The Torch Theatre

St Peter's Road, Milford Haven SA73 2BU
☎ 01646 694192   🖷 01646 698919
info@torchtheatre.co.uk
www.torchtheatre.co.uk
Artistic Director *Peter Doran*

Founded 1977. Stages a mixed programme of in-house and middle-scale touring work. Unsolicited scripts will be read and guidance offered but production unlikely due to restricted funding. Please include s.a.e. for return of script and notes; mark clearly, 'FAO Peter Doran'. Torch Theatre Company productions include: *Dancing at Lughnasa*; *Neville's Island*; *The Woman in Black*; *Abigail's Party*; *Taking Steps*; *Blue Remembered Hills*; *A Prayer for Wings*; *Little Shop of Horrors*; *The Caretaker*; *One Flew Over the Cuckoo's Nest*; *One for the Road*; *Noises Off*; *Dead Funny* plus annual Christmas musicals.

Traverse Theatre

Cambridge Street, Edinburgh EH1 2ED
☎ 0131 228 3223   🖷 0131 229 8443
louise.stephens@traverse.co.uk
www.traverse.co.uk
Artistic Director *Dominic Hill*
Literary Manager *Katherine Mendelsohn*
Literary Assistant *Louise Stephens*

The Traverse is Scotland's only new writing theatre, with a particular commitment to producing new Scottish plays. However, it also has a strong international programme of work in translation and visiting companies. Previous productions include *The People Next Door* Henry Adam; *Dark Earth* David Harrower; *15 Seconds* François Archambarlt, version by Isabel Wright; *Iron* Rona Munro; *Damascus* David

Greig; *Gagarin Way* Gregory Burke; *Perfect Days* Liz Lochhead. Please address hard copies of unsolicited scripts to *Louise Stephens*, Literary Assistant.

Trestle Theatre Company

Trestle Arts Base, Russet Drive, St Albans AL4 0JQ
☎ 01727 850950   🖷 01727 855558
admin@trestle.org.uk
www.trestle.org.uk
Artistic Director *Emily Gray*

Founded 1981. Some devised work, but increasing collaboration with writers to create new writing for touring visual/physical performance work nationally and internationally and also local community projects. No unsolicited scripts.

Tricycle Theatre

269 Kilburn High Road, London NW6 7JR
☎ 020 7372 6611   🖷 020 7328 0795
www.tricycle.co.uk
Artistic Director *Nicolas Kent*

Founded 1980. Presenting an eclectic, culturally diverse programme that reflects the local area of Kilburn – particularly Black, Irish, Jewish, Asian and South African work. Known for Tribunal Plays, including *Guantanamo*, *Bloody Sunday*, *The Stephen Lawrence Inquiry* and *The Colour of Justice*, adapted from the enquiry transcripts by Richard Norton-Taylor and seen by more than 25 million people worldwide. Previous productions include: *Joe Turner's Come and Gone*, *The Piano Lesson* and *Two Trains Runnin'* all by August Wilson; and *Let There Be Love* by Kwame Kwei-Armah. New writing welcome. Looks for a strong narrative drive with popular appeal. Fee: £15 per script. Supplies written reader's report. Can only return scripts if s.a.e. is enclosed with original submission.

Tron Theatre Company

63 Trongate, Glasgow G1 5HB
☎ 0141 552 3748   🖷 0141 552 6657
gregory.thompson@tron.co.uk
www.tron.co.uk
Artistic Director *To be appointed*

Founded 1981. Plays to a broad cross-section of Glasgow and beyond, including international tours. Recent productions: *Pyrenees* David Greig (co-production with Paines Plough); *Half Life* John Mighton (co-production with Necessary Angel Theatre Company, Canada and Perth Rep); *Ubu the King* adapted by David Greig (co-production with Dundee Rep, BITE05, Barbican and the Young Vic as part of 'Young Genius'); *The Patriot* Grae Cleugh and *Antigone* – version by David Levin (both Tron Theatre productions). Interested in ambitious plays by UK and international writers. No unsolicited scripts.

Unicorn Theatre

147 Tooley Street, London SE1 2HZ
℡ 020 7645 0500   ℻ 020 7645 0550
stagedoor@unicorntheatre.com
www.unicorntheatre.com
Artistic Director *Tony Graham*
Associate Artistic Director *Rosamunde Hutt*
Associate Director (Literary) *Carl Miller*
Executive Director *Chris Moxon*

Founded 1947, resident at the Arts Theatre from 1967 to 1999 and now based at the new Unicorn Children's Centre with two theatres. Produces full-length professionally performed plays for children and adults. Recent work includes *Tom's Midnight Garden* by Philippa Pearce, adapt. David Wood; *Yikes* by Bryony Lavery; *Oz* by Patrick Shanahan; *Journey to the River Sea* by Eva Ibbotson, adapt. Carl Miller. Does not produce unsolicited scripts but works with commissioned writers. Writers interested in working with the company should send an e-mail (artistic@unicorntheatre.com), requesting further information.

Upstairs at the Gatehouse

▷ Ovation Productions under Film, TV and Radio Producers

Charles Vance

Hampden House, 2 Weymouth Street, London W1W 5BT
℡ 020 7636 4343   ℻ 020 7636 2323
admin@charlesvance.co.uk
Contact *Charles Vance*

In the market for medium-scale touring productions and summer-season plays. Hardly any new work and no commissions but writing of promise stands a good chance of being passed on to someone who might be interested in it. Occasional try-outs for new work in the Sidmouth repertory theatre. Send s.a.e. for return of mss.

Warehouse Theatre

Dingwall Road (adjacent to East Croydon Station), Croydon CRO 2NF
℡ 020 8681 1257   ℻ 020 8688 6699
info@warehousetheatre.co.uk
www.warehousetheatre.co.uk
Artistic Director *Ted Craig*

South London's new writing theatre seats 90–100 and produces up to six new plays a year. Also co-produces with, and hosts, selected touring companies who share the theatre's commitment to new work. The theatre continues to build upon a tradition of discovering and nurturing new writers through the **International Playwriting Festival** (see entry under *Festivals*). Also runs a vigorous writers' workshop and hosts youth theatre workshops and Saturday morning children's theatre. Scripts will get a guaranteed response if submitted through the International Playwriting Festival.

Watford Palace Theatre

Clarendon Road, Watford WD17 1JZ
℡ 01923 235455   ℻ 01923 819664
melody@watfordpalacetheatre.co.uk
www.watfordpalacetheatre.co.uk
Contact *Assistant Producer*

Reopened in Autumn 2004, following an £8.7 million refurbishment. An important part of artistic policy is developing new work suitable for this 600-seat proscenium arch theatre, which generally means the writer has some professional production experience. 'We run a writers' group for local playwrights, by invitation. We regret we are unable to offer a script reading and reporting service for unsolicited scripts and are unable to return unsolicited mss.'

West Yorkshire Playhouse

Playhouse Square, Leeds LS2 7UP
℡ 0113 213 7800   ℻ 0113 213 7250
alex.chisholm@wyp.org.uk
www.wyp.org.uk
Associate Director *Alex Chisholm*

Committed to working with new writing originating from the Yorkshire and Humberside region. New writing from outside the region is programmed usually where writer or company is already known to the theatre. For more information, call 0113 213 7286 or e-mail to address above. Scripts should be submitted with s.a.e. for return. 'Not all submitted scripts will be read. You are strongly advised to check guidelines on the website *before* submitting.'

Whirligig Theatre

14 Belvedere Drive, Wimbledon, London SW19 7BY
℡ 020 8947 1732   ℻ 020 8879 7648
whirligig-theatre@virgin.net
Contact *David Wood*

Occasional productions and tours to major theatre venues, usually a musical for primary school audiences and weekend family groups. Interested in scripts which exploit the theatrical nature of children's tastes. Previous productions: *The See-Saw Tree*; *The Selfish Shellfish*; *The Gingerbread Man*; *The Old Man of Lochnagar*; *The Ideal Gnome Expedition*; *Save the Human*; *Dreams of Anne Frank*; *Babe, the Sheep-Pig*.

White Bear Theatre Club

138 Kennington Park Road, London SE11 4DJ
www.whitebeartheatre.co.uk

Administration: 3 Dante Road, Kennington, London SE11 4RB ⓣ 020 7793 9193
Artistic Director *Michael Kingsbury*

Founded 1988. Output primarily new work for an audience aged 20–35. Unsolicited scripts welcome, particularly new work with a keen eye on contemporary issues, though not agitprop. *Absolution* by Robert Sherwood was nominated by the Writers' Guild for 'Best Fringe Play' and *Spin* by the same author was the *Time Out* Critics' Choice in 2000. The theatre received the *Time Out* award for Best Fringe Venue in 2001 and a Peter Brook award for best up-and-coming venue. In 2004 *Round the Horne ... Revisited* transferred to The Venue, Leicester Square and ran for 14 months.

Windsor Theatre Royal
▷ Theatre Royal Windsor

The Young Vic

66 The Cut, London SE1 8LZ
ⓣ 020 7922 2800   ⓕ 020 7922 2801
info@youngvic.org
www.youngvic.org
Artistic Director *David Lan*
Executive Director *Kevin Fitzmaurice*
Dramaturg *Ruth Little*

Founded 1970. Re-opened in 2006 having been rebuilt at a cost of £13 million, The Young Vic is a theatre for everyone. Produces revivals of classics – old and new – as well as new plays and annual events that embrace both young people and adults.

UK and Irish Writers' Courses

England

Berkshire

University of Reading

School of Continuing Education, London Road,
Reading RG1 5AQ
℡ 0118 378 8347
Cont-Ed@reading.ac.uk
www.reading.ac.uk/ContEd

An expanding and popular programme of creative
writing courses, including *Life-writing*; *Poetry
Workshop*; *Getting Started*; *Writing Fiction*;
Publishing Your Poetry; *Scriptwriting*; *Playwriting*;
Travel Writing; *Article Writing*; *Comedy Writing*
and *Writing for Radio*. Additional one-off event
includes readings by students of their work and
various Saturday workshops. Tutors include
novelist Leslie Wilson and poets Jane Draycott,
Susan Utting, David Grubb and Paul Bavister.
Fees vary depending on the length of course.
Concessions available.

The Write Coach

2 Rowan Close, Wokingham RG41 4BH
℡ 0118 978 4904
enquiries@thewritecoach.co.uk
www.thewritecoach.co.uk
Contact *Bekki Hill*

Workshops and one-to-one coaching (both
in person and by telephone) to assist both
professional and aspiring writers to become more
successful, break through their blocks, build
confidence, increase motivation and expand
creativity, define direction and find time and
space to write. Free initial consultation available.
See website for details.

Buckinghamshire

Missenden Abbey

Evreham Adult Learning Centre, Swallow Street,
Iver SL0 0HS
℡ 01296 383582
dcevreham@buckscc.gov.uk
www.arca.uk.net/missendenabbey

Residential and non-residential weekend
workshops, Easter and summer school.
Programmes have included *Writing Stories for
Children*; *Short Story Writing*; *Poetry Workshop*;
Writing Comedy for Television; *Life Writing*;
Travel Tales.

National Film & Television School

Beaconsfield Studios, Station Road, Beaconsfield
HP9 1LG
℡ 01494 731425   ℻ 01494 674042
info@nfts.co.uk
www.nfts.co.uk

Two-year programme covering all aspects of
screenwriting from the development of ideas
through to final production. Studying alongside
other filmmaking students allows writers
to have their work tested in workshops and
productions. Unlike a Screenwriting MA set
in an academic institution, this course is set
in a working studio and emphasizes practical
filmmaking and contact with industry personnel
through seminars and pitching sessions. The
course requires and encourages a high level of
dedication and a prolific output. Year One deals
with the basic principles of storytelling, the craft
of screenwriting for film and television and the
collaborative nature of production, and includes
several short writing assignments. Year Two
focuses on longer writing assignments, with
students writing a feature script and TV project
as well as their MA dissertation. Opportunities
may be available to write short fiction or
animation scripts for production. In partnership
with **The Script Factory**, The NFTS also offers
an 18-month, part-time Diploma in *Script
Development*.

Cambridgeshire

National Extension College

Michael Young Centre, Purbeck Road, Cambridge
CB2 8HN
℡ 0800 389 2839   ℻ 01223 400321
info@nec.ac.uk
www.nec.ac.uk/courses
Contact *Customer Relations Adviser*

The National Extension College (NEC) was set up as a charity over 40 years ago to help people of all ages fit learning into their lives. Supports more than 10,000 learners each year on over 100 open learning courses including *Essential Editing*; *Creative Writing* and *Writing Short Stories*. 'You can enrol at any time, study at home, work at your own pace and fit learning into your lives.' The NEC is fully accredited by the Open Distance Learning Quality Council. Contact the NEC for a free copy of the *Guide to Courses* or visit the website.

University of Cambridge Institute of Continuing Education

Madingley Hall, Madingley, Cambridge CB23 8AQ
℡ 01954 280399 📠 01954 280200
registration@cont-ed.cam.ac.uk
www.cont-ed.cam.ac.uk

A wide range of weekend, five-day and week-long creative writing courses for adults is offered by the University at the Institute of Continuing Education's residential headquarters at Madingley Hall. Evening courses are also available in Cambridgeshire and surrounding areas. Details of all courses can be found on the website or phone for a brochure.

Cheshire

Burton Manor

The Village, Burton, Neston CH64 5SJ
℡ 0151 336 5172 📠 0151 336 6586
enquiry@burtonmanor.com
www.burtonmanor.com

Wide variety of short courses, residential and non-residential, on writing and literature, including *Creative Writing Workshop* and *Writing for Pleasure*. Full details in brochure.

University of Chester

Parkgate Road, Chester CH1 4BJ
℡ 01244 512128 📠 01244 511330
a.chantler@chester.ac.uk
www.chester.ac.uk
Contact *Dr Ashley Chantler*

The Department of English offers two programmes: *Creative Writing* (full-time Combined Honours degree over three years) and *Creative Writing* (full-time MA over one year; part-time is possible).

Cornwall

University College Falmouth

Woodlane, Falmouth TR11 4RH
℡ 01326 211077
www.falmouth.ac.uk

www.bloc-online.com (magazine)
Contact *Admissions*

MA *Professional Writing*. An intensive vocational writing programme developing skills in fiction, magazine journalism/features, screenwriting. Students work on an extended writing project and form links with other postgraduate programmes such as television production and broadcast journalism.

Cumbria

Higham Hall College

Bassenthwaite Lake, Cockermouth CA13 9SH
℡ 01768 776276 📠 01768 776013
admin@highamhall.com
www.highamhall.com

Winter and summer residential courses. Programme includes *Creative Writing*; *Writing for Radio and TV*. Brochure available.

Derbyshire

Real Writers

PO Box 170, Chesterfield S40 1FE
℡ 01246 520834 📠 01246 520834
info@real-writers.com
www.real-writers.com

Appraisal service with personal tuition by post or e-mail from working writers. Send s.a.e. or see website for details.

Swanwick – The Writers' Summer School

The Hayes Conference Centre, Nr Alfreton
jeanfsutton@btinternet.com
www.wss.org.uk
Secretary *Jean Sutton*

Offers six-days of informal talks, discussion groups, interactive workshops and quizzes. Competitions and 'a lot of fun'. All levels welcome. Held mid-August from Saturday to Friday morning. Cost (2008) £360–£445 per person, all inclusive. S.a.e. to the Secretary at 4 Home Farm Close, Sandown, Isle of Wight PO36 9Q
(℡ 01983 406759/e-mail as above).

University of Derby

Admissions Office, Kedleston Road, Derby DE22 1GB
℡ 01332 622236 📠 01332 622754
J.Bains@derby.ac.uk
www.derby.ac.uk
Contact *Carl Tighe*

With upwards of 300 students, this is one of the oldest *Creative Writing* operations at degree level in the UK. The subject offers a BA Hons *Creative Writing* with particular strengths in Short Stories, Writing and New Technologies and Storytelling.

All teaching is done through practical workshops led by established and experienced writers. In the final year students produce an independent research project on a subject of their choice. The subject also contributes to the BA Hons Combined Studies.

Devon

Arvon Foundation (Devon)

See entry under **London**

Dartington College of Arts

Totnes TQ9 6EJ
℡ 01803 862224   📠 01803 861666
registry@dartington.ac.uk
www.dartington.ac.uk
Writing Admissions Tutor *Jerome Fletcher*

BA (Hons) *Writing* or *Writing (Contemporary Practices)* or *Writing (Scripted Media)*;*Textual Practices minor award*; *MA Performance Writing*: exploratory approaches to writing as it relates to performance, visual arts, sound arts and contemporary culture. Encourages the interdisciplinary, with minor awards and electives at BA level in arts and cultural management, choreography, music, theatre, visual performance. Contact the Writing Admissions Tutor (see the website).

Exeter Phoenix

Bradninch Place, Gandy Street, Exeter EX4 3LS
℡ 01392 667080   📠 01392 667599
www.exeterphoenix.org.uk

Regular literature events, focusing on performances by living poets and writers, often linked to wider programmes. Tutors in a wide range of writing skills run classes and workshops. 'Uncut Poets' group holds monthly meetings.

Fire in the Head Courses & Writers' Services

PO Box 17, Yelverton PL20 6YF
www.fire-in-the-head.co.uk
Contact *Roselle Angwin*

Comprehensive year-round writing programme in poetry, fiction (short and novel-writing), reflective writing and creative development. Also available, online and snailmail correspondence courses and mentoring.

University of Exeter

Department of English, Queen's Building, Queen's Drive, Exeter EX4 4QH
℡ 01392 269305   📠 01392 264361
soe.pgoffice@ex.ac.uk
www.sall.ex.ac.uk/english

Offers BA (Hons) in *English* with 2nd and 3rd-year options in creative writing, poetry, short

fiction, screenwriting and creative non-fiction. MA *English*; MA *Creative Writing*: poetry, novella, screenwriting, life writing and novel. PhD *Creative Writing*.

Dorset

Bournemouth University

The Media School, Weymouth House, Talbot Campus, Fern Barrow, Poole BH12 5BB
℡ 01202 965553   📠 01202 965099
Programme Administrator *Katrina King*

Three-year, full-time BA(Hons) course in *Scriptwriting for Film and Television*.

Essex

National Council for the Training of Journalists

The New Granary, Station Road, Newport, Saffron Walden CB11 3PL
℡ 01799 544014   📠 01799 544015
info@nctj.com
www.nctj.com

For details of journalism courses, both full-time and via distance learning, please write to the NCTJ enclosing a large s.a.e. or visit the website.

Gloucestershire

Chrysalis – The Poet In You

5 Oxford Terrace, Uplands, Stroud GL5 1TW
℡ 01453 759436/020 7794 8880
jay@ramsay3892.fsnet.co.uk
www.lotusfoundation.org.uk
Contact *Jay Ramsay, BAHons (Oxon), PGDip, UKCP member*

Two-part correspondence course combining poetry and personal development. Offers postal courses, day and weekend workshops (including 'The Sacred Space of the Word'), an on-going poetry group based in London and Gloucestershire and individual therapy related to the participant's creative process. The Chrysalis course itself consists of Part 1, 'for those who feel drawn to reading more poetry as well as wanting to start to write their own', and Part 2, 'a more in-depth course designed for those who are already writing and who want to go more deeply into its process and technique, its background and cultural history'. Both courses combine course notes with in-depth individual correspondence and feedback. Editing and information about publication also provided. Brochure and workshop dates available from the address above.

Wye Valley Arts Centre

Hephzibah Gallery, Llandogo NP25 4TW
☎ 01594 530214/01291 689463   ⓕ 01594 530321
info@wyearts.co.uk
www.wyearts.co.uk

Courses (held at Hephzibah Gallery, Llandogo in the Wye Valley and 'at our house in Cornwall') include *Writing and Poetry* and *Writing Workshop*. All styles and abilities.

Hampshire

Highbury College, Portsmouth

Dovercourt Road, Cosham, Portsmouth PO6 2SA
☎ 023 9238 3131   ⓕ 023 9237 8382
media.journalism@highbury.ack.uk
www.highbury.ac.uk
Course Administrator *Christine Stallard*
(☎ 023 9231 3287)

The 20-week fast-track postgraduate courses include: *Pre-entry Magazine Journalism*; *Pre-entry Newspaper Journalism*; and Diploma in *Broadcasting Journalism*, run under the auspices of the Periodicals Training Council, the National Council for Training of Journalists and the Broadcast Journalism Training Council respectively.

University of Portsmouth

School of Creative Arts, Film and Media, Portsmouth PO1 2EG
☎ 023 9284 5138   ⓕ 023 9284 5152
creative@port.ac.uk
www.port.ac.uk/departments/academic/scafm
Contact *School of Creative Arts, Film and Media*

Offers creative writing and combined creative writing degrees across a wide range of undergraduate programmes and in all genres. BA, MA and PhD study is available. The School is a key partner in a national AHRC Research training project in Creative Writing. Staff include a number of national and/or international award-winning writers, short story writers/novelists, playwrights, screenwriters, new-media writers and poets.

University of Winchester

Winchester SO22 4NR
☎ 01962 841515   ⓕ 01962 842280
course.enquiries@winchester.ac.uk
www.winchester.ac.uk
Contact *Enquiries* (☎ 01962 827234)

The University of Winchester offers a range of courses for budding writers or English enthusiasts both at undergraduate and postgraduate level. Undergraduate Programmes: BA *Creative Writing*; BA *English*; BA *English Literature and Language*; BA *English with American Literature*.

Postgraduate: MA *Creative and Critical Writing*; BA *English: Contemporary Literature*; MA *Writing for Children*.

Winchester Writers' Conference, Festival, Bookfair & Week-long Writing Workshops

University of Winchester, Faculty of Arts, Winchester SO22 4NR
☎ 01962 827238
barbara.large@winchester.ac.uk
www.Writersconference.co.uk
Conference Director *Barbara Large, MBE, FRSA, HFUW*
Honorary Patrons *Dame Beryl Bainbridge, Maureen Lipman, Jacqueline Wilson OBE*

Held on 26th, 27th, 28th June in 2009, followed by week-long writing workshops, 29th June to 3rd July. This Festival of Writing, now in its 29th year, attracts authors, playwrights, poets, literary agents, commissioning editors and book production specialists who give workshops, mini courses, lectures, seminars and one-to-one appointments to help writers harness their creative ideas and to develop their writing, editing and marketing skills. Fifteen writing competitions are attached to the conference. All first-place winners are published in the anthology *The Best of* series. The Bookfair offers delegates a wide choice of exhibits including authors' services, publishers, societies, booksellers, printers and trade associations. See also **Pitstop Refuelling Writers' Weekend Workshops** under *Writers' Circles and Workshops*.

Hertfordshire

Liberato Breakaway Writing Courses

16 Middle King, Braintree CM7 3XY
☎ 01376 551379
liberato@talktalk.net
www.liberato.co.uk
Contact *Maureen Blundell*

Specializes in beginner fiction writers with weekend and week-long courses. Emphasis on individual writing with written ms critiques and one-to-one sessions. Weekends held at Polstead in Suffolk throughout the year. Greek weeks in June and/or September on the Saronic island of Agistri, near Aegina – a relaxing, informal holiday with one-to-one sessions and editing service. B&B accommodation. Also offers postal and e-mail ms critiques on all fiction/autobiography.

Isle of Wight

Annual Writers' Writing Courses & Workshops

F * F Productions, 39 Ranelagh Road, Sandown
PO36 8NT
℡ 01983 407772   🖷 01983 407772
felicity@writeplot.co.uk
www.learnwriting.co.uk
Contact *Felicity Fair Thompson*

Regular residential weeks and weekends on creative writing for beginners and experienced writers – individual advice and workshops. Also offers postal ms critiques and one-to-one advice on fiction and film scripts

Kent

North West Kent College

Oakfield Lane, Dartford DA1 2JT
℡ 01322 629436/0800 074 1447 (Freephone helpline)
🖷 01322 629468
www.nwkcollege.ac.uk
Contact *Neil Nixon*, Head of School, Media & Communications

Two-year, full-time course that explores writing from a number of angles, teaching essential skills, market and academic aspects of the subject. Successful students progress to work or the University of Greenwich, the latter option allowing them to gain a BA(Hons) in Humanities from a further year of study. Staff include scriptwriters, novelists and a book publisher. Students compile a portfolio in the final year.

University of Kent

The Registry, Canterbury CT2 7NZ
℡ 01227 827272
information@kent.ac.uk
www.kent.ac.uk/studying
www.kent.ac.uk/english
Contact *Dr Emma Bainbridge* (℡ 01227 823402)

Diploma and degree level course in English and American Literature with Creative Writing in a vibrant department. Undergraduate certificate courses in *Creative Writing*, *Practical* and *Imaginative Writing*. Also Combined Studies courses in *English and Creative Writing*.

Lancashire

Alston Hall College

Alston Lane, Longridge, Preston PR3 3BP
℡ 01772 784661   🖷 01772 785835
alston.hall@ed.lancscc.gov.uk
www.alstonhall.com

Holds regular day and residential *Creative Writing* workshops. Full colour brochure available.

Edge Hill University

St Helen's Road, Ormskirk L39 4QP
℡ 01695 575171
shepparr@edgehill.ac.uk
www.edgehill.ac.uk
Contact *Professor Robert Sheppard*

Offers a two-year, part-time MA in *Writing Studies*. Combines advanced-level writers' workshops with closely related courses in the poetics of writing and contemporary writing in English. There is also provision for MPhil and PhD-level research in writing and poetics. A full range of creative writing courses is available at undergraduate level, in poetry, fiction and scriptwriting which may be taken as part of a modular BA (contact: *Daniele Pantano* daniele. pantano@edgehill.ac.uk).

Lancaster University

English & Creative Writing, Bowland College, Bailrigg, Lancaster LA1 4YN
℡ 01524 594169   🖷 01524 594247
l.kellett@lancaster.ac.uk
www.lancs.ac.uk/depts/english/crew/index.htm
Contact *Lyn Kellett*

Offers practical graduate and undergraduate courses in writing fiction, poetry and scripts. All based on group workshops – students' work-in-progress is circulated and discussed. Distance learning MA now available. Graduates include Andrew Miller, Justin Hill, Monique Roffey, Alison MacLeod, Jacob Polley.

Leicestershire

Writing School Leicester

c/o Leicester Adult Education College, 2 Wellington Street, Leicester LE1 6HL
℡ 0116 233 4343   🖷 0116 233 4344
val.moore@writingschoolleicester.co.uk
www.writingschoolleicester.co.uk
Contact *Valerie Moore*

Offers a wide range of quality part-time creative writing and journalism courses throughout the year in Leicester and elsewhere. The writing school is staffed by professional writers and offers a mix of critical workshops, one-day and term-length courses and short craft modules. Supports writers through to publication and has strong links with local media.

Lincolnshire

Creuse Writers' Workshop & Retreat

45 Browning Drive, Lincoln LN2 4HF
℡ 01522 880565   ℻ 01522 880565
info@cresuewritersworkshopandretreat.com
www.cresuewritersworkshopandretreat.com

A selection of home study is available with
personal tuition from professional published
writers. One-to-one tutoring. Visit the website for
further information.

London

Arvon Foundation

National Administration: 2nd Floor, 42A
Buckingham Palace Road, London SW1W ORE
℡ 020 7931 7611   ℻ 020 7963 0961
www.arvonfoundation.org
Devon: Totleigh Barton, Sheepwash, Beaworthy
EX21 5NS ℡ 01409 231338 ℻ 01409 231144
totleighbarton@arvonfoundation.org
Yorkshire: Lumb Bank, The Ted Hughes Arvon
Centre, Heptonstall, Hebden Bridge HX7 6DF
℡ 01422 843714 ℻ 01422 843714
lumbbank@arvonfoundation.org
Inverness-shire: Moniack Mhor, Teavarran,
Kiltarlity, Beauly IV4 7HT ℡ 01463 741675
moniackmhor@arvonfoundation.org
Shropshire: The John Osborne Centre, The
Hurst, Clunton, Craven Arms SY7 0JA
℡ 01588 640658 ℻ 01588 640509
thehurst@arvonfoundation.org
Joint Presidents *Terry Hands, Sir Robin
Chichester-Clark*
Chairman *Nigel Pantling*
National Director *Ariane Koek*

Founded 1968. Offers people of any age (over 16)
and any background the opportunity to live and
work with professional writers. Four-and-a-half-
day residential courses are held throughout the
year at Arvon's four centres, covering poetry,
fiction, drama, writing for children, songwriting
and the performing arts. Bursaries towards the
cost of course fees are available for those on
low incomes, the unemployed, students and
pensioners. Runs a biennial international poetry
competition (see entry under *Prizes*).

Birkbeck College, University of London

School of English and Humanities, Malet Street,
London WC1E 7HX
℡ 020 77079 0689
a.taylor@english.bbk.ac.uk
www.bbk.ac.uk/eh/eng
Contact *Anne Marie Taylor*

Taught by published writers, Birkbeck offers
an MA course in *Fiction Writing*. The course
will extend and cultivate existing writing skills,
help develop writing to a professional level and
is supported by masterclasses and readings
from visiting professionals. All classes are held
in the evening. Study part-time over two years
or full-time over one year. Applications must
be supported by a portfolio of creative writing.
Download an application form from the website.

The Central School of Speech and Drama

Embassy Theatre, Eton Avenue, London NW3 3HY
℡ 020 7722 8183   ℻ 020 7722 4132
enquiries@cssd.ac.uk
Contact *Academic Registry*

MA in *Advanced Theatre Practice*. One-year,
full-time course aimed at providing a grounding
in principal areas of professional theatre
practice – *Writing for Performance, Dramaturgy,
Directing, Performance, Puppetry* and *Design*,
with an emphasis on collaboration between the
various strands. *Writing for Performance* students
have the opportunity of working with companies
to create new and innovative work for the theatre.
Prospectus available. Also MA in *Writing for
Stage & Broadcast Media*. Focuses on texts for
the theatre, television, cinema and radio. Course
Leader: *Dymphna Callery*(d.callery@cssd.ac.uk).

City Lit

Keeley Street, Covent Garden, London WC2B 4BA
℡ 020 7492 2652   ℻ 020 7492 2735 (Humanities)
humanities@citylit.ac.uk
www.citylit.ac.uk

The writing school offers a wide range of courses,
with workshops catering for every level of student
– from beginners to previously published authors
– in a variety of disciplines. Classes include
*Ways into Creative Writing; Writing for Children;
Playwriting (stages 1 & 2); Writing Short Stories;
Screenwriting (Stages 1 & 2); Comedy Writing;
Writing Novels; Advanced Critical Workshop.*
Various lengths of course are available, and the
department offers information and advice all year
round, with the exception of August.

City University

Northampton Square, London EC1V 0HB
℡ 020 7040 8268   ℻ 020 7040 8256
ell@city.ac.uk
www.city.ac.uk/ell/cfa/writing_journalism
www.city.ac.uk/journalism

The Courses for Adults evening programme
includes: *Certificate in Novel Writing* (three-term
course); *Short Story Writing; Writing for Children;
Writing Poetry; Writing Television Drama;
Writing Situation Comedy; Food Writing; Feature
Writing; Narrative Non-Fiction; Writing about
Travel; Writing for the Theatre.* Contact the
Education and Lifelong Learning Department at

the address above or e-mail ell@city.ac.uk
(⊕ 020 7040 8268 ⊕ 020 7040 8256).
The University's Journalism Department offers
an MA in *Creative Writing (Non-fiction)* which
focuses on writing and research skills including
how to identify a subject, how to use archives,
exploring existing genres and structuring
material to complete a book. The course offers
a taught component, one-to-one tutorials and
opportunities to hear guest speakers. The course
advisory board, made up of leading non-fiction
writers, provides students with expert advice
and direct contact with the publishing industry.
Contact Programme Director *Julie Wheelwright*,
Postgraduate Admissions, Journalism at the
address above or e-mail julie.wheelwright.1@city.
ac.uk (⊕ 020 7040 8221).

The Complete Creative Writing Course at the Groucho Club

⊕ 020 7249 3711   ⊕ 020 7683 8141
maggie@writingcourses.org.uk
www.writingcourses.org.uk
Contact *Maggie Hamand*

Courses of ten two-hour sessions held at the
Groucho Club in London's Soho, starting in
January, April and September, Monday, Tuesday
or Saturday afternoons. Early evening sessions
offered at nearby venue. Beginners and advanced
courses available. Courses include stimulating
exercises, discussion and weekly homework. Fee:
£225–£265 for whole course.

Fiction Writing Workshops and Tutorials

5 Queen Elizabeth Close, London N16 0HL
⊕ 020 8809 4725
henrietta@writtenwords.net
www.WrittenWords.net
Contact *Henrietta Soames*

Fiction workshops for beginner and advanced
writers led by author Henrietta Soames. 'Lively
discussions, stimulating exercises, valuable
feedback.' Fiction tutorials for students who
require concentrated attention and ongoing
support. Short stories/novels, full-length mss
welcomed.

London College of Communication

Elephant & Castle, London SE1 6SB
⊕ 020 7514 2105
shortcourses@lcc.arts.ac.uk
www.lccarts.co.uk/training

Intensive courses in writing, editing and
journalism. Courses include: *News Writing
Practice*; *Feature Writing*; *How to Write and
Sell Travel Features*; *Brush Up Your Grammar*.
Prospectus and information leaflets available;
telephone or access the website.

London School of Journalism

126 Shirland Road, London W9 2BT
⊕ 020 7289 7777   ⊕ 020 7432 8141
info@lsjournalism.com
www.home-study.com
www.lsj.org
Contact *Student Administration Office*

Distance learning courses with an individual
and personal approach. Students remain
with the same tutor throughout the course.
Options include: *Novel Writing*; *Short Story
Writing*; *Writing for Children*; *Poetry*; *Freelance
Journalism*; *Media Law*; *Improve Your English*;
Cartooning; *Thriller Writing*. Fees vary but range
from £295 for *Enjoying English Literature* to £395
for *Journalism and Newswriting*. Postgraduate
Diploma Courses taught in London (three month
full-time, six month part-time, nine month vening
classes). Online postgraduate diploma course (24
months) also available.

Middlesex University

Trent Park, Bramley Road, London N14 4YZ
⊕ 020 8411 5000   ⊕ 020 8411 6652
admissions@mdx.ac.uk
www.mdx.ac.uk
The UK's longest established writing degree offers
a Single or Combined Honours programme in
Creative and Media Writing (full or part-time).
This modular programme gives an opportunity to
explore creative non-fiction, poetry, prose fiction
and dramatic writing for a wide range of genres
and audiences. Option for work experience
in the media and publishing industries.
Contact Admissions or *James Martin Charlton*
(j.charlton@mdx.ac.uk).
MA *Writing* (full-time, part-time) concentrates
on fiction writing. Includes writing workshops;
critical seminars; lectures and workshops from
established writers; introduction to agents and
publishers. Contact *David Rain* (d.rain@mdx.
ac.uk).
Also offers research degrees M.Phil/PhD in
Creative Writing. Contact: Maggie Butt (m.butt@
mdx.ac.uk). The University has a thriving Writing
Centre running an annual literary festival,
weekly talks, community projects and writers in
residence. 'All of our writing courses take place
in the setting of a beautiful country park, within
easy reach of central London.'

PMA Media Training

The PMA Centre for Media Excellence, 7a
Bayham Street, London NW1 0EY
⊕ 01480 300653/020 7278 0606
⊕ 01480 496022
training@pma-group.com
www.pma-group.com
www.becomeajournalist.co.uk

Contacts *Vicky Chandler, Melanie Gilbert*
Fast-track nine-week Postgraduate Diploma
and more than 150 one and two-day editorial,
PR, design and publishing courses held in
central London. Intensive workshops run by top
journalists, print and online, designers and PR
professionals. All short courses are designed to
lead to the PMA Gold Standard qualification.
Apple, Adobe, Quark and the Periodical
Publishers' Association accreditation. NCTJ, NUJ
and Communicators in Business recommended.
Edexcel Exam Centre. Special discount for
freelancers. See website for dates and fees.

The Poetry School
▷ entry under Organizations of Interest to Poets

Roehampton University
Roehampton Lane, London SW15 5PU
☎ 020 8392 3232   🖷 020 8392 3470
www.roehampton.ac.uk

MA/MRes Creative and Professional Writing
includes modules in: Fiction: how to grow stories;
Writing for a child audience; Poetry: form and
innovation; Creative non-fiction and journalism;
Knowing and subverting the rules: screenwriting
for independent film; and Screenwriting for
independent film.

The Script Factory
The Square, 61 Frith Street, London W1D 3JL
☎ 020 7323 1414
general@scriptfactory.co.uk
www.scriptfactory.co.uk

Established in 1996, The Script Factory is a
screenwriter and script development organization
set up to bridge the gap between writers
and the industry, and to promote excellence
in screenwriting. Specializing in training,
screenings, masterclasses and various services,
The Script Factory operates throughout the
UK and internationally. For the best source of
information on upcoming activities, courses and
latest news, and to join the free mailing list, check
out the website

Soho Theatre Company
▷ entry under Theatre Producers

Travellers' Tales
92 Hillfield Road, London NW6 1QA
info@travellerstales.org
www.travellerstales.org

'Travellers' Tales is Britain's foremost provider of
travel writing and travel photography training for
non-professionals. Offers practical courses with
top professionals.' Tutors include award-winning
authors William Dalrymple and Colin Thubron,
travel editors Simon Calder (*The Independent*),
Lyn Hughes (*Wanderlust*) and Jonathan Lorie

(*Traveller*). Courses include beginners' weekends,
UK masterclasses and creative holidays overseas.

University of Westminster
School of Media, Arts and Design, Harrow
Campus, Watford Road, Harrow HA1 3TP
☎ 020 7911 5903   🖷 020 7911 5955
harrow-admissions@wmin.ac.uk
www.wmin.ac.uk

Courses include part-time MAs available
in *Journalism Studies*; *Film and Television
Studies*; *Screenwriting and Producing*; *Public
Communication and Public Relations*;
Communication and Communication Policy.

Greater Manchester

Manchester Metropolitan University – The Writing School
Department of English, Geoffrey Manton
Building, Rosamond Street West, off Oxford
Road, Manchester M15 6LL
☎ 0161 247 1732/1   🖷 0161 247 6345
a.biswell@mmu.ac.uk (campus course)
h.beck@mmu.ac.uk (online course)
Course Convenors *Andrew Biswell*
(campus address), *Heather Beck* (online)

The Writing School offers three 'routes' for
students to follow: *Novel*, *Poetry*, and *Children's
Writing*. A key feature of the programme is
regular readings, lectures, workshops and
masterclasses by writers, publishers, producers,
booksellers, librarians and agents. Tutors include
Simon Armitage, Heather Beck, Carol Ann
Duffy, Paul Magrs, Michael Symmons Roberts,
Jackie Roy, Nick Royle and Jeffrey Wainwright.
The Novel and Poetry routes are available online
as well as through the campus (contact *Heather
Beck*, Online Convenor).

University of Manchester
English & American Studies, School of Arts,
Histories and Cultures, Mansfield Cooper
Building, Oxford Road, Manchester M13 9PL
☎ 0161 306 1259   🖷 0161 275 5987
englishpg@manchester.ac.uk
www.manchester.ac.uk/english
Course Directors *John McAuliffe,
Patricia Duncker*

Offers a one-year MA in *Creative Writing* (the
novel, short story and poetry) and a PhD.

University of Salford
Postgraduate Admissions, School of Media,
Music & Performance, Adelphi Building, Peru
Street, Salford M3 6EQ
☎ 0161 295 6026   🖷 0161 295 6023
r.humphrey@salford.ac.uk
www.smmp.salford.ac.uk

MA in *Television and Radio Scriptwriting*. Two-year, part-time course taught by professional writers and producers. Also offers masterclasses with leading figures in the radio and television industry.

The Writers Bureau

Sevendale House, 7 Dale Street, Manchester M1 1JB
☎ 0161 228 2362 🖷 0161 236 9440
studentservices@writersbureau.com
www.writersbureau.com

Comprehensive home-study writing course with personal tuition service from professional writers. Fiction, non-fiction, articles, short stories, novels, TV, radio and drama all covered in detail. Trial period, guarantee and no time limits. Writing for children, using the Internet to sell your writing and biographies, memoirs and family history courses also available. ODLQC accredited. Quote Ref. EH08. Free enquiry line: 0800 389 7360.

The Writers Bureau College of Journalism

Address etc. as The Writers Bureau above

Home-study course covering all aspects of journalism. Real-life assignments assessed by qualified tutors with the emphasis on getting into print. Comprises 28 modules and three supplements. Ref: EHJ08. Free enquiry line: 0800 389 7360.

The Writers College

Address etc. as The Writers Bureau above

The Art of Writing Poetry Course from The Writers Bureau sister college. A home-study course with a more `recreational' emphasis. The 60,000-word course has 17 modules and lets you complete twelve written assignments for tutorial evaluation. Quote Ref. EHP08. Free enquiry line: 0800 389 7360.

Merseyside

University of Liverpool

Continuing Education, 126 Mount Pleasant, Liverpool L69 3GR
☎ 0151 794 6900/6952 (24 hours)
🖷 0151 794 2544
conted@liverpool.ac.uk
www.liv.ac.uk/conted
Course Organizer *Dr John Redmond*

Courses include: *Exploring Writing, Scriptwriting* (for radio and TV, and for film and TV), *Comedy Writing, Writing for Women by Women, Writing for the Stage, Writing Science Fiction, Writing for Children, Writing Poetry*. Most courses take place once weekly (evening or daytime) and there is also a series of Saturday courses on aspects of writing. CE courses offer University credits which

can be accumulated for personal development purposes, or towards an award such as the Certificate in Higher Education (*Creative Writing*) – 120 credits. For most courses no previous knowledge is required. Fee concessions available if you receive certain benefits or are retired. Full programme at www.liv.ac.uk/conted or free printed prospectus on request.

Norfolk

University of East Anglia

School of Literature and Creative Writing, Norwich NR4 7TJ
☎ 01603 592154 🖷 01603 250599
pgt.hum@uea.ac.uk
www.uea.ac.uk/creativewriting
Contact *Jack Camplin*, Postgraduate Admissions

UEA has a history of concern with contemporary literary culture. Among its programmes is the MA in *Creative Writing* (founded by Angus Wilson and Malcolm Bradbury in 1970/71). The course has three parallel entry points: Prose fiction, poetry and scriptwriting. A series of weekly workshops provide intensive examination of students' own work, and also draw on aspects of teaching in nineteenth and twentieth century literature, literary theory and film and cultural studies.

Northamptonshire

Knuston Hall Residential College for Adult Education

Irchester, Wellingborough NN29 7EU
☎ 01933 312104 🖷 01933 357596
enquiries@knustonhall.org.uk
www.knustonhall.org.uk

Writing courses have included: *Articles for Magazines*; *My Life in Poems and Stories*; *Writer's Workshops* and *Creative Writing*.

Northumberland

Community Creative Writing

'Sea Winds', 2 St Helens Terrace, Berwick-upon-Tweed TD15 1RJ
☎ 01289 305213
mavismaureen@btinternet.com
Contact/Tutor *Maureen Raper, MBE*

The courses, which are held at the Community Centre College, Berwick-upon-Tweed as well as by distance learning for those unable to travel into class, include: *Creative Writing for Beginners*; *Creative Writing for Intermediates*; *Writing for Children*; *Writing Comedy*; *Writing for Radio and Television*; *Writing Crime and Mystery*.

Nottinghamshire

The Nottingham Trent University

College of Arts, Humanities and Education, Clifton Campus, Nottingham NG11 8NS
☎ 0115 848 8977
acc.postgrad@ntu.ac.uk
www.ntu.ac.uk/postgrad/

MA in *Creative Writing*. Hands-on and workshop-based, the course concentrates primarily on the practice and production of writing. A choice of options from *Fiction, Poetry, Creative Non-Fiction, New Media Writing, Children's Writing* and *Scriptwriting*. Assignments and a dissertation, but no formal exams. There is a full programme of visiting speakers. Study either full-time or part-time.

Oxfordshire

University of Oxford Department for Continuing Education

Rewley House, 1 Wellington Square, Oxford OX1 2JA
☎ 01865 280356   ⊞ 01865 270309
pp@conted.ox.ac.uk
www.conted.ox.ac.uk

Creative writing classes held during the autumn and spring terms. There are also one-week summer school courses, a two-year part-time Diploma and a two-year part-time MSt in *Creative Writing*. Early booking is advised.

Shropshire

Arvon Foundation (Shropshire)

See entry under **London**

Somerset

Ammerdown Conference and Retreat Centre

Ammerdown Park, Radstock, Bath BA3 5SW
☎ 01761 433709   ⊞ 01761 433094
centre@ammerdown.org
www.ammerdown.org
www.ammerdown-conference.co.uk

Courses have included *Creative Journaling* with Ann Beazer, *Writing for Well Being* with Sue Ashby and *Writing the Spirit* with Judy Clinton. En suite residential facilities. Brochure available or full details on the website.

Bath Spa University

Newton Park, Bath BA2 9BN
☎ 01225 875573   ⊞ 01225 875503
r.kerridge@bathspa.ac.uk
www.bathspa.ac.uk

Contact *Elizabeth Rack* (☎ 01225 875603)

MA in *Creative Writing*. A course designed to help writers bring their work to a more publishable, performable and broadcastable level held in weekend workshops. Teaching is by published writers in the novel, poetry, short stories and scriptwriting and includes visits by literary agents and publishers. In recent years, several students from this course have won the Manchester Book Prize and been longlisted for the Man Booker, Orange and Costa book prizes, and have had work produced on BBC Radio. MA in *Writing for Young People*. A course for writers for children of all ages, from the picture-book age through to adolescent and 'crossover' writing, which aims at markets among adults as well as young people. An experienced teaching team helps and encourages students to create a significant body of writing, with practical plans for its place in the real world of publishing.

Dillington House

Ilminster TA19 9DT
☎ 01460 52427   ⊞ 01460 52433
dillington@somerset.gov.uk
www.dillington.com

Dillington House is one of the finest historic houses in Somerset. Provides short residential courses across a wide range of subjects, including writing and literary appreciation. Full details of the programme are available in the free brochure and on the website.

Institute of Copywriting

Overbrook Business Centre, Poolbridge Road, Blackford, Wedmore BS28 4PA
☎ 0800 781 1715   ⊞ 01934 713492
copy@inst.org
www.inst.org/copy

Comprehensive copywriting home-study course, including advice on becoming a self-employed copywriter. Each student has a personal tutor who is an experienced copywriter and who provides detailed feedback on the student's assignments. Other courses include: Diploma in *Creative Writing* and Diploma in *Screenwriting*.

University of Bristol

Department of English, 3/5 Woodland Road, Bristol BS8 1TB
☎ 0117 954 6969
www.bris.ac.uk/english
Contact *Lifelong Learning Organizer*

A variety of short courses and day schools designed for both experienced writers and for absolute beginners, including courses exploring writing for therapeutic purposes. Also offers *Diploma in Creative Writing*. Detailed brochure available.

Staffordshire

Keele University

The Centre for Continuing and Professional Education, Keele University, (Freepost ST1666), Newcastle under Lyme ST5 5BR
☎ 01782 583436

Daytime, evening and weekend courses on literature and creative writing. The 2008 programme includes courses on novel writing, writers workshops and sessions on poetry writing.

Surrey

The Guildford Institute

Guildford Institute, Ward Street, Guildford GU1 4LH
☎ 01483 562142   🖷 01483 451034
www.guildford-institute.org.uk

The Guildford Institute is the venue for various creative writing courses. It also hosts Guildford Writers, a writers' circle meeting on alternate Tuesday evenings from 7.30 pm to 9.30 pm. Members of the group are writing short stories, poetry or novels, and bring their work to read aloud to the group for other members' advice, constructive criticism and general comments. New members are always welcome.

Royal Holloway University of London

Department of Drama and Theatre, Egham Hill, Egham TW20 0EX
☎ 01784 443922   🖷 01784 431018
drama@rhul.ac.uk
www.rhul.ac.uk/Drama
www.rhul.ac.uk/english
Contact *Dan Rebellato*

Three-year BA courses in *Drama and Creative Writing* or *English and Creative Writing*, during which students progressively specialize in playwriting, poetry or fiction. Playwriting can be studied as part of the BA *Drama and Creative Writing* degree and poetic practice is an option in the *English* programme. Playwriting can be studied at postgraduate level in the MA in *Theatre* and there are MAs in *Poetic Practice* and *Screenwriting*.

Sussex

Earnley Concourse

Earnley, Chichester PO20 7JN
☎ 01243 670392   🖷 01243 670832
info@earnley.co.uk
www.earnley.co.uk

Offers a range of residential and non-residential courses throughout the year. Previous programme has included *Creative Writing: Getting Started*. Brochure available.

The University of Chichester

Bishop Otter Campus, College Lane, Chichester PO19 6PE
☎ 01243 816000   🖷 01243 816080
S.Norgate@chi.ac.uk
www.chi.ac.uk
MA Programme Coordinator (Creative Writing) *Stephanie Norgate* (☎ 01243 816296)

MA in *Creative Writing*, both full and part-time. Also, MPhil/PhD in *Creative Writing, Creative/ Critical*. Contact: *Dr Bill Gray* (b.gray@chi.ac.uk).

University of Sussex

Centre for Continuing Education, The Sussex Institute, Essex House, Brighton BN1 9QQ
☎ 01273 606755
cce@sussex.ac.uk
www.sussex.ac.uk/cce
Contact *Sue Roe*

MA in *Creative Writing and Authorship*: a unique opportunity for graduate writers to develop writing practice in the context of the study of cultural and aesthetic issues of authorship, past and present, in workshops, masterclasses and seminars. One year, full-time, two years, part-time. Convenor: *Sue Roe*. Certificate in *Creative Writing*: short fiction, novel and poetry for imaginative writers. Two-years, part-time. Convenor: *Mark Slater*. Both courses include CCE Agents' and Publishers' Day.

Tyne & Wear

University of Sunderland

Centre for Lifelong Learning, 2nd Floor, Bedson Building, Kings Road, Newcastle upon Tyne NE1 7RU
☎ 0191 515 2800   🖷 0191 515 2890
lifelong.learning@sunderland.ac.uk
cll.sunderland.ac.uk

Courses, held in Newcastle and across the North East, include: *Creative Writing: Feature Writing; Screenwriting; Writing and Illustrating for Children* and *Writing From the Inside Out*, a workshop for women. Contact the Centre for Lifelong Learning.

West Midlands

Birmingham City University

School of English, City Campus North, Birmingham B42 2SU
☎ 0121 331 5540   🖷 0121 331 6692

ruth.page@bcu.ac.uk
www.lhds.bcu.ac.uk/english/index.php?page=ma_creative_writing

The *Creative Writing* MA is staffed entirely by established writers and can be studied full or part time. The course has core modules of *Prose Fiction* and *Scriptwriting*, but there is also the opportunity to study within the wider context of reading and literature. With opportunities to get feedback in small groups and from staff, participants receive invaluable editorial advice. (See also **National Academy of Writing**).

National Academy of Writing (based at Birmingham City University)

Birmingham City University, School of English, Perry Barr Campus, Birmingham B42 2SU

☎ 0121 331 5471   ℻ 0121 331 6692

nicola.monaghan@bcu.ac.uk
www.thenationalacademyofwriting.org.uk

Diploma in *Writing*. This course has a strong professional focus and aims to produce working writers. Modules are available in a number of disciplines, including fiction, life writing and screenwriting. Teaching includes taught modules and supervision by Birmingham City University staff, all of whom are established writers. There are also regular masterclasses with Academy patrons. Applications are welcome from all, regardless of previous academic experience. Admission is based on talent and commitment to a career in writing. (See also **Birmingham City University**.)

Starz! – Film and Theatre Performing Arts

Christmas House, Chester Road, Castle Bromwich, Solihull B36 0ET

☎ 0121 749 7147

www.gavinprime.mac.com
Creative Director *Gavin Prime*

Teaching disabled, disadvantaged and mainstream children. The only totally free theatre and film school of this type in England. Students range in age from eight to 23 and work in all ways connected to performing arts. This includes writing and showing them the structure of the business and how to try and get work accepted.

University of Birmingham

Department of Drama and Theatre Arts, The Old Library, 998 Bristol Road, Selly Oak, Birmingham B29 6LQ

☎ 0121 414 5998

drama@contacts.bham.ac.uk
www.drama.bham.ac.uk
Course Convenor *Steve Waters* (s.waters@bham.ac.uk)

The MPhil in *Playwriting Studies*, established by playwright David Edgar in 1989, was the UK's first postgraduate course in playwriting. An intensive course which encourages students to think critically about dramatic writing, assisting them to put these insights into practice in their own plays.

University of Warwick

Open Studies, Centre for Lifelong Learning, Coventry CV4 7AL

☎ 024 7657 3739

k.rainsley@warwick.ac.uk
openstudies@warwick.ac.uk
www.warwick.ac.uk/cll/OpenStudies

Creative writing courses held at the university or in regional centres. Subjects include: *Creative Writing*; *Publishing and Editing*; *Creative Writing for All*; *Ways into Creative Writing* and *Writing for Pleasure*. One-year certificates in *Creative Writing* and *Journalism* are available; individual modules can be taken.

Wiltshire

Marlborough College Summer School

Marlborough SN8 1PA

☎ 01672 892388   ℻ 01672 892476

admin@mcsummerschool.org.uk
www.mcsummerschool.org.uk

Summer School with literature and creative writing included in its programme. Caters for residential and day students. Brochure available giving full details and prices.

Urchfont Manor College

Urchfont, Devizes SN10 4RG

☎ 01380 840495   ℻ 01380 840005

urchfontmanor@wiltshire.gov.uk
www.urchfontmanor.co.uk

Short courses – days, residential weeks and weekends – offered in a varied programme which includes literature and creative writing. 'Beautiful location; historic environment; delicious home cooking.' Send for a brochure.

Yorkshire

Arvon Foundation (Yorkshire)

See entry under **London**

Leeds Metropolitan University

School of Film, Television & Performing Arts, H505, Civic Quarter, Calverley Street, Leeds LS1 3HE

☎ 0113 812 3860

screenwriting@leedsmet.ac.uk
www.leedsmet.ac.uk
Administrator *Chris Pugh*

Offers a Diploma/MA in *Screenwriting (Fiction)*.

Open College of the Arts

Unit 1B, Redbrook Business Park, Wilthorpe Road, Barnsley S75 1JN
℡ 0800 731 2116   🖷 01226 730838
open.arts@ukonline.co.uk
www.oca-uk.com

The OCA correspondence course, *Starting to Write*, offers help and stimulus from experienced writers/tutors. Emphasis is on personal development rather than commercial genre. Subsequent levels available include specialist poetry, fiction, autobiographical and children's writing courses. OCA writing courses are accredited by the Buckinghamshire New University. Prospectus and Guide to Courses available on request.

Sheffield Hallam University

Faculty of Development & Society, Sheffield Hallam University, City Campus, Sheffield S1 1WB
℡ 0114 225 5555   🖷 0114 225 2167
fdsenquiries@shu.ac.uk
www.shu.ac.uk

Offers MA in *Writing* (one-year, full-time; also part-time) and MA *English Studies* (one year full-time, also part-time).

University of Hull

School of Arts and New Media, Scarborough Campus, Filey Road, Scarborough YO11 3AZ
℡ 01723 362392
a.head@hull.ac.uk
www.hull.ac.uk

BA Single Honours in *Theatre and Performance Studies* incorporates opportunities in writing and other media across each level of the programme. The campus hosts the annual National Student Drama Festival which includes the International Student Playscript Competition (details from The National Information Centre for Student Drama; nsdf@hull.ac.uk).

University of Leeds – Lifelong Learning Centre

Leeds LS2 9JT
℡ 0113 343 3212
part-time@leeds.ac.uk
www.leeds.ac.uk/lifelonglearningcentre
Contacts *Rebecca O'Rourke, Pat Owens*

The Lifelong Learning Centre provides accredited part-time courses in creative writing on the Leeds campus during the day and evening. Introductory workshop courses are offered each year with follow-on and advanced courses providing the opportunity to focus on particular aspects of creative writing. These more specialized courses change from year to year and can include poetry, fiction, script and lifewriting or themed work. These courses form part of the Lifelong Learning Centre's Open Studies programme, which is designed to provide adults with opportunities for continuing education and to create routes into higher education. The Centre also offers a dedicated Part-time Joint Honours Degree where students can study Creative Writing in combination with a choice of other subjects including Arabic and Middle Eastern Studies, Business Studies and Theology and Religious Studies. Tutors include: Rosanna Blagg, Ray French, Sophie Nicholls, Rebecca O'Rourke, Rommi Smith, Beccy Stirrup and Adam Strickson.

University of Leeds

School of Performance and Cultural Industries, University of Leeds, Leeds LS2 9JT
℡ 0113 343 8735   🖷 0113 343 8711
enquiries-pci@leeds.ac.uk
www.leeds.ac.uk/paci
Contact *Garry Lyons or Jane Richardson* (Admissions Secretary)

MA in *Writing for Performance and Publication*. This new postgraduate programme (launched in September 2006) is particularly relevant to aspiring writers with professional ambitions in the areas of theatre, film, television and radio drama, while also allowing opportunities to work in prose fiction and other culturally significant genres. Studies are offered over one-year full-time and two-years part-time and the tutors are established authors in their chosen fields. The course director is award winning playwright and screenwriter Garry Lyons (*The Worst Witch, Leah's Trials, Britain's First Suicide Bombers*). The MA is one of a new portfolio of post-graduate degrees to be offered by Leeds University's School of Performance and Cultural Industries and is housed in a brand new theatre complex – stage@leeds – at the heart of the University's city centre campus. Students are not only given the opportunity to work on their own writing projects but are also encouraged to collaborate with colleagues on other MAs such as *Performance Studies*, with a view to seeing their work staged.

University of Sheffield

Institute for Lifelong Learning, 196–198 West Street, Sheffield S1 4ET
℡ 0114 222 7000   🖷 0114 222 7001
www.shef.ac.uk/till

Certificate in *Creative Writing* (Degree Level 1) is part-time and open to all. Courses in poetry, journalism, scriptwriting, comedy, short story writing, travel writing. Brochures and information available from the address above.

Yorkshire Art Circus

School Lane, Glasshoughton, Castleford
WF10 4QH
℡ 01977 550401
admin@artcircus.org.uk
www.artcircus.org.uk
Administrator *Angela Sibbit*

Yorkshire Art Circus is a community arts
organization and a registered charity. Runs
courses, forums and masterclasses that aim to
meet the needs of writers and artists who are
looking to learn more about creative writing, ICT
and visual arts. All tutors are professional writers
and artists who make a living working in the
field of arts that they teach. Also runs outreach
projects, a Writer Development Programme (for
writers who wish to pursue a career in writing),
and sessions for groups on request. An annual
training brochure is available.

Ireland

Dingle Writing Courses

Ballintlea, Ventry, Co. Kerry
℡ 00 353 66 915 9815
info@dinglewritingcourses.ie
www.dinglewritingcourses.ie
Directors *Abigail Joffe, Nicholas McLachlan*

An autumn programme of weekend residential
courses for beginners and experienced writers
alike. Tutored by professional writers the courses
include poetry, fiction, starting to write and
writing for theatre as well as special themed
courses. Past tutors have included Michael
Donaghy, Paul Durcan, Anne Enright, Nuala Ni
Dhomhnaill, Carlo Gébler, Jennifer Johnston,
Paula Meehan. Also organizes tailor-made
courses for schools, writers' groups or students
on a *Creative Writing* programme.

INKwell Writers' Workshops

The Old Post Office, Kilmacanogue, Co. Wicklow
℡ 00 353 1 276 5921/00 353 87 2835382
www.inkwellwriters.ie
Contact *Vanessa O'Loughlin*

Offers a series of one-day (Saturday) intensive
fiction writing workshops with bestselling Irish
writers, held at the Fitzpatrick's Castle Hotel,
Killiney, Co. Dublin. 'Designed to inspire and
guide new and accomplished writers, giving
them invaluable access to authors at the top
of their field.' Also, Pure Fiction Weekend:
19th–21st September 2008, full board, single
accommodation weekend workshop run at
Kippure Lodge located on a 240-acre private
estate in the heart of the Wicklow Mountains.
2008 Workshops: *Start Writing* (6th September):
Julie Parsons and Sarah Webb; *Womens' Fiction*

(4th October): Sinead Moriarty and Sarah
Webb; *Short Stories* (1st November): Martina
Devlin. Workshops held at the beginning of
2008 included: *Writing for Children* with Marita
Conlon McKenna and Maeve Friel; *Romance
with Mills and Boon®* writers Daisy Cummins and
Trish Wylie; *Getting Published* with Agent Sheila
Crowley of A.P. Watt, Patricia Deevy, Editorial
Director, Penguin Ireland, Paula Campbell of
Poolbeg Press and Patricia O'Reilly.

Queen's University of Belfast

School of Education, Belfast BT7 1LN
℡ 028 9097 3323   ℻ 028 9097 1084
openlearning.education@qub.ac.uk
www.qub.ac.uk/edu

Courses have included *Creative Writing* and
Writing for Profit and Pleasure.

University of Dublin (Trinity College)

Graduate Studies Office, Arts Building, Trinity
College, Dublin 2
℡ 00 353 1 896 1166
gradinfo@tcd.ie
www.tcd.ie/owc
Contact *Admissions*

Offers an MPhil *Creative Writing* course. A one-
year, full-time course intended for students who
are seriously committed to writing or prospective
authors.

The Writer's Academy

Carrig-on-Bannow, Co. Wexford
℡ 00 353 51 561789
thewritersacademy@eircom.net
www.thewritersacademy.net

Distance learning freelance article writing and
short story writing courses, critique service and
marketing advice. Personal tuition plus expert
information and support for writers of all abilities.

Scotland

Arvon Foundation (Inverness-shire)

See entry under **London**

Edinburgh University

Office of Lifelong Learning, 11 Buccleuch Place,
Edinburgh EH8 9LW
℡ 0131 650 4400   ℻ 0131 667 6097
oll@ed.ac.uk
www.lifelong.ed.ac.uk

Several writing-orientated courses and summer
schools. Full course information available on the
website.

7:84 Summer School

▷ 7:84 Theatre Company Scotland under Theatre
Producers

University of Dundee

Continuing Education, Nethergate, Dundee
DD1 4HN
www.dundee.ac.uk/learning/conted

Various creative writing courses held at the
University and elsewhere in Tayside. Detailed
course brochure available from end of June.

University of Glasgow

Department of Adult and Continuing Education,
11 Eldon Street, Glasgow G3 6NH
📞 0141 330 1835   📠 0141 330 1821
dace-query@educ.gla.ac.uk
www.gla.ac.uk/adulteducation

Runs writers' workshops and courses at all levels;
all friendly and informal. Daytime and evening
meetings. Tutors are all experienced published
writers. Call 0141 330 1829 for course brochure.

University of St Andrews

School of English, The University, St Andrews
KY16 9AL
📞 01334 462666   📠 01334 462655
english@st-andrews.ac.uk
www.st-andrews.ac.uk

Offers postgraduate study in *Creative Writing*.
Modules in either Writing Fiction or Writing
Poetry. Students submit a dissertation of original
writing – prose fiction of 15,000 words or a
collection of around 30 short poems. Taught by
Professor Douglas Dunn, John Burnside, Kathleen
Jamie, Meaghan Delahunt and Don Paterson.
Also offers PhD in *Creative Writing*.

Wales

Ty Newydd Writers' Centre

Llanystumdwy, Cricieth LL52 0LW
📞 01766 522811   📠 01766 523095
post@tynewydd.org
www.tynewydd.org

Residential writers' centre set up by the Taliesin
Trust with the support of the **Arts Council of
Wales** to encourage and promote writing in
both English and Welsh. Most courses run from
Monday evening to Saturday morning. Each
course has two tutors and a maximum of 16
participants. A wide range of courses for all levels
of experience. Early booking essential. Fees start
at £215 for weekends and £420 for week-long
courses, all inclusive. People on low incomes may
be eligible for a grant or bursary. Course leaflet
available. (See also *Organizations of Interest to
Poets*.)

University of Aberystwyth

Department of English, Hugh Owen Building,
Aberystwyth SY23 3DY
📞 01970 622534/5   📠 01970 622530
www.aber.ac.uk/english

BA in *English and Creative Writing*, a three-year
course taught in part by practising writers:
Director of Creative Writing, novelist and poet,
Jem Poster and acclaimed poets Tiffany Atkinson,
Matthew Francis and Kelly Grovier. Also offers
PhD in *Creative Writing*, and taught MA in
Creative Writing with modules in writing poetry,
writing fiction, research for writers, and writing
and publication.

University of Glamorgan

Department of English, Treforest, Pontypridd
CF37 1DL
📞 01443 483598
www.glam.ac.uk

MPhil in *Writing*: a two-year part-time
Masters degree for writers of fiction and poets.
Established 1993. Contact *Professor Tony Curtis* at
the Faculty of Humanities and Social Sciences.
BA in *Creative and Professional Writing*: a
three-year course (5–6 years part-time) for
undergraduates.
MA in *Scriptwriting (Theatre, Film, TV or
Radio)*: a two-year part-time Masters degree
for scriptwriters (held at the Cardiff School for
Creative and Cultural Industries). Contact: *Wyn
Mason* (📞 01443 668579; wmason@glam.ac.uk).

University of Wales, Bangor

School of English, College Road, Bangor LL57 2DG
📞 01248 382102
els029@bangor.ac.uk
postgrad-english@bangor.ac.uk
www.bangor.ac.uk

The School of English offers: MPhil/PhD *Creative
and Critical Writing*; Diploma/MA *Creative
Studies (Creative Writing)*; BA (Hons) *English
with Creative Writing*; BA (Hons) *English with
Songwriting*; BA (Hons) *English with Journalism*.
The Centre for Creative and Performing
Arts offers *Creative Writing* courses from
undergraduate level to MA and PhD.

John Wilson's Writing Courses

Argoed Hall, Tregaron SY25 6JR
📞 01974 298070
john.wilson@virgin.net

Residential writing intensives ranging from basic
skills to workshops in creative and business
writing and journalism. For advanced writers
an opportunity to understand and develop
the creative process is provided. The courses
normally run from Thursday evening to Sunday
afternoon, with lunch at Argoed Hall and nearby

B&B accommodation, with evening meals (or full board) at nearby hotels. Phone or e-mail for details.

Writers' Holiday at Caerleon

School Bungalow, Church Road, Pontnewydd, Cwmbran NP44 1AT
☎ 01633 489438
writersholiday@lineone.net
www.writersholiday.net
Contact *Anne Hobbs*

Annual six-day comprehensive conference including 12 courses for writers of all standards held in the summer at the University of Wales' Caerleon Campus. Courses, lectures, concert and excursion all included in the fee. Private, single and en-suite, full board accommodation. Courses have included *Writing for Publication*; *Writing Poetry*; *Writing Romantic Fiction* and *Writing for the Radio*.

US Writers' Courses

Arizona

University of Arizona

MFA Program in Creative Writing, Department of English, Modern Languages 445, PO Box 210067, Tucson AZ 85721–0067
℡ 001 520 621 3880
creativewriting@english.arizona.edu
cwp.web.arizona.edu

Established in 1974, the MFA in *Creative Writing* is a two-year programme with concentrations in Poetry, Fiction and Creative Non-Fiction. 'It boasts a faculty of 14 distinguished writers who share an uncommon commitment to teaching and supporting their students in the completion of publishable manuscripts.'

Arkansas

University of Arkansas

Programs in Creative Writing & Translation, Department of English, 333 Kimpel Hall, Fayetteville AR 72701
℡ 001 479 575 4301   ℻ 001 479 575 5919
mfa@uark.edu
www.uark.edu/depts/english/PCWT.html

MFA in *Creative Writing*, offering small, intensive workshops and innovative classes in fiction and poetry.

California

American Film Institute

2021 N. Western Avenue, Los Angeles CA 90027–1657
℡ 001 323 856 7740
admissions@afi.com
www.afi.com
Contact *Admissions Counselor*

Screenwriting at AFI focuses on narrative storytelling in an environment designed to stimulate the world of the professional screenwriter. Screenwriting Fellows in the First Year are immersed in the production process in order to learn how screenplays are visualized.

Initially writing short screenplays – one of which will be the basis for a first year production – writers collaborate with Producing and Directing Fellows to see their work move from page to screen. The remainder of the first year is devoted to the completion of a feature-length screenplay. Second Year Fellows may write a second feature-length screenplay, or develop materials for television, including biopics, television movies and spec scripts for sitcoms and one-hour dramas. They also have the opportunity to work closely with other disciplines by writing a Second Year thesis script.

California Institute of the Arts

MFA Writing Program, School of Critical Studies, 24700 McBean Parkway, Valencia CA 91355–2397
℡ 001 661 255 1050
writing@calarts.edu
www.calarts.edu/~writing
Director, Writing Program *Brighde Mullins*

The two-year MFA *Writing Programme* is rooted in principles of critical thought, experimentation and innovation. It is intended as an alternative to traditional programmes, with a founding premise that distinctions between creative and critical writing should be suspended. Students are encouraged to work across genres and often work in various multi-media forms in addition to their literary production.

Chapman University

Graduate Admissions Office, One University Drive, Orange CA 92866
℡ 001 714 997 6711
www.chapman.edu
Contact *Graduate Admissions*

The programme is a two-year Master of Fine Arts (MFA) degree in *Creative Writing* intended for graduate students who wish to specialize in writing literary fiction. There is an emphasis on fiction but secondary studies include screenwriting, playwriting and poetry.

Saint Mary's College of California

MFA Program in Creative Writing, PO Box 4686, Moraga CA 94575–4686
℡ 001 925 631 4762

writers@stmarys-ca.edu
www.stmarys-ca.edu
Programme Coordinator *Thomas Cooney*

Two-year MFA course in *Creative Writing* with Fiction, Non-Fiction or Poetry. The core of the programme is the Writing Workshop which provides the opportunity for students to work with established writers.

San Francisco State University

Creative Writing Department, College of Humanities, 1600 Holloway Avenue, San Francisco CA 94132
℡ 001 415 338 1891
cwriting@sfsu.edu
www.sfsu.edu/~cwriting

Offers BA and MA in *English:* with a concentration on *Creative Writing*; and MFA in *Creative Writing*. Undergraduate classes include feature writing, short story writing, the craft of poetry and the craft of playwriting while graduate classes include advanced story writing, advanced poetry writing, advanced playwriting, experimental fiction, playwright's theatre workshop and workshops in fiction, poetry and playwriting.

University of Southern California (USC)

School of Cinematic Arts, Attn: Writing for Screen & Television Program, University Park, LUC 301, Los Angeles CA 90089–2211
℡ 001 213 740 3303
www-cntv.usc.edu

Offers a Bachelor of Fine Arts or Master of Fine Arts degree in *Writing for Screen and Television*. There is a strong international presence, having accepted students from over 44 countries.

Colorado

Colorado State University

Creative Writing Program, 1773 Campus Delivery, 359 Eddy Hall, Fort Collins CO 80523–1773
℡ 001 970 491 2403
english@lamar.colostate.edu
www.colostate.edu/Depts/English/programs/mfa.htm
Contact *Assistant to the Director of Creative Writing or Director of Creative Writing*

Three-year Master of Fine Arts (MFA) programme in creative writing with concentrations in fiction or poetry. The programme offers a balance of intimate and intensive writing and translation. Course work culminates in a thesis – a collection of poetry or short stories or a novel. Students have the opportunity to teach introductory creative-writing courses, intern with literary journals

including the *Colorado Review* and be a part of a thriving community.

District of Columbia

American University

Department of Literature, 237 Battelle-Tompkins, 4400 Massachusetts Avenue, N.W., Washington DC 20016–8047
℡ 001 202 885 2971
www.american.edu/cas/lit/mfa-lit.cfm
Contact *Kristin Toburen* (Graduate Programs Assistant)

Established in 1980, the MFA in *Creative Writing* is a 48-semester-hour programme allowing students to concentrate in Poetry, Fiction or Creative Non-Fiction. The programme is allied with the University's MA in Literature which offers numerous courses and master's-level seminars enabling creative writing students to explore both classic texts and contemporary literature.

Florida

Florida International University

English Department, Biscayne Bay Campus, 3000 NE 151st Street, North Miami FL 33181
℡ 001 305 919 5857
crwriting@fiu.edu
www.fiu.edu/crwriting
Director *Les Standiford*

The MFA Program in *Creative Writing* includes writing workshop, literature, form and theory and thesis (most students complete the course in about three years; completion of study within eight years is required). There is no language requirement. Graduate workshops include short fiction, the novel, popular fiction, screenwriting, creative non-fiction and poetry. Great emphasis is placed upon preparation and completion of a book-length thesis. Admission is based primarily on the strength of the applicant's submitted writing sample. Application deadline: January 15.

University of Florida

Program in Creative Writing, 4008 Turlington Hall, PO Box 117310, Gainesville FL 32611–7310
℡ 001 352 392 6650   ℻ 001 352 392 0860
www.english.ufl.edu/crw
Director, Program in Creative Writing *David Leavitt*

MFA@FLA, the graduate program in Creative Writing at the University of Florida, is one of the oldest writing programs in the United States, begun in 1948. 'We require an equal interest in writing and in reading literature. We don't believe in any particular school of writing; we have no wish to foster or found one. All the members of our faculty – four fiction writers and four poets – live in Gainesville and are tenured. We put out *Subtropics*, a new and important literary magazine.' In the autumn of 2008 the programme is changing from a two-year to a three-year format; all students will receive full tuition waivers and teaching assistantships.

Illinois

Chicago State University

MFA Program in Creative Writing, Department of English, Division of Graduate Studies, 9501 South King Drive, LIB338, Chicago IL 60628
☎ 001 773 995 2189
www.csu.edu/GraduateSchool

Offers an MFA with courses in Creative Writing; writing workshops in fiction, poetry, creative non-fiction, playwriting and scriptwriting. Students undertake coursework in African American Literature and non-African American Literature. Students will show competency in the genre of their choice, choosing from fiction, creative non-fiction and poetry. Students may take course work in playwriting and scriptwriting for film and television as electives.

Indiana

University of Notre Dame

Creative Writing Program, Department of English, Notre Dame IN 46556
☎ 001 574 631 7526   ✆ 001 574 631 4795
creativewriting@nd.edu
www.nd.edu/~alcwp

The MFA in *Creative Writing* is a two-year degree programme centred around workshops in poetry and fiction, offering literature courses, translation, a literary publishing course and twelve credits of thesis preparation with an individual faculty adviser.

Purdue University

MFA Program in Creative Writing, Department of English, West Lafayette IN 47907
☎ 001 765 494 3740
www.cla.purdue.edu/mfacw
Program Director *Porter Shreve*

MFA in *Creative Writing* offering 'full funding, generous stipends and editorial experience with the award-winning literary journal *Sycamore Review*'. The three-year programme in either fiction or poetry is small and intensive and includes workshops, literature and craft courses, and thesis tutorials toward a book-length manuscript.

Taylor University

Department of English, 1025 West Rudisill Blvd, Fort Wayne IN 46807–2170
☎ 001 260 744 8647
DNHensley@TaylorU.edu
Contact *Dr Dennis E. Hensley*

Summer Honours Programme one-week seminars in *Professional Writing*; evening courses in *Freelance Writing* and *Fiction Writing*; one year college certificate in *Professional Writing*. Dr Hensley is a professor of English and director of the professional writing major.

Louisiana

Louisiana State University

English Department, 260 Allen Hall, Baton Rouge LA 70803
☎ 001 225 578 4086
www.english.lsu.edu
Director of Creative Writing *James Wilcox*

Master of Fine Arts in *Creative Writing* course with a focus in Poetry, Fiction, Drama, Screenwriting. Includes opportunity to edit literary journals.

Maryland

Goucher College

Welch Center for Graduate Studies, 1021 Dulaney Valley Road, Baltimore MD 21204–2794
☎ 001 410 337 6200
center@goucher.edu
Program Director *Patsy Sims*

The two-year MFA Program in *Creative Nonfiction* is a limited-residency programme that allows students to complete most of the requirements off campus while developing their skill as non-fiction writers under the close supervision of a faculty mentor. Provides instruction in the following areas: narrative non-fiction, literary journalism, memoir, the personal essay, travel/nature/science writing and biography/profiles.

Massachusetts

Boston University

Creative Writing Program, 236 Bay State Road, Boston MA 02215
☎ 001 617 353 2510   ✆ 001 617 353 3653
crwr@bu.edu
www.bu.edu/writing
Director *Leslie Epstein*
Contact *Matthew Yost*

Intensive Master of Fine Arts in *Creative Writing*: Fiction, Poetry or Drama, can be completed in one or two years, depending on a student's needs. The Boston University Creative Writing Program is one of the oldest in the country. Students participate in workshops or seminars concentrating on their chosen specialization and are expected to balance that work with an equal number of graduate literature courses. Internships at the literary journal *AGNI*, assistantships at the Boston Arts Academy, as well as teaching fellowships are available on a limited basis.

Emerson College

120 Boylston Street, Boston MA 01226–4624
℡ 001 617 824 8610   Ⓕ 001 617 824 8614
gradapp@emerson.edu
www.emerson.edu/graduate_admission
Graduate Program Director *Douglas Whynott*

Offers MFA in *Creative Writing* with a focus on writing fiction, non-fiction and poetry. Includes writing workshops, internship opportunities, a student reading series and student teaching opportunities. Publishes literary magazines, *Ploughshares* and *Redivider*. Contact the Graduate Admission Office for more information.

Michigan

Western Michigan University

Department of English, 1903 W. Michigan Avenue, Kalamazoo MI 49008-5331
℡ 001 269 387 2584
english.graduate@wmich.edu
www.wmich.edu/english
Director of Graduate Studies *Jana Schulman*

BA English Major/Minor with *Creative Writing Emphasis* (poetry, fiction, playwriting). The English Major aims at giving students intensive practice in writing and criticism in various genres in a workshop format; for general writing careers or for prospective candidates for the MFA in Creative Writing.
MFA in *Creative Writing* (poetry, fiction, playwriting, non-fiction) For students who wish to become professional writers of poetry, fiction, drama or non-fiction. Qualifies them to teach the craft at college or university level.
PhD in *English, with an Emphasis on Creative Writing* (poetry, fiction, playwriting, non-fiction). Qualifies students to teach creative writing and literature at university level.

Minnesota

Minnesota State University Moorhead

Graduate Studies Office, MSUM, 1104 7th Avenue South, Moorhead MN 56563
℡ 001 218 477 2344
enger@mnstate.edu
www.mnstate.edu/finearts
MFA Coordinator *Lin Enger*

MFA in *Creative Writing*. Students specialize in poetry, fiction, playwriting, screenwriting, or creative non-fiction and are encouraged to experiment in more than one genre. The programme offers the opportunity to take workshops, seminars and tutorials in chosen areas and to work with New Rivers Press (www.newriverspress.com) or with *Red Weather*, the campus literary magazine.

Minnesota State University, Mankato

Department of English, 230 Armstrong Hall, Mankato MN 56001
℡ 001 507 389 2117   Ⓕ 001 507 389 21175362
richard.robbins@mnsu.edu
www.english.mnsu.edu/cw
Programme Director *Richard Robbins*

The MFA programme in *Creative Writing* meets the needs of students who want to strike a balance between the development of individual creative talent and the close study of literature and language. The programme gives appropriate training for careers in freelancing, college-level teaching, editing and publishing and arts administration.

Missouri

University of Missouri-Columbia

Creative Writing Program, Department of English, 107 Tate Hall, Columbia MO 65211–1500
℡ 011 573 884 7773
creativewriting@missouri.edu
creativewriting.missouri.edu
Contact *Sharon Fisher*

Creative Writing MA and PhD in English programmes in fiction, poetry and non-fiction. Access the website for further information.

University of Missouri

MFA Program, Department of English, One University Drive, St Louis MO 63121
℡ 00 314 516 6845
www.umsl.edu/~mfa
Contact *Mary Troy*

The MFA in *Creative Writing* provides opportunities for growth in the writing of fiction and poetry (with some work in non-fiction) as well as practical training in literary editing.

While normally a studio/academic programme mixing the study of literature and criticism with workshops and independent study and editing, the plan of study is flexible and individual. Students ordinarily specialize in one genre, either fiction or poetry, and regular workshops in these forms are at the heart of the programme. Five workshops, at least four in the student's chosen genre, are required for the degree though more may be taken as electives. Students also take from five to nine courses on offer, choosing from graduate courses in literary journal editing; in poetry and fiction form, theory and technique; in literature and literary criticism; in composition theory; and in linguistics. A creative thesis of six hours completes the 390-hour programme.

New York

Adelphi University

MFA Program in Creative Writing, Department of English, PO Box 701, Garden City NY 11530
℡ 001 516 877 4044
mfa@adelphi.edu
academics.adelphi.edu/artsci/creativewriting
Director of Creative Writing *Judith Baumel*

The MFA in *Creative Writing* offers students the opportunity to specialize in three major genres: fiction, poetry and dramatic writing. Its unique Professional Development Practicum introduces students to the professional and practical life of writers across many disciplines and prepares them for careers in writing, teaching and/or more advanced graduate studies through training in creative writing, language and literary studies, research and teaching. Students are required to complete 37 credits in a plan of study that includes writing workshops and literature classes. A student thesis is a degree requirement.

Brooklyn College of the City University of New York

MFA Program in Creative Writing, Department of English, Brooklyn NY 11210–2889
℡ 001 718 951 5914
profmsp@msn.com
academic.brooklyn.cuny.edu/english
Deputy Chair for Graduate Studies *Mark Patkowski*

The MFA in *Creative Writing* is a small, highly personal two-year programme which confers degrees in fiction, poetry and playwriting. Admission is highly competitive. Students must complete 12 courses (four workshops, four tutorials and four literature courses), submit a book-length work in their chosen genre near the end of the second year and pass a comprehensive examination that tests their knowledge and

expertise in literature. The workshops, seminars and one-on-one tutorials with established writers particularly emphasize relationships between instructors and students.

New York University

Lillian Vernon Creative Writers House, 58 West 10th Street, New York NY 10011
℡ 001 212 998 8816
creative.writing@nyu.edu
www.cwp.fas.nyu.edu
Program Coordinator *Allison Brotherton*

Offers MFA in *English and American Literature with a Concentration in Creative Writing* and MA in *Creative Writing* in the genres of either poetry or fiction. Includes writing workshops and craft courses, literary outreach programmes, a public reading series, student readings, special literary seminars and student teaching opportunities. Publishes a literary journal, *Washington Square*. Contact the department via e-mail.

North Carolina

University of North Carolina at Greensboro

MFA Writing Program, 3302 HHRA Building, Greensboro NC 27402
℡ 001 336 334 5459
TLKenned@uncg.edu
www.uncg.edu/eng/mfa
Contact *Terry Kennedy*

MFA in *Creative Writing (Poetry, Fiction)*. One of the oldest of its kind in the country, the MFA Writing Program at Greensboro is a two-year residency with an emphasis on providing students with studio time in which to study the writing of poetry or fiction. Gives a flexibility that permits students to develop their particular talents through small classes in writing, literature and the arts.

University of North Carolina at Wilmington

Department of Creative Writing, 601 S. College Road, Wilmington NC 28403
℡ 001 910 962 3070
mfa@uncw.edu
www.uncw.edu/writers
Contact *MFA Coordinator*

An intensive studio-academic programme in the writing of fiction, poetry and creative non-fiction leading to either the Master of Fine Arts or Bachelor of Fine Arts degree in *Creative Writing*. Courses include workshops in the three genres, special topics and forms courses as well as a range of courses in literature. Course work in publishing and editing is also offered in the Publishing Laboratory, a university press imprint which supports local, regional and national publishing projects. The MFA programme is home to the literary journal, *Ecotone: reimagining place*.

Ohio

Bowling Green State University

Creative Writing Program, English Department, Bowling Green OH 43403
☎ 001 419 372 8370   ℻ 001 419 372 6805
www.bgsu.edu/departments/creative-writing/

Undergraduate BFA programme: a four-year programme which offers comprehensive and rigorous training in the art of writing and develops students' skills in preparation for numerous post-graduate careers.
Graduate MFA programme: a two-year studio/academic composite, mostly work in writing itself in either poetry or fiction. Also offers ten new teaching assistantships each autumn. See the website for details on both programmes.

Oregon

University of Oregon

Creative Writing Program, 5243 University of Oregon, Eugene OR 97403–5243
☎ 001 541 346 3944
www.uoregon.edu
Program Director *Karen J. Ford*

Offers MFA in *Creative Writing* in poetry or fiction. Includes writing workshops, craft courses, individualized tutorials with faculty, public reading series, student readings and student teaching opportunities. Admission is highly competitive.

Philadelphia

Seton Hill University

1 Seton Hill Drive, Greensburg PA 15601
☎ 001 724 830 4600
lynn@setonhill.edu
www.setonhill.edu
Contact *Director of Graduate Studies*

The MA in *Writing Popular Fiction* programme at Seton Hill University allows students to earn a graduate degree by writing fiction that people *actually* read. Students attend week-long residencies in January and June and complete writing projects off-campus, working with a faculty mentor who is a published author in the chosen genre. Students can choose to specialize in science fiction, fantasy, horror, children's literature, romance or mystery.

Rhode Island

Brown University

Literary Arts, Box 1923, Providence RI 02912
☎ 001 401 863 3260   ℻ 001 401 863 1535
Writing@brown.edu
www.brown.edu/cw
Contact *Director of Literary Arts*

Two-year MFA in *Fiction, Poetry, Dramatic Writing and Electronic Writing*. Students take three workshops, four electives and one independent study (through which they complete a thesis project). Application deadline: 15 December.

Tennessee

University of Memphis

MFA Program in Creative Writing, Department of English, 471A Patterson Hall, Memphis TN 38152
☎ 001 901 678 4692
creativewriting@memphis.edu
www.mfainmemphis.com
Contact *MFA Coordinator*

'The Writing Workshop is the premier creative writing program in the South.' Concentrations on fiction, creative non-fiction and poetry. Also features an award-winning national journal, *The Pinch* and the River City Writers Series. Students work in small groups with nationally recognized authors. Students have opportunities to teach creative writing locally and abroad.

Texas

University of Houston

Creative Writing Program, Department of English, Room 229, Roy Cullen Building, University of Houston, Houston TX 77204-3015
☎ 001 713 743 3015   ℻ 001 713 743 3697
cwp@uh.edu
www.uh.edu
Office Coordinator *Shatera Dixon*

Offers MFA in *Creative Writing* and PhD in *Literature and Creative Writing*. Admission is highly competitive; students can apply for admission in the genres of poetry, fiction and non-fiction. Normally admits 20 students a year. The MFA is a three-year degree programme; the PhD normally takes five years. Students take courses both in creative writing and literature. Fellowships and teaching assistantships are available. In collaboration with Inprint, it hosts the Inprint/Brown Reading series and the Inprint Studio series. The programme offers opportunities to collaborate with artists in other disciplines. Publishes a literary journal, *Gulf Coast*.

Virginia

George Mason University

Graduate Creative Writing Program, Department of English, MSN 3E4, 4400 University Drive, Fairfax VA 22030

☎ 001 703 993 1180
writing@gmu.edu
creativewriting.gmu.edu
Contact *Graduate Coordinator*

The MFA in *Creative Writing* is a 48-credit hour programme with flexibility for students to tailor their studies to their interests and craft development. Twelve to 18 hours are in writing seminars in the chosen genre and include a course in the historic forms of the genre. At least 12 hours are in literature and at least three hours are in a genre other than the concentration. Six hours are in thesis. The distribution of the remaining hours is generally up to the students with consultation from the faculty.

Hollins University

MFA in Creative Writing, Hollins University Graduate Center, PO Box 9603, Roanoke VA 24020–1603

☎ 001 540 362 6575   ⨏ 001 540 362 6288
hugrad@hollins.edu
www.hollins.edu/grad
Manager, Graduate Studies *Cathy Koon*

MFA in *Creative Writing*: a small, highly selective two-year full-time programme which allows students to specialize in poetry, fiction or both, while also offering opportunities to study and write screenplays and creative non-fiction. 'The programme is characterized by an individual approach, a lively community of writers and a stimulating combination of challenge and support.' Also three summer-term MA/MFA programmes in *Children's Literature*; *Playwriting*; and *Screenwriting & Film Studies*. Students typically study for three to five six-week terms.

Virginia Commonwealth University

Department of English, PO Box 842005, Richmond VA 23284–2005

☎ 001 804 828 1329
tndidato@vcu.edu
www.has.vcu.edu/eng
Graduate Programs Coordinator *Thom Didato*

Three-year MFA in *Creative Writing* with tracks in fiction and poetry, and workshops in short fiction, the novel, poetry, drama, screenwriting and creative non-fiction. Opportunities to work with *Blackbird*: an online journal of literature and the arts (www.blackbird.vcu.edu), *Stand*

Magazine, the Levis Reading Prize and the First Novelist Award.

Washington

Eastern Washington University

Inland Northwest Center for Writers, Creative Writing, 501 N. Riverpoint Blvd., Suite 425, Spokane WA 99202

☎ 001 509 359 4956
prussell@ewu.edu
www.ewumfa.com
Director *Jonathan Johnson*

The MFA Program is an intensive two-year, pre-professional course of study with an emphasis on the practice of literature as a fine art. Includes course work in the study of literature from the vantage point of its composition and history, but the student's principal work is done in advanced workshops and in the writing of a book-length thesis of publishable quality in fiction, literary non-fiction or poetry. Students have the opportunity to work as interns on several projects: the literary magazine, *Willow Springs*; the community outreach programme, Writers in the Community; and Eastern Washington University Press. The MFA is a terminal degree programme.

Seattle Pacific University

MFA Program in Creative Writing, 3307 Third Avenue West, Seattle WA 98119

☎ 001 206 281 2727
mfa@spu.edu
www.spu.edu/mfa
Contact *Gregory Wolfe*, Program Director

The low-residency MFA at SPU is a two-year programme for apprentice writers who not only want to pursue excellence in the craft of writing but also place their work within the larger context of the Judeo-Christian tradition of faith. Students work closely with a faculty mentor in their chosen genre (poetry, fiction or creative non-fiction) toward the completion of a creative thesis. They also write a critical paper each quarter and read a minimum of 62 books over the course of the programme. Students attend two intensive ten-day residencies per year, one on Whidbey Island in Washington State and one in Santa Fe, New Mexico. Graduating students attend a final residency in which they deliver a public reading and a lecture.

Wyoming

University of Wyoming

MFA in Creative Writing, Department of English
– 3353, 1000 E. University Avenue, Laramie
WY 82071–2000
☎ 001 307 766 2867
cw@uwyo.edu
www.uwyo.edu/creativewriting
Contact *Rachel Ferrell* (Program Assistant)

The two-year Master of Fine Arts in *Creative Writing* is an intensive 40-hour studio degree in Poetry, Fiction or Creative Non-Fiction. Special features include opportunities for interdisciplinary study, supported by a wide range of university departments, and a required professional internship on campus, in the community or further afield, ensuring the acquisition or polishing of 'real-life' writing skills.

Writers' Circles and Workshops

Directory of Writers' Circles, Courses and Workshops

39 Lincoln Way, Harlington LU5 6NG
℡ 01525 873197
diana@writers-circles.com
www.writers-circles.com
Editor *Diana Hayden*

Directory of UK writers' circles, courses and workshops.

Camelford Poetry Workshops

▷ entry under Organizations of Interest to Poets

Carmarthen Writers' Circle

Lower Carfan, Tavernspite, Whitland SA34 0NP
℡ 01994 240441
Contact *Jenny White*

Founded 1989. The Circle meets monthly on the second Monday of the month at the Indoor Bowls Centre, Carmarthen. All levels and genres welcome. Holds occasional workshops.

Chiltern Writers' Group

151 Chartridge Lane, Chesham HP5 2SE
℡ 01494 772308
info@chilternwriters.org
www.chilternwriters.org

Guest speakers, workshops, manuscript critiques and monthly meetings, held at Wendover Library every second Thursday of the month at 8.00 pm. Regular newsletter. SUBSCRIPTION £17 p.a.; £12 (concessions); £3 (non-members meetings).

Coleg Harlech WEA

Harlech, LL46 2PU 01766 781900
info@fc.harlech.ac.uk
www.harlech.ac.uk

Coleg Harlech WEA was formed in 2001 with the merger of Coleg Harlech and the Workers' Educational Association North Wales.

The Cotswold Writers' Circle

Bliss's Cottage, Lower Chedworth, Cheltenham GL54 4AN
℡ 01285 720668
elaine@pandelunt.co.uk
Patron *Katie Fforde*

Membership Secretary *Elaine Lunt*
Competition Secretary *Mrs Anne Brookes*

The Circle meets every Tuesday morning (except second Tuesday of the month) in Cirencester. Activities include workshops with well-known authors and an annual International Open Writing Competition; closing date: 31 January (send s.a.e. for details). Winners published in the Circle's anthology, *Pen Ultimate*.

Creuse Writers' Workshop & Retreat

45 Browning Drive, Lincoln LN2 4HF
℡ 01522 880565   ℻ 01522 880565
info@creusewritersworkshopandretreat
www.creusewritersworkshopandretreat

One week writers' workshops held twice yearly in Creuse, central France. Run by a team of professional writers to encourage creativity in writers of all grades. The week includes guest speakers, fun practical discussions and social events. Opportunities to write exercises, discuss work as a group and one-to-one. Check the website for details.

Cumbrian Literary Group

`Calgarth', The Brow, Flimby, Maryport CA15 8TD
℡ 01900 813444
President *Glyn Matthews*
Secretary *Joyce E. Fisher*

Founded 1946 to provide a meeting place for readers and writers in Cumbria. The Group meets once a month (April to November) in Keswick. Invites speakers to meetings and holds annual competitions for poetry and prose. Publishes *Bookshelf* magazine. Further details from the Secretary at the address above.

'Sean Dorman' Manuscript Society

3 High Road, Britford, Salisbury SP5 4DS
Director *Jan Smith*

Founded 1957. Provides mutual help among writers and aspiring writers in England, Wales and Scotland. By means of circulating manuscript parcels, members receive constructive comment on their own work and read and comment on the work of others. Full details and application forms available on receipt of s.a.e.

Includes FREE online access to **www.thewritershandbook.com**

East Anglian Writers

194 North Walsham Road, Norwich NR6 7QW
chair@eastanglianwriters.org.uk
www.eastanglianwriters.org.uk
Chair *Victoria Manthorpe*

A group for professional writers living in Norfolk, Suffolk, Essex, Cambridgeshire and Bedfordshire. Affiliated to the **Society of Authors**. Aims to help writers socialize and promote their work. Informal social events, speakers' evenings and contact point for professional writers in the area.

éQuinoxe TBC

43 rue Beaubourg, 75003 Paris, France
Ⓣ 00 33 1 47 55 11 89   Ⓕ 00 33 1 48 04 52 58
equinoxetbc@equinoxetbc.fr
www.equinoxetbc.fr
President & Artistic Director *Noëlle Deschamps*

Week-long intensive screenwriting workshops, held twice a year in English and French, open to experienced European screenwriters. Participants are selected by an international jury: candidates must have written a script for a feature film, which is sufficiently developed and ready to go into production. Entry form available on the website.

Euroscript Ltd

64 Hemingford Road, London N1 1DB
Ⓣ 07958 244656
ask@euroscript.co.uk
www.euroscript.co.uk

Script development organization offering creative and editorial input to writers and production companies. Develops screenplays through an intensive consultancy programme, including residential script workshops; also offers a broad programme of short courses, provides script reports, and runs an annual Screen Story Competition. Open to writers of any nationality. E-mail, telephone, or access the website for further information.

Foyle Street Writers Group
▷ See Sunderland City Library and Arts Centre under Library Services

Guildford Writers
▷ The Guildford Institute under UK and Irish Writers' Courses

Historical Novel Folio

187 Baker Street, Alvaston, Derby DE24 8SG
Contact *Mrs Heidi Sullivan*

An independent postal workshop – single folio dealing with any period before World War II. Send s.a.e. for details.

'How to Self-Publish Your Book' Workshops

Faculty of Arts, University of Winchester, Winchester SO22 4NR
Ⓣ 01962 827238
barbara.large@winchester.ac.uk
www.writersconference.co.uk
Contact *Barbara Large, MBE, FRSA, HFUW*

One-day workshops to inform writers who wish to self-publish their books about the digital and lithographic processes. Participants will learn the skills of researching, writing and revising their mss; the use of photographs, maps, diagrams, illustrations, design and formatting; types of paper, fonts, sizes, ISBN numbers and costs, and marketing their books. 2009 workshops: Friday, 8th May and Friday, 18th September.

Janus Writers
▷ See Sunderland City Library and Arts Centre under Library Services

Bernard Kops and Tom Fry Drama Writing Workshops

41B Canfield Gardens, London NW6 3JL
Ⓣ 020 7624 2940/8533 2472
bernardkops@blueyonder.co.uk
Tutors *Bernard Kops, Tom Fry*

Three 11-week terms per year. Small group drama writing workshops for stage, screen and television with actors in attendence. Tutorials also available. Call for details.

New Writing South

9 Jew Street, Brighton BN1 1UT
Ⓣ 01273 735353
admin@newwritingsouth.com
www.newwritingsouth.com
Contact *Chris Taylor*

Works throughout the South East offering resources and development to all creative writers in the region. Also builds partnerships between writers and those able to produce their work. Membership open to professional, emerging and aspiring creative writers in the region and to companies with a professional interest in new writing.

North West Playwrights (NWP)

18 Express Networks, 1 George Leigh Street, Manchester M4 5DL
Ⓣ 0161 237 1978
newplaysnw@hotmail.com
www.newplaysnw.co.uk

Founded in 1982, NWP is the regional development agency for new theatre writing in the north west of England. NWP operates a script-reading service, classes and script development scheme and *The Lowdown* newsletter. Services available to writers in the region only.

The Original Writers Group

Garfield Community Centre, 64 Garfield Rd, London SW11 5PN
info@theoriginalwriter.com
www.theoriginalwriter.com
Contact *Rupert Davies-Cooke*

Currently we have 64 members: poets, novelists, playwrights, scriptwriters, historians, philosophers. It is a rich mix and makes a lively evening. We welcome writers from all walks of life as everyone has a point of view, and the more varied these opinions, the richer the experience. Usually six to eight writers turn up each evening and there is time to discuss three to four pieces. Even if you are just starting out and are just interested in writing, but don't have anything to read, come along and experience the evening; as sometimes the best motivation to writing is to hear how others got there before you.' The group meets at Garfield Community Centre, 64 Garfield Road, London, SW11 5PN every first and third Wednesday of the month 7.00 pm to 9:00 pm; £2 entrance fee. Directions to the Community Centre are available on the website.

Pitstop Refuelling Writers' Weekend Workshops

Faculty of Arts, University of Winchester, Winchester SO22 4NR
℡ 01962 827238
barbara.large@winchester.ac.uk
www.writersconference.co.uk
Director *Barbara Large, MBE, FRSA, HFUW*

Following on from the **Winchester Writers' Conference** in Winchester, these are small-group writing workshops under the guidance of professional writers. Join *Writing Marketable Children's Fiction or Non-Fiction*; *Editing the Final Draft of Your Novel and Marketing It*; *Lifewriting and Research*. 2008 workshops: 24th–26th October; 2009: 13th–15th March and 24th–26th October.

Player–Playwrights

▷ entry under Professional Associations

Scribo

1/31 Hamilton Road, Bournemouth BH1 4EQ
℡ 01202 302533
Contacts *K. & P. Sylvester*

Scribo (established over 20 years ago) is a postal workshop for novelists giving criticism and support via manuscript folios (e.g. fantasy/sci-fi, mainstream, women's fiction, crime). Joining fee: £5. No annual subscription. Send s.a.e. for further details.

Script Yorkshire

Communications Officer: 409 Gateway West, East Street, Leeds LS9 8DZ
members@scriptyorkshire.co.uk
www.scriptyorkshire.co.uk
Membership *Joanna Loveday*

'Script Yorkshire is the region's only support and advocacy organization for scriptwriters across performance and broadcast media. Run democratically by writers for writers, it puts the needs of members at the very core of its activity, developing both their practice and their industry connections. Script Yorkshire aims to empower Yorkshire's writers for performance and broadcast media by giving them the skills and insights they need to develop their own careers.' Send an e-mail or visit the website for information on becoming a member.

Short Story Writers' Folio/Children's Novel Writers' Folio

5 Park Road, Brading, Sandown, Isle of Wight PO36 0HU
℡ 01983 407697
nott.dawn-wortley@tiscali.co.uk
Contact *Mrs Dawn Wortley-Nott*

Postal workshops; members receive constructive criticism of their work and read and offer advice on fellow members' contributions. Send an s.a.e. for further details.

Society of Sussex Authors

Dolphin House, 51 St Nicholas Lane, Lewes BN7 2JZ
℡ 01273 470100   F 01273 470100
sussexbooks@aol.com
Contact *David Arscott*

Founded 1968. Six meetings per year, held in Lewes, plus social events. Membership restricted to Sussex-based writers only.

Southwest Scriptwriters

℡ 0125 541798
info@southwest-scriptwriters.co.uk
www.southwest-scriptwriters.co.uk
Secretary *John Colborn*

Founded 1994 to offer support to regional writers for all media. The group meets at the Naval Volunteer pub opposite the currently dark Bristol Old Vic.

Speakeasy – Milton Keynes Writers' Group

PO Box 5948, Stoke Hammond, Milton Keynes MK10 1GX
℡ 01908 663860
speakeasy@writerbrock.co.uk
www.mkweb.co.uk/speakeasy
Contact *Martin Brocklebank*

Monthly meetings on the first Friday of each month at 7.30 pm. Full and varied programme includes Local Writer Nights where work can be read and peformed and Guest Nights where writers, poets and journalists are invited to speak. Mini-workshops and information nights are also in the programme. Invites entries to Open Creative Writing Competitions. Phone, e-mail or send s.a.e. for details to address above.

Spread the Word

77 Lambeth Walk, London SE11 6DX
℡ 020 7735 3111   🖷 020 7735 2666
info@spreadtheword.org.uk
www.spreadtheword.org.uk
General Manager *Nick Murza*

Supports new writing and writers in London throughout their careers through creative writing workshops, talks and events, cross arts projects, advice sessions and one-to-one surgeries. Also runs creative writing competitions and provides online resources for writers.

Sussex Playwrights' Club

2 Brunswick Mews, Hove BN3 1HD
℡ 01273 730106
www.sussexplaywrights.com
Secretary *Dennis Evans*

Founded 1935. Monthly readings of members' work by experienced actors. Membership open to all. Meetings held at New Venture Theatre, Bedford Place, Brighton.

Ver Poets

℡ 01727 864898
daphne.schiller@virgin.net
www.verpoets.org.uk
Secretary *Daphne Schiller*

Founded 1966. Postal and local members. Meets in St Albans, runs evening meetings and daytime workshops and organizes competitions.

Walton Wordsmiths

27 Braycourt Avenue, Walton on Thames
KT12 2AZ
℡ 01932 702874
wendy@stickler.org.uk
www.waltonwordsmiths.co.uk
Contact *Wendy Hughes*

Founded in 1999 by Wendy Hughes. Supports writers of all grades and capabilities by offering constructive criticism and advice.

Workers' Educational Association

National Office: 3rd Floor, 70 Clifton Street, London EC2A 4HB
℡ 020 7426 3450
national@wea.org.uk
www.wea.org.uk
Founded in 1903, the WEA is a charitable

provider of adult education with students drawn from all walks of life. It runs writing courses and workshops in many parts of the country which are open to everyone. Contact your local WEA office for details of courses in your region.
Eastern Cintra House, 12 Hills Road, Cambridge, B2 1JP eastern@wea.org.uk
East Midlands 39 Mapperley Road, Mapperley Park, Nottingham NG3 5AQ ℡ 0115 962 8400 eastmidlands@wea.org.uk
London 4 Luke Street, London EC2A 4XW
℡ 020 7426 1950 london@wea.org.uk
North East 1st Floor, Unit 6, Metro Riverside Park, Delta Bank Road, Gateshead NE11 9DJ
℡ 0191 461 8100 northeast@wea.org.uk
North West Suite 405–409, The Cotton Exchange, Old Hall Street (Bixteth Street Entrance), Liverpool L3 9JR ℡ 0151 243 5340 northwest@wea.org.uk
Southern Unit 57 Riverside 2, Sir Thomas Longley Road, Rochester ME2 4DP
℡ 01634 298600 southern@wea.org.uk
South West Bradninch Court, Castle Street, Exeter EX4 3PL ℡ 01392 490970 southwest@wea.org.uk
West Midlands 4th Floor, Lancaster House, 67 Newhall Street, Birmingham, B3 1NQ westmidlands@wea.org.uk
Yorkshire and Humber 6 Woodhouse Square, Leeds LS3 1AD ℡ 0113 245 3304 yorkshumber@wea.org.uk

Workers' Educational Association North Wales
▷ Coleg Harlech WEA

Workers' Educational Association Northern Ireland

1–3 Fitzwilliam Street, Belfast BT9 6AW
℡ 028 9032 9718   🖷 028 9023 0306
info@wea-ni.com
www.wea-ni.com

Autonomous from the WEA National Association, WEA Northern Ireland was founded in 1910.

Workers' Educational Association Scotland

Riddles Court, 322 Lawnmarket, Edinburgh EH1 2PG
℡ 0131 226 3456
hq@weascotland.org.uk
www.weascotland.org.uk
Scottish Secretary *Joyce Connon*

WEA Scotland was founded in 1905 and is part of the national Workers' Educational Association.

Workers' Educational Association South Wales

7 Coopers Yard, Curran Road, Cardiff CF10 5NB
℡ 029 2023 5277   🖷 029 2023 3986

weasw@swales.wea.org.uk
www.swales.wea.org.uk

Autonomous from the WEA National Association, WEA South Wales has been active in community education for over 100 years.

Writers in Oxford

www.writersinoxford.org
Membership Secretary *Parichehre Mosteshar-Gharai*

Chair *Frank Egerton*
Treasurer *John Geake*
SUBSCRIPTION £20 p.a.

Founded 1992. Open to published authors, playwrights, poets and journalists. Literary seminars and social functions. Publishes *The Oxford Writer* newsletter.

Yorkshire Playwrights
▷ Script Yorkshire

Bursaries, Fellowships and Grants

Arts Council of Wales ACW's Creative Wales Awards
▷ entry under Arts Councils and Regional Offices

The Authors' Contingency Fund
The Society of Authors, 84 Drayton Gardens, London SW10 9SB
📞 020 7373 6642   📠 020 7373 5768
info@societyofauthors.org
www.societyofauthors.org

This fund makes modest grants to published authors who find themselves in sudden financial difficulties. Contact *Sarah Baxter* at the **Society of Authors** for an information sheet and application form.

The Authors' Foundation
The Society of Authors, 84 Drayton Gardens, London SW10 9SB
📞 020 7373 6642   📠 020 7373 5768
info@societyofauthors.org
www.societyofauthors.org

Grants to novelists, poets and writers of non-fiction in need of funding (in addition to a proper advance) for research, travel or other necessary expenditure. Application by letter to The Authors' Foundation giving details, in confidence, of the advance and royalties, together with the reasons for needing additional funding. Grants are sometimes given even if there is no commitment by a publisher, so long as the applicant has had a book published and the new work will almost certainly be published. About £40,000 is distributed each year. Contact the **Society of Authors** for full entry details. Final entry dates: 30 April and 30 September.

The K. Blundell Trust
The Society of Authors, 84 Drayton Gardens, London SW10 9SB
📞 020 7373 6642   📠 020 7373 5768
info@societyofauthors.org
www.societyofauthors.org

Grants to writers to fund travel, important research or other expenditure. Author must be under 40, has to submit a copy of his/her previous book and the work must 'contribute to the greater understanding of existing social and economic organization'. Application by letter. Contact the **Society of Authors** for full entry details. Final entry dates: 30 April and 30 September.

Alfred Bradley Bursary Award
c/o BBC Radio Drama, Room 2130, New Broadcasting House, Oxford Road, Manchester M60 1SJ
📞 0161 244 4253   📠 0161 244 4248
Contact *Coordinator*

Established 1992. Biennial award in commemoration of the life and work of the distinguished radio producer Alfred Bradley. Aims to encourage and develop new radio writing talent in the BBC North region. There is a change of focus for each award, e.g. previous years have targeted comedy drama, verse drama, etc. Entrants must live in the north of England. The award is given to help writers to pursue a career in writing for radio. Next award launched summer 2008. (Further information can be found on bbc.co.uk/writersroom) Previous winners: Mark Shand, Anthony Cropper, Julia Copus, Michael Stewart, Ben Tagoe, Katie Douglas.
£ Award Up to £6,000 over two years; potential BBC Radio Drama commissions and opportunities for mentoring with producers.

British Academy Small Research Grants
10 Carlton House Terrace, London SW1Y 5AH
📞 020 7969 5217   📠 020 7969 5414
grants@britac.ac.uk
www.britac.ac.uk
Contact *Research Grants Department*

Award to further original academic research at postdoctoral level in the humanities and social sciences. Entrants must be resident in the UK. See website for final entry dates. Three competitions a year.
£ Award £7,500 (maximum).

Cholmondeley Awards
The Society of Authors, 84 Drayton Gardens, London SW10 9SB
📞 020 7373 6642   📠 020 7373 5768
info@societyofauthors.org
www.societyofauthors.org

Founded in 1965 by the late Dowager Marchioness of Cholmondeley. Annual honorary awards to recognize the achievement and distinction of individual poets. 2007 winners: Judith Kazantzis, Robert Nye, Penelope Shuttle.
- £ Awards £6,000 (total).

Olive Cook Prize
▷ Tom-Gallon Trust Award

The Economist/Richard Casement Internship
The Economist, 25 St James's Street, London SW1A 1HG
☎ 020 7830 7000
casement@economist.com
www.economist.com
Contact *Geoffrey Carr*, Science Editor (re. Casement Internship)

For an aspiring journalist to spend three months in the summer writing for *The Economist* about science and technology. Applicants should write a letter of introduction along with an article of approximately 600 words suitable for inclusion in the Science and Technology Section. 'Our aim is more to discover writing talent in a scientist or science student than scientific aptitude in a budding journalist.' Competition details normally announced in the magazine late January or early February and 4–5 weeks allowed for application.

European Jewish Publication Society
PO Box 19948, London N3 3ZJ
☎ 020 8346 1776   ℱ 020 8346 1776
colinshindler@gmel.com
www.ejps.org.uk
Contact *Dr Colin Shindler*

Established in 1995 to help fund the publication of books of European Jewish interest which would otherwise remain unpublished. Helps with the marketing, distribution and promotion of such books. Publishers who may be interested in publishing works of Jewish interest should approach the Society with a proposal and manuscript in the first instance. Books which have been supported recently include: *The History of Zionism* Walter Laqueur; *Photographing the Holocaust* Janina Struk; *The Arab-Israeli Cookbook* Robin Soans; *Daughters of Sarah: An Anthology of Jewish Women Writing in French* eds. Eva Martin Sartori and Madeleine Cottenet-Hage; *Whistleblowers and the Bomb: Vanunu, Israel and Nuclear Secrecy* Yoel Cohen. Also supports the publication of poetry, and translations from and into other European languages.
- £ Grant £3,000 (maximum).

Fulbright Awards
The Fulbright Commission, Fulbright House, 62 Doughty Street, London WC1N 2JZ
☎ 020 7404 6880   ℱ 020 7404 6834
programmes@fulbright.co.uk
www.fulbright.co.uk
Contact *Programme Coordinator*

The Fulbright Commission offers a number of scholarships given at postgraduate level and above, open to any field of study/research to be undertaken in the USA. Length of award is typically an academic year. Application deadline for postgraduate awards is usually early May of the preceding year of study; and mid-March/early April for postdoctoral awards. Full details and application forms are available on the Commission's website. Alternatively, send A4 envelope with sufficient postage for 100g with a covering letter explaining which level of award is of interest.

Tony Godwin Memorial Trust
c/o 38 Lyttelton Court, Lyttelton Road, London N2 0EB
☎ 020 8209 1613
info@tgmt.org.uk
www.tgmt.org.uk
Chairman *Iain Brown*

Biennial award established to commemorate the life of Tony Godwin, a prominent publisher in the 1960s/70s. Open to all young people (under 35 years old) who are UK nationals and working, or intending to work, in publishing. The award provides the recipient with the means to spend at least one month as the guest of an American publishing house in order to learn about international publishing. The recipient is expected to submit a report upon return to the UK. Next award: 2009; final entry date: 31 December 2008. Previous winners: George Lucas (Hodder), Clive Priddle (Fourth Estate), Richard Scrivener (Penguin), Lisa Shakespeare (Weidenfeld & Nicolson), Fiona Stewart (HarperCollins).
- £ Bursary of approx. US$5,000.

Eric Gregory Trust Fund
The Society of Authors, 84 Drayton Gardens, London SW10 9SB
☎ 020 7373 6642   ℱ 020 7373 5768
info@societyofauthors.org
www.societyofauthors.org

Annual awards of varying amounts are made for the encouragement of poets under the age of 30 on the basis of a submitted collection. Open only to British-born subjects resident in the UK. Final entry date: 31 October. Contact the **Society of Authors** for full entry details. 2007 winners: Rachel Curzon, Miriam Gamble, Michael McKimm, Helen Mort, Jack Underwood.
- £ Awards £20,000 (total).

The Guardian Research Fellowship

Nuffield College, Oxford OX1 1NF
☎ 01865 278542   ℻ 01865 278666
Contact *The Academic Administrator*

One-year fellowship endowed by the Scott Trust, owner of *The Guardian*, to give someone working in the media the chance to put their experience into a new perspective, publish the outcome and give a *Guardian* lecture. Applications welcomed from journalists and management members, in newspapers, periodicals or broadcasting. Research or study proposals should be directly related to experience of working in the media. Accommodation and meals in college will be provided and a stipend. Advertised biennially in November.

Hawthornden Literary Institute

Hawthornden International Retreat for Writers, Lasswade EH18 1EG
☎ 0131 440 2180   ℻ 0131 440 1989
Contact *The Administrator*

Established 1982 to provide a peaceful setting where published writers can work in silence. The Castle houses up to six writers at a time, who are known as Hawthornden Fellows. Writers from any part of the world may apply for the fellowships. No monetary assistance is given, nor any contribution to travelling expenses, but once arrived at Hawthornden, the writer is the guest of the Retreat. Applications on forms provided must be made by the end of November for the following calendar year. Previous winners include: Les Murray, Alasdair Gray, Helen Vendler, David Profumo, Hilary Spurling.

Francis Head Bequest

The Society of Authors, 84 Drayton Gardens, London SW10 9SB
☎ 020 7373 6642   ℻ 020 7373 5768
info@societyofauthors.org
www.societyofauthors.org

Provides grants to published British authors over the age of 35 who need financial help during a period of illness, disablement or temporary financial crisis. Contact *Sarah Baxter* at the **Society of Authors** for an information sheet and application form.

Hosking Houses Trust

33 The Square, Clifford Chambers, Stratford-upon-Avon CV37 8HT
sarahhosking@btinternet.com
www.hoskinghouses.co.uk
Contact *The Secretary*

Established in 2001 to offer women writers over the age of 40 a period of financially-protected, domestic peace. Residence of Church Cottage two miles from Stratford-upon-Avon; furnished and equipped; offered for between two months and one year, all bills paid (except BT) also a personal bursary. No duties except to complete personal work, report to the Trustees every three months, to acknowledge the Trust in all work produced during tenure or published as a result of it. Advertised as available: appointed on merit and need to an established, published woman writer on any subject, and who has the legal right to be in the UK.

Jerwood Awards

▷ The Royal Society of Literature/Jerwood Awards

Journalists' Charity

Dickens House, 35 Wathen Road, Dorking RH4 1JY
☎ 01306 887511   ℻ 01306 888212
enquiries@journalistscharity.org.uk
www.journalistscharity.org.uk
Director/Secretary *David Ilott*

Aims to relieve distress among journalists and their dependants. Continuous and/or occasional financial grants; also retirement homes for eligible beneficiaries. Further information available from the Director.

Ralph Lewis Award

University of Sussex Library, Brighton BN1 9QL
☎ 01273 678158   ℻ 01273 873413
a.timoney@sussex.ac.uk
01273 678158

Established 1985. Occasional award set up by Ralph Lewis, a Brighton author and art collector who left money to fund awards for promising manuscripts which would not otherwise be published. The award is given in the form of a grant to a UK-based publisher in respect of a publication of literary works by new authors. No direct applications from writers. Previous winners: **Peterloo Poets; Serpent's Tail**.

The Elizabeth Longford Grants

The Society of Authors, 84 Drayton Gardens, London SW10 9SB
☎ 020 7373 6642   ℻ 020 7373 5768
info@societyofauthors.org
www.societyofauthors.org

Biannual grant, sponsored by Flora Fraser and Peter Soros, to a historical biographer for essential research, travel or other necessary expenditure. Contact the **Society of Authors** for full details. Final entry dates: 30 April and 30 September.
£ Grant £2,500.

The John Masefield Memorial Trust

The Society of Authors, 84 Drayton Gardens, London SW10 9SB
☎ 020 7373 6642   ℻ 020 7373 5768

info@societyofauthors.org
www.societyofauthors.org

This trust makes occasional grants to professional poets (or their immediate dependants) who are faced with sudden financial problems. Contact *Sarah Baxter* at the **Society of Authors** for an information sheet and application form.

The Somerset Maugham Awards

The Society of Authors, 84 Drayton Gardens, London SW10 9SB
☎ 020 7373 6642   📠 020 7373 5768
info@societyofauthors.org
www.societyofauthors.org

Annual awards designed to encourage writers under the age of 35 to travel. Given on the basis of a published work of fiction, non-fiction or poetry. Open only to British-born subjects resident in the UK. Final entry date: 20 December. 2007 winners: Horatio Clare *Running for the Hills* and James Scudamore *The Amnesia Clinic*.
💷 Awards £10,000 (total).

The Airey Neave Trust

PO Box 36800, 40 Bernard Street, London WC1N 1WJ
☎ 020 7833 4440
hanthoc@aol.com
www.theaireyneavetrust.org.uk
Contact *Hannah Scott*

Initiated 1989. Annual research fellowships for up to three years – towards a book or paper – for serious research connected with national and international law, and human freedom. Preferably attached to a particular university in Britain.

North East Literary Fellowship

Arts Council England, North East, Central Square, Forth Street, Newcastle upon Tyne NE1 3PJ
☎ 0191 255 8542   📠 0191 230 1020
www.artscouncil.org.uk
Contact *Literature Officer*

A competitive fellowship in association with the Universities of Durham and Newcastle upon Tyne. Contact **Arts Council England, North East**, for details.

Northern Writers' Awards
▷ New Writing North under Professional Associations and Societies

PAWS (Public Awareness of Science) Drama Fund

The PAWS Office, OMNI Communications, First Floor, 155 Regents Park Road, London NW1 8BB
pawsomni@btconnect.com
www.pawsdrama.com
Contacts *Andrew Millington, Andree Molyneux*

Established 1994. PAWS encourages and supports television drama drawing on science, engineering and technology. It offers a range of activities to bring science and technology to TV drama writers and producers in an easily accessible way. Events: covering the latest science and technology topics and enabling writers to meet scientists and engineers; Contacts & Information Service: aims to put writers and producers in 'one-to-one' contact with specialists who can help them develop their TV drama ideas; New Funding Scheme: rolling funding support primarily intended to help writers and producers who have a recent track record of successful TV drama experience to bring their ideas closer to production. People with lesser experience may also apply. Contact the PAWS Office or check out the website for further information.

Pearson Playwrights' Scheme

c/o Pearson Plc, 80 Strand, London WC2R ORL
playwrightscheme@tiscali.co.uk
www.pearson.com
Administrator *Jack Andrews, MBE*

Awards up to five bursaries to playwrights annually, each worth £7,000. Applicants must be sponsored by a theatre which then submits the play for consideration by a panel. Each award allows the playwright a 12-month attachment. Applications invited via theatres in October each year. A further award of £10,000, for Best Play, is on offer each year to previous bursary winners. E-mail for up-to-date information.

The Charles Pick Fellowship

School of Literature and Creative Writing, University of East Anglia, Norwich NR4 7TJ
☎ 01603 592286   📠 01603 507728
charlespickfellowship@uea.ac.uk
www1.uea.ac.uk/cm/home/schools/hum/lit/awards/pick
Fellowship Administrator *Shawn Alexander*

Founded in 2001 by the Charles Pick Consultancy in memory of the publisher and literary agent who died in 2000, to support a new unpublished writer of fictional or non-fictional prose. This residential Fellowship is awarded annually and lasts six months. There is no interview; candidates will be judged on the quality and promise of their writing, the project they describe and the strength of their referee's report. 2007 winner: Erin Soros. Deadline for applications: 31 January each year.
💷 Award £10,000 plus accommodation on campus.

Peggy Ramsay Foundation

Hanover House, 14 Hanover Square, London W1S 1HP
☎ 020 7667 5000   📠 020 7667 5100

laurence.harbottle@harbottle.com
www.peggyramsayfoundation.org
Contact *G. Laurence Harbottle*

Founded 1992 in accordance with the will of the late Peggy Ramsay, the well-known agent. Grants are made to writers for the stage who have some experience and who need time and resources to make writing possible. Grants are also made for writing projects by organizations connected with the theatre. The Foundation does not support production costs or any project that does not have a direct benefit to playwriting. Writers applying must have had one full-length play professionally produced for adults.

Ⓔ Grants total £150,000 to £200,000 per year.

The Royal Literary Fund

3 Johnson's Court, off Fleet Street, London EC4A 3EA
Ⓣ 020 7353 7150 Ⓕ 020 7353 1350
egunnrlf@globalnet.co.uk
www.rlf.org.uk
Secretary *Eileen Gunn*

Grants and pensions are awarded to published authors of several works in financial need, or to their dependants. Examples of author's works are needed for assessment by Committee. Contact the Secretary for further details and application form.

The Royal Society of Literature/Jerwood Awards

The Royal Society of Literature, Somerset House, Strand, London WC2R 1LA
Ⓣ 020 7845 4676 Ⓕ 020 7845 4679
paulaj@rslit.org
www.rslit.org
Submissions *Paula Johnson*

The awards offer financial assistance to authors engaged in writing their first major commissioned works of non-fiction. Three awards: one of £10,000 and two of £5,000 will be offered annually to writers working on substantial non-fiction projects. Open to UK and Irish writers and writers who have been resident in the UK for at least three years. Applications for the awards should be submitted by mid-October. Further details available on the website.

Scottish Arts Council Creative Scotland Awards

Scottish Arts Council, 12 Manor Place, Edinburgh EH3 7DD
Ⓣ 0131 226 6051/ Help Desk: 0845 603 6000
Ⓕ 0131 225 9833
help.desk@scottisharts.org.uk
www.scottisharts.org.uk

The Scottish Arts Council is currently reviewing the scope and opportunities offered by the Creative Scotland Awards in their present format in order to build on their existing success and to widen the range of potential applicants, as well as developing stronger links with successful applicants. It is anticipated that the Awards will be re-launched during 2008. Further information will be available on the website and through SAC's 'Information Bulletin'.

Scottish Arts Council New Writers' Bursaries
▷ Scottish Book Trust New Writers' Awards

Scottish Arts Council Writers' Bursaries

Scottish Arts Council, 12 Manor Place, Edinburgh EH3 7DD
Ⓣ 0131 226 6051 Ⓕ 0131 225 9833
gavin.wallace@scottisharts.org.uk
www.scottisharts.org.uk
Head of Literature *Dr Gavin Wallace*

Bursaries to enable published writers of literary work, playwrights and storytellers to devote more time to their writing. Around 20 bursaries of up to £15,000 awarded annually; deadline for applications in July and January. Application open to writers based in Scotland.

Scottish Book Trust Mentoring Scheme

Sandeman House, Trunk's Close, 55 High Street, Edinburgh EH1 1SR
Ⓣ 0131 524 0160 Ⓕ 0131 524 0161
www.scottishbooktrust.com
Contact *Caitrin Armstrong* (Writer Development Coordinator)

The mentoring scheme aims to support published writers living and working in Scotland who have a specific project on which they'd like some focused, dedicated support and advice. Successful applicants will be paired with another writer or industry professional who has appropriate experience and will support them as they work together over an intensive period of nine months. Applicants should see the website for full details.

Scottish Book Trust New Writers' Awards

Sandeman House, Trunk's Close, 55 High Street, Edinburgh EH1 1SR
Ⓣ 0131 524 0160 Ⓕ 0131 524 0161
www.scottishbooktrust.com
Contact *Caitrin Armstrong* (Writer Development Coordinator)

(Formerly the Scottish Arts Council New Writers' Bursaries.) Eight awards available for writers living and working in Scotland who have little or no previous publication or production track record. Successful applicants will receive a cash award of £2,000 plus a bespoke professional development package which may include nine months working with a mentor, plus networking opportunities with publishers/agents. The award is open to writers of fiction, literary non-fiction, poets and playwrights who can demonstrate their commitment to writing. A c.v., sample of work and referee will be required. Applicants must see website for full application details.

Laurence Stern Fellowship

Department of Journalism, City University, Northampton Square, London EC1V OHB
① 020 7040 4036
a.r.mckane@city.ac.uk
www.city.ac.uk/journalism
Contact *Anna McKane*

Founded 1980. Awarded each year to a young journalist experienced enough to work on national stories. It gives a chance to work on the national desk of the *Washington Post*. A senior executive from the newspaper selects from a shortlist drawn up in February/March. 2008 winner: Holly Watt of *The Sunday Times*. Full details about how to apply available on the City University website in January.

Tom-Gallon Trust Award and the Olive Cook Prize

The Society of Authors, 84 Drayton Gardens, London SW10 9SB
① 020 7373 6642   ⑤ 020 7373 5768
info@societyofauthors.org
www.societyofauthors.org

An award of £1,000 is made on the basis of a submitted story to fiction writers of limited means who have had at least one short story accepted for publication. Both awards are biennial and are awarded in alternate years. Contact the **Society of Authors** for an entry form. Final entry date: 31 October.

The Betty Trask Awards

The Society of Authors, 84 Drayton Gardens, London SW10 9SB
① 020 7373 6642   ⑤ 020 7373 5768
info@societyofauthors.org
www.societyofauthors.org

These annual awards are for authors who are under 35 and Commonwealth citizens, awarded on the strength of a first novel (published or unpublished) of a traditional or romantic nature. The awards must be used for a period or periods of foreign travel. Final entry date: 31 January. Contact the **Society of Authors** for an entry form. 2007 winners: Will Davis *My Side of the Story*; Adam Foulds *The Truth About These Strange Times*; Cynan Jones *The Long Dry*; Julie Maxwell *You Can Live Forever*; Karen McLeod *In Search of the Missing Eyelash*.
€ Awards £20,000 (total).

The Travelling Scholarships

The Society of Authors, 84 Drayton Gardens, London SW10 9SB
① 020 7373 6642   ⑤ 020 7373 5768
info@societyofauthors.org
www.societyofauthors.org

Annual honorary grants to established British writers. No submissions. 2007 winners: Naomi Alderman, Susan Elderkin, Philip Marsden.
€ Award £1,800 each.

The David T.K. Wong Fellowship

School of Literature and Creative Writing, University of East Anglia, Norwich NR4 7TJ
① 01603 592286   ⑤ 01603 507728
davidtkwongfellowship@uea.ac.uk
www1.uea.ac.uk/cm/home/schools/hum/lit/awards/wong
Fellowship Administrator *Shawn Alexander*

Named for its sponsor, Mr. David Wong (retired Hong Kong businessman, teacher, journalist, senior civil servant and writer of short stories); this unique and generous award was launched in 1997 to enable a fiction writer who wants to write in English about the Far East. This residential Fellowship is awarded annually and lasts one academic year. Unpublished or established published writers are welcome to apply. There is no interview; candidates will be judged on the quality and promise of their writing and the project they describe. 2007 winner: Balli Jaswal. Deadline for applications: 31 January each year.
€ Award £26,000.

Prizes

Academi Book of the Year Awards

Academi, 3rd Floor, Mount Stuart House, Mount Stuart Square, Cardiff CF10 5FQ
℡ 029 2047 2266   🖷 029 2047 0691
post@academi.org
www.academi.org
Contacts *Peter Finch* (Chief Executive), *Lleucu Siencyn* (Deputy)

Annual prizes awarded for works of exceptional literary merit written by Welsh authors (by birth or residence), published in Welsh or English during the previous calendar year. There is one major prize in English, the Welsh Book of the Year Award, and one major prize in Welsh, Gwobr Llyfr y Flwyddyn.

💷 Prizes of £5,000 (1st); £500 (2nd); £250 (3rd); five runners-up prizes of £50 each.

Academi Cardiff International Poetry Competition

Academi, PO Box 438, Cardiff CF10 5YA
℡ 029 2047 2266   🖷 029 2049 2930
post@academi.org
www.academi.org
Contact *Peter Finch (Chief Executive)*

Established 1986. An annual competition supported by Cardiff Council for unpublished poems in English of up to 50 lines. Closing date in January each year.

💷 Prizes total £6,000; £5,000 (1st); £500 (2nd); £250 (3rd); five runners-up receive £50 each.

J.R. Ackerley Prize

English PEN, 6–8 Amwell Street, London EC1R 1UQ
℡ 020 7713 0023   🖷 020 7837 7838
enquiries@englishpen.org
www.englishpen.org

Commemorating the novelist/autobiographer J.R. Ackerley, this prize is awarded for a literary autobiography, written in English and published in the year preceding the award. Entry restricted to nominations from the Ackerley Trustees only ('please do not submit books'). 2007 winner: Brian Thompson *Keeping Mum*.

💷 Prize of £1,000, plus silver pen.

Aldeburgh Poetry Festival Prize
▷ Jerwood Aldeburgh First Collection Prize

Alexander Prize

Royal Historical Society, University College London, Gower Street, London WC1E 6BT
℡ 020 7387 7532   🖷 020 7387 7532
royalhistsoc@ucl.ac.uk
www.royalhistoricalsociety.org/grants.htm
Contact *Administrative Secretary*

Established 1897. Awarded for a published journal article or essay based upon original research. The article/essay must have been published during the period 1 January 2006 to 31 December 2007. Competitors may choose their own subject for the essay. Closing date: 31 December.

💷 Prize of £250.

Hans Christian Andersen Awards

IBBY, Nonnenweg 12, Postfach, CH-4003 Basel, Switzerland
℡ 00 41 61 272 2917   🖷 00 41 61 272 2757
ibby@ibby.org
www.ibby.org
Deputy Director of Administration *Forest Zhang*
Director, Member Services, Communications & New Projects *Liz Page*

The highest international prizes for children's literature: The Hans Christian Andersen Award for Writing, established 1956; The Hans Christian Andersen Award for Illustration, established 1966. Candidates are nominated by National Sections of IBBY (The International Board on Books for Young People). Biennial prizes are awarded, in even-numbered years, to an author and an illustrator whose body of work has made a lasting contribution to children's literature. 2008 winners: Award for Writing: Jürg Schubiger (Switzerland); Award for Illustration: Roberto Innocenti (Italy).

💷 Awards: gold medals.

Angus Book Award

Angus Council Cultural Services, County Buildings, Forfar DD8 3WF
℡ 01307 461460   🖷 01307 462590

cultural.services@angus.gov.uk
www.angus.gov.uk/bookaward
Contact *Moyra Hood,* Educational Resources
Librarian

Established 1995. Designed to try to help
teenagers develop an interest in and enthusiasm
for reading. Eligible books are read and voted on
by third-year schoolchildren in all eight Angus
secondary schools. 2008 winner: Kate Cann
Leaving Poppy.

£ Prize of £500 cheque, plus trophy in the form
of a replica Pictish stone.

Annual Theatre Book Prize
▷ The Society for Theatre Research Annual Theatre Book
Prize

Arts Council Children's Award
▷ Brian Way Award

Arvon Foundation International Poetry Competition

2nd Floor, 42a Buckingham Palace Road, London
SW1W ORE
☏ 020 7931 7611 📠 020 7963 0961
poetry@arvonfoundation.org
www.arvonfoundation.org
Contact *The London Office*

Established 1980. Biennial competition (next
in 2010) for poems written in English and not
previously broadcast or published. There are no
restrictions on the number of lines, themes, age of
entrants or nationality. No limit to the number of
entries. Entry fee: £7 per poem. Previous winners:
Paul Farley *Laws of Gravity*; Don Paterson *A
Private Bottling.*

£ Prize of £5,000 (1st), and £5,000 worth of other
prizes.

Asham Award for Women

Asham Literary Endownment Trust, c/o Town
Hall, High Street, Lewes BN7 2QS
☏ 01273 483159
www.ashamaward.com
Contact *Carole Buchan*

Founded in 1996, the award (named after the
house in Sussex where Virginia Woolf lived) is
a biennial short story competition for women
and is administered by the Asham Literary
Endownment Trust which encourages new
writing through competitions, mentoring,
training and publication. Candidates must be
female, over 18, resident in the UK and must not
have had a novel or complete anthology of work
previously published. Next award (sponsored by
the John. S. Cohen Foundation and Much Ado
Books of Alfriston, East Sussex) was launched in
June 2008; details available on the website.

£ Prize money up to £3,600; 12 winning writers
are published by Bloomsbury alongside specially
commissioned stories by leading professional
writers.

Authors' Club First Novel Award

Authors' Club, 40 Dover Street, London W1S 4NP
☏ 020 7499 8581 📠 020 7409 0913
stella@theauthorsclub.co.uk
Contact *Stella Kane*

Established 1954. This award is made for the most
promising work published in Britain by a British
author, and is presented at a dinner held at The
Arts Club. Entries for the award are accepted
from publishers from September of the year in
question and must be full-length (short stories are
not eligible). For further details/application form,
contact *Stella Kane* at the e-mail address above
or access the website. 2007 winner: Segun Afolabi
Goodbye Lucille.

£ Award of £1,000.

The BA/Nielson BookData Author of the Year Award

Booksellers Association Ltd, 272 Vauxhall Bridge
Road, London SW1V 1BA
☏ 020 7802 0801 📠 020 7802 0803
naomi.gane@booksellers.org.uk
www.booksellers.org.uk
Contact *Naomi Gane*

Founded as part of the BA Annual Conference to
involve authors more closely in the event. Authors
must be British or Irish. Not an award open to
entry but voted on by the BA's membership. 2007
winner: Anthony Horowitz.

£ Award of £1,000, plus trophy.

Banipal Prize
▷ The Translators Association Awards

Verity Bargate Award

Soho Theatre, 21 Dean Street, London W1D 3NE
☏ 020 7287 5060 📠 020 7287 5061
www.sohotheatre.com

The award was set up to commemorate the late
Verity Bargate, co-founder and director of the
Soho Theatre. This national award is presented
biennially for a new and unperformed play (next
submission deadline: July 2009). Access details
on the website. Previous winners include: Matt
Charman, Shan Khan, Fraser Grace, Lyndon
Morgans, Adrian Pagan, Diane Samuels, Judy
Upton and Toby Whithouse.

The BBC National Short Story Prize

Room 316, BBC Henry Wood House, 3–6
Langham Place, London W1B 3DF
www.theshortstory.org.uk
www.bbc.co.uk/radio4

Launched at the 2005 Edinburgh International Book Festival, this annual short story award – the largest for a single story – is funded by the BBC and is administered in partnership with the **Booktrust** and **Scottish Book Trust**. It was formerly known as the National Short Story Prize. Open to authors with a previous record of publication who are UK nationals or residents, aged 18 years or over. Entries can be stories published during 2007 or previously unpublished and must not be more than 8,000 words. Entrants must have a prior record of publication and submit original work that does not infringe the copyright or any other rights of any third party. The stories must be written in English and only one will be accepted per author. Full entry details from the address above or via the website. 2007 winner: Julian Gough *The Orphan & The Mob*.

£ Prizes of £15,000 (overall); £3,000 (runner-up).

BBC Wildlife Magazine Nature Writing Awards

BBC Wildlife Magazine, 14th Floor, Tower House, Fairfax Street, Bristol BS1 3BN
☎ 0117 927 9009   🖷 0117 933 8032
jamesfair@bbcmagazinesbristol.com
www.bbcwildlifemagazine.com
Contact *James Fair*

This competition aims to revive the art of nature writing. Entrants should submit an original 800-word essay on any aspect of nature that interests them. The article must not have been previously published either in print or on the internet. See the March issue and website for entry details and entry form.

£ Award of an Earthwatch expedition and publication in the magazine.

BBC Wildlife Magazine Poet of the Year Awards

BBC Wildlife Magazine, 14th Floor, Tower House, Fairfax Street, Bristol BS1 3BN
☎ 0117 927 9009   🖷 0117 933 8032
wildlifemagazine@bbcmagazinesbristol.com
www.bbcwildlifemagazine.com
Contact *Sophie Stafford*

Annual award for a poem, the subject of which must be the natural world and/or our relationship with it. Entrants may submit one poem only of no more than 50 lines with the entry form which is published in the April issue. Closing date for entries varies from year to year. See the magazine and website for entry details.

£ Poet of the Year Award: a wildlife break, publication in the magazine, plus the possibility of a reading of your poem on Radio 4's *Poetry Please*; runners-up: books and publication in the

magazine; four young poets awards, categories: age 7 and under, 8–11, 12–14, 15–17.

BBC Wildlife Magazine Travel Writing Award

BBC Wildlife Magazine, Bristol Magazines Ltd, 14th Floor, Tower House, Fairfax Street, Bristol BS1 3BN
☎ 0117 314 8375   🖷 0117 934 9008
jamesfair@bbcmagazinesbristol.com
www.bbcwildlifemagazine.com

Awarded to a travel essay that is a true account involving an intimate encounter with wildlife, either local or exotic. The essay should convey a strong impression of the environment and incorporate the idea of travel and discovery. Maximum 800 words. See website and the magazine for entry details. Call for entries appears in the January issue of the magazine.

£ Award winner receives publication in the magazine, plus 'holiday of a lifetime'.

The BBCFour Samuel Johnson Prize for Non-Fiction

Colman Getty, Middlesex House, 28 Windmill Street, London W1T 2JJ
☎ 020 7631 2666   🖷 020 7631 2699
lois@colmangetty.co.uk
Contact *Lois Tucker*

Established 1998. Annual prize sponsored by BBCFour. Eligible categories include the arts, autobiography, biography, business, commerce, current affairs, history, natural history, popular science, religion, sport and travel. Entries submitted by publishers only. 2007 winner: Rajiv Chandrasekaran *Imperial Life in the Emerald City*.

£ Prize of £30,000; £1,000 to each shortlisted author.

David Berry Prize

Royal Historical Society, University College London, Gower Street, London WC1E 6BT
☎ 020 7387 7532   🖷 020 7387 7532
royalhistsoc@ucl.ac.uk
www.royalhistoricalsociety.org/grants.htm
Contact *Administrative Secretary*

Annual award for an essay of not more than 10,000 words on Scottish history. Closing date: 31 October.

£ Prize of £250,

Besterman/McColvin Medal

43 Yardington, Whitchurch SY13 1BL
☎ 01948 663570
millbrookend-awards@yahoo.co.uk
www.cilip.org.uk
Contact *Awards Administrator* (at address above)

ISG (CILIP) Reference Awards. The Besterman McColvin medals are awarded annually for an outstanding reference work first published in the UK during the preceding year. Consists of two categories: printed and electronic. Works eligible for consideration include: encyclopedias, general and special dictionaries; annuals, yearbooks and directories; handbooks and compendia of data; atlases. Nominations are invited from members of **CILIP**, publishers and others. 2006 winners: *Oxford African American Studies Centre*, published by Oxford University Press (electronic category); *Dictionary of Pastellists Before 1800* Neil Jeffares (printed category).

£ Medal and cash prize for each category.

The Biographers' Club Prize

119A Fordwych Road, London NW2 3NJ
☎ 020 8452 4993
anna@annaswan.co.uk
www.biographersclub.co.uk
Contact *Anna Swan*

Established 1999 by literary agent, biographer and founder of the Biographers' Club, Andrew Lownie, to finance and encourage first-time writers researching a biography. Sponsored by the *Daily Mail*. Open to previously unpublished writers, applicants should submit a proposal of no more than 20 pages, including outline, sample chapter, a note on the market for the book, sources consulted, competing books and a c.v. Entry fee: £10 (cheques made payable to the Biographers' Club). 2007 winner: Clare Mulley.

£ Prize of £2,000.

Birdwatch Bird Book of the Year

c/o Birdwatch Magazine, B403A The Chocolate Factory, 5 Clarendon Road, London N22 6XJ
☎ 020 8881 0550
editorial@birdwatch.co.uk
www.birdwatch.co.uk
Contact *Dominic Mitchell*

Established in 1992 to acknowledge excellence in ornithological publishing – an increasingly large market with a high turnover. Annual award. Entries, from publishers, must offer an original and comprehensive treatment of their particular ornithological subject matter and must have a broad appeal to British-based readers. 2007 winner: Hadoram Shirihai *The Birds of Israel*.

Bisto Book of the Year Awards
▷ CBI Bisto Book of the Year Awards

James Tait Black Memorial Prizes

University of Edinburgh, David Hume Tower, George Square, Edinburgh EH8 9JX
☎ 0131 650 3619   ☎ 0131 650 6898
www.englit.ed.ac.uk/jtbinf.htm

Contact *Department of English Literature*

Established 1918 in memory of a partner of the publishing firm of **A.&C. Black Ltd**. Two prizes, one for biography and one for fiction. Closing date for submissions: 1 December. Each prize is awarded for a book first published in Britain in the previous calendar year (1 January–31 December). 2006 winners: Cormac McCarthy *The Road* (fiction); Byron Rogers *The Man Who Went into the West: The Life of R.S. Thomas* (biography).

£ Prizes of £10,000 each.

Blue Peter Children's Book Awards

Booktrust, Book House, 45 East Hill, London SW18 2QZ
☎ 020 8516 2972/8516 2960
www.bbc.co.uk/bluepeter
Contacts *Tarryn McKay, Helen Hayes*

Judged by a panel of adults and children. The three categories are: Book I Couldn't Put Down, Most Fun Story with Pictures and Best Book with Facts. Books must be published in paperback between 1 November and 31 October the following year. Winners are announced in the spring in a special broadcast on Blue Peter. 2007 winners: Book I Couldn't Put Down and winner of Blue Peter Book of the Year 2007: *The Outlaw Varjak Paw* by S.F. Said, illustrated by Dave McKean; Most Fun Story with Pictures: *You're a bad man, Mr Gum* by Andy Stanton, illustrated by David Tazzyman; Best Book with Facts: *Worst Children's Jobs in History* by Tony Robinson.

Boardman Tasker Award

Pound House, Llangennith, Swansea SA3 1JQ
☎ 01792 386215   ☎ 01792 386215
margaretbody@lineone.net
www.boardmantasker.com
Contact *Maggie Body*, Honorary Secretary

Established 1983, this award is given for a work of fiction, non-fiction or poetry, whose central theme is concerned with the mountain environment and which can be said to have made an outstanding contribution to mountain literature. Authors of any nationality are eligible, but the book must have been published or distributed in the UK for the first time between 1 November 2006 and 31 October 2007. Entries from publishers only. 2007 winner: Robert Macfarlane *The Wild Places*.

£ Prize of £3,000 (at Trustees' discretion).

Bollinger Everyman Wodehouse Prize

Colman Getty, 28 Windmill Street, London W1T 2JJ
☎ 020 7631 2666
katy@colmangetty.co.uk
Contact *Katy Macmillan-Scott*

Annual prize, launched in 2000 on the 25th anniversary of the death of P.G. Wodehouse. Presented at the **Guardian Hay Festival**, awarded to the best comic novel published in the preceding twelve months. 2008 winner: Will Self *The Butt*.

The Booker Prize for Fiction
▷ The Man Booker Prize for Fiction

Booktrust Early Years Awards

Booktrust, Book House, 45 East Hill, London SW18 2QZ
☎ 020 8516 2972   📠 020 8516 2978
tarryn@booktrust.org.uk
www.booktrust.org.uk
Contact *Tarryn McKay*

Formerly the Sainsbury's Baby Book Award, established in 1999. Annual awards with three categories: the Baby Book Award, the Best Book for Pre-School Children (up to the age of 5), and an award for the Best Emerging Illustrator. Books to be submitted by publishers only; authors and illustrators must be of British nationality or other nationals who have been resident in the UK for at least five years. 2007 winners: Jess Stockham (Baby Book); Polly Dunbar (Pre-School); Emily Gravett (Best Emerging Illustrator).

💷 Prizes of £2,000 for each category; in addition the Best Emerging Illustrator receives a specially commissioned piece of artwork.

Booktrust Teenage Prize

Book House, 45 East Hill, London SW18 2QZ
☎ 020 8875 4821   📠 020 8516 2978
megan@booktrust.org.uk
www.bookheads.org.uk
Contact *Megan Farr*

Established 2003. Annual prize that recognizes and celebrates the best in teenage fiction. Funded and administered by Booktrust. Open to works of fiction for young adults in the UK, the books to be published between 1 July 2006 and 30 June 2007. Final entry date in March. 2007 winner: Marcus Sedgwick *My Swordhand is Singing*.

💷 Prize of £2,500.

Harry Bowling Prize

c/o 16 St Briac Way, Exmouth EX8 5RN
☎ 01395 279659
marina@marina-oliver.net
www.harrybowlingprize.net
Contacts *Margaret James, Marina Oliver*

Established 2000 in honour of Harry Bowling, 'the king of Cockney sagas' (died 1999). Biennial award (next in 2010), sponsored by **Headline Book Publishing,** to encourage writers of adult fiction set in London. Open to anyone who has not been published previously. Final entry date:

31 March. Entry fee charged; forms from address above (enclose s.a.e.) or via the website. 2006 winner: Jean Fullerton.

💷 Prize winner receives £1,000; runners-up, £100.

The Branford Boase Award

8 Bolderwood Close, Bishopstoke, Eastleigh SO50 8PG
☎ 01962 826658   📠 01962 856615
anne.marley@tiscali.co.uk
www.branfordboaseaward.org.uk
Administrator *Anne Marley*

Established in 2000 in memory of children's novelist, Henrietta Branford and editor and publisher, Wendy Boase. To be awarded annually to encourage and celebrate the most promising novel by a new writer of children's books, while at the same time highlighting the importance of the editor in nurturing new talent. 2007 winner: Siobhan Dowd *Swift Pure Cry* (author); David Fickling, **David Fickling Books**(editor).

💷 Award of a specially commissioned box, carved and inlaid in silver with the Branford Boase Award logo; winning author receives £1,000.

The Bridport Prize

PO Box 6910, Bridport DT6 9BQ
☎ 01308 428333
frances@bridportprize.org.uk
www.bridportprize.org.uk
Contact *Frances Everitt*, Administrator

Annual competition for poetry and short story writing. Unpublished work only, written in English. Winning stories are read by a literary agent, the winning poems are put forward to the **Forward Prize**, and an anthology of winning entries is published. Final entry date: 30 June. Send s.a.e. for entry forms.

💷 Prizes of £5,000, £1,000 & £500 in each category, plus 10 supplementary prizes of £50 each. Winning stories are submitted to the **National Short Story Prize**.

Katharine Briggs Folklore Award

The Folklore Society, c/o The Warburg Institute, Woburn Square, London WC1H 0AB
☎ 020 7862 8564
enquiries@folklore-society.com
www.folklore-society.com
Contact *The Convenor*

Established 1982. An annual award in November for the book, published in Britain and Ireland between 1 June in the previous calendar year and 30 May, which has made the most distinguished non-fiction contribution to folklore studies. Intended to encourage serious research in the field which Katharine Briggs did so much to establish. The term folklore studies is interpreted broadly to include all aspects of traditional and popular culture, narrative, belief, custom and folk arts. 2007 winner: Jack Zipes *Why Fairy Tales Stick*.

£ Prize of £200, plus engraved goblet.

Bristol Festival of Ideas Best Book of Ideas Prize

Bristol Cultural Development, Leigh Court, Abbots Leigh, Bristol BS8 3RA
☎ 07778 932778
ideas@gwebusinesswest.co.uk
www.ideasfestival.co.uk
Contact *Vicky Washington*

This annual award was established to mark the fifth anniversary of the Bristol Festival of Ideas in 2009 and was founded by Andrew Kelly, Director of the Festival, in association with Blackwell and **Arts & Business**. The award is given to the book which is considered to present new, important and challenging ideas, which is rigorously argued, and which is engaging and accessible. Submitted books must be first published in the year prior to each award (2008 for the 2009 award). Final entry date: end October (for books published in November and December proofs and manuscripts can be provided). Submissions by publishers only.

£ Prize of £10,000.

British Association for Applied Linguistics (BAAL) Book Prize

BAAL Publications Secretary, Department of Linguistics and English Language, Lancaster University, Lancaster LA1 4YT
☎ 01524 594642   📠 01524 843085
v.koller@lancaster.ac.uk
www.baal.org.uk
Contact *Veronika Koller*

Annual award made by the British Association for Applied Linguistics to an outstanding book in the field of applied linguistics. Final entry in mid-December . Nominations from publishers only. 2007 winner: Ben Rampton *Language in Late Modernity*.

British Book Awards
▷ Galaxy British Book Awards

British Czech & Slovak Association Prize

The BCSA Prize Administrator, 24 Ferndale, Tunbridge Wells TN2 3NS
☎ 01892 543206
prize@bcsa.co.uk
www.bcsa.co.uk
Contact *Prize Administrator*

The British Czech & Slovak Association offers an annual prize for the best piece of original writing, in English, on the links between Britain and the Czech and Slovak Republics, or describing society in transition in those Republics since the Velvet Revolution in 1989. Entries can be fiction or factual, should not have been previously published and should be up to 2,000 words in length. Submissions are invited from individuals of any age, nationality or educational background. Closing date: 30 June. Entry details for 2008 should be checked with the Prize Administrator. 2007 winners: James Gault *Old Honza's Day Out*; Bethany Croy *Preserving National Identity among Czech Jewish Children in Britain during World War II*.

£ Prize of £300, presented at the BCSA's annual dinner in London; 2nd prize, £100. The winning entry is published in the *British Czech & Slovak Review*.

British Fantasy Awards

c/o 5 Greenbank, Barnt Green, Birmingham B45 8DH
bfsawards@britishfantasysociety.org
www.britishfantasysociety.org
Awards Administrator *David Sutton*

Awarded by the **British Fantasy Society** by members at its annual conference for Best Novel and Best Short Story categories, among others. Not an open competition. Previous winners include: Ramsey Campbell, Dan Simmons, Michael Marshall Smith, Thomas Ligotti.

British Press Awards

Press Gazette, Wilmington Business Information, Paulton House, 8 Shepherdess Walk, London N1 7LB
☎ 020 7324 2382   📠 020 7566 5769
apont@wilmington.co.uk
www.pressgazette.co.uk

'The Oscars of British journalism.' Open to all British national newspapers and news agencies. April event. Run by *Press Gazette*.

British Science Fiction Association Awards

awards@bsfa.co.uk
www.bsfa.co.uk
Awards Administrator *Donna Scott*

Established 1970. Categories for novel, short fiction, non-fiction and artwork. The awards are announced and presented at the British National Science Fiction Convention every Easter. 2007 winners: Ian McDonald *Brasyl* (novel); Ken MacLeod *Lighting Out* (short fiction); Andy Bigwood *Cracked World* (artwork – for cover of *disLocations*); BSFA Fiftieth Anniversary Award – Best Novel of 1958: *Non-Stop* by Brian Aldiss.

British Sports Book Awards

National Sporting Club, Café Royal, 68 Regent Street, London W1B 5EL

☎ 020 7437 0144   🖷 020 7437 5441

david@nationalsportingclub.co.uk

www.nationalsportingclub.co.uk

Chairman *David Willis*

Established 2003. Annual awards presented at a Café Royal lunch in London in March. Four categories: Best Autobiography, Best Biography, Best Illustrator , Best New Writer. Nominations received in November each year by e-mail (address above). Sponsors include Skysports, *The Times*, Llanllyr Spring Water, Littlehampton Book Services and Ladbrokes.

The Caine Prize for African Writing

51a Southwark Street, London SE1 1RU

☎ 020 7378 6234   🖷 020 7378 6235

info@caineprize.com

www.caineprize.com

Administrator *Nick Elam*

Secretary *Jan Hart*

Annual award founded in 1999 in memory of Sir Michael Caine, former chairman of Booker plc, to recognize the worth of African writing in English. Awarded for a short story by an African writer, published in English anywhere in the world. An 'African writer' is someone who was born in Africa, or who is a national of an African country, or whose parents are African, and whose work has reflected African sensibilities. Final entry date: 31 January; submissions by publishers only. 2007 winner: Monica Arac de Nyeko *Jambula Tree*.

💷 Prize of $10,000.

The Calouste Gulbenkian Prize
▷ The Translators Association Awards

James Cameron Award

City University, Department of Journalism, Northampton Square, London EC1V 0HB

☎ 020 7040 8221   🖷 020 7040 8594

H.Stephenson@city.ac.uk

Contact *Hugh Stephenson*

Annual award for journalism to a reporter of any nationality, working for the British media, whose work is judged to have contributed most during the year to the continuance of the Cameron tradition. 2006 winner: Patrick Cockburn.

Canadian Poetry Association (Annual Poetry Contest)

Canadian Poetry Association, 331 Elmwood Dr., Suite 4-212, Moncton, New Brunswick E1A 1X6, Canada

brianstevens@live.ca

www.canadianpoetryassoc.com

Annual contest open only to members of the CPA worldwide. See the website for entry details. All winning poems are published on the CPA website and in *Poemata*.

💷 Three cash prizes.

Cardiff Book of the Year Awards
▷ Academi Book of the Year Awards

Cardiff International Poetry Competition
▷ Academi Cardiff International Poetry Competition

Carey Award

Society of Indexers, Woodbourn Business Centre, 10 Jessell Street, Sheffield S9 3HY

☎ 0114 244 9561   🖷 0114 244 9563

admin@indexers.org.uk

www.indexers.org.uk

Secretary *Judith Menes*

A private award made by the Society to a member who has given outstanding services to indexing. The recipient is selected by the Executive Board with no recommendations considered from outside the Society.

Carnegie Medal
▷ CILIP: The Chartered Institute of Library and Information Professionals Carnegie Medal

CBI Bisto Book of the Year Awards

17 North Great Georges Street, Dublin 1, Republic of Ireland

☎ 00 353 1 872 7475   🖷 00 353 1 872 8486

info@childrensbooksireland.ie

www.childrensbooksireland.ie

Contact *Jenny Murray*

Founded 1990 as Bisto Book of the Decade Awards. This led to the establishment of an annual award made by the Irish Children's Book Trust, later to become Children's Books Ireland. Open to any author or illustrator of children's books born or resident in Ireland; open to English or Irish languages. 2007/8 winners: Siobhan Dowd *The London Eye Mystery* (Bisto Book of the Year); Tom Kelly *The Thing with Finn* (Eilís Dillon Award).

💷 Prizes of €10,000 (Bisto Book of the Year); €3,000 (Eilís Dillon Award); three honour awards of €1,000 each.

Sid Chaplin Short Story Competition

Shildon Town Council, Civic Hall Square, Shildon DL4 1AH

☎ 01388 772563   🖷 01388 775227

Contact *Mrs J.M. Stafford*

Established 1986. Annual short story competition. Maximum 3,000 words; £2.50 entrance fee (juniors free). All stories must be unpublished and not broadcast and/or performed. Application forms available from 1 January; closing date: May.

£ Prizes of £300 (1st); £150 (2nd); £75 (3rd); £30 (Junior).

Chapter One Promotions International Short Story Competition

PO Box 43667, London SE22 9XU
☎ 0845 456 5364
info@chapteronepromotions.com
www.chapteronepromotions.com
Contacts *Johanna Bertie, Preetha Leela*

Established in 2005, the competition is open to new and established writers to submit unpublished short stories of fewer than 2,500 words. Author details on separate sheet. Submissions and payment accepted online. Closing date: 14 January. 2007 winner: Andrea Owen.

£ Prizes of £2,500, £1,000, £500, plus publication of winners, runners-up and highly recommended in anthology.

Chapter One Promotions Open Poetry Competition

PO Box 43667, London SE22 9XU
☎ 0845 456 5364
info@chapteronepromotions.com
www.chapteronepromotions.com
Contacts *Johanna Bertie, Preetha Leela*

Established 2005. Unpublished poems of no more than 30 lines. Judge decides on the best 20 poems which are displayed on the website and the public votes for their favourite poem. The three poems with the most votes win. Closing date: 1 June. Online voting period: 1–15 July. Winning poems appear on the website in August. Submissions and payment accepted online. 2007 winner: Gary Smillie.

£ Prizes £1,000, £500, £250.

Children's Book Circle Eleanor Farjeon Award
▷ Eleanor Farjeon Award

The Children's Laureate

Booktrust, Book House, 45 East Hill, London SW18 2QZ
☎ 020 8516 2976   ⊕ 020 8516 2978
childrenslaureate@booktrust.org.uk
www.childrenslaureate.org

Established 1998. The Laureate is awarded biennially to an eminent British writer or illustrator of children's books both in celebration of a lifetime's achievement and to highlight the role of children's book creators in inspiring, informing and entertaining young readers. 2007–09 winner: Michael Rosen.

£ Award winner receives a medal and £10,000.

CILIP: The Chartered Institute of Library and Information Professionals Carnegie Medal

7 Ridgmount Street, London WC1E 7AE
☎ 020 7255 0650   ⊕ 020 7255 0651
ckg@cilip.org.uk
www.ckg.org.uk

Established 1936. Presented for an outstanding book for children written in English and first published in the UK during the preceding year. Fiction, non-fiction and poetry are all eligible. 2007 winner: Meg Rosoff *Just in Case*.

£ Award winner receives a medal.

CILIP: The Chartered Institute of Library and Information Professionals Kate Greenaway Medal

7 Ridgmount Street, London WC1E 7AE
☎ 020 7255 0650   ⊕ 020 7255 0651
ckg@cilip.org.uk
www.ckg.org.uk

Established 1955. Presented annually for the most distinguished work in the illustration of children's books first published in the UK during the preceding year. 2007: Mini Grey *The Adventures of the Dish and the Spoon*.

£ Award winner receives a medal. The Colin Mears Award (£5,000 cash) is given annually to the winner of the Kate Greenaway Medal.

Arthur C. Clarke Award for Science Fiction

246e Bethnal Green Road, London E2 0AA
clarkeaward@gmail.com
www.clarkeaward.com
Administrator *Tom Hunter*

Established 1986. The Arthur C. Clarke Award is given annually to the best science fiction novel with first UK publication in the previous calendar year. Both hardcover and paperback books qualify. Made possible by a generous donation by the late Arthur C. Clarke, this award is selected by a rotating panel of judges nominated by the **British Science Fiction Association**, the **Science Fiction Foundation** and the website SFCrowsnest.com. 2008 winner: Richard Morgan *Black Man*.

£ Award of £2,008 (award increases by £1 per year), plus trophy.

David Cohen Prize for Literature

Arts Council England, 14 Great Peter Street, London SW1P 3NQ
☎ 020 7973 5325   ⊕ 020 7973 6983
www.artscouncil.org.uk/davidcohenprize

Established 1993 by the Arts Council and awarded biennially, the David Cohen Prize for Literature is one of the most distinguished literary prizes in Britain. It recognizes writers who use the English language and who are citizens of the UK and the Republic of Ireland, encompassing dramatists as well as novelists, poets and essayists. The prize is for a lifetime's achievement and is donated by the David Cohen Family Charitable Trust. Set up in 1980 by David Cohen, general practitioner and son of a property developer, the Trust Arts Council England is providing a further £12,500 (The Clarissa Luard Award) to enable the winner to encourage new work, with the dual aim of encouraging young writers and readers. Previous winners: William Trevor, Dame Muriel Spark, Harold Pinter, V.S. Naipaul, Doris Lessing, Beryl Bainbridge and Thom Gunn. 2007 winner: Derek Mahon.

Ⓔ Award of £40,000, plus £12,500 towards new work.

The Commonwealth Writers Prize

Commonwealth Foundation, Marlborough House, Pall Mall, London SW1Y 5HY
☎ 020 7747 6262
j.sobol@commonwealth.int
www.commonwealthfoundation.com/culturediversity/writersprize/

Established 1987. An annual award to reward and encourage the upsurge of new Commonwealth fiction. Any work of prose or fiction is eligible, i.e. a novel or collection of short stories. No drama or poetry. The work must be first written in English by a citizen of the Commonwealth and be first published in the year before its entry for the prize. Entries must be submitted by the publisher to the region of the writer's Commonwealth citizenship. The four regions are: Africa, Europe and South Asia, South East Asia and South Pacific, Caribbean and Canada. 2008 winners: Lawrence Hill (Canada) *The Book of Negroes* (Best Book Award); Tahmima Anam (Bangladesh) *A Golden Age* (Best First Book Award).

Ⓔ Prizes of £10,000 for Best Book; £5,000 for Best First Book; 8 prizes of £1,000 for each best and best first book in four regions.

The Duff Cooper Prize

54 St Maur Road, London SW6 4DP
artemiscooper@btopenworld.com
Contact *Artemis Cooper*

An annual award for a literary work of biography, history, politics or poetry, published by a recognized publisher (member of the **Publishers Association**) during the previous 12 months. The book must be submitted by the publisher, not the author. Financed by the interest from a trust fund commemorating Duff Cooper, first Viscount Norwich (1890–1954). 2006 winner: William Dalrymple *The Last Mughal*.

Ⓔ Prize of £5,000.

Costa Book Awards

The Booksellers Association, Minster House, 272 Vauxhall Bridge Road, London SW1V 1BA
☎ 020 7802 0802   ᖴ 020 7802 0803
naomi.gane@booksellers.org.uk
www.costabookawards.com
Contact *Naomi Gane*

Established 1971. Formerly the Whitbread Book Awards. The awards celebrate and promote the best contemporary British writing. They are judged in two stages and offer a total of £50,000 prize money. The awards are open to novel, first novel, biography, poetry and children's book, each judged by a panel of three judges. The winner of each award receives £5,000. The Costa Book of the Year (£25,000) is chosen from the category winners. Writers must have lived in Britain and Ireland for three or more years. Submissions received from publishers only. Closing date: early July. Sponsored by Costa. 2007 winners: A.L. Kennedy *Day* (novel & overall winner); Catherine O'Flynn *What Was Lost* (first novel); Simon Sebag Montefiore *Young Stalin* (biography); Jean Sprackland *Tilt* (poetry); Ann Kelley *The Bower Bird* (children's).

The Coventry Inspiration Book Awards

Coventry Schools Library & Resource Service, Central Library, Smithford Way, Coventry CV1 1FY
☎ 02476 832338   ᖴ 02476 832358
sls@coventry.gov.uk
www.myvotescoventry.org
Children & Young People's Categories and management of awards *Joy Court*
Adult Categories *Natalie Morrissey*

Established 2006/07. Annual award inspired by the Family Reading Campaign uniquely has an award for every age-group of children and young people and by genre categories for adults. The awards go live in October during Children's Book Week and every shortlisted title is open for reader's comments and votes. From January each week the two books with the least votes in each category get 'eliminated'. Voting then starts again each week so fans have to keep voting to keep their favourites in. The final winners are announced in February half term. An Awards ceremony is held later in the year to celebrate the winners in each category and to present individual trophies. Books in all C&YP categories should not have won major national awards before. The exception is the 'Our Special Favourites' which by its very nature and the nomination process could not make that distinction.

Rose Mary Crawshay Prize

The British Academy, 10 Carlton House Terrace,
London SW1Y 5AH
☎ 020 7969 5200   📠 020 7969 5300
events@britac.ac.uk
www.britac.ac.uk
Contact *British Academy Chief Executive and Secretary*

Established 1888 by Rose Mary Crawshay, this
prize is given for an historical or critical work to
a woman of any nationality on English literature,
with particular preference for a work on Keats,
Byron or Shelley. The work must have been
published in the preceding three years.

John Creasey Dagger
▷ Crime Writers' Association (The New Blood Dagger for Best First Crime Novel)

Crime Writers' Association (Cartier Diamond Dagger)

PO Box 273, Boreham Wood WD6 2XA
secretary@thecwa.co.uk
www.thecwa.co.uk
Contact *The Secretary*

Established 1986. An annual award for a lifetime's
outstanding contribution to the genre. 2008
winner: Sue Grafton.

Crime Writers' Association (The CWA Gold Dagger for Non-Fiction)

PO Box 273, Boreham Wood WD6 2XA
secretary@thecwa.co.uk
www.thecwa.co.uk
Contact *The Secretary*

Biennial award for the best non-fiction crime
book published during the year. Nominations
from publishers only. 2006 winner: Linda Rhodes,
Lee Shelden and Kathryn Abnett *The Dagenham Murder*.

£ Award winner receives a Dagger, plus cheque
(sum varies).

Crime Writers' Association (The Dagger in the Library)

PO Box 273, Boreham Wood WD6 2XA
secretary@thecwa.co.uk
www.thecwa.co.uk
Contact *The Secretary*

Reinstated 2002. Annual award (sponsored by
Random House) to the author whose work
has given most pleasure to readers. Nominated
and judged by librarians. 2007 winner: Stuart
MacBride.

£ Award winner receives a Dagger, plus cheque.

Crime Writers' Association (The Debut Dagger)

The Debut Dagger Competition, PO Box 165,
Wirral CH31 9BD
debut.dagger@thecwa.co.uk
www.thecwa.co.uk
Contact *The Secretary*

Annual competition (sponsored by **Orion**) for
unpublished writers to submit the first 3,000
words and 500-word outline of a crime novel
(entry fee and form required). For full details
see website or send s.a.e. to address above. 2007
winner: Alan Bradley *The Sweetness At The Bottom Of The Pie*.

£ Award winner receives a Dagger, plus cheque.

Crime Writers' Association (The Ian Fleming Steel Dagger)

PO Box 273, Boreham Wood WD6 2XA
secretary@thecwa.co.uk
www.thecwa.co.uk
Contact *The Secretary*

Founded 2002. Annual award for the best
thriller, adventure or spy novel. Sponsored by Ian
Fleming Publications Ltd to celebrate the best
of contemporary thriller writing. 2007 winner:
Gillian Flynn *Sharp Objects*.

£ Award winner receives a Dagger, plus cheque.

Crime Writers' Association (The Gold and Silver Daggers for Fiction)

These awards have now been superseded by the
Duncan Lawrie Dagger (see entry).

Crime Writers' Association (The New Blood Dagger for Best First Crime Novel)

PO Box 273, Boreham Wood WD6 2XA
secretary@thecwa.co.uk
www.thecwa.co.uk
Contact *The Secretary*

Established in 1973 following the death of crime
writer John Creasey, founder of the **Crime
Writers' Association**. This award, sponsored by
BBC Audio Books, is given annually for the best
crime novel by an author who has not previously
published a full-length work of fiction. Fomerly
known as the John Creasey Dagger. Nominations
from publishers only. 2007 winner: Gillian Flynn
Sharp Objects.

£ Award winner receives a Dagger, plus cheque.

Crime Writers' Association (The CWA Ellis Peters Award)

PO Box 273, Boreham Wood WD6 2XA
secretary@thecwa.co.uk
www.thecwa.co.uk
Contact *The Secretary*

Sponsored by the Estate of Ellis Peters, Time Warner and **Headline**. Established 1999. Annual award for the best historical crime novel. Nominations from publishers only. 2007 winner: Ariana Franklin *Mistress of the Art of Death*.

Ⓔ Award plus cheque.

The Roald Dahl Funny Prize
Booktrust, Book House, 45 East Hill, London SW18 2QZ
☎ 020 8516 2972   🖷 020 8516 2978
tarryn@booktrust.org.uk
www.booktrust.org.uk
Contacts *Tarryn McKay, Helen Hayes*

Established in 2008 by Michael Rosen as part of his work as Children's Laureate, and administered by Booktrust. The prize is to honour those books that simply make us laugh and aims to reward those authors and artists who write and illustrate their books using humour in their stories, poetry and fiction. Books must have been first published in the UK or Ireland between 1 October and 31 September. Authors and illustrators must be British or Irish, or other nationals living in Britain or Ireland at the time of entering. The prize is awarded to the funniest book in two categories: for children aged six and under (picture book) and for children aged 7-14.

Ⓔ Prizes of £2,500 for each category.

Daily Mail First Novel Award
Transworld Publishers, 61–63 Uxbridge Road, London W5 5SA
www.dailymail.co.uk/books

Founded 2007. Open competition for writers aged 16 or over, resident in the UK or Republic of Ireland. The winning novel will be published by Transworld in 2009 with an advance of £30,000.

Hunter Davies Prize
▷ Lakeland Book of the Year Awards

Felix Dennis Prize for Best First Collection
▷ The Forward Prizes for Poetry

George Devine Award
9 Lower Mall, Hammersmith, London W6 9DJ
Contact *Christine Smith*

Annual award for a promising new playwright writing for the stage in memory of George Devine, artistic director of the **Royal Court Theatre**, who died in 1965. The play, which can be of any length, does not need to have been produced. Send two copies of the script, plus outline of work, to *Christine Smith*. Closing date: 1 March. Send s.a.e. for the script to be returned if required. Information leaflet available from January on receipt of s.a.e.

Ⓔ Prize of £10,000.

Dingle Prize
PO Box 3401, Norwich NR7 7JF
☎ 01603 516236
execsec@bshs.org.uk
www.bshs.org.uk

Biennial award made by the British Society for the History of Science to the best book in the history of science (broadly construed) which is accessible to a wide audience of non-specialists. Next award: 2009. Previous winners: Ken Alder *The Measure of All Things*; Deborah Cadbury *The Dinosaur Hunters*; Steven Shapin *The Scientific Revolution*.

Ⓔ Prize of £300.

Dolman Best Travel Book Award
Authors' Club, 40 Dover Street, London W1S 4NP
☎ 020 7408 5092   🖷 020 7409 0913
stella@theauthorsclub.co.uk
Contact *Stella Kane*

William Dolman, a former chairman of the **Authors' Club**, instituted this prize for the most promising book of travel literature. Books published in Great Britain by British writers are eligible and submissions are accepted in February of each year. Inaugural winner: Nicholas Jubber *The Prester Quest*.

Ⓔ Award of £1,000.

Drama Association of Wales Playwriting Competition
The Old Library, Singleton Road, Splott, Cardiff CF24 2ET
☎ 029 2045 2200   🖷 029 2045 2277
aled.daw@virgin.net
Contact *Teresa Hennessy*

Annual competition held to promote the writing of one-act plays in English and Welsh of between 20 and 50 minutes' playing time. Application forms from the address above. Closing date: 31 January.

The John Dryden Competition
School of Literature and Creative Writing, University of East Anglia, Norwich NR4 7TJ
🖷 01603 250599
transcomp@uea.ac.uk
www.bcla.org
Competition Organizer *Dr Jean Boase-Beier*

Established 1983. Annual competition open to unpublished literary translations from all languages. Maximum submission: 25 pages.

Ⓔ Prizes of £350 (1st); £200 (2nd); £100 (3rd); plus publication for all winning entries in the Association's journal. Other entries may receive commendations.

The Duke of Westminster's Medal for Military Literature

Royal United Services Institute for Defence and Security Studies, Whitehall, London SW1A 2ET

☎ 020 7747 2637   📠 020 7321 0943

www.rusi.org

Established in 1997, this annual award is sponsored by the Duke of Westminster and aims to mark a notable and original contribution to the study of international or national security, or the military profession. Work must be in English, by a living author, and have been published as a book, rather than an article, in the preceding or next six months of the closing date for entries. 2006 winner: *The Pursuit of Victory* Professor Roger Knight.

💷 Silver medal, plus £1,000. Winning author is invited to give lecture at RUSI where the Duke of Westminster will present the Medal.

Edge Hill Prize for the Short Story

Dept. of English, Edge Hill University, St Helen's Road, Ormskirk L39 4QP

☎ 01695 584121   📠 01695 579997

coxa@edgehill.ac.uk

www.edgehill.ac.uk

Contact *A. Cox*

Annual award founded in 2007 following a one-day conference on the short story. Its aim is to reward high achievement in writing the short story and to promote the genre. Awarded to the author of a published short story collection from the UK or Ireland. Entries are submitted through publishers and must be published in the previous year.

💷 Prize of £5,000.

T.S. Eliot Prize

Truman State University Press, 100 East Normal Street, Kirksville, MO 63501, USA

☎ 001 660 785 7336   📠 001 660 785 4480

tsup@truman.edu

tsup.truman.edu

Contact *Nancy Rediger*

An annual award, established in 1997 in honour of native Missourian, T.S. Eliot, to publish and promote contemporary English-language poetry regardless of a poet's nationality, reputation, stage in career or publication history. Entry requirements: 60–100 pages of original poetry with $25 fee. Final entry date: 31 October. 2007 winner: Carol V. Davis *Into the Arms of Pushkin: Poems of St Petersburg*.

💷 Prize of $2,000, plus publication.

T.S. Eliot Prize

Poetry Book Society, Fourth Floor, 2 Tavistock Place, London WC1H 9RE

☎ 020 7833 9247   📠 020 7833 9247

info@poetrybooks.co.uk

www.poetrybookshoponline.com

Contact *Chris Holifield*

Established in 1993 to mark the 40th anniversary of the founding of the Poetry Book Society, the T.S. Eliot prize is awarded by PBS in January of each year for the best new single author collection of poetry published in the previous year. Submissions must be made by the publisher and are for a single-author poetry collection published in the preceding calendar year in the UK or Republic of Ireland. No self-published work is eligible. Final entry date: beginning of August. The prize-winners' cheques are given by Mrs Valerie Eliot, who presents them at an award ceremony in central London. T.S. Eliot Prize Readings are held on the evening before the ceremony. The Prize is distinctive in having a Shadowing Scheme, run in association with the English and Media Centre's *emagazine*. 2007 winner: Sean O'Brien *The Drowned Book*.

The Desmond Elliott Prize

The Desmond Elliott Charitable Trust, 84 Godolphin Road, London W12 8JW

☎ 020 8222 6580

ema.manderson@googlemail.com

www.desmondelliottprize.com

Contact *Mrs Emma Manderson*

Established in 2008 in memory of literary agent and publisher, Desmond Elliott, this biennial prize is for a first full-length novel, written in English and published in book form in the UK, by an author who is permanently resident in the UK. Final entry date: 23 November 2009.

💷 Prize of £10,000.

The Encore Award

The Society of Authors, 84 Drayton Gardens, London SW10 9SB

☎ 020 7373 6642   📠 020 7373 5768

info@societyofauthors.org

www.societyofauthors.org

Established 1990. Awarded biennially for the best second published novel or novels of the year. Final entry date: 30 November 2008. Details from the **Society of Authors**. 2007 winner: M.J. Hyland *Carry Me Down*.

💷 Prize of £10,000.

Envoi Poetry Competition

Ty Meirion, Glan yr afon, Tanygrisiau, Blaenau Ffestiniog LL41 3SU

☎ 01766 832112

Contact *Jan Fortune-Wood*

Run by *Envoi* poetry magazine. Competitions are featured regularly, with prizes of £300, plus three annual subscriptions to *Envoi*. Winning poems along with full adjudication report are published. Send s.a.e. to Competition Secretary at the address above.

Geoffrey Faber Memorial Prize

Faber & Faber Ltd, 3 Queen Square, London WC1N 3AU

☎ 020 7465 0045   🖷 020 7465 0034
www.faber.co.uk

Established 1963 as a memorial to the founder and first chairman of **Faber & Faber**, this prize is awarded in alternate years for the volume of verse and the volume of prose fiction published in the UK in the preceding two years, which is judged to be of greatest literary merit. Authors must be under 40 at the time of publication and citizens of the UK, Commonwealth, Republic of Ireland or South Africa. 2007 winner: Edward Docx *Self Help*.

💷 Prize of £1,000.

The Alfred Fagon Award

The Royal Court Theatre, Sloane Square, London SW1W 8AS
www.talawatheatrecompany.co.uk/afa

First presented in 1997. An annual award, in memory of playwright Alfred Fagon, which is open to any playwright of Caribbean descent, resident in the UK, for the best new stage play in English, which need not have been produced (television and radio plays and film scripts not eligible). Two copies of the script (plus s.a.e. for return), together with a brief history of the play and c.v. (including author's Caribbean connection), should be sent to the address above by 31 August. 2007 winner: Allia V. Oswald.

💷 Award of £5,000.

Eleanor Farjeon Award

childrensbookcircle@hotmail.co.uk
www.childrensbookcircle.org.uk

This award, named in memory of the much-loved children's writer, is for distinguished services to children's books by an individual or organization. It may be given to a librarian, teacher, publisher, bookseller, author, artist, reviewer, television producer, or any other person or organization working with or for children through books. Nominations from members of the **Children's Book Circle**. 2007 winner: **Jane Nissen Books**.

Fish Fiction Prizes

Fish Publishing, Durrus, Bantry, Co. Cork, Republic of Ireland
info@fishpublishing.com
www.fishpublishing.com
Contact *Clem Cairns*

Offers several writing contests including the Short Story Prize, an annual award, founded in 1994, which aims to discover, encourage and publish new literary talent. The best 15 stories are published in an anthology. Also, the Fish-Knife Award run in conjunction with the **Crime Writers' Association**. Details of these and other contests can be found on the website.

Sir Banister Fletcher Award

Authors' Club, 40 Dover Street, London W1S 4NP
☎ 020 7408 5092   🖷 020 7409 0913
stella@theauthorsclub.co.uk
Contact *Stella Kane*

Founded in 1954, this award was created by the late Sir Banister Fletcher, former President of the **Authors' Club** and the Royal Institute of British Architects, and is presented for 'the most deserving book on architecture or the arts'. The award is open to titles written by British authors or those resident in the UK and published under a British imprint. 2007 winner: Pelkonen and Albrecht *Eero Saarinen: Shaping the Future*.

💷 Prize of £1,000.

The John Florio Prize
▷ **The Translators Association Awards**

The Paul Foot Award

Private Eye, 6 Carlisle Street, London W1D 3BN
www.private-eye.co.uk
Contact *Digby Halsby* (Midas PR ☎ 020 7590 8909))

Annual award, established in 2005 by *The Guardian* and *Private Eye* for campaigning journalism in memory of former contributor Paul Foot. Submissions can be made by individual journalists, teams of journalists or publications for work appearing between October 2006 and August 2007. Single pieces or entire campaigns are eligible. Final entry date: 12 September 2007. No broadcast material. 'Please do not send original material as no correspondence can be entered into.' 2007 joint winners: David Leigh and Rob Evans (Guardian) and Deborah Wain (Doncaster Free Press).

💷 Award of £5,000 with five runners-up receiving £1,000.

The Forward Prizes for Poetry

Administrator: Colman Getty, 28 Windmill Street, London W1T 2JJ
☎ 020 7631 2666   🖷 020 7631 2699

kate@colmangetty.co.uk
www.forwardartsfoundation.co.uk
Contact *Kate Wright-Morris*

Established 1992. Three awards: the Forward Prize for Best Collection, the Felix Dennis Prize for Best First Collection and the Forward Prize for Best Single Poem in memory of Michael Donaghy, which is not already part of an anthology or collection. All entries must be published in the UK or Republic of Ireland and submitted by poetry publishers (collections) or newspaper and magazine editors (single poems). Individual entries of poets' own work are not accepted. 2007 winners: Sean O'Brien (best collection), Daljit Nagra (best first collection), Alice Oswald (best single poem).

£ Prizes of £10,000 (best collection); £5,000 (best first collection); £1,000 (best single poem).

The Foyle Young Poets of the Year Award

The Poetry Society, 22 Betterton Street, London WC2H 9BX
☎ 020 7420 9880   ℻ 020 7240 4818
fyp@poetrysociety.org.uk
www.foyleyoungpoets.org
Contact *George Ttoouli*

This annual award was founded by the Poetry Society in 1997 in order to help young poets thrive as readers and writers. Since 2001 the Foyle Foundation has sponsored the award which encourages creativity, imagination and literacy amongst young people at school. Two top poets are selected as judges and they choose 100 of the best young poets, out of almost 5,000 poets who enter each year. Poets must be aged 11–17 on or before the closing date of 31st July each year. Entry is free and can be via the web entry form, or by post. Poets can enter as many poems as they like (unlimited), with no restrictions on length, content, style, form or theme.

£ Fifteen overall winners aged 11–17, including three international winners, have their poems printed in the winners' anthology; five 11–14-year-olds win a visit to their school by a leading poet; fifteen 15–17 year-old-winners attend a week-long residential course at one of the prestigious Arvon Centres. The three schools who inspire the most entries receive a special selection of books from **Faber & Faber**, **Bloodaxe Books** and **Salt Publishing**.

The Frogmore Poetry Prize

42 Morehall Avenue, Folkestone CT19 4EF
☎ 07751 251689
www.frogmorepress.co.uk
Contact *Jeremy Page*

Established 1987. Awarded annually and sponsored by the Frogmore Foundation. The winning poem, runners-up and short-listed entries are all published in the magazine. Closing date 31 May. Previous winners: Bill Headdon, John Latham, Diane Brown, Tobias Hill, Mario Petrucci, Gina Wilson, Ross Cogan, Joan Benner, Ann Alexander, Gerald Watts, Katy Darby, David Angel, Howard Wright, Caroline Price, Julie-ann Rowell, Arlene Ang, Peter Marshall.

£ The winner receives 200 guineas and a two-year subscription to the biannual literary magazine, *The Frogmore Papers*.

FT & Goldman Sachs Business Book of the Year Award

Financial Times, One Southwark Bridge, London SE1 9HL
☎ 020 7873 3000   ℻ 020 7873 3072
bookaward@ft.com
www.ft.com/bookaward
Contacts *Lizzie Allen*

Established in 2005 to identify the book that provides the most compelling and enjoyable insight into modern business issues, including management, finance and economics. Titles must be published for the first time in the English language, or in English translation, between 31 October 2007 and 1 November 2008. Submissions by a publisher or *bona fide* imprint which holds English language rights in the book. 2007 winner: William D. Cohan *The Last Tycoons*.

£ Award of £30,000 (winner); shortlisted authors receive £5,000 each.

Galaxy British Book Awards

Publishing News, 7 John Street, London WC1N 2ES
☎ 0870 870 2345   ℻ 0870 870 0385
mailbox@publishingnews.co.uk
www.galaxybritishbookawards.com

Established 1988. Viewed by the book trade as the one to win, 'The Nibbies' are presented annually. The awards are made in various categories. Each winner receives the prestigious Nibbie and the awards are presented to those who have made the most impact in the book trade during the previous year. 2009 winners included: Ian McEwan, J.K. Rowling, Khaled Hosseni, Francesca Simon, Catherine O'Flynn. For further information contact: Merric Davidson, PO Box 60, Cranbrook TN17 2ZR (☎ l/ ℻ 01580 212041; gbba@mdla.co.uk).

Martha Gellhorn Trust Prize

Crosscombe, Town's Lane, Loddiswell TQ7 4QY
☎ 01548 550344   ℻ 01548 550344
sandyandshirlee@tiscali.co.uk

Annual prize for journalism in honour of one of the twentieth century's greatest reporters. Open for journalism published in English, giving 'the view from the ground – a human story that penetrates the established version of events and illuminates an urgent issue buried by prevailing fashions of what makes news'. The subject matter can involve the UK or abroad. Six copies of each entry should be sent to the address above by 31 March 2009. Previous winners include: Hala Jaber, Michael Tierney, Ghaith Abdul-Ahad, Patrick Cockburn, Robert Fisk, Jeremy Harding, Geoffrey Lean.

Ⓔ Prize of £5,000.

The Gladstone History Book Prize

Royal Historical Society, University College London, Gower Street, London WC1E 6BT
Ⓣ 020 7387 7532   Ⓕ 020 7387 7532
royalhistsoc@ucl.ac.uk
www.royalhistoricalsociety.org/grants.htm
Contact *Administrative Secretary*

Established 1998. Annual award for the best new work on any historical subject which is not primarily related to British history, published in the UK in the preceding calendar year. The book must be the author's first (solely written) history book and be an original and scholarly work of historical research. Closing date: 31 December.

Ⓔ Prize of £1,000.

Glenfiddich Food & Drink Awards

c/o Wild Card PR, Brettenham House, 5 Savoy Street, London WC2E 7AE
www.glenfiddich.com/foodanddrink

Established in 1970, the Glenfiddich Food & Drink Awards honour the very best writers and broadcasters who find inspiration at the bar, on the menu and on the wine list. The categories cover everything from authors to bar writers, food writers to television presenters, plus the Glenfiddich Independent Spirit Award and the Glenfiddich Trophy. Winners are presented with a special bottling of Glenfiddich Gran Reserva 21 Year Old Single Malt Scotch Whisky and a cheque for £1,000. From among the category winners the judges select one outstanding candidate to receive The Glenfiddich Trophy (to be held for one year) along with a further £3,000 and a bottle of Glenfiddich 30 Year Old Single Malt Scotch Whisky. (The Awards did not take place in 2008.)

Golden Hearts Awards

▷ Romance Writers of America under Professional Associations and Societies

Golden PEN Award for Lifetime Distinguished Service to Literature

English PEN, 6–8 Amwell Street, London EC1R 1UQ
Ⓣ 020 7713 0023   Ⓕ 020 7837 7838
enquiries@englishpen.org
www.englishpen.org

Awarded to a senior writer, with a distinguished body of work written over many years, who has made a significant and constructive impact on fellow writers, the reading public and the literary world. Nominations by members of English PEN only. Previous winners include: Harold Pinter, Doris Lessing, Michael Frayn, Nina Bawden. 2007 winner: Josephine Pullein-Thompson.

Good Housekeeping Book Awards

The National Magazine Company Ltd, 72 Broadwick Street, London W1F 9EP
Ⓣ 020 7439 5218
natalie.earl@natmags.co.uk
Contact *Natalie Earl*

Annual awards, established in 2007. Categories: Best Fiction, Best Non-Fiction, Best Debut and The Good Housekeeping Reader Award for Best Author. Books are submitted by publishers for the first three categories while readers of the magazine vote for 'Best Author'. 2008 winners: Rose Tremain *The Road Home* (Best Fiction); Sally Brampton *Shoot the Damn Dog* (Best Non-Fiction); Sadie Jones *The Outcast* (Best Debut); Ian McEwan (The Good Housekeeping Reader Award for Best Author).

Ⓔ Prize winners receive a certificate.

Gourmand World Cookbook Awards

Pintor Rosales 36, 8°A, 28008 Madrid, Spain
Ⓣ 00 34 91 541 6768   Ⓕ 00 34 91 541 6821
icr@virtualsw.es
www.cookbookfair.com
Contact *Edouard Cointreau*

Founded in 1995 by Edouard Cointreau to reward those who 'cook with words'. The only world competition for food and wine books in all languages. Annual event with the winners of the worldwide local or regional competitions competing for the 'Best Book in the World' award. 107 countries participated in the 2008 awards with the Awards event taking place in London.

The Winston Graham Historical Prize

Royal Cornwall Museum, River Street, Truro TR1 2SJ
enquiries@royalcornwallmuseum.org
www.royalcornwallmuseum.org
Contact *Prize Administrator*

Established in 2008 the prize has been funded by a generous legacy to the Royal Institution of Cornwall from the author. It will be awarded in its inaugural year for an unpublished work of historical fiction, preferably with a clear connection to Cornwall. Details and conditions of entry can be found on the website or by writing to the Royal Cornwall Museum. Deadline for

shortlist submissions (synopsis and first two chapters): 31 March 2009.
£ Prize 2008/9: £5,000 and potential for publication by **Pan Macmillan**.

Kate Greenaway Medal
▷ CILIP: The Chartered Institute of Library and Information Professionals Kate Greenaway Medal

The Griffin Poetry Prize
6610 Edwards Boulevard, Mississauga, Ontario, Canada L5T 2V6
☎ 001 905 565 5993   🖷 001 905 564 3645
info@griffinpoetryprize.com
www.griffinpoetryprize.com
Contact *Ruth Smith, Manager*

Annual award established in 2000 by Toronto-based entrepreneur, Scott Griffin, for books of poetry written in or translated into English. Trustees include Margaret Atwood and Michael Ondaatje. Submissions from publishers only.
£ Prizes total C$100,000, divided into two categories: International and Canadian.

The Guardian Children's Fiction Award
The Guardian, 119 Farringdon Road, London EC1R 3ER (From Nov. 2008: Kings Place, 90 York Way, London N1 9AG)
☎ 020 7239 9694   🖷 020 7239 9933
Children's Book Editor *Julia Eccleshare*

Established 1967. Annual award for an outstanding work of fiction for children aged seven and over by a British or Commonwealth author, first published in the UK in the year of the award, excluding picture books. No application form necessary. 2007 winner: Jenny Valentine *Finding Violet Park*.
£ Award of £1,500.

The Guardian First Book Award
The Guardian, 119 Farringdon Road, London EC1R 3ER(From Nov. 2008: Kings Place, 90 York Way, London N1 9AG)
☎ 020 7886 9317
Contact *Thea Skelton*

Established 1999. Annual award for first time authors published in English in the UK. All genres of writing eligible, apart from academic, guidebooks, children's, educational, manuals, reprints and TV, radio and film tie-ins. All books must be published between January and December 2008 and have an ISDN number or equivalent. Submissions are only received direct from publishers, not from individual authors. Self-published work is not eligible. 2007 winner: Dinaw Mengestu *Children of the Revolution*.
£ Award of £10,000, plus *Guardian/Observer* advertising package.

Guild of Food Writers Awards
255 Kent House Road, Beckenham BR3 1JQ
☎ 020 8659 0422
awards@gfw.co.uk
www.gfw.co.uk
Contact *Jonathan Woods*

Established 1985. Annual awards in recognition of outstanding achievement in all areas in which food writers work and have influence. Entry is not restricted to members of the Guild. Entry form available from the address above. 2007 winners: Michael Smith Award for Work on British Food: Mark Hix *British Regional Food*; Jeremy Round Award for Best First Book and Food Book of the Year Award: Ruth Cowen *Relish: The Extraordinary Life of Alexis Soyer*; Evelyn Rose Award for Cookery Journalist of the Year: Diana Henry for work published in *The Sunday Telegraph*'s *Stella* magazine; Cookery Book of the Year Award: Skye Gyngell *A Year in My Kitchen*; Food Journalist of the Year Award: Felicity Lawrence for work published in *The Guardian*; Miriam Polunin Award for Work on Healthy Eating: *The Food Programme: Arthritis* presented by Sheila Dillon, producer, Margaret Collins, BBC Radio 4.

Guildford Book Festival First Novel Award
Guildford Book Festival, c/o Tourist Information Centre, 14 Tunsgate, Guildford GU1 3QT
☎ 01483 444334
deputy@guildfordbookfestival.co.uk
www.guildfordbookfestival.co.uk
Contacts *Pamela Thomas, Glenis Pycraft*

This annual award was originally known as The Pendleton May Award. It was initiated to encourage local talent but has now been opened to all writers in the UK. Submissions must be from agents or publishers with accompanying letter (proofs accepted). Books to be published between 1 November and 31 October i.e. between Festivals. Shortlisted authors must be prepared to attend the presentation dinner held during the Festival. 2007 winner: Catherine O'Flynn *What Was Lost*.
£ Prize of £2,500.

Gwobr Llyfr y Flwyddyn
▷ Academi Book of the Year Awards

The Anthony Hecht Poetry Prize
The Waywiser Press, 14 Lyncroft Gardens, Ewell KT17 1UR
☎ 020 8393 7055   🖷 020 8393 7055
waywiserpress@aol.com
www.waywiser-press.com/hechtprize.html
Contact *Philip Hoy*, Managing Editor

Established 2005. Annual prize awarded for an outstanding, unpublished, full collection of poems. Entrants must be at least 18 years of age and may not have published more than one previous collection. Detailed guidelines and entry forms available from July on the website or by sending an A4-sized s.a.e.

£ Prize of £1,750 and publication by Waywiser in the UK and USA.

Hellenic Foundation Prize
▷ The Translators Association Awards

Felicia Hemans Prize for Lyrical Poetry
University of Liverpool, PO Box 147, Liverpool L69 7WZ
☎ 0151 794 2458   🖷 0151 794 2454
wilderc@liv.ac.uk
Contact *The Sub-Dean, Faculty of Arts*

Established 1899. Annual award for published or unpublished verse. Open to past or present members and students of the University of Liverpool. One poem per entrant only. Closing date 1 May.

£ Prize of £30.

The Hessell-Tiltman Prize for History
English PEN, 6–8 Amwell Street, London EC1R 1UQ
☎ 020 7713 0023   🖷 020 7837 7838
enquiries@englishpen.org
www.englishpen.org

Founded 2002. Awarded for a history book covering any period up to the end of the Second World War, written in English (including translations) and aimed at a wide audience. Submissions cannot be made as books are nominated by PEN Literary Foundation Patrons and members of PEN's Executive Committee. 2007 winner: Vic Gattrell *City of Laughter*.

£ Prize of £3,000.

William Hill Sports Book of the Year
Greenside House, Station Road, Wood Green, London N22 7TP
☎ 020 8918 3731   🖷 020 8918 3728
pressoffice@williamhill.co.uk
Contact *Graham Sharpe*

Established 1989. Annual award introduced by Graham Sharpe of bookmakers William Hill. Sponsored by William Hill and thus dubbed the 'Bookie' prize, it is the first, and only, Sports Book of the Year award. Final entry date: September. 2007 winner: Duncan Hamilton *Provided You Don't Kiss Me: 20 Years with Brian Clough*.

£ Prize (reviewed annually) of £22,000 package, including £18,000 cash, hand-bound copy, £2,000 free bet. Runners-up prizes.

HISSAC Annual Short Story Award
20 Lochslin, Balintore, Easter Ross IV20 1UP
☎ 01862 832266   🖷 01862 832266
info@hissac.co.uk
www.hissac.co.uk
Contact *Clio Gray*

Established 2003. Annual short story award organized by HISSAC, the Highlands & Islands Short Story Association, an international network of writers aiming to critique and encourage each other's work and encourage all manner of creative writing, including the short story form. Entry fees: £4 per entry, £10 for three. Maximum wordage: 2,500. No connection to Scotland is needed. Double-spaced, single-sided entries; no children's stories. Final entry date: 31 July 2009.

£ Prizes of £400 (1st); £50 each (2nd & 3rd).

Calvin & Rose G. Hoffman Prize
King's School, Canterbury CT1 2ES
☎ 01227 595544
Contact *The Bursar*

Annual award for distinguished publication on Christopher Marlowe, established by the late Calvin Hoffman, author of *The Man Who was Shakespeare* (1955) as a memorial to himself and his wife. For unpublished works of at least 5,000 words written in English for their scholarly contribution to the study of Christopher Marlowe and his relationship to William Shakespeare. Final entry date: 1 September each year.

David C. Horn Prize
▷ The Yale Drama Series

L. Ron Hubbard's Writers of the Future Contest
PO Box 218, East Grinstead RH19 4GH
Contest Administrator *Andrea Grant-Webb*

Established 1984 by L. Ron Hubbard to encourage new and amateur writers of science fiction and fantasy. Quarterly awards with an annual grand prize. Entrants must submit a short story of up to 10,000 words, or a novelette of fewer than 17,000 words, which must not have been published previously. The contest is open only to those who have not been published professionally. Send s.a.e. for entry form. Previous winners: Roge Gregory, Malcolm Twigg, Tom Brennan, Alan Smale, Janet Barron.

£ Prizes of £500 (1st); £375 (2nd); £250 (3rd) each quarter; Annual Grand Prize: £2,500. All winners are awarded a trip to the annual L. Ron Hubbard Achievement Awards which include a series of professional writers' workshops, and are published in the *L. Ron Hubbard Presents Writers of the Future* anthology.

The Imison Award

The Society of Authors, 84 Drayton Gardens
SW10 9SB
☏ 020 7373 6642   ℻ 020 7373 5768
jhodder@societyofauthors.org
www.societyofauthors.org
Contact *Jo Hodder*

Annual award established 'to perpetuate the
memory of Richard Imison, to acknowledge
the encouragement he gave to writers working
in the medium of radio, and in memory of the
support and friendship he invariably offered
writers in general, and radio writers in particular.'
Administered by the **Society of Authors** and
generally sponsored by the **Peggy Ramsay
Foundation**, the purpose is 'to encourage new
talent and high standards in writing for radio
by selecting the radio drama by a writer new
to radio which, in the opinion of the judges, is
the best of those submitted.' An adaptation for
radio of a piece originally written for the stage,
television or film is not eligible. Any radio drama
first transmitted in the UK between 1 January
and 31 December is eligible, provided the work
is an original piece for radio and it is the first
dramatic work by the writer(s) that has been
broadcast. Submission may be made by any party
to the production in the form of three copies
of both script and recording (non-returnable),
accompanied by a nomination form and 250 word
synopsis and c.v. 2007 winner: Mike Bartlett *Not
Talking*.
£ Prize of £1,500.

The Independent Foreign Fiction Prize

c/o Literature Department, Arts Council England,
14 Great Peter Street, London SW1P 3NQ
☏ 020 7973 5204   ℻ 020 7973 6983
www.artscouncil.org.uk/iffp

The Independent Foreign Fiction Prize celebrates
an exceptional work of fiction by a living author
which has been translated into English from
any other language and published in the United
Kingdom in the last year. 2008 winner: *Omega
Minor* by Paul Verhaeghen, translated by the
author from the Dutch.
£ Prize of £10,000 shared equally between
author and translator.

The International Dundee Book Prize

City of Discovery Campaign, 3 City Square,
Dundee DD1 3BA
☏ 01382 434214   ℻ 01382 434650
book.prize@dundeecity.gov.uk
www.dundeebookprize.com
Contact *Karin Johnston*

'The Dundee International Book Prize has
established itself as the UK's premier prize for
emerging novelists. Its £10,000 cash award
together with publication by **Birlinn Ltd**,
publishers of the Polygon imprint, make the Prize
highly valued by tomorrow's great new writers
seeking to break into the publishing world.' The
next award, in 2009, is for an unpublished novel
on any theme and in any genre. Telephone or
e-mail *Karin Johnston* for further information.
2007 winner: Fiona Dunscombe *The Triple Point
of Water*.

The International IMPAC Dublin Literary Award

Dublin City Library & Archive, 138–144 Pearse
Street, Dublin 2, Republic of Ireland
☏ 00 353 1 674 4802   ℻ 00 353 1 674 4879
literaryaward@dublincity.ie
www.impacdublinaward.ie

Established 1995. Sponsored by Dublin City
Council and US-based productivity improvement
firm, IMPAC, this prize is awarded for a work
of fiction written and published in the English
language or written in a language other than
English and published in English translation.
Initial nominations are made by municipal public
libraries in major and capital cities worldwide,
each library putting forward up to three books
to the international panel of judges in Dublin.
2007 winner: *Out Stealing Horses*, translated from
Norwegian by Anne Born.
£ Prize of €100,000 (if the winning book is in
English translation, the prize is shared €75,000 to
the author and €25,000 to the translator).

International Prize for Arabic Fiction

Abu Dhabi office: Emirates Foundation, Chamber
of Commerce Tower, PO Box 111445, Abu Dhabi,
United Arab Emirates
☏ 00 961 993 5333   ℻ 00 961 993 5333
info@arabicfiction.org
www.arabicfiction.com
Beirut office: IPAF Office, Sahel Alma, PO Box
280 – Jounieh – Lebanon
Prize Administrator *Joumana Haddad*

Launched in Abu Dhabi in April 2007 in
association with the Booker Prize Foundation,
the Emirates Foundation and the Weidenfeld
Institute for Strategic Dialogue. Annual award
which aims to reward excellence in contemporary
Arabic creative writing and to encourage
wider readership of quality Arabic literature
internationally. Awarded specifically for prose
fiction in Arabic (no short stories nor poetry)
with each of the six shortlisted authors receiving
$10,000 with a further $50,000 going to the
winner. Submissions from publishers only. Final
entry date: 31 July. Inaugural winner: Baha Taher
Sunset Oasis.

International Student Playscript Competition
▷ University of Hull under UK and Irish Writers' Courses

Irish Book Awards
Unique Media Ltd, The Gate House, Parker Hill,
Lr. Rathmines Road, Dublin 6, Republic of Ireland
☎ 00 353 1 497 8738   ℻ 00 353 1 497 8739
info@uniquemedia.ie
www.irishbooksawards.ie

Established in 2005 to acknowledge the wealth
of talent in Irish literature. Annual awards
comprising nine categories which are open to all
Irish authors who had a book published in the
previous twelve months. 2008 winners: Anne
Enright *The Gathering* (The Hughes & Hughes
Irish Novel of the Year); Diarmaid Ferriter *Judging
Dev* (The Argosy Irish Non-Fiction Book of
the Year; The Tubridy Show Listeners' Choice
Award; Eason Irish Published Book of the Year);
Anita Notaro *Take a Look at Me Now* (Galaxy
Irish Popular Fiction Book of the Year); *Trevor
Brennan: Heart and Soul* Trevor Brennan with
Gerry Thornley (Energise Sport Irish Sports
Book of the Year); Julia Kelly *With My Lazy
Eye* (International Education Services Irish
Newcomer of the Year Award); Brendan O'Brien
The Story of Ireland and Roddy Doyle *Wilderness*
(The Dublin Airport Authority Irish Children's
Book of the Year – Junior and Senior); William
Trevor (Bob Hughes Lifetime Achievement in
Irish Literature).

Jerwood Aldeburgh First Collection Prize
The Poetry Trust, The Cut, 9 New Cut,
Halesworth IP19 8BY
☎ 01986 835950
info@thepoetrytrust.org
www.thepoetrytrust.org
Contact *Naomi Jaffa*

Funded by Jerwood Charitable Foundation,
the prize is awarded to the author of what in
the opinion of the judges is the best first full
collection of poetry published in book form in the
UK and the Republic of Ireland in the preceding
twelve months. The winner receives £3,000 plus
an invitation to read (fee paid) at the **Aldeburgh
Poetry Festival** the following year. Please refer
to the website for details on how to enter. 2007/8
winner: Tiffany Atkinson *Kink and Particle*.

Jewish Quarterly Literary Prize
PO Box 37645, London NW7 1WB
☎ 020 8343 4675
www.jewishquarterly.org
Administrator *Pam Lewis*

Formerly the H.H. Wingate Prize. Annual
award (for fiction or non-fiction) for a work that
best stimulates an interest in and awareness of
themes of Jewish interest. Books must have been
published in the UK in the year of the award
(written in English originally or in translation) by
an author resident in Britain, the Commonwealth,
Israel, Republic of Ireland or South Africa.
2007 winner: Howard Jacobson *Kalooki Nights*
(fiction).

£ Prizes of £5,000.

Samuel Johnson Prize for Non-Fiction
▷ The BBCFour Samuel Johnson Prize for Non-Fiction

Mary Vaughan Jones Award
Cyngor Llyfrau Cymru (Welsh Books Council),
Castell Brychan, Aberystwyth SY23 2JB
☎ 01970 624151   ℻ 01970 625385
wbc.children@wbc.org.uk
www.wbc.org.uk
Contact *The Administrator*

Triennial award for distinguished services in
the field of children's literature in Wales over a
considerable period of time. 2006 winner: Mair
Wynn Hughes.

£ Award winner receives a silver trophy.

Keats–Shelley Prize
Keats–Shelley Memorial Association, 117 Cheyne
Walk, London SW10 0ES
☎ 020 7352 2180   ℻ 020 7352 6705
harrietcullenuk@yahoo.com
www.keats-shelley.com
Contact *Harriet Cullen*

Established 1998. Annual award to promote the
study and appreciation of Keats and Shelley,
especially in the universities, and of creative
writing inspired by the younger Romantic poets.
Two categories: essay and poem; open to all
ages and nationalities. Recent winners: Richard
Marggraf Turley (poems), Adam Gyngell (essays).

£ Prize of £3,000 distributed between the
winners of the two categories.

Kelpies Prize
Floris Books, 15 Harrison Gardens, Edinburgh
EH11 1SH
☎ 0131 337 2372   ℻ 0131 347 9919
floris@florisbooks.co.uk
www.florisbooks.co.uk/kelpiesprize
Contact *Prize Administrator*

Established 2004. Annual prize to encourage and
reward new Scottish writing for children. The
prize is for an unpublished novel (40–60,000
words) for children aged 9–12, set wholly or
mainly in Scotland. The author does not need to
be Scottish. Application form, guidelines, terms
and conditions are available on the website.

£ Prize of £2,000 and publication in Floris
Books' Kelpies series.

The Petra Kenney Poetry Competition

PO Box 32, Filey YO14 9YG
morgan@kenney.uk.net
www.petrapoetrycompetition.co.uk
Contact *Secretary*

Established 1995. Annual poetry award. Original,
unpublished poems up to 80 lines on any theme.
Closing date: 1 December. Entry fee: £3 per poem.
Send s.a.e. for rules and entry form.

£ Prizes of £1,000 (1st); £500 (2nd); £250 (3rd)
plus inscribed Royal Brierley Crystal Vase; and
three highly commended prizes of £125 each. New
prizes for comic verse: £250 (1st) and Young Poets
(14–18): £250 (1st); £125 (2nd).

Kent & Sussex Poetry Society Open Competition

13 Ruscombe Close, Southborough, Tunbridge
Wells TN4 0SG
☎ 01892 543862
www.kentandsussexpoetrysociety.org
Chairman *Clive R. Eastwood*

Annual competition. Entry fee: £4 per poem,
maximum 40 lines. Full details available on the
website.

£ Prizes total £1,350.

Kiriyama Pacific Rim Book Prize

Pacific Rim Voices, 300 Third Street, Suite 822,
San Francisco, CA 94107, USA
☎ 001 415 777 1628   ᖴ 001 415 777 1646
jeannine@kiriyamaprize.org
www.kiriyamaprize.org
Contact *Jeannine Stronach*, Prize Manager

Founded 1996 with the aim of promoting books
that contribute to greater understanding and
cooperation among the peoples and nations of
the Pacific Rim and South Asia, this annual award
takes its name from the Reverend Seiyu Kiriyama.
Entry details may be obtained from the address
above or from the website. 2008 winners: Lloyd
Jones *Mister Pip* (fiction); Julia Whitty *The Fragile
Edge*.

£ Prizes of $30,000, divided between both
winners.

Kraszna-Krausz Book Awards

37 Eton Avenue, London N12 0BD
☎ 020 7100 2523
awards@kraszna-krausz.org.uk
www.kraszna-krausz.org.uk
Administrator *Jenny Boyce*

Established 1985. Annual award to encourage
and recognize oustanding achievements in
the publishing and writing of books on the
art, practice, history and technology of still
photography and the moving image (film,
television, video and related screen media). Books
in English, distributed in the UK, are eligible.
Entries must be submitted by publishers only.

Lakeland Book of the Year Awards

Cumbria Tourism, Windermere Road, Staveley
LA8 9PL
☎ 01539 825052   ᖴ 01539 825076
slindsay@cumbriatourism.org
www.golakes.co.uk
Contact *Sheila Lindsay*

Established in 1984 by local author and broadcaster
Hunter Davies in conjunction with the Cumbria
Tourist Board, the books entered can be about any
aspect of life in the county of Cumbria, from local
history books and walking guides to novels and
poetry. The contest attracts entries from both new
and established authors. Since the establishment of
the awards they have grown in importance and are
now attracting in the region of 60 entries annually,
all competing for the Hunter Davies Prize for the
Lakeland Book of the Year.

In addition to the Hunter Davies Prize there
are currently five categories including Award
for Guides, Walks and Places; Award for People
and Social History; Award for Arts and Culture;
Award for Heritage and Tradition; and the Best
Illustrated Book. Closing date for entries is
mid-March. The awards are presented at a charity
luncheon in early July.

£ Awards of £100 for each category together
with a framed certificate. Overall winner of the
Hunter Davies Prize for the Lakeland Book of the
Year also receives a cheque for £100 and a framed
certificate.

Lancashire County Library and Information Service Children's Book of the Year Award

Lancashire County Library Headquarters, County
Hall, PO Box 61, Preston PR1 8RJ
☎ 01772 534751   ᖴ 01772 534880
jacob.hope@lcl.lancscc.gov.uk
Award Coordinator *Jake Hope*

Established 1986. Annual award, presented
in June for a work of original fiction suitable
for 12–14-year-olds. The winner is chosen by
13–14-year-old secondary school pupils in
Lancashire. Books must have been published
between 1 September and 31 August in the
previous year of the award and authors must be
UK and Republic of Ireland residents. Final entry
date: 1 September each year. 2007 winner: Robert
Muchamore *Divine Madness*.

£ Prize of £1,000, plus engraved glass decanter.

Lannan Literary Award

Lannan Foundation, 313 Read Street, Santa Fe,
New Mexico 87501, USA
☎ 001 505 986 8160
www.lannan.org

Established 1989. Annual awards given to writers of exceptional poetry, fiction and non-fiction who have made a significant contribution to English-language literature, as well as emerging writers of distinctive literary merit who have demonstrated potential for outstanding future work. On occasion, the Foundation recognizes a writer for lifetime achievement. Candidates for the awards and fellowships are recommended to the Foundation by a network of writers, literary scholars, publishers and editors. Applications or unsolicited nominations for the awards and fellowships are not accepted.

The Duncan Lawrie Dagger (in association with Crime Writers' Association)

PO Box 273, Boreham Wood WD6 2XA
secretary@thecwa.co.uk
www.thecwa.co.uk
Contact *The Secretary*

Established 2006. Sponsored by Duncan Lawrie Bank. An annual award for the best crime fiction published during the year. Nominations from publishers only. 2007 winner: Peter Temple *The Broken Shore*.

Ⓔ Award winner receives a Dagger, plus cheque for £20,000.

The Duncan Lawrie International Dagger (in association with Crime Writers' Association)

PO Box 273, Boreham Wood WD6 2XA
secretary@thecwa.co.uk
www.thecwa.co.uk
Contact *The Secretary*

Established 2006. Sponsored by Duncan Lawrie Bank. An annual award for the best foreign crime fiction translated into English during the year. Nominations from publishers only. 2007 winner: Fred Vargas *Wash This Blood Clean From My Hand* (translator: Sian Reynolds).

Ⓔ Award winner receives a Dagger, plus cheque for £5,000 to the author; £1,000 to the translator.

Le Prince Maurice Prize

Constance Hotels Experience, Poste de Flacq, Mauritius, Indian Ocean
Ⓣ 020 7235 3245   Ⓕ 020 7235 3246
comm@constancehotels.com
www.constancehotels.com
Contact *Kelly Desire*

Founded in 2003 and sponsored by one of Mauritius's five-star resorts, Constance Le Prince Maurice. Annual award designed to celebrate the literary love story. Administered in both the UK and France, it is awarded alternately to an English-speaking and French-speaking writer. The aim is to strengthen the cultural links between Mauritius and Europe. Submissions are made by publishers only; closing date for next English

award is September 2009. Previous winner: Anne Donovan *Buddha Da*.

Ⓔ Prize winner receives a trophy, plus all-expenses-paid two-week 'writer's retreat' at Constance Le Prince Maurice.

Legend Writing Award

39 Emmanuel Road, Hastings TN34 3LB
www.legendwritingaward.com
Contact *Legend Coordinator*

Established 2001. Annual award to encourage new fiction writers resident in the UK. The competition is for short stories of 2,000 words maximum and there is no set theme. Closing date: 31 August. Entry fees: £7 (first entry); £5 (each subsequent entry). Rules/entry form available from website or by sending s.a.e. to address above. The competition is organized and judged by Hastings Writers' Group.

Ⓔ Prizes of £500 (1st); £250 (2nd); £100 (3rd); plus three runners-up prizes of £50.

The Bernard Levin Award

Society of Indexers, Woodbourn Business Centre, 10 Jessell Street, Sheffield S9 3HY
Ⓣ 0114 244 9561   Ⓕ 0114 244 9563
admin@indexers.org.uk
www.indexers.org.uk
Secretary *Judith Menes*

Established in 2000 to celebrate the late Bernard Levin, a journalist and author whose writings showed untiring and eloquent support for indexers and indexing. An occasional award for outstanding services to the Society of Indexers. Last awarded in 2007 to Jill Halliday.

The Astrid Lindgren Memorial Award for Literature (ALMA)

Swedish Arts Council, PO Box 27215, SE-102 53 Stockholm, Sweden
Ⓣ 00 46 8 519 264 00   Ⓕ 00 46 8 519 264 99
literatureaward@alma.se
www.alma.se
Director *Anna Cokorilo*

Established 2002 by the Swedish government in memory of the children's author Astrid Lindgren. Administered by the Swedish Arts Council, it is an international award for children's and young people's literature given annually to one or more recipients, irrespective of language or nationality. Writing, illustrating and storytelling, as well as reading promotion activities may be awarded. Selected organizations worldwide are invited to submit nominations once a year; jury members may also contribute nominations. 2008 winner: Sonya Hartnett.

Ⓔ Award of SEK 5 million (approx. £400,000)

John Llewellyn Rhys Prize

Booktrust, Book House, 45 East Hill, London
SW18 2QZ
☎ 020 8516 2972   📠 020 8516 2978
tarryn@booktrust.org.uk
www.booktrust.org.uk
Contact *Tarryn McKay*

Established 1942. An annual writer's award for a memorable work of fiction, non-fiction, drama or poetry. Entrants must be 35 or under at the time of publication; books must have been published in the UK during the calendar year of the award. The author must be a citizen of the UK or the Commonwealth, writing in English. 2006/7 winner: Sarah Hall, *The Carhullan Army*.

£ Prize of £5,000 (1st); £500 for shortlisted entries.

The Elizabeth Longford Prize for Historical Biography

The Society of Authors, 84 Drayton Gardens, London SW10 9SB
☎ 020 7373 6642   📠 020 7373 5768
info@societyofauthors.org
www.societyofauthors.org

Established in 2003 in memory of Elizabeth Longford and sponsored by Flora Fraser and Peter Soros. Awarded annually for a historical biography published in the year preceding the prize. No unsolicited submissions. 2007 winner: Jessie Childs *Henry VIII's Last Victim – The Life and Times of Henry Howard, Earl of Surrey*.

£ Prize of £5,000.

Longman-History Today Book of the Year Award

c/o History Today, 20 Old Compton Street, London W1D 4TW
☎ 020 7534 8000
www.historytoday.com
Contact *Peter Furtado*

Established 1993. Annual award set up as a joint initiative between the magazine *History Today* and the publisher Longman (**Pearson**) to mark the past links between the two organizations, to encourage new writers, and to promote a wider public understanding of, and enthusiasm for, the study and publication of history. Award for author's first or second book.

£ Prize of £2,000.

The Clarissa Luard Award
▷ David Cohen Prize for Literature

Sir William Lyons Award

The Guild of Motoring Writers, 40 Baring Road, Bournemouth BH6 4DT
☎ 01202 422424   📠 01202 422424

generalsec@gomw.co.uk
www.guildofmotoringwriters.co.uk
Contact *Patricia Lodge*

An annual competitive award, sponsored by Jaguar, to encourage young people in automotive journalism and to foster interests into motoring and the motor industry. Entrance by two essays and interview with Awards Committee. Applicants must be British, aged 17–23 and resident in UK. Final entry date: 1 October.

McColvin Medal
▷ Besterman/McColvin Medal

W.J.M. Mackenzie Book Prize

Political Studies Association, Dept. of Politics, University of Newcastle, Newcastle upon Tyne NE1 7RU
☎ 0191 222 8021   📠 0191 222 3499
psa@ncl.ac.uk
www.psa.ac.uk
PSA Executive Director *Jack Arthurs*

Established 1987. Annual award to best work of political science published in the UK during the previous year. Submissions from publishers only. Final entry date: 31 October. Prizes are judged in the year following publication and awarded the year after that. Previous winner: Iain McLean and Alistair McMillan *State of the Union*.

McKitterick Prize

Society of Authors, 84 Drayton Gardens, London SW10 9SB
☎ 020 7373 6642   📠 020 7373 5768
info@societyofauthors.org
www.societyofauthors.org
Contact *Awards Secretary*

Annual award for a full-length novel in the English language, first published in the UK or unpublished. Open to writers over 40 who have not had any novel published other than the one submitted (excluding works for children). Closing date: 20 December. 2007 winner: Reina James *This Time of Dying*.

£ Prize of £4,000.

Enid McLeod Prize

Franco-British Society, 2 Dovedale Studios, 465 Battersea Park Road, London SW11 4LR
☎ 020 7924 3511
www.francobritishsociety.org.uk
Executive Secretary *Kate Brayn*

Established 1982. Annual award to the author of the work of literature published in the UK which, in the opinion of the judges, has contributed most to Franco-British understanding. Any full-length work written in English by a citizen of the UK Commonwealth is eligible. No English translation of a book written originally in any other language will be considered. Nominations from publishers for books published between 1 January and 31 December of the year of the prize. Closing date: 31 December. 2007 winner: Professor Tim Blanning *The Pursuit of Glory*.

Ⓔ Prize winner receives a cheque.

Macmillan Prize for a Children's Picture Book Illustration

Macmillan Children's Books, 20 New Wharf Road, London N1 9RR
☏ 020 7014 6124   Ⓕ 020 7014 6142
lindsey.evans@macmillan.co.uk
www.panmacmillan.com
Contact *Lindsey Evans*, Macmillan Children's Books

Set up in order to stimulate new work from young illustrators in art schools, and to help them start their professional lives. Fiction or non-fiction. Macmillan have the option to publish any of the prize winners.

Ⓔ Prizes of £1,000 (1st); £500 (2nd); £250 (3rd).

Macmillan Writer's Prize for Africa

Macmillan Education, Between Towns Road, Oxford OX4 3PP
☏ 01865 405700   Ⓕ 01865 405788
www.write4africa.com
Contact *Nicky Price*

Founded in 2001 to encourage and to recognize original writing for children and young people by African writers. Sponsored by **Macmillan Publishers Ltd**, which has publishing companies throughout sub-Saharan Africa. The prize is for previously unpublished works of fiction by African writers, and aims to promote and to celebrate story writing from all over the continent. Two main awards – for children's literature and teenage fiction – and an additional award dedicated to new, previously unpublished writers. See website for entry qualifications. 2007/08 winners: Jayne Bauling *E Eights* (Senior Award); Nnedi Okorafor-Mbachu *Long Juju Man* (Junior); Ekow Kwegyir Bentum *Kwansa and the Bandit Crabs* (New Children's Writer).

Ⓔ Prizes of US$ 5,000 each (Senior and Junior awards); US$3,000 (New Children's Writer).

The Mail on Sunday Novel Competition

Postal box address may change annually (see below)

Annual award established 1983. Judges (2008: Sir John Mortimer, Fay Weldon and Michael Ridpath) look for a story/character that springs to life in the 'tantalizing opening 50–150 words of a novel'. Details of the competition, including the postal box address, are published in *The Mail on Sunday* in July/August. 2007 winner: Jane Forrest.

Ⓔ Awards of £400 book tokens and a writing course at the **Arvon Foundation** (1st); £300 tokens (2nd); £200 tokens (3rd); three further prizes of £150 tokens each.

The Man Asian Literary Prize

22nd Floor, Easy Commercial Building, 253–261 Hennessy Road, Hong Kong
☏ 00 852 2877 8444
info@manasianliteraryprize.org
www.manasianliteraryprize.org
Chairman *Peter Gordon*

Initiated by the Man Group and the Hong Kong Literary Festival in 2007. An annual award which aims to recognize the best of new Asian literature and to bring it to the attention of the world literary community. Open to a single work of Asian fiction in English (but unpublished in English) of no less than 30,000 words by an Asian author residing in an Asian country or territory. Final entry date: end March. Inaugural winner: Jiang Rong *Wolf Totem*.

The Man Booker International Prize

Colman Getty, 28 Windmill Street, London W1T 2JJ
☏ 020 7631 2666   Ⓕ 020 7631 2699
info@colmangetty.co.uk
www.manbookerinternational.com

Established in 2004 to complement **The Man Booker Prize for Fiction** by recognizing one writer's achievement in literature and their significant influence on writers and readers worldwide. Sponsored by the Man Group, this biennial award is given to a living author who has published fiction either originally in English or whose work is generally available in translation into the English language. Submissions for the prize are not invited. Where the winning author's work has been translated into English an additional prize of £15,000 is awarded to the translator. The winning author chooses who the translator's prize should go to and whether it is to be awarded to one translator or divided between several. 2007 winner: Chinua Achebe.

The Man Booker Prize for Fiction

Colman Getty, 28 Windmill Street, London W1T 4JE
☏ 020 7631 2666   Ⓕ 020 7631 2699
info@colmangetty.co.uk
www.themanbookerprize.com
Contact *Lois Tucker (Submissions)*

The Booker Prize for Fiction was originally set up by Booker plc in 1968 to reward merit, raise the stature of the author in the eyes of the public and encourage an interest in contemporary fiction. In April 2002 it was announced that the Man Group had been chosen by the Booker Prize

Foundation as the new sponsor of the Booker Prize. The sponsorship is due to run until 2011. United Kingdom publishers may enter up to two full-length novels, with scheduled publication dates between 1 October 2007 and 30 September 2008. In addition, any title by an author who has previously won the Booker or Man Booker Prize and any title by an author who has been shortlisted in the last ten years may be submitted. 2007 winner: Anne Enright *The Gathering*.

£ The winner receives £50,000. The six shortlisted authors each receive £2,500.

Marsh Award for Children's Literature in Translation

The English-Speaking Union, Dartmouth House, 37 Charles Street, London W1J 5ED
☏ 020 7529 1550   ⨏ 020 7495 6108
education@esu.org
Contact *Elizabeth Stokes*

Established 1995 and sponsored by the Marsh Christian Trust, the award aims to encourage translation of foreign children's books into English. It is a biennial award (next award: 2009), open to British translators of books for 4–16-year-olds, published in the UK by a British publisher. Any category will be considered with the exception of encyclopedias and reference books. No electronic books. 2007 winner: Anthea Bell for her translation of *The Flowing Queen* by Kai Meyer.

£ Prize of £1,000.

Marsh Biography Award

The English-Speaking Union, Dartmouth House, 37 Charles Street, London W1J 5ED
☏ 020 7529 1563   ⨏ 020 7495 6108
katie_brock@esu.org
www.esu.org
Contact *Katie Brock*

A biennial award for the most significant biography published over a two-year period by a British publisher. Next award October 2009. 2007 winner: Maggie Ferguson *George Mackay Brown: A Life*.

£ Award: Membership of the ESU and £4,000, plus a silver trophy presented at a gala dinner.

Colin Mears Award

▷ CILIP: The Chartered Institute of Library and Information Professionals Kate Greenaway Medal

Medical Book Awards

The Society of Authors, 84 Drayton Gardens, London SW10 9SB
☏ 020 7373 6642   ⨏ 020 7373 5768
sbaxter@societyofauthors.org
www.societyofauthors.org
Contact *Sarah Baxter*, Medical Writers Group

Annual awards sponsored by the Royal Society of Medicine. Nine categories for medical textbooks published in the twelve months preceding the deadline. Contact the **Society of Authors** for entry details. Closing date: 20 April.

£ Prizes total £6,500.

The Mercedes-Benz Award for the Montagu of Beaulieu Trophy

Guild of Motoring Writers, 40 Baring Road, Bournemouth BH6 4DT
☏ 01202 422424   ⨏ 01202 422424
generalsec@gomw.co.uk
www.guildofmotoringwriters.co.uk
Contact *Patricia Lodge*

First presented by Lord Montagu on the occasion of the opening of the National Motor Museum at Beaulieu in 1972. Awarded annually to a member of the **Guild of Motoring Writers** who, in the opinion of the nominated jury, has made the greatest contribution to recording in the English language the history of motoring or motor cycling in a published book or article, film, television or radio script, or research manuscript available to the public.

£ Cash prize sponsored by Mercedes-Benz UK.

Mere Literary Festival Open Competition

'Lawrences', Old Hollow, Mere BA12 6EG
☏ 01747 860475
www.merewilts.org.uk
Contact *Mrs Adrienne Howell* (Events Organizer)

Annual open competition which alternates between short stories and poetry. The winners are announced at the **Mere Literary Festival** during the second week of October. The 2008 competition is for short fiction. For further details, including entry fees and form, access the website or contact the address above from 1 March with s.a.e.

£ Cash prizes.

Meyer-Whitworth Award

Playwrights' Studio Scotland, CCA, 350 Sauchiehall Street, Glasgow G2 3JD
☏ 0141 332 4403
info@playwrightsstudio.co.uk
www.playwrightsstudio.co.uk
Contact *Claire Burkitt*

The Meyer-Whitworth Award is the largest annual monetary prize for playwriting in the UK. Funded by the Royal National Theatre Foundation and managed by the Playwrights' Studio, Scotland in association with the UK Playwrights Network, it is intended to help further the careers of UK playwrights who are not yet established. Nominations by directors of professional theatre companies only. Plays must be original work, be in the English language and have been produced professionally in the UK for the first time between 1 November and 30 November of the previous year. Translations are not eligible. Candidates will have had no more than two of their plays professionally produced, including the play submitted. No writer who has previously won the award may reapply and no play that has previously been submitted for the award is eligible. Writers must be resident in the British Isles or Republic of Ireland. Nomination forms available from the Playwrights' Studio, Scotland.

£ Award of £10,000.

MIND Book of the Year

Granta House, 15–19 Broadway, London E15 4BQ
☎ 020 8215 2301   🖷 020 8215 2269
j.bird@mind.org.uk
www.mind.org.uk

Established 1981. Annual award, in memory of Sir Allen Lane, for the author of a book published in the current year (fiction or non-fiction), which furthers public understanding of mental health problems. 2008 winner: Martin Townsend *The Father I Had*.

The Mitchell Prize for Art History/The Eric Mitchell Prize

c/o The Burlington Magazine, 14–16 Duke's Road, London WC1H 9SZ
☎ 020 7388 8157   🖷 020 7388 1230
hopcraft@burlington.org.uk (with 'Mitchell Prize' in subject line)
Chairman *Caroline Elam*
Contact *Alice Hopcraft*

The Mitchell Prize is awarded for a book which has made an outstanding and original contribution to the understanding of the visual arts, and The Eric Mitchell Prize is awarded for the best exhibition catalogue. Both must be on western art, written in English and published in the previous two years. Nominations should be submitted by publishers before the end of April. 2006 winners: Mark P. McDonald *The Print Collection of Ferdinand Columbus* (The Mitchell Prize); Elena Phipps, Johanna Hecht, Christina Esteras Martin *The Colonial Andes: Tapestry & Silverware 1530–1830*, The Metropolitan Museum of Art, New York in association with Yale University Press 2004 (The Eric Mitchell Prize).

Momaya Press Short Story Writing Competition

Momaya Press, Flat 1, 189 Balham High Road, Rear Building, London SW12 9BE
☎ 020 8673 9616
infouk@momayapress.com
www.momayapress.com
Contact *Monisha Saldanha*

Established 2004. Annual short story competition sponsored by Momaya Press. Open to writers in the English language worldwide. Entries, which should not have been published previously, can be in any style and format and on any subject; 2,500 words maximum. Entry fee: £7. Submissions by post or via entry form on the website.

£ Prizes of £110 (1st); £60 (2nd); £30 (3rd); all winners are published in the *Momaya Annual Review*.

Scott Moncrieff Prize
▷ The Translators Association Awards

The Sheridan Morley Prize

c/o Oberon Books, 521 Caledonian Road, London N7 9RH
☎ 020 7607 3637   🖷 020 7607 3629
info@oberonbooks.com
Contact *David Miller*

Established in 2007/08 as a lasting memorial to Sheridan Morley, who during a long and illustrious career as critic, broadcaster and author, wrote more than 30 books. The prize celebrates the art of theatre biography and is awarded every April to the best biography on a theatrical subject, written in the English language and published in the given year. Final entry date: early March. Inaugural winner: *Will & Me. How Shakespeare Took Over My Life* Dominic Dromgoole.

£ Prize of £2,000.

nasen Awards 2008 – Celebrating Inclusive Practice, in association with TES

Nasen House, 4–5 Amber Business Village, Amber Close, Amington, Tamworth B77 4RP
☎ 01827 311 500   🖷 01827 313 005
welcome@nasen.org.uk
www.nasen.org.uk

Organized by nasen in association with the *Times Educational Supplement*. Awards presented in eight categories: Children's Book; Academic Book; Book to Support Teaching and Learning – Teacher Book; Book to Support Teaching and Learning – Pupil Book; Book that supports SEN & Disability Issues; Inclusive Resource for Primary Classrooms; Inclusive Resource for Secondary Classrooms; ICT Accessibility. Deadline: 30 April each year. Eligibility for 2008 Awards: books and resources must have been published in the UK between 22 June 2007 and 30 April 2008. Contact nasen for more information.

£ Prize of £500 for each category.

Melissa Nathan Award for Comedy Romance

Melissa Nathan Foundation and Award, PO Box 56923, London N10 3YU
☎ 020 8671 8424   🖷 020 8883 6694

information@melissanathan.co.uk
www.melissanathan.com
Chair *Andrew Saffron*
Committee *Maggie Phillips*

Established in 2007 and named after bestselling novelist, Melissa Nathan, who died of cancer aged 37, to honour her memory and writers of comedy romance novels. Awarded to the novel that best combines laugh-out-loud comedy with believable, heart-warming romance (varying sub-category awards each year). To be eligible, novels must be published in the UK in the year of the award, including paperback version of hardbacks published the previous year. Final entry date: 31 December. Inaugural winner: Marian Keyes *Anybody Out There?*

£ Prize of £5,000, plus trophy.

National Poetry Anthology

United Press, Admail 3735, London EC1B 1JB
℡ 0870 240 6190   ℻ 0870 240 6191
mail@unitedpress.co.uk
www.unitedpress.co.uk
Contact *Julie Embury*

Free-to-enter annual poetry competition. Organizers United Press allow up to three poems (20 lines and 160 words maximum each) by annual closing date of 30 June. They select around 250 regional winners in the UK. These receive a free copy of the annual book, *The National Poetry Anthology* and vote for the overall winner who receives £1,000 and a trophy to keep for life.

National Poetry Competition

The Poetry Society, 22 Betterton Street, London WC2H 9BX
℡ 020 7420 9895   ℻ 020 7240 4818
marketing@poetrysociety.org.uk
www.poetrysociety.org.uk
Contact *Competition Organizer (WH)*

One of Britain's major open poetry competitions. Closing date: 31 October. Poems on any theme, up to 40 lines. For rules and entry form send s.a.e. to the competition organizer at the address above or enter the competition via the website.

£ Prizes of £5,000 (1st); £1,000 (2nd); £500 (3rd); plus 10 commendations of £50.

New Writing Ventures

www.newwritingpartnership.org.uk/ventures

New Writing Ventures will not be running in 2008. The administrators, Booktrust, reported: 'After three very successful years we are taking a break to evaluate the impact of the programme and to determine its direction in the future. Over the past three years we've worked with 27 great new writers and seen some amazing collections of poetry and novels published. We've seen writers find agents, change genres, discover whole new areas of interest and most importantly, benefit from in-depth mentoring relationships with established writers that will continue to have an impact on their writing careers for some time to come. We hope that you'll keep an eye on the New Writing Ventures website for further news about new developments.'

'The Nibbies'

▷ Galaxy British Book Awards

Nielsen Gold & Platinum Book Awards

Nielsen BookScan, 3rd Floor, Midas House, 62 Goldsworth Road, Woking GU21 6LQ
℡ 01483 712222   ℻ 01483 712220
awards.bookscan@nielsen.com
www.nielsenbookscan.co.uk
Contact *Mo Siewcharran*

Established in 2000 to award sales achieved for actual consumer purchases of a title through UK bookshops. Awarded to any title, in all its editions, that sells more than 500,000 copies (Gold) or one million copies (Platinum) over a period of five consecutive years.

£ Award winner receives commemorative plaque issued by the publisher to the author.

Nobel Prize

The Nobel Foundation, PO Box 5232, 102 45 Stockholm, Sweden
℡ 00 46 8 663 0920   ℻ 00 46 8 660 3847
www.nobel.se
Contact *Information Section*

Awarded yearly for outstanding achievement in physics, chemistry, physiology or medicine, literature and peace. Founded by Alfred Nobel, a chemist who proved his creative ability by inventing dynamite. In general, individuals cannot nominate someone for a Nobel Prize. The rules vary from prize to prize but the following are eligible to do so for Literature: members of the Swedish Academy and of other academies, institutions and societies similar to it in constitution and purpose; professors of literature and of linguistics at universities or colleges; Nobel Laureates in Literature; presidents of authors' organizations which are representative of the literary production in their respective countries. British winners of the literature prize, first granted in 1901, include Rudyard Kipling, John Galsworthy and Winston Churchill. Winners since 1998: José Saramago (Portugal); Günter Grass (Germany); Gao Xingjian (France); V.S. Naipaul (UK); Imre Kertész (Hungary); J.M. Coetzee (South Africa); Elfriede Jelinek (Austria); Harold Pinter (UK); Orhan Pamuk (Turkey). Nobel Laureate in Literature 2007: Doris Lessing (UK).

The Noma Award for Publishing Africa

PO Box 128, Witney OX8 5XU

☎ 01993 775235   ⓕ 01993 709265

maryljay@aol.com

www.nomaaward.org

Contact *Mary Jay*, Secretary to the Managing Committee

Established 1979. Annual award, founded by the late Shoichi Noma, President of Kodansha Ltd, Tokyo. The award is for an outstanding book, published in Africa by an African writer, in three categories: scholarly and academic; literature and creative writing; children's books. Entries, by publishers only, by 28 February for a title published in the previous year. Maximum number of three entries. Previous winners: Hamdi Sakkut *The Arabic Novel: Bibliography and Critical Introduction 1865–1995*; Elinor Sisulu *Walter and Albertina Sisulu. In Our Lifetime*; Werewere-Liking *La mémoire amputée*; Lebogang Mashile *In a Ribbon of Rhythm*; Shimmer Chinodya *Strife*.

ⓔ Prize of US$10,000 and presentation plaque.

Frank O'Connor International Short Story Award

The Munster Literature Centre, Frank O'Connor House, 84 Douglas Street, Cork, Republic of Ireland

☎ 00 353 21 431 2955

munsterlit@eircom.net

www.munsterlit.ie

Contact *Patrick Cotter*

Founded in 2005 to reward and encourage writers and publishers of short stories. Annual competition for a collection of original short stories by a single living author published between September of the previous year and August of the award year. Final entry date: 31 August. Vanity press publications not accepted. 2007 winner: Miranda July.

ⓔ Award of €35,000.

Sean O'Faolain Short Story Prize

The Munster Literature Centre, Frank O'Connor House, 84 Douglas Street, Cork, Republic of Ireland

☎ 00 353 21 431 2955

munsterlit@eircom.net

www.munsterlit.ie

Contact *Patrick Cotter*

Founded 2003. Annual competition for an original, unpublished story of under 3,000 words. Final entry date: 1 July. 2006 winner: Joyce Russell.

ⓔ Awards of €1,500 (1st); €500 (2nd).

C.B. Oldman Prize

Special Libraries & Archives, University of Aberdeen, King's College, Aberdeen AB24 3SW

☎ 01224 274266   ⓕ 01224 273891

r.turbet@abdn.ac.uk

Contact *Richard Turbet*

Established 1989 by the International Association of Music Libraries, UK & Ireland Branch. Annual award for best book of music bibliography, librarianship or reference published the year before last (i.e. books published in 2006 considered for the 2008 prize). Previous winners: Pamela Thompson, Malcolm Lewis, Andrew Ashbee, Michael Talbot, Donald Clarke, John Parkinson, John Wagstaff, Stanley Sadie, William Waterhouse, Richard Turbet, John Gillaspie, David Fallows, Arthur Searle, Graham Johnson, Michael Twyman, John Henderson, David Golby, David Wyn Jones.

ⓔ Prize of £200.

Ondaatje Prize

▷ The Royal Society of Literature Ondaatje Prize

Orange Broadband Prize for Fiction/Orange Broadband Award for New Writers

Booktrust, 45 East Hill, London SW18 2QZ

☎ 020 8516 2972   ⓕ 020 8516 2978

tarryn@booktrust.org.uk

www.orangeprize.co.uk

Contact *Tarryn McKay*

Established 1996. Annual award founded by a group of senior women in publishing to 'create the opportunity for more women to be rewarded for their work and to be better known by the reading public'. Awarded for a full-length novel written in English by a woman of any nationality, and published in the UK between 1 April and 31 March of the following year. 2008 winner: Rose Tremain *The Road Home*..

ⓔ Prize of £30,000 and a work of art (a limited edition bronze figurine known as 'The Bessie' in acknowledgement of anonymous prize endowment).

Established 2005, the **Orange Broadband Award for New Writers** is open to all first works of fiction, written by women of any age or nationality, published in the UK between 1 April and 31 March (short story collections and novellas also eligible). 2008 winner: Joanna Cavenna *Inglorious*..

The Orwell Prize

Blackwell Publishing, 9600 Garsington Road, Oxford OX4 2DQ

☎ 01865 476255   ⓕ 01865 471255

lucie.crowther@oxon.blackwellpublishing.com

www.theorwellprize.co.uk

www.blackwellpublishing.com/poqu
Contact *Lucie Crowther*

Jointly established in 1993 by the George Orwell Memorial Fund and *Political Quarterly* to encourage and reward writing in the spirit of Orwell's 'What I have most wanted to do ... is to make political writing into an art'. Two categories: book or pamphlet; sustained journalism, reporting, comment or blogs (on a theme) in broadcast, online, newspaper, periodicals, features or columns. Submissions by editors, publishers or authors. 2007 winners (for work published in 2007): Johann Hari of *The Independent* (journalism); Raja Shehadeh, *Palestinian Walks* (book); and a Special Award for Writing and Broadcasting was presented to Clive James.

£ Prizes of £1,000 for each category.

The Wilfred Owen Award for Poetry

21 Culverden Avenue, Tunbridge Wells TN4 9RE
☎ 01892 532712   📠 01892 532712
mmccrane@ukonline.co.uk
www.1914-18.co.uk/owen
Contact *Meg Crane*

Established in 1988 by the **Wilfred Owen Association**. Given to a poet whose poetry reflects the spirit of Owen's work in its thinking, expression and inspiration. Applications are not sought; the decision is made by the Association's committee. Previous winners: Seamus Heaney, Christopher Logue, Michael Longley, Harold Pinter, Tony Harrison.

£ Award winner receives a silver and gunmetal work of art, suitably decorated and engraved.

OWPG Awards for Excellence

Outdoor Writers & Photographers Guild, PO Box 520, Bamber Bridge, Preston PR5 8LF
☎ 01772 321243   📠 0870 137 8888
secretary@owpg.org.uk
www.owpg.org.uk
Secretary *Terry Marsh*

Established 1980. Annual awards by the **Outdoor Writers & Photographers Guild** to raise the standard of outdoor writing, journalism, broadcasting and photography. Categories include guidebook, outdoor book, outdoor feature, travel feature, words and pictures, photography. Open to OWPG members only. Final entry date: April.

The Oxford Weidenfeld Translation Prize

St Anne's College, Oxford OX2 6HS
☎ 01865 274820   📠 01865 274895
sandra.madley@st-annes.ox.ac.uk
www.stannes.ox.ac.uk/about/translationprize.html
Contact *The Principals' Secretary*

Established in 1996 by publisher Lord Weidenfeld and now includes St. Anne's, Queen's and New colleges to encourage good translation into English. Annual award to the translator(s) of a work of fiction, poetry or drama written in any living European language. Translations must have been published in the previous calendar year. Submissions from publishers only.

£ Prize of £2,000.

PEN Awards

See **J.R. Ackerley Prize**; **Golden PEN Award for Lifetime Distinguished Service to Literature**; **The Hessell-Tiltman Prize for History**

Samuel Pepys Award

Haremoor House, Faringdon SN7 8PN
Contact *Paul Gray*

Established 2003. Biennial award (closing date: 31 May 2009) given by the Samuel Pepys Award Trust for a book that makes the greatest contribution to the understanding of Samuel Pepys, his times or his contemporaries. 2007 winner: John Adamson *The Noble Revolt*.

£ Award of £2,000, plus commemorative medal.

Peterloo Poets Open Poetry Competition

The Old Chapel, Sand Lane, Calstock PL18 9QX
☎ 01822 833473
publisher@peterloopoets.com
www.peterloopoets.com
Contact *Harry Chambers*

Established 1986. Annual competition for unpublished English language poems of not more than 40 lines. Final entry date: 1 March. Send s.a.e. for rules and entry form. Previous winners: Susan Utting, David Craig, Rodney Pybus, Debjani Chatterjee, Donald Atkinson, Romesh Gunesekera, Anna Crowe, Carol Ann Duffy, Mimi Khalvati, John Lyons, M.R. Peacocke, Alison Pryde, John Weston, Maureen Wilkinson, Chris Woods, Judy Gahagan, Carol Rumens, John Godfrey.

£ Prizes total £4,100; £1,500 (1st); £1,000 (2nd); £500 (3rd); £100 (4th); plus ten prizes of £50; 15–19 age group: five prizes of £100.

Poetry Business Competition

Bank Street Arts, 32–40 Bank Street, Sheffield S1 2DS
edit@poetrybusiness.co.uk
www.poetrybusiness.co.uk
Contact *The Competition Administrator*

Established 1986. Annual award which aims to discover and publish new writers. Entrants should submit a short manuscript of poems. Winners will have their work published by the **Poetry Business** under the Smith/Doorstop imprint. Contact for conditions of entry. Previous winners include: Pauline Stainer, Michael Laskey, Mimi Khalvati, David Morley, Moniza Alvi, Selima Hill, Catherine Smith, Daljit Nagra.

£ Prizes: Publication of full collection; runners-up have pamphlets; 20 complimentary copies. Also cash prize (£1,000) to be shared equally between all winners.

The Poetry Society's National Poetry Competition
▷ National Poetry Competition

The Portico Prize for Literature
The Portico Library, 57 Mosley Street, Manchester M2 3HY
☎ 0161 236 6785   ⊞ 0161 236 6803
librarian@theportico.org.uk
www.theportico.org.uk
Contact *Miss Emma Marigliano*

Established 1985. Administered by the Portico Library in Manchester. Biennial award for a work of fiction or non-fiction published between the two closing dates. Set wholly or mainly in the North West of England, including Cumbria and the High Peak District of Derbyshire. Previous winners include: Anthony Burgess *Any Old Iron*; Jenny Uglow *Elizabeth Gaskell: A Habit of Stories*. 2006 winner: Andrew Biswell *The Real Life of Anthony Burgess*.

Practical Art Book of the Year
PO Box 3, Huntingdon PE28 0QX
☎ 01832 710201   ⊞ 01832 710488
award@acaward.com
www.acaward.com
www.artists-choice.co.uk
Contact *Henry Malt*

Jointly sponsored by **Artists' Choice** book club and *Leisure Painter* and *The Artist* magazines. A short list is drawn up by the editors and the winner voted by readers. Shortlist will be announced in June and the winner in November.

The Premio Valle Inclán
▷ The Translators Association Awards

V.S. Pritchett Memorial Prize
The Royal Society of Literature, Somerset House, Strand, London WC2R 1LA
☎ 020 7845 4676   ⊞ 020 7845 4679
info@rslit.org
www.rslit.org

Established 1999. Awarded for a previously unpublished short story of between 2,000 and 5,000 words. For entry forms, please contact the Secretary from 15 November onwards. Closing date: 31 February. Open to UK and Irish writers.

£ Prize of £1,000.

The Projection Box Essay Awards 2008–2009
12 High Street, Hastings TN34 3EY
☎ 01424 204144   ⊞ 01424 204144
admin@pbawards.co.uk
www.pbawards.co.uk
Contact *Mo Heard*

Founded 2007 by small independent publishers, The Projection Box. Annual open competition which in the current year is to encourage new research and thinking into any historical, artistic or technical aspect of popular optical media. The prize is for an essay in English, not previously published, of between 5,000 and 8,000 words (including notes). Further information available on the website.

£ Prize of £250 plus publication in *Early Popular Visual Culture*.

Prose & Poetry Prizes
The New Writer, PO Box 60, Cranbrook TN17 2ZR
☎ 01580 212626   ⊞ 01580 212041
admin@thenewwriter.com
www.thenewwriter.com
Contact *Merric Davidson*

Established 1997. Annual award. Open to all poets writing in the English language for an original, previously unpublished poem or collection of six to ten poems. Also open to writers of short stories and novellas/serials, features, articles, essays and interviews. Final entry date: 30 November. Full guidelines on the website. Previous winners: Wes Lee, Graham Clifford, Mario Petrucci, Ros Barber, Celia de Fréine, Andrew McGuinness, Katy Darby, David Grubb.

£ Prizes total £2,000, plus publication in annual collection published by *The New Writer*.

Pulitzer Prizes
The Pulitzer Prize Board, 709 Journalism Building, Columbia University, 2950 Broadway, New York, NY 10027, USA
☎ 001 212 854 3841   ⊞ 001 212 854 3342
pulitzer@pulitzer.org
www.pulitzer.org

Awards for journalism in US newspapers, and for published literature, drama and music by American nationals. 2008 winners include: Junot Diaz *The Brief Wondrous Life of Oscar Wao* (fiction); John Matteson *Eden's Outcasts* (biography); *The Washington Post* (public service).

The Red House Children's Book Award
The Federation of Children's Book Groups, 2 Bridge Wood View, Horsforth, Leeds LS18 5PE
☎ 0113 258 8910
andrea@wickave.waitrose.com
www.redhousechildrensbookaward.co.uk
National Coordinator *Marianne Adey*

Established 1980. Three categories. Awarded annually for best book of fiction suitable for children. Unique in that it is judged by the children themselves. 2007 overall winner: Andy Stanton *You're a Bad Man, Mr Gunn!*.

£ Award winners receive silver bowls and trophy, portfolio of letters, drawings and comments from the children who take part in the judging.

Trevor Reese Memorial Prize

Institute of Commonwealth Studies, School of Advanced Studies, University of London, 28 Russell Square, London WC1B 5DS

☏ 020 7862 8853 ℻ 020 7862 8820
troy.rutt@sas.ac.uk
commonwealth.sas.ac.uk/reese.htm

Established in 1979 with the proceeds of contributions to a memorial fund to Dr Trevor Reese, Reader in Commonwealth Studies at the Institute and a distinguished scholar of Imperial History (d.1976). Triennial award (next award 2009 for works published 2006, 2007 and 2008) for a scholarly work, usually by a single author, which has made a wide-ranging, innovative and scholarly contribution in the broadly-defined field of Imperial and Commonwealth History. All correspondence relating to the prize should be marked 'Trevor Reese Memorial Prize'.

£ Prize of £1,000.

Regional Press Awards

Press Gazette, Paulton House, Sheperdess Walk, London N1 7LB

☏ 020 7324 2396 ℻ 020 7566 5769
apont@wilmington.co.uk
www.pressgazette.co.uk

Open to all regional newspapers and regional journalists, whether freelance or staff. June event. Run by the *Press Gazette*.

Renault UK Journalist of the Year Award

Guild of Motoring Writers, 40 Baring Road, Bournemouth BH6 4DT

☏ 01202 422424 ℻ 01202 422424
generalsec@gomw.co.uk
www.guildofmotoringwriters.co.uk
Contact *Patricia Lodge*

Originally the Pierre Dreyfus Award, established in 1977. Awarded annually by Renault UK Ltd in honour of Pierre Dreyfus, president director general of Renault 1955–75, to the member of the **Guild of Motoring Writers** who is judged to have made the most outstanding journalistic effort in any medium during the year. Particular emphasis is placed on initiative and endeavour.

RIBA International Book Awards

c/o 15 Bonhill Street, London EC2P 2EA
☏ 020 7496 8364 ℻ 020 7374 8500

john.morgan@ribabookshops.com
www.ribabookshops.com
Contact *John Morgan*

Founded in 2005 and hosted by RIBA Bookshops and the British Architectural Library (RIBA Trust). Annual open, free competition to recognize and celebrate the best writing on architecture, construction, architectural practice and interior design. Entries must have been published in the year preceding the award year; final entry date: 31 December. Foreign (English language) books, which are encouraged, currently represent around 40 per cent of entries. 2007 winners: RIBA Sir Nikolaus Pevsner Award for Architecture: *Interpreting the Renaissance* Manfredo Tafuri (Yale University Press); RIBA Sir Robert McAlpine Award for Construction: *Stone Conservation* Alison Henry (Donhead Publishing).

£ Money award to author and publishers' trophy (in four categories from 2008).

RITA Awards

▷ Romance Writers of America under Professional Associations and Societies

Romantic Novel of the Year

www.rna-uk.org
Honorary Administrator *Mrs P. Fenton*

Established 1960. Annual award for the best romantic novel of the year, open to non-members as well as members of the **Romantic Novelists' Association**. Novels must be first published in the UK between specified dates. Full details on the RNA website. 2008 winner: Freya North *Pillow Talk*. Contact the organizer for entry form via the website address.

Rossica Translation Prize

Academia Rossica, 151 Kensington High Street, London W8 6SU

☏ 020 7937 5001 ℻ 020 7937 5001
rossica-prize@academia-rossica.org
www.academia-rossica.org
Contacts *Svetlana Adjoubei, Bettina Wrichert*

Founded in 2005 by Academia Rossica (AR), a UK registered charity that supports cultural collaboration between Russia and the West. A biennial prize, which is awarded for the best new literary translation from Russian into English, published anywhere in the world. The aim is to promote the best of Russian literature and a better understanding of Russian culture in English-speaking countries. Four copies of the (published) English translation and three copies of the Russian original should be submitted to Academia Rossica. Final entry date for the 2009 award: 31 December 2008. 2007 winner: Joanne

Turnbull, translator of *7 Stories* by Sigizmund Krzhizhanovsky.

Ⓔ Prizes of £3,000 (translator); £1,000 (publisher).

Royal Economic Society Prize

The Economic Journal, Department of Economics, London Business School, Regent's Park, London NW1 4SA
☎ 020 7000 8413   Ⓕ 020 7000 8401
econjournal@london.edu
www.res.org.uk/society.resprize.asp
Contact *Heather Daly*

Annual award for the best article published in *The Economic Journal*. Open to members of the Royal Economic Society only. Previous winners: Professors Marcos Rangal, Tilman Börgers, Christian Dustmann, Paul Cheshire, Stephen Sheppard.

Ⓔ Prize of £3,000.

Royal Mail Awards for Scottish Children's Books

Scottish Book Trust, Sandeman's House, Trunk's Close, 55 High Street, Edinburgh EH1 1SR
☎ 0131 524 0160   Ⓕ 0131 524 0161
anna.gibbons@scottishbooktrust.com
www.scottishbooktrust.com
Children's Programme Manager *Anna Gibbons*

Awards are given to new and established authors of published books in recognition of high standards of writing for children in three age group categories: younger children (0–7 years); younger readers (8–11), older readers (12–16). There is also a competition for short stories written in Gaelic by children. A shortlist is drawn up by a panel of children's book experts and the winner in each category is selected by children and young people voting for their favourites in schools and libraries across Scotland. Authors should be Scottish or resident in Scotland, but books of particular Scottish interest by other authors are eligible. Final entry date: 31 January; guidelines available on request.

Ⓔ Awards total £15,000.

Royal Society of Literature Awards

The Royal Society of Literature Ondaatje Prize, V.S. Pritchett Memorial Prize and (under *Bursaries, Fellowships and Grants*) The Royal Society of Literature/Jerwood Awards

The Royal Society of Literature Ondaatje Prize

The Royal Society of Literature, Somerset House, Strand, London WC2R 1LA
☎ 020 7845 4676   Ⓕ 020 7845 4679
paulaj@rslit.org
www.rslit.org

Submissions *Paula Johnson*

Administered by the **Royal Society of Literature** and endowed by Sir Christopher Ondaatje, the prize is awarded annually to a book of literary merit, fiction, non-fiction or poetry, best evoking the spirit of a place. All entries must be published within the calendar year 2008 and should be submitted by mid-December. The writer must be a citizen of the UK, Commonwealth or Ireland. Further details available on the website. 2008 winner: Graham Robb *The Discovery of France*.

Ⓔ Prize of £10,000.

The Royal Society Prizes for Science Books

c/o The Royal Society, 6–9 Carlton House Terrace, London SW1Y 5AG
☎ 020 7451 2531   Ⓕ 020 7930 2170
sciencebooks@royalsociety.org
www.royalsociety.org
Contact *Katherine Hardaker*

Annual prizes established in 1988 to celebrate the best in popular science writing. Awarded to books that make science more accessible to readers of all ages and backgrounds. Owned and managed by the Royal Society and supported by the Beecroft Trust. All entries must be written in English and their first publication must have been between 1 January and 31 December; submission by publishers only. Educational textbooks published for professional or specialist audiences are not eligible. 2007 winners: Richard Hammond *Can You Feel the Force?* (Junior Prize); Daniel Gilbert *Stumbling on Happiness* (General Prize).

Ⓔ Prizes total £30,000; £10,000 to each winner; £1,000 to each of the five shortlisted authors in each prize.

Runciman Award

The Anglo-Hellenic League, c/o The Hellenic Centre, 16–18 Paddington Street, London W1U 5AS
☎ 020 7486 9410
Ⓕ 020 7486 4254 (c/o Hellenic Centre)
info@anglohellenicleague.org
www.anglohellenicleague.org
Contact *The Administrator*

Established 1985. Annual award, sponsored by the National Bank of Greece. Founded by the Anglo-Hellenic League to promote Anglo-Greek understanding and friendship, for a work wholly or mainly about some aspect of Greece or the world of Hellenism, published in English in any country of the world in its first edition during 2008. Books published anywhere in the world are eligible. It is a condition of the award that shortlisted books should be available for purchase to readers in the UK at the time of the award ceremony. Named after the late Professor Steven Runciman, former chairman of the Anglo-Hellenic League. No category of writing will be excluded from consideration. Works in translation, with the exception of translations from Greek literature, will not be considered. Final entry date in late January 2009; award presented in May/June. Additional information available on the website.

£ Award of £9,000.

Sainsbury's Baby Book Award
> Booktrust Early Years Awards

The Saltire Literary Awards
Saltire Society, 9 Fountain Close, 22 High Street, Edinburgh EH1 1TF
☎ 0131 556 1836 ℻ 0131 557 1675
saltire@saltiresociety.org.uk
www.saltiresociety.org.uk
Administrator *Kathleen Munro*

Established 1982 and 1988. Annual awards, one for Book of the Year and one for a First Book by an author publishing for the first time. Open to any author of Scottish descent or living in Scotland, or to anyone who has written a book which deals with either the work and life of a Scot or with a Scottish problem, event or situation. Nominations are invited from editors of leading newspapers, magazines and periodicals. 2007 winners: The Faculty of Advocates/Saltire Society Scottish Book of the Year: A.L. Kennedy *Day*; Royal Mail/Saltire Society Scottish First Book of the Year: Mark McNay *Fresh: A Novel*.

£ Prizes of £5,000 (Scottish Book); £1,500 (First Book).

Schlegel–Tieck Prize
> The Translators Association Awards

Science Writer Awards
The Daily Telegraph, 111 Buckingham Palace Road, London SW1W 0DT
☎ 020 7931 2000/0845 094 6367 (Hotline: 10.00 am to 4.00 pm Mon–Fri)
enquiries@science-writer.co.uk
www.science-writer.co.uk
Contact *Roger Highfield*

Established 1987, this award is designed to bridge the gap between science and writing, challenging the writer to come up with a piece of no more than 700 words that is friendly, informative and, above all, understandable. Sponsored by Bayer, the award is open to two age groups: 15–19 and 20–28.

£ Award winners and runners-up receive cash prizes and have the opportunity to have their pieces published on the science pages of *The Daily Telegraph*. The winner in each category also gets a cash prize and a chance to meet the judges, which include Fay Weldon, Sir David Attenborough and Adam Hart-Davis. There is also a prize for schools and a prize for teachers. To check launch date, visit the website, telephone the hotline number or e-mail for further information.

Scottish Arts Council Book of the Year Awards
> Sundial Scottish Arts Council Book of the Year Awards

Scottish Book of the Year
> The Saltire Literary Awards

Scottish History Book of the Year
The Saltire Society, 9 Fountain Close, 22 High Street, Edinburgh EH1 1TF
☎ 0131 556 1836 ℻ 0131 557 1675
saltire@saltiresociety.org.uk
www.saltiresociety.org.uk
Administrator *Kathleen Munro*

Established 1965. Annual award in memory of the late Dr Agnes Mure Mackenzie for a published work of distinguished Scottish historical research of scholarly importance (including intellectual history and the history of science). Editions of texts are not eligible. Nominations are invited and should be sent to the Administrator. 2007 winner: Christopher A. Whatley *The Scots and the Union*.

£ Prize of £1,500.

SES Book Prizes
24 Ireton Grove, Attenborough, Nottingham NG9 6BJ
☎ 0115 925 5959 ℻ 0115 925 5959
grahamlittler@msn.com
www.soc-for-ed-studies.org.uk
Honorary Secretary *Professor Graham H. Littler*
Chair, Book Prize Sub-committee *Dr Ian Davies* (id5@york.ac.uk)

Annual awards given by the Society for Educational Studies for the best books on education published in the UK during the preceding year. Nominations by publishers or by individual authors based in the UK.

£ Prizes of £2,000 (1st); £1,000 (2nd); £400 (highly commended).

Bernard Shaw Translation Prize
> The Translators Association Awards

André Simon Memorial Fund Book Awards
1 Westbourne Gardens, Glasgow G12 9XE
☎ 0141 342 4929
katie@andresimon.co.uk
Contact *Katie Lander*

Established 1978. Awards given annually for the best book on drink, best on food and special commendation in either. 2007 winners: Hugh Fearnley-Whittingstall and Nick Fisher *The River*

Cottage Fish Book; Philip Williamson and David Moore *Wine Behind the Label*.

£ Awards of £2,000 (best book); £1,000 (special commendation); £200 to shortlisted books.

SJA British Sports Journalism Awards
▷ Sports Journalists' Association of Great Britain under Professional Associations and Societies

The Society for Theatre Research Annual Theatre Book Prize

The Society for Theatre Research, PO Box 53971, London SW15 6UL
theatrebookprize@btinternet.com
www.str.org.uk

Established 1997. Annual award for books, in English, of original research into any aspect of the history and technique of British or British-related theatre. Not restricted to authors of British nationality nor books solely from British publishers. Books must be first published in English (no translations) during the calendar year. Play texts and those treating drama as literature are not eligible. Publishers submit books directly to the independent judges and should contact the Book Prize Administrator for further details. 2007 winner: *State of the Nation* Michael Billington.

£ Award of £400.

Sony Radio Academy Awards

Alan Zafer & Associates, 47–48 Chagford Street, London NW1 6EB
☎ 020 7723 0106   🖷 020 7724 6163
secretariat@radioawards.org
www.radioawards.org
Contact *The Secretariat*

Established 1982 in association with the **Society of Authors** and Sony UK. Presented in association with the **Radio Academy**. Annual awards to recognize excellence in radio broadcasting. Entries must have been broadcast in the UK between 1 January and 31 December in the year preceding the award. The categories for the awards are reviewed each year and announced in November.

Southport Writers' Circle Open Short Story Competition

16 Ormond Avenue, Westhead LA40 6HT
☎ 01695 577938
southportwriterscircle@yahoo.co.uk

Founded 2005. Open to previously unpublished work, the story must be entered anonymously. No entry form required; include a cover sheet supplying title of story, name and contact details (send s.a.e. for results if no e-mail). Fee: £3 per story (payable to Southport Writers' Circle). Final entry date: 31 October.

£ Prizes of £150 (1st); £75 (2nd); £25 (3rd); additional £25 Local Prize.

Southport Writers' Circle Poetry Competition

32 Dover Road, Southport PR8 4TB
Contact *Mrs Hilary Tinsley*

For previously unpublished work. Entry fee: £3 per poem. Any subject, any form; maximum 40 lines. Closing date: end April. Poems must be entered anonymously, accompanied by a sealed envelope marked with the title of poem, containing s.a.e. Entries must be typed on A4 paper and be accompanied by the appropriate fee payable to Southport Writers' Circle. No application form is required. Envelopes should be marked 'Poetry Competition'. Postal enquiries only. No calls.

£ Prizes of £200 (1st); £100 (2nd); £50 (3rd); additional £25 Humour Prize.

Bram Stoker Awards for Superior Achievement

Horror Writers Association, 244 Fifth Ave., Suite 2767, New York, NY 10001, USA
hwa@horror.org
www.horror.org

Founded 1988 and named in honour of Bram Stoker, author of *Dracula*. Presented annually by the **Horror Writers Association** (HWA) for works of horror first published in the English language. Works are eligible during their first year of publication. HWA members recommend works for consideration in eight categories: Novel, First Novel, Short Fiction, Long Fiction, Fiction Collection, Poetry Collection, Anthology and Non-fiction. In addition, Lifetime Achievement Stokers are occasionally presented to individuals whose entire body of work has substantially influenced horror.

Strokestown International Poetry Competition

Strokestown International Poetry Festival Office, Strokestown, Co. Roscommon, Republic of Ireland
☎ 00 353 71 963 3759
www.strokestownpoetry.org
Administrator *Melissa Newman*

Annual poetry festival and competition. The festival takes place over the first weekend in May with readings and competitions. A centrepiece of the festival is the Strokestown International Poetry Competition for unpublished poems not exceeding 70 lines. Entry forms available on the website or call the Festival Office. Entry fee: €5.

£ Prizes of €4,000, €2,000 and €1,000 for a poem in English; €4,000, €2,000 and €1,000 for a poem in Irish or Scottish Gaelic.

Sunday Times Writer of the Year Award

The Sunday Times, 1 Pennington Street, London
E98 1ST

☎ 020 7782 5770   🖷 020 7782 5798
www.societyofauthors.org

Established 1987. Annual award to fiction and
non-fiction writers. The panel consists of *Sunday
Times* journalists and critics. Previous winners:
Anthony Burgess, Seamus Heaney, Stephen
Hawking, Ruth Rendell, Muriel Spark, William
Trevor, Martin Amis, Margaret Atwood, Ted
Hughes, Harold Pinter, Tom Wolfe, Robert
Hughes. No applications; prize at the discretion
of the Literary Editor.

Sunday Times Young Writer of the Year Award

The Society of Authors, 84 Drayton Gardens,
London SW10 9SB

☎ 020 7373 6642   🖷 020 7373 5768
info@societyofauthors.org
Contact *Awards Secretary*

Established 1991. Annual award given on the
strength of the promise shown by a full-length
published work of fiction, non-fiction, poetry
or drama. Entrants must be British citizens,
resident in Britain and under the age of 35. The
panel consists of *Sunday Times* journalists and
critics. Closing date: 31 October. The work must
be by one author, in the English language, and
published in Britain. Applications by publishers
via the **Society of Authors**. 2008 winner: Adam
Foulds *The Truth About These Strange Times*.

Sundial Scottish Arts Council Book of the Year Awards

Scottish Arts Council, 12 Manor Place, Edinburgh
EH3 7DD

☎ 0131 226 6051   🖷 0131 225 9833
aly.barr@scottisharts.org.uk
www.scottisharts.org.uk
Literature Officer *Aly Barr*

Sponsored by Sundial Properties Ltd, the Scottish
Arts Council launched these new annual awards
which are the largest literary prizes of their kind
in Scotland and the third biggest in the UK. Given
in recognition of outstanding literary merit in
fiction, poetry and literary non-fiction, there
are four categories of awards: Fiction (including
the short story); Poetry; Literary Non-Fiction;
First Book of the Year. Authors should be
Scottish or resident in Scotland, but books of
Scottish interest by other authors are eligible
for consideration. Submissions are made by
publishers only, on behalf of their authors. Final
closing date: 31 December for books published
in that calendar year. Guidelines are available on
request.

💷 Prize of £5,000 is given to the winning author
of each category and the four winning authors are
considered for the Sundial Scottish Arts Council
Book of the Year, worth an additional £20,000.

Reginald Taylor and Lord Fletcher Essay Prize

Journal of the British Archaeological Association,
Institute of Archaelogy, c/o School of Art History,
St Andrew's University, 9 The Scorel, St Andrew's
KY16 9AR
Contact *Dr Julian Luxford*

A biennial prize, in memory of the late E.
Reginald Taylor and of Lord Fletcher, for the best
unpublished scholarly essay, not exceeding 7,500
words, on a subject of archaeological, art history
or antiquarian interest within the period from the
Roman era to AD 1830. The essay should show
original research on its chosen subject, and the
author will be invited to read the essay before
the Association. The essay may be published in
the journal of the Association if approved by the
Editorial Committee. Closing date for entries is
30 April 2010. All enquiries by post, please. No
phone calls. Send s.a.e. for details.

💷 Prize of £300 and a medal.

Theakstons Old Peculier Prize for the Crime Novel of the Year

Raglan House, Raglan Street, Harrogate HG1 1LE

☎ 01423 562303   🖷 01423 521264
crime@harrogatefestival.org.uk
www.harrogate-festival.org.uk/crime
Festival Manager *Sharon Canavar*
Coordinator *Erica Morris*

Established 2005. Sponsored by Theakstons Old
Peculier in association with Waterstone's, the
award is open to full-length crime or mystery
novels by British authors, published in the UK. It
is the only crime fiction prize to be voted for by
the general public following announcement of a
long list in April. 2007 winner: (announced on the
opening night of the **Theakstons Old Peculier
Harrogate Crime Festival**) Allan Guthrie *Two
Way Split*.

💷 Prize of £3,000 cash.

The Dylan Thomas Prize

The Dylan Thomas Centre, Somerset Place,
Swansea SA1 1RR

☎ 01792 474051   🖷 01792 463993
tim@dylanthomasprize.com
www.dylanthomasprize.com
Contact *Tim J. Prosser*

Founded 2004. An award of £60,000 will be given
to the winner of this prize, which was established
to encourage, promote and reward exciting new
writing in the English-speaking world and to
celebrate the poetry and prose of Dylan Thomas.
Entrants should be the author of a published

book (in English), under the age of 30 (when the work was published), writing within one of the following categories: poetry, novel, collection of short stories by one author, play that has been professionally performed, a broadcast radio play, a professionally produced screenplay that has resulted in a feature-length film. Authors need to be nominated by their publishers, or producers in the case of performance art. Final entry date: 30 April. The award will be presented at the **Dylan Thomas Festival** in Swansea on 10 November 2008. Further entry details can be found on the website.

The Tinniswood Award

The Society of Authors, 84 Drayton Gardens, London SW10 9SB
☎ 020 7373 6642   📠 020 7373 5768
jhodder@societyofauthors.org.uk
www.societyofauthors.net
Contact *Jo Hodder*

Annual award established in 2004 by the **Society of Authors** and the **Writers' Guild of Great Britain** to perpetuate the memory of playwright Peter Tinniswood as well as to celebrate and encourage high standards in radio drama. Producers are invited to send in any radio drama first transmitted within the UK and Northern Ireland during the period 1 January to 31 December 2007. Submissions must come from the producers and are restricted to a maximum of *two entries only* per producer. An adaptation for radio of a piece originally written for any other medium e.g. stage, television, film, novel, poem or a short story will not be eligible. Final entry date: 28 January. 2007 winners: *Not Talking* by Mike Bartlett and *To Be A Pilgrim* by Rachel Joyce.

💷 Prize of £1,500, donated by the **ALCS**.

The Tir Na N-Og Award

Cyngor Llyfrau Cymru (Welsh Books Council), Castell Brychan, Aberystwyth SY23 2JB
☎ 01970 624151   📠 01970 625385
wbc.children@wbc.org.uk
www.wbc.org.uk

An annual award given to the best original book published for children in the year prior to the announcement. There are three categories: Welsh Language Book – Primary Sector; Welsh Language Book – Secondary Sector; Best English Book with an authentic Welsh background.

💷 Awards of £1,000 (each category).

The Translators Association Awards

The Translators Association, 84 Drayton Gardens, London SW10 9SB
☎ 020 7373 6642   📠 020 7373 5768
pjohnson@societyofauthors.org
www.societyofauthors.org

Contact *Awards Secretary*

Various awards for published translations into English from Arabic (Banipal Prize), Dutch and Flemish (The Vondel Translation Prize), French (Scott Moncrieff Prize), German (Schlegel-Tieck Prize), Modern Greek (Hellenic Foundation Prize), Italian (The John Florio Prize), Portuguese (The Calouste Gulbenkian Prize), Spanish (The Premio Valle Inclán) and Swedish (Bernard Shaw Translation Prize). Contact the **Translators Association** for full details.

The Betty Trask Awards
▷ entry under Bursaries, Fellowships and Grants

Sir Peter Ustinov Television Scriptwriting Award

Foundation of the International Academy of Television Arts & Sciences, 888 Seventh Avenue, 5th Floor, New York, NY 10019, USA
☎ 001 212 489 6969   📠 001 212 489 6557
emmys@iemmys.tv
www.iemmys.tv
Contact *Award Administrator*

Established 1998. The late Sir Peter Ustinov gave his name to the Foundation's Television Scriptwriting Award. This annual competition is designed to motivate novice writers worldwide. The scriptwriter cannot be a United States citizen nor resident and must be below 30 years of age. Further details and entry form available on the website.

💷 The award winner will receive US$2,500 and a trip to New York.

Ver Poets Open Competition

181 Sandridge Road, St Albans AL1 4AH
☎ 01727 762601
gillknibbs@yahoo.co.uk
www.verpoets.org.uk
Contact *Gillian Knibbs*

Various competitions are organized by **Ver Poets**, the main one being the annual Open for unpublished poems of no more than 30 lines written in English. Entry fee: £3 per poem. Entry form available from address above and/or website. Two copies of poems typed on A4 white paper. The anthology of winning and selected poems, and the adjudicators' report are normally available from mid-June. Final entry date: 30 April. Back numbers of the anthology are available for £4, post-free; one copy each free to those included.

💷 Prizes of £500 (1st); £300 (2nd); £100 (3rd). Young Writers' Prize (16–21), £100. Also runs High Lights, an open competition for younger poets with an 31 October closing date. Details available from address above.

Vogue Talent Contest

Vogue, Vogue House, Hanover Square, London
W1S 1JU

☎ 020 7152 3003   🖷 020 7408 0559
Contact *Frances Bentley*

Established 1951. Annual award for young writers
and journalists (under 25 on 1 January in the
year of the contest). Final entry date is in April.
Entrants must write three pieces of journalism on
given subjects.

💷 Prizes of £1,000, plus a month's paid work
experience with *Vogue*; £500 (2nd).

The Vondel Translation Prize
▷ The Translators Association Awards

Wadsworth Prize for Business History

School of Management, University of
Southampton, Building 2, Highfield,
Southampton SO17 1BJ
R.A.Edwards@soton.ac.uk
Contact *Dr Roy Edwards*

Now in its 32nd year, the Wadsworth Prize is
awarded annually by the Business Archives
Council to an individual judged to have made
an outstanding contribution to the study of
British business history. Books are nominated by
publishers. 2006 winner: Dr Terry Gourvish with
Mike Anson *The Official History of Britain and
the Channel Tunnel*.

Walford Award

43 Yardington, Whitchurch SY13 1BL
☎ 01948 663570
millbrookend-awards@yahoo.co.uk
www.cilip.org.uk
Contact *Awards Administrator* (at address above)

Part of the ISG (CILIP) Reference Awards.
Awarded to an individual who has made a
sustained and continual contribution to British
bibliography over a period of years. The nominee
need not be resident in the UK. The award is
named after Dr A.J. Walford, a bibliographer
of international repute. 2006 winner: Dr Diana
Dixon. 💷 Cash prize and certificate.

Waterstone's Children's Book Prize

Waterstone's, Capital Court, Capital Interchange
Way, Brentford TW8 0EX
☎ 020 8742 3800   🖷 020 8742 0215
waterstones.com
Contact *Sarah Clarke*

Founded 2005. An annual award which aims to
support new children's authors and introduce them
to a wide audience while introducing Waterstone's
customers to some of the brightest new talent
writing today. Books submitted should be fiction
for children between the ages of 7 and 14 with
a focus on text rather than illustration. Open to
new authors of any nationality with not more than
two previously published fiction titles (adults'
or children's). Publishers must declare any titles
written under another name, including series
fiction. All submitted titles must be available in
paperback for a period of shortlist promotion and
can be the first part of a series or trilogy but stand
alone novel. See website for submission deadline.
2008 winner: Sally Nicholl *Ways to Live Forever*.

💷 Prize of £5,000 and major promotion in
Waterstone's.

The David Watt Prize

Rio Tinto plc, 2 Eastbourne Terrace, London
W2 6LG
☎ 01985 844613   🖷 01985 844002
davidwattprize@riotinto.com
celiabeale@globalnet.co.uk
www.riotinto.com
Contact *The Administrator*

Initiated in 1988 to commemorate the life
and work of David Watt. Annual award, open
to writers currently engaged in writing for
English language newspapers and journals, on
international and national affairs. The winner
is judged as having made 'an outstanding
contribution towards the greater understanding
of national, international or global issues'. Entries
must have been published during the year
preceding the award. Final entry date 31 March.
2007 winner Richard Tomkins of the *Financial
Times* for his article 'Profits of doom'.

💷 Prize of £10,000.

Brian Way Award

c/o Theatre Centre, Shoreditch Town Hall, 380
Old Street, London EC1V 9LT
☎ 020 7729 3066
admin@theatre-centre.co.uk
www.theatre-centre.co.uk
Contact *Theatre Writing Section*

Formerly the Arts Council Children's Award,
founded 2000. An annual award for playwrights
who write for children and young people.
The plays, which must have been produced
professionally, must be at least 45 minutes long.
The playwright must be resident in the UK.
Contact e-mail address/website for full details
and application form. 2008 winner: David Greig
Yellow Moon.

💷 Award of £6,000.

The Harri Webb Prize

10 Heol Don, Whitchurch, Cardiff CF14 2AU
☎ 029 2062 3359
Contact *Professor Meic Stephens*

Established 1995. Annual award to commemorate
the Welsh poet, Harri Webb (1920–94), for a

single poem in any of the categories in which he wrote: ballad, satire, song, polemic or a first collection of poems. The poems are chosen by three adjudicators; no submissions. Previous winner: Grahame Davies.

£ Prizes of £100/£200.

Welsh Book of the Year Award
▷ Academi Book of the Year Awards

The Wheatley Medal
Society of Indexers, Woodbourn Business Centre, 10 Jessell Street, Sheffield S9 3HY
☎ 0114 244 9561   ⊟ 0114 244 9563
admin@indexers.org.uk
www.indexers.org.uk
Marketing Director *Ann Kingdom*

The Wheatley Medal is awarded for an outstanding index and was established by the **Society of Indexers** and the Library Association (now **CILIP**) to highlight the importance of good indexing and confer recognition on the most highly regarded practitioners and their publishers. First awarded in 1963, it has since been presented for indexes to a wide range of publications, from encyclopedias to journals.

£ Cash prize, plus medal and certificate.

Whitbread Book Awards
▷ Costa Book Awards

Whitfield Prize
Royal Historical Society, University College London, Gower Street, London WC1E 6BT
☎ 020 7387 7532   ⊟ 020 7387 7532
royalhistsoc@ucl.ac.uk
www.royalhistoricalsociety.org/grants.htm
Contact *Administrative Secretary*

Established 1977. An annual award for the best new work within a field of British or Irish history, published in the UK in the preceding calendar year. The book must be the author's first (solely written) history book and be an original and scholarly work of historical research. Final entry date: 31 December.

£ Prize of £1,000.

Whiting (John) Award
▷ The Peter Wolff Theatre Trust supports the Whiting Award

Wilkins Memorial Poetry Prize
Birmingham & Midland Institute, 9 Margaret Street, Birmingham B3 3BS
☎ 0121 236 3591   ⊟ 0121 212 4577
www.bmi.org.uk
Administrator *Mr P.A. Fisher*

The structure of the Wilkins Memorial Poetry Prize is currently under discussion. For further information, contact the Administrator.

Winchester Writers' Conference Prizes
Faculty of Arts, University of Winchester, Winchester SO22 4NR
☎ 01962 827238
christian.francis@winchester.ac.uk
www.writersconference.co.uk

Fifteen writing competitions sponsored by major publishers who offer 56 prizes and trophies for entries under categories of: The First Three Pages of the Novel; Poetry; Writing can be Murder; Feature Articles and Scriptwriting; Slim Volume; Small Edition; Short Stories; Shorter Short Story; Haibun; Lifewriting and Retirement. Competitions open for all writers in January; closing date: 30 May. Phone or write for booklet giving details of each competition and the entry blanks or access them on the website.

H.H. Wingate Prize
▷ Jewish Quarterly Literary Prize

Wisden (Cricket) Book of the Year
John Wisden & Co., 13 Old Aylesfield, Froyle Road, Alton GU34 4BY
☎ 01420 83415
almanack@wisdengroup.com
www.wisden.com
Contact *Hugh Chevallier*

Founded 2002. Annual award. The Wisden Book of the Year is chosen exclusively by the Wisden Cricketers' Almanack book reviewer. In each annual *Wisden Almanack*, a different person is commissioned to review all cricket books published in the previous calendar year, and that reviewer selects the Wisden Book of the Year. Final entry date: 1 December. 2006 winner: *Brim Full of Passion* by Wasim Khan, selected by book reviewer Peter Oborne.

The Peter Wolff Theatre Trust supports the Whiting Award
Hampstead Theatre, Eton Avenue, London NW3 3EU
☎ 020 7749 4200
info@hampsteadtheatre.com
www.hampsteadtheatre.com
Contact *Neil Morris* (General Manager)

The original award was founded in 1965 by the Arts Council to commemorate the life and work of the playwright John Whiting (*The Devils*, *A Penny for a Song*). In 2008 it was taken over by a consortium of theatres, headed by Hampstead Theatre and funded by the Peter Wolff Theatre Trust. Annual, open award, for a play that is an original piece of writing (no adaptations or pieces of verbatim theatre), produced by a company in the subsidized sector and must have had its first performance in the award time period. The judges look for a play in which the writing is of special quality, of relevance and importance to contemporary life and of potential value to British theatre. Each year a different member of the consortium will host the award. The 2008 award is hosted by Liverpool Everyman (closing date for submissions: 18 August; award announced in October 2008). For information on the 2009 award, contact Hampsted Theatre.

Ⓔ Prize of £6,000.

Wolfson History Prize

Wolfson Foundation, 8 Queen Anne Street, London W1G 9LD
Ⓣ 020 7323 5730　Ⓕ 020 7323 3241
www.wolfson.org.uk
Contact *Prize Administrator*

The Wolfson History Prize, established in 1972, is awarded annually to promote and encourage standards of excellence in the writing of history for the general reading public. Submissions are made through publishers. 2006 winners (awarded in 2007): Dr Adam Tooze *The Wages of Destruction*; Dr Christopher Clark *Iron Kingdom*; Professor Vic Gatrell *City of Laughter*.

The Writers Bureau Poetry and Short Story Competition

The Writers Bureau, Sevendale House, 7 Dale Street, Manchester M1 1JB
Ⓣ 0161 228 2362　Ⓕ 0161 228 3533
studentservices@writersbureau.com
www.writersbureau.com/competition
Competition Secretary *Angela Cox*

Established 1994. Annual award. Poems should be no longer than 40 lines and short stories no more than 2,000 words. £5 entry fee. Closing date: 30 June 2009.

Ⓔ Prizes in each category: £1,000 (1st); £400 (2nd); £200 (3rd); £100 (4th); £50 (x 6).

Writers' Forum Short Story & Poetry Competitions

See *Writers' Forum* under *Magazines*.

Writers-of-the-Year Competition

▷ writers inc under Organizations of Interest to Poets

The Yale Drama Series/David C. Horn Prize

PO Box 209040, New Haven, CT 06520–9040, USA
www.yalebooks.com/drama

Major new annual playwriting competition intended to support emerging playwrights. Submissions must be original, unpublished full-length plays written in English (translations, musicals and children's plays are not accepted). Only one manuscript may be submitted per year; plays that have had professional productions are not eligible. 2009 submissions accepted between 1 June and 15 August. Further information available on the website.

Ⓔ Prize winner receives the David C. Horn Prize of $10,000, publication by Yale University Press and a staged reading at Yale Rep.

Yeovil Literary Prize

The Octagon Theatre, Hendford, Yeovil BA20 1UX
Ⓣ 01935 422884 (messages only)
www.yeovilprize.co.uk
Contact *Penny Deacon*

Annual competition, established in 2004, to stimulate interest in literary pursuits and to provide funds for the Yeovil Community Arts Association. Categories: novel (send synopsis and opening pages; maximum 15,000 words); short story: (maximum 2,000 words); poetry (maximum 40 lines). Entries, which can be sent by post or online, must be in English and not previously published; submissions from overseas welcome. Entrants must be over 18 years of age. Final entry date: end May. 2007 winners: Rachel Moses (novel); Alison Theaker (short story); Gabriel Griffin (poetry).

Ⓔ Prizes: Novel: £1,000 (1st); £250 (2nd); £100 (3rd). Short Story/Poetry: £500 (1st); £200 (2nd); £100 (3rd).

Yorkshire Post Book Awards

Yorkshire Post Newspapers Ltd, PO Box 168, Wellington Street, Leeds LS1 1RF
Contact *Duncan Hamilton, Deputy Editor*

Annual awards: **Yorkshire Post Novel of the Year Award**; **Yorkshire Post Non-Fiction Book of the Year Award**; **Best Title With a Yorkshire Theme**. Awarded to the book which, in the opinion of the judges, is the best work published in the preceding year.

Ⓔ Prizes of £1,200 (Novel of the Year); £1,200 (Non-Fiction Book of the Year); £500 (Best Title with a Yorkshire Theme) .

YoungMinds Book Award

Youngminds, 48–50 St John Street, London EC1M 4DG
Ⓣ 020 7336 8445　Ⓕ 020 7336 8446
hannah.smith@youngminds.org.uk
www.youngminds.org.uk/bookaward
Contact *Hannah Smith*

Established 2003. This award is given to a published work of literature that throws fresh light on the ways a child takes in and makes sense of the world he or she is growing into – novels, memoirs, diaries, poetry collections which portray something of the unique subtlety of a child's experience. 2007 winner: Suzanne Sojqvist *Still Here With Me*.

Ⓔ Prize of £3,000.

Professional Associations and Societies

ABSW

Wellcome Wolfson Building, 165 Queen's Gate,
London SW7 5HD
℡ 0870 770 3361
absw@absw.org.uk
www.absw.org.uk
Chairman *Ted Nield*
Administrator & EUSJA Board Member *Barbara Drillsma*

ABSW has played a central role in improving
the standards of science journalism in the UK
over the last 40 years. The Association seeks
to improve standards by means of networking,
lectures and organized visits to institutional
laboratories and industrial research centres.
Puts members in touch with major projects
in the field and with experts worldwide. A
member of the European Union of Science
Journalists' Associations, ABSW is able to offer
heavily subsidized places on visits to research
centres in most other European countries, and
hosts reciprocal visits to Britain by European
journalists. Membership open to those who are
considered to be *bona fide* science writers/editors,
or their film/TV/radio equivalents, who earn a
substantial part of their income by promoting
public interest in and understanding of science.
ABSW is hosting the World Conference of
Science Journalists (29th June – 1st July 2009).

Academi (Yr Academi Gymreig)

3rd Floor, Mount Stuart House, Mount Stuart
Square, Cardiff CF10 5FQ
℡ 029 2047 2266   ℻ 029 2049 2930
post@academi.org
www.academi.org

Glyn Jones Centre, Wales Millenium Centre,
Cardiff Bay, Cardiff
North West Wales office: Ty Newydd,
Llanystumdwy, Cricieth, Gwynedd LL52 0LW
℡ 01766 522817 ℻ 01766 523095
llio@academi.org
West Wales office: Dylan Thomas Centre,
Somerset Place, Swansea SA1 1RR
℡ 01792 463980 ℻ 01792 463993
dylanthomas.centre@swansea.gov.uk

Chief Executive *Peter Finch*

Academi is the trading name of Yr Academi
Gymreig, the Welsh National Literature
Promotion Agency and Society for Writers.
Yr Academi Gymreig was founded in 1959 as
an association of Welsh language writers. An
English language section was established in
1968. Membership, for those who have made a
significant contribution to the literature of Wales,
is by invitation. Membership currently stands at
600. The Academi runs courses, competitions
(including the **Cardiff International Poetry
Competition**), conferences, tours by authors,
festivals, and represents the interests of Welsh
writers and Welsh writing both inside Wales
and beyond. Its publications include *Taliesin*, a
quarterly literary journal in the Welsh language,
The Oxford Companion to the Literature of Wales,
The Welsh Academy English-Welsh Dictionary and
a variety of translated works.

The Academi won the franchise from the
Arts Council of Wales to run the Welsh National
Literature Promotion Agency. The new, much
enlarged organization now administers a variety
of schemes including bursaries, the annual **Book
of the Year Award**, critical services, writers'
mentoring, Writers on Tour, Writers Residencies
and a number of literature development projects.
It promotes an annual literary festival alternating
between North and South Wales, runs its own
programme of literary activity and publishes
A470, a bi-monthly literature information
magazine. The Academi was also in receipt of a
lottery grant to publish the first Welsh National
Encyclopedia (published January 2008).

Those with an interest in literature in Wales
can become an associate of the Academi (which
carries a range of benefits). Rates are £15 p.a.
(waged); £7.50 (unwaged).

ALCS
▷ Authors' Licensing & Collecting Society Limited

Alliance of Literary Societies

59 Bryony Road, Selly Oak, Birmingham B29 4BY
℡ 0121 475 1805
l.j.curry@bham.ac.uk
www.allianceofliterarysocieties.org.uk

Chair *Linda J. Curry*

Founded 1974. Aims to help and support its 100+ member societies and, when necessary, to act as a pressure group. Produces an annual journal (*ALSo*).

Arts & Business (A&B)

Nutmeg House, 60 Gainsford Street, Butlers Wharf, London SE1 2NY

☎ 020 7378 8143   🖷 020 7407 7527
head.office@AandB.org.uk
www.AandB.org.uk
Director of Press & Public Affairs *Jonathan Tuchner*
Director of Marketing & Communications *Sebastian Paul*
Media Manager *Sophie Gaskill*

A&B enables business and its people to be more successful by engaging with the arts and to increase resources for the arts from business and its people. It provides a wide range of services to over 400 business members and 1,100 arts organizations through the Development Forum. With the support of its President, HRH The Prince of Wales, it explores and develops new ways for business, the arts and society to interact. Arts & Business has 12 regional offices offering a range of services throughout the UK.

Arvon Foundation
▷ entry under UK Writers' Courses

ASLS
▷ Association for Scottish Literary Studies

Association for Scottish Literary Studies

c/o Department of Scottish Literature, 7 University Gardens, University of Glasgow, Glasgow G12 8QH

☎ 0141 330 5309   🖷 0141 330 5309
office@asls.org.uk
www.asls.org.uk
Contact *Duncan Jones*

Founded 1970. ASLS is an educational charity promoting the languages and literature of Scotland. Publishes works of Scottish literature; essays, monographs and journals; and *Scotnotes*, a series of comprehensive study guides to major Scottish writers. Also produces *New Writing Scotland*, an annual anthology of contemporary poetry and prose in English, Gaelic and Scots (see entry under *Magazines*).

Association of American Correspondents in London

c/o People Magazine, Third Floor, Brettenham House, Lancaster Place, London WC2E 7TL
info@theaacl.co.uk
www.theaacl.co.uk
Secretary/Treasurer *Monique Jessen*

Founded 1919 to serve the professional interests of its member organizations, promote social cooperation among them, and maintain the ethical standards of the profession. (An extra £40 is charged for each department of an organization which requires separate listing in the Association's handbook and a charge of £10 for each full-time editorial staff listed, up to a maximum of £120 regardless of the number listed.)

Association of American Publishers, Inc

71 Fifth Avenue, 2nd Floor, New York, NY 10003, USA

☎ 001 212 255 0200   🖷 001 212 255 7007
www.publishers.org
Also at: 50 F Street, NW, Suite 400, Washington, DC 20001

☎ 001 202 347 3375 🖷 001 202 347 3690

Founded 1970. For information, visit the Association's website.

Association of Authors' Agents (AAA)

LAW, 14 Vernon Street, London W14 0RJ
www.agentsassoc.co.uk
President *Philippa Milne-Smith*

Founded 1974. Membership voluntary. The AAA maintains a code of practice, provides a forum for discussion and represents its members in issues affecting the profession. For a full list of members and a list of frequently asked questions visit the AAA website. The AAA is a voluntary body and unable to operate as an information service to the public.

Association of Authors' Representatives (AAR)

676A 9th Avenue #312, New York, NY 10036, USA
aarinc@mindspring.com
www.aar-online.org
Contact *Administrative Secretary*

Founded in 1991 through the merger of the Society of Authors' Representatives and the Independent Literary Agents Association. Membership of this US organization is restricted to agents of at least two years' operation. Provides information, education and support for its members and works to protect their best interests.

Association of British Editors
▷ Society of Editors

Association of British Science Writers
▷ ABSW

Association of Canadian Publishers

174 Spadina Avenue, Suite 306, Toronto, Ontario M5T 2C2, Canada

☎ 001 416 487 6116   🖷 001 416 487 8815

admin@canbook.org
www.publishers.ca
Executive Director *Carolyn Wood*

Founded 1971. ACP represents around 140 Canadian-owned book publishers country-wide from the literary, general trade, scholarly and education sectors. Aims to encourage the writing, publishing, distribution and promotion of Canadian books and to support the development of a 'strong, independent and vibrant Canadian-owned publishing industry'. The organization's website has information on getting published and links to many of their member publishers' websites. The ACP does not accept manuscripts and cannot provide assistance to authors who wish to find a Canadian publisher.

Association of Christian Writers

23 Moorend Lane, Thame OX9 3BQ
℡ 01844 213673
admin@christianwriters.org.uk
www.christianwriters.org.uk
President *Adrian Plass*
Chairman *Brian Vincent*
Administrator *Rev. Simon Baynes*

Founded in 1971 'to inspire and equip men and women to use their talents and skills with integrity to devise, write and market excellent material which comes from a Christian worldview. In this way we seek to be an influence for good and for God in this generation'. Publishes a quarterly magazine. Runs three training events each year, biennial conference, competitions, postal workshops, area groups, prayer support and manuscript criticism. Charity No. 1069839.

Association of Freelance Editors, Proofreaders & Indexers (Ireland)

11 Clonard Road, Sandyford, Dublin 16, Republic of Ireland
℡ 00 353 1 295 2194
brenda@ohanlonmediaservices.com
poweredititing@eircom.net
www.afepi.ie
Co-chairs *Brenda O'Hanlon* (at address above), *Winifred Power* (28 Parklane Avenue, Abbeyside, Dungarvan, Co Waterford ℡ 00 353 58 48458)

The organization was established in Ireland to protect the interests of its members and to provide information to publishers on freelancers working in the relevant fields. Full membership is restricted to freelancers with experience and/or references (but the association does not test or evaluate the skills of members). A new category of membership, Associate Member, is available for trainees in proofreading/editing who are taking the Publishing Training Centre correspondence courses in Proofreading and Copy-editing.

Association of Freelance Writers

Sevendale House, 7 Dale Street, Manchester M1 1JB
℡ 0161 228 2362   🖷 0161 228 3533
fmn@writersbureau.com
www.freelancemarketnews.com
Contact *Angela Cox*

Founded in 1995 to help and advise new and established freelance writers. Members receive a copy of *Freelance Market News* each month which gives news, views and the latest advice and guidelines about publications at home and abroad. Other benefits include one free appraisal of prose or poetry each year, reduced entry to The **Writers Bureau** writing competition, reduced fees for writing seminars and discounts on books for writers.

Association of Golf Writers

1 Pilgrims Bungalow, Mulberry Hill, Chilham CT4 8AH
℡ 01227 732496   🖷 01227 732496
andyfarrell292@btinternet.com
Honorary Secretary *Andy Farrell*

Founded 1938. Aims to cooperate with golfing bodies to ensure best possible working conditions.

Association of Illustrators

2nd Floor, Back Building, 150 Curtain Road, London EC2A 3AT
℡ 020 7613 4328   🖷 020 7613 4417
info@theaoi.com
www.theaoi.com
Contact *Membership Coordinator*

The AOI is a non-profit-making trade association and members consist primarily of freelance illustrators as well as agents, clients, students and lecturers. As the only body to represent illustrators and campaign for their rights in the UK, the AOI has successfully increased the standing of illustration as a profession and improved the commercial and ethical conditions of employment. Organizes annual events programme and provides an advisory service for members. Publications include: triannual magazine, *Varoom – the journal of illustration and made images*; *The Illustrator's Guide to Law and Business Practice*; *The Illustrator's Guide to Success (not guaranteed)*, client directories (online) and *Images*, the only jury-selected annual of British contemporary illustration.

Association of Independent Libraries

Leeds Library, 18 Commercial Street, Leeds LS1 6AL
℡ 0113 245 3071
enquiries@leedslibrary.co.uk
www.independentlibraries.co.uk

Chairman *Geoffrey Forster*
Contact *Catherine Levy*

Established in 1989 to 'further the advancement, conservation and restoration of a little-known but important living portion of our cultural heritage'. Members include the **London Library**, **Devon & Exeter Institution**, **Linen Hall Library** and **Plymouth Proprietary Library**.

Association of Learned and Professional Society Publishers

Bluebell Lodge, 8 Rickford Road, Nailsea, Bristol BS48 4PY
☎ 01275 856444   🖷 07968 504763
ian.russell@alpsp.org
www.alpsp.org
Chief Executive *Ian Russell*
Chief Operating Officer *Nick Evans*
Editor-in-Chief, Learned Publishing
Sally Morris (editor@alpsp.org)

The Association of Learned and Professional Society Publishers (ALPSP) is the international trade association for not-for-profit publishers and those who work with them. It currently has over 360 members in 36 countries. ALPSP provides representation of its sector, cooperative services such as the ALPSP Learned Journals Collection and ALPSP eBooks Collection, a professional development programme of training courses and seminars and a wealth of information and advice. ALPSP runs an annual international conference and awards programme. See the website for further information.

Association of Scottish Motoring Writers

32 Caledonia Crescent, Ardrossan KA22 8LW
☎ 07775 930686
spark@cfpress.co.uk
Secretary *Stephen Park*

Founded 1961. Aims to co-ordinate the activities of, and provide shared facilities for, motoring writers resident in Scotland. Welcomes applications from active journalists.

Audiobook Publishing Association

c/o 18 Green Lanes, Hatfield AL10 9JT
☎ 07971 280788
info@theapa.net
www.theapa.net
Chair *Ali Muirden*
Administrator *Charlotte McCandlish* (at address above)

Founded 1994. The Audiobook Publishing Association (formerly the Spoken Word Publishing Association) is the UK trade association for the audiobook industry with membership open to all those involved in the publishing of spoken word audio. Publishes annual *APA Resources Directory*, available from the address above, and now also available online at www.theapa.net (click on 'APA Members').

Australian Copyright Council

PO Box 1986, Strawberry Hills, NSW 2012, Australia
☎ 00 612 8815 9777   🖷 00 612 8815 9799
info@copyright.org.au
www.copyright.org.au
Contact *Customer Service*

Founded 1968. The Council's activities and services include a range of publications, organizing and speaking about copyright at seminars, research, consultancies and free legal advice. Aims include assistance for copyright owners to exercise their rights effectively, raising awareness about the importance of copyright and seeking changes to the law of copyright.

Australian Publishers Association

60/89 Jones Street, Ultimo, NSW 2007, Australia
☎ 00 612 9281 9788   🖷 00 612 9281 1073
www.publishers.asn.au
Chief Executive *Maree McCaskill*

Founded 1948. The APA initiates programmes that contribute to the development of publishing in Australia, virgorously protects and furthers the interests of copyright owners, agents and licensees and actively represents members' interests to government and other organizations as appropriate. The Association encourages excellence in writing, editing, design, production, marketing and distribution of published works in Australia, protects freedom of expression and manages members' funds to further the interests of the industry.

Australian Society of Authors

PO Box 1566, Strawberry Hills, NSW 2012, Australia
☎ 00 612 9318 0877   🖷 00 612 9318 0530
asa@asauthors.org
www.asauthors.org
Executive Director *Dr Jeremy Fisher*

Founded in 1963, the Australian Society of Authors is the professional association for Australia's literary creators with 3,000 members in Australia and overseas.

Authors' Club

40 Dover Street, London W1S 4NP
☎ 020 7499 8581   🖷 020 7409 0913
stella@theauthorsclub.co.uk
www.theartsclub.co.uk
Secretary *Stella Kane*

Founded in 1891 by Sir Walter Besant, the Authors' Club welcomes as members writers, agents, publishers, critics, journalists, academics and anyone involved with literature and the written word. Administers the **Authors' Club Best First Novel Award**, the **Dolman Best Travel Book Award** and **Sir Banister Fletcher Award**, and organizes regular talks and dinners with well-known guest speakers. Membership fee: apply to the Secretary.

Authors' Licensing & Collecting Society Limited (ALCS)

The Writers' House, 13 Haydon Street, London EC3N 1DB
☎ 020 7264 5700   🖷 020 7264 5755
alcs@alcs.co.uk
www.alcs.co.uk
Chief Executive *Owen Atkinson*

Founded 1977. The UK collective rights management society for writers and their successors, ALCS is a non-profit organization whose principal purpose is to ensure that hard-to-collect revenues due to writers are efficiently collected and speedily distributed. Established to give assistance to writers through the protection and exploitation of collective rights, ALCS has distributed over £150 million in secondary royalties to writers since its creation.

ALCS represents all types of writer, fiction and non-fiction, including educational, research and academic authors, scriptwriters, playwrights, poets, editors and freelance journalists across the print and broadcast media. On joining, members give ALCS a mandate to administer on their behalf those rights which the law determines must be received or which are best handled collectively. Chief among these are: photocopying, cable retransmission, rental and lending rights (but not British Public Lending Right), off-air recording, electronic rights, the performing right and public reception of broadcasts. The society is a prime resource and a leading authority on copyright matters and writers' collective interests. It maintains a watching brief on all matters affecting copyright both in Britain and abroad, making representations to UK government authorities and the EU. Visit the ALCS website or contact the office for registration forms and further information.

BAPLA (British Association of Picture Libraries and Agencies)

18 Vine Hill, London EC1R 5DZ
☎ 020 7713 1780   🖷 020 7713 1211
enquiries@bapla.org.uk
www.bapla.org.uk

Everything you need to know about finding, buying and selling pictures. Represents over 400 image supplying members.

BFI

21 Stephen Street, London W1T 1LN
☎ 020 7255 1444   🖷 020 7436 0439
www.bfi.org.uk
Chair *Greg Dyke*
Director *Amanda Nevill*

'There's more to discover about film and television through the BFI. Our world-renowned archive, cinemas, festivals, films, publications and learning resources are here to inspire you.' To find out more, visit the website.

The Bibliographical Society

c/o Institute of English Studies, University of London, Senate House, Malet Street, London WC1E 7HU
☎ 020 7862 8679   🖷 020 7862 8720
secretary@bibsoc.org.uk
www.bibsoc.org.uk
President *E. Leedham-Green*
Honorary Secretary *M.L. Ford*

Aims to promote and encourage the study and research of historical, analytical, descriptive and textual bibliography, and the history of printing, publishing, bookselling, bookbinding and collecting; to hold meetings at which papers are read and discussed; to print and publish works concerned with bibliography; to form a bibliographical library. Awards grants and bursaries for bibliographical research. Publishes a quarterly magazine called *The Library*.

Booksellers Association of the UK & Ireland Ltd

Minster House, 272 Vauxhall Bridge Road, London SW1V 1BA
☎ 020 7802 0802   🖷 020 7802 0803
mail@booksellers.org.uk
www.booksellers.org.uk
Chief Executive *Tim Godfray*

Founded 1895. The BA helps 4,400 independent, chain and multiple retail outlets to sell more books, reduce costs and improve efficiency. It represents members' interests to the UK Government, European Commission, publishers, authors and others in the trade as well as offering marketing assistance, conferences, seminars and exhibitions. Together with **The Publishers Association**, coordinates World Book Day. Publishes directories, catalogues, surveys and various other publications connected with the book trade and administers the **Costa Book Awards**.

Booktrust

Book House, 45 East Hill, London SW18 2QZ
☎ 020 8516 2977   🖷 020 8516 2978

query@booktrust.org.uk
www.booktrust.org.uk
Director *Viv Bird*
Head of Promotions *Helen Hayes*

Founded 1925. Booktrust is an independent national charity that encourages readers of all ages and cultures to discover and enjoy reading. The reader is at the heart of everything they do. Their website includes reviews of thousands of books, as well as information about Children's Book Week and the **Children's Laureate**, both of which are administered by Booktrust. Booktrust also administers literary prizes, including the **Orange Broadband Prize for Fiction**, the **John Llewellyn Rhys Prize** and the **Booktrust Teenage Prize** (see entries under *Prizes*), and promotes reading through campaigns such as Get London Reading (www.getlondonreading.com) and Story (www.theshortstory.org.uk). Booktrust runs Bookstart (www.bookstart.org.uk), the acclaimed national scheme that works through locally based organizations to give a free pack of books to babies and guidance materials for parents and carers. They also run two other gifting schemes for older children (Booktime and Booked Up), as well as a scheme for looked after children (the Letterbox Club), and a programme to encourage writing in schools (Everybody Writes).

British Academy of Composers and Songwriters

2nd Floor, British Music House, 25–27 Berners Street, London W1T 3LR
℡ 020 7636 2929   ℻ 020 7636 2212
info@britishacademy.com
www.britishacademy.com
Membership Manager *Vicky Hunt*

The Academy represents the interests of music writers of all genres, providing advice on professional and artistic matters. Publishes bi-monthly magazines and administers the annual Ivor Novello Awards and British Composer Awards.

British Association of Communicators in Business (CiB)

Suite GA2, Oak House, Woodlands Business Park, Linford Wood, Milton Keynes MK14 6EY
℡ 01908 313755   ℻ 01908 313661
enquiries@cib.uk.com
www.cib.uk
Chief Executive *Kathie Jones*

The professional body for in-house, freelance and agency staff involved in internal and corporate communications. Services to members include a business support helpline, an online forum, a monthly e-zine, the CiB Freelance Directory, an annual conference and gala dinner, automatic eligibility to FEIEA, a European network of 5,000 communicators.

British Association of Journalists

89 Fleet Street, London EC4Y 1DH
℡ 020 7353 3003   ℻ 020 7353 2310
office@bajunion.org.uk
www.bajunion.org.uk
General Secretary *Steve Turner*

Founded 1992. Aims to protect and promote the industrial and professional interests of journalists.

British Association of Picture Libraries and Agencies
▷ BAPLA

British Centre for Literary Translation

University of East Anglia, Norwich NR4 7TJ
℡ 01603 592785   ℻ 01603 592737
bclt@uea.ac.uk
www.uea.ac.uk/bclt
Director *Amanda Hopkinson*

Founded 1989, BCLT is funded jointly by Arts Council England and the University of East Anglia. It aims to raise the profile of literary translation in the UK through events, publications, activities and research aimed at professional translators, students and the general reader. Member of the international RECIT literary translation network. Activities include the annual Sebald Lecture, Summer School and www.literarytranslation.com website with the British Council. BCLT has a PhD programme in literary translation as well as an MA in Literary Translation and other units at undergraduate and postgraduate level. Joint sponsor with BCLA of the **John Dryden Translation Prize**. Publishes a journal *In Other Words* and *New Books in German*. Free mailing list.

British Copyright Council

Copyright House, 29–33 Berners Street, London W1T 3AB
℡ 01986 788122   ℻ 01986 788847
secretary@britishcopyright.org
www.britishcopyright.org
Contact *Janet Ibbotson*

Works for the national and international acceptance of copyright and acts as a lobby/watchdog organization on behalf of creators, publishers and performers on copyright and associated matters.

The British Council

10 Spring Gardens, London SW1A 2BN
℡ 020 7389 3194   ℻ 020 7389 3199
arts@britishcouncil.org
www.britishcouncil.org/arts

The British Council promotes Britain abroad. It provides access to British ideas, expertise and experience in education, the English language, literature and the arts, science and technology and governance. Works in 110 countries running a mix of offices, libraries, resource centres and English teaching operations.

British Equestrian Writers' Association

Priory House, Station Road, Swavesey, Cambridge CB24 4QJ
℡ 01954 232084   🖷 01954 231362
gnewsumn@aol.com
Contact *Gillian Newsum*

Founded 1973. Aims to further the interests of equestrian sport and improve, wherever possible, the working conditions of the equestrian press. Membership is by invitation of the committee. Candidates for membership must be nominated and seconded by full members and receive a majority vote of the committee.

British Film Institute
▷ BFI

British Guild of Beer Writers

Woodcote, 2 Jury Road, Dulverton TA22 9DU
℡ 01398 324314
tierneyjones@btinternet.com
www.beerwriters.co.uk
Secretary *Adrian Tierney-Jones*

Founded 1988. Aims to improve standards in beer writing and at the same time extend public knowledge of beers and brewing. Publishes an annual directory of members with details of their publications and their particular areas of expertise; this is then circulated to newspapers, magazines, trade press and broadcasting organizations. Also publishes a monthly newsletter, the *BGBW Newsletter*, edited by Adrian Tierney-Jones. As part of the plan to improve writing standards and to achieve a higher profile for beer, the Guild offers annual awards, including a Travel Bursary, to writers and broadcasters judged to have made the most valuable contribution towards this end in their work.

British Guild of Travel Writers

5 Berwick Courtyard, Berwick St. Leonard, Salisbury SP3 5UA
℡ 01747 820455
secretariat@bgtw.org
www.bgtw.org
Chairman *John Carter*
Secretariat *Judy Robson*

The professional association of travel writers, broadcasters, photographers and editors which aims to serve its members' professional interests by acting as a forum for debate, discussion and 'networking'. The Guild publishes an annual Year Book and has a website giving full details of all its members and useful trade contacts, holds monthly meetings and has a monthly newsletter. Members are required to spend a significant proportion of their working time on travel.

British Science Fiction Association

39 Glyn Avenue, New Barnet EN4 9PJ
bsfamail@gmail.com
www.bsfa.co.uk

Founded originally in 1958 by a group of authors, readers, publishers and booksellers interested in science fiction. With a worldwide membership, the Association aims to promote the reading, writing and publishing of science fiction and to encourage SF fans to maintain contact with each other. Publishes *Matrix* e-zine, media review newsletter via the web with comment and opinions, news of conventions, etc. Contributions from members welcomed; *Vector* quarterly critical journal – reviews of books and magazines; *Focus* biannual magazine with articles, original fiction and letters column. Also offers postal and online writer's workshop. For further information, contact the Membership Secretary *Peter Wilkinson* at the postal address above or via e-mail.

British Society of Comedy Writers

61 Parry Road, Ashmore Park, Wolverhampton WV11 2PS
℡ 01902 722729   🖷 01902 722729
comedy@bscw.co.uk
www.bscw.co.uk
Contact *Ken Rock*

Founded 1999. The Society aims to develop good practice and professionalism among comedy writers while bringing together the best creative professionals, and working to standards of excellence agreed with the light entertainment industry. Offers a network of industry contacts and a range of products, services and training initiatives including specialized workshops, an annual international conference, script assessment service and opportunities to visit international festivals.

British Society of Magazine Editors (BSME)

137 Hale Lane, Edgware HA8 9QP
℡ 020 8906 4664   🖷 020 8959 2137
admin@bsme.com
www.bsme.com
Contact *Gill Branston*

Holds regular industry forums and events as well as an annual awards dinner.

Broadcasting Press Guild

c/o Torin Douglas, Room G690, BBC Television Centre, Wood Lane, London W4 1JA
℡ 020 8624 9052   🖷 020 8624 9096

torin.douglas@bbc.co.uk
www.broadcastingpressguild.org
Membership Secretary *Steve Clarke*
Lunch Secretary & Treasurer *Torin Douglas*

Founded 1974 to promote the professional
interests of journalists specializing in writing or
broadcasting about the media. Organizes monthly
lunches addressed by leading industry figures,
and annual TV and radio awards. Membership by
invitation.

BSME
▷ British Society of Magazine Editors

Bureau of Freelance Photographers
Focus House, 497 Green Lanes, London N13 4BP
☎ 020 8882 3315   ᖴ 020 8886 3933
info@thebfp.com
www.thebfp.com
Membership Secretary *Angela Kidd*

Founded 1965. Assists members in selling their
pictures through monthly *Market Newsletter*, and
offers advisory, legal assistance and other services.

Campaign for Press and Broadcasting Freedom
Second Floor, 23 Orford Road, London E17 9NL
☎ 020 8521 5932
freepress@cpbf.org.uk
www.cpbf.org.uk

Broadly based pressure group working for more
accountable and accessible media in Britain.
Advises on right of reply and takes up the issue of
the portrayal of minorities. Members receive *Free
Press* (bi-monthly), discounts on publications and
news of campaign progress.

Canadian Authors Association
National Office: Box 419, 320 South Shores Road,
Campbellford, Ontario K0L 1L0, Canada
☎ 001 705 653 0323   ᖴ 001 705 653 0593
admin@canauthors.org
www.canauthors.org
Contact *The National Director*

Founded 1921. The CAA was founded to promote
recognition of Canadian writers and their works,
and to foster and develop a climate favourable
to the creative arts. It is Canada's national
association for writers of every kind; for those
actively seeking to become writers and for those
who want to support writers. The Association
has branches across the country providing
support to local members in the form of advice,
local contests, publications and writers' circles.
Publishes *National Newsline* (quarterly) and *The
Canadian Writer's Guide*.

Canadian Federation of Poets
2431 Cyprus Avenue, Burlington, Ontario L7P
1G5, Canada
☎ 001 818-859-7210
info@federationofpoets.com
www.federationofpoets.com
President *Tracy Repchuk*

Founded 2003. Federations and members from
around the world can post messages, 'meet' via
forums and live chats and learn about themselves
and their international colleagues with regard to
poetry. Membership includes poetry workshops,
anthologies, publishing opportunities, online
forums, free book promotion, calendar of events
and free subscription to *Poetry Canada* magazine
(www.poetrycanada.com). Also offers an online
52-week poetry certification programme to
help educate on the various types of poetry and
promote the diversity of poetry.

Canadian Publishers' Council
250 Merton Street, Suite 203, Toronto, Ontario
M4S 1B1, Canada
☎ 001 416 322 7011   ᖴ 001 416 322 6999
pubadmin@pubcouncil.ca
www.pubcouncil.ca
Executive Director, External Relations *Jacqueline
Hushion*

Founded 1910. Trade association of English-
language publishers which represents the
domestic and international interests of member
companies.

Chartered Institute of Journalists
2 Dock Offices, Surrey Quays Road, London
SE16 2XU
☎ 020 7252 1187   ᖴ 020 7232 2302
memberservices@cioj.co.uk
www.cioj.co.uk
General Secretary *Dominic Cooper*

Founded 1884. The Institute is concerned with
professional journalistic standards and with
safeguarding the freedom of the media. It is open
to writers, broadcasters and journalists (including
self-employed) in all media. Affiliate membership
(£133) is available to part-time or occasional
practitioners and to overseas journalists who
can join the Institute's International Division.
Non-employing members also belong to the IOJ
(TU), an independent trade union which protects,
advises and represents them in their employment
or freelance work; negotiates on their behalf
and provides legal assistance and support. The
IOJ (TU) is a certificated independent trade
union which represents members' interests in
the workplace, and the CIOJ is also a constituent
member of the Media Society, the **British
Copyright Council** and the Journalists Copyright

Fund. Editor of the Institute's quarterly magazine is *Andrew Smith*.

Chartered Institute of Linguists (IoL)

Saxon House, 48 Southwark Street, London
SE1 1UN
☏ 020 7940 3100   ⓕ 020 7940 3101
info@iol.org.uk
www.iol.org.uk
Chief Executive *John Hammond*
Director of Communications *Cetty Zambrano*

Founded 1910. Professional association for translators, interpreters and trainers; examining body for languages at degree level and above for vocational purposes; the National Register of Public Service Interpreters is managed by NRPSI Limited, an IoL subsidiary. The Institute's limited company, Language Services Ltd, provides customized assessments of language-oriented requirements, skills, etc.

Children's Book Circle

childrensbookcircle@hotmail.co.uk
www.childrensbookcircle.org.uk
Membership Secretary *Katie Jennings*

The Children's Book Circle provides a discussion forum for anybody involved with children's books. Regular meetings are addressed by a panel of invited speakers and topics focus on current and controversial issues. Administers the **Eleanor Farjeon Award** and the Patrick Hardy lecture.

Children's Books Ireland

17 North Great George's Street, Dublin 1, Republic of Ireland
☏ 00 353 1 872 7475   ⓕ 00 353 1 872 7476
info@childrensbooksireland.ie
www.childrensbooksireland.ie
Contact *Mags Walsh*

Irish national children's books organization. Holds annual conferences as well as other occasional seminars and events. Quarterly magazine, *Inis*. Annual children's book festival in October; **CBI Bisto Book of the Year Award** (see entry under *Prizes*). Partners other organizations to promote and develop ways of bringing young people and books together.

CiB

▷ British Association of Communicators in Business

CILIP in Scotland (CILIPS)

www.slainte.org.uk
Director *Elaine Fulton*

CILIP in Scotland (Chartered Institute of Library and Information Professionals in Scotland) aims to bring together everyone engaged in or interested in library work in Scotland. Its main aims are the promotion of library services and the qualifications and status of librarians.

CILIP Wales

Department of Information Studies, Llanbadarn Fawr, Aberystwyth SY23 3AS
☏ 01970 622174   ⓕ 01970 622190
cilip-wales@aber.ac.uk
www.dil.aber.ac.uk/cilip_w/index.htm
Contact *Mandy Powell*, (Development Officer)

The Chartered Institute of Library and Information Professionals Wales was founded in 2002 as a branch of **CILIP**. Publishes a thrice-yearly journal, *Y Ddolen*.

CILIP: The Chartered Institute of Library and Information Professionals

7 Ridgmount Street, London WC1E 7AE
☏ 020 7255 0500   ⓕ 020 7255 0501
info@cilip.org.uk
www.cilip.org.uk
Chief Executive *Bob McKee*

The leading membership body for library and information professionals, formed in 2002 following the unification of The Library Association and the Institute of Information Scientists. **Facet Publishing** (successor to LA Publishing) produces 25–30 new titles each year and has over 150 titles in print from the LA's back catalogue.

Circle of Wine Writers

5 Ingatestone Hall Cottages, Ingatestone CM4 9NR
administrator@winewriters.org
www.winewriters.org
Administrator *Andrea Warren*

Founded 1960. Membership is open to all those professionally engaged in communicating about wines and spirits, with the exception of people primarily doing so for promotional purposes. Aims to improve the standard of writing, broadcasting and lecturing about wines, spirits and beers; to contribute to the growing knowledge and interest in wine; to promote wines and spirits of quality and to comment adversely on faulty products or dubious practices; to establish and maintain good relations with the news media and the wine trade; to provide members with a strong voice with which to promote their views; to provide a programme of workshops, meetings, talks and tastings.

CLÉ – Irish Book Publishers' Association

25 Denzille Lane, Dublin 2, Republic of Ireland
☏ 00 353 1 639 4868
info@publishingireland.com
www.publishingireland.com
President *Sean Ócearnaigh*
Administrator *Karen Kenny*
Project Manager *Jolly Ronan*

Founded 1970 to promote Irish publishing, protect members' interests and train the industry.

Comhairle nan Leabhraichean/The Gaelic Books Council

22 Mansfield Street, Glasgow G11 5QP
☎ 0141 337 6211 ⓕ 0141 341 0515
brath@gaelicbooks.net
www.gaelicbooks.org
www.ur-sgeul.com
Chairman *Professor Roibeard Ó Maolalaigh*
Director *Ian MacDonald*

Founded 1968 and now a charitable company with its own bookshop. Encourages and promotes Gaelic publishing by giving grants to publishers and writers; providing editorial and word-processing services; retailing Gaelic books; producing a catalogue of all Gaelic books in print and answering enquiries about them; mounting occasional literary evenings and training courses. Stock list on website.

Commercial Radio Companies Association
▷ RadioCentre

The Copyright Licensing Agency Ltd

Saffron House, 6–10 Kirby Street, London EC1N 8TS
☎ 020 7400 3100/0800 085 6644 (new licences)
ⓕ 020 7400 3101
cla@cla.co.uk
www.cla.co.uk
Chief Executive *Kevin Fitzgerald*

CLA is the UK's reproduction rights organization which looks after the interests of authors, publishers and visual creators in the photocopying and scanning of extracts from books, journals, magazines and digital publications. Founded in 1983 by the **Authors' Licensing and Collecting Society Limited (ALCS)** and the **Publishers Licensing Society Ltd (PLS)** to promote and enforce intellectual property rights of UK rightsholders both at home and abroad. CLA works closely with Reproduction Rights Organizations (RROs) from other countries and has an agency agreement with the **Design and Artists Copyright Society (DACS)** which represents artists and illustrators. A not-for-profit organization, CLA licenses the business, education and government sectors and distributes fees collected to authors, publishers and visual creators. CLA has also developed licences which enable digitization of existing print material. Licences allow photocopying and scanning from original print material as well as copying from born digital publications. Scanning and e-mail distribution is only available for UK works at present. Since its inception in 1983, CLA has distributed over £400 million.

Council for British Archaeology

St Mary's House, 66 Bootham, York YO30 7BZ
☎ 01904 671417 ⓕ 01904 671384
info@britarch.ac.uk
www.britarch.ac.uk
Head of Information & Communications *Dan Hull*
Publications Officer *Catrina Appleby*

Founded in 1944 to represent and promote archaeology at all levels. It is an educational charity working throughout the UK to involve people in archaeology and to promote the appreciation and care of the historic environment for the benefit of present and future generations. The CBA publishes a wide range of academic, educational and general works about archaeology, ranging from monographs and handbooks to online articles and the popular magazine, *British Archaeology* (see **CBA Publishing** under *UK Publishers*).

Crime Writers' Association (CWA)

PO Box 273, Boreham Wood WD6 2XA
secretary@thecwa.co.uk
www.thecwa.co.uk
Hon. Secretary *Liz Evans*
Membership Secretary *Rebecca Tope*
(Crossways Cottage, Walterstone HR2 0DX)

Full membership is limited to professional crime writers, but publishers, literary agents, booksellers, etc. who specialize in crime, are eligible for Associate membership. The Association has regional chapters throughout the country, including Scotland. Meetings are held regularly with informative talks frequently given by police, scenes of crime officers, lawyers, etc., and a weekend conference is held annually in different parts of the country. Produces a monthly newsletter for members called *Red Herrings* and presents various annual awards (see entries under *Prizes*).

The Critics' Circle

c/o Catherine Cooper, 69 Marylebone Lane, London W1U 2PH
☎ 020 7224 1410
www.criticscircle.org.uk
President *Marianne Gray*
Honorary General Secretary *William Russell*

Membership by invitation only. Aims to uphold and promote the art of criticism (and the commercial rates of pay thereof) and preserve the interests of its members: professionals involved in criticism of film, drama, music, dance and art.

Cyngor Llyfrau Cymru
▷ Welsh Books Council

DACS
▷ Design and Artists Copyright Society

Data Publishers Association

Queen's House, 28 Kingsway, London WC2B 6JR
☎ 020 7405 0836
info@dpa.org.uk
www.dpa.org.uk
Contact *Sarah Gooch*

Since 1970, the DPA has existed to serve, promote and protect the interests of all companies operating in the directory, data and search publishing sector. Today, the industry contributes well over £1 billion to the UK economy. The DPA membership not only includes mainstream data providers but also a whole host of organizations providing services and products to the data provision and publishing sector, such as printers, software solution providers and contract directory publishers. The DPA's members are predominantly UK companies but it also offers membership to overseas companies with UK interests. It provides its members with opportunities to develop their businesses by providing a complete and diverse range of services, including communications, networking, training and seminars, statistics, representation to government, publicity and promotion, legal support and recognition.

Department for Culture, Media and Sport

2–4 Cockspur Street, London SW1Y 5DH
☎ 020 7211 6000   📠 020 7211 6270
michael.panayi@culture.gsi.gov.uk
Director of Communications *Paddy Feeny*

The Department for Culture, Media and Sport has responsibilities for Government policies relating to the arts, museums and galleries, sport, the 2012 Olympic & Paralympic Games, gambling, broadcasting, Press freedom and regulation, the built historic environment, the film and music industries, tourism and the National Lottery. It funds the **Arts Council**, national museums and galleries, the **British Library**, the **Public Lending Right** and the Royal Commission on Historical Manuscripts. It is responsible within Government for the public library service in England, and for library and information matters generally, where they are not the responsibility of other departments.

Design and Artists Copyright Society (DACS)

33 Great Sutton Street, London EC1V 0DX
☎ 020 7336 8811   📠 020 7336 8822
info@dacs.org.uk
www.dacs.org.uk
Chief Executive *Joanna Cave*

DACS is the UK's copyright and collecting society for artists and visual creators. Represents 26,000 fine artists and their heirs as well as photographers, illustrators, craftspeople, cartoonists, architects, animators and designers.

Drama Association of Wales

The Old Library, Singleton Road, Splott, Cardiff CF24 2ET
☎ 029 2045 2200   📠 029 2045 2277
aled.daw@virgin.net
www.amdram.co.uk/daw
Contact *Teresa Hennessy*

Runs a large playscript lending library; holds an annual playwriting competition (see entry under *Prizes*); offers a script-reading service (£25 full mss; £17.50 one-act play) which usually takes two months from receipt of play to issue of reports. From plays submitted to the reading service, selected scripts are considered for publication of a short run (70–200 copies). Writers receive a percentage of the cover price on sales and a percentage of the performance fee.

Edinburgh Bibliographical Society

Special Collections, Edinburgh University Library, George Square, Edinburgh EH8 9LJ
warrenmcdougall@aol.com
http://mcs.qmuc.ac.uk/EBS/
Secretary *Warren McDougall*

Founded 1890. Organizes lectures on bibliographical topics and visits to libraries. Publishes an annual journal, which is free to members, and other occasional publications.

Educational Publishers Council
▷ The Publishers Association

Electronic Publishers' Forum
▷ The Publishers Association

The English Association

University of Leicester, University Road, Leicester LE1 7RH
☎ 0116 252 3982   📠 0116 252 2301
engassoc@le.ac.uk
www.le.ac.uk/engassoc
Chief Executive *Helen Lucas*
Contact *Julia Hughes*

Founded 1906 to promote understanding and appreciation of the English language and its literatures. Activities include sponsoring a number of publications and organizing lectures and conferences for teachers, plus annual sixth-form conferences. Publications include *The Year's Work in Critical and Cultural Theory*; *English*, *The Use of English*; *English 4–11*; *Essays and Studies* and *The Year's Work in English Studies*.

English PEN

6–8 Amwell Street, London EC1R 1UQ
☎ 020 7713 0023   📠 020 7837 7838

enquiries@englishpen.org
www.englishpen.org
Director *Jonathan Heawood*
English PEN is part of International PEN, a
worldwide association of writers and other
literary professionals which promotes literature,
fights for freedom of expression and speaks
out for writers who are imprisoned or harassed
for having criticized their governments, or for
publishing other unpopular views. Founded in
London in 1921, International PEN now consists
of over 130 centres in almost 100 countries.
PEN originally stood for poets, essayists and
novelists, but membership is now open to all
literary professionals. It is also possible to become
a 'Friend of English PEN'. A programme of talks
and discussions, and other activities such as
social gatherings, is supplemented by mailings,
website and annual congress at one of the centre
countries.

Federation of Entertainment Unions
c/o Equity, Guild House, Upper St Martin's Lane,
London WC2H 9EG
℡ 07973 714206
entertainment.unions@gmail.com
Secretary & Coordinator *Paul Evans*

Plenary meetings six times annually. Additionally,
there are Training, Equalities & European
Committees. Represents the following unions:
British Actors' Equity Association; Broadcasting
Entertainment Cinematograph and Theatre
Union; Musicians' Union; Unite – the union;
Professional Footballers' Association; **National
Union of Journalists**; The Writers' Guild of
Great Britain.

Fellowship of Authors and Artists
PO Box 158, Hertford SG13 8FA
℡ 01992 511697   ℻ 0870 116 3398
www.author-fellowship.co.uk
Contact *Graham Irwin*

Founded in 2000 to promote and encourage the
use of writing and all art forms as a means of
therapy and self healing; to provide a valuable
resource and meeting point for all interested
parties including, but not limited to, writers,
artists, counsellors and healers; to publish as web
pages or e-books any suitable works that may help
to support or promote the aims of the fellowship.

Foreign Press Association in London
11 Carlton House Terrace, London SW1Y 5AJ
℡ 020 7930 0445   ℻ 020 7925 0469
enquiries@foreign-press.org.uk
www.foreign-press.org.uk
Director *Christoph Wyld*

Founded 1888. Non-profit-making service
association for foreign correspondents based
in London, providing a variety of press-related
services. Welcomes enquiries about media-related
venue hire.

The Gaelic Books Council
▷ Comhairle nan Leabhraichean

The Garden Media Guild
Katepwa House, Ashfield Park Avenue, Ross-on-
Wye HR9 5AX
℡ 01989 567393   ℻ 01989 567676
admin@gardenmediaguild.co.uk
www.gardenmediaguild.co.uk
Contact *Gill Hinton*

Founded 1990. Aims to raise the quality of
gardening communication, to help members
operate efficiently and profitably, to improve
liaison between garden communicators and the
horticultural industry. Administers an annual
awards scheme. Operates a mailing service and
organizes press briefing days.

Guidebookwriters.com
27 Camwood, Clayton-le-Woods, Bamber Bridge
PR5 8LA
℡ 01772 321243   ℻ 0870 137 8888
tm@wpu.org.uk
www.guidebookwriters.com
Contact *Terry Marsh*

Guidebookwriters.com exists to promote the
work of its members, all of whom are professional
guidebook writers, to the trade through direct
contact with public relations companies, tourist
boards, editors, publishers and consultants and,
through its website, regular news bulletins and
press releases.

Guild of Agricultural Journalists
Isfield Cottage, Church Road, Crowborough
TN6 1BN
℡ 01892 610628
don.gomery@btinternet.com
www.gaj.org.uk
Honorary General Secretary *Don Gomery*

Founded 1944 to promote a high professional
standard among journalists who specialize
in agriculture, horticulture and allied
subjects. Represents members' interests with
representative bodies in the industry; provides
a forum through meetings and social activities
for members to meet eminent people in the
industry; maintains contact with associations
of agricultural journalists overseas; promotes
schemes for the education of members and for
the provision of suitable entrants into agricultural
journalism.

Guild of Editors
▷ Society of Editors

The Guild of Food Writers

255 Kent House Road, Beckenham BR3 1JQ
☎ 020 8659 0422
gfw@gfw.co.uk
www.gfw.co.uk
Administrator *Jonathan Woods*

Founded 1985. The objects of the Guild are to bring together professional food writers including journalists, broadcasters and authors, to extend the range of members' knowledge and experience by arranging discussions, tastings and visits, and to encourage new writers through competitions and awards. The Guild aims to contribute to the growth of public interest in, and knowledge of, the subject of food and to campaign for improvements in the quality of food.

Guild of Motoring Writers

40 Baring Road, Bournemouth BH6 4DT
☎ 01202 422424   ☒ 01202 422424
generalsec@gomw.co.uk
www.guildofmotoringwriters.co.uk
General Secretary *Patricia Lodge*

Founded 1944. Represents members' interests and provides a forum for members to exchange information.

Horror Writers Association

244 Fifth Avenue, Suite 2767, New York, NY 10001, USA
hwa@horror.org
www.horror.org
www.horror.org/UK (UK Chapter)

Founded 1987. World-wide organization of writers and publishers dedicated to promoting the interests of writers of horror and dark fantasy. Publishes a monthly newsletter, issues e-mail bulletins, gives access to lists of horror agents, reviewers and bookstores; and keys to the 'Members Only' area of the HWA website. Presents the annual **Bram Stoker Awards** (see entry under *Prizes*).

Independent Publishers Guild

PO Box 93, Royston SG8 5GH
☎ 01763 247014   ☒ 01763 246293
info@ipg.uk.com
www.ipg.uk.com
Executive Director *Bridget Shine*

Founded 1962. Actively represents the interests of independent publishers. With over 500 members and steadily growing with combined revenues of well over £500 million, the IPG provides a networking base whereby members receive regular e-newsletters, training courses and seminars covering important areas for its members, from e-commerce to international trade. Runs a collective stand for members at leading international book fairs including Frankfurt, London and most recently at Beijing. Membership is open to new and established independent publishers. Supplier membership is also available. Check the website for further information.

Independent Theatre Council

12 The Leathermarket, Weston Street, London SE1 3ER
☎ 020 7403 1727   ☒ 020 7403 1745
admin@itc-arts.org
www.itc-arts.org
Chief Executive *Charlotte Jones*

The Independent Theatre Council (ITC) is the management association and political voice of around 700 performing arts professionals and organizations. ITC provides its members with legal and management advice, training and professional development, networking, regular newsletters and a comprehensive web resource. Additionally ITC initiates and develops projects to enrich, enhance and raise the profile of the performing arts. Publications available from ITC include *The ITC Practical Guide for Writers and Companies*. ITC negotiates contracts and has established standard agreements for theatre professionals with Equity. The rights and fee structure agreement, reached with **The Writers' Guild** in 1991, was updated in December 2002 and rates of pay are reviewed annually. Contact The Writers' Guild or visit ITC's website for further details.

Institute of Copywriting

Overbrook Business Centre, Poolbridge Road, Blackford, Wedmore BS28 4PA
☎ 0800 781 1715   ☒ 01934 713492
services@inst.org
www.inst.org/copy
Secretary *Lynn Hall*

Founded 1991 to promote copywriters and copywriting (writing publicity material). Maintains a code of practice. Membership is open to students as well as experienced practitioners. Runs training courses (see entry under *UK Writers' Courses*). Has a list of approved copywriters. Answers queries relating to copywriting. Contact the Institute for a free booklet.

Institute of Translation and Interpreting (ITI)

Fortuna House, South Fifth Street, Milton Keynes MK9 2EU
☎ 01908 325250   ☒ 01908 325259
info@iti.org.uk
www.iti.org.uk

Founded 1986. The ITI is the only independent professional association of practising translators and interpreters in the UK. Its aim is to promote the highest standards in the profession. It has strong corporate membership and runs professional development courses and conferences. Membership is open to those with a genuine and proven involvement in translation and interpreting (including students). ITI's bi-monthly *Bulletin* and other publications are available from the Secretariat. ITI's *Directory of Members* offers a free referral service for those in need of a professional translator/interpreter and is available online. ITI is a full and active member of FIT (International Federation of Translators).

International Association of Scientific, Technical & Medical Publishers

3rd Floor, Prama House, 267 Banbury Road, Oxford OX2 7HT
☏ 01865 339321 📠 01865 339325
www.stm-assoc.org

STM is an international association of scientific, technical, medical and scholarly publishers which aims to create a platform for exchanging ideas and information, and to represent the interest of the STM publishing community in the fields of copyright, technology developments and end user/library relations. The membership consists of large and small companies, not-for-profit organizations, learned societies, traditional, primary, secondary publishers and new entrants to global publishing.

IoL
▷ Chartered Institute of Linguists

Irish Book Publishers' Association
▷ CLÉ

Irish Copyright Licensing Agency Ltd

25 Denzille Lane, Dublin 2, Republic of Ireland
☏ 00 353 1 662 4211 📠 00 353 1 662 4213
info@icla.ie
www.icla.ie
Executive Director *Samantha Holman*

Founded 1992 by writers and publishers in Ireland to provide a scheme through which rights holders can give permission, and users of copyright material can obtain permission to copy.

Irish Playwrights and Screenwriters Guild

Art House, Curved Street, Temple Bar, Dublin 2, Republic of Ireland
☏ 00 353 1 670 9970
info@script.ie
www.script.ie
Contact *David Kavanagh*

Founded in 1969 to safeguard the rights of scriptwriters for radio, stage and screen.

Irish Translators' & Interpreters' Association (ITIA)

Irish Writers' Centre, 19 Parnell Square, Dublin 1, Republic of Ireland
☏ 00 353 1 872 1302 📠 00 353 1 872 6282
itiasecretary@eircom.net
www.translatorsassociation.ie
Chairperson *Annette Schiller*
Honorary Secretary *Mary Phelan*
Treasurer *Mirian Watchorn*

Founded 1986. The Association is open to all translators: technical, commercial, literary and cultural, and to all classes of interpreters. It is also for those with an interest in translation such as teachers and third-level students. Information on the profession of interpreting and translation and a register of members are available on the website. Publishes a newsletter, *Translation Ireland* and an eZine, the *ITIA Bulletin*. Organizes certification tests (for legal documents) for professional members.

Irish Writers' Union

Irish Writers' Centre, 19 Parnell Square, Dublin 1, Republic of Ireland
☏ 00 353 1 872 1302
iwu@ireland-writers.com
www.ireland-writers.com
Chairperson *Helen Dwyer*

Founded 1986 to promote the interests and protect the rights of writers in Ireland.

Isle of Man Authors

11 Christian Close, Ballastowell Gardens, Ramsey IM8 2AU
☏ 01624 815634
Secretary *Mrs Beryl Sandwell*

An association of writers living on the Isle of Man, which has links with the **Society of Authors**.

ITC
▷ Independent Theatre Council

ITI
▷ Institute of Translation and Interpreting

IVCA (International Visual Communications Association)

19 Pepper Street, Glengall Bridge, London E14 9RP
☏ 020 7512 0571 📠 020 7512 0591
info@ivca.org
www.ivca.org
Publications and Information Officer *Philip Fey*

The IVCA is a professional association representing the interests of the users and suppliers of visual communications. In particular it pursues the interests of producers, commissioners and manufacturers involved in the non-broadcast and independent facilities industries and also business event companies. It represents all sizes of company and freelance individuals, offering information and advice services, publications, a professional network, special interest groups, a magazine and a variety of events including the UK's Film and Video Communications Festival.

The Library Association
▷ CILIP: The Chartered Institute of Library and Information Professionals

literaturetraining

PO Box 23595, Leith, Edinburgh EH6 7YX
☎ 0131 553 2210
info@literaturetraining.com
www.literaturetraining.com
Director *Philippa Johnston*

The UK's only dedicated provider of free information and advice on professional development for writers and literature professionals. Draws on the expertise of its eight partner organizations: **Apples & Snakes**, Lapidus, the **National Association of Writers in Education**, the **National Association for Literature Development**, **Scottish Book Trust**, **Survivors' Poetry** and **Writernet**. Their online directory contains latest information on courses, workshops, jobs, residencies, submissions, competitions, conferences and events, organizations, books, magazines and funding for professional development. Other services include a fortnightly e-bulletin, an information and help line and a developing range of info sheets and specially commissioned features relating to creative and professional practice. literaturetraining is a partner in the CreativePeopleNetwork.

Llenyddiaeth Cymru Dramor
▷ Welsh Literature Abroad

The MCPS–PRS Alliance

29–33 Berners Street, London W1T 3AB
☎ 020 7580 5544   🖷 020 7306 4455
admissions@mcps-prs-alliance.co.uk
www.mcps-prs-alliance.co.uk

PRS ensures that composers, songwriters and music publishers are paid royalties when their music is used. The music does not need to be published or recorded for eligibility for PRS membership.

Medical Journalists' Association

Fairfield, Cross in Hand, Heathfield TN21 0SH
☎ 01435 868786   🖷 01435 865714
secretary@mja-uk.org
www.mja-uk.org
Chairman *David Delvin*
Honorary Secretary *Philippa Pigache*

Founded 1967. Aims to improve the quality and practice of medical and health journalism and to improve relationships and understanding between medical and health journalists and the health and medical professions. Over 400 members. Regular meetings with senior figures in medicine and medico politics; educational workshops on important subject areas or issues of the day; debates; awards for medical journalists from commercial sponsors, plus the MJA's own annual awards for journalism and books. Publishes a newsletter five times a year.

Medical Writers' Group

The Society of Authors, 84 Drayton Gardens, London SW10 9SB
☎ 020 7373 6642   🖷 020 7373 5768
s.baxter@societyofauthors.org
www.societyofauthors.org
Contact *Sarah Baxter*

Founded 1980. A specialist group within the **Society of Authors** offering advice and help to authors of medical books.

Mystery Writers of America, Inc.

17 East 47th Street, 6th Floor, New York, NY 10017, USA
☎ 001 212 888 8171   🖷 001 212 888 8107
mwa@mysterywriters.org
www.mysterywriters.org
Administrative Manager *Margery Flax*

Founded 1945. Aims to promote and protect the interests of writers of the mystery genre in all media; to educate and inform its membership on matters relating to their profession; to uphold a standard of excellence and raise the profile of this literary form to the world at large. Holds an annual banquet at which the 'Edgars' are awarded (named after Edgar Allan Poe).

National Association for Literature Development

PO Box 49657, London N8 7YZ
☎ 020 7272 8386
director@nald.org
www.nald.org
Contact *Melanie Abrahams*

NALD is the largest membership organization for literature professionals. The only UK body for all those involved in developing writers, readers and literature audiences. Offers advice, training and advocacy via individual and organizational membership.

National Association of Writers Groups

The Arts Centre, Biddick Lane, Washington NE38 2AB
☎ 01262 609228
nawg@tesco.net
www.nawg.co.uk

Secretary *Diane Wilson*

Founded 1995 with the object of furthering the interests of writers' groups and individuals throughout the UK. A registered charity (No. 1059047), NAWG is strictly non-sectarian and non-political. Publishes a bi-monthly newsletter, distributed to member groups; gives free entry to competitions for group anthologies, poetry, short stories, articles, novels and sketches; holds an annual open festival of writing with 21 workshops, seminars, individual surgeries led by professional, high-profile writers. Membership is open to all writers' groups and individual writers – there are no restrictions or qualifications required for joining; 160 groups are affiliated to-date.

National Association of Writers in Education

PO Box 1, Sheriff Hutton, York YO60 7YU
☎ 01653 618429
info@nawe.co.uk
www.nawe.co.uk
Contact *Paul Munden*

Founded 1991. Aims to promote the contribution of living writers to education and to encourage both the practice and the critical appreciation of creative writing. Has over 900 members. Organizes national conferences and training courses. A directory of writers who work in schools, colleges and the community is available online. Publishes a magazine, *Writing in Education*, issued free to members three times per year.

National Campaign for the Arts

1 Kingly Street, London W1B 5PA
☎ 020 7287 3777   ☎ 020 7287 4777
nca@artscampaign.org.uk
www.artscampaign.org.uk
Director *Louise de Winter*

The UK's only independent campaigning organization representing all the arts. The NCA seeks to safeguard, promote and develop the arts and win public and political recognition for their importance as a key element in the national culture. Its members shape its campaign work, provide its mandate and the core funding, through subscriptions that enable the NCA to act independently on their behalf; the organization receives no public subsidy. Members have access to the NCA's information, advice, seminars, conferences and publications. Membership is open to organizations and individuals working in or with an interest in the arts. Literature subscriptions available.

National Centre for Research in Children's Literature (NCRCL)

Bede House, School of Arts, Digby Stuart College, Roehampton University, Roehampton Lane, London SW15 5PH
☎ 020 8392 3000   ☎ 020 8392 3819
g.lathey@roehampton.ac.uk
l.sainsbury@roehampton.ac.uk
www.ncrcl.ac.uk

The National Centre for Research in Children's Literature facilitates and supports research exchange in the field of children's literature. Runs an internationally acclaimed MA Programme, an annual conference in association with the International Board on Books for Young People (IBBY).

National Literacy Trust

68 South Lambeth Road, London SW8 1RL
☎ 020 7587 1842   ☎ 020 7587 1411
contact@literacytrust.org.uk
www.literacytrust.org.uk
Director *Jonathan Douglas*

Founded 1993. Independent registered charity 'dedicated to building a literate nation in which everyone enjoys the skills, confidence and pleasures that literacy can bring.' The only organization concerned with raising literacy standards for all age groups throughout the UK. Maintains an extensive website with literacy news, summaries of key issues, research and examples of practice nationwide; organizes conferences, courses and training events and runs a range of initiatives to turn promising ideas into effective action. These include the National Reading Campaign, funded by the government; Reading Is Fundamental, UK, which provides free books to children; Reading The Game, involving the professional football community and the Talk To Your Baby campaign.

National Union of Journalists

Headland House, 308 Gray's Inn Road, London WC1X 8DP
☎ 020 7278 7916   ☎ 020 7837 8143
info@nuj.org.uk
www.nuj.org.uk
General Secretary *Jeremy Dear*

Represents journalists in all sectors of publishing, print, broadcast and online media. Responsible for wages and conditions agreements which apply across the industry. Provides advice and representation for its members, as well as administering unemployment and other benefits. Publishes various guides and magazines: *On-Line Freelance Directory*; *On-Line Fees Guide*; *The Journalist* and *The Freelance*.

NCRCL
▷ National Centre for Research in Children's Literature

New Playwrights Trust
▷ Writernet

New Producers Alliance (NPA)

Room 7.03 The Tea Building, 56 Shoreditch High Street, London E1 6JJ
℡ 020 7613 0440   ℻ 020 7729 1852
queries@npa.org.uk
www.npa.org.uk

Founded 1993. National membership and training organization for independent new filmmakers, the NPA provides a forum and focus for over 800 members, ranging from students and first timers to highly experienced feature filmmakers and industry affiliates. Services include: producer, writer and director training from 'entrance' to 'advanced' levels; seminars, masterclasses, business breakfasts, themed networking evenings and preview screenings; in addition, the NPA supports its members at film festivals including Cannes, Berlin and Edinburgh with events and industry receptions. Provides free advice services including Legal and Tax & Accounting; the NPA also publishes a monthly newsletter and online directory included in the website for its 1,000 members.

New Writing North

Culture Lab, Grand Assembly Rooms, Newcastle University, King's Walk, Newcastle upon Tyne NE1 7RU
℡ 0191 222 1332   ℻ 0191 222 1372
mail@newwritingnorth.com
www.newwritingnorth.com
Director *Claire Malcolm*
Administrator *Catherine Robson*

New Writing North is the literature development agency for the Arts Council England, North East region and offers many useful services to writers, organizing events, readings and workshops. NWN has a website and mailing list by which members of the public can receive news of literary events, courses and work opportunities. Administers the Northern Writers' Awards, which include tailored development packages (mentoring and financial help) for local writers at different stages of their careers. NWN works to develop theatre writing and creative radio writing projects within the region, as well as a strong education and community-based programme, working with schools, LEAs and both public and private sector organizations.

The Newspaper Society

St Andrew's House, 18–20 St Andrew Street, London EC4A 3AY
℡ 020 7632 7424   ℻ 020 7632 7401
collette_yates@newspapersoc.org.uk
www.newspapersoc.org.uk
Director *David Newell*

The Newspaper Society is the voice of Britain's regional press, a £4 billion sector read by over 40 million adults a week, It promotes the interests of the regional and local press, provides legal advice and lobbying services, holds a series of conferences and seminars, and runs the annual Local Newspaper Week.

NPA
▷ New Producers Alliance

Outdoor Writers & Photographers Guild

PO Box 520, Bamber Bridge, Preston PR5 8LF
℡ 01772 321243   ℻ 0870 137 8888
secretary@owg.org.uk
www.owpg.org.uk
Secretary *Terry Marsh*

Founded 1980 to promote, encourage and assist the development and maintenance of professional standards among creative professionals working in the outdoors. Members include writers, photographers, broadcasters, filmmakers, editors, publishers and illustrators. Publishes a quarterly journal, *Outdoor Focus*. Presents five Awards for Excellence plus other awards in recognition of achievement. Maintains a website of members' details; informal members' e-mail 'chatline' to exchange news and information; sends out regular media news bulletins; organizes press trips.

Pact (Producers Alliance for Cinema and Television)

Procter House, 2nd Floor, 1 Procter Street, London WC1V 6DW
℡ 020 7067 4367   ℻ 020 7067 4377
enquiries@pact.co.uk
www.pact.co.uk
Chief Executive *John McVay*
Chief Operations Officer *Andrew Chowns*

Pact is the UK trade association that represents and promotes the commercial interests of independent feature film, television, children's, animation and interactive media companies. Headquartered in London, Pact has regional representation throughout the UK, in order to support its members. A highly effective lobbying organization, it has regular and constructive dialogues with government, regulators, public agencies and opinion formers on all issues affecting its members and contributes to key public policy debates on the media industry, both in the UK and in Europe. Pact negotiates terms of trade with all public service broadcasters in the UK and supports members in their business dealings with cable and satellite channels. It also lobbies for a properly structured and funded UK film industry and maintains close contact with the **UK Film Council** and other relevant film organizations and government departments.

PEN
▷ English PEN

Performing Right Society
▷ The MCPS–PRS Alliance

The Personal Managers' Association Ltd
1 Summer Road, East Molesey KT8 9LX
☎ 020 8398 9796   ⊞ 020 8398 9796
info@thepma.com
www.thepma.com
Co-chairs *Michelle Kass, Alan Brodie, Nicholas Young*
Secretary *Angela Adler*

An association of artists' and dramatists' agents (membership not open to individuals). Monthly meetings for exchange of information and discussion. Maintains a code of conduct and acts as a lobby when necessary. Applicants screened. A high proportion of play agents are members of the PMA.

The Picture Research Association
c/o 1 Willow Court, off Willow Street, London EC2A 4QB
☎ 020 7739 8544   ⊞ 020 7782 0011
chair@picture-research.org.uk
www.picture-research.org.uk
Chair *Veneta Bullen*
Membership Secretary *Frances Topp*

Founded 1977 as the Society of Picture Researchers & Editors. The Picture Research Association is a professional body for picture researchers, managers, picture editors and all those involved in the research, management and supply of visual material to all forms of the media. The Association's main aims are to promote the interests and specific skills of its members internationally; to promote and maintain professional standards; to bring together those involved in the research and publication of visual material; to provide a forum for the exchange of information and to provide guidance to its members. Free advisory service for members, regular meetings, quarterly magazine, monthly newsletter and Freelance Register.

Player–Playwrights
37 Woodvale Way, London NW11 8SQ
☎ 07804 786470
playerplaywrights@hotmail.co.uk
cstrickett@hotmail.com
www.playerplaywrights.co.uk
Presidents *Laurence Marks, Maurice Gran*
Contact *Christine Strickett* (at the address above)

Founded 1948. A society giving opportunity for writers new to stage, radio, television and film, as well as others finding difficulty in achieving results, to work with writers established in those media. At weekly meetings (7.30 pm–10.00 pm, Mondays, upstairs at the Horse and Groom, 128 Great Portland Street, London W1), members' scripts are read or performed by actor members and afterwards assessed and dissected in general discussion. New writers and new acting members are always welcome.

Playwrights' Studio, Scotland
350 Sauchiehall Street, CCA, Glasgow G2 3JD
☎ 0141 332 4403
info@playwrightsstudio.co.uk
www.playwrightsstudio.co.uk

The Playwrights' Studio, Scotland is a national initiative designed to engage the people of Scotland directly with new playwriting and raise the standard of plays for presentation to the public. Aims to increase awareness of the wealth of creative talent within Scotland, both at home and internationally, and work to develop the art form of theatre, examining and expanding the playwright's role within it. Delivers a wide range of activities which address the development needs of aspiring, emerging and established playwrights, encourage producers' interest in new plays and playwrights, and promote Scotland's plays and playwrights to the widest possible markets.

PLR
▷ Public Lending Right

PLS
▷ Publishers Licensing Society Ltd

PMA
▷ The Personal Managers' Association Ltd

Poetry Book Society
▷ under Organizations of Interest to Poets

Poetry Ireland
▷ under Organizations of Interest to Poets

The Poetry Society
▷ under Organizations of Interest to Poets

PPA The Association for Publishers and Providers of Consumer, Customer & Business Media in the UK
Queen's House, 55–56 Lincoln's Inn Fields, London WC2A 3LJ
☎ 020 7404 4166   ⊞ 020 7404 4167
info1@ppa.co.uk
www.ppa.co.uk

Founded 1913 to promote and protect the interests of its members in particular and the magazine industry as a whole.

Press Complaints Commission

Halton House, 20/23 Holborn, London EC1N 2JD
☎ 020 7831 0022   📠 020 7831 0025
pcc@pcc.org.uk
www.pcc.org.uk
Director *Tim Toulmin*
Information Officer *Tonia Milton*

Founded in 1991 to deal with complaints from members of the public about the editorial content of newspapers and magazines. Administers the editors' Code of Practice covering such areas as accuracy, privacy, harrassment and intrusion into grief. Publications available: *Code of Practice, How to Complain* and *Annual Review*.

Private Libraries Association

Ravelston, South View Road, Pinner HA5 3YD
dchambrs@aol.com
www.pla.org
Honorary Secretary *Stanley Brett* (stanbrett@f2s.com)

Founded in 1956 for the advancement of the education of the public in the study, production and ownership of books. Publishes work concerned with the objects of the Association, particularly those works which are not commercially viable; holds meetings at which papers on cognate subjects are read and discussed; arranges lectures and exhibitions which are also open to non-members.

Producers Alliance for Cinema and Television
▷ Pact

Public Lending Right

Richard House, Sorbonne Close, Stockton-on-Tees TS17 6DA
☎ 01642 604699   📠 01642 615641
authorservices@plr.uk.com
www.plr.uk.com
Registrar *Jim Parker*

Founded 1979. Public Lending Right is funded by the Department for Culture, Media and Sport to recompense authors for books borrowed from public libraries. Authors and books can be registered for PLR only when application is made during the author's lifetime. To qualify, an author must be resident in a European Union member state, or in Iceland, Liechtenstein or Norway, and the book must be printed, bound and put on sale with an ISBN. In 2008, £6.66 million was distributed to over 23,900 authors (£6.81 million in the previous year) with 242 authors reaching the maximum payment threshold of £6,600. The Rate Per Loan was 5.98 pence in 2008. Consult the website for further information.

The Publishers Association

29b Montague Street, London WC1B 5BW
☎ 020 7691 9191   📠 020 7691 9199
mail@publishers.org.uk
www.publishers.org.uk
Chief Executive *Simon Juden*

The national UK trade association for books, learned journals, and electronic publications, with around 200 member companies in the industry. Very much a trade body representing the industry to Government and the European Commission, and providing services to publishers. Publishes the *Directory of Publishing* in association with **Continuum**. Also home of the Educational Publishers Council (school books), PA's International Division, the Academic and Professional Division, the Trade Publishers Council and the Electronic Publishers' Forum.

Publishers' Association of South Africa

PO Box 106, Green Point, Cape Town, South Africa 8051
☎ 00 27 21 425 2721   📠 00 27 21 421 3270
pasa@publishsa.co.za
www.publishsa.co.za
Contact *Dudley Schroeder* (dudley@publishsa.co.za)

Founded in 1992 to represent publishing in South Africa, a key industry sector. With a membership of 162 companies, the Association includes commercial organizations, university presses, one-person privately-owned publishers as well as importers and distributors.

Publishers Licensing Society Ltd

37–41 Gower Street, London WC1E 6HH
☎ 020 7299 7730   📠 020 7299 7780
pls@pls.org.uk
www.pls.org.uk
Chief Executive *Alicia Wise*
Operations Manager *Tom West*

Founded in 1981, the PLS obtains mandates from publishers which grant PLS the authority to license photocopying of pages from published works. Some licences for digitization of printed works are available. PLS aims to maximize revenue from licences for mandating publishers and to expand the range and repertoire of mandated publishers available to licence holders. It supports the **Copyright Licensing Agency Ltd (CLA)** in its efforts to increase the number of legitimate users through the issuing of licences and vigorously pursues any infringements of copyright works belonging to rights' holders.

Publishers' Publicity Circle

65 Airedale Avenue, London W4 2NN
☎ 020 8994 1881

ppc-@lineone.net
www.publisherspublicitycircle.co.uk
Contact *Heather White*

Enables book publicists from both publishing houses and freelance PR agencies to meet and share information regularly. Meetings, held monthly in central London, provide a forum for press journalists, television and radio researchers and producers to meet publicists collectively. A directory of the PPC membership is published each year and distributed to over 2,500 media contacts.

Publishing Scotland

Scottish Book Centre, 137 Dundee Street, Edinburgh EH11 1BG
☎ 0131 228 6866 📠 0131 228 3220
joanlyle@publishingscotland.org
www.publishingscotland.org
Director *Marion Sinclair*
Membership Services and Marketing Manager *Liz Small*
Financial and Office Administrator *Carol Lothian*
Information and Professional Development Administrator *Joan Lyle*

Publishing Scotland, (formerly Scottish Publishers Association) is the trade, training and development body for publishing in Scotland. It represents over 70 Scottish publishers, from multinationals to small presses, and has developed a Network membership from all sections of the book world, including the **Society of Authors in Scotland**. Publishing Scotland also acts as an information centre for the trade and general public. Publishes annual *Directory of Publishing in Scotland*. Attends international book fairs; runs an extensive training programme; carries out industry research.

The Radio Academy

5 Market Place, London W1W 8AE
☎ 020 7927 9920 📠 020 7636 8924
info@radioacademy.org.uk
www.radioacademy.org.uk
Director *Trevor Dann*

The Radio Academy is a registered charity dedicated to the encouragement, recognition and promotion of excellence in UK broadcasting and audio production. Since 1983 the Academy has represented the radio industry to outside bodies including the government and offered neutral ground where everyone from the national networks to individual podcasters is encouraged to discuss the broadcasting, production, marketing and promotion of radio and audio.

RadioCentre

77 Shaftesbury Avenue, London W1D 5DU
☎ 020 7306 2603 📠 020 7306 2500

Chief Executive *Andrew Harrison*
Head of Strategy & Operations *Michael O'Brien*

RadioCentre represents commercial radio stations in the UK. Its role is to maintain and build a strong and successful commercial radio industry, both in terms of listening hours and revenues.

Romance Writers of America

16000 Stuebner Airline Road, Suite 140, Spring, TX 77379, USA
☎ 001 832 717 5200 📠 001 832 717 5201
info@rwanational.org
www.rwanational.org
Professional Relations Coordinator *Carol Ritter*

Founded 1980. RWA, a non-profit association with more than 9,000 members worldwide, provides a service to authors at all stages of their careers as well as to the romance publishing industry and its readers. Anyone pursuing a career in romantic fiction may join RWA. Holds an annual conference, and provides contests for both published and unpublished writers through the RITA Awards and Golden Hearts Awards.

The Romantic Novelists' Association

3 Griffin Road, Thame OX9 3LB
☎ 01844 213947
ianandcatherinejones@btinternet.com
www.rna-uk.org
Contact *Catherine Jones* (Chairman)

Membership is open to published writers of romantic fiction (modern or historical), or those who have had one or more full-length serials published. Associate membership is open to publishers, editors, literary agents, booksellers, librarians and others having a close connection with novel writing and publishing. Membership in the New Writers' Scheme is available to a limited number of writers who have not yet had a full-length novel published. New Writers must submit a manuscript each year. The manuscripts receive a report from experienced published members and the reading fee is paid with the annual subscription of £90 (£95 for non-EU members). Meetings are held in London and the regions with guest speakers. *Romance Matters* is published quarterly and issued free to members. The Association makes three annual awards: **The Romantic Novel of the Year**; Joan Hessayon New Writers Award for the best published novel by a new writer who has gone through the New Writers' Scheme and remains a member, and the Romance Prize for category romances.

Royal Society of Literature

Somerset House, Strand, London WC2R 1LA
☎ 020 7845 4676 📠 020 7845 4679

info@rslit.org
www.rslit.org
Chair *Maggie Gee, FRSL*
Secretary *Maggie Ferguson*

Founded 1820. Membership by application to
the Secretary. Fellowships are conferred by
the Society on the proposal of two Fellows.
Membership benefits include lectures, discussion
meetings and poetry readings. Lecturers have
included Julian Barnes, Michael Holroyd, Don
Paterson, Zadie Smith and Tom Stoppard.
Presents the **Royal Society of Literature
Ondaatje Prize**, the **Royal Society of Literature/
Jerwood Awards** and the **V.S. Pritchett
Memorial Prize**.

Royal Television Society

Kildare House, 3 Dorset Rise, London EC4Y 8EN
℡ 020 7822 2810   ℻ 020 7822 2811
info@rts.org.uk
www.rts.org.uk
Membership Manager *Deborah Halls* (deborah@
rts.org.uk)

Founded 1927. Covers all disciplines involved
in the television industry. Provides a forum for
debate and conferences on technical, social and
cultural aspects of the medium. Presents various
awards including journalism, programmes,
technology and design. Publishes *Television
Magazine* ten times a year for members and
subscribers.

Science Fiction Foundation

75 Rosslyn Avenue, Harold Wood RM3 0RG
www.sf-foundation.org

Founded 1970. The SFF is an international
academic body for the furtherance of science
fiction studies. Publishes a thrice-yearly
magazine, *Foundation* (see entry under
Magazines), which features academic articles
and reviews of new fiction. Publishes a series of
books on critical studies of science fiction, and
arranges conferences, lectures and seminars on
related topics. It also has a reference library (see
entry under *Library Services*), housed at Liverpool
University.

Scottish Book Trust

Sandeman House, Trunk's Close, 55 High Street,
Edinburgh EH1 1SR
℡ 0131 524 0160   ℻ 0131 524 0161
info@scottishbooktrust.com
www.scottishbooktrust.com
Contact *Marc Lambert*

Scottish Book Trust exists to serve readers
and writers in Scotland and works to ensure
that everyone has access to good books and to
related resources and opportunities. Operates
'Live Literature Funding' which funds over

1,200 visits a year by Scottish writers to a variety
of institutions and groups; supports Scottish
writing through a programme of professional
training opportunities for writers; operates
a touring programme for children's authors,
the Scottish Friendly Children's Book Tour;
works internationally promoting Scottish
literature; publishes a variety of resources and
leaflets to support readership; promotes other
readership development opportunities. The Book
Information Service provides free advice and
support to readers and writers and the general
public.

Scottish Daily Newspaper Society

21 Lansdowne Crescent, Edinburgh EH12 5EH
℡ 0131 535 1064   ℻ 0131 535 1063
info@sdns.org.uk
Director *Mr J.B. Raeburn*

Founded 1915. Trade association representing
publishers of Scottish daily and Sunday
newspapers including the major Scottish editions
of UK national newspapers.

Scottish Library Association

▷ CILIP in Scotland (CILIPS)

Scottish Newspaper Publishers Association

48 Palmerston Place, Edinburgh EH12 5DE
℡ 0131 220 4353   ℻ 0131 220 4344
info@snpa.org.uk
www.snpa.org.uk
Director *Mr S. Fairclough*

The representative body for the publishers of
paid-for weekly and associated free newspapers in
Scotland. Represents the interests of the industry
to government, public and other bodies and
provides a range of services including marketing
of *The Scottish Weekly Press* and education and
training. It is an active supporter of the Press
Complaints Commission.

Scottish Print Employers Federation

48 Palmerston Place, Edinburgh EH12 5DE
℡ 0131 220 4353   ℻ 0131 220 4344
info@spef.org.uk
www.spef.org.uk
Director *Mr S. Fairclough*

Founded 1910. Employers' organization and
trade association for the Scottish printing
industry. Represents the interests of the industry
to government, public and other bodies and
provides a range of services including industry
promotion, industrial relations, education,
training and commercial activities. Negotiates a
national wages and conditions agreement with
Unite GPM Sector. The Federation is a member
of Intergraf, the international confederation for
employers' associations in the printing industry,
which represents its interests at European level.

Scottish Screen

2nd Floor, 249 West George Street, Glasgow
G2 4QE
☎ 0141 302 1700   🖷 0141 302 1711
info@scottishscreen.com
www.scottishscreen.com
Contact *Judy Anderson*

Scottish Screen is the national development
agency for the screen industries in Scotland.
'We aim to inspire audiences, support new and
existing talent and businesses, educate young
people, and promote Scotland as a creative place
to make great films, award-winning television and
world renowned digital entertainment.'

Society for Editors and Proofreaders (SfEP)

Riverbank House, 1 Putney Bridge Approach,
London SW6 3JD
☎ 020 7736 3278/7736 0901 (training dept.)
🖷 020 7736 3318
administration@sfep.org.uk
www.sfep.org.uk
Chair *Penny Williams*
Vice-Chair/Company Secretary *Sarah Price*
SUBSCRIPTION £85 p.a. (individuals) plus £27.50
joining fee; corporate membership available.

Founded 1988 in response to the growing
number of freelance editors and their increasing
importance to the publishing industry.
Aims to promote high editorial standards by
disseminating information through advice
and training, and to achieve recognition of
the professional status of its members. The
Society also supports moves towards recognized
standards of training and accreditation for
editors and proofreaders. It launched its own
Accreditation in Proofreading test in 2002.

Society for Technical Communications (STC)

901 N. Stuart Street, Suite 904, Arlington,
Virginia 22203, USA
☎ 001 703 522 4114   🖷 001 703 522 2075
www.stc.org
www.stcuk.org
UK Chapter Membership Manager *Karen Lewis*
(membership@stcuk.org)

Dedicated to advancing the arts and sciences
of technical communication, the STC has more
than 15,000 members worldwide, including
technical writers, editors, graphic designers,
videographers, multimedia artists, Web and
intranet page information designers, translators
and others whose work involves making technical
information available to those who need it. The
UK Chapter, established in 1985, hosts talks and
seminars, runs an annual technical publications
competition, and publishes its own newsletter six
times a year.

The Society of Authors

84 Drayton Gardens, London SW10 9SB
☎ 020 7373 6642   🖷 020 7373 5768
info@societyofauthors.org
www.societyofauthors.org
General Secretary *Mark Le Fanu*
Deputy General Secretary *Kate Pool*

The Society of Authors is an independent
trade union, chaired by Tracy Chevalier, with
some 8,000 members. As well as campaigning
for the profession, it advises members on
negotiations with publishers, agents, broadcasting
organizations, theatre managers and film
companies and on queries or concerns members
may have about the business aspects of their
writing. The Society assists with complaints and
takes action for breach of contract, copyright
infringement, etc. Among the Society's
publications are *The Author* (quarterly) and the
Quick Guides series to various aspects of writing
(all free of charge to members). Other services
include vetting of contracts, emergency funds
for writers, meetings and seminars, and various
special discounts, including money off books,
hotels and insurance. There are groups within
the Society for academic writers, broadcasters,
children's writers and illustrators, educational
writers, medical writers and translators, as well
as some regional groups. Authors under 35 or
over 65, not earning a significant income from
their writing, may apply for lower subscription
rates (after their first year, in the case of over-65s).
Contact the Society for a free booklet and a copy
of *The Author* or visit the website.

The Society of Authors in Scotland

c/o 99A/2 St. Stephen Street, Edinburgh EH3 5AB
☎ 0131 556 9446   🖷 0131 556 9446
anguskonstam@aol.com
www.writersorg.co.uk
Secretary *Brian D. Osborne*

The Scottish branch of the **Society of Authors**,
which organizes business meetings, social and
bookshop events throughout Scotland.

Society of Children's Book Writers & Illustrators

ra@britishscbwi.org
www.britishscbwi.org
British Isles Regional Adviser *Natascha Biebow*

Founded in 1968 by a group of Los Angeles-based
writers, the SCBWI is the only international
professional organization for the exchange of
information between writers, illustrators, editors,
publishers, agents and others involved with
literature for young people. With a membership
of more than 12,000 worldwide, it holds two
annual international conferences plus a number

of regional events, publishes a bi-monthly journal and awards grants for works in progress. The British Isles region sponsors workshops and speaker events to help authors and illustrators develop their craft and market their work. It also facilitates local critique groups and publishes a quarterly regional newsletter. For membership enquiries, visit the website or contact membership@british scbwi.org

Society of Civil and Public Service Writers

17 The Green, Corby Glen, Grantham NG33 4NP
joan@lewis5634.fsnet.co.uk
www.scpsw.co.uk
Membership Secretary *Mrs Joan Lewis*

Founded 1935. Welcomes serving and retired members of the Civil Service, armed forces, Post Office, BT, nursing profession and other public servants who are aspiring or published writers. Offers competitions for short story, article and poetry; postal folios for short story and article; AGM, occasional meetings and luncheon held in London; quarterly magazine, *Civil Service Author*, to which members may submit material; Poetry Workshop (extra £3) offers annual weekend outside London, anthology, newsletter, postal folio, competitions. S.a.e. to the Secretary for details.

Society of Editors

University Centre, Granta Place, Mill Lane, Cambridge CB2 1RU
Ⓣ 01223 304080   Ⓕ 01223 304090
info@societyofeditors.org
www.societyofeditors.org
Executive Director *Bob Satchwell*

Formed by a merger of the Association of British Editors and the Guild of Editors, the Society of Editors has nearly 500 members in national, regional and local newspapers, magazines, broadcasting, new media, journalism education and media law. Campaigns for media freedom and self-regulation. For further information call *Bob Satchwell* (Ⓣ 07860 562815).

Society of Indexers

Woodbourn Business Centre, 10 Jessell Street, Sheffield S9 3HY
Ⓣ 0114 244 9561   Ⓕ 0114 244 9563
admin@indexers.org.uk
www.indexers.org.uk
Secretary *Judith Menes*
Administrator *Wendy Burrow*

Founded 1957. Publishes *The Indexer* (biannual, April and October), a quarterly newsletter, an online directory, *Indexers Available (IA)*, which lists members and their subject expertise. In addition, the Society runs an open-learning course entitled *Training in Indexing* and

recommends rates of pay (currently £18.50–£30 per hour or £2–£5 per page). Awards annually **The Wheatley Medal** and **The Bernard Levin Award** (see entries under *Prizes*).

The Society of Medical Writers

Membership Secretary: 30 Dollis Hill Lane, London NW2 6JE
richard_cutler_novelist@yahoo.co.uk
www.somw.org
Contact *Dr R.C. Cutler*

Now in its 21st year, the Society is open to anyone in any branch of medicine or its allied professions who publishes or aspires to publish their written work of whatever nature – fact or fiction, prose or poetry. The Society aims to improve standards of writing; to encourage literacy; to provide meetings for members for the exchange of views, skills and ideas; to provide education on the preparation, presentation and submission of written material for publication; to advise on sources of assistance with regard to technical, legal and financial aspects of writing; to consider questions of ethics relating to writing and publication; to further developments in the art of writing. Biannual conferences held in the spring and autumn at various locations; details published in *The Writer* or available from the secretary (copy for *The Writer* may be submitted to the editor throughout the year).

Society of Picture Researchers & Editors
▷ The Picture Research Association

Society of Women Writers & Journalists

27 Braycourt Avenue, Walton-on-Thames KT12 2AZ
wendy@stickler.org
www.swwj.co.uk
Membership Secretary *Wendy Hughes*

Founded 1894. For women writers, the SWWJ upholds professional standards and literary achievements through regional group meetings; residential weekends; postal critique service; competitions; outings and major London meetings. For Full members, membership card doubles as a press card. Publishes *The Woman Writer*. Male writers are accepted as Associate Members. A new category has been introduced for Friends of the Society.

Society of Young Publishers

c/o The Bookseller, Endeavour House, 189 Shaftesbury Avenue, London WC2H 8TJ
www.thesyp.org.uk
Contact *Membership Secretary*

Provides facilities whereby members can increase their knowledge and widen their experience of all aspects of publishing, and holds regular social events. Run entirely by volunteers, it is open to those in related occupations, students, and associate membership is available for over-35s. Publishes a bi-monthly newsletter called *InPrint* and holds meetings on the last Wednesday of each month featuring trade professionals. For other events see the website. Please enclose an s.a.e. when writing to the Society.

Sports Journalists' Association of Great Britain

c/o Start2Finish Event Management, Unit 92, Capital Business Centre, 22 Carlton Road, Croydon CR2 0BS
☎ 020 8916 2234   ℻ 020 8916 2235
stevenwdownes@btinternet.com
www.sportsjournalists.co.uk
Secretary *Steven Downes* (80 Southbridge Road, Croydon CR0 1AE ☎ 07710 428562)

Founded in 1948 as the Sports Writers' Association of Great Britain to promote and maintain a high professional standard among journalists who specialize in sport in all its branches – including photography, broadcasting and online – and to serve members' interests. Publishes a quarterly *Bulletin* and a Yearbook, and promotes the annual SJA British Sports Awards in Nov/Dec, and the SJA British Sports Journalism Awards and SJA British Sports Photography Awards every March.

STM
▷ **International Association of Scientific, Technical & Medical Publishers**

Tees Valley Arts

Melrose House, Melrose Street, Middlesbrough TS1 2HZ
☎ 01642 264651   ℻ 01642 264955
info@teesvalleyarts.org.uk
www.teesvalleyarts.org.uk
Executive Director *Rosi Lister*
Director of Programmes *Rowena Sommerville*
Communications Officer *Simon Smith*

TVA is an arts development agency which operates throughout the Tees Valley, encompassing all art forms, working with educational and community organizations and groups, to add value to existing activities and to develop new ones. TVA works in partnership with statutory and voluntary organizations to embed cultural activity in educational and social practice, and in neighbourhood renewal and regeneration initiatives.

Theatre Writers' Union
▷ **The Writers' Guild of Great Britain**

Trade Publishers Council
▷ **The Publishers Association**

The Translators Association

84 Drayton Gardens, London SW10 9SB
☎ 020 7373 6642   ℻ 020 7373 5768
sbaxter@societyofauthors.org
www.societyofauthors.org/translators
Contact *Sarah Baxter*

Founded 1958 as a subsidiary group within the **Society of Authors** to deal exclusively with the special problems of literary translators into the English language. Benefits to members include free legal and general advice and assistance on all business matters relating to translators' work, including the vetting of contracts and advice on rates of remuneration. Membership is normally confined to translators who have had their work published in volume or serial form or produced in this country for stage, television or radio. The Association administers several prizes for translators of published work (see *Prizes*) and maintains an online database of members' details for the use of publishers who are seeking a translator for a particular work.

UK Film Council

10 Little Portland Street, London W1W 7JG
☎ 020 7861 7861   ℻ 020 7861 7863
info@ukfilmcouncil.org.uk
www.ukfilmcouncil.org.uk

The UK's leading film body. Uses lottery money and Government Grant-in-Aid to encourage the development of new talent, skills and creative and technological innovation in UK film; help new and established filmmakers make distinctive British films; support the creation and growth of stable businesses in the film sector; provide access to finance and help the UK film industry to compete in the global marketplace. Also promotes enjoyment and understanding of the cinema and ensures that film's economic and creative and cultural interests are properly represented in public policy making.

Voice of the Listener & Viewer Ltd (VLV)

101 Kings Drive, Gravesend DA12 5BQ
☎ 01474 352835   ℻ 01474 351112
info@vlv.org.uk
www.vlv.org.uk
Chairman *Jocelyn Hay*
Executive Director *Peter Blackman*
Administrative Secretary *Sue Washbrook*

VLV represents the citizen and consumer interests in broadcasting. It is an independent, non-profit-making organization working to ensure independence, quality and diversity in broadcasting. VLV is the only consumer body speaking for listeners and viewers on the full range of broadcasting issues. It is funded by its members and is free from sectarian, commercial and political affiliations. Holds public lectures, seminars and conferences throughout the UK, and has frequent contact with MPs, civil servants, the BBC and independent broadcasters, regulators, academics and other consumer groups. Provides an independent forum where all with an interest in broadcasting can speak on equal terms. Produces a quarterly news bulletin and regular briefings on broadcasting issues. Holds its own archive and those of the former Broadcasting Research Unit (1980–90) and BACTV (British Action for Children's Television). Maintains a panel of speakers, the VLV Forum for Children's Broadcasting, the VLV Forum for Educational Broadcasting, and acts as secretariat for the European Alliance of Listeners' and Viewers' Associations (EURALVA). VLV has responded to all major public enquiries on broadcasting since 1984. The VLV does not handle complaints.

WATCH
▷ Writers, Artists and their Copyright Holders

Welsh Academy
▷ Academi

Welsh Books Council (Cyngor Llyfrau Cymru)
Castell Brychan, Aberystwyth SY23 2JB
☎ 01970 624151   ℻ 01970 625385
castellbrychan@cllc.org.uk
www.cllc.org.uk
www.gwales.com
Director *Gwerfyl Pierce Jones*

Founded 1961 to stimulate interest in books from Wales and to support publishing in Wales. The Council distributes publishing grants and promotes and fosters all aspects of book production in Wales in both Welsh and English. Its Editorial, Design, Marketing and Children's Books departments and wholesale distribution centre offer central services to publishers in Wales. Writers in Welsh and English are welcome to approach the Editorial Department for advice on how to get their manuscripts published.

Welsh Literature Abroad/Llenyddiaeth Cymru Dramor
Canolfan Mercator Centre, University of Wales Aberystwyth, Llanbadarn Campus, Aberystwyth SY23 3AS
☎ 01970 622544   ℻ 01970 621524
post@welshlitabroad.org
www.welshlitabroad.org
Director *Sioned Puw Rowlands*

Founded 2000. Welsh Literature Abroad works to facilitate the translation of Wales' literature. Translation grants are available to publishers. Participates in international book fairs, and works with translators, publishers and festivals abroad to promote the literature and writers of Wales internationally.

Welsh National Literature Promotion Agency
▷ Academi

West Country Writers' Association
6 The Beals, Greenway, Woodbury, Exeter EX5 1LU
☎ 01395 233753
secretarywcwa@aol.com
Secretary *Sue Bury*

Founded 1951 in the interest of published authors living in or writing about the West Country. Annual congress and newsletters. Regional meetings to discuss news and views.

Women in Publishing
3 Gordon Road, London W5 2AD
gill@srowley.dircon.co.uk
www.wipub.org.uk
Liaison Officer *Gill Rowley*

Aims to promote the status of women working in publishing and related trades by helping them to develop their careers. Through WiP there are opportunities for members to learn more about every area of the industry, share information and expertise, give and receive support and take part in a practical forum for the exchange of information about different disciplines and industry sectors. Monthly meetings with speakers provide a forum for discussion. Meetings are held on the second Wednesday of each month. See website for further information. Publishes monthly newsletter, *WiPlash*, and *Women in Publishing Directory*.

Women Writers Network (WWN)
23 Prospect Road, London NW2 2JU
☎ 020 7794 5861
www.womenwriters.org.uk
Membership Secretary *Cathy Smith*

Founded 1985. Provides a forum for the exchange of information, support, career and networking opportunities for working women writers. Meetings, seminars, excursions, newsletter and directory. Full membership includes free admission to monthly meetings and a monthly newsletter. Details from the Membership Secretary at the address above.

Writernet (formerly New Playwrights Trust)
Cabin V, Clarendon Buildings, 25 Horsell Road, London N5 1XL
☎ 020 7609 7474   ℻ 020 7609 7557
info@writernet.org.uk
www.writernet.org.uk
Director *Jonathan Meth*
Chair *Bonnie Greer*
Administrator *Elizabeth Robertson*

Provides writers for all forms of live and recorded performance (at any stage in their career) with a range of services that enable them to pursue their careers more effectively. These include: a network connecting dramatic writers to the industry and to each other; online resources to support dramatic writers and those who work with them; a script-reading service, publications, free monthly e-bulletin and guides. Aims to help writers from all parts of the country and a wide diversity of backgrounds to fulfil their potential both inside and outside the new-writing mainstream.

Writers and Photographers unLimited

27 Camwood, Clayton-le-Woods, Bamber Bridge PR5 8LA
℡ 01772 321243   ℻ 0870 137 8888
tm@wpu.org.uk
www.wpu.org.uk
Contact *Terry Marsh*

An online marketing service for English-language professional writers and photographers worldwide who specialize in all aspects of travel, tourism, the outdoors, adventure sports, food and drink. The USP of WPu is that all its members are full-time professionals. Promotes the work of its members by direct contact with public relations companies, tourist boards, editors, publishers and consultants and, through its website, regular news bulletins and press releases.

The Writers' Guild of Great Britain

15 Britannia Street, London WC1X 9JN
℡ 020 7833 0777   ℻ 020 7833 4777
erik@writersguild.org.uk
www.writersguild.org.uk
President *David Edgar*
Chair *Katharine Way*
General Secretary *Bernie Corbett*
Deputy General Secretary *Anne Hogben*
SUBSCRIPTION Full member: 1% of earnings from professional writing (min. £150, max. £1,500); Candidate member: £100; Student member: £20; Affiliate member (agent or writers' group): £275

Founded in 1959 as the Screenwriters' Guild, the Writers' Guild is a trade union, affiliated to the TUC, representing professional writers in television, radio, theatre, film, books and new media. It negotiates Minimum Terms Agreements governing writers' contracts and covering minimum fees; advances; repeat fees; royalties and residuals; rights; credits; number of drafts; script alterations and the resolution of disputes. The most important MTAs cover BBC TV Drama; BBC Radio Drama; ITV Companies; **Pact** (independent TV and film producers); TAC (Welsh language independent TV producers); Theatrical Management Association; **Independent Theatre Council**; and an agreement covering the **Royal National Theatre**, **Royal Shakespeare Company**, and **Royal Court Theatre**. These agreements are regularly renegotiated and in most cases the

minimum fees are reviewed annually.The Guild advises members on all aspects of their working lives, including contract vetting, legal advice, help with copyright problems and representation in disputes with producers, publishers or other writers. The Guild organizes seminars, discussions and lectures on all aspects of writing. *UK Writer*, the Guild's quarterly magazine, is free to members and contains features about professional writing, Guild news and details of work opportunities, training courses, literary competitions, etc. A detailed e-mail of news, opportunities and other information is sent to members every Friday. Full Membership is open to anyone who has received payment for a piece of written work under a contract with terms not less than those negotiated by the Guild. Writers who do not qualify can join as Candidate Members and those on accredited writing courses or theatre attachments can become Student Members.

Writers, Artists and their Copyright Holders (WATCH)

The Library, The University of Reading, PO Box 223, Whiteknights, Reading RG6 6AE
℡ 0118 378 8783   ℻ 0118 378 6636
www.watch-file.com
Contact *Dr David Sutton*

Founded 1994. Provides an online database of information about the copyright holders of literary authors and artists. The database is available free of charge on the Internet and the web. WATCH is the successor project to the Location Register of English Literary Manuscripts and Letters, and continues to deal with location register enquiries.

Yachting Journalists' Association

36 Church Lane, Lymington SO41 3RB
℡ 01590 673894
sec@yja.co.uk
www.yja.co.uk
Contact *Vice Chairman/Membership Secretary*

To further the interest of yachting, sail and power, and to provide support and assistance to journalists in the field; current membership is just over 260 with 31 from overseas. A handbook, listing details of members and subscribing PR organizations, press facility recommendations, forthcoming events and other useful information, is published in alternate years at a cost to non-members and non-advertisers of £10. The YJA organizes the Yachtsman of the Year and Young Sailor of the Year Awards, presented annually at the beginning of January.

Yr Academi Gymreig
▷ Academi

Literary Societies

Margery Allingham Society
39 Morgan Street, London E3 5AA
www.margeryallingham.org.uk
Contact *Paula Cooze* (Hon. Treasurer)
SUBSCRIPTION £14 p.a.

Founded in 1988 to promote interest in and study of the works of Margery Allingham. The Society publishes two issues of the journal, *The Bottle Street Gazette*, per year. Contributions welcome. Two social events a year. Open membership.

Jane Austen Society
9 Nicola Close, South Croydon CR2 6NA
www.janeaustensociety.org.uk
Secretary *Mrs Maureen Stiller*
Membership Secretary *Mrs Rosemary Culley*
SUBSCRIPTION UK: £7.50 (student for 3 years); £15 (standard annual); £12 (annual, members of JAS branches/groups); £20 (joint); £45 (corporate); £250 (life); Overseas: £18 (annual) or £50/€75 (for 3 years); £50 (corporate); £300/€450 (life) all by credit card or UK banker's order only

Founded in 1940 to promote interest in and enjoyment of Jane Austen's novels and letters. The Society has the following Branches/Groups: Bath and Bristol, Cambridge, Midlands, London, Norfolk, Kent, Hampshire, Isle of Wight, Southern Circle, Wales and Scotland. There are independent Societies in North America, Australia and Japan.

The Baskerville Hounds (The Dartmoor Sherlock Holmes Study Group)
6 Bramham Moor, Hill Head, Fareham PO14 3RU
℡ 01329 667325
baskervillehounds@acd-221b.info
www.acd-221b.info
Chairman *Philip Weller*

Founded 1989. An international Sherlock Holmes society specializing solely in studies of *The Hound of the Baskervilles* and its Dartmoor associations. Organizes many study functions, mostly on Dartmoor. Currently involved in several large-scale research projects. Membership is open only to those who are willing to be actively involved, by correspondence and/or attendance at functions, in the group's study activities.

The BB Society
Secretary: 8 Park Road, Solihull B91 3SU
℡ 0121 704 1002
enquiries@roseworldproductions.com
Chairman *Brian Mutlow*
Secretary *Bryan Holden*
SUBSCRIPTION £10 (individual); £17.50 (family); £50 (corporate); £5 (student); £17.50 (overseas)

Founded in 2000 to bring together devotees of the writer/illustrator BB (Denys Watkins-Pitchford). BB's oeuvre included books on wildfowling, angling, children's fantasy and the countryside. The Society holds two meetings a year and publishes newsletters and an annual journal, *Sky Gypsy*.

The Beckford Society
The Timber Cottage, Crockerton, Warminster BA12 8AX
℡ 01985 213195
sidney.blackmore@btinternet.com
Secretary *Sidney Blackmore*
SUBSCRIPTION £20 (minimum) p.a.

Founded 1995 to promote an interest in the life and works of William Beckford (1760–1844) and his circle. Encourages Beckford studies and scholarship through exhibitions, lectures and publications, including an annual journal, *The Beckford Journal* and occasional newsletters.

Thomas Lovell Beddoes Society
9 Amber Court, Belper DE56 1HG
℡ 01773 828066   ℻ 01773 828066
john@beddoes.demon.co.uk
www.phantomwooer.org
Chairman *John Lovell Beddoes*
Secretary *Christine Hunkinson*
Journal Editor *Shelley Rees* (srees@ossm.edu)

Formed to research the life, times and work of poet Thomas Lovell Beddoes (1803–1849), encourage relevant publications, further the reading and appreciation of his works by a wider public and liaise with other groups and organizations. Publishes an annual journal.

The Adrian Bell Society

3 The Maltings, Church Close, Coltishall
NR12 7DZ
☎ 01603 737168
Secretary *Moya Leighton*

SUBSCRIPTION £5 p.a.

Founded in 1995, the Society aims to promote the study and enjoyment of the writing of Adrian Bell (1901–80); to publish journals twice a year and hold meetings; to help members obtain copies of out of print titles; to assemble a collection of photocopies of his 1500 *Countryman's Notebooks* that appeared in the *Eastern Daily Press* and to index them; to commemorate his career as *The Times* first crossword compiler; to share knowledge of the places and factual background mentioned in his books. The society has 250 members worldwide who keep in touch through the journal. *A Centenary Countryman's Notebook*, published 2001.

Hilaire Belloc Society

1 Hillview, Elsted, Midhurst GU29 0JX
☎ 01730 825575
hilairebelloc1@aol.com
Contact *Dr Grahame Clough*

SUBSCRIPTION £12 p.a. (UK); $20 (US)

Founded in 1996 to promote the life and work of Hilaire Belloc through the publication of newsletters containing rare and previously unpublished material. Organizes walks, social events, a Belloc weekend and a three-day conference. Currently 70 members, of which 30% are from overseas.

Arnold Bennett Society

4 Field End Close, Trentham, Stoke on Trent
ST4 8DA
arnoldbennettscty@btinternet.com
www.arnoldbennettsociety.org.uk
Secretary *Mrs Carol Gorton*

SUBSCRIPTION £12 p.a. (single); £14 (family) plus £2 for membership outside the UK

Aims to promote interest in the life and works of 'Five Towns' author Arnold Bennett and other North Staffordshire writers. Annual dinner. Regular functions and talks in and around Burslem, plus annual seminar at Wedgwood College, Barlaston and annual conference at the Potties Museum & Art Gallery, Stoke on Trent.

E.F. Benson Society

The Old Coach House, High Street, Rye TN31 7JF
☎ 01797 223114
www.efbensonsociety.org
Secretary *Allan Downend*

SUBSCRIPTION £7.50 (UK/Europe); £12.50 (overseas)

Founded in 1985 to promote the life and work of E.F. Benson and the Benson family. Organizes social and literary events, exhibitions, talks and Benson interest walks in Rye. Publishes a quarterly newsletter and annual journal, *The Dodo*, postcards and reprints of E.F. Benson articles and short stories in a series called 'Bensoniana' plus other books of Benson interest. Holds an archive which includes the Seckersen Collection (transcriptions of the Benson collection at the Bodleian Library in Oxford). Also sells second-hand Benson books.

The Betjeman Society

6 St Annes Road, Shrewsbury SY3 6AU
☎ 01743 350372
honsecbetjeman@fsmail.net
www.johnbetjeman.com
Honorary Secretary *Colin Wright*
Chairman *John Heald*

SUBSCRIPTION £10 (individual); £12 (family); £3 (student); £3 additional for overseas members

Aims to promote the study and appreciation of the work and life of Sir John Betjeman. Annual programme includes poetry readings, lectures, discussions, visits to places associated with Betjeman, and various social events. Meetings are held in London and other centres. Regular newsletter and annual journal, *The Betjemanian*.

The Bewick Society

c/o The Hancock Museum, Newcastle upon Tyne
NE2 4PT
☎ 0191 232 6386
nhsn@ncl.ac.uk
www.bewicksociety.org.uk
Membership Secretary *June Holmes*

SUBSCRIPTION £10 p.a. (individual); £12 (family)

Founded in 1988 to promote an interest in the life and work of Thomas Bewick, wood-engraver and naturalist (1753–1828). Organizes related events and meetings, and is associated with the Bewick birthplace museum at Cherryburn, Northumberland.

Birmingham Central Literary Association

23 Arden Grove, Ladywood, Birmingham B16 8HG
☎ 0121 454 9352
bruce@bakerbrum.co.uk.
Contact *The Secretary*

Holds fortnightly meetings in central Birmingham to discuss the lives and work of authors and poets. Holds an annual dinner to celebrate Shakespeare's birthday.

The George Borrow Society

60 Upper Marsh Road, Warminster BA12 9PN
www.clough5.fsnet.co.uk/gb.html
Chairman/Bulletin Editor *Dr Ann M. Ridler*

Membership Secretary *Michael Skillman*
Honorary Treasurer *David Pattinson*

SUBSCRIPTION £15 p.a.

Founded in 1991 to promote knowledge of the life and works of George Borrow (1803–81), traveller, linguist and writer. The Society holds biennial conferences (with published proceedings) and informal intermediate gatherings, all at places associated with Borrow. Publishes the *George Borrow Bulletin* twice yearly, containing scholarly articles, reviews of publications relating to Borrow, reports of past events and news of forthcoming events. Member of the **Alliance of Literary Societies** and corporate associate member of the Centre of East Anglian Studies (CEAS) at the University of East Anglia, Norwich (Borrow's home city for many years).

Elinor Brent-Dyer
▷ Friends of the Chalet School and The New Chalet Club

British Fantasy Society
5 Greenbank, Barnt Green, Birmingham B45 8DH
secretary@britishfantasysociety.org
www.britishfantasysociety.org
Honorary President *Ramsey Campbell*
Chairman *Marie O'Regan*
Secretary *Vicky Cook*

SUBSCRIPTION from £25 p.a.

Founded in 1971 for devotees of fantasy, horror and related fields in literature, art and the cinema. Publishes a regular newsletter with information and reviews of new books and films, plus related fiction and non-fiction magazines. Annual conference at which the **British Fantasy Awards** are presented. These awards are voted on by the membership and are not an open competition.

The Brontë Society
Brontë Parsonage Museum, Haworth, Keighley
BD22 8DR
☎ 01535 640195   🖷 01535 647131
bronte@bronte.org.uk
www.bronte.info
Contact *Membership Officer*

SUBSCRIPTION details from the Membership Officer or the website

Founded 1893. Aims and activities include the preservation of manuscripts and other objects related to or connected with the Brontë family, and the maintenance and development of the museum and library at Haworth. The Society holds regular meetings, lectures and exhibitions; publishes information relating to the family and a triannual *Gazette*. Freelance contributions for either publication should be sent to the Publications Secretary at the address above. Members can receive the journal, *Brontë Studies*, at a reduced subscription.

The Rupert Brooke Society
The Orchard, 45–47 Mill Way, Grantchester
CB3 9ND
☎ 01223 551118   🖷 01223 842331
rbs@callan.co.uk
www.rupertbrooke.com
Contact *Secretary*

SUBSCRIPTION £10 (UK); £15 (overseas)

Founded in 1999 to foster an interest in the work of Rupert Brooke, help preserve places associated with him and to increase the knowledge and appreciation of the village of Grantchester. Members receive a newsletter with information about events, new books and activities.

The Browning Society
38b Victoria Road, London NW6 6PX
☎ 020 7604 4257
v.l.greenaway@btinternet.com
Honorary Secretary *Dr Vicky Greenaway*

SUBSCRIPTION £15 p.a.

Founded in 1969 to promote an interest in the lives and poetry of Robert and Elizabeth Barrett Browning. Meetings are arranged in the London area, one of which occurs in December at Westminster Abbey to commemorate Robert Browning's death.

The John Buchan Society
'The Toft', 37 Waterloo Road, Lanark ML11 7QH
☎ 01555 662103
glennismac2000@yahoo.co.uk
www.johnbuchansociety.co.uk
Secretary *Mrs Glennis McClemont*
Membership Secretary *Diana Durdon*

SUBSCRIPTION £15 (full/overseas); £120 (10 years)

To perpetuate the memory of John Buchan and to promote a wider understanding of his life and works. Holds regular meetings and social gatherings, publishes a journal, and liaises with the John Buchan Centre at Broughton in the Scottish borders.

The Robert Burns World Federation Ltd
Dean Castle Country Park Dower House, Kilmarnock KA3 1XB
☎ 01563 572469   🖷 01563 572469
admin@robertburnsfederation.com
www.worldburnsclub.com
Chief Executive *Shirley Bell*
Office Administrator *Margaret Craig*

SUBSCRIPTION £25 p.a.(individual); £27 (family); £46 (club subscription)

Founded in 1885 to encourage interest in the life and work of Robert Burns and keep alive the old Scottish Tongue. The Society's interests go beyond Burns himself in its commitment to the development of Scottish literature, music and arts in general. Publishes the triannual *Burns Chronicle*.

Randolph Caldecott Society

Clatterwick House, Clatterwick Lane, Little Leigh, Northwich CW8 4RJ

☎ 01606 891303 (day)/781731 (evening)

Honorary Secretary *Kenneth N. Oultram*

SUBSCRIPTION £10–£15 p.a.

Founded in 1983 to promote the life and work of artist/book illustrator Randolph Caldecott. Meetings held in the spring and autumn in Caldecott's birthplace, Chester. Guest speakers, outings, newsletter, exchanges with the Society's American counterpart. (Caldecott died and was buried in St Augustine, Florida.) A medal in his memory is awarded annually in the US for children's book illustration.

Lewis Carroll Society

69 Cromwell Road, Hertford SG13 7DP

☎ 01992 584530

alanwhite@tesco.net

www.lewiscarrollsociety.org.uk

Secretary *Alan White*

SUBSCRIPTION details on the website

Founded in 1969 to bring together people with an interest in Charles Dodgson and promote research into his life and works. Publishes bi-annual journal, *The Carrollian*, featuring scholarly articles and reviews; a newsletter (*Bandersnatch*) which reports on Carrollian events and the Society's activities; and *The Lewis Carroll Review*, a book-reviewing journal. Regular meetings held in London with lectures, talks, outings, etc.

Lewis Carroll Society (Daresbury)

Clatterwick House, Clatterwick Lane, Little Leigh, Northwich CW8 4RJ

☎ 01606 891303 (day)/781731 (evening)

Honorary Secretary *Kenneth N. Oultram*

SUBSCRIPTION £7–10 p.a.

Founded 1970. To promote the life and work of Charles Dodgson, author of the world-famous *Alice's Adventures*. Holds meetings in the spring and autumn in Carroll's birthplace, Daresbury, Cheshire. Guest speakers, a newsletter and theatre visits.

The New Chalet Club

18 Nuns Moor Crescent, Newcastle upon Tyne NE4 9BE

www.newchaletclub.co.uk

Membership Secretary *Mrs Rona S.S. Falconer*

SUBSCRIPTION £6 p.a. (UK under 18); £10 (UK adult & Europe); £12 (RoW)

Founded in 1995 for all those with an interest in the books of Elinor Brent-Dyer. Publishes a quarterly journal and occasional supplements, including *Children's Series Fiction*, and regularly holds local and national meetings.

Friends of the Chalet School

4 Rock Terrace, Coleford, Bath BA3 5NF

☎ 01373 812705

focs@rockterrace.demon.co.uk

www.chaletschool.org.uk

Contacts *Ann Mackie-Hunter, Clarissa Cridland*

SUBSCRIPTION details on the website or on application

Founded in 1989 to promote the works of Elinor Brent-Dyer. The society has members worldwide; publishes four magazines a year and runs a lending library.

The Chesterton Society UK

11 Lawrence Leys, Bloxham, Near Banbury OX15 4NU

☎ 01295 720869/07766 711984 (mobile)

🖷 01295 720869

Honorary Secretary *Rev. Deacon Robert Hughes, KCHS*

SUBSCRIPTION £12.50 p.a.

Founded in 1964 to promote the ideas and writings of G.K. Chesterton.

The Children's Books History Society

26 St Bernards Close, Buckfast, Buckfastleigh TQ11 0EP

☎ 01364 643568

cbhs@abcgarrett.demon.co.uk

Chair/Membership Secretary *Mrs Pat Garrett*

SUBSCRIPTION £10 p.a. (UK/Europe); write for overseas subscription details

Established 1969. Aims to promote an appreciation of children's books and to study their history, bibliography and literary content. Meetings held in London and the provinces, plus a summer meeting to a collection, or to a location with a children's book connection. Three substantial newsletters issued annually, sometimes with an occasional paper. Review copies should be sent to Newsletter co-editor *Mrs Pat Garrett* (address above). The Society constitutes the British branch of the Friends of the Osborne and Lillian H. Smith Collections in Toronto, Canada, and also liaises with **CILIP** (formerly The Library Association). In 1990, the Society established its biennial Harvey Darton Award for a book, published in English, which extends our knowledge of some aspect of British children's literature of the past. 2006 winner: Lawrence Darton for *The Dartons. An Annotated*

Check-List of Children's Books Issued by Two Publishing Houses 1787–1876.

The John Clare Society

9 The Chase, Ely CB6 3DR
☎ 01353 668438
www.johnclare.org.uk human.ntu.ac.uk/clare/clare.html
Honorary Secretary *Miss Sue Holgate*

SUBSCRIPTION £12 (individual); £15 (joint); £10 (retired); £15 (group); £17 (library); £5 (student, full-time); £17 sterling draft/$34/€28 (eurocheque) (overseas)

Founded in 1981 to promote a wider appreciation of the life and works of the poet John Clare (1793–1864). Organizes an annual festival in Helpston in July; arranges exhibitions, poetry readings and conferences; and publishes an annual society journal and quarterly newsletter.

William Cobbett Society

3 Park Terrace, Tillington, Petworth GU28 9AE
☎ 01798 342008
www.williamcobbett.org.uk
Chairman *Barbara Biddell*
Also: Boynell House, Outlands Lane, Curdridge, Southampton SO30 2HR ☎ 01489 782453
Contact *David Chun*

SUBSCRIPTION £8 p.a.

Founded in 1976 to bring together those with an interest in the life and works of William Cobbett (1763–1835) and to extend the interest to a wider public. Society activities include an annual Memorial Lecture; publication of an annual journal (*Cobbett's New Register*) containing articles on various aspects of his life and times; an annual expedition retracing routes taken by Cobbett on his Rural Rides in the 1820s; visits to his birthplace and his tomb in Farnham, Surrey. In association with the Society, the Museum of Farnham holds bound volumes of *Cobbett's Political Register*, a large collection of Cobbett's works, books about Cobbett, and has various Cobbett artefacts on display.

The Friends of Coleridge

87 Richmond Road, Montpelier, Bristol BS6 5EP
☎ 0117 942 6366
gcdd@blueyonder.co.uk
www.friendsofcoleridge.com
Membership Secretary/Website Manager *Paul Cheshire* (74 Wells Road, Bath BA2 3AR; friendscol@btconnect.com)
Editor (Coleridge Bulletin) *Graham Davidson*

SUBSCRIPTION details on the website

Founded in 1987 to advance knowledge about the life, work and times of Samuel Taylor Coleridge and his circle, and to support his Nether Stowey Cottage, with the National Trust, as a centre of Coleridge interest. Holds literary evenings, study weekends and a biennial international academic conference. Publishes *The Coleridge Bulletin* biannually. Short articles on Coleridge-related topics may be sent to the editor at the (Bristol) address above.

Wilkie Collins Society

4 Ernest Gardens, London W4 3QU
☎ 020 8747 0115
paul@wilkiecollins.org
www.wilkiecollins.org
Chairman *Andrew Gasson*
Membership Secretary *Paul Lewis* (at address above)

SUBSCRIPTION £10 (UK/Europe); £18 (RoW; remittance must be made in UK sterling or by PayPal to paul@paullewis.co.uk)

Founded in 1980 to provide information on and promote interest in the life and works of Wilkie Collins, one of the first English novelists to deal with the detection of crime. *The Woman in White* appeared in 1860 and *The Moonstone* in 1868. Publishes newsletters, reprints of Collins' work and an annual academic journal.

The Arthur Conan Doyle Society

PO Box 1360, Ashcroft, British Columbia, Canada V0K 1A0
☎ 001 250 453 2045   🖷 001 250 453 2075
sirhenry@telus.net
www.ash-tree.bc.ca/acdsocy.html
Joint Organizers *Christopher Roden, Barbara Roden*

Founded in 1989 to promote the study and discussion of the life and works of Sir Arthur Conan Doyle. Occasional meetings, functions and visits. Publishes an occasional journal together with reprints of Conan Doyle's writings.

Joseph Conrad Society (UK)

c/o The Polish Social and Cultural Association (POSK), 238–246 King Street, London W6 0RF
theconradian@aol.com
www.josephconradsociety.org
Honorary Secretary *Hugh Epstein*
Treasurer/Editor (*The Conradian*) *Allan Simmons*

SUBSCRIPTION £25 p.a. (individual)

Founded in 1973 to promote the study of the works and life of Joseph Conrad (1857–1924). A scholarly society, supported by the Polish Library at the Polish Cultural Association where a substantial library of Conrad texts and criticism is held by the **Polish Library**. Publishes a journal of Conrad studies (*The Conradian*) biannually and holds an annual International Conference in the first week of July. Awards an essay prize and travel and study grants to scholars on application.

The Dartmoor Sherlock Holmes Study Group
▷ The Baskerville Hounds

The Rhys Davies Trust
10 Heol Don, Whitchurch, Cardiff CF14 2AU
☎ 029 2062 3359
Contact *Professor Meic Stephens*

Founded in 1990 to perpetuate the literary
reputation of the Welsh prose writer, Rhys Davies
(1901–78), and to foster Welsh writing in English.
Organizes competitions in association with
other bodies such as the **Welsh Academy**, puts
up plaques on buildings associated with Welsh
writers, offers grant-aid for book production, etc.

The Walter de la Mare Society
3 Hazelwood House, New River Crescent, London
N13 5RE
☎ 020 8886 1771
www.bluetree.co.uk/wdlmsociety
Honorary President *Professor John Bayley, CBE*
Honorary Secretary & Treasurer *Julie de la Mare*
SUBSCRIPTION £15 p.a.

Founded in 1997 to honour the memory of Walter
de la Mare; to promote the study and deepen
the appreciation of his works; to widen the
readership of his works; to facilitate research by
making available the widest range of contacts and
information about de la Mare; and to encourage
and facilitate new Walter de la Mare publications.
Produces a regular newsletter and organizes
events. Membership information from the
address above.

Warwick Deeping Appreciation Society
23 Merton Road, Enfield EN2 0LS
☎ 020 8367 0263
geoffrey@gillam.fsworld.co.uk
Secretary *Geoffrey Gillam*
SUBSCRIPTION £3.50 p.a.

Open to all with an interest in the life and work
of Warwick Deeping who produced 70 novels
and many short stories. A quarterly newsletter
includes reviews of his books and results of
research into his life.

The Dickens Fellowship
The Charles Dickens Museum, 48 Doughty Street,
London WC1N 2LX
☎ 020 7405 2127
dickens.fellowship@btinternet.com
www.dickens.fellowship.org
Joint Honorary General Secretaries *Mrs Lee Ault,
Mrs Joan Dicks*

Founded 1902. The Society's particular aims and
objectives are: to bring together lovers of Charles
Dickens; to spread the message of Dickens,
his love of humanity ('the keynote of all his
work'); to remedy social injustice for the poor
and oppressed; to assist in the preservation of
material and buildings associated with Dickens.
Annual conference. Publishes journal called *The
Dickensian* (founded 1905 and available at special
rate to members) and organizes a full programme
of lectures, discussions, visits and conducted
walks throughout the year. Branches worldwide.

Dymock Poets
▷ The Friends of the Dymock Poets

Early English Text Society
Lady Margaret Hall, Oxford OX2 6QA
🖷 01865 286581
www.eets.org.uk
Executive Secretary *Professor Vincent Gillespie*
(at address above)
Editorial Secretary *Dr H.L Spencer* (at Exeter
College, Oxford OX1 3DP)
Membership Secretary *Mrs J.M. Watkinson*
(at 12 North End, Durham DH1 4NJ)
SUBSCRIPTION £20 p.a. (UK); $30 (US); $35
(Canada)

Founded 1864. Concerned with the publication
of early English texts. Members receive annual
publications (one or two a year) or may select
titles from the backlist in lieu.

The George Eliot Fellowship
12 Fair Isle Drive, Nuneaton CV10 7LJ
www.george-eliot-fellowship.com
Secretary *Mrs E. Mellor*
SUBSCRIPTION £10 p.a.; £100 (life); concession for
pensioners

Founded 1930. Exists to honour George Eliot
and promote interest in her life and works.
Readings, memorial lecture, birthday luncheon
and functions. Issues a quarterly newsletter and
an annual journal. Awards an annual prize for a
George Eliot essay.

Rev. G. Bramwell Evens
▷ The Romany Society

The John Meade Falkner Society
Greenmantle, Main Street, Kings Newton,
Melbourne DE73 8BX
☎ 01332 865315
www.johnmeadefalknersociety.co.uk
Secretary *Kenneth Hillier*
SUBSCRIPTION £5 (UK); $10 (overseas)

Founded in 1999 to promote the appreciation
and study of John Meade Falkner's life, times and
works. Produces three newsletters a year and an
annual journal.

Folly (Fans of Light Literature for the Young)

21 Warwick Road, Pokesdown, Bournemouth
BH7 6JW
☎ 01202 432562
folly@sims.abel.co.uk
Contact *Mrs Sue Sims*

SUBSCRIPTION £9 p.a. (UK); £12 (Europe); £15
(RoW)

Founded in 1990 to promote interest in a wide
variety of collectable children's authors – with a
bias towards writers of girls' books and school
stories. Publishes three magazines a year.

The Ford Madox Ford Society

c/o Dr Sara Haslam, English Department, The
Open University, Milton Keynes MK7 6AA
☎ 01908 653453   ℻ 01908 653750
s.j.haslam@open.ac.uk
www.rialto.com/fordmadoxford_society
Chairman *Professor Max Saunders*
Treasurer *Dr Sara Haslam*

SUBSCRIPTION £12 or £6 concession

Founded in 1996 to promote the works of Ford
Madox Ford and to increase knowledge of
his writing and impact on writing in the 20th
century. Meets for academic conferences and
more popular events annually. Distributes
International Ford Madox Ford Studies free to
members. Welcomes all with an interest in Ford
and his works. Has members in the UK, USA,
Italy, France and Germany and holds events in as
many different places as possible.

The Franco-Midland Hardware Company

6 Bramham Moor, Hill Head, Fareham PO14 3RU
☎ 01329 667325
fmhc@acd-221b.info
www.acd-221b.info
Chairman *Philip Weller*

Founded 1989. A federation of societies in four
continents. Two core groups, '221B' (Holmesian/
Sherlockian) and 'The Arthur Conan Doyle Study
Group' (Doylean), form the UK base. Membership
is open only to those interested in being actively
involved in the programme of correspondence
studies and/or functions. Arranges certificated
self-study projects and assists group and
individual research programmes, plus scholarly
gatherings.

The Friars' Club

33 Grenville Court, Chorleywood WD3 5PZ
☎ 01923 283795
petermccall.westlodge@btinternet.com
Secretary *Frances-Mary Blake*

SUBSCRIPTION £10 (first year); £8.50 (renewals)

Founded in 1982 to promote the writings of
Frank Richards (Charles Hamilton), author of
the Greyfriars School stories in *The Magnet* and
creator of Billy Bunter. The Club produces a
quarterly magazine for members. Articles and
comments are welcomed.

The Friends of Shandy Hall (The Laurence Sterne Trust)

Shandy Hall, Coxwold, York YO61 4AD
☎ 01347 868465   ℻ 01347 868465
shandyhall@dsl.pipex.com
www.asterisk.org.uk
Curator *Patrick Wildgust*

SUBSCRIPTION £7 (annual); £70 (life)

Promotes interest in the works of Laurence
Sterne and aims to preserve the house in which
they were created (open to the public). Publishes
annual journal, *The Shandean*. An Annual
Memorial Lecture is delivered each summer.
Regular exhibitions in gallery. Bookshop.

The Friends of the Dymock Poets

Valentine's Cottage, Hollybush, Ledbury HR8 2ET
cateluck2003@yahoo.com
www.dymockpoets.co.uk
Chairman *Roy Palmer*
Hon. Secretary *Catharine Lumby*
Membership Secretary *Jeff Cooper*

SUBSCRIPTION £7 (individual); £12 (couple); £3
(student); £12 (society/family)

Founded 1993. Established to foster an interest in
the work of the Dymock Poets – Edward Thomas,
Robert Frost, Wilfrid Gibson, Rupert Brooke,
John Drinkwater, Lascelles Abercrombie; help
preserve places and things associated with them;
keep members informed of literary and other
matters relating to the poets; increase knowledge
and appreciation of the landscape between May
Hill (Gloucestershire) and the Malvern Hills.
Members are offered lectures, poetry readings,
social meetings, newsletters and annual journal,
guided walks in the countryside around Dymock;
links with other literary societies; and annual
event to commemorate the first meeting between
Edward Thomas and Robert Frost on 6 October
1913.

The Gaskell Society

Far Yew Tree House, Chester Road, Tabley,
Knutsford WA16 0HN
☎ 01565 634668
joanleach@aol.com
gaskellsociety.co.uk lang.nagoya-u.ac.jp/
~matsuoka/EG-Society.html
Honorary Secretary *Joan Leach*

SUBSCRIPTION £15 p.a.; £20 (corporate &
overseas)

Founded in 1985 to promote and encourage the study and appreciation of the life and works of Elizabeth Cleghorn Gaskell. Meetings held in Knutsford, Manchester, Bath, London and York; annual journal and biannual newsletter. On alternate years holds either a residential weekend conference or overseas visit.

The Ghost Story Society

PO Box 1360, Ashcroft, British Columbia, Canada V0K 1A0

☎ 001 250 453 2045   ℻ 001 250 453 2075
nebuly@telus.net
www.ash-tree.bc.ca/GSS.html
Joint Organizers *Barbara Roden, Christopher Roden*

SUBSCRIPTION UK: £25 (airmail); $40 (US$); $45 (Canadian$)

Founded 1988. Devoted mainly to supernatural fiction in the literary tradition of M.R. James, Walter de la Mare, Algernon Blackwood, E.F. Benson, A.N.L. Munby, R.H. Malden, etc. Publishes a thrice-yearly journal, *All Hallows*, which includes new fiction in the genre and non-fiction of relevance to the genre.

The Robert Graves Society

3 rue des grands Augustins, Paris 75006, France

☎ 00 33 1 43 29 33 48
dunstanward@yahoo.com
www.gravesiana.org
President *Professor Dunstan Ward*

SUBSCRIPTION £17, €30, $40 p.a.

The Society aims to promote interest in, and research on, the life and works of Robert Graves (1895–1985), author of some 140 books of poetry, fiction, biography, criticism, anthropology, social history, mythology, biblical studies, translation and children's books. Organizes an international conference every two years. The 2008 conference, 'Innovation and Tradition: Robert Graves in the Twentieth Century', will be held at St John's College, Oxford, 9th–13th September. Publishes annually the leading journal of Graves studies, *Gravesiana*, edited by Professor Dunstan Ward. The Society's membership of more than 200 experts and enthusiasts forms the core of an international research community.

Graham Greene Birthplace Trust

17 North Road, Berkhamsted HP4 3DX

☎ 01442 866694
secretary@grahamgreenebt.org
www.grahamgreenebt.org
Acting Secretary *Colin Garrett*

SUBSCRIPTION £10 (UK, £25 for 3 years); £14 (Europe, £33); £18 (RoW, £45)

Founded on 2nd October 1997, the 93rd anniversary of Graham Greene's birth, to promote the appreciation and study of his works. Publishes a quarterly newsletter, occasional papers, videos and compact discs. Organizes the annual four-day Graham Greene Festival during the weekend nearest to the writer's birthday and administrates the Graham Greene Memorial Awards.

The Ivor Gurney Society

65 Coombe Road, Irby, Wirral CH61 4UW

☎ 0151 648 4957
rolfe@rolfjordan.com
www.ivorgurney.org.uk
General Secretary *Rolf Jordan*

SUBSCRIPTION £12.50 (individual); £15 (joint); £12.50 (group)

Founded in 1995 to make Ivor Gurney's music and poetry available to a wider audience by way of performances, readings, conferences, recordings and publications; to enhance and promote informed scholarship in all aspects of his life and work through the publication of a regular newsletter and an annual journal, available free to Society members; and to assist the Gloucester City Library in maintaining the Ivor Gurney collection. Publishes occasional papers including unpublished letters, poems and compositions. Society events are held each spring and autumn when talks and recitals enable members to explore Gurney's verse and music.

Rider Haggard Society

27 Deneholm, Whitley Bay NE25 9AU

☎ 0191 252 4516   ℻ 0191 252 4516
rb27allen@blueyonder.co.uk
www.riderhaggardsociety.org.uk
Contact *Roger Allen*

SUBSCRIPTION £9 p.a. (UK); £12 (overseas)

Founded in 1985 to promote appreciation of the life and works of Sir Henry Rider Haggard, English novelist, 1856–1925. News/books exchange and annual meetings. Vast source for Haggard purchases, research and information.

James Hanley Network

Old School House, George Green Road, George Green, Wexham SL3 6BJ

☎ 01753 578632
gostick@london.com
www.jameshanley.inspiron.co.uk/INDEX.HTM
Network Coordinator *Chris Gostick*

An informal international association founded in 1997 for all those interested in exploring and publicizing the works and contribution to literature of the novelist and dramatist James Hanley (1901–1985). Publishes an annual newsletter. Occasional conferences are planned for the future. All enquiries welcome.

The Thomas Hardy Society

PO Box 1438, Dorchester DT1 1YH
℡ 01305 251501   🖷 01305 251501
info@hardysociety.org
www.hardysociety.org
Honorary Secretary *Mike Nixon*

SUBSCRIPTION £18 (individual); £25 (corporate);
£22.50 (individual, overseas); £30 (corporate,
overseas)

Founded in 1967 to promote the reading and
study of the works and life of Thomas Hardy.
Thrice-yearly journal, events and a biennial
conference. An international organization.

The Hazlitt Society

c/o The Guardian, 119 Farringdon Road, London
EC1R 3ER (From Nov. 2008: King's Place, 90 York
Way, London N1 9AG)
correspondence@williamhazlitt.org
www.williamhazlitt.org
Founding President *Michael Foot*
Contact *Helen Hodgson* (Secretary and
Correspondent)

Established to encourage appreciation of Hazlitt's
work and to promote his values. The first *Hazlitt
Review*, launched in 2008, aims to foster and
sustain the restitution of Hazlitt as a canonical
author and to stimulate new scholarship in his
work.

The Henty Society

205 Icknield Way, Letchworth Garden City
SG6 4TT
Honorary Secretary *David Walmsley*
SUBSCRIPTION £15 p.a. (UK); £18 (overseas)

Founded in 1977 to study the life and work of
George Alfred Henty, and to publish research,
bibliographical data and lesser-known works,
namely short stories. Organizes conferences and
social gatherings in the UK and North America,
and publishes bulletins to members. Published
in 1996: *G.A. Henty (1832–1902) a Bibliographical
Study* by Peter Newbolt (2nd edition with
addenda and corrigenda 2005).

James Hilton Society

49 Beckingthorpe Drive, Bottesford, Nottingham
NG13 0DN
www.jameshiltonsociety.co.uk
Honorary Secretary *J.R. Hammond*
SUBSCRIPTION £10 (UK/EU); £7 concession

Founded in 2000 to promote interest in the life
and work of novelist and scriptwriter James
Hilton (1900–1954). Publishes *The Hiltonian*
(annually) and *The James Hilton Newsletter*
(quarterly) and organizes meetings and
conferences.

Sherlock Holmes Society (The Musgraves)

Hallas Lodge, Greenside Lane, Cullingworth,
Bradford BD13 5AP
℡ 01535 273468
hallaslodge@btinternet.com
Contact *Anne Jordan*

SUBSCRIPTION £17 p.a.; £18 (joint)

Founded in 1987 to promote enjoyment and study
of Sir Arthur Conan Doyle's Sherlock Holmes
through publications and meetings. One of
the largest Sherlock Holmes societies in Great
Britain. Honorary members include Bert Coules,
Edward Hardwicke and Clive Merrison. Past
honorary members: Dame Jean Conan Doyle,
Peter Cushing, Jeremy Brett, Michael Williams
and Richard Lancelyn Green. Open membership.
Lectures, presentations and consultation on
matters relating to Holmes and Conan Doyle
available.

Sherlock Holmes

▷ The Baskervillle Hounds and The Franco-Midland
Hardware Company

Housman Society

80 New Road, Bromsgrove B60 2LA
℡ 01527 874136
info@housman-society.co.uk
www.housman-society.co.uk
Contact *Jim Page*

SUBSCRIPTION £10 (UK); £12.50 (overseas)

Founded in 1973 to promote knowledge and
appreciation of the lives and work of A.E.
Housman and other members of his family, and
to promote the cause of literature and poetry.
Sponsors a lecture at **The Guardian Hay Festival**
each year under the title of 'The Name and
Nature of Poetry'. Publishes an annual journal and
biannual newsletter.

W.W. Jacobs Appreciation Society

3 Roman Road, Southwick BN42 4TP
℡ 01273 596217
Contact *A.R. James*

Founded in 1988 to encourage and promote
the enjoyment of the works of W.W. Jacobs,
and stimulate research into his life and works.
No SUBSCRIPTION charge. Material available
for purchase includes *W.W. Jacobs*, a biography
published in 1999, price £12, post paid, and
WWJ Book Hunter's Field Guide, a narrative
bibliography published in 2001, price £6, post
paid.

Richard Jefferies Society

Pear Tree Cottage, Longcot SN7 7SS
℡ 01793 783040
R.Jefferies_Society@tiscali.co.uk
www.bath.ac.uk/~lissmc/rjeffs.htm

Honorary Secretary *Jean Saunders*
Membership Secretary *Mrs Margaret Evans*

SUBSCRIPTION £7 p.a. (individual); £8 (joint); life membership for those over 60

Founded in 1950 to promote understanding of the work of Richard Jefferies, nature/country writer, novelist and mystic (1848–87). Produces newsletters, reports and an annual journal; organizes talks, discussions and readings. Library and archives. Assists in maintaining the museum in Jefferies' birthplace at Coate near Swindon. Membership applications should be sent to *Margaret Evans*, 23 Hardwell Close, Grove, Nr Wantage OX12 0BN.

Jerome K. Jerome Society

c/o Fraser Wood, Mayo and Pinson, 15/16 Lichfield Street, Walsall WS1 1TS
☎ 01922 629000   🖷 01922 721065
tonygray@jkj.demon.co.uk
www.jeromekjerome.com
Honorary Secretary *Tony Gray*

SUBSCRIPTION £12 p.a. (individual); £8 (student); £15 (family); £30 (corporate)

Founded in 1984 to stimulate interest in Jerome K. Jerome's life and works (1859–1927). One of the Society's principal activities is the support of a small museum in the author's birthplace, Walsall. Meetings, lectures, events and a newsletter, *Idle Thoughts*. Annual dinner in Walsall near Jerome's birth date (2nd May).

The Captain W.E. Johns Appreciation Society

221 Church Road, Northolt UB5 5BE
brian@zhong-ding.com
Contact (Derby meeting) *Brian Woodruff* (at address above; ☎ 07836 680048)
Contact (Twyford) *Joy Tilley* (☎ 01785 240299)

Society for the appreciation of W.E. Johns, creator of Biggles. Meets twice a year in Derby and Twyford, near Reading (see above for contacts). Biannual magazine, *Biggles Flies Again* (Editor *Roger Davies*, ☎ 01722 320761; rda@salisbury75. freeserve.co.uk).

Johnson Society

Johnson Birthplace Museum, Breadmarket Street, Lichfield WS13 6LG
☎ 01543 264972
sjmuseum@lichfield.gov.uk
www.lichfieldrambler.co.uk
Chairman *Mary Baker*

SUBSCRIPTION £10 p.a.; £15 (joint); £100/£150 (life: individual/joint)

Founded in 1910 to encourage the study of the life, works and times of Samuel Johnson (1709–1784) and his contemporaries. The Society is committed to the preservation of the Johnson Birthplace Museum and Johnson memorials.

Johnson Society of London

255 Baring Road, Grove Park, London SE12 0BQ
☎ 020 8851 0173
JSL@nbbl.demon.co.uk
www.nbbl.demon.co.uk/index.html
President *Lord Harmsworth*
Honorary Secretary *Mrs Z.E. O'Donnell*

SUBSCRIPTION £20 p.a.; £25 (joint); £15 (student)

Founded in 1928 to promote the knowledge and appreciation of Dr Samuel Johnson and his works. Publishes an annual journal, *New Rambler* and occasional newsletter, *The New Idler*. Regular meetings from October to April in the meeting room of Wesley's Chapel, City Road, London EC1 on the second Saturday of each month, and a commemoration ceremony around the anniversary of Johnson's death (December) held in Westminster Abbey.

The David Jones Society

22 Gower Road, Sketty, Swansea SA2 9BY
☎ 01792 206144   🖷 01792 475037
anne.price-owen@davidjonesociety.org
www.davidjonesociety.org
Contact *Anne Price-Owen*

SUBSCRIPTION £20 (individual); £35 (corporate)

Founded in 1996, the Society aims to promote and encourage knowledge of the painter-poet David Jones. The Society hosts annual scholarly meetings. The annual SUBSCRIPTION includes a copy of the *David Jones Journal*. Unsolicited material relating to David Jones, art, literature or any of his philosophies may be sent for consideration for publication in the journal.

The Just William Society

Easter Badbea, Dundonnell IV23 2QX
philandpaula@easter-badbea.co.uk
Secretary/Treasurer *Paula Cross*

SUBSCRIPTION £10 p.a. (UK); £12 (overseas); £5 (juvenile/student); £15 (family)

Founded in 1994 to further knowledge of Richmal Crompton's *William* and *Jimmy* books. The *Just William Society Magazines* are published throughout the year and an annual 'William' meeting is held in April.

The Keats–Shelley Memorial Association (Inc)

Registered office: Bedford House, 76A Bedford Street, Leamington Spa CV32 5DT
☎ 01926 427400   🖷 01926 335133
Contact *Honorary Secretary*

SUBSCRIPTION £12 p.a.

Founded in 1903 to promote appreciation of the works of Keats and Shelley, and their circle. One of the Society's main tasks is the preservation of 26 Piazza di Spagna in Rome as a memorial to the British Romantic poets in Italy, particularly Keats and Shelley. Publishes an annual review of Romantic Studies called the *Keats-Shelley Review*, arranges events and lectures for Friends and promotes bursaries and competitive writing on Romantic Studies (see **Keats-Shelley Prize** under *Prizes*). The *Review* is edited by *Stephen Hebron*, c/o Wordsworth Trust, Dove Cottage, Grasmere LA22 9SH

Kent & Sussex Poetry Society

39 Rockington Way, Crowborough TN6 2NJ
☏ 01892 662781
john@kentandsussexpoetrysociety.org
www.kentandsussexpoetrysociety.org
Publicity Secretary *John Arnold*

SUBSCRIPTION £10 p.a. (full); £6 (concessionary; country members living farther afield, senior citizen, under-16, unemployed)

Founded in 1946 to promote the enjoyment of poetry. Monthly meetings are held in Tunbridge Wells, including readings by major poets, a monthly workshop and an annual writing retreat week. Publishes an annual folio of members' work based on Members' Competition, adjudicated and commented upon by a major poet. Runs an annual **Open Poetry Competition** (see entry under *Prizes* and details on website) and Saturday workshops twice a year with leading poets.

The Kilvert Society

30 Bromley Heath Avenue, Downend, Bristol BS16 6JP
☏ 0117 957 2030
www.communigate.co.uk/here/kilvertsociety
Honorary Secretary *Alan Brimson*

SUBSCRIPTION £12 p.a.; £15 (two persons at same address)

Founded in 1948 to foster an interest in the Diary, the diarist and the countryside he loved. Publishes three journals each year; during the summer holds three weekends of walks, commemoration services and talks.

The Kipling Society

6 Clifton Road, London W9 1SS
☏ 020 7286 0194   ⊞ 020 7286 0194
jane@keskar.fsworld.co.uk
www.kipling.org.uk
Honorary Secretary *Jane Keskar*

SUBSCRIPTION £24 p.a. (£22 p.a. by standing order)

Founded 1927. This is a literary society for all who enjoy the prose and verse of Rudyard

Kipling (1865–1936) and are interested in his life and times. The Society's main activities are: maintaining a specialized library at the City University in Islington, London; answering enquiries from the public (schools, publishers, writers and the media); arranging a regular programme of lectures, especially in London and in Sussex, and an annual luncheon with guest speaker; maintaining a small museum and reference at The Grange, in Rottingdean near Brighton; issuing a quarterly journal. Please contact the Secretary by letter, telephone, fax or e-mail for further information.

The Kitley Trust

Woodstock, Litton Dale, Litton SK17 8QL
☏ 01298 871564
stottie2@waitrose.com
Contact *Rosie Ford*

Founded in 1990 by a teacher in Sheffield to promote the art of creative writing, in memory of her mother, Jessie Kitley. Activities include: biannual poetry competitions; a 'Get Poetry' day (distribution of children's poems in shopping malls); annual sponsorship of a writer for a school; campaigns; organizing conferences for writers and teachers of writing. Funds are provided by donations and profits (if any) from competitions.

The Charles Lamb Society

BM ELIA, London WC1N 3XX
Chairman *Nicholas Powell*

SUBSCRIPTION UK: £12 (single); £18 (double); £18 (corporate)

Founded 1935. Publishes studies of the life, works and times of Charles Lamb and his circle. Holds meetings in London five times a year. *The Charles Lamb Bulletin* is published four times a year (Editor: *Professor Richard S. Tomlinson* e-mail: romanticism@insightbb.com).

Lancashire Authors' Association

Heatherslade, 5 Quakerfields, Westhoughton, Bolton BL5 2BJ
☏ 01942 791390
eholt@cwctv.net
General Secretary *Eric Holt*

SUBSCRIPTION £10 p.a.; £13 (joint); £1 (junior)

Founded in 1909 for writers and lovers of Lancashire literature and history. Aims to foster and stimulate interest in Lancashire history and literature as well as in the preservation of the Lancashire dialect. Meets three times a year on Saturday at various locations. Publishes a quarterly journal called *The Record*, which is issued free to members, and holds eight annual competitions (open to members only) for both verse and prose. Comprehensive library with access for research to members.

The Philip Larkin Society

16 Mere View Avenue, Hornsea HU18 1RR
☏ 01964 533324
www.philiplarkin.com
General Secretary *Andrew Eastwood*

SUBSCRIPTION £20 (full rate); £14 (unwaged/senior citizen); £6 (student); £50 (institutions)

Founded in 1995 to promote awareness of the life and work of Philip Larkin (1922–1985) and his literary contemporaries; to bring together all those who admire Larkin's work as a poet, writer and librarian; to bring about publications on all things Larkinesque. Organizes a programme of events ranging from lectures to rambles exploring the countryside of Larkin's schooldays and publishes a biannual journal, *About Larkin*.

The D.H. Lawrence Society

24 Briarwood Avenue, Nottingham NG3 6JQ
☏ 0115 950 3008
Secretary *Ron Faulks*

SUBSCRIPTION £14; £12 (concession); £16 (European); £19 (RoW); £50 (institutions)

Founded in 1974 to increase knowledge and the appreciation of the life and works of D.H. Lawrence. Monthly meetings, addressed by guest speakers, are held in the library at Eastwood (birthplace of DHL). Organizes visits to places of interest in the surrounding countryside, supports the activities of the D.H. Lawrence Centre at Nottingham University, and has close links with DHL Societies worldwide. Publishes two newsletters and one journal each year, free to members.

The Leamington Literary Society

52 Newbold Terrace East, Leamington Spa CV32 4EZ
☏ 01926 425733
Honorary Secretary *Mrs Margaret Watkins*

SUBSCRIPTION £10 p.a.

Founded in 1912 to promote the study and appreciation of literature and the arts. Holds regular meetings every second Tuesday of the month (from September to June) at the Royal Pump Rooms, Leamington Spa. Also smaller groups which meet regularly to study poetry and the modern novel. The Society has published various books of local interest.

Lewes Monday Literary Club

c/o 1c Prince Edward's Road, Lewes BN7 1BJ
☏ 01273 478512
chris.lutrario@btinternet.com
Contact *Christopher Lutrario*

SUBSCRIPTION £20 p.a.; £5 (guest, per meeting)

Founded in 1948 for the promotion and enjoyment of literature. Seven meetings are held during the winter generally on the last Monday of each month (from October to April) at the Pelham House Hotel in Lewes. The Club attracts speakers of the highest quality and a balance between all forms of literature is aimed for. Guests are welcome to attend meetings.

The George MacDonald Society

38 South Vale, Upper Norwood, London SE19 3BD
☏ 020 8653 8768
r.lines878@btinternet.com
www.george-macdonald.com
Contact *Richard Lines*
Membership Enquiries *Roger Bardet* (r.bardet@hotmail.co.uk)

SUBSCRIPTION £10 p.a. (UK); £11 (overseas); US$18 (N. America)

Founded in 1980 to promote interest in the life and works of George MacDonald (1824–1905) and his friends and contemporaries. Organizes conferences, meetings, lectures and outings. Annual newsletter, *Orts* and an academic journal, *North Wind*, published by St Norbert College, De Pere, WA, USA, are distributed free to members. (Registered Charity No: 1024021.)

The Friends of Arthur Machen

Apt 5, 26 Hervey Road, Blackheath, London SE3 8BS
www.machensoc.demon.co.uk
Contact *Jeremy Cantwell*

SUBSCRIPTION £17 p.a. (UK); £20 (Europe/RoW); $36 (US)

Brings together those who appreciate Machen's writings and generally promotes discussion and research. Publishes two journals and two newsletters annually, plus occasional publications for the membership only.

The Marlowe Society

9 Middlefield Gardens, Hurst Green Road, Halesowen B62 9QH
☏ 0121 421 1482   ☏ 0121 421 1482
www.marlowe-society.org
Membership Secretary *Frieda Barker*
Treasurer *Bruce Young*

SUBSCRIPTION £12 p.a. (individual); £7 p.a. (pensioner/student/unwaged); £15 p.a. (overseas); £200 (group); £100 (individual life membership)

Founded 1955. Holds meetings, lectures and discussions, stimulates research into Marlowe's life and works, encourages production of his plays and publishes a biannual newsletter. At a ceremony on 11th July 2002 the Society celebrated their success in establishing a memorial to the playwright and poet in Poets' Corner in Westminster Abbey.

The John Masefield Society

The Frith, Ledbury HR8 1LW
☎ 01531 633800
www2.sas.ac.uk/ies/cmps/Projects/Masefield/
Society/jms1.htm
Chairman *Peter Carter*

SUBSCRIPTION £5 p.a. (individual); £8 (family, institution, library); £10 (overseas); £2.50 (junior, student)

Founded in 1992 to stimulate the appreciation of and interest in the life and works of John Masefield (Poet Laureate 1930–1967). The Society is based in Ledbury, the Herefordshire market town of his birth and holds various public events in addition to publishing a journal and occasional papers.

William Morris Society

Kelmscott House, 26 Upper Mall, Hammersmith, London W6 9TA
☎ 020 8741 3735 📠 020 8748 5207
william.morris@care4free.net
www.morrissociety.org
Curator *Helen Elletson*

SUBSCRIPTION £18 p.a.; £8 (student); £30 (corporate)

Founded in 1955 to promote interest in the life, work and ideas of William Morris (1834–1896), English poet and craftsman. Organizes events and educational programmes. Members receive newsletters and biannual journal. Archive and library open by appointment. Museum open Thursdays and Saturdays, 2.00 pm to 5.00 pm.

The Neil Munro Society

8 Briar Road, Kirkintilloch, Glasgow G66 3SA
☎ 0141 776 4280
brian@bdosborne.fsnet.co.uk
www.neilmunro.co.uk
Secretary *Brian D. Osborne*

SUBSCRIPTION £12 p.a.; £7 (unwaged); £15 (institutional)

Founded in 1996 to encourage interest in the works of Neil Munro (1863–1930), the Scottish novelist, short story writer, poet and journalist. An annual programme of meetings is held in Glasgow and Munro's home-town of Inveraray. Publishes *ParaGraphs*, a twice-yearly magazine, sponsors reprints of Munro's work and is developing a Munro archive.

Violet Needham Society

c/o 19 Ashburnham Place, London SE10 8TZ
☎ 020 8692 4562
richardcheffins@aol.com
violetneedhamsociety.org.uk

Honorary Secretary *R.H.A. Cheffins*

SUBSCRIPTION £7.50 p.a. (UK & Europe); £11 (RoW)

Founded in 1985 to celebrate the work of children's author Violet Needham and stimulate critical awareness of her work. Publishes thrice-yearly *Souvenir*, the Society journal, with an accompanying newsletter; organizes meetings and excursions to places associated with the author and her books. The journal includes articles about other children's writers of the 1940s and 1950s and on ruritanian fiction. Contributions welcome.

The Edith Nesbit Society

26 Strongbow Road, Eltham, London SE9 1DT
www.edithnesbit.co.uk
Contact *Mrs Marion Kennett*

SUBSCRIPTION £7 p.a.; £9 (joint)

Founded in 1996 to celebrate the life and work of Edith Nesbit (1858–1924), best known as the author of *The Railway Children*. The Society's activities include a regular newsletter, booklets, talks and visits to relevant places. 2008 is the 150th anniversary of Edith Nesbit's birth and several special events, including the launch of a children's poetry competition, are planned.

The Wilfred Owen Association

29 Arthur Road, London SW19 7DN
☎ 020 8947 0476
vcedavis@hotmail.com
www.1914-18.co.uk/owen
Chairman *Mrs Meg Crane*

SUBSCRIPTION £10 adult (£8 concessions); £15 couple (£13 concessions); £15 overseas (adult); £20 overseas (couple)

Founded in 1989 to commemorate the life and works of Wilfred Owen by promoting readings, visits, talks and performances relating to Owen and his work. The Association offers practical support for students of literature and future poets through links with education, support for literary foundations and information on historical and literary background material. Membership is international with 380 members. Publishes regular newsletters. Speakers are available.

The Elsie Jeanette Oxenham Appreciation Society

32 Tadfield Road, Romsey SO51 5AJ
☎ 01794 517149
abbeybufo@gmail.com
www.bufobooks.demon.co.uk/abbeylnk.htm
Membership Secretary/Treasurer *Ms Ruth Allen*
Editor (The Abbey Chronicle) *Fiona Dyer*
(fionadyer@btopenworld.com)

SUBSCRIPTION £7.50 p.a.; enquire for overseas rates

Includes FREE online access to **www.thewritershandbook.com**

Founded in 1989 to promote the works of Elsie J. Oxenham. Publishes a newsletter for members, *The Abbey Chronicle*, three times a year. 500+ members.

Thomas Paine Society

14 Park Drive, Forest Hall, Newcastle NE12 9JP
☏ 0191 266 2988
President *The Rt. Hon. Michael Foot*
Honorary Secretary *Stuart Hill* (at address above)
Treasurer *Stuart Wright*
SUBSCRIPTION (minimum) £15 p.a. (UK); $35 (overseas); £5 (unwaged/pensioner/student)

Founded in 1963 to promote the life and work of Thomas Paine, and continues to expound his ideals. Meetings, newsletters, lectures and research assistance. The Society has members worldwide and keeps in touch with American and French Thomas Paine associations. Publishes magazine, *The Journal of Radical History*, twice yearly (Editor, *R.W. Morrell*, 43 Eugene Gardens, Nottingham NG2 3LF) and a newsletter. Holds occasional exhibitions and lectures, including the biannual Thomas Paine Memorial Lecture and the annual Eric Paine Memorial Lecture.

The Beatrix Potter Society

The Lodge, Salisbury Avenue, Harpenden AL5 2PS
☏ 01582 769755
beatrixpottersociety@tiscali.co.uk
www.beatrixpottersociety.org.uk
Membership Secretary *Jenny Akester*
SUBSCRIPTION UK: £20 p.a. (individual); £25 (institution); Overseas: £25 (individual); £30 (institution)

Founded in 1980 to promote the study and appreciation of the life and works of Beatrix Potter (1866–1943). Potter was not only the author of *The Tale of Peter Rabbit* and other classics of children's literature; she was also a landscape and natural history artist, diarist, farmer and conservationist, and was responsible for the preservation of large areas of the Lake District through her gifts to the National Trust. The Society upholds and protects the integrity of the unique work of Potter, her aims and bequests; holds regular talks and meetings in London and elsewhere with visits to places connected with Beatrix Potter. International Study Conferences are held in the UK and the USA. The Society has an active publishing programme. (UK Registered Charity No. 281198.)

Anthony Powell Society

76 Ennismore Avenue, Greenford UB6 0JW
☏ 020 8864 4095   ☏ 020 8864 6109
secretary@anthonypowell.org
www.anthonypowell.org
Patron *John M.A. Powell*

President *Simon Russell Beale*
Honorary Secretary *Dr Keith C. Marshall*
SUBSCRIPTION £20 p.a. (individual); £30 (joint/gold); £12 (student); £100 (organization); £5 p.a. supplement for overseas members

Founded in June 2000 to advance education and interest in the life and works of the English author Anthony Powell (1905–2000). The Society's major activity is a biennial Powell conference, the first three of which were at Eton College (2001), Balliol College, Oxford (2003) and The Wallace Collection, London (2005). The Society organizes events for members, ranging from 'pub meets' to talks and visits to places of Powell interest as well as publishing *Secret Harmonies* (annual journal) and a quarterly *Newsletter*. Member of the **Alliance of Literary Societies**.

The Powys Society

25 Mansfield Road, Taunton TA1 3NJ
☏ 01823 278177
peter_lazare@hotmail.com
www.powys-society.org
Honorary Secretary *Peter Lazare*
SUBSCRIPTION £18.50 (UK); £22 (overseas); £10 (student)

The Society (with a membership of 300) aims to promote public education and recognition of the writings, thought and contribution to the arts of the Powys family; particularly of John Cowper, Theodore and Llewelyn, but also of the other members of the family and their close associates. The Society holds two major collections of Powys published works, letters, manuscripts and memorabilia. Publishes the *Powys Society Newsletter* in April, June and November and *The Powys Journal* in August. Organizes an annual conference as well as lectures and meetings in Powys places.

The J.B. Priestley Society

Eldwick Crag Farm, Otley Road, Bingley BD16 3BB
☏ 01274 563078
reavill@globalnet.co.uk
www.jbpriestley-society.com
President *Tom Priestley*
Chairman *Dr Ken Smith*
Honorary Secretary *R.E.Y. Slater*
Membership Secretary *Tony Reavill*
SUBSCRIPTION £12 (individual); £7 (concession); £18 (group/family)

Founded in 1997 to widen the knowledge and understanding of Priestley's works; promote the study of his life and his social, cultural and political influences; provide members of the Society with lectures, seminars, films, journals and stimulate education projects; promote public performances of his works and the distribution of material associated with him. For further information, contact the Membership Secretary at the address above.

The Barbara Pym Society

St Hilda's College, Oxford OX4 1DY

☎ 01865 276828   ℻ 01865 276820

eileen.roberts@st-hildas.oxford.ac.uk

www.barbara-pym.org

Chairman *Clemence Schultz*

Archivist *Yvonne Cocking*

SUBSCRIPTION £15 p.a. (individual); £22 p.a. (household); £8 p.a. (concession)

Aims to promote interest in and scholarly research into the works of Barbara Pym; to bring together like-minded people to enjoy the exploration of all aspects of her novels and to continue to encourage publishers to keep her novels in print. Annual weekend conference at St. Hilda's College in September, focusing on a theme or a particular novel. Annual North American Conference at Harvard Law School. Annual one-day spring meeting in London, with speaker. Publishes a newsletter, *Green Leaves*, biannually. In addition to reports on conferences and society activities, the newsletter includes scholarly papers and various unpublished items from the Pym archives at the Bodleian Library (i.e. short stories).

The Queen's English Society

The Clergy House, Hide Place, London SW1P 4NJ

☎ 07766 416900

enquiries@queens-english-society.com

www.queens-english-society.com

Honorary Secretary *Mr G.J. Hardwick*

SUBSCRIPTION £15 p.a. (ordinary); £20 (joint)

Founded in 1972 to promote and uphold the use of good English and to encourage the enjoyment of the language. Holds regular meetings and conferences to which speakers are invited, an annual luncheon and publishes a quarterly journal, *Quest*, for which original articles are welcome. (Registered Charity No. 272901.)

Rainbow Poetry Recitals

74 Marina, St Leonards-on-Sea TN38 0BJ

☎ 01424 444072

beyondcloister@hotmail.co.uk

Administrator *Hugh Hellicar*

SUBSCRIPTION £6.50/€10 (individual); £10/€15 (joint/family); £12 (school)

Founded 1994. Regular recitals of poetry and other literature are held in Sussex and London. Seminars in some branches. Research library in Sussex. Quarterly magazine with overseas distribution.

The Arthur Ransome Society Ltd

Abbot Hall Art Gallery & Museum, Kendal LA9 5AL

tarsinfo@arthur-ransome.org

www.arthur-ransome.org

Trustee Chairman *Geraint Lewis*

Company Secretary *Peter Hyland*

SUBSCRIPTION UK: £5 (junior); £10 (student); £15 (adult); £20 (family); £40 (corporate); Overseas: £5 (junior); £10 (student); £20 (adult); £25 (family) Payable in local currency in US, Canada, Australia, Japan and New Zealand

Founded in 1990 to celebrate the life and to promote the works and ideas of Arthur Ransome, author of *Swallows and Amazons* titles for children, biographer of Oscar Wilde, works on the Russian Revolution and extensive articles on fishing. TARS seeks to encourage children and adults to engage in adventurous pursuits, to educate the public about Ransome and his works, and to sponsor research into his literary works and life.

The Romany Society

10 Haslam Street, Bury BL9 6EQ

☎ 0161 764 7078

phil_shelley@btopenworld.com

www.romanysociety.org.uk

Honorary Secretary *John Thorpe* (at address above)

Publications Officer *Phil Shelley*

SUBSCRIPTION £5 (individual); £9 (family); £10 (institution); £3.50 (student/unwaged)

Promotes the life, work and conservation message of the Rev. G. Bramwell Evens – 'Romany' – the first broadcasting naturalist who influenced generations of listeners with his *Out With Romany* radio programmes on the BBC during the 1930s and 1940s and with his series of natural history books. The Romany Society was established originally following Evens' death at the age of 59 in 1943 and ran until 1965. Revived in 1996, the Society holds an annual members' weekend to areas of significance in the Romany story, erects memorial plaques where appropriate, encourages media coverage of Romany and supports young naturalists with the Romany Memorial Grant.

The Followers of Rupert

31 Whiteley, Windsor SL4 5PJ

☎ 01753 865562

followersofrupert@hotmail.com

www.rupertthebear.org.uk

Membership Secretary *Mrs Shirley Reeves*

SUBSCRIPTION UK: £25 (single); £25 (joint); Europe: £30 (single); £32 (joint); RoW: £35 (single); £37 (joint)

Founded 1983. The Society caters for the growing interest in the Rupert Bear stories, past, present and future. Publishes the *Nutwood Newsletter* quarterly which gives up-to-date news of Rupert and information on Society activities. A national get-together of members – the Followers Annual – is held during the summer.

The Ruskin Society of London

20 Parmoor Court, Summerfield, Oxford OX2 7XB
℡ 01865 310987
Honorary Secretary *Miss O.E. Forbes-Madden*

SUBSCRIPTION £15 p.a.

Founded in 1986 to promote interest in John Ruskin (1819–1900) and his contemporaries. All aspects of Ruskinia are introduced. Functions are held in London. Publishes the annual *Ruskin Gazette*, a journal concerned with Ruskin's influence. Affiliated to other literary societies.

The Ruskin Society

49 Hallam Street, London W1W 6JP
℡ 020 7580 1894
cgamble@zen.co.uk
www.lancs.ac.uk/depts/ruskin/links.htm
President *Sir Richard Body*
Chairman *Dr Malcolm Hardman*
Honorary Secretary *Dr Cynthia J. Gamble*

SUBSCRIPTION £10 p.a. (payable on 1st Jan.)

Founded in 1997 to encourage a wider understanding of John Ruskin and his contemporaries. Organizes lectures and events which seek not only to explain to the public at large the nature of Ruskin's theories but also to place these in a modern context.

The Malcolm Saville Society

6 Redcliffe Street, London SW10 9DS
mystery@witchend.com
www.witchend.com
Secretary *George Jasieniecki*

SUBSCRIPTION £10 p.a. (UK); £12 (EU); £16 (RoW)

Founded in 1994 to remember and promote interest in the work of the popular children's author. Regular social activities, booksearch, library, contact directory and four magazines per year.

The Dorothy L. Sayers Society

Rose Cottage, Malthouse Lane, Hurstpierpoint BN6 9JY
℡ 01273 833444   ℻ 01273 835988
www.sayers.org.uk
Contact *Christopher Dean*

SUBSCRIPTION £14 p.a. (UK); £16.50 (Europe); $34 (US)

Founded in 1976 to promote the study of the life, works and thoughts of Dorothy Sayers; to encourage the performance of her plays and publication of her books and books about her; to preserve original material and provide assistance to researchers. Acts as a forum and information centre, providing material for study purposes which would otherwise be unavailable. Annual conventions, usually in August and other meetings. Co-founder of the Dorothy L. Sayers Centre in Witham. Publishes bi-monthly bulletin, annual proceedings and other papers.

The Shaw Society

Flat F, 10 Compton Road, London N1 2PA
℡ 020 7226 4266
shawsociety@blueyonder.co.uk
Honorary Secretary *Alan Knight*

SUBSCRIPTION £15 p.a. (individual); £22 (joint)

Founded in 1941 to promote interest in the life and works of G. Bernard Shaw. Meetings are held on the last Friday of every month (except July, August and December) at Conway Hall, Red Lion Square, London WC1 (6.30 pm for 7.00 pm) at which speakers are invited to talk on some aspect of Shaw's life or works. Monthly playreadings are held on the first Friday of each month (except August). Over 40 years ago the Society began the annual open-air performance held at Shaw's Corner, Ayot St Lawrence in Hertfordshire, on the weekend nearest to Shaw's birthday (26th July). Publishes a biannual magazine, *The Shavian*, with a calendar of Shavian events. (No payment for contributors.)

The Robert Southey Society

1 Lewis Terrace, Abergarwed, Neath SA11 4DL
℡ 01639 711480
Contact *Robert King*

SUBSCRIPTION £10 p.a.

Founded in 1990 to promote the work of Robert Southey. Publishes an annual newsletter and arranges talks on his life and work. Open membership.

The Laurence Sterne Trust
▷ The Friends of Shandy Hall

Robert Louis Stevenson Club

12 Dean Park, Longniddry EH32 0QR
℡ 01875 852976   ℻ 01875 853328
alan@amarchbank.freeserve.co.uk
www.rlsclub.org.uk
Secretary *Alan Marchbank*

SUBSCRIPTION £20 p.a. (individual); £26 p.a. (overseas); £150 (10-year)

Founded in 1920 to foster interest in Robert Louis Stevenson's life and works. The Club organizes an annual lunch and other events. Publishes *RLS Club News* twice a year.

The R.S. Surtees Society

Manor Farm House, Nunney, Near Frome
BA11 4NJ
☎ 01373 836937   🖷 01373 836574
rssurtees@fsmail.net
www.r.s.surteessociety.org
Contact *Orders and Membership Secretary*
(☎ 01373 302155)

Founded in 1979 to republish the works of R.S.
Surtees and others.

The Tennyson Society

Central Library, Free School Lane, Lincoln
LN2 1EZ
☎ 01522 552862   🖷 01522 552858
kathleenjefferson@lincolnshire.gov.uk
www.tennysonsociety.org.uk
Honorary Secretary *Miss K. Jefferson*

SUBSCRIPTION £10 p.a. (individual); £12 (family);
£15 (corporate); £160 (life)

Founded in 1960. An international society with
membership worldwide. Exists to promote the
study and understanding of the life and work of
Alfred, Lord Tennyson. The Society is concerned
with the work of the Tennyson Research Centre,
'probably the most significant collection of
manuscripts, family papers and books in the
world'. Publishes annually the *Tennyson Research
Bulletin*, which contains articles and critical
reviews; and organizes lectures, visits and
seminars. Annual memorial service at Somersby
in Lincolnshire.

The Angela Thirkell Society

54 Belmont Park, London SE13 5BN
☎ 020 8244 9339
penny.aldred@ntlworld.com
www.angelathirkellsociety.com
www.angelathirkell.org (N. American branch)
Honorary Secretary *Mrs. P. Aldred*

SUBSCRIPTION £10 p.a.

Founded in 1980 to honour the memory of
Angela Thirkell as a writer and to make her works
available to new generations. Publishes an annual
journal, holds an AGM in the autumn and a
spring meeting which usually takes the form of
a visit to a location associated with Thirkell. Has
a flourishing North American branch which has
frequent contact with the UK parent society.

The Dylan Thomas Society of Great Britain

Fern Hill, 24 Chapel Street, Mumbles, Swansea
SA3 4NH
☎ 01792 363875
Contact *Mrs Cecily Hughes*

SUBSCRIPTION £5 (individual); £8 (two adults
from same household)

Founded in 1977 to foster an understanding of the
work of Dylan Thomas and to extend members'
awareness of other 20th century writers,
especially Welsh writers in English. Meetings take
place monthly, mainly in Swansea.

The Edward Thomas Fellowship

1 Carfax, Undercliff Drive, St Lawrence, Isle of
Wight PO38 1XG
☎ 01983 853366
colingthornton@btopenworld.com
www.edward-thomas-fellowship.org.uk
Hon. Secretary *Colin G. Thornton*

SUBSCRIPTION £7 p.a. (single); £10 p.a. (joint)

Founded in 1980 to perpetuate and promote the
memory of Edward Thomas and to encourage an
appreciation of his life and work. The Fellowship
holds a commemorative birthday walk on the
Sunday nearest the poet's birthday (3rd March)
and an autumn walk; issues newsletters and holds
various events and seminars.

The Tolkien Society

210 Prestbury Road, Cheltenham GL52 3ER
secretary@tolkiensociety.org
www.tolkiensociety.org
Secretary *Sally Kennett*

SUBSCRIPTION £21 p.a. (UK); £24 (EU); £27
(RoW)

An international organization which aims to
encourage and further interest in the life and
works of the late Professor J.R.R. Tolkien, CBE,
author of *The Hobbit* and *Lord of the Rings*.
Current membership stands at 1,100. Publishes
Mallorn annually and *Amon Hen* bi-monthly.

The Trollope Society

Maritime House, Old Town, Clapham, London
SW4 0JW
☎ 020 7720 6789   🖷 020 7627 2965
info@trollopesociety.org
www.trollopesociety.org
Contact *Pelham Ravenscroft*

Founded in 1987 to study and promote Anthony
Trollope's works. Publishes the complete works of
Trollope's novels and travel books.

Wainwright Society

Kendal Museum, Station Road, Kendal LA9 6BT
☎ 01539 721374   🖷 01539 737976
membership@wainwright.org.uk
www.wainwright.org.uk
Membership Secretary *Morag Clement*

SUBSCRIPTION £10 p.a. (per household, from 1st
Jan.)

Founded in 2002 'to keep alive the things that
Alfred Wainwright (1907–1991) promoted
through his guidebooks (*Pictorial Guides to the*

Lakeland Fells), started 50 years ago, and the many other publications which were the "labour of love" of a large portion of his life'. Produces a newsletter three times per year, organizes walks, events, annual dinner and lecture.

The Walmsley Society

April Cottage, No 1 Brand Road, Hampden Park, Eastbourne BN22 9PX

☎ 01323 506447

walmsley@haughshw.demon.co.uk

Honorary Secretary *Fred Lane*

SUBSCRIPTION £13 p.a.; £15 (family); £11 (student/ senior citizen); £12 (overseas, £20, one year; £35, two years)

Founded in 1985 to promote interest in the writings of Leo Walmsley and to foster an appreciation of the work of his father, the artist Ulric Walmsley. Two annual meetings – one held in Robin Hood's Bay on the East Yorkshire coast, spiritual home of the author Leo Walmsley. Publishes a journal twice-yearly and newsletters, and is involved in other publications which benefit the aims of the Society. A biography of Leo Walmsley is now available.

Sylvia Townsend Warner Society

2 Vicarage Lane, Fordington, Dorchester DT1 1LH

☎ 01305 266028

judithmbond@tiscali.co.uk

www.townsendwarner.com

Contact *Eileen Johnson*

SUBSCRIPTION £10 p.a.; $25 (overseas)

Founded in 2000 to promote a wider readership and better understanding of the writings of Sylvia Townsend Warner.

Mary Webb Society

8 The Knowe, Willaston, Neston CH64 1TA

☎ 0151 327 5843

suehigginbotham@yahoo.co.uk

www.marywebbsociety.co.uk

Secretary *Sue Higginbotham*

SUBSCRIPTION UK: £10 p.a. (individual); £13 p.a. (joint); Overseas: £13 p.a. (individual); £15 p.a. (joint)

Founded 1972. Attracts members from the UK and overseas who are devotees of the literature of Mary Webb and of the beautiful Shropshire countryside of her novels. Publishes biannual journal, organizes summer schools in various locations related to the authoress's life and works. Archives; lectures; tours arranged for individuals and groups.

H.G. Wells Society

Flat 3, 27b Church Road, London NW4 4EB

secretaryhgwellsociety@hotmail.com

www.hgwellsusa.50megs.com/UK/index.html

Hon. General Secretary *Mark Egerton*

SUBSCRIPTION £16 (UK/EU); £19 (RoW); £20 (corporate); £10 (concessions)

Founded in 1960 to promote an interest in and appreciation of the life, work and thought of Herbert George Wells. Publishes *The Wellsian* (annual) and *The H.G. Wells Newsletter* (three issues yearly). Organizes meetings and conferences.

The Oscar Wilde Society

19 South Hill Road, Gravesend DA12 1LA

vanessasalome@blueyonder.co.uk

www.oscarwildesociety.co.uk

Chairman *Donald Mead*

Honorary Secretary *Vanessa Heron*

SUBSCRIPTION £25 p.a. (UK); £30 p.a. (Europe); £35 p.a. (RoW)

Founded in 1990 to promote knowledge, appreciation and study of the life, personality and works of the writer and wit Oscar Wilde. Activities include meetings, lectures, readings and exhibitions, and visits to locations associated with Wilde. Members receive a journal of Oscar Wilde studies, *The Wildean*, twice-yearly and a newsletter journal *Intentions* (six per year).

The Charles Williams Society

35 Broomfield, Stacey Bushes, Milton Keynes MK12 6HA

charles_wms_soc@yahoo.co.uk

www.geocities.com/charles_wms_soc

www.charleswilliamssociety.org.uk

Contact *Honorary Secretary*

SUBSCRIPTION £12.50 p.a.

Founded in 1975 to promote interest in, and provide a means for the exchange of views and information on the life and work of Charles Walter Stansby Williams (1886–1945).

The Henry Williamson Society

7 Monmouth Road, Dorchester DT1 2DE

☎ 01305 264092

zseagull@aol.com

www.henrywilliamson.co.uk

General Secretary *Mrs Sue Cumming*

Membership Secretary *Mrs Margaret Murphy* (16 Doran Drive, Redhill RH1 6AX ☎ 01737 763228 mm@ misterman.freeserve.co.uk)

SUBSCRIPTION £12 p.a.; £15 (family); £5 (student)

Founded in 1980 to encourage, by all appropriate means, a wider readership and deeper understanding of the literary heritage left by the 20th-century English writer Henry Williamson (1895–1977). Publishes annual journal.

The P.G. Wodehouse Society (UK)

26 Radcliffe Road, Croydon CRO 5QE
info@pgwodehousesocitey.org.uk
www.pgwodehousesociety.org.uk
Membership Secretary *Christine Hewitt*

SUBSCRIPTION £15 p.a. (£20 Dec. to March, for
membership extended to May the following year)

Formed in 1997 to promote enjoyment of the
works of P.G. Wodehouse, author, lyricist,
playwright and journalist, creator of Lord
Emsworth, Jeeves & Wooster and many more.
Membership of 1,000+. Publications include the
quarterly journal, *Wooster Sauce* and regular *By
The Way papers*. Meetings and events are held in
London and around the country. The Society is
allied to Wodehouse societies around the world.

The Parson Woodforde Society

22 Gaynor Close, Wymondham NR18 0EA
℡ 01953 604124
mabrayne@supanet.com
www.parsonwoodforde.org.uk
Membership Secretary *Mrs Ann Elliott*

SUBSCRIPTION £12.50 (UK); £25 (overseas)

Founded 1968. Aims to extend and develop
knowledge of James Woodforde's life and the
society in which he lived and to provide the
opportunity for fellow enthusiasts to meet
together in places associated with the diarist.
Publishes a quarterly journal and newsletter. The
Society has produced a complete edition of the
diary of James Woodforde in 17 volumes covering
the period 1759–1802.

The Virginia Woolf Society of Great Britain

Fairhaven, Charnleys Lane, Banks, Southport
PR9 8HJ
stuart.n.clarke@btinternet.com
www.virginiawoolfsociety.co.uk
Contact *Stuart N. Clarke*

SUBSCRIPTION £16 p.a.; £21 (overseas)

Founded in 1998 to promote interest in the life
and work of Virginia Woolf, author, essayist
and diarist. The Society's activities include trips
away, walks, reading groups and talks. Publishes
a literary journal, *Virginia Woolf Bulletin*, three
times a year.

The Wordsworth Trust

Dove Cottage, Grasmere LA22 9SH
℡ 01539 435544   ℻ 01539 463508
enquiries@wordsworth.org.uk
www.wordsworth.org.uk
Literature Officer *Andrew Forster*

Founded in 1880, the Trust is a living memorial
to the life and poetry of William Wordsworth
and his contemporaries. As Centre for British
Romanticism, the Wordsworth Trust, with
its wealth of manuscripts, books, drawings
and pictures, provides 'the full context for
understanding and celebrating a major cultural
moment in history in which Britain played a
profound role'. Research facilities are available
at the Jerwood Centre; contact: *Jeff Cowton*,
Curator. 'Summer Poetry at Grasemere' is a
live poetry event that takes place each year (see
The Wordsworth Trust entry under *Festivals*).
(Registered Charity No. 1066184.)

WW2 HMSO PPBKS Society

3 Roman Road, Southwick BN42 4TP
℡ 01273 596217
Contact *A.R. James*

Founded in 1994 to encourage collectors and
to promote research into HMSO's World War
II series of paperbacks. Most of them were
written by well-known authors, though in many
cases anonymously. No SUBSCRIPTION charge.
Available for purchase: Collectors' Guide (£5);
Handbook, *Informing the People* (£10).

The Yeats Society Sligo

Yeats Memorial Building, Hyde Bridge, Sligo,
Republic of Ireland
℡ 00 353 71 914 2693   ℻ 00 353 71 914 2780
info@yeats-sligo.com
www.yeats-sligo.com
President *Joe Cox*

SUBSCRIPTION €25 (single); €40 (couple)

Founded in 1958 to promote the heritage of W.B.
Yeats and the Yeats family. Attractions include
a permanent Yeats Exhibition; films for public
viewing throughout the year; the annual Yeats
International Summer School in July/August;
the Winter School in January. The Yeats Festival
is also held in the summer, and lectures, poetry
readings, workshops, etc. are held in the autumn
and spring, together with a variety of outings
and other events for members. Visits are made
to schools, and contacts are made with Yeats
scholars and students worldwide.

The Charlotte M. Yonge Fellowship

8 Anchorage Terrace, Durham DH1 3DL
℡ 0191 384 7857
www.cmyf.org.uk
Contact *Dr C.E. Schultze*

SUBSCRIPTION £9 p.a.

Founded in 1996 to provide the opportunity
for all those who enjoy and admire the work of
Charlotte M. Yonge (1823–1901), author of *The
Heir of Radclyffe*, to learn more about her life and
writings. Holds two meetings annually, one in
November in London and the other, the spring
AGM, at venues associated with Yonge's life,
works or interests, or where buildings or gallery
and museum collections relate to Victorian

history, art and society. Publishes *Review* twice-yearly and an occasional *Journal*. A loan collection of Yonge's works enables members to borrow publications that are hard to obtain. The Fellowship's archive is held at St Hugh's College, Oxford and material is available to members and *bona fide* researchers.

Yorkshire Dialect Society

51 Stepney Avenue, Scarborough YO12 5BW
www.ydsociety.org.uk
Secretary *Michael Park*

SUBSCRIPTION £10 p.a.

Founded in 1897 to promote interest in and preserve a record of the Yorkshire dialect. Publishes dialect verse and prose writing. Two journals to members annually. Details of publications are available from YDS at address above.

Francis Brett Young Society

92 Gower Road, Halesowen B62 9BT
☎ 0121 422 8969
www.fbysociety.co.uk
Honorary Secretary *Mrs Jean Hadley*

SUBSCRIPTION £7 p.a. (individual); £10 (couple sharing a journal); £5 (student); £7 (organization/ overseas); £70 (life); £100 (joint, life)

Founded 1979. Aims to provide a forum for those interested in the life and works of English novelist Francis Brett Young and to collate research on him. Promotes lectures, exhibitions and readings; publishes a regular newsletter.

Arts Councils and Regional Offices

Arts Council England

14 Great Peter Street, London SW1P 3NQ
☏ 0845 300 6200 Textphone 020 7973 6564
enquiries@artscouncil.org.uk
www.artscouncil.org.uk
Chairman *Professor Sir Christopher Frayling*
Chief Executive *Alan Davey*
Director of Literature Strategy *Antonia Byatt*

Founded 1946. Arts Council England is the national development agency for the arts in England, distributing public money from government and the National Lottery to artists and arts organizations. ACE works independently and at arm's length from government. Information about Arts Council England funding is available on the website, by e-mail or by contacting the enquiry line on 0845 300 6200.

Arts Council England has 9 regional offices:

Arts Council England, East
Eden House, 48–49 Bateman Street, Cambridge CB2 1LR
☏ 0845 300 6200 ⓕ 0870 242 1271
Textphone 01223 306893

Arts Council England, East Midlands
St Nicholas Court, 25–27 Castle Gate, Nottingham NG1 7AR
☏ 0845 300 6200 ⓕ 0115 950 2467

Arts Council England, London
2 Pear Tree Court, London EC1R 0DS
☏ 0845 300 6200 ⓕ 020 7608 4100
Textphone 020 7973 6564

Arts Council England, North East
Central Square, Forth Street, Newcastle upon Tyne NE1 3PJ
☏ 0845 300 6200 ⓕ 0191 230 1020
Textphone 0191 255 8585

Arts Council England, North West
Manchester House, 22 Bridge Street, Manchester M3 3AB
☏ 0845 300 6200 ⓕ 0161 834 6969
Textphone 0161 834 9131

Arts Council England, South East
Sovereign House, Church Street, Brighton BN1 1RA
☏ 0845 300 6200 ⓕ 0870 242 1257
Textphone 01273 710659

Arts Council England, South West
Senate Court, Southernhay Gardens, Exeter EX1 1UG
☏ 0845 300 6200 ⓕ 01392 498546
Textphone 01392 433503

Arts Council England, West Midlands
82 Granville Street, Birmingham B1 2LH
☏ 0845 300 6200 ⓕ 0121 643 7239
Textphone 0121 643 2815

Arts Council England, Yorkshire
21 Bond Street, Dewsbury WF13 1AX
☏ 0845 300 6200 ⓕ 01924 466522
Textphone 01924 438585

The Arts Council/An Chomhairle Ealaíon

70 Merrion Square, Dublin 2, Republic of Ireland
☏ 00 353 1 618 0200   ⓕ 00 353 1 676 1302
artistsservices@artscouncil.ie
www.artscouncil.ie
Director *Mary Cloake*
Head of Literature *Sarah Bannon*
Literature Specialist (English language)
Bronwen Williams

The Arts Council is the Irish government agency for developing the arts. It provides financial assistance to artists, arts organizations, local authorities and others for artistic purposes. It offers advice and information on the arts to Government and to a wide range of individuals and organizations. As an advocate for the arts and artists, the Arts Council undertakes projects and research, often in new and emerging areas of arts practice, and increasingly in cooperation with partner organizations. Its funding from government for 2008 was €82.3 million.

Of particular interest to individual writers is the Council's free booklet, *Support for Artists 2008*, which describes bursaries, awards and schemes on offer and how to apply for them. Applicants to these awards must have been born in, or be resident in, the Republic of Ireland.

The Arts Council of Northern Ireland

MacNeice House, 77 Malone Road, Belfast
BT9 6AQ
℡ 028 9038 5200   🅕 028 9066 1715
dsmyth@artscouncil-ni.org
www.artscouncil-ni.org
Literature and Language Arts Officer *Damian Smyth*

Funds book production by established publishers, programmes of readings, literary festivals, mentoring services, writers-in-residence schemes, network service providers, literary magazines and periodicals. Annual awards and bursaries for writers are available. Holds information also on various groups associated with regional arts, workshops and courses.

Scottish Arts Council

12 Manor Place, Edinburgh EH3 7DD
℡ 0131 226 6051   🅕 0131 225 9833
help.desk@scottisharts.org.uk
www.scottisharts.org.uk
Acting Chief Executive *Jim Tough*
Head of Literature *Dr Gavin Wallace*
Literature Officers *Aly Barr, Emma Turnbull*
Literature Administrator *Catherine Allan*

Principal channel for government funding of the arts in Scotland. The Scottish Arts Council is funded by the Scottish Government and National Lottery. It aims to develop and improve the knowledge, understanding and practise of the arts, and to increase their accessibility throughout Scotland. It offers grants to artists and arts organizations concerned with the visual arts, crafts, dance and mime, drama, literature, music, festivals and traditional, ethnic and community arts. Scottish Arts Council's support for Scottish-based writers with a track record of publication includes bursaries, writing fellowships and book awards (see entries under *Bursaries, Fellowships and Grants* and *Prizes*). Information offered includes lists of literature awards, literary magazines, agents and publishers.

The Arts Council of Wales/Cyngor Celfyddydau Cymru

9 Museum Place, Cardiff CF10 3NX
℡ 029 2037 6500   🅕 029 2022 1447
www.artswales.org.uk
www.celfcymru.org.uk

The Arts Council of Wales (ACW) is the development body for the arts in Wales. ACW's Creative Wales Awards offer substantial grants for established writers and playwrights. Funds the **Welsh Academy** and **Ty Newydd Writers' Centre** to provide services to individual writers, including bursaries, mentoring, the critical writers services and writers in residence/on tour. Responsibility for promoting the publishing and sales of books from Wales and funding literary magazines rests with the **Welsh Books Council**.

Library Services

Aberdeen Central Library

Rosemount Viaduct, Aberdeen AB25 1GW
☎ 01224 652500   ⅁ 01224 641985
centrallibrary@aberdeencity.gov.uk
www.aberdeencity.gov.uk

OPEN Central Library Lending Services: 9.00 am to 7.00 pm Monday to Thursday; 9.00 am to 5.00 pm Friday & Saturday. Business Information, Community Reference & Local Studies: 9.00 am to 8.00 pm Monday to Thursday; 9.00 am to 5.00 pm Friday & Saturday. Branch library opening hours vary
OPEN ACCESS

General reference and loans. Books, pamphlets, periodicals and newspapers; videos, CDs, DVDs; arts equipment lending service; Internet and Learning Centre for public access; photographs of the Aberdeen area; census records, maps, newspapers; online and remote access databases, patents and standards. The library offers special services to housebound readers.

Armitt Collection, Museum & Library

Rydal Road, Ambleside LA22 9BL
☎ 015394 31212   ⅁ 015394 31313
info@armitt.com
www.armitt.com
OPEN 10.00 am to 5.00 pm Monday to Wednesday; Friday to Sunday (last admission 4.30 pm)
FREE ACCESS (Prior appointment for research)

A small but unique reference library of rare books, manuscripts, pictures, antiquarian prints, maps and museum items, mainly about the Lake District. It includes early guidebooks and topographical works, books and papers relating to Ruskin, H. Martineau, Charlotte Mason and others; fine art including work by W. Green, J.B. Pyne, John Harden, K. Schwitters, and Victorian photographs by Herbert Bell; also a major collection of Beatrix Potter's scientific watercolour drawings and microscope studies. Museum and Exhibition open seven days per week from 10.00 am to 5.00 pm. Entry charge for museum.

The Athenaeum, Liverpool

Church Alley, Liverpool L1 3DD
☎ 0151 709 7770   ⅁ 0151 709 0418
info@theathenaeum.org.uk
www.theathenaeum.org.uk
OPEN 9.00 am to 4.00 pm Monday and Tuesday; 9.00 am to 9.00 pm Wednesday to Friday
ACCESS To club members; researchers by application only

General collection, with books dating from the 15th century, now concentrated mainly on local history with a long run of Liverpool directories and guides. SPECIAL COLLECTIONS Liverpool playbills; William Roscoe; Blanco White; Robert Gladstone; 18th-century plays; 19th-century economic pamphlets; Bibles; Yorkshire and other genealogy. Some original drawings, portraits, topographical material and local maps.

Bank of England Information Centre

Threadneedle Street, London EC2R 8AH
☎ 020 7601 4715   ⅁ 020 7601 4356
informationcentre@bankofengland.co.uk
www.bankofengland.co.uk
OPEN 9.00 am to 5.30 pm Monday to Friday
ACCESS For research workers only by prior arrangement, when material is not readily available elsewhere
50,000 volumes of books and periodicals. 2,000 periodicals taken. UK and overseas coverage of banking, finance and economics. SPECIAL COLLECTIONS Central bank reports; UK 17th–19th-century economic tracts; government reports in the field of banking.

Barbican Library

Barbican Centre, London EC2Y 8DS
☎ 020 7638 0569/7638 0568 (24-hr renewals)
⅁ 020 7638 2249
barbicanlib@cityoflondon.gov.uk
www.cityoflondon.gov.uk/libraries
Librarian *John Lake, BA, MCLIP*
OPEN 9.30 am to 5.30 pm Monday and Wednesday; 9.30 am to 7.30 pm Tuesday and Thursday; 9.30 am to 2.00 pm Friday; 9.30 am to 4.00 pm Saturday
OPEN ACCESS

Situated on Level 2 of the Barbican Centre, this is the City of London's largest lending library. Study facilities are available plus free Internet access. In addition to a large general lending department, the library seeks to reflect the Centre's emphasis on the arts and includes strong collections (including DVDs, videos and CD-ROMs) on painting, sculpture, theatre, cinema and ballet, as well as a large music library with books, scores and CDs (sound recording loans available at a small charge). Also houses the City's main children's library and has special collections on basic skills, materials for young adults, finance, natural resources, conservation, socialism and the history of London. Service available for housebound readers. A literature events programme is organized by the Library which supplements and provides cross-arts planning opportunities with the Barbican Centre artistic programme. Reading groups meet in the library on the first Thursday of every month and a weekly Basic Skills advice service is offered during term time.

Barnsley Public Library

Central Library, Shambles Street, Barnsley S70 2JF
☏ 01226 773930   🖷 01226 773955
barnsleylibraryenquiries@barnsley.gov.uk
www.barnsley.gov.uk/Libraries
OPEN Lending & Reference: 9.30 am to 7.00 pm Monday and Wednesday; 9.30 am to 5.30 pm Tuesday, Thursday, Friday; 9.30 am to 4.00 pm Saturday. Telephone to check hours of other departments and other branch libraries.
OPEN ACCESS

General library, lending and reference. Archive collection of family history and local firms; local studies: coal mining, local authors, Yorkshire and Barnsley; large junior library. (Specialist departments are closed on certain weekday evenings and Saturday afternoons.)

BBC Written Archives Centre

Caversham Park, Reading RG4 8TZ
☏ 0118 948 6281   🖷 0118 946 1145
heritage@bbc.co.uk
www.bbc.co.uk/heritage
Contact *Jacqueline Kavanagh*
OPEN 9.30 am to 5.30 pm Monday to Friday
ACCESS For reference, by appointment only, Wednesday to Friday

Holds the written records of the BBC, including internal papers from 1922 to the 1990s and published material to date. 20th century biography, social history, popular culture and broadcasting. Charges for certain services.

Bedford Central Library

Harpur Street, Bedford MK40 1PG
☏ 01234 350931/270102 (Reference Library)
🖷 01234 342163
www.bedfordshire.gov.uk
OPEN 9.00 am to 6.00 pm Monday, Tuesday, Wednesday, Friday; 9.00 am to 1.00 pm Thursday; 9.00 am to 5.00 pm Saturday
OPEN ACCESS

Lending library with a wide range of stock, including books, music, audiobooks, DVDs and videos; reference and information library, children's library, local history library and Internet facilities.

Belfast Public Libraries: Central Library

Royal Avenue, Belfast BT1 1EA
☏ 028 9050 9150   🖷 028 9033 2819
info.belb@ni-libraries.net
www.belb.org.uk
www.ni-libraries.net
OPEN 9.00 am to 8.00 pm Monday to Thursday; 9.00 am to 5.30 pm Friday; 9.00 am to 4.30 pm Saturday
OPEN ACCESS is available to the Central Lending Library and to the stock of the specialist reference departments within the building.

Over two million volumes for lending and reference. SPECIAL COLLECTIONS United Nations depository; complete British Patent Collection; Northern Ireland Newspaper Library; British and Irish government publications. The Central Library provides the following Reference Departments: General Reference; Belfast, Ulster and Irish Studies; Music. A Learning Gateway includes public Internet facilities. The Lending Library is one of 19 branches along with a range of outreach services to hospitals, care homes and housebound readers.

BFI National Library

21 Stephen Street, London W1T 1LN
☏ 020 7255 1444   🖷 020 7436 2338
library@bfi.org.uk
www.bfi.org.uk
OPEN 10.30 am to 5.30 pm Monday and Friday; 10.30 am to 8.00 pm Tuesday and Thursday; 1.00 pm to 8.00 pm Wednesday; Telephone Enquiry Service operates from 10.00 am to 5.00 pm (closed 1.00 pm to 2.00 pm)
ACCESS For reference only; annual, 5-day and limited day membership available

The world's largest collection of information on film and television including periodicals, cuttings, scripts, related documentation, personal papers. Main library catalogue available via website.

Birmingham and Midland Institute

9 Margaret Street, Birmingham B3 3BS
☏ 0121 236 3591   🖷 0121 212 4577
admin@bmi.org.uk
www.bmi.org.uk

Administrator & General Secretary *Philip Fisher*
ACCESS Members only

Established 1854. Later merged with the
Birmingham Library which was founded in 1779.
The Library specializes in the humanities, with
approximately 100,000 volumes in stock. Founder
member of the **Association of Independent
Libraries**. Meeting-place of many affiliated
societies devoted to poetry and literature.

Birmingham Library Services

Central Library, Chamberlain Square,
Birmingham B3 3HQ
℡ 0121 303 4511
central.library@birmingham.gov.uk
www.birmingham.gov.uk/libraries
OPEN 9.00 am to 8.00 pm Monday to Friday;
9.00 am to 5.00 pm Saturday

Over a million volumes. RESEARCH COLLECTIONS
include the Shakespeare Library; War Poetry
Collection; Parker Collection of Children's
Books and Games; Johnson Collection; Milton
Collection; Cervantes Collections; Early and
Fine Printing Collection (including the William
Ridler Collection of Fine Printing); Joseph
Priestley Collection; Loudon Collection;
Railway Collection; Wingate Bett Transport
Ticket Collection; Labour, Trade Union and Co-
operative Collections. PHOTOGRAPHIC ARCHIVES
Sir John Benjamin Stone; Francis Bedford; Francis
Frith; Warwickshire Photographic Survey;
Boulton and Watt Archive. Also, Charles Parker
Archive; Birmingham Repertory Theatre Archive
and Sir Barry Jackson Library; Local Studies
(Birmingham); Patents Collection; Song Sheets
Collection; Oberammergau Festival Collection.

Bournemouth Library

22 The Triangle, Bournemouth BH2 5RQ
℡ 01202 454848   ℻ 01202 454840
bournemouth@bournemouthlibraries.org.uk
www.bournemouth.gov.uk/libraries
OPEN 10.00 am to 7.00 pm Monday; 9.30 am
to 7.00 pm Tuesday, Thursday, Friday; 9.30 am
to 5.00 pm Wednesday; 10.00 am to 4.00 pm
Saturday
OPEN ACCESS

Main library for Bournemouth with lending,
reference and music departments, plus the
Heritage Zone – local and family history.

Bradford Central Library

Princes Way, Bradford BD1 1NN
℡ 01274 433600   ℻ 01274 433687
public.libraries@bradford.gov.uk
www.bradford.gov.uk/libraries

OPEN 9.00 am to 7.30 pm Monday to Friday; 9.00
am to 5.00 pm Saturday

Wide range of books and media loan services.
Comprehensive reference and information
services, including major local history collections
and specialized business information service.

Brighton Jubilee Library

Jubilee Street, Brighton BN1 1GE
℡ 01273 290800
libraries@brighton-hove.gov.uk
www.citylibraries.info/libraries/jubilee.asp
OPEN 10.00 am to 7.00 pm Monday and Tuesday;
10.00 am to 5.00 pm Wednesday, Friday and
Saturday; 10.00 am to 8.00 pm Thursday; 11.00
am to 4.00 pm Sunday
ACCESS Stock on open access and in onsite store;
material for reference use and lending

Specializations include art and antiques, history
of Brighton, local illustrations, Hebrew and
Oriental literature, natural history, children's
books, World War Two and large bequests of rare
and historical books.

Bristol Central Library

College Green, Bristol BS1 5TL
℡ 0117 903 7200   ℻ 0117 922 1081
www.bristol.gov.uk
OPEN 9.30 am to 7.30 pm Monday, Tuesday and
Thursday; 10.00 am to 5.00 pm Wednesday; 9.30
am to 5.00 pm Friday and Saturday; 1.00 pm to
5.00 pm Sunday
OPEN ACCESS

Lending, reference, art, music, business and local
studies are particularly strong. DVD, video and
CD collections on site. Facilities available: PCs
(large screen with Jaws and Zoomtext), Internet,
printing; videophone on site; black & white and
colour photocopiers.

British Architectural Library

Royal Institute of British Architects, 66 Portland
Place, London W1B 1AD
℡ 020 7580 5533   ℻ 020 7631 1802
info@inst.riba.org
www.architecture.com
OPEN 10.00 am to 8.00 pm Tuesday; 10.00 am to
5.00 pm Wednesday to Friday; 10.00 am to 1.30
pm Saturday; Closed Sunday, Monday and any
Saturday preceding a Bank Holiday; full details on
the website
ACCESS Free to RIBA members and non-members
on proof of identity and address; loans available to
RIBA members only

Collection of books, photographs and periodicals.
All aspects of architecture, current and historical.
Material both technical and aesthetic, covering
related fields including: interior design, landscape
architecture, topography, the construction

industry and applied arts. Brochure available; queries by telephone, letter, e-mail or in person. Charge for research (min. charge £40 + VAT).

RIBA British Architectural Library Drawings & Archives Collections

Drawings and manuscripts can be consulted at the RIBA Study Rooms, Henry Cole Wing, Victoria and Albert Museum: ☏ 020 7307 3708 drawings&archives@inst.riba.org
ACCESS Free to all.

The British Cartoon Archive
▷ entry under Picture Libraries

The British Library

Admission to St Pancras Reading Rooms – British Library Readers' Passes
Everyone is welcome to visit the British Library exhibition galleries or to tour the building. However, to use the reading rooms you will need to apply for a reader's pass, for which identification is required. The British Library issues passes to those who want to use its collections – researchers, innovators and entrepreneurs across all fields of study, and in academic, commerce or personal research – whether or not they are affiliated to a research institution. Two pieces of identification are required (original documents only). One proof of home address, e.g. a utility bill, bank statement or driving licence; and proof of signature, i.e. a bank or credit card, passport, driving licence or national identity card.

It would also be helpful to take with you anything to support your application (such as a student card, business card, professional membership card or details of the items you wish to see).

For further information, visit the website at www.bl.uk or contact the Reader Admissions Office, The British Library, 96 Euston Road, London NW1 2DB ☏ 020 7412 7676 🖷 020 7412 7794; reader-admissions@bl.uk

British Library Asia, Pacific and Africa Collections

96 Euston Road, London NW1 2DB
☏ 020 7412 7873   🖷 020 7412 7641
apac-enquiries@bl.uk
www.bl.uk

OPEN 10.00 am to 5.00 pm Monday; 9.30 am to 5.00 pm Tuesday to Saturday; closed for public holidays
ACCESS By British Library reader's pass
An extensive collection of printed volumes and manuscripts in the languages of Africa, the Near and Middle East and all of Asia, plus records of the East India Company and British government in India until 1947. Also prints, drawings and paintings by British artists of India.

For information on British Library collections and services, visit the website.

British Library Business & IP Centre

96 Euston Road, London NW1 2DB
☏ 020 7412 7454 (free enquiry service)
🖷 020 7412 7453 (free enquiry service)
bipc@bl.uk
www.bl.uk/bipc
OPEN 10.00 am to 8.00 pm Monday; 9.30 am to 8.00 pm Tuesday to Thursday; 9.30 am to 5.00 pm Friday and Saturday; closed for public holidays
ACCESS By British Library reader's pass

The Business & IP Centre holds the most comprehensive collection of business information literature and patent specifications in the UK. Includes market research reports and journals, directories, company annual reports, trade and business journals, up-to-date literature on patents, trade marks, designs and copyright.

British Library Early Printed Collections/ Rare Books and Music Reading Room

96 Euston Road, London NW1 2DB
☏ 020 7412 7564   🖷 020 7412 7691
rare-books@bl.uk
www.bl.uk/collections/early.html
General enquiries about reader services & advance reservations: ☏ 020 7412 7676 🖷 01937 546321 reader-services-enquiries@bl.uk
OPEN 10.00 am to 8.00 pm Monday; 9.30 am to 8.00 pm Tuesday to Thursday; 9.30 am to 5.00 pm Friday and Saturday; closed for public holidays
ACCESS By British Library reader's pass

Early Printed Collections selects, acquires, researches and provides access to material printed in Britain and Western Europe from the 15th to early 20th centuries. The collections are available in the Rare Books and Music Reading Room at St Pancras which also functions as the focus for the British Library's extensive collection of humanities microforms.

Further information about Early Printed Collections can be found at the British Library website.

British Library Humanities Reading Room

96 Euston Road, London NW1 2DB
☏ 020 7412 7676   🖷 020 7412 7794
humanities-enquiries@bl.uk
www.bl.uk/resources/humanities
OPEN 10.00 am to 8.00 pm Monday; 9.30 am to 8.00 pm Tuesday, Wednesday, Thursday; 9.30 am to 5.00 pm Friday and Saturday; closed for public holidays
ACCESS By British Library reader's pass

This reading room is the focus for the Library's modern collections service in the humanities. It is on two levels, Humanities 1 and Humanities 2 and provides access to the Library's comprehensive collections of books and periodicals in all subjects in the humanities and social sciences and in all languages apart from Oriental. These collections are not available for browsing at the shelf. Material is held in closed access storage and needs to be identified and ordered from store using an online catalogue. A selective open access collection on most humanities subjects can be found in Humanities 1 whilst in Humanities 2 there are open access reference works relating to periodicals and theses, to recorded sound and to librarianship and information science.

To access British Library catalogues, go to the website.

British Library Manuscript Collections

96 Euston Road, London NW1 2DB
Ⓣ 020 7412 7513   Ⓕ 020 7412 7745
mss@bl.uk
www.bl.uk
OPEN 10.00 am to 5.00 pm Monday; 9.30 am to 5.00 pm Tuesday to Saturday; closed for public holidays
ACCESS Reading facilities only, by British Library reader's pass; a written letter of recommendation and advance notice is required for certain categories of material

Two useful publications, *Index of Manuscripts in the British Library,* Cambridge 1984–6, 10 vols, and *The British Library: Guide to the Catalogues and Indexes of the Department of Manuscripts* by M.A.E. Nickson, help to guide the researcher through this vast collection of manuscripts dating from Ancient Greece to the present day. Approximately 300,000 mss, charters, papyri and seals are housed here.

For information on British Library collections and services and to access British Library catalogues, including the Manuscripts online catalogue, visit the website.

British Library Map Collections

96 Euston Road, London NW1 2DB
Ⓣ 020 7412 7702   Ⓕ 020 7412 7780
maps@bl.uk
www.bl.uk/collections/maps
OPEN 10.00 am to 5.00 pm Monday; 9.30 am to 5.00 pm Tuesday to Saturday; closed for public holidays
ACCESS By British Library reader's pass

A collection of about 4.5 million maps, charts and globes, manuscript, printed and, increasingly, digital, with particular reference to the history of British cartography. Maps for all parts of the world in a wide range of scales, formats and dates, including the most comprehensive collection of Ordnance Survey maps and plans. SPECIAL COLLECTIONS King George III Topographical

Collection and Maritime Collection, the Crace Collection of maps and plans of London and the cartographic archive of the Ministry of Defence (i.e. GSGS).

For information on British Library collections and services, visit the website. To access main British Library catalogues, go to the website at http://catalogue.bl.uk

British Library Music Collections

96 Euston Road, London NW1 2DB
Ⓣ 020 7412 7772   Ⓕ 020 7412 7751
music-collections@bl.uk
www.bl.uk/collections/music/music.html
OPEN 10.00 am to 8.00 pm Monday; 9.30 am to 8.00 pm Tuesday to Thursday; 9.30 am to 5.00 pm Friday and Saturday; closed for public holidays
ACCESS By British Library reader's pass

SPECIAL COLLECTIONS The Royal Music Library (containing almost all Handel's surviving autograph scores), The Zweig Collection of Music & Literary Mss, The Royal Philharmonic Society Archive and the Paul Hirsch Music Library. Also a large collection (about one and a half million items) of printed music (UK via legal deposit) and about 100,000 items of manuscript music, both British and foreign.

The British Library website contains details of collections and services, and provides access to the catalogues.

British Library Newspapers

Colindale Avenue, London NW9 5HE
Ⓣ 020 7412 7353   Ⓕ 020 7412 7379
newspaper@bl.uk
www.bl.uk/collections/newspapers.html
OPEN 10.00 am to 5.00 pm Monday to Saturday (last newspaper issued 4.15 pm); closed for public holidays
ACCESS By British Library reader's pass or Newspaper Library pass (available from and valid only for Colindale)

Major collections of English provincial, Scottish, Welsh, Irish, Commonwealth and selected overseas foreign newspapers from *c.*1700 are housed here. Some earlier holdings are also available. London newspapers from 1801 and many weekly and fortnightly periodicals are also in stock. (London newspapers pre-dating 1801 are housed at the new library building in St Pancras – 96 Euston Road, NW1 – though many are available at Colindale Avenue on microfilm.) Readers are advised to check availability of material in advance.

For information on British Library Newspapers collections and services, visit the website.

British Library Science, Technology and Business Collections

96 Euston Road, London NW1 2DB
Ⓣ 020 7412 7676 (General Enquiries)
Ⓕ 020 7412 7495
scitech@bl.uk
www.bl.uk
Business enquiries: Ⓣ 020 7412 7454 (Free enquiry service)
OPEN 10.00 am to 8.00 pm Monday; 9.30 am to 8.00 pm Tuesday to Thursday; 9.30 am to 5.00 pm Friday and Saturday; closed for public holidays
ACCESS By British Library reader's pass

Engineering, business information on companies, markets and products, physical science and technologies. See also **British Library Business & IP Centre**.
To access British Library catalogues go to the website at http://catalogue.bl.uk

British Library Social Sciences & Official Publications

96 Euston Road, London NW1 2DB
Ⓣ 020 7412 7676   Ⓕ 020 7412 7794
social-sciences@bl.uk
www.bl.uk/collections/social/social.html
OPEN Reading Room: 10.00 am to 8.00 pm Monday; 9.30 am to 8.00 pm Tuesday to Thursday; 9.30 am to 5.00 pm Friday and Saturday; closed for public holidays
ACCESS By British Library reader's pass

Provides an information service on the social sciences, law, public administration, and current and international affairs, and access to current and historical official publications from all countries and intergovernmental bodies. Material available in the reading room includes House of Commons sessional papers, UK legislation, UK electoral registers, up-to-date reference books on official publications and on the social sciences, a major collection of statistics and a browsing collection of recent social science books and periodicals; other collections are kept in closed stores and items have to be ordered for delivery to the reading room.
To access British Library catalogues, go to the website at http://catalogue.bl.uk

British Library Sound Archive

96 Euston Road, London NW1 2DB
Ⓣ 020 7412 7676   Ⓕ 020 7412 7441
sound-archive@bl.uk
www.bl.uk/soundarchive
OPEN 10.00 am to 8.00 pm Monday; 9.30 am to 8.00 pm Tuesday to Thursday; 9.30 am to 5.00 pm Friday and Saturday; closed for public holidays
Listening service (by appointment)

Northern Listening Service
British Library Document Supply Centre, Boston Spa: 9.15 am to 4.30 pm Monday to Friday
OPEN ACCESS

An archive of over 1,000,000 discs and more than 200,000 hours of tape recordings, including all types of music, oral history, drama, literature, poetry, wildlife, selected BBC broadcasts and BBC Sound Archive material. Produces a twice-yearly newsletter, *Playback*.
For information on British Library Sound Archive collections and services, visit the website.

British Psychological Society Library

c/o Psychology Library, Senate House Library, University of London, Senate House, Malet Street, London WC1E 7HU
Ⓣ 020 7862 8451/8461   Ⓕ 020 7862 8480
enquiries@shl.lon.ac.uk
www.shl.lon.ac.uk
OPEN Term-time: 9.00 am to 9.00 pm Monday to Thursday; 9.00 am to 6.30 pm Friday; 9.30 am to 5.30 pm Saturday (Holidays: 9.00 am to 6.00 pm Monday to Friday; 9.30 am to 5.30 pm Saturday)
ACCESS Members only; non-members £5 day ticket

Reference library, containing the British Psychological Society collection of periodicals – over 140 current titles housed alongside the University of London's collection of books and journals. Largely for academic research. General queries referred to **Swiss Cottage Library** in London which has a good psychology collection.

Benjamin Britten Collection
▷ **Suffolk County Council – Suffolk Libraries**

Bromley Central Library

High Street, Bromley BR1 1EX
Ⓣ 020 8460 9955   Ⓕ 020 8466 7860
central.library@bromley.gov.uk
www.bromley.gov.uk
OPEN 9.30 am to 6.00 pm Monday, Wednesday, Friday; 9.30 am to 8.00 pm Tuesday and Thursday; 9.30 am to 5.00 pm Saturday
OPEN ACCESS

A large selection of fiction and non-fiction books for loan, both adult and children's. Also DVDs, videos, CDs, cassettes, language courses, open learning packs, CD-ROM and Playstation games for hire. Other facilities include CD-ROMs, computer hire, People's Network Internet, local studies library, 'Upfront' teenage section, large reference library with photocopying, fax, microfiche and film facilities, 'Bromley Knowledge' – online community information, reading groups and Local Links (access to council services). Library and Archives catalogues online, 'Bromley Training Truck' – free computer and basic skills training. SPECIALIST COLLECTIONS include: H.G. Wells, Walter de la Mare, Crystal Palace, The Harlow Bequest, and the history and geography of Asia, America, Australasia and the Polar regions.

Bromley House Library

Angel Row, Nottingham NG1 6HL
℡ 0115 947 3134
enquiries@bromleyhouse.org
www.bromleyhouse.org
Librarian *Carol Allison*
OPEN 9.30 am to 5.00 pm Monday to Friday; also first Saturday of each month from 10.00 am to 12.30 pm
ACCESS For members only

Founded 1816 as the Nottingham Subscription Library. Collection of 35,000 books including local history, topography, biography, travel and fiction.

Robert Burns Collection
▷ The Mitchell Library

CAA Library and Information Centre

Aviation House, Gatwick Airport South, Gatwick RH6 0YR
℡ 01293 573725   ℱ 01293 573181
infoservices@caa.co.uk
www.caa.co.uk
OPEN 9.00 am to 5.00 pm Monday to Friday
OPEN ACCESS

A collection of books, reports, directories, statistics and periodicals on most aspects of civil aviation and related subjects.

Cambridge Central Library (Reference Library & Information Service)

7 Lion Yard, Cambridge CB2 3QD
℡ 0845 045 5225   ℱ 01223 712011
your.library@cambridgeshire.gov.uk
www.cambridgeshire.gov.uk/leisure/libraries

The Central Library is closed until late autumn 2008 for major redevelopment. During this period a full range of services is available through the network of 31 libraries in the county (details on the website) and by telephone, post, e-mail and online referral.

Large stock of books, periodicals, newspapers, maps, plus comprehensive collection of directories and annuals covering UK, Europe and the world. Microfilm and fiche reading and printing services. Online access to news and business databases. News databases on CD-ROM; Internet access. Monochrome and colour photocopiers.

Camomile Street Library

12–20 Camomile Street, London EC3A 7EX
℡ 020 7332 1855

camomile@cityoflondon.gov.uk
www.cityoflondon.gov.uk/camomilestlibrary
OPEN 9.30 am to 5.30 pm Monday, Tuesday, Thursday, Friday; 9.30 am to 6.30 pm Wednesday
OPEN ACCESS

City of London Corporation lending library. Wide range of fiction and non-fiction books and language courses on cassette and CD, foreign fiction, paperbacks, maps and guides for travel at home and abroad, children's books, a selection of large print, and collections of DVDs, videos and music CDs, books in Bengali. Spoken word recordings on cassette and CD. Free Internet access.

Cardiff Central Library

John Street, Cardiff CF10 5BA
℡ 029 2038 2116   ℱ 029 2087 1588
rboddy@cardiff.gov.uk
www.cardiff.gov.uk/libraries
OPEN 9.00 am to 6.00 pm Monday, Tuesday, Wednesday, Friday; 9.00 am to 7.00 pm Thursday; 9.00 am to 5.30 pm Saturday

General lending library with the following departments: leisure, music, children's, local studies, information, science and humanities.

Carmarthen Public Library

St Peter's Street, Carmarthen SA31 1LN
℡ 01267 224824   ℱ 01267 221839
wtphillips@carmarthenshire.gov.uk
www.carmarthenshire.gov.uk
OPEN 9.30 am to 7.00 pm Monday, Tuesday, Wednesday, Friday; 9.30 am to 5.00 pm Thursday and Saturday
OPEN ACCESS

Comprehensive range of fiction, non-fiction, children's books and reference works in English and in Welsh. Large local history library. Free Internet access and computer facilities. Large Print books, audiobooks, CDs, CD-ROMs, DVDs and videos available for loan.

Catholic National Library

St Michaels Abbey, Farnborough Road, Farnborough GU14 7NQ
℡ 01252 543818
library@catholic-library.org.uk
www.catholic-library.org.uk
OPEN ACCESS For reference (non-members must sign in; loans restricted to members)

Contains books, many not readily available elsewhere, on theology, religions worldwide, scripture and the history of churches of all denominations.

City Business Library

1 Brewers' Hall Garden, off Aldermanbury Square,
London EC2V 5BX
☎ 020 7332 1812/(3803 textphone)
🖷 020 7600 1185
cbl@cityoflondon.gov.uk
www.cityoflondon.gov.uk/citybusinesslibrary
OPEN 9.30 am to 5.00 pm Monday to Friday
(except public holidays)
OPEN ACCESS

Local authority free public reference library run
by the Corporation of London. Books, directories,
periodicals, newspapers and electronic databases
of current business interest. Provided for anyone
with a business information enquiry and is one
of the leading public resource centres in Britain
in its field. Large directory collection for both the
UK and overseas, plus companies information,
market research reports, management, banking,
insurance, investment and statistics. Free
public Internet access. No academic journals or
textbooks.

City of London Libraries

▷ Barbican Library; Camomile Street Library; City
Business Library; Guildhall Library

Commonwealth Secretariat Library

Marlborough House, Pall Mall, London SW1Y 5HX
☎ 020 7747 6164   🖷 020 7747 6168
library@commonwealth.int
www.thecommonwealth.org
Library Services Manager *David Blake*
OPEN 10.00 am to 4.45 pm Monday to Friday
ACCESS For reference only, by appointment

Extensive reference source concerned with
economy, development, trade, production and
industry of Commonwealth countries; also
human resources including women, youth, health,
management and education. Includes the archives
of the Secretariat.

Coventry Central Library

Smithford Way, Coventry CV1 1FY
☎ 024 7683 2314/2395 (Minicom)
🖷 024 7683 2440
central.library@coventry.gov.uk
www.coventry.gov.uk
OPEN 9.00 am to 8.00 pm Monday to Friday; 9.00
am to 4.30 pm Saturday; 12.00 am to 4.00 pm
Sunday
OPEN ACCESS

Located in the middle of the city's main shopping
centre. Approximately 120,000 items (books,
cassettes, CDs and DVDs) for loan; plus reference
collection of business information and local
history. SPECIAL COLLECTIONS Cycling and
motor industries; George Eliot; Angela Brazil;
Tom Mann Collection (trade union and labour

studies); local newspapers on microfilm from 1740
onwards. Over 300 periodicals taken. 'Peoplelink'
community information database available.

Crace Collection

▷ British Library Map Collections

Department for Environment, Food and Rural Affairs

Nobel House, 17 Smith Square, London SW1P 3JR
☎ 020 7238 3000   🖷 020 7238 6591
DEFRA Helpline 08459 335577 (local call
rate): general contact point which can provide
information on the work of DEFRA, either
directly or by referring callers to appropriate
contacts. Available 9.00 am to 5.00 pm Monday
to Friday (excluding Bank Holidays)
OPEN 9.30 am to 5.00 pm Monday to Friday
ACCESS For reference (but at least 24 hours notice
must be given for intended visits)

Derby Central Library

Wardwick, Derby DE1 1HS
☎ 01332 255398   🖷 01332 369570
www.derby.gov.uk/libraries
OPEN 9.30 am to 7.00 pm Monday, Tuesday,
Thursday, Friday; 9.30 am to 1.00 pm Wednesday;
9.30 am to 4.00 pm Saturday
Local Studies Library
25B Irongate, Derby DE1 3GL
☎ 01332 255393 🖷 01332 255381
OPEN 9.30 am to 7.00 pm Monday and Tuesday;
9.30 am to 5.00 pm Wednesday, Thursday, Friday;
9.30 am to 4.00 pm Saturday
OPEN ACCESS

General library for lending, information and
Children's Services. The Central Library also
houses specialist private libraries: Derbyshire
Archaeological Society; Derby Philatelic Society.
The Local Studies Library houses the largest
multimedia collection of resources in existence
relating to Derby and Derbyshire. The collection
includes mss deeds, family papers, business
records including the Derby Canal Company,
Derby Board of Guardians and the Derby China
Factory. Both libraries offer free Internet access.

Devon & Exeter Institution Library

7 Cathedral Close, Exeter EX1 1EZ
☎ 01392 251017
J.P.Gardner@exeter.ac.uk
www.exeter.ac.uk/library
OPEN 9.30 am to 5.00 pm Monday to Friday
ACCESS Members only (temporary membership
available)

Founded 1813. Under the administration of
Exeter University Library. Contains over 36,000
volumes, including long runs of 19th-century
journals, theology, history, topography, early
science, biography and literature. A large

and growing collection of books, journals, newspapers, prints and maps relating to the South West.

Doncaster Library and Information Services

Central Library, Waterdale, Doncaster DN1 3JE
℡ 01302 734305   🖷 01302 369749
Reference.Library@doncaster.gov.uk
library.doncaster.gov.uk
OPEN 9.00 am to 6.00 pm Monday; 8.30 am to 6.00 pm Tuesday and Friday; 8.30 am to 8.00 pm Wednesday and Thursday; 9.00 am to 5.00 pm Saturday
OPEN ACCESS

Books, newspapers, periodicals, spoken word cassettes/CDs, DVDs. Reading aids unit for people with visual impairment; activities for children during school holidays and regular storytimes. Events programme, including visits by authors and talks, readers' groups, etc. Also reference library and Local Studies library.

Dorchester Library (part of Dorset County Library)

Colliton Park, Dorchester DT1 1XJ
℡ 01305 224440 (lending)/224448 (reference)
🖷 01305 266120
dorchesterreferencelibrary@dorsetcc.gov.uk
www.dorsetforyou.com/libraries
OPEN 10.00 am to 7.00 pm Monday; 9.30 am to 7.00 pm Tuesday, Wednesday, Friday; 9.30 am to 5.00 pm Thursday; 9.00 am to 4.00 pm Saturday
OPEN ACCESS

General lending and reference library, including special collections on Thomas Hardy, the Powys Family and William Barnes. Periodicals, children's library, CD-ROMs, free Internet access. Video lending service.

Dundee Central Library

The Wellgate, Dundee DD1 1DB
℡ 01382 431500   🖷 01382 431558
central.library@dundeecity.gov.uk
www.dundeecity.gov.uk

OPEN Departments of Leisure Reading, Art & Music, Children's, Reference & Information, and Local History Centre: 9.30 am to 6.00 pm Monday, Tuesday, Friday; 10.00 am to 6.00 pm Wednesday; 9.30 am to 8.00 pm Thursday; 9.30 am to 5.00 pm Saturday. Commerce, Science & Business: 9.30 am to 9.00 pm Monday, Tuesday, Thursday, Friday; 10.00 am to 9.00 pm Wednesday; 9.30 am to 5.00 pm Saturday
ACCESS Reference services available to all; lending services to those who live, work, study or were educated within Dundee City

Adult lending, reference and children's services. Art, music, audio, video and DVD lending services. Internet access. Schools service

(Agency). Housebound and mobile services.
SPECIAL COLLECTIONS The Wighton Collection of National Music; The Wighton Heritage Centre; The Wilson Photographic Collection; The Lamb Collection.

Durning-Lawrence Library
▷ Senate House Library

English Nature
▷ Natural England

Equality and Human Rights Commission

Arndale House, Arndale Centre, Manchester M4 3EQ
℡ 0845 604 6610   🖷 0161 829 8110
englandhelpline@equalityhumanrights.com
www.equalityhumanrights.com

The EHRC is open to the public via the Helpline (0845 604 6610) 9.00 am to 5.00 pm Monday, Tuesday, Thursday, Friday; 9.00 am to 7.45 pm Wednesday.

Essex County Council Libraries

Goldlay Gardens, Chelmsford CM2 0EW
℡ 01245 284981   🖷 01245 492780
essexlib@essexcc.gov.uk
www.essexcc.gov.uk

Essex County Council Libraries has 73 static libraries throughout Essex as well as 12 mobile libraries. Services to the public include books, newspapers, periodicals, CDs, cassettes, videos, CD-ROMs and Internet access. Specialist subjects and collections are listed below at the relevant library.

Chelmsford Library
PO Box 882, Market Road, Chelmsford CM1 1LH
℡ 01245 492758 🖷 01245 492536
chelmford.library@essexcc.gov.uk
OPEN 8.30 am to 7.00 pm Monday to Friday; 8.30 am to 5.30 pm Saturday; 12.30 pm to 4.30 pm Sunday

Colchester Library
Trinity Square, Colchester CO1 1JB
℡ 01206 245900 🖷 01206 245901
colchester.library@essexcc.gov.uk
OPEN 8.30 am to 7.30 pm Monday to Friday; 8.30 am to 5.00 pm Saturday; 12.30 pm to 4.30 pm Sunday
Local studies; Castle collection (18th-century subscription library); Cunnington collection; Margaret Lazell collection; Taylor collection.

Harlow Library
The High, Harlow CM20 1HA
℡ 01279 413772 🖷 01279 424612
harlow.library@essexcc.gov.uk
OPEN 9.00 am to 7.00 pm Monday to Friday; 9.00 am to 5.00 pm Saturday; 1.00 pm to 4.00 pm Sunday

Sir John Newson Memorial collection; Maurice Hughes Memorial collection.

Loughton Library
Traps Hill, Loughton IG10 1HD
☎ 020 8502 0181 🖷 020 8508 5041
loughton.library@essexcc.gov.uk
OPEN 9.00 am to 7.00 pm Monday to Friday; 9.30 am to 5.30 pm Saturday; 11.00 am to 3.00 pm Sunday
National Jazz Foundation Archive.

Saffron Walden Library
2 King Street, Saffron Walden CB10 1ES
☎ 01799 523178 🖷 01799 513642
OPEN 9.00 am to 7.00 pm Monday to Friday; 9.00 am to 5.00 pm Saturday; 1.00 pm to 4.00 pm Sunday
Victorian studies collection.

Witham Library
18 Newland Street, Witham CM8 2AQ
☎ 01376 519625 🖷 01376 501913
OPEN 9.00 am to 7.00 pm Monday to Friday; 9.00 am to 5.00 pm Saturday; 1.00 pm to 4.00 pm Sunday
Dorothy L. Sayers and Maskell collections.

Family Records Centre

Part of the **National Archives** (see entry)

The Fawcett Library
▷ The Women's Library

Edward Fitzgerald Collection
▷ Suffolk County Council – Suffolk Libraries

Forestry Commission Library

Forest Research Station, Alice Holt Lodge, Wrecclesham, Farnham GU10 4LH
☎ 01420 22255   🖷 01420 23653
library@forestry.gsi.gov.uk
www.forestry.gov.uk
www.forestresearch.gov.uk
OPEN 9.00 am to 5.00 pm Monday to Thursday; 9.00 am to 4.30 pm Friday
ACCESS By appointment for personal visits

Approximately 20,000 books on forestry and arboriculture, plus 500 current journals. CD-ROMS include TREECD (1939 onwards). Offers a Research Advisory Service for advice and enquiries on forestry (☎ 01420 23000) with a charge for consultations and diagnosis of tree problems exceeding ten minutes.

French Institute Library

Institut français du Royaume Uni, 17 Queensberry Place, London SW7 2DT
☎ 020 7073 1350   🖷 020 7073 1363
library@ambafrance.org.uk
www.institut-francais.org.uk
OPEN 12 noon to 7.00 pm Tuesday to Friday; 12 noon to 6.00 pm Saturday; Children's Library: 2.00 pm to 6.00 pm Tuesday to Friday; 12 noon to 6.00 pm Saturday; closed in August and for one week at Christmas
OPEN ACCESS For reference and consultation (loans restricted to members; leaflet available on demand)

A collection of over 55,000 volumes mainly centred on French cultural interests with special emphasis on French as a Foreign Language, literature and history. Books and periodicals, mainly in French, a few in English and some bilingual. Collection of DVDs, videos, CDs (French music), CD-ROMs, audiobooks; special collections: 'France Libre'. Denis Saurat MSSS, Paris. Recordings of lectures, press-cuttings on French current affairs; Campus France; Internet access to members; Wifi. Group visits on request. Reading group; documentary screenings; talks. Children's library (8,000 documents). Bistro.

The Froebel Archive for Childhood Studies
▷ entry under Useful Websites

John Frost Newspapers

22b Rosemary Avenue, Enfield EN2 0SS
☎ 020 8366 1392/0946   🖷 020 8366 1379
andrew@johnfrostnewspapers.com
www.johnfrostnewspapers.co.uk
Contacts *Andrew Frost, John Frost*

A collection of 80,000 original newspapers (1630 to the present day) and 200,000 press cuttings available, on loan, for research and rostrum/stills work (TV documentaries, book and magazine publishers and audiovisual presentations). Historic events, politics, sports, royalty, crime, wars, personalities, etc., plus many in-depth files.

German Historical Institute Library

17 Bloomsbury Square, London WC1A 2NJ
☎ 020 7309 2050   🖷 020 7309 2055
library@ghil.ac.uk
www.ghil.ac.uk
OPEN 10.00 am to 5.00 pm Monday, Tuesday, Wednesday, Friday; 10.00 am to 8.00 pm Thursday
OPEN ACCESS Visitors must bring proof of address and recent photograph to be issued with a reader's ticket

70,000 volumes; around 200 journals; databases. Devoted primarily to German history from the Middle Ages to the present day, with special emphasis on the 19th and 20th centuries, in particular Germany between 1933 and 1945, the development of the two German states after 1945/49 and German unification after 1989.

Gloucestershire County Council Libraries & Information

Quayside House, Shire Hall, Gloucester GL1 2HY
☎ 08452 305420   🖷 01452 425042

libraryhelp@gloucestershire.gov.uk
www.libraries.gloucestershire.gov.uk/
Head of Libraries & Information *David Paynter*
OPEN ACCESS

The service includes 39 local libraries and five mobile libraries. The website includes library opening hours, mobile library route schedules; the library catalogue; book renewal/reservations facility; and information about news and events.

Goethe-Institut Library

50 Princes Gate, Exhibition Road, London
SW7 2PH
☎ 020 7596 4040/4044   ᖴ 020 7594 0230
library@london.goethe.org
www.goethe.de/london
Librarian *Elisabeth Pyroth*
OPEN 1.00 pm to 6.30 pm Monday to Thursday; 1.00 pm to 5.00 pm Saturday

Library specializing in German literature and books/audiovisual material on German culture and history: 20,000 books (4,000 of them in English), 120 periodicals, 13 newspapers, 3,700 audiovisual media (including 1,200 videos/DVDs), selected press clippings on German affairs from the German and UK press, information service, photocopier, video facility. Also German language teaching material for teachers and students of German. See website for membership details.

Goldsmiths' Library
▷ Senate House Library

Greater London Council History Library
▷ London Metropolitan Archives Library Services

Guildford Institute Library

Ward Street, Guildford GU1 4LH
☎ 01483 562142   ᖴ 01483 451034
guildford-institute@surrey.ac.uk
OPEN 10.00 am to 3.00 pm Tuesday to Friday; 10.30 am to 1.30 pm each first Saturday of the month
OPEN ACCESS to members only but open to enquirers for research purposes

Founded 1834. Some 14,000 volumes of which 7,500 were printed before the First World War. The remaining stock consists of recently published works of fiction and non-fiction. Newspapers and periodicals also available. SPECIAL COLLECTIONS include an almost complete run of the *Illustrated London News* from 1843–1906, a collection of Victorian and early 20th century local history ephemera albums, and about 400 photos and other pictures relating to the Institute's history and the town of Guildford.

Guildhall Library

Aldermanbury, London EC2V 7HH
☎ See below   ᖴ 020 7600 3384

www.cityoflondon.gov.uk/guildhalllibrary
ACCESS For reference (but much of the material is kept in storage areas and is supplied to readers on request; proof of identity is required for consultation of certain categories of stock). Free, *limited* enquiry service available. Also a fee-based service for in-depth research: ☎ 020 7332 1854 ᖴ 020 7600 3384 search.guildhall@cityoflondon. gov.uk www.cityoflondon.gov.uk/search_guildhall

Part of the City of London libraries. Seeks to provide a basic general reference service but its major strength, acknowledged worldwide, is in its historical collections. The library is divided into three sections, each with its own catalogues and enquiry desks. These are: Printed Books; Manuscripts; the Print & Maps Room.

PRINTED BOOKS
☎ 020 7332 1868/1870
printed books.guildhall@cityoflondon.gov.uk
OPEN 9.30 am to 5.00 pm Monday to Saturday; NB closes on Saturdays preceding Bank Holidays; check for details

Strong on all aspects of London history, with wide holdings of English history, topography and genealogy, including local directories, poll books and parish register transcripts. Also good collections of English statutes, law reports, parliamentary debates and journals, and House of Commons papers. Home of several important collections deposited by London institutions: the Marine collection of the Corporation of Lloyd's, the Stock Exchange's historical files of reports and prospectuses, the Clockmakers' Company library and museum, the Gardeners' Company, Fletchers' Company, the Institute of Masters of Wine, International Wine and Food Society and Gresham College.

MANUSCRIPTS
☎ 020 7332 1862/3
manuscripts.guildhall@cityoflondon.gov.uk
www.history.ac.uk/gh/
OPEN 9.30 am to 5.00 pm Monday to Saturday (no requests for records after 4.30 pm; no manuscripts can be produced between 12 noon and 2.00 pm on Saturdays); NB closes on Saturdays preceding Bank Holidays; check for details

The official repository for historical records relating to the City of London (except those of the City of London Corporation itself, which are housed at the **London Metropolitan Archives**). Records date from the 11th century to the present day. They include archives of most of the City's parishes, wards and livery companies, and of many individuals, families, estates, schools, societies and other institutions, notably the Diocese of London and St Paul's Cathedral, as well as the largest collection of business archives in any public repository in the UK. Although mainly of City interest, holdings include material for the London area as a whole and beyond. Many of the records are of national and international significance.

PRINTS & MAPS ROOM

☎ 020 7332 1839

prints&maps@cityoflondon.gov.uk

OPEN 9.30 am to 5.00 pm Monday to Friday

An unrivalled collection of prints and drawings relating to London and the adjacent counties. The emphasis is on topography, but there are strong collections of portraits and satirical prints. The map collection includes maps of the capital from the mid-16th century to the present day and various classes of Ordnance Survey maps. Other material includes photographs, theatre bills and programmes, trade cards, book plates and playing cards as well as a sizeable collection of Old Master prints. Over 30,000 items have been digitally imaged on the 'Collage' website, including topographical prints, some maps, a small number of photographs and all the Guildhall Art Gallery collection. 'Collage' website: collage.cityoflondon.gov.uk

Guille-Alles Library

Market Street, St Peter Port, Guernsey GY1 1HB

☎ 01481 720392   ℻ 01481 712425

ga@library.gg

www.library.gg

OPEN 9.00 am to 5.00 pm Monday, Thursday, Friday, Saturday; 10.00 am to 5.00 pm Tuesday; 9.00 am to 8.00 pm Wednesday

OPEN ACCESS For residents; payment of returnable deposit by visitors. Music CD collection: £10 for two-year subscription.

Lending, reference and information services. Public Internet service.

Herefordshire Libraries

Shirehall, Hereford HR1 2HX

☎ 01432 261644   ℻ 01432 260744

libraries@herefordshire.gov.uk

www.herefordshire.gov.uk/leisure/libraries

OPEN Opening hours vary in the libraries across the county

ACCESS Information and reference services open to anyone; loans to members only (membership criteria: resident, being educated, working in Herefordshire. Proof of identity and address required.) Temporary membership to visitors is also available.

Information service, reference and lending libraries. Non-fiction and fiction for all age groups, including normal and large print, spoken word cassettes, and CDs, music CDs, DVDs, playstation2. Hereford Library houses the largest

reference section and the county local history collection, although some reference material and local history is available at all libraries. Internet access at all libraries. *Special collections* Cidermaking; Beekeeping; Alfred Watkins; John Masefield; Pilley.

University of Hertfordshire Library

College Lane, Hatfield AL10 9AB

☎ 01707 284678   ℻ 01707 284666

www.herts.ac.uk/lis

OPEN See website for term-time and vacation opening hours

ACCESS For reference use of printed collections. Appropriate ID required for Visitor's pass.

For further information please refer to the website.

Highgate Literary and Scientific Institution Library

11 South Grove, London N6 6BS

☎ 020 8340 3343   ℻ 020 8340 5632

librarian@hlsi.net

www.hlsi.net

OPEN 10.00 am to 5.00 pm Tuesday to Friday; 10.00 am to 4.00 pm Saturday (closed Sunday and Monday)

ANNUAL MEMBERSHIP £55 (individual); £90 (household)

25,000 volumes of general fiction and non-fiction, with a children's section and extensive local archives. SPECIAL COLLECTIONS on local history, London, and local poets Samuel Taylor Coleridge and John Betjeman.

Highland Libraries, The Highland Council, Education, Culture and Sport Service

Library Support Unit, 31a Harbour Road, Inverness IV1 1UA

☎ 01463 235713   ℻ 01463 236986

libraries@highland.gov.uk

www.highland.gov.uk

OPEN Library opening hours vary to suit local needs. Contact administration and support services for details (8.00 am to 5.00 pm Monday to Friday)

OPEN ACCESS

Comprehensive range of lending and reference stock: books, pamphlets, periodicals, newspapers, compact discs, audio and video cassettes, maps, census records, genealogical records, photographs, educational materials, etc. Free access to the Internet in all libraries. Highland Libraries provides the public library service throughout the Highlands with a network of 43 static and 12 mobile libraries.

Paul Hirsch Music Library
▷ British Library Music Collections

Holborn Library

32–38 Theobalds Road, London WC1X 8PA

☎ 020 7974 6345

holbornlibrary@camden.gov.uk

www.camden.gov.uk/libraries

OPEN 10.00 am to 7.00 pm Monday and
Thursday; 10.00 am to 6.00 pm Tuesday,
Wednesday and Friday; 10.00 am to 5.00 pm
Saturday

OPEN ACCESS

London Borough of Camden public library.
Includes the London Borough of Camden Local
Studies and Archive Centre.

Holcenberg Jewish Collection
▷ City of Plymouth Library and Information Services

Sherlock Holmes Collection (Westminster)

Marylebone Library, Marylebone Road, London
NW1 5PS

☎ 020 7641 1206   🖷 020 7641 1019

ccooke@westminster.gov.uk

www.westminster.gov.uk/libraries/special/
sherlock.cfm

OPEN 9.30 am to 8.00 pm Monday, Tuesday,
Thursday, Friday; 10.00 am to 8.00 pm
Wednesday; closed Saturday and Sunday (unless
by prior arrangement)

ACCESS By appointment only

Located in Westminster's Marylebone Library.
An extensive collection of material from all over
the world, covering Sherlock Holmes and Sir
Arthur Conan Doyle. Books, pamphlets, journals,
newspaper cuttings and photos, much of which
is otherwise unavailable in this country. Some
background material.

IHR Library
▷ The Institute of Historical Research

Imperial College Central Library
▷ Science Museum Library

Imperial War Museum

Department of Printed Books, Lambeth Road,
London SE1 6HZ

☎ 020 7416 5342   🖷 020 7416 5246

collections@iwm.org.uk

www.iwm.org.uk

OPEN 10.00 am to 5.00 pm Monday to Saturday
(restricted service Saturday; closed on Bank
Holiday Saturdays and two weeks during the year
for annual stock check)

ACCESS For reference (but at least 24 hours'
notice must be given for intended visits)

A large collection of material on British and
Commonwealth 20th-century life with detailed
coverage of the two World Wars and other
conflicts. This collection includes substantial
holdings of European language material. Books,

pamphlets and periodicals, including many
produced for short periods in unlikely wartime
settings; also maps, biographies and privately
printed memoirs, and foreign language material.
Additional research material available in the
departments of Art, Documents, Exhibits and
Firearms and in the Film, Photographs and Sound
Archives. See www.iwmcollections.org.uk for
their online catalogues.

The Institute of Historical Research

Senate House, Malet Street, London WC1E 7HU

☎ 020 7862 8760

IHR.Library@sas.ac.uk

www.history.ac.uk/library

OPEN 9.00 am to 8.45 pm Monday to Friday; 9.30
am to 5.15 pm Saturday

ACCESS For reference only. Non-members are
advised to telephone to check availability of items
before visiting.

Reference collection of printed primary sources,
bibliographies, guides to archives, periodicals and
works covering the history of Western Europe
from the fall of the Roman Empire. Holds a
substantial microform collection of mainly British
material from repositories outside London.

Instituto Cervantes

102 Eaton Square, London SW1W 9AN

☎ 020 7201 0757   🖷 020 7235 0329

biblon@cervantes.es

londres.cervantes.es

OPEN 12 noon to 7.30 pm Monday to Thursday;
12 noon to 6.30 pm Friday; 9.30 am to 2.00 pm
Saturday

OPEN ACCESS For reference and lending

Spanish language and literature, history, art,
philosophy. The library houses a collection of
books, periodicals, videos, DVDs, slides, tapes,
CDs, cassettes, CD-ROMs specializing entirely in
Spain and Latin America.

Italian Institute Library

39 Belgrave Square, London SW1X 8NX

☎ 020 7396 4425   🖷 020 7235 4618

www.icilondon.esteri.it

OPEN 10.00 am to 1.30 pm and 2.30 pm to 5.00
pm Monday to Friday. Term-time: 10.00 am
to 1.30 pm and 2.30 pm to 5.00 pm Monday,
Tuesday, Thursday, Friday; 2.30 pm to 8.00 pm
Wednesday; 10.30 am to 4.30 pm Saturday (first
and third of each month). The Library is now
open to the general public every Wednesday
evening and on the first and third Saturday of
each month during term time. Call 020 7396 4425
to confirm exact opening times.

OPEN ACCESS For reference

A collection of over 30,000 volumes relating to
all aspects of Italian culture, including DVDs and

CDs relating to Italian cinema. Texts are mostly in Italian, with some in English.

Jersey Library

Halkett Place, St Helier, Jersey JE2 4WH
℡ 01534 448700 (Jersey Library)/448701 (Reference)/448702 (Open Learning)/448733 (Branch Library) 🄵 01534 448730
je.library@.gov.je
www.gov.je/library
OPEN 9.30 am to 5.30 pm Monday, Wednesday, Thursday, Friday; 9.30 am to 7.30 pm Tuesday; 9.30 am to 4.00 pm Saturday
OPEN ACCESS

Books, periodicals, newspapers, CDs, DVDs and language packs, cassettes, videos, microfilm, specialized local studies collection, public Internet access. Branch Library at Les Quennevais School, St Brelade. Mobile library and homes services. Open Learning Centre.

Kent County Central Library

Kent Libraries and Archives, Springfield, Maidstone ME14 2LH
℡ 01622 696511 🄵 01622 696494
countycentrallibrary@kent.gov.uk
www.kent.gov.uk/libs
OPEN 9.00 am to 6.00 pm Monday to Friday; 9.00 am to 5.00 pm Saturday
OPEN ACCESS

50,000 volumes available on the floor of the library plus 250,000 volumes of non-fiction, mostly academic, available on request to staff. English literature, poetry, classical literature, drama (including play sets), music (including music sets). Strong, too, in sociology, art history, business information and government publications. Loans to all who live or work in Kent; those who do not may consult stock for reference or arrange loans via their own local library service.

King George III Topographic Collection and Maritime Collection
▷ British Library Map Collections

Leeds Central Library

Calverley Street, Leeds LS1 3AB
℡ 0113 247 8911 🄵 0113 247 8426
enquiry.express@leeds.gov.uk
www.leeds.gov.uk/libraries
OPEN 9.00 am to 8.00 pm Monday, Tuesday, Wednesday; 9.00 am to 5.00 pm Thursday and Friday; 10.00 am to 5.00 pm Saturday; 1.00 pm to 5.00 pm Sunday
ACCESS Level access and lift. Open access to lending libraries and some reference materials. Other reference materials on request. Free public Internet access.

Lending Library ℡ 0113 247 8270
Adult and children's popular fiction and non-fiction for loan, including talking books, large print and books in world languages.

Music Library ℡ 0113 247 8273
Scores, books, DVD and audio.

The Information Centre ℡ 0113 247 8282 or 0113 395 1833
Company and product information, market research, statistics, directories, journals and computer-based information. Highlighted collections covering Law and Rights, Health, Jobs and Careers, Europe Direct, Education and Business. Extensive files of newspapers and periodicals plus all government publications since 1960. Special collections include military history, Judaic, early gardening books.

Art Library ℡ 0113 247 8247
Major collection of material on fine and applied arts.

Local and Family History Library
℡ 0113 247 8290
Extensive collection on Leeds and Yorkshire, including maps, books, pamphlets, local newspapers, illustrations and playbills. Census returns for the whole of Yorkshire also available. International Genealogical Index and parish registers.

Leeds Library and Information Service has an extensive network of branch and mobile libraries.

Leeds Library

18 Commercial Street, Leeds LS1 6AL
℡ 0113 245 3071 🄵 0113 245 1191
OPEN 9.00 am to 5.00 pm Monday to Friday; 9.30 am to 1.00 pm on the first Saturday of each month
ACCESS To members; research use upon application to the librarian

Founded 1768. Contains over 125,000 books and periodicals from the 15th century to the present day. SPECIAL COLLECTIONS include Reformation pamphlets, Civil War tracts, Victorian and Edwardian children's books and fiction, European language material, spiritualism and psychical research, plus local material.

Library of the Religious Society of Friends

Friends House, 173 Euston Road, London NW1 2BJ
℡ 020 7663 1135 🄵 020 7663 1001
library@quaker.org.uk
www.quaker.org.uk/library
OPEN 1.00 pm to 5.00 pm Monday, Tuesday, Thursday, Friday; 10.00 am to 5.00 pm Wednesday
OPEN ACCESS Reader registration required with proof of permanent address

Quaker history, thought and activities from the 17th century onwards. Supporting collections on peace, anti-slavery and other subjects in which Quakers have maintained long-standing interest. Also archives and manuscripts relating to the Society of Friends.

Lincoln Central Library

Free School Lane, Lincoln LN2 1EZ
☎ 01522 782010   🖷 01522 535882
lincolnlibrary@lincolnshire.gov.uk
www.lincolnshire.gov.uk
OPEN 9.00 am to 7.00 pm Monday to Friday; 9.30 am to 4.00 pm Saturday
OPEN ACCESS to the library; appointment required for the Tennyson Research Centre

Lending and reference library. Special collections include Lincolnshire local history (printed and published material, photographs, maps, directories and census data) and the Tennyson Research Centre (contact *Grace Timmins*, grace.timmins@lincolnshire.gov.uk).

Linen Hall Library

17 Donegall Square North, Belfast BT1 5GB
☎ 028 9032 1707   🖷 028 9043 8586
info@linenhall.com
www.linenhall.com
Librarian *John Gray*
OPEN 9.30 am to 5.30 pm Monday to Friday; 9.30 am to 1.00 pm Saturday
OPEN ACCESS For reference (loans restricted to members)

Founded 1788. Contains about 200,000 books. Major Irish and local studies collections, including the Northern Ireland Political Collection relating to the current troubles (c. 250,000 items).

Literary & Philosophical Society of Newcastle upon Tyne

23 Westgate Road, Newcastle upon Tyne NE1 1SE
☎ 0191 232 0192   🖷 0191 261 4494
library@litandphil.org.uk
www.litandphil.org.uk
Librarian *Kay Easson*
OPEN 9.30 am to 7.00 pm Monday, Wednesday, Thursday; 9.30 am to 8.00 pm Tuesday; 9.30 am to 5.00 pm Friday; 9.30 am to 1.00 pm Saturdays
ACCESS Members; research facilities for *bona fide* scholars on application to the Librarian

200-year-old library of 140,000 volumes, periodicals (including 130 current titles), classical music on vinyl recordings and CD, plus a collection of scores. Free public lectures, events and recitals. Recent publications include: *The Reverend William Turner: Dissent and Reform in Georgian Newcastle upon Tyne* Stephen Harbottle; *History of the Literary and Philosophical Society of Newcastle upon Tyne, Vol.*

2 *(1896–1989)* Charles Parish; *Bicentenary Lectures 1993* ed. John Philipson.

Liverpool Libraries and Information Services

William Brown Street, Liverpool L3 8EW
☎ 0151 233 5829   🖷 0151 233 5886
refbt.central.library@liverpool.gov.uk
www.liverpool.gov.uk/libraries
OPEN 9.00 am to 6.00 pm Monday to Friday; 9.00 am to 5.00 pm Saturday; 12 noon to 4.00 pm Sunday
OPEN ACCESS

Learn Direct Centre – UK Online Centre Free broadband internet access.

Humanities Reference Library A total stock in excess of 120,000 volumes and 24,000 maps, plus book plates, prints and autographed letters. SPECIAL COLLECTIONS Walter Crane and Edward Lear illustrations, Kelmscott Press, Audubon's *Birds of America*.

Business and Technology Reference Library Extensive stock dealing with all aspects of science, commerce and technology, including British and European standards and patents and trade directories.

Audio Visual Library Extensive stock relating to all aspects of music. Includes 128,000 volumes and music scores, over 4,000 CDs, 1,000 videos and 800 DVDs.

Record Office and Local History Department Material relating to Liverpool, Merseyside, Lancashire and Cheshire, together with archive material mainly on Liverpool. Proof of name and address required to obtain reader's ticket.

Lending Library Graphic novels, large print collection, children's collections, audio books, study support collections, Reader friendly displays of bestsellers and out of print titles.

University of the Arts London – London College of Communication

Library and Learning Resources, Elephant and Castle, London SE1 6SB
☎ 020 7514 6527   🖷 020 7514 6527
www.lcc.arts.ac.uk
ACCESS Appointment required

Library and Learning Resources provides books, periodicals, slides, CD-ROMs, videos and computer software on all aspects of the art of the book, printing, management, film/photography, graphic arts, plus retailing. SPECIAL COLLECTIONS History and development of published and unpublished scripts and the art of the western book.

The London Library

14 St James's Square, London SW1Y 4LG
Ⓣ 020 7930 7705   Ⓕ 020 7766 4766
membership@londonlibrary.co.uk
www.londonlibrary.co.uk
Librarian *Miss Inez Lynn*
OPEN 9.30 am to 7.30 pm Monday, Tuesday,
Wednesday; 9.30 am to 5.30 pm Thursday, Friday,
Saturday
ACCESS For members only (2008 fees: £375
p.a.; £190 for 16–24-year-olds). Day and weekly
reference tickets available for non-members (£10
and £30).

With over a million books and 8,000 members,
The London Library 'is the most distinguished
private library in the world; probably the largest,
certainly the best loved'. Founded in 1841, it is
a registered charity and wholly independent
of public funding. Its permanent collection
embraces most European languages as well
as English. Its subject range is predominantly
within the humanities, with emphasis on
literature, history, fine and applied art,
architecture, bibliography, philosophy, religion,
and topography and travel. Some 8,000–9,000
titles are added yearly. Over 95 per cent of the
stock is on open shelves to which members
have free access. Members may take out up to
10 volumes at a time; 15 if they live more than
20 miles from the Library and may hold the
books on loan until they are needed by another
member. The comfortable Reading Room has an
annexe for users of personal computers. There are
photocopiers, CD-ROM workstations, free access
to the Internet, and the Library also offers a postal
loans service.

The London Library has recently embarked on
its most ambitious redevelopment project in over
a century; it includes the integration of a new
contiguous building which will increase the total
area of the site by approximately 30%.

Membership is open to all: prospective
members are required to submit a refereed
application form in advance of admission, but
there is at present no waiting list for membership.
The London Library Trust may offer reduced cost
'Carlyle Memberships' to those who are unable
to afford the full annual fee; details on application.

London Metropolitan Archives Library Services

History Library: 40 Northampton Road, London
EC1R OHB
Ⓣ 020 7332 3820   Ⓕ 020 7833 9136
ask.lma@cityoflondon.gov.uk
www.cityoflondon.gov.uk/lma
Contact *The Enquiry Team*
OPEN 9.30 am to 4.45 pm Monday, Wednesday,
Friday; 9.30 am to 7.30 pm Tuesday and Thursday;
open some Saturdays – call for details
ACCESS For reference only

This 100,000 volume library covers all aspects
of the life and development of London, with
strong holdings on the history and organization
of London local government. As the former
Greater London Council History Library, the
collection covers all subjects of London life,
from architecture and biography to theatres
and transport. The collection includes London
directories from 1677 to the present, Acts of
Parliament, statistical returns, several hundred
periodical titles and public reports.

London's Transport Museum Poster Archives
▷ London's Transport Museum Photographic Library
under Picture Libraries

Lord Louis Library

Orchard Street, Newport, Isle of Wight PO30 1LL
Ⓣ 01983 527655/823800 (Reference Library)
Ⓕ 01983 825972
reflib@postmaster.co.uk
www.iwight.com/thelibrary
OPEN 9.00 am to 5.30 pm Monday, Tuesday,
Wednesday and Friday; 10.00 am to 8.00 pm
Thursday; 9.00 am to 5.00 pm Saturday; 10.00 am
to 1.00 pm Sunday
OPEN ACCESS

General adult and junior fiction and non-fiction
collections; local history collection at Library HQ
(Ⓣ 01983 203880 for details). Internet access in all
branches of the library on the Isle of Wight. Also
the county's main reference library.

Manchester Central Library

St Peters Square, Manchester M2 5PD
Ⓣ 0161 234 1900   Ⓕ 0161 234 1963
libraries@manchester.gov.uk
www.manchester.gov.uk/libraries
OPEN 9.00 am to 8.00 pm Monday to Thursday;
9.00 am to 5.00 pm Friday and Saturday.
Commercial Library: 9.00 am to 6.00 pm Monday
to Thursday; 9.00 am to 5.00 pm Friday and
Saturday
OPEN ACCESS

One of the country's leading reference libraries
with extensive collections covering all subjects.
Departments include: Commercial, European,
Science & Humanities, Arts, Music, Local
Studies, Chinese, General Readers, Language &
Literature. Large lending stock and VIP (visually
impaired) service available. Internet access.

Tom Mann Collection
▷ Coventry Central Library

Marylebone Library (Westminster)
▷ Sherlock Holmes Collection

Ministry of Defence Cartographic Archive
▷ British Library Map Collections

The Mitchell Library

North Street, Glasgow G3 7DN
☎ 0141 287 2999   🖷 0141 287 2815
lil@cls.glasgow.gov.uk
www.glasgowlibraries.org
OPEN 9.00 am to 8.00 pm Monday to Thursday;
9.00 am to 5.00 pm Friday and Saturday
OPEN ACCESS

One of Europe's largest public reference libraries
with stock of over 1,200,000 volumes. It
subscribes to 48 newspapers and more than 1,200
periodicals. There are collections in microform,
records, tapes and videos, as well as CD-ROMs,
electronic databases, illustrations, photographs,
postcards, etc.
    The library contains a number of special
collections, e.g. the Robert Burns Collection
(5,000 volumes), the Scottish Poetry Collection
(12,000 items) and the Scottish Drama Collection
(1,650 items).

Morrab Library

Morrab House, Morrab Gardens, Penzance
TR18 4DA
☎ 01736 364474
Librarian *Annabelle Read*
OPEN 10.00 am to 4.00 pm Tuesday to Friday;
10.00 am to 1.00 pm Saturday
ACCESS Non-members may use the library for a
small daily fee but may not borrow books

Formerly known as the Penzance Library. An
independent subscription lending library of over
40,000 volumes covering virtually all subjects
except modern science and technology, with large
collections on history, literature and religion.
There is a comprehensive Cornish collection of
books, newspapers and manuscripts including
the Borlase letters; a West Cornwall photographic
archive; many runs of 18th and 19th-century
periodicals; a collection of over 2,000 books
published before 1800.

The National Archives

Kew, Richmond TW9 4DU
☎ 020 8876 3444
www.nationalarchives.gov.uk
www.nationalarchives.gov.uk/contact/form
OPEN 9.00 am to 5.00 pm Monday and Friday;
9.00 am to 7.00 pm Tuesday and Thursday; 10.00
am to 5.00 pm Wednesday; 9.30 am to 5.00 pm
Saturday (closed public holidays and for annual
stocktaking in early December). From April 2008
the Family Record Centre services have been
incorporated at the Kew site.
ACCESS Reader's ticket required for access to
original documents, available free of charge on
production of proof of identity (UK citizens:
banker's card or driving licence; non-UK:
passport or national identity card). Telephone for
further information.

Over 168 kilometres of shelving house the
national repository of records of central
government in the UK and law courts of England
and Wales, which extend in time from the 11th
to the 20th century. Medieval records and the
records of the State Paper Office from the early
16th to late 18th century, plus the records of the
Privy Council Office and the Lord Chamberlain's
and Lord Steward's departments. Modern
government department records, together with
those of the Copyright Office dating mostly from
the late 18th century. Under the Public Records
Act, records are normally only open to inspection
when they are 30 years old.

National Library of Ireland

Kildare Street, Dublin 2, Republic of Ireland
☎ 00 353 1 603 0200   🖷 00 353 1 676 6690
info@nli.ie
www.nli.ie
OPEN 9.30 am to 9.00 pm Monday, Tuesday,
Wednesday; 9.30 am to 5.00 pm Thursday and
Friday; 9.30 am to 1.00 pm Saturday
ACCESS Passes are issued for genealogical
research or to consult newspapers; Reader's
Ticket required for access to other material.

Collections include books, manuscripts, prints
and drawings, maps, photographs, newspapers,
music, ephemera and genealogical materials.

National Library of Scotland

George IV Bridge, Edinburgh EH1 1EW
☎ 0131 623 3700   🖷 0131 623 3701
enquiries@nls.uk
www.nls.uk
OPEN Main Reading Room: 9.30 am to 8.30 pm
Monday, Tuesday, Thursday, Friday; 10.00 am to
8.30 pm Wednesday; 9.30 am to 1.00 pm Saturday.
Map Library: 9.30 am to 5.00 pm Monday,
Tuesday, Thursday, Friday; 10.00 am to 5.00 pm
Wednesday; 9.30 am to 1.00 pm Saturday
ACCESS to all reading rooms, for research not
easily done elsewhere, by reader's ticket

Collection of over seven million volumes. The
library receives all British and Irish publications.
Large stock of newspapers and periodicals. Many
special collections, including early Scottish books,
theology, polar studies, baking, phrenology and
liturgies. Also large collections of maps, music
and manuscripts including personal archives of
notable Scottish persons.

National Library of Wales

Penglais, Aberystwyth SY23 3BU
☎ 01970 632800   🖷 01970 615709

holi@llgc.org.uk
www.llgc.org.uk
OPEN 9.30 am to 6.00 pm Monday to Friday; 9.30
am to 5.00 pm Saturday (closed Bank Holidays
and the week after Christmas)
ACCESS to reading rooms by reader's ticket,
available on application. Open access to a wide-
ranging exhibition programme

Collection of over four million books and
including large collections of periodicals, maps,
manuscripts and audiovisual material. Particular
emphasis on humanities in printed foreign
material, and on Wales and other Celtic areas in
all collections.

National Meteorological Library and Archive

FitzRoy Road, Exeter EX1 3PB
℡ 01392 884841   🖷 0870 900 5050
metlib@metoffice.gov.uk
www.metoffice.gov.uk
OPEN Library: 8.30 am to 4.30 pm Monday to
Friday. Archive: 10.00 am to 6.00 pm
ACCESS By Visitor's Pass available from the
reception desk; advance notice of a planned visit
is appreciated.

The major repository of most of the important
literature on the subjects of meteorology,
climatology and related sciences from the 16th
century to the present day. The Library houses
a collection of books, journals, articles and
scientific papers, plus published climatological
data from many parts of the world.

The Technical Archive (National Meteorological
Archive, Great Moor House, Sowton Industrial
Estate, Bittern Road, Exeter EX2 7NL ℡ 01392
360987, 🖷 0870 900 5050; metarc@metoffice.
gov.uk) holds the document collection of
meteorological data and charts from England,
Wales and British overseas bases, including ships'
weather logs. Records from Scotland are stored
in Edinburgh and those from Northern Ireland in
Belfast.

National Monuments Record
▷ entry under Picture Libraries

Natural England

1 East Parade, Sheffield S1 2ET
℡ 0114 241 8920   🖷 0114 241 8921
enquiries@naturalengland.org.uk
www.naturalengland.org.uk
Contact *Enquiry Service* (℡ 0845 600 3078
🖷 01733 455103)
OPEN 8.30 am to 5.00 pm Monday to Thursday;
8.30 am to 4.30 pm Friday
ACCESS *Bona fide* students only (telephone the
library for appointment on 01733 455094)

Information on nature conservation, nature
reserves, SSSIs, planning, legislation, countryside,
recreation, stewardship schemes and advice to
farmers.

The Natural History Museum Library

Cromwell Road, London SW7 5BD
℡ 020 7942 5460   🖷 020 7942 5559
library@nhm.ac.uk
www.nhm.ac.uk/library/index.html
OPEN 10.00 am to 4.30 pm Monday to Friday
ACCESS To *bona fide* researchers, by reader's
ticket on presentation of identification (telephone
first to make an appointment)

The library is in five sections: general; botany;
zoology; entomology; earth sciences. The
sub-department of ornithology is housed at
the Zoological Museum, Akeman Street, Tring
HP23 6AP (℡ 020 7942 6156). Resources available
include books, journals, maps, manuscripts,
drawings and photographs covering all aspects
of natural history, including palaeontology and
mineralogy, from the 15th century to the present
day. Also archives and historical collection on the
museum itself.

Norfolk Library & Information Service

Norfolk and Norwich Millennium Library, The
Forum, Millennium Plain, Norwich NR2 1AW
millennium.library@norfolk.gov.uk
www.norfolk.gov.uk/council/departments/lis/
libhome.htm
OPEN Lending Library, Reference and Information
Service and Norfolk Studies: 9.00 am to 8.00 pm
Monday to Friday; 9.00 am to 5.00 pm Saturday.
EXPRESS: 9.00 am to 9.30 pm Monday to Friday;
9.00 am to 8.30 pm Saturday; 10.30 am to 4.30 pm
Sunday
OPEN ACCESS

Reference lending library (stock merged together)
with wide range, including books, recorded
music, music scores, plays and videos. Houses
the 2nd Air Division Memorial Library and has
a strong Norfolk Heritage Library. Extensive
range of reference stock including business
information. Online databases. Public 🖷 and
colour photocopying, free access to the Internet.
EXPRESS (fiction, sound & vision library within
a library): selection of popular fiction, videos,
CDs and DVDs available, with extended opening
hours.

Northamptonshire Libraries & Information Service

PO Box 216, John Dryden House, 8–10 The Lakes,
Northampton NN4 7DD
℡ 01604 237959   🖷 01604 237937
nlis@northamptonshire.gov.uk
kwilkinson@northamptonshire.gov.uk
www.northamptonshire.gov.uk/Leisure/Libraries/
home.htm
http://litnorthants.wordpress.com

Literature Development Officer *Kate Wilkinson*

'Northamptonshire Libraries and Information Service provides its users with much more than just book loans. There are clubs to join, events to attend, skills to learn, local history resources to use, CDs and DVDs to rent – to list but a few services.' In addition, writers and readers are supported by 'Literature Northants'. Run by the Literature Development Officer, Literature Northants provides access to literature opportunities across the county. The LDO is also the key contact for all the latest literary news and events.

Northumberland County Library

Beechfield, Gas House Lane, Morpeth NE61 1TA
☎ 01670 534518/534514
libraries@northumberland.gov.uk
www.northumberlandlibraries.com
OPEN 9.00 am to 7.30 pm Monday, Tuesday, Wednesday, Friday; 9.00 am to 12.30 pm Saturday (closed Thursday)
OPEN ACCESS

Books, periodicals, newspapers, story cassettes, CD-ROMs, DVDs, CDs, videos, Free Internet access, word processing facilities, prints, microforms, vocal scores, playsets. SPECIAL COLLECTIONS Northern Poetry Library: 15,000 volumes of modern poetry (see entry under *Organizations of Interest to Poets*); Cinema: comprehensive collection of about 5,000 volumes covering all aspects of the cinema; Family History.

Nottingham Central Library

Angel Row, Nottingham NG1 6HP
☎ 0115 915 2828   ☏ 0115 915 2840
central_admin.library@nottinghamcity.gov.uk
enquiryline@nottinghamcity.gov.uk
www.nottinghamcity.gov.uk/libraries
OPEN 9.00 am to 7.00 pm Monday to Friday; 9.00 am to 1.00 pm Saturday
OPEN ACCESS

City Centre Library; includes wide range of stock. Strengths: Business collection; Sound and Vision collection; comprehensive Nottinghamshire Local Studies. Drama and music sets for loan to groups. Free Internet access for visitors. Contemporary Art Gallery (not administered by the library service).

Nottingham Subscription Library
▷ Bromley House Library

Office for National Statistics, National Statistics Information and Library Service

Customer Contact Centre, Room 1.015 Office for National Statistics, Cardiff Road, Newport NP10 8XG
☎ 0845 601 3034/01633 812399   ☏ 01633 652747

info@statistics.gov.uk
www.statistics.gov.uk
OPEN 9.00 am to 5.00 pm; appointment required; visitors must supply some form of identification

Also: National Statistics Information and Library Service, Government Buildings, Cardiff Road, Newport NP9 1XG
OPEN as above

Wide range of government statistical publications and access to government Internet-based data. Census statistical data from 1801; population and health data from 1837; government social survey reports from 1941; recent international statistical data (UN, Eurostat, etc.); monograph and periodical collections of statistical methodology. The library in south Wales holds a wide range of government economic and statistical publications.

Orkney Library and Archive

44 Junction Road, Kirkwall KW15 1AG
☎ 01856 873166   ☏ 01856 875260
general.enquiries@orkneylibrary.org.uk
and
archives@orkneylibrary.org.uk
www.orkneylibrary.org.uk
Principal Archivist *Gary Amos*
Principal Librarian *Karen Walker*
OPEN 9.00 am to 7.00 pm Monday to Thursday; 9.00 am to 5.00 pm Friday and Saturday. Archives: 9.00 am to 5.00 pm Monday, Tuesday, Wednesday, Friday; 9.00 am to 7.00 pm Thursday; 9.00 am to 5.00 pm Saturday
OPEN ACCESS

Local studies collection. Archive includes sound and photographic departments.

Oxford Central Library

Westgate, Oxford OX1 1DJ
☎ 01865 815549   ☏ 01865 721694
oxfordcentral.library@oxfordshire.gov.uk
www.oxfordshire.gov.uk
OPEN 9.00 am to 7.00 pm Monday to Thursday; 9.00 am to 5.30 pm Friday and Saturday

General lending and reference library including Oxfordshire Studies. Also periodicals, audio visual materials, music library, children's library and Business Information Point.

PA News Centre

Central Park, New Lane, Leeds LS11 5DZ
☎ 0870 830 6824   ☏ 0870 830 6825
palibrary@pressassociation.co.uk
www.pa.press.net
OPEN 10.00 am to 6.00 pm Monday to Friday
OPEN ACCESS

PA News, the 24-hour national news and information group, offers the PA Digital Library which holds Press Association stories. Research undertaken by in-house staff.

Penzance Library
▷ Morrab Library

City of Plymouth Library and Information Services

Central Library, Drake Circus, Plymouth PL4 8AL
℡ 01752 305923
library@plymouth.gov.uk
www.plymouthlibraries.info
OPEN ACCESS

CENTRAL LIBRARY LENDING DEPARTMENTS:
Lending ℡ 01752 305912 lendlib@plymouth.gov.uk
Children's Department ℡ 01752 305916
childrens.library@plymouth.gov.uk
Music & Drama Department ℡ 01752 305914
music@plymouth.gov.uk

OPEN 9.00 am to 7.00 pm Monday and Friday; 9.00 am to 5.30 pm Tuesday, Wednesday, Thursday; 9.00 am to 5.00 pm Saturday

The Lending departments offer books on all subjects; language courses on cassette and foreign language books; the Holcenberg Jewish Collection; books on music and musicians, drama and theatre; music parts and sets of music parts; play sets; DVDs, videos; song index; cassettes and CDs; public Internet access.

CENTRAL LIBRARY REFERENCE DEPARTMENTS:
Reference ℡ 01752 305907/305908
ref@plymouth.gov.uk
Local Studies & Naval History Department
℡ 01752 305909
localstudies@plymouth.gov.uk

OPEN 9.00 am to 7.00 pm Monday to Friday; 9.00 am to 5.00 pm Saturday

The Reference departments include an extensive collection of Ordnance Survey maps and town guides; community and census information; marketing and statistical information; Patents; books on every aspect of Plymouth; naval history; Mormon Index on microfilm; Baring Gould manuscript of 'Folk Songs of the West'; public Internet access, including some electronic subscriptions.

Plymouth Proprietary Library

Alton Terrace, 111 North Hill, Plymouth PL4 8JY
℡ 01752 660515
Librarian *John R. Smith*
OPEN Monday to Saturday from 9.30 am (closing time varies)
ACCESS To members; visitors by appointment only

Founded 1810. The library contains approximately 17,000 volumes of mainly 19th and 20th

century work. Member of the **Association of Independent Libraries**.

The Poetry Library
▷ entry under Organizations of Interest to Poets

Polish Library POSK

238–246 King Street, London W6 0RF
℡ 020 8741 0474   ℻ 020 8741 7724
polish.library@posk.org
library@polishlibrary.co.uk
www.posk.org
Librarian *Mrs Jadwiga Szmidt*
OPEN 10.00 am to 8.00 pm Monday and Wednesday; 10.00 am to 5.00 pm Friday; 10.00 am to 1.00 pm Saturday (library closed Tuesday and Thursday)
ACCESS For reference to all interested in Polish affairs; limited loans to members and *bona fide* scholars only through inter-library loans.

Books, pamphlets, periodicals, maps, music, photographs on all aspects of Polish history and culture. SPECIAL COLLECTIONS Emigré publications; Joseph Conrad and related works; Polish underground publications; bookplates.

Poole Central Library

Dolphin Centre, Poole BH15 1QE
℡ 01202 262424   ℻ 01202 262442
libraries@poole.gov.uk
www.boroughofpoole.com/libraries
OPEN 9.00 am to 6.00 pm Monday to Friday; 9.00 am to 5.00 pm Saturday
OPEN ACCESS

General lending and reference library, including Healthpoint health information centre, business information, children's library, periodicals and newspapers, Internet suite and meeting room.

Press Association Library
▷ PA News Centre

Public Record Office
▷ The National Archives

Racing Collection (Newmarket)
▷ Suffolk County Council – Suffolk Libraries

Reading Central Library

Abbey Square, Reading RG1 3BQ
℡ 0118 901 5950   ℻ 0118 901 5954
info@readinglibraries.org.uk
www.readinglibraries.org.uk
OPEN 9.00 am to 5.30 pm Monday and Friday; 9.00 am to 7.00 pm Tuesday and Thursday; 9.00 am to 5.00 pm Wednesday; 9.30 am to 5.00 pm Saturday
OPEN ACCESS

Ground Floor: Fiction, audio-visual material, children's library; First Floor: Newspapers, non-fiction – biography, business, careers, cookery,

DIY, engineering, gardening, health, languages, law, mind, body and spirit, pets, science, social studies, sport, transport, travel; Second Floor: LearnDirect, non-fiction – art, music, literature, plays, computing; Third floor: Non-fiction – world history, national history, local history. Magazines and periodicals and Internet access on all floors.

RIBA British Architectural Library Drawings & Archives Collection
▷ British Architectural Library

Richmond Central Reference Library
Old Town Hall, Whittaker Avenue, Richmond TW9 1TP
☎ 020 8940 5529   🖷 020 8940 6899
reference.services@richmond.gov.uk
www.richmond.gov.uk
OPEN 9.30 am to 6.00 pm Monday, Tuesday, Thursday, Friday; 9.30 am to 8.00 pm Wednesday; 9.30 am to 5.00 pm Saturday
OPEN ACCESS

General reference library serving the needs of local residents and organizations. Internet access and online databases for public use. Enquiries received by visit, telephone and e-mail.

Royal Geographical Society (with the Institute of British Geographers)
1 Kensington Gore, London SW7 2AR
☎ 020 7591 3000   🖷 020 7591 3001
enquiries@rgs.org
www.rgs.org
OPEN 10.00 am to 5.00 pm Monday to Friday (except Bank Holidays and the period from Christmas to New Year)

A collection of over 150,000 volumes, dating primarily from the foundation of the Society in 1830 onwards, focusing on the history and geography of places worldwide. For information on the picture library see entry under *Picture Libraries*.

Royal Institute of Philosophy
▷ Senate House Library, University of London

The Royal Philharmonic Society Archive
▷ British Library Music Collections

Royal Society Library
6–9 Carlton House Terrace, London SW1Y 5AG
☎ 020 7451 2606   🖷 020 7930 2170
library@royalsociety.org
royalsociety.org
OPEN 10.00 am to 5.00 pm Monday to Friday
ACCESS Open to all researchers with an interest in the history of science, the Fellowship of the Royal Society and science policy. Researchers are advised to contact the Library in advance of their first visit.

History of science, scientists' biographies, science policy reports, and publications of international scientific unions and national academies from all over the world.

RSA (Royal Society for the Encouragement of Arts, Manufactures & Commerce)
8 John Adam Street, London WC2N 6EZ
☎ 020 7930 5115   🖷 020 7839 5805
general@rsa.org.uk
www.theRSA.org
OPEN Library: 8.30 am to 8.00 pm every weekday. Archive material by appointment
ACCESS to Fellows of RSA; Archive: by appointment to all researchers (contact the Archivist, ☎ 020 7451 6847; archive@rsa.org.uk)

Archives of the Society since 1754. A collection of approximately 15,000 items including minutes of the Society, correspondence, prints, original drawings and international exhibition material and an early library of over 700 volumes.

Royal Society of Medicine Library
1 Wimpole Street, London W1G 0AE
☎ 020 7290 2940   🖷 020 7290 2939
library@rsm.ac.uk
www.rsm.ac.uk
Contact *Director of Information Services*
OPEN 9.00 am to 5.00 pm Monday to Thursday; 9.00 am to 5.30 pm Friday; 10.00 am to 4.30 pm Saturday
ACCESS For reference only, on introduction by Fellow of the Society or temporary membership is available to non-members; £10 per day; £35 per week; £90 per month. Identification required.

Books, periodicals, databases on postgraduate biomedical information. Extensive historical collection dating from the fifteenth century and medical portrait collection.

Eric Frank Russell Archive
▷ Science Fiction Foundation Research Library

St Bride Library
Bride Lane, Fleet Street, London EC4Y 8EE
☎ 020 7353 4660   🖷 020 7353 1547
nigelroache@stbridefoundation.org
elizabethklaiber@stbridefoundation.org
www.stbridefoundation.org
Librarian *Nigel Roache*
Assistant Librarian *Elilzabeth Klaiber*
OPEN 12 noon to 5.30 pm Tuesday and Thursday; 12 noon to 9.00 pm Wednesday
OPEN ACCESS

A public reference library maintained by the St Bride Foundation. Appointments advisable for consultation of special collections. Every aspect of printing and related matters: publishing and bookselling, newspapers and magazines, graphic design, calligraphy and typography, papermaking and bookbinding. One of the world's largest specialist collections in its field, with over 50,000 volumes, over 3,000 periodicals (200 current titles), and extensive collection of drawings, manuscripts, prospectuses, patents and materials for printing and typefounding. Noted for its comprehensive holdings of historical and early technical literature.

Science Fiction Foundation Research Library

Liverpool University Library, PO Box 123, Liverpool L69 3DA
Ⓣ 0151 794 3142   Ⓕ 0151 794 2681
asawyer@liverpool.ac.uk
www.sfhub.ac.uk
www.sf-foundation.com
Contact *Andy Sawyer*
ACCESS For research, by appointment only (telephone first)

This is the largest collection outside the US of English-language science fiction and related material – including autobiographies and critical works. SPECIAL COLLECTIONS Runs of 'pulp' magazines dating back to the 1920s. Foreign-language material (including a large Russian collection), and the papers of the Flat Earth Society. The collection also features a growing range of archive and manuscript material, including material from Stephen Baxter, Ramsey Campbell and John Brunner. The University of Liverpool also holds the Olaf Stapledon, Eric Frank Russell and John Wyndham archives.

Science Museum Library

Imperial College Road, London SW7 5NH
Ⓣ 020 7942 4242   Ⓕ 020 7942 4243
smlinfo@nmsi.ac.uk
www.sciencemuseum.org.uk/library
OPEN 10.00 am to 5.00 pm Monday to Friday
OPEN ACCESS Reference only; no loans

National reference library for the history and public understanding of science and technology, with a large collection of source material. Operates jointly with Imperial College Central Library.

Scottish Drama Collection/Scottish Poetry Collection
▷ **The Mitchell Library**

Scottish Poetry Library
▷ entry under **Organizations of Interest to Poets**

Seckford Collection
▷ **Suffolk County Council – Suffolk Libraries**

2nd Air Division Memorial Library
▷ **Norfolk Library & Information Service**

Senate House Library, University of London

Senate House, Malet Street, London WC1E 7HU
Ⓣ 020 7862 8461/62 (Information Centre)
Ⓕ 020 7862 8480
enquiries@shl.lon.ac.uk (Information Centre)
www.shl.lon.ac.uk
MEMBERSHIP DESK: Ⓣ 020 7862 8439/40
userservices@shl.lon.ac.uk

OPEN Term-time: 9.00 am to 9.00 pm Monday to Thursday; 9.00 am to 6.30 pm Friday; 9.45 am to 5.30 pm Saturday. Vacation: 9.00 am to 6.00 pm Monday to Friday; 9.45 am to 5.30 pm Saturday (closed on Sundays and at certain periods during Bank Holidays)

The Senate House Library is a major academic research library predominantly based across the Humanities and Social Sciences. Housed within its 16 floors are some two million titles including 5,500 current periodicals and a wide range of electronic resources. It contains a number of outstanding research collections which, as well as supporting the scholarly activities of the University, attract researchers from throughout the UK and internationally. These include: English (e.g. the Durning-Lawrence Library and Sterling Collection of first editions); Economic and Social History (the Goldsmiths' Library, containing 70,000 items ranging from 15th to early 19th century); Modern Languages (primarily Romance and Germanic); Palaeography (acclaimed as being the best open access collection in its field in Europe); History (complementary to the **Institute of Historical Research**); Music, Philosophy (acts as the Library of the Royal Institute of Philosophy); Psychology (includes the BPS library); Major area studies collections (Latin-American, including Caribbean; United States and Commonwealth Studies, British Government Publications and maps). The Library has a wide range of Special Collections holdings. Check the website for full details of collections and current access arrangements.

Sheffield Libraries, Archives and Information

Central Library, Surrey Street, Sheffield S1 1XZ
Ⓣ 0114 273 4712   Ⓕ 0114 273 5009
libraries@sheffield.gov.uk
www.sheffield.gov.uk (click on 'In your area')

Central Lending Library
Ⓣ 0114 273 4727 (enquiries)/4729 (book renewals)
OPEN 10.00 am to 8.00 pm Monday; 9.30 am to 5.30 pm Tuesday, Thursday and Friday; 9.30 am to 8.00 pm Wednesday; 9.30 am to 5.30 pm Saturday

Books, talking books, large print, language courses, European fiction, books in cultural languages, play sets. Free Internet access. Writers' Resource Centre, Wednesday evenings, 5.00 pm to 7.00 pm. Proof of signature and a separate proof of address required to join.

Sheffield Archives

52 Shoreham Street, Sheffield S1 4SP
℡ 0114 203 9395 🄕 0114 203 9398
archives@sheffield.gov.uk

OPEN 10.00 am to 5.30 pm Monday; 9.30 am to 5.30 pm Tuesday to Thursday; 9.00 am to 1.00 pm and 2.00 pm to 5.00 pm Saturday (documents should be ordered by 5.00 pm Thursday for Saturday); closed Friday
ACCESS By reader's card

Holds documents relating to Sheffield and South Yorkshire, dating from the 12th century to the present day, including records of the City Council, churches, businesses, landed estates, families and individuals, institutions and societies. Free Internet access.

Arts and Social Sciences and Sports Reference Service

℡ 0114 273 4747/8

OPEN 10.00 am to 8.00 pm Monday; 9.30 am to 5.30 pm Tuesday, Thursday, Friday, Saturday; 9.30 am to 8.00 pm Wednesday
ACCESS Mainly reference but some lending material

A comprehensive collection of books, periodicals and newspapers covering all aspects of the arts (excluding music), sports, social sciences and free Internet access.

Music and Video Service

℡ 0114 273 4733
musicandav.library@sheffield.gov.uk
OPEN As for Central Lending Library above
ACCESS For reference and lending

An extensive range of books, CDs, cassettes, scores, etc. related to music. Also a video cassette and DVD loan service. Free Internet access.

Local Studies Service

℡ 0114 273 4753
localstudies.library@sheffield.gov.uk
OPEN As for Arts & Social Sciences above (except Wednesday 9.30 am to 5.30 pm)
ACCESS For reference

Extensive material covering all aspects of Sheffield and its population, including maps, photos and videos. Free Internet access. Photograph collection available on www.picturesheffield.co.uk

Business, Science and Technology Reference Services

℡ 0114 273 4736/7 or 273 4743
businessandtech.library@sheffield.gov.uk
OPEN As for Arts & Social Sciences above
ACCESS For reference only

Extensive coverage of science and technology as well as commerce and commercial law. British patents and British and worldwide standards with emphasis on metals. Hosts the World Metal Index. The business section holds a large stock of business and trade directories, and reference works with business emphasis. Free Internet access.

Sheffield Information Service

℡ 0114 273 4712 🄕 0114 275 7111
sis@sheffield.gov.uk
OPEN 10.00 am to 5.30 pm Monday; 9.30 am to 5.30 pm Tuesday to Saturday

Full local information service covering all aspects of the Sheffield community. Free Internet access.

Central Children's Library

℡ 0114 273 4734
kidsandteens.library@sheffield.gov.uk
OPEN 10.30 am to 5.00 pm Monday and Friday; 1.00 pm to 5.00 pm Tuesday, Wednesday, Thursday; 9.30 am to 5.30 pm Saturday

Books, spoken word cassettes, DVDs, videos; under-five play area; teenage reference section; readings and promotions; storytime and Babytime sessions. Free Internet access.

Shetland Library

Lower Hillhead, Lerwick ZE1 0EL
℡ 01595 743868 🄕 01595 694430
shetlandlibrary@sic.shetland.gov.uk
www.shetland-library.gov.uk
OPEN 9.30 am to 8.00 pm Monday and Thursday; 9.30 am to 5.00 pm Tuesday, Wednesday, Friday, Saturday

General lending and reference library; extensive local interest collection including complete set of *The Shetland Times*, *The Shetland News* and other local newspapers on microfilm and many old and rare books; audio collection including talking books/newspapers. Junior room for children. Disabled access and Housebound Readers Service (delivery to reader's home). Mobile library services to rural areas. Open Learning Service. Same day photocopying service. Publishing programme of books in dialect, history, literature. Learning centre providing access to learning opportunities including the Internet.

Shoe Lane Library

Hill House, Little New Street, London EC4A 3JR
℡ 020 7583 7178   🄕 020 7353 0884
shoelane@cityoflondon.gov.uk
www.cityoflondon.gov.uk/shoelanelibrary
OPEN 9.00 am to 5.30 pm Monday, Wednesday, Thursday, Friday; 9.00 am to 6.30 pm Tuesday
OPEN ACCESS

Corporation of London general lending library, with a comprehensive stock of 50,000 volumes, most of which are on display. Free Internet access.

Shrewsbury Library and Reference & Information Service
Castlegates, Shrewsbury SY1 2AS
℡ 01743 255300   🖷 01743 255309
shrewsbury.library@shropshire.gov.uk
www.shropshire.gov.uk/library.nsf
OPEN 9.30 am to 5.00 pm Monday, Wednesday, Friday; 9.30 am to 8.00 pm Tuesday and Thursday; 9.00 am to 5.00 pm Saturday; 1.00 pm to 4.00 pm Sunday
OPEN ACCESS

The largest public library in Shropshire. Books, cassettes, CDs, talking books, DVDs, videos, language courses. Open Learning and study centre with public use computers for word processing and Internet access. Music, literature and art book collection for lending and reference. The West Midlands Literary Heritage Collection is housed at Shrewsbury Library.

Spanish Institute Library
▷ Instituto Cervantes

Olaf Stapledon
▷ Science Fiction Foundation Research Library

Sterling Collection
▷ Senate House Library

Suffolk County Council – Suffolk Libraries
Endeavour House, Russell Road, Ipswich IP1 2BX
℡ 01473 584563   🖷 01473 583700
help@suffolklibraries.co.uk
www.suffolk.gov.uk/LeisureAndCulture/Libraries/SuffolkLibrariesDirect
OPEN See website for details of individual libraries or contact Endeavour House. Major libraries open seven days a week; all libraries open Sunday
ACCESS A single library card gives access to the lending service of 43 libraries and resource centres across the county. Loans can be collected and/or returned at any service point. Details on website.

Full range of lending and reference services, free public access to the Internet. Wide range of online services for registered library users through the Suffolk Libraries Direct pages, e.g. self-service reservations, renewals, free access to subscription services. Suffolk InfoLink Plus database gives details of local organizations throughout the county. SPECIAL COLLECTIONS include Suffolk Archives and Local History Collection; Benjamin Britten Collection; Edward Fitzgerald Collection; Seckford Collection and Racing Collection (Newmarket). The Suffolk Infolink service gives details of local groups and societies and is available in libraries throughout the county and on the website.

Sunderland City Library and Arts Centre
28–30 Fawcett Street, Sunderland SR1 1RE
℡ 0191 514 1235   🖷 0191 514 8444
enquiry.desk@sunderland.gov.uk
www.sunderland.gov.uk/libraries
OPEN 9.30 am to 7.30 pm Monday and Wednesday; 9.30 am to 5.00 pm Tuesday, Thursday, Friday; 9.30 am to 4.00 pm Saturday

The city's main library for lending and reference services. Local studies and children's sections, plus Sound and Vision department (CDs, DVDs, CD-ROMs, talking books). Sunderland Public Libraries also maintains community libraries of varying size, offering a range of services, plus mobile libraries. Free Internet access is available in all libraries across the city. A Books on Wheels service is available to housebound readers; the Schools Library Service serves teachers and schools. Two writers' groups meet at the City Library and Arts Centre: Janus Writers, every Wednesday, 1.30 pm to 3.30 pm; Foyle Street Writers Group, every Wednesday, 10.00 am to 12 noon.

Swansea Central Library
The Civic Centre, Oystermouth Road, Swansea SA1 3SN
℡ 01792 636464   🖷 01792 637103
central.library@swansea.gov.uk
www.swansea.gov.uk/libraries
OPEN 8.30 am to 8.00 pm Tuesday to Friday; 10.00 am to 4.00 pm Saturday and Sunday. The Library has a lending service open the same hours, including books, DVDs, videos, music CDs and talking books.
ACCESS For reference only (many local studies items now available on open access; other items must be requested on forms provided)

General reference material; also Welsh Assembly information, statutes, company information, maps, European Community information. Local studies: comprehensive collections on Wales; Swansea & Gower; Dylan Thomas. Local maps, periodicals, illustrations, local newspapers from 1804. B&w and colour photocopying and 🖷 facilities, free computer and Internet access available (library membership required and pre-booking recommended) and microfilm/microfiche copying facility.

Swiss Cottage Central Library
88 Avenue Road, London NW3 3HA
℡ 020 7974 6522   🖷 020 7974 6532
OPEN 10.00 am to 7.00 pm Monday; 10.00 am to 6.00 pm Tuesday, Wednesday, Friday; 10.00 am to 8.00 pm Thursday; 10.00 am to 5.00 pm Saturday; 11.00 am to 4.00 pm Sunday
OPEN ACCESS

Over 300,000 volumes in the lending and reference libraries. Home of the London Borough of Camden's Information and Reference Services.

Tennyson Research Centre
> Lincoln Central Library

Thurrock Communities, Libraries & Cultural Services

Grays Library, Orsett Road, Grays RM17 5DX
☎ 01375 413973   🖷 01375 370806
grays.library@thurrock.gov.uk
www.thurrock.gov.uk/libraries
OPEN 10.00 am to 7.00 pm Monday; 9.00 am to 7.00 pm Tuesday and Thursday; 9.00 am to 5.00 pm Wednesday, Friday, Saturday; branch library opening times vary.
OPEN ACCESS

General library lending and reference through ten libraries and a mobile library. Services include books, magazines, newspapers, audiocassettes, CDs, videos and language courses. Large collection of Thurrock materials. Internet and Microsoft Office.

Truro Library

Union Place, Truro TR1 1EP
☎ 01872 279205 (lending)/272702 (reference)/0845 607 2119 (24hr-renewals)
truro.library@cornwall.gov.uk
www.cornwall.gov.uk/library
OPEN 8.30 am to 6.00 pm Monday, Tuesday, Thursday, Friday; 9.30 am to 6.00 pm Wednesday; 9.00 am to 4.00 pm Saturday

Books, cassettes, CDs, videos, DVDs and pc games for loan through branch or mobile networks. Internet access and IT support and training. Reference collection. SPECIAL COLLECTIONS on local studies.

Vin Mag Archive Ltd

84–90 Digby Road, London E9 6HS
☎ 020 8533 7588   🖷 020 8533 7283
piclib@vinmag.com
www.vinmagarchive.com
Contact *Angela Maguire*

Wholly owned subsidiary of the Vintage Magazine Company. A collection of original printed material – magazines, newspapers, posters, adverts, books, ephemera and movie material. Also photographic collection; see entry under *Picture Libraries*.

West Midlands Literary Heritage Collection
> Shrewsbury Library and Reference & Information Service

Western Isles Libraries

Public Library, 19 Cromwell Street, Stornoway HS1 2DA
☎ 01851 708631   🖷 01851 708676
www.cne-siar.gov.uk
OPEN 10.00 am to 5.00 pm Monday to Wednesday; 10.00 am to 6.00 pm Thursday and Friday; 10.00 am to 5.00 pm Saturday
OPEN ACCESS

General public library stock, plus local history and Gaelic collections including maps, videos, printed music, cassettes and CDs; census records and Council minutes; music collection (CDs and cassettes). Branch libraries on the isles of Barra, Benbecula, Harris and Lewis.

City of Westminster Archives Centre

10 St Ann's Street, London SW1P 2DE
☎ 020 7641 5180/4879 (Minicom)
🖷 020 7641 5179
archives@westminster.gov.uk
www.westminster.gov.uk/archives
OPEN 10.00 am to 7.00 pm Tuesday, Wednesday, Thursday, 10.00 am to 5.00 pm Friday and Saturday (closed Monday)
ACCESS For reference

Comprehensive coverage of the history of Westminster and selective coverage of general London history. 22,000 books, together with a large stock of maps, prints, photographs, local newspapers, theatre programmes and archives.

Westminster Music Library

Victoria Library, 160 Buckingham Palace Road, London SW1W 9UD
☎ 020 7641 1300   🖷 020 7641 4281
musiclibrary@westminster.gov.uk
www.westminster.gov.uk/libraries/special/music
OPEN 11.00 am to 7.00 pm Monday to Friday; 10 am to 5.00 pm Saturday
OPEN ACCESS

Located at Victoria Library, this is the largest public music library in the South of England, with extensive coverage of all aspects of music, including books, periodicals and printed scores. No recorded material, notated only. Lending library includes a small collection of CDs and DVDs.

Westminster Reference Library

35 St Martin's Street, London WC2H 7HP
☎ 020 7641 1300   🖷 020 7641 5247
referencelibrarywc2@westminster.gov.uk
www.westminster.gov.uk/libraries
OPEN: 10.00 am to 8.00 pm Monday to Friday. 10.00 am to 5.00 pm Saturday
ACCESS For reference only, but there are now within the Art & Design and Performing Arts Collections several thousand older books available for loan to library members. Membership open to anyone with proof of a permanent address in Britain.

A general reference library with emphasis on the following: Art & Design – fine and decorative arts, architecture, graphics and design; Performing Arts – theatre, cinema, radio, television and dance; Official Publications – major collection of HMSO publications from 1947, plus parliamentary papers dating back to 1906 (TSO subscription ceased April 2006); Business –online and hard copy directories, market research, industry and company data; Periodicals – long files of many titles. One working day's notice is required for some government documents and some older periodicals. Extensive picture resource library.

The Wiener Library

4 Devonshire Street, London W1W 5BH
☎ 020 7636 7247   ℻ 020 7436 6428
info@wienerlibrary.co.uk
www.wienerlibrary.co.uk
Director *Ben Barkow*
Education and Outreach Coordinator *Katherine Klinger*
OPEN 10.00 am to 5.30 pm Monday to Friday
ACCESS By letter of introduction (readers needing to use the Library for any length of time should become members)

Private library – one of the leading research centres on European history since the First World War, with special reference to the era of totalitarianism and to Jewish affairs. Founded by Dr Alfred Wiener in Amsterdam in 1933, it holds material that is not available elsewhere. Books, periodicals, press archives, documents, pamphlets, leaflets, photo archive, audiovisual material, brochures.

The Wighton Collection of National Music
▷ Dundee Central Library

Vaughan Williams Memorial Library

English Folk Dance and Song Society, Cecil Sharp House, 2 Regent's Park Road, London NW1 7AY
☎ 020 7485 2206 ext. 33   ℻ 020 7284 0523
library@efdss.org
www.efdss.org
Contact *Malcolm Taylor*
OPEN 9.30 am to 5.30 pm Tuesday to Friday; 10.00 am to 4.00 pm 1st and 3rd Saturday (sometimes closed between 1.00 pm and 2.00 pm)
ACCESS For reference to the general public, on payment of a daily fee; EFDSS members may borrow books and use the library free of charge

A multimedia collection: books, periodicals, manuscripts, tapes, records, CDs, films, videos. Mostly British traditional culture and how this has developed around the world. Some foreign language material, and some books in English about foreign cultures. Also, the history of the English Folk Dance and Song Society.

Dr Williams's Library

14 Gordon Square, London WC1H 0AR
☎ 020 7387 3727
enquiries@dwlib.co.uk
www.dwlib.co.uk
OPEN 10.00 am to 5.00 pm Monday, Wednesday, Friday; 10.00 am to 6.30 pm Tuesday and Thursday
OPEN ACCESS to reading room (loans restricted to subscribers). Visitors required to supply identification and register for a visitor's ticket. Requests to see manuscripts and rare books must be made in advance; separate registration required. Annual subscription £10; ministers of religion and undergraduate students £5.

Primarily a library of theology, religion and ecclesiastical history. Also philosophy, history (English and Byzantine). Particularly important for the study of English Nonconformity. Also major manuscript collections. Trustees of Dr Williams's Library manage the Congregational Library on behalf of the Memorial Hall Trustees. The Dr Williams's Centre for Dissenting Studies is a collaboration with Queen Mary, University of London (see www.english.qmul.ac.uk/drwilliams).

Wolverhampton Central Library

Snow Hill, Wolverhampton WV1 3AX
☎ 01902 552025 (lending)/552026 (reference)
℻ 01902 552024
libraries@wolverhampton.gov.uk
www.wolverhampton.gov.uk/libraries
OPEN 9.00 am to 7.00 pm Monday to Thursday; 9.00 am to 5.00 pm Friday and Saturday

General lending and reference libraries, plus children's library and audiovisual library holding cassettes, CDs, videos and music scores. Internet access.

Archives & Local Studies Collection
42–50 Snow Hill, Wolverhampton WV2 4AG
☎ 01902 552480
OPEN 10.00 am to 5.00 pm Monday, Tuesday, Friday, 1st and 3rd Saturday of each month; 10.00 am to 7.00 pm Wednesday; closed Thursday

The Women's Library

London Metropolitan University, Old Castle Street, London E1 7NT
☎ 020 7320 2222   ℻ 020 7320 2333
moreinfo@thewomenslibrary.ac.uk
www.thewomenslibrary.ac.uk
OPEN Reading Room: 9.30 am to 5.00 pm Tuesday to Friday (8.00 pm Thursday); 10.00 am to 4.00 pm Saturday; Exhibition: 9.30 am to 5.30 pm Monday to Friday (8.00 pm Thursday); 10.00 am to 4.00 pm Saturday
OPEN ACCESS

The Women's Library, national research library for women's history, is the UK's oldest and most comprehensive research library on all aspects of women in society, with both historical and contemporary coverage. The Library includes materials on feminism, work, education, health, the family, law, arts, sciences, technology, language, sexuality, fashion and the home. The main emphasis is on Britain but many other countries are represented, especially the Commonwealth and the developing countries. Established in 1926 as the library of the London Society of Women's Service (formerly Suffrage), a non-militant organization led by Millicent Fawcett. In 1953 the Society was renamed after her and the library became the Fawcett Library. Collections include: women's suffrage, work, education, women and the church, the law, sport, art, music, abortion, prostitution. Mostly British materials but some American, Commonwealth and European works. Books, journals, pamphlets, archives, photographs, posters, postcards, audiovisual materials, artefacts, scrapbooks, albums and press cuttings dating mainly from the 19th century although some materials date from the 17th century.

The Library's new building, which opened in 2002, includes a reading room, exhibition space, café, education areas and a conference room, and is the cultural and research centre for anyone interested in women's lives and achievements.

Worcestershire Libraries and Information Service

Cultural Services, Worcestershire County Council, County Hall, Spetchley Road, Worcester WR5 2NP

☎ 01905 766231   📠 01905 766930
libraries@worcestershire.gov.uk
www.worcestershire.gov.uk/libraries
OPEN Opening hours vary in the 22 libraries, the History Centre and mobile libraries covering the county; all full-time libraries open at least one evening a week until 8.00 pm, and on Saturday until 5.30 pm; part-time libraries vary
ACCESS Information and reference services open to anyone; loans to members only (membership criteria: resident, being educated, working, or an elector in the county or neighbouring authorities; temporary membership to visitors. Proof of identity and address required). No charge for membership or for borrowing books.

Information service, and reference and lending libraries. Non-fiction and fiction for all age groups, including normal and large print, spoken word cassettes, sound recordings (CD, cassette), videos, maps, local history, CD-ROMs for reference at main libraries, free public Internet access in all libraries. Joint Libraries Service/ County Record Office History Centre with resources for local and family history. SPECIAL

COLLECTIONS Carpets and Textiles; Needles & Needlemaking; Stuart Period; A.E. Housman.

The Wordsworth Trust (Jerwood Centre)
▷ entry under Literary Societies

John Wyndham Archive
▷ Science Fiction Foundation Research Library

York Central Library

Museum Street, York YO1 7DS
☎ 01904 655631   📠 01904 552835
lending@york.gov.uk
reference.library@york.gov.uk
www.york.gov.uk/leisure/Libraries

Lending Library
OPEN 9.30 am to 8.00 pm Monday, Tuesday, Friday; 9.30 am to 5.30 pm Wednesday and Thursday; 9.30 am to 4.00 pm Saturday (NB closed until 1.00 pm on the first Thursday of each month)

General lending library including CDs, DVDs, audio books, children's storytapes language courses and printed music. Large print books. Photocopying, Internet and ⊞ facilities.

Reference Library
OPEN 9.00 am to 8.00 pm Monday, Tuesday, Wednesday, Friday; 9.00 am to 5.30 pm Thursday; 9.00 am to 4.00 pm Saturday (NB closed until 1.00 pm on the first Thursday of each month)

General reference library; periodicals, newspapers, local newspaper index, EU information, organizations database; local studies library for York and surrounding area; microfilm/ fiche readers for national and local newspapers; census returns and family history resource; general reference collection. Maintains strong links with other local history resource centres, namely the Borthwick Institute, York City Archive and York Minster Library. CD-ROM and Internet facilities. Room 18: IT resource centre available to the public. In addition, 14 branch libraries and one mobile, serving the City of York Council area.

Zoological Society Library

Regent's Park, London NW1 4RY
☎ 020 7449 6293   📠 020 7586 5743
library@zsl.org
www.zsl.org
OPEN 9.30 am to 5.30 pm Monday to Friday
ACCESS To members and staff; non-members require proof of address and photographic ID

160,000 volumes on zoology including 5,000 journals (1,300 current) and a wide range of books on animals and particular habitats. Slide collection available and many historic zoological prints. Library catalogue available online (library. zsl.org).

Picture Libraries

Acme
> Popperfoto.com

actionplus sports images
54–58 Tanner Street, London SE1 3PH
℡ 020 7403 1558   ℻ 020 7403 1526
info@actionplus.co.uk
www.actionplus.co.uk

Founded 1986. Specialist sports and action library providing complete and creative international coverage of over 300 varieties of professional and amateur sport. Elite, junior, professional, amateur, extreme, disabled, minority, offbeat, major and minor events, adults, children, high and low, fast and slow. Millions of images covering over 30 years of sport.

Lesley & Roy Adkins Picture Library
Ten Acre Wood, Whitestone, Exeter EX4 2HW
℡ 01392 811357
mail@adkinshistory.com
www.adkinshistory.com

Colour and some b&w coverage of archaeology, antiquity, heritage and related subjects in the UK, Europe, Egypt and Turkey. Subjects include towns, villages, housing, landscape and countryside, churches, temples, castles, monasteries, art and architecture, gravestones and tombs, inscriptions, naval, nautical and antiquarian views. Images supplied in digital form. No service charge if pictures are used.

The Advertising Archive Limited
45 Lyndale Avenue, London NW2 2QB
℡ 020 7435 6540   ℻ 020 7794 6584
suzanne@advertisingarchives.co.uk
www.advertisingarchives.co.uk
Contacts *Suzanne Viner, Larry Viner*

With over one million images, the largest collection of British and American press ads, TV commercial stills and magazine cover illustrations in Europe. Material spans the period from 1850 to the present day. In-house research; rapid service, competitive rates. On-line database and digital delivery available.

akg-images Ltd, The Arts and History Picture Library
5 Melbray Mews, 158 Hurlingham Road, London SW6 3NS
℡ 020 7610 6103   ℻ 020 7610 6125
enquiries@akg-images.co.uk
www.akg-images.co.uk

250,000 images online with direct access to ten million kept in the Berlin AKG Library. Specializes in art, archaeology, history, topography, music, personalities and film.

Bryan & Cherry Alexander Photography
Higher Cottage, Manston, Sturminster Newton DT10 1EZ
℡ 01258 473006   ℻ 01258 473333
alexander@arcticphoto.co.uk
www.arcticphoto.com
Contact *Cherry Alexander*

Arctic and Antarctic specialists; indigenous peoples, wildlife and science in polar regions; Canada, Greenland, the Scandinavian Arctic, Siberia and Alaska.

Alpine Garden Society
AGS Centre, Avon Bank, Pershore WR10 3JP
℡ 01386 554790   ℻ 01386 554801
ags@alpinegardensociety.net
www.alpinegardensociety.net
Contact *Peter Sheasby*

Over 28,000 colour transparencies (35mm) covering plants in the wild from many parts of the world; particularly strong in plants from mountain and sub-alpine regions, and from Mediterranean climates; South Africa, Australia, Patagonia, California and the Mediterranean. Extensive coverage of show alpines in pots and in gardens and of European orchids. Full slide list available.

Alvey & Towers
The Springboard Centre, Mantle Lane, Coalville LE67 3DW
℡ 01530 450011   ℻ 01530 450011
office@alveyandtowers.com
www.alveyandtowers.com
Contact *Emma Rowen*

Houses one of the country's most comprehensive collections of transport images depicting not only actual transport systems but their surrounding industries as well. Also specialist modern railway image collection.

Andes Press Agency

26 Padbury Court, London E2 7EH

☎ 020 7613 5417   🖷 020 7739 3159

apa@andespressagency.com

www.andespressagency.com

Contacts *Val Baker, Carlos Reyes*

80,000 colour transparencies and 300,000 b&w, specializing in social documentary, world religions, Latin America and Britain.

Heather Angel/Natural Visions

Highways, 6 Vicarage Hill, Farnham GU9 8HG

☎ 01252 716700   🖷 01252 727464

hangel@naturalvisions.co.uk

www.naturalvisions.co.uk

Contact *Heather Angel*

Constantly expanding worldwide natural history, wildlife and landscapes: polar regions, tropical rainforest flora and fauna, all species of plants and animals in natural habitats from Africa, Asia (notably China and Malaysia), Australasia, South America and USA, urban wildlife, pollution, biodiversity, global warming. Also worldwide gardens and cultivated flowers. Can send a digital lightbox to authors with an e-mail. The library then supplies high resolution digital files direct to the publisher.

Aquarius Library

PO Box 5, Hastings TN34 1HR

☎ 01424 721196   🖷 01424 717704

aquarius.lib@clara.net

www.aquariuscollection.com

Contact *David Corkill*

A unique collection of film stills, star portraits and candids covering the entire history of the cinema to the present day. Available as high resolution downloads from the Aquarius Library website containing thousands of film and TV titles and over 100,000 high quality stills. Postal deliveries and free searches also available.

Architectural Association Photo Library

36 Bedford Square, London WC1B 3ES

☎ 020 7887 4066   🖷 020 7414 0782

photolib@aaschool.ac.uk

www.aaschool.ac.uk/photlib

Contacts *Valerie Bennett, Sarah Franklin, Henderson Downing*

100,000 35mm transparencies on architecture, historical and contemporary. Archive of large-format b&w negatives from the 1920s and 1930s.

ArenaPAL

Thompson House, 42–44 Dolben Street, London SE1 0UQ

☎ 020 7403 8542   🖷 020 7403 8561

enquiries@arenapal.com

www.arenapal.com

Contacts *Biddy Hayward, Robert Piwko, Mike Markiewicz*

'Probably the best entertainment images in the world.' Continually updated performing arts image collection covering classical music, opera, theatre, musicals, film, pop, rock, jazz, instruments, festivals, venues, circus, ballet and contemporary dance, props and personalities. Over two million images from late 19th century onwards. Phone, fax or e-mail to make a selection.

ArkReligion.com

57 Burdon Lane, Cheam SM2 7BY

☎ 020 8642 3593   🖷 020 8395 7230

images@artdirectors.co.uk

www.arkreligion.com

Contacts *Helene Rogers, Bob Turner*

Extensive coverage of all religions.

The Art Archive

2 The Quadrant, 135 Salusbury Road, London NW6 6RJ

☎ 020 7624 3500   🖷 020 7624 3355

info@picture-desk.com

www.picture-desk.com

Picture library holding over 80,000 high-resolution images covering fine art, history and ancient civilizations.

Art Directors & Trip Photo Library

57 Burdon Lane, Cheam SM2 7BY

☎ 020 8642 3593   🖷 020 8395 7230

images@artdirectors.co.uk

www.artdirectors.co.uk

Contacts *Helene Rogers, Bob Turner*

Extensive coverage, with over 750,000 images, of all countries, lifestyles, peoples, etc. with detailed coverage of all religions. Backgrounds a speciality.

Aspect Picture Library Ltd

40 Rostrevor Road, London SW6 5AD

☎ 020 7736 1998   🖷 020 7731 7362

Aspect.Ldn@btinternet.com

www.aspect-picture-library.co.uk

Colour and b&w worldwide coverage of countries, events, industry and travel, with large files on art, namely paintings, space, China, the Middle East, French villages, English villages and Ireland.

Australia Pictures

28 Sheen Common Drive, Richmond TW10 5BN

☎ 020 7602 1989   🖷 020 7602 1989

equilibrium.films@virgin.net

Contact *John Miles*

Collection of 4,000 transparencies covering all aspects of Australia: Aboriginal people, paintings, Ayers Rock, Kakadu, Tasmania, underwater, reefs, Arnhem Land, Sydney. Also Africa, Middle East and Asia.

aviation-images.com

42B Queens Road, London SW19 8LR

☎ 020 8944 5225　🖷 020 8944 5335

pictures@aviation-images.com

www.aviation-images.com

Contacts *Mark Wagner, Denise Lalonde*

Two million 'stunning' aviation images, civil and military, archive and modern, from the world's best aviation photographers. Member of **BAPLA** and RAeS.

Barnaby's Library
▷ Mary Evans Picture Library

Barnardo's Photographic and Film Archive

Barnardo's House, Tanners Lane, Barkingside, Ilford IG6 1QG

☎ 020 8498 7345　🖷 020 8498 7090

stephen.pover@barnardos.org.uk

www.barnardos.org.uk

Image Librarian *Stephen Pover*

The Barnardo's image archive is a unique resource of images of social history dating from 1874. Images can be sent either by post or e-mail.

Colin Baxter Photography Limited

Woodlands Industrial Estate, Grantown-on-Spey PH26 3NA

☎ 01479 873999　🖷 01479 873888

sales@colinbaxter.co.uk

www.colinbaxter.co.uk

Contact *Mike Rensner* (Editorial)

Over 50,000 images specializing in Scotland. Also the Lake District, Yorkshire, France, Iceland and a special collection on Charles Rennie Mackintosh's work. Publishes guidebooks plus books, calendars, postcards and greetings cards on landscape, cityscape and natural history containing images which are primarily, but not exclusively, Colin Baxter's. Also publishers of the *Worldlife Library* of natural history books, and a range of books and stationery based on reproductons of the designs of Charles Rennie Mackintosh.

The Photographic Library Beamish, The North of England Open Air Museum

The North of England Open Air Museum, Beamish DH9 0RG

☎ 0191 370 4000　🖷 0191 370 4001

museum@beamish.org.uk

www.beamish.org.uk

www.beamishcollections.co.uk

Comprehensive collection; images relate to the North East of England and cover agricultural, industrial, topography, advertising and shop scenes, people at work and play. On laser disk for rapid searching. Visitors by appointment weekdays.

Francis Bedford
▷ Birmingham Library Services under Library Services

Ivan J. Belcher Colour Picture Library

57 Gibson Close, Abingdon OX14 1XS

☎ 01235 521524　🖷 01235 521524

Extensive colour picture library specializing in top-quality medium-format transparencies depicting the British scene. Particular emphasis on tourist, holiday and heritage locations, including famous cities, towns, picturesque harbours, rivers, canals, castles, cottages, rural scenes and traditions photographed throughout the seasons. Mainly of recent origin and constantly updated.

BFI Stills Sales

BFI, 21 Stephen Street, London W1T 1LN

☎ 020 7957 4797　🖷 020 7323 9260

stills.films@bfi.org.uk

www.bfi.org.uk/collections/stills/index.html

'BFI Stills Sales is the world's most comprehensive collection of film and television images.' The collection captures on and off screen moments, portraits of the world's most famous stars – and those behind the camera who made them famous – as well as publicity posters, set designs, images of studios, cinemas, special events and early film and TV technologies.

Anthony Blake Photo Library

20 Blades Court, Deodar Road, Putney, London SW15 2NU

☎ 020 8877 1123　🖷 020 8877 9787

info@abpl.co.uk

www.abpl.co.uk

'Europe's premier source' of food and wine related images. From the farm and the vineyard to the plate and the bottle. Cooking and kitchens, top chefs and restaurants, country trades and markets, worldwide travel with an extensive Italian section.

Peter Boardman Collection
▷ Chris Bonington Picture Library

Chris Bonington Picture Library

Badger Hill, Hesket Newmarket, Wigton CA7 8LA

☎ 016974 78286　🖷 016974 78238

frances@bonington.com

www.bonington.com

Contact *Frances Daltrey*

Based on the personal collection of climber and author Chris Bonington and his extensive travels and mountaineering achievements; also work by Doug Scott and other climbers, including the Peter Boardman and Joe Tasker Collections. Full coverage of the world's mountains, from British hills to Everest, depicting expedition planning and management stages, the approach march showing inhabitants of the area, flora and fauna, local architecture and climbing action shots on some of the world's highest mountains.

Boulton and Watt Archive
▷ Birmingham Library Services under Library Services

The Bridgeman Art Library
17–19 Garway Road, London W2 4PH
☎ 020 7727 4065   🖷 020 7792 8509
research@bridgeman.co.uk
www.bridgemanart.co.uk
Head of Picture Research *Jenny Page*

Fine art photo archive representing more than 1,000 museums, galleries and picture owners around the world. High res digital files and large-format colour transparencies of paintings, sculptures, prints, manuscripts, antiquities photographs and the decorative arts. From prehistoric sculpture to contemporary paintings. Offices in London, Paris, New York and Berlin. Collections represented include the British Library, National Galleries of Scotland, Boston Museum of Fine Arts, the Wallace Collection, The Courtauld, Detroit Institute of Arts, The Barnes Foundation, the National Army Museum, the National Trust as well as the Giraudon archive.

The British Cartoon Archive
The Templeman Library, University of Kent, Canterbury CT2 7NU
☎ 01227 823127   🖷 01227 823127
N.P.Hiley@kent.ac.uk
J.M.Newton@kent.ac.uk
www.kent.ac.uk/cartoons
Contacts *Dr Nicholas Hiley, Jane Newton*

A national research archive of over 120,000 original cartoons and caricatures by 350 British cartoonists, supported by a library of books, papers, journals, catalogues, cuttings and assorted ephemera. A computer database of 120,000 cartoons provides quick and easy catalogued access via the web. A source for exhibitions and displays as well as a picture library service. Specializes in historical, political and social cartoons from British newspapers and magazines.

British Library Images Online
British Library, 96 Euston Road, London NW1 2DB
Images Online ☎ 020 7412 7614 (Research)/7755 (Reproduction rights) 🖷 020 7412 7771
imagesonline@bl.uk www.imagesonline.bl.uk

Images Online offers instant access to thousands of images from the British Library's historic worldwide collections which include manuscripts, rare books, music scores and maps spanning almost 3,000 years. A collection of 20,000 images including digital images of drawings, illustrations and photographs. Specialities: historical images.

Brooklands Museum Photo Archive
Brooklands Museum, Brooklands Road, Weybridge KT13 0QN
☎ 01932 857381   🖷 01932 855465
info@brooklandsmuseum.com
www.brooklandsmuseum.com
Head of Collections *John Pulford*
General Manager – Operations *Julian Temple*

About 40,000 b&w and colour prints and slides. Subjects include: Brooklands Motor Racing 1907–1939; British aviation and aerospace, 1908 to the present day, particularly BAC, Hawker, Sopwith and Vickers aircraft built at Brooklands.

Hamish Brown MBE, D.Litt, FRSGS Scottish Photographic
26 Kirkcaldy Road, Burntisland KY3 9HQ
☎ 01592 873546   🖷 01592 873546
Contact *Hamish Brown*

Coverage of most topics and areas of Scotland (sites, historic, buildings, landscape, mountains), also travel, mountains, general (50,000 items) and Morocco. Commissions undertaken.

Capital Pictures
85 Randolph Avenue, London W9 1DL
☎ 020 7286 2212
sales@capitalpictures.com
www.capitalpictures.com
Contact *Phil Loftus*

Up-to-date celebrity and entertainment photography from around the world and a film stills collection of all the great films and yet to be released blockbusters. The website is fully searchable.

Cephas Picture Library
A1 Kingsway Business Park, Oldfield Road, Hampton TW12 2HD
☎ 020 8979 8647   🖷 020 8941 4001
pictures@cephas.com
www.cephas.com
Contact *Mick Rock*

The wine industry and vineyards of the world is the subject on which Cephas has made its reputation. '100,000 images make this the most comprehensive and up-to-date archive in Britain.' Almost all wine-producing countries and all aspects of the industry are covered in depth. Spirits, beer and cider also included. A major food and drink collection now also exists, through preparation and cooking, to eating and drinking.

Giles Chapman Library

25 Homefield Road, Sevenoaks TN13 2DU
℡ 01732 452547
chapman.media@virgin.net
www.gileschapman.com
Contact *Giles Chapman*

Some 165,000 colour and b&w images of cars and motoring, from 1945 to the present day. No research fees.

Christian Aid
▷ Still Pictures' Whole Earth Photolibrary

Christie's Images Ltd

1 Langley Lane, London SW8 1TJ
℡ 020 7582 1282   Ｆ 020 7582 5632
imageslondon@christies.com
www.christiesimages.com
Contact *Mark Lynch*

The collection spans 80 categories representing every culture, from the distant past to the present day. An unrivalled collection from Old Masters to Impressionists and Twentieth Century masterpieces. Antiquities, Art Nouveau and Decorative Art, Books and Manuscripts, Contemporary Art and Furniture, Music and Movie Memorabilia, Textiles, Toys. Experienced researchers will search for specific requests.

The Cinema Museum

The Master's House, Old Lambeth Workhouse, 2 Dugard Way, off Renfrew Road, London SE11 4TH
℡ 020 7840 2200   Ｆ 020 7840 2299
martin@cinemamuseum.org.uk
pixdesk@rgapix.com
www.ronaldgrantarchive.com

Colour and b&w coverage (including stills) of the motion picture industry throughout its history, including the Ronald Grant Archive. Smaller collections on theatre, variety, television and popular music.

Classical Collection Ltd

22 Avon, Hockley, Tamworth B77 5QA
℡ 01827 286086/07963 194921   Ｆ 01827 286086
neil@classicalcollection.co.uk
Managing Director *Neil Arthur Williams, MA (Hum), MA (Mus)*

Archive specializing in classical music ephemera, particularly portraits of composers, musicians, conductors and opera singers, comprising old and very rare photographs, postcards, antique prints, cigarette cards, stamps, First Day Covers, concert programmes, Victorian newspapers, etc. Also modern photos of composer references such as museums, statues, busts, paintings, monuments, memorials and graves. Other subjects covered

include ballet, musical instruments, concert halls, opera houses, 'music in art', manuscripts, opera scenes, music-caricatures, bands, orchestras and other music groups. Neil Arthur Williams is a qualified music historian, musicologist and freelance writer of music articles, concert programme notes and CD as well as being a commissionable composer.

John Cleare/Mountain Camera

Hill Cottage, Fonthill Gifford, Salisbury SP3 6QW
℡ 01747 820320   Ｆ 01747 820320
cleare@btinternet.com
www.mountaincamera.com

Colour and b&w coverage of mountains and wild places, climbing, ski-touring, trekking, expeditions, wilderness travel, landscapes, people and geographical features from all continents with Himalaya, Andes, Antarctic, Alps and the British countryside in depth. Topics range from reindeer in Lapland and camels in Australia to whitewater rafting in Utah, ski-mountaineering in China and abstract illustration for poetry and Japanese haiku. Commissions, presentations and consultancy work undertaken. Researchers welcome by appointment. Member of **BAPLA** and the **OWPG**.

Michael Cole Camerawork

The Coach House, 27 The Avenue, Beckenham BR3 5DP
℡ 020 8658 6120   Ｆ 020 8658 6120
mikecole@dircon.co.uk
Contacts *Michael Cole, Derrick Bentley*

Probably the largest and most comprehensive collection of tennis pictures in the world. Over 50 years' coverage of the Wimbledon Championships. M.C.C. incorporates the tennis archives of Le Roye Productions, established in 1945. Picture requests and enquiries by e-mail.

Collections

13 Woodberry Crescent, London N10 1PJ
℡ 020 8883 0083   Ｆ 020 8883 9215
collections@btinternet.com
www.collectionspicturelibrary.co.uk
www.collectionspicturelibrary.com
Contact *Simon Shuel*

Extensive coverage of the British Isles and Ireland, from the Shetlands to the Channel Islands and Connemara to East Anglia, including people, traditional customs, landscapes, buildings old and new – both well known and a bit obscure, 'plus a considerable collection of miscellaneous bits and pieces which defy logical filing'. Images supplied digitally or as original transparencies. Visitors welcome but please call first.

Corbis Images

111 Salusbury Road, London NW6 6RG

☎ 020 7644 7644   🖷 020 7644 7645

sales.uk@corbis.com

www.corbis.com

A unique and comprehensive resource containing more than 65 million images, with over 2.1 million of them available online. The images come from professional photographers, museums, cultural institutions and public and private collections worldwide. Subjects include history, travel, celebrities, events, science, world art and cultures.

Sylvia Cordaiy Photo Library

45 Rotherstone, Devizes SN10 2DD

☎ 01380 728327   🖷 01380 728328

info@sylvia-cordaiy.com

www.sylvia-cordaiy.com

www.alamy.com/sylviacordaiyphotolibrary
(search: 'Sylvia Cordaiy')

Over 180 countries on file from the obscure to main stock images – Africa, North, Central and South America, Asia, Atlantic, Indian and Pacific Ocean islands, Australasia, Europe, polar regions. Covers travel, architecture, ancient civilizations, world heritage sites, people worldwide, environment, wildlife, natural history, Antarctica, domestic pets, livestock, marine biology, veterinary treatment, equestrian, ornithology, flowers. UK files cover cities, towns villages, coastal and rural scenes, London collection. Transport, railways, shipping and aircraft (military and civilian). Aerial photography. Backgrounds and abstracts. Also the Paul Kaye B/W archive. All images available by digital delivery.

Country Images Picture Library

27 Camwood, Clayton-le-Woods, Bamber Bridge PR5 8LA

☎ 01772 321243   🖷 0870 137 8888

tm@wpu.org.uk

www.countryimages.info

Contact *Terry Marsh*

35mm film and digital colour coverage of landscapes and countryside features generally throughout the UK (Cumbria, North Yorkshire, Lancashire, southern Scotland, Isle of Skye, St Kilda and Scottish islands, Wales, Cornwall), France (Alps, Auvergne, Pyrenees, Provence, Charente-Maritime, Hérault, Aube-en-Champagne, Loire valley, Somme), Madeira, the Azores and Australia. Commissions undertaken.

Country Life Picture Library

Blue Fin Building, 110 Southwark Street, London SE1 0SU

☎ 020 3148 4474

picturelibrary@ipcmedia.com

www.clpicturelibrary.co.uk

Contact *Justin Hobson*

Over 150,000 b&w negatives, dating back to 1897, and 100,000 colour transparencies. Country houses, stately homes, churches and town houses in Britain and abroad, interiors of architectural interest (ceilings, fireplaces, furniture, paintings, sculpture), and exteriors showing many landscaped gardens, sporting and social events, crafts, people and animals. Visitors by appointment.

Philip Craven Worldwide Photo-Library

Surrey Studios, 21 Nork Way, Nork, Banstead SM7 1PB

☎ 01737 814646

www.philipcraven.com

Contact *Philip Craven*

Extensive coverage of British scenes, cities, villages, English countryside, gardens, historic buildings and wildlife. Worldwide travel and wildlife subjects on medium and large-format transparencies.

Sue Cunningham Photographic

56 Chatham Road, Kingston upon Thames KT1 3AA

☎ 020 8541 3024   🖷 020 8541 5388

info@scphotographic.com

www.scphotographic.com

Contacts *Patrick Cunningham, Sue Cunningham*

Extensive coverage of many geographical areas: South America (especially Brazil), Eastern Europe from the Baltic to the Balkans, various African countries, Western Europe including the UK. Colour and b&w.

Dale Concannon Collection
▷ Phil Sheldon Golf Picture Library

Dalton–Watson Collection
▷ Ludvigsen Library

The Defence Picture Library

14 Mary Seacole Road, The Millfields, Plymouth PL1 3JY

☎ 01752 312061   🖷 01752 312063

picdesk@defencepictures.com

www.defencepictures.com

Contacts *David Reynolds, Jessica Kelly, James Rowlands*

Specialist source of military photography covering all areas of the UK Armed Forces, supported by a research agency of facts and figures. More than one million images. Visitors welcome by appointment.

Douglas Dickins Photo Library

2 Wessex Gardens, Golders Green, London
NW11 9RT
☎ 020 8455 6221
Sole Proprietor *Douglas Dickins, FRPS*

Worldwide colour and b&w coverage, specializing
in Asia, particularly India, Indonesia and Japan.
In Grandpa's Footsteps, highly illustrated book
of world travel published by The Book Guild in
2000.

Dominic Photography

4B Moore Park Road, London SW6 2JT
☎ 020 7381 0007   🖷 020 7381 0008
Contacts *Zoë Dominic, Catherine Ashmore*

Colour and b&w coverage of the entertainment
world from 1957 onwards: dance, opera, theatre,
ballet, musicals and personalities.

E&E Picture Library – Ecclesiastical and Eccentricities

Beggars Roost, Woolpack Hill, Brabourne Lees,
Ashford TN25 6RR
☎ 01303 812608   🖷 01303 812608
info@picture-library.freeserve.co.uk
www.eeimag
www.heritage-images.com
Contact *Isobel Sinden*

Over 400,000 images in stock. Categories
covered: religions; ancient; architecture; death &
commemorative; festivals; heritage; eccentricities;
places; saints; stained glass; transport; nature.
Website updates continuously; eventually, all
subject areas will be available online. E-mail
image requests.

Patrick Eagar Photography

1 Queensberry Place, Richmond TW9 1NW
☎ 020 8940 9269   🖷 020 8332 1229
info@patrickeagar.com
www.patrickeagar.com

The cricket library consists of Patrick Eagar's
work over the last 30 years with coverage of over
290 Test matches worldwide, unique coverage
of all eight World Cups, countless one-day
internationals and player action portraits of over
2,000 cricketers. The wine library consists of
vineyards, grapes and festivals from Argentina to
New Zealand. France and Australia are specialist
areas. Photographs can be supplied by e-mail, on
CD or DVD.

Ecoscene

Empire Farm, Throop Road, Templecombe
BA8 0HR
☎ 01963 371700
pix@ecoscene.com
www.ecoscene.com
Contact *Sally Morgan*

Expanding colour library of over 80,000
transparencies specializing in all aspects of the
environment: pollution, conservation, recycling,
restoration, wildlife (especially underwater),
habitats, education, industry and agriculture.
All parts of the globe are covered with specialist
collections covering Antarctica, Australia,
North America. Sally Morgan, who runs the
library, is a professional ecologist and expert
source of information on all environmental
topics. Photographic and writing commissions
undertaken. Images delivered by e-mail or
available for download.

Edifice

Cutterne Mill, Evercreech BA4 6LY
☎ 01749 831400
info@edificephoto.com
www.edificephoto.com
Contacts *Philippa Lewis, Gillian Darley*

Colour material on exteriors of all possible types
of building worldwide from grand landmarks
to the small and simple. Specialists in domestic
architecture, garden features, architectural detail,
ornament, historical and period style, building
materials and techniques. Website searchable
online, but expert help and advice happily given
over the phone or by e-mail.

Education Photos

8 Whitemore Road, Guildford GU1 1QT
☎ 01483 511666
johnwalmsley@educationphotos.co.uk
www.educationphotos.co.uk

Formerly the John Walmsley Photo Library.
Specialist library of learning/training/working
subjects. Comprehensive coverage of learning
environments such as playgroups, schools,
colleges and universities. Images reflect a multi-
racial Britain. Commissions undertaken.

English Heritage Photo Library

Kemble Drive, Swindon SN2 2GZ
☎ 01793 414903   🖷 01793 414556
images@english-heritage.org.uk
www.englishheritageimages.com
Contacts *Duncan Brown, Javis Gurr,
Jonathan Butler*

Images of English castles, abbeys, houses,
gardens, Roman remains, ancient monuments,
battlefields, industrial and post-war buildings,
interiors, paintings, artifacts, architectural
details, conservation, archaeology, scenic views,
landscapes. See also **National Monuments
Record** entry.

Mary Evans Picture Library

59 Tranquil Vale, Blackheath, London SE3 0BS
☎ 020 8318 0034   🖷 020 8852 7211

pictures@maryevans.com
www.maryevans.com

Leading historical archive of artwork,
photographs, prints and ephemera illustrating
and documenting all aspects of the past, from
ancient times to the later decades of the 20th
century. Emphasis on everyday life plus extensive
coverage of events, portraits, transport, and
costume. 'Unrivalled material on folklore and
paranormal phenomena.' Notable collections
include the Weimar Archive documenting the
Third Reich, and Barnaby's Library of social
documentary photography from the 1930s to
the 1970s. Own material complemented by
important contributor collections, such as the
Illustrated London News **Picture Library** (see
entry); the archive of the National Magazine
Company; the Harry Price Library of Magical
Literature; Sigmund Freud Copyrights; and the
Women's Library. Arrangements with agencies
in France (Rue des Archives); Italy (De Agostini);
Germany (Interfoto); and Spain (Aisa) have added
thousands of additional photographs and fine
art images to the website. Over 200,000 images
searchable online, with 80,000 available in high-
resolution for immediate download. Brochure on
request. Founder member of **BAPLA**. Compilers
of the *Picture Researcher's Handbook* published
by Pira International.

Eye Ubiquitous

65 Brighton Road, Shoreham BN43 6RE
℡ 01273 440113   ℻ 01273 440116
library@eyeubiquitous.com
www.eyeubiquitous.com

General stock specializing in social documentary
worldwide, and an extensive travel related
collection.

Faces and Places

28 Sheen Common Drive, Richmond TW10 5BN
℡ 020 7602 1989/07930 622964 (mobile)
equilibrium.films@virgin.net
john@facesandplacespix.com
www.facesandplacespix.com

Extensive worldwide library built up over twenty
years which covers 40,000 photographs focusing
mainly on travel, tourism, unusual locations,
indigenous peoples, scenery and underwater
photography; Australia, Tibet, Mali, Yemen and
Iran.

Famous Pictures & Features Agency

13 Harwood Road, London SW6 4QP
℡ 020 7731 9333   ℻ 020 7731 9330
features@famous.uk.com
www.famous.uk.com

'Famous is a long-established celebrity features
and pictures agency that has worked successfully
for many years with writers who value our
integrity and efficiency.' Represents top celebrity
writers from the UK, LA, New York, Europe and
Australia, syndicating their copy worldwide. 'We
are always interested in new showbiz material,
so please contact us if you are a journalist
or freelancer looking to sell your celebrity
interviews, features or gossip stories.'

ffotograff

10 Kyveilog Street, Pontcanna, Cardiff CF11 9JA
℡ 029 2023 6879
ffotograff@easynet.co.uk
www.ffotograff.com
Contact *Patricia Aithie*

Library and agency specializing in travel,
exploration, the arts, architecture, traditional
culture, archaeology and landscape. Based in
Wales but specializing in the Middle and Far East;
Africa, Central and South America; Yemen and
Wales are strong aspects of the library. Churches
and cathedrals of Britain and Crusader castles.

Financial Times Pictures

One Southwark Bridge, London SE1 9HL
℡ 020 7873 3671   ℻ 020 7873 4606
photosynd@ft.com
Contacts *Nicky Burr, Richard Pigden*

Photographs from around the world ranging from
personalities in business, politics and the arts,
people at work and other human interests and
activities. Delivery via ISDN, ftp or e-mail.

Fine Art Photographic Library Ltd

2A Milner Street, London SW3 2PU
℡ 020 7589 3127
info@fineartphotolibrary.com
www.fineartphotolibrary.com

Over 30,000 large-format transparencies, with
a specialist collection of 19th and 20th century
paintings. CD-ROM available.

Firepix International

68 Arkles Lane, Anfield, Liverpool L4 2SP
℡ 0151 260 0111/0777 5930419 (mobile)
℻ 0151 260 0111
info@firepix.com
www.firepix.com
Contact *Tony Myers*

The UK's only fire photo library. 23,000 images of
fire-related subjects, firefighters, fire equipment
manufacturers. Website contains 15 categories
from industrial fire, domestic, digital images and
abstract flame.

Fogden Wildlife Photographs

16 Locheport, North Uist, Western Isles HS6 5EU
℡ 01876 580245   ℻ 01876 580777
susan.fogden@fogdenphotos.co.uk
www.fogdenphotos.com

Contact *Susan Fogden*

Natural history collection, with special reference to rain forests and deserts. 'Emphasis on quality rather than quantity'; growing collection of around 25,000 images.

Food Features

Stream House, West Flexford Lane, Wanborough, Guildford GU3 2JW

☎ 01483 810840 🖷 01483 811587

frontdesk@foodpix.co.uk

www.foodfeatures.net

Contacts *Steve Moss, Alex Barker*

Specialized high-quality food and drink photography, features and tested recipes. Clients' specific requirements can be incorporated into regular shooting schedules. Commissions welcome.

Forest Life Picture Library

231 Corstorphine Road, Edinburgh EH12 7AT

☎ 0131 314 6411

neill.campbell@forestry.gsi.gov.uk

www.forestry.gov.uk/pictures

Contact *Neill Campbell*

The official image bank of the Forestry Commission, the library provides a single source for all aspects of forest and woodland management. The comprehensive subject list includes tree species, scenic landscapes, employment, wildlife, flora and fauna, conservation, sport and leisure.

Werner Forman Archive Ltd

36 Camden Square, London NW1 9XA

☎ 020 7267 1034 🖷 020 7267 6026

wfa@btinternet.com

www.werner-forman-archive.com

Colour and b&w coverage of ancient civilizations, oriental and primitive societies around the world. A number of rare collections. Searchable website.

Formula One Pictures

☎ 00 36 26 322 826/00 36 70 776 9682 (mobile)

jt@f1pictures.com

www.f1pictures.com

Contact *John Townsend*

500,000 35mm colour slides, b&w and colour negatives and digital images of all aspects of Formula One Grand Prix racing from 1980 including driver profiles and portraits.

Robert Forsythe Picture Library

16 Lime Grove, Prudhoe NE42 6PR

☎ 01661 834511 🖷 01661 834511

robert@forsythe.demon.co.uk

www.forsythe.demon.co.uk

Contacts *Robert Forsythe, Fiona Forsythe*

30,000 transparencies of industrial and transport heritage; plus a unique collection of 250,000 items of related publicity ephemera from 1945. Image finding service available. Robert Forsythe is a transport/industrial heritage historian and consultant. Nationwide coverage, particularly strong on northern Britain. A bibliography of published material is available.

The Francis Frith Collection

Frith's Barn, Teffont, Salisbury SP3 5QP

☎ 01722 716376 🖷 01722 716881

julia_skinner@francisfrith.co.uk

www.francisfrith.co.uk

Contact *Julia Skinner*, Managing Editor

Publishers of *Frith's Photographic Memories* series of illustrated local books, all featuring nostalgic photographs from the archive, founded by Frith in 1860. The archive contains over 360,000 images of 7,000 British towns.

John Frost Newspapers

▷ entry under Library Services

Andrew N. Gagg's PHOTO FLORA

Town House Two, Fordbank Court, Henwick Road, Worcester WR2 5PF

☎ 01905 748515

gaggsphotoflora@googlemail.com

www.gagg.f2s.com/PHOTOFLORA/photoflora.html

Contact *Andrew Gagg*

Specialist in British and European wild plants, flowers, ferns, grasses, trees, shrubs, etc. with colour coverage of most British and many European species (rare and common) and habitats; also travel in India, Sri Lanka, Nepal, Egypt and North Africa, China, Mexico, Thailand, Tibet, Vietnam and Cambodia.

Galaxy Picture Library

34 Fennels Way, Flackwell Heath, High Wycombe HP10 9BY

☎ 01628 521338

robin@galaxypix.com

www.galaxypix.com

Contact *Robin Scagell*

Specializes in astronomy, space, telescopes, observatories, the sky, clouds and sunsets. Composites of foregrounds, stars, moon and planets prepared to commission. Editorial service available.

Garden and Wildlife Matters Photo Library

'Marlham', Henley's Down, Battle TN33 9BN

☎ 01424 830566 🖷 01424 830224

gardens@gmpix.com

www.gmpix.com

Contact *Dr John Feltwell*

Collection of 110,000 6x4" and 35mm images. General gardening techniques and design; cottage gardens and USA designer gardens. 10,000 species of garden plants and over 1,000 species of trees. Flowers, wild and house plants, trees and crops. Environmental, ecological and conservation pictures, including sea, air, noise and freshwater pollution, SE Asian and Central and South American rainforests; Eastern Europe, Mediterranean. Recycling, agriculture, forestry, horticulture and oblique aerial habitat shots from Europe and USA. Digital images supplied worldwide.

Garden Picture Library

2nd Floor, Waterside House, 9 Woodfield Road, London W9 2BA

☎ 020 7432 8200   ℻ 020 7432 8201
sales@gardenpicture.com
creative@gardenpicture.com
www.gardenpicture.com
Creative Director *Lee Wheatley*

'Our inspirational images of gardens, plants and gardening lifestyle offer plenty of scope for writers looking for original ideas to write about.' The collection covers all garden related subjects; everything from plant portraits to floral graphics and garden design details to whole landscapes. Brings together the work of over 100 professional photographers from across the gardening globe. Holds approximately 300,000 fully captioned images of which around 65,000 are available digitally. Free in-house picture research can be undertaken on request or searches can be made online via keyword or subject category to create lightboxes which can be e-mailed or downloaded. Visitors to the library are welcome by appointment and copies of promotional literature are available on request.

Ed Geldard Photo Collection

9 Sunderland Bridge Village, Durham DH6 5HB
☎ 0191 378 2592
camera-one@talktalk.net
Contact *Ed Geldard*

Approximately 30,000 colour transparencies and b&w negs, all by Ed Geldard, specializing in mountain landscapes: particularly the mountain regions of the Lake District, and the Yorkshire limestone areas, from valley to summit. Commissions undertaken. Books published: *Wainwright's Tour of the Lake District*; *Wainwright in the Limestone Dales*; *The Lake District*; *Northumberland and the Land of the Prince Bishops*.

Genesis Space History Photo Library

20 Goodwood Park Road, Bideford EX39 2RR
☎ 01237 477883

tim@spaceport.co.uk
www.spaceport.co.uk
Contact *Tim Furniss*

Historical b&w spaceflight collection including rockets, spacecraft, spacemen, Earth, moon and planets. Catalogue available on the website.

Geo Aerial Photography

4 Christian Fields, London SW16 3JZ
☎ 020 8764 6292/0115 981 9418
℻ 020 8764 6292/0115 981 9418
geo.aerial@geo-group.co.uk
www.geo-group.co.uk
Contact *Kelly White*

Established 1990 and now a growing collection of aerial oblique photographs from the UK, Scandinavia, Asia and Africa – landscapes, buildings, industrial sites, etc. Commissions undertaken.

GeoScience Features

6 Orchard Drive, Wye TN25 5AU
☎ 01233 812707   ℻ 01233 812707
gsf@geoscience.uk.com
www.geoscience.uk.com

Fully computerized and comprehensive library containing the world's principal source of volcanic phenomena. Extensive collections, providing scientific detail with technical quality, of rocks, minerals, fossils, microsections of botanical and animal tissues, animals, biology, birds, botany, chemistry, earth science, ecology, environment, geology, geography, habitats, landscapes, macro/microbiology, peoples, sky, weather, wildlife and zoology. Over 500,000 original colour transparencies in medium and 35mm-format. Subject lists and CD-ROM catalogue available. Incorporates the RIDA photolibrary and Landform Slides.

Geoslides Photography

4 Christian Fields, London SW16 3JZ
☎ 0115 981 9418   ℻ 0115 981 9418
geoslides@geo-group.co.uk
www.geo-group.co.uk
Contact *John Douglas*

Established in 1968. Landscape and human interest subjects from the Arctic, Antarctica, Scandinavia, UK, Africa (south of Sahara), Middle East, Asia (south and southeast). Also specialist collections of images from British India (the Raj) and Boer War. Forty years of photography services.

Getty Images

101 Bayham Street, London NW1 0AG
☎ 0800 376 7981/020 7428 6109
www.gettyimages.com
Contact *Sales Department*

'Getty Images is the world's leading imagery company, creating and providing the largest collection of still and moving images to communication professionals around the globe.' From sports and news photography to archival and contemporary imagery. For those who wish to commission photographers to fulfil specific needs, the company maintains a full-service department for custom-shot images.

Martin and Dorothy Grace
Boxwood Cottage, The Row, Lyth, Kendal LA8 8DD
℡ 01539 568569
martin.grace@tiscali.co.uk

Colour coverage of natural history, topography and travel in Britain, the Galapagos Islands, France, southern Spain, Peru, Madagascar, the Falklands, South Georgia and the Antarctic. Specializes in trees, shrubs and wild flowers but also cover ferns, birds and butterflies, habitats, landscapes, ecology. High quality digital stock supplied online.

Ronald Grant Archive
▷ The Cinema Museum

Sally and Richard Greenhill
357 Liverpool Road, London N1 1NL
℡ 020 7607 8549
sr.greenhill@virgin.net
www.srgreenhill.co.uk
Photo Librarian *Denise Lalonde*

Social documentary photography in colour and b&w of working lives: pregnancy and birth, child development, education, work, old people, medical, urban. Also Modern China, 1971 to the present; most London statues. Some material from Borneo, USA, India, Israel, Philippines and Sri Lanka.

V.K. Guy Ltd
Browhead Cottage, Troutbeck LA23 1PG
℡ 015394 33519   ℻ 015394 32971
admin@vkguy.co.uk
www.vkguy.co.uk
Contacts *Vic Guy, Mike Guy*

25,000 5x4" UK and Ireland scenic transparencies from Shetland in the north to the Channel Isles in the south, suitable for tourism brochures, calendars, etc. Can be supplied digitally and browsed through via the online database.

Robert Harding World Imagery
58–59 Great Marlborough Street, London W1F 7JY
℡ 020 7478 4000   ℻ 020 7478 4161
info@robertharding.com
www.robertharding.com

A leading source of stock photography with over two million colour images covering a wide range of subjects – worldwide travel and culture, geography and landscapes, people and lifestyle, architecture. Rights protected and royalty-free images. Visitors welcome; telephone or visit the website.

Dennis Hardley Photography, Scottish Photo Library
Rosslynn, Ceum Dunrigh, Benderloch, Oban PA37 1ST
℡ 01631 720434   ℻ 01631 720434
info@dennishardley.com
www.ineedanimage.com
Contacts *Dennis Hardley, Tony Hardley*

Established 1974. About 30,000 images (6x7", 6x9" format colour transparencies) of Scotland: castles, historic, scenic landscapes, islands, transport, etc. Also English views – Liverpool, Chester, Bath, Weston Super Mare, Sussex, Somerset and Cambridge. All images available to buy and instantly download from the website. Includes Scotland, England and North Wales, covering lifestyle, landscapes, transport, cities, etc.

Jim Henderson Photography
Crooktree, Kincardine O'Neil, Aboyne AB34 4JD
℡ 01339 882149   ℻ 01339 882149
JHende7868@aol.com
www.jimhendersonphotography.com
Contact *Jim Henderson*

Scenic and general activity coverage of north east Scotland (Aberdeenshire) and Highlands for tourist, holiday and activity illustration. Specialist collection of over 300 Aurora Borealis displays from 1989–2008 on Royal Deeside and co-author of *The Aurora* (published 1997). Author and photographer for *Aberdeen* and *Aberdeen & Royal Deeside* titles for Aberdeen Journals. Recent images of ancient Egyptian sites: Cairo through to Abu-Simbel. **BAPLA** and Alamy member. Commissions undertaken.

John Heseltine Archive
Mill Studio, Frogmarsh Mill, South Woodchester GL5 5ET
℡ 01453 873792
john@heseltine.co.uk
www.heseltine.co.uk
Contact *John Heseltine*

Over 200,000 colour transparencies and digital files of landscapes, architecture, food and travel with particular emphasis on Italy, France and the UK.

Christopher Hill Photographic
17 Clarence Street, Belfast BT2 8DY
℡ 028 9024 5038   ℻ 028 9023 1942
sales@scenicireland.com
www.scenicireland.com
Contact *Christopher Hill*

A comprehensive collection of scenic landscapes of Ireland. Every aspect of Irish life is shown, concentrating on the positive. Also large miscellaneous section. The website contains over 7,000 images available to buy online.

Hobbs Golf Collection

5 Winston Way, New Ridley, Stocksfield NE43 7RF
☎ 01661 842933   ℱ 01661 842933
info@hobbsgolfcollection.com
www.hobbsgolfcollection.com
Contact *Margaret Hobbs*

Specialist golf collection: players, courses, art, memorabilia and historical topics (1300–present). 40,000+ images, mainly 35mm colour transparencies and b&w prints. All images can be supplied in digital format. Commissions undertaken. Author of 30 golf books.

David Hoffman Photo Library

c/o BAPLA, 18 Vine Hill, London EC1R 5DZ
☎ 020 8981 5041
info@hoffmanphotos.com
www.hoffmanphotos.com
Contact *David Hoffman*

David Hoffman Photo Library is a social issues focused stock archive built up from journalistic and documentary work over more than 30 years. The collection covers drugs and drug use, policing, disorder, race and racists, riots, youth, prostitution, protest, homelessness, housing, environmental demonstrations and events, waste disposal, alternative energy and pollution. A wide range of reportage and general stock, plus topical issues from flash mobs to SUVs. Primarily UK images, plus Europe, USA, Canada, Venezuela, Mexico and Thailand.

Holt Studios International Ltd
▷ Frank Lane Picture Agency Ltd

Houghton's Horses/Kit Houghton Photography

Radlet Cottage, Spaxton, Bridgwater TA5 1DE
☎ 01278 671362   ℱ 01278 671739
kit@enterprise.net
www.houghtonshorses.com
Contacts *Kit Houghton, Kate Houghton*

Specialist equestrian library of over 300,000 transparencies and digital images on all aspects of the horse world, with images ranging from the romantic to the practical and competition pictures in all equestrian disciplines worldwide. Online picture delivery is now the norm. All transparencies in the process of being scanned. Commissions undertaken.

Chris Howes/Wild Places Photography

PO Box 100, Abergavenny NP7 9WY
☎ 01873 737707
photos@wildplaces.co.uk
Contacts *Chris Howes, Judith Calford*

Expanding collection of over 50,000 colour transparencies, b&w prints and digital photos covering travel, topography and natural history worldwide, plus action sports such as climbing. Supply now fully digital. Specialist areas include caves, caving and mines (with historical coverage using engravings and early photographs), wildlife, landscapes and the environment, including pollution and conservation. Europe (including Britain), Canada, USA, Africa and Australia are all well represented within the collection. Commissions undertaken.

Huntley Film Archive

22 Islington Green, London N1 8DU
☎ 020 7226 9260   ℱ 020 7359 9337
films@huntleyarchives.com
www.huntleyarchives.com
Contact *Amanda Huntley*

Originally a private collection, the library is now a comprehensive archive of rare and vintage documentary film dating from 1895. 50,000 films on all subjects of a documentary nature. Online catalogue available.

Jacqui Hurst

66 Richford Street, Hammersmith, London W6 7HP
☎ 020 8743 2315/07970 781336 (mobile)
jacquihurst@yahoo.co.uk
www.jacquihurstphotography.co.uk
Contact *Jacqui Hurst*

A specialist library of traditional and contemporary applied arts, regional food producers and markets. The photos form illustrated essays of how something is made and finish with a still life of the completed object. The collection is always being extended and a list is available on request. Commissions undertaken.

Hutchison Picture Library

65 Brighton Road, Shoreham-by-Sea BN43 6RE
☎ 01273 440113   ℱ 01273 440113
library@hutchisonpictures.co.uk
www.hutchisonpictures.co.uk

Worldwide contemporary images from the straight-forward to the esoteric and quirky. Over half a million documentary colour photographs on file, covering people, places, customs and faiths, agriculture, industry and transport. Special collections include the environment and climate, family life (including pregnancy and birth), ethnic minorities worldwide (including Disappearing World archive), conventional and alternative medicine, and music around the world. Search service available.

Illustrated London News Picture Library

c/o Mary Evans Picture Library, 59 Tranquil Vale,
Blackheath, London SE3 0BS
℡ 020 7805 5585   🖷 020 8852 7211
iln@maryevans.com
www.ilnpictures.co.uk

Engravings, photographs and illustrations from
1842 to the present day, taken from magazines
published by Illustrated Newspapers: *Illustrated
London News*; *Graphic*; *Sphere*; *Tatler*; *Sketch*;
Illustrated Sporting and Dramatic News;
Illustrated War News 1914–18; *Bystander*;
Britannia & Eve. Social history, London,
Industrial Revolution, wars, travel. Now housed
and managed by **Mary Evans Picture Library**
(see entry). Research and scanning services
available.

Images of Africa Photobank

11 The Windings, Lichfield WS13 7EX
℡ 01543 262898
info@imagesofafrica.co.uk
www.imagesofafrica.co.uk
Contact *David Keith Jones, FRPS*

Images covering Botswana, Chad, Egypt,
Ethiopia, Kenya, Lesotho, Madagasca, Malawi,
Morocco, Namibia, Rwanda, South Africa,
Swaziland, Tanzania, Uganda, Zaire, Zambia,
Zanzibar and Zimbabwe. 'Probably the best
collection of photographs of Kenya in Europe.'
A wide range of topics are covered. Strong on
African wildlife with over 80 species of mammals
including many sequences showing action
and behaviour. Popular animals like lions and
elephants are covered in encyclopædic detail.
More than 100 species of birds and many reptiles
are included. Other strengths include National
Parks and reserves, natural beauty, tourism
facilities, traditional and modern people. Most
work is by David Keith Jones, FRPS; several other
photographers are represented. Many thousands
of these images can now be accessed online. Go
to www.imagesofafrica.co.uk and look for Online
Images.

Imperial War Museum Photograph Archive

All Saints Annexe, Austral Street, London SE11 4SJ
℡ 020 7416 5333   🖷 020 7416 5355
photos@iwm.org.uk
www.iwm.org.uk

A national archive of ten million photographs
illustrating all aspects of 20th and 21st century
conflict. Emphasis on the two World Wars but
includes material from other conflicts involving
Britain and the Commonwealth. Majority of
material is b&w, although holdings of colour
material increase with more recent conflicts.
Visitors welcome by appointment, Monday to
Friday, 10.00 am to 5.00 pm.

International Photobank

PO Box 6554, Dorchester DT1 9BS
℡ 01305 854145   🖷 01305 853065
peter@internationalphotobank.co.uk
www.internationalphotobank.co.uk
Contacts *Peter Baker, Gary Goodwin*

Over 450,000 images, in digital and transparency
format. Colour coverage of travel subjects: places,
people, folklore, events. Digital broadband service
available for newspapers, magazines and other
users. Pictures can also be selected from the
comprehensive website.

Robbie Jack Photography

45 Church Road, Hanwell, London W7 3BD
℡ 020 8567 9616   🖷 020 8567 9616
robbie@robbiejack.com
www.robbiejack.com
Contact *Robbie Jack*

Built up over the last 25 years, the library contains
over 500,000 colour images of the performing
arts – theatre, dance, opera and music. Includes
West End shows, the RSC and Royal National
Theatre productions, English National Opera and
Royal Opera. The dance section contains images
of the Royal Ballet, English National Ballet, the
Rambert Dance Company, plus many foreign
companies. Also holds the largest selection of
colour material from the Edinburgh International
Festival. Researchers are welcome to visit by
appointment. From 2001 began offering colour
transparencies as digital images which are being
added to on a daily basis.

Jayawardene Travel Photo Library

7A Napier Road, Wembley HA0 4UA
℡ 020 8795 3581   🖷 020 8795 4083
jaytravelphotos@aol.com
www.jaytravelphotos.com
Contact *Rohith Jayawardene*

160,000 travel and travel-related images – digital
and original transparencies. Many places
photographed in depth, with more than 800
pictures per country. Over 10,000 images
available online.

Paul Kaye B/W Archive

▷ **Sylvia Cordaiy Photo Library**

David King Collection

90 St Pauls Road, London N1 2QP
℡ 020 7226 0149
davidkingcollection@btopenworld.com
www.davidkingcollection.com
Contact *David King*

250,000 b&w original and copy photographs and
colour transparencies of historical and present-
day images. Russian history and the Soviet Union
from 1900 to the fall of Khrushchev; the lives

of Lenin, Trotsky and Stalin; the Tzars, Russo-Japanese War, 1917 Revolution, World War I, Red Army, the Great Purges, Great Patriotic War, etc. Special collections on China, Eastern Europe, the Weimar Republic, John Heartfield, American labour struggles, Spanish Civil War. Open to qualified researchers by appointment, Monday to Friday, 10.00 am to 6.00 pm. Staff will undertake research; negotiable fee for long projects. David King's latest photographic books: *The Commissar Vanishes*, documenting the falsification of photographs and art in Stalin's Russia, and *Ordinary Citizens*, mugshots of victims shot without trial from the archives of Stalin's secret police. A new display of Soviet posters, 1917–1960, from the David King Collection was held at Tate Modern in the spring of 2008.

The Kobal Collection

2 The Quadrant, 135 Salusbury Road, London NW6 6RJ
☎ 020 7624 3500   🖷 020 7624 3355
info@picture-desk.com
www.picture-desk.com

Picture library holding colour and b&w coverage of Hollywood films, portraits, stills, publicity shots, posters and ephemera.

Kos Picture Source Ltd

Po Box 52854, 7 Spice Court, London SW11 3UU
☎ 020 7801 0044   🖷 020 7801 0055
images@kospictures.com
www.kospictures.com
Contact *Chris Savage*

Specialists in water-related images including international yacht racing and cruising, classic boats and superyachts, and extensive range of watersports. Also worldwide travel including seascapes, beach scenes, underwater photography and the weather.

Ed Lacey Collection
▷ Phil Sheldon Golf Picture Library

Landform Slides
▷ GeoScience Features

Frank Lane Picture Agency Ltd

Pages Green House, Wetheringsett, Stowmarket IP14 5QA
☎ 01728 860789   🖷 01728 860222
pictures@flpa-images.co.uk
www.flpa-images.co.uk

Colour coverage of natural history, environment, pets and weather. Represents Sunset from France, Foto Natura from Holland, Minden Pictures from the US, Holt Studios and works closely with Eric and David Hosking, plus 270 freelance photographers. Website with 110,000 images.

Last Resort Picture Library

Manvers Studios, 12 Ollerton Road, Tuxford, Newark NG22 0LF
☎ 01777 870166   🖷 01777 871739
dick@dmimaging.co.uk
www.dmimaging.co.uk
Contacts *Jo Makin, Dick Makin*

Subject areas include agriculture, architecture, IT, education, food, industry, landscapes, people at work, skiing, trees, flowers, mountains and winter landscapes. Images cover a wide variety of topics, ranging from the everyday to the obscure. 'Bespoke service available through our linked photographic studio. Contact us for details.'

LAT Photographic

Teddington Studios, Broom Road, Teddington TW11 9BE
☎ 020 8267 3000   🖷 020 8267 3001
lat.photo@haymarket.com
www.latphoto.co.uk
Contact *Zoë Schafer*

Motor sport collection of almost 12 million images dating from 1895 to the present day.

Lebrecht Music and Arts

3 Bolton Road, London NW8 0RJ
☎ 020 7625 5341
pictures@lebrecht.co.uk
www.lebrecht.co.uk
Contact *Elbie Lebrecht*

Classical music, opera, jazz, rock, musicians, instruments, performers, writers, artists, architects, sculptors, historical personalities, politicians, society and glamorous showbusiness personalities.

The Erich Lessing Archive of Fine Art & Culture

c/o akg-images Ltd, The Arts and History Picture Library, 5 Melbray Mews, 158 Hurlingham Road, London SW6 3NS
☎ 020 7610 6103   🖷 020 7610 6125
enquiries@akg-images.co.uk
www.akg-images.co.uk

Top quality high resolution scans depicting the contents of many of the world's finest art galleries as well as ancient archaeological and biblical sites available in the UK via the akg-images website.

Library of Art & Design
▷ V&A Images

Lindley Library, Royal Horticultural Society

80 Vincent Square, London SW1P 2PE
☎ 020 7821 3050   🖷 020 7821 3022
Contact *Picture Librarian*

22,000 original drawings and approx. 8,000 books with hand-coloured plates of botanical illustrations. Appointments are necessary to visit the collection of unpublished material; all photography is carried out by their own photographer.

Link Picture Library

41A The Downs, London SW20 8HG
☎ 020 8944 6933
library@linkpicturelibrary.com
www.linkpicturelibrary.com
Contact *Orde Eliason*

100,000 images of South Africa, India, China, Vietnam and Israel. A more general collection of colour transparencies from 100 countries worldwide. Link Picture Library has an international network and can source material not in its file from Japan, Scandinavia, India and South Africa. Original photographic commissions undertaken.

London Aerial Photo Library

Studio D1, Fairoaks Airport, Chobham, Woking GU24 8HU
☎ 01276 855997/855344   🖷 01276 855455
info@londonaerial.co.uk
www.londonaerial.co.uk
www.flightimages.com
Contact *Amanda Campbell*
Librarian *Glyn Tassell*

Extensive collections of aerial imagery. 450,000 images including continually updated photography of London plus excellent coverage of the UK in general. Covers landmarks, industrial/retail properties, sporting venues/football stadiums, conceptual shots. Library searches by expert staff, free of charge. Selection of digital images also available via website. Commissioned photography undertaken.

London Metropolitan Archives

40 Northampton Road, London EC1R 0HB
☎ 020 7332 3820   🖷 020 7833 9136
ask.lma@cityoflondon.gov.uk
www.cityoflondon.gov.uk/lma
Contact *The Enquiry Team*

London Metropolitan Archives (LMA) is the largest local authority record office in the UK. Holds over 32 miles of archives – an enormous amount of information about the capital and its people. These include records of London government, hospitals, charities, businesses and parish churches. Types of record range from books and manuscript documents to photographs, maps and drawings. 'Nearly 900 years of London history can be brought to life at LMA.' There is also a 100,000 volume reference library specializing in London history. See entry under *Library Services*.

London's Transport Museum Photographic Library

39 Wellington Street, London WC2E 7BB
☎ 020 7379 6344   🖷 020 7565 7252
www.ltmcollection.org/photos/index.html
www.ltmcollection.org/posters/index.html

Photographic collection reflecting London's public transport history from the 1860s to the present day. The core of the collection is made up of the old London Transport photographic archive of over 150,000 b&w photographs. Also includes a smaller archive of colour material, as well as historic albums and prints from all periods. Over 16,000 images are now available to view and purchase on-line, with new photographs added regularly. Each image has a brief caption, stating location and date (where known) to assist retrieval. For more specialist enquiries, please make an appointment to view the photographic collection at Acton Depot (contact: dutycurator@ltmuseum.co.uk).

London Transport Museum Poster Archives: for 100 years, since Frank Pick commissioned the first graphic poster for London Underground, the company and its successors have kept copies of everything they produced. In the early 20th century, under Pick's guidance, London Transport commissioned work from the best artists and designers in the country. By the 1980s, when the collection was transferred to the Museum, it contained more than 5,000 printed posters and nearly 1,000 original artworks. The collection has grown steadily ever since and is now a uniquely comprehensive picture of a century of British graphic design.

Ludvigsen Library

Scoles Gate, Hawkedon, Bury St Edmunds IP29 4AU
☎ 01284 789246   🖷 01284 789246
library@ludvigsen.com
www.ludvigsen.com
Contact *Karl Ludvigsen*

Extensive information research facilities for writers and publishers. Approximately 400,000 images (both b&w and many colour transparencies) of automobiles and motorsport, from 1890s through 1980s. Glass plate negatives from the early 1900s; Formula One, Le Mans, motor car shows, vintage, antique and classic cars from all countries. Includes the Dalton-Watson Collection and the work of noted photographers such as John Dugdale, Edward Eves, Peter Keen, Max le Grand, Karl Ludvigsen, Rodolfo Mailander, Ove Nielsen, Stanley Rosenthall and others.

MacQuitty International Photographic Collection

7 Elm Lodge, River Gardens, Stevenage Road, London SW6 6NZ
☎ 020 7385 5606   🖷 020 7385 5606
miranda.macquitty@btinternet.com
Contact *Dr Miranda MacQuitty*

Colour and b&w collection on aspects of life in over 70 countries: dancing, music, religion, death, archaeology, buildings, transport, food, drink, nature. Visitors by appointment.

Magnum Photos Ltd

Ground Floor, 63 Gee Street, London EC1V 3RS
☎ 020 7490 1771   🖷 020 7608 0020
magnum@magnumphotos.co.uk
www.magnumphotos.com
Archive Director *Nick Galvin*

Founded 1947 by Cartier Bresson, George Rodger, Robert Capa and David 'Chim' Seymour. Represents over 70 of the world's leading photo-journalists. Coverage of all major world events from the Spanish Civil War to present day. Also a large collection of travel images and personalities.

The Raymond Mander & Joe Mitchenson Theatre Collection

Jerwood Library of the Performing Arts, Trinity College of Music, King Charles Court, Old Royal Naval College, London SE10 9JF
☎ 020 8305 4426   🖷 020 8305 9426
rmangan@tcm.ac.uk
www.mander-and-mitchenson.co.uk
Contact *Richard Mangan*
Archive Officer/Cataloguer *Kristy Davis*

Enormous collection covering all aspects of the theatre: plays, actors, dramatists, music hall, theatres, singers, composers, etc. Visitors welcome by appointment.

The Roger Mann Collection

Wensley Court, 48 Barton Road, Torquay TQ1 4DW
☎ 01803 323868   🖷 01803 616448
rogermann48bart@aol.com
www.therogermanncollection.co.uk
Contact *R.F. Mann*

Comprehensive collection of cricket photographs covering the period 1750 to 1945. The photographs feature most of the first-class players, teams, Test match action and overseas tours of the period. This collection includes almost 2,000 original match scorecards, cartoons, images, prints, postcards, cigarette cards, letters and the personal memorabilia of many of the best-known players of the time. Also some coverage of the period 1946 to 1970.

S&O Mathews Photography

Little Pitt Place, Brighstone, Isle of Wight PO30 4DZ
☎ 01983 741098
oliver@mathews-photography.com
www.mathews-photography.com

Specialist stock library of fine photographs of botanical and horticultural subjects, plants, plant portraits and plant associations, as well as gardens, including views, details and features.

Institution of Mechanical Engineers

1 Birdcage Walk, London SW1H 9JJ
☎ 020 7973 1274   🖷 020 7222 8762
library@imeche.org
www.imeche.org

Historical images and archives on mechanical engineering. Open 9.15 am to 5.30 pm, Monday to Friday.

Lee Miller Archives

Farley Farm House, Muddles Green, Chiddingly, Near Lewes BN8 6HW
☎ 01825 872691   🖷 01825 872733
www.leemiller.co.uk

The work of Lee Miller (1907–77). As a photo-journalist she covered the war in Europe from early in 1941 to VE Day with further reporting from the Balkans. Collection includes photographic portraits of prominent Surrealist artists: Ernst, Eluard, Miró, Picasso, Penrose, Carrington, Tanning, and others. Surrealist and contemporary art, poets and writers, fashion, the Middle East, Egypt, the Balkans in the 1930s, London during the Blitz, war in Europe and the liberation of Dachau and Buchenwald.

Mirrorpix

22nd Floor, One Canada Square, Canary Wharf, London E14 5AP
☎ 020 7293 3700   🖷 020 7293 0357
desk@mirrorpix.com
www.mirrorpix.com
Account Manager *Mel Knight*

Founded 1903. The photographic collection of the *Daily Mirror*. 200,000 images online, archive of 30 million negatives and prints. Ideal for autobiographical, entertainment, social and political history (domestic and foreign), the arts, culture, fashion, industry, sport and royalty.

Monitor Picture Library

The Forge, Roydon, Harlow CM19 5HH
☎ 01279 792700
sales@monitorpicturelibrary.com
www.monitorpicturelibrary.com

Colour and b&w coverage of 1960s, '70s and '80s personalities and celebrities from: music, entertainment, sport, politics, royals, judicial, commerce etc. Specialist files on Lotus cars. Syndication to international, national and local media.

Motoring Picture Library

National Motor Museum, Beaulieu SO42 7ZN
℡ 01590 614656   🖷 01590 612655
motoring.pictures@beaulieu.co.uk
www.motoringpicturelibrary.com
www.alamy.com/mpl
Contacts *Jonathan Day, Tim Woodcock, Tom Wood*

Three-quarters of a million b&w images, plus over 120,000 colour images covering all forms of motoring history from the 1880s to the present day. Commissions undertaken. Own studio.

Mountain Camera
▷ John Cleare

Moving Image Communications

9 Faversham Reach, Faversham ME14 7LA
℡ 0845 257 2968   🖷 01795 534306
mail@milibrary.com
nathalie@milibrary.com
mike@milibrary.com
www.milibrary.com
Contact *Nathalie Banaigs*

Over 16,000 hours of quality archive and contemporary footage, including: Channel X; TVAM Archive 1983–92; Drummer Films (travel classics, 1950–70); The Freud Archive (1930–39); Film Finders (early cinema); Adrian Brunel Films; Cuban Archives; Stockshots (timelapse, cityscapes, land and seascapes, chroma-key); Space Exploration (NASA); Wild Islands; Flying Pictures; National Trust.

Museum of Childhood
▷ V&A Images

Museum of London Picture Library

150 London Wall, London EC2Y 5HN
℡ 020 7814 5604/5612   🖷 020 7600 1058
picturelib@museumoflondon.org.uk
www.museumoflondon.org.uk
Library Manager *Sean Waterman*
Senior Picture Researcher *Nikki Braunton*
Picture Researcher *Sarah Williams*

The Museum of London picture library holds over 35,000 images illustrating the history of London and the life of its people from prehistoric times to the present day. The images are drawn from the Museum's extensive and unique collections of oil paintings, historic photographs, drawings, prints maps and artefacts including costume, jewellery and ceramics.

The National Archives Image Library

Ruskin Avenue, Kew, Richmond TW9 4DU
℡ 020 8392 5225   🖷 020 8487 1974
image-library@nationalarchives.gov.uk
www.nationalarchives.gov.uk/imagelibrary
Contacts *Paul Johnson, Hugh Alexander*

British and colonial history from the Domesday Book to the 1970s, shown in photography, maps, illuminations, posters, advertisements, textiles and original manuscripts. Approximately 30,000 5x4" and 35mm colour transparencies and b&w negatives. Open 9.00 am to 5.00 pm, Monday to Friday.

National Art Library
▷ V&A Images

National Galleries of Scotland Picture Library

The Scottish National Gallery of Modern Art, 75 Belford Road, Edinburgh EH4 3DR
℡ 0131 624 6258/6260   🖷 0131 623 7135
picture.library@nationalgalleries.org
www.nationalgalleries.org
Contact *Shona Corner*

Over 80,000 b&w and several thousand images in colour of works of art from the Renaissance to present day. Specialist subjects cover fine art (painting, sculpture, drawing), portraits, Scottish, historical, still life, photography and landscape.

National Magazine Company
▷ Mary Evans Picture Library

National Maritime Museum Picture Library

Greenwich, London SE10 9NF
℡ 020 8312 6631/6704   🖷 020 8312 6533
picturelibrary@nmm.ac.uk
www.nmm.ac.uk/picturelibrary
Contacts *David Taylor, Doug McCarthy*

Over 400,000 images from the leading collection of maritime art and artefacts, including oil paintings from the 16th century to present day, prints and drawings, historic photographs, ships plans, models, rare maps and charts, globes, manuscripts and navigation and scientific instruments. Images are available as digital files. Contact for further information or for picture research assistance.

National Media Museum
▷ Science & Society Picture Library

National Monuments Record

English Heritage, Kemble Drive, Great Western Village, Swindon SN2 2GZ
℡ 01793 414600   🖷 01793 414606
www.english-heritage.org.uk/nmr

The National Monuments Record (NMR), the public archive of English Heritage, holds over ten million photographs, plans, drawings, reports, records and publications covering England's archaeology, architecture, social and local history. The NMR is the largest publicly accessible archive in Britain and is the biggest dedicated to the historic environment.

National Portrait Gallery Picture Library

St Martin's Place, London WC2H 0HE
☎ 020 7312 2474/5/6   🖷 020 7312 2464
picturelibrary@npg.org.uk
www.npg.org.uk
Contact *Tom Morgan*

Pictures of brilliant, daring and influential characters who have made British history are available for publication. Images can be searched, viewed and ordered on the website. Copyright clearance is arranged for all images supplied.

National Railway Museum Search Engine – Library & Archive

Leeman Road, York YO26 4XJ
☎ 01904 686235   🖷 01904 611112
nrm.researchcentre@nrm.org.uk
www.nrm.org.uk

The National Railway Museum and the **Science and Society Picture Library** are both part of the National Museum of Science and Industry (NMSI) whose other constituents are the Science Museum in London and the National Media Museum in Bradford (formerly the National Museum of Photography, Film & Television). 1.5 million images, mainly b&w, covering every aspect of railways from 1850s to the present day; over one million engineering drawings; the most comprehensive railway library in the UK; sound and oral history archives; a growing digital video record of today's railway industry; a large collection of British railway posters, graphic art and advertising material from the 1820s onwards. Search engine is open 10.00 am to 5.30 pm, seven days a week. Closed 24th–26th December.

The National Trust Photo Library

Heelis, Kemble Drive, Swindon SN2 2NA
☎ 01793 817700   🖷 01793 817401
photo.library@nationaltrust.org.uk
www.ntpl.org.uk
Contact *Chris Lacey*

The National Trust Photo Library houses a unique collection of contemporary photography which vividly illustrates the rich diversity and historical range of properties in the National Trust's care, throughout England, Wales and Northern Ireland. The Library has an exceptional choice of images, created by commissioned specialist photographers, suitable for a broad range of creative applications. 40,000 images can be searched and ordered online.

Natural History Museum Image Resources

Cromwell Road, London SW7 5BD
☎ 020 7942 5401/5324   🖷 020 7942 5443
nhmpl@nhm.ac.uk
www.nhm.ac.uk/piclib
Contacts *Jennifer Vladimirsky, Sally Jennings*

Pictures from the Museum's collections, including dinosaurs, man's evolution, extinct species and fossil remains. Also pictures of gems, minerals, birds and animals, plants and insect specimens, plus historical artworks depicting the natural world.

Nature Photographers Ltd

West Wit, New Road, Little London, Tadley RG26 5EU
☎ 01256 850661
info@naturephotographers.co.uk
www.naturephotographers.co.uk
Contact *Dr Paul Sterry*

Over 150,000 images, digitized to order, on worldwide natural history and environmental subjects. The library is run by a trained biologist and experienced author on his subject.

Nature Picture Library

5a Great George Street, Bristol BS1 5RR
☎ 0117 911 4675   🖷 0117 911 4699
info@naturepl.com
www.naturepl.com
Contact *Helen Gilks*

A collection of 300,000 nature photos from around the world, including strong coverage of animal portraits and behaviour. Other subjects covered include plants, pets, landscapes and travel, environmental issues and wildlife film-makers at work. Thousands of images can be viewed online and downloaded direct for reproduction.

Peter Newark's Picture Library

3 Barton Buildings, Queen Square, Bath BA1 2JR
☎ 01225 334213   🖷 01225 480554

Over one million images covering world history from ancient times to the present day. Includes an extensive military collection of photographs, paintings and illustrations. Also a special collection on American history covering Colonial times, exploration, social, political and the Wild West and Native-Americans in particular. Subject list available. Telephone, fax or write for further information.

NHPA
▷ Photoshot

Odhams Periodicals Library
▷ Popperfoto.com

Offshoot
▷ Skishoot

Only Horses Picture Agency
27 Greenway Gardens, Greenford UB6 9TU
☎ 020 8578 9047   🖷 020 8575 7244
onlyhorsespics@aol.com
www.onlyhorsespictures.com

Colour and b&w coverage of all aspects of the horse. Foaling, retirement, racing, show jumping, eventing, veterinary, polo, breeds, personalities.

Oxford Picture Library
15 Curtis Yard, North Hinksey Lane, Oxford OX2 0LX
☎ 01865 723404   🖷 01865 725294
opl@cap-ox.com
www.cap-ox.co.uk
Contacts *Chris Andrews, Annabel Matthews*

Specialist collection on Oxford: the city, university and colleges, events, people, spires and shires. Also, the Cotswolds, architecture and landscape from Stratford-upon-Avon to Bath; the Thames and Chilterns, including Henley on Thames and Windsor; Channel Islands, especially Guernsey and Sark. Aerial views of all areas specified above. General collection includes wildlife, trees, plants, clouds, sun, sky, water and teddy bears. Commissions undertaken.

Oxford Scientific (OSF)
4th Floor, 83–84 Long Acre, London WC2E 9NG
☎ 020 7836 5591   🖷 020 7379 4650
enquiries@osf.co.uk
www.osf.co.uk

Collection of 350,000 colour transparencies and digital files of wildlife and natural science images supplied by over 300 photographers worldwide, covering all aspects of wildlife plus landscapes, weather, seasons, plants, pets, environment, anthropology, habitats, science and industry, space, creative textures and backgrounds and geology. Macro and micro photography. UK agents for Animals Animals, USA, Okapia, Germany and Dinodia, India. Research by experienced researchers for specialist and creative briefs. Visits welcome, by appointment.

Panos Pictures
1 Honduras Street, London EC1Y 0TH
☎ 020 7253 1424   🖷 020 7253 2752
pics@panos.co.uk
www.panos.co.uk

Documentary colour and b&w library specializing in Third World and Eastern Europe, with emphasis on environment and development issues. Leaflet available. Fifty per cent of all profits from this library go to the Panos Institute to further its work in international sustainable development.

Charles Parker Archive
▷ Birmingham Library Services under Library Services

Ann & Bury Peerless Picture Library
St David's, 22 King's Avenue, Minnis Bay, Birchington-on-Sea CT7 9QL
☎ 01843 841428   🖷 01843 848321
www.peerlessimages.com
Contacts *Ann Peerless, Bury Peerless*

Specialist collection on world religions: Hinduism, Buddhism, Confucianism, Taoism, Jainism, Christianity, Islam, Sikhism, Zoroastrianism (Parsees of India). Geographical areas covered: India, Afghanistan (Bamiyan Valley of the Buddhas), Pakistan, Bangladesh, Sri Lanka, Cambodia (Angkor), Java (Borobudur), Bali, Malaysia, Thailand, Taiwan, Russia, China, Spain, Poland, Uzbekistan (Samarkand and Bukhara), Vietnam. Basis of collection (35mm colour transparencies), historical, cultural, extensive coverage of art (sculpture and miniature paintings), architecture including Pharaonic Egypt, earlier coverage of significance on Iran, Kenya, Uganda and Zimbabwe.

Photofusion
17A Electric Lane, London SW9 8LA
☎ 020 7733 3500   🖷 020 7738 5509
library@photofusion.org
www.photofusionpictures.org
Contact *Liz Somerville*

Colour and b&w coverage of contemporary social and environmental UK issues including babies and children, disability, education, the elderly, environment, family, health, housing, homelessness, people and work. Brochure available.

The Photolibrary Wales
2 Bro-nant, Church Road, Pentyrch, Cardiff CF15 9QG
☎ 029 2089 0311   🖷 029 2089 0311
info@photolibrarywales.com
www.photolibrarywales.com
Contacts *Steve Benbow, Kate Benbow*

Over 100,000 digital images covering all areas and subjects of Wales. Represents the work of 260 photographers, living and working in Wales.

Photoshot
29–31 Saffron Hill, London EC1N 8SW
☎ 020 7421 6000   🖷 020 7421 6006
www.photoshot.com
www.nhpa.co.uk
www.staystill.com
www.worldpictures.com

Contacts *Charles aylor* (U.P.P.A.); *Emma Hier* (Stay Still); *Tracey Howells* (Starstock); *Colin Finlay* (World Illustrated); *Tim Harris* (NHPA & Woodfall Wild Images); *David Brenes* (World ictures)

Photoshot collections include U.P.P.A. (daily national and international business, political and establishment news); Stay Still (exclusive celebrity portraiture and TV publicity images); Starstock (live images of celebrity and entertainment personalities and events from around the world); World Illustrated (the world in pictures, art, culture, environment and heritage); NHPA (wildlife and nature); World Pictures (comprehensive archive of practical and stylish travel images); Woodfall Wild Images (wildlife, landscape and environmental).

PictureBank Photo Library Ltd

Parman House, 30–36 Fife Road, Kingston upon Thames KT1 1SY
℡ 020 8547 2344   ℻ 020 8974 5652
info@picturebank.co.uk
www.picturebank.co.uk

Over 400,000 images covering people (girls, couples, families, children), travel and scenic (UK and world), moods (sunsets, seascapes, deserts, etc.), industry and technology, environments and general. Commissions undertaken. Visitors welcome. Member of **BAPLA**. New material in digital format welcome.

Pictures Colour Library

10 James Whatman Court, Turkey Mill, Ashford Road, Maidstone ME14 5SS
℡ 01622 609809   ℻ 01622 609806
enquiries@picturescolourlibrary.co.uk
www.picturescolourlibrary.co.uk

Travel and travel-related images depicting lifestyles and cultures, people and places, attitudes and environments from around the world, including a comprehensive section on Great Britain.

Axel Poignant Archive

115 Bedford Court Mansions, Bedford Avenue, London WC1B 3AG
℡ 020 7636 2555   ℻ 020 7636 2555
Rpoignant@aol.com
Contact *Roslyn Poignant*

Anthropological and ethnographic subjects, especially Australia and the South Pacific. Also Scandinavia (early history and mythology), Sicily and England.

H.G. Ponting
▷ Popperfoto.com

Popperfoto.com

The Old Mill, Overstone Farm, Overstone, Northampton NN6 0AB
℡ 01604 670670   ℻ 01604 670635
inquiries@popperfoto.com
www.popperfoto.com

Home to over 14 million images, covering 150 years of photographic history. Renowned for its archival material, a world-famous sports library and stock photography. Popperfoto's credit line includes Bob Thomas Sports Photography, UPI, Acme, INP, Planet, Paul Popper, Exclusive News Agency, Victory Archive, Odhams Periodicals Library, Illustrated, Harris Picture Agency, and H.G. Ponting. Colour from 1940, b&w from 1870 to the present. Major subjects covered worldwide include events, personalities, wars, royalty, sport, politics, transport, crime, history and social conditions. Material available on the same day to clients throughout the world. Nearly 200,000 high-resolution images online available by direct download, e-mail, FTP and CD. Researchers welcome by appointment.

PPL (Photo Agency) Ltd

Bookers Yard, The Street, Walberton, Arundel BN18 0PF
℡ 01243 555561   ℻ 01243 555562
ppl@mistral.co.uk
www.pplmedia.com
Contacts *Barry Pickthall, Richard Johnson*

Two million pictures covering watersports, sub-aqua, business and commerce, travel and tourism; pictures of yesteryear and a fast growing archive on Sussex and the home counties. Pictures available in high resolution directly from the website.

Premaphotos Wildlife

Amberstone, 1 Kirland Road, Bodmin PL30 5JQ
℡ 01208 78258   ℻ 01208 72302
enquiries@premaphotos.com
www.premaphotos.com
Contact *Jean Preston-Mafham*, Library Manager

Natural history worldwide. Subjects include flowering and non-flowering plants, fungi, slime moulds, fruits and seeds, galls, leaf mines, seashore life, mammals, birds, reptiles, amphibians, insects, spiders, habitats, scenery and cultivated cacti. Commissions undertaken. Searchable website. Visitors welcome.

Harry Price Library of Magical Literature
▷ Mary Evans Picture Library

Professional Sport UK Ltd

18–19 Shaftesbury Quay, Hertford SG14 1SF
℡ 01992 505000   ℻ 01992 505020

pictures@prosport.co.uk
www.professionalsport.com

Photographic coverage of tennis, soccer, athletics, golf, cricket, rugby, winter sports and many minor sports. Major international events including the Olympic Games, World Cup soccer and all Grand Slam tennis events. Also news and feature material supplied worldwide. Online photo archive; photo transmission services available for editorial and advertising.

Public Record Office Image Library
▷ The National Archives

Punch

87–135 Brompton Road, London SW1X 7XL
☎ 020 7225 6710   🖷 020 7225 6712
punch.library@harrods.com
www.punch.co.uk
www.punchcartoons.com
Owner *Punch Limited*

'Illustrate your point more effectively with a Punch cartoon.' Choose from half a million images published from 1841 onwards. 'The world's largest cartoon archive' with a vast subject range including social history, politics, fashion, famous faces and much, much more all drawn by some of the worlds' greatest cartoonists such as Sir John Tenniel, E.H. Shepard and Norman Thelwell. 'Search through thousands of the very best' at www.punchcartoons.com

Redferns Music Picture Library

7 Bramley Road, London W10 6SZ
☎ 020 7792 9914   🖷 020 7792 0921
info@redferns.com
www.redferns.com

Picture library covering every aspect of music, from 18th century classical to present day pop. Over one million archived artists and other subjects including musical instruments, recording studios, crowd scenes, festivals, etc. Brochure available. Over 250,000 images available on the website. 'Contact us for free lightbox searches.'

Report Digital

4 Clarence Road, Stratford on Avon CV37 9DL
☎ 01789 262151
info@reportdigital.co.uk
www.reportdigital.co.uk
Contact *John Harris*

A collection of 60,000 images. Subjects available online include work issues and occupations, leisure, economy, health, education, politics, social issues, protest, trades union, environmental issues, culture. An increasing coverage of international issues of globalization, migration and climate change. Pictures are available online;

search, browse and download high resolution images. Member of **BAPLA**.

Retna Pictures Ltd

Units 1a & 1b, Farm Lane Trading Estate, 101 Farm Lane, London SW6 1QJ
☎ 0845 034 0645   🖷 0845 034 0646
info@retna.co.uk
www.retna.co.uk

Established 1978, Retna Pictures Ltd is a leading picture agency with two libraries: celebrity/music and lifestyle. The former specializes in images of international and national celebrities, music from the '60s to the current day, films and personalities. The lifestyle library specializes in people, family life, work, leisure and food. Both libraries are constantly receiving new material from established and up and coming photographers.

Rex Features Ltd

18 Vine Hill, London EC1R 5DZ
☎ 020 7278 7294   🖷 020 7696 0974
enquiries@rexfeatures.com
www.rexfeatures.com
Contact *Sales Manager*

Extensive picture library established in the 1950s. Daily coverage of news, politics, personalities, showbusiness, glamour, humour, art, medicine, science, landscapes, royalty, etc.

RIDA Photolibrary
▷ GeoScience Features

Royal Air Force Museum

Grahame Park Way, Hendon, London NW9 5LL
☎ 020 8205 2266   🖷 020 8200 1751
photographic@rafmuseum.org
www.rafmuseum.org
www.rafmuseumphotos.com

About a quarter of a million images, mostly b&w, with around 1,500 colour in all formats, on the history of aviation. Particularly strong on the activities of the Royal Air Force from the 1870s to 1970s. Researchers are requested to enquire in writing only.

The Royal Collection, Photographic Services

St. James's Palace, London SW1A 1JR
☎ 020 7839 1377   🖷 020 7024 5643
picturelibrary@royalcollection.org.uk
www.royalcollection.org.uk
www.royal.gov.uk
Contacts *Shruti Patel, Karen Lawson*

Photographic material of items in the Royal Collection, particularly oil paintings, drawings and watercolours, works of art, and interiors and exteriors of royal residences. Over 35,000 colour transparencies, plus 25,000 b&w negatives.

Royal Geographical Society Picture Library

1 Kensington Gore, London SW7 2AR

☎ 020 7591 3060   🖶 020 7591 3001

images@rgs.org

www.rgs.org/images

Contacts *Jamie Owen, Joy Wheeler*

A strong source of geographical images, both historical and modern, showing the world through the eyes of photographers and explorers dating from the 1830s to the present day. The RGS Contempory Collection provides up-to-date transparencies from around the world, highlighting aspects of cultural activity, environmental phenomena, anthropology, architectural design, travel, mountaineering and exploration. The RGS Picture Library offers a professional and comprehensive service for both commercial and academic use.

Royal Photographic Society Collection

▷ Science & Society Picture Library

RSPB Images

The Old Dairy, Broadfield Road, Sheffield S8 0XQ

☎ 0114 258 0001   🖶 0114 258 0101

sales@rspb-images.com

www.rspb-images.com

Contact *Naddy Tweed*

RSPB Images represents some of the UK's leading wildlife photographers, and holds stunning images ranging from birds, butterflies, moths, mammals, reptiles and their habitats to most RSPB reserves, plants and abstract subjects. High resolution downloads are now available for many of the images and a free picture research service is also provided.

RSPCA Photolibrary

RSPCA Trading Limited, Wilberforce Way, Southwater, Horsham RH13 9RS

☎ 01403 793 150

pictures@rspcaphotolibrary.com

www.rspcaphotolibrary.com

With over 150,000 colour images, the RSPCA Photolibrary has a comprehensive collection of natural history images whose subjects include mammals, birds, domestic and farm animals, amphibians, insects and the environment, as well as a unique photographic record of the RSPCA's work. Also includes the Wild Images collection. Downloadable website and e-mail lightboxes. No search fees.

Russia and Eastern Images

'Sonning', Cheapside Lane, Denham, Uxbridge UB9 5AE

☎ 01895 833508

easteuropix@btinternet.com

www.easteuropix.com

Architecture, cities, landscapes, people and travel images of Russia and the former Soviet Union. Considerable background knowledge available and Russian language spoken.

Peter Sanders Photography Ltd

24 Meades Lane, Chesham HP5 1ND

☎ 01494 773674/771372   🖶 01494 773674

photos@petersanders.com

www.petersanders.com

Contacts *Peter Sanders, Hafsa Garwatuk*

Specializes in the world of Islam in all its aspects from culture, arts, industry, lifestyles, celebrations, etc. Areas included are north, east and west Africa, the Middle East (including Saudi Arabia), China, Asia, Europe and USA. A continually expanding library.

Science & Society Picture Library

Science Museum, Exhibition Road, London SW7 2DD

☎ 020 7942 4400   🖶 020 7942 4401

piclib@nmsi.ac.uk

www.scienceandsociety.co.uk

Contact *David Thompson*

'The Science & Society Picture Library has one of the widest ranges of photographs, paintings, prints, posters and objects in the world.' The images come from the Science Museum, the **National Railway Museum** and the National Media Museum (formerly the National Museum of Photography, Film & Television) which now includes the Royal Photographic Society Collection. Images are available as high or low resolution files via e-mail, FTP or direct from the website.

Science Museum

▷ Science & Society Picture Library

Science Photo Library

327–329 Harrow Road, London W9 3RB

☎ 020 7432 1100   🖶 020 7286 8668

info@sciencephoto.com

www.sciencephoto.com

Subjects covered include the human body, health and medicine, research, genetics, technology and industry, space exploration and astronomy, earth science, satellite imagery, environment, flowers, plants and gardens, nature and wildlife and the history of science. The whole collection, more than 250,000 images, is available online.

Mick Sharp Photography

Eithinog, Waun, Penisarwaun, Caernarfon LL55 3PW

☎ 01286 872425   🖶 01286 872425

mick.jean@virgin.net

Contacts *Mick Sharp, Jean Williamson*

Archaeology, ancient monuments, buildings, churches, countryside, environment, history, landscape, past cultures and topography. Emphasis on British Isles but material also from other countries. Features the photos of both Mick Sharp and Jean Williamson. Access to other specialist collections on related subjects. Commissions undertaken. Medium format and 35mm colour transparencies plus b&w prints from 5x4 negatives.

Phil Sheldon Golf Picture Library

40 Manor Road, Barnet EN5 2JQ
℡ 020 8440 1986   ℻ 020 8440 9348
gill@philsheldongolfpics.co.uk
info@philsheldongolfpics.co.uk
www.philsheldongolfpics.co.uk

An expanding collection of over 600,000 quality images of the 'world of golf'. In-depth worldwide tournament coverage including every Major championship and Ryder Cup since 1976. Instruction, portraits, trophies and over 400 golf courses from around the world. Also the Dale Concannon collection covering the period 1870 to 1940, the classic 1960s collection by photographer Sidney Harris and the Ed Lacey Collection.

Skishoot–Offshoot

Hall Place, Upper Woodcott, Whitchurch RG28 7PY
℡ 01635 255527   ℻ 01635 255528
pictures@skishoot.co.uk
www.skishoot.co.uk
Contact *Claire Randall*

Skishoot ski and snowboarding picture library has 500,000 images. Offshoot travel library specializes in France. Also produces the ski information website: www.welove2ski.com

The Skyscan Photolibrary

Oak House, Toddington, Cheltenham GL54 5BY
℡ 01242 621357   ℻ 01242 621343
info@skyscan.co.uk
www.skyscan.co.uk
Contact *Brenda Marks*

As well as the Skyscan Photolibrary collection of unique balloon's-eye views of Britain, the library includes the work of photographers from across the aviation spectrum; air to ground, aviation, aerial sports – 'in fact, anything aerial!' Links have been built with photographers across the world; photographs can be handled on an agency basis and held in house, or as a brokerage where the collection stays with the photographer; terms 50/50 for both. Commissioned photography arranged. Enquiries welcome.

Snookerimages (Eric Whitehead Photography)

10 Brow Close, Bowness on Windermere LA23 2HA
℡ 01539 448894
snooker@snookerimages.co.uk
www.snookerimages.co.uk
www.snookerimages-pictures.com
Contact *Eric Whitehead*

Over 30,000 images of snooker from 1982 to the present day.

Solo Syndication Ltd

17–18 Haywards Place, Clerkenwell, London EC1R 0EQ
℡ 020 7566 0360   ℻ 020 7566 0388
Syndication Director *Trevor York*
Sales *Danny Howell, Nick York*
Online transmissions *Geoff Malyon*
(℡ 020 7566 0370)

Three million images from the archives of the *Daily Mail, Mail on Sunday, Evening Standard* and *Evening News*. Hard prints or Mac-to-Mac delivery. 24-hour service.

Sotheby's Picture Library

Level 2, Olympia 2, Hammersmith Road, London W14 8UX
℡ 020 7293 5383   ℻ 020 7293 5062
piclib@sothebys.com
www.sothebys.com
Contact *Joanna Ling*

The library consists of over one million subjects sold at Sotheby's. Images from the 15th to the 20th century. Oils, drawings, watercolours, prints and decorative items. 'Happy to do searches or, alternatively, visitors are welcome by appointment.'

South American Pictures

48 Station Road, Woodbridge IP12 4AT
℡ 01394 383963
info@southamericanpictures.com
www.southamericanpictures.com
www.nonesuchinfo.info (specialist site)
Contact *Marion Morrison*

Colour and b&w images of South/Central America, Cuba, Mexico, New Mexico (USA), Dominican Republic and Haiti, including archaeology and the Amazon. There is an archival section, with pictures and documents from most countries. Now with 40 contributing photographers. All images are supplied digitally.

Starstock
▷ Photoshot

Stay Still
▷ Photoshot

Still Digital

1c Castlehill, Doune FK16 6BU
℡ 01786 842790
info@stillmovingpictures.com
www.stilldigital.co.uk
Contact *John Hutchinson*

Fully searchable and downloadable service for thousands of Scottish images.

Still Pictures' Whole Earth Photolibrary

199 Shooters Hill Road, London SE3 8UL
℡ 020 8858 8307   ℻ 020 8858 2049
info@stillpictures.com
www.stillpictures.com
Contact *Ben Alcraft*

Founded 1970. High profile photo library specializing in environment, the Third World, social issues and nature. Represents 15 leading European and US agencies and over 400 photographers as well as the United Nations Environment Programme (UNEP) archive, the Christian Aid collection and a growing selection of images from Woodfall Wild Images. The website has nearly 250,000 images online, ready for instant download.

Stockfile

5 High Street, Sunningdale SL5 0LX
℡ 01344 872249
info@stockfile.co.uk
www.stockfile.co.uk
Contacts *Jill Behr, Steven Behr*

Specialist cycling collection with emphasis on mountain biking. Expanding adventure sports section covering snow, land, air and water activities.

Stockscotland.com

Croft Studio, Croft Roy, Crammond Brae, Tain IV19 1JG
℡ 01862 892298   ℻ 01862 892298
info@stockscotland.com
www.stockscotland.com
Contact *Hugh Webster*

150,000 Scottish images with 10,000 currently available online.

Sir John Benjamin Stone
▷ Birmingham Library Services under Library Services

Jessica Strang Photo Library

504 Brody House, Strype Street, London E1 7LQ
℡ 020 7247 8982   ℻ 0870 705 9992
jessica@jessicastrang.com
Contact *Jessica Strang*

Approximately 60,000 transparencies covering architecture, interiors (contemporary), gardens, 'obsessive and not just small but tiny, or from almost no space at all', men, women, couples and animals in architecture, and vanishing London details. Recycled ideas for the home.

Joe Tasker Collection
▷ Chris Bonington Picture Library

Tate Images

Top Floor, The Lodge, Tate Britain, Millbank, London SW1P 4RG
℡ 020 7887 8871/8890/4933   ℻ 020 7887 8805
tate.images@tate.org.uk
www.tate-images.com
Contact *Alison Fern*

Tate is one of the world's leading visual arts organizations and its collection encompasses the national collection of historic British art from 1500, including iconic masterpieces by Gainsborough, Constable, Turner, David Hockney and Henry Moore, and the national collection of international modern art which includes works by Dali, Picasso, Matisse, Rothko, Emin, Hirst, Warhol and Andreas Gursky. This vast collection of art imagery is available from the Tate Images website. Clients can browse the entire Tate collection using a new, keyword-based search engine; select images and save their selections to a personal light box; download low-resolution imagery for layout purposes. E-commerce coming soon.

Theatre Museum
▷ V&A Images

Bob Thomas Sports Photography
▷ Popperfoto.com

Thoroughbred Photography Ltd

The Hornbeams, 2 The Street, Worlington IP28 8RU
℡ 01638 713944
mail@thoroughbredphoto.com
www.thoroughbredphoto.com
Contacts *Gill Jones, Laura Green*

Photography by Trevor Jones. Extensive library of high-quality colour images depicting all aspects of thoroughbred horseracing dating from 1987. Major group races, English Classics, studs, stallions, mares and foals, early morning scenes, personalities, jockeys, trainers and prominent owners. Also international work: USA Breeders' Cup, Arc de Triomphe, French Classics, Irish Derby, Dubai World Cup Racing and scenes; and more unusual scenes such as racing on the sands at low tide, Ireland, and on the frozen lake at St Moritz. Visitors by appointment.

TopFoto

PO Box 33, Edenbridge TN8 5PF
℡ 01732 863939   ℻ 01732 860215
admin@topfoto.co.uk
www.topfoto.co.uk

Contact *Alan Smith*

International editorial distributor with over two million pictures available online to download and 12 million in hard copy, representing 40 leading suppliers.

B.M. Totterdell photography

Constable Cottage, Burlings Lane, Knockholt TN14 7PE

☎ 01959 532001   🖷 01959 532001

btrial@btinternet.com

whatvolleyball.co.uk

Contact *Barbara Totterdell*

Specialist volleyball library covering all aspects of the sport.

Tessa Traeger Picture Library

7 Rossetti Studios, 72 Flood Street, London SW3 5TF

☎ 020 7352 3641   🖷 020 7352 4846

info@tessatraeger.com

www.tessatraeger.com

Food, gardens, travel and artists.

Travel Ink Photo Library

The Old Coach House, 14 High Street, Goring on Thames, Nr Reading RG8 9AR

☎ 01491 873011   🖷 01491 875558

info@travel-ink.co.uk

www.travel-ink.co.uk

Contacts *Frances Honnor, Felicity Bazell*

A collection of over 120,000 travel, tourism and lifestyle images, carefully edited and constantly updated, from countries worldwide. Specialist collections from the UK, Greece, France, Far East and Caribbean. The website offers a fully captioned and searchable selection of over 31,000 images and is ideal for picture researchers.

Peter Trenchard Photography

The Studio, West Hill, St Helier, Jersey JE2 3HB

☎ 01534 769933   🖷 01534 789191

peter-trenchard@jerseymail.co.uk

www.peter-trenchard.com

Contact *Peter Trenchard, FBIPP, AMPA, PPA*

Tropix Photo Library

44 Woodbines Avenue, Kingston upon Thames KT1 2AY

☎ 020 8546 0823

info@tropix.co.uk

www.tropix.co.uk

Managing Director *Veronica Birley*

Specializes in images of developing nations: travel and editorial pictures emphasizing the attractive and progressive. Fully searchable, keyworded website. Evocative photos concerning the economies, environment, culture and society of 100+ countries across Africa, Central and South America, Caribbean, Eastern Europe, Middle East, Indian sub-continent, South East Asia, CIS and Far East. Worldwide travel collections also include UK, Europe, North America and Antarctica. Assignment photography available. All photos supplied with detailed captions. Established 1982. **BAPLA** member.

True North

Louper Weir, Ghyll Head, Windermere LA23 3LN

☎ 07941 630420

hurlmere@btinternet.com

www.northpix.co.uk

Contact *John Morrison*

The collection features the life and landscape of the north of England, photographed by John Morrison. No other photographer's work required.

U.P.P.A.
▷ Photoshot

Ulster Museum Picture Library

Botanic Gardens, Belfast BT9 5AB

☎ 028 9038 3000 ext 3114   🖷 028 9038 3103

michelle.ashmore@magni.org.uk

www.ulstermuseum.org.uk

Specialist subjects: art (fine and decorative, late 17th–20th century), particularly Irish art, archaeology, ethnography, treasures from the Armada shipwrecks, geology, botany, zoology, local history and industrial archaeology. The Museum is closed until 2009 for major redevelopment. Until then, contact the Ulster Folk and Transport Museum at Cultra, Holywood, County Down BT18 0EU ☎ 028 9042 8428.

United Nations Environment Programme (UNEP)
▷ Still Pictures' Whole Earth Photolibrary

Universal Pictorial Press & Agency (U.P.P.A.)
▷ Photoshot

UPI
▷ Popperfoto.com

V&A Images

Victoria and Albert Museum, Cromwell Road, South Kensington, London SW7 2RL

☎ 020 7942 2489 (commercial)/2479 (academic)

🖷 020 7942 2482

vaimages@vam.co.uk

www.vandaimages.com

A vast collection of photographs from the world's largest museum of decorative and applied arts, reflecting culture and lifestyle spanning over 1,000 years of history to the present time. Digital delivery of contemporary and historical textiles, costumes and fashions, ceramics, furniture, metalwork, glass, sculpture, toys, and games, design and photographs from around the world. Unique photographs include 1960s fashion by John French, Harry Hammond's behind the scenes pop idols, Houston Rogers theatrical world of the 1930s to 1970s, images of royalty by Lafayette, Cecil Beaton, and the 19th century pioneer photographers. Images from the National Art Library and Library of Art & Design, and from the Theatre Museum and Museum of Childhood are readily available.

Victory Archive
▷ Popperfoto.com

Vin Mag Archive Ltd
84–90 Digby Road, London E9 6HX
☎ 020 8533 7588   🖷 020 8533 7283
piclib@vinmag.com
www.vinmagarchive.com
Contact *Angela Maguire*

Wholly owned subsidiary of the Vintage Magazine Company. 'A unique source of images covering just about any subject, period, theme and location.' Also a collection of half a million movie, TV, celebrity and sports images available online. Contact *Angela Maguire* for code and password. See also entry under *Library Services*.

John Walmsley Photo Library
▷ Education Photos

Warwickshire Photographic Survey
▷ Birmingham Library Services under Library Services

Waterways Photo Library
39 Manor Court Road, Hanwell, London W7 3EJ
☎ 020 8840 1659   🖷 020 8567 0605
watphot39@aol.com
www.waterwaysphotolibrary.com
Contact *Derek Pratt*

A specialist photo library on all aspects of Britain's inland waterways. Top-quality 35mm and medium-format colour transparencies, plus a large collection of b&w and an increasing collection of digital photography. Rivers and canals, bridges, locks, aqueducts, tunnels and waterside buildings. Town and countryside scenes, canal art, waterway holidays, boating, fishing, windmills, watermills, watersports and wildlife.

Weimar Archive
▷ Mary Evans Picture Library

Wellcome Images
183 Euston Road, London NW1 2BE
☎ 020 7611 8348   🖷 020 7611 8577
images@wellcome.ac.uk
http://images.wellcome.ac.uk
Contact *Venita Paul*

Approximately 180,000 images on the history of medicine and human culture worldwide, including religion, astronomy, botany, genetics, landscape and cell biology.

West Cornwall Photographic Archive
▷ Morrab Library under Library Services

Westminster Reference Library
▷ entry under Library Services

Eric Whitehead Photography
10 Brow Close, Bowness on Windermere
LA23 2HA
☎ 01539 448894
snooker@snookerimages.co.uk
www.snookerimages.co.uk
www.snookerimages-pictures.com
Contact *Eric Whitehead*

The agency covers local news events, PR and commercial material, also leading library of snooker images (see **Snookerimages**).

Wild Images
▷ RSPCA Photolibrary

Wild Places Photography
▷ Chris Howes

David Williams Picture Library
50 Burlington Avenue, Glasgow G12 0LH
☎ 0141 339 7823   🖷 0141 337 3031
david@scotland-guide.co.uk

Specializes in travel photography with wide coverage of Scotland, Iceland and Spain. Many other European countries also included plus smaller collections of Western USA and Canada. The main subjects in each country are: cities, towns, villages, 'tourist haunts', buildings of architectural or historical interest, landscapes and natural features. The Scotland and Iceland collections include many pictures depicting physical geography and geology. Photographic commissions and illustrated travel articles undertaken. Catalogue available.

Vaughan Williams Memorial Library
English Folk Dance and Song Society, Cecil Sharp House, 2 Regent's Park Road, London NW1 7AY
☎ 020 7485 2206 ext. 29/33/
☎ direct line: 020 7241 8959   🖷 020 7284 0523
library@efdss.org
www.efdss.org

Mainly b&w coverage of traditional/folk music, dance and customs worldwide, focusing on Britain and other English-speaking nations. Photographs date from the late 19th century to the present day.

The Wilson Photographic Collection
▷ Dundee Central Library under Library Services

Woodfall Wild Images
▷ Photoshot

World Illustrated
▷ Photoshot

World Pictures
▷ Photoshot

Yemen Pictures

28 Sheen Common Drive, Richmond TW10 5BN
☎ 020 7602 1989   ⊞ 020 7602 1989
www.facesandplacespix.com

Large collection (4,000 transparencies) covering all aspects of Yemen – culture, people, architecture, dance, qat, music. Also Africa, Australia, Middle East, and Asia.

York Archaeological Trust Picture Library

47 Aldwark, York YO1 7BX
☎ 01904 663006   ⊞ 01904 663024

ckyriacou@yorkat.co.uk
www.yorkarchaeology.co.uk

Specialist library of rediscovered artifacts, historic buildings and excavations, presented by the creators of the highly acclaimed Jorvik Viking Centre. The main emphasis is on the Roman, Anglo-Saxon, Medieval and Viking periods.

John Robert Young Collection

Paxvobiscum, 16 Greenacres Drive, Ringmer, Lewes BN8 5LX
☎ 01273 814172
pax@freedom255.com
www.johnrobertyoung.com
Contacts *Jennifer Barrett, John Robert Young*

50,000 colour and b&w images. Military (French Foreign Legion, Spanish Foreign Legion, Royal Marines), religion (Christian), travel, China, personalities of the 1960s and '70s, also sixties' fashion.

Press Cuttings Agencies

Cision

Cision House, 16–22 Baltic Street West, London
EC1Y 0UL
☏ 0870 736 0010   ⨎ 020 7689 1164
info.uk@cision.com
www.cision.com

Formerly Romeike, Cision monitors national
and international dailies and Sundays, provincial
papers, consumer magazines, trade and technical
journals, teletext services as well as national radio
and TV networks. Back research, advertising
checking and Internet monitoring, plus analysis
and editorial summary service available.

Durrants

Discovery House, 28–42 Banner Street, London
EC1Y 8QE
☏ 020 7674 0200   ⨎ 020 7674 0222
contact@durrants.co.uk
www.durrants.co.uk

Wide coverage of all print media sectors including
foreign press plus Internet, newswire and
broadcast monitoring. High speed, early morning
press cuttings from the national press by web or
e-mail. Overnight delivery via courier to most
areas or first-class mail. Well presented, laser
printed, A4 cuttings. Rates on application.

International Press-Cutting Bureau

224–236 Walworth Road, London SE17 1JE
☏ 020 7708 2113   ⨎ 020 7701 4489
info@ipcb.co.uk
www.ipcb.co.uk
Contact *Robert Podro*

Covers national, provincial, trade, technical and
magazine press. Cuttings are normally sent twice
weekly by first-class post. Charges start at £75 per
month + £1.10 per cutting.

We Find It (Press Clippings)

40 Galwally Avenue, Belfast BT8 7AJ
☏ 028 9064 6008   ⨎ 028 9064 6008
Contact *Avril Forsythe*

Specializes in Northern Ireland press and
magazines, both national and provincial. Rates on
application.

Festivals

Aberdeen Arts Carnival

Aberdeen Arts Centre, 33 King Street, Aberdeen
AB24 5AA
☏ 01224 635208
www.aberdeenartscentre.org.uk
Venue Manager *Paula Gibson*

Performances by local amateurs and professional
theatre companies. Arts workshops in drama,
music, art, dance and creative writing take place
each summer during the school holidays.

University of Aberdeen Writers Festival
▷ Word

The Aldeburgh Literary Festival

44 High Street, Aldebugh IP15 5AB
☏ 01728 452389   ℻ 01728 452389
johnandmary@aldeburghbookshop.co.uk
www.aldeburghbookshop.co.uk
Festival Organizers *John & Mary James*

Founded 2002. Held on the first weekend
in March. The programme includes literary
workshops, a literary dinner, and talks. Previous
speakers have included Beryl Bainbridge,
Alan Bennett, Craig Brown, Richard Dawkins,
Lady Antonia Fraser, Michael Frayn, Anthony
Horowitz, P.D. James, Doris Lessing, David
Lodge, Ian McEwan, Harold Pinter, Matt Ridley,
Alexander McCall Smith, Claire Tomalin, Salley
Vickers, A.N. Wilson.

Aldeburgh Poetry Festival

The Poetry Trust, The Cut, 9 New Cut,
Halesworth IP19 8BY
☏ 01986 835950   ℻ 01986 874524
info@thepoetrytrust.org
www.thepoetrytrust.org
Director *Naomi Jaffa*

Founded 1989. One of the UK's biggest
celebrations of international contemporary
poetry, held in the small coastal town of
Aldeburgh annually in early November (6th–8th
in 2009). Features readings, performances, craft
talks, public masterclass, workshops, discussions
and a family event. Includes a reading by the
winner of the annual **Jerwood Aldeburgh First
Collection Prize** (see entry under *Prizes*).

Aspects Literature Festival

North Down Borough Council, Town Hall, The
Castle, Bangor BT20 4BT
☏ 028 9127 8032   ℻ 028 9127 1370
gail.prentice@northdown.gov.uk
Festival Coordinator/Arts Officer *Gail Prentice*

Founded 1992. 'Ireland's Premier Literary Festival'
is held at the end of September and celebrates the
richness and diversity of living Irish writers with
occasional special features on past generations. It
draws upon all disciplines – fiction (of all types),
poetry, theatre, non-fiction, cinema, song-writing,
etc. It also includes a day of writing for young
readers and sends writers to visit local schools
during the festival. Highlights of recent festivals
include appearances by Bernard MacLaverty,
Marion Keyes, Alice Taylor, Frank Delaney,
Seamus Heaney, Brian Keenan and Fergal Keane.

Aye Write! Bank of Scotland Book Festival

The Mitchell Library, North Street, Charing
Cross, Glasgow G3 7DN
☏ 0141 287 2999/2876   ℻ 0141 287 2815
lil@csglasgow.org
www.ayewrite.com
Festival Director *Karen Cunningham*
Festival Programmer *Andrew Kelly*

Launched 2005. Annual festival which aims
to increase the use of libraries, encourage a
love of books, reading and writing and an
awareness of Glasgow's writing heritage. The
2008 Festival guests included Kathleen Turner,
Gerry Anderson, Hanif Kureishi, Dame Helena
Kennedy, Louis de Bernières, Edwin Morgan, Will
Self, Alasdair Gray, Joanne Harris. 2009 Festival:
6th–14th March.

Bath Festival of Children's Literature

PO Box 4123, Bath BA1 0FR
☏ 01225 314676   ℻ 01225 445485/463362 (tickets)
info@bathkidslitfest.co.uk
www.bathkidslitfest.co.uk
Festival Directors *John McLay, Gill McLay*

Founded 2007. Annual event celebrating
children's books and reading held 19th–28th
September 2008 (18th–27th in 2009). Over 80
events for readers aged up to 16. Guest authors

have included Jacqueline Wilson, Eoin Colfer, Anthony Horowtiz, Darren Shan, Lauren Child, Louise Rennison, Francesca Simon, Michelle Paver, Charlie Higson, Rick Riordan, Allan Ahlberg. Sponsored by *The Daily Telegraph*.

Bath Literature Festival

Bath Festivals, Abbey Chambers, Kingston Buildings, Bath BA1 1NT
☎ 01225 462231   🖷 01225 445551
info@bathfestivals.org.uk
www.bathlitfest.org.uk
Box Office: Bath Festivals Box Office, 2 Church Street, Abbey Green, Bath BA1 1NL
☎ 01225 463362 boxoffice@bathfestivals.org.uk
Artistic Director *Sarah LeFanu*

Founded 1995. This annual festival (held in March) programmes over 100 different literary events from debates and lectures to readers' groups and workshops in venues throughout the city. In addition there are a number of events for children and young people. Previous featured writers include Martin Amis, Terry Pratchett, Margaret Drabble, Joanna Trollope, A.C. Grayling, Steven Berkoff, Hermione Lee.

BayLit Festival

Academi, Mount Stuart House, Mount Stuart Square, Cardiff CF10 5FQ
☎ 029 2047 2266   🖷 029 2049 2930
post@academi.org
www.academi.org
Chief Executive *Peter Finch*

Annual literature festival held in Cardiff Bay, featuring writers from Wales and beyond. Lectures, readings, performances, workshops and book launches in English and Welsh. Dates and further details available on the Academi website. Writers who appeared in recent festivals include: Ian McMillan, Fay Weldon, Simon Singh, Howard Marks and Will Self.

Belfast Festival at Queen's

8 Fitzwilliam Street, Belfast BT9 6AW
☎ 028 9097 1034   🖷 028 9097 1336
g.farrow@qub.ac.uk
www.belfastfestival.com
Director *Graeme Farrow*

Founded 1964. Annual three-week festival held in October/November (17th–1st in 2008). Organized by Queen's University, the festival covers a wide variety of events, including literature. Programme available in September.

Beverley Literature Festival

Wordquake, Libraries and Information, Council Offices, Skirlaugh HU11 5HN
☎ 01482 392745   🖷 01482 392710
john@bevlit.org
www.beverley-literature-festival.org

Festival Director *John Clarke*

Founded 2002. Annual October festival which has acquired 'a national reputation for commissioning and hosting quality poetry events'. Also covers all other literary genres and mixes readings and performances with author-led readers' groups and creative writing workshops. Held in intimate and historic venues across Beverley.

Birmingham Book Festival

c/o Unit 116, The Custard Factory, Gibb Street, Birmingham B9 4AA
☎ 0121 246 2770   🖷 0121 246 2771
jonathan@birminghambookfestival.org
www.birminghambookfestival.org
Contact *Jonathan Davidson*

Founded 1999. Annual literature festival held during October at venues around Birmingham. Also promotes events throughout the year. Includes performances, lectures, discussion events and workshops.

Book Now – Richmond Literature Festival

The Arts Service, Orleans House Gallery, Riverside, Twickenham TW1 3DJ
artsinfo@richmond.gov.uk
www.richmond.gov.uk/literature

Founded 1992. Annual three-week literature festival held in November, delivered by the Arts Service of Richmond Borough. Plays host to leading writers.

Borders Book Festival

Harmony House, St Mary's Road, Melrose TD6 9LJ
☎ 07929 435575
info@bordersbookfestival.org
www.bordersbookfestival.org
Festival Coordinator *Paula Ogilvie*

Founded in 2004. Annual literary festival held over four days during the third weekend in June. The event brings internationally acclaimed and new authors to the gardens of the National Trust for Scotland's splendid Georgian property, Harmony House, and other locations in Melrose, for lively and entertaining discussions and performances to audiences of all ages and tastes. Past speakers include Ian Rankin, Iain Banks, Michael Palin, Rory Bremner, Alexander McCall Smith, Sarah Raven, Roy Hattersley, Claire Messud, Sheila Hancock, William Dalrymple and A.L. Kennedy. A well-stocked Festival bookshop is open throughout the Festival dates. Free mailing list.

Bournemouth Literary Festival

☎ 01202 417535
info@bournemouthliteraryfestival.co.uk
www.bournemouthliteraryfestival.co.uk

Director & Founder *Lillian Avon*

Founded in 2004, the Festival fuses literature with performing arts. Events throughout the year, in venues all over Bournemouth, with a focus on September and October.

Brighton Children's Book Festival

Brighton Writers' Centre, 49 Grand Parade, Brighton BN2 9QA
① 01273 571700
info@bcbf.org.uk
www.bcbf.org.uk
Festival Director *Laura Atkins*

Founded 2007. Held in Brighton, annually in April, this festival runs over one weekend and includes a number of activities including keynote speeches, workshops, readings and displays for writers and readers, young people and adults.

Brighton Festival

Brighton Dome & Festival Ltd, 12a Pavilion Buildings, Castle Square, Brighton BN1 1EE
① 01273 700747   ⓕ 01273 707505
info@brightonfestival.org
www.brightonfestival.org
Contact *Festival Office*

Founded 1966. For 24 days every May, Brighton hosts England's largest mixed arts festival. Music, dance, theatre, film, opera, literature, comedy and exhibitions. Literary enquiries will be passed to the literature officer. Deadline October for following May.

Bristol Poetry Festival

▷ *Poetry Can under Organizations of Interest to Poets*

Broadstairs Dickens Festival

10 Lanthorne Road, Broadstairs CT10 3NH
① 01843 861827   ⓕ 01843 861827
www.broadstairs.gov.uk/DickensFestival.html
Organizer *Sylvia Hawkes*

Founded in 1937 to commemorate the 100th anniversary of Charles Dickens' first visit to Broadstairs in 1837, which he continued to visit until 1859. The Festival lasts for four days in June and events include a parade, a performance of a Dickens play (*Bleak House* in 2008), melodramas, Dickens readings, a Victorian cricket match, Victorian bathing parties, talks, music hall, four-day Victorian country fair. Costumed Dickensian ladies in crinolines with top-hatted escorts promenade during the Festival.

Buxton Festival

3 The Square, Buxton SK17 6AZ
① 01298 70395   ⓕ 01298 72289
info@buxtonfestival.co.uk
www.buxtonfestival.co.uk
Chief Executive *Glyn Foley*

Founded 1979. Held annually in July, this 19-day opera, literature and music festival includes a varied literary programme, attracting those with a broad interest in well-crafted writing, whether in biography, fiction, politics or personal memoir.

Cambridge Wordfest

① 01223 264404
admin@cambridgewordfest.co.uk
www.cambridgewordfest.co.uk
Festival Director *Cathy Moore*
Festival Patrons *Dame Gillian Beer, Rowan Pelling, Ali Smith*

An annual spring festival for writers as well as readers which takes place at various venues across Cambridge and the surrounding area. A weekend event packed with the best of contemporary writing, poetry, political debate, events for children and writing workshops. Also a one-day winter festival during the last weekend in November.

Canterbury Festival

Christ Church Gate, The Precincts, Canterbury CT1 2EE
① 01227 452853   ⓕ 01227 781830
info@canterburyfestival.co.uk
www.canterburyfestival.co.uk
Festival Director *Rosie Turner*

Canterbury provides a unique location for the largest festival of arts and culture in the region (11th–25th October in 2008), showcasing local, national and international talent in classical and contemporary music, world theatre, opera, comedy, dance, walks and community events, plus an extended programme of literature and debate.

Carnegie Sporting Words Festival

Harrogate International Festivals, Raglan House, Raglan Street, Harrogate HG1 1LE
① 01423 562303   ⓕ 01423 521264

Founded in 2008, the Carnegie Sporting Words Festival (CSWF) takes place in Leeds and Harrogate over the weekend of 2nd–5th October. It is the UK's first festival celebrating the sporting word as expressed through books, newspapers, broadcast media and the web, and will feature sports celebrities, broadcasters, journalists and authors. Events include a Sports Quiz Dinner hosted by BBC's Mark Pougatch; a seminar exploring the future of digital sports media and a Schools Activity Day at the Headingley Carnegie Stadium in partnership with BBC Sport.

Centre for Creative & Performing Arts Spring Literary Festival at UEA

University of East Anglia, School of Literature and Creative Writing, Norwich NR4 7TJ
① 01603 592810

v.striker@uea.ac.uk
www1.uea.ac.uk/cm/home/schools/hum/
booksandwriters/litfest
Contact *Val Striker*

Founded 1993. Annual event held in the spring and summer.

Charleston Festival

The Charleston Trust, Charleston, Nr Firle, Nr Lewes BN8 6LL
☎ 01323 811626
info@charleston.org.uk
www.charleston.org.uk
Festival Programmer *Diana Reich*

Annual literary festival held in May over nine days. Novelists, biographers, travel writers, broadcasters, poets, food writers, actors and artists gather at Charleston, the country home of the Bloomsbury Group. Past speakers include Margaret Atwood, Alan Bennett, Jeanette Winterson, Louis de Bernières, Patti Smith, Paula Rego, Clive James, Ali Smith, Roger McGough, Germaine Greer and Harold Pinter.

Cheltenham Literature Festival
▷ The Times Cheltenham Literature Festival

Chester Literature Festival

55–57 Watergate Row South, Chester CH1 2LE
☎ 01244 409113   ⊞ 01244 401697
info@chesterlitfest.org
www.chesterfestivals.org.uk/literature
Chairman *Bill Hughes*

Founded 1989. Annual festival (6th–24th October in 2008). Events include international and nationally known writers, as well as events by local literary groups. There is a Literary Lunch and a Festival Dinner, events for children, workshops and competitions. Free mailing list.

Children's Books Ireland – Annual Festival
▷ Children's Books Ireland under Professional Associations

Dartington Literary Festival
▷ Ways With Words

Derbyshire Literature Festival

c/o Arts Office, Derbyshire County Council, Cultural & Community Services Department, Alfreton Library, Severn Square, Alfreton DE55 7BQ
☎ 01773 832497   ⊞ 01773 831359
ann.wright@derbyshire.gov.uk
www.derbyshire.gov.uk
Festival Organizer *Ann Wright*

Founded 2000. The festival takes place in June every two years and covers the whole county. The festival programming includes all types of live literature events; performance poetry, theatre, talks, readers' groups, workshops, dramatized readings, signings, storytelling and literary trails, as well as a number of cross-art form events. The festival takes place in libraries and many other community venues including heritage centres, industrial buildings, stately homes, parks and moors, churches and schools and is specifically designed to reach as many different communities and geographical areas in the county as possible.

Dorchester Festival

Dorchester Arts Centre, School Lane, The Grove, Dorchester DT1 1XR
☎ 01305 266926   ⊞ 01305 266143
enquiries@dorchesterarts.org.uk
www.dorchesterarts.org.uk
Artistic Director *Sharon Hayden*

Founded 1996. An annual five-day international festival held over early May Bank Holiday which includes performance, live music and visual arts with associated educational and community projects. Also three days of free events in various venues around the town, including literature and poetry. The theme for the 2008 festival was Cuba and the Culture of South America.

The Daphne du Maurier Festival of Arts & Literature

Restormel Borough Council, 39 Penwinnick Road, St Austell PL25 5DR
☎ 01726 223300   ⊞ 01726 223301
rbc@dumaurierfestival.co.uk
www.dumaurierfestival.co.uk

Founded in 1997, this annual arts and literature festival is held in and around Fowey in Cornwall in May and includes a wide range of professional and community events.

Dumfries and Galloway Arts Festival

Gracefield Arts Centre, 28 Edinburgh Road, Dumfries DG1 1JQ
☎ 01387 260447   ⊞ 01387 260447
info@dgartsfestival.org.uk
www.dgartsfestival.org.uk
Festival Organizer *Mrs Barbara Kelly*

Founded 1980. Annual week-long festival held at the end of May with a variety of events including classical and folk music, theatre, dance, literary events, exhibitions and children's events.

Dundee Literary Festival

The Tower Building, Nethergate, Dundee DD1 4HN
☎ 01382 384768
dundeeliteraryfestival@gmail.com
www.dundeeliteraryfestival
Festival Director *Anna Day*

Founded 2007. Annual literary festival held in June with a mix of internationally renowned writers, local authors, poetry and new writing. Speakers have included Philip Pullman, Jacqueline Wilson and Kirsty Gunn. E-mail to be added to the mailing list.

Durham Book Festival

Durham City Arts, 2 The Cottages, Fowlers Yard, Back Silver Street, Durham DH1 3RA
☎ 0191 375 0763
enquiries@durhamcityarts.org.uk
www.durhamcityarts.org
Festival Coordinators *Becca Pelly-Fry, Kate James*

Founded 1989. Annual 2–3-week festival held in September/October at various locations in the city. Performances and readings plus workshops, cabaret, exhibitions and other events.

Edinburgh International Book Festival

5a Charlotte Square, Edinburgh EH2 4DR
☎ 0131 718 5666   📠 0131 226 5335
admin@edbookfest.co.uk
www.edbookfest.co.uk
Director *Catherine Lockerbie*
Children and Education Programme Director *Sara Grady*
Programme Manager *Roland Gulliver*

Founded 1983. 'The world's largest and most dynamic annual book event' takes place across 17 days in Edinburgh each August. An extensive programme of over 650 events featuring 700 or more authors, showcasing the work of the world's top writers and thinkers to an audience of over 200,000 adults and children. Featuring workshops, readings, lectures and a high profile debate and discussion series, book signings, bookshops and cafes.

Folkestone Literary Festival

The Block, 65–69 Tontine Street, Folkestone CT20 1JR
☎ 01303 245799   📠 01303 223761
info@folkestonelitfest.com
www.folkestonelitfest.com
Festival Director *Ellie Beedham*

Annual festival. Tours, discussions, talks, live performances and readings. 2008 festival: 1st–8th November.

Frome Festival

25 Market Place, Frome BA11 1AH
☎ 01373 453889
office@fromefestival.co.uk
www.fromefestival.co.uk

An annual festival held from the first Friday in July for ten days celebrating all aspects of visual and performing arts and entertainment and with a strong literary element. Features music, film, dance, drama and visual arts plus readings, talks and workshops. Previous literary guests have

included nationally renowned writers as well as local authors.

Graham Greene Festival
▷ Graham Greene Birthplace Trust under Literary Societies

The Guardian Hay Festival

25 Lion Street, Hay-on-Wye HR3 5AD
☎ 0870 787 2848   📠 01497 821066
admin@hayfestival.com
www.hayfestival.com
Festival Director *Peter Florence*

Founded 1988. Annual May festival sponsored by *The Guardian*. Also has a children's programme which began in 2003. The event coincides with half term and caters for young children through to teenage. Guests have included Paul McCartney, Bill Clinton, Salman Rushdie, Toni Morrison, Stephen Fry, Joseph Heller, Carlos Fuentes, Maya Angelou, Amos Oz, Arthur Miller, Jimmy Carter.

Guildford Book Festival

c/o Tourist Information Centre, 14 Tunsgate, Guildford GU1 3QT
☎ 01483 444334
deputy@guildfordbookfestival.co.uk
www.guildfordbookfestival.co.uk
Book Festival Director *Glenis Pycraft*

Patrons: Elizabeth Buchan, Michael Buerk, Michael Rosen, Sandi Toksvig, Fay Weldon, Timothy West and Jacqueline Wilson. Held annually, during October in venues throughout the ancient town of Guildford. The festival includes events with established and new writers, workshops, poetry performances, children's events and the **Guildford Book Festival First Novel Award** (see entry under *Prizes*). This diverse festival aims to involve, instruct and entertain all who care about literature. Guest authors in 2007 included Roy Hattersley, A.A. Gill, Colin Dexter, Jacqueline Wilson, Kate Mosse, Michael Dobbs, Barry Cryer, Julian Clary, Patrick Gale.

Harrogate Crime Writing Festival
▷ Theakstons Old Peculier Harrogate Crime Writing Festival

Harrogate International Festival

Raglan House, Raglan Street, Harrogate HG1 1LE
☎ 01423 562303   📠 01423 521264
info@harrogate-festival.org.uk
www.harrogate-festival.org.uk
Chief Executive *William Culver-Dodds*
Ops Director *Sharon Canavar*

Founded 1966. Annual two-week festival at the end of July and beginning of August. Events include international symphony orchestras, chamber concerts, ballet, celebrity recitals, contemporary dance, opera, drama, jazz and comedy.

Harwich Festival of the Arts

2A Kings Head Street, Harwich CO12 3EG
☎ 01255 503571
anna@rendell-knights.freeserve.co.uk
Contact *Anna Rendell-Knights*

Founded 1980. Annual ten-day summer festival held in June/July. Events include concerts, drama, film, dance, art, exhibitions, historic town walks. The Festival in 2008 commemorated the building of the Harwich Redoubt Fort in 1808.

The Hay Festival
▷ The Guardian Hay Festival

Hebden Bridge Arts Festival

New Oxford House, Albert Street, Hebden Bridge HX7 8AH
☎ 01422 842684
hbfestival@gmail.com
festival@rebeccayorke.co.uk
www.hebdenbridge.co.uk/festival
Administrator *Rebecca Yorke*
Programming *Stephen May*

Founded 1994. Annual arts festival with increasingly strong adult and children's literature events. Previous guest writers include Roger McGough, Benjamin Zephaniah, Anne Fine, Juliet Barker, Jacqueline Wilson, Quentin Blake, Ian McMillan, Adele Geras, George Alagiah.

Humber Mouth – Hull Literature Festival

City Arts Unit, Central Library, Albion Street, Kingston upon Hull HU1 3TF
☎ 01482 616961   ℻ 01482 616827
humbermouth@gmail.com
www.humbermouth.org.uk
Contact *Maggie Hannan*

Founded 1992. Hull's largest festival, held in the summer, and one of the region's liveliest events. Features readings, talks, performances and workshops by writers and artists from around the world and from the city.

Ilkley Literature Festival

The Manor House, 2 Castle Hill, Ilkley LS29 9DT
☎ 01943 601210   ℻ 01943 817079
admin@ilkelyliteraturefestival.org.uk
www.ilkleyliteraturefestival.org.uk
Director *Rachel Feldberg*

Founded 1973. Major literature festival in the north held over 17 days every October. Includes children's literature weekend and Festival fringe. Recent guests include Maya Angelou, Alan Bennett, Sarah Waters, Claire Tomalin, Benjamin Zephaniah, Richard Ford, Carol Ann Duffy, P.D. James, Donna Tartt, Kate Adie, Germaine Greer, Michael Ondaatje, Chimamanda Adichie, Liz Lochhead, Tony Harrison, Ian Rankin. Telephone or e-mail to join free mailing list.

Imagine: Writers and Writing for Children
▷ Royal Festival Hall Literature & Talks

The International Festival of Mountaineering Literature

15 Ednaston Court, Ednaston, Ashbourne DE6 3BA
☎ 01335 360581   ℻ 01333 439148
t.gifford@chi.ac.uk
www.festivalofmountaineeringliterature.co.uk
Director *Terry Gifford*

Founded 1987. Annual one-day festival held at Kendal Mountain Festivals, Kendal, Cumbria. Taking place in November, the Festival celebrates recent books, commissions new writing, gives overviews of national literatures, holds debates of issues, book signings. Announces the winner of the festival writing competition run in conjunction with *Climb* magazine. Write to join free mailing list.

International Playwriting Festival

Warehouse Theatre, Dingwall Road, Croydon CR0 2NF
☎ 020 8681 1257   ℻ 020 8688 6699
info@warehousetheatre.co.uk
www.warehousetheatre.co.uk/ipf.html
Festival Administrator *Rose Marie Vernon*

The Warehouse Theatre Company's International Playwriting Festival will celebrate 23 successful years in 2008. It is held in two parts: the *first* is the competition with entries accepted between January and June from all over the world and judged by a panel of distinguished theatre practitioners. The *second* is the festival itself which showcases the selected plays from the competition. This takes place in November each year. The IPF also showcases the successful plays in Europe in association with its partners Extra Candoni in Italy and Theatro Ena in Cyprus. Rules and entry form on web page above. Previous winners produced at the theatre include: Guy Jenkin *Fighting for the Dunghill*; James Martin Charlton *Fat Souls*; Peter Moffat *Iona Rain*; Dino Mahoney *YoYo*; Dominic McHale *The Resurrectionists*; Philip Edwards *51 Peg*; Roumen Shomov *The Dove*; Maggie Nevill *The Shagaround*; Andrew Shakeshaft *Just Sitting*; Mark Norfolk *Knock Down Ginger*; Des Dillon *Six Black Candles*; Alex Evans *Mother Russia*.

Isle of Man Literature Festival

Isle of Man Arts Council, St Andrew's House, Finch Road, Douglas IM1 2PX
☎ 01624 694598   ℻ 01624 686860
iomartscouncil@dtl.gov.im
www.iomarts.com

Contact *Arts Development Manager*

Founded 1997. Regular programme of literature events throughout the year.

King's Lynn, The Fiction Festival

19 Tuesday Market Place, King's Lynn PE30 1JW
☎ 01553 691661 (office hours) or 761919
🖷 01553 691779
tony.ellis@hawkins-solicitors.com
www.lynnlitfests.com
Contact *Anthony Ellis*

Founded 1989. Annual weekend festival held in March (13th–15th in 2009). Over the weekend there are readings and discussions, attended by guest writers of which there are usually twelve. Previous guests have included Beryl Bainbridge, Louis de Bernière, John Buchan, J.P. Donleavy and D.J. Taylor.

King's Lynn, The Poetry Festival

19 Tuesday Market Place, King's Lynn PE30 1JW
☎ 01553 691661 (office hours) or 761919
🖷 01553 691779
tony.ellis@hawkins-solicitors.com
www.lynnlitfests.com
Contact *Anthony Ellis*

Founded 1985. Annual weekend festival held at the end of September, with guest poets (usually eight). Previous guests have included Clive James, Les Murray, Peter Porter, D.M. Thomas, C.K. Williams and Kit Wright.

King's Sutton Literary Festival

Glebe House, St Rumbold's Drive, King's Sutton, Banbury OX17 3PJ
☎ 01295 811473
info@kslitfest.co.uk
Festival Organizer *Mrs Sara Allday*

Founded 2004. Takes place over a weekend in early March. Celebrates reading in a village that has attracted many world-renowned writers to the festival, such as Margaret Drabble, Andrew Motion, Salley Vickers, Justin Cartwright, David Lodge, Rachel Billington. Takes place under one roof in the modern village centre and café with new and second-hand book sales, a programme involving the village primary school. Proceeds to the Parish Church Restoration Fund.

Knutsford Literature Festival

☎ 01565 722738/07050 183417
knutsfordlitfest@yahoo.co.uk
www.knutsfordlitfest.blogspot.com
Contact *Charlotte Peters Rock*

An annual two-week festival, held at the beginning of October to celebrate writing and performance by distinguished national, international and local authors. Events include readings and discussions, a literary lunch and theatrical performances.

Lambeth Readers & Writers Festival

London Borough of Lambeth, Brixton Library, Brixton Oval, London SW2 1JQ
☎ 020 7926 1105
to'dell@lambeth.gov.uk
www.lambethgov.uk/Libraries
Contact *Tim O'Dell*

Founded in 2001. Held annually in May, the festival brings together internationally recognized, new and local authors for a month of talks, poetry and writing courses focusing on the reading experience. Previous guests have included Peter Ackroyd, Ben Okri, Armando Iannucci, Polly Toynbee, Linton Kwesi Johnson, Doris Lessing, Buchi Emecheta, Jo Brand and Lionel Shriver.

Lancaster LitFest

PO Box 751, Lancaster LA1 9AJ
☎ 01524 62166
all@litfest.org
www.litfest.org

Founded 1978. Regional Literature Development Agency, organizing workshops, readings, residencies, publications. Year-round programme of literature-based events, and annual festival in November featuring a wide range of writers from the UK and overseas. Publishes new Lancashire and Cumbrian prose and poetry writers under the Flax Books imprint.

Ledbury Poetry Festival

Church Lane, Ledbury HR8 1DH
☎ 0845 458 1743
admin@poetry-festival.com
www.poetry-festival.com
Festival Director *Chloe Garner*

Founded 1997. Annual ten-day festival held in June/July. Includes readings, discussions, workshops, exhibitions, music and walks. There is also an extensive year-round community programme and a national poetry competition. Past guests have included James Fenton, Helen Dunmore, Bernard MacLaverty, Andrew Motion, Benjamin Zephaniah, Simon Armitage, Roger McGough, John Hegley. Full programme available in May.

Lewes Live Literature

PO Box 2766, Lewes BN7 2WF
info@leweslivelit.co.uk
www.leweslivelit.co.uk
Contact *Mark Hewitt*

Founded 1995. Literature productions leading to occasional seasons of events throughout the year.

Lichfield Festival

7 The Close, Lichfield WS13 7LD
☎ 01543 306270   ☎ 01543 306274
info@lichfieldfestival.org
www.lichfieldfestival.org
Festival Director *Richard Hawley*

Annual July festival with events taking place in the 13th century Cathedral, the Lichfield Garrick Theatre and other venues. Mainly music but a growing programme of literary events such as poetry, plays and talks. New dedicated literature weekend each autumn; literary dinner plus two days of writer talks.

City of London Festival

12–14 Mason's Avenue, London EC2V 5BB
☎ 020 7796 4949
admin@colf.org
www.colf.org
Director *Ian Ritchie*

Founded 1962. Annual three-week festival held in June and July. Features over fifty classical and popular music events alongside poetry and prose readings, street theatre and open-air extravaganzas, in some of the most outstanding performance spaces in the world.

London Literature Festival

Southbank Centre, Belvedere Road, London SE1 8XX
☎ 0871 663 2500 (box office)
www.southbankcentre.co.uk

Established in 2008, events take place at the Southbank Centre over two weeks in early July. 2008 guests included Wendy Cope, Sebastian Barry, Halima Bashir, Lemn Sissay, James Hopkin, Kate Summerscale, Timothy O'Grady, Rebecca Miller and George Monbiot.

Lowdham Book Festival

The Bookcase, 50 Main Street, Lowdham NG14 7BE
☎ 0115 966 4143
janestreeter@thebookcase.co.uk
www.lowdhambookfestival.co.uk
Contact *Jane Streeter*

Founded 1999. Annual nine-day festival held in June combining a village fête atmosphere with that of a major literature festival. Guests have included Carol Ann Duffy, Polly Toynbee, Ian McMillan, Alan Sillitoe, Jackie Kay, Gary Younge and Carole Blake. Talks on everything from St Kilda to the Jewish roots of rock 'n' roll.

Manchester Literature Festival

3rd Floor, 24 Lever Street, Manchester M1 1DZ
☎ 0161 236 5725
admin@manchesterliteraturefestival.co.uk
www.manchesterliteraturefestival.co.uk
Contact *Administrator*

Held in October (6th–16th in 2008), Manchester Literature Festival offers 'unique and imaginative literature experiences to its audiences' with a programme that features readings by some of the world's finest authors, freshly commissioned work, and a series of cutting-edge events exploring the crossover between new technology and literature.

Mere Literary Festival

Lawrence's, Old Hollow, Mere BA12 6EG
☎ 01747 860475/861211 (Tourist Information)
www.merewilts.org.uk
Contact *Adrienne Howell*

Founded 1997. Annual festival held in the second week of October in aid of registered charity, The Mere & District Linkscheme. Events include readings, quiz, workshop, writer's lunch and talks. Finale is adjudication of the festival's writing competition (see entry under *Prizes*) and presentation of awards.

Arthur Miller Centre Literary Festival at UEA

University of East Anglia, School of American Studies, Norwich NR4 8TJ
☎ 01603 592810
v.striker@uea.ac.uk
www1.uea.ac.uk/cm/home/schools/hum/booksandwriters/litfest
Contact *Val Striker*

Founded 1991. Annual festival, held in the autumn.

National Association of Writers' Groups (NAWG) Open Festival of Writing

The Arts Centre, Washington NE38 2AB
☎ 01262 609228
nawg@tesco.net
www.nawg.co.uk
Festival Administrator *Mike Wilson*

Founded 1995. Annual Festival held at St Aidan's College, University of Durham, 5th–7th September 2008. Three days of creative writing tuition covering poetry, short and long fiction, playwriting, journalism, TV sitcom and many other subjects all led by professional writer-tutors. Workshops, one-to-one tutorials and fringe events. Saturday gala dinner and awards ceremony. Full or part-residential weekend, or single workshops only. Open to all, no qualifications or NAWG membership required.

National Eisteddfod of Wales

40 Parc Ty Glas, Llanishen, Cardiff CF14 5DU
☎ 029 2076 3777   ☎ 029 2076 3737
info@eisteddfod.org.uk
www.eisteddfod.org.uk

The National Eisteddfod, held in August, is the largest arts festival in Wales, attracting over 170,000 visitors during the week-long celebration of more than 800 years of tradition. Competitions, bardic ceremonies and concerts.

National Student Drama Festival
▷ University of Hull under UK Writers' Courses

Northern Children's Book Festival
Schools Library Service, Sandhill Centre, Grindon Lane, Sunderland SR3 4EN
☎ 0191553 8866/7/8   🖷 0191 553 8869
schools.library@sunderland.gov.uk
www.ncbf.org.uk
Secretary *Eleanor Dowley*

Founded 1984. Annual two-week festival during November. Events in schools and libraries for children in the North East region. One Saturday during the festival sees the staging of a large book event hosted by one of the local authorities involved.

Off the Page Literature Series
County Library Support Services, Glaisdale Parkway, Nottingham NG8 4GP
☎ 0115 982 9027   🖷 0115 928 6400
alison.hirst@nottscc.gov.uk
Contact *Alison Hirst*

Author visits aimed at making authors accessible to readers. Programme is available from late summer/early autumn.

Off the Shelf Literature Festival
Central Library, Surrey Street, Sheffield S1 1XZ
☎ 0114 273 4716/4400   🖷 0114 273 5009
offtheshelf@sheffield.gov.uk
www.offtheshelf.org.uk
Festival Organizers *Maria de Souza,
Susan Walker, Lesley Webster*

Founded 1992. Annual two-week festival held during the last fortnight in October. Lively and diverse mix of readings, workshops, children's events, storytelling and competitions. Previous guests have included Michael Frayn, Rabbi Lionel Blue, Germaine Greer, Robert Goddard, Dom Joly, Barry Davies, Michael Wood, Linton Kwesi Johnson.

Oundle Festival of Literature
2 New Road, Oundle PE8 4LA
☎ 01832 273050
enquiries@oundlelitfest.org.uk
www.oundlelitfest.org.uk
Festival Deputy Chair *Liz Dillarstone*

Founded 2002. Annual festival running in the first two weeks of March for adults and children. Features talks by high profile authors and poets, writing workshops, poetry and prose showcases for local writers, children's writing

competitions, school events, drama productions and community readings for young and old.

Oxford Literary Festival
▷ The Sunday Times Oxford Literary Festival

Poetry Otherwise
Emerson College, Forest Row RH18 5JX
☎ 01342 822238   🖷 01342 826055
mail@emerson.org.uk
www.emerson.org.uk
www.poetryotherwise.org
Director *Paul Matthews*

Held in August (17th–23rd in 2008). A summer gathering of poets, writers and all lovers of language. Workshops, readings, talks, conversation, practice in the art of speaking poetry. Contributors have included Andrea Hollander Budy, John Freeman, Ashley Ramsden, Andie Lewenstein, Paul Matthews, Peter Abbs, Katherine Pierpoint, Fiona Owen, Lee Harwood, Jay Ramsay, Mimi Khalvati and Roselle Angwin. 'Wholesome food, wonderful Sussex countryside.'

PULSE International Poetry Festival
THE SOUTH, Brighton Writers' Centre, 49 Grand Parade, Brighton BN2 9QA
☎ 01273 571700
info@thesouth.org.uk
www.thesouth.org.uk
Festival Director *John Davies*

Founded 2007. Annual, two-week poetry festival held in late September and early October. An international festival twinned with CUISLE in Ireland and Apokalypsa in Slovenia, both of which each year feature visiting poets. The festival has a particular focus on translation in poetry. Performances, readings, workshops and fairs.

Purple Patch Poetry Convention
25 Griffiths Road, West Bromwich B71 2EH
ppatch66@hotmail.com
www.purplepatchpoetry.co.uk
Contact *Geoff Stevens*

Held at the Barlow Theatre near Birmingham this event celebrates the strength of small press poets. The Purple Patch Poetry Convention is a three-day event of poetry readings, talks, workshops and discussions. Admission charges are kept very low, payments to participants being low to non-existent. The Convention's aims are quality and an opportunity to participate. Poets at past festivals have included Ray Avery, R.G. Bishop, Tilla Brading, Gerald England, J.F. Haines, Brendan Hawthorne, Martin Holroyd, Mike Hoy, Eamer O'Keeffe, Carolyn King, Paul McDonald, Bob Mee, Les Merton, Michael Newman, Simon Pitt, Andy Robson, Sam Smith and Steve Sneyd. Monthly poetry readings are held at the Barlow

Theatre with open mic, guest poet, music and residents Unleaded Petrels.

Redbridge Book and Media Festival

3rd Floor, Central Library & Museum, Clements Road, Ilford IG1 1EA
☎ 020 8708 2855   ⓕ 020 8708 2431
jacqueline.eggleston@redbridge.org.uk
www.redbridge.gov.uk
Contact *Arts Development Officer*

Founded in 2003, the festival celebrates literature in all its forms, and takes place annually over two weeks in April/May. Features a programme of author readings, book signings, creative writing classes, school-based workshops, community events, performance poetry evenings, competitions, exhibitions and panel debates.

Royal Court Young Writers Programme

Royal Court Theatre, Sloane Square, London SW1W 8AS
☎ 020 7565 5050   ⓕ 020 7565 5001
ywp@royalcourttheatre.com
www.royalcourttheatre.com
Associate Director *Ola Animashawun*

Open to young people up to the age of 25. The YWP focuses on the process of playwriting by running a series of writers' groups throughout the year. Additionally the YWP welcomes unsolicited scripts from all young writers from across the country. 'We are always looking for scripts for development and possible production (not film scripts).'

Runnymede International Literature Festival

Royal Holloway, University of London, Egham TW20 0EX
☎ 01273 571700
info@rfest.org.uk
www.rfest.co.uk
Festival Director *Professor Robert Hampson*

Founded 2006. Annual literary festival held in Runnymede as a partnership beween Royal Holloway, University of London, Runnymede Borough Council and THE SOUTH. Held in February, the festival runs for a core weekend with associated community and education activities. Includes readings, performances, masterclasses and community workshops.

Saffron Walden Literary Festival

Harts, Hart House, Shire Hill, Saffron Walden CB11 3AQ
☎ 01799 523456 (box office)
events@harts1836.co.uk
www.hartsbooks.co.uk
Festival Director *Jo Burch*

Founded 2006. Various events in 2008 run by family-owned Harts Bookshop: 'Sense of Place' Book Festival, held in February; monthly children's events; and main Book Festival held over a fortnight (25th September–5th October) with events and workshops for adults and children. 'Sense of Place' Festival speakers: Nicholas Crane, Andrew Motion, Rob Macfarlane, Polly Evans, Ronald Blythe and Malika's Kitchen.

Scotland's Book Town Festival

Festival Office, County Buildings, Wigtown DG8 9JH
☎ 01988 403222
mail@wigtownbookfestival.com
www.wigtownbookfestival.com
Festival Director *Michael McCreath*
Festival Admin *Anne Barclay*

Founded 1999. Annual Festival held over ten days (last week September/first week October). Author readings, poetry events, workshops, drama, music. Children's events and school outreach programme. Also Spring Festival (founded 2005) held over first May Bank Holiday weekend.

Southwold Festival
▷ Ways with Words

StAnza: Scotland's Poetry Festival

Registered Office: 57 Lade Braes, St Andrews KY16 9DA
☎ 05600 433847
info@stanzapoetry.org
www.stanzapoetry.org
Festival Director *Brian Johnstone* (admin@stanzapoetry.org)
Artistic Director *Eleanor Livingstone* (arts@stanzapoetry.org)
Press & Media Manager *Annie Kelly* (press@stanzapoetry.org)

The only regular festival dedicated to poetry in Scotland, StAnza is international in outlook. Held annually in March in the ancient university town of St Andrews, the festival is an opportunity to hear world class poets reading in exciting and atmospheric venues. The 2008 festival themes were *Poetry & Conflict* and *Sea of Tongues*. The festival features readings, discussions, conversations, performance poetry, poetry in exhibition, workshops and children's poetry. It has a strong showing of poets from across the UK and overseas as well as the festival's signature foreign language readings. Free programmes can be ordered from Fife Contemporary Art & Craft on 01334 474610 or by e-mail to mail@fcac.co.uk

Stratford-upon-Avon Poetry Festival

The Shakespeare Centre, Henley Street, Stratford-upon-Avon CV37 6QW
☎ 01789 204016/292176 (box office)
ⓕ 01789 296083

info@shakespeare.org.uk
www.shakespeare.org.uk
Festival Director *Paul Edmonson*

Sponsored by the Shakespeare Birthplace Trust, this Festival, now in its 55th year, takes place in July and celebrates poetry past and present with special reference to the work of William Shakespeare. Features weekly recitals given by established poets and by actors who present themed evenings of verse. The 2008 Festival included an evening of local children's poetry, a Poetry Mass and local poets' event. Full details are available from late April and can be accessed on the Trust's website.

Strokestown International Poetry Festival
▷ Strokestown International Poetry Competition under Prizes

The Sunday Times Oxford Literary Festival
301 Woodstock Road, Oxford OX2 7NY
☎ 01865 514149
info@sundaytimes-oxfordliteraryfestival.co.uk
www.sundaytimes-oxfordliteraryfestival.co.uk
Directors *Sally Dunsmore, Angela Prysor-Jones*

Founded 1997. Annual week-long festival held in March/April. Authors speaking about their books, covering a wide variety of writing: fiction, poetry, biography, travel, food, gardening, children's art. Previous guests have included William Boyd, Andrew Motion, Beryl Bainbridge, Sophie Grigson, Philip Pullman, Doris Lessing, Seamus Heaney, Richard Dawkins, Zandra Rhodes, Kazuro Ishiguro.

Swindon Festival of Literature
Lower Shaw Farm, Shaw, Swindon SN5 5PJ
☎ 01793 771080
swindonlitfest@lowershawfarm.co.uk
www.swindonfestivalofliterature.co.uk
Festival Director *Matt Holland*

Founded 1994. Annual festival held in May, starting with 'Dawn Chorus' at sunrise. Includes a wide range of authors, speakers, discussions, performances and workshops, plus the Swindon Performance Poetry Slam competition. Guests at the 2008 Festival included Andrew Motion, Ben Okri and John Pilger.

Theakstons Old Peculier Harrogate Crime Writing Festival
Raglan House, Raglan Street, Harrogate HG1 1LE
☎ 01423 562303   ✆ 01423 521264
crime@harrogate-festival.org.uk
www.harrogate-festival.org.uk
Operations Director *Sharon Canavar*
Festival Coordinator *Erica Morris*

Launched 2003. Weekend of events at the end of July each year featuring the best of British and American crime writers. The winner of the **Theakstons Old Peculier Prize for the Crime Novel of the Year** is announced at the Festival (see entry under *Prizes*). Events also include industry 'How to ...' sessions, social events and late night shows. Part of the **Harrogate International Festival**.

Dylan Thomas Festival
Dylan Thomas Centre, Somerset Place, Swansea SA1 1RR
☎ 01792 463980   ✆ 01792 463993
dylanthomas.lit@swansea.gov.uk
www.dylanthomas.com
Contacts *David Woolley, Jo Furber*

Two weeks of performances, talks, lectures, films, music, poetry, exhibitions and celebrity guests held in 2008 from 27th October to 10th November, including the award of the second Dylan Thomas Prize for young writers. The Dylan Thomas Centre also runs a year-round programme of literary events; please e-mail for details.

The Times Cheltenham Literature Festival
109 Bath Road, Cheltenham GL53 7LS
☎ 01242 775861
clair.greenaway@cheltenhamfestivals.com
www.cheltenhamfestivals.com
Festival Manager *Clair Greenaway*

Founded 1949. Annual festival held in October (10th–19th in 2008). The first purely literary festival of its kind, it has over the past decade developed from an essentially local event into the largest and most popular in Europe. A wide range of events including talks and lectures, poetry readings, novelists in conversation, exhibitions, discussions and a large bookshop.

Torbay Weekend Festival of Poetry
6 The Mount, Higher Furzeham, Brixham TQ5 8QY
☎ 01803 851098
pwoxley@aol.com
www.acumen-poetry.co.uk/events
Festival Organizer *Patricia Oxley*

Founded 2001. Held annually at the end of October, Thursday evening until Monday afternoon with workshops, poetry readings, talks and debates. Programme for adults and children. Encourages active participation through workshops, open mike events, etc. Guest poets include internationally known writers as well as local authors.

Ty Newydd Festival
Ty Newydd, Llanystumdwy, Cricieth LL52 0LW
☎ 01766 522811   ✆ 01766 523095
post@tynewydd.org
www.tynewydd.org

Director *Sally Baker*

Biennial, bilingual literature festival held on alternate years. Held in June and run by the National Writers' Centre for Wales, it is located at Ty Newydd Writers' Centre and other venues near Cricieth. Features writers and poets from Wales (working in both English and Welsh) and worldwide. Ty Newydd also holds an annual weekend festival in November devoted solely to Cynghanedd – Welsh strict meter poetry.

Warwick Words

The Court House, Jury Street, Warwick CV34 4EW
☎ 01926 427056
info@warwickwords.co.uk
www.warwickwords.co.uk

Founded 2002. Annual festival featuring both living writers and those who have had connections with Warwick – Tolkien, Larkin and Landor in particular. Also workshops and an education programme. A weekend festival of literature and spoken word for all the family. Events at the Bridge House Theatre, St Mary's Church, the Lord Leycester Hospital and other historic buildings around Warwick over the first weekend in October (2nd–5th October in 2008).

Ways with Words

Droridge Farm, Dartington, Totnes TQ9 6JG
☎ 01803 867373   🖷 01803 863688
admin@wayswithwords.co.uk
www.wayswithwords.co.uk
Festival Director *Kay Dunbar*

Ways With Words runs three major annual literature festivals: at Dartington Hall, Devon for ten days in July, featuring over 150 writers giving talks, readings, interviews, discussions, performances and workshops; Words by the Water, Keswick, Cumbria in March, a ten-day festival as Dartington; and a five-day festival at Southwold, Suffolk in November. Also organizes writing, reading and painting courses in Italy and France.

Wellington Literary Festival

Civic Offices, Larkin Way, Tan Bank, Wellington, Telford TF1 1LX
☎ 01952 567697   🖷 01952 567690
welltowncl@aol.com
www.wellington-shropshire.gov.uk
Contact *Howard Perkins*

Founded 1997. Annual festival held throughout October. Events include storytelling, writers' forum, 'Pints and Poetry', children's poetry competition, story competition, theatre review and guest speakers.

Wells Festival of Literature

25 Chamberlain Street, Wells BA5 2PQ
☎ 01749 670929
www.somersite.co.uk/wellsfest.htm

Founded 1992. Annual week-long festival held in the middle of October. Main venue is the historic, moated Bishop's Palace. A wide range of speakers caters for different tastes in reading. Recent guests include: Sir Roy Strong, Kate Adie, Libby Purves, Melvyn Bragg, Margaret Drabble, William Dalrymple, Timothy West. Short story and poetry competitions and writing workshops are run in conjunction with the Festival.

Wigtown Book Festival
▷ Scotland's Book Town Festival

Winchester Writers' Conference, Bookfair & Workshops
▷ entry under UK and Irish Writers' Courses

Wonderful Words

c/o Redruth Library, Redruth TR15 2QE
☎ 07967 340907
mtwose@cornwall.gov.uk
www.cornwall.gov.uk/library
Festival Organizer *Maureen Twose*

Founded 1994. Biennial literature festival hosted throughout Cornwall. Organized by the Cornwall Library Service, Wonderful Words has grown into a prestigious event for children and adults, offering author talks, writers' workshops, poetry performances, theatre and storytelling. Guest authors have included Doris Lessing, Ruth Rendell, Margaret Drabble, Simon Callow and Dick King-Smith.

Word – University of Aberdeen Writers Festival

Office of External Affairs, University of Aberdeen, King's College, Aberdeen AB24 3FX
☎ 01224 273874
word@abdn.ac.uk
www.abdn.ac.uk/word
Artistic Director *Alan Spence*

Held annually in May, the Festival takes place over six days at the historic King's College Campus and at venues throughout the city. Attracts over 70 authors and over 10,000 visitors to a weekend of readings, lectures, debates, music, art exhibitions and film screenings. Also includes an extended schools' and children's programme and a Festival of Gaelic.

Word Market

PO Box 150, Barrow-in-Furness LA14 3WF
☎ 07812 178193
janice@wordmarket.org.uk
www.wordmarket.org.uk
Project Coordinator *Janice Benson*

Word Market celebrates the written word through a programme of exciting, live events and projects throughout the year. As well as an annual festival, regular workshops, master-classes and performances are held with well-known poets, best-selling authors, publishers, literary agents and theatre directors. Check the website for further details.

Words by the Water
▷ Ways with Words

The Wordsworth Trust

Dove Cottage, Grasmere LA22 9SH
☎ 015394 35544   📠 015394 35748
enquiries@wordsworth.org.uk
www.wordsworth.org.uk
Contact *Literature Officer*
Ongoing contemporary poetry programme held over the summer and one-off events held around the country. Readers include Simon Armitage, Seamus Heaney, Paul Muldoon, Sharon Olds, Philip Pullman and Don Paterson as well as a 'support' reader at most events.
The Trust also operates an artists-in-residence programme throughout the year, giving time and space for new writers to develop their work in this setting: a cottage and stipend are supplied and no demands are made upon the artist. Past artists-in-residence include Henry Shukman, Owen Sheers, Paul Farley, Helen Farish, Jack Mapanje, Rebecca O'Connor, Jacob Polley, Neil Rollinson and John Hartley Williams.

Writenow!
▷ Queen's Theatre, Hornchurch under Theatre Producers

Writing on the Wall

First Floor, Mission Hall, 36 Windsor Street, Liverpool L8 1XE
☎ 0151 703 0020
info@writingonthewall.org.uk
www.writingonthewall.org.uk
Festival Coordinator *Madeline Heneghan*

Annual festival held in Liverpool that works alongside schools, young people, local communities and broader audiences to celebrate writing, diversity, tolerance, story-telling and humour through controversy, inquiry and debate. Performances, readings, workshops, screenings, high profile debates and discussions. Guest speakers have included Noam Chomsky, Irvine Welsh, Howard Marks, Roddy Doyle, Benjamin Zephaniah.

Useful Websites

A2A Access to Archives

www.a2a.org.uk

Database of UK archive catalogues dating from
the eighth century to the present day. Now
contains '10.2 million records relating to 9.35
million items held in 416 record offices and other
repositories'.

AbeBooks

www.abebooks.co.uk

Includes new and second-hand books search
engine and 'The Rare Book Room' for antiquarian,
rare and collectible titles online. Also features
a search engine for textbooks and reference
publications, both new and second-hand.

Academi (Welsh Academy/ Yr Academi Gymreig)

www.academi.org

News of events, publications and funding for
Welsh-based literary events. (See entry under
Professional Associations and Societies.)

Alibris

www.alibris.com

Over 60 million used, new and hard-to-find
books online. Bargain books, text and reference
listings and rare books. ISBN search engine. Also
movies and music.

Alliance of Literary Societies

www.sndc.demon.co.uk/als.htm

Details of societies and events. (See entry under
Professional Associations and Societies.)

Amazon

www.amazon.co.uk

The online shop.

Ancestry

www.ancestry.co.uk

Family history information – databases, articles
and other sources of genealogical data.

Arts Council England

www.artscouncil.org.uk

Includes information on funding applications,
publications and the National Lottery. (See entry
under *Arts Councils and Regional Offices*.)

Arts Council of Northern Ireland

www.artscouncil-ni.org

Information on funding and awards, events and
free E-Newsletter service. (See entry under *Arts
Councils and Regional Offices*.)

Arts Council of Wales/ Cyngor Celfyddydau Cymru

www.acw-ccc.org.uk

Information on publications, grants, council
meetings, the arts in Wales. (See entry under *Arts
Councils and Regional Offices*.)

Arvon Foundation

www.arvonfoundation.org

See entry under *UK and Irish Writers' Courses*.

Ask About Writing

www.askaboutwriting.net

Resource site for writers.

Association for Scottish Literary Studies

www.asls.org.uk

The educational charity promoting the languages
and literature of Scotland. (See entry under
Professional Associations and Societies.)

Association of Authors' Agents (AAA)

www.agentsassoc.co.uk

UK agents' organization including list of
current members. (See entry under *Professional
Associations and Societies*.)

Association of Authors' Representatives (AAR)

www.aar-online.org

US agents' organization including search engine for current members. (See entry under *Professional Associations and Societies.*)

Author-Network

www.author-network.com

Writers' resource site run by Karen Scott.

Author.co.uk

www.author.co.uk

Links to advice and information for writers.

Authorbank

www.authorbank.com

Registration for a fee enables authors to present book ideas to publishers online and access to a free advice service.

Authors' Licensing and Collecting Society Limited (ALCS)

www.alcs.co.uk

Details of membership, news and publications. (See entry under *Professional Associations and Societies.*)

Bartleby.com

www.bartleby.com

An ever-expanding list of books published online for reference, free of charge.

BBC

www.bbc.co.uk

Access to all BBC departments and services.

bibliofind

www.bibliofind.com

Millions of second-hand and rare books, periodicals and ephemera for sale online via Amazon.com.

Bibliomania

www.bibliomania.com

Over 2,000 classic texts, study guides and reference resources available online for free. Also selected books for sale in the Bibliomania shop.

book2book/booktrade.info

www.booktrade.info/index.php

Established by a group of publishers, booksellers, website developers and trade journalists to provide up-to-date news, features and useful information for the book trade.

Booktrust

www.booktrust.org.uk

Book information service, guide to prizes and awards, factsheets on getting published. (See entry under *Professional Associations and Societies.*)

British Association of Picture Libraries and Agencies (BAPLA)

www.bapla.org.uk

Website includes the BAPLA Online Database search facility by category or name. (See entry under *Professional Associations and Societies.*)

British Centre for Literary Translation

www.literarytranslation.com

A joint website with the British Council with workshops by leading translators, contacts and networks, and search engine for translation conferences, seminars and events. (See entry under *Professional Associations and Societies.*)

The British Council

www.britishcouncil.org

Information on the Council's English Language services, education programmes, society and science links. (See entry under *Professional Associations and Societies.*)

British Film Institute (bfi)

www.bfi.org.uk

Information on the services offered by the Institute. (See entry under *Professional Associations and Societies.*)

British Library

www.bl.uk

Reader service enquiries, access to main catalogues, information on collections, links to the various reading rooms and exhibitions. (See related entries under *Library Services.*)

Chapter One Promotions

www.chapteronepromotions.com

Provides services to writers at various stages of their writing career: story and poetry critique services, proof reading, 'Kids Korner'.

Characterization Tool for Novelists

www.synergise.com/p4

The 'P4 Personality Mapping tool' provides novelists with the means to quickly create realistic characters based upon accepted psychological types. Works online and offline

Chartered Institute of Linguists (IoL)

www.iol.org.uk

Discussion forum, news on regional societies, job opportunities, 'Find a Linguist' service and *The*

Linguist magazine. (See entry under *Professional Associations and Societies.*)

CILIP
www.cilip.org.uk

The professional body for librarians and information professionals. (See entry under *Professional Associations and Societies.*)

CILIP in Scotland (CLIPS)
www.slainte.org.uk

(See entry under *Professional Associations and Societies.*)

CILIP Wales
www.dil.aber.ac.uk/cilip_w/index.htm

(See entry under *Professional Associations and Societies.*)

Complete Works of William Shakespeare
www-tech.mit.edu/Shakespeare/works.html

Access to the text of the complete works.

Copyright Licensing Agency Ltd (CLA)
www.cla.co.uk

Copyright information, customer support and information on CLA services. (See entry under *Professional Associations and Societies.*)

Crime Writers' Association (CWA)
www.thecwa.co.uk

Website of the professional crime writers' association. (See entry under *Professional Associations and Societies.*)

Daily Express
www.express.co.uk

Daily Express online.

Daily Mail
www.dailymail.co.uk

Daily Mail online.

Daily Mirror
www.mirror.co.uk

Daily Mirror online.

Daily Telegraph
www.telegraph.co.uk

Daily Telegraph online.

Dictionary of Slang
dictionaryofslang.co.uk

A guide to slang 'from a British perspective'. Research information and search facility.

The Eclectic Writer
www.eclectics.com/writing/writing.html

US website offering a selection of articles giving advice for writers and an online discussion board.

The English Association
www.le.ac.uk/engassoc

News, publications, conference and membership information. (See entry under *Professional Associations and Societies.*)

Federation of Worker Writers and Community Publishers (FWWCP)
myweb.tiscali.co.uk/thefwwcp/Info.htm

Links to members of the FWWCP, the Federation magazine, information on membership. (See entry under *Professional Associations and Societies.*)

Film Angel
www.filmangel.co.uk

Established in conjunction with **Hammerwood Films** to create a shop window for writers and would-be film angels alike. Submitted synopses are displayed for a pre-determined period, for a fee, while would-be angels are invited to finance a production of their choice.

Filmmaker Store
www.filmmakerstore.com

US site giving scriptwriting resources, listings and advice.

Financial Times
www.ft.com

Financial Times online.

Frankfurt Book Fair
www.frankfurt-book-fair.com/en/portal.html

Provides latest news and market analysis of the book business plus information on the annual Book Fair.

The Froebel Archive for Childhood Studies
www.roehampton.ac.uk/froebel/
    froebelarchive/index.html

Collection of children's literature. Supports courses at Roehampton University and is available to *bona fide* researchers. Catalogue available online.

The Good Web Guide Ltd
www.thegoodwebguide.co.uk

Guide to the best websites. Provides thousands of detailed and independent reviews of a wide range of websites.

Google Images
www.images.google.com

Search engine for images.

The Guardian/Observer
www.guardian.co.uk

Website of the *Guardian* and *Observer* newspapers online.

Guide to Grammar and Style
www.andromeda.rutgers.edu/~jlynch/Writing

A guide to grammar and style, organized alphabetically, plus articles and links to other grammatical reference sites.

Hansard
www.parliament.the-stationery-office.co.uk/
pa/cm/cmhansrd.htm

The official record of debates and written answers in the House of Commons. The transcript of each day's business appears at noon on the following weekday.

The Herald
www.theherald.co.uk

Scottish daily broadsheet.

House of Commons Research Library
www.parliament.uk/parliamentary_publications_
and_archives/research_papers.cfm

Gives access to the text of research reports prepared for MPs on a wide range of current issues.

The Independent
www.independent.co.uk

The *Independent* newspaper online.

Independent Northern Publishers
www.northernpublishers.co.uk

Promotes the work of new writers across the region.

IngentaConnect
www.ingentaconnect.com

A comprehensive online UK academic research service. The site offers access to more than 23 million articles, chapters and reports online.

Inpress Books
www.inpressbooks.co.uk

Provides sales, marketing and distribution for small independent publishers.

Institute of Translation and Interpreting (ITI)
www.iti.org.uk/indexMain.html

Website of the professional association of translators and interpreters, with the ITI directory of members, publications, training and membership information. (See entry under *Professional Associations and Societies*.)

Internet Classics Archive
classics.mit.edu

Includes works of classical literature; mostly Greek and Roman with some Chinese and Persian. All are in English translation.

Internet Movie Database (IMDb)
www.imdb.com

Essential resource for film buffs and researchers with search engine for cast lists, screenwriters, directors and producers; film and television news, awards, film preview information, video releases.

The Irish Arts Council/An Chomhairle Ealaíon
www.artscouncil.ie

Monthly e-mail newsletter available giving latest information on grants and awards, news and events, etc. (See entry under *Arts Councils and Regional Offices*.)

Journalism UK
www.journalismuk.co.uk

A website for UK-based print journalists who write for text-based publications. Includes links to newspapers, magazines, e-zines, news sources plus information on training and organizations.

The Library Association
▷ CILIP

Literature North East
www.literaturenortheast.co.uk

Monthly e-newsletter; listings of events and readings in the region; writing courses and local training opportunities.

Literature North West
www.publishingnorthwest.co.uk

Promotional agency for the region's independent presses and literature organizations

literaturetraining
www.literaturetraining.com

Online directory of training and professional development opportunities for UK writers and literature professionals. Includes information on courses, workshops, jobs, residencies, submissions, competitions, organizations and funding for professional development. Also, information sheets and specially commissioned features.

Location Register of 20th-Century English Literary Manuscripts and Letters
www.reading.ac.uk/library/about-us/projects/lib-
location-register.asp

Reference source for the study of English literature. Information about the manuscript holdings of repositories of all sizes, from the

British Library to small-town museums, of literary authors – from major poets to minor science fiction writers.

The Mail on Sunday
www.mailonsunday.co.uk

The Mail on Sunday online.

Mr William Shakespeare and the Internet
shakespeare.palomar.edu

Guide to scholarly Shakespeare resources on the Internet.

National Union of Journalists (NUJ)
www.nuj.org.uk

Represents those journalists who work in all sectors of publishing, print and broadcasting. (See entry under *Professional Associations and Societies.*)

The Never Ending Story
www.TheNeverEndingStory.co.uk

A range of constantly evolving online publications which online users can read and contribute to. Includes stories and poems by well-known writers, personalities and site members (free membership).

New Writers Consultancy
www.new-writers-consultancy.com

Advice for writers and critiques.

New Writing North
www.newwritingnorth.com

Essentially for writers based in the north of England but also a useful source of advice and guidelines. (See entry under *Professional Associations and Societies.*)

New Writing Partnership
www.newwritingpartnership.org.uk

Promotes and supports creative writing in Eastern England.

Newnovelist
www.newnovelist.com

Novel-writing software. Aids research, characterization, structure and plot, and breaks down the process of writing a novel into manageable chunks.

PEN
www.englishpen.org

Website of the English Centre of International PEN. News of events, information on prizes, membership details. (See entry under *Professional Associations and Societies.*)

PlaysOnTheNet
www.playsonthenet.com

Information and help for new playwrights.

Poets and Writers
www.pw.org

A US site containing information and advice for writers.

Producers Alliance for Cinema and Television (pact)
www.pact.co.uk

Publications, training, production companies, membership details. (See entry under *Professional Associations and Societies.*)

The Publishers Association
www.publishers.org.uk

Information about the Association and careers in publishing; also 'Getting Published' pages. (See entry under *Professional Associations and Societies.*)

Publishing North West
▷ Literature North West.

Publishing Scotland
www.publishingscotland.org

Links to websites of members of the Association, information on activities and publications. (See entry under *Professional Associations and Societies.*)

RefDesk.com
refdesk.com

Free facts and statistics on every country in the world plus charts and maps, illustrations and related sources.

Royal Society of Literature
www.rslit.org

Information on lectures, discussions and readings; membership details and prizes. (See entry under *Professional Associations and Societies.*)

The Scotsman
www.scotsman.com

The Scotsman newspaper online.

Scottish Arts Council
www.sac.org.uk

Information on funding and events. (See entry under *Arts Councils and Regional Offices.*)

Scottish Book Trust
www.scottishbooktrust.com

Information on the Trust's activities. (See entry under *Professional Associations and Societies.*)

Scottish Publishers Association
▷ Publishing Scotland

Screenwriters Online
screenwriter.com/insider/news.html

US website, described as the '*only* professional screenwriter's site run by major screenwriters who get their scripts and screenplays made into movies'. Contains screenplay analysis, expert articles and *The Insider Report*.

The SF Hub
www.sfhub.ac.uk/

Science fiction research website created by the University of Liverpool. Includes links to the Science Fiction Foundation collection and the John Wyndham archives.

Shots Magazine
www.shotsmag.co.uk

Electronic magazine of crime and mystery fiction.

Slainte: Information & Libraries Scotland
www.slainte.org.uk

Links to various services and information on librarianship and information management in Scotland. (See entry under *Professional Associations and Societies*.)

Society for Editors and Proofreaders (SfEP)
www.sfep.org.uk

Information about the Society including online directory of members. (See entry under *Professional Associations and Societies*.)

The Society of Authors
www.societyofauthors.org

Includes FAQs for new writers, diary of events, membership details, prizes and awards. (See entry under *Professional Associations and Societies*.)

Society of Indexers
www.indexers.org.uk/

Indexing information for publishers and authors, 'Find an Indexer' pages. Membership information. (See entry under *Professional Associations and Societies*.)

South Bank Centre, London
www.sbc.org.uk

Links to the Royal Festival Hall, Purcell Room, Queen Elizabeth Hall, the Hayward Gallery and Poetry Library; news of literature events.

Story Wizard
www.storywizard.co.uk

Software to enable children to write creative stories.

The Sun
www.thesun.co.uk

The Sun online.

The Times
www.thetimes.co.uk

The Times online.

trAce Online Writing Centre
tracearchive.ntu.ac.uk

Online archive of work published by the trAce Online Writing Centre between 1995–2005. Also holds articles and transcripts of discussions.

UK Children's Books Directory
www.ukchildrensbooks.co.uk

Website created by Steve and Diana Kimpton to 'increase the profile of UK children's books on the Internet'.

The UK Public Libraries Page
dspace.dial.pipex.com/town/square/ac940/
    weblibs.html

Website links to public libraries throughout the UK, compiled by Sheila and Robert Harden.

Webster Dictionary/Thesaurus
www.m-w.com/home.htm

Merriam-Webster Online. Includes a search facility for words in the Webster Dictionary or Webster Thesaurus; word games and daily podcast.

Welsh Academy
▷ Academi

Welsh Books Council (Cyngor Llyfrau Cymru)
www.cllc.org.uk

Information about books from Wales, editorial and design services, publishing grants. (See entry under *Professional Associations and Societies*.)

The Word Pool
www.wordpool.co.uk

Children's book site with information on writing for children and a thriving discussion group for children's writers. Also Word Pool Design (www.wordpooldesign.co.uk): web design for writers, illustrators and publishers.

WordCounter
www.wordcounter.com

Highlights the most frequently used words in a given text. Use as a guide to see what words are overused. Also Political Vocabulary Analysis which measures indications of political leanings in given text.

Write4kids.com

www.write4kids.com

US website for children's writers, whether published or beginners. Includes special reports, articles, advice, news on the latest bestsellers and links to related sites.

Writernet

www.writernet.co.uk

Information, advice and guidance for writers on all aspects of live and recorded performance. (See entry under *Professional Associations and Societies*.)

The Writers' Guild of Great Britain

www.writersguild.org.uk

Information on contracts, copyright, news, writers' resources and industry regulations. (See entry under *Professional Associations and Societies*.)

Writers Nexus International

www.writersnexus.com

Enables writers to place their work online and for publishers, agents and readers to access it quickly and effectively.

Writers' Circles

www.writers-circles.com

Directory of writers' circles, courses and workshops.

Writers, Artists and their Copyright Holders (WATCH)

tyler.hrc.utexas.edu

Database of copyright holders in the UK and North America. (See entry under *Professional Associations and Societies*.)

WritersNet

www.writers.net

Internet directory of writers, editors, publishers and literary agents.

WritersServices

www.WritersServices.com

Established in March 2000 by Chris Holifield, former deputy managing director and publisher at Cassell. Offers factsheets, book reviews, advice, links and other resources for writers including editorial services, contract vetting and self-publishing.

The Writing Centre

www.thewritingcentre.com

A resource site for writers in Cornwall, the South West and beyond. Covers all genres from poetry to business writing, from novels to radio comedy. Helps find mentors, run workshops, seminars and short courses. Details of training and listings of magazines and publishers in the region.

Writing-World.com

writing-world.com

'The online community for readers and writers.'

Miscellany

Apple Coaching (inc. CoachingWriters.co.uk)

8 Feering Road, Billericay CM11 2DR
☎ 01277 632085
eve@applecoaching.com
www.CoachingWriters.co.uk
Contact *Eve Menezes Cunningham*

Helping professional and aspiring writers increase their success and happiness through life coaching, business coaching and NLP. Writing and editorial services also available. Sign up for free e-newsletter on the website. Rates negotiable.

Combrógos

10 Heol Don, Whitchurch, Cardiff CF14 2AU
☎ 029 2062 3359
Contact *Professor Meic Stephens*

Founded 1990. Arts and media research, editorial services, specializing in books about Wales or by Welsh authors. 'Encyclopaedic knowledge of Welsh history, language, literature and culture.'

DOI Registration Agency

3rd Floor, Midas House, 62 Goldsworth Road, Woking
☎ 0870 777 8712   🖷 0870 777 8714
doi.agency@nielsen.com
www.doi.nielsenbookdata.co.uk
Manager, ISBN, SAN & DOI Agencies *Diana Williams*
Senior Manager, Registration Services ISBN, SAN & DOI Agencies *Julian Sowa*

DOIs (Digital Object Identifiers) are used to uniquely identify files or other resources on the Internet. The DOI Registration Agency is responsible for issuing DOI prefixes and numbers and can provide help and advice on maintaining DOIs. Nielsen Book also runs the **ISBN** and **SAN** agencies (see entries).

Jacqueline Edwards

104 Earlsdon Avenue South, Coventry CV5 6DQ
twigsbranches@yahoo.co.uk
Contact *Jacqueline Edwards, MA, LLB(Hons)*

Historical research: family, local and 19th and 20th century legal history. Covers Warwickshire, Gloucestershire, Northamptonshire, Worcestershire and the National Archives, Kew, London.

Fiction Writing Workshops and Tutorials

5 Queen Elizabeth Close, London N16 0HL
☎ 020 8809 4725
henrietta@writtenwords.net
www.WrittenWords.net
Contact *Henrietta Soames*

Fiction workshops for beginner and advanced writers led by author, Henrietta Soames. 'Lively discussion, stimulating exercises, valuable feedback.' Fiction tutorials for students who require concentrated attention and ongoing support. Short stories/novels, full-length mss welcomed.

Susan Grossman – Writing Workshops

☎ 07905 888835
susangrossman@tiscali.co.uk
writeowords@tiscali.co.uk
www.susangrossman.co.uk
Contact *Susan Grossman*

Established travel writer and freelance journalist, Susan Grossman runs journalism workshops for freelancers looking to increase their writing ability and earning potential. Workshops cover how to focus on an idea, write compelling synopses and approach editors with confidence. Selling Freelance Features and Travel Writing workshops are held regularly in central London. One-to-one sessions are also available. See website for details.

ISBN Agency

3rd Floor, Midas House, 62 Goldsworth Road, Guildford
☎ 0870 777 8712   🖷 0870 777 8714
isbn.agency@nielsen.com
www.isbn.nielsenbookdata.co.uk
Manager, ISBN, SAN & DOI Agencies *Diana Williams*
Senior Manager, Registration Services ISBN, SAN & DOI Agencies *Julian Sowa*

ISBNs are product numbers used by all sections of the book trade for ordering and listing purposes. The ISBN Agency is responsible for issuing ISBNs to publishers based in the UK and Republic of Ireland and can provide help and advice on changing from 10 to 13-digits. Nielsen Book also runs the **SAN & DOI** agencies (see entries).

Caroline Landeau

6 Querrin Street, London SW6 2SJ
☎ 07050 600420
Contact *Caroline Landeau*

Experienced research and production – films, multimedia, books, magazines, exhibitions, animation, general interest, art, music, crime, travel, food, film, theatre.

M-Y Books Ltd

187 Ware Road, Hertford SG13 7EQ
☎ 01992 586279
jonathan@m-ybooks.co.uk
www.m-ybooks.co.uk
Contact *Jonathan Miller*

Offers book promotional and marketing services, also provides audio book production and distribution.

Julia McCutchen, Writers' Coach & Professional Publishing Consultant

PO Box 3703, Trowbridge
☎ 01380 871331   ℻ 01380 871331
julia@juliamccutchen.com
www.JuliaMcCutchen.com
Contact *Julia McCutchen*

Julia McCutchen is the author of *The Writer's Journey: From Inspiration to Publication* and has 20 years experience of publishing. Specializes in helping writers who want to write a book for publication by offering individual coaching, courses and classes to provide information, guidance, feedback and support for each stage of the writing journey. Primary areas of expertise include how to write a first-class book proposal and how to approach the right people in the right way for the best possible chances of success. See the website for more information and a range of free resources for writers.

Nielsen BookNet

3rd Floor, Midas House, 62 Goldsworth Road, Woking GU21 6LQ
☎ 0870 777 8710   ℻ 0870 777 8711
sales.booknet@nielsen.com
www.nielsenbooknet.co.uk
Head of BookNet Sales *Stephen Long*

BookNet provides a range of e-commerce services that allows electronic trading between booksellers, publishers/distributors, libraries and other suppliers. Services include BookNet Web for booksellers and publishers/distributors, TeleOrdering and EDI messaging.

Nielsen BookScan

3rd Floor, Midas House, 62 Goldsworth Road, Woking GU21 6LQ
☎ 01483 712222   ℻ 01483 712220
info.bookscan@nielsen.com
www.nielsenbookscan.co.uk
Publisher Account Manager *Reeta Windsor*

BookScan collects transactional data at the point of sale from tills and despatch systems of all the major book retailers in the UK, Ireland, US, South Africa, Spain and Italy (with New Zealand and Denmark due to launch shortly). Each week data is coded and analysed, producing complete market information for retailers, publishers, libraries, agents and the media within 72 hours of the week ending Saturday.

Ormrod Research Services

Weeping Birch, Burwash TN19 7HG
☎ 01435 882541
richardormrod@aol.com
www.writeonservices.co.uk
Contact *Richard Ormrod*

Established 1982. Comprehensive research service: literary, historical, academic, biographical, commercial. Verbal quotations available. Also editing, indexing, ghost-writing and critical reading.

Patent Research

Dachsteinstr. 12a, D–81825 Munich, Germany
☎ 00 49 89 430 7833
Contact *Gerhard Everwyn*

All world, historical patents for researchers, authors, archives, museums and publishers. Rates on application.

Leda Sammarco

☎ 07930 568516
leda.sammarco@btopenworld.com
www.ledasammarco.com
Contact *Leda Sammarco*

Provides coaching service for writers in the area of mind, body and spirit, personal development and self-help. Also copywriting and advice on PR and marketing.

SAN Agency

3rd Floor, Midas House, 62 Goldsworth Road, Guildford 0870 777 8712
☎ 0870 777 8714
san.agency@nielsen.com
www.san.nielsenbookdata.co.uk
Manager, ISBN, SAN & DOI Agencies *Diana Williams*
Senior Manager, Registration Services ISBN, SAN & DOI Agencies *Julian Sowa*

SANs, Standard Address Numbers, are unique identifiers for geographical locations and can be assigned to the addresses of organizations involved in the book selling or publishing industries. The SAN Agency is responsible for managing the scheme on behalf of Book Industry Communication in the UK and Republic of Ireland. Nielsen Book also runs the **ISBN** & **DOI** agencies (see entries).

The United Kingdom Copyright Bureau

110 Trafalgar Road, Portslade BN41 1GS

☎ 01273 277333

info@copyrightbureau.co.uk
www.copyrightbureau.co.uk
Contacts *Ralph de Straet von Kollman, Petra Ginman*

The UKCB provides a secure copyright service at reasonable cost, enabling multiple copyrights to be registered nominally when required. Prices are advertised on the website including the UKCB's solicitors, etc. Copyrights preferred on floppy disk or CD-ROM; manuscripts are not accepted due to storage space.

Index of Entries

A

AA Publishing, 64
A2A Access to Archives, 642
Aard Press, 225
Aardman, 417
Abacus, 64
ABC News Intercontinental
Inc., 394
ABC–CLIO (US), 172
ABC–CLIO (UK), 64
AbeBooks, 642
Abel (Dominick) Literary
Agency, Inc., 280
Aberdeen (University of)
Writers Festival, 629
Aberdeen Arts Carnival, 629
Aberdeen Central Library, 574
Aberystwyth, University of, 467
Abingdon Press, 172
Ableman (Sheila) Literary
Agency, 249
Above The Title, 417
Abrams (Harry N.), Inc., 172
Abraxas Unbound 294
Absolute Press, 64
Absolutely Productions Ltd, 417
Abson Books London, 64
Abstract Images, 417
ABSW, 526
Acacia Productions Ltd, 417
Academi, 219, 526
Academi Book of the Year
Awards, 488
Academi Cardiff International
Poetry Competition, 488
Academic Press, 64
Academy Chicago Publishers,
172
Acair Ltd, 64
Accent Press, 225
Acclaim, 294
Accountancy, 294
Accountancy Age, 294
Ace Books, 172
ACE Tennis Magazine, 294
ACER Press, 191
Ackerley (J.R.) Prize, 488
Acme, 601
Acoustic, 294

Acrobat Television, 417
Actaeon Films Ltd, 418
actionplus sports images, 601
Actors Touring Company, 438
Acumen, 208, 294
Acumen Publishing Limited, 64
Addison Wesley, 64
Adelphi Edizioni SpA, 166
Adelphi University, 473
Adkins (Lesley & Roy) Picture
Library, 601
Adlard Coles Nautical, 64
Adrian Bell Society, The, 553
Advertising Archive Limited,
The, 601
Aeroplane, 295
Aesthetica, 208
Affiliated East West Press Pvt
Ltd, 195
African Books Collective, 64
Age Concern Books, 65
Agency (London) Ltd, The, 249
Agenda Editions, 199
Agenda Poetry Magazine, 208
Agent Research & Evaluation,
Inc. (AR&E), 273
AIR International, 295
Aireings, 208
AirForces Monthly, 295
Airlife Publishing, 65
Aitken Alexander Associates
Ltd, 249
Akadémiai Kiadó, 166
akg-images Ltd, Arts and
History Picture Library, 601
Akros Publications, 65, 199
Alabama Press, University of,
172
Aladdin Books Ltd, 242
Aladdin Paperbacks, 172
Alaska Press, University of, 172
Albion Press Ltd, The, 242
Alcemi, 65
ALCS, 526
Aldeburgh Literary Festival,
The, 629
Aldeburgh Poetry Festival, 629
Aldeburgh Poetry Festival Prize,
488

Alexander (Bryan & Cherry)
Photography, 601
Alexander Prize, 488
Alianza Editorial SA, 169
Alibris, 642
All About Soap, 295
All Out Productions, 418
Allan (Ian) Publishing Ltd, 65
Allardyce, Barnett, Publishers,
225
Allen & Unwin Pty Ltd, 191
Allen (J.A.) & Co., 65
Allen Lane, 65
Allen-Jones (Louise) Literary
Scouts, 274
Alliance of Literary Societies,
526
Allingham (Margery) Society,
552
Allison & Busby, 65
Alloway Publishing Limited, 225
Allyn & Bacon, 65
Alma Books Ltd, 65
Almeida Theatre Company, 438
Alomo Productions, 418
Alpha Books, 172
Alpha Press, 66
Alpine Garden Society, 601
Alston Hall College, 457
Alternative Theatre Company
Ltd, 438
Alton Douglas Books, 66
Altshuler (Miriam) Literary
Agency, 280
Alvey & Towers, 601
AMACOM Books, 172
Amateur Gardening, 295
Amateur Photographer, 295
Amateur Stage, 295
Amazon, 642
Amber Books Ltd, 242
Amber Lane Press Ltd, 66
Ameet Sp. zo.o., 168
American Film Institute, 469
American University, 470
Amistad, 173
Ammerdown Conference and
Retreat Centre, 462
Ampersand Agency Ltd, The,
249

Amsco, 66
An Gúm, 153
Ancestors, 295
Ancestry, 642
Anchor Canada, 193
Andersen (Hans Christian) Awards, 488
Andersen Press Ltd, 66
Anderson (Darley) Literary, TV & Film Agency, 249
Andes Press Agency, 602
Andrews (Chris) Publications, 66
Andrews (Julie) Collection, 173
Angel (Heather)/Natural Visions, 602
Angels' Share, The, 66
Angler's Mail, 295
Anglo-Saxon Books, 225
Anglo/Fortunato Films Ltd, 418
Angus Book Award, 488
Animal Action, 296
Animals and You, 296
Anness Publishing Ltd, 66
Annual Theatre Book Prize, 489
Annual Writers' Writing Courses & Workshops, 457
Anova Books, 66
Anthem Press, 67
Antique Collecting, 296
Antique Collectors' Club, 67
Antiquesnews, 296
Anubis Literary Agency, 250
Anvil, 173
Anvil Books, 153
Anvil Press Poetry Ltd, 67, 199
Apex Advertiser, 296
Apex Publishing Ltd, 67
Apollo, 296
Apollos, 68
Apostrof (Forlaget) ApS, 161
Apple, 68
Apple Coaching, 641
Apples & Snakes, 219
Appletree Press Ltd, 68
Aquamarine, 68
Aquarius, 208
Aquarius Library, 602
Arbour E-Books, 247
Arc Publications, 199
Arc Publications Ltd, 68
Arcadia Books, 68
Arcane, 68
Arche Verlag AG, 171
Architects' Journal, The, 296
Architectural Association Photo Library, 602
Architectural Design, 296
Architectural Press, 68
Architectural Review, The, 296
ARD – Das Erste, 414
Ardis, 68

Areheart (Shaye) Books, 173
Arena, 191, 297
ArenaPAL, 602
Areopagus, 208
Argentum, 68
Argus, The (Brighton), 386
Aris & Phillips, 68
Arizona Press, University of, 173
Arizona, University of, 469
Arkansas Press, University of, 173
Arkansas, University of, 469
ArkReligion.com, 602
Arlington Productions Limited, 418
Armchair Traveller, 68
Armitt Collection, Museum & Library, 574
Armourer Magazine, The, 297
Arnefold Publishing, 68
Arris Publishing Ltd, 68
Arrow, 69
Arrowhead Press, 199
Arscott's (David) Sussex Book Club, 293
Art & Training Films Ltd, 418
Art Archive, The, 602
Art Directors & Tripp Photo Library, 602
Art Monthly, 297
Art Newspaper, The, 297
Art Review, 297
Arte France, 414
Arte Germany, 414
Artech House, 69
Artellus Limited, 250
Arthaud (Éditions), 162
Arthur Miller Centre Literary Festival, 636
Artist, The, 297
Artists' Choice, 293
Arts & Business (A&B), 527
Arts Council Children's Award, 489
Arts Council England, 572
Arts Council of Northern Ireland, The, 573
Arts Council of Wales ACW's Creative Wales Awards, 482
Arts Council of Wales, The/ Cyngor Celfyddydau Cymru, 573
Arts Council, The/An Chomhairle Ealaíon, 572
Arvon Foundation, 458
Arvon Foundation (Devon), 455
Arvon Foundation (Inverness-shire), 466
Arvon Foundation (Shropshire), 462
Arvon Foundation (Yorkshire), 464

Arvon Foundation International Poetry Competition, 489
Aschehoug (H) & Co (W Nygaard), 168
Asham Award for Women, 489
Ashfield Press, 153
Ashford Entertainment Corporation Ltd, The, 418
Ashgate Publishing Ltd, 69
Ashgrove Publishing, 69
Ashwell Publishing Limited, 69
Ask About Writing, 642
ASLS, 527
Aspect Picture Library Ltd, 602
Aspects Literature Festival, 629
Associated Press, 394
Associated Press Limited, 393
Association for Scottish Literary Studies, 527
Association of American Correspondents in London, 527
Association of American Publishers, Inc, 527
Association of Authors' Agents (AAA), 527
Association of Authors' Representatives (AAR, 527
Association of British Editors, 527
Association of British Science Writers, 527
Association of Canadian Publishers, 527
Association of Christian Writers, 528
Association of Freelance Editors, Proofreaders & Indexers (Ireland), 528
Association of Freelance Writers, 528
Association of Golf Writers, 528
Association of Illustrators, 528
Association of Independent Libraries, 528
Association of Learned and Professional Society Publishers, 529
Association of Scottish Motoring Writers, 529
ATC, 438
Athenaeum, Liverpool, The, 574
Atheneum Books for Young Readers, 173
Athletics Weekly, 297
Atlantean Publishing, 199
Atlantic Books, 69
Atlantic Europe Publishing Co. Ltd, 69
Atlantic Monthly Press, 173
Atlantis Ltd, 161
Atom, 70

Atria Books, 173
Atrium, 153
Attic Press, 153
Auckland University Press, 196
Audiobook Publishing
    Association, 529
Aurora Metro, 70
Aurum Press Ltd, 70
Austen (Jane) Society, 552
Austin & Macauley Publishers
    Limited, 70
Australia Pictures, 602
Australian Copyright Council,
    529
Australian Publishers
    Association, 529
Australian Society of Authors,
    529
Authentic Media, 70
Author Literary Agents, 250
Author, The, 298
Author.co.uk, 642
Author-Network, 643
Authorbank, 642
AuthorHouse UK, 70
Authors Online, 70
Authors' Club, 529
Authors' Club First Novel
    Award, 489
Authors' Contingency Fund,
    The, 482
Authors' Foundation, The, 482
Authors' Licensing and
    Collecting Society Limited
    (ALCS), 530
Auto Express, 298
Autocar, 298
Autumn Publishing, 71
Avalon, 418
Avalon Travel, 173
Avery, 173
Aviation News, 298
aviation–images.com, 603
Avon, 71, 173
AVRO (Algemene Omroep
    Vereniging), 415
Award Publications Limited, 71
Awen, 208
Axelrod Agency, The, 280
Aye Write! Bank of Scotland
    Book Festival, 629
Azure, 71

B

BA/Nielson BookData Author
    of the Year Award, 489
Baby Cow Productions, 418
Back Bay Books, 173
Badcock & Rozycki Literary
    Scouts, 274
Badger Publishing Ltd, 71
Baillière Tindall, 71

Baird (Duncan) Publishers, 71
Baker Books, 293
Balance, 298
Baldi (Malaga) Literary Agency,
    280
Baldwin (M.&M.), 225
Balkin Agency, Inc., The, 280
Ballantine Books, 173
Baltimore Sun, The, 394
Banipal Prize, 489
Bank of England Information
    Centre, 574
Banker, The, 298
Banner Books, 173
Bantam Dell Publishing Group,
    173
Bantam/Bantam Press, 71
BAPLA (British Association
    of Picture Libraries and
    Agencies), 530
Barbara Pym Society, The, 566
Barbican Library, 574
Bard, 208
Bard Hair Day, A, 208
Barefoot Books Ltd, 71
Bargate (Verity) Award, 489
Barnaby's Library, 603
Barnardo's Photographic and
    Film Archive, 603
Barnsley Public Library, 575
Barny Books, 225
Barque Press, 199
Barrett (Loretta) Books, Inc.,
    280
Barrington Stoke, 71
Barron's Educational Series,
    Inc., 173
Bartleby.com, 643
Basic Books/Basic Civitas, 173
Baskerville Hounds, The, 552
Bath Chronicle, The, 385
Bath Festival of Children's
    Literature, 629
Bath Literature Festival, 630
Bath Spa University, 462
Batsford (B.T.), 71
Baxter (Colin) Photography
    Limited, 603
BayLit Festival, 630
BB Books, 199
BB Society, The, 552
BBC Active, 71
BBC Asian Network, 399
BBC Audiobooks Ltd, 238
BBC Birmingham, 399
BBC Books, 71
BBC Bristol, 399
BBC Children's Books, 71
BBC Coventry and
    Warwickshire, 401
BBC Drama, Comedy and
    Children's, 396

BBC Essex, 401
*BBC Gardeners' World
    Magazine*, 298
BBC Good Food, 299
BBC Guernsey, 402
BBC Hereford & Worcester, 402
BBC History Magazine, 299
BBC Local Radio, 400
BBC London, 400, 403
BBC Music Magazine, 299
BBC National Short Story Prize,
    The, 489
BBC New Talent, 397
BBC News, 397
BBC Northern Ireland, 398
BBC Oxford, 403
BBC Radio Berkshire, 401
BBC Radio Bristol, 401
BBC Radio Cambridgeshire, 401
BBC Radio Cornwall, 401
BBC Radio Cumbria, 401
BBC Radio Cymru, 401
BBC Radio Derby, 401
BBC Radio Devon, 401
BBC Radio Foyle, 401
BBC Radio Gloucestershire, 402
BBC Radio Humberside, 402
BBC Radio Jersey, 402
BBC Radio Kent, 402
BBC Radio Lancashire, 402
BBC Radio Leeds, 402
BBC Radio Leicester, 402
BBC Radio Lincolnshire, 402
BBC Radio Manchester, 403
BBC Radio Merseyside, 403
BBC Radio Newcastle, 403
BBC Radio Norfolk, 403
BBC Radio Northampton, 403
BBC Radio Nottingham, 403
BBC Radio Orkney, 403
BBC Radio Scotland
    (Dumfries), 403
BBC Radio Scotland (Selkirk),
    404
BBC Radio Sheffield, 404
BBC Radio Shetland, 404
BBC Radio Shropshire, 404
BBC Radio Solent, 404
BBC Radio Stoke, 404
BBC Radio Suffolk, 404
BBC Radio Swindon, 404
BBC Radio Ulster, 405
BBC Radio Wales, 405
BBC Radio Wiltshire, 405
BBC Radio York, 405
BBC Religion, 397
BBC Scotland, 398
BBC Somerset, 404
BBC South East, 400
BBC Southern Counties Radio,
    404
BBC Sport, 397

BBC Tees, 404
BBC Three Counties Radio, 405
BBC Top Gear Magazine, 299
BBC Vision, BBC Audio & Music, 396
BBC Wales, 399
BBC West/BBC South/BBC South West/BBC South East, 400
BBC Wildlife Magazine, 299
BBC Wildlife Magazine Nature Writing Awards, 490
BBC Wildlife Magazine Poet of the Year Awards, 490
BBC Wildlife Magazine Travel Writing Award, 490
BBC WM, 405
BBC World Service, 397
BBC Worldwide, 72
BBC writersroom, 398
BBC Written Archives Centre, 575
BBC Yorkshire/BBC North West/BBC North East & Cumbria, 400
BBCFour Samuel Johnson Prize for Non-Fiction, The, 490
BCA (Book Club Associates), 293
BCS Publishing Ltd, 242
Beacon Press, 173
Beamish, The Photographic Library, 603
Beautiful Books Ltd, 72
Beck (C.H., OHG) Verlag, 164
Beckford Society, The, 552
Beckmann International, 419
Beddoes (Thomas Lovell) Society, 552
Bedford (Francis), 603
Bedford Central Library, 575
Bee World, 299
Belair, 72
Belcher (Ivan J.) Colour Picture Library, 603
Belfast Festival at Queen's, 630
Belfast News Letter, 389
Belfast Public Libraries: Central Library, 575
Belfast Telegraph, 389
Belfond (Éditions), 162
Bell Lomax Moreton Agency, The, 250
Bella, 299
Belli (Lorella) Literary Agency (LBLA), 250
Belloc (Hilaire) Society, 553
Bender Richardson White, 242
Bennett (Arnold) Society, 553
Benson (E.F.) Society, 553
Berg Publishers, 72
Berghahn Books, 72

Berkley, 174
Berlitz Publishing, 72
Bernstein (Meredith) Literary Agency, Inc., 280
Berry ((David) Prize, 490
Bertelsmann (C.), 164
Bertrand Editora Lda, 169
Best, 300
Best of British, 300
Besterman/McColvin Medal, 490
Betjeman Society, The, 553
Between the Lines, 199, 226
Beverley Literature Festival, 630
Bewick Society, The, 553
BeWrite Books, 247
Beyond the Cloister Publications, 199
BFI, 530
BFI National Library, 575
BFI Publishing, 72
BFI Stills Sales, 603
BFP Books, 72
bibliofind, 643
Bibliographical Society, The, 530
Bibliomania, 643
Bibliophile Books, 293
Big Heart Media, 419
Big Issue Cymru, The, 300
Big Issue in Scotland, The, 300
Big Issue in the North, The, 300
Big Issue South West, The, 300
Big Issue, The, 300
Bike, 300
Bingley (Clive) Books, 73
Biographers' Club Prize, The, 491
Biography, 20–24
Bird Life Magazine, 301
Bird Watching, 301
Birds, 301
Birdwatch, 301
Birdwatch Bird Book of the Year, 491
Birkbeck College, University of London, 458
Birlinn Ltd, 73
Birmingham and Midland Institute, 575
Birmingham Book Festival, 630
Birmingham Central Literary Association, 553
Birmingham City University, 463
Birmingham Library Services, 576
Birmingham Mail, 387
Birmingham Post, 387
Birmingham Repertory Theatre, 438
Birmingham, University of, 464
Bison Books, 174

Bisto Book of the Year Awards, 491
Bitter Lemon Press, 73
Bizarre, 301
Black & White Publishing Ltd, 73
Black (A.&C.) Publishers Ltd, 73
Black (James Tait) Memorial Prizes, 491
Black Ace Books, 73
Black Beauty & Hair, 301
Black Lace, 73
Black Spring Press Ltd, 74
Black Static, 302
Black Swan, 74
Black Theatre Co-op, 438
BlackAmber, 74
Blackbirch Press, 174
Blackhall Publishing Ltd, 153
Blackstaff Press Ltd, 74
Blackstone Publishers, 74
Blackwatch Productions Limited, 419
Blackwell Munksgaard, 161
Blackwell Publishing, 74
Blackwell Publishing (US), 174
Blake (Anthony) Photo Library, 603
Blake (John) Publishing Ltd, 74
Blake Friedmann Literary Agency Ltd, 251
Blandford Press, 74
Bleecker Street Associates, Inc., 280
Blinking Eye Publishing, 199
Bliss, 74
Bliss Magazine, 302
Blithe Spirit, 209
Bloodaxe Books Ltd, 74, 200
Bloody Books, 74
Bloomberg News, 394
Bloomsbury Publishing Plc, 74
Blue Butterfly Publishers, 200
Blue Door, 75
Blue Peter Children's Book Awards, 491
Bluechrome Publishing, 200
Bluemoose Books Limited, 226
Blueprint, 302
Blundell (K.) Trust, The, 482
BMM, 75
BNT, 413
Boardman (Peter) Collection, 603
Boardman Tasker Award, 491
Boatswain Press, 75
Bobcat, 75
Bodleian Library, 75
Bodley Head/Bodley Head Children's Books, 75
Bollinger Everyman Wodehouse Prize, 491

Boltneck Publications Limited, 75
Bolton News, The, 382
Bompiani, 166
Bona Broadcasting Limited, 419
Bond Street, 193
Bonington (Chris) Picture Library, 603
Bonnier Books, 170
Bonnier Books (UK), 75
Bonniers (Albert) Förlag, 170
Bonomi (Luigi) Associates Limited (LBA), 251
Book Blocks, 75
Book Bureau Literary Agency, The, 279
Book Castle, The, 226
Book Club Associates, 293
Book Collector, The, 302
Book Guild Publishing, 75
Book House, 242
Book Now – Richmond Literature Festival, 630
Book World Magazine, 302

book2book/booktrade.info, 643
BookBlast Ltd, 251
Booker Prize for Fiction, The, 492
Books 4 Publishing, 247
Bookseeker Agency, 251
Bookseller, The, 302
Booksellers Association of the UK & Ireland Ltd, 530
Booktrust, 530
Booktrust Early Years Awards, 492
Booktrust Teenage Prize, 492
Bootleg Theatre Company, 439
Borchardt (Georges), Inc., 281
Bordas (Éditions), 162
Border Lines Biographies, 76
Borderline Theatre Company, 439
Borders Book Festival, 630
Borgens Forlag A/S, 161
Borrow (George) Society, The, 553
Boston University, 471
Boulevard Books & The Babel Guides, 76
Boulton and Watt Archive, 604
Bound Biographies Limited, 76
Bounty, 76
Bournemouth Library, 576
Bournemouth Literary Festival, 630
Bournemouth University, 455
Bowker (UK) Ltd, 76
Bowles (Claire) Publicity, 276
Bowling (Harry) Prize, 492
Bowling Green State University, 474

Boxing Monthly, 302
Boxtree, 76
Boyars (Marion) Publishers Ltd, 76
Boyds Mills Press, 174
Boyle (Maria) Communications Limited, 276
Boyz, 303
Bra Böcker AB, 170
Bradford Central Library, 576
Bradley (Alfred) Bursary Award, 482
Bradshaw Books, 153, 200
Bradt Travel Guides, 77
Brand Literary Magazine, 303
Brandon/Mount Eagle Publications, 154
Branford Boase Award, The, 492
Brassey's, Inc., 174
Braun (Barbara) Associates, Inc., 281
Brealey (Nicholas) Publishing, 77
Breedon Books Publishing Co. Ltd, The, 77
Breese (Martin) International, 77
Brent-Dyer, Elinor, 554
Brepols Publishers NV, 160
Breslich & Foss Ltd, 242
Brewin Books Ltd, 77
Brides, 303
Bridge Pamphlets, 200
Bridgeman Art Library, The, 604
Bridport Prize, The, 492
Briggs (Katharine) Folklore Award, 492
Bright 'I's, 77
Brighter Pictures, 419
Brighton Children's Book Festival, 631
Brighton Festival, 631
Brighton Jubilee Library, 576
Brilliant Publications, 226
Bristol Central Library, 576
Bristol Classical Press, 77
Bristol Festival of Ideas Best Book of Ideas Prize, 493
Bristol Old Vic Theatre Company, 439
Bristol Phoenix Press, 77
Bristol Poetry Festival, 631
Bristol, University of, 462
British Academic Press, 77
British Academy of Composers and Songwriters, 531
British Academy Small Research Grants, 482
British Academy, The, 77
British Architectural Library, 576

British Association for Applied Linguistics (BAAL) Book Prize, 493
British Association of Communicators in Business, 531
British Association of Journalists, 531
British Association of Picture Libraries and Agencies, 531
British Birds, 303
British Book Awards, 493
British Cartoon Archive, The, 604
British Centre for Literary Translation, 531
British Chess Magazine, 303
British Computer Society, The, 78
British Copyright Council, 531
British Council, The, 531
British Czech & Slovak Association Prize, 493
British Equestrian Writers' Association, 532
British Fantasy Awards, 493
British Fantasy Society, 554
British Film Institute, 532
British Guild of Beer Writers, 532
British Guild of Travel Writers, 532
British Haiku Society, The, 219
British Journalism Review, 303
British Library, 577
British Library Asia, Pacific and Africa Collections, 577
British Library Business & IP Centre, 577
British Library Early Printed Collections/Rare Books and Music Reading Room, 577
British Library Humanities Reading Room, 577
British Library Images Online, 604
British Library Manuscript Collections, 578
British Library Map Collections, 578
British Library Music Collections, 578
British Library Newspapers, 578
British Library Science, Technology and Business Collections, 579
British Library Social Sciences & Official Publications, 579
British Library Sound Archive, 579
British Library, The, 78
British Medical Journal, 303

British Museum Press, The, 78
British Philatelic Bulletin, 304
British Press Awards, 493
British Psychological Society Library, 579
British Railway Modelling, 304
British Science Fiction Association, 532
British Science Fiction Association Awards, 493
British Sky Broadcasting Ltd (BSkyB), 409
British Society of Comedy Writers, 532
British Society of Magazine Editors (BSME), 532
British Sports Book Awards, 494
Britten (Benjamin) Collection, 579
Brittle Star, 209
Broadcast, 304
Broadcasting Press Guild, 532
Broadstairs Dickens Festival, 631
Broadway, 174
Brodie (Alan) Representation Ltd, 251
Brodie (Andrew), 78
Brodie Press, The, 200, 226
Brombergs Bokförlag AB, 170
Bromley Central Library, 579
Bromley House Library, 580
Brontë Society, The, 554
Brooke (Rupert) Society, The, 554
Brooklands Museum Photo Archive, 604
Brooklyn College of the City University of New York, 473
Brookside, 154
Brown (Hamish) Scottish Photographic, 604
Brown (Jenny) Associates, 251
Brown Skin Books, 78
Brown University, 474
Brown Wells and Jacobs Ltd, 242
Brown, Son & Ferguson, Ltd, 78
Browne & Miller Literary Associates, 281
Browne (Pema) Ltd, 281
Browning Society, The, 554
Bruna (A.W.) Uitgevers BV, 167
Bryan (Felicity), 252
Bryntirion Press, 79
Brána a.s., 161
BSME, 533
Buccaneer Films, 419
Buchan (John) Society, The, 554
Buckman Agency, The, 252
Build It, 304

Building Magazine, 304
Bulfinch Press, 174
Bullseye Publications, 200
Bulzoni Editore SRL, 166
Bureau of Freelance Photographers, 533
Burke (Edmund) Publisher, 154
Burkeman (Brie), 252
Burlington Magazine, The, 304
Burning House, 79
Burns & Oates, 79
Burns (Robert) Collection, 580
Burns (Robert) World Federation Ltd, The, 554
Burton (Juliet) Literary Agency, 252
Burton Mail, 385
Burton Manor, 454
Buses, 305
Bush Theatre, 439
Business Brief, 305
Business Education Publishers Ltd, 79
Business Plus, 79, 174
Business Traveller, 305
Business Week, 394
Buster Books, 79
Butterworth Heinemann, 79
Buxton Festival, 631
Buzz Extra, 305
Bykofsky (Sheree) Associates, Inc., 281
BZZTÔH (Uitgeverij) BV, 167

C

CAA Library and Information Centre, 580
Cactus TV, 419
Cadogan Guides, 79
Caedmon, 174
Caine Prize for African Writing, The, 494
Caldecott (Randolph) Society, 555
Calder Publications Ltd, 79
Calder Wood Press, 200
California Institute of the Arts, 469
California Press (University of), 79
California Press, University of, 174
Calkins Creek, 174
Calmann-Lévy (Éditions), 163
Calon TV, 419
Calouste Gulbenkian Prize, The, 494
Cambridge (University of) Institute of Continuing Education, 454
Cambridge Central Library, 580
Cambridge Evening News, 379

Cambridge University Press, 79
Cambridge Wordfest, 631
Cambridgeshire Journal Magazine, The, 305
Camden Press Ltd, 80
Camelford Poetry Workshops, 219
Cameron & Hollis, 243
Cameron (James) Award, 494
Cameron Publicity and Marketing, 276
Caminho (Editorial) SARL, 169
Camomile Street Library, 580
Campaign, 305
Campaign for Press and Broadcasting Freedom, 533
Campbell Books, 80
Campbell Thomson & McLaughlin Ltd, 252
Camping and Caravanning, 305
Canadian Authors Association, 533
Canadian Broadcasting Corporation, 394
Canadian Federation of Poets, 533
Canadian Poetry Association (Annual Poetry Contest), 494
Canadian Publishers' Council, 533
Canadian Scholars' Press, Inc, 193
Canadian Television Network, 394
Canal Sur Televisión, 416
Canal+, 414
Canal+ Espana, 416
Canals and Rivers, 305
Candelabrum Poetry Magazine, 209
Candis, 306
Candle, 80
Cannon's Mouth, 209
Canongate Books Ltd, 80
Canongate US, 174
Canterbury Festival, 631
Canterbury Press, 80
Canterbury University Press, 196
Capall Bann Publishing, 80
Cape (Jonathan)/Jonathan Cape Children's Books, 80
Capel & Land Ltd, 252
Capital Pictures, 604
Capital Radio London, 410
Cappelen Damm AS, 168
Cappelli Editore, 166
Capstone Publishing, 80
Car Mechanics, 306
Caravan Magazine, 306
Carcanet Press Ltd, 80

Cardiff Academic Press, 81
Cardiff Book of the Year
	Awards, 494
Cardiff Central Library, 580
Cardiff International Poetry
	Competition, 494
Carey Award, 494
Carillon, 209
Caring, 195
Carlsen Verlag GmbH, 164
Carlton Publishing Group, 81
Carmarthen Public Library, 580
Carmarthen Writers' Circle, 477
Carnegie Medal, 494
Carnegie Sporting Words
	Festival, 631
Carnival (Films & Theatre) Ltd,
	419
Carolrhoda Books, 174
Carroll & Brown Publishers
	Limited, 81
Carroll (Lewis) Society, 555
Carroll (Lewis) Society
	(Daresbury), 555
Cartwn Cymru, 420
Carvainis (Maria) Agency, Inc.,
	281
Casarotto Ramsay and
	Associates Ltd, 252
Cassell, 81
Cassell Illustrated, 81
Castiglia Literary Agency, 281
Cat World, 306
Catchpole (Celia), 253
Caterer and Hotelkeeper, 306
Cathie (Kyle) Ltd, 81
Catholic Herald, The, 306
Catholic National Library, 580
Catholic Truth Society (CTS),
	81
Catnip Publishing Ltd, 81
Causeway Press, 82
Caxton Press, The, 196
CBA Publishing, 82
CBD Research Ltd, 82
CBI Bisto Book of the Year
	Awards, 494
CBS News, 394
CCV, 82
Celador Films, 420
Celador Productions, 420
Celebs on Sunday, 306
Celtic Films Entertainment Ltd,
	420
Cengage Learning (EMEA)
	Ltd, 82
Centaur Press, 82
Center Street, 174
Central FM Ltd, 410
Central School of Speech and
	Drama, The, 458

Centre for Creative &
	Performing Arts Spring
	Literary Festival at UEA, 631
Century, 82
Cephas Picture Library, 604
CF4K, 82
CHA, 82
Chalet School (Friends of), 555
Chambers Harrap Publishers
	Ltd, 82
Chameleon Television Ltd, 420
Chancery House Press, 83
Chand (S.) & Co Ltd, 195
Channel 4, 405
Channel 4 Books, 83
Channel Television, 406
Channel Television Ltd, 420
Chanticleer Magazine, 209
Chaplin (Sid) Short Story
	Competition, 494
Chapman, 209, 306
Chapman & Vincent, 253
Chapman (Giles) Library, 605
Chapman (Paul) Publishing
	Ltd, 83
Chapman Publishing, 83
Chapman University, 469
Chapter One Promotions, 643
Chapter One Promotions
	International Short Story
	Competition, 495
Chapter One Promotions Open
	Poetry Competition, 495
Characterization Tool for
	Novelists, 643
Charlesbridge Publishing, 174
Charleston Festival, 632
Charlewood Press, 226
Charnwood, 83
Chartered Institute of
	Journalists, 533
Chartered Institute of Linguists
	(IoL), 534
Chartered Institute of Personnel
	and Development (CIPD), 83
Chastleton Travel, 83
Chat, 307
Chatto & Windus, 83
Cheetah Television, 420
Cheetham (Mic) Associates, 253
Chelius (Jane) Literary Agency,
	Inc., 282
Cheltenham Literature Festival,
	632
Chemist & Druggist, 307
Cherrytree Books, 83
Chester (Linda) & Associates,
	282
Chester Chronicle, 379
Chester Literature Festival, 632
Chester, University of, 454
Chesterton Society UK, The, 555

Chicago Press, The University
	of, 174
Chicago State University, 471
Chicago Tribune Press Service,
	394
Chichester, The University of,
	463
Chicken House Publishing, 83
Chilcote (Judith) Agency, 253
Child's Play (International)
	Ltd, 83
Children's Book Circle, 534
Children's Book Circle Eleanor
	Farjeon Award, 495
Children's Books History
	Society, The, 555
Children's Books Ireland, 534
Children's Books Ireland –
	Annual Festival, 632
Children's Film & Television
	Foundation Ltd, The, 420
Children's Laureate, The, 495
Children's Press, 174
Children's Press, The, 154
Chiltern Writers' Group, 477
Chimera, 83
Chivers Audio Books/Chivers
	Children's Audio Books, 238
Choice, 307
Cholmondeley Awards, 482
Chris (Teresa) Literary Agency
	Ltd, 253
Christian Aid, 605
Christian Focus Publications, 83
Christian Heritage, 84
Christie's Images Ltd, 605
Chrome Dreams, 84, 238
*Chronicle and Echo
	(Northampton)*, 384
Chronicle Books LLC, 174
Chrysalis – The Poet In You, 455
Chrysalis Books Group, 84
Chrysalis Press, 226
Church Music Quarterly, 307
Church of England Newspaper,
	307
Church of Ireland Publishing,
	154
Church Times, 307
Churchill Livingstone, 84
Churchwarden Publications
	Ltd, 84
CiB, 534
Cicerone Press, 84
Cico Books, 84
CILIP in Scotland (CILIPS), 534
CILIP Wales, 534
CILIP: The Chartered Institute
	of Library and Information
	Professionals, 534
CILIP: The Chartered Institute
	of Library and Information

Professionals Carnegie Medal, 495
CILIP: The Chartered Institute of Library and Information Professionals Kate Greenaway Medal, 495
Cinema Museum, The, 605
Cinnamon Press, 200
Circle of Wine Writers, 534
Cisco Press, 84
Cision, 628
Citizen, The (Gloucester), 381
Citizens Theatre, 439
City Business Library, 581
City Lit, 458
City of London Libraries, 581
City University, 458
Civilização Editora, 169
Clairview Books Ltd, 84
Clare (John) Society, The, 556
Clark (T&T), 84
Clark (William) Associates, 282
Clarke (Arthur C.) Award for Science Fiction, 495
Clarke (Arthur H.) Company, 175
Clarke (James) & Co., 84
Clarkson Potter, 175
Classic, 84
Classic & Sports Car, 308
Classic Bike, 308
Classic Boat, 308
Classic Cars, 308
Classic FM, 410
Classic Land Rover World, 308
Classic Stitches, 308
Classical Collection Ltd, 605
Classical Guitar, 309
Classical Music, 309
Cleare (John)/Mountain Camera, 605
Clemmey (Mary) Literary Agency, 253
Cleveland Productions, 420
Cliffs Notes, 191
Climber, 309
Closer, 309
Clowes (Jonathan) Ltd, 254
Club International, 309
Clwyd Theatr Cymru, 439
Clyde 1/Clyde 2, 410
CLÉ – Irish Book Publishers' Association, 534
Cló Iar-Chonnachta, 154
CNBC, 394
CNBC Europe, 409
CNN International, 394, 409
CNP Publications, 226
Coach and Bus Week, 309
Cobbett (William) Society, 556
Cochrane (Elspeth) Personal Management, 254

Codron (Michael) Plays Ltd, 439
Cohen (David) Prize for Literature, 495
COI, 421
Coin News, 309
Colbert Macalister PR, 276
Colchester Mercury Theatre Limited, 439
Cole (Michael) Camerawork, 605
Coleg Harlech WEA, 477
Colin (Rosica) Ltd, 254
Coliseum Oldham, The, 439
Collections, 605
Collective Press, The, 200
Collector's Library/Collector's Library Editions in Colour, 85
Collin (Frances) Literary Agent, 282
Collin (Peter) Publishing Ltd, 85
Collingwood O'Hare Entertainment Ltd, 421
Collins, 85
Collins & Brown, 85
Collins (US), 175
Collins (Wilkie) Society, 556
Collins Press, The, 154
Colman Getty, 276
Colorado State University, 470
Colourpoint Books, 85
Columba Press, The, 154
Columbia Press (University of), 85
Columbia University Press, 175
Combrógos, 641
Comedy Unit, The, 421
Comhairle nan Leabhraichean/ The Gaelic Books Council, 535
Commercial Radio Companies Association, 535
Commonwealth Secretariat Library, 581
Commonwealth Writers Prize, The, 496
Communication Ethics, 85
Community Creative Writing, 461
Company, 310
Company Pictures, 421
Compendium Publishing Ltd, 85
Complete Creative Writing Course at the Groucho Club, The, 459
Complete Works of William Shakespeare, 643
Computer Arts, 310
Computer Weekly, 310
Computeractive, 310
Computing, 310

Conan Doyle (Arthur) Society, The, 556
Concordance, 440
Condor, 85
Condé Nast Traveller, 310
Congdon (Don) Associates, Inc., 282
Conrad (Joseph) Society (UK), 556
Conran Octopus, 85
Constable & Robinson Ltd, 85
Construction News, 310
Contact Publishing Ltd, 226
Contact Theatre Company, 440
Contemporary Poetics Research Centre, 220
Continuum International Publishing Group Ltd, The, 85
Conville & Walsh Limited, 254
Conway, 86
Conway-Gordon (Jane) Ltd, 254
Cook (Olive) Prize, 483
Cook (Thomas) Publishing, 86
Cool FM, 410
Coop Traveller, 310
Cooper (Duff) Prize, The, 496
Cooper (Leo), 86
Copyright, 46–51
Copyright Licensing Agency Ltd, The, 535
Corbis Images, 606
Cordaiy (Sylvia) Photo Library, 606
Corgi, 86
Corgi Audio, 238
Corgi Children's Books, 86
Cork University Press, 155
Cosgrove Hall Films, 421
Cosmic Elk, The, 227
Cosmopolitan, 311
Costa Book Awards, 496
Cotler (Joanna) Books, 175
Cotswold Writers' Circle, The, 477
Council for British Archaeology, 535
Counselling at Work, 311
Country Homes and Interiors, 311
Country Life, 311
Country Life Picture Library, 606
Country Living, 311
Country Matters Picture Library, 606
Country Publications Ltd, 86
Country Smallholding, 311
Country Walking, 312
Countryman's Weekly, The, 312
Countryman, The, 312

Countryside Alliance Update, 312
Countryside Books, 86
Courier and Advertiser, The (Dundee), 391
Coventry Central Library, 581
Coventry Inspiration Book Awards, The, 496
Coventry Telegraph, 387
Crace Collection, 581
Craven (Philip) Worldwide Photo-Library, 606
Crawshay (Rose Mary) Prize, 497
Creasey (John) Dagger, 497
Creative Partnership, The, 421
Crescent Moon Publishing and Joe's Press, 227
Cressrelles Publishing Co. Ltd, 86
Creuse Writers' Workshop & Retreat, 458, 477
Crew (Rupert) Ltd, 254
Cricket Ltd, 421
Cricketer International, The, 312
Crime Express, 227
Crime Scene Scotland, 247
Crime Writers' Association (CWA), 535
Crime Writers' Association (Cartier Diamond Dagger), 497
Crime Writers' Association (The CWA Ellis Peters Award), 497
Crime Writers' Association (The CWA Gold Dagger for Non-Fiction), 497
Crime Writers' Association (The Dagger in the Library), 497
Crime Writers' Association (The Debut Dagger), 497
Crime Writers' Association (The Gold and Silver Daggers for Fiction), 497
Crime Writers' Association (The Ian Fleming Steel Dagger), 497
Crime Writers' Association (The New Blood Dagger for Best First Crime Novel), 497
Crimewave, 312
Crimson Publishing, 86
Critics' Circle, The, 535
Crocodile Books, USA, 175
Crombie Jardine Publishing Limited, 87
Crossbridge Books, 227
Crossway, 87
Crown House Publishing, 87
Crown Publishing Group, 175

Crowood Press Ltd, The, 87
Crucible, 312
Crucible Theatre, 440
CRW Publishing Ltd, 87
Crème de la Crime Ltd, 86
CSA Word, 238, 421
CT (Ceska Televize), 413
Cumbria, 312
Cumbrian Literary Group, 477
Cummings (Benjamin), 87
Cunningham (Sue) Photographic, 606
Currach Press, 155
Currency, 175
Currency Press Pty Ltd, 191
Current Accounts, 209
Currey (James) Publishers, 87
Curtis Brown Group Ltd, 254
Curtis Brown Ltd, 282
Custom Publishing, 87
Cutting Edge Productions Ltd, 422
Cycle Sport, 313
Cycling Weekly, 313
Cygnus Books, 293
Cyngor Llyfrau Cymru, 535
CYP, 239

D

D&B Publishing, 87
DaCapo, 175
DACS, 536
Dahl (Roald) Funny Prize, The, 498
Daily Echo (Bournemouth), 380
Daily Express, 372
Daily Mail, 372
Daily Mail First Novel Award, 498
Daily Mirror, 372
Daily Post, 391
Daily Record (Glasgow), 372
Daily Sport, 373
Daily Star, 373
Daily Star Sunday, 373
Daily Telegraph, The, 373
Daish (Judy) Associates Ltd, 255
Dale Concannon Collection, 606
Dalesman, 313
Dalton (Terence) Ltd, 88
Dalton–Watson Collection, 606
Dance Theatre Journal, 313
Dance Today, 313
Dancing Times, The, 313
Dandelion Arts Magazine, 209
Dandy, The, 313
Dartington College of Arts, 455
Dartington Literary Festival, 632
Dartmoor Sherlock Holmes Study Group, The, 557

Darton, Longman & Todd Ltd, 88
Darts World, 314
Data Publishers Association, 536
David & Charles Publishers, 88
David Jones Society, The, 561
Davidson (Caroline) Literary Agency, 255
Davidson (Merric) Literary Agency, 255
Davies (Christopher) Publishers Ltd, 88
Davies (Hunter) Prize, 498
Davies (Rhys) Trust, The, 557
DAW Books, Inc., 175
Day Books, 227
Day by Day, 314
Day Dream Press, 200
Day Five Ltd, 276
Dazed & Confused, 314
de la Mare (Giles) Publishers Ltd, 88
de la Mare (Walter) Society, The, 557
de Wolfe (Felix), 255
Debrett's Ltd, 88
Decanter, 314
Decanto, 209
Dedalus Ltd, 89
Dedalus Press, The, 200
Deeping (Warwick) Appreciation Society, 557
Defence Picture Library, The, 606
Del Rey, 175
Delacorte Press, 175
Delicious, 314
Dell, 175
Delta Books, 175
Dench Arnold Agency, The, 255
Dennis (Felix) Prize for Best First Collection, 498
Denoël (Éditions), 163
Dent (JM), 89
Department for Culture, Media and Sport, 536
Department for Environment, Food and Rural Affairs, 581
Derby Central Library, 581
Derby Evening Telegraph, 380
Derby Playhouse, 440
Derby, University of, 454
Derbyshire Life and Countryside, 314
Derbyshire Literature Festival, 632
Descent, 314
Design and Artists Copyright Society (DACS), 536
Desire, 315
Despatches, 89

Destination France, 315
Deutsch (André) Ltd, 89
Deutscher Taschenbuch Verlag GmbH & Co. KG, 164
Devine (George) Award, 498
Devon & Exeter Institution Library, 581
Devon Life, 315
Diagram Visual Information Ltd, 243
Dial Books for Young Readers, 175
Dial Press, 175
Dickens Fellowship, The, 557
Dickins (Douglas) Photo Library, 607
Dictionary of Slang, 644
Diehard Publishers, 200
Digital One, 410
Digital Press, 89
Dijkstra (Sandra) Literary Agency, 282
Dillington House, 462
Dimensions for Living, 175
Dinas, 89
Dingle Prize, 498
Dingle Writing Courses, 466
Diogenes Verlag AG, 171
Dionysia Press, 201
Director, 315
Directory of Writers' Circles, Courses and Workshops, 477
Disability Now, 315
Discovered Authors, 89
Diva, 315
Diverse Production Limited, 422
Dog World, 316
DogHouse, 201
Dogs Monthly, 316
DOI Registration Agency, 641
Dokorán Ltd, 161
Dolman Best Travel Book Award, 498
Dominic Photography, 607
Donald (John) Publishers Ltd, 89
Doncaster Library and Information Services, 582
Doncaster Star, The, 388
Donhead Publishing Ltd, 89
Dorchester Festival, 632
Dorchester Library, 582
Dorian Literary Agency (DLA), 255
Dorling Kindersley Ltd, 90
Dorman (Sean) Manuscript Society, 477
Dorset Echo, 380
Dorset Life – The Dorset Magazine, 316

DoubleBand Films, 422
Doubleday, 90
Doubleday Broadway Publishing Group, 175
Doubleday Canada, 193
Doubleday Children's Books, 90
Dow Jones Newswires, 393, 394
Downtown Press, 176
Downtown Radio, 410
DR (Danmarks Radio), 413
Dragonby Press, The, 227
Drake A-V Video Ltd, 422
Drake Educational Associates, 90
Drama Association of Wales, 536
Drama Association of Wales Playwriting Competition, 498
Dramatic Lines, 227
Dreadful Night Press, 201
Dream Catcher, 201, 209
Dref Wen, 90
Driftwood Publications, 201
Druid Theatre Company, 440
Dryden (John) Competition, The, 498
du Maurier (Daphne) Festival of Arts & Literature, The, 632
Dublin, University of (Trinity College), 466
Duckworth (Gerald) & Co. Ltd, 90
Duke of Westminster's Medal for Military Literature, The, 499
Dukes, The, 440
Dumfries and Galloway Arts Festival, 632
DUNA TV (Duna Televízió), 414
Dundee Central Library, 582
Dundee Literary Festival, 632
Dundee Repertory Theatre, 440
Dundee, University of, 467
Dunedin Academic Press Ltd, 90
Dunham Literary Inc., 283
Dunne (Thomas) Books, 176
Dunstan (Charles) Communications Ltd, 422
Durham Book Festival, 633
Durning-Lawrence Library, 582
Durrants, 628
Dutton/Dutton's Children's Books, 176
DW-TV (Deutsche Welle), 414
Dymock Poets, 557
Dystel & Goderich Literary Management, 283

E
E&E Picture Library, 607
e-book-downloads-uploads.com, 247
Eady (Toby) Associates Ltd, 256
Eagar (Patrick) Photography, 607
Early English Text Society, 557
Earnley Concourse, 463
Earth Love, 209
East Anglia, University of, 461
East Anglian Daily Times, 385
East Anglian Writers, 478
East Gate Books, 176
Eastern Angles Theatre Company, 441
Eastern Daily Press, 384
Eastern Eye, 316
Eastern Rainbow, 209
Eastern Washington University, 475
Easy Living, 316
Ebury Publishing/Ebury Press, 90
Ecco, 176
Echo Newspapers (Basildon), 381
Eclectic Writer, The, 644
Ecologist, The, 316
Economist Books, 91
Economist, The, 316
Economist/Richard Casement Internship, The, 483
Ecoscene, 607
Eddison Pearson Ltd, 256
Eddison Sadd Editions, 243
Eden, 91
Edge Hill Prize for the Short Story, 499
Edge Hill University, 457
Edge, The, 317
Edgewell Publishing, 227
EDHASA (Editora y Distribuidora Hispano – Americana SA), 169
Edifice, 607
Edinburgh Bibliographical Society, 536
Edinburgh International Book Festival, 633
Edinburgh Review, 317
Edinburgh Royal Lyceum Theatre, 441
Edinburgh University, 466
Edinburgh University Press, 91
Education Photos, 607
Educational Design Services, LLC, 283
Educational Heretics Press, 227
Educational Publishers Council, 536
Edwards (Jacqueline), 641

Edwards Fuglewicz, 256
Eerdmans Publishing Company, 176
Egg Box Publishing, 201
Egmont Press, 91
Egmont Publishing, 91
Eight Hand Gang, The, 220
Eilish Press, 228
Eland Publishing Ltd, 91
Electric Airwaves Ltd (Ladbroke Radio), 422
Electrical Times, 317
Electronic Publishers' Forum, 536
Element, 91
11:9, 91
Elgar (Edward) Publishing Ltd, 91
Eliot (George) Fellowship, The, 557
Eliot (T.S.) Prize, 499
Elle, 317
Elle Decoration, 317
Ellenberg (Ethan) Literary Agency, 283
Elliot Right Way, 91
Elliott & Thompson, 91
Elliott (Desmond) Prize, The, 499
Elm Publications/Training, 92
Elsevier Ltd, 92
Elwin Street Productions, 243
Emblem Editions, 193
Embroidery, 317
Emerson College, 472
Emissary Publishing, 92
Emma Treehouse Ltd, 92
Empire, 317
Empiricus Books, 92
Enable Enterprises, 228
Encore Award, The, 499
Endemol UK Productions, 422
Engine, The, 210
Engineer, The, 317
English Association, The, 536
English Garden, The, 318
English Heritage (Publishing), 92
English Heritage Photo Library, 607
English Nature, 582
English PEN, 536
English Touring Theatre, 441
Enitharmon Press, 92, 201
Envoi, 210
Envoi Poetry Competition, 499
EO (Evangelische Omroep), 415
Eos, 176
Epworth, 93
Equality and Human Rights Commission, 582
Equinox, 210

éQuinoxe TBC, 478
Eratica, 210
Erotic Review Books, 93
Erskine Press, 243
ERT S.A. (Elliniki Radiophonia Tileorassi), 414
ES, 318
Espasa-Calpe (Editorial) SA, 170
Esquire, 318
Essence Press, 201
Essentials, 318
Essex County Council Libraries, 582
Essex Life, 318
ETB-1/ETB-2 (Euskal Telebista S.A.), 416
Etruscan Books, 201
ETV (Eesti Televisioon), 413
EuroCrime, 93
Euromonitor International plc, 93
Europa-América (Publicações) Lda, 169
European Jewish Publication Society, 483
Euroscript Ltd, 478
Evans (Faith) Associates, 256
Evans (Mary) Picture Library, 607
Evans Brothers Ltd, 93
Eve, 318
Evening Chronicle (Newcastle upon Tyne), 386
Evening Express (Aberdeen), 390
Evening Gazette (Colchester), 381
Evening Gazette (Middlesbrough), 386
Evening Leader (Mold), 391
Evening News (Edinburgh), 390
Evening News (Norwich), 384
Evening Post (Bristol), 385
Evening Post Nottingham, 384
Evening Standard (London), 383
Evening Star (Ipswich), 386
Evening Telegraph (Dundee), 390
Evening Telegraph (Kettering), 384
Evening Telegraph (Peterborough), 379
Evening Times (Glasgow), 390
Evens (Rev. G. Bramwell), 557
Eventing, 318
Evergreen, 318
Everyman, 93
Everyman Chess, 93
Everyman's Library, 93
Exeter Northcott, 441
Exeter Phoenix, 455
Exeter Press, University of, 93

Exeter, University of, 455
Exley Publications Ltd, 93
Expert Books, 94
Expert Publications Ltd, 243
Express & Echo (Exeter), 380
Express & Star (Wolverhampton), 387
Eye Ubiquitous, 608
Eye: The international review of graphic design, 318
Eyewitness Guides/Eyewitness Travel Guides, 94

F

Faber & Faber Ltd, 94
Faber & Faber, Inc., 176
Faber (Geoffrey) Memorial Prize, 500
Fabulous, 319
Faces and Places, 608
Facet Publishing, 94
Fagon (Alfred) Award, The, 500
Fairchild Publications of New York, 394
Fairgame Magazine, 319
FaithWords, 176
Fal Publications, 201
FalconGuides, 176
Falkner (John Meade) Society, The, 557
Families First, 319
Family Records Centre, 583
Family Tree Magazine, 319
Famous Pictures & Features Agency, 608
Fancy Fowl, 319
Fand Music Press, 228
Farjeon (Eleanor) Award, 500
Farmar (A.&A.) Ltd, 155
Farmers Weekly, 319
Farnham Film Company Ltd, 422
Farrant (Natasha), 274
Farrar, Straus & Giroux, Inc., 176
Fast Car, 319
Fast Films, 423
Fawcett Library, The, 583
Fayard (Librarie Arthème), 163
Feather Books, 201, 228
Feather's Miscellany, 210
Federation of Entertainment Unions, 537
Feldstein Agency, The, 256
Fellowship of Authors and Artists, 537
Fenn (H.B.) & Company Ltd, 193
Fern HousE-Publishing, 248
Fernhurst Books, 94
Festival Film and Television Ltd, 423
ffotograff, 608

FHM, 320
Fickling (David) Books, 94
Fiction Factory, 423
Fiction Writing Workshops and
    Tutorials, 459, 641
Fidra Books, 228
Field, The, 320
5oconnect, 248
57 Productions, 220
Fig Tree, 94
Film and General Productions
    Ltd, 423
Film Angel, 644
Film Review, 320
Filmmaker Store, 644
Financial Times, 373
Financial Times Pictures, 608
Finborough Theatre, 441
Findhorn Press Ltd, 94
Fine Art Photographic Library
    Ltd, 608
Fire in the Head Courses &
    Writers' Services, 455
Firebird, 176
Firefly Books Ltd, 193
Firefly Publishing, 94
Firehouse Productions, 423
Firepix International, 608
Fireside, 176
Firing Squad, The, 210
First & Best in Education Ltd,
    94
First Avenue Editions, 176
First Film Company Ltd, The,
    423
First Offense, 210
First Writes Theatre Company,
    423
Fischer (S.) Verlag GmbH, 164
Fish Fiction Prizes, 500
Fisher (Anne Louise), 274
Fishing News, 320
Fitch (Laurence) Ltd, 256
Fitzgerald (Edward) Collection,
    583
Fitzgerald Publishing, 95
Fitzhenry & Whiteside Limited,
    193
Fitzjames Press, 95
Five, 406
Five Leaves Publications, 201,
    228
Five Star, 95, 176
5–Spot, 176
Fix, The, 320
Flambard Press, 201
Flaming Arrows, 210
Flammarion Éditions, 163
Flannel, 423
Flarestack Publishing, 201
Flash, 320

Flashback Television Limited,
    423
Fledgling Press, 248
Fletcher (Sir Banister) Award,
    500
Flight International, 320
Flora International, 321
Florida International University,
    470
Florida, University of, 470
Florio (John) Prize, The, 500
Floris Books, 95
Flyleaf Press, 155
FlyPast, 321
FMcM Associates, 276
Focal Press, 95
Focus, 321
Focus Films Ltd, 424
Fogden Wildlife Photographs,
    608
Folens Limited, 95
Folio Society, The, 293
Folkestone Literary Festival, 633
Folly, 558
Font International Literary
    Agency, 279
Food Features, 609
Foot (Paul) Award, The, 500
Football Poets, The, 220
For Dummies, 191
Forbes Magazine, 394
Ford (Ford Madox) Society,
    The, 558
Foreign Press Association in
    London, 537
Forest Life Picture Library, 609
Forestry Commission Library,
    583
Forever, 176
Forlaget Forum, 161
Forman (Werner) Archive Ltd,
    609
Formula One Pictures, 609
Forstater (Mark) Productions
    Ltd, 424
Forsythe (Robert) Picture
    Library, 609
Fort Publishing Ltd, 95
*Fortean Times: The Journal of
    Strange Phenomena*, 321
Forth One/Forth 2, 410
Fortune, 394
Forum (Bokförlaget) AB, 171
Forward Press, 202
Forward Prizes for Poetry, The,
    500
Foster (Jill) Ltd, 257
Foulsham Publishers, 95
*Foundation: the international
    review of science fiction*, 321
Fountain Press, 95
Four Courts Press Ltd, 155

Four O Clock Press, 96
Fourth Estate, 96
Fox & Howard Literary Agency,
    257
Fox News Channel, 394
Foyle Street Writers Group, 478
Foyle Young Poets of the Year
    Award, The, 501
FQ Magazine, 321
France 2/France 3, 414
France 5, 414
France Magazine, 321
Franco-Midland Hardware
    Company, The, 558
Frankfurt Book Fair, 644
Fraser Ross Associates, 257
Frazer (Cathy) PR, 277
Fredericks (Jeanne) Literary
    Agency, Inc., 283
Free Association Books Ltd, 96
Free Press, 176
Freedman (Robert A.) Dramatic
    Agency, Inc., 283
Freelance Market News, 321
Freelance, The, 322
Freeman (W. H.), 96
FremantleMedia Group Ltd,
    424
French (Samuel) Ltd, 96
French (Samuel), Inc., 176
French Institute Library, 583
French Literary Review, The, 210
Friars' Club, The, 558
Friday Project, The, 96
Friends of Coleridge, The, 556
Friends of Shandy Hall, The, 558
Friends of the Dymock Poets,
    The, 558
Frith (Francis) Collection, The,
    609
Froebel Archive for Childhood
    Studies, The, 644
Frogmore Papers, The, 210
Frogmore Poetry Prize, The, 501
Frogmore Press, The, 202
Frome Festival, 633
Frommer's, 191
Front Magazine, 322
Front Street, 176
Frontier Publishing, 228
Frontline Books, 96
Frost (John) Newspapers, 583
FT & Goldman Sachs Business
    Book of the Year Award, 501
Fulbright Awards, 483
Fulton (David) Publishers Ltd,
    96
Funfax, 97
Fusion Press, 97
Futerman, Rose & Associates,
    257

G

Gabriel (Jüri), 257
Gaelic Books Council, The, 537
Gagg's (Andrew N.)
    PHOTOFLORA, 609
Gaia Books, 97
Gaia Communications, 424
Galactic Central Publications,
    228
Galaxy British Book Awards,
    501
Galaxy Picture Library, 609
Gale, 176
Gallery Press, The, 155, 202
Gallimard (Éditions), 163
Galore Park Publishing Ltd, 228
Garbaj, 210
Garden and Wildlife Matters
    Photo Library, 609
Garden Answers, 322
Garden Media Guild, The, 537
Garden News, 322
Garden Picture Library, 610
Garden, The, 322
Gardens Illustrated, 322
Garnet Publishing Ltd, 97
Garzanti Libri SpA, 166
Gaskell Society, The, 558
Gate Theatre Company Ltd, 441
Gay (Noel) Television, 424
Gay Times, 322
Gazette, The (Blackpool), 382
Geddes & Grosset, 97
Geldard (Ed) Photo Collection,
    610
Gelfman Schneider Literary
    Agents, Inc., 283
Gellhorn (Martha) Trust Prize,
    501
Genesis Space History Photo
    Library, 610
Geo Aerial Photography, 610
Geological Society Publishing
    House, The, 97
George Mason University, 475
GeoScience Features, 610
Geoslides Photography, 610
Geringer (Laura) Books, 176
German Historical Institute
    Library, 583
Getty Images, 610
Ghost Story Society, The, 559
Giant Rooster PR, 277
Gibbons (Stanley) Publications,
    97
Gibbons Stamp Monthly, 322
Gibson (Douglas) Books, 193
Gibson Square, 97
Gill & Macmillan, 155
Ginger Productions, 424
Ginn, 98
Girl Talk/Girl Talk Extra, 322

Giunti Editoriale SpA, 166
GL Assessment, 98
Gladstone History Book Prize,
    The, 502
Glamorgan, University of, 467
Glamour, 323
Glasgow, University of, 467
Glass (Eric) Ltd, 257
Glenfiddich Food & Drink
    Awards, 502
Global Tapestry Journal, 210
Globe and Mail, The, 394
Globe Pequot Press, The, 176
Glosa Education Organisation,
    229
Gloucester Publishers Plc, 98
Gloucestershire County Council
    Libraries & Information, 583
Gloucestershire Echo, 381
GMC Publications Ltd, 98
GMTV, 406
Godsfield Press, 98
Godstow Press, 229
Godwin (David) Associates, 258
Godwin (Tony) Memorial
    Trust, 483
Goethe-Institut Library, 584
Gold Dust Magazine, 323
Goldcrest Films International
    Ltd, 424
Golden Hearts Awards, 502
Golden PEN Award for Lifetime
    Distinguished Service to
    Literature, 502
Goldsmith Press, The, 202
Goldsmiths' Library, 584
Golf Monthly, 323
Golf World, 323
Gollancz, 98
Gomer Press/Gwasg Gomer,
    98, 202
Good Film Company Ltd, The,
    424
Good Holiday Magazine, 323
Good Homes Magazine, 323
Good Housekeeping, 323
Good Housekeeping Book
    Awards, 502
Good Motoring, 324
Good News, 324
Good Ski Guide, The, 324
Good Web Guide Ltd, The, 645
Google Images, 645
Gotch Solutions, 277
Gotham, 177
Goucher College, 471
Gourmand World Cookbook
    Awards, 502
Gower, 98
GQ, 324
Grace (Martin and Dorothy),
    611

Gradiva–Publicações S.A., 169
Graeae Theatre Company, 441
Graffeg, 229
Graham & Whiteside, 177
Graham (Winston) Historical
    Prize, The, 502
Graham Maw Christie Literary
    Agency, 258
Graham-Cameron Publishing &
    Illustration, 98
Grand Central Publishing, 177
Granite Productions, 425
Grant (Ronald) Archive, 611
Grant Books, 229
Grant Naylor Productions, 425
Granta, 324
Granta Books, 98
Graphic Universe, 177
Grasset & Fasquelle (Éditions),
    163
Graves (Robert) Society, The,
    559
Grazia, 324
Great Northern Publishing, 229
Great Outdoors, The, 324
Great War (1914–1918), The, 324
Greater London Council
    History Library, 584
Green (Annette) Authors'
    Agency, 258
Green (Christine) Authors'
    Agent, 258
Green (W.) Scotland, 99
Green Arrow Publishing, 202
Green Books, 99
Green Print, 99
Green Umbrella Ltd, 425
Green Willow Books, 177
Greenaway (Kate) Medal, 503
Greenberg (Louise) Books Ltd,
    258
Greenburger (Sanford J.)
    Associates, Inc., 284
Greene & Heaton Ltd, 258
Greene (Graham) Birthplace
    Trust, 559
Greene (Graham) Festival, 633
Greenhaven Press, 177
Greenhill (Sally and Richard),
    611
Greenock Telegraph, 390
Greenwich Village Productions,
    425
Greenwood-Heinemann, 177
Gregory & Company Authors'
    Agents, 258
Gregory (Eric) Trust Fund, 483
Gresham Books Ltd, 99
Griffin, 177
Griffin Poetry Prize, The, 503
Griffin Publishing Associates,
    229

Grijalbo, 170
Grimsby Telegraph, 383
Grolier, 177
Grosset & Dunlap, 177
Grossman (David) Literary
    Agency Ltd, 259
Grossman (Susan) – Writing
    Workshops, 641
Groupo Anaya, 169
Grove/Atlantic Inc., 177
Grow Your Own, 325
Grub Street, 99
GSSE, 229
GT (Gay Times), 325
Guardian Children's Fiction
    Award, The, 503
Guardian First Book Award,
    The, 503
Guardian Hay Festival, The, 633
Guardian Research Fellowship,
    The, 484
Guardian Weekend, 325
Guardian, The, 374
Guernsey Press & Star, 392
Guide to Grammar and Style,
    645
Guidebookwriters.com, 537
Guiding magazine, 325
Guild of Agricultural
    Journalists, 537
Guild of Editors, 537
Guild of Food Writers Awards,
    503
Guild of Food Writers, The, 538
Guild of Motoring Writers, 538
Guildford Book Festival, 633
Guildford Book Festival First
    Novel Award, 503
Guildford Institute Library, 584
Guildford Institute, The, 463
Guildford Writers, 478
Guildhall Library, 584
Guille–Alles Library, 585
Guinness World Records Ltd,
    99
Gulf Professional Press, 99
Gummerus Oy, 162
Gunn O'Connor (Marianne)
    Literary Agency, 279
Gunnmedia Ltd, 259
Gurney (Ivor) Society, The, 559
Gusay (Charlotte) Literary
    Agency, The, 284
Guy (V.K.) Ltd, 611
Gwasg Carreg Gwalch, 99
Gwobr Llyfr y Flwyddyn, 503
Gyldendal, 161
Gyldendal Norsk Forlag, 168

H

H&E Naturist, 325
Hachette Book Group USA, 177

Hachette Books Ireland, 155
Hachette Children's Books, 99
Hachette Digital, 239
Hachette Livre (France), 163
Hachette Livre Australia, 191
Hachette Livre New Zealand
    Ltd, 196
Hachette Livre UK, 100
Haggard (Rider) Society, 559
Haiku Scotland, 210
Hair, 325
Hairflair & Beauty, 325
Halban Publishers, 100
Haldane Mason Ltd, 243
Hale (Robert) Ltd, 101
Hall (G.K.), 177
Hall (Rod) Agency Limited,
    The, 259
Halsgrove, 101
Halstar, 101
Hambledon Continuum, 101
Hambly (Katie) PR, 277
Hamilton (Margaret) Books, 191
Hamish Hamilton, 101
Hamlyn Octopus, 101
Hammersmith Press Ltd, 101
Hammerwood Film
    Productions, 425
Hampstead Theatre, 442
Hanbury Agency, The, 259
Hancock (Roger) Ltd, 259
Handshake, 211
Hanley (James) Network, 559
Hansard, 645
Hanser (Carl) Verlag GmbH &
    Co. KG, 165
HappenStance Press, 202
Happy Cat, 101
Harcourt, 177
Harding (Robert) World
    Imagery, 611
Hardley (Dennis) Photography,
    Scottish Photo Library, 611
Hardy (Thomas) Society, The,
    560
Harlequin, 211
Harlequin Enterprises Ltd, 193
Harlequin Mills & Boon
    Limited, 101
Harmony Books, 177
Harper's Bazaar, 326
HarperCollins AudioBooks, 239
HarperCollins Canada Ltd, 193
HarperCollins Publishers
    (Australia) Pty Ltd, 191
HarperCollins Publishers (New
    Zealand) Ltd, 196
HarperCollins Publishers India
    Ltd, 195
HarperCollins Publishers Ltd,
    102

HarperCollins Publishers, Inc.,
    177
Harpercroft, 202
Harrap, 103
Harriman House Ltd, 103
Harris (Joy) Literary Agency,
    284
Harrogate Crime Writing
    Festival, 633
Harrogate International
    Festival, 633
Harrogate Theatre, 442
Hartlepool Mail, 379
Hartswood Films Ltd, 425
Harvard University Press, 103,
    177
Harvill Secker, 103
Harwich Festival of the Arts,
    634
Harwood (Antony) Limited, 259
Hat Trick Productions Ltd, 425
Haunted Library, 229
Haus Publishing, 103
Hawai'i Press, University of, 177
Hawkins (John) & Associates,
    Inc., 284
Hawthornden Literary Institute,
    484
Hay Festival, The, 634
Haynes Publishing, 103
Hazlitt Society, The, 560
Head (Francis) Bequest, 484
Head to Head Communication
    Ltd, 425
Headland Publications, 202
Headline/Headline Review, 103
Headlong Theatre, 442
Headpress, 230
Health & Fitness Magazine, 326
Healthcare Productions
    Limited, 425
Healthy, 326
Hearing Eye, 202
Heart of Albion Press, 230
Heat, 326
Heath (A.M.) & Co. Ltd, 259
Heath (Rupert) Literary
    Agency, 260
Hebden Bridge Arts Festival,
    634
Hecht (Anthony) Poetry Prize,
    The, 503
Heinemann, 104
Heinemann (William), 104
Heinemann International
    Southern Africa, 197
Helicon Publishing, 104
Hellenic Foundation Prize, 504
Hello!, 326
Helm (Christopher), 104
Helm Information Ltd, 104
Helter Skelter Publishing, 104

Hemans (Felicia) Prize for Lyrical Poetry, 504
Henderson (Jim) Photography, 611
Henderson Publishing, 104
Henry (Ian) Publications Ltd, 104
Henser Literary Agency, 260
Henty Society, The, 560
Herald Express (Torquay), 380
Herald, The (Glasgow), 374
Herald, The (Plymouth), 380
Herbert Press, The, 104
Herefordshire Libraries, 585
Heritage Theatre Ltd, 425
Herman (Jeff) Agency, LLC, The, 284
Hermes House, 104
Hern (Nick) Books, 104
Hertfordshire (University of) Library, 585
Hertfordshire Press, University of, 104
Heseltine (John) Archive, 611
Hesperus Press Limited, 104
Hessell-Tiltman Prize for History, The, 504
Heyne Verlag, 165
hhb agency ltd, 260
Hi-Fi News, 326
Hiddenspring Books, 178
High Life, 326
Higham (David) Associates Ltd, 260
Higham Hall College, 454
Highbury College, Portsmouth, 456
Highgate Literary and Scientific Institution Library, 585
Highland Libraries, The Highland Council, Education, Culture and Sport Service, 585
Hill (Christopher) Photographic, 611
Hill (William) Sports Book of the Year, 504
Hill Street Press LLC, 178
Hilltop Press, 202
Hilton (James) Society, 560
Hind Pocket Books Pvt. Ltd, 195
Hiperión (Ediciónes) SL, 170
Hippocrene Books, Inc., 178
Hippopotamus Press, 202
Hirsch (Paul) Music Library, 585
Hiss & Boo Co. Ltd, The, 442
HISSAC Annual Short Story Award, 504
Historical Novel Folio, 478
History Into Print, 105
History Press Ltd, The, 105
History Today, 327

HMSO, 105
Hobbs Golf Collection, 612
Hobsons Plc, 105
Hodder & Stoughton, 105
Hodder & Stoughton Audiobooks, 239
Hodder Arnold, 105
Hodder Children's Books, 105
Hodder Gibson, 105
Hodder Headline Ltd, 105
Hoffman (Calvin & Rose G.) Prize, 504
Hoffman (David) Photo Library, 612
Hoffmann und Campe Verlag GmbH, 165
Hogenson (Barbara) Agency, Inc., The, 284
Holborn Library, 586
Holcenberg Jewish Collection, 586
Holiday House, Inc., 178
Hollins University, 475
Holmes (Sherlock) 560
Holmes (Sherlock) Collection, 586
Holmes (Sherlock) Society (The Musgraves), 560
Holmes Associates, 425
Holt (Henry) & Company, Inc., 178
Holt (Vanessa) Ltd, 260
Holt McDougal, 178
Holt Studios International Ltd, 612
Holtzbrinck (Georg von) Verlagsgruppe GmbH, 165
Home & Family, 327
Homes & Antiques, 327
Homes & Gardens, 327
Honeyglen Publishing Ltd, 105
Honno Welsh Women's Press, 105, 202
Hordern (Kate) Literary Agency, 260
Horizon Press, 106
Horizonte (Livros) Lda, 169
Horn (David C.) Prize, 504
Horror Writers Association, 538
Horse, 327
Horse and Hound, 327
Horse and Rider, 327
Horsecross Arts Ltd, 442
Horus Editions, 106
Hosking Houses Trust, 484
Hoskins (Valerie) Associates Limited, 260
Hotline, 328
Houghton Mifflin Harcourt Publishing Company, 178
Houghton's Horses/Kit Houghton Photography, 612

Hourglass Productions, 426
House & Garden, 328
House Beautiful, 328
House of Commons Research Library, 645
Housman Society, 560
Houston, University of, 474
How To Books Ltd, 106
'How to Self-Publish Your Book' Workshops, 478
How to Spend It, 328
Howard Books, 178
Howes (Chris)/Wild Places Photography, 612
HPBooks, 178
HQ Poetry Magazine (Haiku Quarterly), 211
HRT (Hrvatska Televizija), 413
Hubbard's (L. Ron) Writers of the Future Contest, 504
Huddersfield Daily Examiner, 388
Hudson Street Press, 178
Huia (NZ) Ltd, 196
Hull Daily Mail, 388
Hull Truck Theatre Company, 442
Hull, University of, 465
Human Horizons, 106
Humanity Books, 178
Humber Mouth – Hull Literature Festival, 634
Hunt (John) Publishing Ltd, 106
Huntley Film Archive, 612
Hurst (Jacqui), 612
Hurst Publishers, Ltd, 106
Hutchinson/Hutchinson Children's Books, 106
Hutchison Picture Library, 612
Hymns Ancient & Modern Ltd, 106
Høst & Søn Publishers Ltd, 161
Hüthig GmbH & Co. KG, 165

I

I*D Books, 203
i-D Magazine, 328
Icon Books Ltd, 106
Icon Films, 426
Idea Generation, 277
Ideal Home, 328
Ignotus Press, 230
IHR Library, 586
IKON (Interkerkelijke Omroep Nederland), 415
Ilex Press Limited, The, 244
Ilkley Literature Festival, 634
Illinois Press, University of, 178
Illustrated London News Picture Library, 613
Image Magazine, 328
Imagenation, 211

Images of Africa Photobank, 613
Imagine: Writers and Writing for Children, 634
Imari Entertainment Ltd, 426
Imison Award, The, 505
Immanion Press, 230
Imperial College Central Library, 586
Imperial War Museum, 586
Imperial War Museum Photograph Archive, 613
Imprint Academic, 106
In Britain, 328
In Pinn, The, 107
Inclement, 211
Income Tax, 52–55
Incomes Data Services, 107
Independent Foreign Fiction Prize, The, 505
Independent Magazine, 329
Independent Music Press, 107
Independent Northern Publishers, 645
Independent on Sunday, 374
Independent Publishers Guild, 538
Independent Talent Group Limited, 261
Independent Theatre Council, 538
Independent Voices, 107
Independent, The, 374
Indepublishing CIA – Consultancy for Independent Authors, 261
Indiana University Press, 179
Infinite Ideas, 107
Infinity Junction, 230
Informa Law, 107
IngentaConnect, 645
Inheritance Tax, 61–62
INKwell Writers' Workshops, 466
Inner Sanctum Publications, 230
Inpress Books, 645
Insiders' Guides, 179
Inspira Group, The, 261
Inspire, 107, 329
Inspired Living, 191
Institute of Copywriting, 462, 538
Institute of Historical Research, The, 586
Institute of Public Administration, 156
Institute of Translation and Interpreting (ITI), 538
Instituto Cervantes, 586
InStyle, 329
Inter-Varsity Press, 107

Intercontinental Literary Agency, 261
Interlink Publishing Group, Inc., 179
International Association of Scientific, Technical & Medical Publishers, 539
International Dundee Book Prize, The, 505
International Festival of Mountaineering Literature, The, 634
International Herald Tribune, 374, 394
International IMPAC Dublin Literary Award, The, 505
International Photobank, 613
International Playwriting Festival, 634
International Press-Cutting Bureau, 628
International Prize for Arabic Fiction, 505
International Scripts, 261
International Student Playscript Competition, 506
Internet Classics Archive, 645
Internet Movie Database (IMDb), 645
Interpreter's House, The, 211
Interzone: Science Fiction & Fantasy, 329
Investors Chronicle, 329
IoL, 539
Iolo, 230
Iota, 211
Iowa Press, University of, 179
Irish Academic Press Ltd, 156
Irish Book Awards, 506
Irish Book Publishers' Association, 539
Irish Book Review, The, 329
Irish Copyright Licensing Agency Ltd, 539
Irish News, The, 389
Irish Pages, 211, 329
Irish Playwrights and Screenwriters Guild, 539
Irish Translators' & Interpreters' Association, 539
Irish Writers' Union, 539
Iron Press, 203
IRRV Valuer, 329
ISBN Agency, 641
Isis Productions, 426
Isis Publishing, 107
Island, 211
Isle of Man Authors, 539
Isle of Man Literature Festival, 634
Isle of Wight Radio, 411
Isolde Films, 426

Italian Institute Library, 586
ITC, 539
Itchy Coo, 107
Ithaca Press, 107
ITI, 539
ITN, 406
ITV Anglia Television, 407
ITV Border Television, 407
ITV Central, 407
ITV Granada, 407
ITV London, 407
ITV Meridian, 407
ITV Network, 406
ITV plc, 406
ITV Productions, 426
ITV Thames Valley, 407
ITV Tyne Tees, 407
ITV Wales, 407
ITV West, 407
ITV West Country, 408
ITV Yorkshire, 408
IVCA, 539
IVP, 108
Ivy Press Limited, The, 244
Ivy Publications, 230

J

J. Garnet Miller, 108
Jacana Books, 191
Jacaranda, 191
Jack (Robbie) Photography, 613
Jacobs (W.W.) Appreciation Society, 560
Jacqui Small, 108
Jade, 330
JAI, 108
Jaico Publishing House, 195
JAM Pictures and Jane Walmsley Productions, 426
James (Barrie) Literary Agency, 261
Jane's Defence Weekly, 330
Jane's Information Group, 108
Janklow & Nesbit (UK) Ltd, 261
Janklow & Nesbit Associates, 284
Janus Publishing Company Ltd, 108
Janus Writers, 478
Jarrold, 108
Jayawardene Travel Photo Library, 613
Jazz Journal International, 330
JCA Literary Agency, Inc., 284
Jefferies (Richard) Society, 560
Jelenkor Kiadó Szolgáltató Kft., 166
Jerome (Jerome K.) Society, 561
Jersey Evening Post, 392
Jersey Library, 587
Jersey Now, 330

Jerwood Aldeburgh First Collection Prize, 506
Jerwood Awards, 484
Jewish Chronicle, 330
Jewish Quarterly, 330
Jewish Quarterly Literary Prize, 506
Jewish Telegraph Group of Newspapers, 331
Johns (Captain W.E.) Appreciation Society, The, 561
Johnson & Alcock, 261
Johnson (Samuel) Prize for Non-Fiction, 506
Johnson Society, 561
Johnson Society of London, 561
Jones (David) Journal, The, 211
Jones (Mary Vaughan) Award, 506
Joseph (Michael), 108
Joseph (Stephen) Theatre, 442
Journal of Apicultural Research, 331
Journal, The, 211
Journal, The (Newcastle upon Tyne), 386
Journalism UK, 645
Journalists' Charity, 484
Jove, 179
JR Books Ltd, 108
Judd (Jane) Literary Agency, 262
Junior Education PLUS, 331
Jupiter Press, The, 230
Just William Society, The, 561
Justice Entertainment Ltd, 426
Justis Publishing Ltd, 248

K

Kahn & Averill, 108
Kangaroo Press, 191
Kansas, University Press of, 179
Kar-Ben Publishing, 179
Karisto Oy, 162
Kass (Michelle) Associates, 262
Katabasis, 203
Kates Hill Press, 203
Kaye (Paul) B/W Archive, 613
Keats–Shelley Memorial Association (Inc), The, 561
Keats–Shelley Prize, 506
Keele University, 463
Kellom Books, 193
Kelly (Frances), 262
Kelpies Prize, 506
Kenilworth Press Ltd, 108
Kenney (Petra) Poetry Competition, The, 507
Kent & Sussex Poetry Society, 562
Kent & Sussex Poetry Society Open Competition, 507

Kent County Central Library, 587
Kent Messenger, 382
Kent State University Press, The, 179
Kent, University of, 457
Kentucky, The University Press of, 179
Kenwright (Bill) Ltd, 443
Kenyon-Deane, 109
Keo Films.com Ltd, 427
Kerrang!, 331
Kettillonia, 203
Kibea Publishing Company, 160
KidHaven Press, 179
Kilvert Society, The, 562
King (David) Collection, 613
King (Laurence) Publishing Ltd, 109
King George III Topographic Collection and Maritime Collection, 587
King's England Press, The, 203
King's Lynn, The Fiction Festival, 634
King's Lynn, The Poetry Festival, 635
King's Sutton Literary Festival, 635
Kingfisher, 109
Kingfisher Television Productions Ltd, 427
Kingsley (Jessica) Publishers Ltd, 109
Kingswood Books, 179
Kipling Society, The, 562
Kirchoff/Wohlberg, Inc., 284
Kiriyama Pacific Rim Book Prize, 507
Kiss József Könyvkiadó, 166
Kitley Trust, The, 562
Kittiwake, 231
Klett (Ernst) Verlag GmbH, 165
Klinger (Harvey), Inc., 284
Kluwer Law International, 109
kmfm, 411
Knight Features, 262
Knopf Canada, 193
Knuston Hall Residential College for Adult Education, 461
Knutsford Literature Festival, 635
Kobal Collection, The, 614
Kogan Page Ltd, 109
Koi Magazine, 331
Kok (Uitgeversmaatschappij J. H.) BV, 168
Komedia, 443
Konner (Linda) Literary Agency, 285

Kops (Bernard) and Tom Fry Drama Writing Workshops, 478
Kos Picture Source Ltd, 614
Kraszna-Krausz Book Awards, 507
Krax, 211
Krieger Publishing Co., 179
KRO (Katholieke Radio Omroep), 415
Kráter Association and Publishing House, 166
KT Publications, 203
Kudos Film and Television Ltd, 427
Kwa-Zulu–Natal Press, University of, 197

L

Lacey (Ed) Collection, 614
Lady, The, 331
Ladybird, 109
Laffont (Robert) Éditions, 163
Lagan Pictures Ltd, 427
Lake District Life, 332
Lakeland Book of the Year Awards, 507
Lamb (Charles) Society, The, 562
Lambeth Readers & Writers Festival, 635
Lampack (Peter) Agency, Inc., 285
Lancashire Authors' Association, 562
Lancashire County Library and Information Service Children's Book of the Year Award, 507
Lancashire Evening Post (Preston), 382
Lancashire Life, 332
Lancashire Telegraph (Blackburn), 382
Lancaster LitFest, 635
Lancaster University, 457
Land Rover World, 332
Landeau (Caroline), 641
Landform Slides, 614
Landmark Publishing Ltd, 109
Landseer Productions Ltd, 427
Lane (Frank) Picture Agency Ltd, 614
Lannan Literary Award, 507
Lannoo, Uitgeverij, 160
Large Print Press, 180
Lark, 180
Larkin (Philip) Society, The, 563
Larousse (Éditions), 163
Larsen (Michael)/Elizabeth Pomada Literary Agency, 285

Last Resort Picture Library, 614
LAT Photographic, 614
Latino Voices, 180
Lattès (Éditions Jean-Claude), 163
Laughing Stock Productions, 239
LAW, 262
Lawrence & Wishart Ltd, 110
Lawrence (D.H.) Society, The, 563
Lawrie (Duncan) Dagger, The, 508
Lawrie (Duncan) International Dagger, The, 508
Lawyer 2B, 332
Lawyer, The, 332
LB Kids, 180
LBA, 262
LBC Radio Ltd, 411
Le Prince Maurice Prize, 508
Leaf Books, 203
Leafe Press, 203
Leamington Literary Society, The, 563
Leamington Spa Courier, 387
Learning Factory, The, 244
Learning Institute, The, 110
Leavitt (Ned) Agency, The, 285
Lebrecht Music and Arts, 614
Ledbury Poetry Festival, 635
Leeds Central Library, 587
Leeds Library, 587
Leeds Metropolitan University, 464
Leeds Poetry Quarterly, 211
Leeds, University of, 465
Leeds, University of (Lifelong Learning Centre), 465
Legend Press Limited, 110
Legend Writing Award, 508
Leicester Mercury, 382
Lemniscaat, 180
Lennard Associates Ltd, 244
Lerner Publishing Group, 180
Lescher & Lescher Ltd, 285
Lessing (Erich) Archive of Fine Art & Culture, The, 614
Lettera, 161
Letterbox Library, 293
Levin (Bernard) Award, The, 508
Levy (Barbara) Literary Agency, 262
Lewes Live Literature, 635
Lewes Monday Literary Club, 563
Lewis (Dewi) Publishing, 110
Lewis (Ralph) Award, 484
Lewis/Lewis Masonic, 110
LexisNexis Butterworths (Pty) Ltd South Africa, 197

LexisNexis Canada Ltd, 193
LexisNexis India, 195
LexisNexis New Zealand, 196
Lexus Ltd, 244
Liberato Breakaway Writing Courses, 456
Library Association Publishing, 110
Library Association, The, 540
Library of Art & Design, 614
Library of the Religious Society of Friends, 587
Library of Wales, 110
Library Theatre Company, 443
Lichfield Festival, 636
Life&Times, 110
Liffey Press Ltd, The, 156
Lilliput Press, The, 156
Lilyville Screen Entertainment Ltd, 427
Limelight Management, 262
Lincoln (Frances) Ltd, 110
Lincoln Central Library, 588
Lincolnshire Echo, 383
Lincolnshire Life, 332
Lincolnshire Poacher, The, 332
Linden Press, 111
Lindgren (Astrid) Memorial Award for Literature, The, 508
Lindhardt og Ringhof Forlag A/S, 162
Lindley Library, Royal Horticultural Society, 614
Lindsey Press, The, 231
Linen Hall Library, 588
Linen Press, The, 231
Linford Romance/Linford Mystery/Linford Western, 111
Link Picture Library, 615
Linkway, 211
Lion Hudson plc, 111
Lionheart Books, 244
List, The, 332
Literary & Philosophical Society of Newcastle upon Tyne, 588
Literary Review, 333
Literature and Psychoanalysis, 212
Literature North East, 645
Literature North West, 645
literaturetraining, 540, 645
Little (Christopher) Literary Agency, The, 263
Little Black Dress, 111
Little Books Ltd, 111
Little Simon, 180
Little Tiger Press, 111
Little, Brown and Company (USA)/Little, Brown Books for Young Readers, 180

Little, Brown Book Group UK, 111
Live Theatre Company, 443
Liverpool Daily Post, 383
Liverpool Echo, 383
Liverpool Libraries and Information Services, 588
Liverpool University Press, 112
Liverpool, University of, 461
Livewire Books for Teenagers, 112
Living France, 333
Llenyddiaeth Cymru Dramor, 540
Llewellyn Publications, 180
Llewellyn Rhys (John) Prize, The, 509
Loaded, 333
Location Register of 20th-Century English Literary Manuscripts and Letters, 645
Loftus Audio Ltd, 427
Logos, 333
London (City of) Festival, 636
London (University of the Arts) – London College of Communication, 588
London Aerial Photo Library, 615
London Bubble Theatre Company, 443
London College of Communication, 459
London Independent Books, 263
London Library, The, 589
London Literature Festival, 636
London Magazine, The, 333
London Metropolitan Archives, 615
London Metropolitan Archives Library Services, 589
London Review of Books, 333
London School of Journalism, 459
London Scientific Films Ltd, 428
London's Transport Museum Photographic Library, 615
London's Transport Museum Poster Archives, 589
Lonely Planet Publications Ltd, 112
Longanesi (Casa Editrice) SpA, 167
Longford (Elizabeth) Grants, The, 484
Longford (Elizabeth) Prize for Historical Biography, The, 509

Longman-History Today Book of the Year Award, 509
Lord (Sterling) Literistic, Inc., 285
Lord Louis Library, 589
Lorenz Books, 112
Los Angeles Times, 395
Lothian Life, 334
Louisiana State University, 471
Louisiana State University Press, 180
Lowdham Book Festival, 636
Lowenstein–Yost Associates Inc., 285
Lownie (Andrew) Literary Agency, 263
LRT (Lietuvos radijas ir televizija), 415
LTV1 & LTV7, 415
Luard (Clarissa) Award, The, 509
Luath Press Ltd, 112
Lucas Alexander Whitley, 263
Lucent Books, 180
Lucida Productions, 428
Luck (Lucy) Associates, 263
Lucky Duck Publishing, 112
Ludvigsen Library, 615
Luis Vives (Grupo Editorial), 170
Lulu.com, 112
Lund Humphries, 112
Lutterworth Press, The, 112
Lutyens and Rubinstein, 263
Luxton Harris Ltd, 263
Lyfrow Trelyspen, 231
Lyons, 180
Lyons (Sir William) Award, 509
Lyric Hammersmith, 443
Lübbe (Verlagsgruppe) GmbH & Co. KG, 165

M
M&M Publishing Services, 244
M-Y Books Ltd, 641
mac, 444
MacBride (Margret) Literary Agency, The, 286
Macdonald & Co., 113
MacDonald (George) Society, The, 563
MacElderry (Margaret K) Books, 180
Machen, The Friends of Arthur, 563
Machine Knitting Monthly, 334
Mackenzie (W.J.M.) Book Prize, 509
MacLehose Press, 113
Macmillan Digital Audio, 239
Macmillan India Ltd, 195

Macmillan Prize for a Children's Picture Book Illustration, 510
Macmillan Publishers Australia Pty Ltd, 191
Macmillan Publishers Ltd, 113
Macmillan Publishers New Zealand, 196
Macmillan Reference USA, 181
Macmillan South Africa, 197
Macmillan Writer's Prize for Africa, 510
MacMullen (Eunice) Ltd, 264
Macquarie Dictionary Publishers, 192
MacQuitty International Photographic Collection, 616
MacUser, 334
Made Simple Books, 114
Magma, 212
Magnard (Éditions), 163
Magnum Photos Ltd, 616
Maia Press, The, 231
Mail on Sunday Novel Competition, The, 510
Mail on Sunday, The, 375
Mainstream Publishing Co. (Edinburgh) Ltd, 114
Malfunction Press (1969), 203
Man Asian Literary Prize, The, 510
Man Booker International Prize, The, 510
Man Booker Prize for Fiction, The, 510
Management Books 2000 Ltd, 114
Management Today, 334
Manchester Central Library, 589
Manchester Evening News, 383
Manchester Library Theatre, 444
Manchester Literature Festival, 636
Manchester Metropolitan University – The Writing School, 460
Manchester University Press, 115
Manchester, University of, 460
Mander (Raymond) & Joe Mitchenson Theatre Collection, The, 616
Mann (Andrew) Ltd, 264
Mann (Carol) Agency, 286
Mann (George) Books, 115
Mann (Roger) Collection, The, 616
Mann (Tom) Collection, 589
Manson (Sarah) Literary Agent, 264
Manson Publishing Ltd, 115

Manus & Associates Literary Agency, Inc., 286
Marchmont Films Ltd, 428
Mare's Nest, 203
marie claire, 334
Marino Books, 156
Mariscat Press, 203
Marjacq Scripts Ltd, 264
Market House Books Ltd, 244
Market News International, 395
Marketing Week, 334
Markings, 212, 334
Marks (Andrea) Public Relations, 277
Marland (Folly), Literary Scout, 274
Marlborough College Summer School, 464
Marlowe Society, The, 563
Marsh Agency Ltd, The, 264
Marsh Award for Children's Literature in Translation, 511
Marsh Biography Award, 511
Marshall (Jane) Productions, 428
Marshall Cavendish Ltd, 115
Martin Books, 115
Marylebone Library (Westminster), 589
Masefield (John) Memorial Trust, The, 484
Masefield (John) Society, The, 564
Maskew Miller Longman, 197
Mason (Kenneth) Publications Ltd, 115
Masque Publishing, 203
Massachusetts Press, University of, 181
Matador, 115
Match, 335
Mathews (S&O) Photography, 616
Maugham (Somerset) Awards, The, 485
Maverick House Publishers, 156
Maverick Television, 428
Max/Max Press, 115
Maxim, 335
Maya Vision International Ltd, 428
Mayfair, 335
Mayfair Times, 335
Mayhew (Kevin) Publishers, 115
Maypole Editions, 231
MBA Literary Agents Ltd, 264
MBP TV, 428
McAra (Duncan), 264
McClelland & Stewart Ltd, 193
McColvin Medal, 509

McCutchen (Julia), Writers' Coach & Professional Publishing Consultant, 641
McFarland & Company, Inc., Publishers, 180
McGraw-Hill Australia Pty Ltd, 192
McGraw-Hill Companies Inc., The, 180
McGraw-Hill Education, 113
McGraw-Hill Education (India), 195
McGraw-Hill New Zealand Pty Ltd, 196
McGraw-Hill Ryerson Ltd, 194
McKernan Literary Agency & Consultancy, 265
McKitterick Prize, 509
McLean (Bill) Personal Management, 264
McLeod (Enid) Prize, 509
MCPS–PRS Alliance, The, 540
Meadowside Children's Books, 115
Mears (Colin) Award, 511
Mechanical Engineers (Institution of), 616
Medal News, 335
Medavia Publishing, 115
Media Week, 335
Medical Book Awards, 511
Medical Journalists' Association, 540
Medical Writers' Group, 540
Medway Messenger, 382
Megalithica Books, 231
Melbourne University Publishing Ltd, 192
Melendez Films, 428
Melody Maker, 335
Melrose Books, 116
Melrose Press Ltd, 116
Memphis, University of, 474
Men's Health, 335
Mendoza Film Productions, 429
Mentor, 116
Mercat Press, 116
Mercedes-Benz Award for the Montagu of Beaulieu Trophy, The, 511
Mercia Cinema Society, 231
Mercier Press Ltd, 156
Mere Literary Festival, 636
Mere Literary Festival Open Competition, 511
Meridian Books, 231
Merlin Press Ltd, The, 116
Merlin Publishing, 157
Merrell Publishers Ltd, 116
Mersey Television, 429
Merseyfilm, 427

Methodist Publishing House, 116
Methuen Drama, 116
Methuen Publishing Ltd, 116
Metro Publishing, 116
Metropolitan Books, 181
Meulenhoff (Uitgeverij J.M.) BV, 168
Mews Books Ltd, 286
Mews Press, 203
Meyer-Whitworth Award, 511
MGA, 277
Michaels (Doris S.) Literary Agency Inc., 286
Michelin Maps & Guides, 116
Michelin Éditions des Voyages, 163
Michigan Press, The University of, 181
Middlesex University, 459
Midland Publishing, 117
Miegunyah Press, 192
Milet Publishing LLC, 181
Millbrook Press, 181
Miller (Lee) Archives, 616
Miller (Neil) Publications, 232
Miller's, 117
Millers Dale Publications, 232
Millivres Prowler Limited, 117
Mills & Boon, 117
Milo Books Limited, 117
MIND Book of the Year, 512
Mindfield, 117
Ministry of Defence Cartographic Archive, 590
MiniWorld Magazine, 336
Minnesota Press, University of, 181
Minnesota State University Moorhead, 472
Minnesota State University, Mankato, 472
Minotaur, 181
Minuit (Les Éditions de) SA, 164
Mirage Publishing, 232
Mirrorpix, 616
Missenden Abbey, 453
Mississippi, University Press of, 181
Missouri Press, University of, 181
Missouri, University of, 472
Missouri-Columbia, University of, 472
Mitchell Beazley, 117
Mitchell Library, The, 590
Mitchell Prize for Art History, The/The Eric Mitchell Prize, 512
Mixmag, 336
Mizz, 336
Mobilise, 336

Mobius, 117
Model Collector, 336
Modern Library, 181
Modern Poetry in Translation, 212
Mohr Books, 232
Mojo, 336
Molino SL (Editorial), 170
Momaya Press Short Story Writing Competition, 512
Monarch Books, 117
Moncrieff (Scott) Prize, 512
Mondadori (Arnoldo) Editore SpA, 167
Moneywise, 337
Monitor Picture Library, 616
Monkey Kettle, 212
Monkey Puzzle Media Ltd, 245
Monomyth, 212
Moodswing, 212
Moonstone Films Ltd, 429
Moore (Christy) Ltd, 265
More, 337
Morgan Kauffman, 117
Morhaim (Howard) Literary Agency, 287
Morley (Sheridan) Prize, The, 512
Morning Star, 375
Morrab Library, 590
Morris (William) Agency (UK) Ltd, 265
Morris (William) Society, 564
Morrison (Henry) Inc., 287
Morrow (William), 181
Mosby, 117
Mother and Baby, 337
Motley (Michael) Ltd, 265
Motor Boat & Yachting, 337
Motor Boats Monthly, 337
Motor Caravan Magazine, 337
Motor Cycle News, 337
Motor Racing Publications, 117
Motorcaravan Motorhome Monthly (MMM), 337
Motoring Picture Library, 617
Mount Eagle, 157
Mountain Camera, 617
Mountain Magic, 338
Moving Image Communications, 617
mph, 117
Mr William Shakespeare and the Internet, 646
Mslexia, 212, 338
MTV 1/MTV 2 (Magyar Televízió), 414
MTV Networks Europe, 409
MTV3 Finland, 414
Mudfog Press, 203
Mulino (Societe editrice il), 167
Munro (Neil) Society, The, 564

Munshiram Manoharlal Publishers Pvt Ltd, 195
Murdoch (Judith) Literary Agency, 265
Murdoch Books UK Ltd, 118
Murray (John) (Publishers) Ltd, 118
Mursia (Gruppo Ugo) Editore SpA, 167
Muscadine Books, 181
Museum of Childhood, 617
Museum of London Picture Library, 617
Music Week, 338
Musical Opinion, 338
My Weekly, 338
My Weekly Pocket Novels, 338
Myriad Editions, 118
Myrmidon Books, 118
Mystery Writers of America, Inc., 540
Móra Könyvkiadö Zrt., 166

N

N.T.C. Touring Theatre Company, 445
NAG Press Ltd, 118
Naggar (Jean V.) Literary Agency, The, 287
NAL, 182
Naouka I Izkoustvo, 161
Narrow Road Company, The, 265
nasen Awards 2008, 512
Nasza Ksiegarnia Publishing House, 168
Nathan (Les Éditions), 164
Nathan (Melissa) Award for Comedy Romance, 512
National Academy of Writing, 464
National Archives Image Library, The, 617
National Archives, The, 118, 590
National Art Library, 617
National Association for Literature Development, 540
National Association of Writers Groups, 540
National Association of Writers in Education, 541
National Association of Writers' Groups (NAWG) Open Festival of Writing, 636
National Campaign for the Arts, 541
National Centre for Research in Children's Literature, 541
National Council for the Training of Journalists, 455
National Eisteddfod of Wales, 636

National Extension College, 453
National Film & Television School, 453
National Galleries of Scotland Picture Library, 617
National Library of Ireland, 157, 590
National Library of Scotland, 590
National Library of Wales, 590
National Literacy Trust, 541
National Magazine Company, 617
National Maritime Museum Picture Library, 617
National Media Museum, 617
National Meteorological Library and Archive, 591
National Monuments Record, 591, 617
National News Press and Photo Agency, 393
National Poetry Anthology, 513
National Poetry Competition, 513
National Portrait Gallery Picture Library, 618
National Public Radio, 395
National Publishing House, 195
National Railway Museum, 618
National Student Drama Festival, 637
National Trust Books, 118
National Trust Magazine, The, 339
National Trust Photo Library, The, 618
National Union of Journalists, 541
Natur och Kultur Bokförlaget, 171
Natural England, 591
Natural History Museum Image Resources, 618
Natural History Museum Library, The, 591
Naturalist, The, 339
Nature, 339
Nature Photographers Ltd, 618
Nature Picture Library, 618
Nautical Data Ltd, 118
Naxos AudioBooks, 239
NB, 339
NBC News, 395
NCRCL, 542
NCRV (Nederlandse Christelijke Radio Vereniging), 415
NCVO Publications, 119
Neave (Airey) Trust, The, 485
Nebraska Press, University of, 182

Need2Know, 232
Needham (Violet) Society, 564
Nelson (B.K.) Literary Agency, 287
Nelson (Thomas) & Sons Ltd, 119
Nelson Education, 194
Nelson Thornes Limited, 119
Neon, 429
Neon Highway, 212
Neptun-Verlag, 171
Nesbit (Edith) Society, The, 564
Nevada Press, University of, 182
Never Bury Poetry, 212
Never Ending Story, The, 646
New Authors Showcase, 248
New Beacon Books Ltd, 119
New Chalet Club, The, 555
New England, University Press of, 182
New Holland Publishers (UK) Ltd, 119
New Hope International, 204
New Humanist, 339
New Internationalist, 339
New Island, 157, 204
New Mexico Press, University of, 182
New Musical Express, 339
New Nation, 339
New Playwrights Trust, 542
New Playwrights' Network, 119
New Producers Alliance, 542
New Scientist, 340
New Seeds, 182
New Shetlander, The, 340
New Star Books Ltd, 194
New Statesman, 340
New Theatre Quarterly, 340
New Vic Theatre, 444
New Welsh Review, 212, 340
New Writer, The, 340
New Writers Consultancy, 646
New Writing North, 542, 646
New Writing Partnership, 646
New Writing Scotland, 341
New Writing South, 478
New Writing Ventures, 513
New York Times, 395
New York University, 473
New!, 341
Newark's (Peter) Picture Library, 618
newbooks, 341
Newman Press, 182
Newnes, 119
Newnovelist, 646
Newpalm Productions, 444
News & Star (Carlisle), 379
News of the World, 375
News, The (Portsmouth), 381
Newspaper Society, The, 542

Newsweek, 395
Nexus, 119
nferNelson, 119
NHPA, 618
Nia, 119
Nibbies, The, 513
Nicholas Enterprises, 245
Nielsen BookData, 119
Nielsen BookNet, 641
Nielsen BookScan, 641
Nielsen Gold & Platinum Book
  Awards, 513
Nightingale Books, 120
Nimax Theatres Limited, 444
96.4 FM The Wave, 412
Nissen (Jane) Books, 232
Nitro, 444
NMS Enterprises Limited –
  Publishing, 120
Noach (Maggie) Literary
  Agency, The, 265
Nobel Prize, 513
Noma Award for Publishing
  Africa, The, 514
Nonsuch, 120
Norfolk Library & Information
  Service, 591
*Norfolk Poets and Writers –
  Tips Newsletter*, 212
Norstedts Förlag, 171
North Carolina at Greensboro,
  University of, 473
North Carolina at Wilmington,
  University of, 473
North East Literary Fellowship,
  485
North Texas Press, University
  of, 182
North West Evening Mail
  (Barrow in Furness), 379
North West Kent College, 457
North West Playwrights
  (NWP), 478
North, The, 212
North-Holland, 120
Northamptonshire Libraries &
  Information Service, 591
Northcote House Publishers
  Ltd, 120
Northern Children's Book
  Festival, 637
Northern Echo (Darlington),
  The, 381
Northern Writers' Awards, 485
NorthSound Radio, 411
Northumberland County
  Library, 592
Northwestern University Press,
  182
Norton (W.W.) & Company
  Ltd, 120

Norton (W.W.) & Company,
  Inc., 182
Norwich Puppet Theatre, 444
NOS (Nederlandse Omroep
  Stichting), 415
Nostalgia Collection, The, 232
Notes from the Underground,
  341
Notre Dame, University of, 471
Nottingham Central Library,
  592
Nottingham Playhouse, 444
Nottingham Playhouse
  Roundabout Theatre in
  Education, 444
Nottingham Subscription
  Library, 592
Nottingham Trent University,
  The, 462
Nottingham University Press,
  120
Now, 341
NPA, 542
NPS (Nederlandse Programma
  Stichting), 415
NRK (Norsk Rikskringkasting),
  415
nthposition, 204
Nuffield Theatre, 445
Number 9 Films, 429
Nurnberg (Andrew) Associates
  Ltd, 265
Nursery Education, 341
Nursing Times, 342
Nuts, 342
NWP, 120
Nyala Publishing, 232
Nye (Alexandra), 266
Nyt Nordisk Forlag Arnold
  Busck A/S, 162

O

O'Brien Press Ltd, The, 157
O'Connor (Frank) International
  Short Story Award, 514
O'Faolain (Sean) Short Story
  Prize, 514
O'Mara (Michael) Books
  Limited, 122
O/Books/ John Hunt Publishing
  Ltd, 120
Oak, 120
Oak Tree Press, 157
Oakhill Publishing Ltd, 240
Oberon Books, 120
Object Permanence, 204
Observer, The, 375
Obsessed With Pipework, 213
Octagon Press Ltd, 121
Octagon Theatre Trust Ltd, 445
Octopus Publishing Group, 121
Odhams Periodicals Library, 618

Odyssey Productions Ltd, 429
Off the Page Literature Series,
  637
Off the Shelf Literature Festival,
  637
Office for National Statistics,
  592
Offshoot, 619
Ohio University Press, 182
OK! Magazine, 342
Oklahoma Press, University
  of, 183
Old Glory, 342
Old Stile Press, The, 204
Old Street Publishing Ltd, 121
Oldham Evening Chronicle, 382
Oldie, The, 342
Oldman (C.B.) Prize, 514
Olive, 342
Olive Branch Press, 183
Olympia Publishers, 122
Olympia Publishing Co. Inc.,
  161
Omnibus Books, 192
Omnibus Press, 122
Omnivision, 429
On Stream Publications Ltd, 157
Once Orange Badge Poetry
  Press, The, 204
*Once Orange Badge Poetry
  Supplement, The*, 213
Ondaatje Prize, 514
One Time Press, The, 204
One World, 183
Oneworld Publications, 122
Online Originals, 248
Only Horses Picture Agency,
  619
Onlywomen Press Ltd, 122
OPC, 123
Open College of the Arts, 465
Open Forum, 183
Open Gate Press, 123
Open University Press, 123
Open Wide Magazine, 213
Opera, 342
Opera Now, 343
Orange Broadband Prize for
  Fiction/Orange Broadband
  Award for New Writers, 514
Orange Tree Theatre, 445
Orbit, 123, 183, 183
Orchard Books, 123
Oregon, University of, 474
Orell Füssli Verlag, 171
ORF(Österreichisher
  Rundfunk)1/ORF2, 413
Organic Gardening, 343
Orient Paperbacks, 195
Original Plus, 204
Original Writers Group, The,
  479

Orion Audio Books, 240
Orion Publishing Group Limited, The, 123
Orkney Library and Archive, 592
Orlando Television Productions Ltd, 429
Ormrod Research Services, 641
Orpheus Books Limited, 245
Orpheus Productions, 429
Orpheus Publishing House, 232
Orwell Prize, The, 514
OS (Office Secretary) Magazine, 343
Oshun, 197
Osprey Publishing Ltd, 123
Otava Publishing Co. Ltd, 162
Other Poetry, 213
Oundle Festival of Literature, 637
Out of Joint, 445
Outdoor Writers & Photographers Guild, 542
Ovation, 429
Owen (Deborah) Ltd, 266
Owen (Peter) Publishers, 123
Owen (Wilfred) Association, The, 564
Owen (Wilfred) Award for Poetry, The, 515
Owl Books, 183
OWPG Awards for Excellence, 515
Oxbow Books, 124
Oxenham (Elsie Jeanette) Appreciation Society, The, 564
Oxford (University of) Department for Continuing Education, 462
Oxford Central Library, 592
Oxford Literary Festival, 637
Oxford Mail, 384
Oxford Picture Library, 619
Oxford Scientific (OSF), 619
Oxford University Press, 124
Oxford University Press (Australia, New Zealand & the Pacific), 192
Oxford University Press India, 196
Oxford University Press New Zealand, 196
Oxford University Press Southern Africa (Pty) Ltd, 197
Oxford University Press, Canada, 194
Oxford Weidenfeld Translation Prize, The, 515

P

PA News Centre, 592
Packard Publishing Limited, 233
Pact, 542
Page (Louise) PR, 277
Paine (Thomas) Society, 565
Paines Plough, 446
Paisley Daily Express, 390
Palgrave, 183
Palgrave Macmillan, 124
Palgrave Macmillan (Australia), 192
Palmer (G.J.) & Sons Ltd, 124
Palores Publishing, 204
Pan, 124
Pan Macmillan Australia Pty Ltd, 192
Panacea Press Limited, 233
Panos Pictures, 619
Paper Moon Productions, 430
Paper Tiger, 124
PaperBooks Ltd, 124
Paragon House, 183
Parallax East Ltd, 430
Parameter, 213
Parapress, 233
Park Home & Holiday Caravan, 343
Parker (Charles) Archive, 619
Parks (Richard) Agency, 287
Parthian, 124, 204
Partners, 204
Partners Annual Poetry Competition, 213
Passion Pictures, 430
Past and Present, 233
Patent Research, 641
Paternoster, 124
Paterson Marsh Ltd, 266
Pathfinder, 124
Pathé Pictures, 430
Paulist Press, 183
Paupers' Press, 233
Pavilion, 124
PAWS (Public Awareness of Science) Drama Fund, 485
Pawsey (John), 266
Payne–Gallway, 125
Payot Libraire, 171
PBS BiH, 413
PC Format, 343
Peace and Freedom, 213
Pearlstine (Maggie) Associates, 266
Pearson, 125
Pearson (Heinemann), 125
Pearson Canada, 194
Pearson Educacion SA, 170
Pearson Education (Benelux), 168
Pearson Education Australia Pty Ltd, 192

Pearson Education New Zealand Ltd, 196
Pearson Playwrights' Scheme, 485
Pearson Scott Foresman, 183
Pearson Television, 430
Peepal Tree Press Ltd, 204
Peer Poetry International, 213
Peerless (Ann & Bury) Picture Library, 619
Pegasus Elliot Mackenzie Publishers Ltd, 125
Pelican Publishing Company, 183
Pelicula Films, 430
PEN, 543
Pen & Sword Books Ltd, 125
PEN Awards, 515
Pen Press Publishing Ltd, 233
Penguin Audiobooks, 240
Penguin Books (NZ) Ltd, 197
Penguin Group (Australia), 192
Penguin Group (Canada), 194
Penguin Group (UK), 125
Penguin Group (USA) Inc., 183
Penguin Ireland, 158
Penguin Longman, 126
Pennant Books Ltd, 126
Penniless Press, 204
Penniless Press, The, 213
Pennine Ink Magazine, 213
Pennine Platform, 213
Pennsylvania Press, University of, 184
Pensions World, 343
Penzance Library, 593
People Magazine, 395
People Management, 343
People's Friend Pocket Novels, 343
People's Friend, The, 344
People, The, 376
Pepper Productions, 430
Pepys (Samuel) Award, 515
Perennial, 184
Perfect Publishers Ltd, 233
Performance Poetry Society, 220
Performing Right Society, 543
Pergamon, 126
Perigee, 184
Period Living, 344
Persephone Books, 126
Perseus Books Group, 184
Personal Managers' Association Ltd, The, 543
Perth Repertory Theatre Ltd, 446
Peterloo Poets, 204
Peterloo Poets Open Poetry Competition, 515

Peters Fraser & Dunlop Group Ltd, 266
PFD, 266
Phaidon Press Limited, 126
Philip's, 126
Phillimore, 127
Philomel Books, 184
Philosopher, The, 344
Phoenix, 127
Photofusion, 619
Photolibrary Wales, The, 619
Photoplay Productions Ltd, 430
Photoshot, 619
Piano, 344
Piatkus Books, 127
Pica Press, 127
Picador, 127
Picador USA, 184
Picard (Alison J.) Literary Agent, 287
Picardy Media & Communication, 430
Piccadilly Press, 127
Pick (Charles) Fellowship, The, 485
Pictorial Presentations, 127
Picture Palace Films Ltd, 430
Picture Postcard Monthly, 344
Picture Research Association, The, 543
PictureBank Photo Library Ltd, 620
Pictures Colour Library, 620
Pigasus Press, 205
Pikestaff Press, 205
Pilot, 344
Pimlico, 127
Pinder Lane & Garon-Brooke Associates Ltd, 287
Pink Paper, 345
Pinwheel, 127
Pipers' Ash Ltd, 205, 233
Pitkin, 127
Pitstop Refuelling Writers' Weekend Workshops, 479
Planet, 213
Planet24 Pictures Ltd, 431
Planeta SA (Editorial), 170
Plantagenet Films Limited, 431
Planète, 414
Platypus PR, 277
Player–Playwrights, 479, 543
Players Press, 184
Playne Books Limited, 245
PlaysOnTheNet, 646
Playwrights Publishing Co., 234
Playwrights' Studio, Scotland, 543
PLR, 543
PLS, 543
Plume, 184
Pluto Press Ltd, 127

Plymouth Library and Information Services, City of, 593
Plymouth Proprietary Library, 593
Plymouth Theatre Royal, 446
PMA, 543
PMA Literary & Film Management, Inc., 287
PMA Media Training, 459
PN Review, 213
Pocket Books, 127
Pocket Books/Pocket Star, 184
Pocket Mountains Ltd, 127
Poems in the Waiting Room, 205
Poet Tree, 213
Poetic Hours, 214
Poetic Licence, 214
Poetry, 25–33
Poetry Archive, The, 220
Poetry Book Society, The, 220
Poetry Business Competition, 515
Poetry Business, The, 221
Poetry Can, 221
Poetry Church, The, 214
Poetry Combination Module, 214
Poetry Cornwall/Bardhonyeth Kernow, 214
Poetry Express, 214
Poetry Ireland News, 214
Poetry Ireland Review/Eigse Eireann, 214
Poetry Ireland/Eigse Eireann, 221
Poetry Library, The, 221
Poetry London, 214
Poetry Nottingham, 214
Poetry Now, 205
Poetry Otherwise, 637
Poetry pf, 221
Poetry Powerhouse Press, 205
Poetry Press Ltd, 205
Poetry Reader, 214
Poetry Review, 214
Poetry Salzburg, 205
Poetry Salzburg Review, 215
Poetry School, The, 222
Poetry Scotland, 215
Poetry Society's National Poetry Competition, The, 516
Poetry Society, The, 222
Poetry Trust, The, 222
Poetry Wales, 215
Poetry Wednesbury, 205
Poets and Writers, 646
Poets Anonymous, 205
Poignant (Axel) Archive, 620
Point Editions, 222
Poisoned Pen Press UK Ltd, 127

Policy Press, The, 128
Polish Library POSK, 593
Polish Scientific Publishers PWN, 169
Politico's Publishing, 128
Polity Press, 128
Polka Theatre, 446
Pollinger Limited, 266
Polygon, 128
Pomegranate Press, 234
Pont Books, 128
Ponting (H.G.), 620
PONY, 345
Poolbeg Press Ltd, 158
Poole Central Library, 593
Pop Universal, 128
Popperfoto.com, 620
Poppy, 184
Porteous (David) Editions, 234
Portfolio, 184
Portico, 128
Portico Prize for Literature, The, 516
Portland Press Ltd, 128
Portobello Books, 128
Portobello Pictures, 431
Portsmouth, University of, 456
post plus, 345
Potomac Books, Inc., 184
Potter (Beatrix) Society, The, 565
Potter (Nicky), 278
Powell (Anthony) Society, 565
Power (Shelley) Literary Agency Ltd, 267
Powys Society, The, 565
Poyser (T&AD), 128
Pozzitive Television, 431
PPA The Association for Publishers and Providers of Consumer, Customer & Business Media in the UK, 543
PPL (Photo Agency) Ltd, 620
PR Week, 345
Practical Art Book of the Year, 516
Practical Boat Owner, 345
Practical Caravan, 345
Practical Fishkeeping, 346
Practical Land Rover World, 346
Practical Parenting, 346
Practical Photography, 346
Practical Wireless, 346
Practical Woodworking, 346
Practitioner, The, 346
Praxis Books, 234
Precious Pearl Press, 205
Prediction, 346
Preface Publishing, 128
Pregnancy & Birth, 347

Pregnancy, Baby & You, 347
Premaphotos Wildlife, 620
Premier Radio, 411
Premio Valle Inclán, The, 516
Premonitions, 215
Prentice Hall/Prentice Hall
   Business, 129
Presence, 215
Presidio Press, 184
Press and Journal, The
   (Aberdeen), 391
Press Association Library, 593
Press Association Ltd, The, 393
Press Complaints Commission,
   544
Press Gazette, 347
Press, (York) The, 388
Presses de la Cité, 164
Presses Universitaires de France
   (PUF), 164
Prestel Publishing Limited, 129
Price (Harry) Library of Magical
   Literature, 620
Price (Mathew) Ltd, 245
Price Stern Sloan, 184
Pride, 347
Priest (Aaron M.) Literary
   Agency, The, 288
Priestley (J.B.) Society, The, 565
Prima, 347
Primary Source Media, 184
Princeton Press (University
   of), 129
Princeton University Press, 184
Prion Books, 129
Pritchett (V.S.) Memorial Prize,
   516
Private Eye, 347
Private Libraries Association,
   544
Producers Alliance for Cinema
   and Television, 544
Professional Practices, 184
Professional Sport UK Ltd, 620
Profile Books, 129
Projection Box Essay Awards
   2008–2009, The, 516
Promenade Enterprises Limited,
   431
Prometheus Books, 184
ProQuest, 129
Prose & Poetry Prizes, 516
Prospect, 347
Protter (Susan Ann) Literary
   Agent, 288
PS Avalon, 205
Psychic News, 348
Psychologies, 348
Psychology Press, 129
Public History, 185
Public Lending Right, 544
Public Record Office, 593

Public Record Office Image
   Library, 621
PublicAffairs, 185
Publicat SA, 169
Publishers Association, The, 544
Publishers Licensing Society
   Ltd, 544
Publishers' Association of South
   Africa, 544
Publishers' Publicity Circle, 544
Publishing House, 129
Publishing News, 348
Publishing Scotland, 545
Publishing Services, 278
Puffin, 129
Puffin Audiobooks, 240
Puffin Books (USA), 185
Pugh (Sonia) PR, 278
Pulitzer Prizes, 516
Pulsar Poetry Magazine, 215
PULSE International Poetry
   Festival, 637
Punch, 621
Puppet State Press, 205
Purdue University, 471
Purple Patch, 215
Purple Patch Poetry
   Convention, 637
Pushkin Press Ltd, 129
Putnam's (G.P.) Sons
   (Children's), 185
PVA Management Limited, 267
Pyr™, 185

Q

Q, 348
Q-News, The Muslim Magazine,
   348
QED Publishing, 129
QQ Press, 205
Quadrille Publishing Ltd, 130
Quantum, 130
Quantum Leap, 215
Quarterly Review, 348
Quartet Books, 130
Quarto Publishing plc, 245
Quartos, 348
QUE Publishing, 130
Queen's English Society, The,
   566
Queen's Theatre, Hornchurch,
   446
Queen's University of Belfast,
   466
Queensland Press, University
   of, 192
QueenSpark Books, 234
Quercus Audiobooks, 240
Quercus Publishing, 130
Quicksilver Books, Literary
   Agents, 288
Quiet Feather, The, 215

Quiller Press, 130
Quince Books Limited, 130
QWF Magazine, 348

R

*RA Magazine (Royal Academy
   of Arts Magazine)*, 349
Rabén och Sjögren Bokförlag,
   171
Racecar Engineering, 349
Racing Collection (Newmarket),
   593
Racing Post, 349
Rack Press, 206
Radcliffe Press, 131
Radcliffe Publishing Ltd, 131
Radiator, The, 215
Radikal Phase Publishing House
   Ltd, 234
Radio Academy, The, 545
Radio Telefís Éireann (RTÉ One
   – RTÉ Two), 414
Radio Times, 349
Radio User, 349
Radio XL 1296 AM, 411
RadioCentre, 545
Ragged Raven Press, 206
RAI (RadioTelevisione Italiana),
   414
Rail, 349
Rail Express, 350
Railway Gazette International,
   350
Railway Magazine, The, 350
Railway Prism, The, 223
Rainbow Poetry News, 215
Rainbow Poetry Recitals, 566
Raines & Raines, 288
Raintree, 131
Rajpal, 196
Ramsay (Peggy) Foundation,
   485
Ramsay (Rosalind) Limited, 274
Randall (Sally) PR, 278
Random Acts of Writing, 350
Random House Audio Books,
   240
Random House Australia, 192
Random House Group Ltd,
   The, 131
Random House Mondadori, 170
Random House New Zealand,
   197
Random House of Canada Ltd,
   194
Random House Publishing
   Group, 185
Ransom Publishing Ltd, 132
Ransome (Arthur) Society Ltd,
   The, 566
Raunchland Publications, 206
Raupo Publishing (NZ) Ltd, 197

Rayo, 185
Razorbill, 185
Read the Music, 215
Reader's Digest, 350
Reader's Digest Association Inc, 185
Reader's Digest Association Ltd, 132, 395
Reader, The, 350
Readers' Review Magazine, 350
Readers' Union Ltd, 293
Reading Central Library, 593
Reading Evening Post, 379
Reading, University of, 453
Reaktion Books, 132
Real Writers, 454
Reality Street Editions, 206
Really Useful Group Ltd, The, 446
Reardon Publishing, 133
Reater, The, 216
Recollections, 133
Record Collector, 351
Recusant, The, 216
Red, 351
Red Arrow Books, 133
Red Candle Press, 206
Red Fox Children's Books, 133
Red House Children's Book Award, The, 516
Red Kite Books, 245
Red Ladder Theatre Company, 446
Red Poets –Y Beirdd Coch, 216
Red Shift Theatre Company, 447
Red Squirrel Publishing, 133
Redbridge Book and Media Festival, 638
Redferns Music Picture Library, 621
Redhammer Management Ltd, 267
Redweather, 431
Reece Halsey North Literary Agency, 288
Reed (Thomas), 133
Reed (William) Directories, 133
Reed Elsevier Group plc, 133
Rees (Helen) Literary Agency, 288
Reese (Trevor) Memorial Prize, 517
RefDesk.com, 646
Reflections, 216
Regency House Publishing Limited, 133
Regional Press Awards, 517
Remember When, 133
Renaissance Films, 431
Renault UK Journalist of the Year Award, 517
Report, 351

Report Digital, 621
Reportage Press, 134
Research Disclosure, 134
Retail Week, 351
Retna Pictures Ltd, 621
Reuters, 393
Reveal, 351
Rex Features Ltd, 621
Rialto Publications, 206
Rialto, The, 216
RIBA British Architectural Library Drawings & Archives Collection, 594
RIBA International Book Awards, 517
Richards (Lisa) Agency, The, 279
Richmond Central Reference Library, 594
Richmond Films & Television Ltd, 431
Richmond House Publishing Company Ltd, 134
Rickshaw Productions Limited, 240
RIDA Photolibrary, 621
Rider, 134
Ridiculusmus, 447
Rigby, 134
Right Way, 134
Rising Stars UK Ltd, 134
RITA Awards, 517
Rittenberg (Ann) Literary Agency, Inc., 288
Rive Gauche Publishing, 206
Riverhead Books, 185
Riverside Publishing Co., The, 185
Rizzoli Editore, 167
RMEP, 134
Robbins (B.J.) Literary Agency, 288
Robérts (L.M.) Limited, 240
Robinson, 134
Robinson Literary Agency Ltd, 267
Robinswood Press, 234
Robson Books, 134
Roc, 185
Rodale, 134
Roehampton University, 460
Rogers, Coleridge & White Ltd, 267
Roghaar (Linda) Literary Agency, LLC, 289
Romance Writers of America, 545
Romantic Novel of the Year, 517
Romantic Novelists' Association, The, 545
Romany Society, The, 566

Rosen Publishing Group, Inc., The, 185
Rosenberg Group, The, 289
Rossica Translation Prize, 517
RotoVision, 134
Roundhouse Group, 134
Roundyhouse, 216
Route Publishing, 206
Routledge, 134
Rowohlt Verlag GmbH, 165
Royal Air Force Museum, 621
Royal Collection Photographic Services, The, 621
Royal Court Theatre, 447
Royal Court Young Writers Programme, 638
Royal Dublin Society, 158
Royal Economic Society Prize, 518
Royal Exchange Theatre Company, 447
Royal Geographical Society, 594
Royal Geographical Society Picture Library, 622
Royal Holloway University of London, 463
Royal Institute of Philosophy, 594
Royal Irish Academy, 158
Royal Literary Fund, The, 486
Royal Lyceum Theatre Company, 447
Royal Mail Awards for Scottish Children's Books, 518
Royal National Theatre, 447
Royal Philharmonic Society Archive, The, 594
Royal Photographic Society Collection, 622
Royal Shakespeare Company, 447
Royal Society Library, 594
Royal Society of Literature, 545
Royal Society of Literature Awards, 518
Royal Society of Literature Ondaatje Prize, The, 518
Royal Society of Literature/ Jerwood Awards, The, 486
Royal Society of Medicine Library, 594
Royal Society Prizes for Science Books, The, 518
Royal Television Society, 546
RS Productions, 431
RSA (Royal Society for the Encouragement of Arts, Manufactures & Commerce), 594
RSPB Images, 622
RSPCA Photolibrary, 622

RTBF (Radio-Télévision Belge de la Communauté Française), 413
RTI SpA (Reti Televisive Italiane), 415
RTP 1 & RTP 2 (Radiotelevisão Portuguesa), 415
RTR (Radio e Televisiun Rumantscha), 416
RTS, 415
RTSI (Radiotelevisione svizzera di lingua Italiana), 416
Rubies In the Darkness, 216
Rugby World, 351
Runciman Award, 518
Runner's World, 351
Running Fitness, 351
Running Press, 185
Runnymede International Literature Festival, 638
Rupert, The Followers of, 566
Rushby-Smith (Uli) Literary Agency, 267
Ruskin Society of London, The, 567
Ruskin Society, The, 567
Russell & Volkening, Inc., 289
Russell (Eric Frank) Archive, 594
Russia and Eastern Images, 622
Rutgers University Press, 185
RYA (Royal Yachting Association), 135
Ryelands, 135
Ryland Peters & Small Limited, 135

S

S&S Libros eñ Espanol, 185
S4C, 408
Sable, 216
Sabras Radio, 411
Saffron Walden Literary Festival, 638
Saga Magazine, 352
Sagalyn Literary Agency, The, 289
Sage Publications, 135
Sailing Today, 352
Sainsbury's Baby Book Award, 519
Sainsbury's Magazine, 352
Saint Mary's College of California, 469
Salamander, 135
Salariya Book Company Ltd, 245
Salford, University of, 460
Salisbury Review, The, 352
Salt Publishing, 206
Saltire Literary Awards, The, 519
SAM Publishing, 135

Sammarco (Leda), 641
SAN Agency, 641
San Francisco State University, 470
Sandberg (Rosemary) Ltd, 267
Sanders (Peter) Photography Ltd, 622
Sanders (Victoria) & Associates LLC, 289
Sands Films, 431
Sandstone Press Ltd, 135
Saqi Books, 135
Sauerländer Verlage AG, 171
Saunders, 136
Saville (Malcolm) Society, The, 567
Savitri Books Ltd, 245
Sawford (Claire) PR, 278
Sayers (Dorothy L.) Society, The, 567
Sayle Literary Agency, The, 268
Sayle Screen Ltd, 268
SB Publications, 136
SBS TV A/S, 413
Scagnetti (Jack) Talent & Literary Agency, 289
Scala Productions Ltd, 431
Scar Tissue, 216
Scarborough Evening News, 388
Scarecrow Press Inc., 186
Sceptre, 136
Schiavone Literary Agency, Inc., 289
Schiller (Heather), 275
Schirmer Reference, 186
Schlegel–Tieck Prize, 519
Scholarly Resources, Inc., 186
Scholastic Australia Pty Limited, 192
Scholastic Book Clubs, 293
Scholastic Canada Ltd, 194
Scholastic Library Publishing, 186
Scholastic Ltd, 136
Schulman (Susan), A Literary Agency, 290
Schønbergske (Det) Forlag, 162
Science & Society Picture Library, 622
Science Fiction Foundation, 546
Science Fiction Foundation Research Library, 595
Science Museum, 622
Science Museum Library, 595
Science Photo Library, 622
Science Press, 192
Science Writer Awards, 519
SCM – Canterbury Press Ltd, 136
Scope Productions Ltd, 432
Scotland in Trust, 352
Scotland Means Business, 352

Scotland on Sunday, 376
Scotland's Book Town Festival, 638
Scots Magazine, The, 352
Scotsman, The, 376
Scottish Arts Council, 573
Scottish Arts Council Book of the Year Awards, 519
Scottish Arts Council Creative Scotland Awards, 486
Scottish Arts Council New Writers' Bursaries, 486
Scottish Arts Council Writers' Bursaries, 486
Scottish Book of the Year, 519
Scottish Book Trust, 546
Scottish Book Trust Mentoring Scheme, 486
Scottish Book Trust New Writers' Awards, 486
Scottish Cultural Press/Scottish Children's Press, 137
Scottish Daily Newspaper Society, 546
Scottish Drama Collection/ Scottish Poetry Collection, 595
Scottish Farmer, The, 352
Scottish Field, 353
Scottish History Book of the Year, 519
Scottish Home & Country, 353
Scottish Library Association, 546
Scottish Newspaper Publishers Association, 546
Scottish Pamphlet Poetry, 223
Scottish Poetry Library, 223
Scottish Print Employers Federation, 546
Scottish Publishers Association, 546
Scottish Screen, 547
Scouting Magazine, 353
Scovil Chichak Galen Literary Agency, Inc., 290
Screen, 353
Screen First Ltd, 432
Screen International, 353
Screen Ventures Ltd, 432
Screenhouse Productions Ltd, 432
Screentrade Magazine, 353
Screenwriters Online, 647
Scribbler!, 216
Scribblers, 246
Scribblers House® LLC Literary Agency, 290
Scribner's (Charles) Sons, 186
Scribner/Scribner Classics, 186
Scribo, 479
Script Factory, The, 460

Script Yorkshire, 479
Scriptor, 216
ScriptWriter Magazine, 354
Scunthorpe Telegraph, 383
Sea Breezes, 354
Seafarer Books Ltd, 137
Seaforth Publishing, 137
Seal Books (Canada), 194
SeaNeverDry Publishing, 137
Search Press Ltd, 137
Seattle Pacific University, 475
Seckford Collection, 595
Second Light, 223
Second Light Publications, 206
2nd Air Division Memorial
    Library, 595
Seix Barral (Editorial) SA, 170
Self Publishing Magazine, The,
    354
Senate House Library,
    University of London, 595
Sentinel, 186
Sentinel (Stoke on Trent), The,
    385
September Films Ltd, 432
Seren, 137, 206
Serif, 235
Serpent's Tail, 138
SES Book Prizes, 519
Seton Hill University, 474
Seuil (Éditions du), 164
Seven Stories Press, 186
7:84 Summer School, 466
7:84 Theatre Company
    Scotland, 448
Seventh Quarry, The, 216
Severn House Publishers, 138
SF DRS (Schweizer Fernsehen),
    416
SF Hub, The, 647
Shakespeare Appreciated, 241
Shambhala Publications, Inc.,
    186
Shared Experience, 448
Sharland Organisation Ltd, The,
    268
Sharp (Mick) Photography, 622
Sharpe (M.E.), Inc., 186
Shaw (Bernard) Translation
    Prize, 519
Shaw Society, The, 567
She Magazine, 354
Shearsman, 216
Shearsman Books, 206
Sheffield Hallam University, 465
Sheffield Libraries, Archives
    and Information, 595
Sheffield, University of, 465
Sheil Land Associates Ltd, 268
Sheldon (Caroline) Literary
    Agency Ltd, 268

Sheldon (Phil) Golf Picture
    Library, 623
Sheldon Press, 138
Sheldrake Press, 138
Shell Like, 432
Shepheard-Walwyn (Publishers)
    Ltd, 138
Sherman Cymru, 448
Shetland Library, 596
Shetland Times Ltd, The, 138
Shields Gazette (South Shields),
    386
Ships Monthly, 354
Shire Publications Ltd, 138
Shoe Lane Library, 596
Shoestring Press, 207
Shoot, 355
Shooting and Conservation, 355
Shooting Gazette, The, 355
*Shooting Times & Country
    Magazine*, 355
*ShOp: A Magazine Of Poetry,
    The,* 217
Short Books, 139
Short Story Writers' Folio/
    Children's Novel Writers'
    Folio, 479
Shots Magazine, 647
Shout Magazine, 355
Shout!, 355
Show of Strength, 448
Showing World, 355
Showrunner, 432
Shrewsbury Library and
    Reference & Information
    Service, 597
Shropshire Magazine, 355
Shropshire Star (Telford), 385
Shuter & Shooter (Pty) Ltd, 197
Sianco Cyf, 432
SIC (Sociedade Independente
    de Comunicacão), 415
Sidgwick & Jackson, 139
Siegel (Rosalie), International
    Literary Agency, Inc., 290
Sight & Sound, 355
Sigma Press, 139
Silent Sound Films Ltd, 432
Silver Link, 235
Simmonds (Dorie) Agency, 269
Simmons (Jeffrey), 269
Simon & Schuster (Australia)
    Pty Ltd, 192
Simon & Schuster Adult
    Publishing Group, 187
Simon & Schuster Audio, 241
Simon & Schuster Children's
    Publishing, 187
Simon & Schuster UK Limited,
    139
Simon (André) Memorial Fund
    Book Awards, 519

Simon (Richard Scott) Ltd, 269
Simon Pulse/Simon Spotlight,
    187
Sinclair-Stevenson, 269
Singing Dragon, 139
Sixties Press, 207
SJA British Sports Journalism
    Awards, 520
*Skier and Snowboarder
    Magazine, The,* 356
Skirt! Books, 187
Skishoot–Offshoot, 623
Skyline Productions, 433
Skymag, 356
Skyscan Photolibrary, The, 623
Sleeping Bear Press, 187
*Slightly Foxed: The Real Reader's
    Quarterly*, 356
Slim at Home, 356
Sluka (Petra), 275
Smallholder, 356
SmartLife International, 356
SmartPass Ltd, 241
SMG Productions & Ginger
    Productions, 433
Smith (Robert) Literary Agency
    Ltd, 269
Smith Gryphon Ltd, 139
Smiths Knoll, 217
Smoke, 217
Smokestack Books, 207
Smythe (Colin) Ltd, 139
Snell (Michael) Literary Agency,
    290
Snooker Scene, 357
Snookerimages (Eric Whitehead
    Photography), 623
Snowbooks Ltd, 139
Society for Editors and
    Proofreaders (SfEP), 547
Society for Promoting Christian
    Knowledge (SPCK), 140
Society for Technical
    Communications (STC), 547
Society for Theatre Research
    Annual Theatre Book Prize,
    The, 520
Society of Authors in Scotland,
    The, 547
Society of Authors, The, 547
Society of Children's Book
    Writers & Illustrators, 547
Society of Civil and Public
    Service Writers, 548
Society of Editors, 548
Society of Indexers, 548
Society of Medical Writers,
    The, 548
Society of Picture Researchers
    & Editors, 548
Society of Sussex Authors, 479

Society of Women Writers & Journalists, 548
Society of Young Publishers, 548
Società Editrice Internazionale – SEI, 167
Soho Theatre Company, 448
Solo Syndication Ltd, 393, 623
Somerset Life, 357
Somethin Else, 433
Sony Radio Academy Awards, 520
Sonzogno, 167
Sophia Books, 140
Sotheby's Picture Library, 623
Soundings Audio Books, 241
South, 217
South American Pictures, 623
South Bank Centre, London, 647
South Wales Argus, 391
South Wales Echo, 391
South Wales Evening Post, 391
South Yorkshire Sport, 393
Southern (Jane), 275
Southern California, University of, 470
Southern Daily Echo, The, 381
Southern Illinois University Press, 187
Southey Society (Robert), The, 567
Southport Writers' Circle Open Short Story Competition, 520
Southport Writers' Circle Poetry Competition, 520
Southwater, 140
Southwest Scriptwriters, 479
Southwold Festival, 638
Southwood Books, 140
Southword, 217
Souvenir Press Ltd, 140
Space Press News and Pictures, 393
Spacelink Books, 235
Spanish Institute Library, 597
SPAudiobooks, 241
SPCK, 140
Speak, 187
Speakeasy – Milton Keynes Writers' Group, 479
Spear's Wealth Management Survey, 357
Special Interest Model Books Ltd, 140
Speciality Food, 357
Specific Films, 433
Spectacular Diseases, 207
Spectator, The, 357
Spectra, 187
Spectrum, 357

Spectrum (Uitgeverij Het) BV, 168
Spectrum Literary Agency, 290
Spectrum Radio, 411
Speechmark Publishing Ltd, 140
Spellmount, 140
Sperling e Kupfer Editori SpA, 167
Sphere, 141
Sphinx, 217
Sphinx Theatre Company, 449
Spice Factory (UK) Ltd, 433
Spiel Unlimited, 223
Spire FM, 411
Spitzer (Philip G.) Literary Agency, Inc., 291
'Spoken' Image Ltd, 433
Spoken Word Publishing Association, 548
Sports Journalists' Association of Great Britain, 549
SportsBooks Limited, 141
Sportsman's Press, The, 141
Spread the Word, 480
Springboard, 217
Springboard Press, 187
Springer GmbH, 165
Springer Science+Business Media BV, 168
Springer-Verlag GmbH, 160
Springer-Verlag London Ltd, 141
Springhill, 141
Spruce, 141
Square Dog Radio LLP, 433
Square Peg, 141
SRG SSR idée suisse (Swiss Broadcasting Corp.), 416
St Andrews, University of, 467
St Bride Library, 594
St James Press, 185
St James Publishing, 234
St Martin's Press LLC, 186
St Pauls Publishing, 135
Stackpole Books, 187
Stadia, 141
Staffordshire Life, 357
Stage, 358
Stainer & Bell Ltd, 141
Stamp Lover, 358
Stamp Magazine, 358
Standaard Uitgeverij, 160
Standen Literary Agency, The, 269
Stanford University Press, 187
StAnza: Scotland's Poetry Festival, 638
Stapledon (Olaf), 597
Star, The (Sheffield), 388
Starstock, 623
Starz! – Film and Theatre Performing Arts, 464

State University of New York Press, 187
Stationery Office, The, 141
Staveacre (Tony) Productions, 433
Stay Still, 623
Steam Railway Magazine, 358
Steel (Elaine), 269
Stein (Abner), 270
Steinberg (Micheline) Associates, 270
Steiner (Rudolf) Press, 141
Stella Magazine, 358
Stenlake Publishing Limited, 235
Sterling Collection, 597
Sterling Publishing Co. Inc., 188
Stern (Lawrence) Fellowship, 487
Sterne (Laurence) Trust, The, 567
Stevenson (Robert Louis) Club, 567
Stewart (Shirley) Literary Agency, 270
Still Digital, 624
Still Pictures' Whole Earth Photolibrary, 624
Stimola Literary Studio, LLC, 291
Stimulus Books, 188
Stinging Fly Press, 207
Stinging Fly, The, 217
Stirling Film & TV Productions Limited, 433
STM, 549
Stockfile, 624
Stockscotland.com, 624
Stoker (Bram) Awards, 520
Stone (Sir John Benjamin), 624
Stone Flower Limited, 235
Stonecastle Graphics Ltd, 246
Story Wizard, 647
Story-Tellers, The, 141
Strad, The, 358
Straight Forward Film & Television Productions Ltd, 434
Straightline Publishing Ltd, 141
Strang (Jessica) Photo Library, 624
Stratford-upon-Avon Poetry Festival, 638
Strebor Books, 188
Strokestown International Poetry Competition, 520
Strokestown International Poetry Festival, 639
Struik Publishers (Pty) Ltd, 198
Student Magazine, The, 217
Studymates Limited, 142
Stuff, 358

Stuhlmann (Barbara W.) Author's Representative, 291
stv, 408
Such (Sarah) Literary Agency, 270
Suffolk and Norfolk Life, 358
Suffolk County Council – Suffolk Libraries, 597
Sugar Magazine, 359
Sugarco Edizioni Srl, 167
Suhrkamp Verlag, 165
Summersdale Publishers Ltd, 142
Sun, The, 376
Sunday Express, 376
Sunday Herald, 377
Sunday Life, 389
Sunday Mail (Glasgow*)*, 377
Sunday Mercury (Birmingham), 387
Sunday Mirror, 377
Sunday Post (Dundee), 377
Sunday Sport, 377
Sunday Sun (Newcastle upon Tyne), 386
Sunday Telegraph, The, 377
Sunday Times Oxford Literary Festival, The, 639
Sunday Times Writer of the Year Award, 521
Sunday Times Young Writer of the Year Award, 521
Sunday Times, The, 378
Sunderland City Library and Arts Centre, 597
Sunderland Echo, 386
Sunderland, University of, 463
Sundial Scottish Arts Council Book of the Year Awards, 521
Sunflower Literary Agency, 270
Sunrise Radio (Yorkshire), 411
Sunset + Vine Productions Ltd, 434
SuperBike Magazine, 359
Superscript, 235
Supplement, The, 217
Surrey Life, 359
Surtees (R.S.) Society, The, 568
Survivors' Poetry, 223
Susijn Agency Ltd, The, 270
Sussex Academic Press, 142
Sussex Life, 359
Sussex Playwrights' Club, 480
Sussex, University of, 463
Sutton, 142
SVT (Sveriges Television), 416
Swallow Press, 188
Swan Hill Press, 142
Swansea Central Library, 597
Swansea Sound 1170 MW/96.4 FM The Wave, 411

Swanwick – The Writers' Summer School, 454
Sweet & Maxwell Group, 142
Swimming Times Magazine, 359
Swindon Advertiser, 387
Swindon Festival of Literature, 639
Swiss Cottage Central Library, 597
Sylph Editions, 235
Syracuse University Press, 188

T

Table Ronde (Les Editions de la), 164
Table Top Productions, 434
Tablet, The, 359
TAFT Group, The, 188
Take a Break, 359
Take It Easy, 360
Talawa Theatre Company Ltd, 449
Talese (Nan A.), 188
Taliesin, 217
Talk 107, 412
talkbackTHAMES, 434
Talking Pen, 207
Talkingheads Production Ltd, 434
talkSPORT, 410
Tamarind Books, 142
Tammi Publishers, 162
Tandem TV & Film Ltd, 434
Tango Books Ltd, 142
Tarcher (Jeremy P.), 188
Tarquin Publications, 235
Tartarus Press, 235
Taschen GmbH, 165
Taschen UK, 143
Tasker (Joe) Collection, 624
Tate Etc, 360
Tate Images, 624
Tatler, 360
Tauris (I.B.) & Co. Ltd, 143
Tax Credits, 62–63
Taxation, 52–63
Taylor & Francis, 143
Taylor (Reginald) and Lord Fletcher Essay Prize, 521
Taylor Made Broadcasts Ltd, 434
Taylor University, 471
Teach Yourself, 143
Teacher, The, 360
Teal (Patricia) Literary Agency, 291
Tears in the Fence, 217
Tees Valley Arts, 549
Tegan (Katherine) Books, 188
Telegram, 143
Telegraph & Argus (Bradford), 388

Telegraph Books, 143
Telemagination Ltd, 434
Teletext Ltd, 408
Templar Poetry, 207
Templar Publishing, 143, 246
Temple Lodge Publishing Ltd, 144
Temple University Press, 188
Tempus, 144
Tennessee (University of) Press, 188
Tennyson Agency, The, 270
Tennyson Research Centre, 598
Tennyson Society, The, 568
10th Muse, 217
Tern Television Productions Ltd, 434
Testimony Films, 435
Texas Press, University of, 188
TG4 (Teilefís na Gaelige), 414
TGO (The Great Outdoors), 360
Thames and Hudson Ltd, 144
Tharpa Publications, 235
That's Life!, 360
Theakstons Old Peculier Harrogate Crime Writing Festival, 639
Theakstons Old Peculier Prize for the Crime Novel of the Year, 521
Theatre Absolute, 449
Theatre Museum, 624
Theatre Royal Plymouth & Drum Theatre Plymouth, 449
Theatre Royal Stratford East, 449
Theatre Royal Windsor, 449
Theatre Workshop Edinburgh, 450
Theatre Writers' Union, 549
Third Alternative, The, 360
Third Half, The, 217
Third Millennium Publishing, 144
Thirkell (Angela) Society, The, 568
This England, 360
Thoemmes, 144
Thomas (Bob) Sports Photography, 624
Thomas (Dylan) Festival, 639
Thomas (Dylan) Prize, The, 521
Thomas (Dylan) Society of Great Britain, The, 568
Thomas (Edward) Fellowship, The, 568
Thomson Gale, 188
Thomson Learning, 144
Thomson Round Hall (Ireland), 144
Thorndike Press, 189

Thornes (Stanley) (Publishers) Ltd, 144
Thoroughbred Photography Ltd, 624
Thorpe (F.A.) Publishing, 144
Thorsons, 144
Three Rivers Press, 189
Threshold Editions, 189
Thurrock Communities, Libraries & Cultural Services, 598
Tiden Norsk Forlag, 168
Tiderne Skifter Forlag A/S, 162
Tiger Aspect Productions, 435
Tigress Productions Ltd, 435
Time, 360, 395
Time Haiku, 218
Time Out, 361
Time Out Guides, 144
Times Books, 144
Times Books (USA), 189
Times Cheltenham Literature Festival, The, 639
Times Educational Supplement Scotland, The, 361
Times Educational Supplement, The, 361
Times Higher Education, The, 361
Times Literary Supplement, The, 361
Times, The, 378
Timewell Press, 144
Tindal Street Press Ltd, 236
Tinniswood Award, The, 522
Tir Na N-Og Award, The, 522
Titan Books, 145
Tlön Books Publishing Ltd, 236
Todariana Editrice, 167
Today's Golfer, 362
Tolkien Society, The, 568
Tom-Gallon Trust Award and the Olive Cook Prize, 487
Top of the Pops Magazine, 362
TopFoto, 624
Tor, 145
Torbay Weekend Festival of Poetry, 639
Torch Theatre, The, 450
Toronto Press, Inc., University of, 194
Total Coarse Fishing, 362
Total Film, 362
Total Flyfisher, 362
Total TV Guide, 362
Total Vauxhall, 362
Totterdell (B.M.) photography, 625
Toucan Books Ltd, 246
Touch Productions Ltd, 435
Touchstone, 189
Touchstone Books Ltd, 246

trAce Online Writing Centre, 647
Trade Publishers Council, 549
Traeger (Tessa) Picture Library, 625
Trail, 363
Transaction Publishers Ltd, 189
Transatlantic Films Production and Distribution Company, 435
Transita Ltd, 145
Translators Association Awards, The, 522
Translators Association, The, 549
Transworld Publishers Ltd, 145
Trask (Betty) Awards, The, 487
Travel Channel, 409
Travel Ink Photo Library, 625
Travel Publishing Ltd, 145
Traveller, 363
Travellers' Tales, 460
Travelling Scholarships, The, 487
Traverse Theatre, 450
Treimel (S©ott) NY, 291
Trenchard (Peter) Photography, 625
Trentham Books Ltd, 145
Trestle Theatre Company, 450
Trevor (Lavinia) Literary Agency, 271
Tribune, 363
Tricycle Theatre, 450
Trident Press Ltd, 145
TriQuarterly, 189
Trollope Society, The, 568
Tron Theatre Company, 450
Tropix Photo Library, 625
Trotman, 146
Troubador Publishing Ltd, 146
Trout Fisherman, 363
True North, 625
Truman Talley Books, 189
Trumpeter, 189
Truro Library, 598
TSO (The Stationery Office), 146
TSR (Télévision Suisse Romande), 416
Tuckwell Press, 146
Tundra Books, 194
Turnbull (Jane), 271
Tusquets Editores, 170
TV 2 Norge, 415
TV Choice Ltd, 435
TV Slovenija, 416
TV–2 Danmark, 413
TV3 (Viasat), 413
TV4, 416
TVC (Televisió de Catalunya S.A.), 416

TVE-1/ TVE-2 (Television Española), 416
TVI (Televisão Independente), 415
TVmyworld, 271
TVP (Telewizja Polska S.A.), 415
TVR (The Romanian Television Corporation), 415
TVTimes, 363
TVV (TVV-Televisio Valenciana), 416
Twayne Publishers, 189
Twelve, 189
Twentieth Century Fox Film Limited, 435
Twenty First Century Publishers Ltd, 146
20/20, 146
Twenty Twenty Television, 436
20x20 magazine, 363
Twenty-First Century Books, 189
Two Dogs, 198
Two Heads, 146
2M Communications Ltd, 291
Two Ravens Press Ltd, 146
Twofour Broadcast, 436
Ty Newydd Festival, 639
Ty Newydd Writers' Centre, 224, 467
Tyburn Film Productions Limited, 436
Tyndale House Publishers, Inc., 189
Tír Eolas, 158

U

U.P.P.A., 625
Ueberreuter (Verlag Carl) GmbH, 160
UK Children's Books Directory, 647
UK Film and TV Production Company Plc, The, 436
UK Film Council, 549
UK Public Libraries Page, The, 647
Ullstein Buchverlag GmbH, 165
Ulster Business, 364
Ulster Museum Picture Library, 625
Ulster Tatler, 364
Ultimate Advertiser, 364
Ulverscroft, 147
Uncut, 364
Understanding, 218
Unicorn Theatre, 451
Unieboek (Uitgeverij) BV, 168
United Agents Limited, 271
United Kingdom Copyright Bureau, The, 641

United Nations Environment Programme (UNEP), 625
Universal Pictorial Press & Agency (U.P.P.A.), 625
Universe, The, 364
Universitaire Pers Leuven, 160
University College Dublin Press, 158
University College Falmouth, 454
University Press of America, Inc., 189
University Presses of California, Columbia & Princeton Ltd, 147
UPI, 625
Upstairs at the Gatehouse, 451
Urchfont Manor College, 464
Usborne Publishing Ltd, 147
Ustinov (Sir Peter) Television Scriptwriting Award, 522
UTV, 408
UWA Press, 193
UXL, 189

V

V&A Images, 625
Vacation Work, 147
Value Added Tax, 57–61
Vance (Charles), 451
Vane Women Press, 207
Vanguard Press, 147, 189
Vegan, The, 364
Ventura, 147
Ver Poets, 480
Ver Poets Open Competition, 522
Vera Productions, 436
Verbo (Editorial) SA, 169
Veritas Publications, 159
Vermilion, 147
Verso, 147
Victor (Ed) Ltd, 271
Victoria University Press, 197
Victory Archive, 626
Video Enterprises, 436
Viking, 147
Viking/Viking Children's Books, 189
Villard, 189
Vin Mag Archive Ltd, 598, 626
Vintage, 147
Vintage Canada, 194
VIP Broadcasting, 436
Virago Press, 147
Virgin Books, 147
Virgin Radio, 410
Virginia Commonwealth University, 475
Virginia Press, University of, 189
Vision, 147, 189
Vision Books Pvt. Ltd, 196

Vision On, 148
Vital Spark, The, 148
Vlaamse Televisie Maatschappij (VTM), 413
Vogue, 364
Vogue Talent Contest, 523
Voice of America, 395
Voice of the Listener & Viewer Ltd (VLV), 549
Voice, The, 365
Volkswagen Golf+, 365
Vondel Translation Prize, The, 523
Voyage, 414
Voyager, 148
VPRO, 415
VRT (Vlaamse Radio- en Televisieomroep), 413
Vuibert (Librairie), 164

W

W.I. Life, 365
Wade & Doherty Literary Agency Ltd, 271
Wadsworth Prize for Business History, 523
Wainwright Society, 568
Wakefield Historical Publications, 236
Walch Education, 189
Wales (University of), Bangor, 467
Wales Literary Agency, Inc., 291
Wales on Sunday, 391
Wales Press, University of, 148
Walford Award, 523
Walk Magazine, 365
Walker & Co., 190
Walker Books Ltd, 148
Walker Large Print, 190
Wall Street Journal, 395
Wall to Wall, 436
Wallflower Press, 148
Wallpaper, 365
Walmsley (Jane) Productions, 436
Walmsley (John) Photo Library, 626
Walmsley Society, The, 569
Walsh Bros. Limited, 436
Walton Wordsmiths, 480
War Cry, The, 365
Ward Lock, 148
Ward Lock Educational Co. Ltd, 148
Ware (Cecily) Literary Agents, 272
Ware (John A.) Literary Agency, 291
Warehouse Theatre, 451
Warne (Frederick), 148, 190

Warner (Sylvia Townsend) Society, 569
Warwick Words, 640
Warwick, University of, 464
Warwickshire Photographic Survey, 626
Wasafiri, 218
Washington Post, 395
Washington Square Press, 190
Washington State University Press, 190
WATCH, 550
Waterloo Press, 207
Waterstone's Children's Book Prize, 523
Waterways Photo Library, 626
Waterways World, 365
Watford Palace Theatre, 451
Watling Street Publishing Ltd, 236
Watson, Little Ltd, 272
Watt (A.P.) Ltd, 272
Watt (David) Prize, The, 523
Watts (Franklin), 148
Watts, Franklin (US), 190
Way (Brian) Award, 523
Wayland, 148
Ways with Words, 640
Waywiser Press, 207, 236
We Find It (Press Clippings), 628
Weatherhill, 190
Web User, 366
Webb (Harri) Prize, The, 523
Webb (Mary) Society, 569
Webb (Wendy) Books, 207
Webster Dictionary/Thesaurus, 647
Wedding, 366
Weekly News, 366
Weidenfeld & Nicolson, 148
Weil (Wendy) Agency, Inc., The, 291
Weiland (Paul) Film Company, 437
Weimar Archive, 626
Weinberger (Josef) Plays, 272
Weiner (Cherry) Literary Agency, 292
WEKA Holding GmbH & Co. KG, 166
Welch (John), Literary Consultant & Agent, 272
Wellcome Images, 626
Wellington Literary Festival, 640
Wellness Central, 190
Wells (H.G.) Society, 569
Wells Festival of Literature, 640
Welsh Academy, 550
Welsh Book of the Year Award, 524

Welsh Books Council (Cyngor Llyfrau Cymru), 550
Welsh Country Magazine, 366
Welsh Literature Abroad/ Llenyddiaeth Cymru Dramor, 550
Welsh National Literature Promotion Agency, 550
West (David) Children's Books, 246
West Cornwall Photographic Archive, 626
West Country Writers' Association, 550
West House Books, 207
West Midlands Literary Heritage Collection, 598
West Yorkshire Playhouse, 451
Western Daily Press (Bristol), 385
Western Isles Libraries, 598
Western Mail, The (Cardiff), 391
Western Michigan University, 472
Western Morning News, 380
Western Writers' Centre, The, 224
Westminster (City of) Archives Centre, 598
Westminster Music Library, 598
Westminster Reference Library, 598, 626
Westminster, University of, 460
Westview, 190
Weyr (Rhoda) Agency, 292
Wharncliffe Books, 148
Wharton, 149
What Car?, 366
What Hi-Fi? Sound & Vision, 366
What Investment, 366
What Satellite and Digital TV, 367
What's New in Building, 367
Wheatley Medal, The, 524
Wheeler Publishing, 190
Which Caravan, 367
Which? Books, 149
Whirligig Theatre, 451
Whistledown Productions Ltd, 437
Whitbread Book Awards, 524
Whitchurch Books Ltd, 236
White (Eve) Literary Agent, 272
White Bear Theatre Club, 451
White Ladder, 149
Whitehead (Eric) Photography, 626
Whitfield Prize, 524
Whiting (John) Award, 524
Whittet Books Ltd, 149
Whittles Publishing, 236

Wiener (Dinah) Ltd, 272
Wiener Library, The, 599
Wighton Collection of National Music, The, 599
Wigtown Book Festival, 640
Wild Goose Publications, 149
Wild Images, 626
Wild Places Photography, 626
Wild Times, 367
Wild Women Press, 207
Wilde (Oscar) Society, The, 569
Wiley (John) & Sons Australia Ltd, 193
Wiley (John) & Sons Canada Ltd, 194
Wiley (John) & Sons, Inc., 190
Wiley Europe Ltd, 149
Wiley Nautical, 149
Wiley-Blackwell, 149
Wilkins Memorial Poetry Prize, 524
William Ernest, 236
Williams (Bridget) Books Ltd, 197
Williams (Charles) Society, The, 569
Williams (David) Picture Library, 626
Williams (Jonathan) Literary Agency, 279
Williams (Vaughan) Memorial Library, 599, 626
Williams's (Dr) Library, 599
Williamson (Henry) Society, The, 569
Willow Bank Publishers Ltd, 236
Wilson (Neil) Publishing Ltd, 149
Wilson (Philip) Publishers Ltd, 150
Wilson Photographic Collection, The, 626
Wilson's (John) Writing Courses, 467
Wimbledon Publishing Company, 150
Winchester (University of), 456
Winchester Writers' Conference Prizes, 524
Winchester Writers' Conference, Festival, Bookfair & Week-long Writing Workshops, 456
Windgather Press, 150
Windsor Theatre Royal, 452
Wine & Spirit, 367
Wingate (H.H.) Prize, 524
Wingbeat, 367
Wingedchariot Press, 150
Wisconsin Press, The University of, 190

Wisden (Cricket) Book of the Year, 524
Wisden Cricketer, The, 367
Wise Buddah Creative Ltd, 437
Wise Publications, 150
WIT Press, 150
Witan Books, 237
Wits University Press, 198
Witzend Productions, 437
Wizard Books, 150
Wodehouse Society (P.G.), The, 570
Wolff (Oswald) Books, 150
Wolff (Peter) Theatre Trust supports the Whiting Award, The, 524
Wolfhound Press, 159
Wolfson History Prize, 525
Wolverhampton Central Library, 599
Woman, 368
Woman and Home, 368
Woman's Own, 368
Woman's Weekly, 368
Woman's Weekly Fiction Special, 368
Women & Golf, 368
Women in Publishing, 550
Women Writers Network (WWN), 550
Women's Library, The, 599
Women's Press, 194
Women's Press, The, 150
Wonderful Words, 640
Wong (David T.K.) Fellowship, The, 487
Woodfall Wild Images, 627
Woodforde (Parson) Society, The, 570
Woodhead Publishing Ltd, 150
Woodworker, The, 368
Woolf (Virginia) Society of Great Britain, The, 570
Worcester News, 388
Worcestershire Libraries and Information Service, 600
Word – University of Aberdeen Writers Festival, 640
Word Market, 640
Word Pool, The, 648
WordCounter, 648
Words by the Water, 641
Wordsmith, 218
Wordsong, 190
Wordsworth Editions Ltd, 151
Wordsworth Trust (Jerwood Centre), The, 600
Wordsworth Trust, The, 570, 641
Workers' Educational Association, 480

Workers' Educational Association North Wales, 480
Workers' Educational Association Northern Ireland, 480
Workers' Educational Association Scotland, 480
Workers' Educational Association South Wales, 480
Working Partners Ltd, 246
Working Title Films Ltd, 437
World Fishing, 369
World Illustrated, 627
World of Interiors, The, 369
World Pictures, 627
World Ski Guide, 369
World Soccer, 369
Worple Press, 207, 237
Wortman Productions UK, 437
WPC Education, 151
WrightBooks, 193
Write Coach, The, 453
Write4kids.com, 648
Writenow!, 641
Writer's Academy, The, 466
Writernet, 550, 648
Writers and Photographers unLimited, 551
Writers Bureau College of Journalism, 461
Writers Bureau Poetry and Short Story Competition, The, 525
Writers Bureau, The, 461
Writers College, The, 461
Writers House, LLC, 292
Writers in Oxford, 481
writers inc, 224, 551
Writers Nexus International, 648
Writers' Circles, 648
Writers' Forum incorporating *World Wide Writers*, 369
Writers' Forum Short Story & Poetry Competitions, 525
Writers' Guild of Great Britain, The, 551
Writers' Holiday at Caerleon, 468

Writers' News/Writing Magazine, 369
Writers, Artists and their Copyright Holders (WATCH), 551
Writers-of-the-Year Competition, 525
WritersNet, 648
WritersServices, 648
Writing Centre, The, 648
Writing on the Wall, 641
Writing School Leicester, 457
Writing-World.com, 648
WSOY (Werner Söderström Osakeyhtiö), 162
WW2 HMSO PPBKS Society, 570
www.ABCtales.com, 247
Wye Valley Arts Centre, 456
Wylie Agency (UK) Ltd, The, 273
Wyndham (John) Archive, 600
Wyoming, University of, 476
Wählströms (B.) Bokförlag AB, 171

X

X Press, The, 151

Y

Y Lolfa Cyf, 151
Yachting Journalists' Association, 551
Yachting Monthly, 369
Yachting World, 370
Yale Drama Series/David C. Horn Prize, The, 525
Yale University Press (London), 151
Yeats Society Sligo, The, 570
Yellow Jersey Press, 151
Yemen Pictures, 627
Yen Press, 190
Yeovil Literary Prize, 525
YLE FST, 414
YLE TV1, 414
YLE TV2, 414
Yonge Fellowship (Charlotte M.), The, 570
York Archaeological Trust Picture Library, 627

York Central Library, 600
York Notes, 151
Yorkshire Art Circus, 466
Yorkshire Dialect Society, 571
Yorkshire Evening Post, 389
Yorkshire Playwrights, 481
Yorkshire Post, 389
Yorkshire Post Book Awards, 525
Yorkshire Women's Life Magazine, 370
You & Your Wedding, 370
You – The Mail on Sunday Magazine, 370
Young (Francis Brett) Society, 571
Young (John Robert) Collection, 627
Young Picador, 151
Young Vic, The, 452
Young Voices, 370
Young Writer, 370
YoungMinds Book Award, 525
Your Cat Magazine, 370
Your Dog Magazine, 370
Your Hair, 371
Your Horse, 371
Yours Magazine, 371
Yr Academi Gymreig, 551

Z

Zambezi Publishing Ltd, 151
Zannier-Betts (Sylvie), 275
ZDF (Zweites Deutsches Fernsehen), 414
Zebra, 198
Zeckendorf (Susan) Associates, Inc., 292
Zed Books Ltd, 151
Zed20, 218
Zeppotron, 437
Zero to Ten, 152
Zest, 371
Zondervan, 190
ZOO, 371
Zoological Society Library, 600
Zsolnay (Paul), Verlag GmbH, 160
Zweig Collection of Music & Literary Mss, The, 600
Zymurgy Publishing, 237

Subject Index

Academic: Literary Agents (UK)
Kelly (Frances), 262

Academic: Magazines
New Scientist, 340

Academic: Prizes
Gladstone History Book Prize, The, 502
Scottish History Book of the Year, 519
Taylor (Reginald) and Lord Fletcher Essay Prize, 521
Whitfield Prize, 524

Academic: Professional Associations
Association of Learned and Professional Society Publishers, 529
International Association of Scientific, Technical & Medical Publishers, 539

Academic: Publishers (European)
Akadémiai Kiadó, 166
Blackwell Munksgaard, 161
Brepols Publishers NV, 160
Gradiva–Publicações S.A., 169
Hanser (Carl) Verlag GmbH & Co. KG, 165
Hüthig GmbH & Co. KG, 165
Nyt Nordisk Forlag Arnold Busck A/S, 162
Presses Universitaires de France (PUF), 164
Springer GmbH, 165
Springer-Verlag GmbH, 160
Universitaire Pers Leuven, 160

Academic: Publishers (International)
Allen & Unwin Pty Ltd, 191
Auckland University Press, 196
Canadian Scholars' Press, Inc, 193

Canterbury University Press, 196
Kwa-Zulu–Natal Press, University of, 197
Macmillan Publishers Australia Pty Ltd, 191
Macmillan Publishers New Zealand, 196
Melbourne University Publishing Ltd, 192
Munshiram Manoharlal Publishers Pvt Ltd, 195
Oxford University Press (Australia, New Zealand & the Pacific), 192
Oxford University Press India, 196
Oxford University Press Southern Africa (Pty) Ltd, 197
Oxford University Press, Canada, 194
Queensland Press, University of, 192
Toronto Press, Inc., University of, 194
UWA Press, 193
Victoria University Press, 197
Wits University Press, 198

Academic: Publishers (Ireland)
Cork University Press, 155
Four Courts Press Ltd, 155
Institute of Public Administration, 156
Irish Academic Press Ltd, 156
National Library of Ireland, 157
On Stream Publications Ltd, 157
Royal Irish Academy, 158
University College Dublin Press, 158

Academic: Publishers (UK)
ABC-CLIO (UK), 64
Acumen Publishing Limited, 64
Ashgate Publishing Ltd, 69
Authentic Media, 70
AuthorHouse UK, 70
Berg Publishers, 72

Berghahn Books, 72
Bodleian Library, 75
Cambridge University Press, 79
Carcanet Press Ltd, 80
Cardiff Academic Press, 81
CBA Publishing, 82
Clarke (James) & Co., 84
Continuum International Publishing Group Ltd, The, 85
Crown House Publishing, 87
Currey (James) Publishers, 87
Discovered Authors, 89
Donald (John) Publishers Ltd, 89
Duckworth (Gerald) & Co. Ltd, 90
Dunedin Academic Press Ltd, 90
Edinburgh University Press, 91
Elsevier Ltd, 92
Exeter Press, University of, 93
Freeman (W. H.), 96
Garnet Publishing Ltd, 97
Geological Society Publishing House, The, 97
Gloucester Publishers Plc, 98
Hammersmith Press Ltd, 101
Harvard University Press, 103
Helm Information Ltd, 104
Hertfordshire Press, University of, 104
Imprint Academic, 106
Kingsley (Jessica) Publishers Ltd, 109
Liverpool University Press, 112
Macmillan Publishers Ltd, 113
Manchester University Press, 115
McGraw-Hill Education, 113
NMS Enterprises Limited – Publishing, 120
Norton (W.W.) & Company Ltd, 120
Open University Press, 123
Oxbow Books, 124
Oxford University Press, 124
Penguin Group (UK), 125
Phaidon Press Limited, 126
Pluto Press Ltd, 127

Portland Press Ltd, 128
Routledge, 134
Sage Publications, 135
Saqi Books, 135
Society for Promoting Christian Knowledge (SPCK), 140
Souvenir Press Ltd, 140
Springer-Verlag London Ltd, 141
Studymates Limited, 142
Sussex Academic Press, 142
Sweet & Maxwell Group, 142
Tauris (I.B.) & Co. Ltd, 143
Taylor & Francis, 143
Trentham Books Ltd, 145
Troubador Publishing Ltd, 146
University Presses of California, Columbia & Princeton Ltd, 147
Wales Press, University of, 148
Wallflower Press, 148
Wimbledon Publishing Company, 150
WIT Press, 150
Yale University Press (London), 151

Academic: Publishers (US)
ABC–CLIO (US), 172
Abingdon Press, 172
Alabama Press, University of, 172
Arizona Press, University of, 173
Arkansas Press, University of, 173
California Press, University of, 174
Chicago Press, The University of, 174
Columbia University Press, 175
Harvard University Press, 177
Hawai'i Press, University of, 177
Illinois Press, University of, 178
Indiana University Press, 179
Iowa Press, University of, 179
Kansas, University Press of, 179
Kent State University Press, The, 179
Kentucky, The University Press of, 179
Louisiana State University Press, 180
Massachusetts Press, University of, 181
McFarland & Company, Inc., Publishers, 180
McGraw-Hill Companies Inc., The, 180
Michigan Press, The University of, 181
Minnesota Press, University of, 181

Mississippi, University Press of, 181
Missouri Press, University of, 181
Nebraska Press, University of, 182
Nevada Press, University of, 182
New England, University Press of, 182
New Mexico Press, University of, 182
Northwestern University Press, 182
Ohio University Press, 182
Oklahoma Press, University of, 183
Paragon House, 183
Pennsylvania Press, University of, 184
Perseus Books Group, 184
Princeton University Press, 184
Rutgers University Press, 185
Scarecrow Press Inc., 186
Sharpe (M.E.), Inc., 186
Southern Illinois University Press, 187
St Martin's Press LLC, 186
Stanford University Press, 187
State University of New York Press, 187
Syracuse University Press, 188
Temple University Press, 188
Tennessee (University of) Press, 188
Texas Press, University of, 188
Transaction Publishers Ltd, 189
University Press of America, Inc., 189
Virginia Press, University of, 189
Wiley (John) & Sons, Inc., 190
Wisconsin Press, The University of, 190

Academic: Small Presses
Eilish Press, 228
Packard Publishing Limited, 233

Academic: UK Packagers
Diagram Visual Information Ltd, 243
Erskine Press, 243

Accountancy/Taxation: Magazines
Accountancy, 294
Accountancy Age, 294

Accountancy/Taxation: Publishers (European)
Orell Füssli Verlag, 171

Accountancy/Taxation: Publishers (International)
LexisNexis Butterworths (Pty) Ltd South Africa, 197
Vision Books Pvt. Ltd, 196

Accountancy/Taxation: Publishers (Ireland)
Blackhall Publishing Ltd, 153

Accountancy/Taxation: Publishers (UK)
McGraw-Hill Education, 113

Adventure & Exploration: Film, TV and Radio Producers
Lucida Productions, 428

Adventure & Exploration: Literary Agents (UK)
Feldstein Agency, The, 256
London Independent Books, 263

Adventure & Exploration: Literary Agents (US)
Gusay (Charlotte) Literary Agency, The, 284
Lampack (Peter) Agency, Inc., 285
Russell & Volkening, Inc., 289

Adventure & Exploration: Magazines
Descent, 314

Adventure & Exploration: Picture Libraries
Cleare (John)/Mountain Camera, 605
Royal Geographical Society Picture Library, 622

Adventure & Exploration: Prizes
Crime Writers' Association (The Ian Fleming Steel Dagger), 497

Adventure & Exploration: Publishers (European)
Flammarion Éditions, 163
Longanesi (Casa Editrice) SpA, 167

Adventure & Exploration: Publishers (International)
Allen & Unwin Pty Ltd, 191

Adventure & Exploration: Publishers (UK)
Birlinn Ltd, 73
Random House Group Ltd, The, 131
Sigma Press, 139

Adventure & Exploration: UK Packagers
Erskine Press, 243

Advertising: Film, TV and Radio Producers
Art & Training Films Ltd, 418
Channel Television Ltd, 420
Creative Partnership, The, 421
Firehouse Productions, 423
Good Film Company Ltd, The, 424
Mendoza Film Productions, 429
Passion Pictures, 430
Shell Like, 432
Weiland (Paul) Film Company, 437
Wise Buddah Creative Ltd, 437

Advertising: Magazines
Campaign, 305

Advertising: Picture Libraries
Advertising Archive Limited, The, 601
Beamish, The Photographic Library, 603

Advertising: Publishers (UK)
RotoVision, 134

Africa: Library Services
British Library Asia, Pacific and Africa Collections, 577

Africa: Literary Agents (UK)
Eady (Toby) Associates Ltd, 256

Africa: Literary Agents (US)
Gusay (Charlotte) Literary Agency, The, 284
Russell & Volkening, Inc., 289

Africa: Picture Libraries
Angel (Heather)/Natural Visions, 602
Australia Pictures, 602
British Library Images Online, 604
Cordaiy (Sylvia) Photo Library, 606
Cunningham (Sue) Photographic, 606
ffotograff, 608

Geo Aerial Photography, 610
Geoslides Photography, 610
Howes (Chris)/Wild Places Photography, 612
Images of Africa Photobank, 613
Link Picture Library, 615
Peerless (Ann & Bury) Picture Library, 619
Sanders (Peter) Photography Ltd, 622
Tropix Photo Library, 625
Yemen Pictures, 627

Africa: Poetry Magazines
Sable, 216

Africa: Prizes
Caine Prize for African Writing, The, 494
Macmillan Writer's Prize for Africa, 510
Noma Award for Publishing Africa, The, 514

Africa: Publishers (UK)
African Books Collective, 64
Currey (James) Publishers, 87
Edinburgh University Press, 91
Evans Brothers Ltd, 93

Africa: Publishers (US)
Indiana University Press, 179
Perseus Books Group, 184

Africa: Small Presses
Serif, 235

Agriculture & Farming: Literary Agents (US)
Gusay (Charlotte) Literary Agency, The, 284

Agriculture & Farming: Magazines
Country Smallholding, 311
Farmers Weekly, 319
Scottish Farmer, The, 352
Smallholder, 356

Agriculture & Farming: Picture Libraries
Beamish, The Photographic Library, 603
Ecoscene, 607
Garden and Wildlife Matters Photo Library, 609
Hutchison Picture Library, 612
Last Resort Picture Library, 614

Agriculture & Farming: Professional Associations
Guild of Agricultural Journalists, 537

Agriculture & Farming: Publishers (Ireland)
Royal Dublin Society, 158

Agriculture & Farming: Publishers (UK)
Crowood Press Ltd, The, 87
Manson Publishing Ltd, 115
Nottingham University Press, 120

Agriculture & Farming: Publishers (US)
Blackwell Publishing (US), 174

Agriculture & Farming: Small Presses
Packard Publishing Limited, 233Packard Publishing Limited,

Animal Care: Literary Agents (US)
Gusay (Charlotte) Literary Agency, The, 284
Snell (Michael) Literary Agency, 290

Animal Care: Magazines
Animal Action, 296
Animals and You, 296
Cat World, 306
Dog World, 316
Dogs Monthly, 316

Animal Care: Picture Libraries
RSPCA Photolibrary, 622

Animal Care: Publishers (UK)
Souvenir Press Ltd, 140

Animal Care: Small Presses
Parapress, 233

Animal Rights: Literary Agents (US)
Gusay (Charlotte) Literary Agency, The, 284

Animation: Film, TV and Radio Producers
Aardman, 417
Baby Cow Productions, 418
Blackwatch Productions Limited, 419

Calon TV, 419
Cartwn Cymru, 420
Collingwood O'Hare
    Entertainment Ltd, 421
Cosgrove Hall Films, 421
Fast Films, 423
Melendez Films, 428
Screen First Ltd, 432
Telemagination Ltd, 434

Animation: Literary Agents (UK)
Hoskins (Valerie) Associates
    Limited, 260
Steinberg (Micheline)
    Associates, 270

Animation: Miscellany
Landeau (Caroline), 641

Animation: Professional Associations
Design and Artists Copyright
    Society (DACS), 536
Pact, 542

Anthropology: Literary Agents (US)
Gusay (Charlotte) Literary
    Agency, The, 284
Russell & Volkening, Inc., 289

Anthropology: Picture Libraries
Natural History Museum Image
    Resources, 618
Oxford Scientific (OSF), 619
Poignant (Axel) Archive, 620
Royal Geographical Society
    Picture Library, 622

Anthropology: Publishers (European)
Adelphi Edizioni SpA, 166
Beck (C.H., OHG) Verlag, 164
Universitaire Pers Leuven, 160

Anthropology: Publishers (International)
Munshiram Manoharlal
    Publishers Pvt Ltd, 195

Anthropology: Publishers (UK)
Boyars (Marion) Publishers
    Ltd, 76
Currey (James) Publishers, 87
Harvard University Press, 103
Hurst Publishers, Ltd, 106
Polity Press, 128
Routledge, 134

Anthropology: Publishers (US)
Alabama Press, University of,
    172
Alaska Press, University of, 172
Indiana University Press, 179
Minnesota Press, University
    of, 181
Oklahoma Press, University
    of, 183
Texas Press, University of, 188

Antiquarian: Picture Libraries
Christie's Images Ltd, 605

Antiquarian: Prizes
Taylor (Reginald) and Lord
    Fletcher Essay Prize, 521

Antiquarian: Publishers (UK)
Souvenir Press Ltd, 140

Antiques: Library Services
Brighton Jubilee Library, 576

Antiques: Literary Agents (UK)
Chris (Teresa) Literary Agency
    Ltd, 253
Limelight Management, 262

Antiques: Magazines
Antiquesnews, 296
Apollo, 296
Homes & Antiques, 327

Antiques: PR Consultants
Frazer (Cathy) PR, 277

Antiques: Picture Libraries
Bridgeman Art Library, The,
    604

Antiques: Publishers (UK)
Antique Collectors' Club, 67
Octopus Publishing Group, 121
Remember When, 133
Souvenir Press Ltd, 140
Wilson (Philip) Publishers Ltd,
    150

Archaeology: Library Services
Derby Central Library, 581

Archaeology: Literary Agents (UK)
Davidson (Caroline) Literary
    Agency, 255
McAra (Duncan), 264

Archaeology: Literary Agents (US)
Braun (Barbara) Associates,
    Inc., 281
Gusay (Charlotte) Literary
    Agency, The, 284

Archaeology: Magazines
Art Newspaper, The, 297
BBC History Magazine, 299
History Today, 327

Archaeology: Picture Libraries
Adkins (Lesley & Roy) Picture
    Library, 601
akg-images Ltd, Arts and
    History Picture Library, 601
English Heritage Photo Library,
    607
ffotograff, 608
Lessing (Erich) Archive of Fine
    Art & Culture, The, 614
MacQuitty International
    Photographic Collection, 616
National Monuments Record,
    617
Sharp (Mick) Photography, 622
South American Pictures, 623
Ulster Museum Picture Library,
    625
York Archaeological Trust
    Picture Library, 627

Archaeology: Prizes
Taylor (Reginald) and Lord
    Fletcher Essay Prize, 521

Archaeology: Professional Associations
Council for British Archaeology,
    535

Archaeology: Publishers (European)
Adelphi Edizioni SpA, 166
Beck (C.H., OHG) Verlag, 164
Brepols Publishers NV, 160
Giunti Editoriale SpA, 166
Longanesi (Casa Editrice) SpA,
    167
Universitaire Pers Leuven, 160

Archaeology: Publishers (International)
Melbourne University
    Publishing Ltd, 192
Munshiram Manoharlal
    Publishers Pvt Ltd, 195

Archaeology: Publishers (Ireland)
Tír Eolas, 158

Archaeology: Publishers (UK)
British Academy, The, 77
British Museum Press, The, 78
CBA Publishing, 82
Donald (John) Publishers Ltd, 89
English Heritage (Publishing), 92
Exeter Press, University of, 93
NMS Enterprises Limited – Publishing, 120
Oxbow Books, 124
Polity Press, 128
Routledge, 134
Scottish Cultural Press/Scottish Children's Press, 137
Souvenir Press Ltd, 140
Thames and Hudson Ltd, 144
Trident Press Ltd, 145

Archaeology: Publishers (US)
Alabama Press, University of, 172

Architecture: Library Services
British Architectural Library, 576
London Library, The, 589
Westminster Reference Library, 598

Architecture: Literary Agents (UK)
Davidson (Caroline) Literary Agency, 255
McAra (Duncan), 264

Architecture: Literary Agents (US)
Braun (Barbara) Associates, Inc., 281
Gusay (Charlotte) Literary Agency, The, 284
Russell & Volkening, Inc., 289

Architecture: Magazines
Apollo, 296
Architects' Journal, The, 296
Architectural Design, 296
Architectural Review, The, 296
Blueprint, 302
Build It, 304
Building Magazine, 304
Country Life, 311

RA Magazine (Royal Academy of Arts Magazine), 349
Wallpaper, 365

Architecture: PR Consultants
Sawford (Claire) PR, 278

Architecture: Picture Libraries
Architectural Association Photo Library, 602
Cordaiy (Sylvia) Photo Library, 606
E&E Picture Library, 607
Edifice, 607
English Heritage Photo Library, 607
ffotograff, 608
Harding (Robert) World Imagery, 611
Heseltine (John) Archive, 611
Last Resort Picture Library, 614
National Monuments Record, 617
National Trust Photo Library, The, 618
Oxford Picture Library, 619
Peerless (Ann & Bury) Picture Library, 619
Royal Geographical Society Picture Library, 622
Strang (Jessica) Photo Library, 624

Architecture: Prizes
Fletcher (Sir Banister) Award, 500
RIBA International Book Awards, 517

Architecture: Professional Associations
Design and Artists Copyright Society (DACS), 536

Architecture: Publishers (European)
Adelphi Edizioni SpA, 166
Brepols Publishers NV, 160
Brána a.s., 161
Flammarion Éditions, 163
Lannoo, Uitgeverij, 160
Nyt Nordisk Forlag Arnold Busck A/S, 162
Orell Füssli Verlag, 171
Springer GmbH, 165
Springer-Verlag GmbH, 160
Taschen GmbH, 165
Universitaire Pers Leuven, 160

Architecture: Publishers (International)
Melbourne University Publishing Ltd, 192
Munshiram Manoharlal Publishers Pvt Ltd, 195
Oxford University Press India, 196
Queensland Press, University of, 192

Architecture: Publishers (UK)
Anova Books, 66
Antique Collectors' Club, 67
de la Mare (Giles) Publishers Ltd, 88
Donald (John) Publishers Ltd, 89
Donhead Publishing Ltd, 89
English Heritage (Publishing), 92
Garnet Publishing Ltd, 97
GMC Publications Ltd, 98
King (Laurence) Publishing Ltd, 109
Lincoln (Frances) Ltd, 110
Merrell Publishers Ltd, 116
Octopus Publishing Group, 121
Phaidon Press Limited, 126
Prestel Publishing Limited, 129
Quiller Press, 130
Reaktion Books, 132
Taschen UK, 143
Tauris (I.B.) & Co. Ltd, 143
Taylor & Francis, 143
Thames and Hudson Ltd, 144
WIT Press, 150

Architecture: Publishers (US)
Mississippi, University Press of, 181
Pelican Publishing Company, 183
Texas Press, University of, 188
Virginia Press, University of, 189

Architecture: UK Packagers
Cameron & Hollis, 243

Art: Book Clubs
Artists' Choice, 293

Art: Library Services
Armitt Collection, Museum & Library, 574
Brighton Jubilee Library, 576
Bristol Central Library, 576
Leeds Central Library, 587
London Library, The, 589

Westminster Reference Library, 598

Ulster Museum Picture Library, 625

Art: Literary Agents (UK)
Davidson (Caroline) Literary Agency, 255
Kelly (Frances), 262
McAra (Duncan), 264

Art: Literary Agents (US)
Gusay (Charlotte) Literary Agency, The, 284
Russell & Volkening, Inc., 289

Art: Magazines
Antiquesnews, 296
Apollo, 296
Art Monthly, 297
Art Newspaper, The, 297
Art Review, 297
Artist, The, 297
Edinburgh Review, 317
RA Magazine (Royal Academy of Arts Magazine), 349
Tate Etc, 360

Art: Miscellany
Landeau (Caroline), 641

Art: PR Consultants
Sawford (Claire) PR, 278

Art: Picture Libraries
akg-images Ltd, Arts and History Picture Library, 601
Art Archive, The, 602
Aspect Picture Library Ltd, 602
Bridgeman Art Library, The, 604
Christie's Images Ltd, 605
Corbis Images, 606
Evans (Mary) Picture Library, 607
Fine Art Photographic Library Ltd, 608
Lessing (Erich) Archive of Fine Art & Culture, The, 614
Miller (Lee) Archives, 616
National Galleries of Scotland Picture Library, 617
National Portrait Gallery Picture Library, 618
Peerless (Ann & Bury) Picture Library, 619
Photoshot, 619
Rex Features Ltd, 621
Royal Collection Photographic Services, The, 621
Sotheby's Picture Library, 623
Tate Images, 624
Traeger (Tessa) Picture Library, 625

Art: Poetry Magazines
Aesthetica, 208
Eratica, 210
Student Magazine, The, 217

Art: Prizes
Fletcher (Sir Banister) Award, 500
Practical Art Book of the Year, 516

Art: Professional Associations
Association of Illustrators, 528
Design and Artists Copyright Society (DACS), 536

Art: Publishers (European)
Adelphi Edizioni SpA, 166
Alianza Editorial SA, 169
Beck (C.H., OHG) Verlag, 164
Bertelsmann (C.), 164
Bompiani, 166
Brepols Publishers NV, 160
Brána a.s., 161
Caminho (Editorial) SARL, 169
Civilização Editora, 169
Denoël (Éditions), 163
Diogenes Verlag AG, 171
Flammarion Éditions, 163
Gallimard (Éditions), 163
Giunti Editoriale SpA, 166
Gyldendal, 161
Hoffmann und Campe Verlag GmbH, 165
Kibea Publishing Company, 160
Lannoo, Uitgeverij, 160
Larousse (Éditions), 163
Longanesi (Casa Editrice) SpA, 167
Lübbe (Verlagsgruppe) GmbH & Co. KG, 165
Mondadori (Arnoldo) Editore SpA, 167
Mursia (Gruppo Ugo) Editore SpA, 167
Nyt Nordisk Forlag Arnold Busck A/S, 162
Orell Füssli Verlag, 171
Random House Mondadori, 170
Springer-Verlag GmbH, 160
Taschen GmbH, 165
Tiderne Skifter Forlag A/S, 162
Tusquets Editores, 170
Universitaire Pers Leuven, 160

Art: Publishers (International)
Melbourne University Publishing Ltd, 192
Munshiram Manoharlal Publishers Pvt Ltd, 195
Queensland Press, University of, 192
Random House New Zealand, 197
Struik Publishers (Pty) Ltd, 198

Art: Publishers (Ireland)
Merlin Publishing, 157

Art: Publishers (UK)
Anova Books, 66
Antique Collectors' Club, 67
Arris Publishing Ltd, 68
Ashgate Publishing Ltd, 69
Black (A.&C.) Publishers Ltd, 73
David & Charles Publishers, 88
de la Mare (Giles) Publishers Ltd, 88
Enitharmon Press, 92
Garnet Publishing Ltd, 97
GMC Publications Ltd, 98
Halsgrove, 101
HarperCollins Publishers Ltd, 102
Harvard University Press, 103
King (Laurence) Publishing Ltd, 109
Lincoln (Frances) Ltd, 110
Mainstream Publishing Co. (Edinburgh) Ltd, 114
Merrell Publishers Ltd, 116
New Holland Publishers (UK) Ltd, 119
NMS Enterprises Limited – Publishing, 120
Phaidon Press Limited, 126
Prestel Publishing Limited, 129
Search Press Ltd, 137
Seren, 137
Taschen UK, 143
Thames and Hudson Ltd, 144
Third Millennium Publishing, 144
Wilson (Philip) Publishers Ltd, 150

Art: Publishers (US)
Abrams (Harry N.), Inc., 172
Academy Chicago Publishers, 172
Barron's Educational Series, Inc., 173
McFarland & Company, Inc., Publishers, 180
Minnesota Press, University of, 181

Mississippi, University Press of, 181
Pelican Publishing Company, 183
Perseus Books Group, 184

Art: Small Presses
Allardyce, Barnett, Publishers, 225
Crescent Moon Publishing and Joe's Press, 227
Sylph Editions, 235
Tlön Books Publishing Ltd, 236
Worple Press, 237

Art: UK Packagers
Cameron & Hollis, 243
Ilex Press Limited, The, 244
Ivy Press Limited, The, 244
Quarto Publishing plc, 245
Touchstone Books Ltd, 246

Art History: Library Services
Kent County Central Library, 587

Art History: Literary Agents (UK)
Artellus Limited, 250
Chris (Teresa) Literary Agency Ltd, 253
Davidson (Caroline) Literary Agency, 255

Art History: Literary Agents (US)
Braun (Barbara) Associates, Inc., 281
Gusay (Charlotte) Literary Agency, The, 284
Russell & Volkening, Inc., 289

Art History: Magazines
Art Monthly, 297
Burlington Magazine, The, 304

Art History: Prizes
Mitchell Prize for Art History, The/The Eric Mitchell Prize, 512
Taylor (Reginald) and Lord Fletcher Essay Prize, 521

Art History: Publishers (European)
Brepols Publishers NV, 160
Denoël (Éditions), 163
Orell Füssli Verlag, 171

Art History: Publishers (International)
Munshiram Manoharlal Publishers Pvt Ltd, 195

Art History: Publishers (UK)
Ashgate Publishing Ltd, 69
British Museum Press, The, 78
Gibson Square, 97
Manchester University Press, 115
Phaidon Press Limited, 126
Reaktion Books, 132
Wilson (Philip) Publishers Ltd, 150
Yale University Press (London), 151

Arts & Entertainment: Bursaries, Fellowships and Grants
Fulbright Awards, 483

Arts & Entertainment: Film, TV and Radio Producers
Diverse Production Limited, 422
Endemol UK Productions, 422
Greenwich Village Productions, 425
Landseer Productions Ltd, 427
Lucida Productions, 428
Redweather, 431
Screen First Ltd, 432
Silent Sound Films Ltd, 432
Staveacre (Tony) Productions, 433

Arts & Entertainment: Library Services
Barbican Library, 574
Manchester Central Library, 589
Sheffield Libraries, Archives and Information, 595
Westminster Reference Library, 598

Arts & Entertainment: Literary Agents (UK)
Heath (Rupert) Literary Agency, 260
hhb agency ltd, 260
Power (Shelley) Literary Agency Ltd, 267
Sinclair-Stevenson, 269
Smith (Robert) Literary Agency Ltd, 269

Arts & Entertainment: Literary Agents (US)
Gusay (Charlotte) Literary Agency, The, 284
Russell & Volkening, Inc., 289

Arts & Entertainment: Magazines
Chapman, 306
Country Life, 311
Day by Day, 314
Edge, The, 317
i-D Magazine, 328
Jersey Now, 330
List, The, 332
London Magazine, The, 333
Markings, 334
New Shetlander, The, 340
New Statesman, 340
Time Out, 361
Times Literary Supplement, The, 361
Tribune, 363

Arts & Entertainment: PR Consultants
FMcM Associates, 276
Idea Generation, 277
Randall (Sally) PR, 278

Arts & Entertainment: Picture Libraries
ArenaPAL, 602
Dominic Photography, 607
ffotograff, 608
Financial Times Pictures, 608
Jack (Robbie) Photography, 613
Mirrorpix, 616
Monitor Picture Library, 616

Arts & Entertainment: Poetry Magazines
Planet, 213

Arts & Entertainment: Prizes
BBCFour Samuel Johnson Prize for Non-Fiction, The, 490

Arts & Entertainment: Professional Associations
Arts & Business (A&B), 527
Critics' Circle, The, 535
Federation of Entertainment Unions, 537
National Campaign for the Arts, 541
Tees Valley Arts, 549

Arts & Entertainment: Publishers (European)
Dokorán Ltd, 161

Payot Libraire, 171
Universitaire Pers Leuven, 160

Arts & Entertainment: Publishers (International)
Currency Press Pty Ltd, 191
Oxford University Press India, 196

Arts & Entertainment: Publishers (UK)
Carlton Publishing Group, 81
Clairview Books Ltd, 84
Icon Books Ltd, 106
Liverpool University Press, 112
Owen (Peter) Publishers, 123
Phaidon Press Limited, 126
ProQuest, 129

Arts & Entertainment: Publishers (US)
Faber & Faber, Inc., 176
McFarland & Company, Inc., Publishers, 180
Mississippi, University Press of, 181
Players Press, 184
Scarecrow Press Inc., 186

Arts & Entertainment: Writers' Courses (UK and Ireland)
Arvon Foundation, 458

Asia/Asian Interest: Library Services
British Library Asia, Pacific and Africa Collections, 577
Bromley Central Library, 579

Asia/Asian Interest: Literary Agents (US)
Gusay (Charlotte) Literary Agency, The, 284
Russell & Volkening, Inc., 289

Asia/Asian Interest: Magazines
Eastern Eye, 316

Asia/Asian Interest: Picture Libraries
Angel (Heather)/Natural Visions, 602
Australia Pictures, 602
British Library Images Online, 604
Cordaiy (Sylvia) Photo Library, 606
Dickins (Douglas) Photo Library, 607

Garden and Wildlife Matters Photo Library, 609
Geo Aerial Photography, 610
Geoslides Photography, 610
Sanders (Peter) Photography Ltd, 622
Tropix Photo Library, 625
Yemen Pictures, 627

Asia/Asian Interest: Poetry Magazines
Sable, 216

Asia/Asian Interest: Prizes
Man Asian Literary Prize, The, 510

Asia/Asian Interest: Publishers (International)
Munshiram Manoharlal Publishers Pvt Ltd, 195

Asia/Asian Interest: Publishers (UK)
Reaktion Books, 132
Routledge, 134
Saqi Books, 135

Asia/Asian Interest: Publishers (US)
Columbia University Press, 175
Hawai'i Press, University of, 177
Indiana University Press, 179
Sharpe (M.E.), Inc., 186
State University of New York Press, 187

Asia/Asian Interest: Television and Radio
BBC Asian Network, 399
Radio XL 1296 AM, 411
Sabras Radio, 411
Sunrise Radio (Yorkshire), 411

Asia/Asian Interest: Theatre Producers
Theatre Royal Stratford East, 449
Tricycle Theatre, 450

Astrology: Magazines
Prediction, 346

Astrology: Publishers (European)
BZZTÔH (Uitgeverij) BV, 167

Astrology: Publishers (International)
HarperCollins Publishers India Ltd, 195

Astrology: Publishers (UK)
Foulsham Publishers, 95

Astrology: Publishers (US)
Llewellyn Publications, 180

Astronomy: Literary Agents (US)
Gusay (Charlotte) Literary Agency, The, 284

Astronomy: Picture Libraries
Galaxy Picture Library, 609
Science Photo Library, 622
Wellcome Images, 626

Astronomy: Publishers (European)
Springer GmbH, 165

Astronomy: Publishers (UK)
Octopus Publishing Group, 121

Atlases: Prizes
Besterman/McColvin Medal, 490

Atlases: Publishers (European)
Cappelen Damm AS, 168
Espasa-Calpe (Editorial) SA, 170
Fayard (Librarie Arthème), 163
Michelin Éditions des Voyages, 163

Atlases: Publishers (International)
Wiley (John) & Sons Australia Ltd, 193

Atlases: Publishers (UK)
AA Publishing, 64
Allan (Ian) Publishing Ltd, 65
Dorling Kindersley Ltd, 90
HarperCollins Publishers Ltd, 102
Michelin Maps & Guides, 116
Myriad Editions, 118
Octopus Publishing Group, 121

Audio Books: Film, TV and Radio Producers
Gaia Communications, 424
Loftus Audio Ltd, 427

Audio Books: Literary Agents (UK)
Chris (Teresa) Literary Agency Ltd, 253

Audio Books: Literary Agents (US)
Gusay (Charlotte) Literary Agency, The, 284

Audio Books: Miscellany
M-Y Books Ltd, 641

Audio Books: Organizations of Interest to Poets
57 Productions, 220

Audio Books: Professional Associations
Audiobook Publishing Association, 529

Audio Books: Publishers (European)
Hoffmann und Campe Verlag GmbH, 165
Lettera, 161
Lübbe (Verlagsgruppe) GmbH & Co. KG, 165
Payot Libraire, 171

Audio Books: Publishers (International)
Penguin Group (Canada), 194

Audio Books: Publishers (Ireland)
Cló Iar-Chonnachta, 154

Audio Books: Publishers (UK)
Berlitz Publishing, 72
Bloomsbury Publishing Plc, 74
Dorling Kindersley Ltd, 90
Hachette Children's Books, 99
Hachette Livre UK, 100
HarperCollins Publishers Ltd, 102
Isis Publishing, 107
Little, Brown Book Group UK, 111
Macmillan Publishers Ltd, 113
Penguin Group (UK), 125
Quercus Publishing, 130
Random House Group Ltd, The, 131
Wild Goose Publications, 149

Audio Books: Publishers (US)
Hachette Book Group USA, 177
HarperCollins Publishers, Inc., 177
Simon & Schuster Children's Publishing, 187
Tyndale House Publishers, Inc., 189

Zondervan, 190

Audio Books: Small Presses
Feather Books, 228

Australia: Library Services
Bromley Central Library, 579

Australia: Literary Agents (US)
Gusay (Charlotte) Literary Agency, The, 284

Australia: Picture Libraries
Angel (Heather)/Natural Visions, 602
Australia Pictures, 602
Cordaiy (Sylvia) Photo Library, 606
Country Matters Picture Library, 606
Eagar (Patrick) Photography, 607
Ecoscene, 607
Faces and Places, 608
Geoslides Photography, 610
Howes (Chris)/Wild Places Photography, 612
Poignant (Axel) Archive, 620
Yemen Pictures, 627

Australia: Prizes
Kiriyama Pacific Rim Book Prize, 507

Australia: Professional Associations
Australian Publishers Association, 529
Australian Society of Authors, 529

Autobiography: Audio Books
Corgi Audio, 238
Hachette Digital, 239
Macmillan Digital Audio, 239
Orion Audio Books, 240
Penguin Audiobooks, 240

Autobiography: Electronic Publishing and Other Services
Fledgling Press, 248
www.ABCtales.com, 247

Autobiography: Literary Agents (Ireland)
Williams (Jonathan) Literary Agency, 279

Autobiography: Literary Agents (UK)
Ableman (Sheila) Literary Agency, 249
Anderson (Darley) Literary, TV & Film Agency, 249
Chilcote (Judith) Agency, 253
Feldstein Agency, The, 256
Heath (Rupert) Literary Agency, 260
hhb agency ltd, 260
Lownie (Andrew) Literary Agency, 263
Simmons (Jeffrey), 269
Wiener (Dinah) Ltd, 272

Autobiography: Literary Agents (US)
Gusay (Charlotte) Literary Agency, The, 284
Nelson (B.K.) Literary Agency, 287
Rees (Helen) Literary Agency, 288
Russell & Volkening, Inc., 289
Schiavone Literary Agency, Inc., 289

Autobiography: PR Consultants
Publishing Services, 278

Autobiography: Prizes
Ackerley (J.R.) Prize, 488
BBCFour Samuel Johnson Prize for Non-Fiction, The, 490

Autobiography: Publishers (European)
Adelphi Edizioni SpA, 166
Bertelsmann (C.), 164
Bompiani, 166
Tusquets Editores, 170

Autobiography: Publishers (International)
Hachette Livre Australia, 191
HarperCollins Publishers India Ltd, 195
Munshiram Manoharlal Publishers Pvt Ltd, 195

Autobiography: Publishers (Ireland)
Gill & Macmillan, 155
Lilliput Press, The, 156
Maverick House Publishers, 156

Autobiography: Publishers (UK)
Apex Publishing Ltd, 67

Austin & Macauley Publishers
Limited, 70
Bound Biographies Limited, 76
Boyars (Marion) Publishers
Ltd, 76
Calder Publications Ltd, 79
Carlton Publishing Group, 81
Hachette Livre UK, 100
Halban Publishers, 100
Hale (Robert) Ltd, 101
HarperCollins Publishers Ltd,
102
Honno Welsh Women's Press,
105
Little, Brown Book Group UK,
111
Macmillan Publishers Ltd, 113
Mainstream Publishing Co.
(Edinburgh) Ltd, 114
Pegasus Elliot Mackenzie
Publishers Ltd, 125
Pen & Sword Books Ltd, 125
Random House Group Ltd,
The, 131
Simon & Schuster UK Limited,
139
Souvenir Press Ltd, 140

**Autobiography: Publishers
(US)**
Houghton Mifflin Harcourt
Publishing Company, 178

Autobiography: Small Presses
Miller (Neil) Publications, 232
Mirage Publishing, 232
Parapress, 233

**Autobiography: Writers'
Courses (UK and Ireland)**
Liberato Breakaway Writing
Courses, 456
Open College of the Arts, 465

Aviation: Book Clubs
BCA (Book Club Associates),
293

Aviation: Library Services
CAA Library and Information
Centre, 580
Norfolk Library & Information
Service, 591

**Aviation: Literary Agents
(UK)**
Welch (John), Literary
Consultant & Agent, 272

**Aviation: Literary Agents
(US)**
Gusay (Charlotte) Literary
Agency, The, 284

Aviation: Magazines
Aeroplane, 295
AIR International, 295
AirForces Monthly, 295
Aviation News, 298
Flight International, 320
FlyPast, 321
Pilot, 344

Aviation: Picture Libraries
aviation–images.com, 603
Brooklands Museum Photo
Archive, 604
Cordaiy (Sylvia) Photo Library,
606
Defence Picture Library, The,
606
Royal Air Force Museum, 621
Skyscan Photolibrary, The, 623

Aviation: Publishers (UK)
Allan (Ian) Publishing Ltd, 65
Anova Books, 66
Ashgate Publishing Ltd, 69
Countryside Books, 86
Crowood Press Ltd, The, 87
Grub Street, 99
Haynes Publishing, 103
Jane's Information Group, 108
Midland Publishing, 117
Osprey Publishing Ltd, 123
Pen & Sword Books Ltd, 125
Special Interest Model Books
Ltd, 140

Aviation: Small Presses
Stenlake Publishing Limited,
235

Aviation: UK Packagers
Amber Books Ltd, 242

Bees: Library Services
Herefordshire Libraries, 585

Bees: Magazines
Buzz Extra, 305
Journal of Apicultural Research,
331

**Bibles/Biblical Studies:
Literary Agents (US)**
Gusay (Charlotte) Literary
Agency, The, 284

**Bibles/Biblical Studies:
Publishers (UK)**
Cambridge University Press, 79
Continuum International
Publishing Group Ltd, The,
85
Darton, Longman & Todd Ltd,
88
Epworth, 93
Hachette Livre UK, 100
Inter-Varsity Press, 107
Oxford University Press, 124
Society for Promoting Christian
Knowledge (SPCK), 140

**Bibles/Biblical Studies:
Publishers (US)**
Eerdmans Publishing Company,
176
Tyndale House Publishers, Inc.,
189
Zondervan, 190

**Bibliography: Library
Services**
London Library, The, 589

Bibliography: Magazines
Book Collector, The, 302

Bibliography: Prizes
Oldman (C.B.) Prize, 514
Walford Award, 523

**Bibliography: Professional
Associations**
Bibliographical Society, The, 530
Edinburgh Bibliographical
Society, 536

**Bibliography: Publishers
(European)**
Brepols Publishers NV, 160
Forlaget Forum, 161
Hachette Livre (France), 163

**Bibliography: Publishers
(UK)**
Bowker (UK) Ltd, 76
British Library, The, 78
Facet Publishing, 94
Nielsen BookData, 119

Bibliography: Small Presses
Dragonby Press, The, 227
Galactic Central Publications,
228

**Bilingual Reference:
Publishers (European)**
Bertrand Editora Lda, 169

BZZTÔH (Uitgeverij) BV, 167
Hachette Livre (France), 163
Hiperión (Ediciónes) SL, 170
Larousse (Éditions), 163
Magnard (Éditions), 163
Random House Mondadori, 170

Bilingual Reference: Publishers (UK)
Chambers Harrap Publishers Ltd, 82
HarperCollins Publishers Ltd, 102
Oxford University Press, 124

Bilingual Reference: UK Packagers
Lexus Ltd, 244

Biochemistry: Library Services
Royal Society of Medicine Library, 594

Biochemistry: Publishers (UK)
Freeman (W. H.), 96
Portland Press Ltd, 128

Biography: Audio Books
Chrome Dreams, 238
CSA Word, 238

Biography: Bursaries, Fellowships and Grants
Longford (Elizabeth) Grants, The, 484

Biography: Festivals
Buxton Festival, 631

Biography: Library Services
Bromley House Library, 580

Biography: Literary Agents (Ireland)
Gunn O'Connor (Marianne) Literary Agency, 279
Williams (Jonathan) Literary Agency, 279

Biography: Literary Agents (UK)
Ableman (Sheila) Literary Agency, 249
Ampersand Agency Ltd, The, 249
Anderson (Darley) Literary, TV & Film Agency, 249
Artellus Limited, 250

Bell Lomax Moreton Agency, The, 250
Brown (Jenny) Associates, 251
Bryan (Felicity), 252
Chris (Teresa) Literary Agency Ltd, 253
Cochrane (Elspeth) Personal Management, 254
Davidson (Caroline) Literary Agency, 255
Edwards Fuglewicz, 256
Feldstein Agency, The, 256
Fox & Howard Literary Agency, 257
Futerman, Rose & Associates, 257
Godwin (David) Associates, 258
Green (Annette) Authors' Agency, 258
Heath (Rupert) Literary Agency, 260
hhb agency ltd, 260
Higham (David) Associates Ltd, 260
Indepublishing CIA – Consultancy for Independent Authors, 261
Johnson & Alcock, 261
Judd (Jane) Literary Agency, 262
Kelly (Frances), 262
Knight Features, 262
Lownie (Andrew) Literary Agency, 263
Luxton Harris Ltd, 263
McAra (Duncan), 264
Pawsey (John), 266
Sayle Literary Agency, The, 268
Sheil Land Associates Ltd, 268
Simmonds (Dorie) Agency, 269
Simmons (Jeffrey), 269
Sinclair-Stevenson, 269
Smith (Robert) Literary Agency Ltd, 269
Turnbull (Jane), 271
United Agents Limited, 271
Wiener (Dinah) Ltd, 272

Biography: Literary Agents (US)
Altshuler (Miriam) Literary Agency, 280
Bernstein (Meredith) Literary Agency, Inc., 280
Bleecker Street Associates, Inc., 280
Borchardt (Georges), Inc., 281
Braun (Barbara) Associates, Inc., 281
Carvainis (Maria) Agency, Inc., 281
Castiglia Literary Agency, 281

Dijkstra (Sandra) Literary Agency, 282
Dystel & Goderich Literary Management, 283
Ellenberg (Ethan) Literary Agency, 283
Gusay (Charlotte) Literary Agency, The, 284
Lampack (Peter) Agency, Inc., 285
Lescher & Lescher Ltd, 285
Michaels (Doris S.) Literary Agency Inc., 286
Naggar (Jean V.) Literary Agency, The, 287
Nelson (B.K.) Literary Agency, 287
Parks (Richard) Agency, 287
Protter (Susan Ann) Literary Agent, 288
Quicksilver Books, Literary Agents, 288
Rees (Helen) Literary Agency, 288
Rittenberg (Ann) Literary Agency, Inc., 288
Russell & Volkening, Inc., 289
Schiavone Literary Agency, Inc., 289
Schulman (Susan), A Literary Agency, 290
Scribblers House® LLC Literary Agency, 290
Ware (John A.) Literary Agency, 291
Writers House, LLC, 292
Zeckendorf (Susan) Associates, Inc., 292

Biography: PR Consultants
Publishing Services, 278
Sawford (Claire) PR, 278

Biography: Poetry Magazines
Dream Catcher, 209

Biography: Poetry Presses
Dream Catcher, 201
Seren, 206

Biography: Prizes
BBCFour Samuel Johnson Prize for Non-Fiction, The, 490
Biographers' Club Prize, The, 491
Black (James Tait) Memorial Prizes, 491
Cooper (Duff) Prize, The, 496
Costa Book Awards, 496
Longford (Elizabeth) Prize for Historical Biography, The, 509

Marsh Biography Award, 511
Morley (Sheridan) Prize, The, 512

Biography: Publishers (European)
Adelphi Edizioni SpA, 166
Arche Verlag AG, 171
Bertelsmann (C.), 164
Bompiani, 166
Borgens Forlag A/S, 161
BZZTÔH (Uitgeverij) BV, 167
Calmann-Lévy (Éditions), 163
Caminho (Editorial) SARL, 169
Civilização Editora, 169
Denoël (Éditions), 163
Deutscher Taschenbuch Verlag GmbH & Co. KG, 164
EDHASA (Editora y Distribuidora Hispano – Americana SA), 169
Espasa-Calpe (Editorial) SA, 170
Fayard (Librarie Arthème), 163
Fischer (S.) Verlag GmbH, 164
Flammarion Éditions, 163
Gallimard (Éditions), 163
Garzanti Libri SpA, 166
Grasset & Fasquelle (Éditions), 163
Gyldendal, 161
Heyne Verlag, 165
Hoffmann und Campe Verlag GmbH, 165
Høst & Søn Publishers Ltd, 161
Lannoo, Uitgeverij, 160
Longanesi (Casa Editrice) SpA, 167
Lübbe (Verlagsgruppe) GmbH & Co. KG, 165
Mondadori (Arnoldo) Editore SpA, 167
Mursia (Gruppo Ugo) Editore SpA, 167
Presses de la Cité, 164
Publicat SA, 169
Random House Mondadori, 170
Seuil (Éditions du), 164
Sperling e Kupfer Editori SpA, 167
Sugarco Edizioni Srl, 167
Suhrkamp Verlag, 165
Table Ronde (Les Editions de la), 164
Tusquets Editores, 170
Ullstein Buchverlage GmbH, 165
Zsolnay (Paul), Verlag GmbH, 160

Biography: Publishers (International)
Allen & Unwin Pty Ltd, 191

Auckland University Press, 196
Currency Press Pty Ltd, 191
Fitzhenry & Whiteside Limited, 193
Hachette Livre Australia, 191
Hachette Livre New Zealand Ltd, 196
HarperCollins Publishers India Ltd, 195
Melbourne University Publishing Ltd, 192
Munshiram Manoharlal Publishers Pvt Ltd, 195
Oxford University Press India, 196
Pan Macmillan Australia Pty Ltd, 192
Queensland Press, University of, 192
Raupo Publishing (NZ) Ltd, 197
Simon & Schuster (Australia) Pty Ltd, 192
Struik Publishers (Pty) Ltd, 198
Victoria University Press, 197
Wits University Press, 198

Biography: Publishers (Ireland)
Anvil Books, 153
Brandon/Mount Eagle Publications, 154
Currach Press, 155
Gill & Macmillan, 155
Hachette Books Ireland, 155
Lilliput Press, The, 156
Maverick House Publishers, 156
Mercier Press Ltd, 156
Merlin Publishing, 157
O'Brien Press Ltd, The, 157
Poolbeg Press Ltd, 158

Biography: Publishers (UK)
Alma Books Ltd, 65
Anova Books, 66
Apex Publishing Ltd, 67
Arris Publishing Ltd, 68
Atlantic Books, 69
Aurora Metro, 70
Aurum Press Ltd, 70
Austin & Macauley Publishers Limited, 70
Authentic Media, 70
AuthorHouse UK, 70
Black Ace Books, 73
Black Spring Press Ltd, 74
Blackstone Publishers, 74
Book Guild Publishing, 75
Bound Biographies Limited, 76
Boyars (Marion) Publishers Ltd, 76
Brewin Books Ltd, 77
Calder Publications Ltd, 79

Carcanet Press Ltd, 80
Carlton Publishing Group, 81
Constable & Robinson Ltd, 85
Davies (Christopher) Publishers Ltd, 88
de la Mare (Giles) Publishers Ltd, 88
Debrett's Ltd, 88
Elliott & Thompson, 91
Faber & Faber Ltd, 94
Gibson Square, 97
Hachette Livre UK, 100
Halban Publishers, 100
Hale (Robert) Ltd, 101
Halsgrove, 101
HarperCollins Publishers Ltd, 102
Haus Publishing, 103
Honeyglen Publishing Ltd, 105
Independent Music Press, 107
JR Books Ltd, 108
Lewis (Dewi) Publishing, 110
Little Books Ltd, 111
Little, Brown Book Group UK, 111
Luath Press Ltd, 112
Macmillan Publishers Ltd, 113
Mainstream Publishing Co. (Edinburgh) Ltd, 114
Melrose Books, 116
Melrose Press Ltd, 116
Mercat Press, 116
O'Mara (Michael) Books Limited, 122
Octagon Press Ltd, 121
Octopus Publishing Group, 121
Omnibus Press, 122
Owen (Peter) Publishers, 123
Pegasus Elliot Mackenzie Publishers Ltd, 125
Pen & Sword Books Ltd, 125
Pennant Books Ltd, 126
Piatkus Books, 127
Profile Books, 129
Quartet Books, 130
Quiller Press, 130
Random House Group Ltd, The, 131
Seren, 137
Shepheard-Walwyn (Publishers) Ltd, 138
Short Books, 139
Sigma Press, 139
Simon & Schuster UK Limited, 139
Souvenir Press Ltd, 140
Tauris (I.B.) & Co. Ltd, 143
Thames and Hudson Ltd, 144
Wilson (Neil) Publishing Ltd, 149
Yale University Press (London), 151

Biography: Publishers (US)
Eerdmans Publishing Company, 176
Holt (Henry) & Company, Inc., 178
Houghton Mifflin Harcourt Publishing Company, 178
Louisiana State University Press, 180
Massachusetts Press, University of, 181
New Mexico Press, University of, 182
Northwestern University Press, 182
Pelican Publishing Company, 183
Perseus Books Group, 184
Potomac Books, Inc., 184
PublicAffairs, 185
St Martin's Press LLC, 186
Washington State University Press, 190

Biography: Small Presses
Accent Press, 225
Chrysalis Press, 226
Mirage Publishing, 232
Nyala Publishing, 232
Parapress, 233
Pipers' Ash Ltd, 233
Witan Books, 237

Biography: UK Packagers
Erskine Press, 243
Savitri Books Ltd, 245

Biography: Writers' Courses (UK and Ireland)
National Academy of Writing, 464
Writers Bureau, The, 461

Biography: Writers' Courses (US)
Goucher College, 471

Biology/Bioscience: Literary Agents (US)
Gusay (Charlotte) Literary Agency, The, 284
Russell & Volkening, Inc., 289

Biology/Bioscience: Magazines
Naturalist, The, 339

Biology/Bioscience: Picture Libraries
GeoScience Features, 610

Biology/Bioscience: Publishers (European)
Adelphi Edizioni SpA, 166
Springer-Verlag GmbH, 160

Biology/Bioscience: Publishers (International)
Canterbury University Press, 196

Biology/Bioscience: Publishers (UK)
Freeman (W. H.), 96
Harvard University Press, 103
Taylor & Francis, 143

Biotechnology: Literary Agents (US)
Gusay (Charlotte) Literary Agency, The, 284
Russell & Volkening, Inc., 289

Biotechnology: Publishers (UK)
Taylor & Francis, 143

Birds: Library Services
Natural History Museum Library, The, 591

Birds: Literary Agents (US)
Gusay (Charlotte) Literary Agency, The, 284

Birds: Magazines
Bird Life Magazine, 301
Bird Watching, 301
Birds, 301
Birdwatch, 301
British Birds, 303
Wild Times, 367
Wingbeat, 367

Birds: Picture Libraries
Cordaiy (Sylvia) Photo Library, 606
GeoScience Features, 610
Grace (Martin and Dorothy), 611
Natural History Museum Image Resources, 618
Premaphotos Wildlife, 620
RSPB Images, 622
RSPCA Photolibrary, 622

Birds: Prizes
Birdwatch Bird Book of the Year, 491

Birds: Publishers (UK)
Black (A.&C.) Publishers Ltd, 73

Birds: Publishers (US)
Texas Press, University of, 188

Black Writing/Issues: Literary Agents (US)
Gusay (Charlotte) Literary Agency, The, 284
Russell & Volkening, Inc., 289

Black Writing/Issues: Magazines
Black Beauty & Hair, 301
New Nation, 339
Pride, 347
Voice, The, 365

Black Writing/Issues: Poetry Presses
Peepal Tree Press Ltd, 204

Black Writing/Issues: Prizes
Fagon (Alfred) Award, The, 500

Black Writing/Issues: Publishers (UK)
Brown Skin Books, 78
New Beacon Books Ltd, 119
X Press, The, 151

Black Writing/Issues: Publishers (US)
Indiana University Press, 179
Massachusetts Press, University of, 181
Mississippi, University Press of, 181
Virginia Press, University of, 189

Black Writing/Issues: Theatre Producers
Nitro, 444
Talawa Theatre Company Ltd, 449
Theatre Royal Stratford East, 449
Tricycle Theatre, 450

Blindness: Magazines
NB, 339

Book Arts & Book Trade: Library Services
London (University of the Arts) – London College of Communication, 588
St Bride Library, 594

Book Arts & Book Trade: Literary Agents (US)
Gusay (Charlotte) Literary Agency, The, 284

Russell & Volkening, Inc., 289

Book Arts & Book Trade: Magazines
Book Collector, The, 302
Book World Magazine, 302
Bookseller, The, 302
Irish Book Review, The, 329
Logos, 333
Publishing News, 348
Self Publishing Magazine, The, 354
Times Literary Supplement, The, 361

Book Arts & Book Trade: Miscellany
DOI Registration Agency, 641
ISBN Agency, 641
Nielsen BookNet, 641
Nielsen BookScan, 641
SAN Agency, 641

Book Arts & Book Trade: Prizes
Galaxy British Book Awards, 501

Book Arts & Book Trade: Professional Associations
Bibliographical Society, The, 530
Booksellers Association of the UK & Ireland Ltd, 530
Booktrust, 530
CLÉ – Irish Book Publishers' Association, 534
Independent Publishers Guild, 538
Publishers Association, The, 544
Publishing Scotland, 545
Scottish Book Trust, 546
Scottish Print Employers Federation, 546
Society for Editors and Proofreaders (SfEP), 547
Society of Authors in Scotland, The, 547
Society of Authors, The, 547
Society of Young Publishers, 548
Welsh Books Council (Cyngor Llyfrau Cymru), 550
Women in Publishing, 550
Writers' Guild of Great Britain, The, 551

Book Arts & Book Trade: Publishers (UK)
British Library, The, 78
Nielsen BookData, 119

Book Arts & Book Trade: Writers' Circles and Workshops
'How to Self-Publish Your Book' Workshops, 478

Botany: Library Services
Natural History Museum Library, The, 591

Botany: Literary Agents (US)
Gusay (Charlotte) Literary Agency, The, 284

Botany: Picture Libraries
Alpine Garden Society, 601
Angel (Heather)/Natural Visions, 602
Cordaiy (Sylvia) Photo Library, 606
Forest Life Picture Library, 609
Gagg's (Andrew N.) PHOTOFLORA, 609
Garden and Wildlife Matters Photo Library, 609
GeoScience Features, 610
Grace (Martin and Dorothy), 611
Lindley Library, Royal Horticultural Society, 614
Mathews (S&O) Photography, 616
Natural History Museum Image Resources, 618
Nature Picture Library, 618
Oxford Picture Library, 619
Oxford Scientific (OSF), 619
Premaphotos Wildlife, 620
Science Photo Library, 622
Ulster Museum Picture Library, 625
Wellcome Images, 626

Botany: Publishers (European)
Flammarion Éditions, 163

Britain: Literary Agents (UK)
Feldstein Agency, The, 256

Britain: Literary Agents (US)
Gusay (Charlotte) Literary Agency, The, 284

Britain: Magazines
Best of British, 300
Evergreen, 318
In Britain, 328
National Trust Magazine, The, 339
This England, 360

Britain: Picture Libraries
Andes Press Agency, 602
Belcher (Ivan J.) Colour Picture Library, 603
Collections, 605
Country Matters Picture Library, 606
Craven (Philip) Worldwide Photo-Library, 606
Cunningham (Sue) Photographic, 606
English Heritage Photo Library, 607
Frith (Francis) Collection, The, 609
Geoslides Photography, 610
Guy (V.K.) Ltd, 611
Heseltine (John) Archive, 611
Howes (Chris)/Wild Places Photography, 612
London Aerial Photo Library, 615
National Monuments Record, 617
National Trust Photo Library, The, 618
Oxford Picture Library, 619
Pictures Colour Library, 620
Poignant (Axel) Archive, 620
Sharp (Mick) Photography, 622
Skyscan Photolibrary, The, 623
Travel Ink Photo Library, 625

Britain: Professional Associations
British Council, The, 531

Britain: Publishers (UK)
Travel Publishing Ltd, 145

Broadcasting: Bursaries, Fellowships and Grants
Bradley (Alfred) Bursary Award, 482

Broadcasting: Library Services
BBC Written Archives Centre, 575

Broadcasting: Literary Agents (US)
Gusay (Charlotte) Literary Agency, The, 284
Russell & Volkening, Inc., 289

Broadcasting: Magazines
Broadcast, 304
Press Gazette, 347

Broadcasting: Professional Associations
Broadcasting Press Guild, 532
Campaign for Press and Broadcasting Freedom, 533
Chartered Institute of Journalists, 533
RadioCentre, 545
Voice of the Listener & Viewer Ltd (VLV), 549

Broadcasting: Writers' Courses (UK and Ireland)
Highbury College, Portsmouth, 456

Building: Library Services
British Architectural Library, 576

Building: Literary Agents (US)
Russell & Volkening, Inc., 289

Building: Magazines
Build It, 304
Building Magazine, 304
Construction News, 310
What's New in Building, 367

Building: Picture Libraries
Edifice, 607

Building: Publishers (UK)
Donhead Publishing Ltd, 89

Business & Commerce: Audio Books
Simon & Schuster Audio, 241

Business & Commerce: Film, TV and Radio Producers
Acacia Productions Ltd, 417
Acrobat Television, 417
Art & Training Films Ltd, 418
Buccaneer Films, 419
Channel Television Ltd, 420
Cricket Ltd, 421
Cutting Edge Productions Ltd, 422
Drake A-V Video Ltd, 422
Dunstan (Charles) Communications Ltd, 422
Electric Airwaves Ltd (Ladbroke Radio), 422
Firehouse Productions, 423
Gaia Communications, 424
Greenwich Village Productions, 425
Imari Entertainment Ltd, 426
Lagan Pictures Ltd, 427

London Scientific Films Ltd, 428
Omnivision, 429
Orpheus Productions, 429
Ovation, 429
Paper Moon Productions, 430
Redweather, 431
Scope Productions Ltd, 432
'Spoken' Image Ltd, 433
Stirling Film & TV Productions Limited, 433
Tern Television Productions Ltd, 434
Twofour Broadcast, 436
Video Enterprises, 436
Wortman Productions UK, 437

Business & Commerce: Library Services
Belfast Public Libraries: Central Library, 575
Bradford Central Library, 576
Bristol Central Library, 576
British Library Business & IP Centre, 577
British Library Science, Technology and Business Collections, 579
City Business Library, 581
Guildhall Library, 584
Kent County Central Library, 587
Leeds Central Library, 587
Liverpool Libraries and Information Services, 588
Nottingham Central Library, 592
Sheffield Libraries, Archives and Information, 595
Westminster Reference Library, 598

Business & Commerce: Literary Agents (UK)
Bell Lomax Moreton Agency, The, 250
Feldstein Agency, The, 256
Fox & Howard Literary Agency, 257
Graham Maw Christie Literary Agency, 258
hhb agency ltd, 260
Kelly (Frances), 262
Knight Features, 262
Power (Shelley) Literary Agency Ltd, 267
Sheil Land Associates Ltd, 268

Business & Commerce: Literary Agents (US)
Bleecker Street Associates, Inc., 280

Browne (Pema) Ltd, 281
Carvainis (Maria) Agency, Inc., 281
Castiglia Literary Agency, 281
Dijkstra (Sandra) Literary Agency, 282
Fredericks (Jeanne) Literary Agency, Inc., 283
Gusay (Charlotte) Literary Agency, The, 284
Herman (Jeff) Agency, LLC, The, 284
Leavitt (Ned) Agency, The, 285
Lowenstein–Yost Associates Inc., 285
MacBride (Margret) Literary Agency, The, 286
Nelson (B.K.) Literary Agency, 287
Rees (Helen) Literary Agency, 288
Schulman (Susan), A Literary Agency, 290
Scribblers House® LLC Literary Agency, 290
Snell (Michael) Literary Agency, 290
Writers House, LLC, 292

Business & Commerce: Magazines
Business Brief, 305
Business Traveller, 305
Director, 315
Investors Chronicle, 329
Management Today, 334
Ulster Business, 364

Business & Commerce: News Agencies
Dow Jones Newswires, 393

Business & Commerce: Picture Libraries
Financial Times Pictures, 608
Photoshot, 619
PPL (Photo Agency) Ltd, 620

Business & Commerce: Prizes
BBCFour Samuel Johnson Prize for Non-Fiction, The, 490
FT & Goldman Sachs Business Book of the Year Award, 501
Wadsworth Prize for Business History, 523

Business & Commerce: Professional Associations
Arts & Business (A&B), 527

Business & Commerce: Publishers (European)
Deutscher Taschenbuch Verlag GmbH & Co. KG, 164
Fischer (S.) Verlag GmbH, 164
Orell Füssli Verlag, 171
Payot Libraire, 171
Pearson Education (Benelux), 168
Ullstein Buchverlage GmbH, 165
Unieboek (Uitgeverij) BV, 168

Business & Commerce: Publishers (International)
Hachette Livre New Zealand Ltd, 196
HarperCollins Publishers (New Zealand) Ltd, 196
LexisNexis Butterworths (Pty) Ltd South Africa, 197
LexisNexis India, 195
Melbourne University Publishing Ltd, 192
Random House of Canada Ltd, 194
Vision Books Pvt. Ltd, 196
Wits University Press, 198

Business & Commerce: Publishers (Ireland)
Blackhall Publishing Ltd, 153
Farmar (A.&A.) Ltd, 155
O'Brien Press Ltd, The, 157
Oak Tree Press, 157

Business & Commerce: Publishers (UK)
Ashgate Publishing Ltd, 69
Austin & Macauley Publishers Limited, 70
Black (A.&C.) Publishers Ltd, 73
Boltneck Publications Limited, 75
Brealey (Nicholas) Publishing, 77
Business Education Publishers Ltd, 79
Capstone Publishing, 80
Crimson Publishing, 86
Crown House Publishing, 87
Discovered Authors, 89
Elgar (Edward) Publishing Ltd, 91
Elm Publications/Training, 92
Euromonitor International plc, 93
Foulsham Publishers, 95
Hachette Livre UK, 100
Harriman House Ltd, 103
Infinite Ideas, 107
Informa Law, 107

Kogan Page Ltd, 109
Management Books 2000 Ltd, 114
McGraw-Hill Education, 113
Piatkus Books, 127
Profile Books, 129
Quiller Press, 130
Random House Group Ltd, The, 131
Reed Elsevier Group plc, 133
Routledge, 134
Simon & Schuster UK Limited, 139
Souvenir Press Ltd, 140
Studymates Limited, 142
TSO (The Stationery Office), 146
Zambezi Publishing Ltd, 151

Business & Commerce: Publishers (US)
AMACOM Books, 172
Barron's Educational Series, Inc., 173
Crown Publishing Group, 175
HarperCollins Publishers, Inc., 177
McGraw-Hill Companies Inc., The, 180
Pelican Publishing Company, 183
Stanford University Press, 187

Business & Commerce: US Media Contacts in the UK
Bloomberg News, 394
Business Week, 394

Business & Commerce: Writers' Courses (UK and Ireland)
Wilson's (John) Writing Courses, 467

Calligraphy & Handwriting: Publishers (UK)
Shepheard-Walwyn (Publishers) Ltd, 138

Canada: Literary Agents (US)
Gusay (Charlotte) Literary Agency, The, 284

Canada: Picture Libraries
Howes (Chris)/Wild Places Photography, 612
Williams (David) Picture Library, 626

Canada: Prizes
Kiriyama Pacific Rim Book Prize, 507

Canada: Professional Associations
Association of Canadian Publishers, 527
Canadian Authors Association, 533
Canadian Federation of Poets, 533
Canadian Publishers' Council, 533

Careers & Employment: Library Services
Leeds Central Library, 587

Careers & Employment: Literary Agents (US)
Gusay (Charlotte) Literary Agency, The, 284

Careers & Employment: Magazines
OS (Office Secretary) Magazine, 343

Careers & Employment: Publishers (International)
Vision Books Pvt. Ltd, 196

Careers & Employment: Publishers (UK)
Crimson Publishing, 86
Hobsons Plc, 105
Kogan Page Ltd, 109
Learning Institute, The, 110
NCVO Publications, 119
Trotman, 146

Careers & Employment: Publishers (US)
AMACOM Books, 172

Caribbean: Library Services
Senate House Library, University of London, 595

Caribbean: Literary Agents (US)
Gusay (Charlotte) Literary Agency, The, 284

Caribbean: Picture Libraries
Travel Ink Photo Library, 625
Tropix Photo Library, 625

Caribbean: Poetry Magazines
Sable, 216

Caribbean: Poetry Presses
Peepal Tree Press Ltd, 204

Caribbean: Prizes
Fagon (Alfred) Award, The, 500

Caribbean: Publishers (UK)
Evans Brothers Ltd, 93

Caribbean: Publishers (US)
Virginia Press, University of, 189

Cartoons/Comics: Literary Agents (UK)
Anderson (Darley) Literary, TV & Film Agency, 249
Knight Features, 262

Cartoons/Comics: Literary Agents (US)
Russell & Volkening, Inc., 289

Cartoons/Comics: Magazines
BBC History Magazine, 299
Dandy, The, 313
Disability Now, 315
Horse and Hound, 327
Notes from the Underground, 341
Private Eye, 347
Tribune, 363

Cartoons/Comics: News Agencies
Solo Syndication Ltd, 393

Cartoons/Comics: Picture Libraries
British Cartoon Archive, The, 604
Punch, 621

Cartoons/Comics: Poetry Magazines
Quiet Feather, The, 215
Scar Tissue, 216

Cartoons/Comics: Professional Associations
Design and Artists Copyright Society (DACS), 536

Cartoons/Comics: Publishers (UK)
Luath Press Ltd, 112

Cartoons/Comics: Writers' Courses (UK and Ireland)
London School of Journalism, 459

CD-Rom: Electronic Publishing and Other Services
Justis Publishing Ltd, 248

CD-Rom: Film, TV and Radio Producers
'Spoken' Image Ltd, 433

CD-Rom: Publishers (European)
Beck (C.H., OHG) Verlag, 164
Brepols Publishers NV, 160
Bruna (A.W.) Uitgevers BV, 167
Espasa-Calpe (Editorial) SA, 170
Hanser (Carl) Verlag GmbH & Co. KG, 165
Hüthig GmbH & Co. KG, 165
Springer GmbH, 165
Springer Science+Business Media BV, 168

CD-Rom: Publishers (International)
Affiliated East West Press Pvt Ltd, 195
Chand (S.) & Co Ltd, 195
LexisNexis Butterworths (Pty) Ltd South Africa, 197
LexisNexis Canada Ltd, 193
LexisNexis India, 195

CD-Rom: Publishers (Ireland)
National Library of Ireland, 157

CD-Rom: Publishers (UK)
ABC-CLIO (UK), 64
Berlitz Publishing, 72
Debrett's Ltd, 88
Elsevier Ltd, 92
HarperCollins Publishers Ltd, 102
Helicon Publishing, 104
Hobsons Plc, 105
Jane's Information Group, 108
Kluwer Law International, 109
Macmillan Publishers Ltd, 113
Nelson Thornes Limited, 119
Pearson (Heinemann), 125
ProQuest, 129
Radcliffe Publishing Ltd, 131
Reed Elsevier Group plc, 133
Sweet & Maxwell Group, 142

Celtic: Library Services
National Library of Wales, 590

Celtic: Literary Agents (UK)
Feldstein Agency, The, 256

Celtic: Publishers (Ireland)
O'Brien Press Ltd, The, 157

Celtic: Publishers (UK)
Capall Bann Publishing, 80
Davies (Christopher) Publishers Ltd, 88
Floris Books, 95
Wales Press, University of, 148
Y Lolfa Cyf, 151

Celtic: Small Presses
Lyfrow Trelyspen, 231

Ceramics: Picture Libraries
V&A Images, 625

Ceramics: Publishers (UK)
Black (A.&C.) Publishers Ltd, 73

Chemistry: Picture Libraries
GeoScience Features, 610

Chemistry: Publishers (European)
Springer-Verlag GmbH, 160

Chemistry: Publishers (UK)
Freeman (W. H.), 96
Springer-Verlag London Ltd, 141

Chemistry: Publishers (US)
Krieger Publishing Co., 179

Children: Audio Books
BBC Audiobooks Ltd, 238
CSA Word, 238
CYP, 239
HarperCollins AudioBooks, 239
Macmillan Digital Audio, 239
Naxos AudioBooks, 239
Oakhill Publishing Ltd, 240
Orion Audio Books, 240
Penguin Audiobooks, 240
Roberts (L.M.) Limited, 240

Children: Book Clubs
BCA (Book Club Associates), 293
Folio Society, The, 293
Letterbox Library, 293

Children: Electronic Publishing and Other Services
Online Originals, 248

Children: Festivals
Brighton Children's Book Festival, 631
Guardian Hay Festival, The, 633

Hebden Bridge Arts Festival, 634
Ilkley Literature Festival, 634
London Literature Festival, 636
Northern Children's Book Festival, 637
Oundle Festival of Literature, 637
Scotland's Book Town Festival, 638
Word – University of Aberdeen Writers Festival, 640

Children: Film, TV and Radio Producers
Calon TV, 419
Children's Film & Television Foundation Ltd, The, 420
Collingwood O'Hare Entertainment Ltd, 421
Comedy Unit, The, 421
Cosgrove Hall Films, 421
Endemol UK Productions, 422
Farnham Film Company Ltd, 422
Screen First Ltd, 432
Sianco Cyf, 432
SMG Productions & Ginger Productions, 433
Sunset + Vine Productions Ltd, 434

Children: Library Services
Barbican Library, 574
Birmingham Library Services, 576
Brighton Jubilee Library, 576
Cardiff Central Library, 580

Children: Literary Agents (Ireland)
Gunn O'Connor (Marianne) Literary Agency, 279

Children: Literary Agents (UK)
Agency (London) Ltd, The, 249
Anderson (Darley) Literary, TV & Film Agency, 249
Bell Lomax Moreton Agency, The, 250
Brown (Jenny) Associates, 251
Catchpole (Celia), 253
Chris (Teresa) Literary Agency Ltd, 253
Conville & Walsh Limited, 254
Curtis Brown Group Ltd, 254
Eddison Pearson Ltd, 256
Fitch (Laurence) Ltd, 256
Fraser Ross Associates, 257
Green (Annette) Authors' Agency, 258

Heath (A.M.) & Co. Ltd, 259
Higham (David) Associates Ltd, 260
Holt (Vanessa) Ltd, 260
Inspira Group, The, 261
Johnson & Alcock, 261
LAW, 262
London Independent Books, 263
MacMullen (Eunice) Ltd, 264
Mann (Andrew) Ltd, 264
Manson (Sarah) Literary Agent, 264
Motley (Michael) Ltd, 265
Noach (Maggie) Literary Agency, The, 265
PFD, 266
Rogers, Coleridge & White Ltd, 267
Rushby-Smith (Uli) Literary Agency, 267
Sandberg (Rosemary) Ltd, 267
Sheldon (Caroline) Literary Agency Ltd, 268
Simmonds (Dorie) Agency, 269
Standen Literary Agency, The, 269
Such (Sarah) Literary Agency, 270
United Agents Limited, 271
Victor (Ed) Ltd, 271
Wade & Doherty Literary Agency Ltd, 271
Ware (Cecily) Literary Agents, 272
Watson, Little Ltd, 272
Watt (A.P.) Ltd, 272
White (Eve) Literary Agent, 272

Children: Literary Agents (US)
Browne (Pema) Ltd, 281
Dijkstra (Sandra) Literary Agency, 282
Dunham Literary Inc., 283
Ellenberg (Ethan) Literary Agency, 283
Kirchoff/Wohlberg, Inc., 284
Lescher & Lescher Ltd, 285
Mews Books Ltd, 286
Picard (Alison J.) Literary Agent, 287
Russell & Volkening, Inc., 289
Stimola Literary Studio, LLC, 291
Treimel (S©ott) NY, 291
Writers House, LLC, 292

Children: Literary Societies
Children's Books History Society, The, 555
Folly, 558

Needham (Violet) Society, 564
Nesbit (Edith) Society, The, 564
Oxenham (Elsie Jeanette) Appreciation Society, The, 564
Potter (Beatrix) Society, The, 565
Saville (Malcolm) Society, The, 567

Children: Magazines
Young Writer, 370

Children: PR Consultants
Frazer (Cathy) PR, 277
Gotch Solutions, 277

Children: Picture Libraries
Photofusion, 619
PictureBank Photo Library Ltd, 620

Children: Poetry Magazines
Scribbler!, 216

Children: Poetry Presses
King's England Press, The, 203

Children: Prizes
Andersen (Hans Christian) Awards, 488
Blue Peter Children's Book Awards, 491
Branford Boase Award, The, 492
CBI Bisto Book of the Year Awards, 494
Children's Laureate, The, 495
CILIP: The Chartered Institute of Library and Information Professionals Carnegie Medal, 495
CILIP: The Chartered Institute of Library and Information Professionals Kate Greenaway Medal, 495
Costa Book Awards, 496
Dahl (Roald) Funny Prize, The, 498
Farjeon (Eleanor) Award, 500
Guardian Children's Fiction Award, The, 503
Irish Book Awards, 506
Jones (Mary Vaughan) Award, 506
Kelpies Prize, 506
Lancashire County Library and Information Service Children's Book of the Year Award, 507
Lindgren (Astrid) Memorial Award for Literature, The, 508

Macmillan Prize for a Children's Picture Book Illustration, 510
Macmillan Writer's Prize for Africa, 510
Marsh Award for Children's Literature in Translation, 511
Red House Children's Book Award, The, 516
Royal Mail Awards for Scottish Children's Books, 518
Tir Na N-Og Award, The, 522
Waterstone's Children's Book Prize, 523
Way (Brian) Award, 523
YoungMinds Book Award, 525

Children: Professional Associations

Booktrust, 530
Children's Book Circle, 534
Children's Books Ireland, 534
National Centre for Research in Children's Literature, 541
Society of Children's Book Writers & Illustrators, 547

Children: Publishers (European)

Ameet Sp. zo.o., 168
Aschehoug (H) & Co (W Nygaard), 168
Bertrand Editora Lda, 169
Bonnier Books, 170
Borgens Forlag A/S, 161
Caminho (Editorial) SARL, 169
Cappelen Damm AS, 168
Cappelli Editore, 166
Carlsen Verlag GmbH, 164
Civilização Editora, 169
Deutscher Taschenbuch Verlag GmbH & Co. KG, 164
Diogenes Verlag AG, 171
Espasa-Calpe (Editorial) SA, 170
Europa-América (Publicações) Lda, 169
Fischer (S.) Verlag GmbH, 164
Flammarion Éditions, 163
Forlaget Forum, 161
Gallimard (Éditions), 163
Giunti Editoriale SpA, 166
Gradiva–Publicações S.A., 169
Groupo Anaya, 169
Gyldendal, 161
Gyldendal Norsk Forlag, 168
Hachette Livre (France), 163
Hanser (Carl) Verlag GmbH & Co. KG, 165
Hiperión (Ediciónes) SL, 170
Horizonte (Livros) Lda, 169
Høst & Søn Publishers Ltd, 161
Kibea Publishing Company, 160

Kok (Uitgeversmaatschappij J. H.) BV, 168
Lannoo, Uitgeverij, 160
Larousse (Éditions), 163
Magnard (Éditions), 163
Molino SL (Editorial), 170
Mondadori (Arnoldo) Editore SpA, 167
Mursia (Gruppo Ugo) Editore SpA, 167
Móra Könyvkiadó Zrt., 166
Nasza Ksiegarnia Publishing House, 168
Nathan (Les Éditions), 164
Neptun-Verlag, 171
Olympia Publishing Co. Inc., 161
Otava Publishing Co. Ltd, 162
Payot Libraire, 171
Publicat SA, 169
Rabén och Sjögren Bokförlag, 171
Random House Mondadori, 170
Rizzoli Editore, 167
Società Editrice Internazionale – SEI, 167
Spectrum (Uitgeverij Het) BV, 168
Standaard Uitgeverij, 160
Tammi Publishers, 162
Ueberreuter (Verlag Carl) GmbH, 160
Unieboek (Uitgeverij) BV, 168
Verbo (Editorial) SA, 169
WSOY (Werner Söderström Osakeyhtiö), 162
Wählströms (B.) Bokförlag AB, 171

Children: Publishers (International)

Allen & Unwin Pty Ltd, 191
Firefly Books Ltd, 193
Fitzhenry & Whiteside Limited, 193
Hachette Livre Australia, 191
Hachette Livre New Zealand Ltd, 196
HarperCollins Canada Ltd, 193
HarperCollins Publishers (Australia) Pty Ltd, 191
HarperCollins Publishers (New Zealand) Ltd, 196
HarperCollins Publishers India Ltd, 195
Huia (NZ) Ltd, 196
Kwa-Zulu–Natal Press, University of, 197
Macmillan Publishers Australia Pty Ltd, 191
National Publishing House, 195

Pan Macmillan Australia Pty Ltd, 192
Penguin Books (NZ) Ltd, 197
Penguin Group (Australia), 192
Queensland Press, University of, 192
Random House Australia, 192
Random House New Zealand, 197
Random House of Canada Ltd, 194
Raupo Publishing (NZ) Ltd, 197
Scholastic Australia Pty Limited, 192
Scholastic Canada Ltd, 194
Simon & Schuster (Australia) Pty Ltd, 192
Tundra Books, 194
UWA Press, 193

Children: Publishers (Ireland)

An Gúm, 153
Anvil Books, 153
Bradshaw Books, 153
Cló Iar-Chonnachta, 154
Collins Press, The, 154
Mercier Press Ltd, 156
O'Brien Press Ltd, The, 157
Poolbeg Press Ltd, 158

Children: Publishers (UK)

AA Publishing, 64
Andersen Press Ltd, 66
Anness Publishing Ltd, 66
Anova Books, 66
Aurora Metro, 70
Austin & Macauley Publishers Limited, 70
Autumn Publishing, 71
Award Publications Limited, 71
Barefoot Books Ltd, 71
Barrington Stoke, 71
Black (A.&C.) Publishers Ltd, 73
Blackstone Publishers, 74
Bloomsbury Publishing Plc, 74
Book Guild Publishing, 75
British Museum Press, The, 78
Carlton Publishing Group, 81
Catnip Publishing Ltd, 81
Chicken House Publishing, 83
Child's Play (International) Ltd, 83
Christian Focus Publications, 83
Compendium Publishing Ltd, 85
CRW Publishing Ltd, 87
Discovered Authors, 89
Dorling Kindersley Ltd, 90
Dref Wen, 90
Egmont Press, 91
Egmont Publishing, 91
Emma Treehouse Ltd, 92

Evans Brothers Ltd, 93
Everyman's Library, 93
Faber & Faber Ltd, 94
Floris Books, 95
Geddes & Grosset, 97
Gomer Press/Gwasg Gomer, 98
Graham-Cameron Publishing & Illustration, 98
Hachette Children's Books, 99
Hachette Livre UK, 100
HarperCollins Publishers Ltd, 102
Haynes Publishing, 103
Icon Books Ltd, 106
Lincoln (Frances) Ltd, 110
Lion Hudson plc, 111
Little Tiger Press, 111
Lutterworth Press, The, 112
Macmillan Publishers Ltd, 113
Mayhew (Kevin) Publishers, 115
Meadowside Children's Books, 115
Melrose Books, 116
NMS Enterprises Limited – Publishing, 120
O'Mara (Michael) Books Limited, 122
O/Books/ John Hunt Publishing Ltd, 120
Olympia Publishers, 122
Onlywomen Press Ltd, 122
Orion Publishing Group Limited, The, 123
Oxford University Press, 124
Pegasus Elliot Mackenzie Publishers Ltd, 125
Penguin Group (UK), 125
Phaidon Press Limited, 126
Piccadilly Press, 127
Pinwheel, 127
Prestel Publishing Limited, 129
QED Publishing, 129
Quercus Publishing, 130
Random House Group Ltd, The, 131
Ransom Publishing Ltd, 132
Ryland Peters & Small Limited, 135
Scholastic Ltd, 136
Scottish Cultural Press/Scottish Children's Press, 137
Short Books, 139
Simon & Schuster UK Limited, 139
Souvenir Press Ltd, 140
Tango Books Ltd, 142
Templar Publishing, 143
Troubador Publishing Ltd, 146
Usborne Publishing Ltd, 147
Walker Books Ltd, 148
Wingedchariot Press, 150
Wordsworth Editions Ltd, 151

X Press, The, 151

Children: Publishers (US)
Bantam Dell Publishing Group, 173
Barron's Educational Series, Inc., 173
Boyds Mills Press, 174
Carolrhoda Books, 174
Charlesbridge Publishing, 174
Chronicle Books LLC, 174
Dial Books for Young Readers, 175
Eerdmans Publishing Company, 176
Farrar, Straus & Giroux, Inc., 176
Hachette Book Group USA, 177
HarperCollins Publishers, Inc., 177
Holiday House, Inc., 178
Holt (Henry) & Company, Inc., 178
Houghton Mifflin Harcourt Publishing Company, 178
Interlink Publishing Group, Inc., 179
Lerner Publishing Group, 180
Milet Publishing LLC, 181
Paulist Press, 183
Pelican Publishing Company, 183
Penguin Group (USA) Inc., 183
Perseus Books Group, 184
Prometheus Books, 184
Putnam's (G.P.) Sons (Children's), 185
Reader's Digest Association Inc, 185
Scholastic Library Publishing, 186
Simon & Schuster Children's Publishing, 187
Sterling Publishing Co. Inc., 188
Tyndale House Publishers, Inc., 189
Walker & Co., 190

Children: Small Presses
Bluemoose Books Limited, 226
Eilish Press, 228
Feather Books, 228
Fidra Books, 228
Galore Park Publishing Ltd, 228
Ivy Publications, 230
Nissen (Jane) Books, 232
Pipers' Ash Ltd, 233
Robinswood Press, 234
Tlön Books Publishing Ltd, 236
Watling Street Publishing Ltd, 236

Children: Television and Radio
BBC Drama, Comedy and Children's, 396

Children: UK Packagers
Aladdin Books Ltd, 242
Albion Press Ltd, The, 242
Amber Books Ltd, 242
Bender Richardson White, 242
Breslich & Foss Ltd, 242
Brown Wells and Jacobs Ltd, 242
Cameron & Hollis, 243
Haldane Mason Ltd, 243
Lionheart Books, 244
Monkey Puzzle Media Ltd, 245
Orpheus Books Limited, 245
Playne Books Limited, 245
Price (Mathew) Ltd, 245
Salariya Book Company Ltd, 245
Stonecastle Graphics Ltd, 246
Toucan Books Ltd, 246
West (David) Children's Books, 246
Working Partners Ltd, 246

Children: Writers' Circles and Workshops
Short Story Writers' Folio/ Children's Novel Writers' Folio, 479

Children: Writers' Courses (UK and Ireland)
Arvon Foundation, 458
Bath Spa University, 462
City Lit, 458
City University, 458
Community Creative Writing, 461
Liverpool, University of, 461
London School of Journalism, 459
Manchester Metropolitan University – The Writing School, 460
Missenden Abbey, 453
Nottingham Trent University, The, 462
Open College of the Arts, 465
Roehampton University, 460
Sunderland, University of, 463
Winchester (University of), 456
Writers Bureau, The, 461

Children: Writers' Courses (US)
Hollins University, 475
Seton Hill University, 474

China: Library Services
Manchester Central Library, 589

China: Literary Agents (UK)
Eady (Toby) Associates Ltd, 256

China: Literary Agents (US)
Gusay (Charlotte) Literary
    Agency, The, 284
Russell & Volkening, Inc., 289

China: Picture Libraries
Angel (Heather)/Natural
    Visions, 602
Aspect Picture Library Ltd, 602
Gagg's (Andrew N.)
    PHOTOFLORA, 609
Greenhill (Sally and Richard),
    611
King (David) Collection, 613
Link Picture Library, 615
Peerless (Ann & Bury) Picture
    Library, 619
Sanders (Peter) Photography
    Ltd, 622
Young (John Robert) Collection,
    627

China: Prizes
Kiriyama Pacific Rim Book
    Prize, 507

Cinema: Festivals
Aspects Literature Festival, 629

Cinema: Library Services
Barbican Library, 574
BFI National Library, 575
Northumberland County
    Library, 592
Vin Mag Archive Ltd, 598
Westminster Reference Library,
    598

**Cinema: Literary Agents
(UK)**
Casarotto Ramsay and
    Associates Ltd, 252
Hoskins (Valerie) Associates
    Limited, 260
London Independent Books,
    263
Simmons (Jeffrey), 269

Cinema: Literary Agents (US)
Gusay (Charlotte) Literary
    Agency, The, 284

Cinema: Magazines
Empire, 317
Film Review, 320

Screentrade Magazine, 353
Sight & Sound, 355
Total Film, 362

Cinema: Picture Libraries
Aquarius Library, 602
BFI Stills Sales, 603
Christie's Images Ltd, 605
Cinema Museum, The, 605
Kobal Collection, The, 614
Moving Image
    Communications, 617

**Cinema: Professional
Associations**
New Producers Alliance, 542
Pact, 542

**Cinema: Publishers
(European)**
Adelphi Edizioni SpA, 166
Larousse (Éditions), 163
Suhrkamp Verlag, 165
Tusquets Editores, 170

**Cinema: Publishers
(International)**
HarperCollins Publishers India
    Ltd, 195

Cinema: Publishers (UK)
Austin & Macauley Publishers
    Limited, 70
BFI Publishing, 72
Boyars (Marion) Publishers
    Ltd, 76
Faber & Faber Ltd, 94
Hern (Nick) Books, 104
Little, Brown Book Group UK,
    111
Wallflower Press, 148

Cinema: Publishers (US)
McFarland & Company, Inc.,
    Publishers, 180
Minnesota Press, University
    of, 181
Players Press, 184
Scarecrow Press Inc., 186

Cinema: Small Presses
Crescent Moon Publishing and
    Joe's Press, 227
Mercia Cinema Society, 231

**Cinema: Writers' Courses
(UK and Ireland)**
Central School of Speech and
    Drama, The, 458

**Cinema: Writers' Courses
(US)**
Southern California, University
    of, 470

**Classical Studies: Literary
Agents (US)**
Gusay (Charlotte) Literary
    Agency, The, 284

**Classical Studies: Publishers
(UK)**
Duckworth (Gerald) & Co. Ltd,
    90
Exeter Press, University of, 93
Harvard University Press, 103
Routledge, 134

**Classical Studies: Publishers
(US)**
Michigan Press, The University
    of, 181
Oklahoma Press, University
    of, 183
Texas Press, University of, 188

Collecting: Library Services
Derby Central Library, 581

Collecting: Magazines
Apollo, 296
Book Collector, The, 302
Book World Magazine, 302
British Philatelic Bulletin, 304
Coin News, 309
Gibbons Stamp Monthly, 322
Homes & Antiques, 327
Medal News, 335
Model Collector, 336
Picture Postcard Monthly, 344
Record Collector, 351
Stamp Lover, 358
Stamp Magazine, 358

Collecting: Picture Libraries
Christie's Images Ltd, 605

Collecting: Publishers (UK)
Anova Books, 66
Antique Collectors' Club, 67
Gibbons (Stanley) Publications,
    97
Quiller Press, 130
Souvenir Press Ltd, 140
Wilson (Philip) Publishers Ltd,
    150

Collecting: Publishers (US)
Pelican Publishing Company,
    183

Collecting: UK Packagers
Cameron & Hollis, 243

Commonwealth: Library Services
Commonwealth Secretariat Library, 581
Senate House Library, University of London, 595

Commonwealth: Prizes
Commonwealth Writers Prize, The, 496
Reese (Trevor) Memorial Prize, 517

Computing: Magazines
Computer Arts, 310
Computer Weekly, 310
Computeractive, 310
Computing, 310
MacUser, 334
PC Format, 343
Web User, 366

Computing: Publishers (European)
Bruna (A.W.) Uitgevers BV, 167
Hanser (Carl) Verlag GmbH & Co. KG, 165
Pearson Education (Benelux), 168
Springer GmbH, 165
Springer Science+Business Media BV, 168
Springer-Verlag GmbH, 160

Computing: Publishers (International)
Chand (S.) & Co Ltd, 195
Jaico Publishing House, 195

Computing: Publishers (UK)
Artech House, 69
British Computer Society, The, 78
Business Education Publishers Ltd, 79
Haynes Publishing, 103
McGraw-Hill Education, 113
Reader's Digest Association Ltd, 132
Springer-Verlag London Ltd, 141

Computing: Publishers (US)
AMACOM Books, 172
Barron's Educational Series, Inc., 173
McGraw-Hill Companies Inc., The, 180

Cookery: Literary Agents (Ireland)
Williams (Jonathan) Literary Agency, 279

Cookery: Literary Agents (UK)
Anderson (Darley) Literary, TV & Film Agency, 249
Chris (Teresa) Literary Agency Ltd, 253
Feldstein Agency, The, 256
Graham Maw Christie Literary Agency, 258
hhb agency ltd, 260
Limelight Management, 262
Sheil Land Associates Ltd, 268
Smith (Robert) Literary Agency Ltd, 269
Wiener (Dinah) Ltd, 272

Cookery: Literary Agents (US)
Dystel & Goderich Literary Management, 283
Ellenberg (Ethan) Literary Agency, 283
Gusay (Charlotte) Literary Agency, The, 284
Lescher & Lescher Ltd, 285
Mews Books Ltd, 286
Parks (Richard) Agency, 287
Quicksilver Books, Literary Agents, 288

Cookery: Magazines
BBC Good Food, 299
Country Smallholding, 311
Delicious, 314
Easy Living, 316
Grow Your Own, 325
Ideal Home, 328
Olive, 342
Sainsbury's Magazine, 352
Speciality Food, 357
That's Life!, 360
W.I. Life, 365

Cookery: Picture Libraries
Blake (Anthony) Photo Library, 603
Cephas Picture Library, 604
Food Features, 609

Cookery: Prizes
Glenfiddich Food & Drink Awards, 502
Gourmand World Cookbook Awards, 502
Guild of Food Writers Awards, 503

Cookery: Publishers (European)
BZZTÔH (Uitgeverij) BV, 167
Civilização Editora, 169
Flammarion Éditions, 163
Garzanti Libri SpA, 166
Giunti Editoriale SpA, 166
Heyne Verlag, 165
Lannoo, Uitgeverij, 160
Larousse (Éditions), 163
Nyt Nordisk Forlag Arnold Busck A/S, 162
Tusquets Editores, 170

Cookery: Publishers (International)
Hachette Livre New Zealand Ltd, 196
HarperCollins Canada Ltd, 193
HarperCollins Publishers (New Zealand) Ltd, 196
HarperCollins Publishers India Ltd, 195
Random House New Zealand, 197
Raupo Publishing (NZ) Ltd, 197

Cookery: Publishers (Ireland)
Gill & Macmillan, 155
On Stream Publications Ltd, 157

Cookery: Publishers (UK)
Anness Publishing Ltd, 66
Anova Books, 66
Ashgrove Publishing, 69
Aurora Metro, 70
Austin & Macauley Publishers Limited, 70
Black & White Publishing Ltd, 73
Blackstaff Press Ltd, 74
Bonnier Books (UK), 75
Cathie (Kyle) Ltd, 81
Davies (Christopher) Publishers Ltd, 88
Dorling Kindersley Ltd, 90
Foulsham Publishers, 95
Garnet Publishing Ltd, 97
GMC Publications Ltd, 98
Grub Street, 99
Hachette Livre UK, 100
HarperCollins Publishers Ltd, 102
Murdoch Books UK Ltd, 118
New Holland Publishers (UK) Ltd, 119
Octopus Publishing Group, 121
Phaidon Press Limited, 126
Quadrille Publishing Ltd, 130
Quiller Press, 130
Random House Group Ltd, The, 131

Reader's Digest Association Ltd, 132
Ryland Peters & Small Limited, 135
Sheldrake Press, 138
Simon & Schuster UK Limited, 139
Souvenir Press Ltd, 140
Sportsman's Press, The, 141
Swan Hill Press, 142
Telegraph Books, 143
Wilson (Neil) Publishing Ltd, 149

Cookery: Publishers (US)
Barron's Educational Series, Inc., 173
Crown Publishing Group, 175
HarperCollins Publishers, Inc., 177
Hippocrene Books, Inc., 178
Interlink Publishing Group, Inc., 179
Pelican Publishing Company, 183
Reader's Digest Association Inc, 185
Washington State University Press, 190

Cookery: Small Presses
Serif, 235

Cookery: UK Packagers
Amber Books Ltd, 242
Quarto Publishing plc, 245

Countryside/Landscape: Book Clubs
Readers' Union Ltd, 293

Countryside/Landscape: Film, TV and Radio Producers
MBP TV, 428

Countryside/Landscape: Library Services
Natural England, 591

Countryside/Landscape: Literary Agents (US)
Gusay (Charlotte) Literary Agency, The, 284

Countryside/Landscape: Literary Societies
Kilvert Society, The, 562

Countryside/Landscape: Publishers (UK)
Sportsman's Press, The, 141

Countryside/Landscape: Magazines
Country Life, 311
Country Living, 311
Country Smallholding, 311
Countryman, The, 312
Cumbria, 312
Field, The, 320
National Trust Magazine, The, 339
Scottish Home & Country, 353
TGO (The Great Outdoors), 360
This England, 360

Countryside/Landscape: News Agencies
Space Press News and Pictures, 393

Countryside/Landscape: Picture Libraries
Cleare (John)/Mountain Camera, 605
Cordaiy (Sylvia) Photo Library, 606
Country Life Picture Library, 606
Country Matters Picture Library, 606
Craven (Philip) Worldwide Photo-Library, 606
English Heritage Photo Library, 607
Forest Life Picture Library, 609
Grace (Martin and Dorothy), 611
Hardley (Dennis) Photography, Scottish Photo Library, 611
Heseltine (John) Archive, 611
Howes (Chris)/Wild Places Photography, 612
Nature Picture Library, 618
Oxford Picture Library, 619
Oxford Scientific (OSF), 619
Photoshot, 619
PictureBank Photo Library Ltd, 620
Rex Features Ltd, 621
Sharp (Mick) Photography, 622
Skyscan Photolibrary, The, 623
Wellcome Images, 626
Williams (David) Picture Library, 626

Countryside/Landscape: Prizes
Lakeland Book of the Year Awards, 507

Countryside/Landscape: Publishers (UK)
Sportsman's Press, The, 141

Swan Hill Press, 142
Whittet Books Ltd, 149

Countryside/Landscape: Small Presses
Graffeg, 229

Crafts & Decorative Arts: Book Clubs
Readers' Union Ltd, 293

Crafts & Decorative Arts: Literary Agents (UK)
Chris (Teresa) Literary Agency Ltd, 253
Limelight Management, 262

Crafts & Decorative Arts: Literary Agents (US)
Bernstein (Meredith) Literary Agency, Inc., 280
Gusay (Charlotte) Literary Agency, The, 284

Crafts & Decorative Arts: Magazines
Classic Stitches, 308
Country Smallholding, 311
Embroidery, 317
Homes & Antiques, 327
Machine Knitting Monthly, 334
Scottish Farmer, The, 352
W.I. Life, 365

Crafts & Decorative Arts: Picture Libraries
Country Life Picture Library, 606

Crafts & Decorative Arts: Publishers (European)
Flammarion Éditions, 163
Giunti Editoriale SpA, 166
Høst & Søn Publishers Ltd, 161

Crafts & Decorative Arts: Publishers (International)
Simon & Schuster (Australia) Pty Ltd, 192

Crafts & Decorative Arts: Publishers (Ireland)
O'Brien Press Ltd, The, 157

Crafts & Decorative Arts: Publishers (UK)
Anness Publishing Ltd, 66
Anova Books, 66
Cico Books, 84
Crowood Press Ltd, The, 87
David & Charles Publishers, 88

Dorling Kindersley Ltd, 90
Floris Books, 95
GMC Publications Ltd, 98
HarperCollins Publishers Ltd, 102
Murdoch Books UK Ltd, 118
New Holland Publishers (UK) Ltd, 119
Octopus Publishing Group, 121
Quadrille Publishing Ltd, 130
Search Press Ltd, 137
Shetland Times Ltd, The, 138
Souvenir Press Ltd, 140

Crafts & Decorative Arts: Publishers (US)
Stackpole Books, 187
Sterling Publishing Co. Inc., 188

Crafts & Decorative Arts: Small Presses
Alloway Publishing Limited, 225
Porteous (David) Editions, 234

Crafts & Decorative Arts: UK Packagers
Breslich & Foss Ltd, 242
Quarto Publishing plc, 245

Cricket: Magazines
Day by Day, 314
Wisden Cricketer, The, 367

Cricket: Picture Libraries
Eagar (Patrick) Photography, 607
Mann (Roger) Collection, The, 616
Professional Sport UK Ltd, 620

Cricket: Prizes
Wisden (Cricket) Book of the Year, 524

Crime: Audio Books
Orion Audio Books, 240
Soundings Audio Books, 241

Crime: Electronic Publishing and Other Services
Crime Scene Scotland, 247

Crime: Festivals
Theakstons Old Peculier Harrogate Crime Writing Festival, 639

Crime: Literary Agents (Ireland)
Book Bureau Literary Agency, The, 279

Crime: Literary Agents (UK)
Ampersand Agency Ltd, The, 249
Anderson (Darley) Literary, TV & Film Agency, 249
Artellus Limited, 250
Bonomi (Luigi) Associates Limited (LBA), 251
Brown (Jenny) Associates, 251
Burton (Juliet) Literary Agency, 252
Cheetham (Mic) Associates, 253
Chris (Teresa) Literary Agency Ltd, 253
Dorian Literary Agency (DLA), 255
Feldstein Agency, The, 256
Gregory & Company Authors' Agents, 258
Holt (Vanessa) Ltd, 260
Indepublishing CIA – Consultancy for Independent Authors, 261
Inspira Group, The, 261
Judd (Jane) Literary Agency, 262
London Independent Books, 263
Motley (Michael) Ltd, 265
Power (Shelley) Literary Agency Ltd, 267
Sayle Literary Agency, The, 268
Sheil Land Associates Ltd, 268
Simmons (Jeffrey), 269
Smith (Robert) Literary Agency Ltd, 269
Wade & Doherty Literary Agency Ltd, 271
Watson, Little Ltd, 272

Crime: Literary Agents (US)
Gusay (Charlotte) Literary Agency, The, 284
Manus & Associates Literary Agency, Inc., 286
Morrison (Henry) Inc., 287
PMA Literary & Film Management, Inc., 287
Quicksilver Books, Literary Agents, 288
Russell & Volkening, Inc., 289

Crime: Magazines
Crimewave, 312

Crime: Miscellany
Landeau (Caroline), 641

Crime: Picture Libraries
Popperfoto.com, 620

Crime: Prizes
Crime Writers' Association (Cartier Diamond Dagger), 497
Crime Writers' Association (The CWA Ellis Peters Award), 497
Crime Writers' Association (The CWA Gold Dagger for Non-Fiction), 497
Crime Writers' Association (The Dagger in the Library), 497
Crime Writers' Association (The Debut Dagger), 497
Crime Writers' Association (The Gold and Silver Daggers for Fiction), 497
Crime Writers' Association (The New Blood Dagger for Best First Crime Novel), 497
Lawrie (Duncan) Dagger, The, 508
Lawrie (Duncan) International Dagger, The, 508
Theakstons Old Peculier Prize for the Crime Novel of the Year, 521

Crime: Professional Associations
Crime Writers' Association (CWA), 535

Crime: Publishers (European)
Garzanti Libri SpA, 166
Karisto Oy, 162
Kiss József Könyvkiadó, 166
Publicat SA, 169

Crime: Publishers (International)
Allen & Unwin Pty Ltd, 191
HarperCollins Publishers India Ltd, 195
Pan Macmillan Australia Pty Ltd, 192

Crime: Publishers (Ireland)
Merlin Publishing, 157
O'Brien Press Ltd, The, 157

Crime: Publishers (UK)
Allison & Busby, 65
Apex Publishing Ltd, 67
Atlantic Books, 69
Austin & Macauley Publishers Limited, 70
Brown Skin Books, 78
Constable & Robinson Ltd, 85
Crème de la Crime Ltd, 86

Fort Publishing Ltd, 95
Hachette Livre UK, 100
HarperCollins Publishers Ltd, 102
Little, Brown Book Group UK, 111
Macmillan Publishers Ltd, 113
Milo Books Limited, 117
Old Street Publishing Ltd, 121
Olympia Publishers, 122
Orion Publishing Group Limited, The, 123
Pegasus Elliot Mackenzie Publishers Ltd, 125
Pennant Books Ltd, 126
Poisoned Pen Press UK Ltd, 127
Polity Press, 128
Polygon, 128
Quercus Publishing, 130
Random House Group Ltd, The, 131
Severn House Publishers, 138
Souvenir Press Ltd, 140
Transworld Publishers Ltd, 145
Twenty First Century Publishers Ltd, 146
Wilson (Neil) Publishing Ltd, 149

Crime: Publishers (US)
St Martin's Press LLC, 186

Crime: Small Presses
Five Leaves Publications, 228

Crime: UK Packagers
Amber Books Ltd, 242

Crime: Writers' Circles and Workshops
Scribo, 479

Crime: Writers' Courses (UK and Ireland)
Community Creative Writing, 461

Current Affairs: Film, TV and Radio Producers
All Out Productions, 418
Hourglass Productions, 426
Lucida Productions, 428
Moonstone Films Ltd, 429

Current Affairs: Library Services
PA News Centre, 592

Current Affairs: Literary Agents (UK)
Artellus Limited, 250

Bryan (Felicity), 252
Chilcote (Judith) Agency, 253
Davidson (Caroline) Literary Agency, 255
Feldstein Agency, The, 256
Futerman, Rose & Associates, 257
Green (Annette) Authors' Agency, 258
Heath (Rupert) Literary Agency, 260
Higham (David) Associates Ltd, 260
Johnson & Alcock, 261
Lownie (Andrew) Literary Agency, 263
Pawsey (John), 266
Sayle Literary Agency, The, 268
Sinclair-Stevenson, 269
Smith (Robert) Literary Agency Ltd, 269
Turnbull (Jane), 271

Current Affairs: Literary Agents (US)
Bernstein (Meredith) Literary Agency, Inc., 280
Bleecker Street Associates, Inc., 280
Dijkstra (Sandra) Literary Agency, 282
Dystel & Goderich Literary Management, 283
Ellenberg (Ethan) Literary Agency, 283
Gusay (Charlotte) Literary Agency, The, 284
Lescher & Lescher Ltd, 285
Michaels (Doris S.) Literary Agency Inc., 286
Protter (Susan Ann) Literary Agent, 288
Russell & Volkening, Inc., 289
Ware (John A.) Literary Agency, 291

Current Affairs: Magazines
BBC History Magazine, 299
New Statesman, 340
Private Eye, 347
Prospect, 347
Time, 360
Tribune, 363

Current Affairs: PR Consultants
Publishing Services, 278

Current Affairs: Picture Libraries
Getty Images, 610

Hoffman (David) Photo Library, 612
Magnum Photos Ltd, 616
Popperfoto.com, 620
Rex Features Ltd, 621

Current Affairs: Poetry Magazines
Planet, 213

Current Affairs: Prizes
BBCFour Samuel Johnson Prize for Non-Fiction, The, 490

Current Affairs: Publishers (International)
Hachette Livre Australia, 191
HarperCollins Publishers India Ltd, 195
Queensland Press, University of, 192
Vision Books Pvt. Ltd, 196

Current Affairs: Publishers (Ireland)
Gill & Macmillan, 155
Hachette Books Ireland, 155
Maverick House Publishers, 156
Mercier Press Ltd, 156

Current Affairs: Publishers (UK)
Atlantic Books, 69
Aurum Press Ltd, 70
Austin & Macauley Publishers Limited, 70
AuthorHouse UK, 70
Carlton Publishing Group, 81
Clairview Books Ltd, 84
Constable & Robinson Ltd, 85
Gibson Square, 97
HarperCollins Publishers Ltd, 102
Lawrence & Wishart Ltd, 110
Mainstream Publishing Co. (Edinburgh) Ltd, 114
Oneworld Publications, 122
Pluto Press Ltd, 127
Profile Books, 129
Random House Group Ltd, The, 131
Tauris (I.B.) & Co. Ltd, 143
Temple Lodge Publishing Ltd, 144
Yale University Press (London), 151

Current Affairs: Publishers (US)
Globe Pequot Press, The, 176
Perseus Books Group, 184
PublicAffairs, 185

Current Affairs: Television and Radio
BBC News, 397

Dance: Library Services
Barbican Library, 574
Westminster Reference Library, 598
Williams (Vaughan) Memorial Library, 599

Dance: Literary Agents (US)
Gusay (Charlotte) Literary Agency, The, 284

Dance: Magazines
Dance Theatre Journal, 313
Dance Today, 313
Dancing Times, The, 313
Mixmag, 336

Dance: Picture Libraries
ArenaPAL, 602
Classical Collection Ltd, 605
Dominic Photography, 607
Jack (Robbie) Photography, 613
MacQuitty International Photographic Collection, 616
Williams (Vaughan) Memorial Library, 626

Dance: Publishers (European)
Arche Verlag AG, 171
Beck (C.H., OHG) Verlag, 164
Fayard (Librarie Arthème), 163
Hoffmann und Campe Verlag GmbH, 165
Larousse (Éditions), 163

Dance: Publishers (International)
Currency Press Pty Ltd, 191
Munshiram Manoharlal Publishers Pvt Ltd, 195

Dance: Publishers (UK)
Luath Press Ltd, 112
Northcote House Publishers Ltd, 120
Oberon Books, 120

Defence: Magazines
Jane's Defence Weekly, 330

Defence: Picture Libraries
aviation–images.com, 603
Defence Picture Library, The, 606

Defence: Publishers (UK)
Jane's Information Group, 108

Defence: Publishers (US)
Potomac Books, Inc., 184Potomac Books, Inc.,

Dentistry: Publishers (US)
Blackwell Publishing (US), 174

Design: Library Services
Westminster Reference Library, 598

Design: Literary Agents (UK)
Davidson (Caroline) Literary Agency, 255
Johnson & Alcock, 261

Design: Literary Agents (US)
Gusay (Charlotte) Literary Agency, The, 284

Design: Magazines
Architectural Review, The, 296
Blueprint, 302
Eye: The international review of graphic design, 318
Wallpaper, 365

Design: PR Consultants
Sawford (Claire) PR, 278

Design: Professional Associations
Design and Artists Copyright Society (DACS), 536

Design: Publishers (European)
Flammarion Éditions, 163
Nyt Nordisk Forlag Arnold Busck A/S, 162

Design: Publishers (UK)
Black (A.&C.) Publishers Ltd, 73
Carlton Publishing Group, 81
King (Laurence) Publishing Ltd, 109
Manchester University Press, 115
Merrell Publishers Ltd, 116
Octopus Publishing Group, 121
Phaidon Press Limited, 126
Prestel Publishing Limited, 129
Reaktion Books, 132
RotoVision, 134
Tauris (I.B.) & Co. Ltd, 143
Thames and Hudson Ltd, 144

Design: Publishers (US)
Crown Publishing Group, 175

Design: UK Packagers
Cameron & Hollis, 243
Ilex Press Limited, The, 244
Quarto Publishing plc, 245

Dictionaries: Prizes
Besterman/McColvin Medal, 490

Dictionaries: Publishers (European)
Akadémiai Kiadó, 166
Beck (C.H., OHG) Verlag, 164
Bertrand Editora Lda, 169
Bordas (Éditions), 162
Cappelen Damm AS, 168
Civilização Editora, 169
Deutscher Taschenbuch Verlag GmbH & Co. KG, 164
Espasa-Calpe (Editorial) SA, 170
Garzanti Libri SpA, 166
Giunti Editoriale SpA, 166
Gyldendal Norsk Forlag, 168
Klett (Ernst) Verlag GmbH, 165
Larousse (Éditions), 163
Lettera, 161
Naouka I Izkoustvo, 161
Presses Universitaires de France (PUF), 164
Random House Mondadori, 170
Società Editrice Internazionale – SEI, 167
Spectrum (Uitgeverij Het) BV, 168
Verbo (Editorial) SA, 169
WSOY (Werner Söderström Osakeyhtiö), 162

Dictionaries: Publishers (International)
Chand (S.) & Co Ltd, 195
HarperCollins Publishers India Ltd, 195
Macmillan India Ltd, 195
Macmillan Publishers Australia Pty Ltd, 191
Munshiram Manoharlal Publishers Pvt Ltd, 195
Oxford University Press India, 196
Wiley (John) & Sons Australia Ltd, 193

Dictionaries: Publishers (UK)
Black (A.&C.) Publishers Ltd, 73
Chambers Harrap Publishers Ltd, 82
HarperCollins Publishers Ltd, 102
Macmillan Publishers Ltd, 113
Oxford University Press, 124

Dictionaries: Publishers (US)
Hippocrene Books, Inc., 178

Dictionaries: UK Packagers
Lexus Ltd, 244
Market House Books Ltd, 244

Directories: Prizes
Besterman/McColvin Medal, 490

Directories: Professional Associations
Data Publishers Association, 536

Directories: Publishers (European)
Beck (C.H., OHG) Verlag, 164
Bordas (Éditions), 162
Cappelen Damm AS, 168
Denoël (Éditions), 163
Deutscher Taschenbuch Verlag GmbH & Co. KG, 164
Europa-América (Publicações) Lda, 169
Fayard (Librarie Arthème), 163
Gradiva–Publicações S.A., 169
Gyldendal, 161
Gyldendal Norsk Forlag, 168
Hachette Livre (France), 163
Kok (Uitgeversmaatschappij J. H.) BV, 168
Larousse (Éditions), 163
Mondadori (Arnoldo) Editore SpA, 167
Mursia (Gruppo Ugo) Editore SpA, 167
Pearson Education (Benelux), 168
Schønbergske (Det) Forlag, 162

Directories: Publishers (International)
Currency Press Pty Ltd, 191
Nelson Education, 194

Directories: Publishers (UK)
CBD Research Ltd, 82
Euromonitor International plc, 93
Hobsons Plc, 105
Jane's Information Group, 108
Kogan Page Ltd, 109
NCVO Publications, 119
Nielsen BookData, 119
Reed (William) Directories, 133
Richmond House Publishing Company Ltd, 134
Straightline Publishing Ltd, 141

Disability: Film, TV and Radio Producers
Redweather, 431

Disability: Magazines
Disability Now, 315
Mobilise, 336

Disability: Picture Libraries
Photofusion, 619

Disability: Poetry Magazines
Once Orange Badge Poetry Supplement, The, 213

Disability: Publishers (UK)
Souvenir Press Ltd, 140

Disability: Small Presses
Enable Enterprises, 228

Disability: Theatre Producers
Graeae Theatre Company, 441

Disability: Writers' Courses (UK and Ireland)
Starz! – Film and Theatre Performing Arts, 464

DIY: Magazines
House Beautiful, 328

DIY: Publishers (International)
Allen & Unwin Pty Ltd, 191

DIY: Publishers (UK)
Crowood Press Ltd, The, 87
Foulsham Publishers, 95
GMC Publications Ltd, 98
HarperCollins Publishers Ltd, 102
Haynes Publishing, 103
Murdoch Books UK Ltd, 118
New Holland Publishers (UK) Ltd, 119
Quiller Press, 130
Reader's Digest Association Ltd, 132

DIY: Publishers (US)
Reader's Digest Association Inc, 185

Documentary: Film, TV and Radio Producers
Above The Title, 417
Abstract Images, 417
Acacia Productions Ltd, 417
All Out Productions, 418
Art & Training Films Ltd, 418

Ashford Entertainment Corporation Ltd, The, 418
Avalon, 418
Beckmann International, 419
Big Heart Media, 419
Blackwatch Productions Limited, 419
Bona Broadcasting Limited, 419
Chameleon Television Ltd, 420
CSA Word, 421
Cutting Edge Productions Ltd, 422
Diverse Production Limited, 422
DoubleBand Films, 422
Dunstan (Charles) Communications Ltd, 422
Electric Airwaves Ltd (Ladbroke Radio), 422
Endemol UK Productions, 422
Fiction Factory, 423
Flannel, 423
Flashback Television Limited, 423
Gaia Communications, 424
Granite Productions, 425
Green Umbrella Ltd, 425
Hartswood Films Ltd, 425
Healthcare Productions Limited, 425
Holmes Associates, 425
Hourglass Productions, 426
Icon Films, 426
Imari Entertainment Ltd, 426
Isis Productions, 426
Isolde Films, 426
JAM Pictures and Jane Walmsley Productions, 426
Keo Films.com Ltd, 427
Lagan Pictures Ltd, 427
Landseer Productions Ltd, 427
Lilyville Screen Entertainment Ltd, 427
Loftus Audio Ltd, 427
London Scientific Films Ltd, 428
Lucida Productions, 428
Maverick Television, 428
Maya Vision International Ltd, 428
Moonstone Films Ltd, 429
Neon, 429
Odyssey Productions Ltd, 429
Omnivision, 429
Orlando Television Productions Ltd, 429
Orpheus Productions, 429
Paper Moon Productions, 430
Passion Pictures, 430
Pelicula Films, 430
Photoplay Productions Ltd, 430
Planet24 Pictures Ltd, 431

Plantagenet Films Limited, 431
Redweather, 431
RS Productions, 431
Screen First Ltd, 432
Screen Ventures Ltd, 432
Screenhouse Productions Ltd, 432
September Films Ltd, 432
Sianco Cyf, 432
Silent Sound Films Ltd, 432
Skyline Productions, 433
SMG Productions & Ginger Productions, 433
Somethin Else, 433
'Spoken' Image Ltd, 433
Square Dog Radio LLP, 433
Staveacre (Tony) Productions, 433
Stirling Film & TV Productions Limited, 433
Straight Forward Film & Television Productions Ltd, 434
talkbackTHAMES, 434
Tandem TV & Film Ltd, 434
Testimony Films, 435
Tiger Aspect Productions, 435
Tigress Productions Ltd, 435
Touch Productions Ltd, 435
Transatlantic Films Production and Distribution Company, 435
Twenty Twenty Television, 436
Twofour Broadcast, 436
VIP Broadcasting, 436
Wall to Wall, 436
Walsh Bros. Limited, 436
Whistledown Productions Ltd, 437
Wise Buddah Creative Ltd, 437
Wortman Productions UK, 437

Documentary: Literary Agents (UK)
London Independent Books, 263
PFD, 266
Robinson Literary Agency Ltd, 267

Documentary: Literary Agents (US)
Gusay (Charlotte) Literary Agency, The, 284

Documentary: Picture Libraries
Huntley Film Archive, 612

Documentary: Television and Radio
BBC Radio Kent, 402

Dogs: Literary Agents (US)
Gusay (Charlotte) Literary Agency, The, 284
Snell (Michael) Literary Agency, 290

Dogs: Magazines
Dog World, 316
Dogs Monthly, 316
Your Dog Magazine, 370

Dogs: Publishers (UK)
Crowood Press Ltd, The, 87

Drama/Plays: Audio Books
BBC Audiobooks Ltd, 238
HarperCollins AudioBooks, 239
Naxos AudioBooks, 239
Roberts (L.M.) Limited, 240
SmartPass Ltd, 241

Drama/Plays: Bursaries, Fellowships and Grants
PAWS (Public Awareness of Science) Drama Fund, 485
Pearson Playwrights' Scheme, 485
Ramsay (Peggy) Foundation, 485
Scottish Arts Council Writers' Bursaries, 486
Scottish Book Trust New Writers' Awards, 486

Drama/Plays: Electronic Publishing and Other Services
Online Originals, 248

Drama/Plays: Festivals
Aspects Literature Festival, 629
International Playwriting Festival, 634
National Association of Writers' Groups (NAWG) Open Festival of Writing, 636
Royal Court Young Writers Programme, 638

Drama/Plays: Film, TV and Radio Producers
Above The Title, 417
Abstract Images, 417
Acrobat Television, 417
Actaeon Films Ltd, 418
Anglo/Fortunato Films Ltd, 418
Arlington Productions Limited, 418
Art & Training Films Ltd, 418
Avalon, 418
Big Heart Media, 419
Blackwatch Productions Limited, 419
Bona Broadcasting Limited, 419
Carnival (Films & Theatre) Ltd, 419
Celador Films, 420
Celtic Films Entertainment Ltd, 420
Chameleon Television Ltd, 420
Company Pictures, 421
CSA Word, 421
DoubleBand Films, 422
Electric Airwaves Ltd (Ladbroke Radio), 422
Endemol UK Productions, 422
Festival Film and Television Ltd, 423
Fiction Factory, 423
Film and General Productions Ltd, 423
First Writes Theatre Company, 423
Flannel, 423
Flashback Television Limited, 423
Focus Films Ltd, 424
FremantleMedia Group Ltd, 424
Goldcrest Films International Ltd, 424
Hammerwood Film Productions, 425
Hartswood Films Ltd, 425
Hat Trick Productions Ltd, 425
Healthcare Productions Limited, 425
Holmes Associates, 425
Hourglass Productions, 426
Imari Entertainment Ltd, 426
ITV Productions, 426
JAM Pictures and Jane Walmsley Productions, 426
Kudos Film and Television Ltd, 427
Lagan Pictures Ltd, 427
Landseer Productions Ltd, 427
Lilyville Screen Entertainment Ltd, 427
Loftus Audio Ltd, 427
Lucida Productions, 428
Maverick Television, 428
Maya Vision International Ltd, 428
Merseyfilm, 427
Neon, 429
Number 9 Films, 429
Parallax East Ltd, 430
Passion Pictures, 430
Picture Palace Films Ltd, 430
Plantagenet Films Limited, 431
Portobello Pictures, 431

Promenade Enterprises Limited, 431
Renaissance Films, 431
Richmond Films & Television Ltd, 431
RS Productions, 431
Sands Films, 431
Scala Productions Ltd, 431
Screen First Ltd, 432
Screen Ventures Ltd, 432
Sianco Cyf, 432
Skyline Productions, 433
SMG Productions & Ginger Productions, 433
Spice Factory (UK) Ltd, 433
Staveacre (Tony) Productions, 433
Table Top Productions, 434
talkbackTHAMES, 434
Talkingheads Production Ltd, 434
Tern Television Productions Ltd, 434
Tiger Aspect Productions, 435
Twenty Twenty Television, 436
Wall to Wall, 436
Walsh Bros. Limited, 436
Working Title Films Ltd, 437
Wortman Productions UK, 437

Drama/Plays: Library Services
Birmingham Library Services, 576
British Library Sound Archive, 579
Mitchell Library, The, 590
Plymouth Library and Information Services, City of, 593

Drama/Plays: Literary Agents (UK)
Agency (London) Ltd, The, 249
Burkeman (Brie), 252
Casarotto Ramsay and Associates Ltd, 252
Cochrane (Elspeth) Personal Management, 254
Curtis Brown Group Ltd, 254
Daish (Judy) Associates Ltd, 255
de Wolfe (Felix), 255
Dench Arnold Agency, The, 255
Fitch (Laurence) Ltd, 256
Glass (Eric) Ltd, 257
Hall (Rod) Agency Limited, The, 259
Hancock (Roger) Ltd, 259
Higham (David) Associates Ltd, 260
Independent Talent Group Limited, 261

International Scripts, 261
Mann (Andrew) Ltd, 264
McLean (Bill) Personal Management, 264
Narrow Road Company, The, 265
PFD, 266
Sayle Screen Ltd, 268
Sharland Organisation Ltd, The, 268
Sheil Land Associates Ltd, 268
Steinberg (Micheline) Associates, 270
Tennyson Agency, The, 270
TVmyworld, 271
United Agents Limited, 271
Ware (Cecily) Literary Agents, 272
Weinberger (Josef) Plays, 272

Drama/Plays: Literary Agents (US)
Freedman (Robert A.) Dramatic Agency, Inc., 283
Gusay (Charlotte) Literary Agency, The, 284
Hogenson (Barbara) Agency, Inc., The, 284

Drama/Plays: Literary Societies
Marlowe Society, The, 563
Shaw Society, The, 567

Drama/Plays: Magazines
Gold Dust Magazine, 323
New Writing Scotland, 341

Drama/Plays: Poetry Magazines
Understanding, 218

Drama/Plays: Poetry Presses
Diehard Publishers, 200
Feather Books, 201
Gallery Press, The, 202
Penniless Press, 204

Drama/Plays: Prizes
Bargate (Verity) Award, 489
Cohen (David) Prize for Literature, 495
Devine (George) Award, 498
Drama Association of Wales Playwriting Competition, 498
Fagon (Alfred) Award, The, 500
Imison Award, The, 505
Llewellyn Rhys (John) Prize, The, 509
Meyer-Whitworth Award, 511
Pulitzer Prizes, 516

Sunday Times Young Writer of the Year Award, 521
Thomas (Dylan) Prize, The, 521
Way (Brian) Award, 523
Wolff (Peter) Theatre Trust supports the Whiting Award, The, 524
Yale Drama Series/David C. Horn Prize, The, 525

Drama/Plays: Professional Associations
Drama Association of Wales, 536
New Writing North, 542
Playwrights' Studio, Scotland, 543
Writernet, 550

Drama/Plays: Publishers (European)
Larousse (Éditions), 163
Tiderne Skifter Forlag A/S, 162

Drama/Plays: Publishers (International)
Currency Press Pty Ltd, 191
Munshiram Manoharlal Publishers Pvt Ltd, 195

Drama/Plays: Publishers (Ireland)
Cló Iar-Chonnachta, 154
Gallery Press, The, 155
New Island, 157

Drama/Plays: Publishers (UK)
Amber Lane Press Ltd, 66
Aurora Metro, 70
Black (A.&C.) Publishers Ltd, 73
Boyars (Marion) Publishers Ltd, 76
Brown, Son & Ferguson, Ltd, 78
Calder Publications Ltd, 79
Chapman Publishing, 83
Cressrelles Publishing Co. Ltd, 86
Faber & Faber Ltd, 94
French (Samuel) Ltd, 96
Hern (Nick) Books, 104
Northcote House Publishers Ltd, 120
Oberon Books, 120
Parthian, 124
Penguin Group (UK), 125
Seren, 137
Smythe (Colin) Ltd, 139

Drama/Plays: Publishers (US)
French (Samuel), Inc., 176
Players Press, 184

Drama/Plays: Small Presses
Dramatic Lines, 227
Feather Books, 228
Griffin Publishing Associates, 229
Maypole Editions, 231
Pipers' Ash Ltd, 233
Playwrights Publishing Co., 234

Drama/Plays: Television and Radio
BBC Drama, Comedy and Children's, 396
BBC Radio Cumbria, 401
BBC Radio Cymru, 401

Drama/Plays: Writers' Circles and Workshops
Kops (Bernard) and Tom Fry Drama Writing Workshops, 478
New Writing South, 478
North West Playwrights (NWP), 478
Original Writers Group, The, 479
Script Yorkshire, 479
Southwest Scriptwriters, 479
Sussex Playwrights' Club, 480

Drama/Plays: Writers' Courses (UK and Ireland)
Arvon Foundation, 458
Birmingham, University of, 464
Central School of Speech and Drama, The, 458
City Lit, 458
City University, 458
Dartington College of Arts, 455
Dingle Writing Courses, 466
Glamorgan, University of, 467
Hull, University of, 465
Lancaster University, 457
Leeds, University of, 465
Liverpool, University of, 461
Middlesex University, 459
Nottingham Trent University, The, 462
Reading, University of, 453
Royal Holloway University of London, 463
Writers Bureau, The, 461

Drama/Plays: Writers' Courses (US)
Adelphi University, 473
Boston University, 471
Brooklyn College of the City University of New York, 473
Brown University, 474
Chapman University, 469
Chicago State University, 471

Hollins University, 475
Louisiana State University, 471
Minnesota State University Moorhead, 472
San Francisco State University, 470
Virginia Commonwealth University, 475
Western Michigan University, 472

Eastern Europe: Literary Agents (US)
Gusay (Charlotte) Literary Agency, The, 284
Russell & Volkening, Inc., 289

Eastern Europe: Picture Libraries
Cunningham (Sue) Photographic, 606
Garden and Wildlife Matters Photo Library, 609
King (David) Collection, 613
Miller (Lee) Archives, 616
Panos Pictures, 619
Russia and Eastern Images, 622
Tropix Photo Library, 625

Eastern Europe: Prizes
British Czech & Slovak Association Prize, 493

Eastern Europe: Publishers (US)
Indiana University Press, 179
Sharpe (M.E.), Inc., 186

Ecology: Literary Agents (US)
Gusay (Charlotte) Literary Agency, The, 284
Quicksilver Books, Literary Agents, 288

Ecology: Magazines
Ecologist, The, 316
Quarterly Review, 348

Ecology: Picture Libraries
Ecoscene, 607
Garden and Wildlife Matters Photo Library, 609
GeoScience Features, 610
Grace (Martin and Dorothy), 611

Ecology: Publishers (Ireland)
Lilliput Press, The, 156
Tír Eolas, 158

Ecology: Publishers (UK)
Octopus Publishing Group, 121

Ecology: Publishers (US)
Holt (Henry) & Company, Inc., 178
Indiana University Press, 179
Krieger Publishing Co., 179

Ecology: Small Presses
Packard Publishing Limited, 233

Economics: Bursaries, Fellowships and Grants
Blundell (K.) Trust, The, 482

Economics: Library Services
Bank of England Information Centre, 574
Senate House Library, University of London, 595

Economics: Literary Agents (US)
Gusay (Charlotte) Literary Agency, The, 284
Russell & Volkening, Inc., 289

Economics: Magazines
Economist, The, 316

Economics: Picture Libraries
Report Digital, 621

Economics: Prizes
FT & Goldman Sachs Business Book of the Year Award, 501
Royal Economic Society Prize, 518

Economics: Publishers (European)
Adelphi Edizioni SpA, 166
Akadémiai Kiadó, 166
Beck (C.H., OHG) Verlag, 164
Calmann-Lévy (Éditions), 163
Caminho (Editorial) SARL, 169
Civilização Editora, 169
Deutscher Taschenbuch Verlag GmbH & Co. KG, 164
Hanser (Carl) Verlag GmbH & Co. KG, 165
Mulino (Societe editrice il), 167
Pearson Education (Benelux), 168
Presses Universitaires de France (PUF), 164
Random House Mondadori, 170
Springer GmbH, 165
Springer Science+Business Media BV, 168

Ullstein Buchverlage GmbH,
165
Universitaire Pers Leuven, 160

**Economics: Publishers
(International)**
Kwa-Zulu–Natal Press,
University of, 197
LexisNexis Butterworths (Pty)
Ltd South Africa, 197
Melbourne University
Publishing Ltd, 192
Wits University Press, 198

**Economics: Publishers
(Ireland)**
Institute of Public
Administration, 156

Economics: Publishers (UK)
Brealey (Nicholas) Publishing,
77
Business Education Publishers
Ltd, 79
Currey (James) Publishers, 87
Elgar (Edward) Publishing Ltd,
91
Freeman (W. H.), 96
Harriman House Ltd, 103
Harvard University Press, 103
Lawrence & Wishart Ltd, 110
Macmillan Publishers Ltd, 113
Manchester University Press,
115
McGraw-Hill Education, 113
Polity Press, 128
Random House Group Ltd,
The, 131
Routledge, 134
Shepheard-Walwyn (Publishers)
Ltd, 138
Verso, 147
Wimbledon Publishing
Company, 150

Economics: Publishers (US)
Massachusetts Press, University
of, 181
Michigan Press, The University
of, 181
Stanford University Press, 187

**Economics: US Media
Contacts in the UK**
Forbes Magazine, 394

Education: Audio Books
SmartPass Ltd, 241

Education: Book Clubs
Baker Books, 293

**Education: Film, TV and
Radio Producers**
Buccaneer Films, 419
Cleveland Productions, 420
Drake A-V Video Ltd, 422
Greenwich Village Productions,
425
Picardy Media &
Communication, 430
Sianco Cyf, 432
TV Choice Ltd, 435

Education: Library Services
Leeds Central Library, 587

**Education: Literary Agents
(US)**
Educational Design Services,
LLC, 283
Gusay (Charlotte) Literary
Agency, The, 284

Education: Magazines
Junior Education PLUS, 331
Nursery Education, 341
Report, 351
Teacher, The, 360
*Times Educational Supplement
Scotland, The*, 361
*Times Educational Supplement,
The*, 361
Times Higher Education, The,
361

Education: Picture Libraries
Education Photos, 607
Greenhill (Sally and Richard),
611
Last Resort Picture Library, 614
Photofusion, 619
Report Digital, 621

Education: Prizes
nasen Awards 2008, 512
SES Book Prizes, 519

**Education: Professional
Associations**
literaturetraining, 540
National Association of Writers
in Education, 541

**Education: Publishers
(European)**
Akadémiai Kiadó, 166
Aschehoug (H) & Co (W
Nygaard), 168
Blackwell Munksgaard, 161
Bordas (Éditions), 162
Bulzoni Editore SRL, 166

Deutscher Taschenbuch Verlag
GmbH & Co. KG, 164
Espasa-Calpe (Editorial) SA, 170
Giunti Editoriale SpA, 166
Groupo Anaya, 169
Gyldendal, 161
Hachette Livre (France), 163
Horizonte (Livros) Lda, 169
Klett (Ernst) Verlag GmbH, 165
Lettera, 161
Luis Vives (Grupo Editorial),
170
Magnard (Éditions), 163
Mondadori (Arnoldo) Editore
SpA, 167
Mursia (Gruppo Ugo) Editore
SpA, 167
Nathan (Les Éditions), 164
Natur och Kultur Bokförlaget,
171
Nyt Nordisk Forlag Arnold
Busck A/S, 162
Otava Publishing Co. Ltd, 162
Payot Libraire, 171
Pearson Educacion SA, 170
Pearson Education (Benelux),
168
Polish Scientific Publishers
PWN, 169
Publicat SA, 169
Sauerländer Verlage AG, 171
Società Editrice Internazionale
– SEI, 167
Standaard Uitgeverij, 160
Tammi Publishers, 162
Verbo (Editorial) SA, 169
Vuibert (Librairie), 164
WSOY (Werner Söderström
Osakeyhtiö), 162

**Education: Publishers
(International)**
ACER Press, 191
Chand (S.) & Co Ltd, 195
Heinemann International
Southern Africa, 197
Macmillan India Ltd, 195
Macmillan Publishers Australia
Pty Ltd, 191
Macmillan Publishers New
Zealand, 196
Macmillan South Africa, 197
Maskew Miller Longman, 197
McGraw-Hill Australia Pty Ltd,
192
McGraw-Hill Education (India),
195
McGraw-Hill Ryerson Ltd, 194
Melbourne University
Publishing Ltd, 192
Nelson Education, 194

Oxford University Press (Australia, New Zealand & the Pacific), 192
Oxford University Press India, 196
Oxford University Press Southern Africa (Pty) Ltd, 197
Oxford University Press, Canada, 194
Pearson Canada, 194
Pearson Education Australia Pty Ltd, 192
Pearson Education New Zealand Ltd, 196
Scholastic Australia Pty Limited, 192
Scholastic Canada Ltd, 194
Science Press, 192
Shuter & Shooter (Pty) Ltd, 197
Wiley (John) & Sons Australia Ltd, 193

Education: Publishers (Ireland)
An Gúm, 153
Gill & Macmillan, 155

Education: Publishers (UK)
Atlantic Europe Publishing Co. Ltd, 69
Austin & Macauley Publishers Limited, 70
Badger Publishing Ltd, 71
Barrington Stoke, 71
Black (A.&C.) Publishers Ltd, 73
Business Education Publishers Ltd, 79
Cambridge University Press, 79
CBA Publishing, 82
Cengage Learning (EMEA) Ltd, 82
Continuum International Publishing Group Ltd, The, 85
Crimson Publishing, 86
Crown House Publishing, 87
Drake Educational Associates, 90
Elm Publications/Training, 92
Evans Brothers Ltd, 93
First & Best in Education Ltd, 94
Folens Limited, 95
Freeman (W. H.), 96
Fulton (David) Publishers Ltd, 96
GL Assessment, 98
Gomer Press/Gwasg Gomer, 98
Graham-Cameron Publishing & Illustration, 98
Hachette Livre UK, 100

HarperCollins Publishers Ltd, 102
Hertfordshire Press, University of, 104
Hodder Gibson, 105
Hymns Ancient & Modern Ltd, 106
Lawrence & Wishart Ltd, 110
Macmillan Publishers Ltd, 113
Marshall Cavendish Ltd, 115
McGraw-Hill Education, 113
Nelson Thornes Limited, 119
Northcote House Publishers Ltd, 120
Open University Press, 123
Oxford University Press, 124
Pearson, 125
Pearson (Heinemann), 125
Pegasus Elliot Mackenzie Publishers Ltd, 125
QED Publishing, 129
Ransom Publishing Ltd, 132
Reed Elsevier Group plc, 133
Rising Stars UK Ltd, 134
Routledge, 134
Sage Publications, 135
Sandstone Press Ltd, 135
Scholastic Ltd, 136
Souvenir Press Ltd, 140
Speechmark Publishing Ltd, 140
Studymates Limited, 142
Trentham Books Ltd, 145
Trotman, 146
TSO (The Stationery Office), 146
Ward Lock Educational Co. Ltd, 148
Wiley-Blackwell, 149
Wimbledon Publishing Company, 150

Education: Publishers (US)
ABC–CLIO (US), 172
Charlesbridge Publishing, 174
Krieger Publishing Co., 179
Lerner Publishing Group, 180
McGraw-Hill Companies Inc., The, 180
Pearson Scott Foresman, 183
Prometheus Books, 184
Scholastic Library Publishing, 186
Southern Illinois University Press, 187
State University of New York Press, 187
Thomson Gale, 188
Walch Education, 189
Wiley (John) & Sons, Inc., 190

Education: Small Presses
Brilliant Publications, 226

Educational Heretics Press, 227
GSSE, 229
Ivy Publications, 230
Robinswood Press, 234
St James Publishing, 234
Witan Books, 237

Education: Writers' Circles and Workshops
Coleg Harlech WEA, 477
Workers' Educational Association, 480
Workers' Educational Association Northern Ireland, 480
Workers' Educational Association Scotland, 480
Workers' Educational Association South Wales, 480

Electronics: Magazines
Radio User, 349

Electronics: Publishers (UK)
Artech House, 69
Helicon Publishing, 104
Special Interest Model Books Ltd, 140

Encyclopedias: Prizes
Besterman/McColvin Medal, 490

Encyclopedias: Publishers (European)
Beck (C.H., OHG) Verlag, 164
Bertrand Editora Lda, 169
Bordas (Éditions), 162
Brepols Publishers NV, 160
Brána a.s., 161
Cappelen Damm AS, 168
Civilização Editora, 169
Deutscher Taschenbuch Verlag GmbH & Co. KG, 164
Espasa-Calpe (Editorial) SA, 170
Garzanti Libri SpA, 166
Gyldendal Norsk Forlag, 168
Hachette Livre (France), 163
Kibea Publishing Company, 160
Klett (Ernst) Verlag GmbH, 165
Kok (Uitgeversmaatschappij J. H.) BV, 168
Larousse (Éditions), 163
Presses Universitaires de France (PUF), 164
Società Editrice Internazionale – SEI, 167
Spectrum (Uitgeverij Het) BV, 168
Verbo (Editorial) SA, 169

WSOY (Werner Söderström Osakeyhtiö), 162

Encyclopedias: Publishers (International)
Chand (S.) & Co Ltd, 195
Macmillan India Ltd, 195
Wiley (John) & Sons Australia Ltd, 193

Encyclopedias: Publishers (UK)
Cambridge University Press, 79
Helicon Publishing, 104
Macmillan Publishers Ltd, 113
Octopus Publishing Group, 121

Encyclopedias: Publishers (US)
Scholastic Library Publishing, 186
Thomson Gale, 188
Wiley (John) & Sons, Inc., 190

Encyclopedias: UK Packagers
Amber Books Ltd, 242
Market House Books Ltd, 244

Engineering: Bursaries, Fellowships and Grants
PAWS (Public Awareness of Science) Drama Fund, 485

Engineering: Library Services
British Library Science, Technology and Business Collections, 579

Engineering: Literary Agents (US)
Russell & Volkening, Inc., 289

Engineering: Magazines
Construction News, 310
Engineer, The, 317

Engineering: Picture Libraries
Mechanical Engineers (Institution of), 616

Engineering: Publishers (European)
Hachette Livre (France), 163
Hanser (Carl) Verlag GmbH & Co. KG, 165
Springer GmbH, 165
Springer Science+Business Media BV, 168

Engineering: Publishers (International)
Chand (S.) & Co Ltd, 195
Jaico Publishing House, 195

Engineering: Publishers (UK)
Cambridge University Press, 79
McGraw-Hill Education, 113
Nottingham University Press, 120
Special Interest Model Books Ltd, 140
Springer-Verlag London Ltd, 141
WIT Press, 150
Woodhead Publishing Ltd, 150

Engineering: Publishers (US)
Krieger Publishing Co., 179
McGraw-Hill Companies Inc., The, 180

Engineering: Small Presses
Whittles Publishing, 236

English Language Teaching (ELT): Professional Associations
English Association, The, 536

English Language Teaching (ELT): Publishers (International)
Macmillan Publishers Australia Pty Ltd, 191
Maskew Miller Longman, 197

English Language Teaching (ELT): Publishers (UK)
Cambridge University Press, 79
Macmillan Publishers Ltd, 113
Oxford University Press, 124
Pearson, 125

English Language Teaching (ELT): Publishers (US)
Michigan Press, The University of, 181

Entomological Studies: Library Services
Natural History Museum Library, The, 591

Entomological Studies: Picture Libraries
Natural History Museum Image Resources, 618

Entomological Studies: Publishers (UK)
Fitzgerald Publishing, 95

Environment/Conservation: Library Services
Natural England, 591

Environment/Conservation: Literary Agents (US)
Fredericks (Jeanne) Literary Agency, Inc., 283
Gusay (Charlotte) Literary Agency, The, 284
Quicksilver Books, Literary Agents, 288

Environment/Conservation: Magazines
BBC Wildlife Magazine, 299
Ecologist, The, 316
Naturalist, The, 339
New Internationalist, 339
W.I. Life, 365

Environment/Conservation: PR Consultants
Publishing Services, 278

Environment/Conservation: Picture Libraries
Cordaiy (Sylvia) Photo Library, 606
Ecoscene, 607
Forest Life Picture Library, 609
Garden and Wildlife Matters Photo Library, 609
GeoScience Features, 610
Hoffman (David) Photo Library, 612
Howes (Chris)/Wild Places Photography, 612
Hutchison Picture Library, 612
Lane (Frank) Picture Agency Ltd, 614
Nature Photographers Ltd, 618
Nature Picture Library, 618
Oxford Scientific (OSF), 619
Panos Pictures, 619
Photofusion, 619
Photoshot, 619
Report Digital, 621
Royal Geographical Society Picture Library, 622
RSPCA Photolibrary, 622
Science Photo Library, 622
Sharp (Mick) Photography, 622
Still Pictures' Whole Earth Photolibrary, 624

Environment/Conservation: Poetry Magazines
Earth Love, 209
Island, 211
Peace and Freedom, 213

Planet, 213

Environment/Conservation: Prizes
BBC Wildlife Magazine Poet of the Year Awards, 490

Environment/Conservation: Publishers (European)
Høst & Søn Publishers Ltd, 161

Environment/Conservation: Publishers (International)
Munshiram Manoharlal Publishers Pvt Ltd, 195

Environment/Conservation: Publishers (UK)
Austin & Macauley Publishers Limited, 70
Capall Bann Publishing, 80
Elgar (Edward) Publishing Ltd, 91
Green Books, 99
Luath Press Ltd, 112
Octopus Publishing Group, 121
Open Gate Press, 123
TSO (The Stationery Office), 146
Zed Books Ltd, 151

Environment/Conservation: Publishers (US)
Indiana University Press, 179
Kansas, University Press of, 179
Louisiana State University Press, 180
Nebraska Press, University of, 182
Texas Press, University of, 188
Washington State University Press, 190

Environment/Conservation: Small Presses
Witan Books, 237Witan Books,

Equal Opportunities: Library Services
Equality and Human Rights Commission, 582

Equal Opportunities: Small Presses
Enable Enterprises, 228

Eroticism: Literary Agents (UK)
London Independent Books, 263
Sunflower Literary Agency, 270

Eroticism: Literary Agents (US)
Gusay (Charlotte) Literary Agency, The, 284

Eroticism: Magazines
Desire, 315
Jade, 330
Mayfair, 335

Eroticism: Publishers (European)
Taschen GmbH, 165

Eroticism: Publishers (UK)
Blackstone Publishers, 74
Brown Skin Books, 78
Erotic Review Books, 93
Pegasus Elliot Mackenzie Publishers Ltd, 125
Random House Group Ltd, The, 131
SeaNeverDry Publishing, 137

Eroticism: Small Presses
Accent Press, 225
Great Northern Publishing, 229

Essays: Literary Agents (UK)
Feldstein Agency, The, 256

Essays: Literary Agents (US)
Gusay (Charlotte) Literary Agency, The, 284

Essays: Magazines
Abraxas Unbound, 294
Edinburgh Review, 317
Irish Pages, 329
New Shetlander, The, 340
Reader, The, 350
20x20 magazine, 363

Essays: Poetry Magazines
Agenda Poetry Magazine, 208
Irish Pages, 211
Penniless Press, The, 213
Quiet Feather, The, 215
Radiator, The, 215
Scriptor, 216
Stinging Fly, The, 217

Essays: Poetry Presses
Seren, 206

Essays: Prizes
Alexander Prize, 488
Cohen (David) Prize for Literature, 495
Keats–Shelley Prize, 506
Prose & Poetry Prizes, 516

Taylor (Reginald) and Lord Fletcher Essay Prize, 521

Essays: Publishers (European)
Arche Verlag AG, 171
Beck (C.H., OHG) Verlag, 164
Deutscher Taschenbuch Verlag GmbH & Co. KG, 164
Diogenes Verlag AG, 171
EDHASA (Editora y Distribuidora Hispano – Americana SA), 169
Espasa-Calpe (Editorial) SA, 170
Flammarion Éditions, 163
Garzanti Libri SpA, 166
Giunti Editoriale SpA, 166
Grasset & Fasquelle (Éditions), 163
Hiperión (Ediciónes) SL, 170
Longanesi (Casa Editrice) SpA, 167
Minuit (Les Éditions de) SA, 164
Naouka I Izkoustvo, 161
Random House Mondadori, 170
Seuil (Éditions du), 164
Tiderne Skifter Forlag A/S, 162
Tusquets Editores, 170

Essays: Publishers (International)
Queensland Press, University of, 192
UWA Press, 193
Victoria University Press, 197

Essays: Publishers (Ireland)
Lilliput Press, The, 156

Essays: Publishers (UK)
Akros Publications, 65
Pushkin Press Ltd, 129
Random House Group Ltd, The, 131

Essays: Small Presses
Paupers' Press, 233
Sylph Editions, 235

Etiquette: Literary Agents (UK)
Graham Maw Christie Literary Agency, 258

Etiquette: Literary Agents (US)
Gusay (Charlotte) Literary Agency, The, 284

Etiquette: Publishers (UK)
Debrett's Ltd, 88

Europe/EU: Library Services
Leeds Central Library, 587
Manchester Central Library, 589
Office for National Statistics,
    592
Swansea Central Library, 597
Wiener Library, The, 599
York Central Library, 600

**Europe/EU: Literary Agents
(US)**
Gusay (Charlotte) Literary
    Agency, The, 284
Siegel (Rosalie), International
    Literary Agency, Inc., 290

Europe/EU: Picture Libraries
Cordaiy (Sylvia) Photo Library,
    606
Cunningham (Sue)
    Photographic, 606
Howes (Chris)/Wild Places
    Photography, 612
Sanders (Peter) Photography
    Ltd, 622
Williams (David) Picture
    Library, 626

Europe/EU: Publishers (UK)
Wales Press, University of, 148

**Family Reference: Literary
Agents (UK)**
Graham Maw Christie Literary
    Agency, 258

**Family Reference: Literary
Agents (US)**
Gusay (Charlotte) Literary
    Agency, The, 284

Family Reference: Magazines
Candis, 306
Families First, 319
Weekly News, 366

**Family Reference: Picture
Libraries**
Photofusion, 619
PictureBank Photo Library Ltd,
    620
Retna Pictures Ltd, 621

**Far East: Bursaries,
Fellowships and Grants**
Wong (David T.K.) Fellowship,
    The, 487

#

**Far East: Literary Agents
(US)**
Gusay (Charlotte) Literary
    Agency, The, 284
Russell & Volkening, Inc., 289

Far East: Picture Libraries
Angel (Heather)/Natural
    Visions, 602
Dickins (Douglas) Photo
    Library, 607
ffotograff, 608
Gagg's (Andrew N.)
    PHOTOFLORA, 609
Greenhill (Sally and Richard),
    611
Peerless (Ann & Bury) Picture
    Library, 619
Travel Ink Photo Library, 625
Tropix Photo Library, 625

Far East: Publishers (UK)
Octagon Press Ltd, 121

**Fashion & Costume: Literary
Agents (UK)**
Anderson (Darley) Literary, TV
    & Film Agency, 249

**Fashion & Costume: Literary
Agents (US)**
Gusay (Charlotte) Literary
    Agency, The, 284

**Fashion & Costume:
Magazines**
Black Beauty & Hair, 301
Dazed & Confused, 314
Easy Living, 316
Elle, 317
Glamour, 323
Harper's Bazaar, 326
Heat, 326
i-D Magazine, 328
InStyle, 329
Jersey Now, 330
Loaded, 333
marie claire, 334
Maxim, 335
Mixmag, 336
More, 337
New!, 341
Shout Magazine, 355
SmartLife International, 356
Vogue, 364
Wallpaper, 365
Your Hair, 371

**Fashion & Costume: PR
Consultants**
Sawford (Claire) PR, 278

**Fashion & Costume: Picture
Libraries**
Miller (Lee) Archives, 616
Mirrorpix, 616
V&A Images, 625

**Fashion & Costume:
Publishers (UK)**
Anova Books, 66
Antique Collectors' Club, 67
Berg Publishers, 72
Carlton Publishing Group, 81
King (Laurence) Publishing
    Ltd, 109
Prestel Publishing Limited, 129
Thames and Hudson Ltd, 144

**Feminism & Women's
Studies: Library Services**
Women's Library, The, 599

**Feminism & Women's
Studies: Literary Agents (US)**
Bernstein (Meredith) Literary
    Agency, Inc., 280
Carvainis (Maria) Agency, Inc.,
    281
Castiglia Literary Agency, 281
Gusay (Charlotte) Literary
    Agency, The, 284
Lowenstein–Yost Associates
    Inc., 285
Michaels (Doris S.) Literary
    Agency Inc., 286
Russell & Volkening, Inc., 289
Snell (Michael) Literary Agency,
    290

**Feminism & Women's
Studies: Magazines**
New Internationalist, 339

**Feminism & Women's
Studies: Picture Libraries**
Evans (Mary) Picture Library,
    607

**Feminism & Women's
Studies: Publishers
(International)**
Williams (Bridget) Books Ltd,
    197

**Feminism & Women's
Studies: Publishers (Ireland)**
Bradshaw Books, 153

**Feminism & Women's
Studies: Publishers (UK)**
Boyars (Marion) Publishers
    Ltd, 76

Free Association Books Ltd, 96
Onlywomen Press Ltd, 122
Polity Press, 128
Souvenir Press Ltd, 140
Verso, 147
Virago Press, 147
Women's Press, The, 150
Zed Books Ltd, 151

Feminism & Women's Studies: Publishers (US)
Illinois Press, University of, 178
Indiana University Press, 179
Massachusetts Press, University of, 181
McFarland & Company, Inc., Publishers, 180
Mississippi, University Press of, 181
North Texas Press, University of, 182
Syracuse University Press, 188

Feminism & Women's Studies: Small Presses
Crescent Moon Publishing and Joe's Press, 227

Feminism & Women's Studies: Theatre Producers
Sphinx Theatre Company, 449

Fiction: Audio Books
BBC Audiobooks Ltd, 238
Corgi Audio, 238
CSA Word, 238
Hachette Digital, 239
HarperCollins AudioBooks, 239
Hodder & Stoughton Audiobooks, 239
Macmillan Digital Audio, 239
Naxos AudioBooks, 239
Oakhill Publishing Ltd, 240
Orion Audio Books, 240
Penguin Audiobooks, 240
Random House Audio Books, 240
Roberts (L.M.) Limited, 240
Simon & Schuster Audio, 241
Soundings Audio Books, 241

Fiction: Book Clubs
BCA (Book Club Associates), 293
Folio Society, The, 293

Fiction: Bursaries, Fellowships and Grants
Maugham (Somerset) Awards, The, 485
Pick (Charles) Fellowship, The, 485

Scottish Arts Council Writers' Bursaries, 486
Scottish Book Trust New Writers' Awards, 486
Tom-Gallon Trust Award and the Olive Cook Prize, 487
Trask (Betty) Awards, The, 487
Wong (David T.K.) Fellowship, The, 487

Fiction: Festivals
Buxton Festival, 631
King's Lynn, The Fiction Festival, 634
London Literature Festival, 636
National Association of Writers' Groups (NAWG) Open Festival of Writing, 636

Fiction: Literary Agents (Ireland)
Book Bureau Literary Agency, The, 279
Font International Literary Agency, 279
Gunn O'Connor (Marianne) Literary Agency, 279
Richards (Lisa) Agency, The, 279
Williams (Jonathan) Literary Agency, 279

Fiction: Literary Agents (UK)
Aitken Alexander Associates Ltd, 249
Ampersand Agency Ltd, The, 249
Anderson (Darley) Literary, TV & Film Agency, 249
Anubis Literary Agency, 250
Artellus Limited, 250
Author Literary Agents, 250
Bell Lomax Moreton Agency, The, 250
Belli (Lorella) Literary Agency (LBLA), 250
Blake Friedmann Literary Agency Ltd, 251
Bonomi (Luigi) Associates Limited (LBA), 251
BookBlast Ltd, 251
Bookseeker Agency, 251
Brown (Jenny) Associates, 251
Bryan (Felicity), 252
Burkeman (Brie), 252
Burton (Juliet) Literary Agency, 252
Campbell Thomson & McLaughlin Ltd, 252
Capel & Land Ltd, 252
Cheetham (Mic) Associates, 253
Chilcote (Judith) Agency, 253

Chris (Teresa) Literary Agency Ltd, 253
Clemmey (Mary) Literary Agency, 253
Clowes (Jonathan) Ltd, 254
Cochrane (Elspeth) Personal Management, 254
Conville & Walsh Limited, 254
Conway-Gordon (Jane) Ltd, 254
Crew (Rupert) Ltd, 254
Curtis Brown Group Ltd, 254
de Wolfe (Felix), 255
Dorian Literary Agency (DLA), 255
Eady (Toby) Associates Ltd, 256
Eddison Pearson Ltd, 256
Edwards Fuglewicz, 256
Feldstein Agency, The, 256
Fraser Ross Associates, 257
Futerman, Rose & Associates, 257
Gabriel (Jüri), 257
Glass (Eric) Ltd, 257
Godwin (David) Associates, 258
Green (Annette) Authors' Agency, 258
Green (Christine) Authors' Agent, 258
Greene & Heaton Ltd, 258
Gregory & Company Authors' Agents, 258
Grossman (David) Literary Agency Ltd, 259
Gunnmedia Ltd, 259
Hanbury Agency, The, 259
Harwood (Antony) Limited, 259
Heath (A.M.) & Co. Ltd, 259
Heath (Rupert) Literary Agency, 260
Henser Literary Agency, 260
Higham (David) Associates Ltd, 260
Holt (Vanessa) Ltd, 260
Hordern (Kate) Literary Agency, 260
Indepublishing CIA – Consultancy for Independent Authors, 261
Inspira Group, The, 261
International Scripts, 261
Janklow & Nesbit (UK) Ltd, 261
Johnson & Alcock, 261
Judd (Jane) Literary Agency, 262
LAW, 262
Levy (Barbara) Literary Agency, 262
Limelight Management, 262
Little (Christopher) Literary Agency, The, 263
London Independent Books, 263
Luck (Lucy) Associates, 263

Lutyens and Rubinstein, 263
Mann (Andrew) Ltd, 264
Marjacq Scripts Ltd, 264
Marsh Agency Ltd, The, 264
MBA Literary Agents Ltd, 264
McAra (Duncan), 264
McKernan Literary Agency &
    Consultancy, 265
Morris (William) Agency (UK)
    Ltd, 265
Motley (Michael) Ltd, 265
Murdoch (Judith) Literary
    Agency, 265
Noach (Maggie) Literary
    Agency, The, 265
Nye (Alexandra), 266
PFD, 266
Power (Shelley) Literary Agency
    Ltd, 267
Robinson Literary Agency Ltd,
    267
Rogers, Coleridge & White Ltd,
    267
Rushby-Smith (Uli) Literary
    Agency, 267
Sayle Literary Agency, The, 268
Sheil Land Associates Ltd, 268
Sheldon (Caroline) Literary
    Agency Ltd, 268
Simmonds (Dorie) Agency, 269
Simmons (Jeffrey), 269
Sinclair-Stevenson, 269
Standen Literary Agency, The,
    269
Stein (Abner), 270
Stewart (Shirley) Literary
    Agency, 270
Such (Sarah) Literary Agency,
    270
Sunflower Literary Agency, 270
Trevor (Lavinia) Literary
    Agency, 271
Turnbull (Jane), 271
United Agents Limited, 271
Victor (Ed) Ltd, 271
Wade & Doherty Literary
    Agency Ltd, 271
Watson, Little Ltd, 272
Watt (A.P.) Ltd, 272
White (Eve) Literary Agent, 272
Wiener (Dinah) Ltd, 272
Wylie Agency (UK) Ltd, The,
    273

Fiction: Literary Agents (US)
Altshuler (Miriam) Literary
    Agency, 280
Axelrod Agency, The, 280
Baldi (Malaga) Literary Agency,
    280
Barrett (Loretta) Books, Inc.,
    280

Bernstein (Meredith) Literary
    Agency, Inc., 280
Bleecker Street Associates, Inc.,
    280
Borchardt (Georges), Inc., 281
Braun (Barbara) Associates,
    Inc., 281
Browne & Miller Literary
    Associates, 281
Browne (Pema) Ltd, 281
Bykofsky (Sheree) Associates,
    Inc., 281
Carvainis (Maria) Agency, Inc.,
    281
Castiglia Literary Agency, 281
Chelius (Jane) Literary Agency,
    Inc., 282
Chester (Linda) & Associates,
    282
Clark (William) Associates, 282
Collin (Frances) Literary Agent,
    282
Congdon (Don) Associates,
    Inc., 282
Curtis Brown Ltd, 282
Dijkstra (Sandra) Literary
    Agency, 282
Dystel & Goderich Literary
    Management, 283
Ellenberg (Ethan) Literary
    Agency, 283
Gelfman Schneider Literary
    Agents, Inc., 283
Greenburger (Sanford J.)
    Associates, Inc., 284
Gusay (Charlotte) Literary
    Agency, The, 284
Harris (Joy) Literary Agency,
    284
Janklow & Nesbit Associates,
    284
JCA Literary Agency, Inc., 284
Klinger (Harvey), Inc., 284
Lampack (Peter) Agency, Inc.,
    285
Larsen (Michael)/Elizabeth
    Pomada Literary Agency,
    285
Leavitt (Ned) Agency, The, 285
Lescher & Lescher Ltd, 285
Lord (Sterling) Literistic, Inc.,
    285
Lowenstein–Yost Associates
    Inc., 285
MacBride (Margret) Literary
    Agency, The, 286
Mann (Carol) Agency, 286
Manus & Associates Literary
    Agency, Inc., 286
Mews Books Ltd, 286
Morhaim (Howard) Literary
    Agency, 287

Morrison (Henry) Inc., 287
Naggar (Jean V.) Literary
    Agency, The, 287
Nelson (B.K.) Literary Agency,
    287
Parks (Richard) Agency, 287
Picard (Alison J.) Literary
    Agent, 287
Pinder Lane & Garon-Brooke
    Associates Ltd, 287
PMA Literary & Film
    Management, Inc., 287
Priest (Aaron M.) Literary
    Agency, The, 288
Protter (Susan Ann) Literary
    Agent, 288
Quicksilver Books, Literary
    Agents, 288
Reece Halsey North Literary
    Agency, 288
Rees (Helen) Literary Agency,
    288
Rittenberg (Ann) Literary
    Agency, Inc., 288
Robbins (B.J.) Literary Agency,
    288
Roghaar (Linda) Literary
    Agency, LLC, 289
Rosenberg Group, The, 289
Russell & Volkening, Inc., 289
Sagalyn Literary Agency, The,
    289
Sanders (Victoria) & Associates
    LLC, 289
Scagnetti (Jack) Talent &
    Literary Agency, 289
Schiavone Literary Agency,
    Inc., 289
Schulman (Susan), A Literary
    Agency, 290
Scovil Chichak Galen Literary
    Agency, Inc., 290
Siegel (Rosalie), International
    Literary Agency, Inc., 290
Spitzer (Philip G.) Literary
    Agency, Inc., 291
Teal (Patricia) Literary Agency,
    291
Wales Literary Agency, Inc., 291
Weil (Wendy) Agency, Inc.,
    The, 291
Weiner (Cherry) Literary
    Agency, 292
Writers House, LLC, 292
Zeckendorf (Susan) Associates,
    Inc., 292

Fiction: Magazines
Black Static, 302
Chat, 307
Crimewave, 312
Edge, The, 317

Edinburgh Review, 317
Gold Dust Magazine, 323
Granta, 324
Mslexia, 338
My Weekly, 338
New Writer, The, 340
People's Friend, The, 344
QWF Magazine, 348
Reader's Digest, 350
Take a Break, 359
20x20 magazine, 363
Weekly News, 366
Woman's Weekly, 368

Fiction: Miscellany
Fiction Writing Workshops and Tutorials, 641

Fiction: PR Consultants
Page (Louise) PR, 277
Randall (Sally) PR, 278

Fiction: Poetry Magazines
Aesthetica, 208
Awen, 208
Penniless Press, The, 213
Southword, 217
Third Half, The, 217

Fiction: Poetry Presses
Bluechrome Publishing, 200
Cinnamon Press, 200
Dionysia Press, 201
Dream Catcher, 201
Egg Box Publishing, 201
Feather Books, 201
Five Leaves Publications, 201
Honno Welsh Women's Press, 202
Iron Press, 203
Kates Hill Press, 203
nthposition, 204
Penniless Press, 204
Route Publishing, 206
Seren, 206

Fiction: Prizes
Authors' Club First Novel Award, 489
Black (James Tait) Memorial Prizes, 491
Boardman Tasker Award, 491
Bowling (Harry) Prize, 492
Cohen (David) Prize for Literature, 495
Commonwealth Writers Prize, The, 496
Costa Book Awards, 496
Daily Mail First Novel Award, 498
Elliott (Desmond) Prize, The, 499

Encore Award, The, 499
Faber (Geoffrey) Memorial Prize, 500
Good Housekeeping Book Awards, 502
Graham (Winston) Historical Prize, The, 502
Guildford Book Festival First Novel Award, 503
Independent Foreign Fiction Prize, The, 505
International Dundee Book Prize, The, 505
International Prize for Arabic Fiction, 505
Irish Book Awards, 506
Jewish Quarterly Literary Prize, 506
Lakeland Book of the Year Awards, 507
Lannan Literary Award, 507
Llewellyn Rhys (John) Prize, The, 509
Macmillan Writer's Prize for Africa, 510
Mail on Sunday Novel Competition, The, 510
Man Asian Literary Prize, The, 510
Man Booker International Prize, The, 510
Man Booker Prize for Fiction, The, 510
McKitterick Prize, 509
Nathan (Melissa) Award for Comedy Romance, 512
Orange Broadband Prize for Fiction/Orange Broadband Award for New Writers, 514
Portico Prize for Literature, The, 516
Sunday Times Writer of the Year Award, 521
Sunday Times Young Writer of the Year Award, 521
Sundial Scottish Arts Council Book of the Year Awards, 521
Thomas (Dylan) Prize, The, 521
Winchester Writers' Conference Prizes, 524
Yeovil Literary Prize, 525
Yorkshire Post Book Awards, 525

Fiction: Publishers (European)
Alianza Editorial SA, 169
Arche Verlag AG, 171
Aschehoug (H) & Co (W Nygaard), 168
Atlantis Ltd, 161
Belfond (Éditions), 162

Bertelsmann (C.), 164
Bompiani, 166
Bonnier Books, 170
Bonniers (Albert) Förlag, 170
Borgens Forlag A/S, 161
Bra Böcker AB, 170
Brombergs Bokförlag AB, 170
Bruna (A.W.) Uitgevers BV, 167
Brána a.s., 161
BZZTÔH (Uitgeverij) BV, 167
Calmann-Lévy (Éditions), 163
Caminho (Editorial) SARL, 169
Cappelen Damm AS, 168
Civilização Editora, 169
Denoël (Éditions), 163
Deutscher Taschenbuch Verlag GmbH & Co. KG, 164
Diogenes Verlag AG, 171
Dokorán Ltd, 161
EDHASA (Editora y Distribuidora Hispano – Americana SA), 169
Espasa-Calpe (Editorial) SA, 170
Europa-América (Publicações) Lda, 169
Fayard (Librairie Arthème), 163
Fischer (S.) Verlag GmbH, 164
Flammarion Éditions, 163
Forlaget Forum, 161
Forum (Bokförlaget) AB, 171
Gallimard (Éditions), 163
Garzanti Libri SpA, 166
Giunti Editoriale SpA, 166
Gradiva–Publicações S.A., 169
Grasset & Fasquelle (Éditions), 163
Grijalbo, 170
Gummerus Oy, 162
Gyldendal, 161
Gyldendal Norsk Forlag, 168
Hachette Livre (France), 163
Heyne Verlag, 165
Hoffmann und Campe Verlag GmbH, 165
Høst & Søn Publishers Ltd, 161
Jelenkor Kiadó Szolgáltató Kft., 166
Karisto Oy, 162
Kibea Publishing Company, 160
Kiss József Könyvkiadó, 166
Kok (Uitgeversmaatschappij J. H.) BV, 168
Laffont (Robert) Éditions, 163
Lattès (Éditions Jean-Claude), 163
Lettera, 161
Lindhardt og Ringhof Forlag A/S, 162
Longanesi (Casa Editrice) SpA, 167
Lübbe (Verlagsgruppe) GmbH & Co. KG, 165

Meulenhoff (Uitgeverij J.M.)
BV, 168
Minuit (Les Éditions de) SA, 164
Mondadori (Arnoldo) Editore
SpA, 167
Mursia (Gruppo Ugo) Editore
SpA, 167
Móra Könyvkiadö Zrt., 166
Natur och Kultur Bokförlaget,
171
Norstedts Förlag, 171
Nyt Nordisk Forlag Arnold
Busck A/S, 162
Olympia Publishing Co. Inc.,
161
Otava Publishing Co. Ltd, 162
Payot Libraire, 171
Planeta SA (Editorial), 170
Presses de la Cité, 164
Publicat SA, 169
Rizzoli Editore, 167
Schønbergske (Det) Forlag, 162
Seuil (Éditions du), 164
Sonzogno, 167
Sperling e Kupfer Editori SpA,
167
Standaard Uitgeverij, 160
Sugarco Edizioni Srl, 167
Suhrkamp Verlag, 165
Table Ronde (Les Editions de
la), 164
Tammi Publishers, 162
Tiden Norsk Forlag, 168
Tiderne Skifter Forlag A/S, 162
Todariana Editrice, 167
Tusquets Editores, 170
Ullstein Buchverlage GmbH,
165
Unieboek (Uitgeverij) BV, 168
WSOY (Werner Söderström
Osakeyhtiö), 162
Wählströms (B.) Bokförlag AB,
171
Zsolnay (Paul), Verlag GmbH,
160

Fiction: Publishers (International)
Allen & Unwin Pty Ltd, 191
Hachette Livre Australia, 191
Hachette Livre New Zealand
Ltd, 196
HarperCollins Canada Ltd, 193
HarperCollins Publishers
(Australia) Pty Ltd, 191
HarperCollins Publishers (New
Zealand) Ltd, 196
HarperCollins Publishers India
Ltd, 195
Hind Pocket Books Pvt. Ltd, 195
Huia (NZ) Ltd, 196

Macmillan Publishers Australia
Pty Ltd, 191
Macmillan Publishers New
Zealand, 196
McClelland & Stewart Ltd, 193
New Star Books Ltd, 194
Orient Paperbacks, 195
Pan Macmillan Australia Pty
Ltd, 192
Pearson Canada, 194
Penguin Books (NZ) Ltd, 197
Penguin Group (Australia), 192
Penguin Group (Canada), 194
Queensland Press, University
of, 192
Random House Australia, 192
Random House New Zealand,
197
Random House of Canada Ltd,
194
Raupo Publishing (NZ) Ltd, 197
Simon & Schuster (Australia)
Pty Ltd, 192
Struik Publishers (Pty) Ltd, 198
Victoria University Press, 197

Fiction: Publishers (Ireland)
Brandon/Mount Eagle
Publications, 154
Cló Iar-Chonnachta, 154
Hachette Books Ireland, 155
Lilliput Press, The, 156
New Island, 157
Penguin Ireland, 158
Poolbeg Press Ltd, 158

Fiction: Publishers (UK)
Allison & Busby, 65
Alma Books Ltd, 65
Arcadia Books, 68
Atlantic Books, 69
Aurora Metro, 70
Austin & Macauley Publishers
Limited, 70
AuthorHouse UK, 70
Beautiful Books Ltd, 72
Bitter Lemon Press, 73
Black & White Publishing Ltd,
73
Black Ace Books, 73
Black Spring Press Ltd, 74
Blackstaff Press Ltd, 74
Blackstone Publishers, 74
Bloomsbury Publishing Plc, 74
Boltneck Publications Limited,
75
Book Guild Publishing, 75
Boulevard Books & The Babel
Guides, 76
Bound Biographies Limited, 76
Boyars (Marion) Publishers
Ltd, 76

Brown Skin Books, 78
Canongate Books Ltd, 80
Carcanet Press Ltd, 80
Dedalus Ltd, 89
Discovered Authors, 89
Enitharmon Press, 92
Faber & Faber Ltd, 94
Garnet Publishing Ltd, 97
Gomer Press/Gwasg Gomer, 98
Hachette Livre UK, 100
Hale (Robert) Ltd, 101
Harlequin Mills & Boon
Limited, 101
HarperCollins Publishers Ltd,
102
Hesperus Press Limited, 104
Honeyglen Publishing Ltd, 105
Honno Welsh Women's Press,
105
Isis Publishing, 107
Janus Publishing Company Ltd,
108
Legend Press Limited, 110
Lewis (Dewi) Publishing, 110
Little, Brown Book Group UK,
111
Luath Press Ltd, 112
Macmillan Publishers Ltd, 113
Melrose Books, 116
Mercat Press, 116
Methuen Publishing Ltd, 116
Monarch Books, 117
Murray (John) (Publishers) Ltd,
118
Myriad Editions, 118
Myrmidon Books, 118
New Beacon Books Ltd, 119
Old Street Publishing Ltd, 121
Olympia Publishers, 122
Onlywomen Press Ltd, 122
Orion Publishing Group
Limited, The, 123
Parthian, 124
Pegasus Elliot Mackenzie
Publishers Ltd, 125
Penguin Group (UK), 125
Persephone Books, 126
Piatkus Books, 127
Polygon, 128
Publishing House, 129
Quercus Publishing, 130
Quince Books Limited, 130
Random House Group Ltd,
The, 131
Reportage Press, 134
SeaNeverDry Publishing, 137
Seren, 137
Serpent's Tail, 138
Severn House Publishers, 138
Short Books, 139
Simon & Schuster UK Limited,
139

Snowbooks Ltd, 139
Souvenir Press Ltd, 140
Telegram, 143
Thorpe (F.A.) Publishing, 144
Timewell Press, 144
Transita Ltd, 145
Transworld Publishers Ltd, 145
Troubador Publishing Ltd, 146
Twenty First Century Publishers
    Ltd, 146
Virago Press, 147
Wilson (Neil) Publishing Ltd,
    149
Women's Press, The, 150
X Press, The, 151
Y Lolfa Cyf, 151

Fiction: Publishers (US)

Academy Chicago Publishers,
    172
Bantam Dell Publishing Group,
    173
California Press, University of,
    174
Charlesbridge Publishing, 174
Crown Publishing Group, 175
Doubleday Broadway
    Publishing Group, 175
Farrar, Straus & Giroux, Inc.,
    176
Grove/Atlantic Inc., 177
Hachette Book Group USA, 177
HarperCollins Publishers, Inc.,
    177
Hill Street Press LLC, 178
Holt (Henry) & Company, Inc.,
    178
Interlink Publishing Group,
    Inc., 179
Llewellyn Publications, 180
Louisiana State University
    Press, 180
Massachusetts Press, University
    of, 181
Michigan Press, The University
    of, 181
Nevada Press, University of, 182
New Mexico Press, University
    of, 182
Northwestern University Press,
    182
Norton (W.W.) & Company,
    Inc., 182
Penguin Group (USA) Inc., 183
Perseus Books Group, 184
Random House Publishing
    Group, 185
Shambhala Publications, Inc.,
    186
Simon & Schuster Adult
    Publishing Group, 187
St Martin's Press LLC, 186

Tyndale House Publishers, Inc.,
    189

Fiction: Small Presses

Accent Press, 225
Barny Books, 225
Bluemoose Books Limited, 226
Chrysalis Press, 226
Contact Publishing Ltd, 226
Crescent Moon Publishing and
    Joe's Press, 227
Five Leaves Publications, 228
Great Northern Publishing, 229
Griffin Publishing Associates,
    229
Ignotus Press, 230
Immanion Press, 230
Ivy Publications, 230
Maia Press, The, 231
Maypole Editions, 231
Miller (Neil) Publications, 232
Pipers' Ash Ltd, 233
Stone Flower Limited, 235
Superscript, 235
Tartarus Press, 235
Tindal Street Press Ltd, 236
Tlön Books Publishing Ltd, 236
Waywiser Press, 236
Willow Bank Publishers Ltd,
    236

Fiction: UK Packagers

Working Partners Ltd, 246

Fiction: Writers' Circles and Workshops

Original Writers Group, The,
    479
Scribo, 479

Fiction: Writers' Courses (UK and Ireland)

Aberystwyth, University of, 467
Annual Writers' Writing
    Courses & Workshops, 457
Arvon Foundation, 458
Bath Spa University, 462
Birkbeck College, University of
    London, 458
Birmingham City University,
    463
City Lit, 458
City University, 458
Dingle Writing Courses, 466
East Anglia, University of, 461
Edge Hill University, 457
Exeter, University of, 455
Fiction Writing Workshops and
    Tutorials, 459
Fire in the Head Courses &
    Writers' Services, 455
Glamorgan, University of, 467

INKwell Writers' Workshops,
    466
Keele University, 463
Knuston Hall Residential
    College for Adult Education,
    461
Lancaster University, 457
Leeds, University of, 465
Liberato Breakaway Writing
    Courses, 456
London School of Journalism,
    459
Manchester Metropolitan
    University – The Writing
    School, 460
Manchester, University of, 460
Middlesex University, 459
National Academy of Writing,
    464
Nottingham Trent University,
    The, 462
Open College of the Arts, 465
Reading, University of, 453
Roehampton University, 460
Sheffield, University of, 465
St Andrews, University of, 467
Sussex, University of, 463
University College Falmouth,
    454
Winchester (University of), 456
Writers Bureau, The, 461

Fiction: Writers' Courses (US)

Adelphi University, 473
American University, 470
Arizona, University of, 469
Arkansas, University of, 469
Boston University, 471
Bowling Green State University,
    474
Brooklyn College of the City
    University of New York, 473
Brown University, 474
Chapman University, 469
Chicago State University, 471
Colorado State University, 470
Eastern Washington University,
    475
Emerson College, 472
Florida International University,
    470
George Mason University, 475
Hollins University, 475
Houston, University of, 474
Louisiana State University, 471
Memphis, University of, 474
Minnesota State University
    Moorhead, 472
Missouri, University of, 472
Missouri-Columbia, University
    of, 472

New York University, 473
North Carolina at Greensboro, University of, 473
North Carolina at Wilmington, University of, 473
Notre Dame, University of, 471
Oregon, University of, 474
Purdue University, 471
Saint Mary's College of California, 469
San Francisco State University, 470
Seattle Pacific University, 475
Seton Hill University, 474
Taylor University, 471
Virginia Commonwealth University, 475
Western Michigan University, 472
Wyoming, University of, 476

Film: Library Services
BFI National Library, 575
Vin Mag Archive Ltd, 598
Westminster Reference Library, 598

Film: Literary Agents (UK)
Agency (London) Ltd, The, 249
Anderson (Darley) Literary, TV & Film Agency, 249
Blake Friedmann Literary Agency Ltd, 251
Brodie (Alan) Representation Ltd, 251
Burkeman (Brie), 252
Capel & Land Ltd, 252
Casarotto Ramsay and Associates Ltd, 252
Clowes (Jonathan) Ltd, 254
Cochrane (Elspeth) Personal Management, 254
Colin (Rosica) Ltd, 254
Curtis Brown Group Ltd, 254
Daish (Judy) Associates Ltd, 255
Dench Arnold Agency, The, 255
Fitch (Laurence) Ltd, 256
Foster (Jill) Ltd, 257
Futerman, Rose & Associates, 257
Glass (Eric) Ltd, 257
Green (Annette) Authors' Agency, 258
Hall (Rod) Agency Limited, The, 259
Hancock (Roger) Ltd, 259
Henser Literary Agency, 260
Hoskins (Valerie) Associates Limited, 260
Independent Talent Group Limited, 261
International Scripts, 261

Johnson & Alcock, 261
Kass (Michelle) Associates, 262
London Independent Books, 263
Mann (Andrew) Ltd, 264
Marjacq Scripts Ltd, 264
MBA Literary Agents Ltd, 264
McLean (Bill) Personal Management, 264
Morris (William) Agency (UK) Ltd, 265
Narrow Road Company, The, 265
PFD, 266
Power (Shelley) Literary Agency Ltd, 267
Sayle Screen Ltd, 268
Sharland Organisation Ltd, The, 268
Sheil Land Associates Ltd, 268
Steel (Elaine), 269
Steinberg (Micheline) Associates, 270
Tennyson Agency, The, 270
TVmyworld, 271
United Agents Limited, 271
Ware (Cecily) Literary Agents, 272
Watt (A.P.) Ltd, 272

Film: Literary Agents (US)
Curtis Brown Ltd, 282
Freedman (Robert A.) Dramatic Agency, Inc., 283
Gusay (Charlotte) Literary Agency, The, 284
Hawkins (John) & Associates, Inc., 284
Russell & Volkening, Inc., 289
Scagnetti (Jack) Talent & Literary Agency, 289

Film: Magazines
Black Static, 302
Dazed & Confused, 314
Edge, The, 317
Empire, 317
Film Review, 320
Gold Dust Magazine, 323
Heat, 326
i-D Magazine, 328
List, The, 332
Screen, 353
Screen International, 353
Screentrade Magazine, 353
ScriptWriter Magazine, 354
Sight & Sound, 355
Total Film, 362
Uncut, 364

Film: Miscellany
Landeau (Caroline), 641

Film: Picture Libraries
akg-images Ltd, Arts and History Picture Library, 601
Aquarius Library, 602
BFI Stills Sales, 603
Capital Pictures, 604
Cinema Museum, The, 605
Huntley Film Archive, 612
Kobal Collection, The, 614
Retna Pictures Ltd, 621
Vin Mag Archive Ltd, 626

Film: Poetry Magazines
Aesthetica, 208

Film: Prizes
Kraszna-Krausz Book Awards, 507
Projection Box Essay Awards 2008–2009, The, 516
Thomas (Dylan) Prize, The, 521

Film: Professional Associations
BFI, 530
Irish Playwrights and Screenwriters Guild, 539
IVCA, 539
New Producers Alliance, 542
Pact, 542
Player–Playwrights, 543
Scottish Screen, 547
UK Film Council, 549
Writers' Guild of Great Britain, The, 551

Film: Publishers (European)
Heyne Verlag, 165
Kiss József Könyvkiadó, 166
Suhrkamp Verlag, 165

Film: Publishers (International)
Currency Press Pty Ltd, 191

Film: Publishers (Ireland)
Merlin Publishing, 157

Film: Publishers (UK)
Anova Books, 66
Aurum Press Ltd, 70
BFI Publishing, 72
Boyars (Marion) Publishers Ltd, 76
Continuum International Publishing Group Ltd, The, 85
Edinburgh University Press, 91
Exeter Press, University of, 93
Faber & Faber Ltd, 94
GMC Publications Ltd, 98

HarperCollins Publishers Ltd, 102
Harvard University Press, 103
Helter Skelter Publishing, 104
Hern (Nick) Books, 104
Macmillan Publishers Ltd, 113
Manchester University Press, 115
Random House Group Ltd, The, 131
Reaktion Books, 132
RotoVision, 134
Tauris (I.B.) & Co. Ltd, 143
Titan Books, 145
Wallflower Press, 148

Film: Publishers (US)
Faber & Faber, Inc., 176
Illinois Press, University of, 178
Indiana University Press, 179
Minnesota Press, University of, 181
Perseus Books Group, 184
Players Press, 184

Film: Small Presses
Crescent Moon Publishing and Joe's Press, 227
Headpress, 230
Ivy Publications, 230
Maypole Editions, 231

Film: UK Packagers
Cameron & Hollis, 243

Film: Writers' Circles and Workshops
éQuinoxe TBC, 478
Kops (Bernard) and Tom Fry Drama Writing Workshops, 478
Original Writers Group, The, 479
Script Yorkshire, 479
Southwest Scriptwriters, 479

Film: Writers' Courses (UK and Ireland)
Annual Writers' Writing Courses & Workshops, 457
Birmingham City University, 463
Bournemouth University, 455
Central School of Speech and Drama, The, 458
City Lit, 458
East Anglia, University of, 461
Edge Hill University, 457
Exeter, University of, 455
Glamorgan, University of, 467
Institute of Copywriting, 462

Leeds Metropolitan University, 464
Leeds, University of, 465
Liverpool, University of, 461
National Academy of Writing, 464
National Film & Television School, 453
Roehampton University, 460
Royal Holloway University of London, 463
Script Factory, The, 460
Sheffield, University of, 465
Starz! – Film and Theatre Performing Arts, 464
Sunderland, University of, 463
University College Falmouth, 454
Westminster, University of, 460

Film: Writers' Courses (US)
American Film Institute, 469
Chapman University, 469
Chicago State University, 471
Florida International University, 470
Hollins University, 475
Louisiana State University, 471
Minnesota State University Moorhead, 472
Southern California, University of, 470
Virginia Commonwealth University, 475

Finance: Library Services
Bank of England Information Centre, 574
Barbican Library, 574
City Business Library, 581
Guildhall Library, 584

Finance: Literary Agents (UK)
Anderson (Darley) Literary, TV & Film Agency, 249
Kelly (Frances), 262

Finance: Literary Agents (US)
Gusay (Charlotte) Literary Agency, The, 284
Konner (Linda) Literary Agency, 285
Lowenstein–Yost Associates Inc., 285
Scribblers House® LLC Literary Agency, 290
Writers House, LLC, 292

Finance: Magazines
Accountancy, 294
Accountancy Age, 294

Banker, The, 298
Director, 315
Economist, The, 316
Investors Chronicle, 329
Moneywise, 337
Pensions World, 343
Spear's Wealth Management Survey, 357
What Investment, 366

Finance: News Agencies
Dow Jones Newswires, 393

Finance: Prizes
FT & Goldman Sachs Business Book of the Year Award, 501

Finance: Publishers (European)
BZZTÔH (Uitgeverij) BV, 167

Finance: Publishers (International)
Jaico Publishing House, 195
LexisNexis Butterworths (Pty) Ltd South Africa, 197
Melbourne University Publishing Ltd, 192
Vision Books Pvt. Ltd, 196

Finance: Publishers (Ireland)
Blackhall Publishing Ltd, 153

Finance: Publishers (UK)
Harriman House Ltd, 103
Kogan Page Ltd, 109
McGraw-Hill Education, 113
NCVO Publications, 119
Random House Group Ltd, The, 131
Telegraph Books, 143
Which? Books, 149
Woodhead Publishing Ltd, 150
Zambezi Publishing Ltd, 151

Finance: Publishers (US)
AMACOM Books, 172

Finance: US Media Contacts in the UK
Dow Jones Newswires, 394
Wall Street Journal, 395

Firearms: Library Services
Imperial War Museum, 586

Firearms: Magazines
Shooting and Conservation, 355
Shooting Gazette, The, 355
Shooting Times & Country Magazine, 355

Firearms: Publishers (UK)
Crowood Press Ltd, The, 87
Sportsman's Press, The, 141
Swan Hill Press, 142

Fish/Fishing: Book Clubs
Readers' Union Ltd, 293

Fish/Fishing: Magazines
Angler's Mail, 295
Fishing News, 320
Koi Magazine, 331
Practical Fishkeeping, 346
Shooting Times & Country Magazine, 355
Total Coarse Fishing, 362
Total Flyfisher, 362
Trout Fisherman, 363
World Fishing, 369

Fish/Fishing: Picture Libraries
Waterways Photo Library, 626

Fish/Fishing: Publishers (European)
Karisto Oy, 162

Fish/Fishing: Publishers (UK)
Crowood Press Ltd, The, 87
Sportsman's Press, The, 141
Swan Hill Press, 142

Fish/Fishing: Publishers (US)
Stackpole Books, 187

Folklore & Tradition: Library Services
Williams (Vaughan) Memorial Library, 599

Folklore & Tradition: Literary Agents (US)
Ware (John A.) Literary Agency, 291

Folklore & Tradition: Magazines
Evergreen, 318
Scots Magazine, The, 352
This England, 360

Folklore & Tradition: Picture Libraries
Evans (Mary) Picture Library, 607
International Photobank, 613
Williams (Vaughan) Memorial Library, 626

Folklore & Tradition: Poetry Presses
King's England Press, The, 203

Folklore & Tradition: Prizes
Briggs (Katharine) Folklore Award, 492

Folklore & Tradition: Publishers (Ireland)
Anvil Books, 153
Mercier Press Ltd, 156
Tír Eolas, 158

Folklore & Tradition: Publishers (UK)
Birlinn Ltd, 73
Capall Bann Publishing, 80
Country Publications Ltd, 86
Gwasg Carreg Gwalch, 99
Luath Press Ltd, 112
Octagon Press Ltd, 121

Folklore & Tradition: Publishers (US)
Indiana University Press, 179
Mississippi, University Press of, 181

Folklore & Tradition: Small Presses
Heart of Albion Press, 230

Food & Drink: Literary Agents (Ireland)
Williams (Jonathan) Literary Agency, 279

Food & Drink: Literary Agents (UK)
Anderson (Darley) Literary, TV & Film Agency, 249
Chris (Teresa) Literary Agency Ltd, 253
Feldstein Agency, The, 256
Graham Maw Christie Literary Agency, 258
hhb agency ltd, 260
Kelly (Frances), 262
Limelight Management, 262
Sheil Land Associates Ltd, 268
Smith (Robert) Literary Agency Ltd, 269
Wiener (Dinah) Ltd, 272

Food & Drink: Literary Agents (US)
Dystel & Goderich Literary Management, 283
Ellenberg (Ethan) Literary Agency, 283

Gusay (Charlotte) Literary Agency, The, 284
Lescher & Lescher Ltd, 285
Mews Books Ltd, 286
Parks (Richard) Agency, 287
Quicksilver Books, Literary Agents, 288

Food & Drink: Magazines
BBC Good Food, 299
Country Smallholding, 311
Decanter, 314
Delicious, 314
Easy Living, 316
Grow Your Own, 325
House & Garden, 328
House Beautiful, 328
Ideal Home, 328
Jersey Now, 330
Olive, 342
Sainsbury's Magazine, 352
Speciality Food, 357
That's Life!, 360
Vogue, 364
W.I. Life, 365
Wine & Spirit, 367

Food & Drink: Miscellany
Landeau (Caroline), 641

Food & Drink: PR Consultants
Frazer (Cathy) PR, 277

Food & Drink: Picture Libraries
Blake (Anthony) Photo Library, 603
Cephas Picture Library, 604
Food Features, 609
Heseltine (John) Archive, 611
Hurst (Jacqui), 612
Last Resort Picture Library, 614
MacQuitty International Photographic Collection, 616
Retna Pictures Ltd, 621
Traeger (Tessa) Picture Library, 625

Food & Drink: Prizes
Glenfiddich Food & Drink Awards, 502
Gourmand World Cookbook Awards, 502
Guild of Food Writers Awards, 503
Simon (André) Memorial Fund Book Awards, 519

Food & Drink: Professional Associations
Guild of Food Writers, The, 538

Food & Drink: Publishers (European)
BZZTÔH (Uitgeverij) BV, 167
Civilização Editora, 169
Flammarion Éditions, 163
Garzanti Libri SpA, 166
Giunti Editoriale SpA, 166
Heyne Verlag, 165
Lannoo, Uitgeverij, 160
Larousse (Éditions), 163
Nyt Nordisk Forlag Arnold Busck A/S, 162
Tusquets Editores, 170

Food & Drink: Publishers (International)
Hachette Livre New Zealand Ltd, 196
HarperCollins Canada Ltd, 193
HarperCollins Publishers (New Zealand) Ltd, 196
HarperCollins Publishers India Ltd, 195
Random House New Zealand, 197
Raupo Publishing (NZ) Ltd, 197

Food & Drink: Publishers (Ireland)
Farmar (A.&A.) Ltd, 155
Gill & Macmillan, 155
Hachette Books Ireland, 155
O'Brien Press Ltd, The, 157
On Stream Publications Ltd, 157

Food & Drink: Publishers (UK)
Absolute Press, 64
Anness Publishing Ltd, 66
Anova Books, 66
Ashgrove Publishing, 69
Aurora Metro, 70
Austin & Macauley Publishers Limited, 70
Black & White Publishing Ltd, 73
Blackstaff Press Ltd, 74
Bonnier Books (UK), 75
Cathie (Kyle) Ltd, 81
Davies (Christopher) Publishers Ltd, 88
Dorling Kindersley Ltd, 90
Foulsham Publishers, 95
Garnet Publishing Ltd, 97
GMC Publications Ltd, 98
Grub Street, 99
Hachette Livre UK, 100
HarperCollins Publishers Ltd, 102
Luath Press Ltd, 112
Murdoch Books UK Ltd, 118

New Holland Publishers (UK) Ltd, 119
Octopus Publishing Group, 121
Phaidon Press Limited, 126
Quadrille Publishing Ltd, 130
Quiller Press, 130
Random House Group Ltd, The, 131
Reader's Digest Association Ltd, 132
Reed (William) Directories, 133
Ryland Peters & Small Limited, 135
Sheldrake Press, 138
Simon & Schuster UK Limited, 139
Souvenir Press Ltd, 140
Sportsman's Press, The, 141
Swan Hill Press, 142
Telegraph Books, 143
Travel Publishing Ltd, 145
Wilson (Neil) Publishing Ltd, 149

Food & Drink: Publishers (US)
Barron's Educational Series, Inc., 173
Crown Publishing Group, 175
HarperCollins Publishers, Inc., 177
Hippocrene Books, Inc., 178
Interlink Publishing Group, Inc., 179
Pelican Publishing Company, 183
Reader's Digest Association Inc, 185
Washington State University Press, 190

Food & Drink: Small Presses
Graffeg, 229
Serif, 235

Food & Drink: UK Packagers
Amber Books Ltd, 242
Quarto Publishing plc, 245

Food & Drink: Writers' Courses (UK and Ireland)
City University, 458

Football: Literary Agents (UK)
Feldstein Agency, The, 256

Football: Magazines
Fairgame Magazine, 319
Match, 335
Shoot, 355
World Soccer, 369

Football: Organizations of Interest to Poets
Football Poets, The, 220

Football: Picture Libraries
Professional Sport UK Ltd, 620

Football: Publishers (UK)
Breedon Books Publishing Co. Ltd, The, 77

France: Library Services
French Institute Library, 583

France: Literary Agents (US)
Gusay (Charlotte) Literary Agency, The, 284
Siegel (Rosalie), International Literary Agency, Inc., 290

France: Magazines
Destination France, 315
France Magazine, 321
Living France, 333

France: Picture Libraries
Aspect Picture Library Ltd, 602
Baxter (Colin) Photography Limited, 603
Country Matters Picture Library, 606
Eagar (Patrick) Photography, 607
Grace (Martin and Dorothy), 611
Heseltine (John) Archive, 611
Skishoot–Offshoot, 623
Travel Ink Photo Library, 625

France: Poetry Magazines
French Literary Review, The, 210

France: Prizes
McLeod (Enid) Prize, 509

Freelancing: Magazines
Freelance Market News, 321

Freelancing: Professional Associations
Society for Editors and Proofreaders (SfEP), 547

Gaelic: Festivals
Word – University of Aberdeen Writers Festival, 640

Gaelic: Library Services
Western Isles Libraries, 598

Gaelic: Magazines
New Writing Scotland, 341

Gaelic: Organizations of Interest to Poets
Scottish Poetry Library, 223
Western Writers' Centre, The, 224

Gaelic: Poetry Presses
Etruscan Books, 201

Gaelic: Prizes
Royal Mail Awards for Scottish Children's Books, 518
Strokestown International Poetry Competition, 520

Gaelic: Professional Associations
Comhairle nan Leabhraichean/ The Gaelic Books Council, 535

Gaelic: Publishers (UK)
Acair Ltd, 64
Birlinn Ltd, 73

Games: Magazines
British Chess Magazine, 303
Darts World, 314
Snooker Scene, 357

Games: Picture Libraries
V&A Images, 625

Games: Publishers (European)
Larousse (Éditions), 163

Games: Publishers (UK)
Anova Books, 66
D&B Publishing, 87
Gloucester Publishers Plc, 98
GMC Publications Ltd, 98
Telegraph Books, 143

Games: Publishers (US)
Sterling Publishing Co. Inc., 188

Games: Small Presses
Jupiter Press, The, 230

Gardening & Horticulture: Book Clubs
Readers' Union Ltd, 293

Gardening & Horticulture: Library Services
Leeds Central Library, 587

Gardening & Horticulture: Literary Agents (Ireland)
Williams (Jonathan) Literary Agency, 279

Gardening & Horticulture: Literary Agents (UK)
Anderson (Darley) Literary, TV & Film Agency, 249
Chris (Teresa) Literary Agency Ltd, 253
Davidson (Caroline) Literary Agency, 255
Limelight Management, 262
Sheil Land Associates Ltd, 268

Gardening & Horticulture: Literary Agents (US)
Fredericks (Jeanne) Literary Agency, Inc., 283
Gusay (Charlotte) Literary Agency, The, 284
Parks (Richard) Agency, 287

Gardening & Horticulture: Magazines
Amateur Gardening, 295
BBC Gardeners' World Magazine, 298
Country Living, 311
English Garden, The, 318
Flora International, 321
Garden Answers, 322
Garden News, 322
Garden, The, 322
Gardens Illustrated, 322
Grow Your Own, 325
Homes & Gardens, 327
House & Garden, 328
House Beautiful, 328
Ideal Home, 328
Jersey Now, 330
Organic Gardening, 343
Sainsbury's Magazine, 352
Vogue, 364
W.I. Life, 365

Gardening & Horticulture: Picture Libraries
Angel (Heather)/Natural Visions, 602
Country Life Picture Library, 606
Craven (Philip) Worldwide Photo-Library, 606
Edifice, 607
Garden and Wildlife Matters Photo Library, 609
Garden Picture Library, 610
Mathews (S&O) Photography, 616
Science Photo Library, 622
Strang (Jessica) Photo Library, 624
Traeger (Tessa) Picture Library, 625

Gardening & Horticulture: Professional Associations
Garden Media Guild, The, 537
Guild of Agricultural Journalists, 537

Gardening & Horticulture: Publishers (European)
Flammarion Éditions, 163
Lannoo, Uitgeverij, 160
Larousse (Éditions), 163
Nyt Nordisk Forlag Arnold Busck A/S, 162

Gardening & Horticulture: Publishers (International)
HarperCollins Publishers (New Zealand) Ltd, 196
HarperCollins Publishers India Ltd, 195
Random House New Zealand, 197
Simon & Schuster (Australia) Pty Ltd, 192

Gardening & Horticulture: Publishers (UK)
Anness Publishing Ltd, 66
Anova Books, 66
Antique Collectors' Club, 67
Austin & Macauley Publishers Limited, 70
Cathie (Kyle) Ltd, 81
Crowood Press Ltd, The, 87
Dorling Kindersley Ltd, 90
GMC Publications Ltd, 98
HarperCollins Publishers Ltd, 102
Lincoln (Frances) Ltd, 110
Merrell Publishers Ltd, 116
Murdoch Books UK Ltd, 118
New Holland Publishers (UK) Ltd, 119
Octopus Publishing Group, 121
Quadrille Publishing Ltd, 130
Quiller Press, 130
Reader's Digest Association Ltd, 132
Ryland Peters & Small Limited, 135
Souvenir Press Ltd, 140
Telegraph Books, 143
Thames and Hudson Ltd, 144
Transworld Publishers Ltd, 145
Whittet Books Ltd, 149

Gardening & Horticulture: Publishers (US)
Crown Publishing Group, 175
Llewellyn Publications, 180
Reader's Digest Association Inc, 185
Sterling Publishing Co. Inc., 188

Gardening & Horticulture: Small Presses
Packard Publishing Limited, 233

Gardening & Horticulture: UK Packagers
Breslich & Foss Ltd, 242
Expert Publications Ltd, 243
Quarto Publishing plc, 245

Gay/Lesbian: Literary Agents (US)
Gusay (Charlotte) Literary Agency, The, 284

Gay/Lesbian: Magazines
Boyz, 303
Diva, 315
GT (Gay Times), 325
Pink Paper, 345
Shout!, 355

Gay/Lesbian: Publishers (UK)
Blackstone Publishers, 74
Millivres Prowler Limited, 117
Onlywomen Press Ltd, 122

Gay/Lesbian: Publishers (US)
Minnesota Press, University of, 181

Genealogy: Library Services
Athenaeum, Liverpool, The, 574
Barnsley Public Library, 575
Bournemouth Library, 576
Guildhall Library, 584
Leeds Central Library, 587
National Library of Ireland, 590
Northumberland County Library, 592
Worcestershire Libraries and Information Service, 600
York Central Library, 600

Genealogy: Magazines
Family Tree Magazine, 319

Genealogy: Miscellany
Edwards (Jacqueline), 641

Genealogy: Publishers (Ireland)
Flyleaf Press, 155

Genealogy: Publishers (UK)
Brewin Books Ltd, 77
Countryside Books, 86
National Archives, The, 118

Genealogy: Writers' Courses (UK and Ireland)
Writers Bureau, The, 461

Geography: Library Services
Royal Geographical Society, 594

Geography: Literary Agents (US)
Gusay (Charlotte) Literary Agency, The, 284

Geography: Picture Libraries
Cleare (John)/Mountain Camera, 605
GeoScience Features, 610
Harding (Robert) World Imagery, 611
Royal Geographical Society Picture Library, 622
Williams (David) Picture Library, 626

Geography: Publishers (European)
Presses Universitaires de France (PUF), 164
Publicat SA, 169
Springer GmbH, 165

Geography: Publishers (International)
Munshiram Manoharlal Publishers Pvt Ltd, 195

Geography: Publishers (UK)
Atlantic Europe Publishing Co. Ltd, 69
Reaktion Books, 132
Routledge, 134
Trident Press Ltd, 145

Geography: Publishers (US)
Louisiana State University Press, 180
Minnesota Press, University of, 181
Syracuse University Press, 188
Texas Press, University of, 188

Geography: Small Presses
Witan Books, 237

Geology: Literary Agents (US)
Gusay (Charlotte) Literary Agency, The, 284

Geology: Picture Libraries
GeoScience Features, 610
Oxford Scientific (OSF), 619
Ulster Museum Picture Library, 625
Williams (David) Picture Library, 626

Geology: Publishers (European)
Universitaire Pers Leuven, 160

Geology: Publishers (UK)
Geological Society Publishing House, The, 97
NMS Enterprises Limited – Publishing, 120

Germany: Library Services
German Historical Institute Library, 583
Goethe-Institut Library, 584

Germany: Literary Agents (US)
Gusay (Charlotte) Literary Agency, The, 284

Germany: Picture Libraries
King (David) Collection, 613

Global Warming: Literary Agents (US)
Gusay (Charlotte) Literary Agency, The, 284

Golf: Literary Agents (UK)
Feldstein Agency, The, 256

Golf: Magazines
Golf Monthly, 323
Golf World, 323
Today's Golfer, 362
Women & Golf, 368

Golf: Picture Libraries
Hobbs Golf Collection, 612
Professional Sport UK Ltd, 620
Sheldon (Phil) Golf Picture Library, 623

Golf: Professional Associations
Association of Golf Writers, 528

Golf: Small Presses
Grant Books, 229

Government: Film, TV and Radio Producers
COI, 421

Government: Library Services
Bank of England Information Centre, 574
Belfast Public Libraries: Central Library, 575
British Library Social Sciences & Official Publications, 579
Guildhall Library, 584
Kent County Central Library, 587
Leeds Central Library, 587
London Metropolitan Archives Library Services, 589
National Archives, The, 590
Office for National Statistics, 592
Senate House Library, University of London, 595
Swansea Central Library, 597
Westminster Reference Library, 598

Government: Literary Agents (US)
Gusay (Charlotte) Literary Agency, The, 284

Government: Publishers (European)
Alianza Editorial SA, 169
Bertelsmann (C.), 164
Denoël (Éditions), 163
Deutscher Taschenbuch Verlag GmbH & Co. KG, 164
Hachette Livre (France), 163
Mulino (Societe editrice il), 167

Government: Publishers (International)
Auckland University Press, 196
Jaico Publishing House, 195
LexisNexis India, 195

Government: Publishers (UK)
Business Education Publishers Ltd, 79
National Archives, The, 118
TSO (The Stationery Office), 146

Greece: Literary Agents (US)
Gusay (Charlotte) Literary Agency, The, 284

Greece: Picture Libraries
Travel Ink Photo Library, 625

Greece: Prizes
Runciman Award, 518

Guidebooks: Literary Agents (UK)
Feldstein Agency, The, 256

Guidebooks: Literary Agents (US)
Gusay (Charlotte) Literary Agency, The, 284

Guidebooks: Professional Associations
Guidebookwriters.com, 537

Guidebooks: Publishers (European)
Random House Mondadori, 170

Guidebooks: Publishers (Ireland)
Gill & Macmillan, 155
On Stream Publications Ltd, 157
Tír Eolas, 158

Guidebooks: Publishers (UK)
AA Publishing, 64
Andrews (Chris) Publications, 66
Berlitz Publishing, 72
Birlinn Ltd, 73
Bradt Travel Guides, 77
Cicerone Press, 84
Cook (Thomas) Publishing, 86
Country Publications Ltd, 86
Donald (John) Publishers Ltd, 89
Dorling Kindersley Ltd, 90
Everyman's Library, 93
Foulsham Publishers, 95
Gwasg Carreg Gwalch, 99
HarperCollins Publishers Ltd, 102
Landmark Publishing Ltd, 109
Lonely Planet Publications Ltd, 112
Luath Press Ltd, 112
Mercat Press, 116
Michelin Maps & Guides, 116
New Holland Publishers (UK) Ltd, 119
Quiller Press, 130

Random House Group Ltd, The, 131
Reardon Publishing, 133
SB Publications, 136
Telegraph Books, 143
Travel Publishing Ltd, 145

Guidebooks: Publishers (US)
Globe Pequot Press, The, 176
Perseus Books Group, 184

Guidebooks: Small Presses
Crescent Moon Publishing and Joe's Press, 227
Kittiwake, 231
Meridian Books, 231

Health & Beauty: Literary Agents (Ireland)
Gunn O'Connor (Marianne) Literary Agency, 279

Health & Beauty: Literary Agents (UK)
Anderson (Darley) Literary, TV & Film Agency, 249
Chris (Teresa) Literary Agency Ltd, 253
Davidson (Caroline) Literary Agency, 255
Graham Maw Christie Literary Agency, 258
Limelight Management, 262

Health & Beauty: Literary Agents (US)
Gusay (Charlotte) Literary Agency, The, 284

Health & Beauty: Magazines
Black Beauty & Hair, 301
Easy Living, 316
Jersey Now, 330
Sainsbury's Magazine, 352
Zest, 371

Health & Beauty: Publishers (European)
Giunti Editoriale SpA, 166

Health & Beauty: Publishers (International)
Raupo Publishing (NZ) Ltd, 197

Health & Beauty: Publishers (UK)
Cathie (Kyle) Ltd, 81
Piatkus Books, 127
Souvenir Press Ltd, 140Souvenir Press Ltd,

Health & Nutrition: Book Clubs
Cygnus Books, 293

Health & Nutrition: Film, TV and Radio Producers
Healthcare Productions Limited, 425
Hourglass Productions, 426
Orlando Television Productions Ltd, 429

Health & Nutrition: Library Services
Leeds Central Library, 587

Health & Nutrition: Literary Agents (UK)
Anderson (Darley) Literary, TV & Film Agency, 249
Bonomi (Luigi) Associates Limited (LBA), 251
Chilcote (Judith) Agency, 253
Chris (Teresa) Literary Agency Ltd, 253
Davidson (Caroline) Literary Agency, 255
Fox & Howard Literary Agency, 257
Graham Maw Christie Literary Agency, 258
Johnson & Alcock, 261
Judd (Jane) Literary Agency, 262
Limelight Management, 262

Health & Nutrition: Literary Agents (US)
Bernstein (Meredith) Literary Agency, Inc., 280
Bleecker Street Associates, Inc., 280
Carvainis (Maria) Agency, Inc., 281
Castiglia Literary Agency, 281
Dijkstra (Sandra) Literary Agency, 282
Ellenberg (Ethan) Literary Agency, 283
Fredericks (Jeanne) Literary Agency, Inc., 283
Gusay (Charlotte) Literary Agency, The, 284
Klinger (Harvey), Inc., 284
Konner (Linda) Literary Agency, 285
Leavitt (Ned) Agency, The, 285
Lowenstein–Yost Associates Inc., 285
Mews Books Ltd, 286
Michaels (Doris S.) Literary Agency Inc., 286

Protter (Susan Ann) Literary Agent, 288
Quicksilver Books, Literary Agents, 288
Rees (Helen) Literary Agency, 288
Schulman (Susan), A Literary Agency, 290
Scribblers House® LLC Literary Agency, 290
Snell (Michael) Literary Agency, 290

Health & Nutrition: Magazines
Candis, 306
Health & Fitness Magazine, 326
Healthy, 326
Men's Health, 335
That's Life!, 360
Zest, 371

Health & Nutrition: PR Consultants
Frazer (Cathy) PR, 277

Health & Nutrition: Picture Libraries
Photofusion, 619
Report Digital, 621

Health & Nutrition: Publishers (European)
Brána a.s., 161
BZZTÔH (Uitgeverij) BV, 167
Deutscher Taschenbuch Verlag GmbH & Co. KG, 164
Karisto Oy, 162
Kibea Publishing Company, 160
Lannoo, Uitgeverij, 160
Larousse (Éditions), 163
Nyt Nordisk Forlag Arnold Busck A/S, 162
Spectrum (Uitgeverij Het) BV, 168
Sperling e Kupfer Editori SpA, 167
Ueberreuter (Verlag Carl) GmbH, 160
Ullstein Buchverlage GmbH, 165

Health & Nutrition: Publishers (International)
Allen & Unwin Pty Ltd, 191
Hachette Livre Australia, 191
Hind Pocket Books Pvt. Ltd, 195
Jaico Publishing House, 195
Orient Paperbacks, 195
Simon & Schuster (Australia) Pty Ltd, 192
Struik Publishers (Pty) Ltd, 198

Health & Nutrition: Publishers (Ireland)
On Stream Publications Ltd, 157

Health & Nutrition: Publishers (UK)
Anness Publishing Ltd, 66
Ashgrove Publishing, 69
Austin & Macauley Publishers Limited, 70
AuthorHouse UK, 70
Baird (Duncan) Publishers, 71
Carlton Publishing Group, 81
Carroll & Brown Publishers Limited, 81
Clairview Books Ltd, 84
Constable & Robinson Ltd, 85
Dorling Kindersley Ltd, 90
Foulsham Publishers, 95
Fulton (David) Publishers Ltd, 96
Hammersmith Press Ltd, 101
Little Books Ltd, 111
Mainstream Publishing Co. (Edinburgh) Ltd, 114
Mason (Kenneth) Publications Ltd, 115
New Holland Publishers (UK) Ltd, 119
Octopus Publishing Group, 121
Open University Press, 123
Publishing House, 129
Quadrille Publishing Ltd, 130
Random House Group Ltd, The, 131
Reader's Digest Association Ltd, 132
Simon & Schuster UK Limited, 139
Society for Promoting Christian Knowledge (SPCK), 140
Speechmark Publishing Ltd, 140
Telegraph Books, 143
Temple Lodge Publishing Ltd, 144
Women's Press, The, 150

Health & Nutrition: Publishers (US)
Avery, 173
Bantam Dell Publishing Group, 173
Barron's Educational Series, Inc., 173
Holt (Henry) & Company, Inc., 178
Llewellyn Publications, 180
Reader's Digest Association Inc, 185
Rosen Publishing Group, Inc., The, 185
Seven Stories Press, 186

Shambhala Publications, Inc., 186

Health & Nutrition: Small Presses
Accent Press, 225
Need2Know, 232

Health & Nutrition: UK Packagers
Breslich & Foss Ltd, 242
Ivy Press Limited, The, 244
Stonecastle Graphics Ltd, 246

Heritage: Literary Agents (UK)
Feldstein Agency, The, 256

Heritage: Magazines
Homes & Antiques, 327
Jersey Now, 330
National Trust Magazine, The, 339
Scotland in Trust, 352
This England, 360

Heritage: Picture Libraries
Adkins (Lesley & Roy) Picture Library, 601
Belcher (Ivan J.) Colour Picture Library, 603
Cordaiy (Sylvia) Photo Library, 606
Country Life Picture Library, 606
Craven (Philip) Worldwide Photo-Library, 606
E&E Picture Library, 607
English Heritage Photo Library, 607
ffotograff, 608
Forsythe (Robert) Picture Library, 609
Hardley (Dennis) Photography, Scottish Photo Library, 611
National Monuments Record, 617
National Trust Photo Library, The, 618
Photoshot, 619

Heritage: Publishers (International)
Wits University Press, 198

Heritage: Publishers (UK)
Anova Books, 66
Breedon Books Publishing Co. Ltd, The, 77
Donhead Publishing Ltd, 89
Sigma Press, 139
Telegraph Books, 143

Third Millennium Publishing, 144

History: Book Clubs
BCA (Book Club Associates), 293
Folio Society, The, 293

History: Film, TV and Radio Producers
Diverse Production Limited, 422
Moonstone Films Ltd, 429

History: Library Services
Guildhall Library, 584
Institute of Historical Research, The, 586
London Library, The, 589
Morrab Library, 590
Senate House Library, University of London, 595
Williams's (Dr) Library, 599

History: Literary Agents (Ireland)
Williams (Jonathan) Literary Agency, 279

History: Literary Agents (UK)
Ableman (Sheila) Literary Agency, 249
Ampersand Agency Ltd, The, 249
Artellus Limited, 250
Bonomi (Luigi) Associates Limited (LBA), 251
Brown (Jenny) Associates, 251
Bryan (Felicity), 252
Chris (Teresa) Literary Agency Ltd, 253
Davidson (Caroline) Literary Agency, 255
Edwards Fuglewicz, 256
Feldstein Agency, The, 256
Fox & Howard Literary Agency, 257
Graham Maw Christie Literary Agency, 258
Green (Annette) Authors' Agency, 258
Heath (Rupert) Literary Agency, 260
hhb agency ltd, 260
Higham (David) Associates Ltd, 260
Johnson & Alcock, 261
Kelly (Frances), 262
Knight Features, 262
Lownie (Andrew) Literary Agency, 263

Luxton Harris Ltd, 263
Nye (Alexandra), 266
Sayle Literary Agency, The, 268
Sheil Land Associates Ltd, 268
Simmons (Jeffrey), 269
Sinclair-Stevenson, 269
Turnbull (Jane), 271
Watson, Little Ltd, 272
Welch (John), Literary Consultant & Agent, 272

History: Literary Agents (US)
Barrett (Loretta) Books, Inc., 280
Bleecker Street Associates, Inc., 280
Dijkstra (Sandra) Literary Agency, 282
Dystel & Goderich Literary Management, 283
Ellenberg (Ethan) Literary Agency, 283
Gusay (Charlotte) Literary Agency, The, 284
Lescher & Lescher Ltd, 285
Lowenstein–Yost Associates Inc., 285
Manus & Associates Literary Agency, Inc., 286
Michaels (Doris S.) Literary Agency Inc., 286
Parks (Richard) Agency, 287
Protter (Susan Ann) Literary Agent, 288
Rees (Helen) Literary Agency, 288
Rittenberg (Ann) Literary Agency, Inc., 288
Russell & Volkening, Inc., 289
Scribblers House® LLC Literary Agency, 290
Siegel (Rosalie), International Literary Agency, Inc., 290
Ware (John A.) Literary Agency, 291
Writers House, LLC, 292

History: Magazines
BBC History Magazine, 299
Edinburgh Review, 317
History Today, 327
New Humanist, 339

History: Miscellany
Edwards (Jacqueline), 641

History: PR Consultants
Publishing Services, 278

History: Picture Libraries
Adkins (Lesley & Roy) Picture Library, 601

akg-images Ltd, Arts and History Picture Library, 601
Art Archive, The, 602
Barnardo's Photographic and Film Archive, 603
Corbis Images, 606
Evans (Mary) Picture Library, 607
King (David) Collection, 613
Mirrorpix, 616
National Archives Image Library, The, 617
National Galleries of Scotland Picture Library, 617
National Monuments Record, 617
Newark's (Peter) Picture Library, 618
Popperfoto.com, 620
Sharp (Mick) Photography, 622

History: Poetry Presses
King's England Press, The, 203

History: Prizes
Alexander Prize, 488
BBCFour Samuel Johnson Prize for Non-Fiction, The, 490
Berry ((David) Prize, 490
Cooper (Duff) Prize, The, 496
Gladstone History Book Prize, The, 502
Hessell-Tiltman Prize for History, The, 504
Longman-History Today Book of the Year Award, 509
Scottish History Book of the Year, 519
Whitfield Prize, 524
Wolfson History Prize, 525

History: Publishers (European)
Adelphi Edizioni SpA, 166
Alianza Editorial SA, 169
Beck (C.H., OHG) Verlag, 164
Borgens Forlag A/S, 161
Brepols Publishers NV, 160
Bruna (A.W.) Uitgevers BV, 167
Calmann-Lévy (Éditions), 163
Caminho (Editorial) SARL, 169
Civilização Editora, 169
Denoël (Éditions), 163
Deutscher Taschenbuch Verlag GmbH & Co. KG, 164
Dokorán Ltd, 161
EDHASA (Editora y Distribuidora Hispano – Americana SA), 169
Fayard (Librarie Arthème), 163
Fischer (S.) Verlag GmbH, 164
Flammarion Éditions, 163

Forlaget Forum, 161
Gallimard (Éditions), 163
Garzanti Libri SpA, 166
Giunti Editoriale SpA, 166
Gyldendal, 161
Heyne Verlag, 165
Hoffmann und Campe Verlag GmbH, 165
Høst & Søn Publishers Ltd, 161
Karisto Oy, 162
Kibea Publishing Company, 160
Kiss József Kónyvkiadó, 166
Kok (Uitgeversmaatschappij J. H.) BV, 168
Lannoo, Uitgeverij, 160
Larousse (Éditions), 163
Longanesi (Casa Editrice) SpA, 167
Lübbe (Verlagsgruppe) GmbH & Co. KG, 165
Mondadori (Arnoldo) Editore SpA, 167
Mulino (Societe editrice il), 167
Mursia (Gruppo Ugo) Editore SpA, 167
Presses de la Cité, 164
Presses Universitaires de France (PUF), 164
Publicat SA, 169
Seuil (Éditions du), 164
Sperling e Kupfer Editori SpA, 167
Sugarco Edizioni Srl, 167
Table Ronde (Les Editions de la), 164
Tiderne Skifter Forlag A/S, 162
Tusquets Editores, 170
Ueberreuter (Verlag Carl) GmbH, 160
Universitaire Pers Leuven, 160
Verbo (Editorial) SA, 169
Zsolnay (Paul), Verlag GmbH, 160

History: Publishers (International)
Auckland University Press, 196
Canterbury University Press, 196
Fitzhenry & Whiteside Limited, 193
Hachette Livre Australia, 191
Huia (NZ) Ltd, 196
Jaico Publishing House, 195
Melbourne University Publishing Ltd, 192
Munshiram Manoharlal Publishers Pvt Ltd, 195
National Publishing House, 195
New Star Books Ltd, 194
Pan Macmillan Australia Pty Ltd, 192

Queensland Press, University of, 192
Random House of Canada Ltd, 194
Raupo Publishing (NZ) Ltd, 197
UWA Press, 193
Victoria University Press, 197
Williams (Bridget) Books Ltd, 197
Wits University Press, 198

History: Publishers (Ireland)
Anvil Books, 153
Burke (Edmund) Publisher, 154
Cork University Press, 155
Currach Press, 155
Farmar (A.&A.) Ltd, 155
Gill & Macmillan, 155
Institute of Public Administration, 156
Lilliput Press, The, 156
Mercier Press Ltd, 156
Merlin Publishing, 157
O'Brien Press Ltd, The, 157
Poolbeg Press Ltd, 158

History: Publishers (UK)
ABC-CLIO (UK), 64
Alma Books Ltd, 65
Apex Publishing Ltd, 67
Arris Publishing Ltd, 68
Ashgate Publishing Ltd, 69
Atlantic Books, 69
AuthorHouse UK, 70
Berg Publishers, 72
Birlinn Ltd, 73
Black Ace Books, 73
Blackstaff Press Ltd, 74
Blackstone Publishers, 74
Brewin Books Ltd, 77
British Academy, The, 77
British Museum Press, The, 78
Carlton Publishing Group, 81
Clarke (James) & Co., 84
Compendium Publishing Ltd, 85
Constable & Robinson Ltd, 85
Continuum International Publishing Group Ltd, The, 85
Country Publications Ltd, 86
Currey (James) Publishers, 87
Davies (Christopher) Publishers Ltd, 88
de la Mare (Giles) Publishers Ltd, 88
Donald (John) Publishers Ltd, 89
Edinburgh University Press, 91
English Heritage (Publishing), 92
Exeter Press, University of, 93
Floris Books, 95

Fort Publishing Ltd, 95
Garnet Publishing Ltd, 97
Gibson Square, 97
Gwasg Carreg Gwalch, 99
Hachette Livre UK, 100
Halban Publishers, 100
HarperCollins Publishers Ltd, 102
Harvard University Press, 103
History Press Ltd, The, 105
Honeyglen Publishing Ltd, 105
Hurst Publishers, Ltd, 106
Janus Publishing Company Ltd, 108
JR Books Ltd, 108
Lawrence & Wishart Ltd, 110
Little Books Ltd, 111
Little, Brown Book Group UK, 111
Luath Press Ltd, 112
Macmillan Publishers Ltd, 113
Mainstream Publishing Co. (Edinburgh) Ltd, 114
Manchester University Press, 115
Merlin Press Ltd, The, 116
Murdoch Books UK Ltd, 118
National Archives, The, 118
New Beacon Books Ltd, 119
New Holland Publishers (UK) Ltd, 119
NMS Enterprises Limited – Publishing, 120
Olympia Publishers, 122
Oneworld Publications, 122
Orion Publishing Group Limited, The, 123
Owen (Peter) Publishers, 123
Oxbow Books, 124
Pegasus Elliot Mackenzie Publishers Ltd, 125
Pen & Sword Books Ltd, 125
Polity Press, 128
Profile Books, 129
Quartet Books, 130
Random House Group Ltd, The, 131
Reader's Digest Association Ltd, 132
Reaktion Books, 132
Routledge, 134
Seren, 137
Sheldrake Press, 138
Short Books, 139
Simon & Schuster UK Limited, 139
Smythe (Colin) Ltd, 139
Souvenir Press Ltd, 140
Spellmount, 140
Tauris (I.B.) & Co. Ltd, 143
Thames and Hudson Ltd, 144
Trident Press Ltd, 145

Verso, 147
Wales Press, University of, 148
Wilson (Neil) Publishing Ltd, 149
Wimbledon Publishing Company, 150
Yale University Press (London), 151

History: Publishers (US)
ABC–CLIO (US), 172
Academy Chicago Publishers, 172
Alabama Press, University of, 172
Alaska Press, University of, 172
Arkansas Press, University of, 173
Holt (Henry) & Company, Inc., 178
Houghton Mifflin Harcourt Publishing Company, 178
Indiana University Press, 179
Kansas, University Press of, 179
Kent State University Press, The, 179
Krieger Publishing Co., 179
Louisiana State University Press, 180
Massachusetts Press, University of, 181
McFarland & Company, Inc., Publishers, 180
Michigan Press, The University of, 181
Mississippi, University Press of, 181
Missouri Press, University of, 181
North Texas Press, University of, 182
Paragon House, 183
Pelican Publishing Company, 183
Perseus Books Group, 184
Potomac Books, Inc., 184
PublicAffairs, 185
St Martin's Press LLC, 186
Stackpole Books, 187
Stanford University Press, 187
Sterling Publishing Co. Inc., 188
Washington State University Press, 190

History: Small Presses
Anglo-Saxon Books, 225
Cosmic Elk, The, 227
Ivy Publications, 230
Parapress, 233
Serif, 235
Witan Books, 237

History: UK Packagers
Amber Books Ltd, 242
Quarto Publishing plc, 245
Savitri Books Ltd, 245

History: Writers' Circles and Workshops
Historical Novel Folio, 478

Home Interests: Literary Agents (US)
Gusay (Charlotte) Literary Agency, The, 284

Home Interests: Magazines
Country Homes and Interiors, 311
Elle Decoration, 317
Good Homes Magazine, 323
Good Housekeeping, 323
Homes & Antiques, 327
Homes & Gardens, 327
House & Garden, 328
House Beautiful, 328
Ideal Home, 328
Period Living, 344
Sainsbury's Magazine, 352
That's Life!, 360
Vogue, 364
Woman and Home, 368

Home Interests: Publishers (European)
Larousse (Éditions), 163

Home Interests: Publishers (International)
HarperCollins Publishers India Ltd, 195

Home Interests: Publishers (UK)
Cathie (Kyle) Ltd, 81
Haynes Publishing, 103
Murdoch Books UK Ltd, 118
Ryland Peters & Small Limited, 135

Home Interests: Publishers (US)
Reader's Digest Association Inc, 185
Sterling Publishing Co. Inc., 188

Horses: Book Clubs
Readers' Union Ltd, 293

Horses: Film, TV and Radio Producers
MBP TV, 428

Horses: Literary Agents (UK)
London Independent Books, 263
Motley (Michael) Ltd, 265

Horses: Literary Agents (US)
Snell (Michael) Literary Agency, 290

Horses: Magazines
Horse, 327
Horse and Hound, 327
Horse and Rider, 327
PONY, 345
Showing World, 355
Your Horse, 371

Horses: Picture Libraries
Cordaiy (Sylvia) Photo Library, 606
Houghton's Horses/Kit Houghton Photography, 612
Only Horses Picture Agency, 619
Thoroughbred Photography Ltd, 624

Horses: Professional Associations
British Equestrian Writers' Association, 532

Horses: Publishers (UK)
Allen (J.A.) & Co., 65
Crowood Press Ltd, The, 87
David & Charles Publishers, 88
Kenilworth Press Ltd, 108
Sportsman's Press, The, 141
Swan Hill Press, 142
Whittet Books Ltd, 149

Housing: Picture Libraries
Hoffman (David) Photo Library, 612
Photofusion, 619

How To: Literary Agents (UK)
Graham Maw Christie Literary Agency, 258

How To: Literary Agents (US)
Browne (Pema) Ltd, 281
Chelius (Jane) Literary Agency, Inc., 282
Nelson (B.K.) Literary Agency, 287
Scribblers House® LLC Literary Agency, 290
Snell (Michael) Literary Agency, 290

Writers House, LLC, 292

How To: Publishers (European)
Gyldendal, 161
Heyne Verlag, 165
Lübbe (Verlagsgruppe) GmbH & Co. KG, 165
Mondadori (Arnoldo) Editore SpA, 167
Sugarco Edizioni Srl, 167

How To: Publishers (UK)
Constable & Robinson Ltd, 85
How To Books Ltd, 106

How To: Publishers (US)
Globe Pequot Press, The, 176
Llewellyn Publications, 180
Sterling Publishing Co. Inc., 188

How To: Small Presses
Pomegranate Press, 234

How To: UK Packagers
Quarto Publishing plc, 245Quarto Publishing plc,

Human Rights: Literary Agents (US)
Gusay (Charlotte) Literary Agency, The, 284

Human Rights: Magazines
New Humanist, 339

Human Rights: Professional Associations
English PEN, 536

Human Rights: Small Presses
Eilish Press, 228

Humanities: Bursaries, Fellowships and Grants
British Academy Small Research Grants, 482

Humanities: Library Services
Birmingham and Midland Institute, 575
British Library Humanities Reading Room, 577
Cardiff Central Library, 580
Liverpool Libraries and Information Services, 588
London Library, The, 589
Manchester Central Library, 589
National Library of Wales, 590
Senate House Library, University of London, 595

Women's Library, The, 599

Humanities: Literary Agents (UK)
Feldstein Agency, The, 256

Humanities: Literary Agents (US)
Gusay (Charlotte) Literary Agency, The, 284

Humanities: Publishers (European)
Brepols Publishers NV, 160
Springer Science+Business Media BV, 168

Humanities: Publishers (International)
Munshiram Manoharlal Publishers Pvt Ltd, 195

Humanities: Publishers (Ireland)
Four Courts Press Ltd, 155
Irish Academic Press Ltd, 156

Humanities: Publishers (UK)
Ashgate Publishing Ltd, 69
Berg Publishers, 72
British Academy, The, 77
Continuum International Publishing Group Ltd, The, 85
Manchester University Press, 115
Routledge, 134
Sage Publications, 135
Wiley-Blackwell, 149

Humanities: Publishers (US)
Harvard University Press, 177
Kentucky, The University Press of, 179
Krieger Publishing Co., 179
Missouri Press, University of, 181
Rutgers University Press, 185
Sharpe (M.E.), Inc., 186
Stanford University Press, 187
State University of New York Press, 187
Texas Press, University of, 188
Virginia Press, University of, 189

Humanities: Small Presses
Superscript, 235

Humour: Audio Books
BBC Audiobooks Ltd, 238
Corgi Audio, 238

Hachette Digital, 239
Laughing Stock Productions, 239
Orion Audio Books, 240

Humour: Festivals
National Association of Writers' Groups (NAWG) Open Festival of Writing, 636

Humour: Film, TV and Radio Producers
Above The Title, 417
Absolutely Productions Ltd, 417
Acrobat Television, 417
Anglo/Fortunato Films Ltd, 418
Avalon, 418
Baby Cow Productions, 418
Celador Productions, 420
Comedy Unit, The, 421
Endemol UK Productions, 422
Fast Films, 423
Flannel, 423
Gay (Noel) Television, 424
Hartswood Films Ltd, 425
Hat Trick Productions Ltd, 425
Mendoza Film Productions, 429
Pozzitive Television, 431
Richmond Films & Television Ltd, 431
Staveacre (Tony) Productions, 433
talkbackTHAMES, 434
Tiger Aspect Productions, 435
Vera Productions, 436

Humour: Literary Agents (Ireland)
Williams (Jonathan) Literary Agency, 279

Humour: Literary Agents (UK)
Anderson (Darley) Literary, TV & Film Agency, 249
Clowes (Jonathan) Ltd, 254
Feldstein Agency, The, 256
Graham Maw Christie Literary Agency, 258
Hancock (Roger) Ltd, 259
Inspira Group, The, 261
Judd (Jane) Literary Agency, 262
Motley (Michael) Ltd, 265
Sheil Land Associates Ltd, 268
Sheldon (Caroline) Literary Agency Ltd, 268
Turnbull (Jane), 271
Ware (Cecily) Literary Agents, 272

Humour: Literary Agents (US)
Gusay (Charlotte) Literary Agency, The, 284
Michaels (Doris S.) Literary Agency Inc., 286

Humour: Magazines
Club International, 309
Dandy, The, 313
Horse and Hound, 327
Jewish Telegraph Group of Newspapers, 331
Loaded, 333
Mayfair, 335
Oldie, The, 342
Private Eye, 347
Reader's Digest, 350
Total Flyfisher, 362

Humour: Picture Libraries
Rex Features Ltd, 621

Humour: Prizes
Bollinger Everyman Wodehouse Prize, 491
Dahl (Roald) Funny Prize, The, 498
Nathan (Melissa) Award for Comedy Romance, 512

Humour: Professional Associations
British Society of Comedy Writers, 532

Humour: Publishers (European)
Borgens Forlag A/S, 161
Calmann-Lévy (Éditions), 163
Forlaget Forum, 161
Heyne Verlag, 165
Standaard Uitgeverij, 160

Humour: Publishers (International)
Allen & Unwin Pty Ltd, 191
Hachette Livre New Zealand Ltd, 196
Orient Paperbacks, 195
Struik Publishers (Pty) Ltd, 198

Humour: Publishers (Ireland)
Hachette Books Ireland, 155
O'Brien Press Ltd, The, 157

Humour: Publishers (UK)
Anova Books, 66
Apex Publishing Ltd, 67
Austin & Macauley Publishers Limited, 70

AuthorHouse UK, 70
Birlinn Ltd, 73
Black & White Publishing Ltd, 73
Blackstaff Press Ltd, 74
Blackstone Publishers, 74
Capstone Publishing, 80
Carlton Publishing Group, 81
Country Publications Ltd, 86
Crombie Jardine Publishing Limited, 87
Emissary Publishing, 92
Exley Publications Ltd, 93
Hachette Livre UK, 100
HarperCollins Publishers Ltd, 102
Henry (Ian) Publications Ltd, 104
JR Books Ltd, 108
Little, Brown Book Group UK, 111
Macmillan Publishers Ltd, 113
Methuen Publishing Ltd, 116
Myriad Editions, 118
New Holland Publishers (UK) Ltd, 119
O'Mara (Michael) Books Limited, 122
Old Street Publishing Ltd, 121
Olympia Publishers, 122
Pegasus Elliot Mackenzie Publishers Ltd, 125
Publishing House, 129
Quiller Press, 130
Random House Group Ltd, The, 131
SeaNeverDry Publishing, 137
Short Books, 139
Simon & Schuster UK Limited, 139
Souvenir Press Ltd, 140
Summersdale Publishers Ltd, 142
Telegraph Books, 143
Wilson (Neil) Publishing Ltd, 149

Humour: Publishers (US)
Perseus Books Group, 184

Humour: Small Presses
Accent Press, 225
Miller (Neil) Publications, 232
Stone Flower Limited, 235

Humour: Television and Radio
BBC Drama, Comedy and Children's, 396
BBC Radio Cymru, 401

Humour: Theatre Producers
Theatre Royal Windsor, 449

Humour: UK Packagers
Elwin Street Productions, 243
Ivy Press Limited, The, 244

Humour: Writers' Courses (UK and Ireland)
City Lit, 458
City University, 458
Community Creative Writing, 461
Liverpool, University of, 461
Missenden Abbey, 453
Reading, University of, 453
Sheffield, University of, 465

Illustrated & Fine Editions: Literary Agents (UK)
Chapman & Vincent, 253
Kelly (Frances), 262

Illustrated & Fine Editions: Literary Agents (US)
Gusay (Charlotte) Literary Agency, The, 284

Illustrated & Fine Editions: PR Consultants
Idea Generation, 277
Pugh (Sonia) PR, 278

Illustrated & Fine Editions: Publishers (European)
Arthaud (Éditions), 162
Beck (C.H., OHG) Verlag, 164
Bompiani, 166
Giunti Editoriale SpA, 166
Gradiva–Publicações S.A., 169
Hoffmann und Campe Verlag GmbH, 165
Kibea Publishing Company, 160
Random House Mondadori, 170
Seuil (Éditions du), 164
Spectrum (Uitgeverij Het) BV, 168

Illustrated & Fine Editions: Publishers (International)
HarperCollins Publishers (Australia) Pty Ltd, 191
Random House Australia, 192
Struik Publishers (Pty) Ltd, 198

Illustrated & Fine Editions: Publishers (UK)
Anness Publishing Ltd, 66
Anova Books, 66
Aurum Press Ltd, 70
Baird (Duncan) Publishers, 71

BFP Books, 72
Blackstaff Press Ltd, 74
British Library, The, 78
Cico Books, 84
Compendium Publishing Ltd, 85
David & Charles Publishers, 88
Dorling Kindersley Ltd, 90
Graham-Cameron Publishing & Illustration, 98
HarperCollins Publishers Ltd, 102
King (Laurence) Publishing Ltd, 109
Lincoln (Frances) Ltd, 110
Lion Hudson plc, 111
Little, Brown Book Group UK, 111
Mainstream Publishing Co. (Edinburgh) Ltd, 114
Octopus Publishing Group, 121
Orion Publishing Group Limited, The, 123
Osprey Publishing Ltd, 123
Prestel Publishing Limited, 129
Quadrille Publishing Ltd, 130
Ryland Peters & Small Limited, 135
Saqi Books, 135
Sheldrake Press, 138
Shepheard-Walwyn (Publishers) Ltd, 138
Souvenir Press Ltd, 140
Tauris (I.B.) & Co. Ltd, 143
Templar Publishing, 143
Thames and Hudson Ltd, 144
Third Millennium Publishing, 144

Illustrated & Fine Editions: Publishers (US)
Abrams (Harry N.), Inc., 172
Chronicle Books LLC, 174
Crown Publishing Group, 175
Perseus Books Group, 184
Sterling Publishing Co. Inc., 188

Illustrated & Fine Editions: Small Presses
Porteous (David) Editions, 234
Zymurgy Publishing, 237

Illustrated & Fine Editions: UK Packagers
Aladdin Books Ltd, 242
Albion Press Ltd, The, 242
Bender Richardson White, 242
Eddison Sadd Editions, 243
Elwin Street Productions, 243
Haldane Mason Ltd, 243
Ivy Press Limited, The, 244
Lionheart Books, 244
Playne Books Limited, 245

Quarto Publishing plc, 245
Salariya Book Company Ltd, 245
Savitri Books Ltd, 245
Stonecastle Graphics Ltd, 246
Toucan Books Ltd, 246
Touchstone Books Ltd, 246
West (David) Children's Books, 246West (David) Children's Books,

Indexing: Miscellany
Ormrod Research Services, 641

Indexing: Prizes
Carey Award, 494

Indexing: Professional Associations
Society of Indexers, 548

India: Library Services
British Library Asia, Pacific and Africa Collections, 577

India: Literary Agents (UK)
Eady (Toby) Associates Ltd, 256

India: Literary Agents (US)
Gusay (Charlotte) Literary Agency, The, 284

India: Picture Libraries
British Library Images Online, 604
Dickins (Douglas) Photo Library, 607
Gagg's (Andrew N.) PHOTOFLORA, 609
Geoslides Photography, 610
Greenhill (Sally and Richard), 611
Link Picture Library, 615
Peerless (Ann & Bury) Picture Library, 619
Tropix Photo Library, 625

Industrial Relations: Magazines
People Management, 343

Industrial Relations: Professional Associations
Federation of Entertainment Unions, 537
Irish Writers' Union, 539
National Union of Journalists, 541
Pact, 542
Scottish Print Employers Federation, 546

Writers' Guild of Great Britain, The, 551

Industrial Relations: Publishers (UK)
Kogan Page Ltd, 109

Industry: Picture Libraries
Aspect Picture Library Ltd, 602
Beamish, The Photographic Library, 603
Ecoscene, 607
Forsythe (Robert) Picture Library, 609
Hutchison Picture Library, 612
Last Resort Picture Library, 614
Mirrorpix, 616
Oxford Scientific (OSF), 619
PictureBank Photo Library Ltd, 620
Science & Society Picture Library, 622
Science Photo Library, 622

Industry: Publishers (UK)
Souvenir Press Ltd, 140

Industry: Small Presses
Stenlake Publishing Limited, 235

Interior Design: Library Services
British Architectural Library, 576

Interior Design: Literary Agents (UK)
Limelight Management, 262

Interior Design: Literary Agents (US)
Gusay (Charlotte) Literary Agency, The, 284

Interior Design: Magazines
Blueprint, 302
Country Homes and Interiors, 311
Country Living, 311
Good Homes Magazine, 323
House & Garden, 328
House Beautiful, 328
Ideal Home, 328
Period Living, 344
SmartLife International, 356
World of Interiors, The, 369

Interior Design: Picture Libraries
Country Life Picture Library, 606
Strang (Jessica) Photo Library, 624

Interior Design: Prizes
RIBA International Book Awards, 517

Interior Design: Publishers (European)
Lannoo, Uitgeverij, 160
Taschen GmbH, 165

Interior Design: Publishers (UK)
Anness Publishing Ltd, 66
Anova Books, 66
Cathie (Kyle) Ltd, 81
Cico Books, 84
King (Laurence) Publishing Ltd, 109
Lincoln (Frances) Ltd, 110
Murdoch Books UK Ltd, 118
New Holland Publishers (UK) Ltd, 119
Octopus Publishing Group, 121
Quadrille Publishing Ltd, 130

Interior Design: UK Packagers
Breslich & Foss Ltd, 242Breslich & Foss Ltd,

Ireland/Irish Interest: Arts Councils and Regional Offices
Arts Council of Northern Ireland, The, 573
Arts Council, The/An Chomhairle Ealaíon, 572

Ireland/Irish Interest: Film, TV and Radio Producers
Lagan Pictures Ltd, 427
Straight Forward Film & Television Productions Ltd, 434

Ireland/Irish Interest: Library Services
Belfast Public Libraries: Central Library, 575
Linen Hall Library, 588
National Library of Ireland, 590

Ireland/Irish Interest: Literary Agents (UK)
Anderson (Darley) Literary, TV & Film Agency, 249
Feldstein Agency, The, 256

Ireland/Irish Interest: Literary Agents (US)
Gusay (Charlotte) Literary Agency, The, 284

Ireland/Irish Interest: Magazines
Irish Book Review, The, 329
Irish Pages, 329

Ireland/Irish Interest: Organizations of Interest to Poets
Poetry Ireland/Eigse Eireann, 221
Western Writers' Centre, The, 224

Ireland/Irish Interest: Picture Libraries
Aspect Picture Library Ltd, 602
Collections, 605
Hill (Christopher) Photographic, 611
National Trust Photo Library, The, 618
Ulster Museum Picture Library, 625

Ireland/Irish Interest: Poetry Magazines
Irish Pages, 211
Poetry Ireland News, 214
Poetry Ireland Review/Eigse Eireann, 214
ShOp: A Magazine Of Poetry, The, 217
Stinging Fly, The, 217

Ireland/Irish Interest: Poetry Presses
Dedalus Press, The, 200
DogHouse, 201
Gallery Press, The, 202
Stinging Fly Press, 207

Ireland/Irish Interest: Press Cuttings Agencies
We Find It (Press Clippings), 628

Ireland/Irish Interest: Prizes
Irish Book Awards, 506

Ireland/Irish Interest: Professional Associations
Association of Freelance Editors, Proofreaders & Indexers (Ireland), 528
Children's Books Ireland, 534
CLÉ – Irish Book Publishers' Association, 534
Irish Copyright Licensing Agency Ltd, 539
Irish Playwrights and Screenwriters Guild, 539
Irish Translators' & Interpreters' Association, 539
Irish Writers' Union, 539

Ireland/Irish Interest: Publishers (Ireland)
Anvil Books, 153
Ashfield Press, 153
Brandon/Mount Eagle Publications, 154
Burke (Edmund) Publisher, 154
Cló Iar-Chonnachta, 154
Collins Press, The, 154
Currach Press, 155
Farmar (A.&A.) Ltd, 155
Gill & Macmillan, 155
Liffey Press Ltd, The, 156
Mercier Press Ltd, 156
New Island, 157
Poolbeg Press Ltd, 158
Royal Irish Academy, 158

Ireland/Irish Interest: Publishers (UK)
Appletree Press Ltd, 68
Blackstaff Press Ltd, 74
Colourpoint Books, 85
Smythe (Colin) Ltd, 139
Travel Publishing Ltd, 145

Ireland/Irish Interest: Publishers (US)
Syracuse University Press, 188

Ireland/Irish Interest: Small Presses
Serif, 235

Ireland/Irish Interest: Theatre Producers
Druid Theatre Company, 440
Tricycle Theatre, 450

Islam: Literary Agents (UK)
London Independent Books, 263

Islam: Literary Agents (US)
Gusay (Charlotte) Literary Agency, The, 284

Islam: Magazines
Q-News, The Muslim Magazine, 348

Islam: Picture Libraries
Peerless (Ann & Bury) Picture Library, 619
Sanders (Peter) Photography Ltd, 622

Islam: Publishers (UK)
Edinburgh University Press, 91
Garnet Publishing Ltd, 97
Oneworld Publications, 122

Italy: Library Services
Italian Institute Library, 586

Italy: Literary Agents (UK)
Belli (Lorella) Literary Agency (LBLA), 250

Italy: Literary Agents (US)
Gusay (Charlotte) Literary Agency, The, 284

Italy: Picture Libraries
Blake (Anthony) Photo Library, 603
Heseltine (John) Archive, 611

Japan: Literary Agents (UK)
Eady (Toby) Associates Ltd, 256
Henser Literary Agency, 260

Japan: Literary Agents (US)
Gusay (Charlotte) Literary Agency, The, 284

Japan: Picture Libraries
Dickins (Douglas) Photo Library, 607
Link Picture Library, 615

Japan: Prizes
Kiriyama Pacific Rim Book Prize, 507

Jewish Interest: Bursaries, Fellowships and Grants
European Jewish Publication Society, 483

Jewish Interest: Library Services
Plymouth Library and Information Services, City of, 593
Wiener Library, The, 599

Jewish Interest: Literary Agents (US)
Gusay (Charlotte) Literary Agency, The, 284

Jewish Interest: Magazines
Jewish Chronicle, 330
Jewish Quarterly, 330
Jewish Telegraph Group of Newspapers, 331

Jewish Interest: Poetry Presses
Five Leaves Publications, 201

Jewish Interest: Prizes
Jewish Quarterly Literary Prize, 506

Jewish Interest: Publishers (UK)
Halban Publishers, 100

Jewish Interest: Publishers (US)
Indiana University Press, 179
Nebraska Press, University of, 182
Paragon House, 183
Syracuse University Press, 188

Jewish Interest: Small Presses
Five Leaves Publications, 228

Journalism & Reportage: Bursaries, Fellowships and Grants
Economist/Richard Casement Internship, The, 483
Guardian Research Fellowship, The, 484
Journalists' Charity, 484
Stern (Lawrence) Fellowship, 487

Journalism & Reportage: Festivals
National Association of Writers' Groups (NAWG) Open Festival of Writing, 636

Journalism & Reportage: Library Services
British Library Newspapers, 578

Frost (John) Newspapers, 583

Journalism & Reportage: Literary Agents (UK)
Belli (Lorella) Literary Agency (LBLA), 250
Feldstein Agency, The, 256
Judd (Jane) Literary Agency, 262

Journalism & Reportage: Literary Agents (US)
Gusay (Charlotte) Literary Agency, The, 284
Ware (John A.) Literary Agency, 291
Weil (Wendy) Agency, Inc., The, 291

Journalism & Reportage: Magazines
British Journalism Review, 303
Press Gazette, 347

Journalism & Reportage: Picture Libraries
Getty Images, 610

Journalism & Reportage: Poetry Magazines
Irish Pages, 211

Journalism & Reportage: Prizes
British Press Awards, 493
Cameron (James) Award, 494
Foot (Paul) Award, The, 500
Gellhorn (Martha) Trust Prize, 501
Lyons (Sir William) Award, 509
Pulitzer Prizes, 516
Regional Press Awards, 517
Renault UK Journalist of the Year Award, 517
Vogue Talent Contest, 523
Watt (David) Prize, The, 523
Winchester Writers' Conference Prizes, 524

Journalism & Reportage: Professional Associations
ABSW, 526
British Association of Journalists, 531
British Society of Magazine Editors (BSME), 532
Campaign for Press and Broadcasting Freedom, 533
Chartered Institute of Journalists, 533
Foreign Press Association in London, 537

Guild of Agricultural Journalists, 537
Medical Journalists' Association, 540
National Union of Journalists, 541
Newspaper Society, The, 542
PPA The Association for Publishers and Providers of Consumer, Customer & Business Media in the UK, 543
Press Complaints Commission, 544
Scottish Daily Newspaper Society, 546
Scottish Newspaper Publishers Association, 546
Society of Editors, 548
Society of Women Writers & Journalists, 548
Sports Journalists' Association of Great Britain, 549

Journalism & Reportage: Publishers (European)
Longanesi (Casa Editrice) SpA, 167

Journalism & Reportage: Publishers (International)
Random House of Canada Ltd, 194

Journalism & Reportage: Publishers (US)
Missouri Press, University of, 181
PublicAffairs, 185

Journalism & Reportage: Writers' Courses (UK and Ireland)
Highbury College, Portsmouth, 456
London College of Communication, 459
London School of Journalism, 459
National Council for the Training of Journalists, 455
PMA Media Training, 459
Roehampton University, 460
Sheffield, University of, 465
University College Falmouth, 454
Wales (University of), Bangor, 467
Warwick, University of, 464
Westminster, University of, 460
Wilson's (John) Writing Courses, 467

Writer's Academy, The, 466
Writers Bureau College of Journalism, 461
Writing School Leicester, 457

Journalism & Reportage: Writers' Courses (US)
Goucher College, 471Goucher College, Goucher College, Goucher College,

Journals: Publishers (European)
Akadémiai Kiadó, 166
Beck (C.H., OHG) Verlag, 164
Blackwell Munksgaard, 161
Brepols Publishers NV, 160
Giunti Editoriale SpA, 166
Gyldendal Norsk Forlag, 168
Hanser (Carl) Verlag GmbH & Co. KG, 165
Hoffmann und Campe Verlag GmbH, 165
Hüthig GmbH & Co. KG, 165
Mondadori (Arnoldo) Editore SpA, 167
Mulino (Societe editrice il), 167
Springer GmbH, 165
Springer Science+Business Media BV, 168
Springer-Verlag GmbH, 160

Journals: Publishers (International)
LexisNexis Canada Ltd, 193
Toronto Press, Inc., University of, 194

Journals: Publishers (Ireland)
Cork University Press, 155
Institute of Public Administration, 156
Royal Dublin Society, 158
Royal Irish Academy, 158

Journals: Publishers (UK)
Authentic Media, 70
BBC Worldwide, 72
Berg Publishers, 72
Berghahn Books, 72
Cambridge University Press, 79
CBA Publishing, 82
Country Publications Ltd, 86
Donhead Publishing Ltd, 89
Edinburgh University Press, 91
Elsevier Ltd, 92
GMC Publications Ltd, 98
Imprint Academic, 106
Informa Law, 107
Jane's Information Group, 108
Kluwer Law International, 109
Liverpool University Press, 112

Macmillan Publishers Ltd, 113
Millivres Prowler Limited, 117
mph, 117
Oxford University Press, 124
Psychology Press, 129
Radcliffe Publishing Ltd, 131
Reed Elsevier Group plc, 133
Routledge, 134
Sage Publications, 135
Scholastic Ltd, 136
Shetland Times Ltd, The, 138
Springer-Verlag London Ltd, 141
Straightline Publishing Ltd, 141
Sweet & Maxwell Group, 142
Trentham Books Ltd, 145

Journals: Publishers (US)
California Press, University of, 174
Prometheus Books, 184
Reader's Digest Association Inc, 185
Sharpe (M.E.), Inc., 186
Texas Press, University of, 188
Transaction Publishers Ltd, 189
Wiley (John) & Sons, Inc., 190

Journals: Small Presses
Crescent Moon Publishing and Joe's Press, 227
Great Northern Publishing, 229
Headpress, 230
Mercia Cinema Society, 231

Languages: Library Services
Manchester Central Library, 589
Senate House Library, University of London, 595

Languages: Literary Agents (US)
Gusay (Charlotte) Literary Agency, The, 284
Schulman (Susan), A Literary Agency, 290
Scribblers House® LLC Literary Agency, 290

Languages: Literary Societies
Queen's English Society, The, 566

Languages: PR Consultants
Frazer (Cathy) PR, 277

Languages: Professional Associations
Chartered Institute of Linguists (IoL), 534

Languages: Publishers (European)
Akadémiai Kiadó, 166
Atlantis Ltd, 161
Beck (C.H., OHG) Verlag, 164
Brepols Publishers NV, 160
Brána a.s., 161
Giunti Editoriale SpA, 166
Hachette Livre (France), 163
Larousse (Éditions), 163
Lettera, 161
Mulino (Societe editrice il), 167
Naouka I Izkoustvo, 161
Payot Libraire, 171
Spectrum (Uitgeverij Het) BV, 168
Todariana Editrice, 167

Languages: Publishers (International)
Melbourne University Publishing Ltd, 192
Munshiram Manoharlal Publishers Pvt Ltd, 195

Languages: Publishers (UK)
Berlitz Publishing, 72
Chambers Harrap Publishers Ltd, 82
Drake Educational Associates, 90
Duckworth (Gerald) & Co. Ltd, 90
HarperCollins Publishers Ltd, 102
Routledge, 134

Languages: Publishers (US)
Barron's Educational Series, Inc., 173
Hippocrene Books, Inc., 178
Texas Press, University of, 188

Languages: Small Presses
Glosa Education Organisation, 229
Packard Publishing Limited, 233

Languages: UK Packagers
Lexus Ltd, 244

Large Print: Publishers (UK)
Isis Publishing, 107
Severn House Publishers, 138
Thorpe (F.A.) Publishing, 144

Latin America: Library Services
Instituto Cervantes, 586
Senate House Library, University of London, 595

Latin America: Literary Agents (US)
Gusay (Charlotte) Literary Agency, The, 284

Latin America: Picture Libraries
Andes Press Agency, 602
ffotograff, 608
Garden and Wildlife Matters Photo Library, 609
South American Pictures, 623
Tropix Photo Library, 625

Latin America: Poetry Presses
Katabasis, 203

Latin America: Publishers (UK)
Evans Brothers Ltd, 93

Latin America: Publishers (US)
Alabama Press, University of, 172
Arizona Press, University of, 173
Louisiana State University Press, 180
Northwestern University Press, 182
Texas Press, University of, 188Texas Press, University of,

Law: Bursaries, Fellowships and Grants
Neave (Airey) Trust, The, 485

Law: Electronic Publishing and Other Services
Justis Publishing Ltd, 248

Law: Library Services
British Library Social Sciences & Official Publications, 579
Guildhall Library, 584
Leeds Central Library, 587
National Archives, The, 590

Law: Literary Agents (UK)
Simmons (Jeffrey), 269

Law: Literary Agents (US)
Lescher & Lescher Ltd, 285
Schulman (Susan), A Literary Agency, 290

Law: Magazines
Lawyer 2B, 332
Lawyer, The, 332

Law: Publishers (European)
Beck (C.H., OHG) Verlag, 164
Mulino (Societe editrice il), 167
Presses Universitaires de France
(PUF), 164
Springer GmbH, 165
Springer-Verlag GmbH, 160
Universitaire Pers Leuven, 160

**Law: Publishers
(International)**
Jaico Publishing House, 195
LexisNexis Butterworths (Pty)
Ltd South Africa, 197
LexisNexis Canada Ltd, 193
LexisNexis India, 195
LexisNexis New Zealand, 196
Melbourne University
Publishing Ltd, 192
Munshiram Manoharlal
Publishers Pvt Ltd, 195

Law: Publishers (Ireland)
Blackhall Publishing Ltd, 153
Institute of Public
Administration, 156

Law: Publishers (UK)
Ashgate Publishing Ltd, 69
Business Education Publishers
Ltd, 79
Cambridge University Press, 79
Elgar (Edward) Publishing Ltd,
91
Informa Law, 107
Kluwer Law International, 109
Manchester University Press,
115
Nottingham University Press,
120
Reed Elsevier Group plc, 133
Routledge, 134
Sweet & Maxwell Group, 142
Trentham Books Ltd, 145
TSO (The Stationery Office),
146
Which? Books, 149

Law: Publishers (US)
Barron's Educational Series,
Inc., 173
Kansas, University Press of, 179
Michigan Press, The University
of, 181
Stanford University Press,
187Stanford University
Press,

**Leisure & Hobbies: Book
Clubs**
Readers' Union Ltd, 293

**Leisure & Hobbies: Literary
Agents (Ireland)**
Williams (Jonathan) Literary
Agency, 279

**Leisure & Hobbies: Literary
Agents (UK)**
Feldstein Agency, The, 256
Watson, Little Ltd, 272

**Leisure & Hobbies:
Magazines**
Good Holiday Magazine, 323
Jersey Now, 330
Motor Caravan Magazine, 337
*Motorcaravan Motorhome
Monthly (MMM)*, 337

**Leisure & Hobbies: Picture
Libraries**
actionplus sports images, 601
Forest Life Picture Library, 609
Henderson (Jim) Photography,
611
Report Digital, 621
Retna Pictures Ltd, 621
Waterways Photo Library, 626

**Leisure & Hobbies:
Publishers (European)**
Aschehoug (H) & Co (W
Nygaard), 168
Giunti Editoriale SpA, 166
Nathan (Les Éditions), 164
Nyt Nordisk Forlag Arnold
Busck A/S, 162
Olympia Publishing Co. Inc.,
161
Orell Füssli Verlag, 171

**Leisure & Hobbies:
Publishers (UK)**
AA Publishing, 64
Carlton Publishing Group, 81
Crowood Press Ltd, The, 87
Gloucester Publishers Plc, 98
GMC Publications Ltd, 98
HarperCollins Publishers Ltd,
102
Haynes Publishing, 103
Sigma Press, 139
Souvenir Press Ltd, 140
Special Interest Model Books
Ltd, 140
Transworld Publishers Ltd, 145

**Leisure & Hobbies:
Publishers (US)**
Barron's Educational Series,
Inc., 173
Globe Pequot Press, The, 176

Sterling Publishing Co. Inc., 188

**Leisure & Hobbies: Small
Presses**
Porteous (David) Editions, 234

**Leisure & Hobbies: UK
Packagers**
Stonecastle Graphics Ltd, 246

Library Science: Prizes
Oldman (C.B.) Prize, 514

**Library Science: Professional
Associations**
CILIP in Scotland (CILIPS), 534
CILIP Wales, 534
CILIP: The Chartered Institute
of Library and Information
Professionals, 534

**Library Science: Publishers
(UK)**
Facet Publishing, 94

**Library Science: Publishers
(US)**
McFarland & Company, Inc.,
Publishers, 180
Scarecrow Press Inc., 186

**Lifestyle: Electronic
Publishing and Other
Services**
50connect, 248

**Lifestyle: Film, TV and Radio
Producers**
Stirling Film & TV Productions
Limited, 433
Straight Forward Film &
Television Productions Ltd,
434
talkbackTHAMES, 434
Twenty Twenty Television, 436

**Lifestyle: Literary Agents
(UK)**
Bonomi (Luigi) Associates
Limited (LBA), 251
Chris (Teresa) Literary Agency
Ltd, 253
Feldstein Agency, The, 256
Graham Maw Christie Literary
Agency, 258
Inspira Group, The, 261
Johnson & Alcock, 261
Smith (Robert) Literary Agency
Ltd, 269

Lifestyle: Literary Agents (US)
Gusay (Charlotte) Literary
Agency, The, 284

Lifestyle: Magazines
Healthy, 326
Ideal Home, 328
InStyle, 329
Jersey Now, 330
Loaded, 333
Lothian Life, 334
Maxim, 335
Men's Health, 335
Reveal, 351
Scottish Field, 353
Skymag, 356
SmartLife International, 356
Time Out, 361
Wallpaper, 365
ZOO, 371

Lifestyle: Picture Libraries
Harding (Robert) World
Imagery, 611
Hardley (Dennis) Photography,
Scottish Photo Library, 611
Retna Pictures Ltd, 621
Travel Ink Photo Library, 625

Lifestyle: Publishers (European)
BZZTÔH (Uitgeverij) BV, 167
Flammarion Éditions, 163
Kibea Publishing Company, 160
Payot Libraire, 171
Unieboek (Uitgeverij) BV, 168

Lifestyle: Publishers (International)
Hachette Livre Australia, 191
Struik Publishers (Pty) Ltd, 198

Lifestyle: Publishers (UK)
Anness Publishing Ltd, 66
Anova Books, 66
Aurum Press Ltd, 70
Baird (Duncan) Publishers, 71
Bonnier Books (UK), 75
Capstone Publishing, 80
Cico Books, 84
Dorling Kindersley Ltd, 90
Foulsham Publishers, 95
HarperCollins Publishers Ltd,
102
Harriman House Ltd, 103
Infinite Ideas, 107
Octopus Publishing Group, 121
Piatkus Books, 127
Random House Group Ltd,
The, 131

Ryland Peters & Small Limited,
135

Lifestyle: Publishers (US)
Globe Pequot Press, The, 176

Lifestyle: UK Packagers
Amber Books Ltd, 242
Ivy Press Limited, The, 244
Quarto Publishing plc, 245
Stonecastle Graphics Ltd, 246

Light Entertainment: Film, TV and Radio Producers
Avalon, 418
Celador Productions, 420
FremantleMedia Group Ltd,
424
Gay (Noel) Television, 424
Hat Trick Productions Ltd, 425
Tiger Aspect Productions, 435

Light Entertainment: Literary Agents (UK)
Hancock (Roger) Ltd, 259
Narrow Road Company, The,
265

Linguistics: Literary Agents (US)
Gusay (Charlotte) Literary
Agency, The, 284

Linguistics: Prizes
British Association for Applied
Linguistics (BAAL) Book
Prize, 493

Linguistics: Professional Associations
Chartered Institute of Linguists
(IoL), 534

Linguistics: Publishers (European)
Adelphi Edizioni SpA, 166
Atlantis Ltd, 161
Beck (C.H., OHG) Verlag, 164
Brána a.s., 161
Caminho (Editorial) SARL, 169
Giunti Editoriale SpA, 166
Hachette Livre (France), 163
Larousse (Éditions), 163
Mulino (Societe editrice il), 167
Presses Universitaires de France
(PUF), 164
Springer GmbH, 165
Todariana Editrice, 167

Linguistics: Publishers (International)
Melbourne University
Publishing Ltd, 192
Munshiram Manoharlal
Publishers Pvt Ltd, 195

Linguistics: Publishers (UK)
Continuum International
Publishing Group Ltd, The,
85
Edinburgh University Press, 91

Literary Criticism: Literary Agents (Ireland)
Williams (Jonathan) Literary
Agency, 279

Literary Criticism: Literary Agents (UK)
Feldstein Agency, The, 256

Literary Criticism: Literary Agents (US)
Gusay (Charlotte) Literary
Agency, The, 284

Literary Criticism: Magazines
Literary Review, 333
New Shetlander, The, 340

Literary Criticism: Poetry Presses
Akros Publications, 199
Poetry Salzburg, 205
Seren, 206

Literary Criticism: Prizes
Crawshay (Rose Mary) Prize,
497

Literary Criticism: Publishers (European)
Adelphi Edizioni SpA, 166
Arche Verlag AG, 171
Beck (C.H., OHG) Verlag, 164
BZZTÔH (Uitgeverij) BV, 167
Deutscher Taschenbuch Verlag
GmbH & Co. KG, 164
Diogenes Verlag AG, 171
EDHASA (Editora y
Distribuidora Hispano –
Americana SA), 169
Flammarion Éditions, 163
Garzanti Libri SpA, 166
Gradiva–Publicações S.A., 169
Grasset & Fasquelle (Éditions),
163
Minuit (Les Éditions de) SA, 164
Mulino (Societe editrice il), 167
Seuil (Éditions du), 164

Tiderne Skifter Forlag A/S, 162
Todariana Editrice, 167

Literary Criticism: Publishers (International)
Auckland University Press, 196
Melbourne University Publishing Ltd, 192
Queensland Press, University of, 192
UWA Press, 193

Literary Criticism: Publishers (Ireland)
Gill & Macmillan, 155
Lilliput Press, The, 156

Literary Criticism: Publishers (UK)
Black Spring Press Ltd, 74
Bloodaxe Books Ltd, 74
Boyars (Marion) Publishers Ltd, 76
Calder Publications Ltd, 79
Currey (James) Publishers, 87
Edinburgh University Press, 91
Enitharmon Press, 92
Harvard University Press, 103
Northcote House Publishers Ltd, 120
Onlywomen Press Ltd, 122
Owen (Peter) Publishers, 123
Routledge, 134
Seren, 137
Smythe (Colin) Ltd, 139
Women's Press, The, 150

Literary Criticism: Publishers (US)
Arkansas Press, University of, 173
Faber & Faber, Inc., 176
Louisiana State University Press, 180
Massachusetts Press, University of, 181
Michigan Press, The University of, 181
Missouri Press, University of, 181

Literary Criticism: Small Presses
Chrysalis Press, 226
Crescent Moon Publishing and Joe's Press, 227
Ivy Publications, 230
Parapress, 233
Paupers' Press, 233Paupers' Press,

Literature: Audio Books
SmartPass Ltd, 241

Literature: Book Clubs
Folio Society, The, 293

Literature: Bursaries, Fellowships and Grants
Scottish Arts Council Writers' Bursaries, 486

Literature: Electronic Publishing and Other Services
Online Originals, 248

Literature: Festivals
Aldeburgh Literary Festival, The, 629
Arthur Miller Centre Literary Festival, 636
Aspects Literature Festival, 629
Bath Literature Festival, 630
BayLit Festival, 630
Belfast Festival at Queen's, 630
Beverley Literature Festival, 630
Book Now – Richmond Literature Festival, 630
Borders Book Festival, 630
Bournemouth Literary Festival, 630
Brighton Festival, 631
Buxton Festival, 631
Cambridge Wordfest, 631
Canterbury Festival, 631
Centre for Creative & Performing Arts Spring Literary Festival at UEA, 631
Charleston Festival, 632
Chester Literature Festival, 632
Derbyshire Literature Festival, 632
Dorchester Festival, 632
du Maurier (Daphne) Festival of Arts & Literature, The, 632
Dundee Literary Festival, 632
Durham Book Festival, 633
Edinburgh International Book Festival, 633
Folkestone Literary Festival, 633
Frome Festival, 633
Guardian Hay Festival, The, 633
Guildford Book Festival, 633
Hebden Bridge Arts Festival, 634
Humber Mouth – Hull Literature Festival, 634
Ilkley Literature Festival, 634
Isle of Man Literature Festival, 634
King's Lynn, The Fiction Festival, 634

Knutsford Literature Festival, 635
Lambeth Readers & Writers Festival, 635
Lancaster LitFest, 635
Lewes Live Literature, 635
Lichfield Festival, 636
London Literature Festival, 636
Lowdham Book Festival, 636
Manchester Literature Festival, 636
Mere Literary Festival, 636
Off the Shelf Literature Festival, 637
Oundle Festival of Literature, 637
Redbridge Book and Media Festival, 638
Runnymede International Literature Festival, 638
Sunday Times Oxford Literary Festival, The, 639
Swindon Festival of Literature, 639
Thomas (Dylan) Festival, 639
Times Cheltenham Literature Festival, The, 639
Ty Newydd Festival, 639
Warwick Words, 640
Ways with Words, 640
Wellington Literary Festival, 640
Wells Festival of Literature, 640
Wonderful Words, 640

Literature: Literary Agents (Ireland)
Book Bureau Literary Agency, The, 279
Gunn O'Connor (Marianne) Literary Agency, 279
Williams (Jonathan) Literary Agency, 279

Literature: Literary Agents (UK)
Ampersand Agency Ltd, The, 249
Artellus Limited, 250
Belli (Lorella) Literary Agency (LBLA), 250
Blake Friedmann Literary Agency Ltd, 251
Bonomi (Luigi) Associates Limited (LBA), 251
Brown (Jenny) Associates, 251
Burkeman (Brie), 252
Cheetham (Mic) Associates, 253
Chris (Teresa) Literary Agency Ltd, 253
Conville & Walsh Limited, 254
Curtis Brown Group Ltd, 254

Davidson (Caroline) Literary Agency, 255
de Wolfe (Felix), 255
Edwards Fuglewicz, 256
Feldstein Agency, The, 256
Fraser Ross Associates, 257
Godwin (David) Associates, 258
Green (Annette) Authors' Agency, 258
Green (Christine) Authors' Agent, 258
Greenberg (Louise) Books Ltd, 258
Greene & Heaton Ltd, 258
Gregory & Company Authors' Agents, 258
Heath (Rupert) Literary Agency, 260
Holt (Vanessa) Ltd, 260
Hordern (Kate) Literary Agency, 260
Janklow & Nesbit (UK) Ltd, 261
Johnson & Alcock, 261
Judd (Jane) Literary Agency, 262
Kass (Michelle) Associates, 262
LAW, 262
Limelight Management, 262
Little (Christopher) Literary Agency, The, 263
London Independent Books, 263
Lutyens and Rubinstein, 263
Marjacq Scripts Ltd, 264
McAra (Duncan), 264
McKernan Literary Agency & Consultancy, 265
Murdoch (Judith) Literary Agency, 265
Nye (Alexandra), 266
Rushby-Smith (Uli) Literary Agency, 267
Sayle Literary Agency, The, 268
Sheil Land Associates Ltd, 268
Sheldon (Caroline) Literary Agency Ltd, 268
Such (Sarah) Literary Agency, 270
Sunflower Literary Agency, 270
Susijn Agency Ltd, The, 270
Trevor (Lavinia) Literary Agency, 271
Watson, Little Ltd, 272

Literature: Literary Agents (US)
Altshuler (Miriam) Literary Agency, 280
Bernstein (Meredith) Literary Agency, Inc., 280
Bleecker Street Associates, Inc., 280

Braun (Barbara) Associates, Inc., 281
Browne & Miller Literary Associates, 281
Carvainis (Maria) Agency, Inc., 281
Castiglia Literary Agency, 281
Chelius (Jane) Literary Agency, Inc., 282
Chester (Linda) & Associates, 282
Clark (William) Associates, 282
Dunham Literary Inc., 283
Dystel & Goderich Literary Management, 283
Ellenberg (Ethan) Literary Agency, 283
Gusay (Charlotte) Literary Agency, The, 284
Hogenson (Barbara) Agency, Inc., The, 284
Janklow & Nesbit Associates, 284
Klinger (Harvey), Inc., 284
Lampack (Peter) Agency, Inc., 285
Larsen (Michael)/Elizabeth Pomada Literary Agency, 285
Leavitt (Ned) Agency, The, 285
Lescher & Lescher Ltd, 285
Mann (Carol) Agency, 286
Manus & Associates Literary Agency, Inc., 286
Michaels (Doris S.) Literary Agency Inc., 286
Naggar (Jean V.) Literary Agency, The, 287
Parks (Richard) Agency, 287
Picard (Alison J.) Literary Agent, 287
Priest (Aaron M.) Literary Agency, The, 288
Quicksilver Books, Literary Agents, 288
Reece Halsey North Literary Agency, 288
Rittenberg (Ann) Literary Agency, Inc., 288
Robbins (B.J.) Literary Agency, 288
Russell & Volkening, Inc., 289
Snell (Michael) Literary Agency, 290
Ware (John A.) Literary Agency, 291
Weil (Wendy) Agency, Inc., The, 291
Writers House, LLC, 292

Literature: Magazines
Chapman, 306

Edinburgh Review, 317
Granta, 324
Irish Pages, 329
Literary Review, 333
London Magazine, The, 333
London Review of Books, 333
New Humanist, 339
New Statesman, 340
New Welsh Review, 340
Spectator, The, 357
Times Literary Supplement, The, 361

Literature: PR Consultants
FMcM Associates, 276

Literature: Poetry Presses
Bluechrome Publishing, 200
Flambard Press, 201
Hearing Eye, 202

Literature: Prizes
Costa Book Awards, 496
Crawshay (Rose Mary) Prize, 497
International IMPAC Dublin Literary Award, The, 505
Lannan Literary Award, 507
Man Booker International Prize, The, 510
McLeod (Enid) Prize, 509
Nobel Prize, 513
Pulitzer Prizes, 516
Saltire Literary Awards, The, 519

Literature: Professional Associations
Authors' Club, 529
English Association, The, 536
Royal Society of Literature, 545

Literature: Publishers (European)
Adelphi Edizioni SpA, 166
Arche Verlag AG, 171
Atlantis Ltd, 161
Beck (C.H., OHG) Verlag, 164
Brepols Publishers NV, 160
Brombergs Bokförlag AB, 170
Brána a.s., 161
BZZTÔH (Uitgeverij) BV, 167
Deutscher Taschenbuch Verlag GmbH & Co. KG, 164
Diogenes Verlag AG, 171
EDHASA (Editora y Distribuidora Hispano – Americana SA), 169
Espasa-Calpe (Editorial) SA, 170
Fischer (S.) Verlag GmbH, 164
Flammarion Éditions, 163
Garzanti Libri SpA, 166
Giunti Editoriale SpA, 166

Gradiva–Publicações S.A., 169
Grasset & Fasquelle (Éditions), 163
Groupo Anaya, 169
Hanser (Carl) Verlag GmbH & Co. KG, 165
Hiperión (Ediciónes) SL, 170
Kráter Association and Publishing House, 166
Larousse (Éditions), 163
Meulenhoff (Uitgeverij J.M.) BV, 168
Minuit (Les Éditions de) SA, 164
Mulino (Societe editrice il), 167
Presses Universitaires de France (PUF), 164
Random House Mondadori, 170
Seix Barral (Editorial) SA, 170
Seuil (Éditions du), 164
Tiderne Skifter Forlag A/S, 162
Todariana Editrice, 167
Tusquets Editores, 170
Ullstein Buchverlage GmbH, 165

Literature: Publishers (International)
Auckland University Press, 196
Hachette Livre Australia, 191
HarperCollins Canada Ltd, 193
HarperCollins Publishers (New Zealand) Ltd, 196
HarperCollins Publishers India Ltd, 195
Hind Pocket Books Pvt. Ltd, 195
Jaico Publishing House, 195
Kwa-Zulu–Natal Press, University of, 197
Melbourne University Publishing Ltd, 192
Munshiram Manoharlal Publishers Pvt Ltd, 195
National Publishing House, 195
New Star Books Ltd, 194
Orient Paperbacks, 195
Pan Macmillan Australia Pty Ltd, 192
Queensland Press, University of, 192
Random House Australia, 192
Random House of Canada Ltd, 194
UWA Press, 193
Victoria University Press, 197

Literature: Publishers (Ireland)
Lilliput Press, The, 156
Penguin Ireland, 158
Poolbeg Press Ltd, 158

Literature: Publishers (UK)
Arcadia Books, 68
Ashgate Publishing Ltd, 69
Atlantic Books, 69
Bloodaxe Books Ltd, 74
Bloomsbury Publishing Plc, 74
Boyars (Marion) Publishers Ltd, 76
Calder Publications Ltd, 79
Canongate Books Ltd, 80
Carlton Publishing Group, 81
Clarke (James) & Co., 84
Continuum International Publishing Group Ltd, The, 85
CRW Publishing Ltd, 87
Davies (Christopher) Publishers Ltd, 88
Dedalus Ltd, 89
Duckworth (Gerald) & Co. Ltd, 90
Elliott & Thompson, 91
Everyman's Library, 93
Exeter Press, University of, 93
Granta Books, 98
Hachette Livre UK, 100
Halban Publishers, 100
HarperCollins Publishers Ltd, 102
Hertfordshire Press, University of, 104
Little, Brown Book Group UK, 111
Macmillan Publishers Ltd, 113
Manchester University Press, 115
Melrose Books, 116
Octagon Press Ltd, 121
Old Street Publishing Ltd, 121
Open Gate Press, 123
Owen (Peter) Publishers, 123
Penguin Group (UK), 125
Polity Press, 128
Polygon, 128
Portobello Books, 128
ProQuest, 129
Pushkin Press Ltd, 129
Quartet Books, 130
Random House Group Ltd, The, 131
Serpent's Tail, 138
Simon & Schuster UK Limited, 139
Smythe (Colin) Ltd, 139
Telegram, 143
Timewell Press, 144
Two Ravens Press Ltd, 146
Wales Press, University of, 148
Wimbledon Publishing Company, 150
Wordsworth Editions Ltd, 151

Literature: Publishers (US)
Farrar, Straus & Giroux, Inc., 176
HarperCollins Publishers, Inc., 177
Houghton Mifflin Harcourt Publishing Company, 178
Kent State University Press, The, 179
Minnesota Press, University of, 181
Mississippi, University Press of, 181
Nebraska Press, University of, 182
Nevada Press, University of, 182
Seven Stories Press, 186
Stanford University Press, 187
Texas Press, University of, 188
Virginia Press, University of, 189

Literature: Small Presses
Allardyce, Barnett, Publishers, 225
Crescent Moon Publishing and Joe's Press, 227
Frontier Publishing, 228
Linen Press, The, 231
Superscript, 235

Literature: Writers' Courses (UK and Ireland)
Arvon Foundation, 458

Literature: Writers' Courses (US)
Chapman University, 469Chapman University, Chapman University,

Local Interest/History: Book Clubs
Arscott's (David) Sussex Book Club, 293

Local Interest/History: Library Services
Armitt Collection, Museum & Library, 574
Athenaeum, Liverpool, The, 574
Barnsley Public Library, 575
Bedford Central Library, 575
Belfast Public Libraries: Central Library, 575
Birmingham Library Services, 576
Bournemouth Library, 576
Bradford Central Library, 576
Brighton Jubilee Library, 576
Bristol Central Library, 576
Bromley Central Library, 579

Bromley House Library, 580
Cardiff Central Library, 580
Carmarthen Public Library, 580
Coventry Central Library, 581
Derby Central Library, 581
Devon & Exeter Institution
    Library, 581
Doncaster Library and
    Information Services, 582
Essex County Council Libraries,
    582
Guildford Institute Library, 584
Herefordshire Libraries, 585
Highgate Literary and Scientific
    Institution Library, 585
Holborn Library, 586
Jersey Library, 587
Leeds Central Library, 587
Lincoln Central Library, 588
Linen Hall Library, 588
Liverpool Libraries and
    Information Services, 588
London Metropolitan Archives
    Library Services, 589
Lord Louis Library, 589
Manchester Central Library, 589
Morrab Library, 590
Norfolk Library & Information
    Service, 591
Nottingham Central Library,
    592
Orkney Library and Archive,
    592
Oxford Central Library, 592
Plymouth Library and
    Information Services, City
    of, 593
Sheffield Libraries, Archives
    and Information, 595
Shetland Library, 596
Suffolk County Council –
    Suffolk Libraries, 597
Sunderland City Library and
    Arts Centre, 597
Swansea Central Library, 597
Truro Library, 598
Western Isles Libraries, 598
Wolverhampton Central
    Library, 599
Worcestershire Libraries and
    Information Service, 600
York Central Library, 600

Local Interest/History:
Literary Agents (UK)
Feldstein Agency, The, 256

Local Interest/History:
Literary Agents (US)
Gusay (Charlotte) Literary
    Agency, The, 284

Local Interest/History:
Magazines
Cambridgeshire Journal
    Magazine, The, 305
Cumbria, 312
Dalesman, 313
Derbyshire Life and
    Countryside, 314
Devon Life, 315
Dorset Life – The Dorset
    Magazine, 316
Essex Life, 318
Image Magazine, 328
Lake District Life, 332
Lancashire Life, 332
Lincolnshire Life, 332
Lincolnshire Poacher, The, 332
Lothian Life, 334
Shropshire Magazine, 355
Somerset Life, 357
Staffordshire Life, 357
Suffolk and Norfolk Life, 358
Surrey Life, 359
Sussex Life, 359
Ulster Tatler, 364

Local Interest/History:
Miscellany
Edwards (Jacqueline), 641

Local Interest/History:
Picture Libraries
Beamish, The Photographic
    Library, 603
National Monuments Record,
    617
Ulster Museum Picture Library,
    625

Local Interest/History:
Poetry Presses
Harpercroft, 202
I*D Books, 203

Local Interest/History: Prizes
Lakeland Book of the Year
    Awards, 507
Portico Prize for Literature,
    The, 516

Local Interest/History:
Publishers (Ireland)
On Stream Publications Ltd, 157

Local Interest/History:
Publishers (UK)
Akros Publications, 65
Birlinn Ltd, 73
Breedon Books Publishing Co.
    Ltd, The, 77
Brewin Books Ltd, 77

Countryside Books, 86
Donald (John) Publishers Ltd,
    89
Fort Publishing Ltd, 95
Halsgrove, 101
Henry (Ian) Publications Ltd,
    104
Hertfordshire Press, University
    of, 104
History Press Ltd, The, 105
Landmark Publishing Ltd, 109
Reardon Publishing, 133
SB Publications, 136
Scottish Cultural Press/Scottish
    Children's Press, 137
Shetland Times Ltd, The, 138
Sigma Press, 139
Straightline Publishing Ltd, 141
Wharncliffe Books, 148

Local Interest/History:
Publishers (US)
Iowa Press, University of, 179

Local Interest/History: Small
Presses
Alloway Publishing Limited, 225
Baldwin (M.&M.), 225
Book Castle, The, 226
Charlewood Press, 226
Heart of Albion Press, 230
Millers Dale Publications, 232
Pipers' Ash Ltd, 233
Pomegranate Press, 234
QueenSpark Books, 234
Stenlake Publishing Limited,
    235
Wakefield Historical
    Publications, 236
Whitchurch Books Ltd, 236

Local Interest/History:
Television and Radio
BBC Three Counties Radio, 405

London: Library Services
Barbican Library, 574
British Library Map Collections,
    578
Guildhall Library, 584
Highgate Literary and Scientific
    Institution Library, 585
London Metropolitan Archives
    Library Services, 589
Westminster (City of) Archives
    Centre, 598

London: Literary Agents
(UK)
Graham Maw Christie Literary
    Agency, 258

London: Literary Agents (US)
Gusay (Charlotte) Literary
    Agency, The, 284

London: Magazines
Mayfair Times, 335
Time Out, 361

London: Picture Libraries
Cordaiy (Sylvia) Photo Library,
    606
Greenhill (Sally and Richard),
    611
Illustrated London News
    Picture Library, 613
London Aerial Photo Library,
    615
London Metropolitan Archives,
    615
London's Transport Museum
    Photographic Library, 615
Miller (Lee) Archives, 616
Museum of London Picture
    Library, 617
Strang (Jessica) Photo Library,
    624

London: Prizes
Bowling (Harry) Prize, 492

London: Small Presses
Watling Street Publishing Ltd,
    236

Magic: Literary Agents (UK)
London Independent Books,
    263

Magic: Publishers (UK)
Breese (Martin) International,
    77
Souvenir Press Ltd, 140

Magic: Publishers (US)
Llewellyn Publications, 180

Maps: Library Services
British Library Map Collections,
    578
Guildhall Library, 584
London Library, The, 589
National Library of Scotland,
    590

Maps: Picture Libraries
British Library Images Online,
    604
National Archives Image
    Library, The, 617

Maps: Publishers (European)
Cappelen Damm AS, 168
Espasa-Calpe (Editorial) SA, 170
Fayard (Librarie Arthème), 163
Michelin Éditions des Voyages,
    163
Standaard Uitgeverij, 160

**Maps: Publishers
(International)**
Wiley (John) & Sons Australia
    Ltd, 193

Maps: Publishers (Ireland)
Burke (Edmund) Publisher, 154

Maps: Publishers (UK)
AA Publishing, 64
Allan (Ian) Publishing Ltd, 65
Cook (Thomas) Publishing, 86
Dorling Kindersley Ltd, 90
HarperCollins Publishers Ltd,
    102
Michelin Maps & Guides, 116
Octopus Publishing Group, 121

**Maritime/Nautical: Library
Services**
British Library Map Collections,
    578
Plymouth Library and
    Information Services, City
    of, 593

**Maritime/Nautical: Literary
Agents (UK)**
London Independent Books,
    263
Welch (John), Literary
    Consultant & Agent, 272

**Maritime/Nautical:
Magazines**
Canals and Rivers, 305
Classic Boat, 308
Jersey Now, 330
Motor Boat & Yachting, 337
Motor Boats Monthly, 337
Practical Boat Owner, 345
Sailing Today, 352
Sea Breezes, 354
Ships Monthly, 354
Waterways World, 365
Yachting Monthly, 369
Yachting World, 370

**Maritime/Nautical: Picture
Libraries**
Adkins (Lesley & Roy) Picture
    Library, 601

Cordaiy (Sylvia) Photo Library,
    606
Ecoscene, 607
Faces and Places, 608
Kos Picture Source Ltd, 614
Moving Image
    Communications, 617
National Maritime Museum
    Picture Library, 617
Waterways Photo Library, 626

**Maritime/Nautical:
Professional Associations**
Yachting Journalists'
    Association, 551

**Maritime/Nautical:
Publishers (European)**
Longanesi (Casa Editrice) SpA,
    167
Mursia (Gruppo Ugo) Editore
    SpA, 167

**Maritime/Nautical:
Publishers (International)**
Melbourne University
    Publishing Ltd, 192
UWA Press, 193

**Maritime/Nautical:
Publishers (UK)**
Allan (Ian) Publishing Ltd, 65
Anova Books, 66
Black (A.&C.) Publishers Ltd, 73
Brown, Son & Ferguson, Ltd, 78
Crowood Press Ltd, The, 87
Exeter Press, University of, 93
Haynes Publishing, 103
Jane's Information Group, 108
Mason (Kenneth) Publications
    Ltd, 115
Nautical Data Ltd, 118
Olympia Publishers, 122
Pen & Sword Books Ltd, 125
RYA (Royal Yachting
    Association), 135
Seafarer Books Ltd, 137
Seaforth Publishing, 137
Shetland Times Ltd, The, 138
Trident Press Ltd, 145
Wiley Nautical, 149

**Maritime/Nautical:
Publishers (US)**
Washington State University
    Press, 190

**Maritime/Nautical: Small
Presses**
Nostalgia Collection, The, 232
Parapress, 233

Stenlake Publishing Limited, 235
Whittles Publishing, 236

Maritime/Nautical: UK Packagers
Amber Books Ltd, 242

Marketing/PR: Film, TV and Radio Producers
Channel Television Ltd, 420
Creative Partnership, The, 421
Cricket Ltd, 421
Picardy Media & Communication, 430
Tandem TV & Film Ltd, 434
Video Enterprises, 436

Marketing/PR: Magazines
Marketing Week, 334
PR Week, 345

Marketing/PR: Miscellany
M-Y Books Ltd, 641

Marketing/PR: Professional Associations
Institute of Copywriting, 538

Marketing/PR: Publishers (UK)
First & Best in Education Ltd, 94
Kogan Page Ltd, 109

Marketing/PR: Publishers (US)
AMACOM Books, 172AMACOM Books,

Mathematics & Statistics: Publishers (European)
Adelphi Edizioni SpA, 166
Alianza Editorial SA, 169
Dokorán Ltd, 161
Springer GmbH, 165
Springer Science+Business Media BV, 168

Mathematics & Statistics: Publishers (UK)
Austin & Macauley Publishers Limited, 70
Freeman (W. H.), 96
Springer-Verlag London Ltd, 141

Mathematics & Statistics: Publishers (US)
Krieger Publishing Co., 179

Media & Cultural Studies: Literary Agents (UK)
Feldstein Agency, The, 256

Media & Cultural Studies: Literary Agents (US)
Gusay (Charlotte) Literary Agency, The, 284

Media & Cultural Studies: Magazines
Media Week, 335

Media & Cultural Studies: Professional Associations
Campaign for Press and Broadcasting Freedom, 533
Chartered Institute of Journalists, 533

Media & Cultural Studies: Publishers (UK)
Edinburgh University Press, 91
Manchester University Press, 115
Penguin Group (UK), 125
Polity Press, 128
Routledge, 134
Tauris (I.B.) & Co. Ltd, 143

Media & Cultural Studies: Publishers (US)
Louisiana State University Press, 180
Michigan Press, The University of, 181
Minnesota Press, University of, 181
Seven Stories Press, 186

Media & Cultural Studies: Small Presses
Crescent Moon Publishing and Joe's Press, 227Crescent Moon Publishing and Joe's Press,

Medical: Library Services
Royal Society of Medicine Library, 594

Medical: Literary Agents (UK)
Davidson (Caroline) Literary Agency, 255

Medical: Literary Agents (US)
Carvainis (Maria) Agency, Inc., 281
Mews Books Ltd, 286

Parks (Richard) Agency, 287
Scribblers House® LLC Literary Agency, 290
Ware (John A.) Literary Agency, 291

Medical: Magazines
Balance, 298
British Medical Journal, 303
Practitioner, The, 346

Medical: Picture Libraries
Hutchison Picture Library, 612
Rex Features Ltd, 621
Science Photo Library, 622
Wellcome Images, 626

Medical: Prizes
Medical Book Awards, 511

Medical: Professional Associations
International Association of Scientific, Technical & Medical Publishers, 539
Medical Journalists' Association, 540
Medical Writers' Group, 540
Society of Medical Writers, The, 548

Medical: Publishers (European)
Adelphi Edizioni SpA, 166
Brána a.s., 161
Flammarion Éditions, 163
Gyldendal, 161
Kibea Publishing Company, 160
Larousse (Éditions), 163
Mondadori (Arnoldo) Editore SpA, 167
Nyt Nordisk Forlag Arnold Busck A/S, 162
Springer GmbH, 165
Springer Science+Business Media BV, 168
Springer-Verlag GmbH, 160

Medical: Publishers (International)
Macmillan Publishers Australia Pty Ltd, 191
McGraw-Hill Australia Pty Ltd, 192
Munshiram Manoharlal Publishers Pvt Ltd, 195

Medical: Publishers (UK)
Cambridge University Press, 79
Elsevier Ltd, 92
Hammersmith Press Ltd, 101
Manson Publishing Ltd, 115

Nottingham University Press, 120

Polity Press, 128

Portland Press Ltd, 128

ProQuest, 129

Radcliffe Publishing Ltd, 131

Sage Publications, 135

Society for Promoting Christian Knowledge (SPCK), 140

Souvenir Press Ltd, 140

Springer-Verlag London Ltd, 141

Wiley-Blackwell, 149

Medical: Publishers (US)
Wiley (John) & Sons, Inc., 190

Memoirs: Book Clubs
Folio Society, The, 293

Memoirs: Festivals
Buxton Festival, 631

Memoirs: Literary Agents (UK)
Ampersand Agency Ltd, The, 249

BookBlast Ltd, 251

Feldstein Agency, The, 256

Graham Maw Christie Literary Agency, 258

hhb agency ltd, 260

Johnson & Alcock, 261

Sheil Land Associates Ltd, 268

Sheldon (Caroline) Literary Agency Ltd, 268

Smith (Robert) Literary Agency Ltd, 269

Memoirs: Literary Agents (US)
Altshuler (Miriam) Literary Agency, 280

Bernstein (Meredith) Literary Agency, Inc., 280

Carvainis (Maria) Agency, Inc., 281

Castiglia Literary Agency, 281

Dijkstra (Sandra) Literary Agency, 282

Gusay (Charlotte) Literary Agency, The, 284

Lescher & Lescher Ltd, 285

Lowenstein–Yost Associates Inc., 285

Manus & Associates Literary Agency, Inc., 286

Michaels (Doris S.) Literary Agency Inc., 286

Naggar (Jean V.) Literary Agency, The, 287

Rittenberg (Ann) Literary Agency, Inc., 288

Schiavone Literary Agency, Inc., 289

Schulman (Susan), A Literary Agency, 290

Scribblers House® LLC Literary Agency, 290

Weil (Wendy) Agency, Inc., The, 291

Memoirs: Magazines
Granta, 324

Irish Pages, 329

London Magazine, The, 333

Memoirs: Poetry Magazines
Sable, 216

Memoirs: Publishers (European)
Bompiani, 166

Garzanti Libri SpA, 166

Random House Mondadori, 170

Tusquets Editores, 170

Ullstein Buchverlage GmbH, 165

Memoirs: Publishers (International)
Hachette Livre Australia, 191

HarperCollins Publishers India Ltd, 195

Oxford University Press India, 196

Queensland Press, University of, 192

Random House of Canada Ltd, 194

Wits University Press, 198

Memoirs: Publishers (Ireland)
Brandon/Mount Eagle Publications, 154

Hachette Books Ireland, 155

New Island, 157

Memoirs: Publishers (UK)
Atlantic Books, 69

Austin & Macauley Publishers Limited, 70

AuthorHouse UK, 70

Black & White Publishing Ltd, 73

Blackstone Publishers, 74

Book Guild Publishing, 75

Halban Publishers, 100

HarperCollins Publishers Ltd, 102

Janus Publishing Company Ltd, 108

Olympia Publishers, 122

Pegasus Elliot Mackenzie Publishers Ltd, 125

Random House Group Ltd, The, 131

Memoirs: Small Presses
Linen Press, The, 231

Miller (Neil) Publications, 232

Nostalgia Collection, The, 232

Memoirs: Writers' Courses (UK and Ireland)
Writers Bureau, The, 461

Memoirs: Writers' Courses (US)
Goucher College, 471

Men's Interests: Literary Agents (US)
Gusay (Charlotte) Literary Agency, The, 284

Men's Interests: Magazines
Arena, 297

Club International, 309

Esquire, 318

FHM, 320

Front Magazine, 322

GQ, 324

Loaded, 333

Maxim, 335

Mayfair, 335

Men's Health, 335

Nuts, 342

ZOO, 371

Men's Interests: Publishers (International)
Struik Publishers (Pty) Ltd, 198

Metaphysics: Publishers (US)
Llewellyn Publications, 180

Mexico: Literary Agents (US)
Gusay (Charlotte) Literary Agency, The, 284

Mexico: Picture Libraries
Gagg's (Andrew N.) PHOTOFLORA, 609

South American Pictures, 623

Mexico: Publishers (US)
Texas Press, University of, 188

Middle East: Library Services
British Library Asia, Pacific and Africa Collections, 577

British Library Asia, Pacific and Africa Collections, 577

Middle East: Literary Agents (UK)
Eady (Toby) Associates Ltd, 256

Middle East: Literary Agents (US)
Gusay (Charlotte) Literary Agency, The, 284

Middle East: Magazines
Jewish Quarterly, 330

Middle East: Picture Libraries
Aspect Picture Library Ltd, 602
Australia Pictures, 602
British Library Images Online, 604
Faces and Places, 608
ffotograff, 608
Geoslides Photography, 610
Henderson (Jim) Photography, 611
Miller (Lee) Archives, 616
Peerless (Ann & Bury) Picture Library, 619
Sanders (Peter) Photography Ltd, 622
Tropix Photo Library, 625
Yemen Pictures, 627

Middle East: Prizes
International Prize for Arabic Fiction, 505

Middle East: Publishers (UK)
Garnet Publishing Ltd, 97
Routledge, 134
Saqi Books, 135
Tauris (I.B.) & Co. Ltd, 143

Middle East: Publishers (US)
Indiana University Press, 179
Interlink Publishing Group, Inc., 179
Syracuse University Press, 188
Texas Press, University of, 188

Military: Book Clubs
BCA (Book Club Associates), 293

Military: Library Services
Leeds Central Library, 587

Military: Literary Agents (UK)
Feldstein Agency, The, 256
Lownie (Andrew) Literary Agency, 263
McAra (Duncan), 264

Sheil Land Associates Ltd, 268
Wade & Doherty Literary Agency Ltd, 271
Welch (John), Literary Consultant & Agent, 272

Military: Magazines
AIR International, 295
AirForces Monthly, 295
Armourer Magazine, The, 297
Aviation News, 298
Flight International, 320
FlyPast, 321
Jane's Defence Weekly, 330

Military: Picture Libraries
aviation–images.com, 603
Defence Picture Library, The, 606
Imperial War Museum Photograph Archive, 613
Newark's (Peter) Picture Library, 618
Royal Air Force Museum, 621
Young (John Robert) Collection, 627

Military: Prizes
Duke of Westminster's Medal for Military Literature, The, 499

Military: Publishers (International)
Kwa-Zulu–Natal Press, University of, 197
Melbourne University Publishing Ltd, 192
Queensland Press, University of, 192
Simon & Schuster (Australia) Pty Ltd, 192

Military: Publishers (UK)
Allan (Ian) Publishing Ltd, 65
Anness Publishing Ltd, 66
Anova Books, 66
Aurum Press Ltd, 70
Birlinn Ltd, 73
Brewin Books Ltd, 77
Compendium Publishing Ltd, 85
Constable & Robinson Ltd, 85
Countryside Books, 86
Crowood Press Ltd, The, 87
David & Charles Publishers, 88
Frontline Books, 96
Grub Street, 99
HarperCollins Publishers Ltd, 102
Haynes Publishing, 103
Macmillan Publishers Ltd, 113
National Archives, The, 118

Olympia Publishers, 122
Orion Publishing Group Limited, The, 123
Osprey Publishing Ltd, 123
Pen & Sword Books Ltd, 125
Routledge, 134
Souvenir Press Ltd, 140
Spellmount, 140

Military: Publishers (US)
Indiana University Press, 179
Kansas, University Press of, 179
Nebraska Press, University of, 182
North Texas Press, University of, 182
Potomac Books, Inc., 184
Stackpole Books, 187

Military: Small Presses
Great Northern Publishing, 229
Parapress, 233
Whittles Publishing, 236

Military: UK Packagers
Amber Books Ltd, 242

Mineralogy/Mining: Library Services
Barnsley Public Library, 575
Natural History Museum Library, The, 591

Mineralogy/Mining: Magazines
Descent, 314

Mineralogy/Mining: Picture Libraries
GeoScience Features, 610
Howes (Chris)/Wild Places Photography, 612
Natural History Museum Image Resources, 618

Motorcycling: Magazines
Bike, 300
Classic Bike, 308
Motor Cycle News, 337
SuperBike Magazine, 359

Motorcycling: Prizes
Mercedes-Benz Award for the Montagu of Beaulieu Trophy, The, 511

Motorcycling: Publishers (UK)
Haynes Publishing, 103
Merrell Publishers Ltd, 116

Motoring & Cars: Library Services
Coventry Central Library, 581

Motoring & Cars: Literary Agents (UK)
Knight Features, 262

Motoring & Cars: Magazines
Auto Express, 298
Autocar, 298
BBC Top Gear Magazine, 299
Car Mechanics, 306
Classic & Sports Car, 308
Classic Cars, 308
Fast Car, 319
Good Motoring, 324
Jersey Now, 330
Land Rover World, 332
Maxim, 335
MiniWorld Magazine, 336
Racecar Engineering, 349
SmartLife International, 356
Total Vauxhall, 362
Volkswagen Golf+, 365
What Car?, 366

Motoring & Cars: Picture Libraries
Brooklands Museum Photo Archive, 604
Chapman (Giles) Library, 605
Formula One Pictures, 609
LAT Photographic, 614
Ludvigsen Library, 615
Monitor Picture Library, 616
Motoring Picture Library, 617

Motoring & Cars: Prizes
Lyons (Sir William) Award, 509
Mercedes-Benz Award for the Montagu of Beaulieu Trophy, The, 511
Renault UK Journalist of the Year Award, 517

Motoring & Cars: Professional Associations
Association of Scottish Motoring Writers, 529
Guild of Motoring Writers, 538

Motoring & Cars: Publishers (UK)
AA Publishing, 64
Crowood Press Ltd, The, 87
Haynes Publishing, 103
Merrell Publishers Ltd, 116
Motor Racing Publications, 117

Motoring & Cars: UK Packagers
Stonecastle Graphics Ltd, 246
Touchstone Books Ltd, 246Touchstone Books Ltd,

Mountaineering: Festivals
International Festival of Mountaineering Literature, The, 634

Mountaineering: Magazines
Climber, 309

Mountaineering: Picture Libraries
Bonington (Chris) Picture Library, 603
Brown (Hamish) Scottish Photographic, 604
Cleare (John)/Mountain Camera, 605
Geldard (Ed) Photo Collection, 610
Last Resort Picture Library, 614
Royal Geographical Society Picture Library, 622

Mountaineering: Prizes
Boardman Tasker Award, 491

Mountaineering: Publishers (UK)
Pocket Mountains Ltd, 127

Multimedia: Film, TV and Radio Producers
Orlando Television Productions Ltd, 429

Multimedia: Literary Agents (UK)
Curtis Brown Group Ltd, 254
Sharland Organisation Ltd, The, 268

Multimedia: Magazines
Empire, 317

Multimedia: Miscellany
Landeau (Caroline), 641

Multimedia: Professional Associations
Society for Technical Communications (STC), 547

Multimedia: Publishers (European)
Brepols Publishers NV, 160
Giunti Editoriale SpA, 166

Groupo Anaya, 169
Hüthig GmbH & Co. KG, 165
Otava Publishing Co. Ltd, 162
Springer GmbH, 165
Springer Science+Business Media BV, 168

Multimedia: Publishers (International)
Affiliated East West Press Pvt Ltd, 195
LexisNexis Butterworths (Pty) Ltd South Africa, 197
LexisNexis Canada Ltd, 193
LexisNexis India, 195
LexisNexis New Zealand, 196
Macmillan Publishers Australia Pty Ltd, 191
Scholastic Australia Pty Limited, 192

Multimedia: Publishers (UK)
ABC-CLIO (UK), 64
Chartered Institute of Personnel and Development (CIPD), 83
Debrett's Ltd, 88
Elm Publications/Training, 92
Elsevier Ltd, 92
Euromonitor International plc, 93
Hachette Livre UK, 100
HarperCollins Publishers Ltd, 102
Helicon Publishing, 104
Hobsons Plc, 105
Jane's Information Group, 108
Kogan Page Ltd, 109
Macmillan Publishers Ltd, 113
Nelson Thornes Limited, 119
Oxford University Press, 124
Pearson (Heinemann), 125
Pegasus Elliot Mackenzie Publishers Ltd, 125
Penguin Group (UK), 125
ProQuest, 129
Psychology Press, 129
Radcliffe Publishing Ltd, 131
Reed (William) Directories, 133
Reed Elsevier Group plc, 133
Rising Stars UK Ltd, 134
Sweet & Maxwell Group, 142

Multimedia: Publishers (US)
ABC–CLIO (US), 172
HarperCollins Publishers, Inc., 177
Scholastic Library Publishing, 186
Thomson Gale, 188
Wiley (John) & Sons, Inc., 190
Zondervan, 190

Multimedia: Writers' Courses (US)
California Institute of the Arts, 469

Music: Audio Books
Chrome Dreams, 238
CYP, 239

Music: Film, TV and Radio Producers
Diverse Production Limited, 422
Electric Airwaves Ltd (Ladbroke Radio), 422
Holmes Associates, 425
Isis Productions, 426
Landseer Productions Ltd, 427
Lucida Productions, 428
Passion Pictures, 430
Pelicula Films, 430
Screen Ventures Ltd, 432
Staveacre (Tony) Productions, 433
Sunset + Vine Productions Ltd, 434
VIP Broadcasting, 436

Music: Library Services
Barbican Library, 574
Belfast Public Libraries: Central Library, 575
Bristol Central Library, 576
British Library Music Collections, 578
British Library Sound Archive, 579
Dundee Central Library, 582
Essex County Council Libraries, 582
Leeds Central Library, 587
Liverpool Libraries and Information Services, 588
Manchester Central Library, 589
National Library of Scotland, 590
Plymouth Library and Information Services, City of, 593
Senate House Library, University of London, 595
Sheffield Libraries, Archives and Information, 595
Westminster Music Library, 598
Williams (Vaughan) Memorial Library, 599

Music: Literary Agents (Ireland)
Williams (Jonathan) Literary Agency, 279

Music: Literary Agents (UK)
Brown (Jenny) Associates, 251
Feldstein Agency, The, 256
Green (Annette) Authors' Agency, 258
Johnson & Alcock, 261

Music: Literary Agents (US)
Gusay (Charlotte) Literary Agency, The, 284
Michaels (Doris S.) Literary Agency Inc., 286

Music: Magazines
Acoustic, 294
BBC Music Magazine, 299
Church Music Quarterly, 307
Classical Guitar, 309
Classical Music, 309
Dazed & Confused, 314
Hi-Fi News, 326
i-D Magazine, 328
Jazz Journal International, 330
Kerrang!, 331
List, The, 332
Loaded, 333
Mixmag, 336
Mojo, 336
Music Week, 338
Musical Opinion, 338
New Musical Express, 339
Piano, 344
Q, 348
Record Collector, 351
Scots Magazine, The, 352
Strad, The, 358
Top of the Pops Magazine, 362
Uncut, 364

Music: Miscellany
Landeau (Caroline), 641

Music: Picture Libraries
akg-images Ltd, Arts and History Picture Library, 601
ArenaPAL, 602
British Library Images Online, 604
Christie's Images Ltd, 605
Cinema Museum, The, 605
Classical Collection Ltd, 605
Hutchison Picture Library, 612
Jack (Robbie) Photography, 613
Lebrecht Music and Arts, 614
MacQuitty International Photographic Collection, 616
Mander (Raymond) & Joe Mitchenson Theatre Collection, The, 616
Monitor Picture Library, 616
Redferns Music Picture Library, 621

Retna Pictures Ltd, 621
Williams (Vaughan) Memorial Library, 626

Music: Poetry Magazines
Aesthetica, 208
Eratica, 210
Read the Music, 215

Music: Prizes
Oldman (C.B.) Prize, 514
Pulitzer Prizes, 516

Music: Professional Associations
British Academy of Composers and Songwriters, 531
MCPS–PRS Alliance, The, 540

Music: Publishers (European)
Adelphi Edizioni SpA, 166
Arche Verlag AG, 171
Beck (C.H., OHG) Verlag, 164
Brepols Publishers NV, 160
Caminho (Editorial) SARL, 169
Fayard (Librarie Arthème), 163
Gallimard (Éditions), 163
Giunti Editoriale SpA, 166
Gyldendal, 161
Hoffmann und Campe Verlag GmbH, 165
Larousse (Éditions), 163
Mondadori (Arnoldo) Editore SpA, 167

Music: Publishers (International)
Currency Press Pty Ltd, 191
Munshiram Manoharlal Publishers Pvt Ltd, 195

Music: Publishers (Ireland)
An Gúm, 153
Currach Press, 155
Merlin Publishing, 157
O'Brien Press Ltd, The, 157

Music: Publishers (UK)
Anova Books, 66
Arc Publications Ltd, 68
Ashgate Publishing Ltd, 69
Aurum Press Ltd, 70
Austin & Macauley Publishers Limited, 70
Black (A.&C.) Publishers Ltd, 73
Boyars (Marion) Publishers Ltd, 76
Calder Publications Ltd, 79
Carlton Publishing Group, 81
Continuum International Publishing Group Ltd, The, 85

de la Mare (Giles) Publishers
Ltd, 88
Faber & Faber Ltd, 94
Harvard University Press, 103
Helter Skelter Publishing, 104
Independent Music Press, 107
JR Books Ltd, 108
Kahn & Averill, 108
Luath Press Ltd, 112
Macmillan Publishers Ltd, 113
Mayhew (Kevin) Publishers, 115
Omnibus Press, 122
Oxford University Press, 124
Phaidon Press Limited, 126
Quartet Books, 130
Random House Group Ltd,
The, 131
Routledge, 134
Shetland Times Ltd, The, 138
Souvenir Press Ltd, 140
Stainer & Bell Ltd, 141
Usborne Publishing Ltd, 147
Wild Goose Publications, 149
Yale University Press (London),
151

Music: Publishers (US)
Faber & Faber, Inc., 176
Illinois Press, University of, 178
Indiana University Press, 179
Louisiana State University
Press, 180
McFarland & Company, Inc.,
Publishers, 180
Michigan Press, The University
of, 181
Nebraska Press, University of,
182
North Texas Press, University
of, 182
Perseus Books Group, 184
Players Press, 184
Scarecrow Press Inc., 186

Music: Small Presses
Allardyce, Barnett, Publishers,
225
Fand Music Press, 228
Parapress, 233
Witan Books, 237

Mysteries: Book Clubs
BCA (Book Club Associates),
293

**Mysteries: Literary Agents
(UK)**
Chris (Teresa) Literary Agency
Ltd, 253
Feldstein Agency, The, 256
Henser Literary Agency, 260

**Mysteries: Literary Agents
(US)**
Axelrod Agency, The, 280
Bernstein (Meredith) Literary
Agency, Inc., 280
Bleecker Street Associates, Inc.,
280
Braun (Barbara) Associates,
Inc., 281
Carvainis (Maria) Agency, Inc.,
281
Dijkstra (Sandra) Literary
Agency, 282
Gusay (Charlotte) Literary
Agency, The, 284
Lampack (Peter) Agency, Inc.,
285
Lescher & Lescher Ltd, 285
Lowenstein–Yost Associates
Inc., 285
Parks (Richard) Agency, 287
Picard (Alison J.) Literary
Agent, 287
Protter (Susan Ann) Literary
Agent, 288
Quicksilver Books, Literary
Agents, 288
Russell & Volkening, Inc., 289
Schulman (Susan), A Literary
Agency, 290
Spectrum Literary Agency, 290
Ware (John A.) Literary Agency,
291
Weiner (Cherry) Literary
Agency, 292

Mysteries: Prizes
Theakstons Old Peculier Prize
for the Crime Novel of the
Year, 521

**Mysteries: Professional
Associations**
Mystery Writers of America,
Inc., 540

**Mysteries: Publishers
(European)**
Bruna (A.W.) Uitgevers BV, 167
BZZTÔH (Uitgeverij) BV, 167
Forlaget Forum, 161
Heyne Verlag, 165
Mondadori (Arnoldo) Editore
SpA, 167
Presses de la Cité, 164
Sonzogno, 167
Ullstein Buchverlage GmbH,
165
Unieboek (Uitgeverij) BV, 168

Mysteries: Publishers (UK)
Blackstone Publishers, 74

Poisoned Pen Press UK Ltd, 127
Thorpe (F.A.) Publishing, 144

Mysteries: Publishers (US)
Academy Chicago Publishers,
172
Bantam Dell Publishing Group,
173
St Martin's Press LLC, 186

Mysteries: Small Presses
Feather Books, 228
Miller (Neil) Publications, 232

**Mysteries: Writers' Courses
(UK and Ireland)**
Community Creative Writing,
461

**Mysteries: Writers' Courses
(US)**
Seton Hill University, 474

**Mythology: Literary Agents
(UK)**
London Independent Books,
263

**Mythology: Literary Agents
(US)**
Gusay (Charlotte) Literary
Agency, The, 284

Mythology: Publishers (UK)
Austin & Macauley Publishers
Limited, 70
Thames and Hudson Ltd, 144

Mythology: Small Presses
Heart of Albion Press, 230

Naturalism: Magazines
H&E Naturist, 325

**Nature/Natural History:
Film, TV and Radio
Producers**
Green Umbrella Ltd, 425

**Nature/Natural History:
Library Services**
Brighton Jubilee Library, 576
Natural History Museum
Library, The, 591

**Nature/Natural History:
Literary Agents (UK)**
Davidson (Caroline) Literary
Agency, 255

Nature/Natural History: Literary Agents (US)
Fredericks (Jeanne) Literary Agency, Inc., 283
Gusay (Charlotte) Literary Agency, The, 284
Ware (John A.) Literary Agency, 291

Nature/Natural History: Literary Societies
Adrian Bell Society, The, 553
Jefferies (Richard) Society, 560
Romany Society, The, 566

Nature/Natural History: Magazines
Naturalist, The, 339
Scots Magazine, The, 352

Nature/Natural History: Picture Libraries
Angel (Heather)/Natural Visions, 602
Baxter (Colin) Photography Limited, 603
Cordaiy (Sylvia) Photo Library, 606
E&E Picture Library, 607
Fogden Wildlife Photographs, 608
Grace (Martin and Dorothy), 611
Howes (Chris)/Wild Places Photography, 612
Lane (Frank) Picture Agency Ltd, 614
MacQuitty International Photographic Collection, 616
Natural History Museum Image Resources, 618
Nature Photographers Ltd, 618
Nature Picture Library, 618
Oxford Scientific (OSF), 619
Photoshot, 619
Premaphotos Wildlife, 620
RSPB Images, 622
RSPCA Photolibrary, 622
Science Photo Library, 622
Still Pictures' Whole Earth Photolibrary, 624

Nature/Natural History: Prizes
BBC Wildlife Magazine Nature Writing Awards, 490
BBC Wildlife Magazine Poet of the Year Awards, 490
BBCFour Samuel Johnson Prize for Non-Fiction, The, 490

Nature/Natural History: Publishers (European)
Dokorán Ltd, 161
Fischer (S.) Verlag GmbH, 164
Flammarion Éditions, 163
Larousse (Éditions), 163
Nyt Nordisk Forlag Arnold Busck A/S, 162

Nature/Natural History: Publishers (International)
Allen & Unwin Pty Ltd, 191
Canterbury University Press, 196
Melbourne University Publishing Ltd, 192
Oxford University Press India, 196
Random House New Zealand, 197
Raupo Publishing (NZ) Ltd, 197
Struik Publishers (Pty) Ltd, 198
UWA Press, 193

Nature/Natural History: Publishers (UK)
Black (A.&C.) Publishers Ltd, 73
Blackstaff Press Ltd, 74
Crowood Press Ltd, The, 87
Dorling Kindersley Ltd, 90
Fitzgerald Publishing, 95
Fountain Press, 95
Freeman (W. H.), 96
HarperCollins Publishers Ltd, 102
Little Books Ltd, 111
New Holland Publishers (UK) Ltd, 119
NMS Enterprises Limited – Publishing, 120
Reader's Digest Association Ltd, 132
Reaktion Books, 132
Shetland Times Ltd, The, 138
Souvenir Press Ltd, 140
Trident Press Ltd, 145
Whittet Books Ltd, 149

Nature/Natural History: Publishers (US)
Abrams (Harry N.), Inc., 172
Alaska Press, University of, 172
Charlesbridge Publishing, 174
Nevada Press, University of, 182
Oklahoma Press, University of, 183
Stackpole Books, 187
Texas Press, University of, 188
Washington State University Press, 190

Nature/Natural History: Small Presses
Packard Publishing Limited, 233
Whittles Publishing, 236

Nature/Natural History: Television and Radio
BBC Bristol, 399

Nature/Natural History: UK Packagers
Brown Wells and Jacobs Ltd, 242

Nature/Natural History: Writers' Courses (US)
Goucher College, 471

Netherlands: Literary Agents (US)
Gusay (Charlotte) Literary Agency, The, 284

New Age: Book Clubs
Cygnus Books, 293

New Age: Literary Agents (UK)
Fox & Howard Literary Agency, 257

New Age: Literary Agents (US)
Bleecker Street Associates, Inc., 280
Quicksilver Books, Literary Agents, 288
Writers House, LLC, 292

New Age: Magazines
Prediction, 346

New Age: Miscellany
Sammarco (Leda), 641

New Age: Publishers (International)
Allen & Unwin Pty Ltd, 191
HarperCollins Publishers India Ltd, 195
Orient Paperbacks, 195

New Age: Publishers (Ireland)
Mercier Press Ltd, 156

New Age: Publishers (UK)
Ashgrove Publishing, 69
Baird (Duncan) Publishers, 71
Capall Bann Publishing, 80

Carroll & Brown Publishers
Limited, 81
Cathie (Kyle) Ltd, 81
Cico Books, 84
Clairview Books Ltd, 84
Crown House Publishing, 87
Findhorn Press Ltd, 94
Foulsham Publishers, 95
HarperCollins Publishers Ltd,
102
Janus Publishing Company Ltd,
108
O/Books/ John Hunt Publishing
Ltd, 120
Octopus Publishing Group, 121
Olympia Publishers, 122
Piatkus Books, 127
Simon & Schuster UK Limited,
139
Temple Lodge Publishing Ltd,
144
Wild Goose Publications, 149
Zambezi Publishing Ltd, 151

New Age: Publishers (US)
Avery, 173
Crown Publishing Group, 175
Llewellyn Publications, 180

New Age: Small Presses
Contact Publishing Ltd, 226
Ignotus Press, 230
Jupiter Press, The, 230
Mirage Publishing, 232

New Age: UK Packagers
Haldane Mason Ltd, 243

New Zealand: Prizes
Kiriyama Pacific Rim Book
Prize, 507

**North America: Literary
Agents (US)**
Gusay (Charlotte) Literary
Agency, The, 284

**North America: Picture
Libraries**
Ecoscene, 607
Tropix Photo Library, 625

Nursing: Magazines
Nursing Times, 342

**Nursing: Publishers
(European)**
Flammarion Éditions, 163
Larousse (Éditions), 163
Nyt Nordisk Forlag Arnold
Busck A/S, 162

Springer-Verlag GmbH, 160

**Occult/Mysticism: Literary
Societies**
Ghost Story Society, The, 559

Occult/Mysticism: Magazines
*Fortean Times: The Journal of
Strange Phenomena*, 321
Prediction, 346
Psychic News, 348

**Occult/Mysticism: Publishers
(European)**
BZZTÔH (Uitgeverij) BV, 167
Heyne Verlag, 165

**Occult/Mysticism: Publishers
(UK)**
Janus Publishing Company Ltd,
108
Souvenir Press Ltd, 140
Wordsworth Editions Ltd, 151

**Occult/Mysticism: Publishers
(US)**
Llewellyn Publications, 180

**Occult/Mysticism: Small
Presses**
Haunted Library, 229
Headpress, 230
Orpheus Publishing House, 232

Opera: Magazines
Opera, 342
Opera Now, 343

Opera: Picture Libraries
ArenaPAL, 602
Classical Collection Ltd, 605
Dominic Photography, 607
Jack (Robbie) Photography, 613
Lebrecht Music and Arts, 614

Opera: Publishers (UK)
Calder Publications Ltd, 79

**Oriental Studies: Library
Services**
British Library Asia, Pacific and
Africa Collections, 577

**Oriental Studies: Literary
Agents (UK)**
Henser Literary Agency, 260

**Oriental Studies: Picture
Libraries**
British Library Images Online,
604

**Oriental Studies: Publishers
(UK)**
Octagon Press Ltd, 121

**Outdoor & Country Pursuits:
Literary Agents (US)**
Gusay (Charlotte) Literary
Agency, The, 284

**Outdoor & Country Pursuits:
Literary Societies**
Wainwright Society, 568

**Outdoor & Country Pursuits:
Magazines**
Camping and Caravanning, 305
Caravan Magazine, 306
Climber, 309
Country Life, 311
Country Walking, 312
Countryman's Weekly, The, 312
Countryside Alliance Update,
312
Field, The, 320
Motor Caravan Magazine, 337
*Motorcaravan Motorhome
Monthly (MMM)*, 337
Park Home & Holiday Caravan,
343
Practical Caravan, 345
Shooting and Conservation, 355
Shooting Gazette, The, 355
*Shooting Times & Country
Magazine*, 355
TGO (The Great Outdoors), 360
Trail, 363
Walk Magazine, 365
Which Caravan, 367

**Outdoor & Country Pursuits:
Picture Libraries**
Howes (Chris)/Wild Places
Photography, 612

**Outdoor & Country Pursuits:
Prizes**
OWPG Awards for Excellence,
515

**Outdoor & Country Pursuits:
Professional Associations**
Outdoor Writers &
Photographers Guild, 542

**Outdoor & Country Pursuits:
Publishers (UK)**
Cicerone Press, 84
Country Publications Ltd, 86
Countryside Books, 86
Crowood Press Ltd, The, 87
Lincoln (Frances) Ltd, 110

Luath Press Ltd, 112
New Holland Publishers (UK) Ltd, 119
Pocket Mountains Ltd, 127
Reardon Publishing, 133
Sigma Press, 139
Sportsman's Press, The, 141
Swan Hill Press, 142
Wilson (Neil) Publishing Ltd, 149

Outdoor & Country Pursuits: Publishers (US)
Globe Pequot Press, The, 176
Stackpole Books, 187

Outdoor & Country Pursuits: Small Presses
Meridian Books, 231
Whittles Publishing, 236

Palaeontology: Library Services
Natural History Museum Library, The, 591
Senate House Library, University of London, 595

Palaeontology: Literary Agents (US)
Gusay (Charlotte) Literary Agency, The, 284

Palaeontology: Publishers (UK)
Freeman (W. H.), 96

Palaeontology: Publishers (US)
Indiana University Press, 179

Pantomime: Literary Agents (UK)
Narrow Road Company, The, 265

Pantomime: Theatre Producers
Hiss & Boo Co. Ltd, The, 442

Parenting: Literary Agents (UK)
Bonomi (Luigi) Associates Limited (LBA), 251

Parenting: Literary Agents (US)
Bleecker Street Associates, Inc., 280
Castiglia Literary Agency, 281

Gusay (Charlotte) Literary Agency, The, 284
Konner (Linda) Literary Agency, 285
Lowenstein–Yost Associates Inc., 285
Parks (Richard) Agency, 287
Protter (Susan Ann) Literary Agent, 288
Scribblers House® LLC Literary Agency, 290
Snell (Michael) Literary Agency, 290

Parenting: Magazines
FQ Magazine, 321
Mother and Baby, 337
Practical Parenting, 346
Pregnancy, Baby & You, 347
That's Life!, 360

Parenting: PR Consultants
Frazer (Cathy) PR, 277

Parenting: Picture Libraries
Greenhill (Sally and Richard), 611
Photofusion, 619

Parenting: Publishers (European)
Deutscher Taschenbuch Verlag GmbH & Co. KG, 164
Karisto Oy, 162
Larousse (Éditions), 163
Spectrum (Uitgeverij Het) BV, 168

Parenting: Publishers (International)
ACER Press, 191
Simon & Schuster (Australia) Pty Ltd, 192
Struik Publishers (Pty) Ltd, 198

Parenting: Publishers (UK)
Carroll & Brown Publishers Limited, 81
Crimson Publishing, 86
Lincoln (Frances) Ltd, 110
Octopus Publishing Group, 121
Piccadilly Press, 127
Random House Group Ltd, The, 131
Ryland Peters & Small Limited, 135

Parenting: UK Packagers
Elwin Street Productions, 243

Pets: Literary Agents (US)
Gusay (Charlotte) Literary Agency, The, 284
Snell (Michael) Literary Agency, 290

Pets: Magazines
Animal Action, 296
Animals and You, 296
Cat World, 306
Your Cat Magazine, 370

Pets: Picture Libraries
Cordaiy (Sylvia) Photo Library, 606
Lane (Frank) Picture Agency Ltd, 614
Nature Picture Library, 618
Oxford Scientific (OSF), 619
RSPCA Photolibrary, 622

Pets: Publishers (European)
Karisto Oy, 162

Pets: Publishers (UK)
Apex Publishing Ltd, 67
HarperCollins Publishers Ltd, 102
New Holland Publishers (UK) Ltd, 119
Octopus Publishing Group, 121
Whittet Books Ltd, 149

Pets: Publishers (US)
Barron's Educational Series, Inc., 173
Sterling Publishing Co. Inc., 188

Philosophy: Library Services
London Library, The, 589
Senate House Library, University of London, 595
Williams's (Dr) Library, 599

Philosophy: Literary Agents (UK)
Feldstein Agency, The, 256

Philosophy: Literary Agents (US)
Gusay (Charlotte) Literary Agency, The, 284

Philosophy: Magazines
New Humanist, 339
Philosopher, The, 344

Philosophy: Publishers (European)
Adelphi Edizioni SpA, 166
Alianza Editorial SA, 169

Beck (C.H., OHG) Verlag, 164
Bompiani, 166
Brepols Publishers NV, 160
Bruna (A.W.) Uitgevers BV, 167
Calmann-Lévy (Éditions), 163
Caminho (Editorial) SARL, 169
Denoël (Éditions), 163
Deutscher Taschenbuch Verlag GmbH & Co. KG, 164
Fayard (Librairie Arthème), 163
Gallimard (Éditions), 163
Garzanti Libri SpA, 166
Grasset & Fasquelle (Éditions), 163
Gyldendal, 161
Hoffmann und Campe Verlag GmbH, 165
Jelenkor Kiadó Szolgáltató Kft., 166
Kibea Publishing Company, 160
Longanesi (Casa Editrice) SpA, 167
Minuit (Les Éditions de) SA, 164
Mondadori (Arnoldo) Editore SpA, 167
Mulino (Societe editrice il), 167
Mursia (Gruppo Ugo) Editore SpA, 167
Naouka I Izkoustvo, 161
Nyt Nordisk Forlag Arnold Busck A/S, 162
Presses Universitaires de France (PUF), 164
Springer GmbH, 165
Sugarco Edizioni Srl, 167
Suhrkamp Verlag, 165
Tusquets Editores, 170
Universitaire Pers Leuven, 160

Philosophy: Publishers (International)
HarperCollins Publishers India Ltd, 195
Jaico Publishing House, 195
Melbourne University Publishing Ltd, 192
Munshiram Manoharlal Publishers Pvt Ltd, 195

Philosophy: Publishers (UK)
Acumen Publishing Limited, 64
Black Ace Books, 73
Boyars (Marion) Publishers Ltd, 76
British Academy, The, 77
Continuum International Publishing Group Ltd, The, 85
Edinburgh University Press, 91
Epworth, 93
Gibson Square, 97
Harvard University Press, 103

Honeyglen Publishing Ltd, 105
Icon Books Ltd, 106
Imprint Academic, 106
Janus Publishing Company Ltd, 108
Macmillan Publishers Ltd, 113
Merlin Press Ltd, The, 116
Octagon Press Ltd, 121
Oneworld Publications, 122
Open Gate Press, 123
Polity Press, 128
Random House Group Ltd, The, 131
Routledge, 134
Shepheard-Walwyn (Publishers) Ltd, 138
Souvenir Press Ltd, 140
Steiner (Rudolf) Press, 141
Verso, 147
Wales Press, University of, 148

Philosophy: Publishers (US)
Indiana University Press, 179
Massachusetts Press, University of, 181
Northwestern University Press, 182
Paragon House, 183
Prometheus Books, 184
Shambhala Publications, Inc., 186
State University of New York Press, 187

Philosophy: Small Presses
Godstow Press, 229
Ivy Publications, 230
Orpheus Publishing House, 232
Paupers' Press, 233
Pipers' Ash Ltd, 233

Photography: Book Clubs
Readers' Union Ltd, 293

Photography: Library Services
Morrab Library, 590

Photography: Literary Agents (Ireland)
Williams (Jonathan) Literary Agency, 279

Photography: Literary Agents (US)
Gusay (Charlotte) Literary Agency, The, 284

Photography: Magazines
Amateur Photographer, 295
Granta, 324
Practical Photography, 346

20x20 magazine, 363

Photography: PR Consultants
Idea Generation, 277

Photography: Poetry Magazines
Dream Catcher, 209
Quiet Feather, The, 215

Photography: Prizes
Kraszna-Krausz Book Awards, 507

Photography: Professional Associations
BAPLA (British Association of Picture Libraries and Agencies), 530
Bureau of Freelance Photographers, 533
Design and Artists Copyright Society (DACS), 536
Writers and Photographers unLimited, 551

Photography: Publishers (European)
Adelphi Edizioni SpA, 166
Arthaud (Éditions), 162
Caminho (Editorial) SARL, 169
Lannoo, Uitgeverij, 160
Orell Füssli Verlag, 171
Taschen GmbH, 165
Tiderne Skifter Forlag A/S, 162

Photography: Publishers (International)
Fitzhenry & Whiteside Limited, 193

Photography: Publishers (Ireland)
On Stream Publications Ltd, 157

Photography: Publishers (UK)
Anness Publishing Ltd, 66
Anova Books, 66
Aurum Press Ltd, 70
BFP Books, 72
Breedon Books Publishing Co. Ltd, The, 77
Constable & Robinson Ltd, 85
Countryside Books, 86
David & Charles Publishers, 88
Enitharmon Press, 92
Fountain Press, 95
Garnet Publishing Ltd, 97
GMC Publications Ltd, 98
Halsgrove, 101

Lewis (Dewi) Publishing, 110
Mainstream Publishing Co. (Edinburgh) Ltd, 114
Mercat Press, 116
Merrell Publishers Ltd, 116
Octopus Publishing Group, 121
Phaidon Press Limited, 126
Prestel Publishing Limited, 129
Quartet Books, 130
Random House Group Ltd, The, 131
Reaktion Books, 132
RotoVision, 134
Taschen UK, 143
Thames and Hudson Ltd, 144
Wharncliffe Books, 148

Photography: Publishers (US)
Abrams (Harry N.), Inc., 172
Barron's Educational Series, Inc., 173
Mississippi, University Press of, 181
Perseus Books Group, 184
Sterling Publishing Co. Inc., 188

Photography: Small Presses
Frontier Publishing, 228
Graffeg, 229
Sylph Editions, 235
Tlön Books Publishing Ltd, 236

Photography: UK Packagers
Ilex Press Limited, The, 244

Phrenology: Library Services
National Library of Scotland, 590

Physics: Literary Agents (US)
Gusay (Charlotte) Literary Agency, The, 284

Physics: Publishers (European)
Dokorán Ltd, 161
Springer Science+Business Media BV, 168

Physics: Publishers (UK)
Freeman (W. H.), 96

Poetry: Audio Books
CSA Word, 238
Hachette Digital, 239
HarperCollins AudioBooks, 239
Naxos AudioBooks, 239
Orion Audio Books, 240
SmartPass Ltd, 241

Poetry: Bursaries, Fellowships and Grants
Cholmondeley Awards, 482
Gregory (Eric) Trust Fund, 483
Masefield (John) Memorial Trust, The, 484
Maugham (Somerset) Awards, The, 485
Scottish Book Trust New Writers' Awards, 486

Poetry: Electronic Publishing and Other Services
Arbour E-Books, 247
Fledgling Press, 248
www.ABCtales.com, 247

Poetry: Festivals
Aldeburgh Poetry Festival, 629
Aspects Literature Festival, 629
Beverley Literature Festival, 630
Cambridge Wordfest, 631
Dorchester Festival, 632
Dundee Literary Festival, 632
Guildford Book Festival, 633
King's Lynn, The Poetry Festival, 635
Lambeth Readers & Writers Festival, 635
Ledbury Poetry Festival, 635
Lichfield Festival, 636
London (City of) Festival, 636
London Literature Festival, 636
National Association of Writers' Groups (NAWG) Open Festival of Writing, 636
Oundle Festival of Literature, 637
Poetry Otherwise, 637
PULSE International Poetry Festival, 637
Purple Patch Poetry Convention, 637
Redbridge Book and Media Festival, 638
Scotland's Book Town Festival, 638
StAnza: Scotland's Poetry Festival, 638
Stratford-upon-Avon Poetry Festival, 638
Sunday Times Oxford Literary Festival, The, 639
Swindon Festival of Literature, 639
Thomas (Dylan) Festival, 639
Times Cheltenham Literature Festival, The, 639
Torbay Weekend Festival of Poetry, 639
Ty Newydd Festival, 639

Wellington Literary Festival, 640
Wells Festival of Literature, 640
Wonderful Words, 640
Wordsworth Trust, The, 641

Poetry: Library Services
Birmingham Library Services, 576
Highgate Literary and Scientific Institution Library, 585
Lincoln Central Library, 588
Mitchell Library, The, 590
Northumberland County Library, 592

Poetry: Literary Agents (Ireland)
Williams (Jonathan) Literary Agency, 279

Poetry: Literary Agents (UK)
Bookseeker Agency, 251
Eddison Pearson Ltd, 256

Poetry: Literary Societies
Beddoes (Thomas Lovell) Society, 552
Betjeman Society, The, 553
Browning Society, The, 554
Burns (Robert) World Federation Ltd, The, 554
Clare (John) Society, The, 556
de la Mare (Walter) Society, The, 557
Friends of Coleridge, The, 556
Graves (Robert) Society, The, 559
Housman Society, 560
Keats–Shelley Memorial Association (Inc), The, 561
Kent & Sussex Poetry Society, 562
Kitley Trust, The, 562
Larkin (Philip) Society, The, 563
Masefield (John) Society, The, 564
Owen (Wilfred) Association, The, 564
Rainbow Poetry Recitals, 566
Tennyson Society, The, 568
Wordsworth Trust, The, 570
Yeats Society Sligo, The, 570

Poetry: Magazines
Abraxas Unbound, 294
Chapman, 306
Day by Day, 314
Edinburgh Review, 317
Evergreen, 318
Gold Dust Magazine, 323
Irish Pages, 329

Literary Review, 333
London Magazine, The, 333
London Review of Books, 333
Markings, 334
Mslexia, 338
New Shetlander, The, 340
New Writer, The, 340
New Writing Scotland, 341
Reader, The, 350
20x20 magazine, 363
Writers' Forum incorporating
    World Wide Writers, 369

Poetry: Prizes
Academi Cardiff International
    Poetry Competition, 488
Arvon Foundation International
    Poetry Competition, 489
BBC Wildlife Magazine Poet of
    the Year Awards, 490
Boardman Tasker Award, 491
Bridport Prize, The, 492
Canadian Poetry Association
    (Annual Poetry Contest),
    494
Chapter One Promotions Open
    Poetry Competition, 495
Cohen (David) Prize for
    Literature, 495
Cooper (Duff) Prize, The, 496
Costa Book Awards, 496
Eliot (T.S.) Prize, 499
Eliot (T.S.) Prize, 499
Envoi Poetry Competition, 499
Faber (Geoffrey) Memorial
    Prize, 500
Forward Prizes for Poetry, The,
    500
Foyle Young Poets of the Year
    Award, The, 501
Frogmore Poetry Prize, The, 501
Griffin Poetry Prize, The, 503
Hecht (Anthony) Poetry Prize,
    The, 503
Hemans (Felicia) Prize for
    Lyrical Poetry, 504
Jerwood Aldeburgh First
    Collection Prize, 506
Keats–Shelley Prize, 506
Kenney (Petra) Poetry
    Competition, The, 507
Kent & Sussex Poetry Society
    Open Competition, 507
Lakeland Book of the Year
    Awards, 507
Lannan Literary Award, 507
Llewellyn Rhys (John) Prize,
    The, 509
Mere Literary Festival Open
    Competition, 511
National Poetry Anthology, 513

National Poetry Competition,
    513
Owen (Wilfred) Award for
    Poetry, The, 515
Peterloo Poets Open Poetry
    Competition, 515
Poetry Business Competition,
    515
Prose & Poetry Prizes, 516
Southport Writers' Circle
    Poetry Competition, 520
Strokestown International
    Poetry Competition, 520
Sunday Times Young Writer of
    the Year Award, 521
Sundial Scottish Arts Council
    Book of the Year Awards, 521
Thomas (Dylan) Prize, The, 521
Ver Poets Open Competition,
    522
Webb (Harri) Prize, The, 523
Wilkins Memorial Poetry Prize,
    524
Winchester Writers' Conference
    Prizes, 524
Writers Bureau Poetry and
    Short Story Competition,
    The, 525
Yeovil Literary Prize, 525

Poetry: Professional
Associations
Canadian Federation of Poets,
    533

Poetry: Publishers
(European)
Alianza Editorial SA, 169
Arche Verlag AG, 171
Atlantis Ltd, 161
Deutscher Taschenbuch Verlag
    GmbH & Co. KG, 164
Gallimard (Éditions), 163
Garzanti Libri SpA, 166
Gyldendal, 161
Hiperión (Ediciónes) SL, 170
Hoffmann und Campe Verlag
    GmbH, 165
Jelenkor Kiadó Szolgáltató Kft.,
    166
Kráter Association and
    Publishing House, 166
Mondadori (Arnoldo) Editore
    SpA, 167
Móra Könyvkiadö Zrt., 166
Random House Mondadori, 170
Seuil (Éditions du), 164
Standaard Uitgeverij, 160
Suhrkamp Verlag, 165
Tiderne Skifter Forlag A/S, 162
Todariana Editrice, 167
Tusquets Editores, 170

Zsolnay (Paul), Verlag GmbH,
    160

Poetry: Publishers
(International)
Auckland University Press, 196
Fitzhenry & Whiteside Limited,
    193
HarperCollins Publishers India
    Ltd, 195
Kwa-Zulu–Natal Press,
    University of, 197
McClelland & Stewart Ltd, 193
New Star Books Ltd, 194
Queensland Press, University
    of, 192
Victoria University Press, 197

Poetry: Publishers (Ireland)
Bradshaw Books, 153
Cló Iar-Chonnachta, 154
Gallery Press, The, 155
Lilliput Press, The, 156
New Island, 157

Poetry: Publishers (UK)
Akros Publications, 65
Anvil Press Poetry Ltd, 67
Arc Publications Ltd, 68
Blackstaff Press Ltd, 74
Bloodaxe Books Ltd, 74
Bound Biographies Limited, 76
Calder Publications Ltd, 79
Carcanet Press Ltd, 80
Cathie (Kyle) Ltd, 81
Chapman Publishing, 83
CRW Publishing Ltd, 87
Enitharmon Press, 92
Everyman's Library, 93
Faber & Faber Ltd, 94
Janus Publishing Company Ltd,
    108
Luath Press Ltd, 112
Macmillan Publishers Ltd, 113
New Beacon Books Ltd, 119
NMS Enterprises Limited –
    Publishing, 120
Onlywomen Press Ltd, 122
Parthian, 124
Pegasus Elliot Mackenzie
    Publishers Ltd, 125
Penguin Group (UK), 125
Polygon, 128
Random House Group Ltd,
    The, 131
Scottish Cultural Press/Scottish
    Children's Press, 137
Seren, 137
Souvenir Press Ltd, 140
Troubador Publishing Ltd, 146
Two Ravens Press Ltd, 146
Wordsworth Editions Ltd, 151

Poetry: Publishers (US)
Arkansas Press, University of, 173
Boyds Mills Press, 174
Farrar, Straus & Giroux, Inc., 176
Iowa Press, University of, 179
Louisiana State University Press, 180
Massachusetts Press, University of, 181
Northwestern University Press, 182

Poetry: Small Presses
Between the Lines, 226
Brodie Press, The, 226
Crescent Moon Publishing and Joe's Press, 227
Edgewell Publishing, 227
Fand Music Press, 228
Feather Books, 228
Five Leaves Publications, 228
Griffin Publishing Associates, 229
Maypole Editions, 231
Pipers' Ash Ltd, 233
Tlön Books Publishing Ltd, 236
Waywiser Press, 236
Worple Press, 237

Poetry: Television and Radio
BBC Somerset, 404
Radio XL 1296 AM, 411

Poetry: Writers' Circles and Workshops
Original Writers Group, The, 479
Ver Poets, 480

Poetry: Writers' Courses (UK and Ireland)
Aberystwyth, University of, 467
Arvon Foundation, 458
Bath Spa University, 462
Chrysalis – The Poet In You, 455
City Lit, 458
Dingle Writing Courses, 466
East Anglia, University of, 461
Edge Hill University, 457
Exeter Phoenix, 455
Exeter, University of, 455
Fire in the Head Courses & Writers' Services, 455
Glamorgan, University of, 467
Keele University, 463
Knuston Hall Residential College for Adult Education, 461
Lancaster University, 457
Liverpool, University of, 461

London School of Journalism, 459
Manchester Metropolitan University – The Writing School, 460
Manchester, University of, 460
Middlesex University, 459
Missenden Abbey, 453
Nottingham Trent University, The, 462
Open College of the Arts, 465
Reading, University of, 453
Roehampton University, 460
Royal Holloway University of London, 463
Sheffield, University of, 465
St Andrews, University of, 467
Sussex, University of, 463
Writers College, The, 461
Writers' Holiday at Caerleon, 468
Wye Valley Arts Centre, 456

Poetry: Writers' Courses (US)
Adelphi University, 473
American University, 470
Arizona, University of, 469
Arkansas, University of, 469
Boston University, 471
Bowling Green State University, 474
Brooklyn College of the City University of New York, 473
Brown University, 474
Chapman University, 469
Chicago State University, 471
Colorado State University, 470
Eastern Washington University, 475
Emerson College, 472
Florida International University, 470
George Mason University, 475
Hollins University, 475
Houston, University of, 474
Louisiana State University, 471
Memphis, University of, 474
Minnesota State University Moorhead, 472
Missouri, University of, 472
Missouri-Columbia, University of, 472
New York University, 473
North Carolina at Greensboro, University of, 473
North Carolina at Wilmington, University of, 473
Notre Dame, University of, 471
Oregon, University of, 474
Purdue University, 471
Saint Mary's College of California, 469

San Francisco State University, 470
Seattle Pacific University, 475
Virginia Commonwealth University, 475
Western Michigan University, 472
Wyoming, University of, 476Wyoming, University of, Wyoming, University of, Wyoming, University of,

Poland: Library Services
Polish Library POSK, 593

Poland: Literary Agents (US)
Gusay (Charlotte) Literary Agency, The, 284

Poland: Picture Libraries
Peerless (Ann & Bury) Picture Library, 619

Poland: Publishers (UK)
Reardon Publishing, 133

Polar Studies: Library Services
Bromley Central Library, 579
National Library of Scotland, 590

Polar Studies: Picture Libraries
Alexander (Bryan & Cherry) Photography, 601
Angel (Heather)/Natural Visions, 602
Cleare (John)/Mountain Camera, 605
Cordaiy (Sylvia) Photo Library, 606
Ecoscene, 607
Geoslides Photography, 610
Grace (Martin and Dorothy), 611
Tropix Photo Library, 625

Polar Studies: Publishers (US)
Alaska Press, University of, 172

Polar Studies: UK Packagers
Erskine Press, 243

Politics: Festivals
Buxton Festival, 631

Politics: Literary Agents (Ireland)
Williams (Jonathan) Literary Agency, 279

Politics: Literary Agents (UK)
Feldstein Agency, The, 256
hhb agency ltd, 260
Sheil Land Associates Ltd, 268
Simmons (Jeffrey), 269

Politics: Literary Agents (US)
Dystel & Goderich Literary Management, 283
Gusay (Charlotte) Literary Agency, The, 284
Lescher & Lescher Ltd, 285
Lowenstein–Yost Associates Inc., 285
Nelson (B.K.) Literary Agency, 287
Russell & Volkening, Inc., 289
Scribblers House® LLC Literary Agency, 290

Politics: Magazines
Chapman, 306
Edinburgh Review, 317
London Review of Books, 333
New Internationalist, 339
New Shetlander, The, 340
New Statesman, 340
Quarterly Review, 348
Salisbury Review, The, 352
Spectator, The, 357
Tribune, 363

Politics: PR Consultants
Publishing Services, 278

Politics: Picture Libraries
British Cartoon Archive, The, 604
Evans (Mary) Picture Library, 607
Financial Times Pictures, 608
King (David) Collection, 613
Lebrecht Music and Arts, 614
Mirrorpix, 616
Monitor Picture Library, 616
Photoshot, 619
Popperfoto.com, 620
Report Digital, 621
Rex Features Ltd, 621

Politics: Poetry Presses
nthposition, 204

Politics: Prizes
Cooper (Duff) Prize, The, 496

Mackenzie (W.J.M.) Book Prize, 509
Orwell Prize, The, 514
Watt (David) Prize, The, 523

Politics: Publishers (European)
Adelphi Edizioni SpA, 166
Alianza Editorial SA, 169
Bertelsmann (C.), 164
Caminho (Editorial) SARL, 169
Civilização Editora, 169
Denoël (Éditions), 163
Deutscher Taschenbuch Verlag GmbH & Co. KG, 164
Fischer (S.) Verlag GmbH, 164
Kiss József Kónyvkiadó, 166
Mulino (Societe editrice il), 167
Orell Füssli Verlag, 171
Presses Universitaires de France (PUF), 164
Tiderne Skifter Forlag A/S, 162
Ueberreuter (Verlag Carl) GmbH, 160
Ullstein Buchverlage GmbH, 165
Universitaire Pers Leuven, 160

Politics: Publishers (International)
Auckland University Press, 196
HarperCollins Publishers India Ltd, 195
Jaico Publishing House, 195
Melbourne University Publishing Ltd, 192
Wits University Press, 198

Politics: Publishers (Ireland)
Currach Press, 155
Hachette Books Ireland, 155
Institute of Public Administration, 156
Maverick House Publishers, 156
Mercier Press Ltd, 156
O'Brien Press Ltd, The, 157

Politics: Publishers (UK)
Arris Publishing Ltd, 68
Atlantic Books, 69
Austin & Macauley Publishers Limited, 70
AuthorHouse UK, 70
Blackstaff Press Ltd, 74
Calder Publications Ltd, 79
Continuum International Publishing Group Ltd, The, 85
Currey (James) Publishers, 87
Edinburgh University Press, 91
Faber & Faber Ltd, 94
Garnet Publishing Ltd, 97

Gibson Square, 97
Green Books, 99
Harriman House Ltd, 103
Harvard University Press, 103
Hurst Publishers, Ltd, 106
Icon Books Ltd, 106
Imprint Academic, 106
JR Books Ltd, 108
Lawrence & Wishart Ltd, 110
Luath Press Ltd, 112
Macmillan Publishers Ltd, 113
Mainstream Publishing Co. (Edinburgh) Ltd, 114
Manchester University Press, 115
Merlin Press Ltd, The, 116
Monarch Books, 117
New Beacon Books Ltd, 119
Oneworld Publications, 122
Open Gate Press, 123
Pluto Press Ltd, 127
Politico's Publishing, 128
Polity Press, 128
Profile Books, 129
Publishing House, 129
Quartet Books, 130
Random House Group Ltd, The, 131
Routledge, 134
Short Books, 139
Simon & Schuster UK Limited, 139
Tauris (I.B.) & Co. Ltd, 143
Verso, 147
Wild Goose Publications, 149
Wimbledon Publishing Company, 150
Yale University Press (London), 151

Politics: Publishers (US)
Arkansas Press, University of, 173
Holt (Henry) & Company, Inc., 178
Indiana University Press, 179
Interlink Publishing Group, Inc., 179
Kansas, University Press of, 179
Louisiana State University Press, 180
Michigan Press, The University of, 181
Minnesota Press, University of, 181
Oklahoma Press, University of, 183
Paragon House, 183
Perseus Books Group, 184
Seven Stories Press, 186
St Martin's Press LLC, 186
Syracuse University Press, 188

Washington State University Press, 190

Politics: Small Presses
Five Leaves Publications, 228
Witan Books, 237

Politics: Theatre Producers
7:84 Theatre Company Scotland, 448

Pollution: Picture Libraries
Angel (Heather)/Natural Visions, 602
Ecoscene, 607
Garden and Wildlife Matters Photo Library, 609
Hoffman (David) Photo Library, 612
Howes (Chris)/Wild Places Photography, 612Howes (Chris)/Wild Places Photography,

Professional: Publishers (European)
Beck (C.H., OHG) Verlag, 164
Blackwell Munksgaard, 161
Borgens Forlag A/S, 161
Hüthig GmbH & Co. KG, 165
Payot Libraire, 171
Polish Scientific Publishers PWN, 169
WEKA Holding GmbH & Co. KG, 166

Professional: Publishers (International)
LexisNexis Butterworths (Pty) Ltd South Africa, 197
LexisNexis New Zealand, 196
Macmillan Publishers Australia Pty Ltd, 191
McGraw-Hill Australia Pty Ltd, 192
McGraw-Hill Ryerson Ltd, 194
Nelson Education, 194
Vision Books Pvt. Ltd, 196
Wiley (John) & Sons Canada Ltd, 194

Professional: Publishers (Ireland)
Blackhall Publishing Ltd, 153
Institute of Public Administration, 156
Oak Tree Press, 157

Professional: Publishers (UK)
Ashgate Publishing Ltd, 69
Brealey (Nicholas) Publishing, 77

Cambridge University Press, 79
Chartered Institute of Personnel and Development (CIPD), 83
Crimson Publishing, 86
Elsevier Ltd, 92
Harriman House Ltd, 103
Informa Law, 107
Kogan Page Ltd, 109
Macmillan Publishers Ltd, 113
Management Books 2000 Ltd, 114
Open University Press, 123
Psychology Press, 129
Reed Elsevier Group plc, 133
Routledge, 134
Sweet & Maxwell Group, 142
Trentham Books Ltd, 145
TSO (The Stationery Office), 146
Wiley Europe Ltd, 149

Professional: Publishers (US)
Abingdon Press, 172
McGraw-Hill Companies Inc., The, 180
Norton (W.W.) & Company, Inc., 182
Pennsylvania Press, University of, 184
Prometheus Books, 184
Stanford University Press, 187
Wiley (John) & Sons, Inc., 190

Professional: Small Presses
Packard Publishing Limited, 233

Proofreading Facilities: Professional Associations
Society for Editors and Proofreaders (SfEP), 547

Psychology: Library Services
British Psychological Society Library, 579
Senate House Library, University of London, 595

Psychology: Literary Agents (UK)
Anderson (Darley) Literary, TV & Film Agency, 249
Fox & Howard Literary Agency, 257
Graham Maw Christie Literary Agency, 258
Henser Literary Agency, 260
London Independent Books, 263
Paterson Marsh Ltd, 266
Watson, Little Ltd, 272

Psychology: Literary Agents (US)
Altshuler (Miriam) Literary Agency, 280
Bleecker Street Associates, Inc., 280
Braun (Barbara) Associates, Inc., 281
Castiglia Literary Agency, 281
Dijkstra (Sandra) Literary Agency, 282
Gusay (Charlotte) Literary Agency, The, 284
Herman (Jeff) Agency, LLC, The, 284
Klinger (Harvey), Inc., 284
Konner (Linda) Literary Agency, 285
Lowenstein–Yost Associates Inc., 285
Naggar (Jean V.) Literary Agency, The, 287
Parks (Richard) Agency, 287
Quicksilver Books, Literary Agents, 288
Schulman (Susan), A Literary Agency, 290
Scribblers House® LLC Literary Agency, 290
Snell (Michael) Literary Agency, 290
Ware (John A.) Literary Agency, 291

Psychology: Poetry Magazines
Leeds Poetry Quarterly, 211
Literature and Psychoanalysis, 212

Psychology: Prizes
MIND Book of the Year, 512

Psychology: Publishers (European)
Adelphi Edizioni SpA, 166
Apostrof (Forlaget) ApS, 161
Borgens Forlag A/S, 161
Bruna (A.W.) Uitgevers BV, 167
Calmann-Lévy (Éditions), 163
Caminho (Editorial) SARL, 169
Denoël (Éditions), 163
Deutscher Taschenbuch Verlag GmbH & Co. KG, 164
Fischer (S.) Verlag GmbH, 164
Giunti Editoriale SpA, 166
Gyldendal, 161
Heyne Verlag, 165
Hoffmann und Campe Verlag GmbH, 165
Kibea Publishing Company, 160

Kok (Uitgeversmaatschappij J. H.) BV, 168
Lannoo, Uitgeverij, 160
Larousse (Éditions), 163
Mondadori (Arnoldo) Editore SpA, 167
Mulino (Societe editrice il), 167
Naouka I Izkoustvo, 161
Nyt Nordisk Forlag Arnold Busck A/S, 162
Orell Füssli Verlag, 171
Presses Universitaires de France (PUF), 164
Springer GmbH, 165
Springer-Verlag GmbH, 160
Suhrkamp Verlag, 165
Table Ronde (Les Editions de la), 164
Tiderne Skifter Forlag A/S, 162
Todariana Editrice, 167
Universitaire Pers Leuven, 160

Psychology: Publishers (International)
ACER Press, 191
Allen & Unwin Pty Ltd, 191
HarperCollins Publishers India Ltd, 195
Melbourne University Publishing Ltd, 192

Psychology: Publishers (UK)
Black Ace Books, 73
Boyars (Marion) Publishers Ltd, 76
Brealey (Nicholas) Publishing, 77
Constable & Robinson Ltd, 85
Crown House Publishing, 87
Free Association Books Ltd, 96
Gibson Square, 97
Harvard University Press, 103
Icon Books Ltd, 106
Imprint Academic, 106
Kingsley (Jessica) Publishers Ltd, 109
Macmillan Publishers Ltd, 113
Monarch Books, 117
Octagon Press Ltd, 121
Oneworld Publications, 122
Open University Press, 123
Piatkus Books, 127
Polity Press, 128
Psychology Press, 129
Random House Group Ltd, The, 131
Society for Promoting Christian Knowledge (SPCK), 140
Souvenir Press Ltd, 140
Women's Press, The, 150

Psychology: Publishers (US)
Holt (Henry) & Company, Inc., 178
Paragon House, 183
Perseus Books Group, 184

Puzzles, Quizzes & Crosswords: Book Clubs
Readers' Union Ltd, 293

Puzzles, Quizzes & Crosswords: Literary Agents (UK)
Knight Features, 262

Puzzles, Quizzes & Crosswords: Magazines
BBC History Magazine, 299
That's Life!, 360

Puzzles, Quizzes & Crosswords: Publishers (International)
HarperCollins Publishers India Ltd, 195
Orient Paperbacks, 195

Puzzles, Quizzes & Crosswords: Publishers (UK)
Apex Publishing Ltd, 67
Carlton Publishing Group, 81
Chambers Harrap Publishers Ltd, 82
GMC Publications Ltd, 98
Telegraph Books, 143
Usborne Publishing Ltd, 147

Puzzles, Quizzes & Crosswords: Publishers (US)
Sterling Publishing Co. Inc., 188

Radio: Bursaries, Fellowships and Grants
Bradley (Alfred) Bursary Award, 482

Radio: Library Services
BBC Written Archives Centre, 575
Westminster Reference Library, 598

Radio: Literary Agents (UK)
Agency (London) Ltd, The, 249
Blake Friedmann Literary Agency Ltd, 251
Casarotto Ramsay and Associates Ltd, 252
Colin (Rosica) Ltd, 254
Curtis Brown Group Ltd, 254
Daish (Judy) Associates Ltd, 255

Fitch (Laurence) Ltd, 256
Foster (Jill) Ltd, 257
Glass (Eric) Ltd, 257
Henser Literary Agency, 260
Hoskins (Valerie) Associates Limited, 260
Independent Talent Group Limited, 261
Mann (Andrew) Ltd, 264
MBA Literary Agents Ltd, 264
McLean (Bill) Personal Management, 264
Narrow Road Company, The, 265
Sayle Screen Ltd, 268
Sharland Organisation Ltd, The, 268
Sheil Land Associates Ltd, 268
Steinberg (Micheline) Associates, 270
Tennyson Agency, The, 270
United Agents Limited, 271
Ware (Cecily) Literary Agents, 272

Radio: Magazines
Heat, 326
Practical Wireless, 346
Radio Times, 349
Radio User, 349
TVTimes, 363

Radio: Prizes
Imison Award, The, 505
Sony Radio Academy Awards, 520
Thomas (Dylan) Prize, The, 521

Radio: Professional Associations
Irish Playwrights and Screenwriters Guild, 539
Player–Playwrights, 543
Radio Academy, The, 545
RadioCentre, 545
Writers' Guild of Great Britain, The, 551

Radio: Writers' Circles and Workshops
Script Yorkshire, 479
Southwest Scriptwriters, 479

Radio: Writers' Courses (UK and Ireland)
Central School of Speech and Drama, The, 458
Community Creative Writing, 461
Glamorgan, University of, 467
Higham Hall College, 454
Leeds, University of, 465

Liverpool, University of, 461
Reading, University of, 453
Salford, University of, 460
Writers Bureau, The, 461
Writers' Holiday at Caerleon, 468

Railways: Book Clubs
BCA (Book Club Associates), 293

Railways: Library Services
Birmingham Library Services, 576

Railways: Magazines
British Railway Modelling, 304
Rail, 349
Rail Express, 350
Railway Gazette International, 350
Railway Magazine, The, 350
Steam Railway Magazine, 358

Railways: Organizations of Interest to Poets
Railway Prism, The, 223

Railways: Picture Libraries
Alvey & Towers, 601
Cordaiy (Sylvia) Photo Library, 606
London's Transport Museum Photographic Library, 615
National Railway Museum, 618
Science & Society Picture Library, 622

Railways: Publishers (Ireland)
On Stream Publications Ltd, 157

Railways: Publishers (UK)
Allan (Ian) Publishing Ltd, 65
Haynes Publishing, 103
SB Publications, 136

Railways: Publishers (US)
Washington State University Press, 190

Railways: Small Presses
Nostalgia Collection, The, 232
Stenlake Publishing Limited, 235

Reference: Literary Agents (Ireland)
Williams (Jonathan) Literary Agency, 279

Reference: Literary Agents (UK)
Davidson (Caroline) Literary Agency, 255
Fox & Howard Literary Agency, 257
Graham Maw Christie Literary Agency, 258
Lownie (Andrew) Literary Agency, 263
Motley (Michael) Ltd, 265

Reference: Literary Agents (US)
Browne (Pema) Ltd, 281
Fredericks (Jeanne) Literary Agency, Inc., 283
Gusay (Charlotte) Literary Agency, The, 284
Herman (Jeff) Agency, LLC, The, 284
Protter (Susan Ann) Literary Agent, 288
Snell (Michael) Literary Agency, 290

Reference: PR Consultants
Publishing Services, 278

Reference: Prizes
Besterman/McColvin Medal, 490
Oldman (C.B.) Prize, 514

Reference: Publishers (European)
Akadémiai Kiadó, 166
Aschehoug (H) & Co (W Nygaard), 168
Beck (C.H., OHG) Verlag, 164
Bordas (Éditions), 162
Brepols Publishers NV, 160
Cappelen Damm AS, 168
Cappelli Editore, 166
Denoël (Éditions), 163
Deutscher Taschenbuch Verlag GmbH & Co. KG, 164
Espasa-Calpe (Editorial) SA, 170
Europa-América (Publicações) Lda, 169
Fayard (Librarie Arthème), 163
Fischer (S.) Verlag GmbH, 164
Flammarion Éditions, 163
Garzanti Libri SpA, 166
Giunti Editoriale SpA, 166
Gradiva–Publicações S.A., 169
Groupo Anaya, 169
Gyldendal, 161
Gyldendal Norsk Forlag, 168
Hachette Livre (France), 163
Kok (Uitgeversmaatschappij J. H.) BV, 168

Larousse (Éditions), 163
Lettera, 161
Mondadori (Arnoldo) Editore SpA, 167
Mursia (Gruppo Ugo) Editore SpA, 167
Nathan (Les Éditions), 164
Nyt Nordisk Forlag Arnold Busck A/S, 162
Payot Libraire, 171
Pearson Education (Benelux), 168
Polish Scientific Publishers PWN, 169
Presses Universitaires de France (PUF), 164
Random House Mondadori, 170
Schønbergske (Det) Forlag, 162
Spectrum (Uitgeverij Het) BV, 168
Springer Science+Business Media BV, 168
Unieboek (Uitgeverij) BV, 168

Reference: Publishers (International)
Chand (S.) & Co Ltd, 195
Currency Press Pty Ltd, 191
Fitzhenry & Whiteside Limited, 193
HarperCollins Canada Ltd, 193
HarperCollins Publishers (New Zealand) Ltd, 196
HarperCollins Publishers India Ltd, 195
Jaico Publishing House, 195
Macmillan India Ltd, 195
Macmillan Publishers Australia Pty Ltd, 191
McGraw-Hill Australia Pty Ltd, 192
Melbourne University Publishing Ltd, 192
Nelson Education, 194
Orient Paperbacks, 195
Oxford University Press India, 196
Queensland Press, University of, 192
Raupo Publishing (NZ) Ltd, 197
Struik Publishers (Pty) Ltd, 198
UWA Press, 193
Wiley (John) & Sons Australia Ltd, 193
Wiley (John) & Sons Canada Ltd, 194

Reference: Publishers (Ireland)
O'Brien Press Ltd, The, 157

Reference: Publishers (UK)
ABC-CLIO (UK), 64
Allan (Ian) Publishing Ltd, 65
Anness Publishing Ltd, 66
Anova Books, 66
Atlantic Books, 69
Austin & Macauley Publishers
    Limited, 70
Baird (Duncan) Publishers, 71
Birlinn Ltd, 73
Black (A.&C.) Publishers Ltd, 73
Blackstone Publishers, 74
Bloomsbury Publishing Plc, 74
Bowker (UK) Ltd, 76
British Library, The, 78
Cambridge University Press, 79
Cathie (Kyle) Ltd, 81
CBD Research Ltd, 82
Chambers Harrap Publishers
    Ltd, 82
Clarke (James) & Co., 84
Compendium Publishing Ltd, 85
Debrett's Ltd, 88
Dorling Kindersley Ltd, 90
Elsevier Ltd, 92
Euromonitor International plc,
    93
Facet Publishing, 94
Foulsham Publishers, 95
Geddes & Grosset, 97
Gibbons (Stanley) Publications,
    97
GMC Publications Ltd, 98
Guinness World Records Ltd,
    99
Hachette Children's Books, 99
Hachette Livre UK, 100
HarperCollins Publishers Ltd,
    102
Harvard University Press, 103
Helicon Publishing, 104
Kogan Page Ltd, 109
Lion Hudson plc, 111
Little, Brown Book Group UK,
    111
Macmillan Publishers Ltd, 113
Manson Publishing Ltd, 115
Melrose Press Ltd, 116
New Holland Publishers (UK)
    Ltd, 119
Nielsen BookData, 119
Octopus Publishing Group, 121
Oxford University Press, 124
Penguin Group (UK), 125
ProQuest, 129
QED Publishing, 129
Quiller Press, 130
Random House Group Ltd,
    The, 131
Reed Elsevier Group plc, 133
Routledge, 134

SCM – Canterbury Press Ltd,
    136
Sweet & Maxwell Group, 142
Taylor & Francis, 143
Telegraph Books, 143
Which? Books, 149
Whittet Books Ltd, 149
Wiley Europe Ltd, 149
Wilson (Neil) Publishing Ltd,
    149
Wordsworth Editions Ltd, 151

Reference: Publishers (US)
ABC–CLIO (US), 172
Abingdon Press, 172
Blackwell Publishing (US), 174
Columbia University Press, 175
Eerdmans Publishing Company,
    176
HarperCollins Publishers, Inc.,
    177
Hippocrene Books, Inc., 178
McFarland & Company, Inc.,
    Publishers, 180
Paragon House, 183
Pennsylvania Press, University
    of, 184
Reader's Digest Association
    Inc, 185
Rosen Publishing Group, Inc.,
    The, 185
Scarecrow Press Inc., 186
Scholastic Library Publishing,
    186
St Martin's Press LLC, 186
Sterling Publishing Co. Inc., 188
Thomson Gale, 188

Reference: Small Presses
Tartarus Press, 235

Reference: UK Packagers
Aladdin Books Ltd, 242
Amber Books Ltd, 242
Bender Richardson White, 242
Diagram Visual Information
    Ltd, 243
Elwin Street Productions, 243
Ilex Press Limited, The, 244
Lexus Ltd, 244
Market House Books Ltd, 244
West (David) Children's Books,
    246

**Religion: Film, TV and Radio
Producers**
Diverse Production Limited,
    422

Religion: Library Services
Catholic National Library, 580
London Library, The, 589

Morrab Library, 590
Williams's (Dr) Library, 599

**Religion: Literary Agents
(UK)**
Anderson (Darley) Literary, TV
    & Film Agency, 249

**Religion: Literary Agents
(US)**
Gusay (Charlotte) Literary
    Agency, The, 284

Religion: Magazines
Catholic Herald, The, 306
Church of England Newspaper,
    307
Church Times, 307
Good News, 324
Inspire, 329
New Humanist, 339
Tablet, The, 359
Universe, The, 364
War Cry, The, 365

Religion: Picture Libraries
Andes Press Agency, 602
ArkReligion.com, 602
Art Directors & Tripp Photo
    Library, 602
E&E Picture Library, 607
MacQuitty International
    Photographic Collection, 616
Peerless (Ann & Bury) Picture
    Library, 619
Wellcome Images, 626
Young (John Robert) Collection,
    627

Religion: Poetry Magazines
Areopagus, 208
Poetry Church, The, 214

Religion: Poetry Presses
Blue Butterfly Publishers, 200
Feather Books, 201

Religion: Prizes
BBCFour Samuel Johnson Prize
    for Non-Fiction, The, 490

**Religion: Professional
Associations**
Association of Christian
    Writers, 528

**Religion: Publishers
(European)**
Adelphi Edizioni SpA, 166
Atlantis Ltd, 161
Caminho (Editorial) SARL, 169

Cappelen Damm AS, 168
Civilização Editora, 169
Deutscher Taschenbuch Verlag GmbH & Co. KG, 164
Fayard (Librarie Arthème), 163
Hiperión (Ediciónes) SL, 170
Kibea Publishing Company, 160
Kok (Uitgeversmaatschappij J. H.) BV, 168
Lannoo, Uitgeverij, 160
Larousse (Éditions), 163
Mondadori (Arnoldo) Editore SpA, 167
Mulino (Societe editrice il), 167
Mursia (Gruppo Ugo) Editore SpA, 167
Nyt Nordisk Forlag Arnold Busck A/S, 162
Publicat SA, 169
Table Ronde (Les Editions de la), 164

Religion: Publishers (International)
HarperCollins Canada Ltd, 193
HarperCollins Publishers (New Zealand) Ltd, 196
HarperCollins Publishers India Ltd, 195
Jaico Publishing House, 195
Melbourne University Publishing Ltd, 192
Munshiram Manoharlal Publishers Pvt Ltd, 195
Orient Paperbacks, 195
Vision Books Pvt. Ltd, 196

Religion: Publishers (Ireland)
Church of Ireland Publishing, 154
Columba Press, The, 154
Veritas Publications, 159

Religion: Publishers (UK)
Apex Publishing Ltd, 67
Ashgate Publishing Ltd, 69
Atlantic Europe Publishing Co. Ltd, 69
Austin & Macauley Publishers Limited, 70
Authentic Media, 70
Baird (Duncan) Publishers, 71
Bryntirion Press, 79
Catholic Truth Society (CTS), 81
Christian Focus Publications, 83
Continuum International Publishing Group Ltd, The, 85
Darton, Longman & Todd Ltd, 88
Edinburgh University Press, 91

Epworth, 93
Floris Books, 95
Garnet Publishing Ltd, 97
Gresham Books Ltd, 99
Hachette Livre UK, 100
Hurst Publishers, Ltd, 106
Hymns Ancient & Modern Ltd, 106
Icon Books Ltd, 106
Inspire, 107
Inter-Varsity Press, 107
Janus Publishing Company Ltd, 108
Lion Hudson plc, 111
Lutterworth Press, The, 112
Mayhew (Kevin) Publishers, 115
Melrose Books, 116
Monarch Books, 117
mph, 117
O/Books/ John Hunt Publishing Ltd, 120
Oneworld Publications, 122
Open Gate Press, 123
Polity Press, 128
Random House Group Ltd, The, 131
Routledge, 134
SCM – Canterbury Press Ltd, 136
Society for Promoting Christian Knowledge (SPCK), 140
Souvenir Press Ltd, 140
St Pauls Publishing, 135
Stainer & Bell Ltd, 141
Tauris (I.B.) & Co. Ltd, 143
Wild Goose Publications, 149
Yale University Press (London), 151

Religion: Publishers (US)
Abingdon Press, 172
Crown Publishing Group, 175
Eerdmans Publishing Company, 176
HarperCollins Publishers, Inc., 177
Illinois Press, University of, 178
Indiana University Press, 179
Paragon House, 183
Paulist Press, 183
Shambhala Publications, Inc., 186
State University of New York Press, 187
Syracuse University Press, 188
Tyndale House Publishers, Inc., 189
Zondervan, 190

Religion: Small Presses
Crossbridge Books, 227
Feather Books, 228

Lindsey Press, The, 231
Orpheus Publishing House, 232
Tharpa Publications, 235

Religion: Television and Radio
BBC Religion, 397
Premier Radio, 411

Research: Miscellany
Combrógos, 641
Landeau (Caroline), 641
Ormrod Research Services, 641
Patent Research, 641

Romantic Fiction: Audio Books
Soundings Audio Books, 241

Romantic Fiction: Literary Agents (UK)
Anderson (Darley) Literary, TV & Film Agency, 249
Chris (Teresa) Literary Agency Ltd, 253
Dorian Literary Agency (DLA), 255
Feldstein Agency, The, 256
Gregory & Company Authors' Agents, 258
London Independent Books, 263
Murdoch (Judith) Literary Agency, 265
Sheil Land Associates Ltd, 268
Sheldon (Caroline) Literary Agency Ltd, 268

Romantic Fiction: Literary Agents (US)
Axelrod Agency, The, 280
Browne (Pema) Ltd, 281
Ellenberg (Ethan) Literary Agency, 283
Lowenstein–Yost Associates Inc., 285
Picard (Alison J.) Literary Agent, 287
Rosenberg Group, The, 289
Spectrum Literary Agency, 290
Teal (Patricia) Literary Agency, 291
Weiner (Cherry) Literary Agency, 292

Romantic Fiction: Magazines
My Weekly Pocket Novels, 338
People's Friend Pocket Novels, 343

Romantic Fiction: Prizes
Nathan (Melissa) Award for
Comedy Romance, 512
Romantic Novel of the Year, 517

**Romantic Fiction:
Professional Associations**
Romance Writers of America,
545
Romantic Novelists'
Association, The, 545

**Romantic Fiction: Publishers
(European)**
BZZTÔH (Uitgeverij) BV, 167
Heyne Verlag, 165
Mondadori (Arnoldo) Editore
SpA, 167
Presses de la Cité, 164
Unieboek (Uitgeverij) BV, 168

**Romantic Fiction: Publishers
(International)**
Harlequin Enterprises Ltd, 193

**Romantic Fiction: Publishers
(UK)**
Allison & Busby, 65
Austin & Macauley Publishers
Limited, 70
Blackstone Publishers, 74
Harlequin Mills & Boon
Limited, 101
Macmillan Publishers Ltd, 113
Severn House Publishers, 138
Thorpe (F.A.) Publishing, 144

**Romantic Fiction: Publishers
(US)**
Bantam Dell Publishing Group,
173

**Romantic Fiction: Writers'
Circles and Workshops**
Scribo, 479

**Romantic Fiction: Writers'
Courses (UK and Ireland)**
Writers' Holiday at Caerleon,
468

**Romantic Fiction: Writers'
Courses (US)**
Seton Hill University, 474

Royal Family: Magazines
Hello!, 326
Weekly News, 366

**Royal Family: Picture
Libraries**
Mirrorpix, 616
Monitor Picture Library, 616
Popperfoto.com, 620
Rex Features Ltd, 621
Royal Collection Photographic
Services, The, 621
V&A Images, 625

**Royal Family: Publishers
(UK)**
O'Mara (Michael) Books
Limited, 122

**Russia & Soviet Studies:
Literary Agents (US)**
Gusay (Charlotte) Literary
Agency, The, 284

**Russia & Soviet Studies:
Picture Libraries**
King (David) Collection, 613
Peerless (Ann & Bury) Picture
Library, 619
Russia and Eastern Images, 622

**Russia & Soviet Studies:
Prizes**
Rossica Translation Prize, 517

**Russia & Soviet Studies:
Publishers (US)**
Indiana University Press,
179Indiana University Press,

**Scandinavia: Literary Agents
(US)**
Gusay (Charlotte) Literary
Agency, The, 284

Scandinavia: Magazines
New Shetlander, The, 340

**Scandinavia: Picture
Libraries**
Geo Aerial Photography, 610
Geoslides Photography, 610
Link Picture Library, 615
Poignant (Axel) Archive, 620

Science: Audio Books
Orion Audio Books, 240

**Science: Bursaries,
Fellowships and Grants**
Economist/Richard Casement
Internship, The, 483
Fulbright Awards, 483
PAWS (Public Awareness of
Science) Drama Fund, 485

**Science: Film, TV and Radio
Producers**
Green Umbrella Ltd, 425
Moonstone Films Ltd, 429
Orlando Television Productions
Ltd, 429
Screenhouse Productions Ltd,
432

Science: Library Services
British Library Science,
Technology and Business
Collections, 579
Cardiff Central Library, 580
Devon & Exeter Institution
Library, 581
Manchester Central Library, 589
Royal Society Library, 594
Science Museum Library, 595
Sheffield Libraries, Archives
and Information, 595

Science: Literary Agents (UK)
Ableman (Sheila) Literary
Agency, 249
Anderson (Darley) Literary, TV
& Film Agency, 249
Artellus Limited, 250
Bonomi (Luigi) Associates
Limited (LBA), 251
Bryan (Felicity), 252
Davidson (Caroline) Literary
Agency, 255
Green (Annette) Authors'
Agency, 258
Heath (Rupert) Literary
Agency, 260
Sheil Land Associates Ltd, 268
Trevor (Lavinia) Literary
Agency, 271
Watson, Little Ltd, 272
Wiener (Dinah) Ltd, 272

Science: Literary Agents (US)
Castiglia Literary Agency, 281
Dijkstra (Sandra) Literary
Agency, 282
Ellenberg (Ethan) Literary
Agency, 283
Fredericks (Jeanne) Literary
Agency, Inc., 283
Gusay (Charlotte) Literary
Agency, The, 284
Klinger (Harvey), Inc., 284
Lowenstein–Yost Associates
Inc., 285
Manus & Associates Literary
Agency, Inc., 286
Naggar (Jean V.) Literary
Agency, The, 287
Protter (Susan Ann) Literary
Agent, 288

Russell & Volkening, Inc., 289
Snell (Michael) Literary Agency, 290
Ware (John A.) Literary Agency, 291
Writers House, LLC, 292

Science: Magazines
Focus, 321
Nature, 339
New Humanist, 339
New Scientist, 340

Science: Picture Libraries
Corbis Images, 606
Oxford Scientific (OSF), 619
Rex Features Ltd, 621
Science & Society Picture Library, 622
Science Photo Library, 622

Science: Prizes
BBCFour Samuel Johnson Prize for Non-Fiction, The, 490
Dingle Prize, 498
Royal Society Prizes for Science Books, The, 518
Science Writer Awards, 519

Science: Professional Associations
ABSW, 526

Science: Publishers (European)
Adelphi Edizioni SpA, 166
Alianza Editorial SA, 169
Aschehoug (H) & Co (W Nygaard), 168
Bruna (A.W.) Uitgevers BV, 167
Fayard (Librarie Arthème), 163
Flammarion Éditions, 163
Giunti Editoriale SpA, 166
Gyldendal, 161
Hachette Livre (France), 163
Hoffmann und Campe Verlag GmbH, 165
Kok (Uitgeversmaatschappij J. H.) BV, 168
Larousse (Éditions), 163
Longanesi (Casa Editrice) SpA, 167
Mondadori (Arnoldo) Editore SpA, 167
Mursia (Gruppo Ugo) Editore SpA, 167
Olympia Publishing Co. Inc., 161
Random House Mondadori, 170
Spectrum (Uitgeverij Het) BV, 168

Sperling e Kupfer Editori SpA, 167
Springer GmbH, 165
Springer Science+Business Media BV, 168
Springer-Verlag GmbH, 160
Suhrkamp Verlag, 165
Universitaire Pers Leuven, 160
Verbo (Editorial) SA, 169

Science: Publishers (International)
HarperCollins Publishers India Ltd, 195
Melbourne University Publishing Ltd, 192
National Publishing House, 195
Wits University Press, 198

Science: Publishers (Ireland)
Royal Dublin Society, 158

Science: Publishers (UK)
Atlantic Books, 69
Atlantic Europe Publishing Co. Ltd, 69
Cambridge University Press, 79
Carlton Publishing Group, 81
Floris Books, 95
Hachette Livre UK, 100
HarperCollins Publishers Ltd, 102
Harvard University Press, 103
Icon Books Ltd, 106
Janus Publishing Company Ltd, 108
Little, Brown Book Group UK, 111
Macmillan Publishers Ltd, 113
Manson Publishing Ltd, 115
Melrose Books, 116
NMS Enterprises Limited – Publishing, 120
Nottingham University Press, 120
Oneworld Publications, 122
Profile Books, 129
ProQuest, 129
Random House Group Ltd, The, 131
Sage Publications, 135
Simon & Schuster UK Limited, 139
Springer-Verlag London Ltd, 141
Woodhead Publishing Ltd, 150
Yale University Press (London), 151

Science: Publishers (US)
Avery, 173
Charlesbridge Publishing, 174
Harvard University Press, 177

Holt (Henry) & Company, Inc., 178
Krieger Publishing Co., 179
McGraw-Hill Companies Inc., The, 180
Perseus Books Group, 184
Prometheus Books, 184

Science: Small Presses
Cosmic Elk, The, 227
Ivy Publications, 230

Science: UK Packagers
Brown Wells and Jacobs Ltd, 242

Science: Writers' Courses (US)
Goucher College, 471

Science Fiction/Fantasy: Audio Books
BBC Audiobooks Ltd, 238

Science Fiction/Fantasy: Book Clubs
BCA (Book Club Associates), 293

Science Fiction/Fantasy: Library Services
Science Fiction Foundation Research Library, 595

Science Fiction/Fantasy: Literary Agents (UK)
Anubis Literary Agency, 250
Artellus Limited, 250
Cheetham (Mic) Associates, 253
Dorian Literary Agency (DLA), 255
Feldstein Agency, The, 256
Henser Literary Agency, 260
Inspira Group, The, 261
London Independent Books, 263
Sheil Land Associates Ltd, 268

Science Fiction/Fantasy: Literary Agents (US)
Gusay (Charlotte) Literary Agency, The, 284
Morrison (Henry) Inc., 287
Parks (Richard) Agency, 287
Protter (Susan Ann) Literary Agent, 288
Spectrum Literary Agency, 290
Weiner (Cherry) Literary Agency, 292

**Science Fiction/Fantasy:
Literary Societies**
British Fantasy Society, 554

**Science Fiction/Fantasy:
Magazines**
Black Static, 302
*Interzone: Science Fiction &
Fantasy*, 329

**Science Fiction/Fantasy:
Organizations of Interest to
Poets**
Eight Hand Gang, The, 220

**Science Fiction/Fantasy:
Poetry Magazines**
Handshake, 211
Premonitions, 215
Scar Tissue, 216

**Science Fiction/Fantasy:
Poetry Presses**
Hilltop Press, 202
Malfunction Press (1969), 203
Pigasus Press, 205

**Science Fiction/Fantasy:
Prizes**
British Fantasy Awards, 493
British Science Fiction
    Association Awards, 493
Clarke (Arthur C.) Award for
    Science Fiction, 495
Hubbard's (L. Ron) Writers of
    the Future Contest, 504

**Science Fiction/Fantasy:
Professional Associations**
British Science Fiction
    Association, 532
Science Fiction Foundation, 546

**Science Fiction/Fantasy:
Publishers (European)**
Calmann-Lévy (Éditions), 163
Denoël (Éditions), 163
Flammarion Éditions, 163
Gyldendal Norsk Forlag, 168
Heyne Verlag, 165
Olympia Publishing Co. Inc.,
    161
Presses de la Cité, 164
Todariana Editrice, 167

**Science Fiction/Fantasy:
Publishers (International)**
HarperCollins Publishers India
    Ltd, 195

**Science Fiction/Fantasy:
Publishers (UK)**
Allison & Busby, 65
Anova Books, 66
Blackstone Publishers, 74
Hachette Livre UK, 100
HarperCollins Publishers Ltd,
    102
Little, Brown Book Group UK,
    111
Macmillan Publishers Ltd, 113
Melrose Books, 116
Olympia Publishers, 122
Orion Publishing Group
    Limited, The, 123
Pegasus Elliot Mackenzie
    Publishers Ltd, 125
Severn House Publishers, 138

**Science Fiction/Fantasy:
Publishers (US)**
Bantam Dell Publishing Group,
    173
DAW Books, Inc., 175
Prometheus Books, 184
Random House Publishing
    Group, 185

**Science Fiction/Fantasy:
Small Presses**
Galactic Central Publications,
    228
Griffin Publishing Associates,
    229
Immanion Press, 230
Miller (Neil) Publications, 232
Pipers' Ash Ltd, 233

**Science Fiction/Fantasy:
Writers' Circles and
Workshops**
Scribo, 479

**Science Fiction/Fantasy:
Writers' Courses (UK and
Ireland)**
Liverpool, University of, 461

**Science Fiction/Fantasy:
Writers' Courses (US)**
Seton Hill University, 474

**Scientific & Technical:
Literary Agents (US)**
Mews Books Ltd, 286

**Scientific & Technical:
Magazines**
London Review of Books, 333

**Scientific & Technical:
Professional Associations**
International Association of
    Scientific, Technical &
    Medical Publishers, 539

**Scientific & Technical:
Publishers (European)**
Akadémiai Kiadó, 166
Polish Scientific Publishers
    PWN, 169
Presses Universitaires de France
    (PUF), 164
Springer-Verlag GmbH, 160

**Scientific & Technical:
Publishers (International)**
Macmillan Publishers Australia
    Pty Ltd, 191

**Scientific & Technical:
Publishers (UK)**
Elsevier Ltd, 92
Reed Elsevier Group plc, 133
Taylor & Francis, 143
Wiley Europe Ltd, 149
Wiley-Blackwell, 149
WIT Press, 150

**Scientific & Technical:
Publishers (US)**
California Press, University of,
    174
Prometheus Books, 184
Wiley (John) & Sons, Inc., 190

**Scotland/Scottish Interest:
Arts Councils and Regional
Offices**
Scottish Arts Council, 573

**Scotland/Scottish Interest:
Bursaries, Fellowships and
Grants**
Scottish Arts Council Creative
    Scotland Awards, 486
Scottish Arts Council Writers'
    Bursaries, 486
Scottish Book Trust New
    Writers' Awards, 486

**Scotland/Scottish Interest:
Library Services**
Mitchell Library, The, 590
National Library of Scotland,
    590

**Scotland/Scottish Interest:
Literary Agents (UK)**
McAra (Duncan), 264

Scotland/Scottish Interest: Literary Agents (US)
Gusay (Charlotte) Literary Agency, The, 284

Scotland/Scottish Interest: Literary Societies
Burns (Robert) World Federation Ltd, The, 554

Scotland/Scottish Interest: Magazines
Big Issue in Scotland, The, 300
Chapman, 306
Edinburgh Review, 317
List, The, 332
Lothian Life, 334
New Shetlander, The, 340
New Writing Scotland, 341
Scotland in Trust, 352
Scots Magazine, The, 352
Scottish Farmer, The, 352
Scottish Field, 353
Scottish Home & Country, 353
Times Educational Supplement Scotland, The, 361

Scotland/Scottish Interest: Organizations of Interest to Poets
Scottish Pamphlet Poetry, 223
Scottish Poetry Library, 223

Scotland/Scottish Interest: Picture Libraries
Baxter (Colin) Photography Limited, 603
Brown (Hamish) Scottish Photographic, 604
Collections, 605
Country Matters Picture Library, 606
Hardley (Dennis) Photography, Scottish Photo Library, 611
Henderson (Jim) Photography, 611
National Galleries of Scotland Picture Library, 617
Still Digital, 624
Stockscotland.com, 624
Williams (David) Picture Library, 626

Scotland/Scottish Interest: Poetry Magazines
Poetry Reader, 214
Poetry Scotland, 215

Scotland/Scottish Interest: Poetry Presses
Akros Publications, 199

Diehard Publishers, 200

Scotland/Scottish Interest: Prizes
Berry ((David) Prize, 490
Kelpies Prize, 506
Saltire Literary Awards, The, 519
Scottish History Book of the Year, 519
Sundial Scottish Arts Council Book of the Year Awards, 521

Scotland/Scottish Interest: Professional Associations
Association for Scottish Literary Studies, 527
Association of Scottish Motoring Writers, 529
CILIP in Scotland (CILIPS), 534
Publishing Scotland, 545
Scottish Book Trust, 546
Scottish Daily Newspaper Society, 546
Scottish Newspaper Publishers Association, 546
Scottish Print Employers Federation, 546
Scottish Screen, 547
Society of Authors in Scotland, The, 547

Scotland/Scottish Interest: Publishers (UK)
Acair Ltd, 64
Appletree Press Ltd, 68
Birlinn Ltd, 73
Black Ace Books, 73
Brown, Son & Ferguson, Ltd, 78
Canongate Books Ltd, 80
Chapman Publishing, 83
Edinburgh University Press, 91
Floris Books, 95
Luath Press Ltd, 112
Mercat Press, 116
NMS Enterprises Limited – Publishing, 120
Scottish Cultural Press/Scottish Children's Press, 137
Wilson (Neil) Publishing Ltd, 149

Scotland/Scottish Interest: Theatre Producers
Dundee Repertory Theatre, 440
Royal Lyceum Theatre Company, 447
7:84 Theatre Company Scotland, 448
Theatre Workshop Edinburgh, 450
Traverse Theatre, 450

Scouting & Guiding: Magazines
Guiding magazine, 325
Scouting Magazine, 353

Self Help: Audio Books
BBC Audiobooks Ltd, 238
HarperCollins AudioBooks, 239
Simon & Schuster Audio, 241

Self Help: Literary Agents (UK)
Graham Maw Christie Literary Agency, 258
Kelly (Frances), 262
Power (Shelley) Literary Agency Ltd, 267
Turnbull (Jane), 271
Watson, Little Ltd, 272

Self Help: Literary Agents (US)
Dijkstra (Sandra) Literary Agency, 282
Fredericks (Jeanne) Literary Agency, Inc., 283
Gusay (Charlotte) Literary Agency, The, 284
Konner (Linda) Literary Agency, 285
Manus & Associates Literary Agency, Inc., 286
Michaels (Doris S.) Literary Agency Inc., 286
Naggar (Jean V.) Literary Agency, The, 287
Nelson (B.K.) Literary Agency, 287
Parks (Richard) Agency, 287
Quicksilver Books, Literary Agents, 288
Schulman (Susan), A Literary Agency, 290
Scribblers House® LLC Literary Agency, 290
Snell (Michael) Literary Agency, 290
Wales Literary Agency, Inc., 291
Zeckendorf (Susan) Associates, Inc., 292

Self Help: Miscellany
Sammarco (Leda), 641

Self Help: Publishers (European)
BZZTÔH (Uitgeverij) BV, 167
Grijalbo, 170
Lannoo, Uitgeverij, 160
Larousse (Éditions), 163
Random House Mondadori, 170

Self Help: Publishers (International)
Allen & Unwin Pty Ltd, 191
Hachette Livre Australia, 191
Hind Pocket Books Pvt. Ltd, 195
Jaico Publishing House, 195
Orient Paperbacks, 195
Pan Macmillan Australia Pty Ltd, 192
Raupo Publishing (NZ) Ltd, 197
Struik Publishers (Pty) Ltd, 198

Self Help: Publishers (UK)
Apex Publishing Ltd, 67
Austin & Macauley Publishers Limited, 70
Brealey (Nicholas) Publishing, 77
Constable & Robinson Ltd, 85
Foulsham Publishers, 95
Hachette Livre UK, 100
Infinite Ideas, 107
JR Books Ltd, 108
Olympia Publishers, 122
Oneworld Publications, 122
Pegasus Elliot Mackenzie Publishers Ltd, 125
Piatkus Books, 127
Random House Group Ltd, The, 131
Simon & Schuster UK Limited, 139
Society for Promoting Christian Knowledge (SPCK), 140
Summersdale Publishers Ltd, 142
Zambezi Publishing Ltd, 151

Self Help: Publishers (US)
Bantam Dell Publishing Group, 173
Llewellyn Publications, 180
Rosen Publishing Group, Inc., The, 185
Shambhala Publications, Inc., 186
St Martin's Press LLC, 186

Self Help: Small Presses
Contact Publishing Ltd, 226
Need2Know, 232
Parapress, 233

Self Help: UK Packagers
Elwin Street Productions, 243
Ivy Press Limited, The, 244

Sexuality/Sex: Literary Agents (UK)
Graham Maw Christie Literary Agency, 258

Sexuality/Sex: Literary Agents (US)
Scribblers House® LLC Literary Agency, 290
Snell (Michael) Literary Agency, 290

Sexuality/Sex: Magazines
Maxim, 335
Men's Health, 335

Sexuality/Sex: Publishers (UK)
Free Association Books Ltd, 96

Short Stories: Audio Books
CSA Word, 238

Short Stories: Bursaries, Fellowships and Grants
Tom-Gallon Trust Award and the Olive Cook Prize, 487

Short Stories: Electronic Publishing and Other Services
Arbour E-Books, 247
www.ABCtales.com, 247

Short Stories: Festivals
National Association of Writers' Groups (NAWG) Open Festival of Writing, 636

Short Stories: Literary Agents (UK)
Feldstein Agency, The, 256

Short Stories: Magazines
Best, 300
Chapman, 306
Crimewave, 312
Darts World, 314
Fix, The, 320
Flash, 320
Irish Pages, 329
London Magazine, The, 333
Markings, 334
New Shetlander, The, 340
New Writing Scotland, 341
Notes from the Underground, 341
People's Friend, The, 344
QWF Magazine, 348
Random Acts of Writing, 350
Reader, The, 350
Woman's Weekly, 368
Woman's Weekly Fiction Special, 368
Writers' Forum incorporating World Wide Writers, 369

Yours Magazine, 371

Short Stories: Poetry Magazines
Brittle Star, 209
Current Accounts, 209
Engine, The, 210
Irish Pages, 211
Krax, 211
Open Wide Magazine, 213
Quiet Feather, The, 215
Scriptor, 216
Springboard, 217
Stinging Fly, The, 217
Understanding, 218

Short Stories: Poetry Presses
Dionysia Press, 201
DogHouse, 201
Dream Catcher, 201
Forward Press, 202
Honno Welsh Women's Press, 202
I*D Books, 203
Mudfog Press, 203
Penniless Press, 204

Short Stories: Prizes
BBC National Short Story Prize, The, 489
Bridport Prize, The, 492
Chaplin (Sid) Short Story Competition, 494
Chapter One Promotions International Short Story Competition, 495
Commonwealth Writers Prize, The, 496
Edge Hill Prize for the Short Story, 499
Fish Fiction Prizes, 500
HISSAC Annual Short Story Award, 504
Hubbard's (L. Ron) Writers of the Future Contest, 504
Legend Writing Award, 508
Mere Literary Festival Open Competition, 511
Momaya Press Short Story Writing Competition, 512
O'Connor (Frank) International Short Story Award, 514
O'Faolain (Sean) Short Story Prize, 514
Orange Broadband Prize for Fiction/Orange Broadband Award for New Writers, 514
Pritchett (V.S.) Memorial Prize, 516
Prose & Poetry Prizes, 516

Southport Writers' Circle Open Short Story Competition, 520
Sundial Scottish Arts Council Book of the Year Awards, 521
Thomas (Dylan) Prize, The, 521
Winchester Writers' Conference Prizes, 524
Writers Bureau Poetry and Short Story Competition, The, 525
Yeovil Literary Prize, 525

Short Stories: Publishers (International)
HarperCollins Publishers India Ltd, 195
Queensland Press, University of, 192

Short Stories: Publishers (Ireland)
Poolbeg Press Ltd, 158

Short Stories: Publishers (UK)
Chapman Publishing, 83
Honno Welsh Women's Press, 105
Parthian, 124

Short Stories: Publishers (US)
Iowa Press, University of, 179
Missouri Press, University of, 181

Short Stories: Small Presses
Edgewell Publishing, 227
Fand Music Press, 228
Miller (Neil) Publications, 232
Pipers' Ash Ltd, 233
Tartarus Press, 235
Tindal Street Press Ltd, 236

Short Stories: Television and Radio
BBC Radio Cumbria, 401
BBC Somerset, 404
Downtown Radio, 410

Short Stories: Writers' Circles and Workshops
Short Story Writers' Folio/ Children's Novel Writers' Folio, 479

Short Stories: Writers' Courses (UK and Ireland)
Bath Spa University, 462
City Lit, 458
City University, 458

Derby, University of, 454
Exeter, University of, 455
Fiction Writing Workshops and Tutorials, 459
Fire in the Head Courses & Writers' Services, 455
INKwell Writers' Workshops, 466
London School of Journalism, 459
Manchester, University of, 460
Missenden Abbey, 453
National Extension College, 453
Sheffield, University of, 465
Sussex, University of, 463
Writer's Academy, The, 466
Writers Bureau, The, 461

Short Stories: Writers' Courses (US)
Colorado State University, 470
Florida International University, 470
San Francisco State University, 470

Show Business: Literary Agents (UK)
Anderson (Darley) Literary, TV & Film Agency, 249
Chilcote (Judith) Agency, 253
Futerman, Rose & Associates, 257
London Independent Books, 263
Lownie (Andrew) Literary Agency, 263

Show Business: Literary Agents (US)
Dystel & Goderich Literary Management, 283
Konner (Linda) Literary Agency, 285
Nelson (B.K.) Literary Agency, 287
Schiavone Literary Agency, Inc., 289

Show Business: Magazines
Closer, 309
Glamour, 323
Heat, 326
Hello!, 326
Hotline, 328
InStyle, 329
New!, 341
Now, 341
OK! Magazine, 342
Reveal, 351
Weekly News, 366

Show Business: Picture Libraries
BFI Stills Sales, 603
Capital Pictures, 604
Corbis Images, 606
Famous Pictures & Features Agency, 608
Kobal Collection, The, 614
Lebrecht Music and Arts, 614
Monitor Picture Library, 616
Photoshot, 619
Retna Pictures Ltd, 621
Rex Features Ltd, 621
Vin Mag Archive Ltd, 626

Show Business: Publishers (UK)
HarperCollins Publishers Ltd, 102
Lewis (Dewi) Publishing, 110
Macmillan Publishers Ltd, 113
Random House Group Ltd, The, 131Random House Group Ltd, The,

Social Work: Publishers (UK)
Kingsley (Jessica) Publishers Ltd, 109

Socialism: Literary Agents (US)
Gusay (Charlotte) Literary Agency, The, 284

Sociology/Social Sciences: Bursaries, Fellowships and Grants
Blundell (K.) Trust, The, 482
British Academy Small Research Grants, 482

Sociology/Social Sciences: Library Services
British Library Humanities Reading Room, 577
British Library Social Sciences & Official Publications, 579
Kent County Central Library, 587
Senate House Library, University of London, 595
Sheffield Libraries, Archives and Information, 595
Women's Library, The, 599

Sociology/Social Sciences: Literary Agents (UK)
Feldstein Agency, The, 256

**Sociology/Social Sciences:
Literary Agents (US)**
Bleecker Street Associates, Inc.,
280
Lowenstein–Yost Associates
Inc., 285
Michaels (Doris S.) Literary
Agency Inc., 286
Schulman (Susan), A Literary
Agency, 290

**Sociology/Social Sciences:
Magazines**
Big Issue Cymru, The, 300
Big Issue in Scotland, The, 300
Big Issue in the North, The, 300
Big Issue South West, The, 300
Big Issue, The, 300

**Sociology/Social Sciences:
Picture Libraries**
Hoffman (David) Photo Library,
612
Illustrated London News
Picture Library, 613
Photofusion, 619
Popperfoto.com, 620
Report Digital, 621
Still Pictures' Whole Earth
Photolibrary, 624

**Sociology/Social Sciences:
Publishers (European)**
Adelphi Edizioni SpA, 166
Alianza Editorial SA, 169
Atlantis Ltd, 161
Beck (C.H., OHG) Verlag, 164
Bruna (A.W.) Uitgevers BV, 167
Brána a.s., 161
Calmann-Lévy (Éditions), 163
Civilização Editora, 169
Dokorán Ltd, 161
Fayard (Librarie Arthème), 163
Gyldendal, 161
Hoffmann und Campe Verlag
GmbH, 165
Kok (Uitgeversmaatschappij J.
H.) BV, 168
Minuit (Les Éditions de) SA, 164
Mulino (Societe editrice il), 167
Mursia (Gruppo Ugo) Editore
SpA, 167
Presses Universitaires de France
(PUF), 164
Random House Mondadori, 170
Springer Science+Business
Media BV, 168
Todariana Editrice, 167
Tusquets Editores, 170
Universitaire Pers Leuven, 160

**Sociology/Social Sciences:
Publishers (International)**
Allen & Unwin Pty Ltd, 191
Auckland University Press, 196
Jaico Publishing House, 195
Kwa-Zulu–Natal Press,
University of, 197
Melbourne University
Publishing Ltd, 192
Munshiram Manoharlal
Publishers Pvt Ltd, 195
New Star Books Ltd, 194

**Sociology/Social Sciences:
Publishers (Ireland)**
Blackhall Publishing Ltd, 153
Veritas Publications, 159

**Sociology/Social Sciences:
Publishers (UK)**
ABC-CLIO (UK), 64
Ashgate Publishing Ltd, 69
Austin & Macauley Publishers
Limited, 70
Berg Publishers, 72
Boyars (Marion) Publishers
Ltd, 76
British Academy, The, 77
Calder Publications Ltd, 79
Camden Press Ltd, 80
Continuum International
Publishing Group Ltd, The,
85
Currey (James) Publishers, 87
Floris Books, 95
Free Association Books Ltd, 96
Harvard University Press, 103
Liverpool University Press, 112
Manchester University Press,
115
McGraw-Hill Education, 113
Open Gate Press, 123
Open University Press, 123
Policy Press, The, 128
Polity Press, 128
Routledge, 134
Sage Publications, 135
Souvenir Press Ltd, 140
Trentham Books Ltd, 145
Verso, 147
Wiley-Blackwell, 149

**Sociology/Social Sciences:
Publishers (US)**
ABC–CLIO (US), 172
Charlesbridge Publishing, 174
Eerdmans Publishing Company,
176
Harvard University Press, 177
Kentucky, The University Press
of, 179
Minnesota Press, University
of, 181
Perseus Books Group, 184
Rutgers University Press, 185
Sharpe (M.E.), Inc., 186
Stanford University Press, 187
State University of New York
Press, 187
Texas Press, University of, 188
Transaction Publishers Ltd, 189
Virginia Press, University of, 189

**Sociology/Social Sciences:
Small Presses**
Superscript, 235

Sound: Library Services
British Library Sound Archive,
579
Imperial War Museum, 586

**South America: Literary
Agents (US)**
Gusay (Charlotte) Literary
Agency, The, 284

**South America: Picture
Libraries**
Angel (Heather)/Natural
Visions, 602
Cordaiy (Sylvia) Photo Library,
606
Cunningham (Sue)
Photographic, 606
ffotograff, 608
Garden and Wildlife Matters
Photo Library, 609
Grace (Martin and Dorothy),
611
South American Pictures, 623
Tropix Photo Library, 625

**Space & Cosmology:
Magazines**
Flight International, 320

**Space & Cosmology: Picture
Libraries**
Aspect Picture Library Ltd, 602
Galaxy Picture Library, 609
Genesis Space History Photo
Library, 610
Moving Image
Communications, 617
Oxford Scientific (OSF), 619
Science Photo Library, 622

**Space & Cosmology:
Publishers (US)**
Krieger Publishing Co., 179

Spain & Hispanic Interest: Library Services
Instituto Cervantes, 586

Spain & Hispanic Interest: Literary Agents (US)
Gusay (Charlotte) Literary Agency, The, 284

Spain & Hispanic Interest: Picture Libraries
Grace (Martin and Dorothy), 611
Peerless (Ann & Bury) Picture Library, 619
Williams (David) Picture Library, 626

Spain & Hispanic Interest: Publishers (UK)
Oxbow Books, 124

Spain & Hispanic Interest: Publishers (US)
Nevada Press, University of, 182

Spain & Hispanic Interest: Small Presses
Five Leaves Publications, 228

Sport & Fitness: Book Clubs
Readers' Union Ltd, 293

Sport & Fitness: Festivals
Carnegie Sporting Words Festival, 631

Sport & Fitness: Film, TV and Radio Producers
Buccaneer Films, 419
Cleveland Productions, 420
Omnivision, 429
Stirling Film & TV Productions Limited, 433
Sunset + Vine Productions Ltd, 434

Sport & Fitness: Library Services
Sheffield Libraries, Archives and Information, 595

Sport & Fitness: Literary Agents (Ireland)
Williams (Jonathan) Literary Agency, 279

Sport & Fitness: Literary Agents (UK)
Anderson (Darley) Literary, TV & Film Agency, 249

Bell Lomax Moreton Agency, The, 250
Brown (Jenny) Associates, 251
Luxton Harris Ltd, 263
Pawsey (John), 266
Simmons (Jeffrey), 269

Sport & Fitness: Literary Agents (US)
Bleecker Street Associates, Inc., 280
Gusay (Charlotte) Literary Agency, The, 284
Michaels (Doris S.) Literary Agency Inc., 286
Rosenberg Group, The, 289
Ware (John A.) Literary Agency, 291

Sport & Fitness: Magazines
ACE Tennis Magazine, 294
Athletics Weekly, 297
Boxing Monthly, 302
Cycle Sport, 313
Cycling Weekly, 313
Good Ski Guide, The, 324
i-D Magazine, 328
Loaded, 333
Racing Post, 349
Rugby World, 351
Runner's World, 351
Running Fitness, 351
Skier and Snowboarder Magazine, The, 356
Swimming Times Magazine, 359
World Ski Guide, 369

Sport & Fitness: News Agencies
South Yorkshire Sport, 393

Sport & Fitness: PR Consultants
Frazer (Cathy) PR, 277

Sport & Fitness: Picture Libraries
actionplus sports images, 601
Cole (Michael) Camerawork, 605
Country Life Picture Library, 606
Eagar (Patrick) Photography, 607
Forest Life Picture Library, 609
Getty Images, 610
Howes (Chris)/Wild Places Photography, 612
London Aerial Photo Library, 615
Mirrorpix, 616
Monitor Picture Library, 616

Popperfoto.com, 620
PPL (Photo Agency) Ltd, 620
Professional Sport UK Ltd, 620
Skishoot–Offshoot, 623
Stockfile, 624
Totterdell (B.M.) photography, 625
Vin Mag Archive Ltd, 626

Sport & Fitness: Prizes
BBCFour Samuel Johnson Prize for Non-Fiction, The, 490
British Sports Book Awards, 494
Hill (William) Sports Book of the Year, 504
Irish Book Awards, 506

Sport & Fitness: Professional Associations
Sports Journalists' Association of Great Britain, 549

Sport & Fitness: Publishers (European)
Brána a.s., 161
BZZTÔH (Uitgeverij) BV, 167
Calmann-Lévy (Éditions), 163
Larousse (Éditions), 163
Mursia (Gruppo Ugo) Editore SpA, 167
Olympia Publishing Co. Inc., 161
Spectrum (Uitgeverij Het) BV, 168
Sperling e Kupfer Editori SpA, 167
Table Ronde (Les Editions de la), 164

Sport & Fitness: Publishers (International)
Fenn (H.B.) & Company Ltd, 193
Fitzhenry & Whiteside Limited, 193
Hachette Livre Australia, 191
Hachette Livre New Zealand Ltd, 196
HarperCollins Publishers India Ltd, 195
Melbourne University Publishing Ltd, 192

Sport & Fitness: Publishers (Ireland)
Currach Press, 155
Hachette Books Ireland, 155
Maverick House Publishers, 156
O'Brien Press Ltd, The, 157

Sport & Fitness: Publishers (UK)

Anova Books, 66
Apex Publishing Ltd, 67
Ashgrove Publishing, 69
Aurum Press Ltd, 70
Austin & Macauley Publishers Limited, 70
AuthorHouse UK, 70
Black & White Publishing Ltd, 73
Black (A.&C.) Publishers Ltd, 73
Blackstaff Press Ltd, 74
Breedon Books Publishing Co. Ltd, The, 77
Carlton Publishing Group, 81
Crowood Press Ltd, The, 87
Davies (Christopher) Publishers Ltd, 88
Fort Publishing Ltd, 95
Hachette Livre UK, 100
HarperCollins Publishers Ltd, 102
JR Books Ltd, 108
Lewis (Dewi) Publishing, 110
Little, Brown Book Group UK, 111
Luath Press Ltd, 112
Macmillan Publishers Ltd, 113
Mainstream Publishing Co. (Edinburgh) Ltd, 114
Methuen Publishing Ltd, 116
Milo Books Limited, 117
New Holland Publishers (UK) Ltd, 119
Octopus Publishing Group, 121
Pennant Books Ltd, 126
Quiller Press, 130
Random House Group Ltd, The, 131
Routledge, 134
RYA (Royal Yachting Association), 135
Simon & Schuster UK Limited, 139
Souvenir Press Ltd, 140
SportsBooks Limited, 141
Transworld Publishers Ltd, 145
Wiley Nautical, 149

Sport & Fitness: Publishers (US)

Barron's Educational Series, Inc., 173
McFarland & Company, Inc., Publishers, 180
Nebraska Press, University of, 182
Perseus Books Group, 184
Potomac Books, Inc., 184
Stackpole Books, 187
Syracuse University Press, 188

Sport & Fitness: Small Presses

Witan Books, 237

Sport & Fitness: Television and Radio

BBC Sport, 397
talkSPORT, 410

Sport & Fitness: UK Packagers

Amber Books Ltd, 242
Stonecastle Graphics Ltd, 246

Technology: Bursaries, Fellowships and Grants

Economist/Richard Casement Internship, The, 483
PAWS (Public Awareness of Science) Drama Fund, 485

Technology: Library Services

British Library Science, Technology and Business Collections, 579
Science Museum Library, 595
Sheffield Libraries, Archives and Information, 595

Technology: Magazines

Engineer, The, 317
Stuff, 358

Technology: Picture Libraries

PictureBank Photo Library Ltd, 620
Science & Society Picture Library, 622
Science Photo Library, 622

Technology: Professional Associations

International Association of Scientific, Technical & Medical Publishers, 539
Society for Technical Communications (STC), 547

Technology: Publishers (European)

Hanser (Carl) Verlag GmbH & Co. KG, 165
Larousse (Éditions), 163
Nathan (Les Éditions), 164
Pearson Education (Benelux), 168
Random House Mondadori, 170
Springer GmbH, 165
Universitaire Pers Leuven, 160

Technology: Publishers (International)

Macmillan Publishers Australia Pty Ltd, 191
Melbourne University Publishing Ltd, 192

Technology: Publishers (Ireland)

Royal Dublin Society, 158

Technology: Publishers (UK)

Artech House, 69
Atlantic Europe Publishing Co. Ltd, 69
NMS Enterprises Limited – Publishing, 120
ProQuest, 129
Sage Publications, 135
Springer-Verlag London Ltd, 141
Taylor & Francis, 143
Wiley-Blackwell, 149
Woodhead Publishing Ltd, 150

Technology: Publishers (US)

Krieger Publishing Co., 179

Teenagers: Book Clubs

Letterbox Library, 293

Teenagers: Literary Agents (UK)

Bonomi (Luigi) Associates Limited (LBA), 251
Conville & Walsh Limited, 254
Fitch (Laurence) Ltd, 256
Futerman, Rose & Associates, 257
Green (Annette) Authors' Agency, 258
London Independent Books, 263
MacMullen (Eunice) Ltd, 264
Manson (Sarah) Literary Agent, 264

Teenagers: Literary Agents (US)

Altshuler (Miriam) Literary Agency, 280
Braun (Barbara) Associates, Inc., 281
Browne (Pema) Ltd, 281
Carvainis (Maria) Agency, Inc., 281
Clark (William) Associates, 282
Dunham Literary Inc., 283
Gusay (Charlotte) Literary Agency, The, 284
Kirchoff/Wohlberg, Inc., 284

Lowenstein–Yost Associates Inc., 285
Manus & Associates Literary Agency, Inc., 286
Mews Books Ltd, 286
Morhaim (Howard) Literary Agency, 287
Picard (Alison J.) Literary Agent, 287
Stimola Literary Studio, LLC, 291
Treimel (S©ott) NY, 291
Writers House, LLC, 292

Teenagers: Magazines
Bliss Magazine, 302
Mizz, 336
Shout Magazine, 355
Sugar Magazine, 359
Top of the Pops Magazine, 362
Young Voices, 370

Teenagers: Poetry Magazines
Wordsmith, 218

Teenagers: Prizes
Angus Book Award, 488
Booktrust Teenage Prize, 492
Lindgren (Astrid) Memorial Award for Literature, The, 508

Teenagers: Publishers (European)
Bertrand Editora Lda, 169
Borgens Forlag A/S, 161
Caminho (Editorial) SARL, 169
Cappelen Damm AS, 168
Cappelli Editore, 166
Civilização Editora, 169
Deutscher Taschenbuch Verlag GmbH & Co. KG, 164
Fischer (S.) Verlag GmbH, 164
Forlaget Forum, 161
Gallimard (Éditions), 163
Gradiva–Publicações S.A., 169
Groupo Anaya, 169
Gyldendal, 161
Horizonte (Livros) Lda, 169
Karisto Oy, 162
Kok (Uitgeversmaatschappij J. H.) BV, 168
Lannoo, Uitgeverij, 160
Larousse (Éditions), 163
Magnard (Éditions), 163
Molino SL (Editorial), 170
Mondadori (Arnoldo) Editore SpA, 167
Mursia (Gruppo Ugo) Editore SpA, 167
Móra Könyvkiadö Zrt., 166

Nasza Ksiegarnia Publishing House, 168
Nathan (Les Éditions), 164
Otava Publishing Co. Ltd, 162
Rabén och Sjögren Bokförlag, 171
Random House Mondadori, 170
Rizzoli Editore, 167
Sperling e Kupfer Editori SpA, 167
Standaard Uitgeverij, 160
Tammi Publishers, 162
Ueberreuter (Verlag Carl) GmbH, 160
Unieboek (Uitgeverij) BV, 168
Verbo (Editorial) SA, 169
WSOY (Werner Söderström Osakeyhtiö), 162
Wählströms (B.) Bokförlag AB, 171

Teenagers: Publishers (International)
Firefly Books Ltd, 193
Fitzhenry & Whiteside Limited, 193
Queensland Press, University of, 192
Random House of Canada Ltd, 194
Tundra Books, 194

Teenagers: Publishers (Ireland)
An Gúm, 153
Cló Iar-Chonnachta, 154
O'Brien Press Ltd, The, 157

Teenagers: Publishers (UK)
Barrington Stoke, 71
Blackstone Publishers, 74
Janus Publishing Company Ltd, 108
Little, Brown Book Group UK, 111
Meadowside Children's Books, 115
Usborne Publishing Ltd, 147
Women's Press, The, 150

Teenagers: Publishers (US)
Boyds Mills Press, 174
Dial Books for Young Readers, 175
Hachette Book Group USA, 177
HarperCollins Publishers, Inc., 177
Holiday House, Inc., 178
Putnam's (G.P.) Sons (Children's), 185
Rosen Publishing Group, Inc., The, 185

Scholastic Library Publishing, 186
Simon & Schuster Children's Publishing, 187
Walker & Co., 190

Teenagers: UK Packagers
Working Partners Ltd, 246

Telecommunications: Publishers (UK)
Artech House, 69

Television: Bursaries, Fellowships and Grants
PAWS (Public Awareness of Science) Drama Fund, 485

Television: Library Services
BBC Written Archives Centre, 575
BFI National Library, 575
Westminster Reference Library, 598

Television: Literary Agents (UK)
Ableman (Sheila) Literary Agency, 249
Agency (London) Ltd, The, 249
Anderson (Darley) Literary, TV & Film Agency, 249
Blake Friedmann Literary Agency Ltd, 251
Brodie (Alan) Representation Ltd, 251
Capel & Land Ltd, 252
Casarotto Ramsay and Associates Ltd, 252
Chilcote (Judith) Agency, 253
Clowes (Jonathan) Ltd, 254
Colin (Rosica) Ltd, 254
Curtis Brown Group Ltd, 254
Daish (Judy) Associates Ltd, 255
Dench Arnold Agency, The, 255
Fitch (Laurence) Ltd, 256
Foster (Jill) Ltd, 257
Futerman, Rose & Associates, 257
Glass (Eric) Ltd, 257
Graham Maw Christie Literary Agency, 258
Hall (Rod) Agency Limited, The, 259
Hancock (Roger) Ltd, 259
Henser Literary Agency, 260
hhb agency ltd, 260
Hoskins (Valerie) Associates Limited, 260
Independent Talent Group Limited, 261
International Scripts, 261

Kass (Michelle) Associates, 262
London Independent Books, 263
Mann (Andrew) Ltd, 264
Marjacq Scripts Ltd, 264
MBA Literary Agents Ltd, 264
McLean (Bill) Personal Management, 264
Morris (William) Agency (UK) Ltd, 265
Narrow Road Company, The, 265
PFD, 266
Robinson Literary Agency Ltd, 267
Sayle Screen Ltd, 268
Sharland Organisation Ltd, The, 268
Sheil Land Associates Ltd, 268
Steinberg (Micheline) Associates, 270
Tennyson Agency, The, 270
TVmyworld, 271
United Agents Limited, 271
Ware (Cecily) Literary Agents, 272
Watt (A.P.) Ltd, 272

Television: Literary Agents (US)
Curtis Brown Ltd, 282
Freedman (Robert A.) Dramatic Agency, Inc., 283
Gusay (Charlotte) Literary Agency, The, 284
Hawkins (John) & Associates, Inc., 284
Scagnetti (Jack) Talent & Literary Agency, 289

Television: Magazines
All About Soap, 295
Empire, 317
Heat, 326
Radio Times, 349
Screen, 353
ScriptWriter Magazine, 354
Skymag, 356
Stage, 358
Total TV Guide, 362
TVTimes, 363
Weekly News, 366
What Satellite and Digital TV, 367

Television: Picture Libraries
Aquarius Library, 602
BFI Stills Sales, 603
Cinema Museum, The, 605
Photoshot, 619
Vin Mag Archive Ltd, 626

Television: Prizes
Kraszna-Krausz Book Awards, 507
Ustinov (Sir Peter) Television Scriptwriting Award, 522

Television: Professional Associations
BFI, 530
Irish Playwrights and Screenwriters Guild, 539
Pact, 542
Player–Playwrights, 543
Royal Television Society, 546
Scottish Screen, 547
Writers' Guild of Great Britain, The, 551

Television: Publishers (UK)
Austin & Macauley Publishers Limited, 70
BFI Publishing, 72
GMC Publications Ltd, 98
Hachette Livre UK, 100
HarperCollins Publishers Ltd, 102
Macmillan Publishers Ltd, 113
Manchester University Press, 115
Random House Group Ltd, The, 131
Titan Books, 145
Trident Press Ltd, 145

Television: Publishers (US)
Players Press, 184
Syracuse University Press, 188

Television: Writers' Circles and Workshops
Kops (Bernard) and Tom Fry Drama Writing Workshops, 478
Script Yorkshire, 479
Southwest Scriptwriters, 479

Television: Writers' Courses (UK and Ireland)
Bournemouth University, 455
Central School of Speech and Drama, The, 458
City University, 458
Community Creative Writing, 461
Glamorgan, University of, 467
Higham Hall College, 454
Leeds Metropolitan University, 464
Leeds, University of, 465
Liverpool, University of, 461
Missenden Abbey, 453

National Film & Television School, 453
Salford, University of, 460
Westminster, University of, 460
Writers Bureau, The, 461

Television: Writers' Courses (US)
American Film Institute, 469
Chicago State University, 471
Southern California, University of, 470

Textbooks: Literary Agents (US)
Educational Design Services, LLC, 283
Gusay (Charlotte) Literary Agency, The, 284
Herman (Jeff) Agency, LLC, The, 284
Rosenberg Group, The, 289

Textbooks: Publishers (European)
Aschehoug (H) & Co (W Nygaard), 168
Beck (C.H., OHG) Verlag, 164
Bulzoni Editore SRL, 166
Cappelen Damm AS, 168
Cappelli Editore, 166
Europa-América (Publicações) Lda, 169
Garzanti Libri SpA, 166
Giunti Editoriale SpA, 166
Gyldendal, 161
Gyldendal Norsk Forlag, 168
Hachette Livre (France), 163
Hanser (Carl) Verlag GmbH & Co. KG, 165
Horizonte (Livros) Lda, 169
Hüthig GmbH & Co. KG, 165
Klett (Ernst) Verlag GmbH, 165
Kok (Uitgeversmaatschappij J. H.) BV, 168
Larousse (Éditions), 163
Lettera, 161
Magnard (Éditions), 163
Mondadori (Arnoldo) Editore SpA, 167
Mulino (Societe editrice il), 167
Mursia (Gruppo Ugo) Editore SpA, 167
Naouka I Izkoustvo, 161
Natur och Kultur Bokförlaget, 171
Nyt Nordisk Forlag Arnold Busck A/S, 162
Olympia Publishing Co. Inc., 161
Otava Publishing Co. Ltd, 162

Pearson Education (Benelux), 168
Presses Universitaires de France (PUF), 164
Schønbergske (Det) Forlag, 162
Società Editrice Internazionale – SEI, 167
Springer-Verlag GmbH, 160
Vuibert (Librairie), 164
WSOY (Werner Söderström Osakeyhtiö), 162

Textbooks: Publishers (International)
Affiliated East West Press Pvt Ltd, 195
Chand (S.) & Co Ltd, 195
Jaico Publishing House, 195
LexisNexis New Zealand, 196
McGraw-Hill Australia Pty Ltd, 192
National Publishing House, 195
Nelson Education, 194
Queensland Press, University of, 192
Wiley (John) & Sons Australia Ltd, 193
Wiley (John) & Sons Canada Ltd, 194

Textbooks: Publishers (Ireland)
An Gúm, 153
Cork University Press, 155
Gill & Macmillan, 155

Textbooks: Publishers (UK)
Atlantic Europe Publishing Co. Ltd, 69
Cambridge University Press, 79
Chartered Institute of Personnel and Development (CIPD), 83
Colourpoint Books, 85
Donald (John) Publishers Ltd, 89
Elm Publications/Training, 92
Freeman (W. H.), 96
Hachette Livre UK, 100
HarperCollins Publishers Ltd, 102
Hodder Gibson, 105
Macmillan Publishers Ltd, 113
Manchester University Press, 115
Oxford University Press, 124
Pearson, 125
Pearson (Heinemann), 125
Psychology Press, 129
Reed Elsevier Group plc, 133
Routledge, 134
Ward Lock Educational Co. Ltd, 148

Wiley Europe Ltd, 149

Textbooks: Publishers (US)
Blackwell Publishing (US), 174
Houghton Mifflin Harcourt Publishing Company, 178
Michigan Press, The University of, 181
Norton (W.W.) & Company, Inc., 182
Pennsylvania Press, University of, 184
Perseus Books Group, 184
St Martin's Press LLC, 186
University Press of America, Inc., 189

Textbooks: Small Presses
Galore Park Publishing Ltd, 228
St James Publishing, 234

Textiles: Library Services
Worcestershire Libraries and Information Service, 600

Textiles: Magazines
Classic Stitches, 308
Embroidery, 317

Textiles: Picture Libraries
Christie's Images Ltd, 605
National Archives Image Library, The, 617
V&A Images, 625

Textiles: Publishers (UK)
King (Laurence) Publishing Ltd, 109
Woodhead Publishing Ltd, 150

Theatre: Bursaries, Fellowships and Grants
Pearson Playwrights' Scheme, 485
Ramsay (Peggy) Foundation, 485
Scottish Arts Council Writers' Bursaries, 486

Theatre: Festivals
Aspects Literature Festival, 629
Derbyshire Literature Festival, 632
International Playwriting Festival, 634
Royal Court Young Writers Programme, 638

Theatre: Film, TV and Radio Producers
Abstract Images, 417

Carnival (Films & Theatre) Ltd, 419
Heritage Theatre Ltd, 425
JAM Pictures and Jane Walmsley Productions, 426

Theatre: Library Services
Barbican Library, 574
Birmingham Library Services, 576
Westminster Reference Library, 598

Theatre: Literary Agents (UK)
Agency (London) Ltd, The, 249
Brodie (Alan) Representation Ltd, 251
Burkeman (Brie), 252
Colin (Rosica) Ltd, 254
Curtis Brown Group Ltd, 254
Daish (Judy) Associates Ltd, 255
Fitch (Laurence) Ltd, 256
Foster (Jill) Ltd, 257
Glass (Eric) Ltd, 257
Hall (Rod) Agency Limited, The, 259
Hancock (Roger) Ltd, 259
Henser Literary Agency, 260
Independent Talent Group Limited, 261
Mann (Andrew) Ltd, 264
MBA Literary Agents Ltd, 264
McLean (Bill) Personal Management, 264
Narrow Road Company, The, 265
Sayle Screen Ltd, 268
Sheil Land Associates Ltd, 268
Simmons (Jeffrey), 269
Steinberg (Micheline) Associates, 270
Tennyson Agency, The, 270
United Agents Limited, 271
Weinberger (Josef) Plays, 272

Theatre: Literary Agents (US)
Curtis Brown Ltd, 282
Freedman (Robert A.) Dramatic Agency, Inc., 283
Hogenson (Barbara) Agency, Inc., The, 284

Theatre: Magazines
Amateur Stage, 295
Chapman, 306
List, The, 332
New Theatre Quarterly, 340
Stage, 358

Theatre: Miscellany
Landeau (Caroline), 641

Theatre: Picture Libraries
ArenaPAL, 602
Cinema Museum, The, 605
Dominic Photography, 607
Jack (Robbie) Photography, 613
Mander (Raymond) & Joe
    Mitchenson Theatre
    Collection, The, 616
V&A Images, 625

Theatre: Poetry Magazines
Aesthetica, 208

Theatre: Prizes
Bargate (Verity) Award, 489
Devine (George) Award, 498
Drama Association of Wales
    Playwriting Competition,
    498
Fagon (Alfred) Award, The, 500
Meyer-Whitworth Award, 511
Morley (Sheridan) Prize, The,
    512
Society for Theatre Research
    Annual Theatre Book Prize,
    The, 520
Way (Brian) Award, 523
Wolff (Peter) Theatre Trust
    supports the Whiting
    Award, The, 524
Yale Drama Series/David C.
    Horn Prize, The, 525

**Theatre: Professional
Associations**
Independent Theatre Council,
    538
Irish Playwrights and
    Screenwriters Guild, 539
Personal Managers' Association
    Ltd, The, 543
Player–Playwrights, 543
Playwrights' Studio, Scotland,
    543
Writers' Guild of Great Britain,
    The, 551

**Theatre: Publishers
(European)**
Adelphi Edizioni SpA, 166
Civilização Editora, 169
Kiss József Kónyvkiadó, 166
Suhrkamp Verlag, 165
Tusquets Editores, 170

**Theatre: Publishers
(International)**
Currency Press Pty Ltd, 191
Munshiram Manoharlal
    Publishers Pvt Ltd, 195

Theatre: Publishers (UK)
Amber Lane Press Ltd, 66
Austin & Macauley Publishers
    Limited, 70
Black (A.&C.) Publishers Ltd, 73
Boyars (Marion) Publishers
    Ltd, 76
Cressrelles Publishing Co. Ltd,
    86
Crowood Press Ltd, The, 87
Hern (Nick) Books, 104
Manchester University Press,
    115
Oberon Books, 120
Richmond House Publishing
    Company Ltd, 134
Routledge, 134
Souvenir Press Ltd, 140

Theatre: Publishers (US)
Faber & Faber, Inc., 176
French (Samuel), Inc., 176
Michigan Press, The University
    of, 181
Northwestern University Press,
    182
Players Press, 184
Scarecrow Press Inc., 186

Theatre: Small Presses
Dramatic Lines, 227
Iolo, 230

**Theatre: Writers' Circles and
Workshops**
Kops (Bernard) and Tom Fry
    Drama Writing Workshops,
    478
New Writing South, 478
North West Playwrights
    (NWP), 478
Script Yorkshire, 479
Southwest Scriptwriters, 479
Sussex Playwrights' Club, 480

**Theatre: Writers' Courses
(UK and Ireland)**
Central School of Speech and
    Drama, The, 458
City University, 458
Dartington College of Arts, 455
Dingle Writing Courses, 466
Glamorgan, University of, 467
Hull, University of, 465
Leeds, University of, 465
Liverpool, University of, 461
Royal Holloway University of
    London, 463
Starz! – Film and Theatre
    Performing Arts, 464

**Theatre-in-Education:
Literary Agents (UK)**
Casarotto Ramsay and
    Associates Ltd, 252
Sharland Organisation Ltd, The,
    268

**Theatre-in-Education:
Theatre Producers**
Nottingham Playhouse
    Roundabout Theatre in
    Education, 444

**Theatre: Children's: Literary
Agents (UK)**
Casarotto Ramsay and
    Associates Ltd, 252
Narrow Road Company, The,
    265
Sharland Organisation Ltd, The,
    268

**Theatre: Children's: Theatre
Producers**
Borderline Theatre Company,
    439
Lyric Hammersmith, 443
Norwich Puppet Theatre, 444
Nottingham Playhouse
    Roundabout Theatre in
    Education, 444
Polka Theatre, 446
Theatre Workshop Edinburgh,
    450
Unicorn Theatre, 451
Whirligig Theatre, 451

**Theatre: Musicals: Literary
Agents (UK)**
Agency (London) Ltd, The, 249
Casarotto Ramsay and
    Associates Ltd, 252
Fitch (Laurence) Ltd, 256
Narrow Road Company, The,
    265
Sharland Organisation Ltd, The,
    268
Tennyson Agency, The, 270

**Theatre: Musicals: Picture
Libraries**
ArenaPAL, 602
Dominic Photography, 607

**Theatre: Musicals: Publishers
(US)**
Players Press, 184

**Theatre: Musicals: Small
Presses**
Dramatic Lines, 227

Theatre: Musicals: Theatre Producers
Concordance, 440
Finborough Theatre, 441
Horsecross Arts Ltd, 442
Newpalm Productions, 444
Nitro, 444
Nuffield Theatre, 445
Queen's Theatre, Hornchurch, 446
Really Useful Group Ltd, The, 446
Theatre Royal Plymouth & Drum Theatre Plymouth, 449
Theatre Royal Stratford East, 449
Whirligig Theatre, 451

Theatre: New Writing: Literary Agents (UK)
Agency (London) Ltd, The, 249
Casarotto Ramsay and Associates Ltd, 252
Narrow Road Company, The, 265
Sharland Organisation Ltd, The, 268
Steinberg (Micheline) Associates, 270
Tennyson Agency, The, 270
Weinberger (Josef) Plays, 272

Theatre: New Writing: Literary Agents (US)
Gusay (Charlotte) Literary Agency, The, 284

Theatre: New Writing: Theatre Producers
Alternative Theatre Company Ltd, 438
ATC, 438
Birmingham Repertory Theatre, 438
Bootleg Theatre Company, 439
Borderline Theatre Company, 439
Bristol Old Vic Theatre Company, 439
Clwyd Theatr Cymru, 439
Colchester Mercury Theatre Limited, 439
Coliseum Oldham, The, 439
Concordance, 440
Contact Theatre Company, 440
Crucible Theatre, 440
Druid Theatre Company, 440
Dundee Repertory Theatre, 440
Eastern Angles Theatre Company, 441
Exeter Northcott, 441

Finborough Theatre, 441
Hampstead Theatre, 442
Headlong Theatre, 442
Hull Truck Theatre Company, 442
Joseph (Stephen) Theatre, 442
Kenwright (Bill) Ltd, 443
Komedia, 443
Library Theatre Company, 443
Live Theatre Company, 443
London Bubble Theatre Company, 443
N.T.C. Touring Theatre Company, 445
New Vic Theatre, 444
Nitro, 444
Nottingham Playhouse Roundabout Theatre in Education, 444
Nuffield Theatre, 445
Octagon Theatre Trust Ltd, 445
Orange Tree Theatre, 445
Out of Joint, 445
Paines Plough, 446
Polka Theatre, 446
Queen's Theatre, Hornchurch, 446
Red Ladder Theatre Company, 446
Red Shift Theatre Company, 447
Ridiculusmus, 447
Royal Court Theatre, 447
Royal Exchange Theatre Company, 447
Royal Lyceum Theatre Company, 447
7:84 Theatre Company Scotland, 448
Sherman Cymru, 448
Show of Strength, 448
Soho Theatre Company, 448
Talawa Theatre Company Ltd, 449
Theatre Absolute, 449
Theatre Royal Plymouth & Drum Theatre Plymouth, 449
Theatre Royal Stratford East, 449
Theatre Royal Windsor, 449
Theatre Workshop Edinburgh, 450
Traverse Theatre, 450
Trestle Theatre Company, 450
Tricycle Theatre, 450
Tron Theatre Company, 450
Unicorn Theatre, 451
Warehouse Theatre, 451
Watford Palace Theatre, 451
West Yorkshire Playhouse, 451
White Bear Theatre Club, 451
Young Vic, The, 452

Theatre: West End: Literary Agents (UK)
Agency (London) Ltd, The, 249
Casarotto Ramsay and Associates Ltd, 252
Fitch (Laurence) Ltd, 256
Narrow Road Company, The, 265
Sharland Organisation Ltd, The, 268
Steinberg (Micheline) Associates, 270
Tennyson Agency, The, 270
Weinberger (Josef) Plays, 272

Theatre: West End: Theatre Producers
Codron (Michael) Plays Ltd, 439
Kenwright (Bill) Ltd, 443
Nimax Theatres Limited, 444
Really Useful Group Ltd, The, 446

Theology: Library Services
Catholic National Library, 580
Devon & Exeter Institution Library, 581
National Library of Scotland, 590
Williams's (Dr) Library, 599

Theology: Publishers (European)
Beck (C.H., OHG) Verlag, 164

Theology: Publishers (Ireland)
Church of Ireland Publishing, 154

Theology: Publishers (UK)
Ashgate Publishing Ltd, 69
Authentic Media, 70
Clarke (James) & Co., 84
Continuum International Publishing Group Ltd, The, 85
Epworth, 93
Janus Publishing Company Ltd, 108
Monarch Books, 117
Polity Press, 128
SCM – Canterbury Press Ltd, 136
Society for Promoting Christian Knowledge (SPCK), 140
St Pauls Publishing, 135

Theology: Publishers (US)
Eerdmans Publishing Company, 176

Paulist Press, 183
Tyndale House Publishers, Inc.,
189

Theology: Small Presses
Lindsey Press, The, 231

**Third Age: Electronic
Publishing and Other
Services**
50connect, 248

Third Age: Magazines
Choice, 307
Oldie, The, 342
Pensions World, 343
Saga Magazine, 352
Yours Magazine, 371

Third Age: Picture Libraries
Photofusion, 619

Third Age: Publishers (UK)
Age Concern Books, 65

**Third World: Literary Agents
(US)**
Gusay (Charlotte) Literary
Agency, The, 284

Third World: Magazines
Catholic Herald, The, 306
New Internationalist, 339

**Third World: Picture
Libraries**
Panos Pictures, 619
Still Pictures' Whole Earth
Photolibrary, 624

**Third World: Poetry
Magazines**
Poetic Hours, 214

Third World: Publishers (UK)
Currey (James) Publishers, 87
Zed Books Ltd, 151

**Thrillers & Suspense: Audio
Books**
Orion Audio Books, 240

**Thrillers & Suspense: Book
Clubs**
BCA (Book Club Associates),
293

**Thrillers & Suspense: Film,
TV and Radio Producers**
Anglo/Fortunato Films Ltd, 418

**Thrillers & Suspense:
Literary Agents (Ireland)**
Book Bureau Literary Agency,
The, 279

**Thrillers & Suspense:
Literary Agents (UK)**
Ampersand Agency Ltd, The,
249
Anderson (Darley) Literary, TV
& Film Agency, 249
Blake Friedmann Literary
Agency Ltd, 251
Bonomi (Luigi) Associates
Limited (LBA), 251
Dorian Literary Agency (DLA),
255
Feldstein Agency, The, 256
Fitch (Laurence) Ltd, 256
Gregory & Company Authors'
Agents, 258
Judd (Jane) Literary Agency, 262
Motley (Michael) Ltd, 265
Sheil Land Associates Ltd, 268
Sheldon (Caroline) Literary
Agency Ltd, 268
Sunflower Literary Agency, 270

**Thrillers & Suspense:
Literary Agents (US)**
Bleecker Street Associates, Inc.,
280
Carvainis (Maria) Agency, Inc.,
281
Dijkstra (Sandra) Literary
Agency, 282
Ellenberg (Ethan) Literary
Agency, 283
Gusay (Charlotte) Literary
Agency, The, 284
Lampack (Peter) Agency, Inc.,
285
Lescher & Lescher Ltd, 285
Lowenstein–Yost Associates
Inc., 285
Parks (Richard) Agency, 287
Picard (Alison J.) Literary
Agent, 287
Protter (Susan Ann) Literary
Agent, 288
Quicksilver Books, Literary
Agents, 288
Rittenberg (Ann) Literary
Agency, Inc., 288
Russell & Volkening, Inc., 289
Schulman (Susan), A Literary
Agency, 290
Snell (Michael) Literary Agency,
290
Spectrum Literary Agency, 290
Spitzer (Philip G.) Literary
Agency, Inc., 291

Ware (John A.) Literary Agency,
291
Writers House, LLC, 292

Thrillers & Suspense: Prizes
Crime Writers' Association
(The Ian Fleming Steel
Dagger), 497

**Thrillers & Suspense:
Publishers (European)**
Borgens Forlag A/S, 161
Bruna (A.W.) Uitgevers BV, 167
BZZTÔH (Uitgeverij) BV, 167
Denoël (Éditions), 163
Deutscher Taschenbuch Verlag
GmbH & Co. KG, 164
Flammarion Éditions, 163
Grasset & Fasquelle (Éditions),
163
Grijalbo, 170
Karisto Oy, 162
Longanesi (Casa Editrice) SpA,
167
Random House Mondadori, 170
Sperling e Kupfer Editori SpA,
167
Unieboek (Uitgeverij) BV, 168

**Thrillers & Suspense:
Publishers (International)**
Allen & Unwin Pty Ltd, 191
HarperCollins Publishers India
Ltd, 195
Pan Macmillan Australia Pty
Ltd, 192

**Thrillers & Suspense:
Publishers (UK)**
Bitter Lemon Press, 73
Blackstone Publishers, 74
Brown Skin Books, 78
Hachette Livre UK, 100
HarperCollins Publishers Ltd,
102
Macmillan Publishers Ltd, 113
Old Street Publishing Ltd, 121
Random House Group Ltd,
The, 131
Twenty First Century Publishers
Ltd, 146

**Thrillers & Suspense: Theatre
Producers**
Hiss & Boo Co. Ltd, The, 442
Theatre Royal Windsor, 449

**Thrillers & Suspense:
Writers' Courses (UK and
Ireland)**
London School of Journalism,
459

Trade Unions: Library Services
Birmingham Library Services, 576
Coventry Central Library, 581

Trade Unions: Magazines
Tribune, 363

Trade Unions: Picture Libraries
Report Digital, 621

Training: Film, TV and Radio Producers
Acrobat Television, 417
Channel Television Ltd, 420
Picardy Media & Communication, 430
Tandem TV & Film Ltd, 434
Video Enterprises, 436

Training: Picture Libraries
Education Photos, 607

Training: Publishers (European)
Nathan (Les Éditions), 164

Training: Publishers (UK)
Ashgate Publishing Ltd, 69
Chartered Institute of Personnel and Development (CIPD), 83
Crown House Publishing, 87
Elm Publications/Training, 92
Hobsons Plc, 105
Kogan Page Ltd, 109
Radcliffe Publishing Ltd, 131
Trotman, 146

Training: Publishers (US)
AMACOM Books, 172

Training: Small Presses
GSSE, 229

Translations: Literary Agents (UK)
Eady (Toby) Associates Ltd, 256
Henser Literary Agency, 260

Translations: Literary Agents (US)
Gusay (Charlotte) Literary Agency, The, 284

Translations: Magazines
Abraxas Unbound, 294
Irish Pages, 329

Translations: Poetry Magazines
Journal, The, 211
Modern Poetry in Translation, 212
Poetry Salzburg Review, 215

Translations: Poetry Presses
Anvil Press Poetry Ltd, 199
Dedalus Press, The, 200
Dionysia Press, 201
Egg Box Publishing, 201
Poetry Salzburg, 205
Spectacular Diseases, 207

Translations: Prizes
Dryden (John) Competition, The, 498
Independent Foreign Fiction Prize, The, 505
International IMPAC Dublin Literary Award, The, 505
Lawrie (Duncan) International Dagger, The, 508
Marsh Award for Children's Literature in Translation, 511
Oxford Weidenfeld Translation Prize, The, 515
Rossica Translation Prize, 517
Translators Association Awards, The, 522

Translations: Professional Associations
British Centre for Literary Translation, 531
Chartered Institute of Linguists (IoL), 534
Institute of Translation and Interpreting (ITI), 538
Irish Translators' & Interpreters' Association, 539
Translators Association, The, 549
Welsh Literature Abroad/ Llenyddiaeth Cymru Dramor, 550

Translations: Publishers (European)
Adelphi Edizioni SpA, 166
Aschehoug (H) & Co (W Nygaard), 168
BZZTÔH (Uitgeverij) BV, 167
Deutscher Taschenbuch Verlag GmbH & Co. KG, 164
Dokorán Ltd, 161
Forum (Bokförlaget) AB, 171
Gradiva–Publicações S.A., 169
Grasset & Fasquelle (Éditions), 163

Hanser (Carl) Verlag GmbH & Co. KG, 165
Hiperión (Ediciónes) SL, 170
Hüthig GmbH & Co. KG, 165
Kiss József Kónyvkiadó, 166
Meulenhoff (Uitgeverij J.M.) BV, 168
Naouka I Izkoustvo, 161
Otava Publishing Co. Ltd, 162
Seix Barral (Editorial) SA, 170
Seuil (Éditions du), 164
Springer-Verlag GmbH, 160
Tammi Publishers, 162
Tiden Norsk Forlag, 168
Tiderne Skifter Forlag A/S, 162
WSOY (Werner Söderström Osakeyhtiö), 162

Translations: Publishers (Ireland)
Cló Iar-Chonnachta, 154

Translations: Publishers (UK)
Anvil Press Poetry Ltd, 67
Arc Publications Ltd, 68
Arcadia Books, 68
Aurora Metro, 70
Bitter Lemon Press, 73
Boulevard Books & The Babel Guides, 76
Carcanet Press Ltd, 80
Dedalus Ltd, 89
Garnet Publishing Ltd, 97
Hesperus Press Limited, 104
Oberon Books, 120
Old Street Publishing Ltd, 121
Owen (Peter) Publishers, 123
Oxbow Books, 124
Parthian, 124
Pushkin Press Ltd, 129
Quartet Books, 130
Random House Group Ltd, The, 131
Seren, 137
Serpent's Tail, 138
Troubador Publishing Ltd, 146
Wingedchariot Press, 150

Translations: Publishers (US)
California Press, University of, 174
Columbia University Press, 175
Michigan Press, The University of, 181
Seven Stories Press, 186
Sharpe (M.E.), Inc., 186
Syracuse University Press, 188

Translations: Small Presses
Pipers' Ash Ltd, 233

Translations: Theatre Producers
ATC, 438
Dundee Repertory Theatre, 440
Gate Theatre Company Ltd, 441
Lyric Hammersmith, 443
Traverse Theatre, 450

Transport: Library Services
Birmingham Library Services, 576

Transport: Magazines
Buses, 305
Old Glory, 342

Transport: Picture Libraries
Alvey & Towers, 601
Cordaiy (Sylvia) Photo Library, 606
E&E Picture Library, 607
Evans (Mary) Picture Library, 607
Forsythe (Robert) Picture Library, 609
Hardley (Dennis) Photography, Scottish Photo Library, 611
Hutchison Picture Library, 612
London's Transport Museum Photographic Library, 615
MacQuitty International Photographic Collection, 616
National Railway Museum, 618
Popperfoto.com, 620
Science & Society Picture Library, 622

Transport: Publishers (European)
Springer GmbH, 165

Transport: Publishers (UK)
Allan (Ian) Publishing Ltd, 65
Anness Publishing Ltd, 66
Anova Books, 66
Brewin Books Ltd, 77
Colourpoint Books, 85
Compendium Publishing Ltd, 85
Countryside Books, 86
Haynes Publishing, 103
Henry (Ian) Publications Ltd, 104
Jane's Information Group, 108
Kogan Page Ltd, 109
TSO (The Stationery Office), 146

Transport: Small Presses
Nostalgia Collection, The, 232

Transport: UK Packagers
Amber Books Ltd, 242

Travel: Audio Books
Corgi Audio, 238
CSA Word, 238

Travel: Bursaries, Fellowships and Grants
Maugham (Somerset) Awards, The, 485
Travelling Scholarships, The, 487

Travel: Library Services
Bromley House Library, 580
London Library, The, 589

Travel: Literary Agents (Ireland)
Williams (Jonathan) Literary Agency, 279

Travel: Literary Agents (UK)
BookBlast Ltd, 251
Feldstein Agency, The, 256
Indepublishing CIA – Consultancy for Independent Authors, 261
Judd (Jane) Literary Agency, 262
London Independent Books, 263
McAra (Duncan), 264
Sayle Literary Agency, The, 268
Sheil Land Associates Ltd, 268
Sinclair-Stevenson, 269

Travel: Literary Agents (US)
Gusay (Charlotte) Literary Agency, The, 284

Travel: Magazines
Business Traveller, 305
Condé Nast Traveller, 310
Coop Traveller, 310
Good Holiday Magazine, 323
Good Motoring, 324
Hotline, 328
House & Garden, 328
In Britain, 328
Jersey Now, 330
Lady, The, 331
Loaded, 333
Motorcaravan Motorhome Monthly (MMM), 337
Mountain Magic, 338
Practical Caravan, 345
Sainsbury's Magazine, 352
SmartLife International, 356
Traveller, 363
Vogue, 364

Wallpaper, 365
Which Caravan, 367

Travel: Miscellany
Grossman (Susan) – Writing Workshops, 641
Landeau (Caroline), 641

Travel: News Agencies
Space Press News and Pictures, 393

Travel: Picture Libraries
Aspect Picture Library Ltd, 602
Brown (Hamish) Scottish Photographic, 604
Cleare (John)/Mountain Camera, 605
Corbis Images, 606
Cordaiy (Sylvia) Photo Library, 606
Craven (Philip) Worldwide Photo-Library, 606
Eye Ubiquitous, 608
Faces and Places, 608
ffotograff, 608
Gagg's (Andrew N.) PHOTOFLORA, 609
Harding (Robert) World Imagery, 611
Heseltine (John) Archive, 611
Howes (Chris)/Wild Places Photography, 612
Illustrated London News Picture Library, 613
International Photobank, 613
Jayawardene Travel Photo Library, 613
Kos Picture Source Ltd, 614
Magnum Photos Ltd, 616
Moving Image Communications, 617
Nature Picture Library, 618
Photoshot, 619
PictureBank Photo Library Ltd, 620
Pictures Colour Library, 620
PPL (Photo Agency) Ltd, 620
Royal Geographical Society Picture Library, 622
Skishoot–Offshoot, 623
Traeger (Tessa) Picture Library, 625
Travel Ink Photo Library, 625
Tropix Photo Library, 625
Williams (David) Picture Library, 626
Young (John Robert) Collection, 627

Travel: Poetry Magazines
Quiet Feather, The, 215

Sable, 216

Travel: Poetry Presses
nthposition, 204

Travel: Prizes
BBC Wildlife Magazine Travel
    Writing Award, 490
BBCFour Samuel Johnson Prize
    for Non-Fiction, The, 490
Dolman Best Travel Book
    Award, 498
Royal Society of Literature
    Ondaatje Prize, The, 518

**Travel: Professional
Associations**
British Guild of Travel Writers,
    532

Travel: Publishers (European)
Arche Verlag AG, 171
Arthaud (Éditions), 162
BZZTÔH (Uitgeverij) BV, 167
Deutscher Taschenbuch Verlag
    GmbH & Co. KG, 164
Lannoo, Uitgeverij, 160
Michelin Éditions des Voyages,
    163
Nyt Nordisk Forlag Arnold
    Busck A/S, 162
Payot Libraire, 171
Spectrum (Uitgeverij Het) BV,
    168
Todariana Editrice, 167
Unieboek (Uitgeverij) BV, 168

**Travel: Publishers
(International)**
Allen & Unwin Pty Ltd, 191
Hachette Livre Australia, 191
HarperCollins Publishers India
    Ltd, 195
Melbourne University
    Publishing Ltd, 192
Munshiram Manoharlal
    Publishers Pvt Ltd, 195
Pan Macmillan Australia Pty
    Ltd, 192
Raupo Publishing (NZ) Ltd, 197
Struik Publishers (Pty) Ltd, 198

Travel: Publishers (Ireland)
O'Brien Press Ltd, The, 157

Travel: Publishers (UK)
AA Publishing, 64
Andrews (Chris) Publications,
    66
Anova Books, 66
Apex Publishing Ltd, 67
Arris Publishing Ltd, 68

Aurum Press Ltd, 70
AuthorHouse UK, 70
Berlitz Publishing, 72
Bradt Travel Guides, 77
Brealey (Nicholas) Publishing,
    77
Business Education Publishers
    Ltd, 79
Constable & Robinson Ltd, 85
Cook (Thomas) Publishing, 86
Crimson Publishing, 86
de la Mare (Giles) Publishers
    Ltd, 88
Dorling Kindersley Ltd, 90
Eland Publishing Ltd, 91
Everyman's Library, 93
Foulsham Publishers, 95
Garnet Publishing Ltd, 97
Gibson Square, 97
Hachette Livre UK, 100
HarperCollins Publishers Ltd,
    102
Haus Publishing, 103
Landmark Publishing Ltd, 109
Little, Brown Book Group UK,
    111
Lonely Planet Publications Ltd,
    112
Melrose Books, 116
Methuen Publishing Ltd, 116
Michelin Maps & Guides, 116
Murdoch Books UK Ltd, 118
New Holland Publishers (UK)
    Ltd, 119
Octagon Press Ltd, 121
Pegasus Elliot Mackenzie
    Publishers Ltd, 125
Quiller Press, 130
Random House Group Ltd,
    The, 131
Reader's Digest Association
    Ltd, 132
Reaktion Books, 132
SB Publications, 136
Sheldrake Press, 138
Simon & Schuster UK Limited,
    139
Summersdale Publishers Ltd,
    142
Tauris (I.B.) & Co. Ltd, 143
Thames and Hudson Ltd, 144
Travel Publishing Ltd, 145
Trident Press Ltd, 145
Wilson (Neil) Publishing Ltd,
    149

Travel: Publishers (US)
Globe Pequot Press, The, 176
Interlink Publishing Group,
    Inc., 179
Pelican Publishing Company,
    183

St Martin's Press LLC, 186

Travel: Small Presses
Crescent Moon Publishing and
    Joe's Press, 227
Frontier Publishing, 228
Graffeg, 229
Ivy Publications, 230
Nyala Publishing, 232
Serif, 235

Travel: Television and Radio
Travel Channel, 409

Travel: UK Packagers
Savitri Books Ltd, 245

**Travel: Writers' Courses (UK
and Ireland)**
City University, 458
London College of
    Communication, 459
Missenden Abbey, 453
Reading, University of, 453
Sheffield, University of, 465
Travellers' Tales, 460

Travel: Writers' Courses (US)
Goucher College, 471

Tropics: Picture Libraries
Angel (Heather)/Natural
    Visions, 602
Fogden Wildlife Photographs,
    608
Garden and Wildlife Matters
    Photo Library, 609

USA: Library Services
Bromley Central Library, 579
Senate House Library,
    University of London, 595

USA: Literary Agents (US)
Siegel (Rosalie), International
    Literary Agency, Inc., 290
Ware (John A.) Literary Agency,
    291

USA: Picture Libraries
Angel (Heather)/Natural
    Visions, 602
Cordaiy (Sylvia) Photo Library,
    606
Garden and Wildlife Matters
    Photo Library, 609
Greenhill (Sally and Richard),
    611
Howes (Chris)/Wild Places
    Photography, 612

Newark's (Peter) Picture
Library, 618
Sanders (Peter) Photography
Ltd, 622
Williams (David) Picture
Library, 626

USA: Prizes
Pulitzer Prizes, 516

**USA: Professional
Associations**
Association of American
Correspondents in London,
527
Association of American
Publishers, Inc, 527
Association of Authors'
Representatives (AAR, 527
Mystery Writers of America,
Inc., 540
Romance Writers of America,
545

USA: Publishers (UK)
Anova Books, 66

USA: Publishers (US)
Alabama Press, University of,
172
Illinois Press, University of, 178
Kansas, University Press of, 179
Kent State University Press,
The, 179
Louisiana State University
Press, 180
Mississippi, University Press
of, 181
Nebraska Press, University of,
182
Nevada Press, University of, 182
Oklahoma Press, University
of, 183
Tennessee (University of) Press,
188
Texas Press, University of, 188
Virginia Press, University of, 189
Washington State University
Press, 190

**Vegetarian: Literary Agents
(US)**
Gusay (Charlotte) Literary
Agency, The, 284

Vegetarian: Magazines
Vegan, The, 364

**Veterinary: Literary Agents
(US)**
Gusay (Charlotte) Literary
Agency, The, 284

Veterinary: Picture Libraries
Cordaiy (Sylvia) Photo Library,
606
Only Horses Picture Agency,
619

Veterinary: Publishers (UK)
Manson Publishing Ltd, 115

Veterinary: Publishers (US)
Blackwell Publishing (US), 174

Video: Magazines
Empire, 317
Sight & Sound, 355

Video: Prizes
Kraszna-Krausz Book Awards,
507

**Video: Professional
Associations**
BFI, 530
IVCA, 539

Video: Publishers (European)
Lettera, 161

**Video: Publishers
(International)**
Currency Press Pty Ltd, 191

Video: Publishers (UK)
Fitzgerald Publishing, 95
Omnibus Press, 122
Phaidon Press Limited, 126

Video: Publishers (US)
Reader's Digest Association
Inc, 185
Tyndale House Publishers, Inc.,
189
Zondervan, 190

**Virtual Reality: Literary
Agents (UK)**
London Independent Books,
263

**Volcanoes: Literary Agents
(UK)**
London Independent Books,
263

Volcanoes: Picture Libraries
GeoScience Features, 610

**Voluntary Sector: Publishers
(UK)**
NCVO Publications, 119

**Wales/Welsh Interest: Arts
Councils and Regional
Offices**
Arts Council of Wales, The/
Cyngor Celfyddydau Cymru,
573

**Wales/Welsh Interest:
Festivals**
Ty Newydd Festival, 639

**Wales/Welsh Interest:
Library Services**
Carmarthen Public Library, 580
National Library of Wales, 590
Swansea Central Library, 597

**Wales/Welsh Interest:
Literary Agents (US)**
Gusay (Charlotte) Literary
Agency, The, 284

**Wales/Welsh Interest:
Literary Societies**
Davies (Rhys) Trust, The, 557
Thomas (Dylan) Society of
Great Britain, The, 568

**Wales/Welsh Interest:
Magazines**
Big Issue Cymru, The, 300
New Welsh Review, 340
Welsh Country Magazine, 366

**Wales/Welsh Interest:
Miscellany**
Combrógos, 641

**Wales/Welsh Interest:
Picture Libraries**
Country Matters Picture
Library, 606
ffotograff, 608
Hardley (Dennis) Photography,
Scottish Photo Library, 611
National Trust Photo Library,
The, 618
Photolibrary Wales, The, 619

**Wales/Welsh Interest: Poetry
Magazines**
New Welsh Review, 212
Planet, 213
Poetry Wales, 215
Red Poets –Y Beirdd Coch, 216
Taliesin, 217

**Wales/Welsh Interest: Poetry
Presses**
Cinnamon Press, 200

Gomer Press/Gwasg Gomer, 202
Honno Welsh Women's Press, 202
Parthian, 204
Rack Press, 206

Wales/Welsh Interest: Prizes
Academi Book of the Year Awards, 488
Drama Association of Wales Playwriting Competition, 498
Jones (Mary Vaughan) Award, 506
Tir Na N-Og Award, The, 522

Wales/Welsh Interest: Professional Associations
Academi, 526
CILIP Wales, 534
Welsh Books Council (Cyngor Llyfrau Cymru), 550
Welsh Literature Abroad/ Llenyddiaeth Cymru Dramor, 550

Wales/Welsh Interest: Publishers (UK)
Bryntirion Press, 79
Davies (Christopher) Publishers Ltd, 88
Dref Wen, 90
Gomer Press/Gwasg Gomer, 98
Gwasg Carreg Gwalch, 99
Honno Welsh Women's Press, 105
Parthian, 124
Seren, 137
Wales Press, University of, 148
Y Lolfa Cyf, 151

Wales/Welsh Interest: Small Presses
Graffeg, 229
Iolo, 230

Wales/Welsh Interest: Theatre Producers
Clwyd Theatr Cymru, 439
Sherman Cymru, 448

Wales/Welsh Interest: Writers' Courses (UK and Ireland)
Ty Newydd Writers' Centre, 467

War: Library Services
Brighton Jubilee Library, 576
German Historical Institute Library, 583
Imperial War Museum, 586

War: Literary Agents (UK)
Feldstein Agency, The, 256

War: Magazines
Armourer Magazine, The, 297
Great War (1914–1918), The, 324

War: Picture Libraries
Defence Picture Library, The, 606
Geoslides Photography, 610
Illustrated London News Picture Library, 613
Imperial War Museum Photograph Archive, 613
King (David) Collection, 613
Magnum Photos Ltd, 616
Miller (Lee) Archives, 616
Popperfoto.com, 620

War: Publishers (UK)
Austin & Macauley Publishers Limited, 70
Blackstone Publishers, 74
Osprey Publishing Ltd, 123
Pegasus Elliot Mackenzie Publishers Ltd, 125
Pen & Sword Books Ltd, 125

War: Small Presses
Miller (Neil) Publications, 232

War: UK Packagers
Erskine Press, 243

Westerns: Literary Agents (US)
Weiner (Cherry) Literary Agency, 292

Westerns: Publishers (UK)
Thorpe (F.A.) Publishing, 144

Wildlife: Film, TV and Radio Producers
Tigress Productions Ltd, 435

Wildlife: Library Services
British Library Sound Archive, 579

Wildlife: Magazines
BBC Wildlife Magazine, 299
Bird Life Magazine, 301
Birds, 301
Country Life, 311
Country Living, 311
Wild Times, 367
Wingbeat, 367

Wildlife: Picture Libraries
Angel (Heather)/Natural Visions, 602
Cordaiy (Sylvia) Photo Library, 606
Craven (Philip) Worldwide Photo-Library, 606
Ecoscene, 607
Forest Life Picture Library, 609
GeoScience Features, 610
Howes (Chris)/Wild Places Photography, 612
Images of Africa Photobank, 613
Nature Photographers Ltd, 618
Oxford Picture Library, 619
Oxford Scientific (OSF), 619
Photoshot, 619
Premaphotos Wildlife, 620
RSPB Images, 622
RSPCA Photolibrary, 622
Science Photo Library, 622
Waterways Photo Library, 626

Wildlife: Prizes
BBC Wildlife Magazine Poet of the Year Awards, 490
BBC Wildlife Magazine Travel Writing Award, 490

Wildlife: Publishers (UK)
Pocket Mountains Ltd, 127

Wine & Spirits: Library Services
Guildhall Library, 584

Wine & Spirits: Literary Agents (UK)
Feldstein Agency, The, 256
Kelly (Frances), 262
Limelight Management, 262

Wine & Spirits: Literary Agents (US)
Lescher & Lescher Ltd, 285

Wine & Spirits: Magazines
Decanter, 314
House & Garden, 328
Wine & Spirit, 367

Wine & Spirits: PR Consultants
Frazer (Cathy) PR, 277

Wine & Spirits: Picture Libraries
Blake (Anthony) Photo Library, 603
Cephas Picture Library, 604

Eagar (Patrick) Photography, 607

Wine & Spirits: Prizes
Glenfiddich Food & Drink Awards, 502
Gourmand World Cookbook Awards, 502
Simon (André) Memorial Fund Book Awards, 519

Wine & Spirits: Professional Associations
Circle of Wine Writers, 534

Wine & Spirits: Publishers (European)
Flammarion Éditions, 163

Wine & Spirits: Publishers (Ireland)
Farmar (A.&A.) Ltd, 155
On Stream Publications Ltd, 157

Wine & Spirits: Publishers (UK)
Absolute Press, 64
Grub Street, 99
Hachette Livre UK, 100
HarperCollins Publishers Ltd, 102
Quiller Press, 130
Special Interest Model Books Ltd, 140

Women's Interests: Literary Agents (UK)
Feldstein Agency, The, 256
Graham Maw Christie Literary Agency, 258
Judd (Jane) Literary Agency, 262

Women's Interests: Literary Agents (US)
Bleecker Street Associates, Inc., 280
Gusay (Charlotte) Literary Agency, The, 284

Women's Interests: Magazines
Bella, 299
Best, 300
Black Beauty & Hair, 301
Chat, 307
Company, 310
Cosmopolitan, 311
Easy Living, 316
Elle, 317
Essentials, 318
Eve, 318
Glamour, 323

Grazia, 324
Harper's Bazaar, 326
InStyle, 329
Lady, The, 331
marie claire, 334
More, 337
Mslexia, 338
My Weekly, 338
Prima, 347
Psychologies, 348
Red, 351
Reveal, 351
She Magazine, 354
Take a Break, 359
Vogue, 364
W.I. Life, 365
Wedding, 366
Weekly News, 366
Woman, 368
Woman and Home, 368
Woman's Own, 368
Woman's Weekly, 368
Yorkshire Women's Life Magazine, 370

Women's Interests: Professional Associations
Society of Women Writers & Journalists, 548

Women's Interests: Publishers (Ireland)
Mercier Press Ltd, 156

Women's Interests: Publishers (UK)
Persephone Books, 126

Woodwork: Magazines
Practical Woodworking, 346
Woodworker, The, 368

Woodwork: Publishers (UK)
GMC Publications Ltd, 98

Woodwork: Publishers (US)
Sterling Publishing Co. Inc., 188

World Affairs: Library Services
PA News Centre, 592

World Affairs: Literary Agents (UK)
Feldstein Agency, The, 256
Simmons (Jeffrey), 269

World Affairs: Literary Agents (US)
Gusay (Charlotte) Literary Agency, The, 284

World Affairs: Magazines
Day by Day, 314
Economist, The, 316
New Internationalist, 339
Prospect, 347
Time, 360
Tribune, 363

World Affairs: Picture Libraries
Getty Images, 610
Magnum Photos Ltd, 616
Photoshot, 619
Popperfoto.com, 620

World Affairs: Prizes
Watt (David) Prize, The, 523

World Affairs: Publishers (UK)
Garnet Publishing Ltd, 97
Luath Press Ltd, 112
Macmillan Publishers Ltd, 113
Mainstream Publishing Co. (Edinburgh) Ltd, 114
Pluto Press Ltd, 127
Tauris (I.B.) & Co. Ltd, 143
Wimbledon Publishing Company, 150
Zed Books Ltd, 151

World Affairs: Publishers (US)
Paragon House, 183
Potomac Books, Inc., 184
Syracuse University Press, 188

World Affairs: Television and Radio
BBC News, 397

Zoology: Library Services
Natural History Museum Library, The, 591
Zoological Society Library, 600

Zoology: Picture Libraries
GeoScience Features, 610
Ulster Museum Picture Library, 625

Zoology: Publishers (International)
Melbourne University Publishing Ltd, 192

Zoology: Publishers (UK)
Freeman (W. H.), 96

Remember to use your FREE access to The Writer's Handbook Online

1. Visit **www.thewritershandbook.com**

2. Register using your unique online access code (this can be found on the sticker on the cover of your book)

3. Explore the great Writer's Resources on offer!

Register with the site and you will have access to:

- The entire directory of *The Writer's Handbook* – accessible by browse and quick searches

- Extra resource material – this includes a list of useful websites, top tips and FAQs for Writers

- Book industry newsfeeds

- Publishing insights – includes a glossary of publishing and advice from an editor

- Extra articles from previous editions of *The Writer's Handbook* which offer a unique insight into how the book industry has changed over the years

- A blog written by industry insiders and creative writing gurus

www.thewritershandbook.com